Physics Galaxy

58 Papers

2017-2021 (Offline & Online)

JEE MAIN

Chapter-Wise Solutions

Physics

Ashish Arora

First edition : Dec, 2000

Second edition : May, 2016

Revised Edition : April, 2022

Title : Physics Galaxy 2022 : JEE Main Physics - Chapter-Wise Solutions (2017-2021)

Language : English

Author's Name : Ashish Arora

Typeset & Published by :

Career Launcher Infrastructure (P) Ltd.

A-45, Mohan Cooperative Industrial Area, Near Mohan Estate Metro Station, New Delhi - 110044

Marketed by :

G.K. Publications (P) Ltd.

Plot No. 9A, Sector-27A, Mathura Road, Faridabad, Haryana-121003

ISBN : **978-93-92837-06-7**

Printer's Details: HT Media Ltd, Greater Noida

For product information :

Visit ***www.gkpublications.com*** or email to ***gkp@gkpublications.com***

Dedicated

to

My Parents, Son, Daughter

and

My beloved wife

In his teaching career since 1992 Ashish Arora personally mentored more than 10000 IITians and students who reached global heights in various career and profession chosen. It is his helping attitude toward students with which all his students remember him in life for his contribution in their success and keep connections with him live. Below is the list of some of the successful students in IITJEE, NEET & International Olympiad personally mentored by him.

NAVNEET LOIWAL	*International GOLD Medal in IPhO-2000 at LONDON*, Also secured **AIR-4** in **IIT JEE 2000** PROUD FOR INDIA : Navneet Loiwal was the first Indian Student who won first International GOLD Medal for our country in International Physics Olympiad.
DUNGRA RAM CHOUDHARY	**AIR-1** in **IIT JEE 2002**
HARSHIT CHOPRA	*National Gold Medal in INPhO-2002* and got **AIR-2** in **IIT JEE-2002**
KUNTAL LOYA	A Girl Student got position **AIR-8** in **IIT JEE 2002**
LUV KUMAR	*National Gold Medal in INPhO-2003* and got **AIR-3** in **IIT JEE-2003**
RAJHANS SAMDANI	*National Gold Medal in INPhO-2003* and got **AIR-5** in **IIT JEE-2003**
SHANTANU BHARDWAJ	*International SILVER Medal in IPhO-2002 at INDONESIA*
SHALEEN HARLALKA	*International GOLD Medal in IPhO-2003 at CHINA* and got **AIR-46** in **IIT JEE-2003**
TARUN GUPTA	*National GOLD Medal in INPhO-2005*
APEKSHA KHANDELWAL	*National GOLD Medal in INPhO-2005*
ABHINAV SINHA	*Hon'ble Mension Award in APhO-2006* at *KAZAKHSTAN*
RAMAN SHARMA	*International GOLD Medal in IPhO-2007* at *IRAN* and got **AIR-20** in **IIT JEE-2007**
PRATYUSH PANDEY	*International SILVER Medal in IPhO-2007* at *IRAN* and got **AIR-85 in IIT JEE-2007**
GARVIT JUNIWAL	*International GOLD Medal in IPhO-2008* at *VIETNAM* and got **AIR-10** in **IIT JEE-2008**
ANKIT PARASHAR	*National GOLD Medal in INPhO-2008*
HEMANT NOVAL	*National GOLD Medal in INPhO-2008* and got **AIR-25** in **IIT JEE-2008**
ABHISHEK MITRUKA	*National GOLD Medal in INPhO-2009*
SARTHAK KALANI	*National GOLD Medal in INPhO-2009*
ASTHA AGARWAL	*International SILVER Medal in IJSO-2009* at *AZERBAIJAN*
RAHUL GURNANI	*International SILVER Medal in IJSO-2009* at *AZERBAIJAN*
AYUSH SINGHAL	*International SILVER Medal in IJSO-2009* at *AZERBAIJAN*
MEHUL KUMAR	*International SILVER Medal in IPhO-2010* at *CROATIA* and got **AIR-19** in **IIT JEE-2010**
ABHIROOP BHATNAGAR	*National GOLD Medal in INPhO-2010*
AYUSH SHARMA	*International Double GOLD Medal in IJSO-2010* at *NIGERIA*
AASTHA AGRAWAL	*Hon'ble Mension Award in APhO-2011* at *ISRAEL* and got **AIR-93** in **IIT JEE 2011**
ABHISHEK BANSAL	*National GOLD Medal in INPhO-2011*
SAMYAK DAGA	*National GOLD Medal in INPhO-2011*
SHREY GOYAL	*National GOLD Medal in INPhO-2012* and secured **AIR-24** in **IIT JEE 2012**
RAHUL GURNANI	*National GOLD Medal in INPhO-2012*
JASPREET SINGH JHEETA	*National GOLD Medal in INPhO-2012*
DIVYANSHU MUND	*National GOLD Medal in INPhO-2012*
SHESHANSH AGARWAL	*International SILVER Medal in IAO-2012 at KOREA*
SWATI GUPTA	*International SILVER Medal in IJSO-2012 at IRAN*
PRATYUSH RAJPUT	*International SILVER Medal in IJSO-2012 at IRAN*
SHESHANSH AGARWAL	*International BRONZE Medal in IOAA-2013 at GREECE*
SHESHANSH AGARWAL	*International GOLD Medal in IOAA-2014 at ROMANIA*
SHESHANSH AGARWAL	*International SILVER Medal in IPhO-2015 at INDIA* and secured **AIR-58** in **JEE(Advanced)-2015**
VIDUSHI VARSHNEY	*International SILVER Medal in IJSO-2015 at SOUTH KOREA*
AMAN BANSAL	**AIR-1 in JEE Advanced 2016**
KUNAL GOYAL	**AIR-3 in JEE Advanced 2016**
GOURAV DIDWANIA	**AIR-9 in JEE Advanced 2016**
DIVYANSH GARG	*International SILVER Medal in IPhO-2016 at SWITZERLAND*
NALIN KHANDELWAL	**AIR-1 in NEET 2019**
MRIDUL AGARWAL	**AIR-1 in JEE Advanced 2021**

ABOUT THE AUTHOR

The complexities of Physics have given nightmares to many, but the homegrown genius of Jaipur- Ashish Arora has helped several students to live their dreams by decoding it.

Newton Law of Gravitation and Faraday's Magnetic force of attraction apply perfectly well with this unassuming genius. A Pied Piper of students, his webportal https://www.physicsgalaxy.com, The world's largest encyclopedia of video lectures on high school Physics possesses strong gravitational pull and magnetic attraction for students who want to make it big in life.

Ashish Arora, gifted with rare ability to train masterminds, has mentored over 10,000 IITians in his past 24 years of teaching sojourn including lots of students made it to Top 100 in IIT-JEE/JEE(Advance) including AIR-1 and many in Top-10. Apart from that, he has also groomed hundreds of students for cracking International Physics Olympiad. No wonder his student Navneet Loiwal brought laurel to the country by becoming the first Indian to win a Gold medal at the 2000 - International Physics Olympiad in London (UK).

His special ability to simplify the toughest of the Physics theorems and applications rates him as one among the best Physics teachers in the world. With this, Arora simply defies the logic that perfection comes with age. Even at 18 when he started teaching Physics while pursuing engineering, he was as engaging as he is now. Experience, besides graying his hair, has just widened his horizon.

Now after encountering all tribes of students - some brilliant and some not-so-intelligent - this celebrated teacher has embarked upon a noble mission to make the entire galaxy of Physics inform of his webportal PHYSICSGALAXY.COM to serve and help global students in the subject. Today students from 183 countries are connected with this webportal. On any topic of physics students can post their queries in comment section in Physics Galaxy Youtube channel on which many Physics experts with Ashish Arora reply to several queries posted online by students.

Dedicated to global students of middle and high school level, his website *www.physicsgalaxy.com* also has teaching sessions dubbed in American accent and subtitles in 87 languages. For students in India preparing for JEE & NEET, his online courses will be available soon on PHYSICSGALAXY.COM.

In 2007 he started authoring Physics Galaxy Youtube Channel which has now become World's Largest Youtube Channel for online physics lectures on high school physics and preparation of JEE & NEET. Daily about 1.5 lakh+ video lectures are being watched on the channel by students preparing for JEE, NEET & High school Physics.

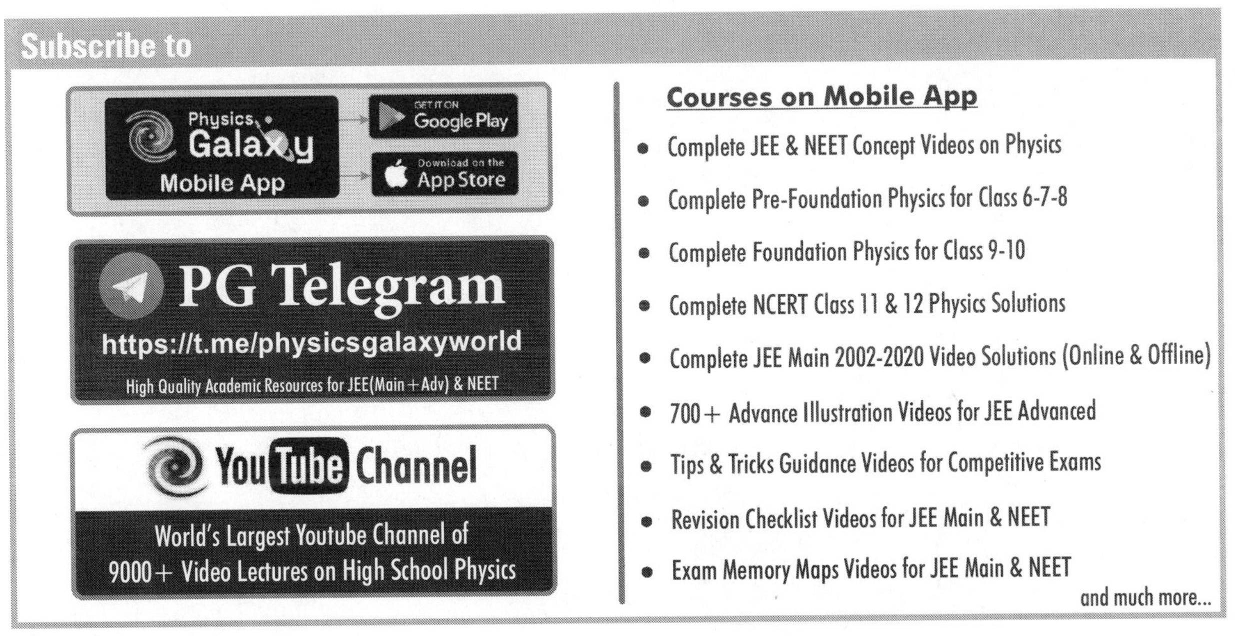

ABOUT THE AUTHOR

FOREWORD

I am extremely happy to share my views about this book written by Er. Ashish Arora. He has contributed immensely to the teaching-learning of Physics though many innovative ways. His approach has always been student centric and this contribution is yet another example of the same.

As the title states "Physics Galaxy JEE Main Solutions" the book provides valuable information and covers the necessary components by offering solution to all questions asked in old JEE Main (all Online & Offline 2017-2021) with detailed explanations. All questions are categorized into 30 topics based on weightage of questions asked in the past 5 years (2017-2021) however in this book total 58 papers of all online and offline shifts of JEE Main are included with detailed comprehensive solutions and upto 2019 all solutions to questions are linked with video solutions available on YouTube channel of Physics Galaxy. This is to indicate and promote understanding of the methods or reasons for specific/particular solutions. Aspirants can also scan the respective QR code given in the book to see its video explanation as and when required. All solutions of the questions are based on the conceptual analysis covered in the popular Physics Galaxy book series (5 volumes) so that aspirants following the book series will be able to relate and recall all the explanations already done.

I am sure the book promotes the concept of deeper learning in the domain of Physics. My best wishes to the many aspirants who will certainly find this book a very valuable tool to enhance thorough understanding of the subject.

April, 2022

Prof. Sandeep Sancheti
Ph. D. (U.K.), B.Tech. FIETE, MIEEE
VC, SRMIST, Chennai &
President, AIU, New Delhi

PREFACE

For a science student, preparing for JEE Main or NEET, Physics is the most important subject, unlike to other subjects it requires logical reasoning and high imagination of brain. Without improving the level of physics it is very difficult to achieve a goal in the present age of competitions. To score better, one does not require hard working at least in physics. It just requires a simple understanding and approach to think a physical situation. Actually physics is the surrounding of our everyday life. If you wish to make the concepts of physics strong, you should try to understand core concepts of physics in practical approach rather than theoretical.

This book "*PHYSICS GALAXY JEE Main Solutions*" is designed in a totally different and friendly approach to develop the physics concepts psychologically. The book covers detailed solutions to all the questions previously asked in JEE Main all online and offline papers from 2017 to 2021. From the analysis of 2021 all four attempts papers it is clearly seen that all these papers held were designed on the basis of past 5 year papers. More than 60% questions are based on concepts asked in previous years which were matching in style with the questions of these old papers, called derived PYQs.

All the questions of these online and offline papers are divided in 30 topics for topic level understanding of questions asked in JEE Main. In my classroom teaching also I found that topic level understanding of previous year papers of exam helps students a lot in concept visualization from the point of view of preparation of final exam as well as for the purpose of last moment revision of all topics.

For every question asked in these old JEE Main papers, video solutions are recorded and made available for students on YouTube channel of Physics Galaxy as well as these video solutions can be accessed through Physics Galaxy mobile app available on iOS and Android play store. While following the book if students face any doubt in understanding the solution of the specific question then they can scan the QR codes given at the end of the book for all the questions to directly start the video solution of that question.

I don't have words for my best friend my wife Anuja for always being together with me to complete this book in the unique style and format along with the whole series of Physics Galaxy book series and lecture notes series.

I would like to pay my gratitude to Sh. Dayashankar Prajapati in assisting me to complete the task in Design Labs of PHYSICSGALAXY.COM and presenting the book in totally new format of second edition.

At last but the most important person, my father who has devoted his valuable time to finally present the book in such a format and a simple language, thanks is a very small word for his dedication in this book.

In this first edition of book I have tried my best to make this book error free but owing to the nature of work, inadvertently, there is possibility of errors left untouched. I shall be grateful to the readers, if they point out me regarding errors and oblige me by giving their valuable and constructive suggestions via emails for further improvement of the book.

Date : April, 2022

Ashish Arora

PHYSICSGALAXY.COM
B-80, Model Town, Malviya Nagar, Jaipur-302017
e-mails: ashisharora@physicsgalaxy.com

CONTENTS

CHAPTER

UNITS AND DIMENSIONS

1.1 MEASUREMENTS AND DIMENSIONAL ANALYSIS

1. Time (T), velocity (C) and angular momentum (h) are chosen as fundamental quantities instead of mass, length and time. In terms of these, the dimensions of mass would be : **[2017 (Online)]**

(A) $[M] = [T^{-1} C^{-2} h]$ (B) $[M] = [T^{-1} C^{2} h]$
(C) $[M] = [T^{-1} C^{-2} h^{-1}]$ (D) $[M] = [T C^{-2} h]$

2. The characteristic distance at which quantum gravitational effects are significant, the Planck length, can be determined from a suitable combination of the fundamental physical constants G, h and c. Which of the following correctly gives the Planck length ? **[2018 (Online)]**

(A) G^2hc (B) $\left(\frac{Gh}{c^3}\right)^{\frac{1}{2}}$
(C) $G^{\frac{1}{2}}h^2c$ (D) Gh^2c^3

3. Expression for time in terms of G (universal gravitational constant), h (Planck constant) and c (speed of light) is proportional to : **[9 Jan. 2019 (Evening)]**

(A) $\sqrt{\frac{Gh}{c^3}}$ (B) $\sqrt{\frac{hc^5}{G}}$
(C) $\sqrt{\frac{c^3}{Gh}}$ (D) $\sqrt{\frac{Gh}{c^5}}$

4. The density of a material in SI units is 128 kg m^{-3}. In certain units in which the unit of length is 25 cm and the unit of mass is 50 g, the numerical value of density of the material is : **[10 Jan. 2019 (Morning)]**

(A) 410 (B) 640
(C) 16 (D) 40

5. The force of interaction between two atoms is given by

$$F = \alpha\beta \exp\left(-\frac{x^2}{\alpha kt}\right);$$ where x is the distance, k is the Boltzmann constant and T is temperature and α and β are two constants. The dimension of β is : **[11 Jan. 2019 (Morning)]**

(A) $M^2L^2T^{-2}$ (B) M^2LT^{-4}
(C) $M^0L^2T^{-4}$ (D) MLT^{-2}

6. If speed (V), acceleration (A) and force (F) are considered as fundamental units, the dimension of Young's modulus will be : **[11 Jan. 2019 (Evening)]**

(A) $V^{-2}A^2F^2$ (B) $V^{-4}A^2F$
(C) $V^{-4}A^{-2}F$ (D) $V^{-2}A^2F^{-2}$

7. Let l, r, c and v represent inductance, resistance, capacitance and voltage, respectively. The dimension of $\frac{l}{rcv}$ in SI units will be : **[12 Jan. 2019 (Evening)]**

(A) [LTA] (B) $[LA^{-2}]$
(C) $[A^{-1}]$ (D) $[LT^2]$

8. If surface tension (S), Moment of inertia (I) and Planck's constant (h), were to be taken as the fundamental units, the dimensional formula for linear momentum would be : **[8 April 2019 (Evening)]**

(A) $S^{3/2}I^{1/2}h^0$ (B) $S^{1/2}I^{1/2}h^0$
(C) $S^{1/2}I^{1/2}h^{-1}$ (D) $S^{1/2}I^{3/2}h^{-1}$

9. In the formula $X = 5YZ^2$, X and Z have dimensions of capacitance and magnetic field, respectively. What are the dimensions of Y in SI units ? **[10 April 2019 (Evening)]**

(A) $[M^{-2} L^{-2} T^6 A^3]$ (B) $[M^{-1} L^{-2} T^4 A^2]$
(C) $[M^{-3} L^{-2} T^8 A^4]$ (D) $[M^{-2} L^0 T^{-4} A^{-2}]$

10. A quantity f is given by $f = \sqrt{\frac{hc^5}{G}}$ where c is speed of light, G universal gravitational constant and h is the Planck's constant. Dimension of f is that of : **[9 Jan. 2020 (Morning)]**

(A) Momentum (B) Area
(C) Energy (D) Volume

11. If speed V, area A and force F are chosen as fundamental units, then the dimension of Young's modulus will be : **[2 Sept. 2020 (Morning)]**

(A) $FA^{-1}V^0$ (B) FA^2V^{-1}
(C) FA^2V^{-3} (D) FA^2V^{-2}

12. If momentum (P), area (A) and time (T) are taken to be the fundamental quantities then the dimensional formula for energy is : **[2 Sept. 2020 (Evening)]**

(A) $[PA^{-1}T^{-2}]$ (B) $[PA_1/2T^{-1}]$
(C) $[P_2AT^{-2}]$ (D) $[P_1/2AT^{-1}]$

13. Dimensional formula for thermal conductivity is (here K denotes the temperature) : **[4 Sept. 2020 (Morning)]**

(A) $MLT^{-3}K$ (B) $MLT^{-2}K$
(C) $MLT^{-2}K^{-2}$ (D) $MLT^{-3}K^{-1}$

14. A quantity x is given by (IFv^2/WL^4) in terms of moment of inertia I, force F, velocity v, work W and Length L. The dimensional formula for x is same as that of : **[4 Sept. 2020 (Evening)]**

(A) Planck's constant (B) Force constant
(C) Energy density (D) Coefficient of viscosity

15. The quantities $x = \frac{1}{\sqrt{\mu_0 \in_0}}$, $y = \frac{E}{B}$ and $z = \frac{l}{CR}$ are defined where C-capacitance, R-Resistance, l-length, E-Electric field, B-magnetic field and $\in_0$ & μ_0 free are permittivity and permeability of free space respectively. Then : **[5 Sept. 2020 (Evening)]**

(A) Only x and y have the same dimension
(B) x, y and z have the same dimension
(C) Only x and z have the same dimension
(D) Only y and z have the same dimension

16. The workdone by a gas molecule in an isolated system is given by, $W = \alpha\beta^2 e^{-\frac{x^2}{\alpha kT}}$, where x is the displacement, k is the Boltzmann constant and T is the temperature. α and β are constants. Then the dimensions of β will be : **[24 Feb. 2021 (Morning)]**

(A) $[M L T^{-2}]$ (B) $[M^2 L T^2]$
(C) $[M^0 L T^0]$ (D) $[M L^2 T^{-2}]$

17. Match List-I with List-II : **[25 Feb. 2021 (Morning)]**

List-I	List-II
(a) h (Planck's constant)	(i) $[M L T^{-1}]$
(b) E (kinetic energy)	(ii) $[M L^2 T^{-1}]$
(c) V (electric potential)	(iii) $[M L^2 T^{-2}]$
(d) P (linear momentum)	(iv) $[M L^2 I^{-1} T^{-3}]$

(A) (a)→(ii), (b)→(iii), (c)→(iv), (d)→(i)
(B) (a)→(iii), (b)→(ii), (c)→(iv), (d)→(i)
(C) (a)→(i), (b)→(ii), (c)→(iv), (d)→(iii)
(D) (a)→(iii), (b)→(iv), (c)→(ii), (d)→(i)

18. If e is the electronic charge, c is the speed of light in free space and h is Planck's constant, the quantity $\frac{1}{4\pi\varepsilon_0}\frac{|e|^2}{\hbar c}$ has dimensions of : **[25 Feb. 2021 (Evening)]**

(A) $[L C^{-1}]$ (B) $[M L T^0]$
(C) $[M^0 L^0 T^0]$ (D) $[M L T^{-1}]$

19. In a typical combustion engine the workdone by a gas molecule is given by $W = \alpha^2\beta e^{\frac{-\beta x^2}{kT}}$, where x is the displacement, k is the Boltzmann constant and T is the temperature. If α and β are constants, dimensions of α will be : **[26 Feb. 2021 (Morning)]**

(A) $[M^2 L T^{-1}]$ (B) $[M L T^{-2}]$
(C) $[M L T^{-1}]$ (D) $[M^0 L T^0]$

20. If C and V represent capacity and voltage respectively then what are the dimensions of λ where $C/V = \lambda$? **[26 Feb. 2021 (Evening)]**

(A) $[M^{-2} L^{-3} I^2 T^6]$ (B) $[M^{-1} L^{-3} I^{-2} T^{-7}]$
(C) $[M^{-3} L^{-4} I^3 T^7]$ (D) $[M^{-2} L^{-4} I^3 T^7]$

21. One main scale division of a vernier callipers is 'a' cm and n^{th} division of the vernier scale coincide with $(n-1)^{th}$ division of the main scale. The least count of the callipers in mm is : **[16 March 2021 (Morning)]**

(A) $\frac{10na}{(n-1)}$ (B) $\frac{10a}{(n-1)}$

(C) $\left(\frac{n-1}{10n}\right)a$ (D) $\frac{10a}{n}$

22. Suppose you have taken a dilute solution of oleic acid in such a way that its concentration becomes 0.01 cm^3 of oleic acid per cm^3 of the solution. Then you make a thin film of this solution (monomolecular thickness) of area 4 cm^2 by considering 100 spherical drops of radius $\left(\frac{3}{40\pi}\right)^{\frac{1}{3}} \times 10^{-3}$ cm. Then the thickness of oleic acid layer will be $x \times 10^{-14}$ m.
Where x is__________. **[17 March 2021 (Evening)]**

23. The entropy of any system is given by

$$S = \alpha^2\beta \ln\left[\frac{\mu kR}{J\beta^2} + 3\right]$$

where α and β are the constants. μ, J, k and R are no. of moles, mechanical equivalent of heat, Boltzmann constant and gas constant respectively. [Take $S = \frac{dQ}{T}$] Choose the incorrect options from the following : **[20 July 2021 (Morning)]**

(A) S and α have different dimensions
(B) S, β, k and μR have the same dimensions
(C) α and J have the same dimensions
(D) α and k have the same dimensions

24. If time (t), velocity (v), and angular momentum (l) are taken as the fundamental units. Then the dimension of mass (m) in terms of t, v, and l is : **[20 July 2021 (Evening)]**

(A) $[t^{-1} v^{-2} l^{1}]$ (B) $[t^{-2} v^{-2} l^{1}]$
(C) $[t^{-1} v^{2} l^{1}]$ (D) $[t^{-1} v^{1} t^{-2}]$

25. The force is given in terms of time t and displacement x by the equation $F = A \cos Bx + C \sin Dt$

The dimensional formula of $\frac{AD}{B}$ is : **[25 July 2021 (Evening)]**

(A) $[ML^2T^{-3}]$ (B) $[M^2L^2T^{-3}]$
(C) $[M^0LT^{-1}]$ (D) $[M^1L^1T^{-2}]$

26. Match List-I with List-II : **[27 July 2021 (Evening)]**

List-I	List-II
(a) Capacitance, C	(i) $M^1L^1 T^{-3}A^{-1}$
(b) Permittivity of free space, ε_0	(ii) $M^{-1}L^{-3} T^4A^2$
(c) Permeability of free space, μ_0	(iii) $M^{-1}L^{-2} T^4A^2$
(d) Electric field, E	(iv) $M_1L^1T^{-2}A^{-2}$

Choose the correct answer from the options given below :

(A) (a) → (iii), (b) → (iv), (c) → (ii), (d) → (i)
(B) (a) → (iii), (b) → (ii), (c) → (iv), (d) → (i)
(C) (a) → (iv), (b) → (ii), (c) → (iii), (d) → (i)
(D) (a) → (iv), (b) → (iii), (c) → (ii), (d) → (i)

27. In a Screw Gauge, fifth division of the circular scale coincides with the reference line when the ratchet is closed. There are 50 divisions on the circular scale, and the main scale moves by 0.5 mm on a complete rotation. For a particular observation the reading on the main scale is 5 mm and the 20th division of the circular scale coincides with reference line. Calculate the true reading : **[26 Aug. 2021 (Morning)]**

(A) 5.00 mm (B) 5.20 mm
(C) 5.25 mm (D) 5.15 mm

28. If E, L, M and G denote the quantities as energy, angular momentum, mass and constant of gravitation respectively, then the dimension of P in the formula $P = EL^2M^{-5}G^{-2}$ are : **[26 Aug. 2021 (Morning)]**

(A) $[M^1L^1T^{-2}]$ (B) $[M^{-1}L^{-1}T^{2}]$
(C) $[M^0L^1T^0]$ (D) $[M^0L^0T^0]$

29. Match List-I with List-II :

List-I	List-II
(a) Magnetic Induction	(i) $ML^2T^{-2}A^{-1}$
(b) Magnetic Flux	(ii) $M^0L^{-1}A$
(c) Magnetic Permeability	(iii) $MT^{-2}A^{-1}$
(d) Magnetization	(iv) $MLT^{-2}A^{-2}$

Choose the most appropriate answer form the options given below : **[26 Aug. 2021 (Evening)]**

(A) (a) → (iii), (b) → (ii), (c) → (iv), (d) → (i)
(B) (a) → (iii), (b) → (i), (c) → (iv), (d) → (ii)
(C) (a) → (ii), (b) → (iv), (c) → (i), (d) → (iii)
(D) (a) → (ii), (b) → (i), (c) → (iv), (d) → (iii)

30. Which of the following is not a dimensionless quantity ? **[27 Aug. 2021 (Morning)]**

(A) Permeability of free space (μ_0)
(B) Relative magnetic permeability (μ_r)
(C) Power factor
(D) Quality factor

31. If E and H represents the intensity of electric field and magnetising field respectively, then the unit of E/H will be : **[27 Aug. 2021 (Morning)]**

(A) Newton (B) mho
(C) ohm (D) Joule

32. Match List-I with List-II :

List-I	List-II
(a) R_H (Rydberg constant)	(i) kg m^{-1} s^{-1}
(b) h (Planck's constant)	(ii) kg m^{-2} s^{-1}
(c) μ_B (Magnetic filed energy density	(iii) m^{-1}
(d) η (coefficient of viscosity)	(iv) kg m^{-1} s^{-2}

Choose the most appropriate answer from the options given below : **[27 Aug. 2021 (Evening)]**

(A) (a) → (iv), (b) → (ii), (c) → (i), (d) → (iii)
(B) (a) → (ii), (b) → (iii), (c) → (iv), (d) → (i)
(C) (a) → (iii), (b) → (ii), (c) → (iv), (d) → (i)
(D) (a) → (iii), (b) → (ii), (c) → (i), (d) → (iv)

33. If force (F), length (L) and time (T) are taken as the fundamental quantities. Then what will be the dimension of density : **[27 Aug. 2021 (Evening)]**

(A) $[FL^{-4}T^2]$ (B) $[FL^{-3}T^3]$
(C) $[FL^{-3}T^2]$ (D) $[FL^{-5}T^2]$

34. Match List-I with List-II : **[31 Aug. 2021 (Morning)]**

List-I	List-II
(a) Torque	(i) MLT^{-1}
(b) Impulse	(ii) MT^{-2}
(c) Tension	(iii) ML^2T^{-2}
(d) Surface Tension	(iv) MLT^{-2}

Choose the most appropriate answer from the option given below :

(A) (a)-(iii), (b)-(iv), (c)-(i), (d)-(ii)
(B) (a)-(ii), (b)-(i), (c)-(iv), (d)-(iii)
(C) (a)-(i), (b)-(iii), (c)-(iv), (d)-(ii)
(D) (a)-(iii), (b)-(i), (c)-(iv), (d)-(ii)

35. Which of the following equations is dimensionally incorrect ?

Where t = time, h = height, s = surface tension, θ = angle, ρ = density, a, r = radius, g = acceleration due to gravity, v = volume, p = pressure, W = work done, Γ = torque, $\in$ = permittivity, E = electric field, J = current density, L = length :

[31 Aug. 2021 (Morning)]

(A) $J = \in \dfrac{\partial E}{\partial t}$ (B) $v = \dfrac{\pi p a^4}{8\eta L}$

(C) $h = \dfrac{2s\cos\theta}{\rho r g}$ (D) $W = \Gamma\theta$

36. If velocity $[V]$ time $[T]$ and force $[F]$ are chosen as the base quantities, the dimensions of the mass will be :

[31 Aug. 2021 (Evening)]

(A) $[FVT^{-1}]$ (B) $[FT^{-1}V^{-1}]$

(C) $[FT^2 V]$ (D) $[FTV^{-1}]$

37. The diameter of a spherical bob is measured using a vernier callipers 9 divisions of the main scale, in the vernier callipers, are equal to 10 divisions of vernier scale. One main scale division is 1 mm. The main scale reading is 10 mm and 8th division of vernier scale was found to coincide exactly with one of the main scale division. If the given vernier callipers has positive zero error of 0.004 cm, then the radius of the bob is_____ $\times 10^{-2}$ cm :

[31 Aug. 2021 (Evening)]

* * * * *

CHAPTER

2 MOTION IN A STRAIGHT LINE

2.1 MOTION IN ONE DIMENSION

1. A body is thrown vertically upwards. Which one of the following graphs correctly represent the velocity *vs* time ? **[2017]**

(A) 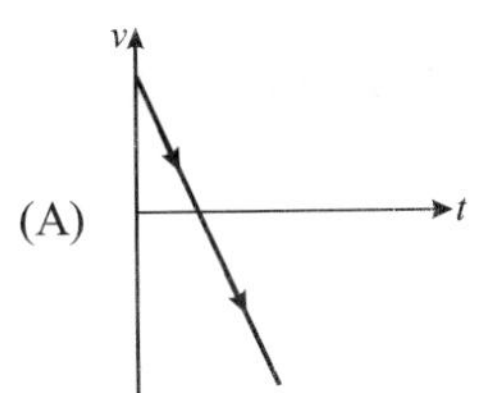

(B) 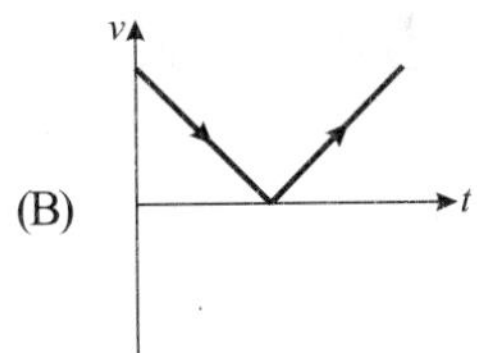

(C) 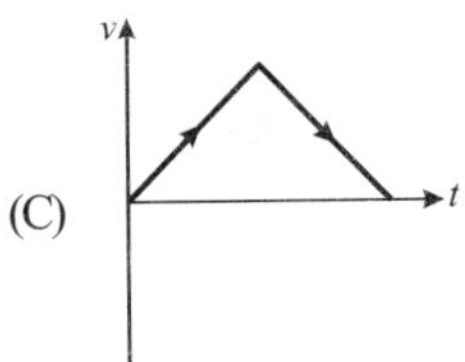

(D) 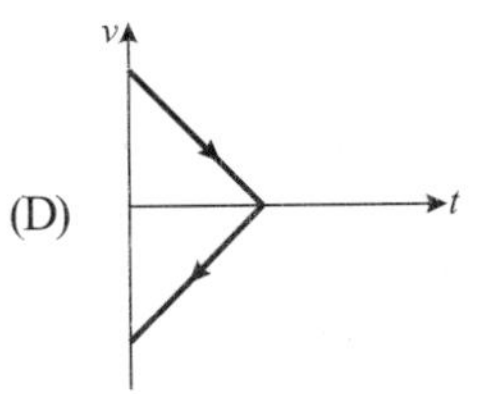

2. Which graph corresponds to an object moving with a constant negative acceleration and a positive velocity ? **[2017 (Online)]**

(A) 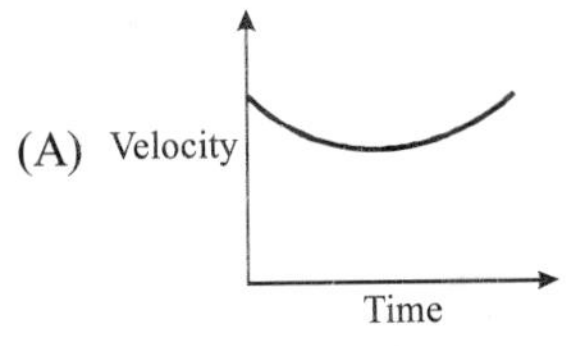

(B) 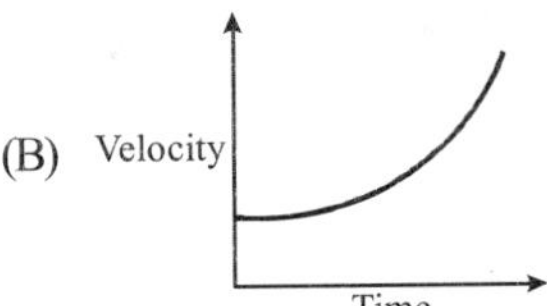

(C) 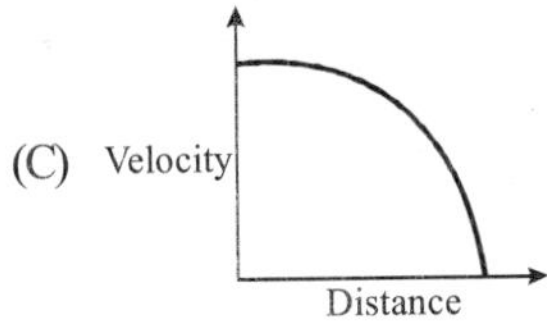

(D) 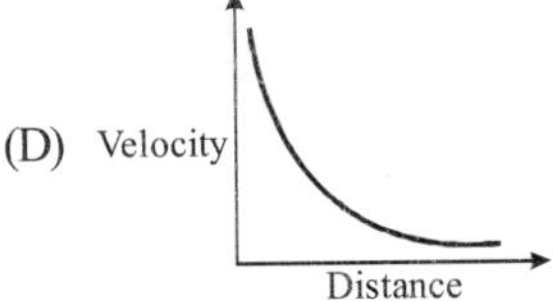

3. All the graphs below are intended to represent the same motion. One of them does it incorrectly. Pick it up : **[2018]**

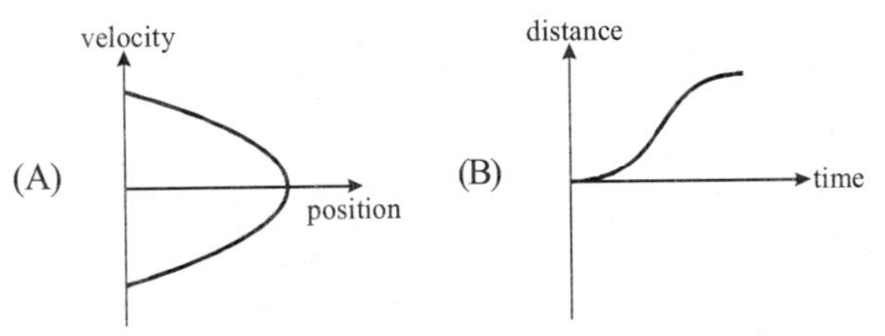

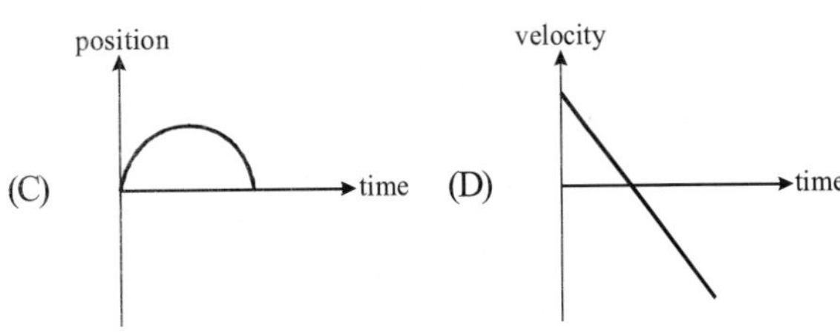

4. An automobile, travelling at 40 km/h, can be stopped at a distance of 40 m by applying brakes. If the same automobile is travelling at 80 km/h, the minimum stopping distance, in metres, is (assume no skidding) : **[2018 (Online)]**

(A) 75 m (B) 160 m
(C) 100 m (D) 150 m

5. The velocity-time graph of a car and a scooter are shown in the figure-(i) the difference between the distance travelled by the car and the scooter in 15 s and (ii) the time at which the car will catch up with the scooter are, respectively : **[2018 (Online)]**

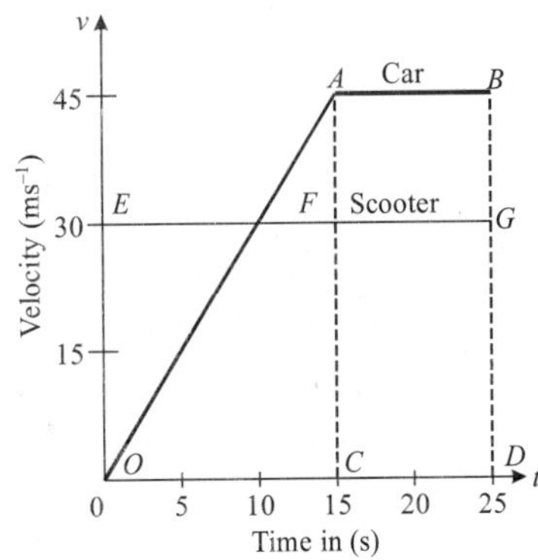

(A) 337.5 m and 25 s (B) 225.5 m and 10 s
(C) 112.5 m and 22.5 s (D) 112.5 m and 15 s

6. A man in a car at location Q on a straight highway is moving with speed v. He decides to reach a point P in a field at a distance d from highway (point M) as shown in the figure. Speed of the car in the field is half to that on the highway. What should be the distance RM, so that the time taken to reach P is minimum? **[2018 (Online)]**

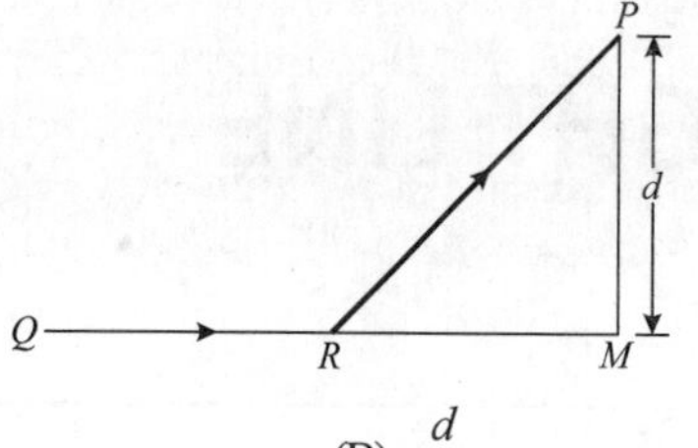

(A) $\frac{d}{\sqrt{3}}$ (B) $\frac{d}{2}$

(C) $\frac{d}{\sqrt{2}}$ (D) d

7. In a car race on straight road, car A takes a time t less than car B at the finish and passes finishing point with a speed 'v' more than that of car B. Both the cars start from rest and travel with constant acceleration a_1 and a_2 respectively. Then 'v' is equal to : **[9 Jan. 2019 (Evening)]**

(A) $\frac{a_1 + a_2}{2}t$ (B) $\sqrt{2a_1a_2}\,t$

(C) $\frac{2a_1a_2}{a_1 + a_2}t$ (D) $\sqrt{a_1a_2}\,t$

8. A particle starts from the origin at time $t = 0$ and moves along the positive x-axis. The graph of velocity with respect to time is shown in figure. What is the position of the particle at time $t = 5$s ? **[10 Jan. 2019 (Evening)]**

(A) 6m (B) 9m

(C) 3m (D) 10m

9. A particle moves from the point $(2.0\hat{i} + 4.0\hat{j})\,m$, at $t = 0$, with an initial velocity $(5.0\hat{i} + 4.0\hat{j})\,\text{ms}^{-1}$. It is acted upon by a constant force which produces a constant acceleration $(4.0\hat{i} + 4.0\hat{j})\,\text{ms}^{-2}$. What is the distance of the particle from the origin at time 2s ? **[11 Jan. 2019 (Evening)]**

(A) $20\sqrt{2}$ m (B) $10\sqrt{2}$ m

(C) 5m (D) 15m

10. A passenger train of length 60 m travels at a speed of 80 km/hr. Another freight train of length 120 m travels at a speed of 30 km/hr. The ratio of times taken by the passenger train to completely cross the freight train when : (i) they are moving in the same direction, and (ii) in the opposite directions is : **[12 Jan. 2019 (Morning)]**

(A) $\frac{5}{2}$ (B) $\frac{25}{11}$

(C) $\frac{3}{2}$ (D) $\frac{11}{5}$

11. A particle starts from origin O from rest and moves with a uniform acceleration along the positive x-axis. Identify all figures that correctly represent the motion qualitatively. (a = acceleration, v = velocity, x = displacement, t = time) : **[8 April 2019 (Evening)]**

(1) 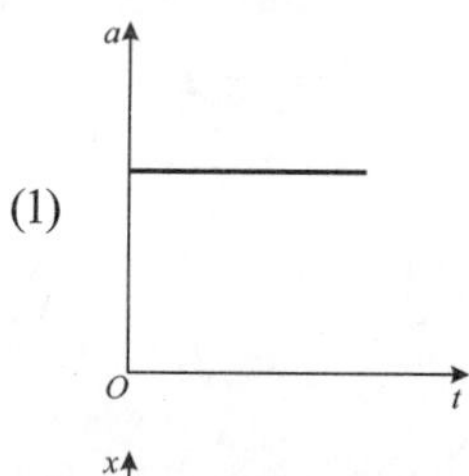

(2) 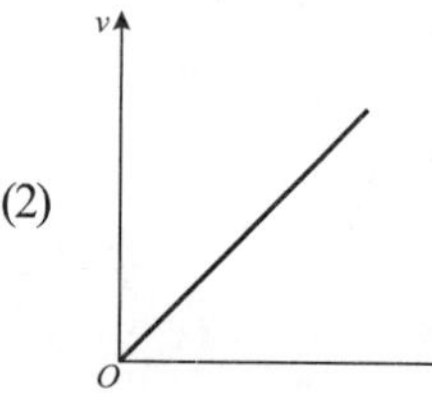

(3) 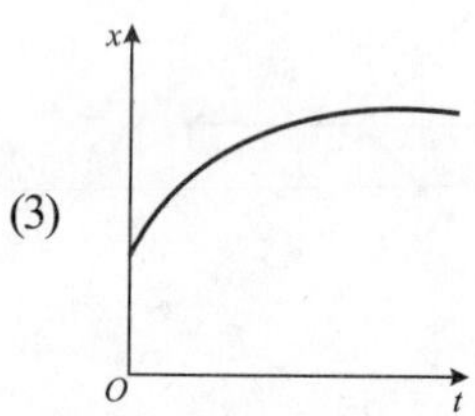

(4) 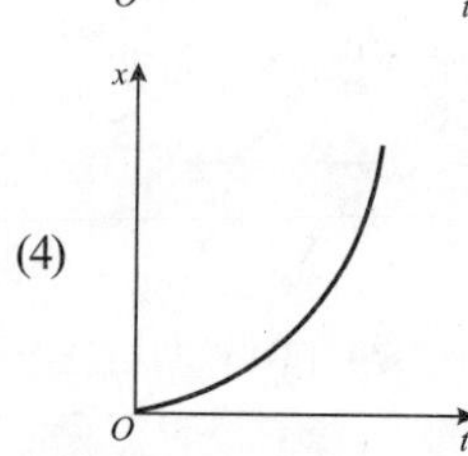

(A) (1), (2), (3) (B) (1)

(C) (1), (2), (4) (D) (2), (3)

12. Let $|\vec{A}_1| = 3$, $|\vec{A}_2| = 5$ and $|\vec{A}_1 + \vec{A}_2| = 5$. The value of $(2\vec{A}_1 + 3\vec{A}_2)\cdot(3\vec{A}_1 - 2\vec{A}_2)$ is : **[8 April 2019 (Evening)]**

(A) -112.5 (B) -106.5

(C) -118.5 (D) -99.5

13. A ball is thrown vertically up (taken as + z-axis) from the ground. The correct momentum-height (p-h) diagram is : **[9 April 2019 (Morning)]**

(A) 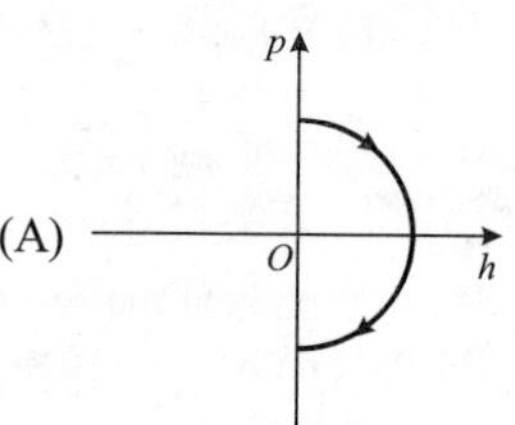

(B) 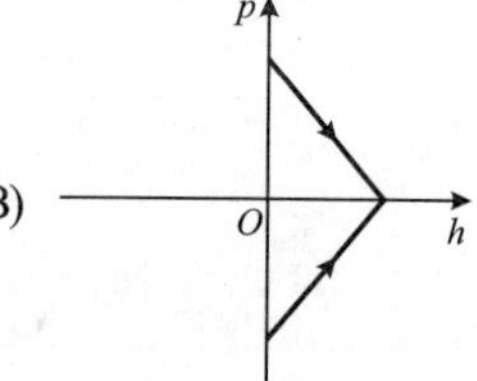

(C) 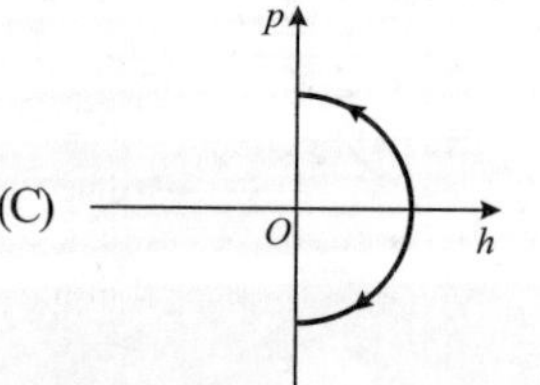

(D) 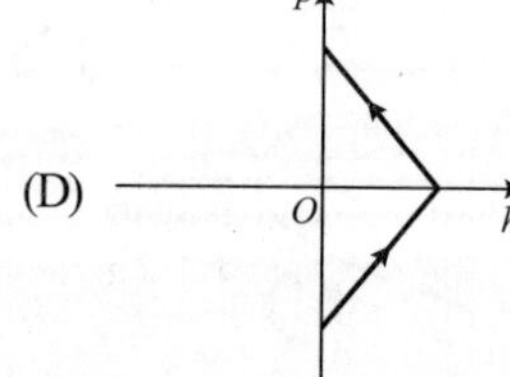

14. The position of a particle as a function of time t, is given by $x(t) = at + bt^2 - ct^3$ where a, b and c are constants. When the particle attains zero acceleration, then its velocity will be : **[9 April 2019 (Evening)]**

(A) $a+\frac{b^2}{4c}$ (B) $a+\frac{b^2}{c}$

(C) $a+\frac{b^2}{2c}$ (D) $a+\frac{b^2}{3c}$

15. A particle is moving with speed $v = b\sqrt{x}$ along positive x-axis. Calculate the speed of the particle at time $t = \tau$ (assume that the particle is at origin at $t = 0$) : **[12 April 2019 (Evening)]**

(A) $\frac{b^2\tau}{4}$ (B) $\frac{b^2\tau}{2}$

(C) $b^2\tau$ (D) $\frac{b^2\tau}{\sqrt{2}}$

16. A ball is dropped from the top of a 100 m high tower on a planet. In the last $\frac{1}{2}s$ before hitting the ground, it covers a distance of 19 m. Acceleration due to gravity (in ms^{-2}) near the surface on that planet is ______. **[8 Jan. 2020 (Evening)]**

17. The distance x covered by a particle in one dimensional motion varies with time t as $x^2 = at^2 + 2bt + c$. If the acceleration of the particle depends on x as x^{-n}, where n is an integer, the value of n is ______. **[9 Jan. 2020 (Morning)]**

18. The speed verses time graph for a particle is shown in the figure. The distance travelled (in m) by the particle during the time interval $t = 0$ to $t = 5$ s will be ______ : **[4 Sept. 2020 (Evening)]**

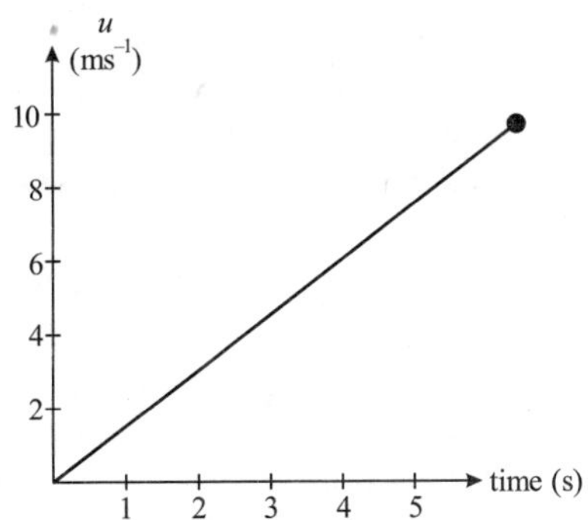

19. A balloon is moving up in air vertically above a point A on the ground. When it is at a height h_1, a girl standing at a distance d (point B) from A (see figure) sees it at an angle 45° with respect to the vertical. When the balloon climbs up a further height h_2, it is seen at an angle 60° with respect to the vertical if the girl moves further by a distance 2.464 d (point C). Then the height h_2 is (given tan 30° = 0.5774) : **[5 Sept. 2020 (Morning)]**

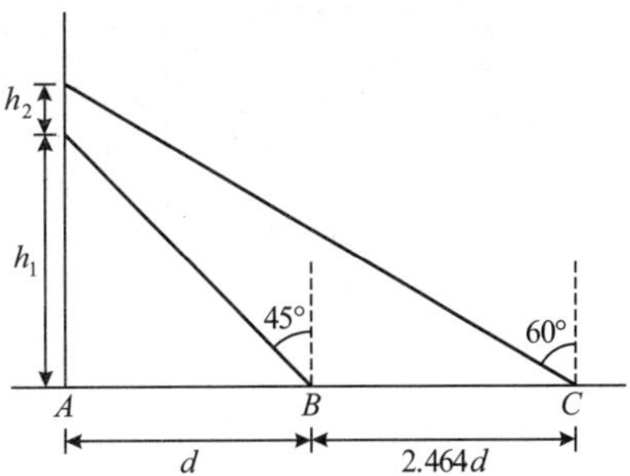

(A) d (B) 0.732 d

(C) 1.464 d (D) 0.464 d

20. A helicopter rises from rest on the ground vertically upwards with a constant acceleration g. A food packet is dropped from the helicopter when it is a height h. The time taken by the packet to reach the ground is close to [g is the acceleration due to gravity] : **[5 Sept. 2020 (Morning)]**

(A) $t=\sqrt{\frac{2h}{3g}}$ (B) $t=1.8\sqrt{\frac{h}{g}}$

(C) $t=3.4\sqrt{\left(\frac{h}{g}\right)}$ (D) $t=\frac{2}{3}\sqrt{\left(\frac{h}{g}\right)}$

21. The velocity (v) and time (t) graph of a body in a straight line motion is shown in the figure. The point S is at 4.333 seconds. The total distance covered by the body in 6s is : **[5 Sept. 2020 (Evening)]**

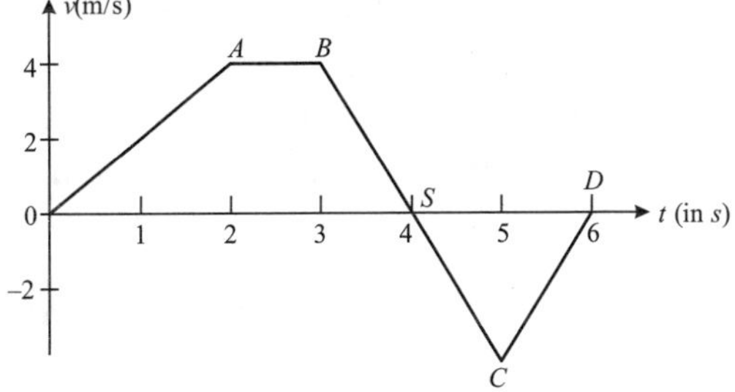

(A) 12m (B) $\frac{49}{4}$ m

(C) 11m (D) $\frac{37}{3}$ m

22. If the velocity-time graph has the shape AMB, what would be the shape of the corresponding acceleration-time graph ?

[24 Feb. 2021 (Morning)]

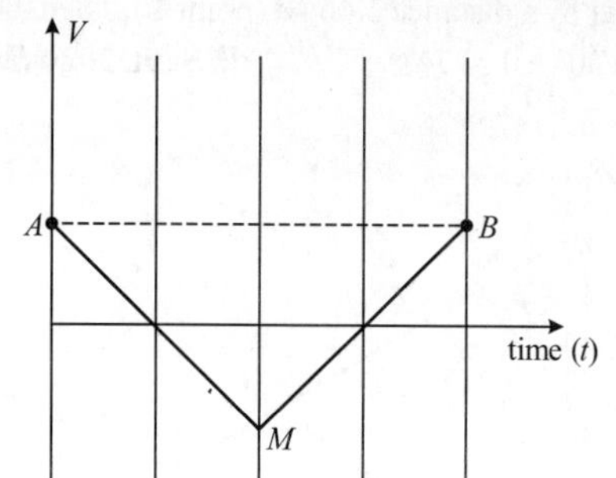

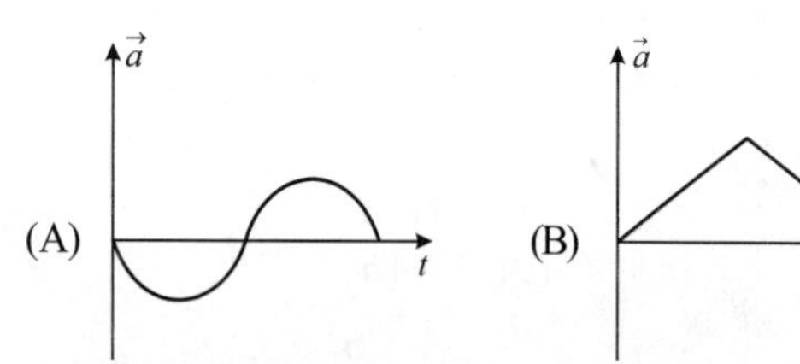

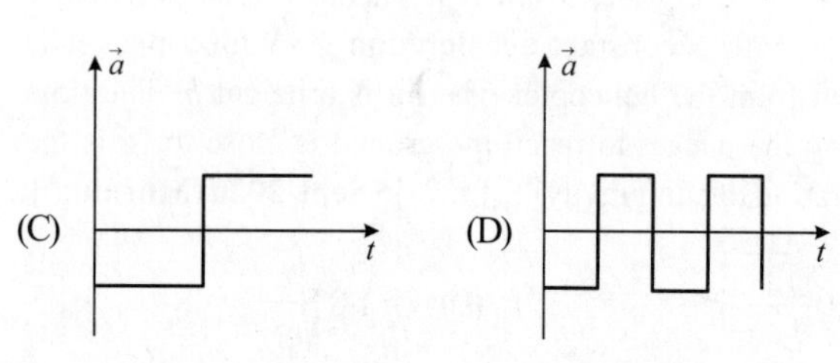

23. An engine of a train, moving with uniform acceleration, passes the signal-post with velocity u and the last compartment with velocity v. The velocity with which middle point of the train passes the signal post is : **[25 Feb. 2021 (Morning)]**

(A) $\sqrt{\dfrac{v^2+u^2}{2}}$ (B) $\sqrt{\dfrac{v^2-u^2}{2}}$

(C) $\dfrac{v-u}{2}$ (D) $\dfrac{u+v}{2}$

24. A stone is dropped from the top of a building. When it crosses a point 5 m below the top, another stone starts to fall from a point 25 m below the top. Both stones reach the bottom of building simultaneously. The height of the building is :

[25 Feb. 2021 (Evening)]

(A) 35 m (B) 45 m

(C) 50 m (D) 25 m

25. A scooter accelerates from rest for time t_1 at constant rate a_1 and then retards at constant rate a_2 for time t_2 and comes to rest. The correct value of $\dfrac{t_1}{t_2}$ will be : **[26 Feb. 2021 (Evening)]**

(A) $\dfrac{a_1}{a_2}$ (B) $\dfrac{a_2}{a_1}$

(C) $\dfrac{a_1+a_2}{a_2}$ (D) $\dfrac{a_1+a_2}{a_1}$

26. The velocity-displacement graph describing the motion of a bicycle is shown in the figure. **[16 March 2021 (Morning)]**

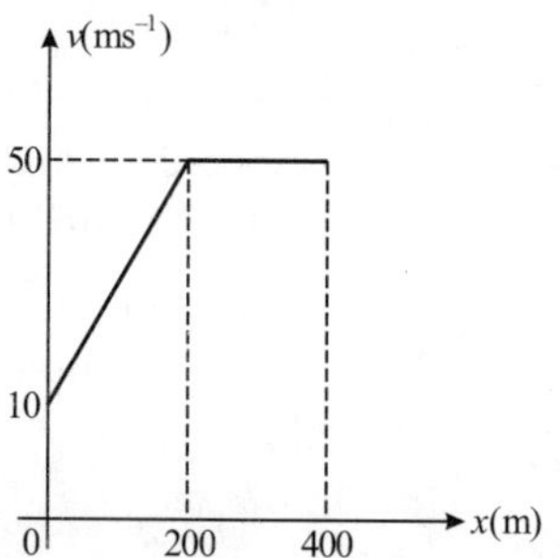

The acceleration-displacement graph of the bicycle's motion is best described by :

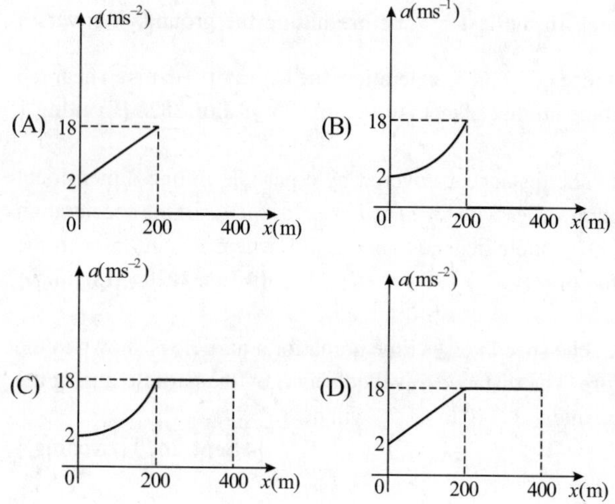

27. A body of mass 2 kg moves under a force of $(2\hat{i}+3\hat{j}+5\hat{k})$N. It starts from rest and was at the origin initially. After 4 s, its new coordinates are (8, b, 20). The value of b is ______.

(Round off to the Nearest Integer) **[16 March 2021 (Evening)]**

28. A car accelerates from rest at a constant rate α for some time after which it decelerates at a constant rate β to come to rest. If the total time elapsed is t seconds, the total distance travelled is : **[17 March 2021 (Morning)]**

(A) $\dfrac{4\alpha\beta}{(\alpha+\beta)}t^2$ (B) $\dfrac{2\alpha\beta}{(\alpha+\beta)}t^2$

(C) $\dfrac{\alpha\beta}{2(\alpha+\beta)}t^2$ (D) $\dfrac{\alpha\beta}{4(\alpha+\beta)}t^2$

29. The velocity of a particle is $v = v_0 + gt + Ft^2$. Its position is $x = 0$ at $t = 0$; then its displacement after time ($t = 1$) is :
[17 March 2021 (Evening)]

(A) $v_0 + g + F$ (B) $v_0 + \frac{g}{2} + \frac{F}{3}$

(C) $v_0 + \frac{g}{2} + F$ (D) $v_0 + 2g + 3F$

30. The position, velocity and acceleration of a particle moving with a constant acceleration can be represented by :
[18 March 2021 (Morning)]

(A)
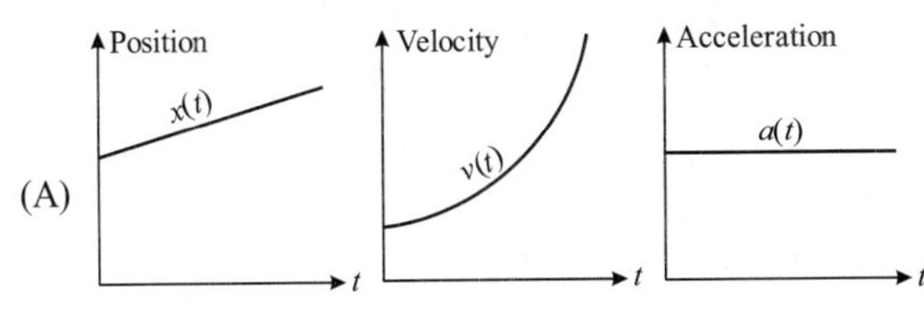

(B)
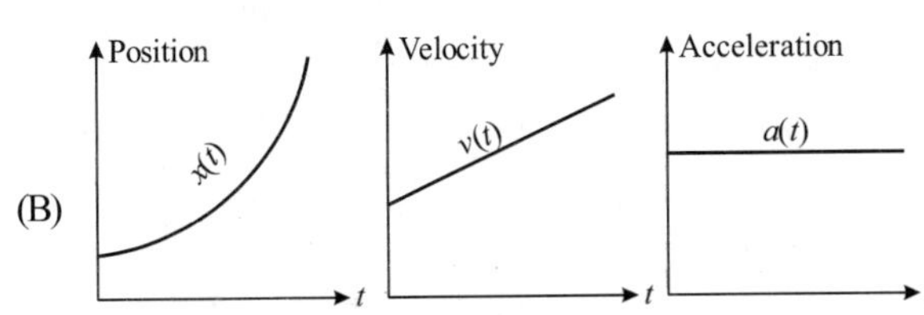

(C)
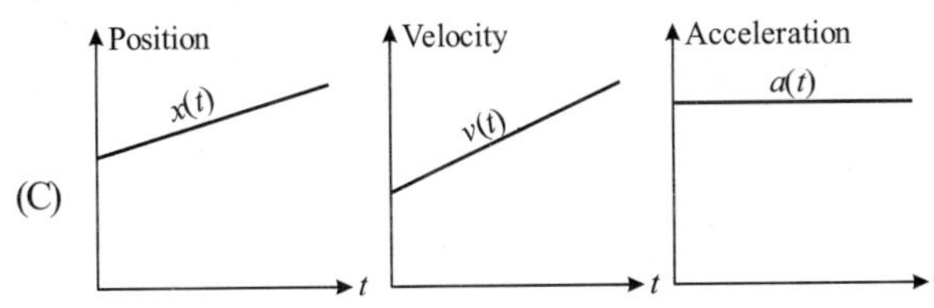

(D)
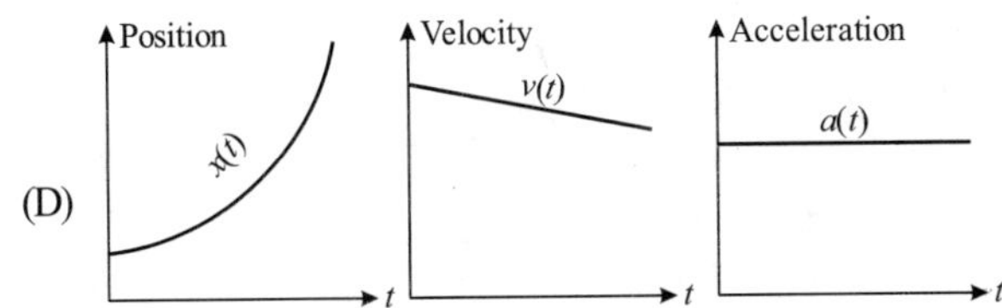

31. The velocity-displacement graph of a particle is shown in the figure.

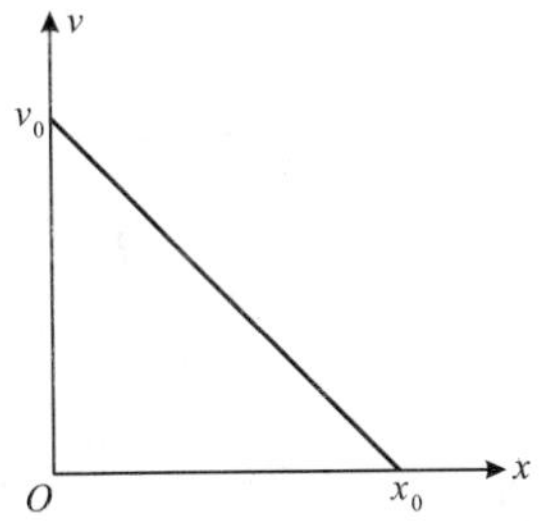

The acceleration-displacement graph of the same particle is represented by :
[18 March 2021 (Evening)]

(A)
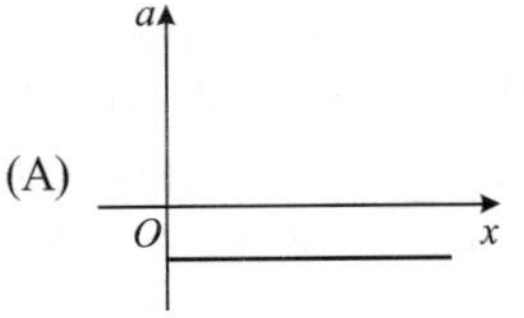

(B)
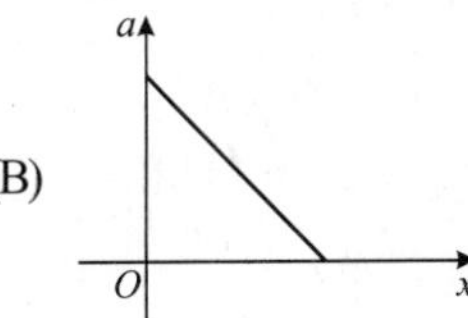

(C)
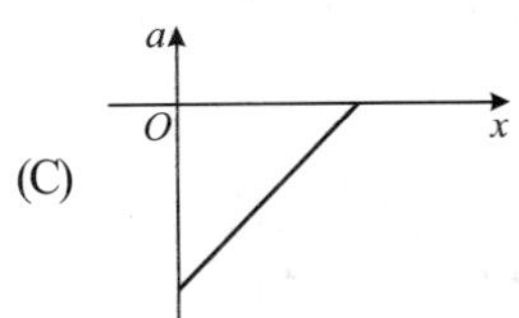

(D)
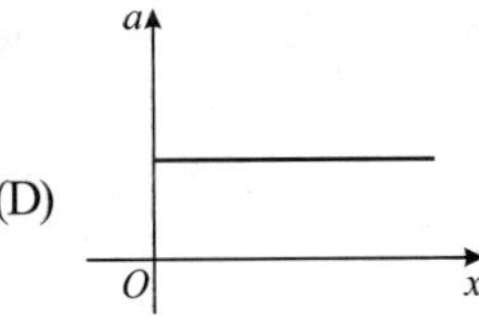

32. Water droplets are coming from an open tap at a particular rate. The spacing between a droplets observed at 4th second after its fall to the next droplet is 34.3 m. At what rate the droplets are coming from the tap ? (Take $g = 9.8$ m/s^2)
[25 July 2021 (Morning)]

(A) 1 drop/second (B) 1 drop/7 seconds
(C) 2 drops/second (D) 3 drops/2 seconds

33. The relation between time t and distance x for a moving body is given as $t = mx^2 + nx$, where m and n are constants. The retardation of the motion is : (Where v stands for velocity)
[25 July 2021 (Evening)]

(A) $2\,mnv^3$ (B) $2\,nv^3$
(C) $2\,mv^3$ (D) $2\,n^2v^3$

34. The instantaneous velocity of a particle moving in a straight line is given as $v = \alpha t + \beta t^2$, where α and β are constants. The distance travelled by the particle between 1 s and 2 s is :
[25 July 2021 (Evening)]

(A) $\frac{\alpha}{2} + \frac{\beta}{3}$ (B) $\frac{3}{2}\alpha + \frac{7}{3}\beta$

(C) $\frac{3}{2}\alpha + \frac{7}{2}\beta$ (D) $3\alpha + 7\beta$

35. A balloon was moving upwards with a uniform velocity of 10 m/s. An object of finite mass is dropped from the balloon when it was at a height of 75 m from the ground level. The height of the balloon from the ground when object strikes the ground was around. (Takes the value of g as 10 m/s^2)
[25 July 2021 (Evening)]

(A) 125 m (B) 200 m
(C) 300 m (D) 250 m

36. A ball is thrown up with a certain velocity so that it reaches a height h. Find the ratio of the two different times of the ball reaching $\frac{h}{3}$ in both the directions : **[27 July 2021 (Morning)]**

(A) $\frac{\sqrt{2}-1}{\sqrt{2}+1}$ (B) $\frac{1}{3}$

(C) $\frac{\sqrt{3}-\sqrt{2}}{\sqrt{3}+\sqrt{2}}$ (D) $\frac{\sqrt{3}-1}{\sqrt{3}+1}$

37. Two spherical balls having equal masses with radius of 5 cm each are thrown upwards along the same vertical direction at an interval of 3s with the same initial velocity of 35 m/s, then these balls collide at a height of_____ m. (Take g = 10 m/s^2)

[26 Aug. 2021 (Morning)]

38. If the velocity of a body related to displacement x is given by $v = \sqrt{5000 + 24x}$ m/s, then calculate the acceleration of the body in m/s^2. **[27 Aug. 2021 (Morning)]**

39. Water drops are falling from a nozzle of a shower onto the floor from a height of 9.8 m. The drops fall at a regular interval of time. When the first drop strikes the floor, at that instant, the third drop begins to fall. Locate the position of second drop from the floor when the first drop strikes the floor :

[27 Aug. 2021 (Evening)]

(A) 2.94 m (B) 4.18 m

(C) 2.45 m (D) 7.35 m

40. A particle is moving with constant acceleration 'a'. Following graph shows v^2 versus x (displacement) plot. The acceleration of the particle is _____ m/s^2 :

[31 Aug. 2021 (Evening)]

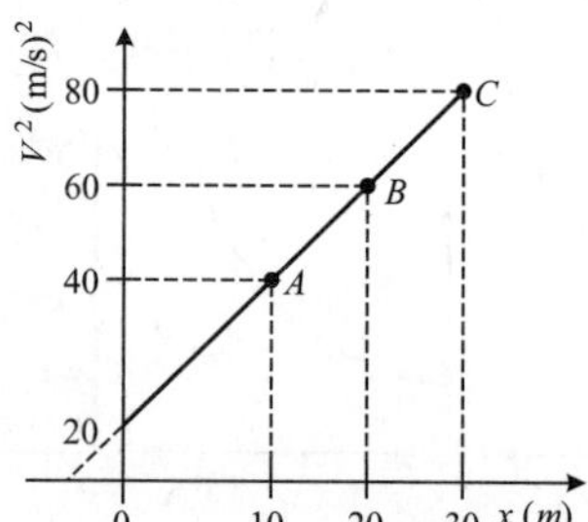

* * * * *

CHAPTER

3 MOTION IN A PLANE

3.1 MOTION IN TWO DIMENSIONS

1. Let $\vec{A} = (\hat{i}+\hat{j})$ and $\vec{B} = (2\hat{i}-\hat{j})$. The magnitude of a coplanar vector $\vec{C}$ such that $\vec{A}\cdot\vec{C} = \vec{B}\cdot\vec{C} = \vec{A}\cdot\vec{B}$ is given by: **[2018 (Online)]**

(A) $\sqrt{\frac{5}{9}}$ (B) $\sqrt{\frac{10}{9}}$

(C) $\sqrt{\frac{20}{9}}$ (D) $\sqrt{\frac{9}{12}}$

2. A particle is moving with a velocity $\vec{v} = K(y\hat{i}+x\hat{j})$, where K is a constant. The general equation for its path is : **[9 Jan. 2019 (Morning)]**

(A) xy = constant (B) $y^2 = x^2$ + constant

(C) $y = x^2$ + constant (D) $y^2 = x$ + constant

3. The position co-ordinates of a particle moving in a 3-D coordinate system is given by

$x = a\cos\omega t$

$y = a\sin\omega t$

and $z = a\omega t$

The speed of the particle is : **[9 Jan. 2019 (Evening)]**

(A) $a\omega$ (B) $\sqrt{3}\,a\omega$

(C) $\sqrt{2}\,a\omega$ (D) $2a\omega$

4. Two guns A and B can fire bullets at speeds 1 km/s and 2 km/s respectively. From a point on a horizontal ground, they are fired in all possible directions. The ratio of maximum areas covered by the bullets fired by the two guns, on the ground is : **[10 Jan. 2019 (Morning)]**

(A) 1 : 2 (B) 1 : 4

(C) 1 : 8 (D) 1 : 16

5. In the cube of side 'a' shown in the figure, the vector from the central point of the face $ABOD$ to the central point of the face $BEFO$ will be : **[10 Jan. 2019 (Morning)]**

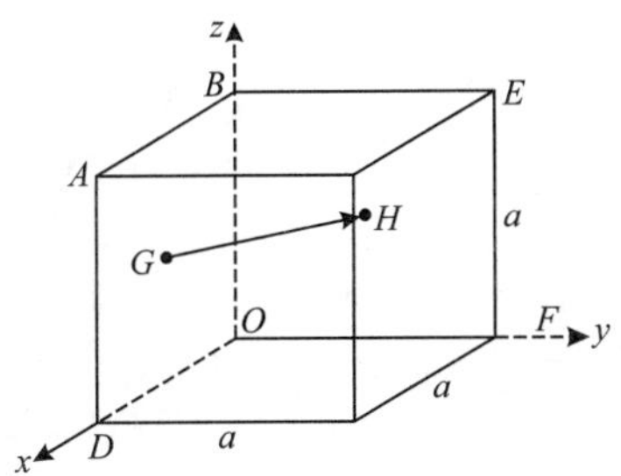

(A) $\frac{1}{2}a(\hat{i}-\hat{k})$ (B) $\frac{1}{2}a(\hat{j}-\hat{i})$

(C) $\frac{1}{2}a(\hat{k}-\hat{i})$ (D) $\frac{1}{2}a(\hat{j}-\hat{k})$

6. Two forces P and Q of magnitude $2F$ and $3F$, respectively, are at an angle θ with each other. If the force Q is doubled, then their resultant also gets doubled. Then, the angle is : **[10 Jan. 2019 (Evening)]**

(A) 30° (B) 60°

(C) 90° (D) 120°

7. Two vectors $\vec{A}$ and $\vec{B}$ have equal magnitudes. The magnitude of $(\vec{A}+\vec{B})$ is 'n' times the magnitude of $(\vec{A}-\vec{B})$. The angle between $\vec{A}$ and $\vec{B}$ is : **[10 Jan. 2019 (Evening)]**

(A) $\sin^{-1}\left[\frac{n^2-1}{n^2+1}\right]$ (B) $\cos^{-1}\left[\frac{n-1}{n+1}\right]$

(C) $\cos^{-1}\left[\frac{n^2-1}{n^2+1}\right]$ (D) $\sin^{-1}\left[\frac{n-1}{n+1}\right]$

8. A body is projected at $t = 0$ with a velocity 10 ms^{-1} at an angle of 60° with the horizontal. The radius of curvature of its trajectory at t = 1s is R. Neglecting air resistance and taking acceleration due to gravity g = 10 ms^{-2}, the value of R is : **[11 Jan. 2019 (Morning)]**

(A) 2.5 m (B) 10.3 m

(C) 2.8 m (D) 5.1 m

9. The position vector of a particle changes with time according to the relation $\vec{r}(t) = 15t^2\hat{i}+(4-20t^2)\hat{j}$. What is the magnitude of the acceleration at $t = 1$? **[9 April 2019 (Evening)]**

(A) 40 (B) 100

(C) 25 (D) 50

10. A plane is inclined at an angle α = 30° with a respect to the horizontal. A particle is projected with a speed u = 2 ms^{-1} from the base of the plane, making an angle θ = 15° with respect to the plane as shown in the figure. The distance from the base, at which the particle hits the plane is close to :
(Take g = 10 ms^{-2}) **[10 April 2019 (Evening)]**

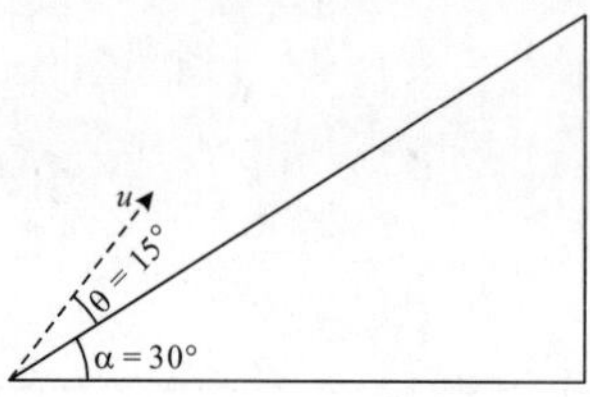

(A) 14 cm (B) 20 cm
(C) 18 cm (D) 26 cm

11. The trajectory of a projectile near the surface of the earth is given as $y = 2x - 9x^2$. If it were launched at an angle θ_0 with speed v_0 then ($g = 10\ \text{ms}^{-2}$) : **[12 April 2019 (Morning)]**

(A) $\theta_0 = \cos^{-1}\left(\frac{1}{\sqrt{5}}\right)$ and $v_0 = \frac{5}{3}\text{ms}^{-1}$

(B) $\theta_0 = \sin^{-1}\left(\frac{1}{\sqrt{5}}\right)$ and $v_0 = \frac{5}{3}\text{ms}^{-1}$

(C) $\theta_0 = \sin^{-1}\left(\frac{2}{\sqrt{5}}\right)$ and $v_0 = \frac{3}{5}\text{ms}^{-1}$

(D) $\theta_0 = \cos^{-1}\left(\frac{2}{\sqrt{5}}\right)$ and $v_0 = \frac{3}{5}\text{ms}^{-1}$

12. A shell is fired from a fixed artillery gun with an initial speed u such that it hits the target on the ground at a distance R from it. If t_1 and t_2 are the values of the time taken by it to hit the target in two possible ways, the product $t_1 t_2$ is :
[12 April 2019 (Morning)]

(A) R/g (B) $R/4g$
(C) $2R/g$ (D) $R/2g$

13. Two particles are projected from the same point with the same speed u such that they have the same range R, but different maximum heights, h_1 and h_2. Which of the following is correct?
[12 April 2019 (Evening)]

(A) $R^2 = 2h_1h_2$ (B) $R^2 = 16h_1h_2$
(C) $R^2 = 4h_1h_2$ (D) $R^2 = h_1h_2$

14. A particle is moving along the x-axis with its coordinate with the time 't' given be $x(t) = 10 + 8t - 3t^2$. Another particle is moving the y-axis with its coordinate as a function of time given by $y(t) = 5 - 8t^3$. At $t = 1$s, the speed of the second particle as measured in the frame of the first particle is given as $\sqrt{v}$. Then v (in m/s) is _____. **[8 Jan. 2020 (Morning)]**

15. A particle moves such that its position vector $\vec{r}(t) = \cos\omega t\hat{i} + \sin\omega t\hat{j}$ where ω is a constant and t is time. Then which of the following statements is true for the velocity $\vec{v}(t)$ and acceleration $\vec{a}(t)$ of the particle : **[8 Jan. 2020 (Evening)]**

(A) $\vec{v}$ is perpendicular to $\vec{r}$ and $\vec{a}$ is directed towards the origin

(B) $\vec{v}$ and $\vec{a}$ both are parallel to $\vec{r}$

(C) $\vec{v}$ and $\vec{a}$ both are perpendicular to $\vec{r}$

(D) $\vec{v}$ is perpendicular to $\vec{r}$ and $\vec{a}$ is directed away from the origin

16. A particle starts from the origin at $t = 0$ with an initial velocity of $3.0\,\hat{i}$ m/s and moves in the x-y plane with a constant acceleration $(6.0\,\hat{i} + 4.0\,\hat{j})$ m/s². The x-coordinate of the particle at the instant when its y-coordinate is 32 m is D meters. The value of D is : **[9 Jan. 2020 (Evening)]**

(A) 50 (B) 32
(C) 60 (D) 40

17. Starting from the origin at time $t = 0$, with initial velocity $5\hat{j}$ ms^{-1}, a particle moves in the x-y plane with a constant acceleration of $(10\hat{i} + 4\hat{j})$ ms^{-2}. At time t, its coordinates are (20 m, y_0m). The values of t and y_0, are respectively :
[4 Sept. 2020 (Morning)]

(A) 4 s and 52 m (B) 2 s and 24 m
(C) 2 s and 18 m (D) 5 s and 25 m

18. In an octagon $ABCDEFGH$ of equal side, what is the sum of $\overrightarrow{AB} + \overrightarrow{AC} + \overrightarrow{AD} + \overrightarrow{AE} + \overrightarrow{AF} + \overrightarrow{AG} + \overrightarrow{AH}$, if, $\overrightarrow{AO} = 2\hat{i} + 3\hat{j} - 4\hat{k}$: **[25 Feb. 2021 (Morning)]**

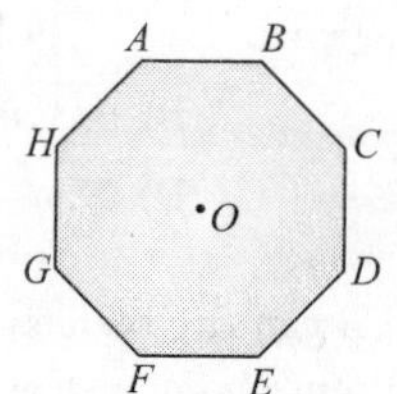

(A) $-16\hat{i} - 24\hat{j} + 32\hat{k}$ (B) $16\hat{i} + 24\hat{j} - 32\hat{k}$
(C) $16\hat{i} + 24\hat{j} + 32\hat{k}$ (D) $16\hat{i} - 24\hat{j} + 32\hat{k}$

19. An electron with kinetic energy K_1 enters between parallel plates of a capacitor at an angle 'α' with the plates. It leaves the plates at angle 'β' with kinetic energy K_2. Then the ratio of kinetic energies $K_1 : K_2$ will be : **[25 Feb. 2021 (Evening)]**

(A) $\frac{\cos\beta}{\sin\alpha}$ (B) $\frac{\cos\beta}{\cos\alpha}$

(C) $\frac{\sin^2\beta}{\cos^2\alpha}$ (D) $\frac{\cos^2\beta}{\cos^2\alpha}$

20. If $\vec{P}\times\vec{Q} = \vec{Q}\times\vec{P}$, the angle between $\vec{P}$ and $\vec{Q}$ is $\theta (0° < \theta < 360°)$. The value of '$\theta$' will be _____°.
[25 Feb. 2021 (Evening)]

21. The trajectory of a projectile in a vertical plane is $y = \alpha x - \beta x^2$, where α and β are constants and x & y are respectively the horizontal and vertical distances of the projectile from the point of projection. The angle of projection θ and the maximum height attained H are respectively given by : **[26 Feb. 2021 (Evening)]**

(A) $\tan^{-1}\alpha, \frac{4\alpha^2}{\beta}$ (B) $\tan^{-1}\left(\frac{\beta}{\alpha}\right), \frac{\alpha^2}{\beta}$

(C) $\tan^{-1}\alpha, \frac{\alpha^2}{4\beta}$ (D) $\tan^{-1}\beta, \frac{\alpha^2}{2\beta}$

22. Consider a frame that is made up of two thin massless rods AB and AC as shown in the figure. A vertical force $\vec{P}$ of magnitude 100 N is applied at point A of the frame.

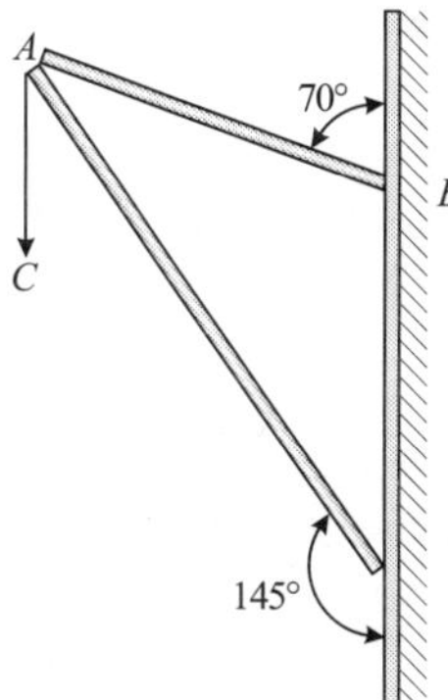

Suppose the force is $\vec{P}$ resolved parallel to the arms AB and AC of the frame. The magnitude of the resolved component along the arm AC is xN. The value of x, to the nearest integer, is _____. [**Given** : sin(35°) = 0.573, cos(35°) = 0.819; sin(110°) = 0.939, cos(110°) = –0.342] **[16 March 2021 (Morning)]**

23. A mosquito is moving with a velocity $\vec{v} = 0.5t^2\hat{i} + 3t\hat{j} + 9\hat{k}$ m/s and accelerating in uniform conditions. What will be the direction of mosquito after 2s ? **[16 March 2021 (Evening)]**

(A) $\tan^{-1}\left(\frac{2}{3}\right)$ from x-axis (B) $\tan^{-1}\left(\frac{2}{3}\right)$ from y-axis

(C) $\tan^{-1}\left(\frac{5}{2}\right)$ from y-axis (D) $\tan^{-1}\left(\frac{5}{2}\right)$ from x-axis

24. A person is swimming with a speed of 10 m/s at an angle of 120° with the flow and reaches to a point directly opposite on the other side of the river. The speed of the flow is 'x' m/s. The value of 'x' to the nearest integer is _____.

[18 March 2021 (Morning)]

25. The projectile motion of a particle of mass 5 g is shown in the figure.

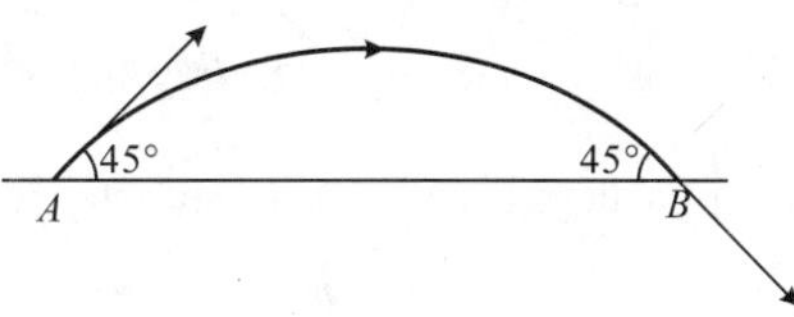

The initial velocity of the particle is $5\sqrt{2}\,\text{ms}^{-1}$ and the air resistance is assumed to be negligible. The magnitude of the change in momentum between the points A and B is $x \times 10^{-2}$ kgms^{-1}. The value of x, to the nearest integer, is _____.

[18 March 2021 (Evening)]

26. If $\vec{A}$ and $\vec{B}$ are two vectors satisfying the relation

$$\vec{A}\cdot\vec{B} = |\vec{A}\times\vec{B}|$$

Then the value of $|\vec{A}-\vec{B}|$ will be :

[20 July 2021 (Morning)]

(A) $\sqrt{A^2+B^2+2AB}$ (B) $\sqrt{A^2+B^2-\sqrt{2}AB}$

(C) $\sqrt{A^2+B^2}$ (D) $\sqrt{A^2+B^2+\sqrt{2}AB}$

27. Two vectors $\vec{P}$ and $\vec{Q}$ have equal magnitudes. If the magnitude of $\vec{P}+\vec{Q}$ is n times the magnitude of $\vec{P}-\vec{Q}$, then angle between $\vec{P}$ and $\vec{Q}$ is : **[20 July 2021 (Evening)]**

(A) $\cos^{-1}\left(\frac{n-1}{n+1}\right)$ (B) $\sin^{-1}\left(\frac{n-1}{n+1}\right)$

(C) $\cos^{-1}\left(\frac{n^2-1}{n^2+1}\right)$ (D) $\sin^{-1}\left(\frac{n^2-1}{n^2+1}\right)$

28. What will be the projection of vector $\vec{A} = \hat{i}+\hat{j}+\hat{k}$ on vector $\vec{B} = \hat{i}+\hat{j}$? **[22 July 2021 (Evening)]**

(A) $2(\hat{i}+\hat{j}+\hat{k})$ (B) $\sqrt{2}(\hat{i}+\hat{j})$

(C) $(\hat{i}+\hat{j})$ (D) $\sqrt{2}(\hat{i}+\hat{j}+\hat{k})$

29. Three particles P, Q and R moving along the vectors $\vec{A} = \hat{i}+\hat{j}$, $\vec{B} = \hat{j}+\hat{k}$ and $\vec{C} = -\hat{i}+\hat{j}$ respectively. They strike on a point and start to move in different directions. Now particle P is moving normal to the plane which contains vector $\vec{A}$ and $\vec{B}$. Similarly particle Q is moving normal to the plane which contains vector $\vec{A}$ and $\vec{C}$. The angle between the direction of motion of

P and Q is $\cos^{-1}\left(\frac{1}{\sqrt{x}}\right)$. Then the value of x is _____.

[22 July 2021 (Evening)]

30. Match List-I with List-II : **[25 July 2021 (Morning)]**

(a) $\vec{C}-\vec{A}-\vec{B}=0$ (i)

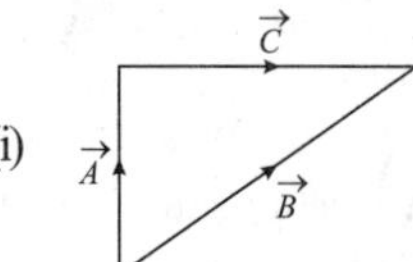

(b) $\vec{A}-\vec{C}-\vec{B}=0$ (ii)

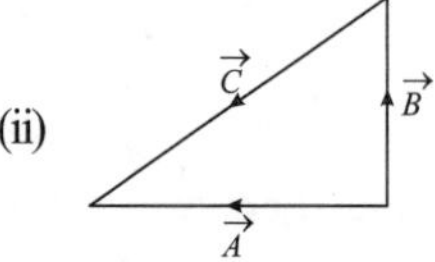

(c) $\vec{B}-\vec{A}-\vec{C}=0$ (iii)

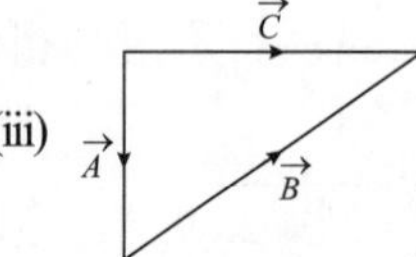

(d) $\vec{A}+\vec{B}=-\vec{C}$ (iv)

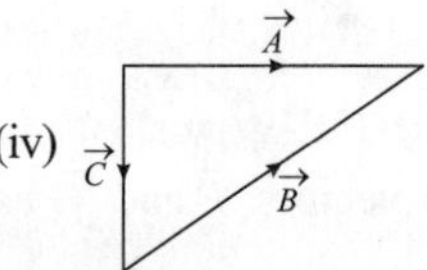

Choose the correct answer from the options given below :

(A) (a)→(iii), (b)→(ii), (c)→(iv), (d)→(i)

(B) (a)→(iv), (b)→(iii), (c)→(i), (d)→(ii)

(C) (a)→(i), (b)→(iv), (c)→(ii), (d)→(iii)

(D) (a)→(iv), (b)→(i), (c)→(iii), (d)→(ii)

31. Two vectors $\vec{X}$ and $\vec{Y}$ have equal magnitude. The magnitude of $(\vec{X}+\vec{Y})$ is n times the magnitude of $(\vec{X}+\vec{Y})$. The angle between $\vec{X}$ and $\vec{Y}$ is : **[25 July 2021 (Evening)]**

(A) $\cos^{-1}\left(\frac{n^2+1}{-n^2-1}\right)$ (B) $\cos^{-1}\left(\frac{n^2-1}{-n^2-1}\right)$

(C) $\cos^{-1}\left(\frac{-n^2-1}{n^2-1}\right)$ (D) $\cos^{-1}\left(\frac{n^2+1}{n^2-1}\right)$

32. A force $\vec{F}=(40\hat{i}+10\hat{j})N$ acts N acts on a body of mass 5 kg. If the body starts from rest, its position vector $\vec{r}$ at time $t = 10$ s, will be : **[25 July 2021 (Evening)]**

(A) $(100\hat{i}+100\hat{j})$ (B) $(400\hat{i}+400\hat{j})$m

(C) $(100\hat{i}+400\hat{j})$m (D) $(400\hat{i}+100\hat{j})$m

33. Assertion-A : If A, B, C, D are four points on a semi-circular arc with centre at 'O' such that $|\overrightarrow{AB}| = |\overrightarrow{BC}| = |\overrightarrow{CD}|$, then $\overrightarrow{AB}+\overrightarrow{AC}+\overrightarrow{AD} = 4\overrightarrow{AO}+\overrightarrow{OB}+\overrightarrow{OC}$

Reason-R : Polygon law of vector addition yields

$$\overrightarrow{AB}+\overrightarrow{BC}+\overrightarrow{CD} = \overrightarrow{AD} = 2\overrightarrow{AO}$$

[27 July 2021 (Morning)]

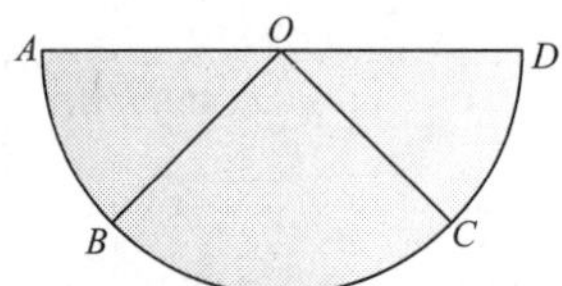

In the light of the above statements, choose the most appropriate answer from the options given below :

(A) A is not correct but R is correct

(B) Both A and R are correct but R is not the correct explanation of A

(C) A is correct but R is not correct

(D) Both A and R are correct and R is the correct explanation of A

34. A swimmer wants to cross a river from point A to point B. Line AB makes an angle of 30° with the flow of river. Magnitude of velocity of the swimmer is same as that of the river. The angle θ with the line AB should be _____°, so that the swimmer reaches point B. **[27 July 2021 (Evening)]**

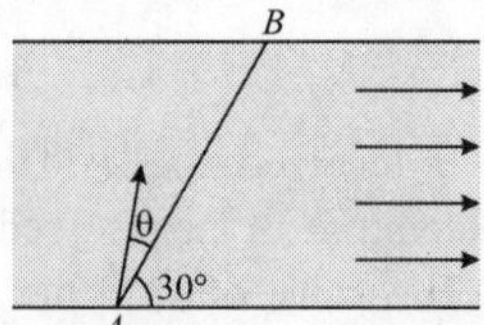

35. The magnitude of vector $\overrightarrow{OA}$, $\overrightarrow{OB}$ and $\overrightarrow{OC}$ in the given figure are equal. The direction of $\overrightarrow{OA}+\overrightarrow{OB}-\overrightarrow{OC}$ with x-axis will be : **[26 Aug. 2021 (Morning)]**

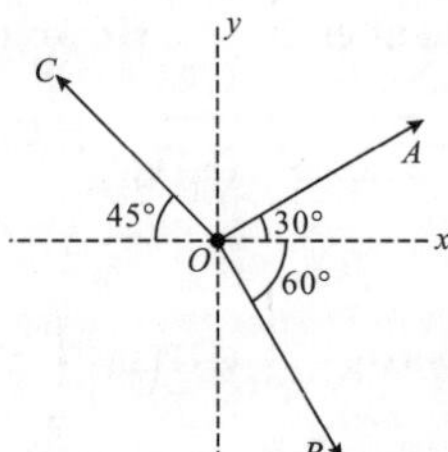

(A) $\tan^{-1}\frac{(1+\sqrt{3}-\sqrt{2})}{(1-\sqrt{3}-\sqrt{2})}$ (B) $\tan^{-1}\frac{(1-\sqrt{3}-\sqrt{2})}{(1+\sqrt{3}+\sqrt{2})}$

(C) $\tan^{-1}\frac{(\sqrt{3}-1-\sqrt{2})}{(1+\sqrt{3}+\sqrt{2})}$ (D) $\tan^{-1}\frac{(\sqrt{3}-1+\sqrt{2})}{(1+\sqrt{3}-\sqrt{2})}$

36. The angle between vector $(\vec{A})$ and $(\vec{A}-\vec{B})$ is : **[26 Aug. 2021 (Evening)]**

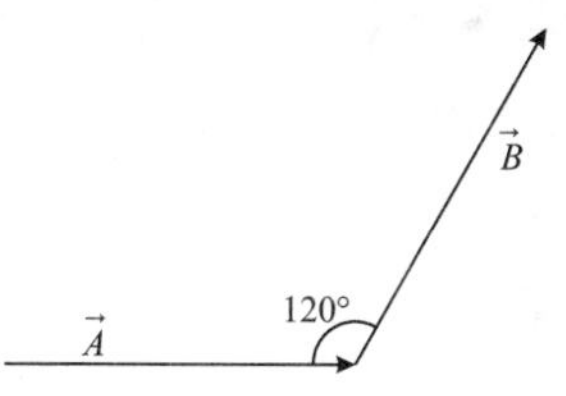

(A) $\tan^{-1}\left(\frac{A}{0.7B}\right)$ (B) $\tan^{-1}\left(\frac{-\frac{B}{2}}{A-B\frac{\sqrt{3}}{2}}\right)$

(C) $\tan^{-1}\left(\frac{\sqrt{3}B}{2A-B}\right)$ (D) $\tan^{-1}\left(\frac{B\cos\theta}{A-B\sin\theta}\right)$

37. A bomb is dropped by a fighter plane flying horizontally. To an observer sitting in the plane, the trajectory of the bomb is a : **[26 Aug. 2021 (Evening)]**

(A) Hyperbola
(B) Straight line vertically down the plane
(C) Parabola in a direction opposite to the motion of plane
(D) Parabola in the direction of motion of plane

38. The resultant of these forces $\overrightarrow{OP}$, $\overrightarrow{OQ}$, $\overrightarrow{OR}$, $\overrightarrow{OS}$ and $\overrightarrow{OT}$ is approximately N. [Take $\sqrt{3}=1.7$, $\sqrt{2}=1.4$, Given $\hat{i}$ and $\hat{j}$ unit vectors along x, y-axis]: **[27 Aug. 2021 (Morning)]**

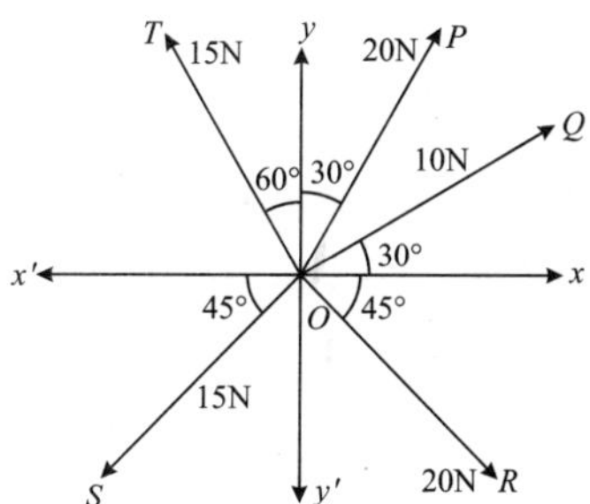

(A) $9.25\hat{i}+5\hat{j}$ (B) $3\hat{i}+15\hat{j}$

(C) $-1.5\hat{i}-15.5\hat{j}$ (D) $2.5\hat{i}-14.5\hat{j}$

39. A player kicks a football with an initial speed of 25 ms^{-1} at an angle of 45° from the ground. What are the maximum height and the time taken by the football to reach at the highest point during motion? (Take $g = 10\ ms^{-2}$) : **[27 Aug. 2021 (Evening)]**

(A) $h_{max} = 15.625$ m T = 1.77 s (B) $h_{max} = 3.54$ m T = 0.125 s
(C) $h_{max} = 10$ m T = 2.5 s (D) $h_{max} = 15.625$ m T = 3.54 s

40. A helicopter is flying horizontally with a speed v at an altitude h has to drop a food packet for a man on the ground. What is the distance of helicopter from the man when the food packet is dropped ? **[31 Aug. 2021 (Morning)]**

(A) $\sqrt{\frac{2ghv^2+1}{h^2}}$ (B) $\sqrt{2ghv^2+h^2}$

(C) $\sqrt{\frac{2gh}{v^2}+h^2}$ (D) $\tan^{-1}\frac{(\sqrt{3}-1+\sqrt{2})}{(1+\sqrt{3}-\sqrt{2})}$

41. A car is moving on a plane inclined at 30° to be horizontal with an acceleration of 10 ms^{-2} parallel to the plane upward. A bob is suspended by a string from the roof of of car. The angle in degrees which the string makes with the vertical is _____ : **[31 Aug. 2021 (Morning)]**

42. Statement-I : If three forces $\vec{F}_1, \vec{F}_2$ and $\vec{F}_3$ are represented by three sides of triangle and $\vec{F}_1+\vec{F}_2=-\vec{F}_3$, then three forces are concurrent forces and satisfy the condition for equilibrium :

Statement-II : A triangle made up of three forces $\vec{F}_1, \vec{F}_2$ and $\vec{F}_3$ as its sides taken in the same order, satisfy the condition for translatory equilibrium. In the light of the above statements, choose the most appropriate answer from the options given below : **[31 Aug. 2021 (Evening)]**

(A) Both Statement-I and Statement-II are false
(B) Both Statement-I and Statement-II are true
(C) Statement-I is false but Statement-II is true
(D) Statement-I is true but Statement-II is false

43. Statement-I : Two forces $(\vec{P}+\vec{Q})$ and $(\vec{P}-\vec{Q})$ where $\vec{P}\perp\vec{Q}$, when act at an angle θ_1 each other, the magnitude of their resultant is $\sqrt{3(P^2+Q^2)}$, when they act at an angle θ_2, then magnitude of their resultant becomes $\sqrt{2(P^2+Q^2)}$. This is possible only when $\theta_1 < \theta_2$.

Statement-II : In the situation given above.
$\theta_1 = 60°$ and $\theta_2 = 90°$

In the light of the above statement, choose the most appropriate answer from the options given below : **[31 Aug. 2021 (Evening)]**

(A) Statement-I is true but Statement-II is false
(B) Both Statement-I and Statement-II are true
(C) Statement-I is false but Statement-II is true
(D) Both Statement-I and Statement-II are false

44. The ranges and heights for two projectiles projected with the same initial velocity at angles 42° and 48° with the horizontal are R_1, R_2 and H_1, H_2 respectively. Choose the correct option :

[01 Sep. 2021 (Evening)]

(A) $R_1 > R_2$ and $H_1 = H_2$ (B) $R_1 = R_2$ and $H_1 < H_2$

(C) $R_1 < R_2$ and $H_1 < H_2$ (D) $R_1 = R_2$ and $H_1 = H_2$

* * * * *

CHAPTER

3 MOTION IN A PLANE

3.2 RELATIVE MOTION

1. A car is standing 200 m behind a bus, which is also at rest. The two start moving at the same instant but with different forward accelerations. The bus has acceleration 2 m/s^2 and the car has acceleration 4 m/s^2. The car will catch up with the bus after a time of : **[2017 (Online)]**

(A) $\sqrt{100}$ s (B) $\sqrt{120}$ s

(C) $10\sqrt{2}$ s (D) 15 s

2. Ship A is sailing towards north-east with velocity $\vec{v} = 30\hat{i} + 50\hat{j}$ km/hr where $\hat{i}$ points east and $\hat{j}$, north. Ship B is at a distance of 80 km east and 150 km north of Ship A and is sailing towards west at 10 km/hr. A will be at minimum distance from B in : **[8 April 2019 (Morning)]**

(A) 4.2 hrs. (B) 2.2 hrs.

(C) 3.2 hrs. (D) 2.6 hrs.

3. The stream of a river is flowing with a speed of 2 km/h. A swimmer can swim at a speed of 4 km/h. What should be the direction of the swimmer with respect to the flow of the river to cross the river straight ? **[9 April 2019 (Morning)]**

(A) 60° (B) 150°

(C) 90° (D) 120°

4. Train A and train B are running on parallel tracks in the opposite directions with speeds of 36 km/hour and 72 km/hour, respectively. A person is walking in train A in the direction opposite to its motion with a speed of 1.8 km/hr. Speed (in ms^{-1}) of this person as observed from train B will be close to : (take the distance between the tracks as negligible) : **[2 Sept. 2020 (Morning)]**

(A) 30.5 ms^{-1} (B) 29.5 ms^{-1}

(C) 31.5 ms^{-1} (D) 28.5 ms^{-1}

5. When a car is at rest, its driver sees rain drops falling on it vertically. When driving the car with speed v, he sees that rain drops are coming at an angle 60° from the horizontal. On further increasing the speed of the car to $(1 + \beta)v$, this angle changes to 45°. The value of b is close to : **[6 Sept. 2020 (Evening)]**

(A) 0.41 (B) 0.50

(C) 0.37 (D) 0.73

6. A swimmer can swim with velocity of 12 km/h in still water. Water flowing in a river has velocity 6 km/h. The direction with respect to the direction of flow of river water he should swim in order to reach the point on the other bank just opposite to his starting point is _____°. (Round off to the Nearest Integer) (find the angle in degree) **[16 March 2021 (Evening)]**

7. A butterfly is flying with a velocity $4\sqrt{2}$ m/s in North-East direction. Wind is slowly blowing at 1 m/s from North to South. The resultant displacement of the butterfly in 3 seconds is : **[20 July 2021 (Morning)]**

(A) $12\sqrt{2}$ m (B) 15 m

(C) 3 m (D) 20 m

8. A boy reaches the airport and finds that the escalator is not working. He walks up the stationary escalator in time t_1. If he remains stationary on a moving escalator then the escalator takes him up in time t_2. The time taken by him to walk up on the moving escalator will be : **[20 July 2021 (Evening)]**

(A) $t_2 - t_1$ (B) $\frac{t_1 t_2}{t_2 - t_1}$

(C) $\frac{t_1 t_2}{t_2 + t_1}$ (D) $\frac{t_1 + t_2}{2}$

* * * * *

CHAPTER

4 LAWS OF MOTION

4.1 NEWTON'S LAWS OF MOTION

1. A mass of 10 kg is suspended vertically by a rope from the roof. When a horizontal force is applied on the rope at some point, the rope deviated at an angle of 45° at the roof point. If the suspended mass is at equilibrium, the magnitude of the force applied is ($g = 10\ \text{ms}^{-2}$) : **[9 Jan. 2019 (Evening)]**

(A) 200 N (B) 100 N
(C) 140 N (D) 70 N

2. A particle of mass m is moving in a straight line with momentum p. Starting at time $t = 0$, a force $F = kt$ acts in the same direction on the moving particle during time interval T so that its momentum changes from p to $3p$. Here k is a constant. The value of T is : **[11 Jan. 2019 (Evening)]**

(A) $2\sqrt{\frac{p}{k}}$ (B) $\sqrt{\frac{2p}{k}}$
(C) $\sqrt{\frac{2k}{p}}$ (D) $2\sqrt{\frac{k}{p}}$

3. A mass of 10 kg is suspended by a rope of length 4 m, from the ceiling. A force F is applied horizontally at the mid-point of the rope such that the top half of the rope makes an angle of 45° with the vertical. Then F equals: (Take $g = 10\ \text{ms}^{-2}$ and the rope to be massless) : **[7 Jan. 2020 (Evening)]**

(A) 100 N (B) 90 N
(C) 75 N (D) 70 N

4. The sum of two forces $\vec{P}$ and $\vec{Q}$ is $\vec{R}$ such that $|\vec{R}| = |\vec{P}|$. The angle θ (in degrees) that the resultant of $2\vec{P}$ and $\vec{Q}$ will make with $\vec{Q}$ is,_____. **[7 Jan. 2020 (Evening)]**

5. A small ball of mass is thrown upward with velocity u from the ground. The ball experiences a resistive force mkv^2 where v is its speed. The maximum height attained by the ball is : **[4 Sept. 2020 (Evening)]**

(A) $\frac{1}{2k}\tan^{-1}\frac{ku^2}{g}en$ (B) $\frac{1}{2k}\ln\left(1+\frac{ku^2}{g}\right)$
(C) $\frac{1}{k}\tan^{-1}\frac{ku^2}{2g}$ (D) $\frac{1}{k}\ln\left(1+\frac{ku^2}{2g}\right)$

6. A spaceship in space sweeps stationary interplanetary dust. As a result, its mass increases at a rate $\frac{dM(t)}{dt} = bv^2(t)$, where $v(t)$ is its instantaneous velocity. The instantaneous acceleration of the satellite is : **[5 Sept. 2020 (Evening)]**

(A) $-\frac{2bv^3}{M(t)}$ (B) $-\frac{bv^3}{2M(t)}$
(C) $-bv^3(t)$ (D) $-\frac{bv^3}{M(t)}$

7. A particle is projected with velocity v_0 along x-axis. A damping force is acting on the particle which is proportional to the square of the distance from the origin i.e. $ma = -\alpha x^2$. The distance at which the particle stops : **[24 Feb. 2021 (Evening)]**

(A) $\left(\frac{2v_0^2}{3\alpha}\right)^{1/2}$ (B) $\left(\frac{3v_0^2}{2\alpha}\right)^{1/3}$
(C) $\left(\frac{2v_0}{3\alpha}\right)^{1/3}$ (D) $\left(\frac{3v_0^2}{2\alpha}\right)^{1/2}$

8. A person standing on a spring balance inside a stationary lift measures 60 kg. The weight of that person if the lift descends with uniform downward acceleration of 1.8 m/s² will be_____ N. [$g = 10$ m/s²]. **[26 Feb. 2021 (Morning)]**

9. A boy pushes a box of mass 2 kg with a force $\vec{F} = (20\hat{i} + 10\hat{j})$N on a frictionless surface. If the box was initially at rest, then _____ m is displacement along the x-axis after 10 s. **[26 Feb. 2021 (Morning)]**

10. A modern grand-prix racing car of mass m is travelling on a flat track in a circular arc of radius R with a speed v. If the coefficient of static friction between the tyres and the track is μ_s, then the magnitude of negative lift F_L acting downwards on the car is : (Assume forces on the four tyres are identical and g = acceleration due to gravity) **[17 March 2021 (Morning)]**

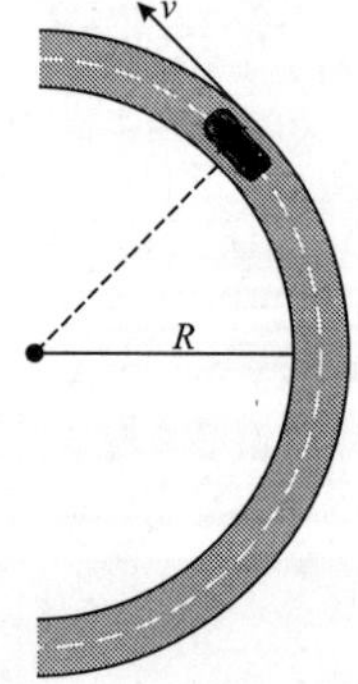

(A) $m\left(\frac{v^2}{\mu_s R}+g\right)$ (B) $m\left(\frac{v^2}{\mu_s R}-g\right)$

(C) $m\left(g-\frac{v^2}{\mu_s R}\right)$ (D) $-m\left(g+\frac{v^2}{\mu_s R}\right)$

11. A boy of mass 4 kg is standing on a piece of wood having mass 5 kg . If the coefficient of friction between the wood and the floor is 0.5, the maximum force that the boy can exert on the rope so that the piece of wood does not move from its place is ______ N. (Round off to the Nearest Integer) [Take $g = 10\ ms^{-2}$]

[17 March 2021 (Evening)]

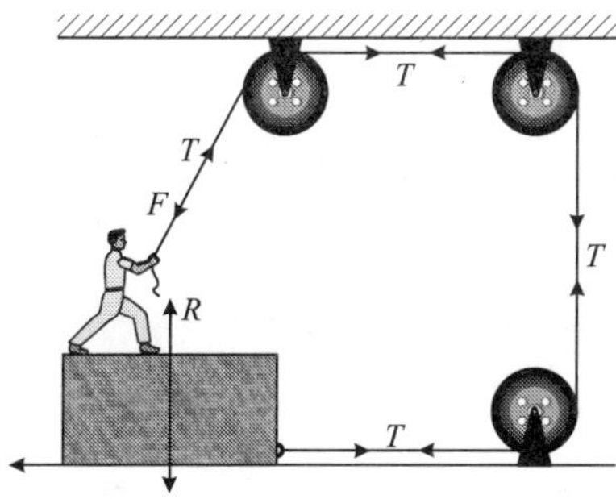

12. A bullet of mass 0.1 kg is fired on a wooden block to pierce through it, but it stops after moving a distance of 50 cm into it. If the velocity of bullet before hitting the wood is 10 m/s and it slows down with uniform deceleration, then the magnitude of effective retarding force on the bullet is 'x' N. The value of 'x' to the nearest integer is ______. **[18 March 2021 (Morning)]**

13. A steel block of 10 kg rests on horizontal floor as shown. When three iron cylinders are placed on its as shown, the block and cylinders go down with an acceleration 0.2 m/s^2. Then normal reaction R' by the floor if mass of the iron cylinders are equal and of 20 kg each, is ______ N.

(Take $g = 10$ m/s^2 and $\mu_s = 0.2$) **[20 July 2021 (Morning)]**

```
            3
          1 | 2
 ______   ______
              a = 0.2 m/s²
```

(A) 686 (B) 714
(C) 684 (D) 716

14. A particle of mass M originally at rest is subjected to a force whose direction is constant but magnitude varies with time according to the relation

$$F = F_0\left[1-\left(\frac{t-T}{T}\right)^2\right]$$

Where F_0 and T are constants. The force acts only for the time interval $2T$. The velocity v of the particle after time $2T$ is :

[27 July 2021 (Evening)]

(A) $\frac{2F_0T}{M}$ (B) $\frac{F_0T}{2M}$

(C) $\frac{4F_0T}{3M}$ (D) $\frac{F_0T}{3M}$

15. The initial mass of a rocket is 1000 kg. Calculate at what rate the fuel should be burnt so that rocket is given an acceleration of 20 ms^{-2}. The gases come out at a relative speed of 500 ms^{-1} with respect to the rocket : [Use $g = 10$ m/s^2]

[26 Aug. 2021 (Morning)]

(A) 6.0×10^2 kg s^{-1} (B) 500 kg s^{-1}
(C) 10 kg s^{-1} (D) 60 kg s^{-1}

16. Two blocks of masses 3 kg are connected by a metal wire going over a smooth pulley. The breaking stress of the metal is $\frac{24}{\pi} \times 10^2$ Nm^{-2}. What is the minimum radius of the wire ? (take $g = 10$ ms^{-2}) : **[26 Aug. 2021 (Evening)]**

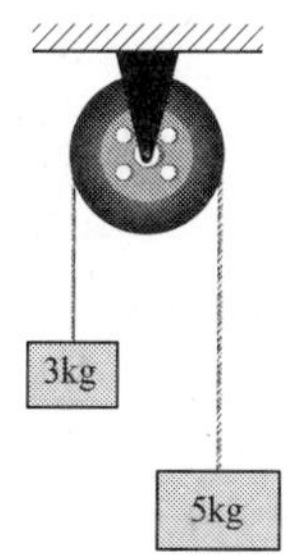

(A) 1250 cm (B) 1.25 cm
(C) 125 cm (D) 12.5 cm

17. The boxes of masses 2 kg and 8 kg are connected by a massless string passing over smooth pulleys. Calculate the time taken by box of mass 8 kg to strike the ground starting from rest. (Use $g = 10$ m/s^2) : **[27 Aug. 2021 (Evening)]**

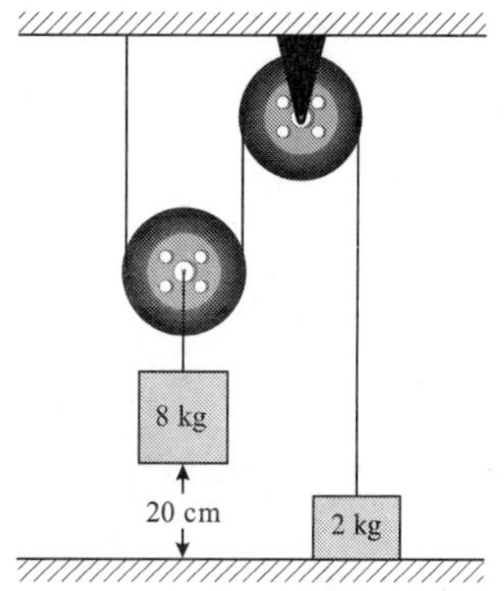

(A) 0.25 s (B) 0.34 s
(C) 0.2 s (D) 0.4 s

18. An object of mass m is being moved with a constant velocity under the action of an applied force of 2 N along a frictionless surface with following surface profile. The correct applied force vs distance graph will be : **[01 Sep. 2021 (Evening)]**

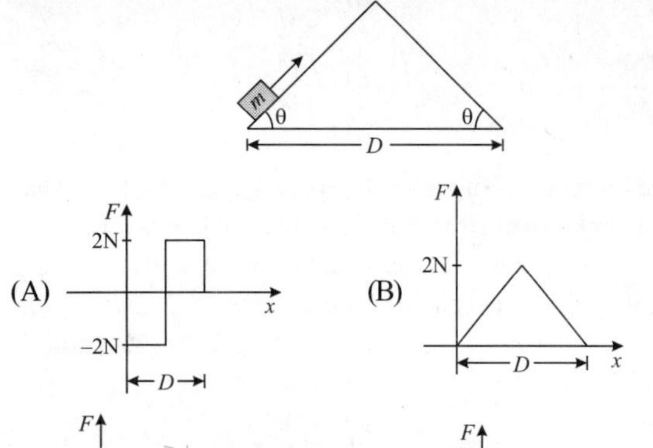

(C) F, D, x, 2N

(D) F, 2N, x, –2N, D

19. A block of mass m slides on the wooden wedge, which in turn slides backwards on the horizontal surface. The acceleration of the block with respect to the wedge is : Given $m = 8$ kg, $M = 16$ kg Assume all the surfaces shown in the figure to be frictionless : **[01 Sep. 2021 (Evening)]**

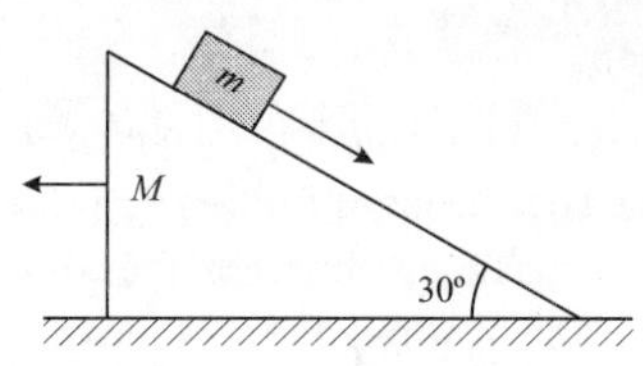

(A) $\frac{4}{3}g$ (B) $\frac{6}{5}g$

(C) $\frac{3}{5}g$ (D) $\frac{2}{3}g$

* * * * *

CHAPTER

4 LAWS OF MOTION

4.2 FRICTION

1. The masses $m_1 = 5$ kg and $m_2 = 10$ kg, connected by an inextensible string over a friction less pulley, are moving as shown in the figure. The coefficient of friction of horizontal surface is 0.15. The minimum weight m that should be put on top of m_2 to stop the motion is : **[2018]**

(A) 18.3 kg
(B) 27.3 kg
(C) 43.3 kg
(D) 10.3 kg

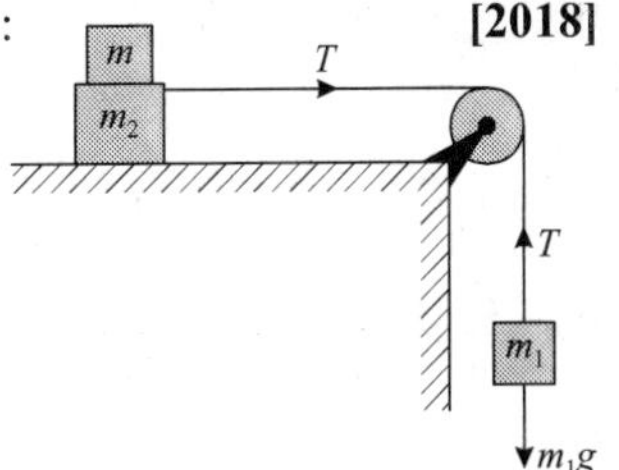

2. A given object taken n times more time to slide down a 45° rough inclined plane as it takes to slide down a perfectly smooth 45° incline. The coefficient of kinetic friction between the object and the incline is : **[2018 (Online)]**

(A) $\sqrt{1-\frac{1}{n^2}}$ (B) $1-\frac{1}{n^2}$

(C) $\frac{1}{2-n^2}$ (D) $\sqrt{\frac{1}{1-n^2}}$

3. A body of mass 2 kg slide down with an acceleration of 3 m/s^2 on a rough inclined plane having a slope of 30°. The external force required to take the same body up the plane with the same acceleration will be : ($g = 10$ m/s^2) : **[2018 (Online)]**

(A) 4 N (B) 14 N
(C) 6 N (D) 20 N

4. A block of mass 10 kg is kept on a rough inclined plane as shown in the figure. A force of 3 N is applied on the block. The coefficient of static friction between the plane and the block is 0.6. What should be the minimum value of force P, such that the block does not move downward ? [Take $g = 10$ ms^{-2}] **[9 Jan. 2019 (Morning)]**

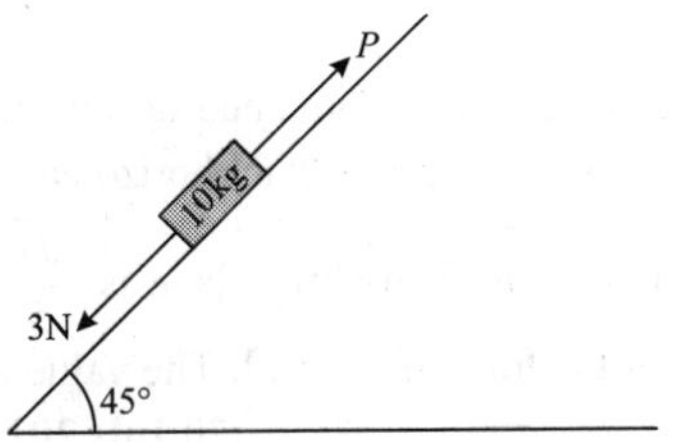

(A) 32 N (B) 25 N
(C) 23 N (D) 18 N

5. A block kept on a rough inclined plane, as shown in the figure, remains at rest upto a maximum force 2 N down the inclined plane. The maximum external force up the inclined plane that does not move the block is 10 N. The coefficient of static friction between the block and the plane is :[Take $g = 10$ m/s^2] **[12 Jan. 2019 (Evening)]**

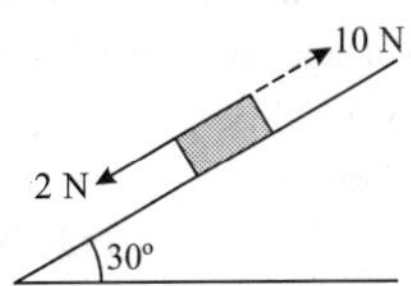

(A) $\frac{2}{3}$ (B) $\frac{\sqrt{3}}{2}$

(C) $\frac{\sqrt{3}}{4}$ (D) $\frac{1}{2}$

6. Two blocks A and B of masses $m_A = 1$ kg and $m_B = 3$ kg are kept on the table as shown in figure. The coefficient of friction between A and B is 0.2 and between B and the surface of the table is also 0.2. The maximum force F that can be applied on B horizontally, so that the block A does not slide over the block B is : (Take $g = 10$ m/s^2) : **[10 April 2019 (Evening)]**

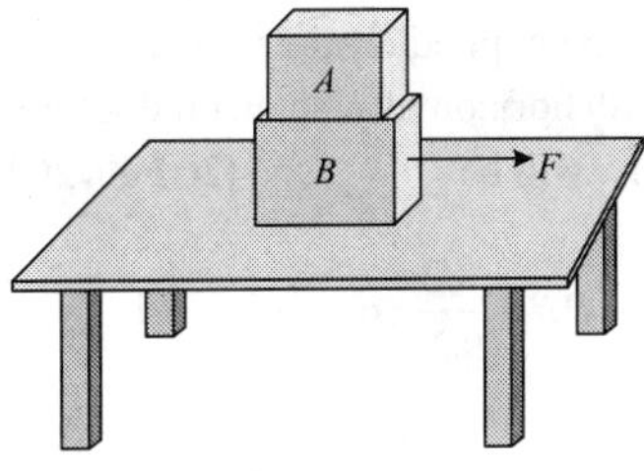

(A) 16 N (B) 40 N
(C) 12 N (D) 8 N

7. A block of mass 5 kg is (i) pushed in case (A) and (ii) pulled in case (B), by a force $F = 20$ N, making an angle of 30° with the horizontal, as shown in the figures. The coefficient of friction between the block and floor is $\mu = 0.2$. The difference between the accelerations of the block, in case (B) and case (A) will be : ($g = 10$ ms^{-2}) : **[12 April 2019 (Evening)]**

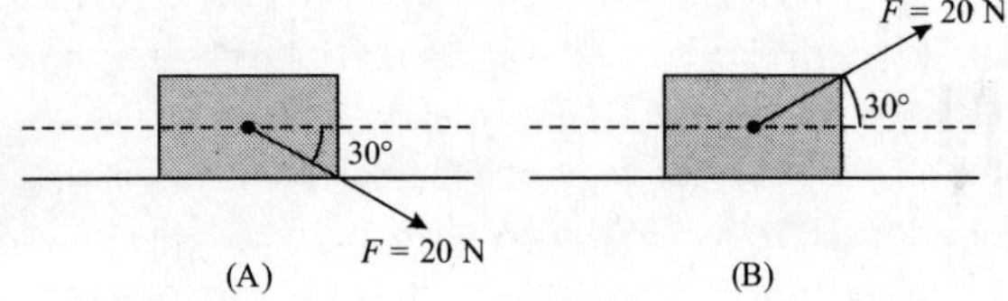

(A) 0 ms^{-2} (B) 0.8 ms^{-2}
(C) 0.4 ms^{-2} (D) 3.2 ms^{-2}

8. An insect is at the bottom of a hemispherical ditch of radius 1 m. It crawls up the ditch but starts slipping after it is at height h from the bottom. If the coefficient of friction between the ground and the insect is 0.75, then h is : ($g = 10 \text{ ms}^{-2}$)

[6 Sept. 2020 (Morning)]

(A) 0.80 m (B) 0.60 m
(C) 0.45 m (D) 0.20 m

9. The coefficient of static friction between a wooden block of mass 0.5 kg and a vertical rough wall is 0.2. The magnitude of horizontal force that should be applied on the block to keep it adhere to the wall will be_____ N. [$g = 10 \text{ ms}^{-2}$]

[24 Feb. 2021 (Morning)]

10. An inclined plane is bent in such a way that the vertical cross-section is given by $y = \frac{x^2}{4}$ where y is in vertical and x in horizontal direction. If the upper surface of this curved plane is rough with coefficient of friction $\mu = 0.5$, the maximum height in cm at which a stationary block will not slip downward is _____ cm.

[24 Feb. 2021 (Morning)]

11. As shown in the figure, a block of mass $\sqrt{3}$ kg is kept on a horizontal rough surface of coefficient of friction $\frac{1}{3\sqrt{3}}$. The critical force to be applied on the vertical surface as shown at an angle 60° with horizontal such that it does not move, will be x. The value of x will be ________. **[26 Feb. 2021 (Morning)]**

[$g = 10 \text{ m/s}^2$; $\sin 60° = \frac{\sqrt{3}}{2}$; $\cos 60° = \frac{1}{2}$]

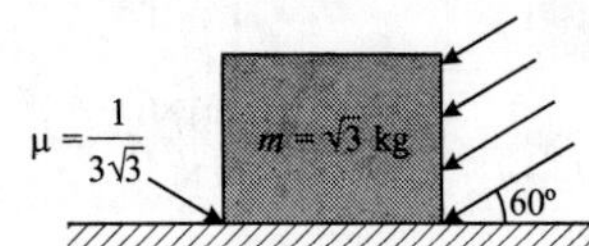

12. A block of mass m slides along a floor while a force of magnitude F is applied to it at an angle θ as shown in figure. The coefficient of kinetic friction is μ_k. Then, the block's acceleration 'a' is given by : (g is acceleration due to gravity)

[16 March 2021 (Morning)]

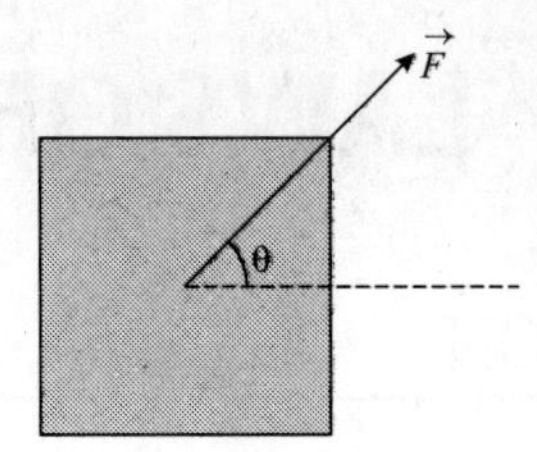

(A) $-\frac{F}{m}\cos\theta - \mu_k\left(g - \frac{F}{m}\sin\theta\right)$

(B) $\frac{F}{m}\cos\theta - \mu_k\left(g - \frac{F}{m}\sin\theta\right)$

(C) $\frac{F}{m}\cos\theta - \mu_k\left(g + \frac{F}{m}\sin\theta\right)$

(D) $\frac{F}{m}\cos\theta - \mu_k\left(g - \frac{F}{m}\sin\theta\right)$

13. Two blocks ($m = 0.5$ kg and $M = 4.5$ kg) are arranged on a horizontal frictionless table as shown in figure. The coefficient of static friction between the two blocks is $\frac{3}{7}$. Then the maximum horizontal force that can be applied on the larger block so that the blocks move together is_____ N. (Round off to the Nearest Integer) [Take g as 9.8 ms^{-2}] **[17 March 2021 (Morning)]**

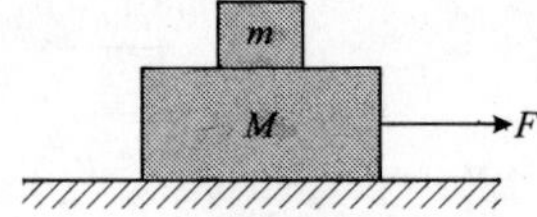

14. A body of mass 1 kg rests on a horizontal floor with which it has a coefficient of static friction $\frac{1}{\sqrt{3}}$. It is desired to make the body move by applying the minimum possible force F N. The value of F will be _____. (Round off to the Nearest Integer) [Take $g = 10 \text{ ms}^{-2}$] **[17 March 2021 (Evening)]**

15. The normal reaction 'N' for a vehicle of 800 kg mass, negotiating a turn on a 30° banked road at maximum possible speed without skidding is _____ $\times 10^3 \text{ kg m/s}^2$. (Given $\cos 30° = 0.87$, $\mu s = 0.2$) **[20 July 2021 (Morning)]**

(A) 7.2 (B) 12.4
(C) 10.2 (D) 6.98

16. A body of mass 'm' is launched up on a rough inclined plane making and angle of 30° with the horizontal. The coefficient of friction between the body and plane is $\frac{\sqrt{x}}{5}$ if the time of ascent is half of the time of descent. The value of x is _____.

[20 July 2021 (Evening)]

17. The coefficient of static friction between two blocks is 0.5 and the table is smooth. The maximum horizontal force that can be applied to move the blocks together is ______ N. (take $g = 10\ \text{ms}^{-2}$) **[26 Aug. 2021 (Evening)]**

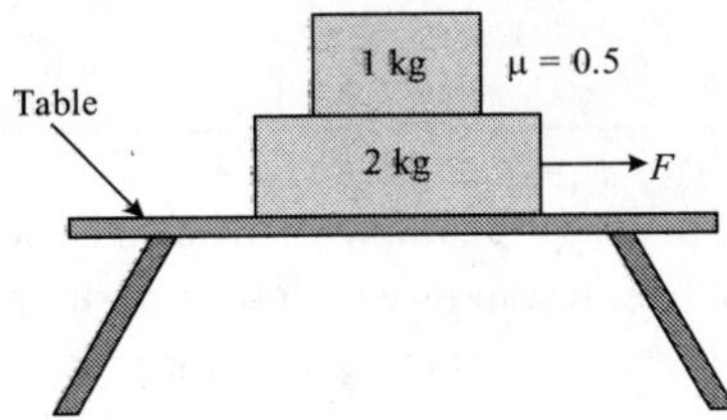

18. When a body slides down from rest along a smooth incline plane making an angle of 30° with the horizontal, it takes time T. When the same body slides down the rest along a rough inclined plane making the same angle and through the same distance, it takes time αT, where α is a constant greater than 1. The co-efficient of friction between the body and the rough plane is

$\frac{1}{\sqrt{x}}\left(\frac{\alpha^2 - 1}{\alpha^2}\right)$ where $x =$ ______. **[01 Sep. 2021 (Evening)]**

* * * * *

CHAPTER 5

WORK ENERGY POWER & APPLICATIONS

5.1 WORK ENERGY POWER

1. A time dependent force $F = 6t$ acts on a particle of mass 1 kg. If the particle starts from rest, the work done by the force during the first 1 second will be : **[2017]**
(A) 9 J (B) 18 J
(C) 4.5 J (D) 22 J

2. A body of mass $m = 10^{-2}$ kg is moving in a medium and experiences a frictional force $F = -kv^2$. Its initial speed is $v_0 = 10\ ms^{-1}$. If, after 10s, its energy is $\frac{1}{8}mv_0^2$, the value of k will be : **[2017]**
(A) 10^{-4} kg m^{-1} (B) 10^{-1} kg m^{-1} s^{-1}
(C) 10^{-3} kg m^{-1} (D) 10^{-3} kg s^{-1}

3. A particle is moving in a circular path of radius a under the action of an attractive potential $U = -\frac{k}{2r^2}$. Its total energy is : **[2018]**
(A) $-\frac{k}{4a^2}$ (B) $\frac{k}{2a^2}$
(C) Zero (D) $-\frac{3}{2}\frac{k}{a^2}$

4. Two particles of the same mass m are moving in circular orbits because of force, given by $F(r) = \frac{-16}{r} - r^3$.
The first particle is at a distance $r = 1$, and the second, at $r = 4$. The best estimate for the ratio of kinetic energies of the first and the second particle is closest to : **[2018 (Online)]**
(A) 10^{-1} (B) 6×10^{-2}
(C) 6×10^2 (D) 3×10^{-3}

5. A body of mass m starts moving from rest along x-axis so that its velocity varies as $v = a\sqrt{s}$ where a is a constant and s is the distance covered by the body. The total work done by all the forces acting on the body in the first t second after the start of the motion is : **[2018 (Online)]**
(A) $\frac{1}{8}ma^4t^2$ (B) $4\,ma^4t^2$
(C) $8\,ma^4t^2$ (D) $\frac{1}{4}ma^4t^2$

6. A block of mass m, lying on a smooth horizontal surface, is attached to a spring (of negligible mass) of spring constant k. The other end of the spring is fixed, as shown in the figure. The block is initially at rest in its equilibrium position. If now the block is pulled with a constant force F, the maximum speed of the block is : **[9 Jan. 2019 (Morning)]**

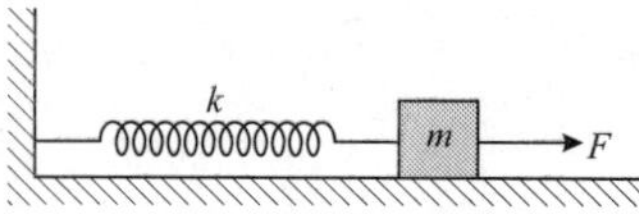

(A) $\frac{\pi F}{\sqrt{mk}}$ (B) $\frac{2F}{\sqrt{mk}}$
(C) $\frac{F}{\sqrt{mk}}$ (D) $\frac{F}{\pi\sqrt{mk}}$

7. A force acts on a 2 kg object so that its position is given as a function of time as $x = 3t^2 + 5$. What is the work done by this force in first 5 seconds ? **[9 Jan. 2019 (Evening)]**
(A) 850 J (B) 900 J
(C) 950 J (D) 875 J

8. A block of mass m is kept on a platform which starts from rest with constant acceleration $g/2$ upward, as shown in figure. Work done bv normal reaction on block in time t is : **[10 Jan. 2019 (Morning)]**

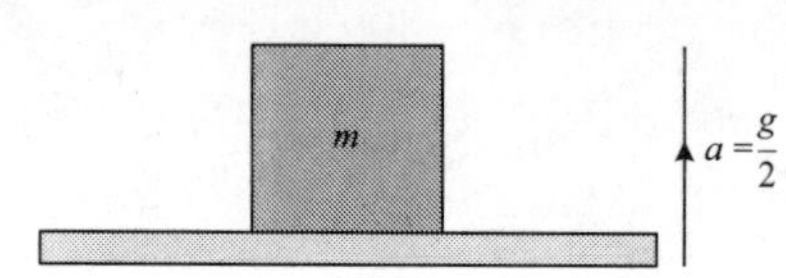

(A) 0 (B) $\frac{3mg^2t^2}{8}$
(C) $-\frac{mg^2t^2}{8}$ (D) $\frac{mg^2t^2}{8}$

9. A particle which is experiencing a force, given by $\vec{F} = 3\vec{i} - 12\vec{j}$, undergoes a displacement of $\vec{d} = 4\vec{i}$. If the particle had a kinetic energy of 3 J at the beginning of the displacement, what is its kinetic energy at the end of the displacement ? **[10 Jan. 2019 (Evening)]**
(A) 15 J (B) 10 J
(C) 12 J (D) 9 J

10. A body of mass 1 kg falls freely from a height of 100 m on a platform of mass 3 kg which is mounted on a spring having spring constant $k = 1.25 \times 10^6$ N/m. The body sticks to the platform and the spring's maximum compression is found to be x. Given that $g = 10$ ms^{-2}, the value of x will be close to :

[11 Jan. 2019 (Morning)]

(A) 4 cm (B) 8 cm
(C) 80 cm (D) 40 cm

11. A particle moves in one dimension from rest under the influence of a force that varies with the distance travelled by the particle as shown in the figure. The kinetic energy of the particle after it has travelled 3 m is : **[8 April 2019 (Morning)]**

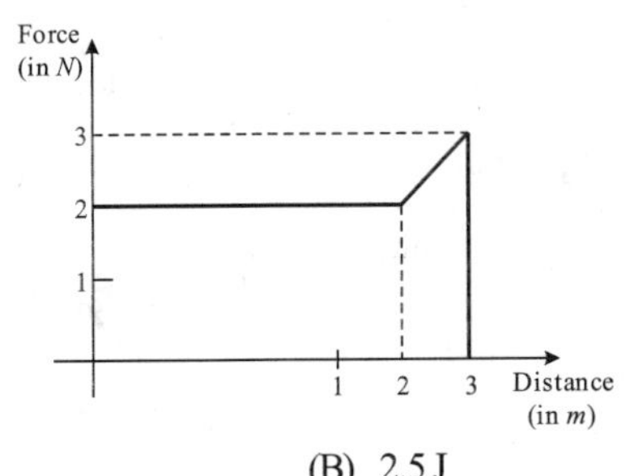

(A) 6.5 J (B) 2.5 J
(C) 4 J (D) 5 J

12. A uniform cable of mass 'M' and length 'L' is placed on a horizontal surface such that its $\left(\frac{1}{n}\right)^{th}$ part is hanging below the edge of the surface. To lift the hanging part of the cable upto the surface, the work done should be :

[9 April 2019 (Morning)]

(A) $\frac{MgL}{n^2}$ (B) $\frac{MgL}{2n^2}$
(C) $\frac{2MgL}{n^2}$ (D) $nMgL$

13. A ball is thrown upward with an initial velocity V_0 from the surface of the earth. The motion of the ball is affected by a drag force equal to $m\gamma\upsilon^2$ (where m is mass of the ball, υ is its instantaneous velocity and γ is a constant). Time taken by the ball to rise to its zenith is : **[10 April 2019 (Morning)]**

(A) $\frac{1}{\sqrt{\gamma g}}\sin^{-1}\left(\sqrt{\frac{\gamma}{g}}V_0\right)$ (B) $\frac{1}{\sqrt{\gamma g}}\tan^{-1}\left(\sqrt{\frac{\gamma}{g}}V_0\right)$

(C) $\frac{1}{\sqrt{2\gamma g}}\tan^{-1}\left(\sqrt{\frac{2\gamma}{g}}V_0\right)$ (D) $\frac{1}{\sqrt{\gamma g}}\ln\left(1+\sqrt{\frac{\gamma}{g}}V_0\right)$

14. A spring whose unstretched length is l has a force constant k. The spring is cut into two pieces of unstretched lengths l_1 and l_2 where, $l_1 = nl_2$ and n is an integer. The ratio k_1/k_2 of the corresponding force constants, will be :

[12 April 2019 (Evening)]

(A) $\frac{1}{n^2}$ (B) n^2

(C) $\frac{1}{n}$ (D) n

15. A 60 HP electric motor lifts an elevator having a maximum total load capacity of 2000 kg. If the frictional force on the elevator is 4000 N, the speed of the elevator at full load is close to: (1 HP = 746 W, $g = 10$ ms^{-2}) : **[7 Jan. 2020 (Morning)]**

(A) 1.7 ms^{-1} (B) 2.0 ms^{-1}
(C) 1.9 ms^{-1} (D) 1.5 ms^{-1}

16. A particle ($m = 1$ kg) slides down a frictionless track (AOC) starting from rest at a point A (height 2 m). After reaching C, the particle continues to move freely in air as a projectile. When it reaching its highest point P (height 1 m), the kinetic energy of the particle (in J) is : (Figure drawn is schematic and not to scale; take $g = 10$ ms^{-2})_____ : **[7 Jan. 2020 (Morning)]**

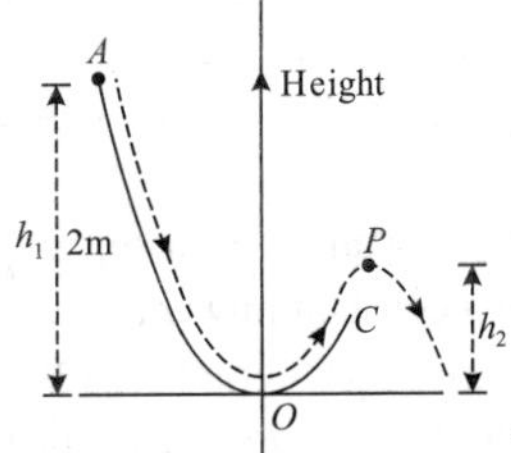

17. An elevator in a building can carry a maximum of 10 persons, with the average mass of each person being 68 kg. The mass of the elevator itself is 920 kg and it moves with a constant speed 3 m/s. The frictional force opposing the motion is 6000 N. If the elevator is moving up with its full capacity, the power delivered by the motor to the elevator ($g = 10$ m/s^2) must be at least :

[7 Jan. 2020 (Evening)]

(A) 56300 W (B) 48000 W
(C) 66000 W (D) 62360 W

18. Consider a force $\vec{F} = -x\hat{i} + y\hat{j}$. The work done by this force in moving a particle from point $A(1, 0)$ to $B(0, 1)$ along the line segment is : (all quantities are in SI units)

[9 Jan. 2020 (Morning)]

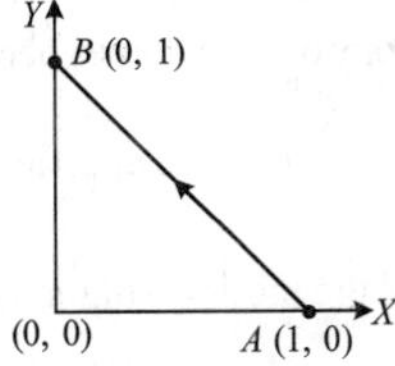

(A) $\frac{3}{2}$ (B) 1

(C) 2 (D) $\frac{1}{2}$

19. A small block starts slipping down from a point B on an inclined plane AB, which is making an angle θ with the horizontal section BC is smooth and the remaining section CA is rough with a coefficient of friction μ. It is found that the block comes to rest as it reaches the bottom (point A) of the inclined plane. If $BC = 2AC$, the coefficient of friction is given by $\mu = k\tan\theta$). The value of k is : **[2 Sept. 2020 (Morning)]**

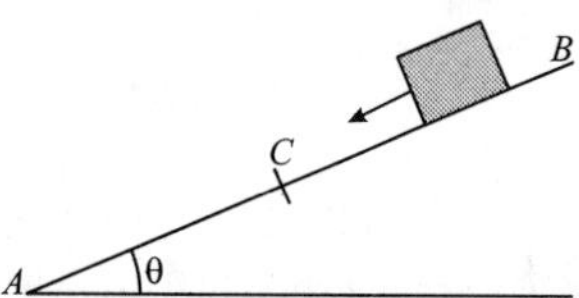

20. A cricket ball of mass 0.15 kg is thrown vertically up by a bowling machine so that it rises to a maximum height of 20 m after leaving the machine. If the part pushing the ball applies a constant force F on the ball and moves horizontally a distance of 0.2 m while launching the ball, the value of F(in N) is ($g = 10\ ms^{-2}$) _____: **[3 Sept. 2020 (Morning)]**

21. A particle is moving unidirectionally on a horizontal plane under the action of a constant power supplying energy source. The displacement (s)-time (t) graph that describes the motion of the particle is (graphs are drawn schematically and are not to scale) : **[3 Sept. 2020 (Evening)]**

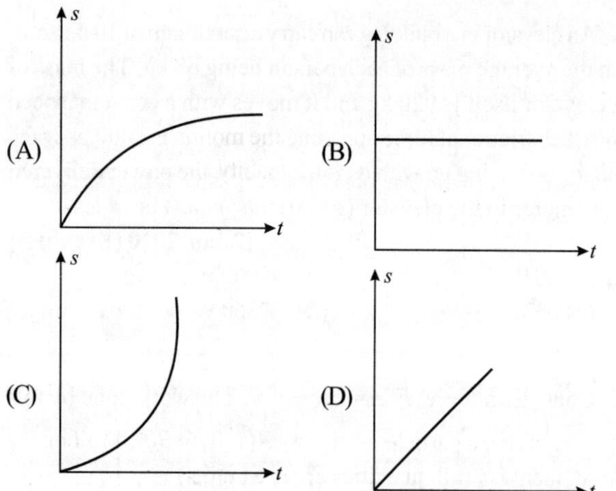

22. A block starts moving up an inclined plane of inclination 30° with an initial velocity of v_0. It comes back to its initial position with velocity $\frac{v_0}{2}$. The value of the coefficient of kinetic friction between the block and the inclined plane is close to $\frac{I}{1000}$. The nearest integer to I is _____. **[3 Sept. 2020 (Evening)]**

23. A person pushes a box on a rough horizontal platform surface. He applies a force of 200 N over a distance of 15 m. Thereafter, he gets progressively tired and his applied force reduces linearly with distance of 100 N. The total distance through which the box has been moved is 30 m. What is the work done by the person during the total movement of the box? **[4 Sept. 2020 (Evening)]**

(A) 5690 J (B) 5250 J

(C) 3280 J (D) 2780 J

24. A body of mass 2 kg is driven by an engine delivering a constant power 1 J/s. The body starts from rest and moves in a straight line. After 9 seconds, the body has moved a distance (in m) _____. **[5 Sept. 2020 (Evening)]**

25. If the potential energy between two molecules is given by $U = -\frac{A}{r^6} + \frac{B}{r^{12}}$, then at equilibrium, separation between molecules, and the potential energy are : **[6 Sept. 2020 (Morning)]**

(A) $\left(\frac{B}{A}\right)^{1/6}, 0$ (B) $\left(\frac{B}{2A}\right)^{1/6}, -\frac{A^2}{2B}$

(C) $\left(\frac{2B}{A}\right)^{1/6}, -\frac{A^2}{4B}$ (D) $\left(\frac{2B}{A}\right)^{1/6}, -\frac{A^2}{2B}$

26. The potential energy (U) of a diatomic molecule is a function dependent on r (interatomic distance) as

$$U = \frac{\alpha}{r^{10}} - \frac{\beta}{r^5} - 3$$

where, α and β are positive constants. The equilibrium distance between two atoms will be $\left(\frac{2\alpha}{\beta}\right)^{a/b}$, where $a =$ _____. **[25 Feb. 2021 (Morning)]**

27. A boy is rolling a 0.5 kg ball on the frictionless floor with the speed of 20 ms^{-1}. The ball gets deflected by an obstacle on the way. After deflection it moves with 5% of its initial kinetic energy. What is the speed of the ball now ? **[17 March 2021 (Morning)]**

(A) 19.0 ms^{-1} (B) 4.47 ms^{-1}

(C) 14.41 ms^{-1} (D) 1.00 ms^{-1}

28. Two identical blocks A and B each of mass m resting on the smooth horizontal floor are connected by a light spring of natural length L and spring constant K. A third block C of mass m moving with a speed v along the line joining A and B collides with A. The maximum compression in the spring is : **[17 March 2021 (Evening)]**

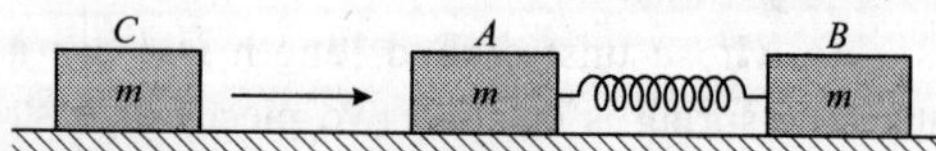

(A) $v\sqrt{\frac{M}{2K}}$ (B) $\sqrt{\frac{mv}{2K}}$

(C) $\sqrt{\frac{mv}{K}}$ (D) $\sqrt{\frac{m}{2K}}$

29. A particle of mass m moves in a circular orbit in a central potential field $U(r) = U_0 r^4$. If Bohr's quantization conditions are applied, radii of possible orbitals r_n vary with $n^{1/\alpha}$, where α is_____ . **[17 March 2021 (Evening)]**

30. A constant power delivering machine has towed a box, which was initially at rest, along a horizontal straight line. The distance moved by the box in time 't' is proportional to : **[18 March 2021 (Morning)]**

(A) $t^{2/3}$ (B) $t^{3/2}$

(3) t (D) $t^{1/2}$

31. As shown in the figure, a particle of mass 10 kg is placed at a point A. When the particle is slightly displaced to its right, it starts moving and reaches the point B. The speed of the particle at B is x m/s. (Take $g = 10$ m/s^2)
The value of 'x' to the nearest integer is_____. **[18 March 2021 (Morning)]**

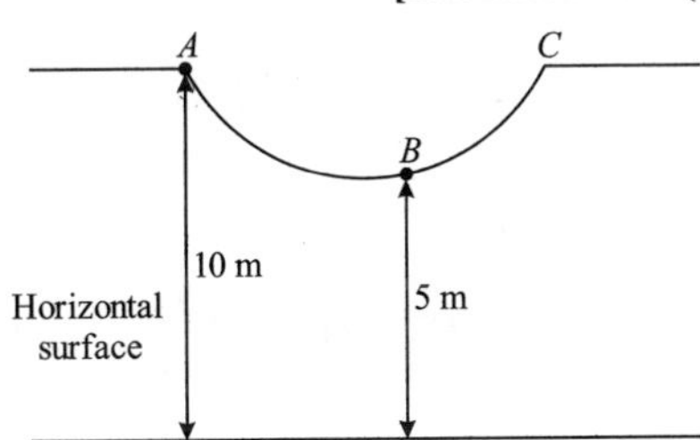

32. A ball of mass 4 kg, moving with a velocity of 10 ms^{-1}, collides with a spring of length 8 m and force constant 100 Nm^{-1}. The length of the compressed spring is x m. The value of x, to the nearest integer, is_____. **[18 March 2021 (Evening)]**

33. If the Kinetic energy of a moving body becomes four times its initial Kinetic energy, then the percentage change in its momentum will be : **[20 July 2021 (Evening)]**

(A) 400 % (B) 100 %

(C) 300 % (D) 200 %

34. A body at rest is moved along a horizontal straight line by a machine delivering a constant power. The distance moved by the body in time 't' is proportional to : **[20 July 2021 (Evening)]**

(A) $t^{1/2}$ (B) $t^{1/4}$

(C) $t^{3/2}$ (D) $t^{3/4}$

35. A porter lifts a suitcase of mass 80 kg and at the destination lowers it down by a distance of 80 cm with a constant velocity. Calculate the work done by the porter in lowering the suitcase. (take $g = 9.8$ ms^{-2}) **[22 July 2021 (Evening)]**

(A) 784.0 J (B) -62700.0 J

(C) 627.2 J (D) -627.2 J

36. A force of $F = (5y + 20)\,\hat{j}$ N acts on a particle. The work done by this force when the particle is moved from $y = 10$ m to $y = 10$ m is_____ . **[25 July 2021 (Evening)]**

37. An automobile of mass 'm' accelerates starting from origin and initially at rest, while the engine supplies constant power P. The position is given as a function of time by : **[27 July 2021 (Evening)]**

(A) $\left(\frac{8P}{9m}\right)^{1/2} t^{3/2}$ (B) $\left(\frac{9P}{8m}\right)^{1/2} t^{3/2}$

(C) $\left(\frac{9m}{8P}\right)^{1/2} t^{3/2}$ (D) 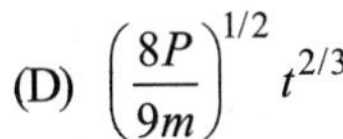 $\left(\frac{8P}{9m}\right)^{1/2} t^{2/3}$

38. Given below is the plot of a potential energy function $U(x)$ for a system, in which a particle is in one dimensional motion, while a conservative force $F(x)$ acts on it. Suppose that **[27 July 2021 (Evening)]**

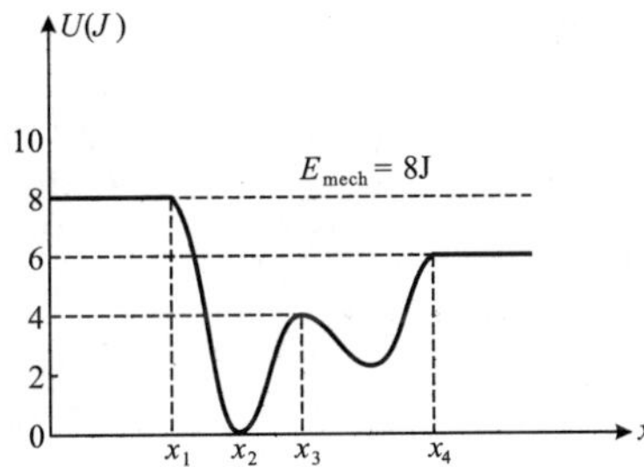

(A) At $x = x_3$, $K.E. = 4$

(B) At $x > x_4$, $K.E.$ is constant throughout the region

(C) At $x < x_1$, $K.E.$ is smallest and the particle is moving at the slowest speed

(D) At $x = x_2$, $K.E.$ is greatest and the particle is moving at the fastest speed

39. A uniform chain of length 3 meter and mass 3 kg overhangs a smooth table with 2 meter laying on the table. If k is the kinetic energy of the chain in joule as it completely slips off the table, then the value of k is_____ . **[26 Aug. 2021 (Morning)]**

40. Two persons A and B perform same amount of work in moving a body through a certain distance d with application of forces acting at angles 45° and 60° with the direction of displacement respectively. The ratio of force applied by person A to the force applied by person B is $\frac{1}{\sqrt{x}}$. The value of x is _____ . **[27 Aug. 2021 (Morning)]**

41. A block moving horizontally on a smooth surface with a speed of 40 ms^{-1} splits into two equal parts. If one of the parts moves at 60 ms^{-1} in the same direction, them the fractional change in the kinetic energy will be $x : 4$ where $x =$ _____.

[31 Aug. 2021 (Morning)]

42. A block moving horizontally on a smooth surface with a speed of 40 m/s splits into two parts with masses in the ratio of 1 : 2. If the smaller part moves at 60 m/s in the same direction, then the fractional change in kinetic energy is :

[31 Aug. 2021 (Evening)]

(A) $\frac{1}{8}$ (B) $\frac{2}{3}$

(C) $\frac{1}{3}$ (D) $\frac{1}{4}$

43. A body of mass m dropped from a height h reaches the ground with a speed of $0.8\sqrt{gh}$. The value of work done by the air - friction is : **[01 Sep. 2021 (Evening)]**

(A) -0.68 mgh (B) 0.64 mgh

(C) mgh (D) 1.64 mgh

44. An engine is attached to a wagon through a shock absorber of length 1.5 m. The system with a total mass of 40000 kg is moving with a speed of 72 kmh^{-1} when the brakes are applied to bring it to rest. In the process of the system being brought to rest, the spring of the shock absorber gest compressed by 1.0 m. If 90% of energy of the wagon is lost due of friction the spring constant is _____ $\times 10^5$ N/m. **[01 Sep. 2021 (Evening)]**

* * * * *

CHAPTER

5 WORK ENERGY POWER & APPLICATIONS

5.2 CIRCULAR MOTION

1. A conical pendulum of length 1 m making an angle $\theta = 45°$ w.r.t. Z-axis and moves in a circle in the XY plane. The radius of the circle is 0.4 m and its centre is vertically below O. The speed of the pendulum, in its circular path, will be : [Take $g = 10\ \text{ms}^{-2}$] **[2017 (Online)]**

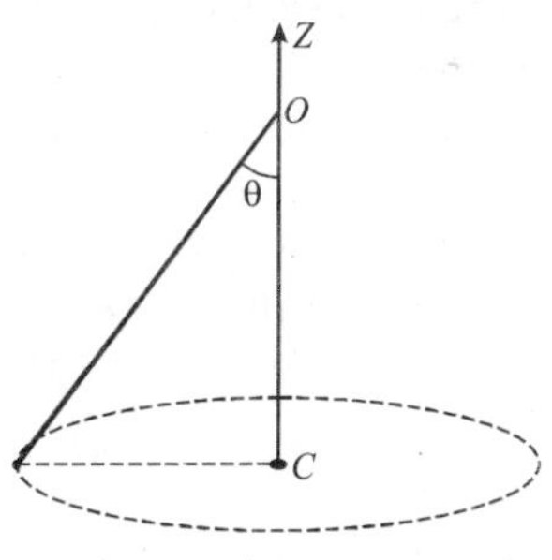

(A) 0.4 m/s (B) 4 m/s
(C) 0.2 m/s (D) 2 m/s

2. A disc rotates about its axis of symmetry in a horizontal plane at a steady rate of 3.5 revolutions per second. A coin placed at a distance of 1.25 cm from the axis of rotation remains at rest on the disc. The coefficient of friction between the coin and the disc is ($g = 10\ \text{m/s}^2$) : **[2018 (Online)]**

(A) 0.5 (B) 0.7
(C) 0.3 (D) 0.6

3. A particle is moving along a circular path with a constant speed of 10 ms^{-1}. What is the magnitude of the change in velocity of the particle, when it moves through an angle of 60° around the centre of the circle? **[11 Jan. 2019 (Morning)]**

(A) Zero (B) 10 m/s
(C) $10\sqrt{3}$ m/s (D) $10\sqrt{2}$ m/s

4. Two particles A, B are moving on two concentric circles of radii R_1 and R_2 with equal angular speed ω. At $t = 0$, their positions and direction of motion are shown in the figure. The relative velocity $\vec{\upsilon}_A - \vec{\upsilon}_B$ at $t = \pi/2\omega$ is given by : **[12 Jan. 2019 (Evening)]**

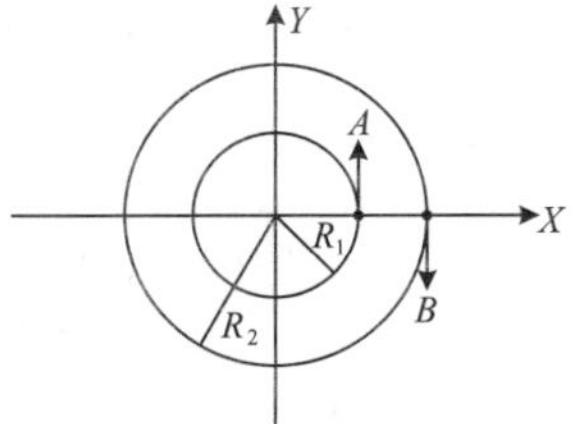

(A) $-\omega(R_1 + R_2)\hat{i}$ (B) $\omega(R_1 + R_2)\hat{i}$
(C) $\omega(R_1 - R_2)\hat{i}$ (D) $\omega(R_2 - R_1)\hat{i}$

5. A smooth wire of length $2\pi r$ is bent into a circle and kept in a vertical plane. A bead can slide smoothly on the wire. When the circle is rotating with angular speed ω about the vertical diameter AB, as shown in figure, the bead is at rest with respect to the circular ring at position P as shown. Then the value of ω^2 is equal to : **[12 April 2019 (Evening)]**

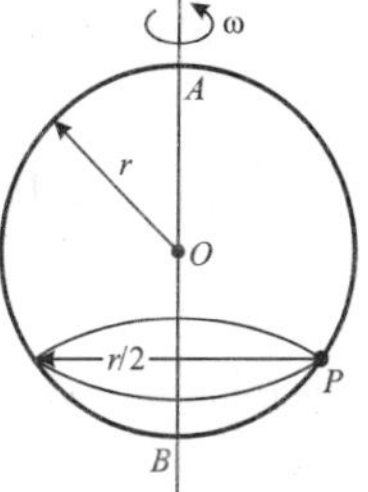

(A) $(g\sqrt{3})/r$ (B) $\dfrac{\sqrt{3}g}{2r}$
(C) $2g/r$ (D) $2g/(r\sqrt{3})$

6. A particle of mass m is fixed to one end of a light spring having force constant k and unstretched length l. The other end is fixed. The system is given an angular speed ω about the fixed end of the spring such that it rotates in a circle in gravity free space. Then the stretch in the spring is : **[8 Jan. 2020 (Morning)]**

(A) $\dfrac{ml\omega^2}{k + m\omega^2}$ (B) $\dfrac{ml\omega^2}{k - m\omega^2}$
(C) $\dfrac{ml\omega^2}{k - \omega m}$ (D) $\dfrac{ml\omega^2}{k + m\omega}$

7. A spring mass system (mass m, spring constant k and natural length l) rest in equilibrium on a horizontal disc. The free end of the spring is fixed at the centre of the disc. If the disc together with spring mass system, rotates about it's axis with an angular velocity ω, ($k >> m\omega^2$) the relative change in the length of the spring is best given by the option : **[9 Jan. 2020 (Evening)]**

(A) $\frac{2m\omega^2}{k}$ (B) $\frac{m\omega^2}{3k}$

(C) $\sqrt{\frac{2}{3}}\left(\frac{m\omega^2}{k}\right)$ (D) $\frac{m\omega^2}{k}$

8. A bead of mass m stays at point $P(a, b)$ on a wire bent in the shape of a parabola $y = 4Cx^2$ and rotating with angular speed ω (see figure). The value of ω is (neglect friction) : **[2 Sept. 2020 (Morning)]**

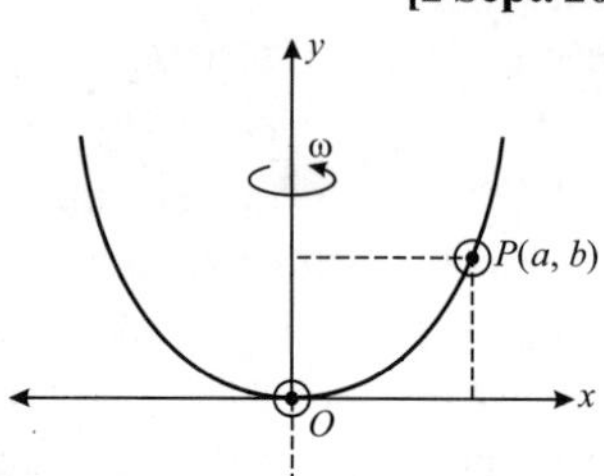

(A) $\sqrt{\frac{2gC}{ab}}$ (B) $2\sqrt{2gC}$

(C) $\sqrt{\frac{2g}{C}}$ (D) $2\sqrt{gC}$

9. A cylindrical vessel containing a liquid is rotated about its axis so that the liquid rises at its sides as shown in the figure. The radius of vessel is 5 cm and the angular speed of rotation is ω rad s^{-1}. The difference in the height, h (in cm) of liquid at the centre of vessel and at the side will be : **[2 Sept. 2020 (Morning)]**

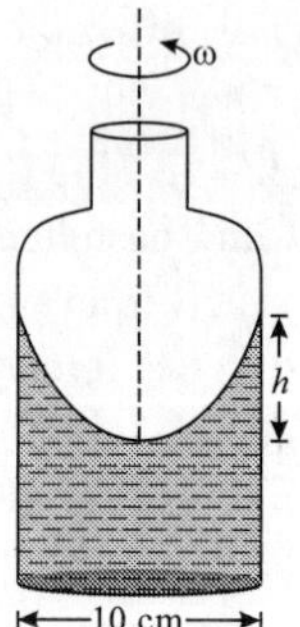

(A) $\frac{25\omega^2}{2g}$ (B) $\frac{2\omega^2}{5g}$

(C) $\frac{5\omega^2}{2g}$ (D) $\frac{2\omega^2}{25g}$

10. A clock has a continuously moving second's hand of 0.1 m length. The average acceleration of the tip of the hand (in units of ms^{-2}) is of the order of : **[6 Sept. 2020 (Morning)]**

(A) 10^{-3} (B) 10^{-2}

(C) 10^{-4} (D) 10^{-1}

11. A small bob tied at one end of a thin string of length 1 m is describing a vertical circle so that the maximum and minimum tension in the string are in the ratio 5 : 1. The velocity of the bob at the highest position is _____ m/s. (Take $g = 10$ m/s^2). **[25 Feb. 2021 (Morning)]**

12. A particle is moving with uniform speed along the circumference of a circle of radius R under the action of a central fictitious force F which is inversely proportional to R^3. Its time period of revolution will be given by : **[26 Feb. 2021 (Morning)]**

(A) $T \propto R^2$ (B) $T \propto R^{3/2}$

(C) $T \propto R^{5/2}$ (D) $T \propto R^{4/3}$

13. A block of 200 g mass moves with a uniform speed in a horizontal circular groove, with vertical side walls of radius 20 cm. If the block takes 40 s to complete one round, the normal force by the side walls of the groove is : **[16 March 2021 (Morning)]**

(A) 0.0314 N (B) 9.859×10^{-2} N

(C) 6.28×10^{-3} N (D) 9.859×10^{-4} N

14. Statement-I : A cyclist is moving on an unbanked road with a speed of 7 kmh^{-1} and takes a sharp circular turn along a path of radius of 2m without reducing the speed. The static friction coefficient is 0.2. The cyclist will not slip and pass the curve ($g = 9.8$ m/s^2)

Statement-II : If the road is banked at an angle of 45°, cyclist can cross the curve of 2m radius with the speed of 18.5 kmh^{-1} without slipping.

In the light of the above statements, choose the correct answer from the options given below. **[16 March 2021 (Evening)]**

(A) Statement-I is incorrect and statement-II is correct

(B) Statement-I is correct and statement-II is incorrect

(C) Both statement-I and statement-II are false

(D) Both statement-I and statement-II are true

15. The angular speed of truck wheel is increased from 900 rpm to 2460 rpm in 26 seconds. The number of revolutions by the truck engine during this time is _____.
(Assuming the acceleration to be uniform). **[17 March 2021 (Morning)]**

16. A particle of mass m moves in a circular orbit under the central potential field, $U(r) = \frac{-C}{r}$, where C is a positive constant. The correct radius – velocity graph of the particle's motion is :

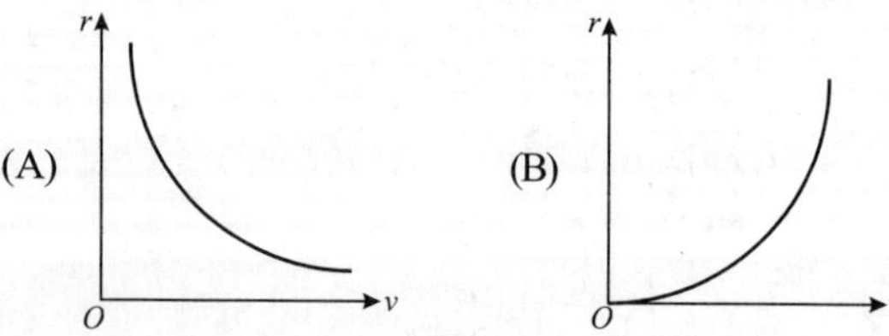

(C) 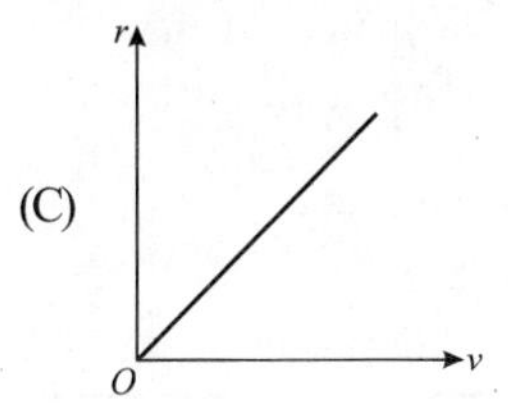

(D) 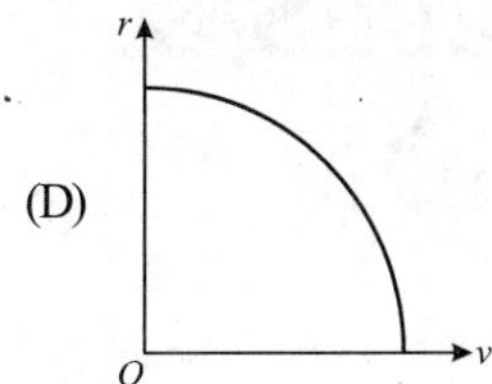

17. A body rotating with an angular speed of 600 rpm is uniformly accelerated to 1800 rpm in 10 s. The number of rotations made in the process is _____ .

[20 July 2021 (Evening)]

18. A pendulum bob has a speed of 3 m/s at its lowest position. The pendulum is 50 cm long. The speed of bob, when the length makes an angle of 60° to the vertical will be ($g = 10\ m/s^2$) _____ m/s : **[25 July 2021 (Morning)]**

19. A small block slides down from the top of hemisphere of radius $R = 3$ m as shown in the figure. The height h at which the block will lose contact with the surface of the sphere is _____ (Assume there is no friction between the block and the hemisphere) : **[27 July 2021 (Evening)]**

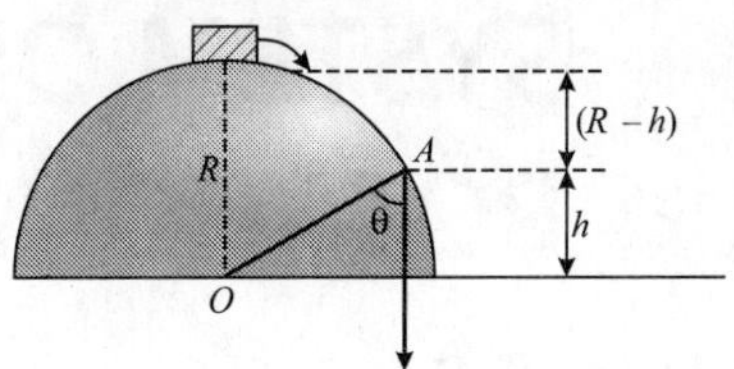

20. Particle of mass m is suspended from a celling through a string of length L. The particle moves in a horizontal circle of radius r such that $r = \frac{L}{\sqrt{2}}$. The speed of particle will be :

[26 Aug. 2021 (Evening)]

(A) $\sqrt{\frac{rg}{2}}$ (B) $2\sqrt{rg}$

(C) $\sqrt{2rg}$ (D) $\sqrt{rg}$

21. A huge circular arc of length 4.4 LY subtends an angle 4s at the centre of circle. How long it would take for a body to complete 4 revolution if its speed is 8 AU per second ?

Given : 1 light year = 9.46×10^{15} m : **[27 Aug. 2021 (Morning)]**

(A) 4.1×10^8 s (B) 4.5×10^{10} s

(C) 7.2×10^8 s (D) 3.5×10^6 s

* * * * *

CHAPTER 6

SYSTEM OF PARTICLES & ROTATIONAL MOTION

6.1 SYSTEM OF PARTICLES

1. An object is dropped from a height h from the ground. Every time it hits the ground it looses 50% of its kinetic energy. The total distance covered as $t \to \infty$ is : **[2017 (Online)]**

(A) $3h$ (B) ∞

(C) $\frac{5}{3}h$ (D) $\frac{8}{3}h$

2. Two particles A and B of equal mass M are moving with the same speed v as shown in the figure. They collide completely inelastically and move as a single particle C. The angle θ that the path of C makes with the X-axis is given by : **[2017 (Online)]**

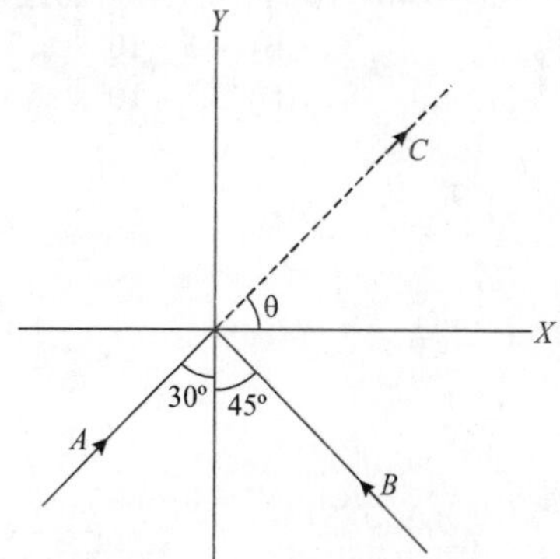

(A) $\tan\theta = \frac{\sqrt{3}+\sqrt{2}}{1-\sqrt{2}}$ (B) $\tan\theta = \frac{\sqrt{3}-\sqrt{2}}{1-\sqrt{2}}$

(C) $\tan\theta = \frac{1-\sqrt{2}}{\sqrt{2}(1+\sqrt{3})}$ (D) $\tan\theta = \frac{1-\sqrt{3}}{1+\sqrt{2}}$

3. In a collinear collision, a particle with an initial speed v_0 strikes a stationary particle of the same mass. If the final total kinetic energy is 50% greater than the original kinetic energy, the magnitude of the relative velocity between the two particles, after collision, is : **[2018]**

(A) $\frac{v_0}{4}$ (B) $\sqrt{2}v_0$

(C) $\frac{v_0}{2}$ (D) $\frac{v_0}{\sqrt{2}}$

4. The mass of a hydrogen molecule is 3.32×10^{-27} kg. If 10^{23} hydrogen molecules strike, per second, a fixed wall of area 2 cm^2 at an angle of 45° to the normal, and rebound elastically with a speed of 10^3 m/s, then the pressure on the wall is nearly : **[2018]**

(A) 2.35×10^3 N/m^2 (B) 4.70×10^3 N/m^2

(C) 2.35×10^2 N/m^2 (D) 4.70×10^2 N/m^2

5. It is found that if a neutron suffers an elastic collinear collision with deuterium at rest, fractional loss of its energy is P_d; while for its similar collision with carbon nucleus at rest, fractional loss of energy is P_c. The values of P_d and P_c are respectively : **[2018]**

(A) (·89, ·28) (B) (·28, ·89)

(C) (0, 0) (D) (0, 1)

6. A proton of mass m collides elastically with a particle of unknown mass at rest. After the collision, the proton and the unknown particle are seen moving at an angle of 90° with respect to each other. The mass of unknown particle is : **[2018 (Online)]**

(A) $\frac{m}{\sqrt{3}}$ (B) $\frac{m}{2}$

(C) 2m (D) m

7. Three blocks A, B and C are lying on a smooth horizontal surface, as shown in the figure. A and B have equal masses, m while C has mass M. Block A is given an initial speed v towards B due to which it collides with B perfectly inelastically. The combined mass collides with C, also perfectly inelastically $\frac{5}{6}$th of the initial kinetic energy is lost in whole process. What is value of M/m ? **[9 Jan. 2019 (Morning)]**

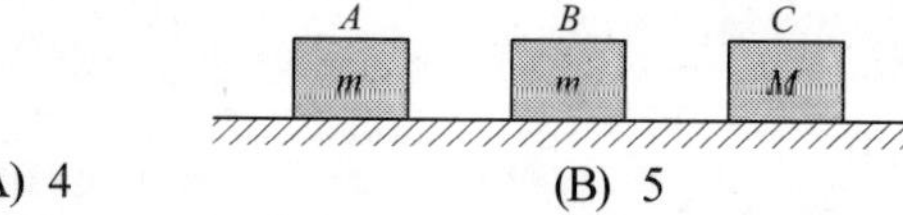

(A) 4 (B) 5

(C) 3 (D) 2

8. A piece of wood of mass 0.03 kg is dropped from the top of a 100 m height building. At the same time, a bullet of mass 0.02 kg is fired vertically upward, with a velocity 100 ms^{-1}, from the ground. The bullet gets embedded in the wood. Then the maximum height to which the combined system reaches above the top of the building before falling below is : ($g = 10$ ms^{-2}) : **[10 Jan. 2019 (Morning)]**

(A) 30m (B) 10m

(C) 40m (D) 20m

9. A simple pendulum, made of a string of length l and a bob of mass m, is released from a small angle θ_0. It strikes a block of mass M, kept on a horizontal surface at its lowest point of oscillations, elastically. It bounces back and goes up to an angle θ_1. Then M is given by : **[12 Jan. 2019 (Morning)]**

(A) $\frac{m}{2}\left(\frac{\theta_0-\theta_1}{\theta_0+\theta_1}\right)$ (B) $\frac{m}{2}\left(\frac{\theta_0+\theta_1}{\theta_0-\theta_1}\right)$

(C) $m\left(\frac{\theta_0+\theta_1}{\theta_0-\theta_1}\right)$ (D) $m\left(\frac{\theta_0-\theta_1}{\theta_0+\theta_1}\right)$

10. The position vector of the centre of mass $\vec{r}$ cm of an symmetric uniform bar of negligible area of cross-section as shown in figure is : **[12 Jan. 2019 (Morning)]**

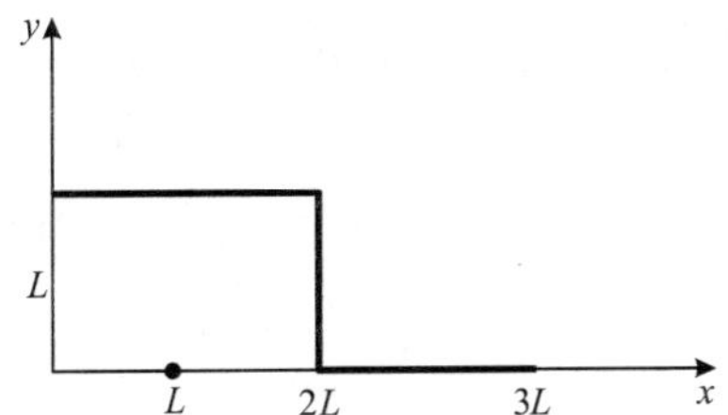

(A) $\vec{r}$ cm $=\frac{13}{8}L\hat{x}+\frac{5}{8}L\hat{y}$ (B) $\vec{r}$ cm $=\frac{11}{8}L\,\hat{x}+\frac{3}{8}L\hat{y}$

(C) $\vec{r}$ cm $=\frac{3}{8}L\,\hat{x}+\frac{11}{8}L\hat{y}$ (D) $\vec{r}$ cm $=\frac{5}{8}L\,\hat{x}+\frac{13}{8}L\hat{y}$

11. An alpha-particle of mass m suffers 1-dimensional elastic collision with a nucleus at rest of unknown mass. It is scattered directly backwards losing, 64% of its initial kinetic energy. The mass of the nucleus is : **[12 Jan. 2019 (Evening)]**

(A) 4m (B) 3.5m
(C) 2m (D) 1.5m

12. Four particles A, B, C and D with masses $m_A = m, m_B = 2m$, $m_C = 3m$ and $m_D = 4m$ are at the corners of a square. They have accelerations of equal magnitude with directions as shown. The acceleration of the centre of mass of the particles is : **[8 April 2019 (Morning)]**

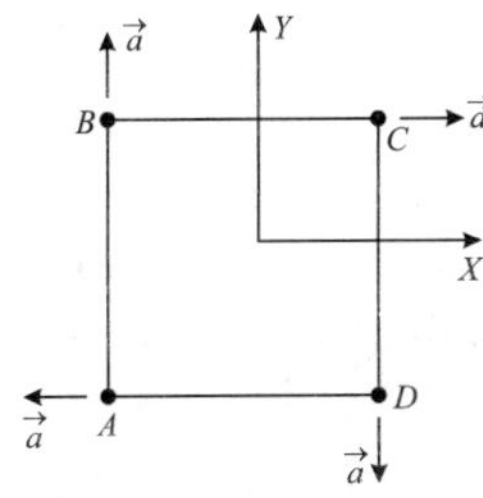

(A) $\frac{a}{5}(\hat{i}-\hat{j})$ (B) $\frac{a}{5}(\hat{i}+\hat{j})$

(C) Zero (D) $a(\hat{i}+\hat{j})$

13. A uniform rectangular thin sheet $ABCD$ of mass M has length a and breadth b, as shown in the figure. If the shaded portion $HBGO$ is cut-off, the coordinates of the centre of mass of the remaining portion will be : **[8 April 2019 (Evening)]**

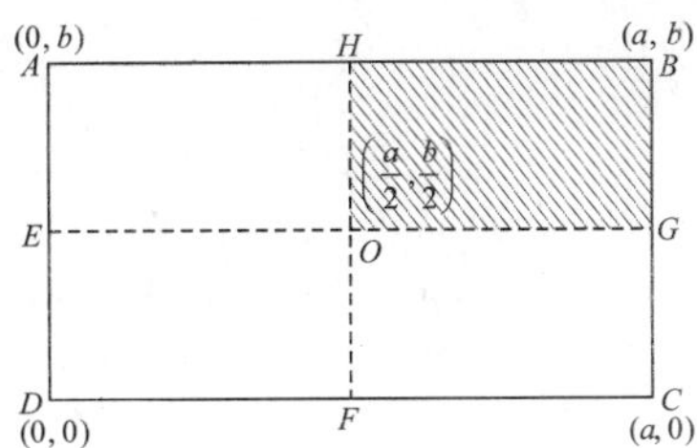

(A) $\left(\frac{2a}{3},\frac{2b}{3}\right)$ (B) $\left(\frac{5a}{3},\frac{5b}{3}\right)$

(C) $\left(\frac{3a}{4},\frac{3b}{4}\right)$ (D) $\left(\frac{5a}{12},\frac{5b}{12}\right)$

14. A body of mass m_1 moving with an unknown velocity of $v_1\hat{i}$, undergoes a collinear collision with a body of mass m_2 moving with a velocity $v_2\hat{i}$. After collision, m_1 and m_2 move with velocities of $v_3\hat{i}$ and $v_4\hat{i}$, respectively. If $m_2 = 0.5\,m_1$ and $v_3 = 0.5v_1$, then v_1 is : **[8 April 2019 (Evening)]**

(A) $v_4-\frac{v_2}{4}$ (B) $v_4-\frac{v_2}{2}$

(C) v_4-v_2 (D) v_4+v_2

15. A body of mass 2 kg makes an elastic collision with a second body at rest and continues to move in the original direction but with one fourth of its original speed. What is the mass of the second body ? **[9 April 2019 (Morning)]**

(A) 1.8 kg (B) 1.2 kg
(C) 1.5 kg (D) 1.0 kg

16. A particle of mass 'm' is moving with speed '$2v$' and collides with a mass '$2m$' moving with speed 'v' in the same direction. After collision, the first mass is stopped completely while the second one splits into two particles each of mass 'm', which move at angle 45° with respect to the original direction. The speed of each of the moving particle will be : **[9 April 2019 (Evening)]**

(A) $v/(2\sqrt{2})$ (B) $2\sqrt{2}v$

(C) $\sqrt{2}v$ (D) $v/\sqrt{2}$

17. A wedge of mass $M = 4m$ lies on a frictionless plane. A particle of mass m approaches the wedge with speed v. There is no friction between the particle and the plane or between the particle and the wedge. The maximum height climbed by the particle on the wedge is given by : **[9 April 2019 (Evening)]**

(A) $\frac{2v^2}{7g}$ (B) $\frac{v^2}{g}$

(C) $\frac{2v^2}{5g}$ (D) $\frac{v^2}{2g}$

18. Two particles, of masses M and $2M$, moving, as shown, with speeds of 10 m/s and 5 m/s, collide elastically at the origin. After the collision, they move along the indicated directions with speeds υ_1 and υ_2, respectively. The values of υ_1 and υ_2 are nearly : **[10 April 2019 (Morning)]**

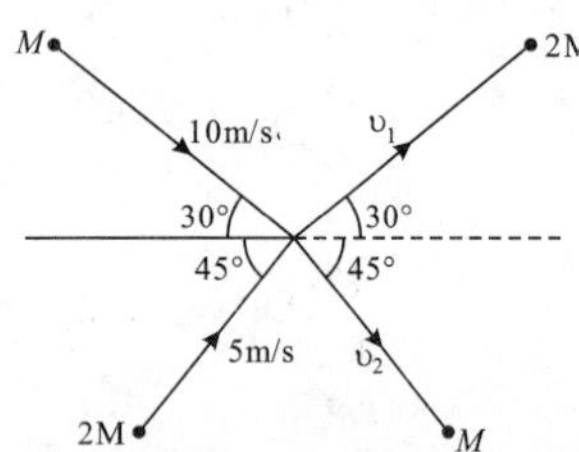

(A) 3.2 m/s and 6.3 m/s (B) 3.2 m/s and 12.6 m/s
(C) 6.5 m/s and 6.3 m/s (D) 6.5 m/s and 3.2 m/s

19. A bullet of mass 20 g has an initial speed of 1 ms^{-1}, just before it starts penetrating a mud wall of thickness 20 cm. If the wall offers a mean resistance of 2.5×10^{-2} N, the speed of the bullet after emerging from the other side of the wall is close to:

[10 April 2019 (Evening)]

(A) 0.4 ms^{-1} (B) 0.1 ms^{-1}
(3) 0.3 ms^{-1} (D) 0.7 ms^{-1}

20. A person of mass M is, sitting on a swing of length L and swinging with an angular amplitude θ_0. If the person stands up when the swing passes through its lowest point, the work done by him, assuming that his centre of mass moves by a distance $l(l << L)$, is close to : **[12 April 2019 (Morning)]**

(A) Mgl (B) $Mgl(1+\theta_0^2)$

(C) $Mgl(1-\theta_0^2)$ (D) $Mgl\left(1+\frac{\theta_0^2}{2}\right)$

21. A man (mass = 50 kg) and his son (mass = 20 kg) are standing on a frictionless surface facing each other. The man pushes his son so that he starts moving at a speed of 0.70 ms^{-1} with respect to the man. The speed of the man with respect to the surface is:

[12 April 2019 (Morning)]

(A) 0.20 ms^{-1} (B) 0.14 ms^{-1}
(C) 0.47 ms^{-1} (D) 0.28 ms^{-1}

22. Three particles of masses 50 g, 100 g and 150 g are placed at the vertices of an equilateral triangle of side 1m (as shown in the figure). The (x, y) coordinates of the centre of mass will be:

[12 April 2019 (Evening)]

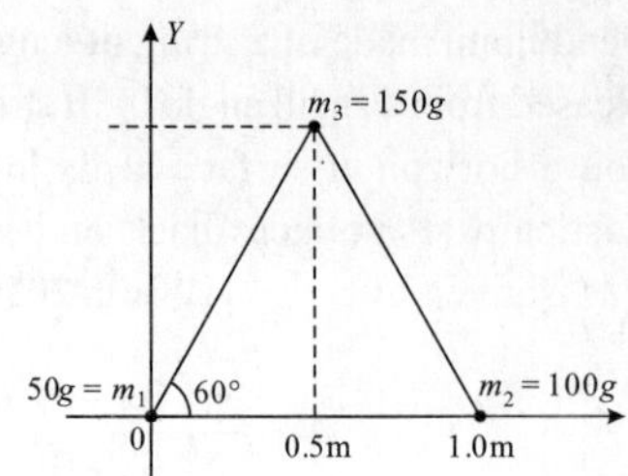

(A) $\left(\frac{7}{12}m, \frac{\sqrt{3}}{8}m\right)$ (B) $\left(\frac{\sqrt{3}}{4}m, \frac{5}{12}m\right)$

(C) $\left(\frac{7}{12}m, \frac{\sqrt{3}}{4}m\right)$ (D) $\left(\frac{\sqrt{3}}{8}m, \frac{7}{12}m\right)$

23. Three point particles of masses 1.0 kg, 1.5 kg and 2.5 kg are placed at three corners of a right angle triangle of sides 4.0 cm, 3.0 cm and 5.0 cm as shown in the figure. The center of mass of the system is at a point : **[7 Jan. 2020 (Morning)]**

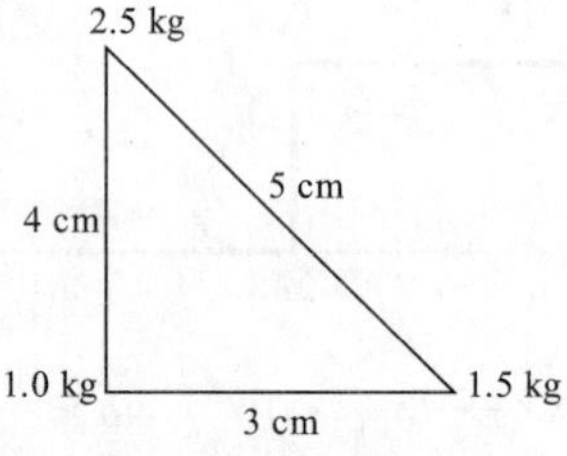

(A) 1.5 cm right and 1.2 cm above 1 kg mass
(B) 0.9 cm right and 2.0 cm above 1 kg mass
(C) 0.6 cm right and 2.0 cm above 1 kg mass
(D) 2.0 cm right and 0.9 cm above 1 kg mass

24. The coordinates of centre of mass of a uniform flag shaped lamina (thin flat plate) of mass 4 kg. (The coordinates of the same are shown in figure) are : **[8 Jan. 2020 (Morning)]**

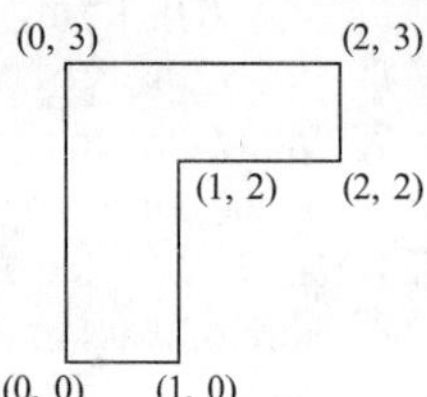

(A) (1.25 m, 1.50 m) (B) (1 m, 1.75 m)
(C) (0.75 m, 0.75 m) (D) (0.75 m, 1.75 m)

25. A body A, of mass m = 0.1 kg has an initial velocity of $3\hat{i}ms^{-1}$. It collides elastically with another body, B of the same mass which has an initial velocity of $5\hat{j}ms^{-1}$. After collision, A moves with a velocity $\vec{v} = 4(\hat{i}+\hat{j})$. The energy of B after collision is written as $\frac{x}{10}J$. The value of x is _____.

[8 Jan. 2020 (Morning)]

26. As shown in figure, when a spherical cavity (centred at O) of radius 1 is cut out of a uniform sphere of radius R (centred at C), the centre of mass of remaining (shaded) part of sphere is at G, i.e, on the surface of the cavity. R can be determined by the equation : **[8 Jan. 2020 (Evening)]**

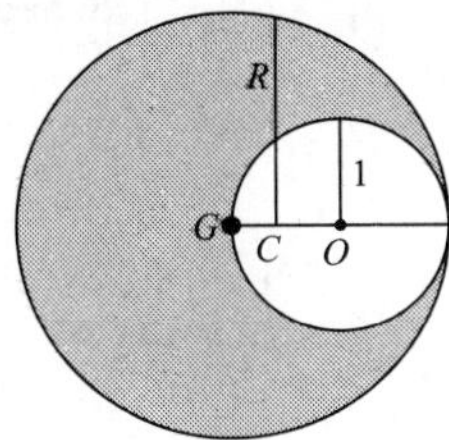

(A) $(R^2-R+1)(2-R)=1$ (B) $(R^2+R-1)(2-R)=1$
(C) $(R^2+R+1)(2-R)=1$ (D) $(R^2-R-1)(2-R)=1$

27. A particle of mass m is dropped from a height h above the ground. At the same time another particle of the same mass is thrown vertically upwards from the ground with a speed of $\sqrt{2gh}$. If they collide head-on completely inelastically, the time taken for the combined mass to reach the ground, in units of $\sqrt{\frac{h}{g}}$ is : **[8 Jan. 2020 (Evening)]**

(A) $\frac{1}{2}$ (B) $\sqrt{\frac{1}{2}}$

(C) $\sqrt{\frac{3}{4}}$ (D) $\sqrt{\frac{3}{2}}$

28. Two particles of equal mass m have respective initial velocities $u\hat{i}$ and $u\left(\frac{\hat{i}+\hat{j}}{2}\right)$. They collide completely inelastically. The energy lost in the process is : **[9 Jan. 2020 (Morning)]**

(A) $\frac{3}{4}$mu^2 (B) $\frac{1}{8}$mu^2

(C) $\sqrt{\frac{2}{3}}$mu^2 (D) $\frac{1}{3}$mu^2

29. A rod of length L has non-uniform linear mass density given by $\rho(x) = a + b\left(\frac{x}{L}\right)^2$, where a and b are constants and $0 \le x \le L$. The value of x for the centre of mass of the rod is at : **[9 Jan. 2020 (Evening)]**

(A) $\frac{4}{3}\left(\frac{a+b}{2a+3b}\right)L$ (B) $\frac{3}{2}\left(\frac{a+b}{2a+b}\right)L$

(C) $\frac{3}{2}\left(\frac{2a+b}{3a+b}\right)L$ (D) $\frac{3}{4}\left(\frac{2a+b}{3a+b}\right)L$

30. A particle of mass m is projected with a speed u from the ground at an angle $\theta = \frac{\pi}{3}$ w.r.t. horizontal (x-axis). When it has reached its maximum height, it collides completely inelastically with another particle of the same mass and velocity $u\hat{i}$. The horizontal distance covered by the combined mass before reaching the ground is : **[9 Jan. 2020 (Evening)]**

(A) $\frac{3\sqrt{2}}{4}\frac{u^2}{g}$ (B) $2\sqrt{2}\frac{u^2}{g}$

(C) $\frac{3\sqrt{3}}{8}\frac{u^2}{g}$ (D) $\frac{5}{8}\frac{u^2}{g}$

31. A particle of mass m with an initial velocity $u\hat{i}$ collides perfectly elastically with a mass $3m$ at rest. It moves with a velocity $v\hat{j}$ after collision, then, v is given by : **[2 Sept. 2020 (Morning)]**

(A) $v=\sqrt{\frac{2}{3}}u$ (B) $v=\frac{1}{\sqrt{6}}u$

(C) $v=\frac{u}{\sqrt{3}}$ (D) $v=\frac{u}{\sqrt{2}}$

32. A particle of mass m is moving along the x-axis with initial velocity $u\hat{i}$. It collides elastically with a particle of mass 10 m at rest and then moves with half its initial kinetic energy (see figure). If $\sin\theta_1 = \sqrt{n}\sin\theta_2$ then value of n is _____: **[2 Sept. 2020 (Evening)]**

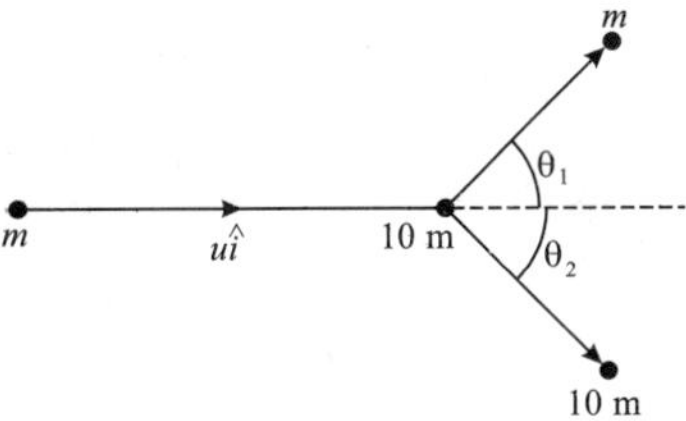

33. A square shaped hole of side $l = \frac{a}{2}$ is carved out at a distance $d = \frac{a}{2}$ from the centre 'O' of a uniform circular disk of radius a. If the distance of the centre of mass of the remaining portion from O is $-\frac{a}{X}$, value of X (to the nearest integer) is _____: **[2 Sept. 2020 (Evening)]**

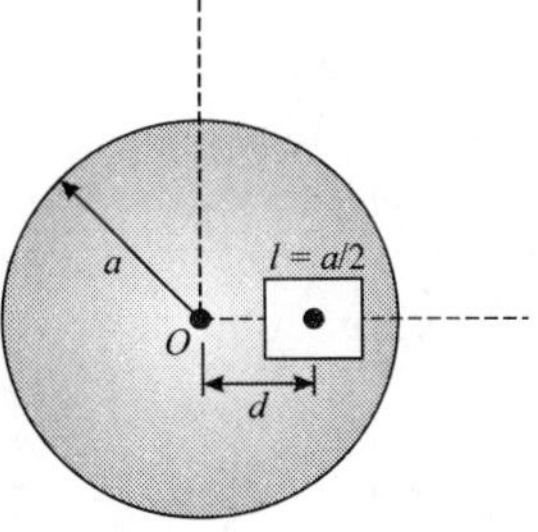

34. A block of mass 1.9 kg is at rest at the edge of a table, of height 1 m. A bullet of mass 0.1 kg collides with the block and sticks to it. If the velocity of the bullet is 20 m/s in the horizontal direction just before the collision then the kinetic energy just before the combined system strikes the floor, is [Take $g = 10$ m/s^2. Assume there is no rotational motion and loss of energy after the collision is negligible.] : **[3 Sept. 2020 (Evening)]**

(A) 21 J (B) 23 J
(C) 19 J (D) 20 J

35. A Tennis ball is released from a height h and after freely falling on a wooden floor it rebounds and reaches height $\frac{h}{2}$. The velocity versus height of the ball during its motion may be represented graphically by. (Graph are drawn schematically and on not to scale) : **[4 Sept. 2020 (Morning)]**

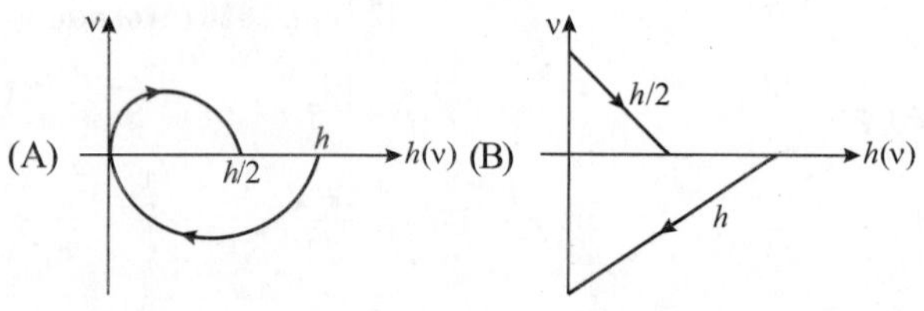

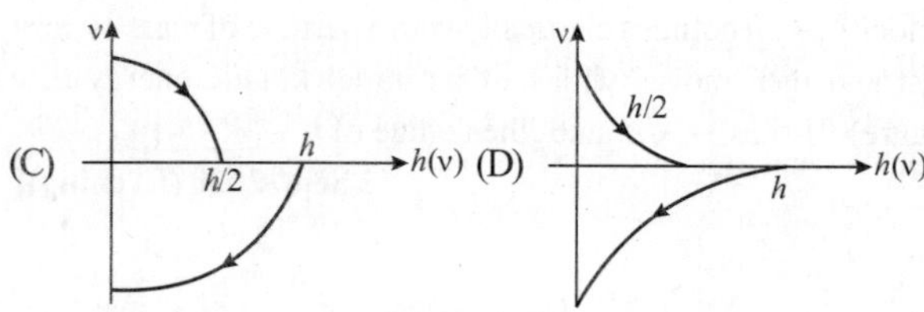

36. Blocks of masses m, $2m$, $4m$ and $8m$ are arranged in a line on a frictionless floor. Another block of mass m, moving with speed v along the same line (see figure) collides with mass m in perfectly inelastic manner. All the subsequent collisions are also perfectly inelastic. By the time the last block of mass $8m$ starts moving the total energy loss is p% of the original energy. Value of 'p' is close to : **[4 Sept. 2020 (Morning)]**

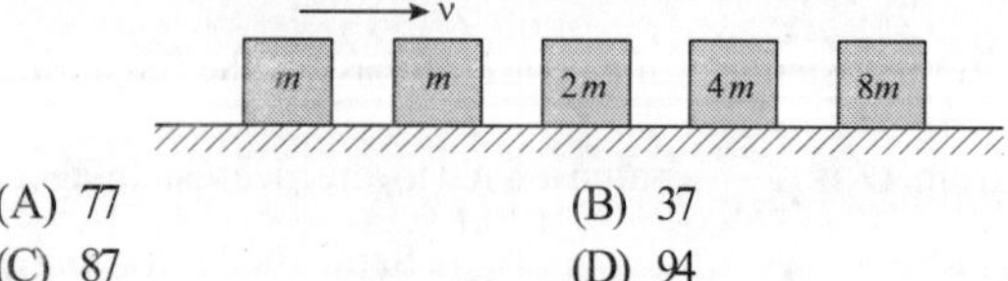

(A) 77 (B) 37
(C) 87 (D) 94

37. Two bodies of the same mass are moving with the same speed, but in different directions in a plane. They have a completely inelastic collision and move together thereafter with a final speed which is half of their initial speed. The angle between the initial velocities of the two bodies (in degree) is _____: **[6 Sept. 2020 (Morning)]**

38. Particle A of mass m_1 moving with velocity $(\sqrt{3}\hat{i} + \hat{j})$ ms^{-1} collides with another particle B of mass m_2 which is at rest initially. Let $\vec{V_1}$ and $\vec{V_2}$ be the velocities of particles A and B after collision respectively. If $m_1 = 2m_2$ and after collision $\vec{V_1} = (\hat{i} + \sqrt{3}\hat{j})$ms^{-1}, the angle between $\vec{V_1}$ and $\vec{V_2}$ is : **[6 Sept. 2020 (Evening)]**

(A) 60° (B) 15°
(C) –45° (D) 105°

39. The centre of mass of a solid hemisphere of radius 8 cm is X cm from the centre of the flat surface. Then value of x is : **[6 Sept. 2020 (Evening)]**

40. A ball with a speed of 9 m/s collides with another identical ball at rest. After the collision, the direction of each ball makes an angle of 30° with the original direction. The ratio of velocities of the balls after collision is $x : y$, where x is _____. **[24 Feb. 2021 (Morning)]**

41. A circular hole of radius $\left(\frac{a}{2}\right)$ is cut out of a circular disc of radius 'a' as shown in figure. The centroid of the remaining circular portion with respect to point 'O' will be : **[24 Feb. 2021 (Evening)]**

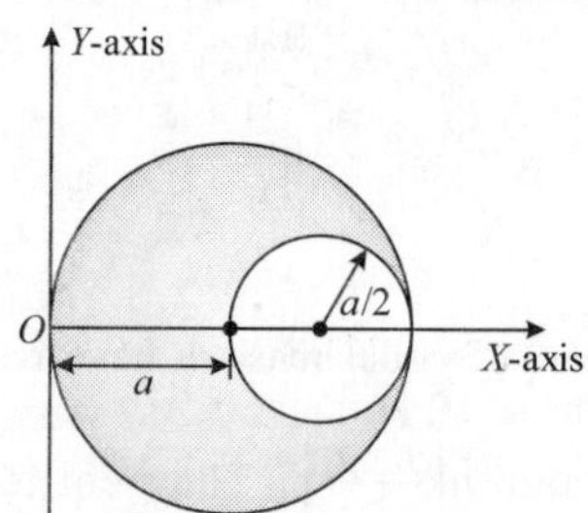

(A) $\frac{5}{6}a$ (B) $\frac{2}{3}a$
(C) $\frac{1}{6}a$ (D) $\frac{10}{11}a$

42. Two solids A and B of mass 1 kg and 2 kg respectively are moving with equal linear momentum. The ratio of their kinetic energies (K.E.)$_A$: (K.E.)$_B$ will be $\frac{A}{1}$, so the value of A will be _____. **[24 Feb. 2021 (Evening)]**

43. Two particles having masses 4g and 16g respectively are moving with equal kinetic energies. The ratio of the magnitudes of their linear momentum is $n : 2$. The value of n will be _____. **[25 Feb. 2021 (Evening)]**

44. Given below are two statements : one is labelled as **Assertion A** and the other is labelled as **Reason R**

Assertion-A : Body 'P' having mass M moving with speed 'u' has head-on collision elastically with another body 'Q' having mass 'm' initially at rest. If $m << M$, body 'Q' will have a maximum speed equal to '$2u$' after collision.

Reason-R : During elastic collision, the momentum and kinetic energy are both conserved

In the light of the above statement, choose the most appropriate answer from the options given below :

[26 Feb. 2021 (Morning)]

(A) **A** is not correct but **R** is correct

(B) Both **A** and **R** are correct and **R** is the correct explanation of **A**

(C) **A** is correct but **R** is not correct

(D) Both **A** and **R** are correct but **R** is NOT the correct explanation of **A**

45. Two masses A and B, each of mass M are fixed together by a massless spring. A force acts on the mass B as shown in figure. If the mass A starts moving away from mass B with acceleration 'a', then the acceleration of mass B will be :

[26 Feb. 2021 (Evening)]

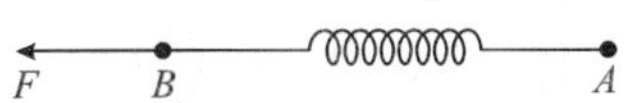

(A) $\frac{MF}{F + Ma}$ (B) $\frac{F + Ma}{M}$

(C) $\frac{F - Ma}{M}$ (D) $\frac{Ma - F}{M}$

46. A ball of mass 10 kg moving with a velocity $10\sqrt{3}\,ms^{-1}$ along X-axis, hits another ball of mass 20 kg which is at rest. After collision, the first ball comes to rest and the second one disintegrates into two equal pieces. One of the pieces starts moving along Y-axis at a speed of 10 m/s. The second piece starts moving at a speed of 20 m/s at an angle θ (degree) with respect to the X-axis.

The configuration of pieces after collision is shown in the figure. The value of θ to the nearest integer is _____.

[16 March 2021 (Morning)]

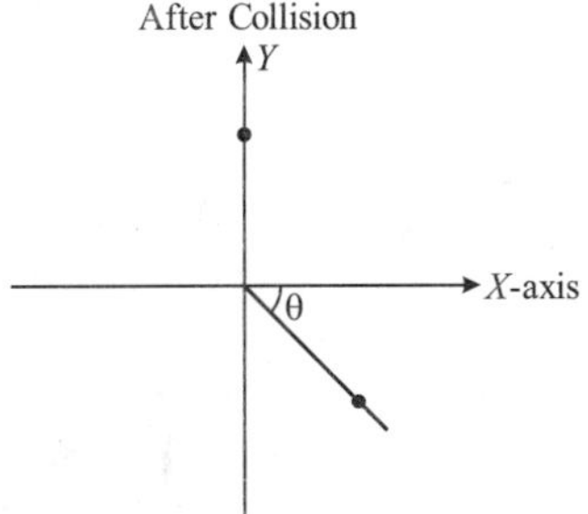

47. A large block of wood of mass $M = 5.99$ kg is hanging from two long massless cords. A bullet of mass $m = 10$g is fired into the block and gets embedded in it. The (block + bullet) then swing upwards, their centre of mass rising a vertical distance $h = 9.8$ cm before the (block + bullet) pendulum comes momentarily to rest at the end of its arc. The speed of the bullet just before collision is : (Take $g = 9.8\,ms^{-2}$)

[16 March 2021 (Evening)]

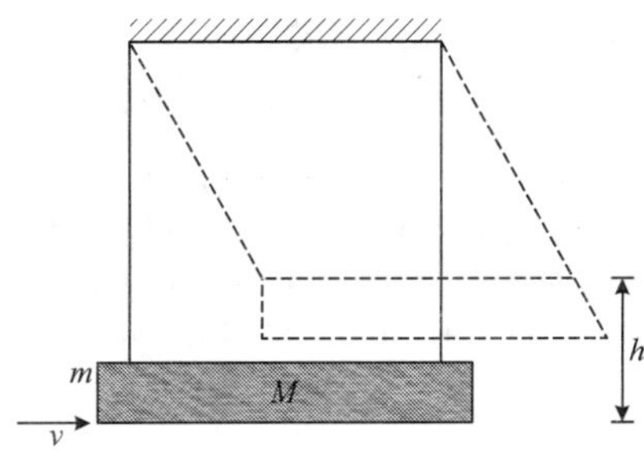

(A) 841.4 m/s (B) 811.4 m/s

(C) 831.4 m/s (D) 821.4 m/s

48. A rubber ball is released from a height of 5 m above the floor. It bounces back repeatedly, always rising to $\frac{81}{100}$ of the height through which it falls. Find the average speed of the ball. (Take g = 10 ms^{-2}) **[17 March, 2021 (Evening)]**

(A) 3.0 ms^{-1} (B) 3.50 ms^{-1}

(C) 2.0 ms^{-1} (D) 2.50 ms^{-1}

49. The disc of mass M with uniform surface mass density σ is shown in the figure. The centre of mass of the quarter disc (the shaded area) is at the position $\frac{x}{3}\frac{a}{\pi}, \frac{x}{3}\frac{a}{\pi}$ where x is _____.

(Round off to the Nearest Integer)

[a is an area as shown in the figure] **[17 March 2021 (Evening)]**

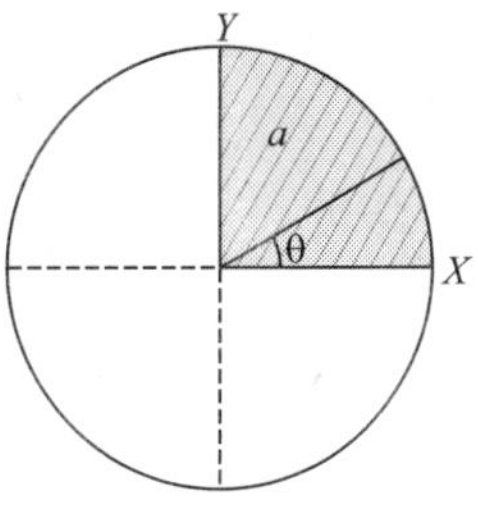

50. A ball of mass 10 kg moving with a velocity $10\sqrt{3}$ m/s along the x-axis, hits another ball of mass 20 kg which is at rest. After the collision, first ball comes to rest while the second ball disintegrates into two equal pieces. One piece starts moving along y-axis with a speed of 10 m/s. The second piece starts moving at an angle of 30° with respect to the x-axis. The velocity of the ball moving at 30° with x-axis is x m/s. The configuration of pieces after collision is shown in the figure below. The value of x to the nearest integer is _____.

[18 March 2021 (Morning)]

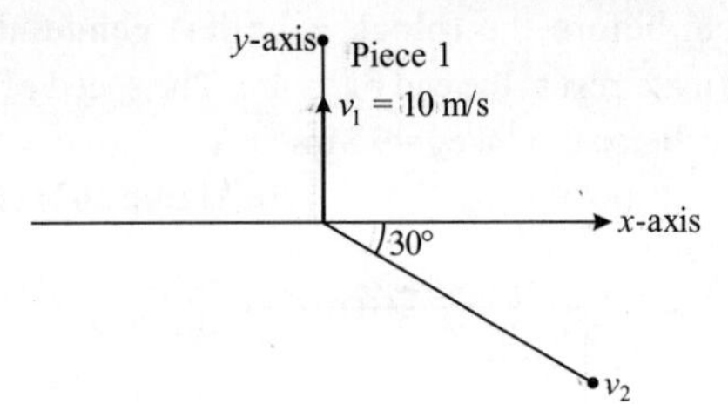

51. An object of mass m_1 collides with another object of mass m_2, which is at rest. After the collision the objects move with equal speeds in opposite direction. The ratio of the masses $m_2 : m_1$ is : **[18 March 2021 (Evening)]**

(A) 3 : 1 (B) 2 : 1

(C) 1 : 2 (D) 1 : 1

52. A bullet of 4 g mass is fired from a gun of mass 4 kg. If the bullet moves the muzzle speed of 50 ms^{-1}, the impulse imparted to the gun and velocity of recoil of gun are :

[22 July 2021 (Evening)]

(A) 0.4 kg ms^{-1}, 0.1 ms^{-1} (B) 0.2 kg ms^{-1}, 0.1 ms^{-1}

(C) 0.2 kg ms^{-1}, 0.05 ms^{-1} (D) 0.4 kg ms^{-1}, 0.05 ms^{-1}

53. The position of the centre of mass of a uniform semi-circular wire of radius 'R' placed in x-y plane with its centre at the origin and the line joining its ends as x-axis is given by $\left(0, \frac{xR}{\pi}\right)$. Then, the value of $|x|$ is _____. **[22 July 2021 (Evening)]**

54. Three objects A, B and C are kept in a straight line on a frictionless horizontal surface. The masses of A, B and C are m, $2m$ and $2m$ respectively. A moves towards B with a speed of 9 m/s and makes an elastic collision with it. Thereafter B makes a completely inelastic collision with C. All motions occur along same straight line. The final speed of C is :

[27 July 2021 (Morning)]

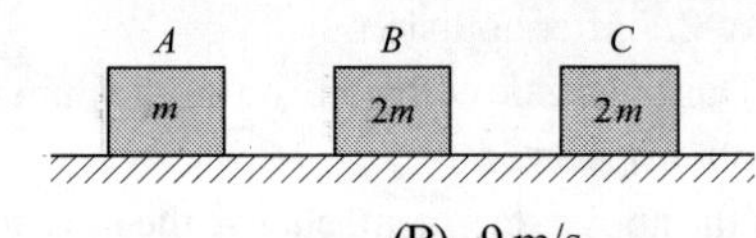

(A) 3 m/s (B) 9 m/s

(C) 4 m/s (D) 6 m/s

55. A bullet of 10g, moving with velocity v, collides head-on with the stationary bob of a pendulum and recoils with velocity 100 m/s. The length of the pendulum is 0.5 m and mass of the bob is 1 kg. The minimum value of v = _____ m/s so that the pendulum describes a circle. (Assume the string to be inextensible and $g = 10$ m/s^2). **[27 Aug. 2021 (Evening)]**

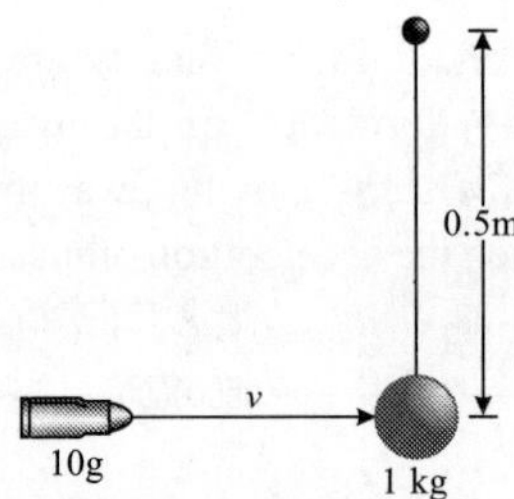

56. A body of mass M moving at speed V_0 collides elastically with another body of same mass m at rest. After the collision, the two masses move at angles θ_1 and θ_2 with respect to the initial direction of motion of the body of mass M. The largest possible value of the ratio M/m, for which the angles θ_1 and θ_2 will be equal, is : **[31 Aug. 2021 (Morning)]**

(A) 1 (B) 4

(C) 2 (D) 3

* * * * *

CHAPTER 6

SYSTEM OF PARTICLES & ROTATIONAL MOTION

6.2 RIGID BODY DYNAMICS

1. The moment of inertia of a uniform cylinder of length l and radius R about its perpendicular bisector is I. What is the ratio l/R such that the moment of inertia is minimum ? **[2017]**

(A) $\frac{\sqrt{3}}{2}$ (B) 1

(C) $\frac{3}{\sqrt{2}}$ (D) $\sqrt{\frac{3}{2}}$

2. A slender uniform rod of mass M and length l is pivoted at one end so that it can rotate in a vertical plane (see figure.) There is negligible friction at the pivot. The free end is held vertically above the pivot and then released. The angular acceleration of the rod when it make an angle θ with the vertical is : **[2017]**

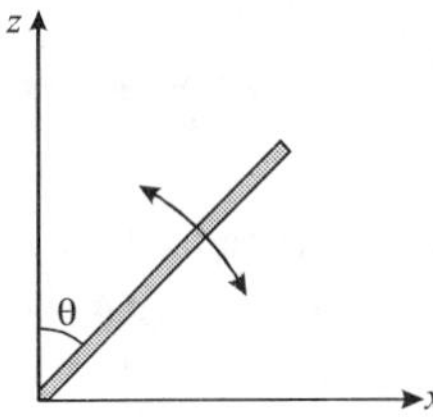

(A) $\frac{3g}{3l}\sin\theta$ (B) $\frac{3g}{3l}\cos\theta$

(C) $\frac{2g}{3l}\cos\theta$ (D) $\frac{3g}{2l}\sin\theta$

3. A circular hole of radius $R/4$ is made in a thin uniform disc having mass M and radius R, as shown in figure. The moment of inertia of the remaining portion of the disc about an axis passing through the point O and perpendicular to the plane of the disc is : **[2017 (Online)]**

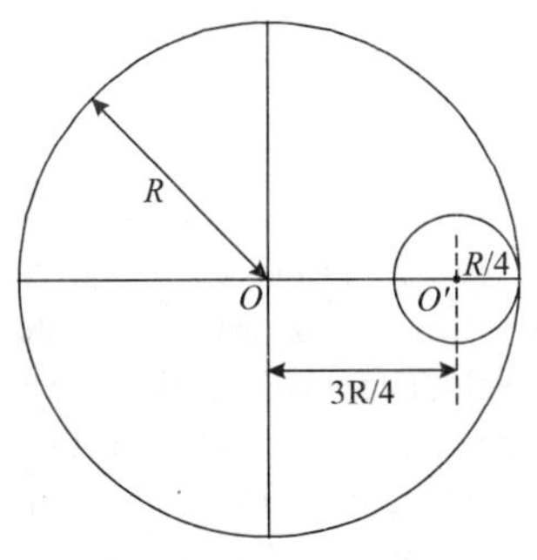

(A) $\frac{219MR^2}{256}$ (B) $\frac{237MR^2}{512}$

(C) $\frac{19MR^2}{512}$ (D) $\frac{197MR^2}{256}$

4. Moment of inertia of an equilateral triangular lamina ABC, about the axis passing through its centre O and perpendicular to its plane is I_0 as shown in the figure. A cavity DEF is cut out from the lamina, where D, E, F are the mid points of the sides. Moment of inertia of the remaining part of lamina about the same axis is : **[2017 (Online)]**

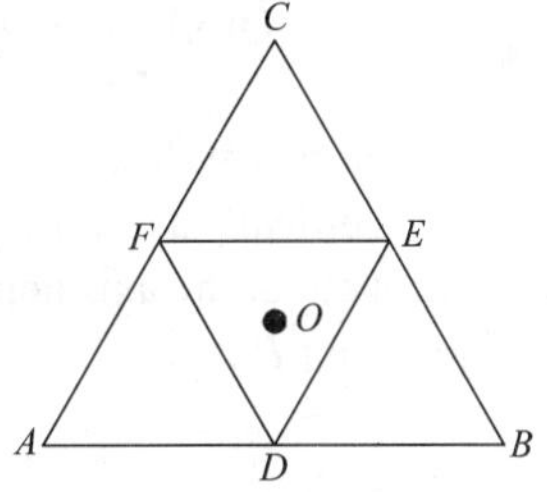

(A) $\frac{7}{8}I_0$ (B) $\frac{15}{16}I_0$

(C) $\frac{3I_0}{4}$ (D) $\frac{31I_0}{32}$

5. The machine as shown has 2 rods of length 1 m connected by a pivot at the top. The end of one rod is connected to the floor by a stationary pivot and the end of the other rod has a roller that rolls along the floor in a slot. As the roller goes back and forth, a 2 kg weight moves up and down. If the roller is moving towards right at a constant speed, the weight moves up with a : **[2017 (Online)]**

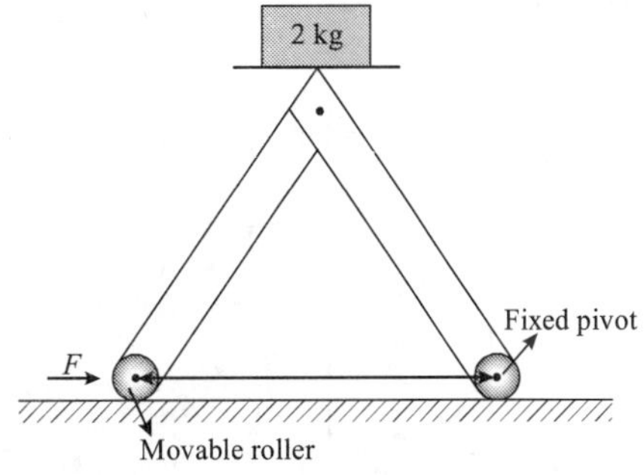

(A) Constant speed (B) Decreasing speed

(C) Increasing speed

(D) Speed which is 3/4th of that of the roller when the weight is 0.4 m above the ground

6. In a physical balance working on the principle of moments, when 5 mg weight is placed on the left pan, the beam becomes horizontal. Both the empty pans of the balance are of equal mass. Which of the following statements is correct ? **[2017 (Online)]**

(A) Left arm is longer than the right arm

(B) Both the arms are of same length

(C) Left arm is shorter than the right arm

(D) Every object that is weighed using this balance appears lighter than its actual weight

7. A particle is moving with a uniform speed in a circular orbit of radius R in a central force inversely proportional to the nth power of R. If the period of rotation of the particle is T, then : **[2018]**

(A) $T \propto R^{3/2}$ for any n (B) $T \propto R^{n/2+1}$

(C) $T \propto R^{(n+1)/2}$ (D) $T \propto R^{n/2}$

8. Seven identical circular planar disks, each of mass M and radius R are welded symmetrically as shown. The moment of inertia of the arrangement about the axis normal to the plane and passing through the point P is : **[2018]**

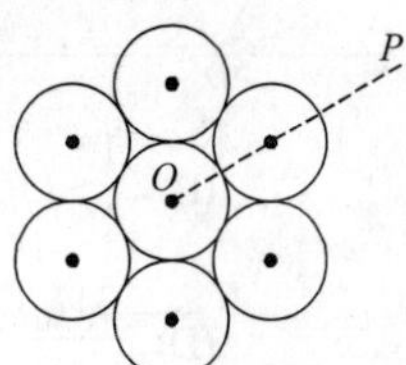

(A) $\frac{19}{2}MR^2$ (B) $\frac{55}{2}MR^2$

(C) $\frac{73}{2}MR^2$ (D) $\frac{181}{2}MR^2$

9. From a uniform circular disc of radius R and mass $9M$, a small disc of radius $R/3$ is removed as shown in the figure. The moment of inertia of the remaining disc about an axis perpendicular to the plane of the disc and passing, through centre of disc is : **[2018]**

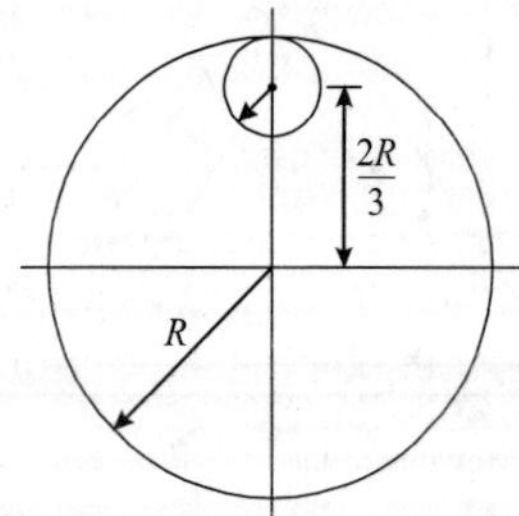

(A) $4\,MR^2$ (B) $\frac{40}{9}MR^2$

(C) $10\,MR^2$ (D) $\frac{37}{9}MR^2$

10. A force of 40N acts on a point B a the end of an L-shaped object, as shown in the figure. The angle θ that will produce maximum moment of the force about point A is given by : **[2018 (Online)]**

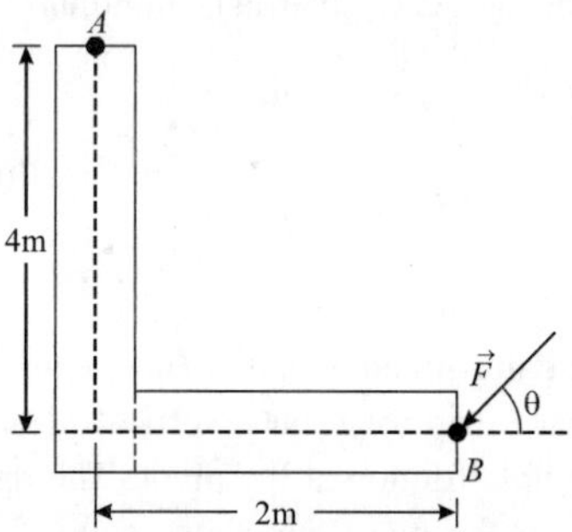

(A) $\tan\theta = \frac{1}{4}$ (B) $\tan\theta = 2$

(C) $\tan\theta = \frac{1}{2}$ (D) $\tan\theta = 4$

11. A uniform rod AB is suspended from a point X, at a variable distance from x from A, as shown. To make the rod horizontal, a mass m is suspended from its end A. A set of (m, x) values is recorded. The appropriate variable that give a straight line, when plotted, are : **[2018 (Online)]**

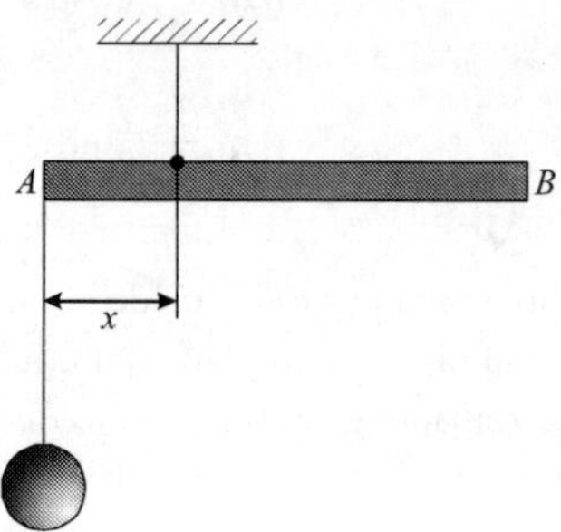

(A) $m, \frac{1}{x}$ (B) $m, \frac{1}{x^2}$

(C) m, x (D) m, x^2

12. A thin uniform bar of length L and mass $8m$ lies on a smooth horizontal table. Two point masses m and $2m$ moving in the same horizontal plane from opposite sides of the bar with speeds $2v$ and v respectively. The masses stick to the bar after collision at a distance $L/3$ and $L/6$ respectively from the centre of the bar. If the bar starts rotating about its center of mass as a result of collision, the angular speed of the bar will be : **[2018 (Online)]**

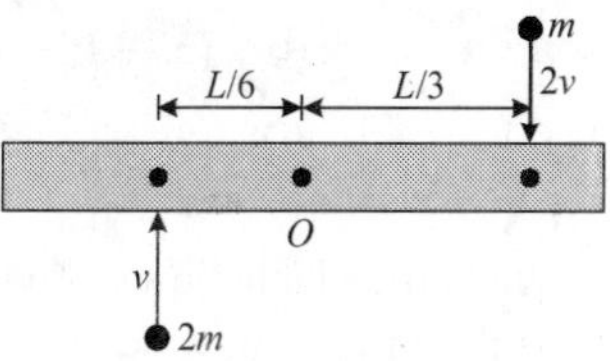

(A) $\frac{v}{6L}$ (B) $\frac{6v}{5L}$

(C) $\frac{3v}{5L}$ (D) $\frac{v}{5L}$

13. A thin circular disc is in the xy plane as shown in the figure. The ratio of its moment of inertia about z and z' axes will be : **[2018 (Online)]**

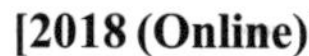

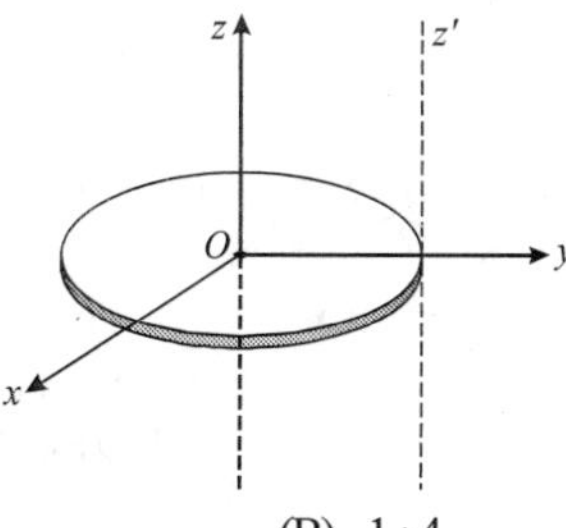

(A) 1 : 2 (B) 1 : 4

(C) 1 : 3 (D) 1 : 5

14. A thin rod MN, free the rotate in the vertical plane about the fixed end N, is held horizontal. When the end M is released the speed of this end, when the rod makes an angle α with the horizontal, will be proportional to (see figure) : **[2018 (Online)]**

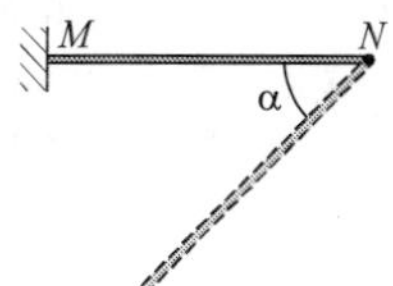

(A) $\sqrt{\cos\alpha}$ (B) $\cos\alpha$

(C) $\sin\alpha$ (D) $\sqrt{\sin\alpha}$

15. An L-shaped object, made of thin rods of uniform mass density, is suspended with a string as shown in figure. If $AB = BC$, and the angle made by AB with downward vertical is θ, then : **[9 Jan. 2019 (Morning)]**

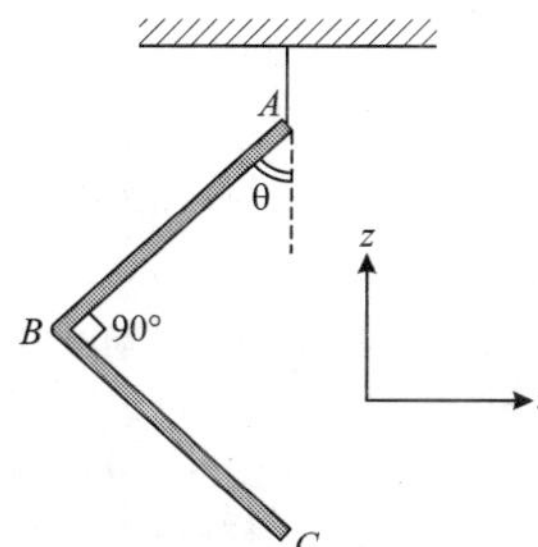

(A) $\tan\theta = \frac{2}{\sqrt{3}}$ (B) $\tan\theta = \frac{1}{3}$

(C) $\tan\theta = \frac{1}{2}$ (D) $\tan\theta = \frac{1}{2\sqrt{3}}$

16. A rod of length 50 cm is pivoted at one end. It is raised such that it makes an angle of 30° from the horizontal as shown and released from rest. Its angular speed when it passes through the horizontal (in rad s^{-1}) will be ($g = 10\ ms^{-2}$) : **[9 Jan. 2019 (Evening)]**

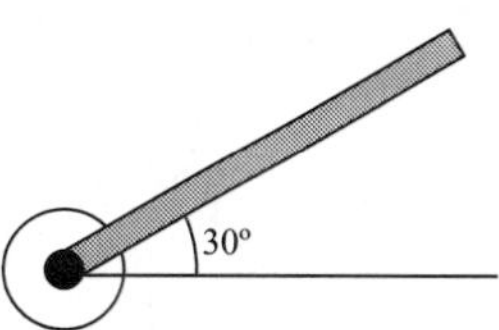

(A) $\sqrt{30}$ (B) $\sqrt{\frac{30}{2}}$

(C) $\frac{\sqrt{30}}{2}$ (D) $\frac{\sqrt{20}}{3}$

17. To mop-clean a floor, a cleaning machine presses a circular mop of radius R vertically down with a total force F and rotates it with a constant angular speed about its axis. If the force F is distributed uniformly over the mop and if coefficient of friction between the mop and the floor is μ, the torque, applied by the machine on the mop is : **[10 Jan. 2019 (Morning)]**

(A) $\frac{2}{3}\mu FR$ (B) $\mu FR/3$

(C) $\mu FR/2$ (D) $\mu FR/6$

18. A homogeneous solid cylindrical roller of radius R and mass M is pulled on a cricket pitch by a horizontal force. Assuming rolling without slipping, angular acceleration of the cylinder is: **[10 Jan. 2019 (Morning)]**

(A) $\frac{3F}{2mR}$ (B) $\frac{F}{3mR}$

(C) $\frac{2F}{3mR}$ (D) $\frac{F}{2mR}$

19. A rigid massless rod of length $3l$ has two masses attached at each end as shown in the figure. The rod is pivoted at point P on the horizontal axis (see figure). When released from initial horizontal position, its instantaneous angular acceleration will be : **[10 Jan. 2019 (Evening)]**

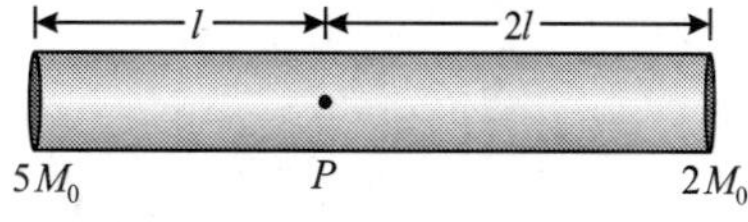

(A) $\frac{g}{2l}$ (B) $\frac{7g}{3l}$

(C) $\frac{g}{13l}$ (D) $\frac{g}{3l}$

20. Two identical spherical balls of mass M and radius R each are stuck on two ends of a rod of length $2R$ and mass M (see figure). The moment of inertia of the system about the axis passing perpendicularly through the centre of the rod is :

[10 Jan. 2019 (Evening)]

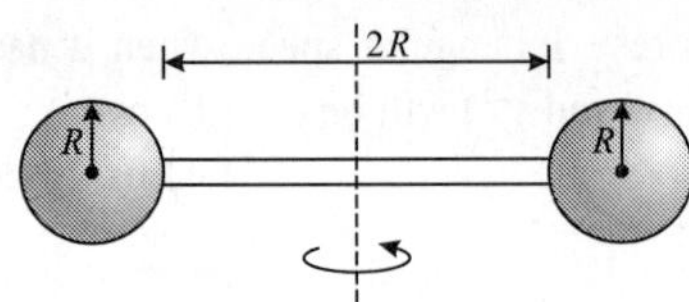

(A) $\frac{152}{15}MR^2$ (B) $\frac{17}{15}MR^2$

(C) $\frac{137}{15}MR^2$ (D) $\frac{209}{15}MR^2$

21. An equilateral triangle ABC is cut from a thin solid sheet of wood. (see figure) D, E and F are the mid-points of its sides as shown and G is the centre of the triangle. The moment of inertia of the triangle about an axis passing through G and perpendicular to the plane of the triangle is I_0. It the smaller triangle DEF is removed from ABC, the moment of inertia of the remaining figure about the same axis is I. Then :

[11 Jan. 2019 (Morning)]

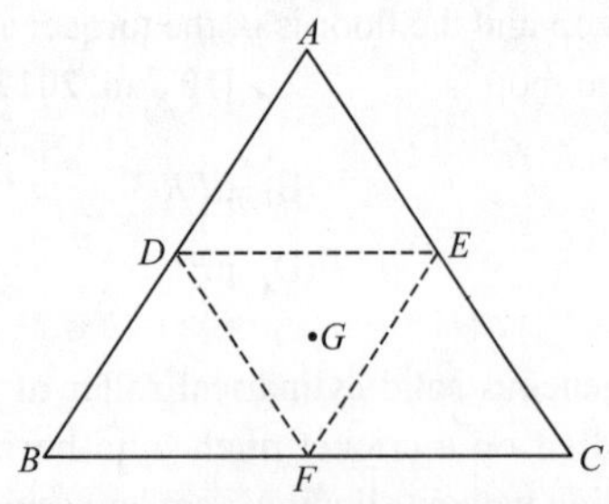

(A) $I = \frac{9}{16}I_0$ (B) $I = \frac{3}{4}I_0$

(C) $I = \frac{I_0}{4}$ (D) $I = \frac{15}{16}I_0$

22. A slab is subjected to two forces $\vec{F_1}$ and $\vec{F_2}$ of same magnitude F as shown in the figure. Force $\vec{F_2}$ is in XY-plane while force F_1 acts along z-axis at the point $(2\vec{i} + 3\vec{j})$. The moment of these forces about point O will be :

[11 Jan. 2019 (Morning)]

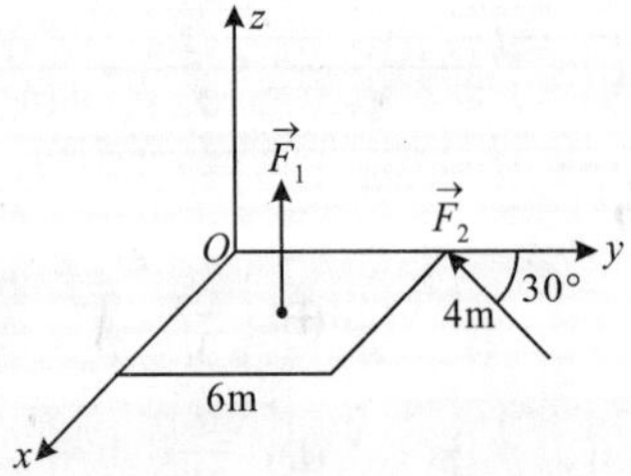

(A) $(3\hat{i} - 2\hat{j} - 3\hat{k})F$ (B) $(3\hat{i} + 2\hat{j} + 3\hat{k})F$

(C) $(3\hat{i} + 2\hat{j} - 3\hat{k})F$ (D) $(3\hat{i} - 2\hat{j} + 3\hat{k})F$

23. A string is wound around a hollow cylinder of mass 5 kg and radius 0.5 m. If the string is now pulled with a horizontal force of 40 N, and the cylinder is rolling without slipping on a horizontal surface (see figure), then the angular acceleration of the cylinder will be (Neglect the mass and thickness of the string) :

[11 Jan. 2019 (Evening)]

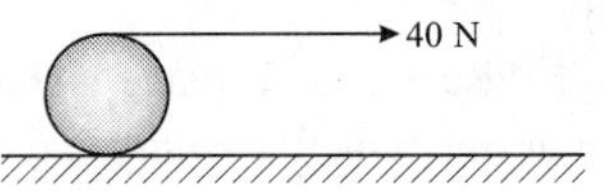

(A) 12 rad/s^2 (B) 16 rad/s^2

(C) 10 rad/s^2 (D) 20 rad/s^2

24. The magnitude of torque on a particle of mass 1 kg is 2.5 Nm about the origin. If the force acting on it is 1 N, and the distance of the particle from the origin is 5 m, the angle between the force and the position vector is (in radians) :

[11 Jan. 2019 (Evening)]

(A) $\frac{\pi}{8}$ (B) $\frac{\pi}{6}$

(C) $\frac{\pi}{4}$ (D) $\frac{\pi}{3}$

25. A circular disc D_1 of mass M and radius R has two identical discs D_2 and D_3 of the same mass M and radius R attached rigidly at its opposite ends (see figure). The moment of inertia of the system about the axis OO', passing through the centre of D_1, as shown in the figure, will be: **[11 Jan. 2019 (Evening)]**

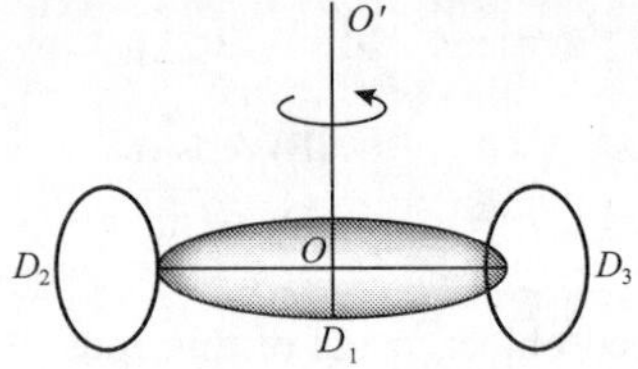

(A) $3MR^2$ (B) $\frac{2}{3}MR^2$

(C) MR^2 (D) $\frac{4}{5}MR^2$

26. Let the moment of inertia of a hollow cylinder of length 30 cm (inner radius 10 cm and outer radius 20 cm), about its axis be I. The radius of a thin cylinder of the same mass such that its moment of inertia about its axis is also I, is :

[12 Jan. 2019 (Morning)]

(A) 12 cm (B) 18 cm

(C) 16 cm (D) 14 cm

27. A particle of mass 20g is released with an initial velocity 5 m/s along the curve from the point A, as shown in the figure. The point A is at height h from point B. The particle slides along the frictionless surface. When the particle reaches point B, its angular momentum about O will be : [Take $g = 10$ m/s^2]
[12 Jan. 2019 (Evening)]

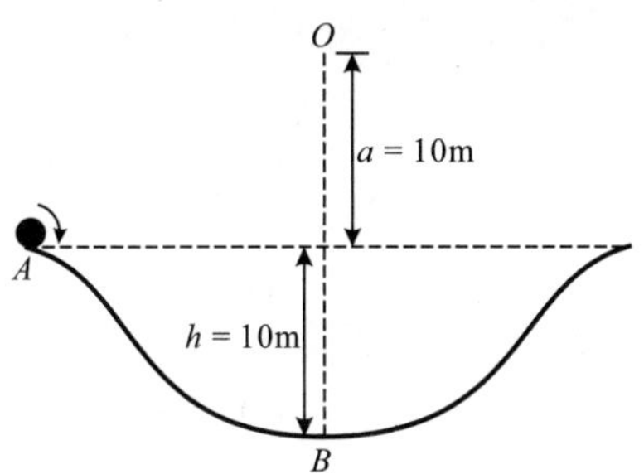

(A) 8 kg-m^2/s (B) 6 kg-m^2/s
(C) 3 kg-m^2/s (D) 2 kg-m^2/s

28. The moment of inertia of a solid sphere, about an axis parallel to its diameter and at a distance of x from it, is $I(x)'$. Which one of the graphs represents the variation of $I(x)$ with x correctly?
[12 Jan. 2019 (Evening)]

(A) (B)

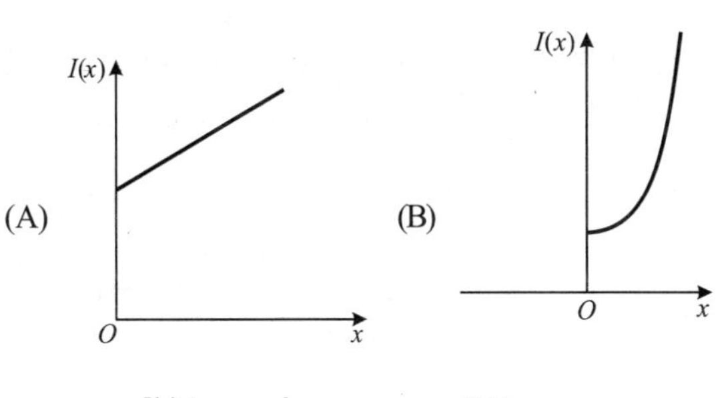

(C) (D)

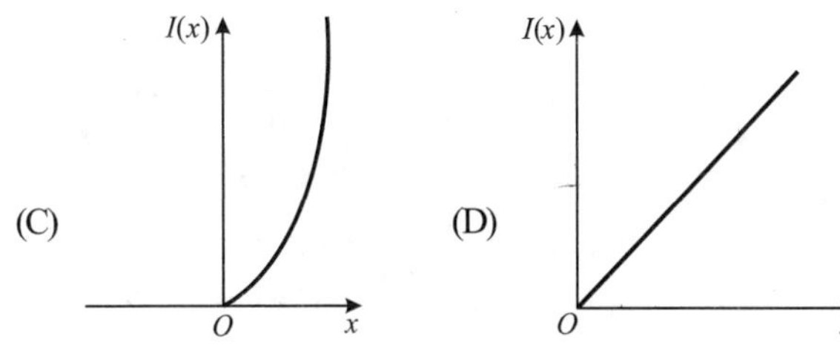

29. A thin circular plate of mass M and radius R has its density varying as $\rho(r) = \rho_0 r$ with ρ_0 as constant and r is the distance from its centre. The moment of Inertia of the circular plate about an axis perpendicular to the plate and passing through its edge is $I = aMR^2$. The value of the coefficient a is :
[8 April 2019 (Morning)]

(A) $\frac{3}{2}$ (B) $\frac{1}{2}$

(C) $\frac{3}{5}$ (D) $\frac{8}{5}$

30. A solid sphere and solid cylinder of identical radii approach an incline with the same linear velocity (see figure). Both roll without slipping all throughout. The two climb maximum heights h_{sph} and h_{cyl} on the incline. The ratio $\frac{h_{sph}}{h_{cyl}}$ is given by :
[8 April 2019 (Evening)]

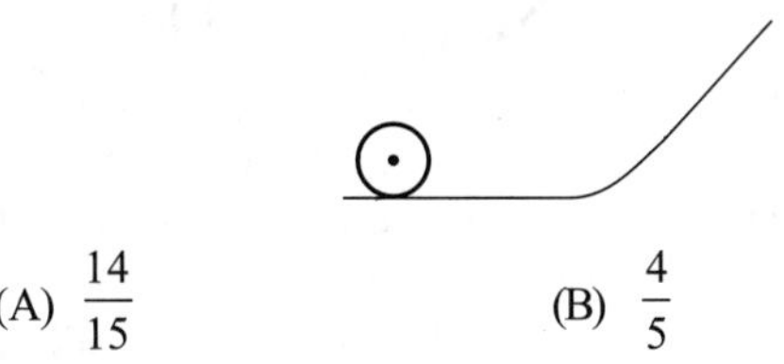

(A) $\frac{14}{15}$ (B) $\frac{4}{5}$

(C) 1 (D) $\frac{2}{\sqrt{5}}$

31. A rectangular solid box of length 0.3 m is held horizontally, with one of its sides on the edge of a platform of height 5 m. When released, it slips off the table in a very short time $\tau = 0.01$s, remaining essentially horizontal. The angle by which it would rotate when it hits the ground will be (in radians) close to :
[8 April 2019 (Evening)]

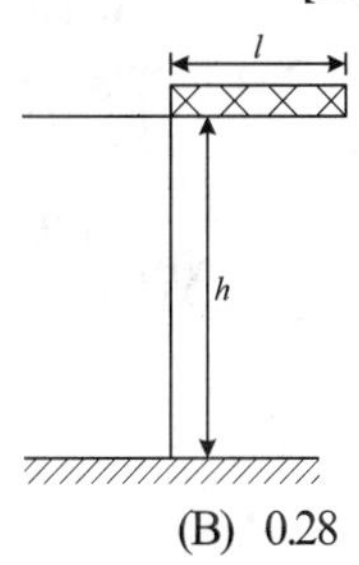

(A) 0.02 (B) 0.28
(C) 0.5 (D) 0.3

32. The following bodies are made to roll up (without slipping) the same inclined plane from a horizontal plane : (i) a ring of radius R, (ii) a solid cylinder of radius $\frac{R}{2}$ and (iii) a solid sphere of radius $\frac{R}{4}$. If in each case, the speed of the centre of mass at the bottom of the incline is same, the ratio of the maximum heights they climb is :
[9 April 2019 (Morning)]

(A) 4 : 3 : 2 (B) 14 : 15 : 20
(C) 10 : 15 : 7 (D) 2 : 3 : 4

33. A stationary horizontal disc is free to rotate about its axis. When a torque is applied on it, its kinetic energy as a function of θ, where θ is the angle by which it has rotated, is given as $k\theta^2$. If its moment of inertia is I then the angular acceleration of the disc is :
[9 April 2019 (Morning)]

(A) $\frac{k}{2I}\theta$ (B) $\frac{k}{I}\theta$

(C) $\frac{k}{4I}\theta$ (D) $\frac{2k}{I}\theta$

34. Moment of inertia of a body about a given axis is 1.5 kg m^2. Initially the body is at rest. In order to produce a rotational kinetic energy of 1200 J, the angular acceleration of 20 rad/s^2 must be applied about the axis for a duration of :

[9 April 2019 (Evening)]

(A) 2s (B) 5s

(C) 2.5s (D) 3s

35. A thin smooth rod of length L and mass M is rotating freely with angular speed ω_0 about an axis perpendicular to the rod and passing through its center. Two beads of mass m and negligible size are at the center of the rod initially. The beads are free to slide along the rod. The angular speed of the system, when the beads reach the opposite ends of the rod, will be :

[9 April 2019 (Evening)]

(A) $\dfrac{M\omega_0}{M+3m}$ (B) $\dfrac{M\omega_0}{M+m}$

(C) $\dfrac{M\omega_0}{M+2m}$ (D) $\dfrac{M\omega_0}{M+6m}$

36. A thin disc of mass M and radius R has mass per unit area $\sigma(r) = kr^2$ where r is the distance from its centre. Its moment of inertia about an axis going through its centre of mass and perpendicular to its plane is : **[10 April 2019 (Morning)]**

(A) $\dfrac{MR^2}{6}$ (B) $\dfrac{MR^2}{3}$

(C) $\dfrac{2MR^2}{3}$ (D) $\dfrac{MR^2}{2}$

37. Two coaxial discs, having moments of inertia I_1 and $\dfrac{I_1}{2}$, are rotating with respective angular velocities ω_1 and $\dfrac{\omega_1}{2}$, about their common axis. They are brought in contact with each other and thereafter they rotate with a common angular velocity. If E_f and E_i are the final and initial total energies, then $(E_f - E_i)$ is :

[10 April 2019 (Morning)]

(A) $\dfrac{I_1\omega_1^2}{12}$ (B) $\dfrac{3}{8}I_1\omega_1^2$

(C) $\dfrac{I_1\omega_1^2}{6}$ (D) $\dfrac{I_1\omega_1^2}{24}$

38. A particle of mass m is moving along a trajectory given by

$x = x_0 + a\cos\omega_1 t$

$y = y_0 + b\sin\omega_2 t$

The torque, acting on the particle about the origin, at $t = 0$ is :

[10 April 2019 (Morning)]

(A) $m(-x_0 b + y_0 a)\,\omega_1^2\hat{k}$ (B) $+my_0 a\,\omega_1^2\hat{k}$

(C) $-m(x_0 b\omega_2^2 - y_0 a\omega_1^2)\hat{k}$ (D) Zero

39. A metal coin of mass $5g$ and radius 1cm is fixed to a thin stick AB of negligible mass as shown in the figure. The system is initially at rest. The constant torque, that will make the system rotate about AB at 25 rotations per second in 5 s, is close to :

[10 April 2019 (Evening)]

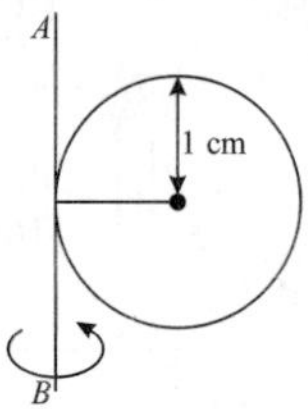

(A) 4.0×10^{-6} Nm (B) 2.0×10^{-5} Nm

(C) 1.6×10^{-5} Nm (D) 7.9×10^{-6} Nm

40. The time dependence of the position of a particle of mass $m = 2$ is given by $\vec{r}(t) = 2t\hat{i} - 3t^2\hat{j}$. Its angular momentum, with respect to the origin, at time $t = 2$ is : **[10 April 2019 (Evening)]**

(A) $36\hat{k}$ (B) $-34(\hat{k} - \hat{i})$

(C) $48(\hat{i} + \hat{j})$ (D) $-48\hat{k}$

41. A solid sphere of mass M and radius R is divided into two unequal parts. The first part has a mass of $\dfrac{7M}{8}$ and is converted into a uniform disc of radius $2R$. The second part is converted into a uniform solid sphere. Let I_1 be the moment of inertia of the disc about its axis and I_2 be the moment of inertia of the new sphere about its axis. The ratio I_1/I_2 is given by :

[10 April 2019 (Evening)]

(A) 185 (B) 65

(C) 285 (D) 140

42. A circular disc of radius b has a hole of radius a at its centre (see figure). If the mass per unit area of the disc varies as $\left(\dfrac{\sigma_0}{r}\right)$, then the radius of gyration of the disc about its axis passing through the centre is : **[12 April 2019 (Morning)]**

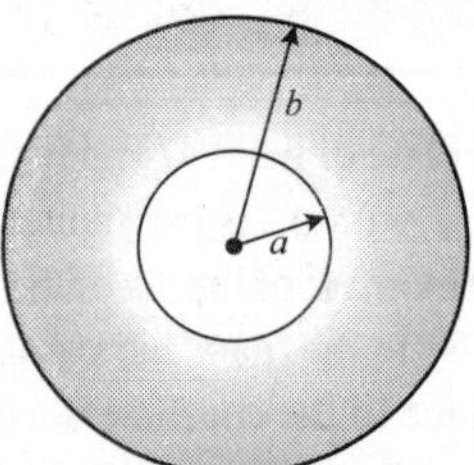

(A) $\dfrac{a+b}{2}$ (B) $\dfrac{a+b}{3}$

(C) $\sqrt{\dfrac{a^2+b^2+ab}{2}}$ (D) $\sqrt{\dfrac{a^2+b^2+ab}{3}}$

43. A uniform rod of length l is being rotated in a horizontal plane with a constant angular speed about an axis passing through one of its ends. If the tension generated in the rod due to rotation is $T(x)$ at a distance x from the axis, then which of the following graphs depicts it most closely ? **[12 April 2019 (Morning)]**

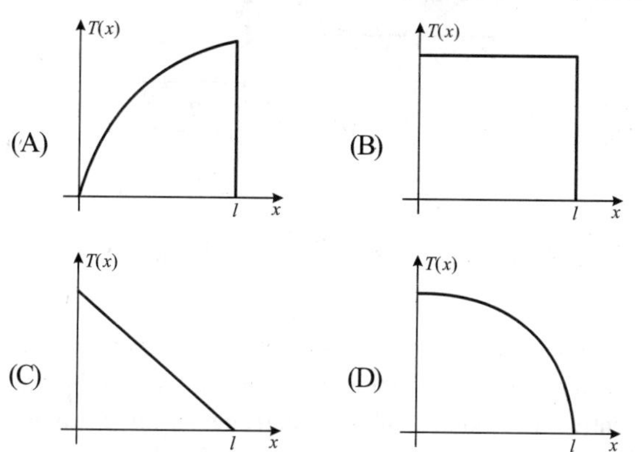

44. As shown in the figure, a bob of mass m is tied by a massless string whose other end portion is wound on a fly wheel (disc) of radius r and mass m. When released from rest the bob starts falling vertically. When it has covered a distance of h, the angular speed of the wheel will be : **[7 Jan. 2020 (Morning)]**

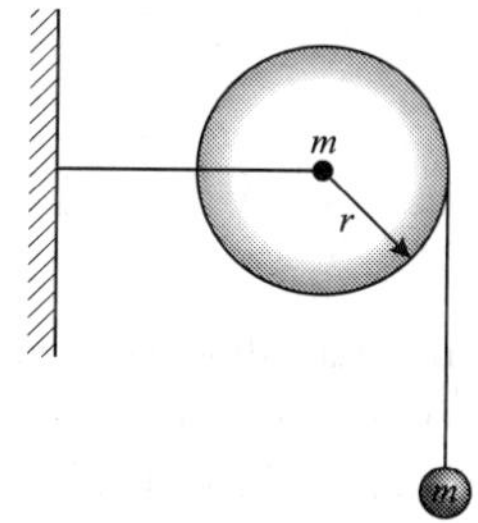

(A) $\frac{1}{r}\sqrt{\frac{2gh}{3}}$ (B) $r\sqrt{\frac{3}{4gh}}$

(C) $\frac{1}{r}\sqrt{\frac{4gh}{3}}$ (D) $r\sqrt{\frac{3}{2gh}}$

45. The radius of gyration of a uniform rod of length l, about an axis passing through a point $l/4$ away from the centre of the rod, and perpendicular to it, is : **[7 Jan. 2020 (Morning)]**

(A) $\frac{1}{8}l$ (B) $\sqrt{\frac{7}{48}}l$

(C) $\sqrt{\frac{3}{8}}l$ (D) $\frac{1}{4}l$

46. Mass per unit area of a circular disc of radius a depends on the distance r from its centre as $\sigma(r) = A + Br$. The moment of inertia of the disc about the axis, perpendicular to the plane and passing through its centre is : **[7 Jan. 2020 (Evening)]**

(A) $2\pi a^4\left(\frac{A}{4}+\frac{aB}{5}\right)$ (B) $\pi a^4\left(\frac{A}{4}+\frac{aB}{5}\right)$

(C) $2\pi a^4\left(\frac{aA}{4}+\frac{B}{5}\right)$ (D) $2\pi a^4\left(\frac{A}{4}+\frac{B}{5}\right)$

47. Consider a uniform cubical box of side a on a rough floor that is to be moved by applying minimum possible force F at a point b above its centre of mass (see figure). If the coefficient of friction is $\mu = 0.4$, the maximum possible value of $100 \times \frac{b}{a}$ for a box not to topple before moving is ________. **[7 Jan. 2020 (Evening)]**

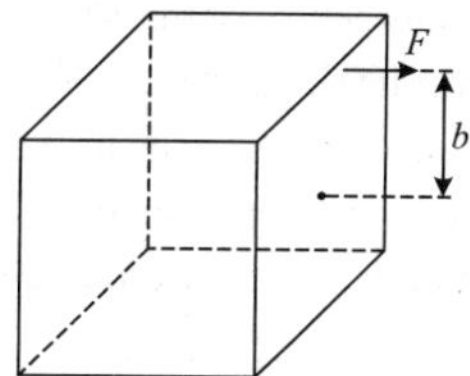

48. Consider a uniform rod of mass $M = 4m$ and length l pivoted about its centre. A mass m moving with velocity v making angle $\theta = \pi/4$ to the rod's long axis collides with one end of the rod and sticks to it. The angular speed of the rod-mass system just after the collision is : **[8 Jan. 2020 (Morning)]**

(A) $\frac{3}{7\sqrt{2}}\frac{v}{l}$ (B) $\frac{3\sqrt{2}}{7}\frac{v}{l}$

(C) $\frac{4}{7}\frac{v}{l}$ (D) $\frac{3}{7}\frac{v}{l}$

49. A uniform sphere of mass 500 g rolls without slipping on a plane horizontal surface with its centre moving at a speed of 5.00 cm/s. Its kinetic energy is : **[8 Jan. 2020 (Evening)]**

(A) 8.75×10^{-4} J (B) 8.75×10^{-3} J

(C) 6.25×10^{-4} J (D) 1.13×10^{-3} J

50. Three solid spheres each of mass m and diameter d are stuck together such that the lines connecting the centres form an equilateral triangle of side of length d. The ratio I_0/I_A of moment of inertia I_0 of the system about an axis passing the centroid and about center of any of the spheres I_A and perpendicular to the plane of the triangle is : **[9 Jan. 2020 (Morning)]**

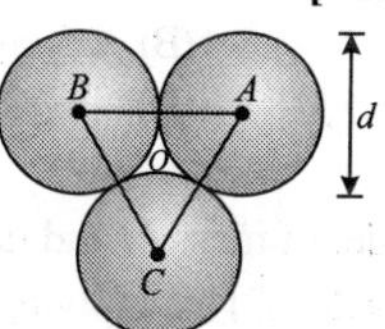

(A) $\frac{13}{23}$ (B) $\frac{15}{13}$

(C) $\frac{23}{13}$ (D) $\frac{13}{15}$

51. One end of a straight uniform 1 m long bar is pivoted on horizontal table. It is released from rest when it makes an angle 30° from the horizontal (see figure). Its angular speed when it hits the table is given as $\sqrt{n}\,s^{-1}$, where n is an integer. The value of n is _____ . **[9 Jan. 2020 (Morning)]**

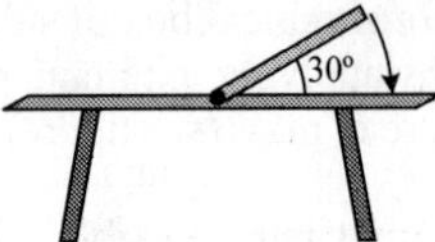

52. A uniformly thick wheel with moment of inertia I and radius R is free to rotate about its centre of mass (see figure). A massless string is wrapped over its rim and two blocks of masses m_1 and m_2 ($m_1 > m_2$) are attached to the ends of the string. The system is released from rest. The angular speed of the wheel when m_1 descents by a distance h is : **[9 Jan. 2020 (Evening)]**

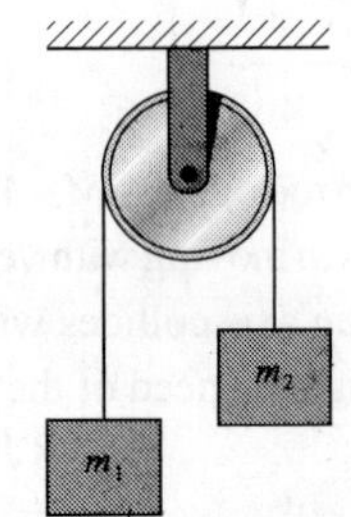

(A) $\left[\dfrac{m_1+m_2}{(m_1+m_2)R^2+I}\right]^{\frac{1}{2}} gh$ (B) $\left[\dfrac{2(m_1-m_2)gh}{(m_1+m_2)R^2+I}\right]^{\frac{1}{2}}$

(C) $\left[\dfrac{2(m_1+m_2)gh}{(m_1+m_2)R^2+I}\right]^{\frac{1}{2}}$ (D) $\left[\dfrac{(m_1-m_2)}{(m_1+m_2)R^2+I}\right]^{\frac{1}{2}} gh$

53. Shown in the figure is rigid and uniform one meter long rod AB held in horizontal position by two strings tied to its ends and attached to the ceiling. The rod is of mass 'm' and has another weight of mass $2m$ hung at a distance of 75 cm from A. The tension in the string at A is : **[2 Sept. 2020 (Morning)]**

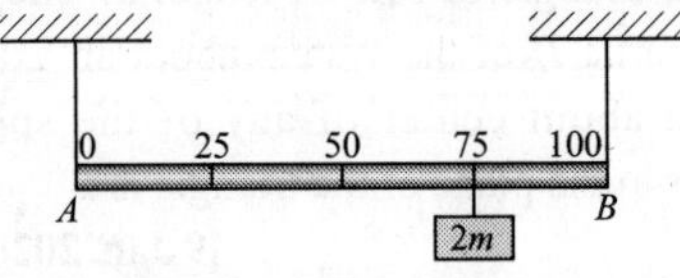

(A) $2mg$ (B) $0.5mg$
(C) $0.75mg$ (D) $1mg$

54. A uniform cylinder of mass M and radius R is to be pulled over a step of height a ($a < R$) by applying a force F at its centre 'O' perpendicular to the plane through the axes of the cylinder on the edge of the step (see figure). The minimum value of F required is : **[2 Sept. 2020 (Morning)]**

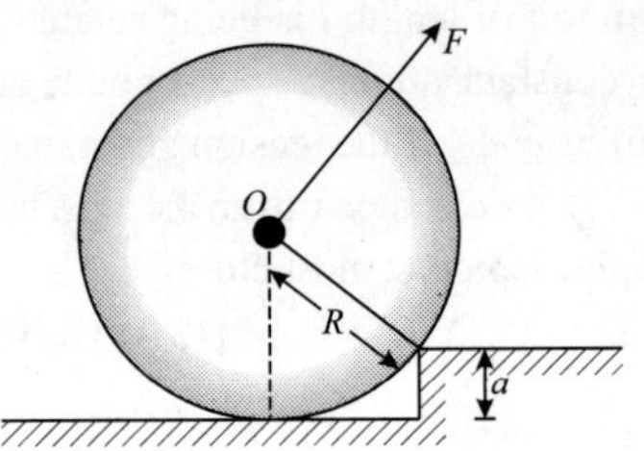

(A) $Mg\sqrt{1-\dfrac{a^2}{R^2}}$ (B) $Mg\sqrt{\left(\dfrac{R}{R-a}\right)^2-1}$

(C) $Mg\dfrac{a}{R}$ (D) $Mg\sqrt{1-\left(\dfrac{R-a}{R}\right)^2}$

55. Two uniform circular discs are rotating independently in the same direction around their common axis passing through their centres. The moment of inertia and angular velocity of the first disc are 0.1 kg-m^2 and 10 rad s^{-1} respectively while those for the second one are 0.2 kg-m^2 and 5 rad s^{-1} respectively. At some instant they get stuck together and start rotating as a single system about their common axis with some angular speed. The Kinetic energy of the combined system is : **[2 Sept. 2020 (Evening)]**

(A) $\dfrac{10}{3}$ J (B) $\dfrac{2}{3}$ J

(C) $\dfrac{5}{3}$ J (D) $\dfrac{20}{3}$ J

56. Moment of inertia of a cylinder of mass M, length L and radius R about an axis passing through its centre and perpendicular to the axis of the cylinder is given as

$$I = M\left(\frac{R^2}{4}+\frac{L^2}{12}\right)$$

If such a cylinder is to be made for a given mass of material, the ratio L/R for it to have minimum possible I is : **[3 Sept. 2020 (Morning)]**

(A) $\sqrt{\dfrac{2}{3}}$ (B) $\dfrac{3}{2}$

(C) $\sqrt{\dfrac{3}{2}}$ (D) $\dfrac{2}{3}$

57. A block of mass $m = 1$ kg slides with velocity $v = 6$ m/s on a frictionless horizontal surface and collides with a uniform vertical rod and sticks to it as shown. The rod is pivoted about O and swings as a result of the collision making angle θ before momentarily coming to rest. If the rod has mass $M = 2$ kg, and length $l = 1$ m, the value of θ is approximately : (Take $g = 10$ m/s^2) : **[3 Sept. 2020 (Morning)]**

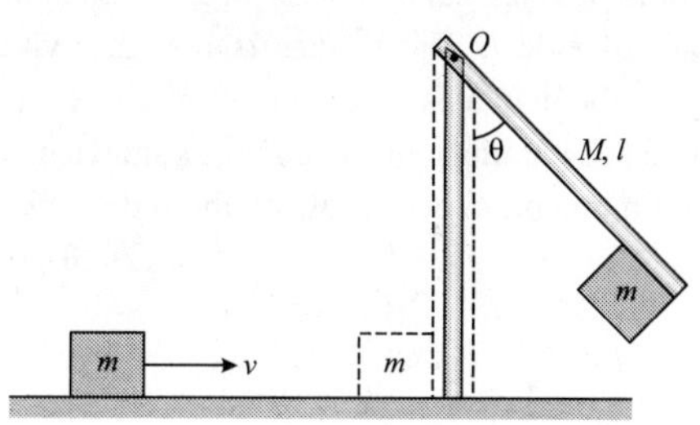

(A) 69°
(B) 63°
(C) 55°
(D) 49°

58. A person of 80 kg mass is standing on the rim of a circular platform of mass 200 kg rotating about its axis as 5 revolutions per minute (rpm). The person now starts moving towards the centre of the platform. What will be the rotational speed (in rpm) of the platform when the person reaches its centre _____ :

[3 Sept. 2020 (Morning)]

59. A uniform rod of length 'l' is pivoted at one of its ends on a vertical shaft of negligible radius. When the shaft rotates at angular speed ω the rod makes an angle θ with it (see figure). To find θ equate the rate of change of angular momentum (direction going into the paper) $\frac{ml^2}{12}\omega^2 \sin\theta\cos\theta$ about the centre of mass (CM) to the torque provided by the horizontal and vertical forces F_H and F_V about the CM. The value of θ is then such that : **[3 Sept. 2020 (Evening)]**

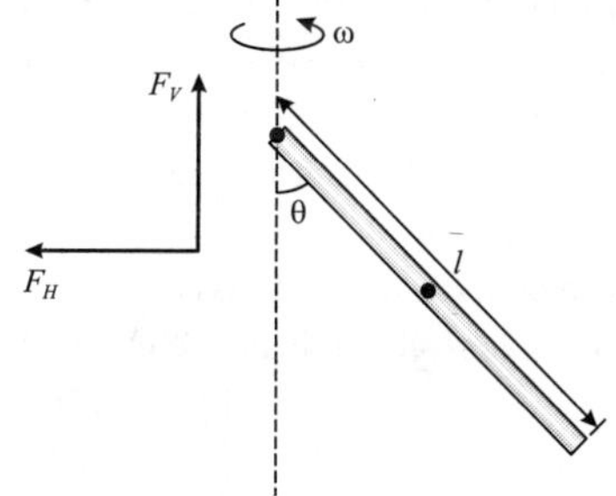

(A) $\cos\theta = \frac{g}{2l\omega^2}$
(B) $\cos\theta = \frac{3g}{2l\omega^2}$
(C) $\cos\theta = \frac{2g}{3l\omega^2}$
(D) $\cos\theta = \frac{g}{l\omega^2}$

60. A massless equilateral triangle EFG of side 'a' (As shown in figure) has three particles of mass m situated at its vertices. The moment of inertia of the system about the line EX perpendicular to EG in the plane of EFG is $\frac{N}{20}ma^2$ where N is an integer. The value of N is _____ :

[3 Sept. 2020 (Evening)]

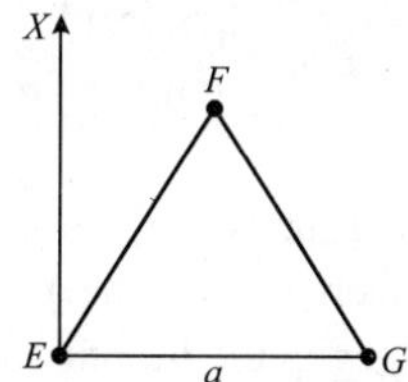

61. ABC is a plane lamina of the shape of an equilateral triangle. D, E are mid points of AB, AC and G is the centroid of the lamina. Moment of inertia of the lamina about an axis passing through G and perpendicular to the plane ABC is I_0. If part ADE is removed, the moment of inertia of the remaining part about the same axis is $\frac{NI_0}{16}$ where N is an integer. Value of N is _____:

[4 Sept. 2020 (Morning)]

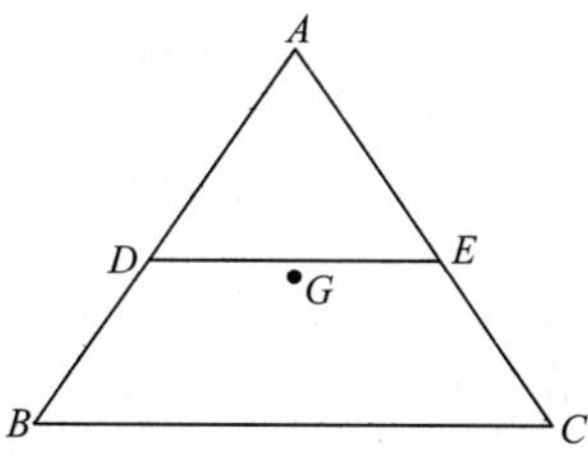

62. A circular disc of mass M and radius R is rotating about its axis with angular speed ω_1. If another stationary disc having radius $\frac{R}{2}$ and same mass M is dropped co-axially on to the rotating disc. Gradually both discs attain constant angular speed ω_2. The energy lost in the process is p% of the initial energy. Value of p is _____ : **[4 Sept. 2020 (Morning)]**

63. Consider two uniform discs of the same thickness and different radii $R_1 = R$ and $R_2 = \alpha R$ made of the same material. If the ratio of their moments of inertia I_1 and I_2, respectively, about their axes is $I_1 : I_2 = 1 : 16$ then the value of α is :

[4 Sept. 2020 (Evening)]

(A) $\sqrt{2}$
(B) 2
(C) 4
(D) $2\sqrt{2}$

64. For a uniform rectangular sheet shown in the figure, the ratio of moments of inertia about the axes perpendicular to the sheet and passing through O (the centre of mass) and O' (corner point) is : **[4 Sept. 2020 (Evening)]**

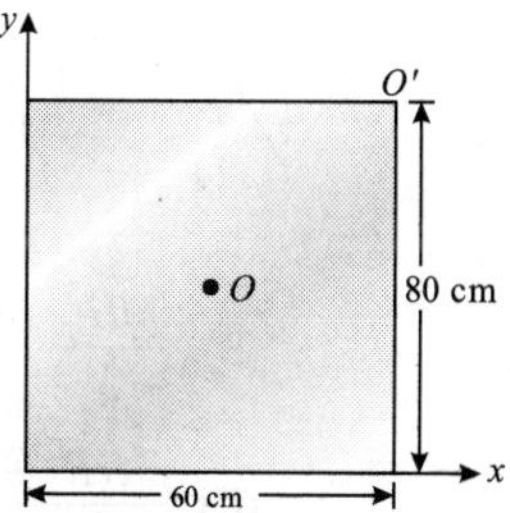

(A) 1/2 (B) 2/3
(C) 1/8 (D) 1/4

65. A wheel is rotating freely with an angular speed ω on a shaft. The moment of inertia of the wheel is I and the moment of inertia of the shaft is negligible. Another wheel of moment of inertia $3I$ initially at rest is suddenly coupled to the same shaft. The resultant fractional loss in the kinetic energy of the system is : **[5 Sept. 2020 (Morning)]**

(A) 0 (B) $\frac{1}{4}$
(C) $\frac{3}{4}$ (D) $\frac{5}{6}$

66. A force $\vec{F} = (\hat{i}+2\hat{j}+3\hat{k})$ N acts at a point $(4\hat{i}+3\hat{j}-\hat{k})m$. Then the magnitude of torque about the point $(\hat{i}+2\hat{j}+\hat{k})m$ will be $\sqrt{x}$ N-m. The value of x is _____: **[5 Sept. 2020 (Morning)]**

67. A thin rod of mass 0.9 kg and length 1 m is suspended, at rest, from one end so that it can freely oscillate in the vertical plane. A particle of move 0.1 kg moving in a straight line with velocity 80 m/s hits the rod at its bottom most point and sticks to it (see figure). The angular speed (in rad/s) of the rod immediately after the collision will be _____ : **[5 Sept. 2020 (Evening)]**

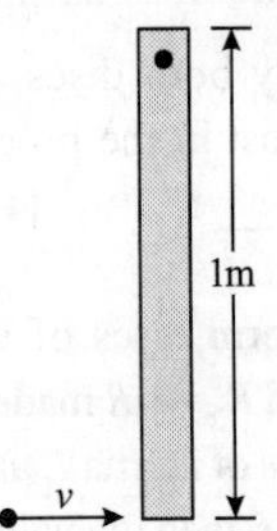

68. Shown in the figure is a hollow icecream cone (it is open at the top). If its mass is M, radius of its top, R and height, H, then its moment of inertia about its axis is : **[6 Sept. 2020 (Morning)]**

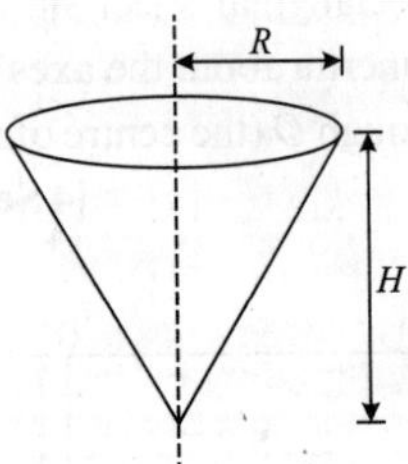

(A) $\frac{MR^2}{2}$ (B) $\frac{MH^2}{3}$
(C) $\frac{MR^2}{3}$ (D) $\frac{M(R^2+H^2)}{4}$

69. Four point masses, each of mass m, are fixed at the corners of a square of side l. The square is rotating with angular frequency ω, about an axis passing through one of the corners of the square and parallel to its diagonal, as shown in the figure. The angular momentum of the square about this axis is : **[6 Sept. 2020 (Morning)]**

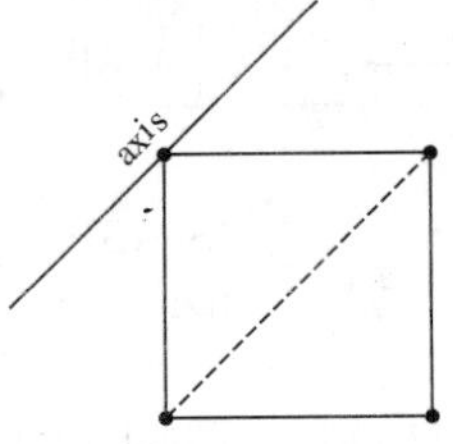

(A) $2ml^2\omega$ (B) $3ml^2\omega$
(C) $ml^2\omega$ (D) $4ml^2\omega$

70. The linear mass density of a thin rod AB of length L varies from A to B as $\lambda(x)=\lambda_0\left(1+\frac{x}{L}\right)$, where x is the distance from A. If M is the mass of the rod then its moment of inertia about an axis passing through A and perpendicular to the rod is : **[6 Sept. 2020 (Evening)]**

(A) $\frac{5}{12}ML^2$ (B) $\frac{3}{7}ML^2$
(C) $\frac{2}{5}ML^2$ (D) $\frac{7}{18}ML^2$

71. Moment of inertia (M.I.) of four bodies, having same mass and radius, are reported as;
I_1 = M.I. of thin circular ring about its diameter,
I_2 = M.I. of circular disc about an axis perpendicular to disc and going through the centre,
I_3 = M.I. of solid cylinder about its axis and
I_4 = M.I. of solid sphere about its diameter
Then : **[24 Feb. 2021 (Morning)]**

(A) $I_1+I_3<I_2+I_4$ (B) $I_1+I_2=I_3+\frac{5}{2}I_4$
(C) $I_1=I_2=I_3>I_4$ (D) $I_1=I_2=I_3<I_4$

72. A uniform thin bar of mass 6 kg and length 24 meter is bent to make an equilateral hexagon. The moment of inertia about an axis passing through the centre of mass and perpendicular to the plane of hexagon is _____ $\times 10^{-1}$ kg-m^2. **[24 Feb. 2021 (Evening)]**

73. A sphere of radius 'a' and mass 'm' rolls along a horizontal plane with constant speed v_0. It encounters an inclined plane at angle θ and climbs upward. Assuming that it rolls without slipping, how far up the sphere will travel ? **[25 Feb. 2021 (Evening)]**

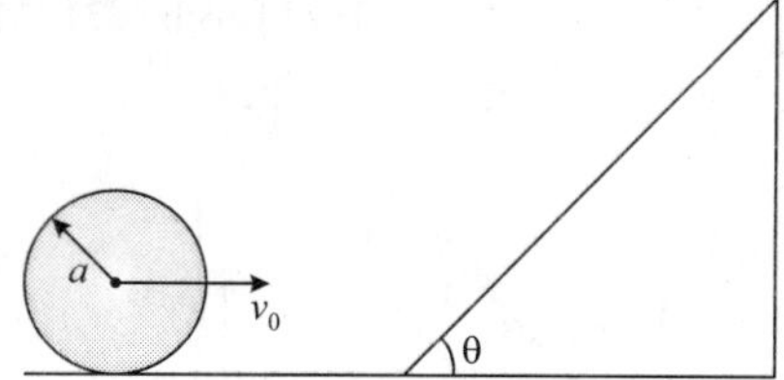

(A) $\frac{v_0^2}{5g\sin\theta}$ (B) $\frac{v_0^2}{2g\sin\theta}$

(C) $\frac{2}{5}\frac{v_0^2}{g\sin\theta}$ (D) $\frac{10v_0^2}{7g\sin\theta}$

74. Four identical solid spheres each of mass 'm' and radius 'a' are placed with their centres on the four corners of a square of side 'b'. The moment of inertia of the system about one side of square where the axis of rotation is parallel to the plane of the square is : **[26 Feb. 2021 (Morning)]**

(A) $\frac{4}{5}ma^2$ (B) $\frac{4}{5}ma^2 + 2mb^2$

(C) $\frac{8}{5}ma^2 + 2mb^2$ (D) $\frac{8}{5}ma^2 + mb^2$

75. A cord is wound round the circumference of wheel of radius r. The axis of the wheel is horizontal and the moment of inertia about it is I. A weight mg is attached to the cord at the end. The weight falls from rest. After falling through a distance 'h', the square of angular velocity of wheel will be : **[26 Feb. 2021 (Evening)]**

(A) $2gh$ (B) $\frac{2mgh}{I+2mr^2}$

(C) $\frac{2gh}{I+mr^2}$ (D) $\frac{2mgh}{I+mr^2}$

76. The maximum and minimum distances of a comet from the Sun are 1.6×10^{12} m and 8.0×10^{10} m respectively. If the speed of the comet at the nearest point is 6×10^4 ms^{-1}, the speed at the farthest point is : **[16 March 2021 (Morning)]**

(A) 1.5×10^3 m/s (B) 6.0×10^3 m/s
(C) 3.0×10^3 m/s (D) 4.5×10^3 m/s

77. Four equal masses, m each are placed at the corners of a square of length (l) as shown in the figure. The moment of inertia of the system about an axis passing through A and parallel to DB would be : **[16 March 2021 (Morning)]**

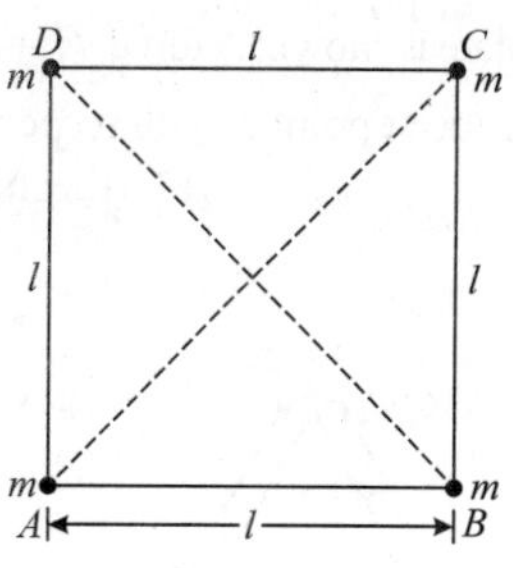

(A) ml^2 (B) $2ml^2$

(C) $3ml^2$ (D) $\sqrt{3}ml^2$

78. Consider a 20 kg uniform circular disk of radius 0.2 m. It is pin supported at its center and is at rest initially. The disk is acted upon by a constant force $F = 20$ N through a massless string wrapped around its periphery as shown in the figure.

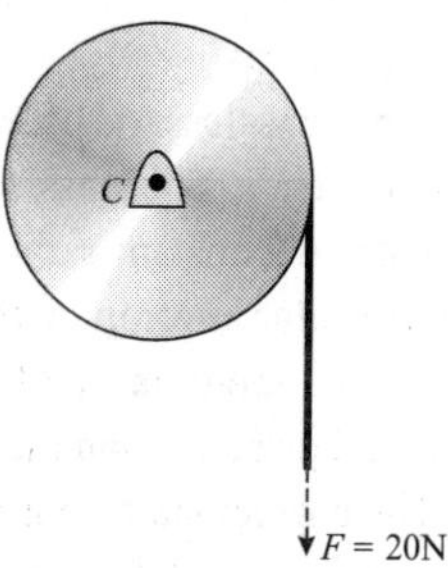

Suppose the disk makes n number of revolutions to attain an angular speed of 50 rad s^{-1}. The value of n, to the nearest integer, is ______. [Given : In one complete revolution, the disk rotates by 6.28 rad] **[16 March 2021 (Morning)]**

79. A force $\vec{F} = 4\hat{i} + 3\hat{j} + 4\hat{k}$ is applied on an intersection point of $x = 2$ plane and x-axis. The magnitude of torque of this force about a point (2, 3, 4) is ______. (Round off to the Nearest Integer) **[16 March 2021 (Evening)]**

80. A solid disc of radius 'a' and mass 'm' rolls down without slipping on an inclined plane making an angle θ with the horizontal. The acceleration of the disc will be $\frac{2}{b}g\sin\theta$ where b is ______. (Round off to the Nearest Integer)
(g = acceleration due to gravity)
(θ = angle as shown in figure) **[16 March, 2021 (Evening)]**

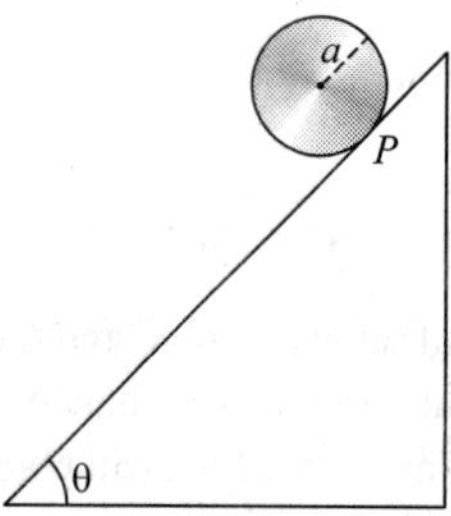

81. A triangular plate is shown. A force $\vec{F} = 4\hat{i} - 3\hat{j}$ is applied at point P. The torque at point P with respect to point 'O' and 'Q' are : **[17 March 2021 (Morning)]**

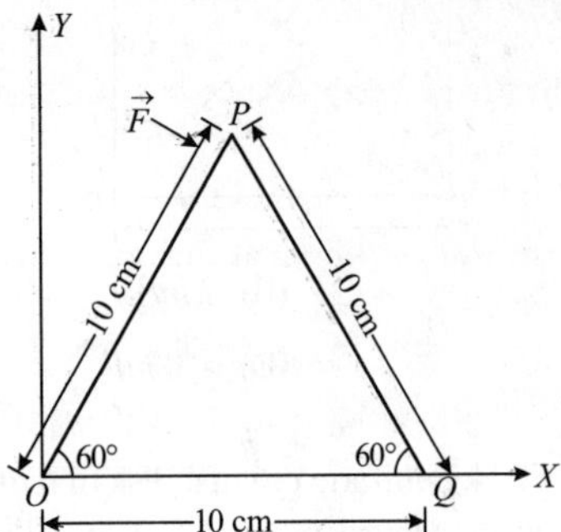

(A) $-15-20\sqrt{3}, 15-20\sqrt{3}$ (B) $15+20\sqrt{3}, 15-20\sqrt{3}$
(C) $15-20\sqrt{3}, 15+20\sqrt{3}$ (D) $-15+20\sqrt{3}, 15+20\sqrt{3}$

82. A mass M hangs on a massless rod of length l which rotates at a constant angular frequency. The mass M moves with steady speed in a circular path of constant radius. Assume that the system is in steady circular motion with constant angular velocity ω. The angular momentum of M about point A is L_A which lies in the positive z direction and the angular momentum of M about B is L_B. The correct statement for this system is : **[17 March 2021 (Morning)]**

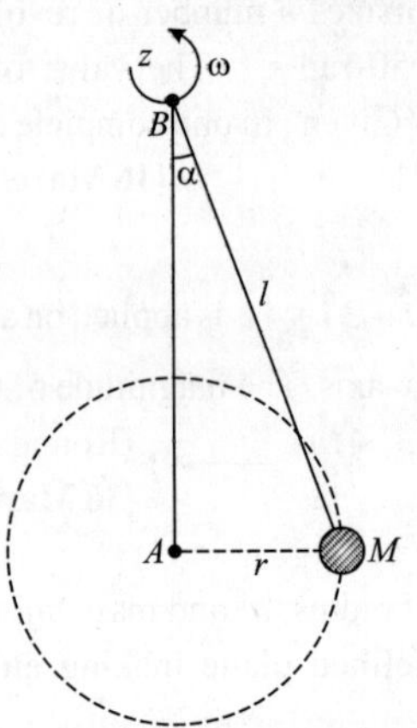

(A) L_A and L_B are both constant in magnitude and direction
(B) L_B is constant in direction with varying magnitude
(C) L_B is constant, both in magnitude and direction
(D) L_A is constant, both in magnitude and direction

83. The following bodies,
(1) a ring
(2) a disc
(3) a solid cylinder
(4) a solid sphere, of same mass 'm' and radius 'R' are allowed to roll down without slipping simultaneously from the top of the inclined plane. The body which will reach first at the bottom of the inclined plane is _____ .

[Mark the body as per their respective numbering given in the question] **[17 March 2021 (Morning)]**

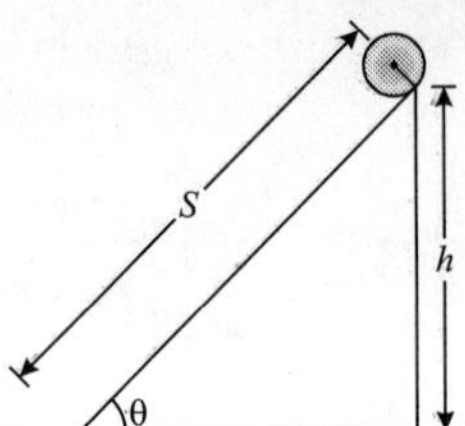

84. A sphere of mass 2kg and radius 0.5 m is rolling with an initial speed of 1 ms^{-1} goes up an inclined plane which makes an angle of 30° with the horizontal plane, without slipping. How long will the sphere take to return to the starting point A ? **[17 March 2021 (Evening)]**

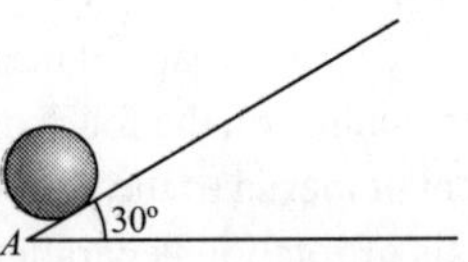

(A) 0.60 s (B) 0.52 s
(C) 0.57 s (D) 0.80 s

85. A thin circular ring of mass M and radius r is rotating about its axis with an angular speed ω. Two particles having mass m each are now attached at diametrically opposite points. The angular speed of the ring will become : **[18 March 2021 (Morning)]**

(A) $\omega\frac{M}{M+m}$ (B) $\omega\frac{M+2m}{M}$

(C) $\omega\frac{M}{M+2m}$ (D) $\omega\frac{M-2m}{M+2m}$

86. Consider a uniform wire of mass M and length L. It is bent into a semicircle. Its moment of inertia about a line perpendicular to the plane of the wire passing through the centre is : **[18 March 2021 (Evening)]**

(A) $\frac{1}{4}\frac{ML^2}{\pi^2}$ (B) $\frac{2}{5}\frac{ML^2}{\pi^2}$

(C) $\frac{ML^2}{\pi^2}$ (D) $\frac{1}{2}\frac{ML^2}{\pi^2}$

87. The angular momentum of a planet of mass M moving around the sun in an elliptical orbit is $\vec{L}$. The magnitude of the areal velocity of the planet is : **[18 March 2021 (Evening)]**

(A) $\frac{4L}{M}$ (B) $\frac{L}{M}$

(C) $\frac{2L}{M}$ (4) $\frac{L}{2M}$

88. A solid cylinder of mass m is wrapped with an inextensible light string and, is placed on a rough inclined plane as shown in the figure. The frictional force acting between the cylinder and the inclined plane is : **[18 March 2021 (Evening)]**

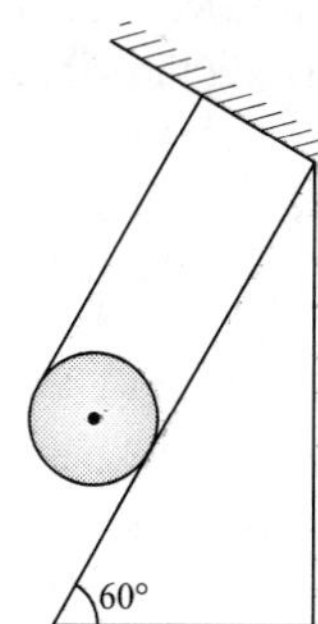

[The coefficient of static friction, μ_s, is 0.4]

(A) $\frac{7}{2}$mg (B) 5 mg

(C) $\frac{mg}{5}$ (D) 0

89. A rod of mass M and length L is lying on a horizontal frictionless surface. A particle of mass 'm' travelling along the surface hits at one end of the rod with a velocity 'u' in a direction perpendicular to the rod. The collision is completely elastic. After collision, particle comes to rest. The ratio masses $\left(\frac{m}{M}\right)$ is $\frac{1}{x}$. The value of 'x' will be _____ . **[20 July 2021 (Morning)]**

90. A circular disc reaches from top to bottom of an inclined plane of length L. When it slips down the plane, it takes time t_1. When it rolls down the plane, it takes time t_2. The value of $\frac{t_2}{t_1}$ is $\sqrt{\frac{3}{x}}$. Then value of x will be _____ .

[20 July 2021 (Morning)]

91. A body rolls down an inclined plane without slipping. The kinetic energy of rotation is 50% of its translational kinetic energy. The body is : **[20 July 2021 (Evening)]**

(A) Hollow cylinder (B) Solid sphere

(C) Sold cylinder (D) Ring

92. Two bodies, a ring and a solid cylinder of same material are rolling down without slipping and inclined plane. The radii of the bodies are same. The ratio of velocity of the centre of mass at the bottom of the inclined plane of the ring to that of the cylinder is $\frac{\sqrt{x}}{2}$. Then, the value of x is _____ .

[20 July 2021 (Evening)]

93. Consider a situation in which a ring, a solid cylinder and a solid sphere roll down on the same inclined plane without slipping, Assume that they start rolling from rest and having identical diameter. The correct statement for this situation is :

[22 July 2021 (Evening)]

(A) The sphere has the greatest and the ring has least velocity of the centre of mass at the bottom of the inclined plane.

(B) The ring has the greatest and the cylinder has the least velocity of the centre of mass at the bottom of the inclined plane.

(C) All of them will have same velocity

(D) The cylinder has the greatest and the sphere has the least velocity of the centre of mass at the bottom of the inclined plane

94. The centre of a wheel rolling on a plane surface moves with a speed V_0. A particle on the rim of the wheel at the same level as the centre will be moving at speed $\sqrt{x}V_0$. Then the value of x is _____ . **[22 July 2021 (Evening)]**

95. Given below are two statements : one is labelled as **Assertion *A*** and the other is labelled as **Reason *R***.

Assertion *A* : Moment of inertia of a circular disc of mass 'M' and radius 'R' about X, Y axes (passing through its plane) and Z-axis which is perpendicular to its plane were found to be I_x, I_Y & I_Z respectively. The respective radii of gyration about all the three axes will be the same.

Reason *R* : A rigid body making rotational motion has fixed mass and shape.

In the light of the above statements, choose the most appropriate answer from the options given below :

[25 July 2021 (Morning)]

(A) Both ***A*** and ***R*** are correct and ***R*** is the correct explanation of A

(B) Both ***A*** and ***R*** are correct but ***R*** is *NOT* the correct explanation of ***A***

(C) ***A*** is not correct but ***R*** is correct

(D) ***A*** is correct hut ***R*** is not correct

96. The minimum and maximum distances of a planet revolving around the Sun are x_1 and x_2. If the minimum speed of the planet on its trajectory is v_0 then its maximum speed will be :

[25 July 2021 (Morning)]

(A) $\frac{v_0 x_2^2}{x_1^2}$ (B) $\frac{v_0 x_2}{x_1}$

(C) $\frac{v_0 x_1}{x_2}$ (D) $\frac{v_0 x_1^2}{x_2^2}$

97. Two billiard balls of equal mass 30 g strike a wall with normally same speed of 108 kmph but at different angles. If the balls get reflected with the same speed then the ratio of the magnitude of impulses imparted to ball A and ball B by the wall along 'X' direction is : **[25 July 2021 (Morning)]**

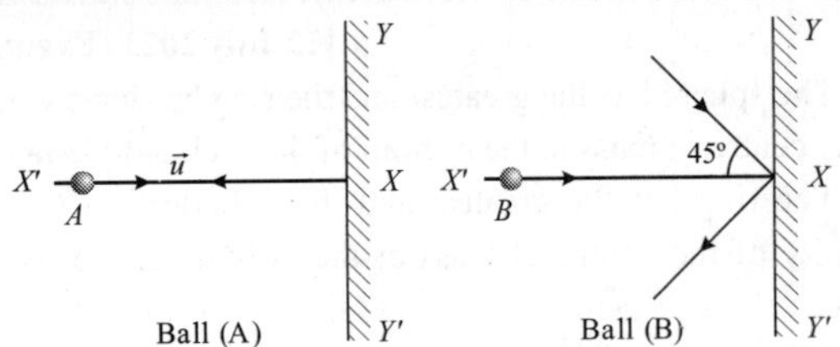

(A) $\sqrt{2}:1$ (B) $1:\sqrt{2}$
(C) $2:1$ (D) $1:1$

98. A body of mass 2 kg moving with a speed of 4 m/s, makes an elastic collision with another body at rest and continues to move in the original direction but with one fourth of its initial speed. The speed of the two body centre of mass is $\frac{x}{10}$ m/s. Then the value of x is ______ : **[25 July 2021 (Morning)]**

99. A particle of mass m is moving in time t on a trajectory given by $\vec{r} = 10\alpha t^2\,\hat{i} + 5\beta(t-5)\,\hat{j}$ Where a and b are dimensional constants. The angular momentum of the particle about origin becomes the same as it was for $t = 0$ at time $t =$ ______ seconds : **[25 July 2021 (Morning)]**

100. A solid disc of radius 20 cm and mass 10 kg is rotating with an angular velocity of 600 rpm, about an axis normal to its circular plane and passing through its centre of mass. The retarding torque required to bring the disc at rest in 10 s is ______ $\pi \times 10^{-1}$ Nm : **[25 July 2021 (Evening)]**

101. The figure shows two solid discs with radius R and r respectively. If mass per unit area is same for both, what is the ratio of MI of bigger disc around axis AB (which is ⊥ to the plane) of the disc and passing through its centre) to MI of smaller disc around one of its diameters lying on its plane ? Given M is the mass of the larger disc. (MI stands for moment of inertia) : **[27 July 2021 (Morning)]**

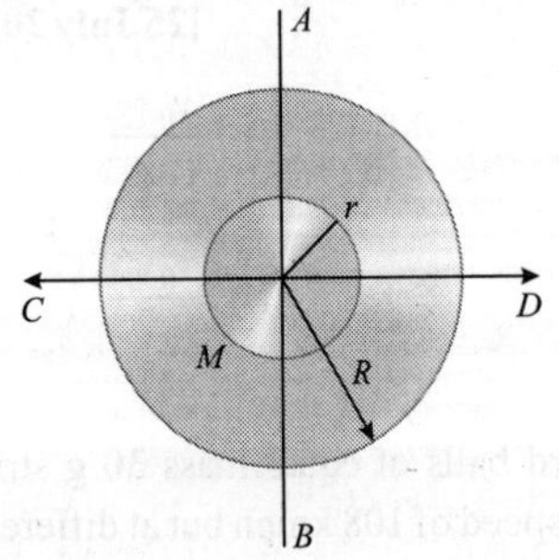

(A) $2R^4 : r^4$ (B) $2R^2 : r^2$
(C) $R^2 : r^2$ (D) $2r^4 : R^4$

102.

List-I	**List-II**
(a) MI of the rod (length L, Mass M, about an axis ⊥ to the rod passing through the mid point)	(i) $8\,ML^2/3$
(b) MI of the rod (length L, Mass $2M$, about an axis ⊥ to the rod passing through one of its end)	(ii) $ML^2/3$
(c) MI of the rod (length $2L$, Mass M, about an axis ⊥ to the rod passing through its midpoint)	(iii) $ML^2/12$
(d) MI of the rod (length $2L$, mass $2M$, about an axis ⊥ to the rod passing through one of its end	(iv) $2\,ML^2/3$

[27 July 2021 (Morning)]

(A) (a)→(iii), (b)→(iv), (c)→(i), (d)→(ii)
(B) (a)→(ii), (b)→(i), (c)→(iii), (d)→(iv)
(C) (a)→(ii), (b)→(i), (c)→(iii), (d)→(iv)
(D) (a)→(iii), (b)→(iv), (c)→(ii), (d)→(i)

103. In the given figure, two wheels P and Q are connected by a belt B. The radius of P is three times as that of Q. In case of same rotational kinetic energy, the ratio of rotational inertias $\left(\frac{I_1}{I_2}\right)$ will be $x : 1$. The value of x will be ______. **[27 July 2021 (Evening)]**

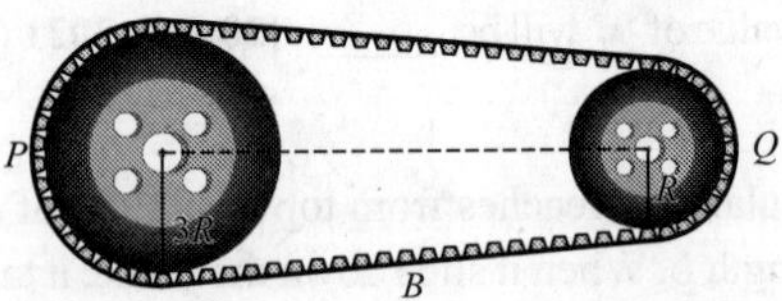

104. Consider a badminton racket with length scales as shown in the figure.

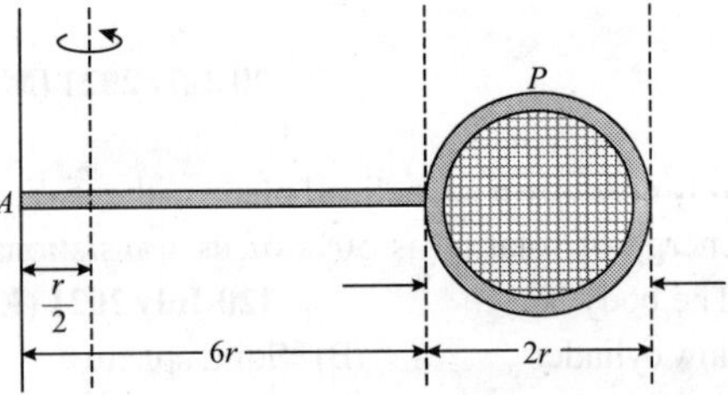

If the mass of the linear and circular portions of the badminton racket are same (M) and the mass of the threads are negligible, the moment of inertia of the racket about an axis perpendicular to the handle and in the plane of the ring at, $\frac{r}{2}$ distance from the end A of the handle will be ______ Mr^2 : **[26 Aug. 2021 (Morning)]**

105. The solid cylinder of length 80 cm and mass M has a radius of 20 cm. Calculate the density of the material used if the moment of inertia of the cylinder about an axis CD parallel of AB as shown in figure is 2.7 kg m^2 : **[26 Aug. 2021 (Evening)]**

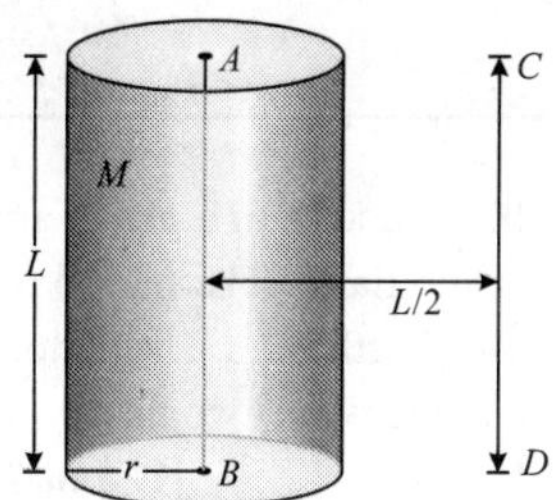

(A) 7.5×10^2 kg/m^3 (B) 7.5×10^1 kg/m^3
(C) 14.9 kg/m^3 (D) 1.49×10^2 kg/m^3

106. Moment of inertia of a square plate of side l about the axis passing through one of the corner and perpendicular to the plane of square plate is given by : **[27 Aug. 2021 (Morning)]**

(A) $\frac{2}{3}Ml^2$ (B) $\frac{Ml^2}{12}$

(C) $\frac{Ml^2}{6}$ (D) Ml^2

107. Two discs have moments of intertia I_1 and I_2 about their respective axes perpendicular to the plane and passing through the centre. They are rotating with angular speeds, ω_1 and ω_2 respectively and are brought into contact face to face with their axes of rotation coaxial. The loss in kinetic energy of the system in the process is given by : **[27 Aug. 2021 (Evening)]**

(A) $\frac{I_1I_2}{(I_1+I_2)}(\omega_1-\omega_2)^2$ (B) $\frac{(\omega_1-\omega_2)^2}{2(I_1+I_2)}$

(C) $\frac{I_1I_2}{2(I_1+I_2)}(\omega_1-\omega_2)^2$ (D) $\frac{(I_1-I_2)^2\omega_1\omega_2}{2(I_1+I_2)}$

108. Angular momentum of a single particle moving with constant speed along circular path :**[31 Aug. 2021 (Morning)]**
(A) Changes in magnitude but remains same in the direction
(B) Is zero
(C) Remains same in magnitude and direction
(D) Remains same in magnitude but changes in the direction

109. A system consists of two identical spheres each of mass 1.5 kg and radius 50 cm at the ends of a light rod. The distance between the centres of the two spheres is 5 m. What will be the moment of inertia of the system about an axis perpendicular to the passing through its midpoint ? **[31 Aug. 2021 (Evening)]**
(A) 1.905×10^5 kgm^2 (B) 19.05 kgm^2
(C) 18.75 kgm^2 (D) 18.45×10^5 kgm^2

110. A 2 kg steel rod of length 0.6 m is clamped on a table vertically at its lower end and is free to rotate in vertical plane. The upper end is pushed so that the rod falls under gravity. Ignoring the friction due to clamping at its lower end, the speed of the free end when it passes through its lowest position is _____ ms^{-1}. (Take $g = 10$ ms^{-2}) **[01 Sep. 2021 (Evening)]**

* * * * *

CHAPTER

7 GRAVITATION

7.1 GRAVITATIONAL FIELD

1. If the Earth has no rotational motion, the weight of a person on the equator is W. Determine the speed with which the earth would have to rotate about its axis so that the person at the equator will weight 3/4 W. Radius of the Earth is 6400 km and $g = 10\ \text{m\s}^2$: **[2017 (Online)]**

(A) 1.1×10^{-3} rad/s (B) 0.83×10^{-3} rad/s
(C) 0.63×10^{-3} rad/s (D) 0.28×10^{-3} rad/s

2. The mass density of a spherical body is given by $\rho(r) = \frac{k}{r}$ for $r \leq R$ and $\rho(r) = 0$ for $r > R$, where r is the distance from the centre. The correct graph that describes qualitatively the acceleration a, of a test particle as a function of r is : **[2017 (Online)]**

(A) (B) (C) (D)

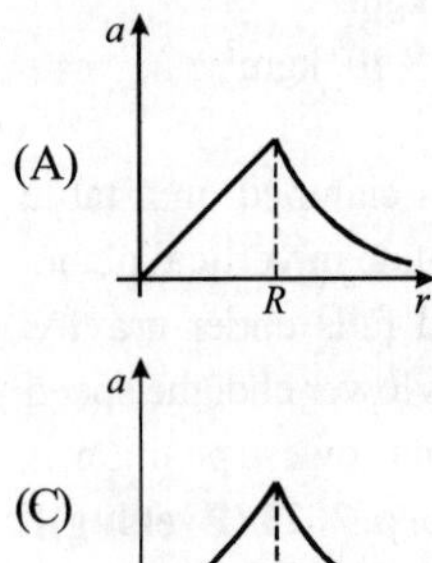

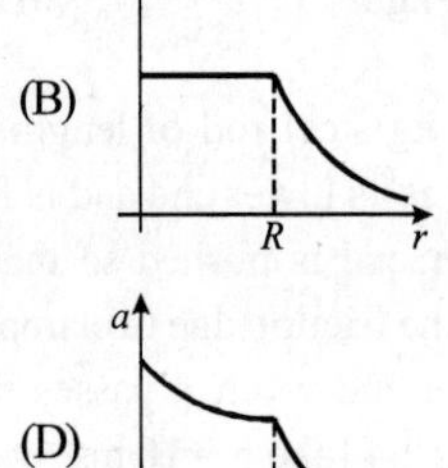

3 Take the mean distance of the moon and the sun from the earth to be 0.4×10^6 km and 150×10^6 km respectively. Their masses are 8×10^{22} kg and 2×10^{30} kg respectively. The radius of the earth is 6400 km. Let ΔF_1 be the difference in the forces exerted by the moon at the nearest and farthest points on the earth and ΔF_2 be the difference in the force exerted by the sun at the nearest and farthest points on the earth. Then, the number closest to $\frac{\Delta F_1}{\Delta F_2}$ is : **[2018 (Online)]**

(A) 2 (B) 6
(C) 10^{-2} (D) 0.6

4. Suppose that the angular velocity of rotation of earth is increased. Then, as a consequence : **[2018 (Online)]**

(A) There will be no change in weight anywhere on the earth
(B) Weight of the object, everywhere on the earth, will decrease
(C) Weight of the object, everywhere on the earth, will increase
(D) Except at poles, weight of the object on the earth will decrease

5. The mass and the diameter of a planet are three times the respective values for the Earth. The period of oscillation of a simple pendulum on the Earth is 2 s. The period of oscillation of the same pendulum on the planet would be : **[11 Jan. 2019 (Evening)]**

(A) $\frac{2}{\sqrt{3}}$ s (B) $2\sqrt{3}$ s
(C) $\frac{\sqrt{3}}{2}$ s (D) $\frac{3}{2}$ s

6. A straight rod of length L extends from $x = a$ to $x = L + a$. The gravitational force exerted on a point mass 'm' at $x = 0$, if the mass per unit length of the rod is $A + Bx^2$, is given by : **[12 Jan. 2019 (Morning)]**

(A) $Gm\left[A\left(\frac{1}{a+L} - \frac{1}{a}\right) - BL\right]$

(B) $Gm\left[A\left(\frac{1}{a} - \frac{1}{a+L}\right) + BL\right]$

(C) $Gm\left[A\left(\frac{1}{a+L} - \frac{1}{a}\right) + BL\right]$

(D) $Gm\left[A\left(\frac{1}{a} - \frac{1}{a+L}\right) - BL\right]$

7. Four identical particles of mass M are located at the corners of a square of side a. What should be their speed if each of them revolves under the influence of other's gravitational field in a circular orbit circumscribing the square ? **[8 April 2019 (Morning)]**

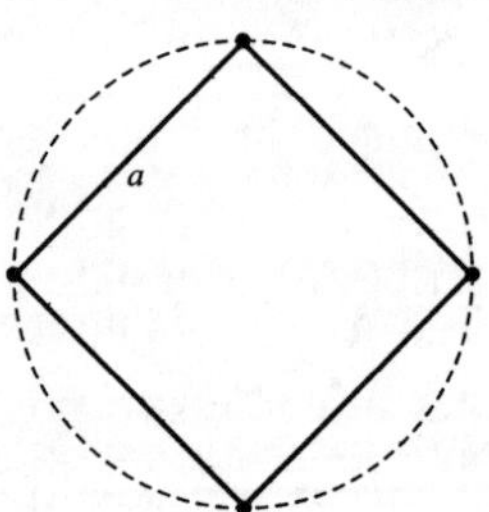

(A) $1.21\sqrt{\frac{GM}{a}}$ (B) $1.41\sqrt{\frac{GM}{a}}$
(C) $1.16\sqrt{\frac{GM}{a}}$ (D) $1.35\sqrt{\frac{GM}{a}}$

8. A rocket has to be launched from earth in such a way that it never returns. If E is the minimum energy delivered by the rocket launcher, what should be the minimum energy that the launcher should have if the same rocket is to be launched from the surface of the moon ? Assume that the density of the earth and the moon are equal and that the earth's volume is 64 times the volume of the moon : **[8 April 2019 (Evening)]**

(A) $\frac{E}{4}$ (B) $\frac{E}{16}$

(C) $\frac{E}{32}$ (D) $\frac{E}{64}$

9. A solid sphere of mass 'M' and radius 'a' is surrounded by a uniform concentric spherical shell of thickness $2a$ and mass $2M$. The gravitational field at distance '$3a$' from the centre will be : **[9 April 2019 (Morning)]**

(A) $\frac{2GM}{9a^2}$ (B) $\frac{GM}{3a^2}$

(C) $\frac{GM}{9a^2}$ (D) $\frac{2GM}{3a^2}$

10. The value of acceleration due to gravity at Earth's surface is 9.8 ms^{-2}. The altitude above its surface at which the acceleration due to gravity decreases to 4.9 ms^{-2}, is close to : (Radius of earth = 6.4×10^6 m) **[10 April 2019 (Morning)]**

(A) 1.6×10^6 m (B) 6.4×10^6 m

(C) 9.0×10^6 m (D) 2.6×10^6 m

11. A box weighs 196 N on a spring balance at the north pole. Its weight recorded on the same balance if it is shifted to the equator is close to (Take $g = 10$ ms^{-2} at the north pole and the radius of the earth = 6400 km) : **[7 Jan. 2020 (Evening)]**

(A) 195.66 N (B) 194.66 N

(C) 194.32 N (D) 195.32 N

12. Consider two solid spheres of radii $R_1 = 1$m, $R_2 = 2$m and masses M_1 and M_2, respectively. The gravitational field due to sphere (1) and (2) are shown. The value of $\frac{M_1}{M_2}$ is : **[8 Jan. 2020 (Morning)]**

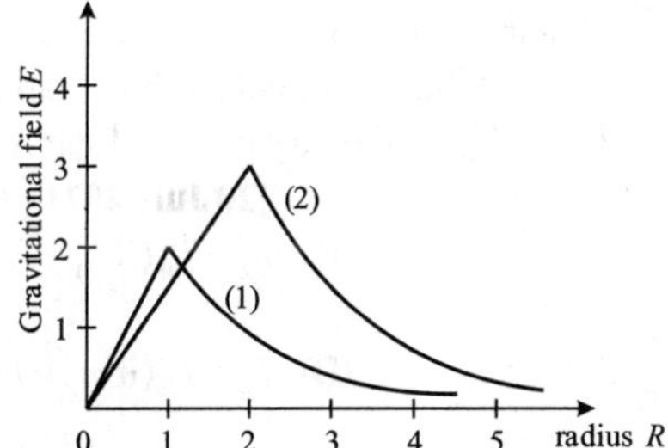

(A) $\frac{1}{2}$ (B) $\frac{2}{3}$

(C) $\frac{1}{3}$ (D) $\frac{1}{6}$

13. The height 'h' at which the weight of a body will be the same as that at the same depth 'h' from the surface of the earth is (Radius of the earth is R and effect of the rotation of the earth is neglected) : **[2 Sept. 2020 (Evening)]**

(A) $\frac{\sqrt{5}R - R}{2}$ (B) $\frac{\sqrt{5}}{2}R - R$

(C) $\frac{R}{2}$ (D) $\frac{\sqrt{3}R - R}{2}$

14. The mass density of a planet of radius R varies with the distance r from its centre as $\rho(r) = \rho_0\left(1 - \frac{r^2}{R^2}\right)$. Then the gravitational field is maximum at : **[3 Sept. 2020 (Evening)]**

(A) $r = \frac{1}{\sqrt{3}}R$ (B) $r = \sqrt{\frac{5}{9}}R$

(C) $r = \sqrt{\frac{3}{4}}R$ (D) $r = R$

15. On the x-axis and a distance x from the origin, the gravitational field due to a mass distribution is given by $\frac{Ax}{(x^2 + a^2)^{3/2}}$ in the x-direction. The magnitude of gravitational potential on the x-axis at a distance x, taking its value to be zero at infinity, is : **[4 Sept. 2020 (Morning)]**

(A) $\frac{A}{(x^2 + a^2)^{1/2}}$ (B) $\frac{A}{(x^2 + a^2)^{3/2}}$

(C) $A(x^2 + a^2)^{3/2}$ (D) $A(x^2 + a^2)^{1/2}$

16. The value of the acceleration due to gravity is g_1 at a height $h = R/2$ (R = radius of the earth) from the surface of the earth. It is again equal to g_1 at a depth d below the surface of the earth. The ratio (d/R) equals : **[5 Sept. 2020 (Morning)]**

(A) $\frac{7}{9}$ (B) $\frac{4}{9}$

(C) $\frac{1}{3}$ (D) $\frac{5}{9}$

17. The acceleration due to gravity on the earth's surface at the poles is g and angular velocity of the earth about the axis passing through the pole is ω. An object is weighed at the equator and at a height h above the poles by using a spring balance. If the weights are found to be same, then h is : ($h << R$, where R is the radius of the earth) : **[5 Sept. 2020 (Evening)]**

(A) $\frac{R^2\omega^2}{8g}$ (B) $\frac{R^2\omega^2}{4g}$

(C) $\frac{R^2\omega^2}{g}$ (D) $\frac{R^2\omega^2}{2g}$

18. A body weighs 49 N on a spring balance at the north pole. What will be its weight recorded on the same weighing machine, if it is shifted to the equator ? **[24 Feb. 2021 (Evening)]**

[Use $g = \frac{GM}{R^2} = 9.8\ ms^{-2}$ and radius of earth, $R = 6400$ km.]

(A) 49.17 N (B) 49.83 N
(C) 49 N (D) 48.83 N

19. A solid sphere of radius R gravitationally attracts a particle placed at 3R from its centre with a force F_1. Now a spherical cavity of radius $\left(\frac{R}{2}\right)$ is made in the sphere (as shown in figure) and the force becomes F_2. The value of $F_1 : F_2$ is : **[25 Feb. 2021 (Morning)]**

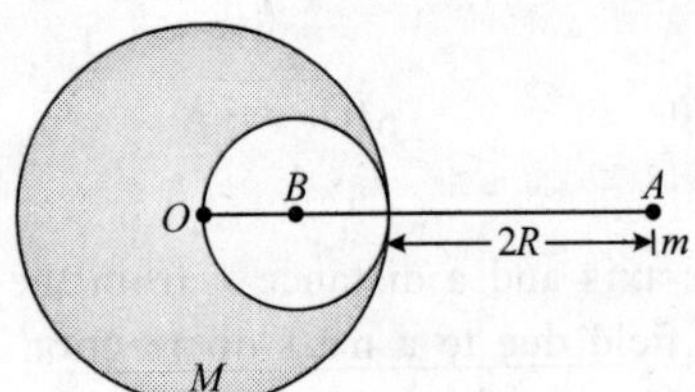

(A) 25 : 36 (B) 50 : 41
(C) 41 : 50 (D) 36 : 36

20. Find the gravitational force of attraction between the ring and sphere as shown in the diagram, where the plane of the ring is perpendicular to the line joining the centres. If $\sqrt{8}R$ is the distance between the centres of a ring (of mass 'm') and a sphere (mass 'M') where both have equal radius 'R'. **[26 Feb. 2021 (Morning)]**

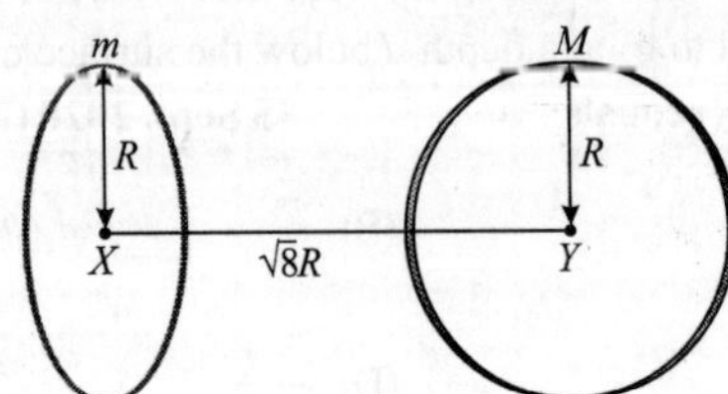

(A) $\frac{1}{3\sqrt{8}} \cdot \frac{GMm}{R^2}$ (B) $\frac{\sqrt{8}}{9} \cdot \frac{GmM}{R}$

(C) $\frac{\sqrt{8}}{27} \cdot \frac{GmM}{R^2}$ (D) $\frac{2\sqrt{2}}{3} \cdot \frac{GMm}{R^2}$

21. In the reported figure of earth, the value of acceleration due to gravity is same at point A and C but it is smaller than that of its value at point B (surface of the earth). The value of $OA : AB$ will be $x : y$. The value of x is ______. **[26 Feb. 2021 (Evening)]**

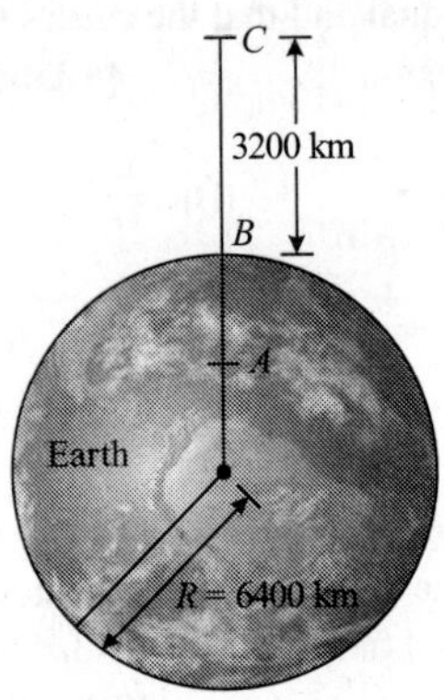

22. If one wants to remove all the mass of the earth to infinity in order to break it up completely. The amount of energy that needs to be supplied will be $\frac{x}{5}\frac{GM^2}{R}$ where x is ______ (Round off to the Nearest Integer)
(M is the mass of earth, R is the radius of earth, G is the gravitational constant) **[16 March 2021 (Evening)]**

23. A person whose mass is 100 kg travels from Earth to Mars in a spaceship. Neglect all other objects in sky and take acceleration due to gravity on the surface of the Earth and Mars as 10 m/s^2 and 4 m/s^2 respectively. Identify from the below figures, the curve that fits best for the weight of the passenger as a function of time : **[20 July 2021 (Morning)]**

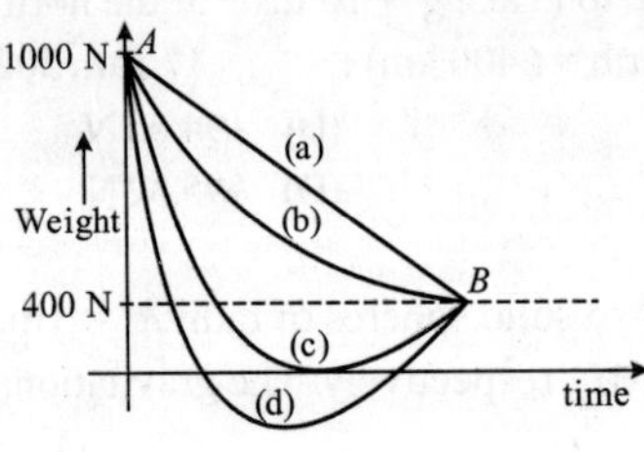

(A) (c) (B) (d)
(C) (b) (D) (a)

24. Consider a binary star system of star A and star B with masses m_A and m_B revolving in a circular orbit of radii r_A and r_B, respectively. If T_A and T_B are the time period of star A and star B, respectively, then : **[20 July 2021 (Evening)]**

(A) $T_A = T_B$ (B) $T_A > T_B$ (if $r_A > r_B$)

(C) $\frac{T_A}{T_B} = \left(\frac{r_A}{r_B}\right)^{3/2}$ (D) $T_A > T_B$ (if $m_A > m_B$)

25. A body is projected vertically upwards from the surface of earth with a velocity sufficient enough to carry to infinity. The time taken by it to reach height h is_____ s :

[22 July 2021 (Evening)]

(A) $\sqrt{\frac{2R_e}{g}}\left[\left(1+\frac{h}{R_e}\right)^{\frac{3}{2}}-1\right]$ (B) $\sqrt{\frac{R_e}{2g}}\left[\left(1+\frac{h}{R_e}\right)^{\frac{3}{2}}-1\right]$

(C) $\frac{1}{3}\sqrt{\frac{R_e}{2g}}\left[\left(1+\frac{h}{R_e}\right)^{\frac{3}{2}}-1\right]$ (D) $\frac{1}{3}\sqrt{\frac{2R_e}{g}}\left[\left(1+\frac{h}{R_e}\right)^{\frac{3}{2}}-1\right]$

26. Consider a planet in some solar system which has a mass double the mass of earth and density equal to the average density of earth. If the weight of an object on earth is W, the weight of the same object on that planet will be :

[25 July 2021 (Evening)]

(A) 2 W (B) $\sqrt{2}$ W

(C) W (D) $2\,2^{\frac{1}{3}}$ W

27. Suppose two planets (spherical in shape) of radii R and $2R$, but mass M and $9M$ respectively have a centre to centre separation $8R$ as shown in the figure. A satellite of mass 'm' is projected from the surface of the planet of mass 'M' directly towards the centre of the second planet. The minimum speed 'v' required for the satellite to reach the surface of the second planet is $\sqrt{\frac{a}{7}\frac{GM}{R}}$ then the value of 'a' is _____ .

[Given : the two planets are fixed in their position]

[27 July 2021 (Morning)]

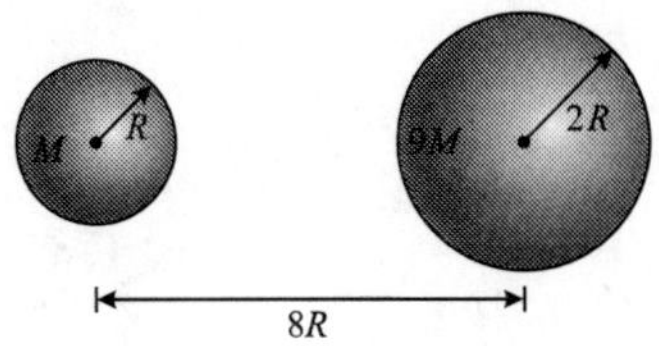

28. Two identical particles of mass 1 kg each to round a circle of radius R, under the action of their mutual gravitational attraction. The angular speed of each particle is :

[27 July 2021 (Evening)]

(A) $\frac{1}{2R}\sqrt{\frac{1}{G}}$ (B) $\sqrt{\frac{2G}{R^3}}$

(C) $\frac{1}{2}\sqrt{\frac{G}{R^3}}$ (D) $\sqrt{\frac{G}{2R^3}}$

29. The planet Mars has two moons, if one of them has a period 7 hours, 30 minutes and an orbital radius of 9.0×10^3 km. Find the mass of Mars. (Given $\frac{4\pi^2}{G} = 6 \times 10^{11}\ \text{N}^{-1}\ \text{m}^{-2}\ \text{kg}^2$) :

[27 July 2021 (Evening)]

(A) 7.02×10^{25} kg (B) 2.25×10^{21} kg
(C) 5.96×10^{19} kg (D) 6.00×10^{23} kg

30. Inside a uniform spherical shell :

[26 Aug. 2021 (Morning)]

(a) The gravitational field is zero
(b) The gravitational field is non-zero but uniform
(c) The gravitational field is same everywhere
(d) The gravitational potential is same everywhere
(e) All of the above

Choose the most appropriate answer from the options given below :

(A) (b), (c), and (d) only (B) (e) only
(C) (a), (c) and (d) only (D) (a), (b) and (c) only

31. A body of mass $(2M)$ splits into four masses $[m, M-m, m, M-m]$ which are rearranged to form a square as shown in the figure. The ratio of M/m for which, the gravitational potential energy of the system becomes maximum is $x : 1$. The value of x is _____ .

[27 Aug. 2021 (Morning)]

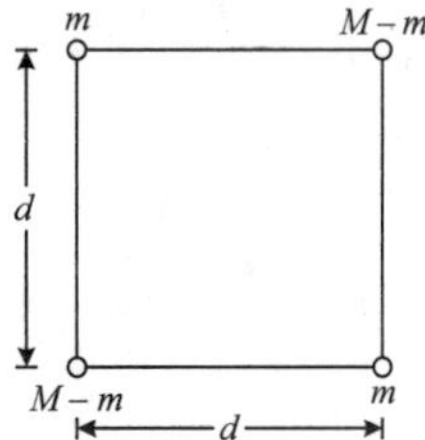

32. A mass of 50 kg is placed at the centre of a uniform spherical shell of mass 100 kg and radius 50 m. If the gravitational potential at a point, 25 m from the centre is V kg/m. The value of V is :

[27 Aug. 2021 (Evening)]

(A) $+2G$ (B) $-20G$
(C) $-4G$ (D) $-60G$

33. If R_E be the radius of Earth, then the ratio between the acceleration due to gravity at a depth r below and a height r above the earth surface is : (Given : $r < R_E$)

[31 Aug. 2021 (Evening)]

(A) $1+\frac{r}{R_E}-\frac{r^2}{R_E^2}-\frac{r^3}{R_E^3}$ (B) $1+\frac{r}{R_E}-\frac{r^2}{R_E^2}+\frac{r^3}{R_E^3}$

(C) $1-\frac{r}{R_E}-\frac{r^2}{R_E^2}-\frac{r^3}{R_E^3}$ (D) $1+\frac{r}{R_E}+\frac{r^2}{R_E^2}+\frac{r^3}{R_E^3}$

34. Four particle of mass M, move along a circle of radius R under the action of their mutual gravitational attraction as shown in figure. The speed of each particle is :

[01 Sep. 2021 (Evening)]

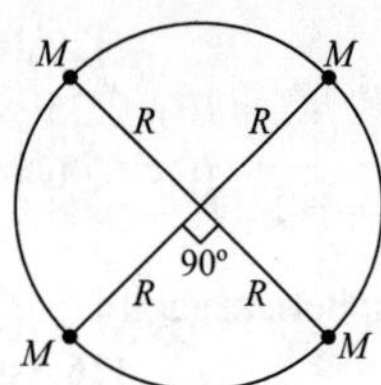

(A) $\frac{1}{2}\sqrt{\frac{GM}{R(2\sqrt{2}+1)}}$ (B) $\frac{1}{2}\sqrt{\frac{GM}{R}}(2\sqrt{2}-1)$

(C) $\frac{1}{2}\sqrt{\frac{GM}{R}}(2\sqrt{2}+1)$ (D) $\sqrt{\frac{GM}{R}}$

* * * * *

CHAPTER

7 GRAVITATION

7.2 SATELLITE MOTION

1. A body of mass m is moving in a circular orbit of radius R about a planet of mass M. At some instant, it splits into two equal masses. The first mass moves in a circular orbit of radius $\frac{R}{2}$, and the other mass, in a circular orbit of radius $\frac{3R}{2}$. The difference between the final and initial total energies is : **[2018 (Online)]**

(A) $-\frac{GMm}{2R}$ (B) $+\frac{GMm}{6R}$

(C) $-\frac{GMm}{6R}$ (D) $\frac{GMm}{2R}$

2. If the angular momentum of a planet of mass m, moving around the Sun in a circular orbit is L, about the center of the Sun, its areal velocity is : **[9 Jan. 2019 (Morning)]**

(A) $\frac{4L}{m}$ (B) $\frac{L}{m}$

(C) $\frac{L}{2m}$ (D) $\frac{2L}{m}$

3. The energy required to take a satellite to a height 'h' above Earth surface (radius of Earth = 6.4×10^3 km) is E_1 and kinetic energy required for the satellite to be in a circular orbit at this height is E_2. The value of h for which E_1 and E_2 are equal, is : **[9 Jan. 2019 (Evening)]**

(A) 1.28×10^4 km (B) 6.4×10^3 km

(C) 3.2×10^3 km (D) 1.6×10^3 km

4. A satellite is moving with a constant speed v in circular orbit around the earth. An object of mass 'm' is ejected from the satellite such that it just escapes from the gravitational pull of the earth. At the time of ejection, the kinetic energy of the object is : **[10 Jan. 2019 (Morning)]**

(A) $\frac{3}{2}mv^2$ (B) mv^2

(C) $2mv^2$ (D) $\frac{1}{2}mv^2$

5. Two stars of masses 3×10^{31} kg each, and at distance 2×10^{11} m rotate in a plane about their common centre of mass O. A meteorite passes through O moving perpendicular to the star's rotation plane. In order to escape from the gravitational field of this double star, the minimum speed that meteorite should have at O is :
(Take Gravitational constant $G = 6.67 \times 10^{-11}$ Nm^2 kg^{-2}) **[10 Jan. 2019 (Evening)]**

(A) 1.4×10^5 m/s (B) 24×10^4 m/s

(C) 3.8×10^4 m/s (D) 2.8×10^5 m/s

6. A satellite is revolving in a circular orbit at a height h from the earth surface, such that $h << R$ where R is the radius of the earth. Assuming that the effect of earth's atmosphere can be neglected then minimum increase in the speed required so that the satellite could escape from the gravitational field of earth is : **[11 Jan. 2019 (Morning)]**

(A) $\sqrt{gR}(\sqrt{2}-1)$ (B) $\sqrt{2gR}$

(C) $\sqrt{gR}$ (D) $\sqrt{\frac{gR}{2}}$

7. A satellite of mass M is in a circular orbit of radius R about the centre of the earth. A meteorite of the same mass, falling towards the earth, collides with the satellite completely inelastically. The speeds of the satellite and the meteorite are the same, just before the collision. The subsequent motion of the combined body will be : **[12 Jan. 2019 (Morning)]**

(A) In a circular orbit of a different radius

(B) In the same circular orbit of radius R

(C) In an elliptical orbit

(D) Such that it escapes to infinity

8. Two satellites, A and B, have masses m and $2m$ respectively. A is in a circular orbit of radius R, and B is in a circular orbit of radius $2R$ around the earth. The ratio of their kinetic energies, T_A/T_B, is : **[12 Jan. 2019 (Evening)]**

(A) 2 (B) $\sqrt{\frac{1}{2}}$

(C) 1 (D) $\frac{1}{2}$

9. A test particle is moving in a circular orbit in the gravitational field produced by a mass density $\rho(r) = \frac{K}{r^2}$. Identify the correct relation between the radius R of the particle's orbit and its period T: **[9 April 2019 (Evening)]**

(A) T/R^2 is a constant (B) TR is a constant

(C) T^2/R^3 is a constant (D) T/R is a constant

10. A spaceship orbits around a planet at a height of 20 km from its surface. Assuming that only gravitational field of the planet acts on the spaceship, what will be the number of complete revolutions made by the spaceship in 24 hours around the planet ?
[Given : Mass of planet = 8×10^{22} kg;
Radius of planet = 2×10^6 m,
Gravitational constant $G = 6.67 \times 10^{-11}$ Nm²/kg²]
[10 April 2019 (Evening)]

(A) 9 (B) 11
(C) 13 (D) 17

11. The ratio of the weights of a body on the Earth's surface to that on the surface of a planet is 9 : 4. The mass of the planet is 1/9th of that of the Earth. If 'R' is the radius of the Earth, what is the radius of the planet ? (Take the planets to have the same mass density) : **[12 April 2019 (Evening)]**

(A) $\frac{R}{3}$ (B) $\frac{R}{2}$
(C) $\frac{R}{4}$ (D) $\frac{R}{9}$

12. A satellite of mass m is launched vertically upwards with an initial speed u from the surface of the earth. After it reaches height R(R = radius of the earth), it ejects a rocket of mass $m/10$ so that subsequently the satellite moves in a circular orbit. The kinetic energy of the rocket is (G is the gravitational constant; M is the mass of the earth) : **[7 Jan. 2020 (Morning)]**

(A) $\frac{m}{20}\left(u - \sqrt{\frac{2GM}{3R}}\right)^2$ (B) $5m\left(u^2 - \frac{119}{200}\frac{GM}{R}\right)$

(C) $\frac{3m}{8}\left(u + \sqrt{\frac{5GM}{6R}}\right)^2$ (D) $\frac{m}{20}\left(u^2 + \frac{113}{200}\frac{GM}{R}\right)$

13. An asteroid is moving directly towards the centre of the earth. When at a distance of $10R$ (R is the radius of the earth) from the earths centre, it has a speed of 12 km/s. Neglecting the effect of earths atmosphere, what will be the speed of the asteroid when it hits the surface of the earth (escape velocity from the earth is 11.2 km/s) ? Give your answer to the nearest integer in kilometer/s ____. **[8 Jan. 2020 (Evening)]**

14. A body A of mass m is moving in a circular orbit of radius R about a planet. Another body B of mass $\frac{m}{2}$ collides with A with a velocity which is half $\left(\frac{\vec{v}}{2}\right)$ the instantaneous velocity $\vec{v}$ of A. The collision is completely inelastic. Then, the combined body : **[9 Jan. 2020 (Morning)]**

(A) Starts moving in an elliptical orbit around the planet
(B) Continues to move in a circular orbit
(C) Falls vertically downwards towards the planet
(D) Escapes from the Planet's Gravitational field.

15. Planet A has mass M and radius R. Planet B has half the mass and half the radius of Planet A. If the escape velocities from the Planets A and B are v_A and v_B, respectively, then $\frac{v_A}{v_B} = \frac{n}{4}$. The value of n is : **[9 Jan. 2020 (Evening)]**

(A) 4 (B) 1
(C) 2 (D) 3

16. The mass density of a spherical galaxy varies as $\frac{K}{r}$ over a large distance 'r' from its centre. In that region, a small star is in a circular orbit of radius R. Then the period of revolution, T depends on R as : **[2 Sept. 2020 (Morning)]**

(A) $T \propto R$ (B) $T^2 \propto \frac{1}{R^3}$
(C) $T^2 \propto R$ (D) $T^2 \propto R^3$

17. A satellite is moving in a low nearly circular orbit around the earth. Its radius is roughly equal to that of the earth's radius R_e. By firing rockets attached to it, its speed is instantaneously increased in the direction of its motion so that is become $\sqrt{\frac{3}{2}}$ times larger. Due to this the farthest distance from the centre of the earth that the satellite reaches is R, value of R is : **[3 Sept. 2020 (Morning)]**

(A) $4R_e$ (B) $3R_e$
(C) $2R_e$ (D) $2.5R_e$

18. A body is moving in a low circular orbit about a planet of mass M and radius R. The radius of the orbit can be taken to be R itself. Then the ratio of the speed of this body in the orbit to the escape velocity from the planet is : **[4 Sept. 2020 (Evening)]**

(A) 1 (B) 2
(C) $\frac{1}{\sqrt{2}}$ (D) $\sqrt{2}$

19. A satellite is in an elliptical orbit around a planet P. It is observed that the velocity of the satellite when it is farthest from the planet is 6 times less than that when it is closest to the planet. The ratio of distances between the satellite and the planet at closest and farthest points is : **[6 Sept. 2020 (Morning)]**

(A) 1 : 6 (B) 3 : 4
(C) 1 : 3 (D) 1 : 2

20. Two planets have masses M and $16\,M$ and their radii are a and $2a$, respectively. The separation between the centres of the planets is $10a$. A body of mass m is fired from the surface of the larger planet towards the smaller planet along the line joining their centres. For the body to be able to reach at the surface of smaller planet, the minimum firing speed needed is : **[6 Sept. 2020 (Evening)]**

(A) $\sqrt{\frac{GM^2}{ma}}$ (B) $\frac{3}{2}\sqrt{\frac{5GM}{a}}$

(C) $4\sqrt{\frac{GM}{a}}$ (D) $2\sqrt{\frac{GM}{a}}$

21. Two stars of masses m and $2m$ at a distance d rotate about their common centre of mass in free space. The period of revolution is : **[24 Feb. 2021 (Morning)]**

(A) $\frac{1}{2\pi}\sqrt{\frac{3Gm}{d^3}}$ (B) $2\pi\sqrt{\frac{d^3}{3Gm}}$

(C) $2\pi\sqrt{\frac{3Gm}{d^3}}$ (D) $\frac{1}{2\pi}\sqrt{\frac{d^3}{3Gm}}$

22. Consider two satellites S_1 and S_2 with periods of revolution 1 hr and 8 hr respectively revolving around a planet in circular orbits. The ratio of angular velocity of satellite S_1 to the angular velocity of satellite S_2 is : **[24 Feb. 2021 (Morning)]**

(A) 2 : 1 (B) 1 : 4
(C) 1 : 8 (D) 8 : 1

23 Four identical particles of equal masses 1 kg made to move the circumference of a circle of radius 1 m under the action of their own mutual gravitational attraction. The speed of each particle will be : **[24 Feb. 2021 (Morning)]**

(A) $\sqrt{G(1+2\sqrt{2})}$ (B) $\sqrt{\frac{G}{2}(1+2\sqrt{2})}$

(C) $\sqrt{\frac{(1+2\sqrt{2})G}{2}}$ (D) $\sqrt{\frac{G}{2}(2\sqrt{2}-1)}$

24. Given below are two statements : one is labelled as **Assertion-A** and the other is labelled as **Reason-R**

Assertion-A : The escape velocities of planet A and B are same. But A and B are of unequal mass

Reason-R : The product of their mass and radius must be same. $M_1R_1 = M_2R_2$

In the light of the above statements, choose the most appropriate answer from the options given below : **[25 Feb. 2021 (Morning)]**

(A) Both **A** and **R** are correct but **R** is NOT the correct explanation of **A**
(B) Both **A** and **R** are correct and **R** is the correct explanation of **A**
(C) **A** is correct but **R** is not correct
(D) **A** is not correct but **R** is correct

25. Two satellites A and B of masses 200 kg and 400 kg are revolving round the earth at height of 600 km and 1600 km respectively. If T_A and T_B are the time periods of A and B respectively then the value of $T_B - T_A$: **[25 Feb. 2021 (Morning)]**

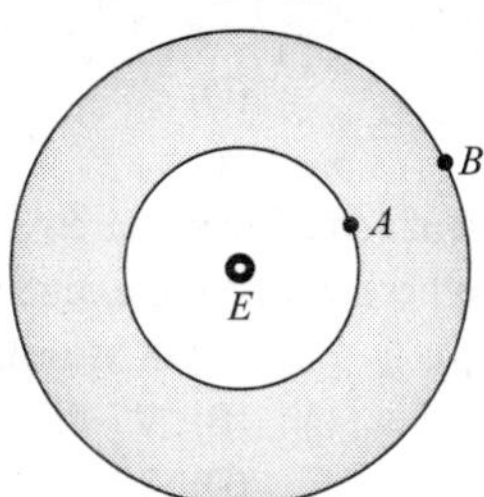

[Given : radius of earth = 6400 km, mass of earth = 6×10^{24} kg] :

(A) 4.24×10^2 s (B) 3.33×10^2 s
(C) 1.33×10^3 s (D) 4.24×10^3 s

26. The initial velocity v_i required to project a body vertically upward from the surface of the earth to reach a height of 10R, where R is the radius of the earth, may be described in terms of escape velocity v_e such that $v_i = \sqrt{\frac{x}{y}} \times v_e$. The value of x will be _____. **[25 Feb. 2021 (Evening)]**

27. A planet revolving in elliptical orbit has : **[26 Feb. 2021 (Morning)]**

(1) A constant velocity of revolution
(2) Has the least velocity when it is nearest to the sun
(3) Its areal velocity is directly proportional to its velocity
(4) Areal velocity is inversely proportional to its velocity
(5) To follow a trajectory such that the areal velocity is constant

Choose the correct answer from the options given below :

(A) 3 only (B) 5 only
(C) 4 only (D) 1 only

28. Assume that a tunnel is dug along a chord of the earth, at a perpendicular distance ($R/2$) from the earth's centre, where 'R' is the radius of the Earth. The wall of the tunnel is frictionless. If a particle is released in this tunnel, it will execute a simple harmonic motion with a time period : **[26 Feb. 2021 (Morning)]**

(A) $2\pi\sqrt{\frac{R}{g}}$ (B) $\frac{2\pi R}{g}$

(C) $\frac{g}{2\pi R}$ (D) $\frac{1}{2\pi}\sqrt{\frac{g}{R}}$

29. The radius in kilometer to which the present radius of earth (R = 6400 km) to be compressed so that the escape velocity is increased 10 time is _____. **[17 March 2021 (Morning)]**

30. A geostationary satellite is orbiting around an arbitary planet 'P' at a height of 11R above the surface of 'P', R being the radius of 'P'. The time period of another satellite in hours at a height of $2R$ from the surface of 'P' is _____. 'P' has the time period of 24 hours. **[17 March 2021 (Evening)]**

(A) $6\sqrt{2}$ (B) $\frac{6}{\sqrt{2}}$

(C) 3 (D) 5

31. The time period of a satellite in a circular orbit of radius R is T. The period of another satellite in a circular orbit of radius $9R$ is : **[18 March 2021 (Morning)]**

(A) 9 T (B) 27 T

(C) 12 T (D) 3 T

32. A satellite is launched into a circular orbit of radius R around earth, while a second satellite is launched into a circular orbit of radius $1.02R$. The percentage difference in the time periods of the two satellites is : **[20 July 2021 (Evening)]**

(A) 1.5 (B) 0.7

(C) 3.0 (D) 2.0

33. The masses and radii of the earth and moon are (M_1, R_1) and (M_2, R_2) respectively. Their centers are at a distance r apart. Find the minimum escape velocity for a particle of mas m to be projected from the middle of these two masses : **[31 Aug. 2021 (Morning)]**

(A) $V=\frac{1}{2}\sqrt{\frac{2GM(M_1+M_2)}{r}}$ (B) $V=\sqrt{\frac{4G(M_1+M_2)}{r}}$

(C) $V=\frac{\sqrt{2G}(M_1+M_2)}{r}$ (D) $V=\frac{1}{2}\sqrt{\frac{4GM(M_1+M_2)}{r}}$

34. Two satellites revolve around a planet in coplanar circular orbits in anticlockwise direction Their period of revolutions are 1 hour and 8 hours respectively. The radius of the orbit of nearer satellite is 2×10^3 km. The angular speed of the farther satellite as observed from the nearer satellite at the instant when both the satellites are closest is $\frac{\pi}{x}$ rad h^{-1} where x is _____.

[01 Sep. 2021 (Evening)]

* * * * *

CHAPTER

8 MECHANICAL PROPERTIES OF SOLIDS

8.1 ELASTICITY

1. A load of mass M kg is suspended from a steel wire of length 2 m and radius 1.0 mm in Searle's apparatus experiment. The increase in length produced in the wire is 4.0 mm. Now the load is fully immersed in a liquid of relative density 2. The relative density of the material of load is 8. The new value of increase in length of the steel wire is : **[12 Jan. 2019 (Evening)]**

(A) 4.0 mm (B) 3.0 mm
(C) 5.0 mm (D) Zero

2. A boy's catapult is made of rubber cord which is 42 cm long, with 6 mm diameter of cross-section and of negligible mass. The boy keeps a stone weighing 0.02 kg on it and stretches the cord by 20 cm by applying a constant force. When released, the stone flies off with a velocity of 20 ms^{-1}. Neglect the change in the area of cross-section of the cord while stretched. The Young's modulus of rubber is closest to :

[8 April 2019 (Morning)]

(A) $10^4\ Nm^{-2}$ (B) $10^8\ Nm^{-2}$
(C) $10^6\ Nm^{-2}$ (D) $10^3\ Nm^{-2}$

3. A steel wire having a radius of 2.0 mm, carrying a load of 4 kg, is hanging from a ceiling. Given that $g = 3.1\ \pi\ ms^{-2}$, what will be the tensile stress that would be developed in the wire ?

[8 April 2019 (Morning)]

(A) $4.8 \times 10^6\ Nm^{-2}$ (B) $5.2 \times 10^6\ Nm^{-2}$
(C) $6.2 \times 10^6\ Nm^{-2}$ (D) $3.1 \times 10^6\ Nm^{-2}$

4. Young's moduli of two wires A and B are in the ratio 7 : 4. Wire A is 2 m long and has radius R. Wire B is 1.5 m long and has radius 2 mm. If the two wires stretch by the same length for a given load, then the value of R is close to :

[8 April 2019 (Evening)]

(A) 1.9 mm (B) 1.7 mm
(C) 1.5 mm (D) 1.3 mm

5. The elastic limit of brass is 379 MPa. What should be the minimum diameter of a brass rod if it is to support a 400 N load without exceeding its elastic limit? **[10 April 2019 (Evening)]**

(A) 1.16 mm (B) 0.90 mm
(C) 1.36 mm (D) 1.00 mm

6. In an experiment, bras and steel wires of length 1 m each with areas of cross section 1 mm^2 are used. The wires are connected in series and one end of the combined wire is connected to a rigid support and other end is subjected to elongation. The stress required to produce a net elongation of 0.2 mm is :

(Given, the Young's Modulus for steel and brass are respectively, $120 \times 10^9\ N/m^2$ and $60 \times 10^9\ N/m^2$) **[10 April 2019 (Evening)]**

(A) $0.2 \times 10^6\ N/m^2$ (B) $4.0 \times 10^6\ N/m^2$
(C) $1.8 \times 10^6\ N/m^2$ (D) $1.2 \times 10^6\ N/m^2$

7. At 40°C, a brass wire of 1 mm radius is hung from the ceiling. A small mass, M is hung from the free end of the wire. When the wire is cooled down from 40°C to 20°C it regains its original length of 0.2 m. The value of M is close to : (Coefficient of linear expansion and Young's modulus of brass are 10^{-5}/°C and $10^{11}\ N/m^2$, respectively; $g = 10\ ms^{-2}$): **[12 April 2019 (Morning)]**

(A) 1.5 kg (B) 9 kg
(C) 0.9 kg (D) 0.5 kg

8. A uniform cylindrical rod of length L and radius r, is made from a material whose Young's modulus of Elasticity equals Y. When this rod is heated by temperature T and simultaneously subjected to a net longitudinal compressional force F, its length remains unchanged. The coefficient of volume expansion, of the material of the rod, is (nearly) equals to :

[12 April 2019 (Evening)]

(A) $F/(3\pi r^2 YT)$ (B) $3F/(\pi r^2 YT)$
(C) $6F/(\pi r^2 YT)$ (D) $9F/(\pi r^2 YT)$

9. A body of mass $m = 10$ kg is attached to one end of a wire of length 0.3 m. The maximum angular speed (in rad s^{-1}) with which it can be rotated about its other end in space station is (Breaking stress of wire = $4.8 \times 10^7\ Nm^{-2}$ and area of cross-section of the wire = $10^{-2}\ cm^2$) is : **[9 Jan. 2020 (Morning)]**

10. Two steel wires having same length are suspended from a ceiling under the same load. If the ratio of their energy stored per unit volume is 1 : 4, the ratio of their diameters is :

[9 Jan. 2020 (Evening)]

(A) $1 : \sqrt{2}$ (B) 1 : 2
(C) 2 : 1 (D) $\sqrt{2} : 1$

11. A wire of density $9 \times 10^3\ kg\ cm^{-3}$ is stretched between two clamps 1 m apart. The resulting strain in the wire is 4.9×10^{-4}. The lowest frequency of the transverse vibrations in the wire is (Young's modulus of wire $Y = 9 \times 10^{10}\ Nm^{-2}$), (to the nearest integer). **[2 Sept. 2020 (Evening)]**

12. A cube of metal is subjected to a hydrostatic pressure of $4GPa$. The percentage change in the length of the side of the cube is close to : (Given bulk modulus of metal, $B = 8 \times 10^{10}$ Pa) : **[4 Sept. 2020 (Evening)]**

(A) 0.6 (B) 1.67
(C) 5 (D) 20

13. An object of mass m is suspended at the end of a massless wire of length L and area of cross-section, A. Young modulus of the material of the wire is Y. If the mass is pulled down slightly its frequency of oscillation along the vertical direction is : **[6 Sept. 2020 (Morning)]**

(A) $f = \frac{1}{2\pi}\sqrt{\frac{YA}{mL}}$ (B) $f = \frac{1}{2\pi}\sqrt{\frac{YL}{mA}}$
(C) $f = \frac{1}{2\pi}\sqrt{\frac{mA}{YL}}$ (D) $f = \frac{1}{2\pi}\sqrt{\frac{mL}{YA}}$

14. If Y, K and η are the values of Young's modulus, bulk modulus and modulus of rigidity of any material respectively. Choose the correct relation for these parameters : **[24 Feb. 2021 (Morning)]**

(A) $K = \frac{Y\eta}{9\eta - 3Y}$ N/m^2 (B) $Y = \frac{9Y\eta}{9K - \eta}$ N/m^2
(C) $\eta = \frac{3YK}{9K + Y}$ N/m^2 (D) $Y = \frac{9K\eta}{2\eta + 3K}$ N/m^2

15. A uniform metallic wire is elongated by 0.04 m when subjected to a linear force F. The elongation, if its length and diameter is doubled and subjected to the same force will be _____ cm. **[24 Feb. 2021 (Evening)]**

16. The normal density of a material is ρ and its bulk modulus of elasticity is K. The magnitude of increase in density of material, when a pressure P is applied uniformly on all sides, will be : **[26 Feb. 2021 (Morning)]**

(A) $\frac{\rho P}{K}$ (B) $\frac{K}{\rho P}$
(C) $\frac{\rho K}{P}$ (D) $\frac{PK}{\rho}$

17. The length of metallic wire is l_1 when tension in it is T_1. It is l_2 when the tension is T_2. The original length of the wire will be : **[26 Feb. 2021 (Evening)]**

(A) $\frac{l_1 + l_2}{2}$ (B) $\frac{T_2 l_1 + T_1 l_2}{T_1 + T_2}$
(C) $\frac{T_1 l_1 - T_2 l_2}{T_2 - T_1}$ (D) $\frac{T_2 l_1 - T_1 l_2}{T_2 - T_1}$

18. In order to determine the Young's Modulus of a wire of radius 0.2 cm (measured using a scale of least count = 0.001 cm) and length 1 m (measured using a scale of least count = 1 mm), a weight of mass 1 kg (measured using a scale of least count = 1 g) was hanged to get the elongation of 0.5 cm (measured using a scale of least count 0.001 cm). What will be the fractional error in the value of Young's Modulus determined by this experiment ? **[16 March 2021 (Evening)]**

(A) 0.14% (B) 0.9%
(C) 9% (D) 1.4%

19. An object is located at 2 km beneath the surface of the water. If the fractional compression $\frac{\Delta V}{V}$ is 1.36%, the ratio of hydraulic stress to the corresponding hydraulic strain will be _____. **[17 March 2021 (Evening)]**
[Given : density of water is 1000 kg m^{-3} and $g = 9.8$ ms^{-2}]

(A) 1.96×10^7 Nm^{-2} (B) 1.44×10^7 Nm^{-2}
(C) 2.26×10^9 Nm^{-2} (D) 1.44×10^9 Nm^{-2}

20. In Young's double slit arrangement, slits are separated by a gap of 0.5 mm, and the screen is placed at a distance of 0.5 m from them. The distance between the first and the third bright fringe formed when the slits are illuminated by a monochromatic light of 5890 Å is : **[18 March 2021 (Morning)]**

(A) 1178×10^{-9} m (B) 1178×10^{-6} m
(C) 1178×10^{-12} m (D) 5890×10^{-7} m

21. Two separate wires A and B are stretched by 2 mm and 4 mm respectively, when they are subjected to a force of 2 N. Assume that both the wires are made up of same material and the radius of wire B is 4 times that of the radius of wire A. The length of the wires A and B are in the ratio of $a : b$. Then a/b can be expressed as $1/x$ where x is _____. **[18 March 2021 (Morning)]**

22. For an adiabatic expansion of an ideal gas, the fractional change in its pressure is equal to (where γ is the ratio of specific heats) : **[18 March 2021 (Evening)]**

(A) $-\gamma\frac{dV}{V}$ (B) $-\gamma\frac{V}{dV}$
(C) $-\frac{1}{\gamma}\frac{dV}{V}$ (D) $\frac{dV}{V}$

23. The value of tension in a long thin metal wire has been changed from T_1 to T_2. The lengths of the metal wire at two different values of tension T_1 and T_2 are l_1 and l_2 respectively. The actual length of the metal wire is : **[20 July 2021 (Morning)]**

(A) $\frac{T_1 l_1 - T_2 l_2}{T_1 - T_2}$ (B) $\frac{l_1 + l_2}{2}$
(C) $\frac{T_1 l_2 + T_2 l_1}{T_1 - T_2}$ (D) $\sqrt{T_1 T_2 l_1 l_2}$

24. The length of a metal wire is l_1, when the tension in it is T_1 and is l_2 when the tension is T_2. The natural length of the wire is : **[20 July 2021 (Evening)]**

(A) $\sqrt{l_1 l_2}$ (B) $\frac{l_1+l_2}{2}$

(C) $\frac{l_1T_2-l_2T_1}{T_2-T_1}$ (D) $\frac{l_1T_2+l_2T_1}{T_2+T_1}$

25. The area of cross-section of a railway track is 0.01 m^2. The temperature variation is 10°C. Coefficient of linear expansion of material of track 10^{-5}/°C. The energy stored per meter in the track is_____J/m. (Young's modulus of material of track is 10^{11} Nm^{-2})_____. **[22 July 2021 (Evening)]**

26. Two wires of same length and radius are joined end to end and loaded. The Young's modulii of the materials of the two wires are Y_1 and Y_2. The combination behaves as a single wire then its Young's modulus is : **[25 July 2021 (Morning)]**

(A) $Y=\frac{Y_1Y_2}{Y_1+Y_2}$ (B) $Y=\frac{Y_1Y_2}{2(Y_1+Y_2)}$

(C) $Y=\frac{2Y_1Y_2}{3(Y_1+Y_2)}$ (D) $Y=\frac{2Y_1Y_2}{Y_1+Y_2}$

27. In Young's double slit experiment, if the source of light changes from orange to blue then : **[27 July 2021 (Morning)]**
(A) The distance between consecutive fringes will decrease
(B) The distance between consecutive fringes will increase
(C) The intensity of the minima will increase
(D) The central bright fringe will become a dark fringe

28. A stone of mass 20 g is projected from a rubber catapult of length 0.1 m and area of cross section 10^{-6} m^2 stretched by an amount 0.04 m. The velocity of the projected stone is _____ m/s (Young's modulus of rubber = 0.5×10^9 N/m^2). **[27 July 2021 (Morning)]**

29. Wires W_1 and W_2 are made of same material having the breaking stress of 1.25×10^9 N/m^2. W_1 and W_2 have cross-sectional area of 8×10^{-7} m^2 and 4×10^{-7} m^2, respectively. Masses of 20 kg and 10 kg hang from them as shown in the figure. The maximum mass that can be placed in the pan without breaking the wires is_____kg (Use $g = 10$ m/s^2). **[27 Aug. 2021 (Evening)]**

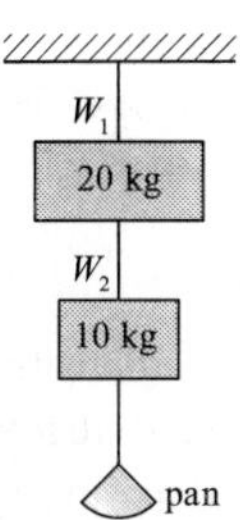

30. A uniform heavy rod of weight 10 kg ms^{-2}, cross-sectional area 100 cm^2 and length 20 cm is hanging from a fixed support. Young modulus of the material of the rod is 2×10^{11} Nm^{-2}. Neglecting the lateral contraction, find the elongation of rod due to its own weight : **[31 Aug. 2021 (Morning)]**
(A) 4×10^{-8} m (B) 5×10^{-10} m
(C) 2×10^{-9} m (D) 5×10^{-8} m

31. When a rubber ball is taken to a depth of_____ m in deep sea, its volume decrease by 0.5%. (The bulk modulus of rubber = 9.8×10^8 Nm^{-2}, Density of sea water = 10^3 kgm^{-3}, $g = 9.8$ m/s^2) **[31 Aug. 2021 (Morning)]**

32. Four identical hollow cylindrical columns of mild steel support a big structure of mass 50×10^3 kg. The inner and outer radii of each column are 50 cm and 100 cm respectively. Assuming uniform local distribution, calculate the compression strain of each column. (use $Y = 2.0 \times 10^{11}$ Pa, $g = 9.8$ m/s^2) : **[31 Aug. 2021 (Evening)]**
(A) 1.87×10^{-3} (B) 60×10^{-8}
(C) 7.07×10^{-4} (D) 2.60×10^{-7}

33. In a Young's double slit experiment, the slits are separated by 0.3 mm and the screen is 1.5 m away from the plane of slits. Distance between fourth bright fringes on both sides of central bright fringe is 2.4 cm. The frequency of light used is _____ $\times 10^{14}$ Hz: **[31 Aug. 2021 (Evening)]**

34. A student determined Young's Modulus of elasticity using the formula $Y = \frac{MgL^3}{4bd^3\delta}$. The value of g is taken to be 9.8 m/s^2, without any significant error, his observation are as following.

Physical Quantity	Least count of the Equipment Used for measurement	Observed Value
Mass (M)	1 g	2 kg
Length of bar (L)	1 mm	1 m
Breadth of bar (b)	0.1 mm	4 cm
Thickness of bar (d)	0.01 mm	0.4 cm
Depression (δ)	0.01 mm	5 mm

Then the fractional error in the measurement of Y is : **[01 Sep. 2021 (Evening)]**
(A) 0.083 (B) 0.0155
(C) 0.155 (D) 0.0083

* * * * *

CHAPTER 9

MECHANICAL PROPERTIES OF FLUIDS

9.1 FLUID STATICS

1. A wooden block floating in a bucket of water has 4/5 of its volume submerged. When certain amount of an oil is poured into the bucket, it is found that the block is just under the oil surface with half of its volume under water and half in oil. The density of oil relative to that of water is : **[9 April 2019 (Evening)]**

(A) 0.5 (B) 0.7
(C) 0.6 (D) 0.8

2. A submarine experiences a pressure of 5.05×10^6 Pa at a depth of d_1 in a sea. When it goes further to a depth of d_2, it experiences a pressure of 8.08×10^6 Pa. Then $d_2 - d_1$ is approximately (density of water = 10^3 kg/m^3 and acceleration due to gravity = 10 ms^{-2}) : **[10 April 2019 (Evening)]**

(A) 500 m (B) 400 m
(C) 300 m (D) 600 m

3. A cubical block of side 0.5 m floats on water with 30% of its volume under water. What is the maximum weight that can be put on the block without fully submerging it under water? (Take density of water = 10^3 kg/m^3) : **[10 April 2019 (Evening)]**

(A) 65.4 kg (B) 87.5 kg
(C) 30.1 kg (D) 46.3 kg

4. Consider a solid sphere of radius R and mass density $\rho(r) = \rho_0\left(1 - \frac{r^2}{R^2}\right), 0 < r \le R$. The minimum density of a liquid in which it will float is : **[8 Jan. 2020 (Morning)]**

(A) $\frac{\rho_0}{5}$ (B) $\frac{\rho_0}{3}$
(C) $\frac{2\rho_0}{3}$ (D) $\frac{2\rho_0}{5}$

5. Two liquids of densities ρ_1 an $\rho_2(\rho_2 = 2\rho_1)$ are filled up behind a square wall of side 10 m as shown in figure. Each liquid has a height of 5 m. The ratio of the forces due to these liquids exerted on upper part MN to that at the lower part NO is (Assume that the liquids are not mixing) : **[8 Jan. 2020 (Evening)]**

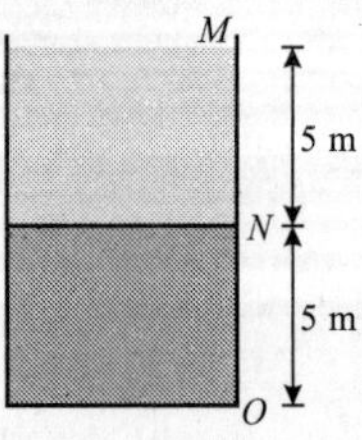

(A) 1/4 (B) 2/3
(C) 1/3 (D) 1/2

6. A air bubble of radius 1 cm in water has an upward acceleration 9.8 cm s^{-2}. The density of water is 1 gm cm^{-3} and water offers negligible drag force on the bubble. The mass of the bubble is (g = 980 cm/s^2) : **[4 Sept. 2020 (Morning)]**

(A) 3.15 gm (B) 4.51 gm
(C) 4.15 gm (D) 1.52 gm

7. Two identical cylindrical vessels are kept on the ground and each contain the same liquid of density d. The area of the base of both vessels is S but the height of liquid in one vessel is x_1 and in the other, x_2. When both cylinders are connected through a pipe of negligible volume very close to the bottom, the liquid flows from one vessel to the other until it comes to equilibrium at a new height. The change in energy of the system in the process is : **[4 Sept. 2020 (Evening)]**

(A) $gdS(x_2 + x_1)^2$ (B) $\frac{3}{4}gdS(x_2 - x_1)^2$
(C) $\frac{1}{4}gdS(x_2 - x_1)^2$ (D) $gdS(x_2^2 + x_1^2)$

8. A hollow spherical shell at outer radius R floats just submerged under the water surface. The inner radius of the shell is r. If the specific gravity of the shell material is $\frac{27}{8}$ w.r.t. water, the value of r is : **[5 Sept. 2020 Morning]**

(A) $\frac{4}{9}R$ (B) $\frac{8}{9}R$
(C) $\frac{1}{3}R$ (D) $\frac{2}{3}R$

9. A hydraulic press can lift 100 kg when a mass 'm' is placed on the smaller piston. It can lift _____ kg when the diameter of the larger piston is increased by 4 times and that of the smaller piston is decreased by 4 times keeping the same mass 'm' on the smaller piston. **[24 Feb. 2021 (Morning)]**

10. The pressure acting on a submarine is 3×10^5 Pa at a certain depth. If the depth is doubled, the percentage increase in the pressure acting on the submarine would be : (Assume that atmospheric pressure is 1×10^5 Pa density of water is 10^3 kg m^{-3}, $g = 10$ ms^{-2}) **[16 March 2021 (Morning)]**

(A) $\frac{200}{3}\%$ (B) $\frac{200}{5}\%$

(C) $\frac{5}{200}\%$ (D) $\frac{3}{200}\%$

11. A raindrop with radius $R = 0.2$ mm falls from a cloud at a height $h = 2000$ m above the ground. Assume that the drop is spherical throughout is fall and the force of buoyance may be neglected, then the terminal speed attained by the raindrop is : (Density of water $f_w = 1000$ kg m^{-3} and density of air $f_a = 1.2$ kg m^{-3}, $g = 10$ m/s^2, Coefficient of viscosity of air $= 1.8 \times 10^{-5}$ Nsm^{-2}) : **[27 July 2021 (Evening)]**

(A) 4.94 ms^{-1} (B) 250.6 ms^{-1}

(C) 43.56 ms^{-1} (D) 14.4 ms^{-1}

12. Two narrow bores of diameter 5.0 mm and 8.0 mm are joined together to form a *U*-shaped tube open at both ends. If this U-tube contains water, what is the difference in level of two limbs of the tube? [Take surface tension of water $T = 7.3 \times 10^{-2}$ Nm^{-1}, angle of contact = 0, $g = 10$ ms^{-2} and density of water $= 1.0 \times 10^3$ kg m^{-3}] **[26 Aug. 2021 (Morning)]**

(A) 4.97 mm (B) 2.19 mm

(C) 5.34 mm (D) 3.62 mm

13. In Millikan's oil drop experiment, what is viscous force acting on a uncharged drop of radius 2×10^{-5} m and density 1.2×10^3 kgm^{-3}? Take viscosity of liquid $= 1.8 \times 10^{-5}$ Nsm^{-2}. (Neglect buoyancy due to air) : **[27 Aug. 2021 (Morning)]**

(A) 5.8×10^{-10} N (B) 1.8×10^{-10} N

(C) 3.8×10^{-11} N (D) 3.9×10^{-10} N

* * * * *

CHAPTER 9

MECHANICAL PROPERTIES OF FLUIDS

9.2 FLUID DYNAMICS

1. The top of a water tank is open to air and its water level is maintained. It is giving out 0.74 m^3 water per minute through a circular opening of 2 cm radius in its wall. The depth of the centre of the opening from the level of water in the tank is close to : **[9 Jan. 2019 (Evening)]**

(A) 9.6 m (B) 4.8 m
(C) 2.9 m (D) 6.0 m

2. Water flows into a large tank with flat bottom at the rate of 10^{-4} m^3s^{-1}. Water is also leaking out of a hole of area 1 cm^2 at its bottom. If the height of the water in the tank remains steady, then this height is : **[10 Jan. 2019 (Morning)]**

(A) 4 cm (B) 2.9 cm
(C) 1.7 cm (D) 5.1 cm

3. A liquid of density ρ is coming out of a hose pipe of radius a with horizontal speed v and hits a mesh. 50% of the liquid passes through the mesh unaffected. 25% looses all of its momentum and 25% comes back with the same speed. The resultant pressure on the mesh will be : **[11 Jan. 2019 (Morning)]**

(A) ρv^2 (B) $\frac{3}{4}\rho v^2$

(C) $\frac{1}{2}\rho v^2$ (D) $\frac{1}{4}\rho v^2$

4. A long cylindrical vessel is half filled with a liquid. When the vessel is rotated about its own vertical axis, the liquid rises up near the wall. If the radius of vessel is 5 cm and its rotational speed is 2 rotations per second, then the difference in the heights between the centre and the sides, in cm, will be : **[12 Jan. 2019 (Evening)]**

(A) 1.2 (B) 0.1
(C) 2.0 (D) 0.4

5. Water from a tap emerges vertically downwards with an initial speed of 1.0 ms^{-1}. The cross-sectional area of the tap is 10^{-4} m^2. Assume that the pressure is constant throughout the stream of water and that the flow is streamlined. The cross-sectional area of the stream, 0.15 m below the tap would be : (Take $g = 10$ ms^{-2}) **[10 April 2019 (Evening)]**

(A) 1×10^{-5} m^2 (B) 5×10^{-5} m^2
(C) 2×10^{-5} m^2 (D) 5×10^{-4} m^2

6. An ideal fluid flows (laminar flow) through a pipe of non-uniform diameter. The maximum and minimum diameters of the pipes are 6.4 cm and 4.8 cm, respectively. The ratio of the minimum and the maximum velocities of fluid in this pipe is : **[7 Jan. 2020 (Evening)]**

(A) $\frac{\sqrt{3}}{2}$ (B) $\frac{3}{4}$

(C) $\frac{81}{256}$ (D) $\frac{9}{16}$

7. Water flows in a horizontal tube (see figure). The pressure of water changes by 700 Nm^{-2} between A and B where the area of cross section are 40 cm^2 and 20 cm^2, respectively. Find the rate of flow of water through the tube. (Density of water = 1000 kgm^{-3}) **[9 Jan. 2020 (Morning)]**

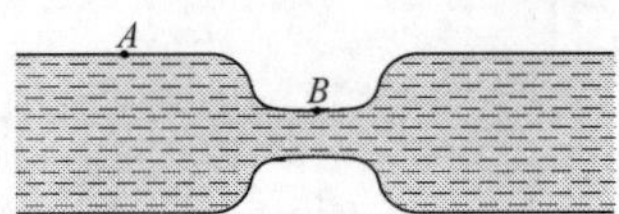

(A) 1810 cm^3/s (B) 3020 cm^3/s
(C) 2720 cm^3/s (D) 2420 cm^3/s

8. A fluid is flowing through a horizontal pipe of varying cross-section, with speed v ms^{-1} at a point where the pressure is P Pascal. At another point where pressure is $P/2$ Pascal its speed is V ms^{-1}. If the density of the fluid is ρ kgm^{-3} and the flow is streamline, then V is equal to : **[6 Sept. 2020 (Evening)]**

(A) $\sqrt{\frac{P}{2\rho}+v^2}$ (B) $\sqrt{\frac{P}{\rho}+v^2}$

(C) $\sqrt{\frac{2P}{\rho}+v^2}$ (D) $\sqrt{\frac{P}{\rho}+v}$

9. Consider a water tank as shown in the figure. It's cross-sectional area is 0.4 m^2. The tank has an opening B near the bottom whose cross-section area is 1 cm^2. A load of 24 kg is applied on the water at the top when the height of the water level is 40 cm above the bottom, the velocity of water coming out the opening B is v ms^{-1}. The value of v, to the nearest integer, is_____. [Take value of g to be 10 ms^{-2}]

[18 March 2021 (Evening)]

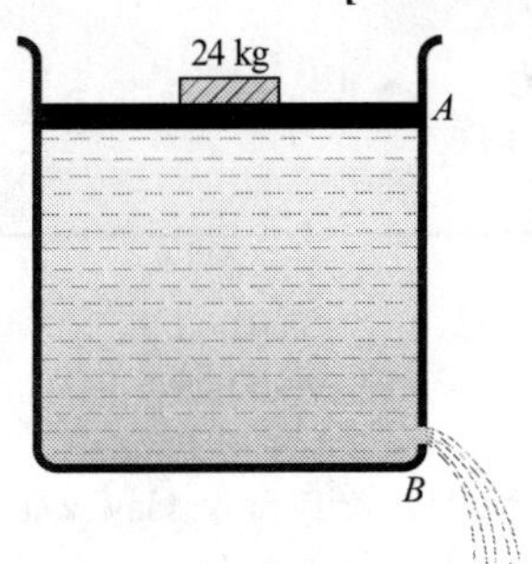

10. A light cylindrical vessel is kept on a horizontal surface. Area of base is A. A hole of cross sectional area 'a' is made just at the bottom side. The minimum coefficient of friction necessary to prevent sliding the vessel due to the impact force of the emerging liquid is ($a << A$) : **[27 July 2021 (Morning)]**

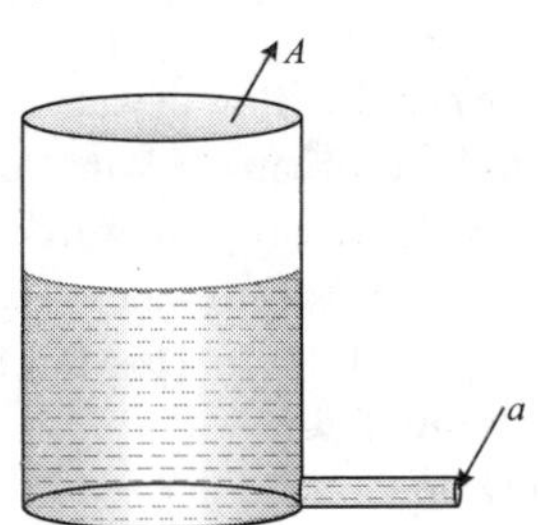

(A) None of these

(B) $\frac{a}{A}$

(C) $\frac{2a}{A}$

(D) $\frac{A}{2a}$

11. The water is filled upto height of 12 m in a tank having vertical sidewalls. A hole is made in one of the walls at a depth 'h' below the water level. The value of 'h' for which the emerging stream of water strikes the ground at the maximum range is _____ m. **[27 July 2021 (Evening)]**

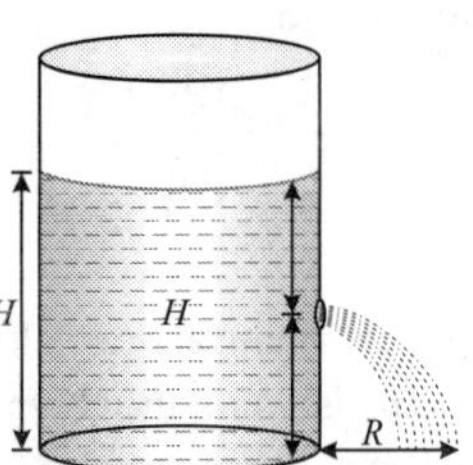

* * * * *

CHAPTER

9 MECHANICAL PROPERTIES OF FLUIDS

9.3 VISCOSITY

1. Two tubes of radii r_1 and r_2, and lengths l_1 and l_2, respectively, are connected in series and a liquid flows through each of them in streamline conditions. P_1 and P_2 are pressure differences across the two tubes. If P_2 is $4P_1$ and l_2 is $l_1/4$, then the radius r_2 will be equal to : **[2017 (Online)]**

(A) r_1 (B) $2r_1$

(C) $4r_1$ (D) $\frac{r_1}{2}$

2. Water from a pipe is coming at a rate of 100 litres per minute. If the radius of the pipe is 5 cm, the Reynolds number for the flow is of the order of : (density of water = 1000 kg/m^3, coefficient of viscosity of water = 1 mPas) : **[8 April 2019 (Morning)]**

(A) 10^6 (B) 10^3

(C) 10^4 (D) 10^2

3. A solid sphere, of radius R acquires a terminal velocity v_1 when falling (due to gravity) through a viscous fluid having a coefficient of viscosity η. The sphere is broken into 27 identical solid spheres. If each of these spheres acquires a terminal velocity v_2, when falling through the same fluid, the ratio (v_1/v_2) equals : **[12 April 2019 (Evening)]**

(A) 1/27 (B) 1/9

(C) 27 (D) 9

4. In an experiment to verify Stokes law, a small spherical ball of radius r and density ρ falls under gravity through a distance h in air before entering a tank of water. If the terminal velocity of the ball inside water is same as its velocity just before entering the water surface, then the value of h is proportional to (ignore viscosity of air) : **[5 Sept. 2020 (Evening)]**

(A) r (B) r^4

(C) r^3 (D) r^2

5. What will be the nature of flow of water from a circular tap, when its flow rate increased from 0.18 L/min to 0.48 L/min ? The radius of the tap and viscosity of water are 0.5 cm and 10^{-3} Pa s, respectively. (Density of water : 10^3 kg/m^3) **[16 March 2021 (Evening)]**

(A) Unsteady to steady flow

(B) Remains steady flow

(C) Remains turbulent flow

(D) Steady flow to unsteady flow

* * * * *

CHAPTER 9

MECHANICAL PROPERTIES OF FLUIDS

9.4 SURFACE TENSION

1. A soap bubble, blown by a mechanical pump at the mouth of a tube, increases in volume, with time, at a constant rate. The graph that correctly depicts the time dependence of pressure inside the bubble is given by : **[12 Jan. 2019 (Evening)]**

(A) 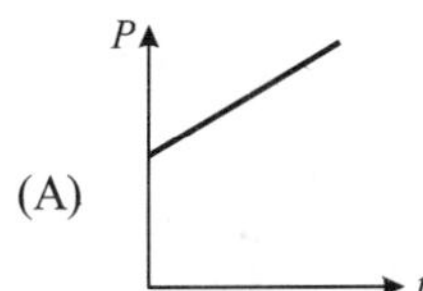

(B) 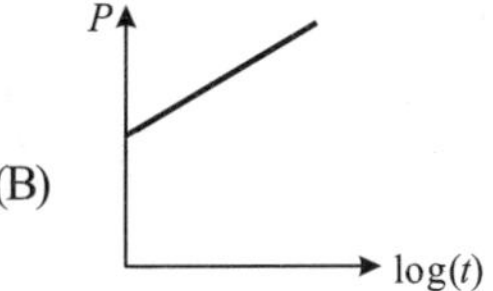

(C) 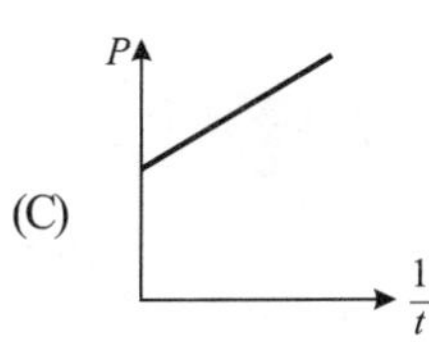

(D) 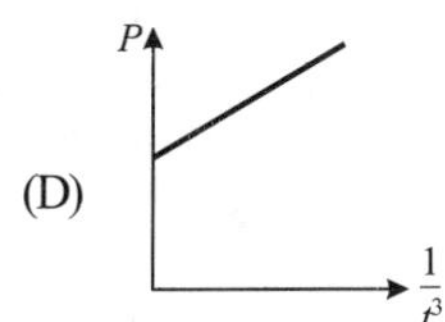

2. If 'M' is the mass of water that rises in a capillary tube of radius 'r', then mass of water which will rise in a capillary tube of radius '$2r$' is : **[9 April 2019 (Morning)]**

(A) $4M$ (B) M
(C) $2M$ (D) $M/2$

3. The ratio of surface tensions of mercury and water is given to be 7.5 while the ratio of their densities is 13.6. Their contact angles, with glass, are close to 135° and 0°, respectively. It is observed that mercury gets depressed by an amount h in a capillary tube of radius r_1, while water rises by the same amount h in a capillary tube of radius r_2. The ratio, (r_1/r_2), is then close to : **[10 April 2019 (Morning)]**

(A) 2/3 (B) 3/5
(C) 2/5 (D) 4/5

4. A small spherical droplet of density d is floating exactly half immersed in a liquid of density ρ and surface tension T. The radius of the droplet is (take note that the surface tension applies an upward force on the droplet) : **[9 Jan. 2020 (Evening)]**

(A) $r = \sqrt{\frac{2T}{3(d+\rho)g}}$ (B) $r = \sqrt{\frac{3T}{(2d-\rho)g}}$

(C) $r = \sqrt{\frac{T}{(d-\rho)g}}$ (D) $r = \sqrt{\frac{T}{(d+\rho)g}}$

5. A capillary tube made of glass of radius 0.15 mm is dipped vertically in a beaker filled with methylene iodide surface tension $= 0.05\ Nm^{-1}$, density $= 667\ kgm^{-3}$) which rises to height h in the tube. It is observed that the two tangents drawn from liquid-glass interfaces (from opposite sides of the capillary) make an angle of 60° with one another. Then h is close to ($g = 10\ ms^{-2}$) : **[2 Sept. 2020 (Evening)]**

(A) 0.137 m (B) 0.172 m
(C) 0.087 m (D) 0.049 m

6. Pressure inside two soap bubbles are 1.01 and 1.02 atmosphere, respectively. The ratio of their volumes is : **[3 Sept. 2020 (Morning)]**

(A) 8 : 1 (B) 0.8 : 1
(C) 2 : 1 (D) 4 : 1

7. When a long glass capillary tube of radius 0.015 cm is dipped in a liquid, the liquid rises to a height of 15 cm within it. If the contact angle between the liquid and glass to close to 0°, the surface tension of the liquid, in milli Newton m^{-1}, is [$\rho_{(liquid)} = 900\ kgm^{-3}, g = 10\ ms^{-2}$] (Give answer in closest integer) _____ : **[3 Sept. 2020 (Morning)]**

8. When two soap bubbles of radii a and b ($b > a$) coalesce, the radius of curvature of common surface is : **[17 March 2021 (Morning)]**

(A) $\frac{ab}{b-a}$ (B) $\frac{a+b}{ab}$

(C) $\frac{b-a}{ab}$ (D) $\frac{ab}{a+b}$

9. Two small drops of mercury each of radius R coalesce to form a single large drop. The ratio of total surface energy before and after the change is : **[20 July 2021 (Evening)]**

(A) $1 : 2^{1/3}$ (B) $2^{3/1} : 1$
(C) $2 : 1$ (D) $1 : 2$

10. Two spherical soap bubbles of radii r_1 and r_2 in vacuum combine under isothermal conditions. The resulting bubble has a radius equal to : **[25 July 2021 (Evening)]**

(A) $\frac{r_1 + r_2}{2}$ (B) $\sqrt{r_1 r_2}$

(C) $\sqrt{r_1^2 + r_2^2}$ (D) $\frac{r_1 r_2}{r_1 + r_2}$

11. A soap bubble of radius 3 cm is formed inside the another soap bubble of radius 6 cm. The radius of an equivalent soap bubble which has the same excess pressure as inside the smaller bubble with respect to the atmospheric pressure is_____cm :

[26 Aug. 2021 (Morning)]

* * * * *

CHAPTER

10 THERMAL PROPERTIES OF MATTER

10.1 THERMAL EXPANSION

1. An external pressure P is applied on a cube at 0°C so that it is equally compressed from all sides. K is the bulk modulus of the material of the cube and α is its coefficient of linear expansion. Suppose we want to bring the cube to its original size by heating. The temperature should be raised by : **[2017]**

(A) $\frac{3\alpha}{PK}$ (B) $3PK\alpha$

(C) $\frac{P}{3\alpha K}$ (D) $\frac{P}{\alpha K}$

2. A compressive force, F is applied at the two ends of a long thin steel rod. It is heated, simultaneously, such that its temperature increases by ΔT. The net change in its length is zero. Let l be the length of the rod, A its area of cross-section, Y its Young's modulus, and α its coefficient of linear expansion. Then, F is equal to : **[2017 (Online)]**

(A) $l^2 Y\alpha\Delta T$ (B) $lA\,Y\alpha\Delta T$

(C) $AY\alpha\Delta T$ (D) $\frac{AY}{\alpha\Delta T}$

3. A steel rail of length 5 m and area of cross-section 40 cm^2 is prevented from expanding along its length while the temperature rises by 10°C. If coefficient of linear expansion and Young's modulus of steel are $1.2 \times 10^{-5}\,K^{-1}$ and $2 \times 10^{11}\,Nm^{-2}$ respectively, the force developed in the rail is approximately : **[2017 (Online)]**

(A) 2×10^7 N (B) 1×10^5 N

(C) 2×10^9 N (D) 3×10^{-5} N

4. A rod, of length L at room temperature and uniform area of cross section A, is made of a metal having coefficient of linear expansion α/°C^{-1}. It is observed that an external compressive force F, is applied on each of its ends, prevents any change in the length of the rod, when its temperature rises by ΔT K. Young's modulus Y, for this metal is : **[9 Jan. 2019 (Morning)]**

(A) $\frac{F}{2A\alpha\Delta T}$ (B) $\frac{F}{A\alpha(\Delta T-273)}$

(C) $\frac{F}{A\alpha\Delta T}$ (D) $\frac{2F}{A\alpha\Delta T}$

5. Two rods A and B of identical dimensions are at temperature 30°C. If A is heated upto 180°C and B upto T°C, then the new lengths are the same. If the ratio of the coefficients of linear expansion of A and B is 4 : 3, then the value of T is : **[11 Jan. 2019 (Evening)]**

(A) 270°C (B) 230°C

(C) 250°C (D) 200°C

6. A thermometer graduated according to a linear scale reads a value x_0 when in contact with boiling water, and $x_0/3$ when in contact with ice. What is the temperature of an object in °C, if this thermometer is in the contact with the object reads $x_0/2$? **[11 Jan. 2019 (Evening)]**

(A) 35 (B) 25

(C) 60 (D) 40

7. A non-isotropic solid metal cube has coefficients of linear expansion as :
5×10^{-5}/°C along the x-axis and 5×10^{-6}/°C along the y and the z-axis. If the coefficient of volume expansion of the solid is $C \times 10^{-16}$/°C then the value of C is ______. **[7 Jan. 2020 (Morning)]**

8. A leak proof cylinder of length 1 m, made of a metal which has very low coefficient of expansion is floating vertically in water at 0°C such that its height above the water surface is 20 cm. When the temperature of water is increased to 4°C, the height of the cylinder above the water surface becomes 21 cm. The density of water at $T = 4$°C, relative to the density at $T = 0$°C is close to : **[8 Jan. 2020 (Morning)]**

(A) 1.01 (B) 1.04

(C) 1.03 (D) 1.26

9. When the temperature of a metal wire is increased from 0°C to 10°C, its length increases by 0.02 %. The percentage change in its mass density will be closest to : **[2 Sept. 2020 (Evening)]**

(A) 0.008 (B) 0.06

(C) 0.8 (D) 2.3

10. A bakelite beaker has volume capacity of 500 cc at 30°C. When it is partially filled with V_m volume (at 30°) of mercury, it is found that the unfilled volume of the beaker remains constant as temperature is varied. If $\gamma_{(beaker)} = 6 \times 10^{-6}$ °C^{-1} and $\gamma_{(mercury)} = 1.5 \times 10^{-4}$ °C^{-1}, where γ is the coefficient of volume expansion, then V_m (in cc) is close to______: **[3 Sept. 2020 (Morning)]**

11. Two different wires having lengths L_1 and L_2, and respective temperature coefficient of linear expansion α_1 and α_2, are joined end-to-end. Then the effective temperature coefficient of linear expansion is : **[5 Sept. 2020 (Evening)]**

(A) $4\frac{\alpha_1\alpha_2}{\alpha_1+\alpha_2}\frac{L_2L_1}{(L_2+L_1)^2}$ (B) $2\sqrt{\alpha_1\alpha_2}$

(C) $\frac{\alpha_1+\alpha_2}{2}$ (D) $\frac{\alpha_1L_1+\alpha_2L_2}{L_1+L_2}$

12. Each side of a box made of metal sheet in cubic shape is 'a' at room temperature 'T', the coefficient of linear expansion of the metal sheet is 'α'. The metal sheet is heated uniformly, by a small temperature ΔT, so that its new temperature is $T + \Delta T$. Calculate the increase in the volume of the metal box : **[24 Feb. 2021 (Morning)]**

(A) $3a^3\alpha\Delta T$ (B) $4\pi a^3\alpha\Delta T$

(C) $4a^3\alpha\Delta T$ (D) $\frac{4}{3}\pi a^3\alpha\Delta T$

13. Given below are two statements : one is labelled as **Assertion-A** and the other is labelled as **Reason-R**

Assertion-A : When a rod lying freely on a smooth surface is heated, no thermal stress is developed in it.

Reason-R : On heating, the length of the rod increases

In the light of the above statements, choose the correct answer from the options given below : **[25 Feb. 2021 (Morning)]**

(A) Both **A** and **R** are true and **R** is the correct explanation of **A**

(B) **A** is false but **R** is true

(C) **A** is true but **R** is false

(D) Both **A** and **R** are true but **R** is NOT the correct explanation of **A**

14. A bimetallic strip consists of metals A and B. It is mounted rigidly as shown. The metal A has higher coefficient of expansion compared to that of metal B. When the bimetallic strip is placed in a cold both, it will : **[16 March 2021 (Evening)]**

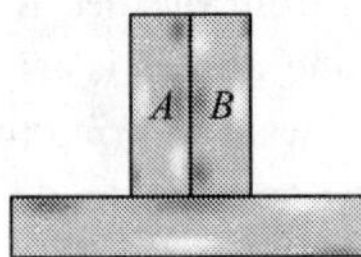

(A) Bend towards the right (B) Not bend but shrink

(C) Neither bend nor shrink (D) Bend towards the left

15. The resistance of a conductor at 15°C is 16 Ω and at 100°C is 20 Ω. What will be the temperature coefficient of resistance of the conductor ? **[27 July 2021 (Evening)]**

(A) 0.010°C^{-1} (B) 0.033°C^{-1}

(C) 0.003°C^{-1} (D) 0.042°C^{-1}

16. A steel rod with $y = 2.0 \times 10^{11}$ Nm^{-2} and $\alpha = 10^{-5}$ °C^{-1} of length 4 m and area of cross-section 10 cm^2 is heated from 0°C to 400°C without being allowed to extend. The tension produced in the rod is $x \times 10^5$ N where the value of x is : **[01 Sep. 2021 (Evening)]**

* * * * *

CHAPTER

10 THERMAL PROPERTIES OF MATTER

10.2 CALORIMETRY

1. A copper ball of mass 100 gm is at a temperature *T*. It is dropped in a copper calorimeter of mass 100 gm, filled with 170 gm of water at room temperature. Subsequently, the temperature of the system is found to be 75°C. *T* is given by (Given : room temperature = 30°C, specific heat of copper = 0.1 cal/gm°C) : **[2017]**

(A) 1250°C` (B) 825°C
(C) 800°C (D) 885°C

2. In an experiment a sphere of aluminium of mass 0.20 kg is heated upto 150°C. Immediately, it is put into water of volume 150 cc at 27°C kept in a calorimeter of water equivalent to 0.025 kg. Final temperature of the system is 40°C. The specific heat of aluminium is : [Take 4.2 Joule = 1 calorie] **[2017 (Online)]**

(A) 378 J/kg – °C (B) 315 J/kg – °C
(C) 476 J/kg – °C (D) 434 J/kg – °C

3. An unknown metal of mass 192 g heated to a temperature of 100°C was immersed into a brass calorimeter of mass 128 g containing 240 g of water a temperature of 8.4°C Calculate the specific heat of the unknown metal if water temperature stabilizes at 21.5°C (Specific heat of brass is 394 J kg^{-1} K^{-1}) :
[10 Jan. 2019 (Evening)]

(A) 1232 J kg^{-1} K^{-1} (B) 458 J kg^{-1} K^{-1}
(C) 654 J kg^{-1} K^{-1} (D) 916 J kg^{-1} K^{-1}

4. Ice at –20°C is added to 50 g of water at 40°C. When the temperature of the mixture reaches 0°C, it is found that 20 g of ice is still unmelted. The amount of ice added to the water was close to **[11 Jan. 2019 (Morning)]**
(Specific heat of water = 4.2 J/g/°C)
Specific heat of Ice = 2.1 J/g/°C
Heat of fusion of water at 0°C = 334 J/g) :

(A) 50 g (B) 40 g
(C) 60 g (D) 100 g

5. When 100 g of a liquid *A* at 100°C is added to 50 g of a liquid *B* at temperature 75°C, the temperature of the mixture becomes 90°C. The temperature of the mixture, if 100 g of liquid *A* at 100°C is added to 50 g of liquid *B* at 50°C, will be :
[11 Jan. 2019 (Evening)]

(A) 80°C (B) 60°C
(C) 70°C (D) 85°C

6. A metal ball of mass 0.1 kg is heated upto 500°C and dropped into a vessel of heat capacity 800 JK^{-1} and containing 0.5 kg water. The initial temperature of water and vessel is 30°C. What is the approximate percentage increment in the temperature of the water ? [Specific Heat Capacities of water and metal are, respectively, 4200 $JKg^{-1}K^{-1}$ and 400 $JKg^{-1}K^{-1}$] :
[11 Jan. 2019 (Evening)]

(A) 30% (B) 20%
(C) 25% (D) 15%

7. A thermally insulated vessel contains 150 g of water at 0°C. Then the air from the vessel is pumped out adiabatically. A fraction of water turns into ice and the rest evaporates at 0°C itself. The mass of evaporated water will be closest to :
(Latent heat of vaporization of water = 2.10×10^6 J kg^{-1} and Latent heat of Fusion of water = 3.36×10^5 J kg^{-1}) :
[8 April 2019 (Morning)]

(A) 130 g (B) 35 g
(C) 20 g (D) 150 g

8. When M_1 gram of ice at –10°C (specific heat = 0.5 cal g^{-1} °C^{-1}) is added to M_2 gram of water at 50°C, finally no ice is left and the water is at 0°C. The value of latent heat of ice, in cal g^{-1} is : **[12 April 2019 (Morning)]**

(A) $\frac{5M_1}{M_2} - 50$ (B) $\frac{50M_2}{M_1}$

(C) $\frac{50M_2}{M_1} - 5$ (D) $\frac{5M_2}{M_1} - 5$

9. *M* grams of steam at 100°C is mixed with 200 g of ice at its melting point in a thermally insulated container. If it produces liquid water at 40°C [heat of vaporization of water is 540 cal/g and heat of fusion of ice is 80 cal/g], the value of *M* is______.
[7 Jan. 2020 (Evening)]

10. Three containers C_1, C_2 and C_3 have water at different temperatures. The table below shows the final temperature *T* when different amounts of water (given in litres) are taken from each containers and mixed (assume no loss of heat during the process). **[8 Jan. 2020 (Evening)]**

C_1	C_2	C_3	T
1l	2l	–	60°C
–	1l	2l	30°C
2l	–	1l	60°C
1l	1l	1l	θ

The value of θ (in °C to the nearest integer) is _____.

11. A calorimeter of water equivalent 20 g contains 180 g of water at 25ºC. 'm' grams of steam at 100°C is mixed in it till the temperature of the mixture is 31ºC. The value of 'm' is close to (Latent heat of water = 540 cal g^{-1}, specific heat of water = 1 cal g^{-1} $°C^{-1}$) : **[3 Sept. 2020 (Evening)]**

(A) 2.6 (B) 2
(C) 4 (D) 3.2

12. The specific heat of water = 4200 J kg^{-1} K^{-1} and the latent heat of ice = 3.4×10^5 J kg^{-1}. 100 grams of ice at 0ºC is placed in 200 g of water at 25ºC. The amount of ice that will melt as the temperature of water reaches 0ºC is close to (in grams) : **[4 Sept. 2020 (Morning)]**

(A) 61.7 (B) 63.8
(C) 69.3 (D) 64.6

13. A bullet of mass 5 g, travelling with a speed of 210 m/s, strikes a fixed wooden target. One half of its kinetic energy is converted into heat in the bullet while the other half is converted into heat in the wood. The rise of temperature of the bullet if the specific heat of its material is 0.030 cal/(g–ºC).
(1 cal = 4.2×10^7 ergs) close to : **[5 Sept. 2020 (Morning)]**

(A) 83.3 ºC (B) 87.5 ºC
(C) 119.2 ºC (D) 38.4 ºC

14. A large number of water drops, each of radius r, combine to have a drop of radius R. If the surface tension is T and mechanical equivalent of heat is J, the rise in heat energy per unit volume will be : **[26 Feb. 2021 (Morning)]**

(A) $\frac{2T}{rJ}$ (B) $\frac{3T}{rJ}$

(C) $\frac{3T}{J}\left(\frac{1}{r}-\frac{1}{R}\right)$ (D) $\frac{2T}{J}\left(\frac{1}{r}-\frac{1}{R}\right)$

15. The temperature of equal masses of three different liquids x, y and z are 10ºC, 20ºC and 30ºC respectively. The temperature of mixture when x is mixed with y is 16°C and that when y is mixed with z is 26ºC. The temperature of mixture when x and z are mixed will be : **[26 Aug. 2021 (Evening)]**

(A) 20.28ºC (B) 25.62ºC
(C) 23.84ºC (D) 28.32ºC

16. The height of victoria falls is 63 m. What is the difference in temperature of water at the top and at the bottom of fall? [Given 1 cal = 4.2 J and specific heat of water = 1 cal g^{-1}° C^{-1}] : **[27 Aug. 2021 (Evening)]**

(A) 14.76 ºC (B) 1.476ºC
(C) 0.147ºC (D) 0.014ºC

17. Due to cold weather a 1 m water pipe of cross-sectional area 1 cm^2 is filled with ice at – 10ºC. Resistance heating is used to melt the ice. Current of 0.5 A is passed through 4 kΩ resistance. Assuming that all the heat produced is used for melting, what is the minimum time required ?
(Given latent heat of fusion for water/ice = 3.33×10^5 J kg^{-1}, specific heat of ice = 2×10^3 J kg^{-1} and density of ice = 10^3 kg/m^3) **[01 Sep 2021 Evening]**

(A) 0.353 s (B) 3.53 s
(C) 70.6 s (D) 35.3 s

* * * * *

CHAPTER

10 THERMAL PROPERTIES OF MATTER

10.3 HEAT TRANSFER

1. Temperature difference of 120°C is maintained between two ends of a uniform rod AB of length $2L$. Another bent rod PQ, of same cross-section as AB and length $\frac{3L}{2}$, is connected across AB (See figure). In steady state, temperature difference between P and Q will be close to : **[9 Jan. 2019 (Morning)]**

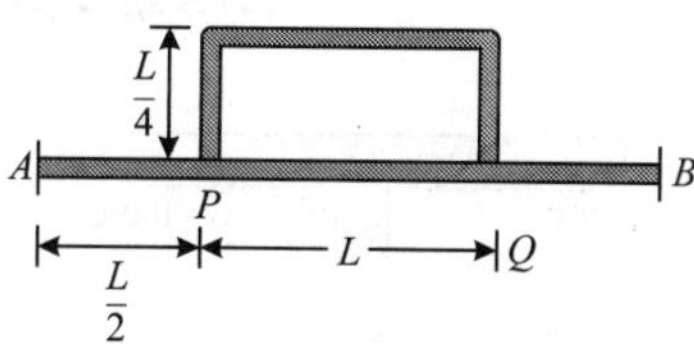

(A) 60°C (B) 75°C
(C) 35°C (D) 45°C

2. A heat source at $T = 10^3$ K is connected to another heat reservoir at $T = 10^2$ K by a copper slab which is 1 m thick. Given that the thermal conductivity of copper is 0.1 WK^{-1} m^{-1}, the energy flux through it in the steady state is :

[10 Jan. 2019 (Morning)]

(A) 90 Wm^{-2} (B) 200 Wm^{-2}
(C) 65 Wm^{-2} (D) 120 Wm^{-2}

3. A cylinder of radius R is surrounded by a cylindrical shell of inner radius R and outer radius $2R$. The thermal conductivity of the material of the inner cylinder is K_1 and that of the outer cylinder is K_2. Assuming no loss of heat, the effective thermal conductivity of the system for heat flowing along the length of the cylinder is : **[12 Jan. 2019 (Morning)]**

(A) $K_1 + K_2$ (B) $\frac{K_1 + K_2}{2}$

(C) $\frac{2K_1 + 3K_2}{5}$ (D) $\frac{K_1 + 3K_2}{4}$

4. Two identical beakers A and B contain equal volumes of two different liquids at 60°C each and left to cool down. Liquid in A has density of 8×10^2 kg/m^3 and specific heat of 2000 J kg^{-1} K^{-1} while liquid in B has density of 10^3 kg m^{-3} and specific heat of 4000 J kg^{-1} K^{-1}. Which of the following best describes their temperature versus time graph schematically? (assume the emissivity of both the beakers to be the same) :

[8 April 2019 (Morning)]

(A)
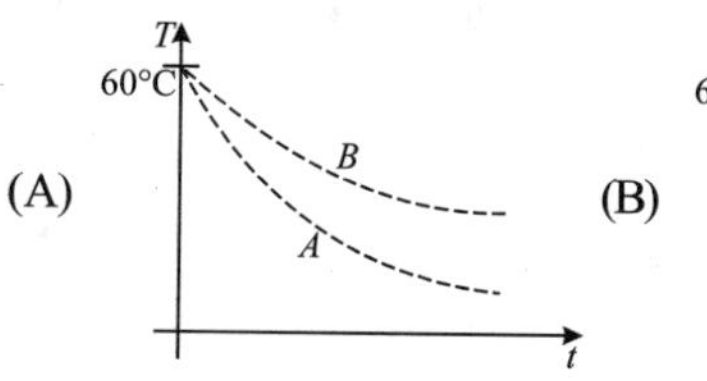

(B)
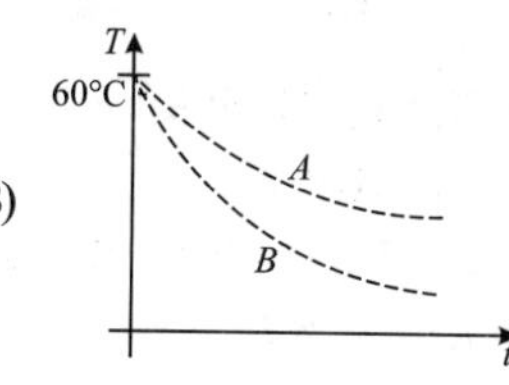

(C)
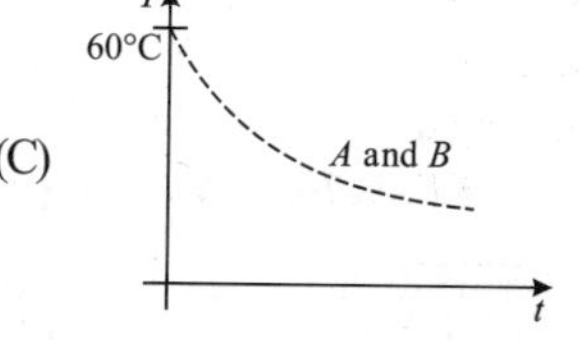

(D)
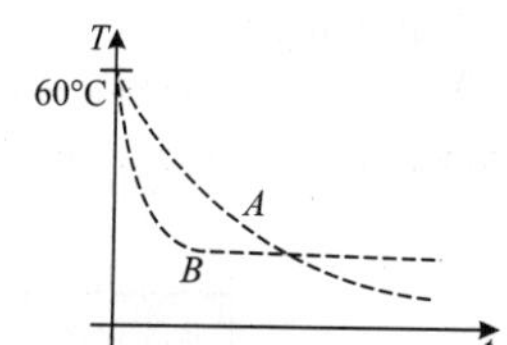

5. Two materials having coefficients of thermal conductivity '$3K$' and 'K' and thickness 'd' and '$3d$', respectively, are joined to form a slab as shown in the figure. The temperatures of the outer surfaces are 'θ_2' and 'θ_1' respectively, ($\theta_2 > \theta_1$). The temperature at the interface is : **[9 April 2019 (Evening)]**

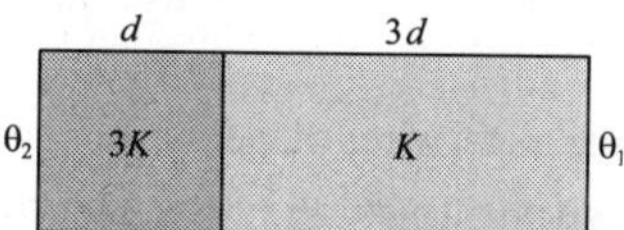

(A) $\frac{\theta_2 + \theta_1}{2}$ (B) $\frac{\theta_1}{10} + \frac{9\theta_2}{10}$

(C) $\frac{\theta_1}{3} + \frac{2\theta_2}{3}$ (D) $\frac{\theta_1}{6} + \frac{5\theta_2}{6}$

6. A metallic sphere cools from 50°C to 40°C in 300 s. If atmospheric temperature around is 20°C, then the sphere's temperature after the next 5 minutes will be close to :

[3 Sept. 2020 (Evening)]

(A) 33°C (B) 35°C
(C) 31°C (D) 28°C

7. Amount of solar energy received on the earth's surface per unit area per unit time is defined a solar constant. Dimension of solar constant is : **[3 Sept. 2020 (Evening)]**

(A) ML^2T^{-2} (B) MLT^{-2}
(C) $M^2L^0T^{-1}$ (D) ML^0T^{-3}

8. Three rods of identical cross-section and lengths are made of three different materials of thermal conductivity K_1, K_2, and K_3, respectively. They are joined together at their ends to make a long rod (see figure). One end of the long rod is maintained at 100°C and the other at 0°C (see figure). If the joints of the rod are at 70°C and 20°C in steady state and there is no loss of energy from the surface of the rod, the correct relationship between K_1, K_2 and K_3 is : **[6 Sept. 2020 (Evening)]**

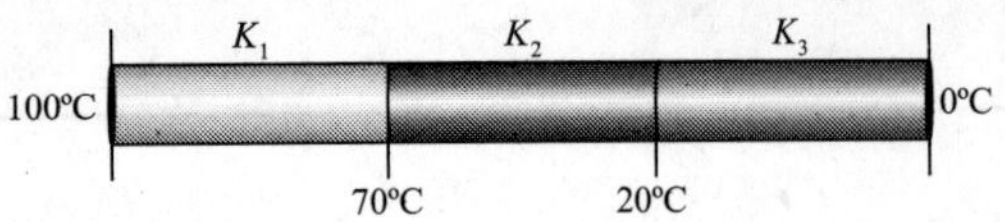

(A) $K_1 : K_3 = 2 : 3; K_2 : K_3 = 2 : 5$
(B) $K_1 < K_2 < K_3$
(C) $K_1 : K_2 = 5 : 2; K_1 : K_3 = 3 : 5$
(D) $K_1 > K_2 > K_3$

9. The temperature θ at the junction of two insulating sheets, having thermal resistances R_1 and R_2 as well as top and bottom temperatures θ_1 and θ_2 (as shown in figure) is given by : **[26 Feb. 2021 (Morning)]**

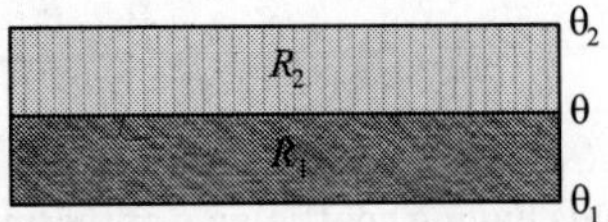

(A) $\dfrac{\theta_1 R_2 + \theta_2 R_1}{R_1 + R_2}$ (B) $\dfrac{\theta_2 R_2 - \theta_1 R_1}{R_2 - R_1}$
(C) $\dfrac{\theta_1 R_1 + \theta_2 R_2}{R_1 + R_2}$ (D) $\dfrac{\theta_1 R_1 - \theta_2 R_1}{R_2 - R_1}$

10. Two identical metal wires of thermal conductivities K_1 and K_2 respectively are connected in series. The effective thermal conductivity of the combination is : **[17 March 2021 (Morning)]**

(A) $\dfrac{2K_1 K_2}{K_1 + K_2}$ (B) $\dfrac{K_1 + K_2}{2K_1 K_2}$
(C) $\dfrac{K_1 + K_2}{K_1 K_2}$ (D) $\dfrac{K_1 K_2}{K_1 + K_2}$

11. In 5 minutes, a body cools from 75°C to 65°C at room temperature of 25°C. The temperature of body at the end of next 5 minutes is ______°C. **[22 July 2021 (Evening)]**

12. A body takes 4 min to cool from 61°C to 59°C. If the temperature of the surroundings is 30°C, the time taken by the body to cool from 51°C to 49°C is : **[27 July 2021 (Morning)]**
(A) 3 min (B) 8 min
(C) 6 min (D) 4 min

13. A rod *CD* of thermal resistance 10.0 KW^{-1} is joined at the middle of an identical rod *AB* as shown in figure. The end *A*, *B* and *D* are maintained at 200°C, 100°C and 125°C respectively. The heat current in *CD* is *P* watt. The value of *P* is ______. **[27 Aug. 2021 (Morning)]**

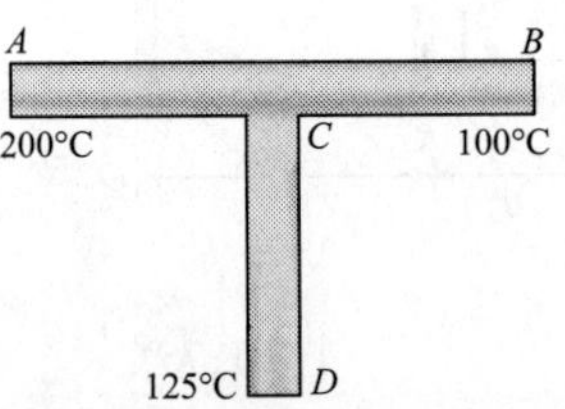

14. Two thin metallic spherical shells of radii r_1 and r_2 ($r_1 < r_2$) are placed their centres coinciding. A material of thermal conductivity *K* is filled in the space between the shells. The inner shell is maintained at temperature θ_1 and the outer shell at temperature θ_2 ($\theta_1 < \theta_2$). The rate at which heat flows radially through the material is : **[31 Aug. 2021 (Evening)]**

(A) $\dfrac{K(\theta_2 - \theta_1)(r_2 - r_1)}{4\pi r_1 r_2}$ (B) $\dfrac{\pi r_1 r_2 (\theta_2 - \theta_1)}{r_2 - r_1}$
(C) $\dfrac{K(\theta_2 - \theta_1)}{r_2 - r_1}$ (D) $\dfrac{4\pi K r_1 r_2 (\theta_2 - \theta_1)}{r_2 - r_1}$

* * * * *

CHAPTER
11 THERMODYNAMICS

11.1 LAWS OF THERMODYNAMICS

1. An engine operates by taking n moles of an ideal gas through the cycle $ABCDA$ shown in figure. The thermal efficiency of the engine is : (Take $C_v = 1.5$ R, where R is gas constant) : **[2017 (Online)]**

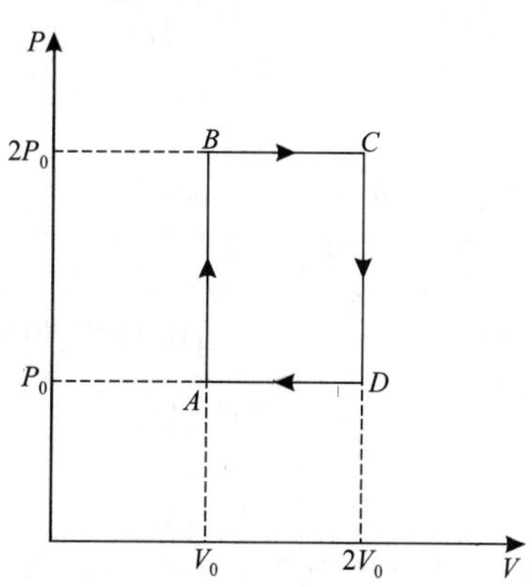

(A) 0.24 (B) 0.15
(C) 0.32 (D) 0.08

2. A gas can be taken from A to B via two different processes ACB and ADB. When path ACB is used 60 J of heat flows into the system and 30 J of work is done by the system. If path ADB is used work done by the system is 10 J. The heat Flow into the system in path ADB is : **[9 Jan. 2019 (Morning)]**

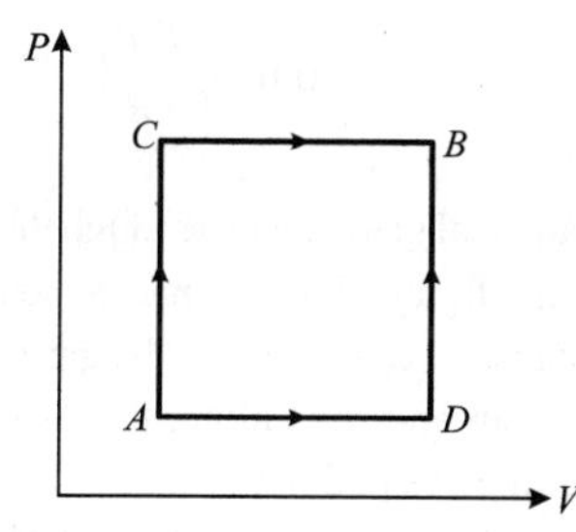

(A) 80 J (B) 20 J
(C) 100 J (D) 40 J

3. A 15g mass of nitrogen gas is enclosed in a vessel at a temperature 27°C. Amount of heat transferred to the gas, so that rms velocity of molecules is doubled, is about :
[Take $R = 8.3$ J/K mole] **[9 Jan. 2019 (Evening)]**

(A) 10 kJ (B) 0.9 kJ
(C) 6 kJ (D) 14 kJ

4. Two Carnot engines A and B are operated in series. The first one, A, receives heat at T_1 (= 600 K) and rejects to a reservoir at temperature T_2. The second engine B receives heat rejected by the first engine and, in turn, rejects to a heat reservoir at T_3 (= 400 K). Calculate the temperature T_2 if the work outputs of the two engines are equal : **[9 Jan. 2019 (Evening)]**

(A) 400 K (B) 600 K
(C) 500 K (D) 300 K

5. Three Carnot engines operate in series between a heat source at a temperature T_1 and a heat sink at temperature T_4 (see figure). There are two other reservoirs at temperature T_2, and T_3, as shown, with $T_2 > T_2 > T_3 > T_4$. The three engines are equally efficient if : **[10 Jan. 2019 (Morning)]**

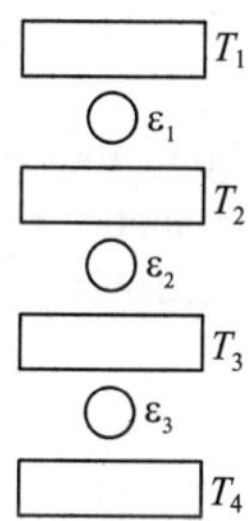

(A) $T_2 = (T_1^2 T_4)^{1/3}; T_3 = (T_1 T_4^2)^{1/3}$

(B) $T_2 = (T_1 T_4^2)^{1/3}; T_3 = (T_1^2 T_4)^{1/3}$

(C) $T_2 = (T_1^3 T_4)^{1/4}; T_3 = (T_1 T_4^3)^{1/4}$

(D) $T_2 = (T_1 T_4)^{1/2}; T_3 = (T_1^2 T_4)^{1/3}$

6. Half mole of an ideal monatomic gas is heated at constant pressure of 1 atm from 20 °C to 90°C. Work done by gas is close to : (Gas constant $R = 8.31$ J /mol.K) **[10 Jan. 2019 (Evening)]**

(A) 73 J (B) 291 J
(C) 581 J (D) 146 J

7. A rigid diatomic ideal gas undergoes an adiabatic process at room temperature. The relation between temperature and volume of this process is TV^x = constant, then x is : **[11 Jan. 2019 (Morning)]**

(A) $\frac{5}{3}$ (B) $\frac{2}{5}$

(C) $\frac{2}{3}$ (D) $\frac{3}{5}$

8. In a process, temperature and volume of one mole of an ideal monoatomic gas are varied according to the relation $VT=K$, where K is a constant. In this process the temperature of the gas is increased by ΔT. The amount of heat absorbed by gas is (R is gas constant) : **[11 Jan. 2019 (Evening)]**

(A) $\frac{1}{2}R\Delta T$ (B) $\frac{3}{2}R\Delta T$

(C) $\frac{1}{2}KR\Delta T$ (D) $\frac{2K}{3}\Delta T$

9. For the given cyclic process CAB as shown for a gas, the work done is : **[12 Jan. 2019 (Morning)]**

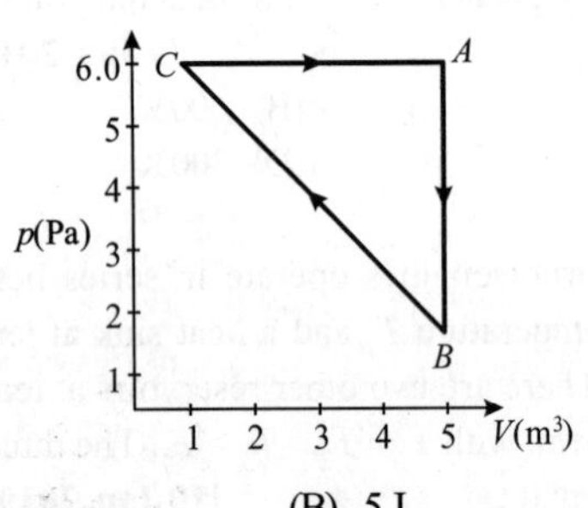

(A) 1 J (B) 5 J

(C) 10 J (D) 30 J

10. The given diagram shows four processes i.e., isochoric, isobaric, isothermal and adiabatic. The correct assignment of the processes, in the same order is given by : **[8 April 2019 (Evening)]**

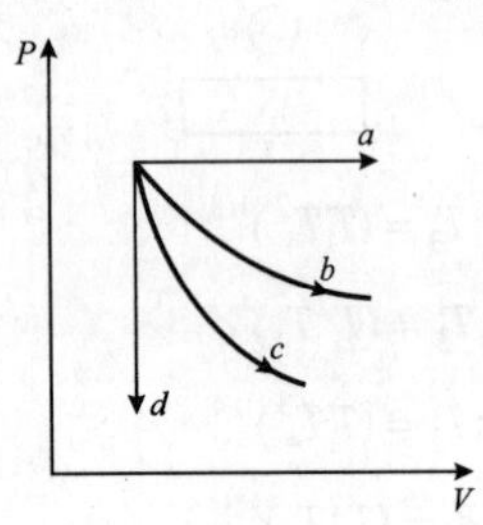

(A) $d\,a\,c\,b$ (B) $a\,d\,c\,b$

(C) $a\,d\,b\,c$ (D) $d\,a\,b\,c$

11. Following figure shows two processes A and B for a gas. If ΔQ_A and ΔQ_B are the amount of heat absorbed by the system in two cases, and ΔU_A and ΔU_B are changes in internal energies, respectively, then : **[9 April 2019 (Morning)]**

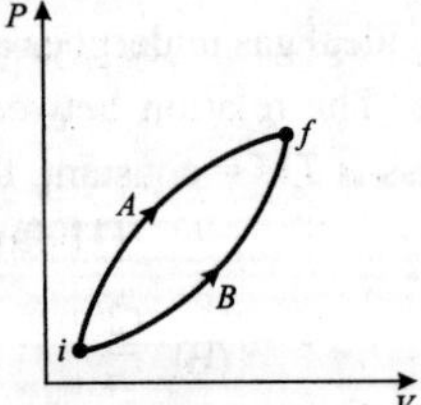

(A) $\Delta Q_A=\Delta Q_B; \Delta U_A=\Delta U_B$ (B) $\Delta Q_A>\Delta Q_B; \Delta U_A=\Delta U_B$

(C) $\Delta Q_A>\Delta Q_B; \Delta U_A>\Delta U_B$ (D) $\Delta Q_A<\Delta Q_B; \Delta U_A<\Delta U_B$

12. A cylinder with fixed capacity of 67.2 lit contains helium gas at STP. The amount of heat needed to raise the temperature of the gas by 20°C is : [Given that $R = 8.31$ J mol^{-1} K^{-1}] : **[10 April 2019 (Morning)]**

(A) 748 J (B) 374 J

(C) 350 J (D) 700 J

13. n moles of an ideal gas with constant volume heat capacity C_V undergo an isobaric expansion by certain volume. The ratio of the work done in the process, to the heat supplied is : **[10 April 2019 (Morning)]**

(A) $\frac{4nR}{C_V - nR}$ (B) $\frac{nR}{C_V - nR}$

(C) $\frac{nR}{C_V + nR}$ (D) $\frac{4nR}{C_V + nR}$

14. When heat Q is supplied to a diatomic gas of rigid molecules, at constant volume its temperature increases by ΔT. The heat required to produce the same change in temperature, at a constant pressure is : **[10 April 2019 (Evening)]**

(A) $\frac{7}{5}Q$ (B) $\frac{3}{2}Q$

(C) $\frac{5}{3}Q$ (D) $\frac{2}{3}Q$

15. One mole of an ideal gas passes through a process where pressure and volume obey the relation $P = P_o\left[1-\frac{1}{2}\left(\frac{V_0}{V}\right)^2\right]$. Here P_0 and V_0 are constants. Calculate the change in the temperature of the gas if its volume changes from V_0 to $2V_0$: **[10 April 2019 (Evening)]**

(A) $\frac{1}{2}\frac{P_oV_o}{R}$ (B) $\frac{3}{4}\frac{P_oV_o}{R}$

(C) $\frac{5}{4}\frac{P_oV_o}{R}$ (D) $\frac{1}{4}\frac{P_oV_o}{R}$

16. A sample of an ideal gas is taken through the cyclic process $abca$ as shown in the figure. The change in the internal energy of the gas along the path ca is – 180 J. The gas absorbs 250 J of heat along the path ab and 60 J along the path bc. The work done by the gas along the path abc is : **[12 April 2019 (Morning)]**

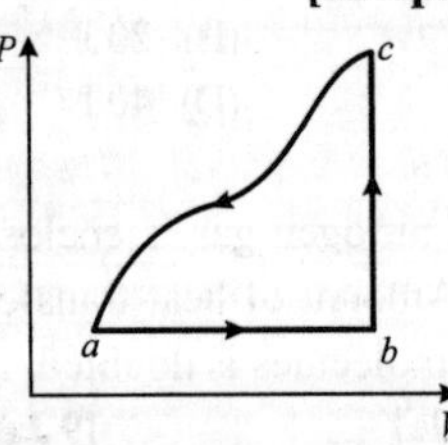

(A) 100 J (B) 120 J

(C) 140 J (D) 130 J

17. A Carnot engine has an efficiency of 1/6. When the temperature of the sink is reduced by 62°C, its efficiency is doubled. The temperatures of the source and the sink are, respectively : **[12 April 2019 (Evening)]**

(A) 124°C, 62°C (B) 37°C, 99°C
(C) 62°C, 124°C (D) 99°C, 37°C

18. A diatomic gas with rigid molecules does 10 J of work when expanded at constant pressure. What would be the heat energy absorbed by the gas, in this process ? **[12 April 2019 (Evening)]**

(A) 35 J (B) 40 J
(C) 25 J (D) 30 J

19. A litre of dry air at STP expands adiabatically to a volume of 3 litres. If $\gamma = 1.40$, the work done by air is : $(3^{1.4} = 4.6555)$ [Take air to be an ideal gas] : **[7 Jan. 2020 (Morning)]**

(A) 90.5 J (B) 48 J
(C) 60.7 J (D) 100.8 J

20. Two moles of an ideal gas with $\frac{C_P}{C_V} = \frac{5}{3}$ are mixed with 3 moles of another ideal gas with $\frac{C_P}{C_V} = \frac{4}{3}$. The value of $\frac{C_P}{C_V}$ for the mixture is : **[7 Jan. 2020 (Morning)]**

(A) 1.50 (B) 1.42
(C) 1.45 (D) 1.47

21. A Carnot engine operates between two reservoirs of temperatures 900 K and 300 K. The engine performs 1200 J of work per cycle. The heat energy (in J) delivered by the engine to the low temperature reservoir, in a cycle, is______. **[7 Jan. 2020 (Morning)]**

22. Two infinite planes each with uniform surface charge density $+\sigma$ are kept in such a way that the angle between them is 30°. The electric field in the region shown between them is given by : **[7 Jan. 2020 (Morning)]**

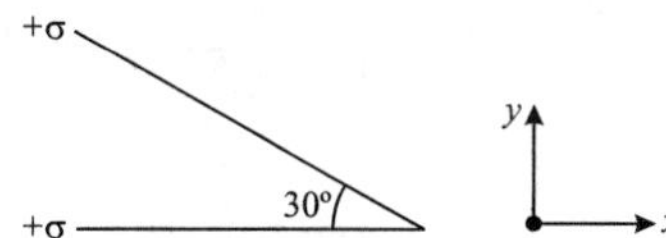

(A) $\frac{\sigma}{\varepsilon_0}\left[\left(1+\frac{\sqrt{3}}{2}\right)\hat{y}+\frac{\hat{x}}{2}\right]$ (B) $\frac{\sigma}{2\varepsilon_0}\left[\left(1-\frac{\sqrt{3}}{2}\right)\hat{y}-\frac{\hat{x}}{2}\right]$

(C) $\frac{\sigma}{2\varepsilon_0}\left[(1+\sqrt{3})\hat{y}+\frac{\hat{x}}{2}\right]$ (D) $\frac{\sigma}{2\varepsilon_0}\left[(1+\sqrt{3})\hat{y}-\frac{\hat{x}}{2}\right]$

23. Two ideal Carnot engines operate in cascade (all heat given up by one engine is used by the other engine to produce work) between temperatures, T_1 and T_2. The temperature of the hot reservoir of the first engine is T_1 and the temperature of the cold reservoir of the second engine is T_2. T is temperature of the sink of first engine which is also the source for the second engine. How is T related to T_1 and T_2, if both the engines perform equal amount of work ? **[7 Jan. 2020 (Evening)]**

(A) $T = \frac{2T_1T_2}{T_1+T_2}$ (B) $T = \sqrt{T_1T_2}$

(C) $T = \frac{T_1+T_2}{2}$ (D) $T = 0$

24. Under an adiabatic process, the volume of an ideal gas gets doubled. Consequently the mean collision time between the gas molecule changes from τ_1 to τ_2. If $\frac{C_p}{C_v} = \gamma$ for this gas then a good estimate for $\frac{\tau_2}{\tau_1}$ is given by : **[7 Jan. 2020 (Evening)]**

(A) $\left(\frac{1}{2}\right)^{\frac{\gamma+1}{2}}$ (B) 2

(C) $\frac{1}{2}$ (D) $\left(\frac{1}{2}\right)^{\gamma}$

25. A thermodynamic cycle $xyzx$ is shown on a V-T diagram. The P-V diagram that best describes this cycle is : (Diagrams are schematic and not to scale) : **[8 Jan. 2020 (Morning)]**

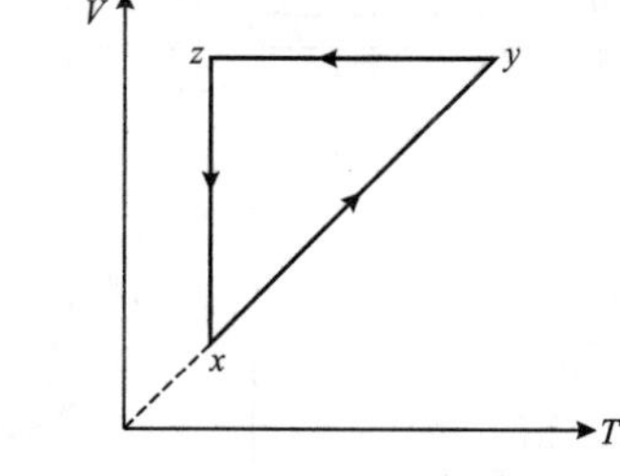

(A)
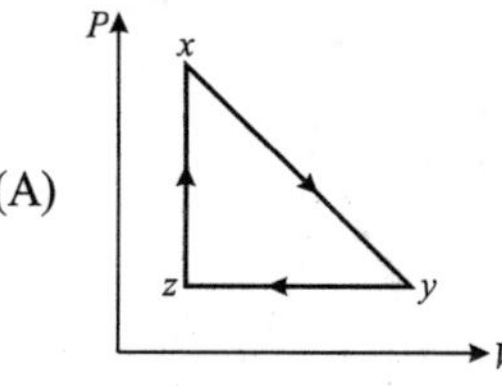

(B)
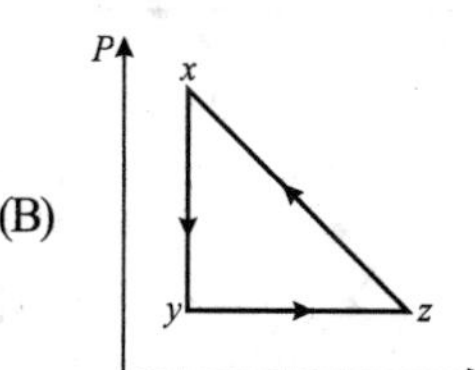

(C)
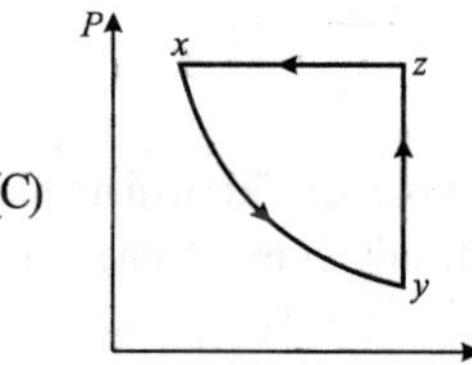

(D)
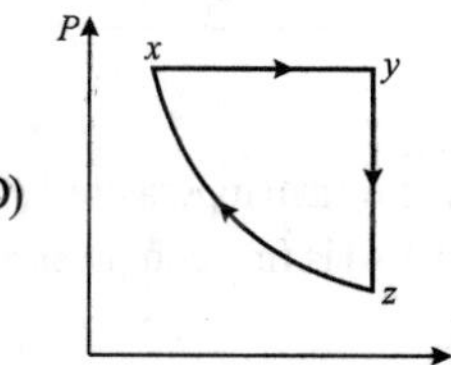

26. A Carnot engine having an efficiency of 1/10 is being used as a refrigerator. If the work done on the refrigerator is 10 J, the amount of heat absorbed from the reservoir at lower temperature is : **[8 Jan. 2020 (Evening)]**

(A) 99 J (B) 100 J
(C) 90 J (D) 1 J

27. Consider a mixture of n moles of helium gas and $2n$ moles of oxygen gas (molecules taken to be rigid) as an ideal gas. Its C_P/C_V value will be : **[8 Jan. 2020 (Evening)]**

(A) 67/45 (B) 19/13
(C) 23/15 (D) 40/27

28. Consider two ideal diatomic gases A and B at some temperature T. Molecules of the gas A are rigid, and have a mass m. Molecules of the gas B have an additional vibrational mode, and have a mass $m/4$. The ratio of the specific heats (C_V^A and C_V^B) of gas A and B, respectively is : **[9 Jan. 2020 (Morning)]**

(A) 7:9 (B) 5:7
(C) 3:5 (D) 5:9

29. Which of the following is an equivalent cyclic process corresponding to the thermodynamic cyclic given in the figure ? Where, $1 \rightarrow 2$ is adiabatic. (Graphs are schematic and are not to scale) **[9 Jan. 2020 (Morning)]**

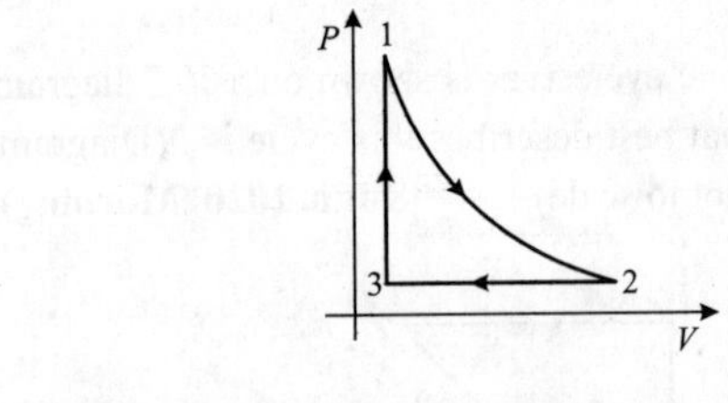

(A)
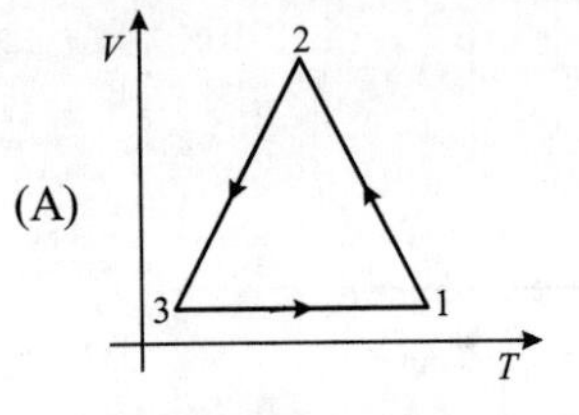

(B)
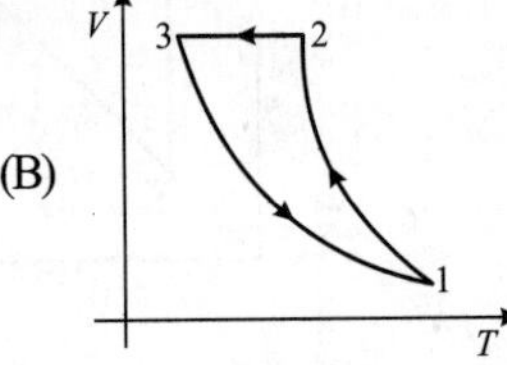

(C)
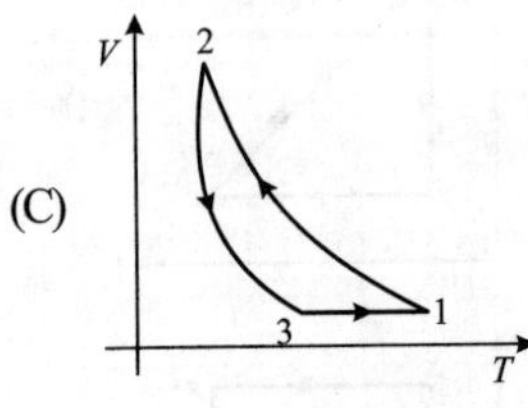

(D)
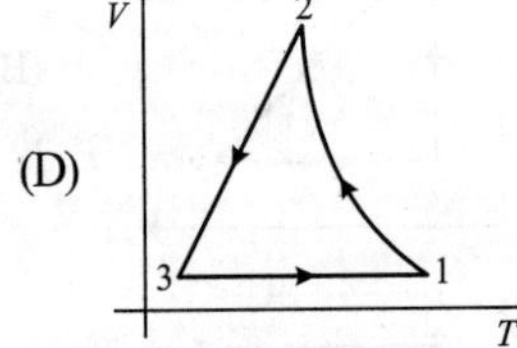

30. Starling at temperature 300 K, one mole of an ideal diatomic gas ($\gamma = 1.4$) is first compressed adiabatically from volume V_1 to $V_2 = \frac{V_1}{16}$. It is then allowed to expand isobarically to volume $2V_2$. If all the processes are the quasi-static then the final temperature of the gas (in °K) is (to the nearest integer) _____. **[9 Jan. 2020 (Evening)]**

31. A gas mixture consists of 3 moles of oxygen and 5 moles of argon at temperature T. Assuming the gases to be ideal and the oxygen bond to be rigid, the total internal energy (in units of RT) of the mixture is : **[2 Sept. 2020 (Morning)]**

(A) 11 (B) 15
(C) 20 (D) 13

32. An engine takes in 5 moles of air at 20°C and 1 atm, and compresses it adiabatically to 1/10th of the original volume. Assuming air to be a diatomic ideal gas made up of rigid molecules, the change in its internal energy during this process comes out to be X kJ. The value of X to the nearest integer is : **[2 Sept. 2020 (Morning)]**

33. A heat engine is involved with exchange of heat of 1915 J, –40 J, +125 J and QJ, during one cycle achieving an efficiency of 50.0%. The value of Q is : **[2 Sept. 2020 (Evening)]**

(A) 640 J (B) 400 J
(C) 980 J (D) 40 J

34. A balloon filled with helium (32°C and 1.7 atm.) bursts. Immediately afterwards the expansion of helium can be considered as : **[3 Sept. 2020 (Morning)]**

(A) Irreversible isothermal (B) Irreversible adiabatic
(C) Reversible adiabatic (D) Reversible isothermal

35. Consider a gas of triatomic molecules. The molecules are assumed to be triangular and made of massless rigid rods whose vertices are occupied by atoms. The internal energy of a mole of the gas at temperature T is : **[3 Sept. 2020 (Morning)]**

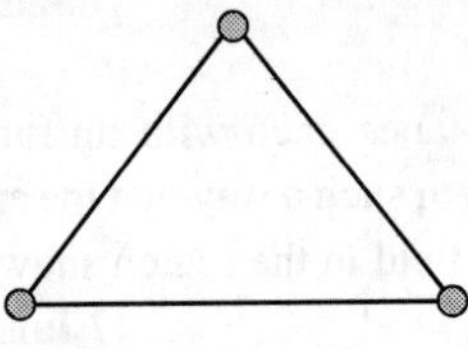

(A) $\frac{9}{2}RT$ (B) $\frac{3}{2}RT$
(C) $\frac{5}{2}RT$ (D) $3RT$

36. If minimum possible work is done by a refrigerator in converting 100 grams of water at 0°C to ice, how much heat (in calories) is released to the surrounding at temperature 27°C (Latent heat of ice = 80 Cal/gram) to the nearest integer ? **[3 Sept. 2020 (Evening)]**

37. A closed vessel contains 0.1 mole of a monoatomic ideal gas at 200 K. If 0.05 mole of the same gas at 400 K is added to it, the final equilibrium temperature (in K) of the gas in the vessel will be closed to _____. **[4 Sept. 2020 (Morning)]**

38. Match the thermodynamic processes taking place in a system with the correct conditions. In the table, ΔQ is the heat supplied, ΔW is the work done and ΔU is change in internal energy of the system : **[4 Sept. 2020 (Evening)]**

Process	Condition
(I) Adiabatic	(1) $\Delta W = 0$
(II) Isothermal	(2) $\Delta Q = 0$
(III) Isochoric	(3) $\Delta U \neq 0, \Delta W \neq 0, \Delta Q \neq 0$
(IV) Isobaric	(4) $\Delta U = 0$

(A) I-2, II-4, III-1, IV-3 (B) I-2, II-1, III-4, IV-3
(C) I-4, II-1, III-2, IV-3 (D) I-1, II-2, III-3, IV-4

39. The change in the magnitude of the volume of an ideal gas when a small additional pressure ΔP is applied at a constant temperature, is the same as the change when the temperature is reduced by a small quantity ΔT at constant pressure. The initial temperature and pressure of the gas were 300 K and 2 atm respectively. If $|\Delta T| = C|\Delta P|$ then value of C in (K/atm) is _____. **[4 Sept. 2020 (Evening)]**

40. Three different processes that can occur in an ideal monoatomic gas are shown in the P vs V diagram. The paths are labelled as $A \to B$, $A \to C$ and $A \to D$. The change in internal energies during these process are taken as EAB, EAC and EAD and the workdone as WAB, WAC and WAD. The correct relation between these parameters are : **[5 Sept. 2020 (Morning)]**

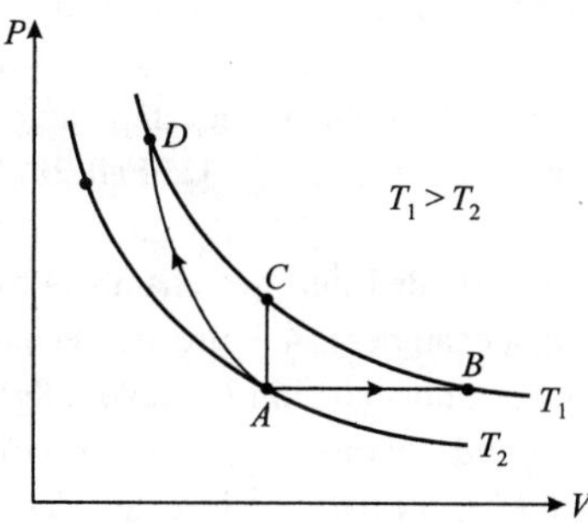

(A) $EAB = EAC = EAD$, $WAB > 0$, $WAC = 0$, $WAD > 0$
(B) $EAB < EAC < EAD$, $WAB > 0$, $WAC > WAD$
(C) $EAB = EAC < EAD$, $WAB > 0$, $WAC = 0$, $WAD < 0$
(D) $EAB > EAC > EAD$, $WAB < WAC < WAD$

41. Number of molecules in a volume of 4 cm^3 of a perfect monoatomic gas at some temperature T and at a pressure of 2 cm of mercury is close to ? (Given, mean kinetic energy of a molecule (at T) is 4×10^{-14} erg, $g = 980$ cm/s^2, density of mercury = 13.6 g/cm^3) : **[5 Sept. 2020 (Morning)]**

(A) 5.8×10^{18} (B) 5.8×10^{16}
(C) 4.0×10^{18} (D) 4.0×10^{16}

42. In an adiabatic process, the density of a diatomic gas becomes 32 times its initial value. The final pressure of the gas is found to be n times the initial pressure. The value of n is : **[5 Sept. 2020 (Evening)]**

(A) 326 (B) $\frac{1}{32}$
(C) 32 (D) 128

43. Molecules of an ideal gas are known to have three translational degrees of freedom and two rotational degrees of freedom. The gas is maintained at a temperature of T. The total internal energy, U of a mole of this gas, and the value of $\gamma\left(=\frac{C_P}{C_v}\right)$ given, respectively, by : **[6 Sept. 2020 (Morning)]**

(A) $U = \frac{5}{2}RT$ and $\gamma = \frac{6}{5}$ (B) $U = 5RT$ and $\gamma = \frac{7}{5}$
(C) $U = 5RT$ and $\gamma = \frac{6}{5}$ (D) $U = \frac{5}{2}RT$ and $\gamma = \frac{7}{5}$

44. An engine operates by taking a monatomic ideal gas through the cycle shown in the figure. The percentage efficiency of the engine is close to : **[6 Sept. 2020 (Evening)]**

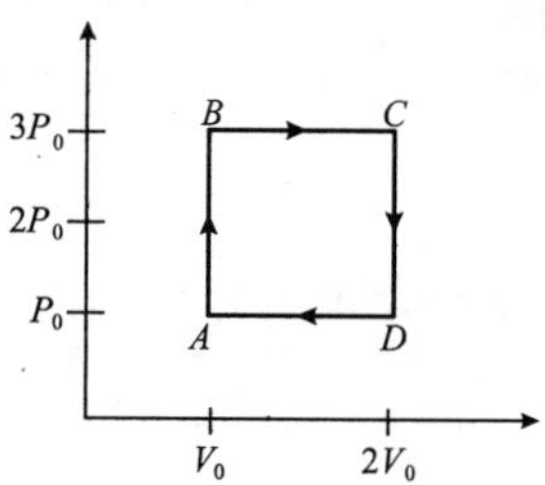

45. Match List-I with List-II. **[24 Feb. 2021 (Morning)]**

List-I	List-II
(a) Isothermal	(i) Pressure constant
(b) Isochoric	(ii) Temperature constant
(c) Adiabatic	(iii) Volume constant
(d) Isobaric	(iv) Heat content is constant

Choose the correct answer from the options given below

(A) (a)→(ii), (b)→(iv), (c)→(iii), (d)→(i)
(B) (a)→(i), (b)→(iii), (c)→(ii), (d)→(iv)
(C) (a)→(iii), (b)→(ii), (c)→(i), (d)→(iv)
(D) (a)→(ii), (b)→(iii), (c)→(iv), (d)→(i)

46. n mole of a perfect gas undergoes a cyclic process $ABCA$ (see figure) consisting of the following processes. **[24 Feb. 2021 (Morning)]**

$A \to B$: Isothermal expansion at temperature T so that the volume is doubled from V_1 to $V_2 = 2V_1$ and pressure changes from P_1 to P_2

$B \to C$: Isobaric compression at pressure P_2 to initial volume V_1

$C \to A$: Isochoric change leading to change of pressure from P_2 to P_1

Total workdone in the complete cycle $ABCA$ is :

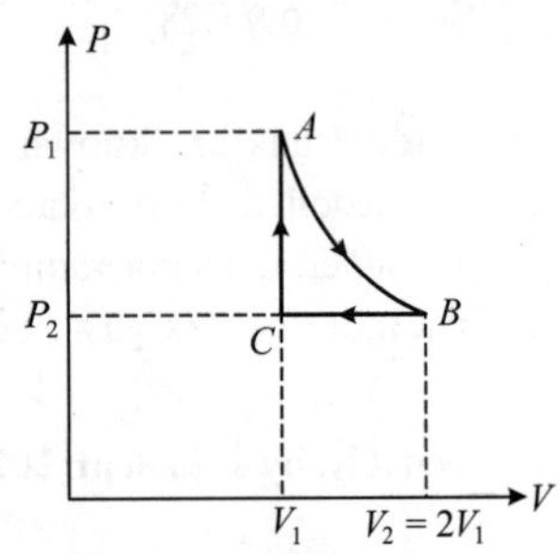

(A) $nRT\left(\ln 2 - \frac{1}{2}\right)$ (B) $nRT\left(\ln 2 + \frac{1}{2}\right)$

(C) $nRT \ln 2$ (D) 0

47. If one mole of an ideal gas at (P_1, V_1) is allowed to expand reversibly and isothermally (A to B) its pressure is reduced to one-half of the original pressure (see figure). This is followed by a constant volume cooling till its pressure is reduced to one-fourth of the initial value ($B \to C$). Then its is restored to its initial state by a reversible adiabatic compression (C to A). The net workdone by the gas is equal to : **[24 Feb. 2021 (Evening)]**

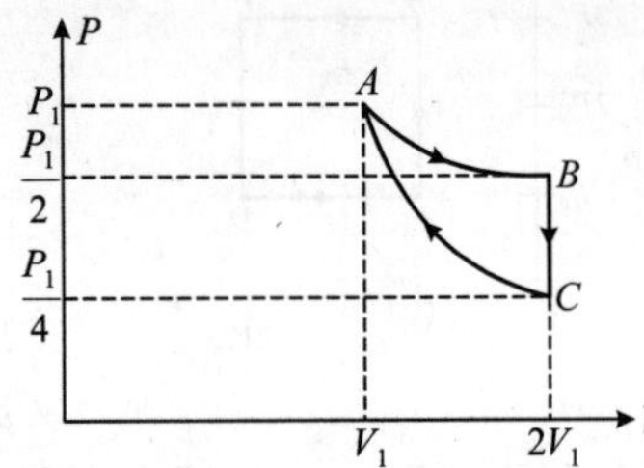

(A) $-\frac{RT}{2(\gamma - 1)}$ (B) 0

(C) $RT \ln 2$ (D) $RT\left(\ln 2 - \frac{1}{2(\gamma - 1)}\right)$

48. A diatomic gas, having $C_P = \frac{7}{2}R$ and $C_V = \frac{5}{2}R$, is heated at constant pressure. The ratio $dU : dQ : dW$: **[25 Feb. 2021 (Morning)]**

(A) 3 : 5 : 2 (B) 5 : 7 : 2

(C) 3 : 7 : 2 (D) 5 : 7 : 3

49. A monoatomic gas of mass 4.0 u is kept in an insulated container. Container is moving with velocity 30 m/s. If container is suddenly stopped then change in temperature of the gas (R = gas constant) is $\frac{x}{3R}$. Value of x is _____. **[25 Feb. 2021 (Morning)]**

50. In a certain thermodynamical process, the pressure of a gas depends on its volume as kV^3. The work done when the temperature changes from 100°C to 300°C will be _____ nR, where n denotes number of moles of a gas. **[25 Feb. 2021 (Morning)]**

51. Thermodynamic process is shown below on a P-V diagram for one mole of an ideal gas. If $V_2 = 2V_1$ then the ratio of temperature T_2/T_1 is : **[25 Feb. 2021 (Evening)]**

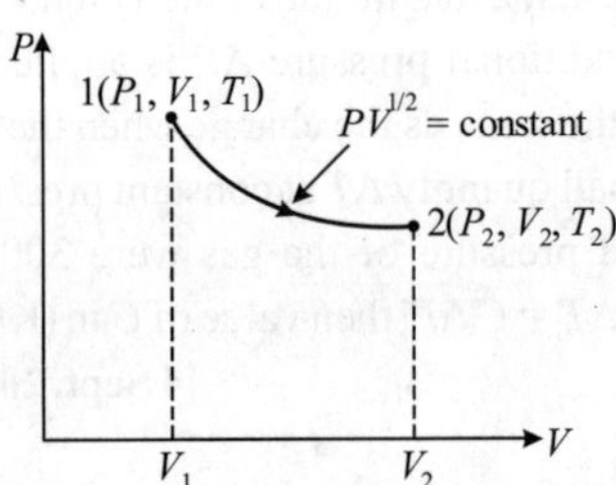

(A) 2 (B) $\sqrt{2}$

(C) $\frac{1}{2}$ (D) $\frac{1}{\sqrt{2}}$

52. A reversible heat engine converts one-fourth of the heat input into work. When the temperature of the sink is reduced by 52 K, its efficiency is doubled. The temperature in Kelvin of the source will be _____. **[25 Feb. 2021 (Evening)]**

53. A container is divided into two chambers by a partition. The volume of first chamber is 4.5 litre and second chamber is 5.5 litre. The first chamber contain 3.0 moles of gas at pressure 2.0 atom and second chamber contain 4.0 moles of gas at pressure 3.0 atm. After the partition is removed and the mixture attains equilibrium, then, the common equilibrium pressure existing in the mixture is $x \times 10^{-1}$ atm. Value of x is _____. **[26 Feb. 2021 (Morning)]**

54. The internal energy (U), pressure (P) and volume (V) of an ideal gas are related as $U = 3PV + 4$. The gas is : **[26 Feb. 2021 (Evening)]**

(A) Either monoatomic or diatomic

(B) Polyatomic only

(C) Diatomic only

(D) Monoatomic only

55. 1 mole of rigid diatomic gas performs a work of $\frac{Q}{5}$ when heat Q is supplied to it. The molar heat capacity of the gas during this transformation is $\frac{xR}{8}$. The value of x is ______.
[R = universal gas constant] **[26 Feb. 2021 (Evening)]**

56. The volume V of a given mass of monoatomic gas changes with temperature T according to the relation $V = KT^{2/3}$. The workdone when temperature changes by 90 K will be xR. The value of x is _____. **[26 Feb. 2021 (Evening)]**
[R = universal gas constant]

57. In thermodynamics, heat and work are : **[16 March 2021 (Morning)]**
(A) Path functions
(B) Intensive thermodynamic state variables
(C) Extensive thermodynamic state variables
(D) Point functions

58. For an ideal heat engine, the temperature of the source is 127°C. In order to have 60% efficiency the temperature of the sink should be ______°C. (Round off to the Nearest Integer) **[16 March 2021 (Evening)]**

59. A polyatomic ideal gas has 24 vibrational modes. What is the value of γ ? **[17 March 2021 (Morning)]**
(A) 1.03 (B) 1.30
(C) 1.37 (D) 10.3

60. A Carnot's engine working between 400 K and 800 K has a work output of 1200 J per cycle. The amount of heat energy supplied to the engine from the source in each cycle is : **[17 March 2021 (Morning)]**
(A) 3200 J (B) 1800 J
(C) 1600 J (D) 2400 J

61. Which one is the correct option for the two different thermodynamic processes ? **[17 March 2021 (Evening)]**

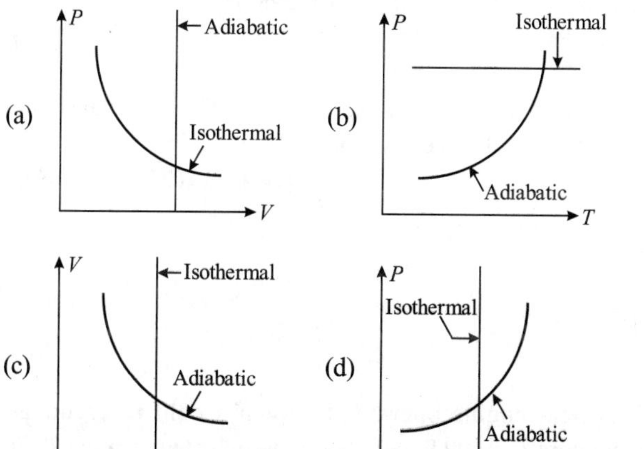

(A) (c) and (a) (B) (c) and (d)
(C) (a) only (D) (b) and (c)

62. The P-V diagram of a diatomic ideal gas system going under cyclic process as shown in figure. The work done during an adiabatic process CD is (use $\gamma = 1.4$) : **[18 March 2021 (Morning)]**

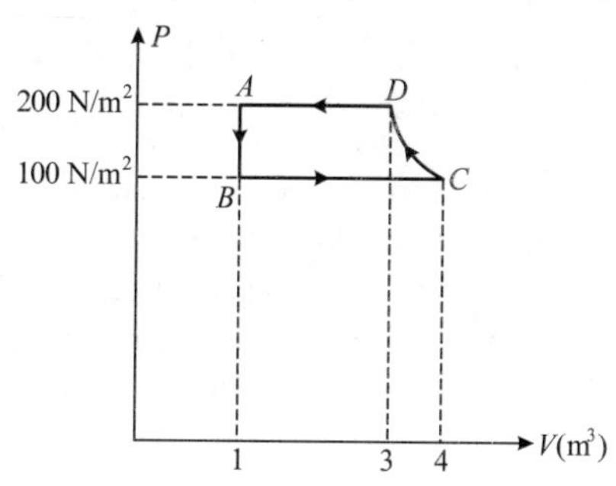

(A) −500 J (B) −400 J
(C) 400 J (D) 200 J

63. The amount of heat needed to raise the temperature of 4 moles of a rigid diatomic gas from 0°C to 50°C when no work is done is ______. (R is the universal gas constant) : **[20 July 2021 (Morning)]**
(A) 750 R (B) 500 R
(C) 250 R (D) 175 R

64. In the reported figure, heat energy absorbed by a system in going through a cyclic process is ______ πJ. **[20 July 2021 (Morning)]**

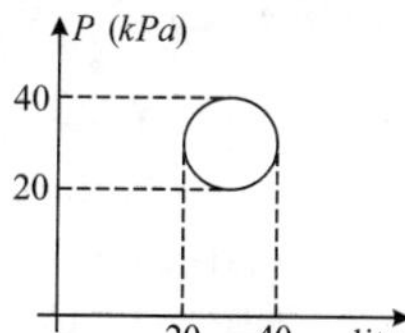

65. One mole of an ideal gas at 27°C is taken from A to B as shown in the given PV indicator diagram. The work done by the system will be _____ $\times 10^{-1}$ J. [Given : R = 8.3 J/mole K, ln2 = 0.6931] (Round off to the nearest integer) **[20 July 2021 (Evening)]**

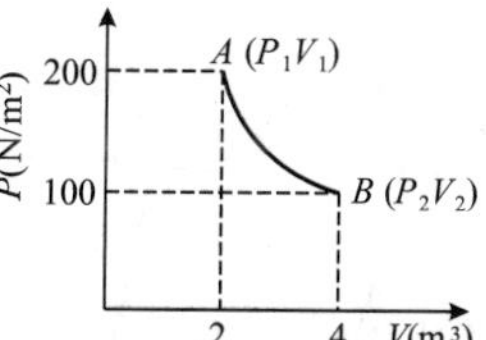

66. A monoatomic ideal gas, initially at temperature T_1 is enclosed in a cylinder fitted with a frictionless piston. The gas is allowed to expand adiabatically to temperature T_2 by releasing the piston suddenly. If l_1 and l_2 are the lengths of the gas column, before and after the expansion respectively, then the value of $\frac{T_2}{T_1}$ will be : **[25 July 2021 (Morning)]**

(A) $\frac{l_2}{l_1}$ (B) $\frac{l_1}{l_2}$

(C) $\left(\frac{l_2}{l_1}\right)^{\frac{2}{3}}$ (D) $\left(\frac{l_1}{l_2}\right)^{\frac{2}{3}}$

67. Two different metal bodies A and B of equal mass are heated at a uniform rate under similar conditions. The variation of temperature of the bodies is graphically represented as shown in the figure. The ratio of specific heat capacities is : **[25 July 2021 (Morning)]**

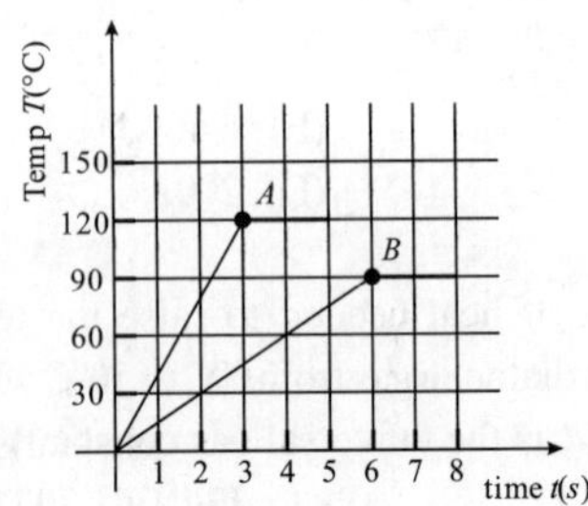

(A) $\frac{3}{4}$ (B) $\frac{4}{3}$

(C) $\frac{8}{3}$ (D) $\frac{3}{8}$

68. A heat engine has an efficiency of $\frac{1}{6}$. When the temperature of sink is reduced by 62°C, its efficiency get doubled. The temperature of the source is : **[25 July 2021 (Evening)]**

(A) 99° (B) 124 °C

(C) 62 °C (D) 37 °C

69. In the reported figure, there is a cycle process $ABCDA$ on a sample of 1 mol of a diatomic gas. The temperature of the gas during the process $A \to B$ and $C \to D$ are T_1 and T_2 $(T_1 > T_2)$ respectively

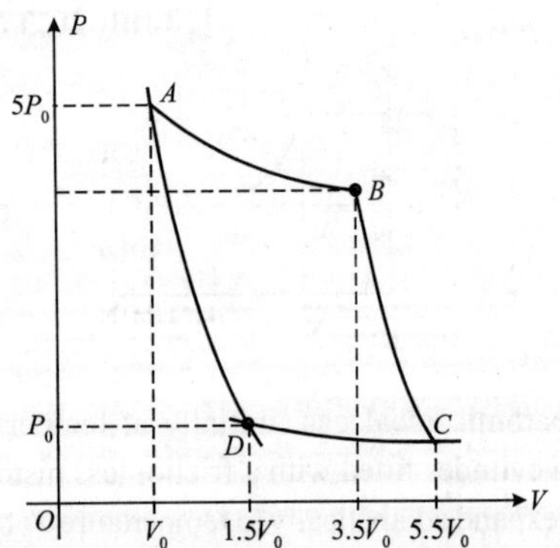

Choose the correct option out of the following for work done if processes BC and DA are adiabatic : **[27 July 2021 (Morning)]**

(A) $W_{BC} + W_{DA} > 0$ (B) $W_{AB} < W_{CD}$

(C) $W_{AB} = W_{DC}$ (D) $W_{AD} = W_{BC}$

70. One mole of an ideal gas is taken through an adiabatic process where the temperature rises from 27°C to 37°C. If the ideal gas is composed of polyatomic molecule that has 4 vibrational modes, which of the following is true ? (R = 8.314 J mol^{-1} K^{-1}) **[27 July 2021 (Evening)]**

(A) Work done by the gas is close to 582 J

(B) Work done on the gas is close to 332 J

(C) Work done on the gas is close to 582 J

(D) Work done by the gas is close to 332 J

71. Two Carnot engines A and B operate in series such that engine A absorbs heat at T_1 and rejects heat to a sink at temperature T. Engine B absorbs half of the heat rejected by Engine A and rejects heat to the sink at T_3, when workdone in both the cases is equal, the value of T is : **[27 July 2021 (Evening)]**

(A) $\frac{3}{2}T_1 + \frac{1}{3}T_3$ (B) $\frac{1}{3}T_1 + \frac{2}{3}T_3$

(C) $\frac{2}{3}T_1 + \frac{3}{2}T_3$ (D) $\frac{2}{3}T_1 + \frac{1}{3}T_3$

72. An electric appliance supplies 6000 J/min heat to the system. If the system delivers a power of 90 W. How long it would take to increase the internal energy by 2.5×10^3 J ? **[26 Aug. 2021 (Morning)]**

(A) 4.1×10^1 s (B) 2.4×10^3 s

(C) 2.5×10^2 s (D) 2.5×10^1 s

73. A refrigerator consumes an average 35 W power to operate between temperatures – 10°C to 25°C. If there is no loss of energy then how much average heat per second does it transfer? **[26 Aug. 2021 (Evening)]**

(A) 263 J/s (B) 298 J/s

(C) 350 J/s (D) 35 J/s

74. An ideal gas is expanding such that PT^3 = constant. The coefficient of volume expansion of the gas is : **[27 Aug. 2021 (Morning)]**

(A) $\frac{1}{T}$ (B) $\frac{3}{T}$

(C) $\frac{4}{T}$ (D) $\frac{2}{T}$

75. A heat engine operates between a cold reservoir at temperature T_2 = 400 K and a hot reservoir at temperature T_1. It takes 300 J of heat from the hot reservoir and delivers 240 J of heat to the cold reservoir in a cycle. The minimum temperature of the hot reservoir has to be_____*K*. **[27 Aug. 2021 (Evening)]**

76. A reversible engine has an efficiency of $\frac{1}{4}$. If the temperature of the sink is reduced by 58°C, its efficiency becomes double. Calculate the temperature of the sink :

[31 Aug. 2021 (Morning)]

(A) 180.4 °C (B) 382 °C
(C) 174 °C (D) 280 °C

77. A sample of gas with $\gamma = 1.5$ is taken through an adiabatic process in which the volume is compressed from 1200 cm^3 to 300 cm^3. If the initial pressure is 200 kPa. The absolute value of the work done by the gas in the process = _____ J :

[31 Aug. 2021 (Evening)]

78. The temperature of 3.00 mol of an ideal diatomic gas is increased by 40.0°C without changing the pressure of the gas. The molecules in the gas rotate but do not oscillate. If the ratio of change in internal energy of the gas to the amount of work done by the gas is $x/10$. Then the value of x (round off to the nearest) is _____. (Given $R = 8.31$ J mol^{-1} K^{-1})

[01 Sep. 2021 (Evening)]

* * * * *

CHAPTER

12 KINETIC THEORY

12.1 KINETIC THEORY OF GASES

1. C_p and C_v are specific heats at constant pressure and constant volume respectively. It is observed that

$C_p - C_v = a$ for hydrogen gas

$C_p - C_v = b$ for nitrogen gas

The correct relation between a and b is : **[2017]**

(A) $a = 14b$ (B) $a = 28b$

(C) $a = \frac{1}{14}b$ (D) $a = b$

2. The temperature of an open room of volume 30 m^3 increases from 17°C to 27°C due to sunshine. The atmospheric pressure in the room remains 1×10^5 *Pa*. If n_i and n_f are the number of molecules in the room before and after heating, the $n_f - n_i$ will be : **[2017]**

(A) 2.5×10^{25} (B) -2.5×10^{25}

(C) -1.61×10^{23} (D) 1.38×10^{23}

3. An ideal gas has molecules with 5 degrees of freedom. The ratio of specific heats at constant pressure (C_p) and at constant volume (C_v) is : **[2017 (Online)]**

(A) 6 (B) $\frac{7}{2}$

(C) $\frac{5}{2}$ (D) $\frac{7}{5}$

4. *N* moles of a diatomic gas in a cylinder are at a temperature *T*. Heat is supplied to the cylinder such that the temperature remains constant but *n* moles of the diatomic gas get converted into monoatomic gas. What is the change in the total kinetic energy of the gas ? **[2017 (Online)]**

(A) $\frac{1}{2}nRT$ (B) 0

(C) $\frac{3}{2}nRT$ (D) $\frac{5}{2}nRT$

5. For the *P-V* diagram given for an ideal gas, out of the following which one correctly represents the *T-P* diagram? **[2017 (Online)]**

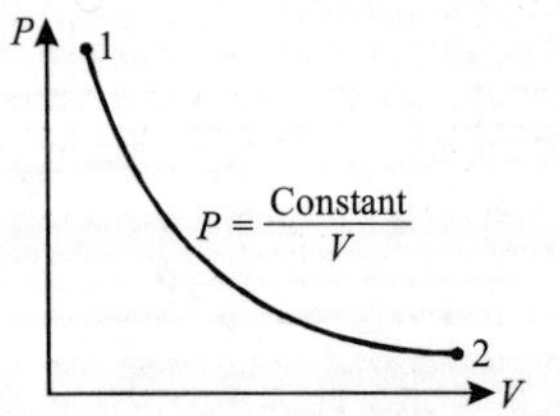

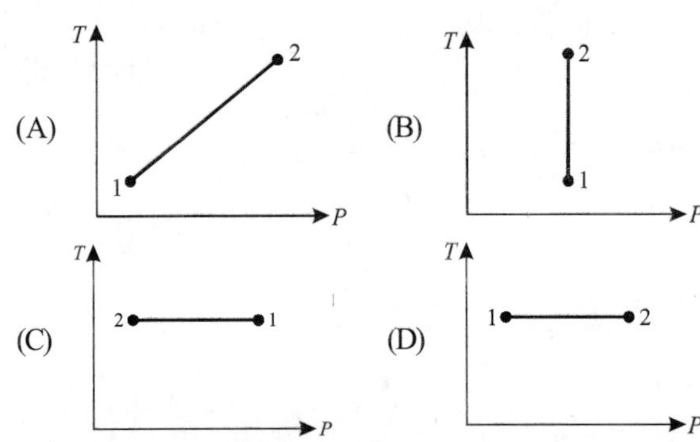

6. A mixture of 2 moles of helium gas (atomic mass = 4 *u*), and 1 mole of argon gas (atomic mass = 40*u*) is kept at 300 K in a container. The ratio of their rms speeds $\left[\frac{V_{rms}(\text{helium})}{V_{rms}(\text{argon})}\right]$, is close to : **[9 Jan. 2019 (Morning)]**

(A) 2.24 (B) 0.45

(C) 0.32 (D) 3.16

7. Two kg of a monoatomic gas is at a pressure of 4×10^4 N/m^2. The density of the gas is 8 kg/m^3. What is the order of energy of the gas due to its thermal motion ? **[10 Jan. 2019 (Evening)]**

(A) 10^3 J (B) 10^5 J

(C) 10^6 J (D) 10^4 J

8. The gas mixture consists of 3 moles of oxygen and 5 moles of argon at temperature *T*. Considering only translational and rotational modes, the total internal energy of the system is : **[11 Jan. 2019 (Morning)]**

(A) 12*RT* (B) 20*RT*

(C) 15*RT* (D) 4*RT*

9. An ideal gas occupies a volume of 2 m^3 at a pressure of 3×10^6 Pa. The energy of the gas is : **[12 Jan. 2019 (Morning)]**

(A) 3×10^2 (B) 10^8 J

(C) 6×10^4 J (D) 9×10^6 J

10. A vertical closed cylinder is separated into two parts by a frictionless piston of mass m and of negligible thickness. The piston is free to move along the length of the cylinder. The length of the cylinder above the piston is l_1, and that below the piston is l_2, such that $l_1 > l_2$. Each part of the cylinder contains *n* moles of an ideal gas at equal temperature *T*. If the piston is stationary, its mass, m, will be given by : (R is universal gas constant and g is the acceleration due to gravity) :

[12 Jan. 2019 (Evening)]

(A) $\frac{nRT}{g}\left[\frac{1}{l_2}+\frac{1}{l_1}\right]$ (B) $\frac{nRT}{g}\left[\frac{l_1-l_2}{l_1l_2}\right]$

(C) $\frac{RT}{g}\left[\frac{2l_1+l_2}{l_1l_2}\right]$ (D) $\frac{RT}{ng}\left[\frac{l_1-3l_2}{l_1l_2}\right]$

11. An ideal gas is enclosed in a cylinder at pressure of 2 atm and temperature, 300 K. The mean time between two successive collisions is 6×10^{-8} s. If the pressure is doubled and temperature is increased to 500 K, the mean time between two successive collisions will be close to : **[12 Jan. 2019 (Evening)]**

(A) 4×10^{-8} s (B) 3×10^{-6} s

(C) 2×10^{-7} s (D) 0.5×10^{-8} s

12. If 10^{22} gas molecules each of mass 10^{-26} kg collide with a surface (perpendicular to it) elastically per second over an area 1 m^2 with a speed 10^4 m/s, the pressure exerted by the gas molecules will be of the order of: **[8 April 2019 (Morning)]**

(A) 10^8 N/m^2 (B) 10^4 N/m^2

(C) 10^3 N/m^2 (D) 10^{16} N/m^2

13. The temperature, at which the root mean square velocity of hydrogen molecules equals their escape velocity from the earth, is closest to :

[Boltzmann Constant $k_B = 1.38 \times 10^{-23}$ J/K
Avogadro Number $N_A = 6.02 \times 10^{26}$ /kg
Radius of Earth : 6.4×10^6 m
Gravitational acceleration on Earth = 10 ms^{-2}]

[8 April 2019 (Evening)]

(A) 650 K (B) 3×10^5 K

(C) 10^4 K (D) 800 K

14. An HCl molecule has rotational, translational and vibrational motions. If the rms velocity of HCl molecules in its gaseous phase is $\bar{v}$, m is its mass and k_B is Boltzmann constant, then its temperature will be : **[9 April 2019 (Morning)]**

(A) $\frac{m\bar{v}^2}{6k_B}$ (B) $\frac{m\bar{v}^2}{5k_B}$

(C) $\frac{m\bar{v}^2}{3k_B}$ (D) $\frac{m\bar{v}^2}{7k_B}$

15. For a given gas at 1 atm pressure, rms speed of the molecule is 200 m/s at 127°C. At 2 atm pressure and at 227°C, the rms speed of the molecules will be : **[9 April 2019 (Morning)]**

(A) 80 m/s (B) $100\sqrt{5}$ m/s

(C) $80\sqrt{5}$ m/s (D) 100 m/s

16. The specific heats, C_P and C_V of a gas of diatomic molecules, A, are given (in units of J mol^{-1} K^{-1}) by 29 and 22, respectively. Another gas of diatomic molecules, B, has the corresponding values 30 and 21. If they are treated as ideal gases, then : **[9 April 2019 (Evening)]**

(A) A has one vibrational mode and B has two

(B) Both A and B have a vibrational mode each

(C) A is rigid but B has a vibrational mode

(D) A has a vibrational mode but B has none

17. A 25×10^{-3} m^3 volume cylinder is filled with 1 mol of O_2 gas at room temperature (300 K). The molecular diameter of O_2, and its root mean square speed, are found to be 0.3 nm, and 200 m/s, respectively. What is the average collision rate (per second) for an O_2 molecule ? **[10 April 2019 (Morning)]**

(A) $\sim 10^{11}$ (B) $\sim 10^{13}$

(C) $\sim 10^{10}$ (D) $\sim 10^{12}$

18. Two moles of helium gas is mixed with three moles of hydrogen molecules (taken to be rigid). What is the molar specific heat of mixture at constant volume ? (R = 8.3 J/mol K):

[12 April 2019 (Morning)]

(A) 21.6 J/mol K (B) 19.7 J/mol K

(C) 17.4 J/mol K (D) 15.7 J/mol K

19. The number density of molecules of a gas depends on their distance r from the origin as, $n(r) = n_0 e^{-\alpha r^4}$. Then the total number of molecules is proportional to :

[12 April 2019 (Evening)]

(A) $n_0\alpha^{1/4}$ (B) $n_0\alpha^{-3}$

(C) $n_0\alpha^{-3/4}$ (D) $\sqrt{n_0}\,\alpha^{1/2}$

20. The plot that depicts the behavior of the mean free time t (time between two successive collisions) for the molecules of an ideal gas, as a function of temperature (T), qualitatively, is : (Graphs are schematic and not drawn to scale) :

[8 Jan. 2020 (Morning)]

(A) 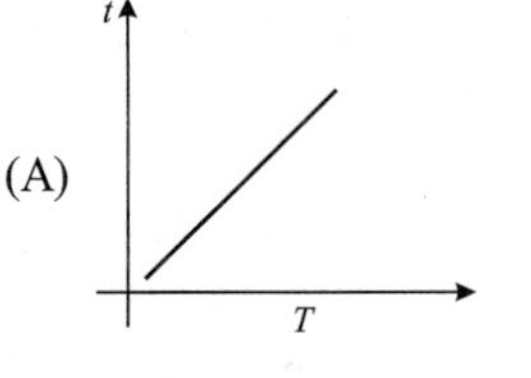

(B) 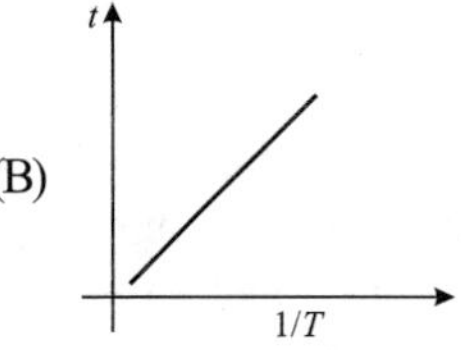

(C) 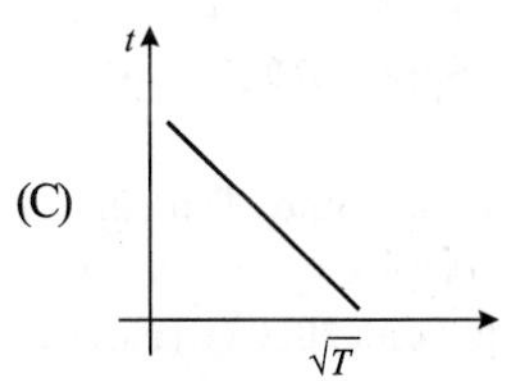

(D) 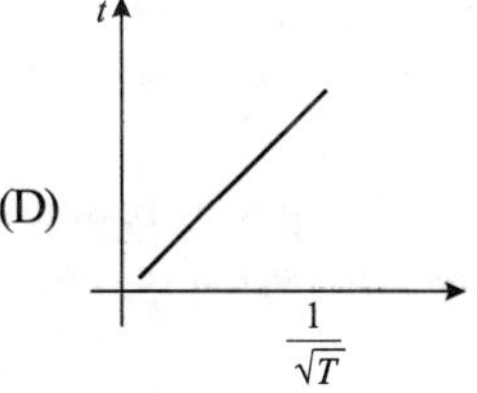

21. Two gases-argon (atomic radius 0.07 nm, atomic weight 40) and xenon (atomic radius 0.1 nm, atomic weight 140) have the same number density and are at the same temperature. The ratio of their respective mean free times is closest to : **[9 Jan. 2020 (Evening)]**

(A) 3.67 (B) 4.67
(C) 1.83 (D) 2.3

22. An ideal gas in a closed container is slowly heated. As its temperature increases, which of the following statements are true : **[2 Sept. 2020 (Evening)]**

(i) The mean free path of the molecules decreases
(ii) The mean collision time between the molecules decreases
(iii) The mean free path remains unchanged
(iv) The mean collision time remains unchanged

(A) (iii) and (iv) (B) (i) and (ii)
(C) (i) and (iv) (D) (ii) and (iii)

23. To raise the temperature of a certain mass of gas by 50°C at a constant pressure, 160 calories of heat is required. When the same mass of gas is cooled by 100°C at constant volume, 240 calories of heat is released. How many degrees of freedom does each molecule of this gas have (assume gas to be ideal) ? **[3 Sept. 2020 (Evening)]**

(A) 5 (B) 3
(C) 6 (D) 7

24. Match the C_P/C_V ratio for ideal gases with different type of molecules : **[4 Sept. 2020 (Morning)]**

Molecular type	C_P/C_V
(A) Monoatomic	(I) 7/5
(B) Diatomic rigid molecules	(II) 9/7
(C) Diatomic non-rigid molecules	(III) 4/3
(D) Triatomic rigid molecules	(IV) 5/3

(A) A-IV, B-I, C-II, D-III (B) A-IV, B-II, C-I, D-III
(C) A-III, B-IV, C-II, D-I (D) A-II, B-III, C-I, D-IV

25. Nitrogen gas is at 300°C temperature. The temperature (in K) at which the rms speed of a H_2 molecule would be equal to the rms speed of a nitrogen molecule, is ______. (Molar mass of N_2 gas 28g) : **[5 Sept. 2020 (Evening)]**

26. Initially a gas of diatomic molecules is contained in a cylinder of volume V_1 at a pressure P_1 and temperature 250 K. Assuming that 25% of the molecules get dissociated causing a change in number of moles. The pressure of the resulting gas at temperature 2000 K, when contained in a volume $2V_1$ is given by P_2. The ratio P_2/P_1 is : **[6 Sept. 2020 (Morning)]**

27. In a dilute gas at pressure P and temperature T, the mean time between successive collisions of a molecule varies with T as : **[6 Sept. 2020 (Evening)]**

(A) $\sqrt{T}$ (B) $\frac{1}{T}$
(C) $\frac{1}{\sqrt{T}}$ (D) T

28. On the basis of kinetic theory of gases, the gas exerts pressure because its molecules : **[24 Feb. 2021 (Evening)]**

(A) Continuously stick to the walls of container
(B) Are attracted by the walls of container
(C) Continuously lose their energy till it reaches wall
(D) Suffer change in momentum when impinge on the walls of container

29. The root mean square speed of molecules of a given mass of a gas at 27°C and 1 atmosphere pressure is 200 ms^{-1}. The root mean square speed of molecules of the gas at 127°C and 2 atmosphere pressure is $\frac{x}{\sqrt{3}}$ ms^{-1}. The value of x will be______. **[24 Feb. 2021 (Evening)]**

30. Given below are two statements : **[25 Feb. 2021 (Evening)]**

Statement-I : In a diatomic molecule, the rotational energy at a given temperature obeys Maxwell's distribution

Statement-II : In a diatomic molecule, the rotational energy at a given temperature equals the translational kinetic energy for each molecule

In the light of the above statements, choose the correct answer from the options given below :

(A) Both Statement-I and Statement-II are false
(B) Statement-I is false but Statement-II is true
(C) Statement-I is true but Statement-II is false
(D) Both Statement-I and Statement-II are true

31. The volume V of an enclosure contains a mixture of three gases, 16 g of oxygen, 28 g of nitrogen and 44 g of carbon dioxide at absolute temperature T. Consider R as universal gas constant. The pressure of the mixture of gases is : **[16 March 2021 (Morning)]**

(A) $\frac{88RT}{V}$ (B) $\frac{3RT}{V}$
(C) $\frac{5}{2}\frac{RT}{V}$ (D) $\frac{4RT}{V}$

32. Calculate the value of mean free path (λ) for oxygen molecules at temperature 27°C and pressure 1.01×10^5 Pa. Assume the molecular diameter 0.3 nm and the gas is ideal. ($k = 1.38 \times 10^{-23}$ JK^{-1}) **[16 March 2021 (Evening)]**

(A) 58 nm (B) 32 nm
(C) 86 nm (D) 102 nm

33. Two ideal polyatomic gases at temperatures T_1 and T_2 are mixed so that there is no loss of energy. If F_1 and F_2, m_1 and m_2, n_1 and n_2 be the degrees of freedom, masses, number of molecules of the first and second gas respectively, the temperature of mixture of these two gases is : **[17 March 2021 (Morning)]**

(A) $\frac{n_1T_1+n_2T_2}{n_1+n_2}$ (B) $\frac{n_1F_1T_1+n_2F_2T_2}{n_1F_1+n_2F_2}$

(C) $\frac{n_1F_1T_1+n_2F_2T_2}{F_1+F_2}$ (D) $\frac{n_1F_1T_1+n_2F_2T_2}{n_1+n_2}$

34. If one mole of the polyatomic gas is having two vibrational modes and β is the ratio of molar specific heats for polyatomic gas $\left(\beta=\frac{C_P}{C_V}\right)$ then the value of β is : **[17 March 2021 (Evening)]**

(A) 1.02 (B) 1.2
(C) 1.25 (D) 1.35

35. What will be the average value of energy along one degree of freedom for an ideal gas in thermal equilibrium at a temperature T? (k_B is Boltzmann constant) **[18 March 2021 (Morning)]**

(A) $\frac{1}{2}k_BT$ (B) $\frac{2}{3}k_BT$

(C) $\frac{3}{2}k_BT$ (D) k_BT

36. An ideal gas in a cylinder is separated by a piston in such a way that the entropy of one part is S_1 and that of the other part is S_2. Given that $S_1 > S_2$. If the piston is removed then the total entropy of the system will be : **[18 March 2021 (Evening)]**

(A) $S_1 \times S_2$ (B) $S_1 - S_2$

(C) $\frac{S_1}{S_2}$ (D) $S_1 + S_2$

37. Consider a sample of oxygen behaving like an ideal gas. At 300 K, the ratio of root mean square (rms) velocity to the average velocity of gas molecule would be :
(Molecular weight of oxygen is 32 g/mol; $R = 8.3$ J K^{-1} mol^{-1})
[18 March 2021 (Evening)]

(A) $\sqrt{\frac{3}{3}}$ (B) $\sqrt{\frac{8}{3}}$

(C) $\sqrt{\frac{3\pi}{8}}$ (D) $\sqrt{\frac{8\pi}{3}}$

38. Consider a mixture of gas molecule of types A, B and C having masses $m_A < m_B < m_C$. The ratio of their root mean square speeds at normal temperature and pressure is : **[20 July 2021 (Morning)]**

(A) $\frac{1}{v_A} > \frac{1}{v_B} > \frac{1}{v_C}$ (B) $v_A = v_B = v_C = 0$

(C) $v_A = v_B \neq v_C$ (D) $\frac{1}{v_A} < \frac{1}{v_B} < \frac{1}{v_C}$

39. The correct relation between the degrees of freedom f and the ratio of specific heat γ is : **[20 July 2021 (Evening)]**

(A) $f = \frac{\gamma+1}{2}$ (B) $f = \frac{1}{\gamma+1}$

(C) $f = \frac{2}{\gamma+1}$ (D) $f = \frac{2}{\gamma-1}$

40. Which of the following graphs represent the behaviour of an ideal gas ? Symbols have their usual meaning : **[20 July 2021 (Evening)]**

(A) 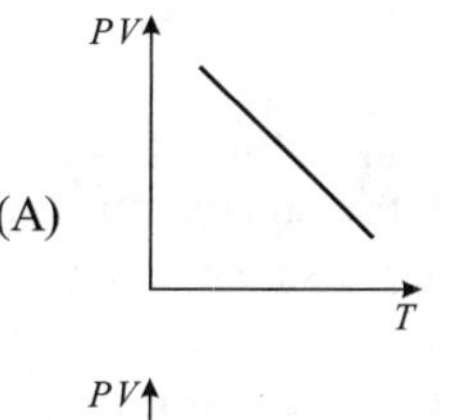

(B) 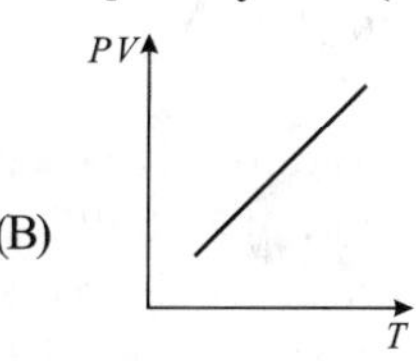

(C) 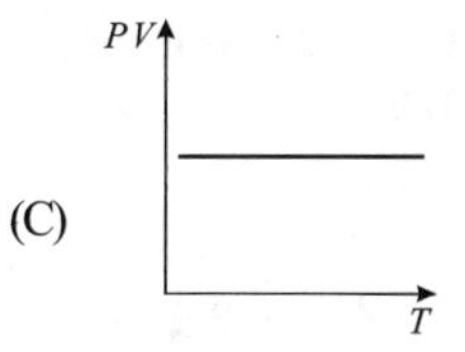

(D) 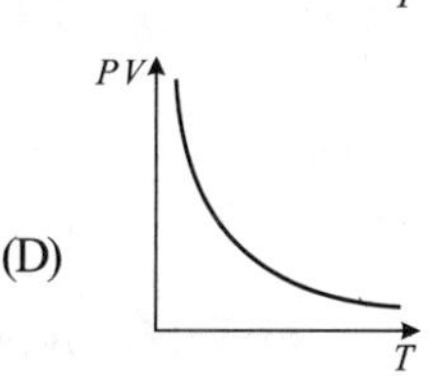

41. What will be the average kinetic energy for a monoatomic gas molecule in thermal equilibrium at temperature T? **[22 July 2021 (Evening)]**

(A) $\frac{1}{2}k_BT$ (B) $\frac{3}{2}k_BT$

(C) $\frac{2}{3}k_BT$ (D) k_BT

42. For a gas $C_P - C_V = R$ in a state P and $C_P - C_V = 1.10\,R$ in a state Q, T_P and T_Q are the temperature in two different state P and Q respectively. Then : **[25 July 2021 (Morning)]**

(A) $T_P > T_Q$ (B) $T_P = T_Q$
(C) $T_P = 0.9\,T_Q$ (D) $T_P < T_Q$

43. A system consists of two types of gas molecules A and B having same number density 2×10^{25} /m^3. The diameter of A and B are 10 Å and 5 Å respectively. They suffer collision at room temperature. The ratio of average distance covered by the molecule A to that of B between two successive collision is ______ $\times 10^{-2}$: **[25 July 2021 (Evening)]**

44. The number of molecules in one litre of an ideal gas at 300 K and 2 atmospheric pressure with mean kinetic energy 2×10^{-9} J per molecule is : **[27 July 2021 (Morning)]**

(A) 3×10^{11} (B) 6×10^{11}

(C) 0.75×10^{11} (D) 1.5×10^{11}

45. The rms speed of the molecules of Hydrogen, Oxygen and Carbon dioxide at the same temperature are V_H, V_O and V_C respectively then : **[26 Aug. 2021 (Morning)]**

(A) $V_H > V_O > V_C$ (B) $V_H = V_O > V_C$

(C) $V_C > V_O > V_H$ (D) $V_H = V_O = V_C$

46. A cylindrical container of volume 4.0×10^{-3} m^3 contains one mole of hydrogen and two moles of carbon dioxide. Assume the temperature of the mixture is 400 K. The pressure of the mixture of gases is : (Take gas constant as 8.3 J mol^{-1} K^{-1}] : **[26 Aug. 2021 (Evening)]**

(A) 249×10^1 Pa (B) 24.9 Pa

(C) 24.9×10^5 Pa (D) 24.9×10^3 Pa

47. A balloon carries a total load of 185 kg at normal pressure and temperature of 27°C. What load will the balloon carry on rising to a height at which the barometric pressure is 45 cm of Hg and the temperature is – 7°C. Assuming constant volume? **[27 Aug. 2021 (Morning)]**

(A) 2.14.15 kg (B) 219.07 kg

(C) 181.46 kg (D) 123.54 kg

48. If the *RMS* speed of oxygen molecules at 0°C is 160 m/s. Find the rms speed of hydrogen molecules at 0°C. **[27 Aug. 2021 (Evening)]**

(A) 40 m/s (B) 80 m/s

(C) 640 m/s (D) 332 m/s

49. For an ideal gas the instantaneous change in pressure p with volume V is given by the equation $dp/dV = -ap$. If $p = p_0$ at $V = 0$ is the given boundary condition, then the maximum temperature one mole of gas can attain is : (Here R is the gas constant) **[31 Aug. 2021 (Morning)]**

(A) $\frac{ap_0}{eR}$ (B) $\frac{p_0}{aeR}$

(C) Infinity (D) 0°C

50. A mixture of hydrogen and oxygen has volume 500 cm^3, temperature 300 K, pressure 400 kPa and mass 0.76 g. The ratio of masses of oxygen to hydrogen will be : **[31 Aug. 2021 (Evening)]**

(A) 3 : 8 (B) 16 : 3

(C) 3 : 16 (D) 8 : 3

51. Consider two separate ideal gases of electrons and protons having same number of particles. The temperature of both the gases are same. The ratio of the uncertainty in determining the position of an electron to that of a proton is proportional to : **[31 Aug. 2021 (Evening)]**

(A) $\sqrt{\frac{m_p}{m_e}}$ (B) $\sqrt{\frac{m_e}{m_p}}$

(C) $\frac{m_p}{m_e}$ (D) $\left(\frac{m_p}{m_e}\right)^{3/2}$

52. The average translational kinetic energy of N_2 gas molecules at ______ °C becomes equal to the kinetic energy of an electron accelerated from rest through a potential difference of 0.1 volt. (Given $k = 1.38 \times 10^{-23}$ J/K) (Fill the nearest integer) **[01 Sep. 2021 (Evening)]**

* * * * *

CHAPTER

13 OSCILLATIONS

13.1 SIMPLE HARMONIC MOTION

1. A particle is executing simple harmonic motion with a time period T. At time $t = 0$, it is at its position of equilibrium. The kinetic energy-time graph of the particle will look like : **[2017]**

(A) 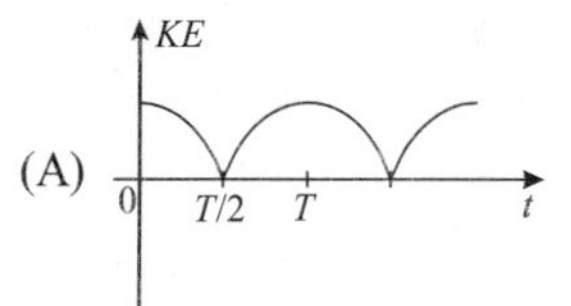

(B) 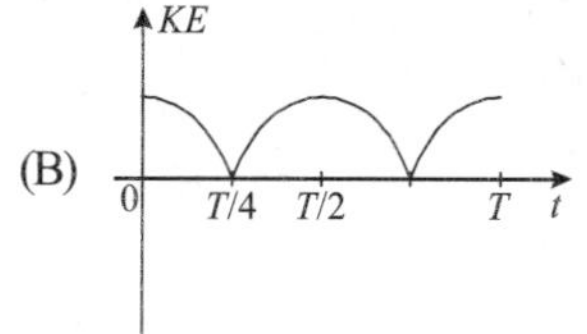

(C) 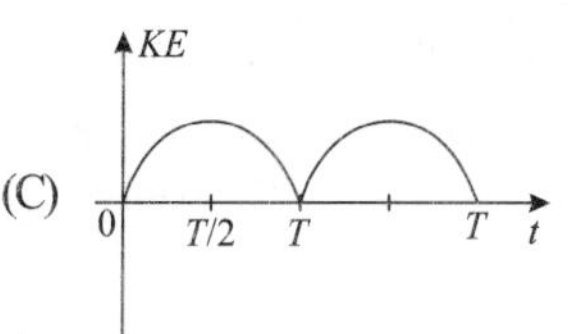

(D) 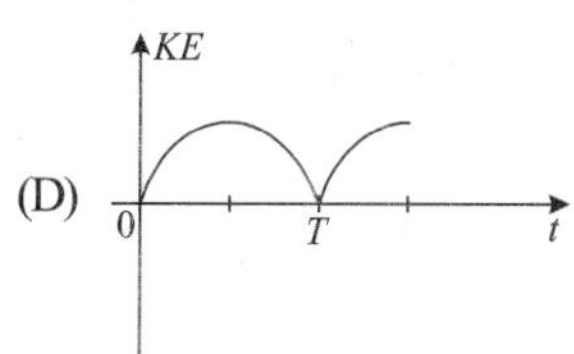

2. A 1 kg block attached to a spring vibrates with a frequency of 1 Hz on a frictionless horizontal table. Two springs identical to the original spring are attached in parallel to an 8 kg block placed on the same table. So, the frequency of vibration of the 8 kg block is : **[2017 (Online)]**

(A) $\frac{1}{4}$Hz (B) $\frac{1}{2\sqrt{2}}$Hz

(C) $\frac{1}{2}$Hz (D) 2 Hz

3. In an experiment to determine the period of a simple pendulum of length 1 m, it is attached to different spherical bobs of radii r_1 and r_2. The two spherical bobs have uniform mass distribution. If the relative difference in the periods, is found to be 5×10^{-4} s, the difference in radii, $|r_1 - r_2|$ is best given : **[2017 (Online)]**

(A) 1 cm (B) 0.1 cm

(C) 0.5 cm (D) 0.01 cm

4. A block of mass 0.1 kg is connected to an elastic spring of spring constant 640 Nm^{-1} and oscillates in a medium of constant 10^{-2} kg s^{-1}. The system dissipates its energy gradually. The time taken for its mechanical energy of vibration to drop to half of its initial value, is closest to : **[2017 (Online)]**

(A) 2 s (B) 3.5 s

(C) 5 s (D) 7 s

5. The ratio of maximum acceleration to maximum velocity in a simple harmonic motion is 10 s^{-1}. At, $t = 0$ the displacement is 5 m. What is the maximum acceleration ? The initial phase is $\pi/4$: **[2017 (Online)]**

(A) 500 m/s^2 (B) $500\sqrt{2}$ m/s^2

(C) 750 m/s^2 (D) $750\sqrt{2}$ m/s^2

6. A silver atom in a solid oscillates in simple harmonic motion in some direction with a frequency of 10^{12}/sec. What is the force constant of the bonds connecting one atom with the other? (Mole wt. of silver = 108 and Avogadro number = 6.02×10^{23} gm mole^{-1}) : **[2018]**

(A) 6.4 N/m (B) 7.1 N/m

(C) 2.2 N/m (D) 5.5 N/m

7. Two simple harmonic motions, as shown, are at right angles. They are combined to form Lissajous figures.

$x(t) = A \sin(at + \delta)$

$y(t) = B \sin(bt)$

Identify the correct match below **[2018 (Online)]**

(A) Parameters : $A = B, a = 2b; \delta = \frac{\pi}{2}$; Curve: Circle

(B) Parameters : $A = B, a = b; \delta = \frac{\pi}{2}$; Curve: Line

(C) Parameters : $A \neq B, a = b; \delta = \frac{\pi}{2}$; Curve: Ellipse

(D) Parameters : $A \neq B, a = b; \delta = 0$; Curve: Parabola

8. A particle executes simple harmonic motion and is located at $x = a$, b and c at times t_0, $2t_0$ and $3t_0$ respectively. The frequency of the oscillation is : **[2018 (Online)]**

(A) $\frac{1}{2\pi t_0}\cos^{-1}\left(\frac{a+b}{2c}\right)$ (B) $\frac{1}{2\pi t_0}\cos^{-1}\left(\frac{a+b}{3c}\right)$

(C) $\frac{1}{2\pi t_0}\cos^{-1}\left(\frac{2a+3c}{b}\right)$ (D) $\frac{1}{2\pi t_0}\cos^{-1}\left(\frac{a+c}{2b}\right)$

9. An oscillator of mass M is at rest in its equilibrium position in a potential $V = \frac{1}{2}k(x - X)^2$. A particle of mass m comes from right with speed u and collides completely inelastically with M and sticks to it. This process repeats every time the oscillator crosses its equilibrium position. The amplitude of oscillations after 13 collisions is : $[M=10, m=5, u=1, k=1]$ **[2018 (Online)]**

(A) $\frac{1}{2}$ (B) $\frac{1}{\sqrt{3}}$

(C) $\frac{2}{3}$ (D) $\sqrt{\frac{3}{5}}$

10. Two masses m and $\frac{m}{2}$ are connected at the two ends of a massless rigid rod of length l. The rod is suspended by a thin wire of torsional constant k at the centre of mass of the rod-mass system (see figure). Because of torsional constant k, the restoring torque is $\tau = k\theta$ for angular displacement θ. If the rod is rotated by θ_0 and released, the tension in it when it passes through its mean position will be : **[9 Jan. 2019 (Morning)]**

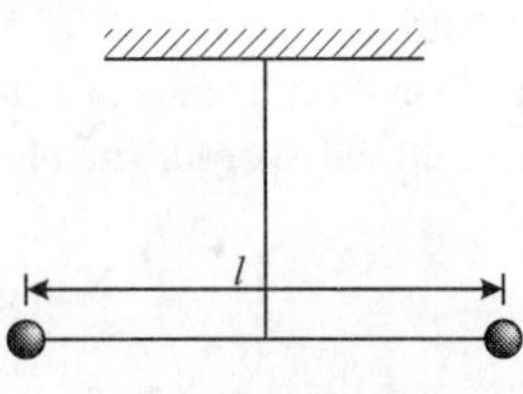

(A) $\frac{3k\theta_0^2}{l}$ (B) $\frac{k\theta_0^2}{2l}$

(C) $\frac{2k\theta_0^2}{l}$ (D) $\frac{k\theta_0^2}{l}$

11. A rod of mass M and length $2L$ is suspended at its middle by a wire. It exhibits torsional oscillations; If two masses each of m are attached at distance $L/2$ from its centre on both sides, it reduces the oscillation frequency by 20%. The value of ratio m/M is close to : **[9 Jan. 2019 (Evening)]**

(A) 0.17 (B) 0.37

(C) 0.57 (D) 0.77

12. A particle is executing simple harmonic motion (SHM) of amplitude A, along the x-axis, about $x = 0$. When its potential Energy (PE) equals kinetic energy (KE), the position of the particle will be : **[9 Jan. 2019 (Evening)]**

(A) $\frac{A}{2}$ (B) $\frac{A}{2\sqrt{2}}$

(C) $\frac{A}{\sqrt{2}}$ (D) A

13. A particle executes simple harmonic motion with an amplitude of 5 cm. When the particle is at 4 cm from the mean position, the magnitude of its velocity in SI units is equal to that of its acceleration. Then, its periodic time in seconds is : **[10 Jan. 2019 (Evening)]**

(A) $\frac{7}{3}\pi$ (B) $\frac{3}{8}\pi$

(C) $\frac{4\pi}{3}$ (D) $\frac{8\pi}{3}$

14. A cylindrical plastic bottle of negligible mass is filled with 310 ml of water and left floating in a pond with still water. If pressed downward slightly and released, it starts performing simple harmonic motion at angular frequency ω. If the radius of the bottle is 2.5 cm then ω close to :

(Density of water $= 10^3$ kg/m^3) : **[10 Jan. 2019 (Evening)]**

(A) 5.00 rad s^{-1} (B) 1.25 rad s^{-1}

(C) 3.75 rad s^{-1} (D) 2.50 rad s^{-1}

15. A particle undergoing simple harmonic motion has time dependent displacement given by $x(t)\ A\sin\frac{\pi t}{90}$. The ratio of potential to kinetic energy of this particle at $t = 210$s will be : **[11 Jan. 2019 (Morning)]**

(A) 2 (B) $\frac{1}{9}$

(C) 3 (D) 1

16. A simple pendulum of length 1 m is oscillating with an angular frequency 10 rad/s. The support of the pendulum starts oscillating up and down with a small angular frequency of 1 rad/s and an amplitude of 10^{-2} m. The relative change in the angular frequency of the pendulum is best given by : **[11 Jan. 2019 (Evening)]**

(A) 10^{-3} rad/s (B) 10^{-1} rad/s

(C) 1 rad/s (D) 10^{-5} rad/s

17. A pendulum is executing simple harmonic motion and its maximum kinetic energy is K_1. If the length of the pendulum is doubled and it performs simple harmonic motion with the same amplitude as in the first case, its maximum kinetic energy is K_2. Then : **[11 Jan. 2019 (Evening)]**

(A) $K_2 = \frac{K_1}{4}$ (B) $K_2 = \frac{K_1}{2}$

(C) $K_2 = 2K_1$ (D) $K_2 = K_1$

18. Two light identical springs of spring constant k are attached horizontally at the two ends of a uniform horizontal rod AB of length l and mass m. The rod is pivoted at its centre O and can rotate freely in horizontal plane. The other ends of the two springs are fixed to rigid supports as shown in figure. The rod is gently pushed through a small angle and released. The frequency of resulting oscillation is : **[12 Jan. 2019 (Morning)]**

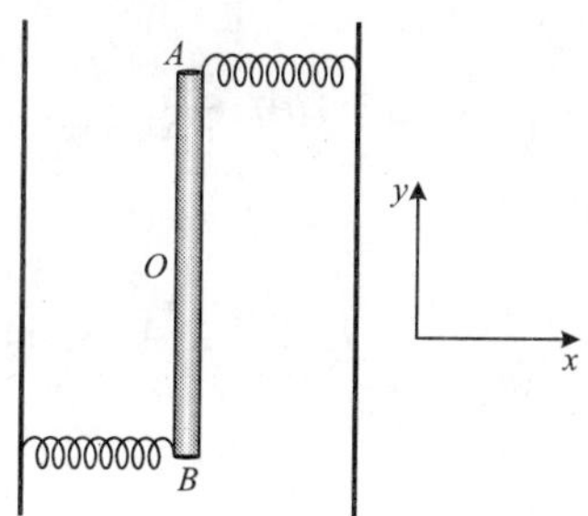

(A) $\frac{1}{2\pi}\sqrt{\frac{6k}{m}}$ (B) $\frac{1}{2\pi}\sqrt{\frac{2k}{m}}$

(C) $\frac{1}{2\pi}\sqrt{\frac{k}{m}}$ (D) $\frac{1}{2\pi}\sqrt{\frac{3k}{m}}$

19. A simple harmonic motion is represented by :

$y = 5(\sin 3\pi t + \sqrt{3}\cos 3\pi t)$ cm

The amplitude and time period of the motion are :

[12 Jan. 2019 (Evening)]

(A) 5 cm, $\frac{3}{2}$ s (B) 5 cm, $\frac{2}{3}$ s

(C) 10 cm, $\frac{3}{2}$ s (D) 10 cm, $\frac{2}{3}$ s

20. A damped harmonic oscillator has a frequency of 5 oscillations per second. The amplitude drops to half its value for every 10 oscillations. The time it will take to drop to 1/1000 of the original amplitude is close to : **[8 April 2019 (Evening)]**

(A) 100 s (B) 20 s

(C) 10 s (D) 50 s

21. A simple pendulum oscillating in air has period T. The bob of the pendulum is completely immersed in a non-viscous liquid. The density of the liquid is $\frac{1}{16}^{\text{th}}$ of the material of the bob. If the bob is inside liquid all the time, its period of oscillation in this liquid is : **[9 April 2019 (Morning)]**

(A) $4T\sqrt{\frac{1}{15}}$ (B) $2T\sqrt{\frac{1}{10}}$

(C) $4T\sqrt{\frac{1}{14}}$ (D) $2T\sqrt{\frac{1}{14}}$

22. A massless spring (k = 800 N/m), attached with a mass (500 g) is completely immersed in 1 kg of water. The spring is stretched by 2 cm and released so that it starts vibrating. What would be the order of magnitude of the change in the temperature of water when the vibrations stop completely ? (Assume that the water container and spring receive negligible heat and specific heat of mass = 400 J/kg K, specific heat of water = 4184 J/kg K) : **[9 April 2019 (Evening)]**

(A) 10^{-3} K (B) 10^{-4} K

(C) 10^{-1} K (D) 10^{-5} K

23. The displacement of a damped harmonic oscillator is given by $x(t) = e^{-0.1t}\cos(10\pi t + \phi)$. Here t is in seconds. The time taken for its amplitude of vibration to drop to half of its initial value is close to : **[10 April 2019 (Morning)]**

(A) 13 s (B) 7 s

(C) 27 s (D) 4 s

24. The displacement time graph of a particle executing SHM is given in figure : (sketch is schematic and not to scale)

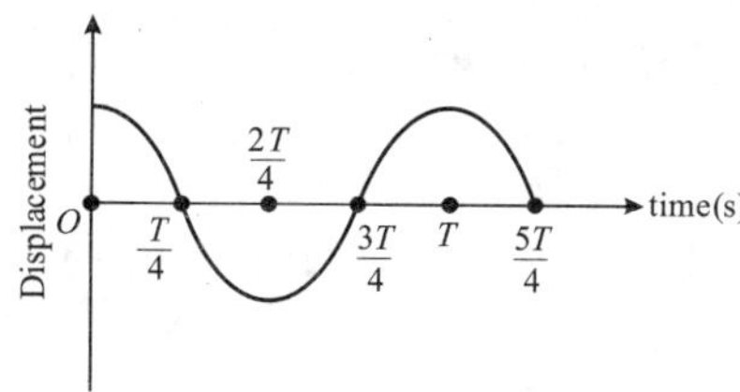

Which of the following statements is/are true for this motion : **[2 sept. 2020 (Evening)]**

(i) The force is zero at $t = \frac{3T}{4}$

(ii) The acceleration is maximum at $t = T$

(iii) The speed is maximum at $t = \frac{T}{4}$

(iv) The potential energy is equal to kinetic energy of the oscillation at $t = \frac{T}{2}$

(A) (i), (ii) and (iv) (B) (ii), (iii) and (iv)

(C) (i) and (iv) (D) (i), (ii) and (iii)

25. A block of mass m attached to massless spring is performing oscillatory motion of amplitude A on a frictionless horizontal plane. If half of the mass of the block breaks off when it is passing through its equilibrium point, the amplitude of oscillation for the remaining system become fA. The value of f is : **[3 Sept. 2020 (Evening)]**

(A) $\frac{1}{2}$ (B) $\sqrt{2}$

(C) 1 (D) $\frac{1}{\sqrt{2}}$

26. A ring is hung on a nail. It can oscillate, without slipping or sliding (i) in its plane with a time period T_1 and, (ii) back and forth in a direction perpendicular to its plane, with a period T_2. The ratio T_1/T_2 will be : **[5 Sept. 2020 (Evening)]**

(A) $\frac{2}{\sqrt{3}}$ (B) $\frac{\sqrt{2}}{3}$

(C) $\frac{2}{3}$ (D) $\frac{3}{\sqrt{2}}$

27. When a particle of mass m is attached to a vertical spring of spring constant k and released, its motion is described by $y(t) = y_0 \sin^2 \omega t$, where '$y$' is measured from the lower end of unstretched spring. Then ω is : **[6 Sept. 2020 (Evening)]**

(A) $\sqrt{\frac{g}{y_0}}$ (B) $\sqrt{\frac{g}{2y_0}}$

(C) $\frac{1}{2}\sqrt{\frac{g}{y_0}}$ (D) $\sqrt{\frac{2g}{y_0}}$

28. In the given figure, a mass M is attached to a horizontal spring which is fixed on one side to a rigid support. The spring constant of the spring is k. The mass oscillates on a frictionless surface with time period T and amplitude A. When the mass is in equilibrium position, as shown in the figure, another mass m is gently fixed upon it. The new amplitude of oscillation will be : **[24 Feb. 2021 (Morning)]**

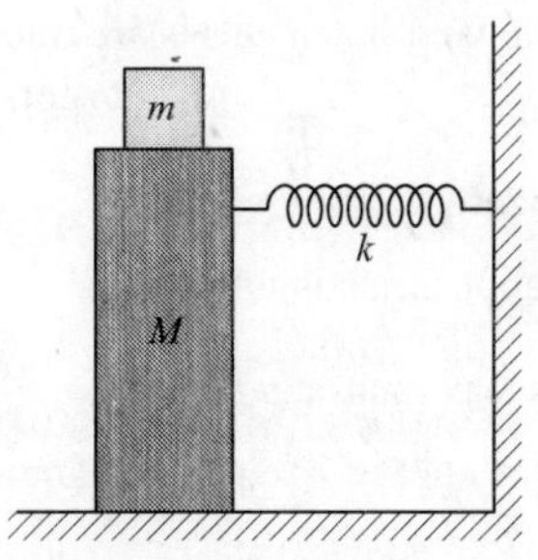

(A) $A\sqrt{\frac{M}{M+m}}$ (B) $A\sqrt{\frac{M-m}{M}}$

(C) $A\sqrt{\frac{M}{M-m}}$ (D) $A\sqrt{\frac{M+m}{M}}$

29. When a particle executes SHM, then nature of graphical representation of velocity as a function of displacement is : **[24 Feb. 2021 (Evening)]**

(A) Straight line (B) Circular
(C) Elliptical (D) Parabolic

30. In the given figure, a body of mass M is held between two massless springs, on a smooth inclined plane. The free ends of the springs are attached to firm supports. If each spring has spring constant k, the frequency of oscillation of given body is : **[24 Feb. 2021 (Evening)]**

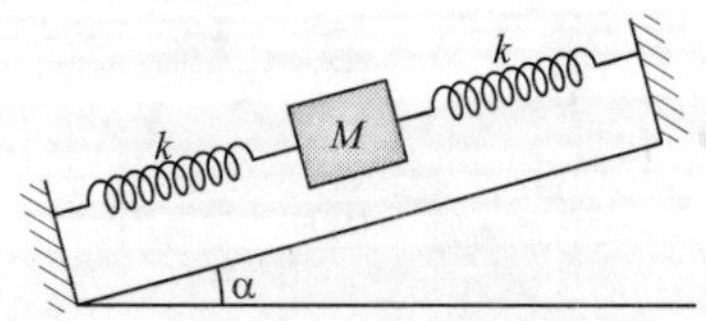

(A) $\frac{1}{2\pi}\sqrt{\frac{k}{Mg \sin \alpha}}$ (B) $\frac{1}{2\pi}\sqrt{\frac{k}{2M}}$

(C) $\frac{1}{2\pi}\sqrt{\frac{2k}{M}}$ (D) $\frac{1}{2\pi}\sqrt{\frac{2k}{Mg \sin \alpha}}$

31. The period of oscillation of a simple pendulum is $T = 2\pi\sqrt{\frac{L}{g}}$.

Measured value of L is 1.0 m from meter scale having a minimum division of 1 mm and time of one complete oscillation is 1.95 s measured from stopwatch of 0.01 s resolution. The percentage error in the determination of g will be : **[24 Feb. 2021 (Evening)]**

(A) 1.03% (B) 1.33%
(C) 1.13% (D) 1.30%

32. If the time period of a two meter long simple pendulum is 2 s, the acceleration due to gravity at the place where pendulum is executing SHM is : **[25 Feb. 2021 (Morning)]**

(A) π^2 ms^{-2} (B) 9.8 ms^{-2}
(C) 16 m/s^2 (D) $2\pi^2$ ms^{-2}

33. $Y = A \sin(\omega t + \phi_0)$ is the time-displacement equation of a SHM. At $t = 0$ the displacement of the particle is $Y = \frac{A}{2}$ and it is moving along negative x-direction. Then the initial phase angle ϕ_0 will be : **[25 Feb. 2021 (Evening)]**

(A) $\frac{2\pi}{3}$ (B) $\frac{\pi}{3}$

(C) $\frac{\pi}{6}$ (D) $\frac{5\pi}{6}$

34. Two identical springs of spring constant $2k$ are attached to a block of mass m and to fixed support (see figure). When the mass is displaced from equilibrium position on either side, it executes simple harmonic motion. The time period of oscillations of this system is : **[25 Feb. 2021 (Evening)]**

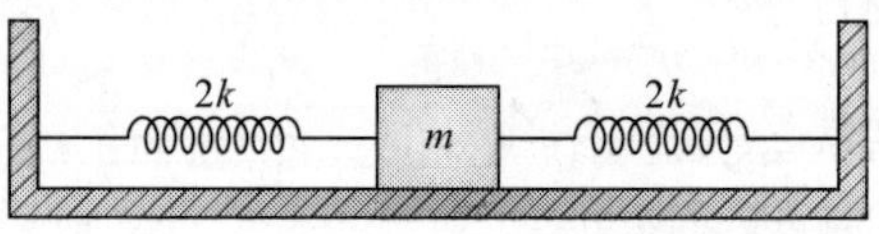

(A) $2\pi\sqrt{\frac{m}{2k}}$ (B) $\pi\sqrt{\frac{m}{2k}}$

(C) $\pi\sqrt{\frac{m}{k}}$ (D) $2\pi\sqrt{\frac{m}{k}}$

35. The point A moves with a uniform speed along the circumference of a circle of radius 0.36 m and covers 30° in 0.1 s. The perpendicular projection P from A on the diameter MN represents the simple harmonic motion of P. The restoration force per unit mass when P touches M will be :

[25 Feb. 2021 (Evening)]

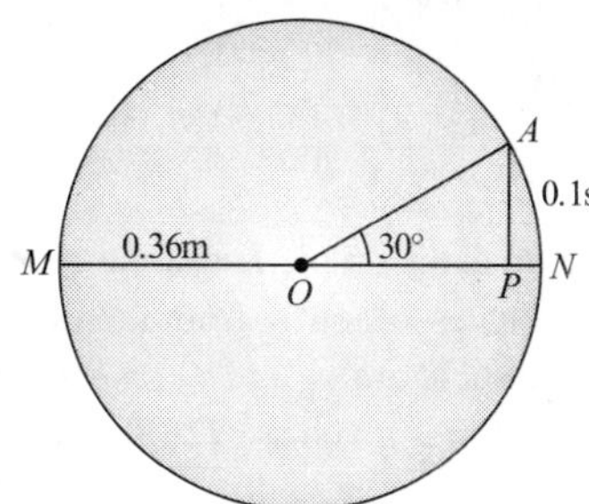

(A) 9.87 N (B) 0.49 N
(C) 100 N (D) 50 N

36. If two similar springs each of spring constant K_1 are joined in series, the new spring constant and time period would be changed by a factor : **[26 Feb. 2021 (Morning)]**

(A) $\frac{1}{4}, \sqrt{2}$ (B) $\frac{1}{2}, \sqrt{2}$

(C) $\frac{1}{4}, 2\sqrt{2}$ (D) $\frac{1}{2}, 2\sqrt{2}$

37. A particle executes SHM, the graph of velocity as a function of displacement is : **[26 Feb. 2021 (Evening)]**

(A) An ellipse (B) A parabola
(C) A helix (D) A circle

38. Given below are two statements :

Statement-I : A second's pendulum has a time period of 1 second

Statement-II : It takes precisely one second to move between the two extreme positions

In the light of the above statements, choose the correct answer from the options given below :

(A) Both Statement-I and Statement-II are false
(B) Statement-I is true but Statement-II is false
(C) Statement-I is false but Statement-II is true
(D) Both Statement-I and Statement-II are true

[26 Feb. 2021 (Evening)]

39. A particle executes SHM with amplitude a and time period T. The displacement of the particle when its speed is half of maximum speed is $\frac{\sqrt{xa}}{2}$. The value of x is ______.

[26 Feb. 2021 (Evening)]

40. Time period of a simple pendulum is T. The time taken to complete $\frac{5}{8}$ oscillations starting from mean position is $\frac{\alpha}{\beta}T$. The value of α is ______. **[26 Feb. 2021 (Evening)]**

41. Time period of a simple pendulum is T inside a lift when the lift is stationary. If the lift moves upwards with an acceleration $g/2$, the time period of pendulum will be :

[16 March 2021 (Morning)]

(A) $\sqrt{3}T$ (B) $\frac{T}{\sqrt{3}}$

(C) $\sqrt{\frac{3}{2}}T$ (D) $\sqrt{\frac{2}{3}}T$

42. Amplitude of a mass-spring system, which is executing simple harmonic motion decreases with time. If mass = 500 g, Decay constant = 20 g/s then how much time is required for the amplitude of the system to drop to half of its initial value ? ($\ln 2 = 0.693$) **[16 March 2021 (Evening)]**

(A) 34.65 s (B) 17.32 s
(C) 0.034 s (D) 15.01 s

43. For what value of displacement the kinetic energy and potential energy of a simple harmonic oscillation become equal ?

[17 March 2021 (Morning)]

(1) $x = 0$ (B) $x = \pm A$

(C) $x = \pm \frac{A}{\sqrt{2}}$ (D) $x = \frac{A}{2}$

44. Consider two identical springs each of spring constant k and negligible mass compared to the mass M as shown. Figure-1 shows one of them and figure-2 shows their series combination. The ratios of time period of oscillation of the two SHM is $\frac{T_b}{T_a} = \sqrt{x}$, where value of x is ______.

(Round off to the Nearest Integer) **[17 March 2021 (Morning)]**

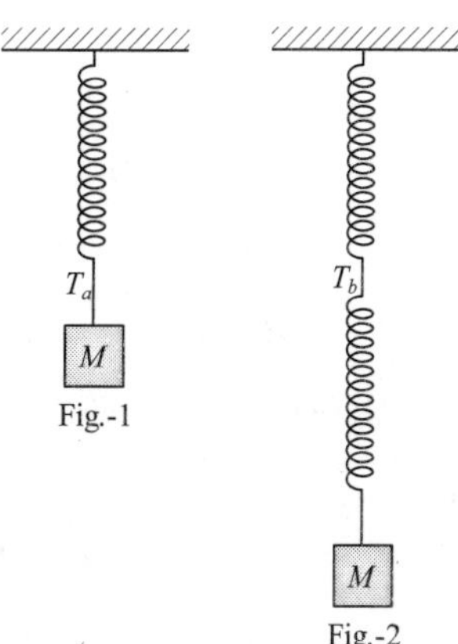

45. A block of mass 1 kg attached to a spring is made to oscillate with an initial amplitude of 12 cm. After 2 minutes the amplitude decreases to 6 cm. Determine the value of the damping constant for this motion. (Take ln 2 = 0.693) **[17 March 2021 (Evening)]**

(A) $0.69 \times 10^2\ kg\ s^{-1}$ (B) $3.3 \times 10^2\ kg\ s^{-1}$

(C) $1.16 \times 10^2\ kg\ s^{-1}$ (D) $5.7 \times 10^{-3}\ kg\ s^{-1}$

46. Two particles A and B of equal masses are suspended from two massless springs of spring constants K_1 and K_2 respectively. If the maximum velocities during oscillations are equal, the ratio of the amplitude of A and B is : **[17 March 2021 (Evening)]**

(A) $\frac{K_2}{K_1}$ (B) $\frac{K_1}{K_2}$

(C) $\sqrt{\frac{K_1}{K_2}}$ (D) $\sqrt{\frac{K_2}{K_1}}$

47. A particle performs simple harmonic motion with a period of 2 second. The time taken by the particle to cover a displacement equal to half of its amplitude from the mean position is $\frac{1}{a}$ s. The value of a to the nearest integer is ______. **[18 March 2021 (Morning)]**

48. If the angular velocity of earth's spin is increased such that the bodies at the equator start floating, the duration of the day would be approximately :

(Take : $g = 10\ ms^{-2}$, the radius of earth, $R = 6400 \times 10^3$ m, Take $\pi = 3.14$) **[18 March 2021 (Evening)]**

(A) 60 minutes (B) Does not change

(C) 1200 minutes (D) 84 minutes

49. The function of time representing a simple harmonic motion with a period of π/ω is : **[18 March 2021 (Evening)]**

(A) $\sin(\omega t) + \cos(\omega)$

(B) $\cos(\omega t) + \cos(2\omega t) + \cos(3\omega t)$

(C) $\sin^2(\omega t)$

(D) $3\cos\left(\frac{\pi}{4} - 2\omega t\right)$

50. In a spring gun having spring constant 100 N/m a small ball B of mass 100 g is put in its barrel (as shown in figure) by compressing the spring through 0.05 m. There should be a box placed at a distance d on the ground so that the ball falls in it. If the ball leaves the gun horizontally at a height of 2 m above the ground. The value of d is ______ m. ($g = 10\ m/s^2$) **[20 July 2021 (Morning)]**

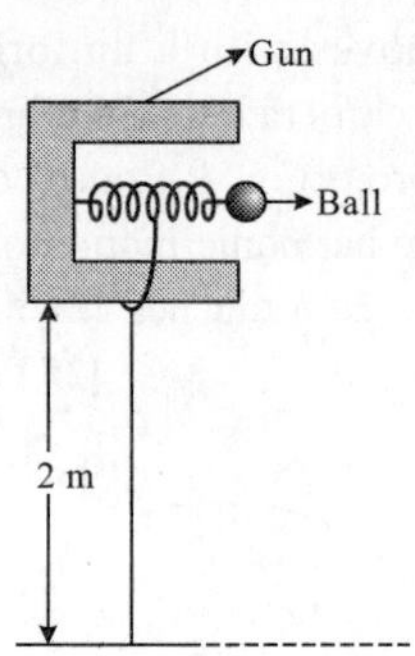

51. A particle is making simple harmonic motion along, the X-axis. If at a distances x_1 and x_2 from the mean position the velocities of the particle are v_1 and v_2 respectively. The time period of its oscillation is given as : **[20 July 2021 (Evening)]**

(A) $T = 2\pi\sqrt{\frac{x_2^2 + x_1^2}{v_1^2 - v_2^2}}$ (B) $T = 2\pi\sqrt{\frac{x_2^2 + x_1^2}{v_1^2 + v_2^2}}$

(C) $T = 2\pi\sqrt{\frac{x_2^2 - x_1^2}{v_1^2 + v_2^2}}$ (D) $T = 2\pi\sqrt{\frac{x_2^2 - x_1^2}{v_1^2 - v_2^2}}$

52. T_0 is the time period of a simple pendulum at a place. If the length of the pendulum is reduced to $\frac{1}{16}$ times of its initial value, the modified time period is : **[22 July 2021 (Evening)]**

(A) T_0 (B) $\frac{1}{4}T_0$

(C) $8\pi T_0$ (D) $4T_0$

53. The motion of a mass on a spring, with spring constant K is as shown in figure. The equation of motion is given by

$$x(t) = A\sin\omega t + B\cos\omega t\ \ \omega = \sqrt{\frac{K}{m}}$$

Suppose that at time $t = 0$, the position of mass is $x(0)$ and velocity $v(0)$, then its displacement can also be represented as $x(t) = C\cos(\omega t - \phi)$, where C and ϕ are : **[22 July 2021 (Evening)]**

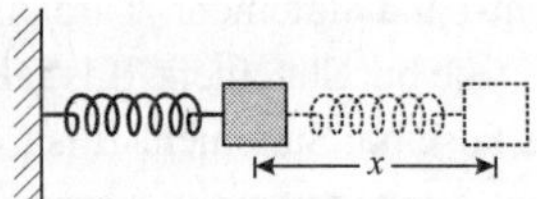

(A) $C = \sqrt{\frac{v(0)^2}{\omega^2} + x(0)^2}$, $\phi = \tan^{-1}\left(\frac{v(0)}{x(0)\omega}\right)$

(B) $C = \sqrt{\frac{2v(0)^2}{\omega^2} + x(0)^2}$, $\phi = \tan^{-1}\left(\frac{x(0)\omega}{2v(0)}\right)$

(C) $C=\sqrt{\dfrac{2v(0)^2}{\omega^2}+x(0)^2}$, $\phi=\tan^{-1}\left(\dfrac{v(0)}{x(0)\omega}\right)$

(D) $C=\sqrt{\dfrac{v(0)^2}{\omega^2}+x(0)^2}$, $\phi=\tan^{-1}\left(\dfrac{x(0)\omega}{v(0)}\right)$

54. In the reported figure, two bodies A and B of masses 200 g and 800 g are attached with the system of springs. Springs are kept in a stretched position with some extension when the system is released. The horizontal surface is assumed to be frictionless. The angular frequency will be ______ rad/s when $k = 20$ N/m : **[25 July 2021 (Morning)]**

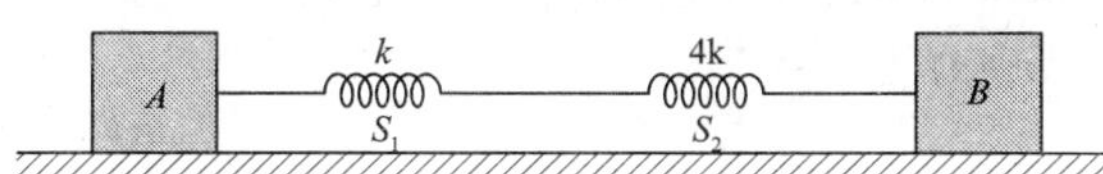

55. In a simple harmonic oscillation, what fraction of total mechanical energy is in the form of kinetic energy, when the particle is midway between mean and extreme position : **[25 July 2021 (Evening)]**

(A) $\dfrac{3}{4}$ (B) $\dfrac{1}{4}$

(C) $\dfrac{1}{2}$ (D) $\dfrac{1}{3}$

56. A particle starts executing simple harmonic motion (SHM) of amplitude a and total energy E. At any instant, its kinetic energy is $\dfrac{3E}{4}$ then its displacement y is given by : **[27 July 2021 (Morning)]**

(A) $y=\dfrac{a}{2}$ (B) $y=a$

(C) $y=\dfrac{a\sqrt{3}}{2}$ (D) $y=\dfrac{a}{\sqrt{2}}$

57. In a uniform magnetic field, the magnetic needle has a magnetic moment 9.85×10^{-2} A/m^2 and moment of inertia 5×10^{-6} kg m^2. If it performs 10 complete oscillations in 5 seconds then the magnitude of the magnetic field is ______ mT. [Take π^2 as 9.85]. **[27 July 2021 (Morning)]**

58. An object of mass 0.5 kg is executing simple harmonic motion. Its amplitude is 5 cm and time period T is 0.2 s. What will be the potential energy of the object at an instant $t = \dfrac{T}{4}$ s starting from mean position. Assume that the initial phase of the oscillation is zero : **[27 July 2021 (Evening)]**

(A) 6.2×10^3 J (B) 1.2×10^3 J

(C) 0.62 J (D) 6.2×10^{-3} J

59. A particle executes simple harmonic motion represented by displacement function as $x(t) = A \sin(\omega t + \phi)$. If the position and velocity of the particle at $t = 0$ s and $x = 2$ cm is $2\ \omega$ cm s^{-1} respectively, then its amplitude is $x\sqrt{2}$ cm where the value of x is ______ **[27 July 2021 (Evening)]**

60. If the length of the pendulum in pendulum clock increases by 0.1%, then the error in time per day is : **[26 Aug. 2021 (Evening)]**

(A) 43.2 s (B) 8.64 s

(C) 4.32 s (D) 86.4 s

61. Two simple harmonic motions are represented by the equations $x_1 = 5\sin\left(2\pi t + \dfrac{\pi}{4}\right)$ and $x_2 = 5\sqrt{2}(\sin 2\pi t + \cos 2\pi t)$. The amplitude of second motion is ______ times the amplitude in first motion. **[26 Aug. 2021 (Evening)]**

62. The variation of displacement with time of a particle executing free simple harmonic motion is shown in the figure below

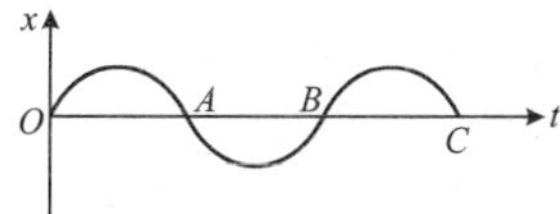

The potential energy curve of the particle with time is represented as: **[27 Aug. 2021 (Morning)]**

(A)
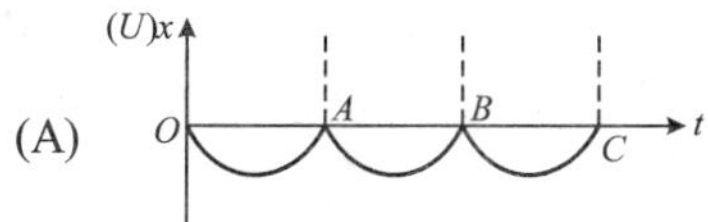

(B)
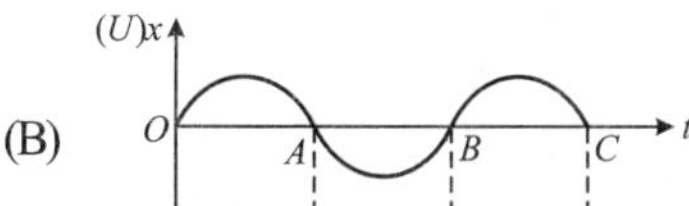

(C)
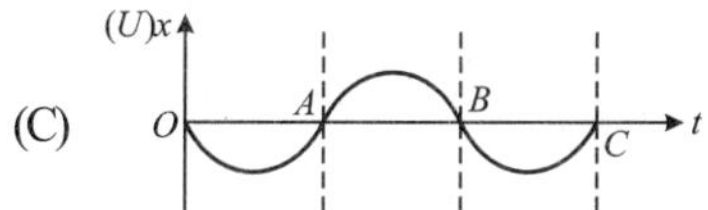

(D)
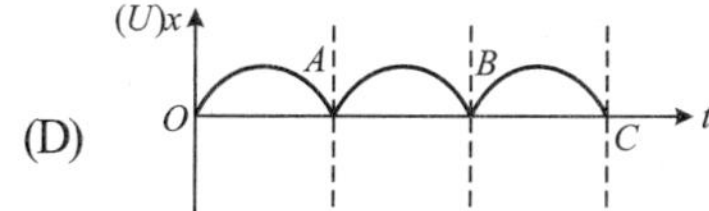

63. Two simple harmonic motion, are represented by the equations

$$y_1 = 10\sin\left(3\pi t + \frac{\pi}{3}\right);\ y_2 = 5(\sin 3\pi t + \sqrt{3}\cos 3\pi t)$$

Ratio of amplitude of y_1 to $y_2 = x : 1$. The value of x is ______. **[27 Aug. 2021 (Evening)]**

64. A particles of mass 1 kg is hanging from a spring of force constant 100 Nm^{-1}. The mass is pulled slightly downward and released so that it executes free simple harmonic motion with time period T. The time when the kinetic energy and potential energy of the system will become equal, is $\frac{T}{x}$. The value of x is_______. **[31 Aug. 2021 (Morning)]**

65. A bob of mass m suspended by a thread of length l undergoes simple harmonic oscillations with time period T. If the bob is immersed in a liquid that has density $\frac{1}{4}$ times that of the bob and the length of the thread is increased by 1/3rd of the original length, then the time period of the simple harmonic oscillations will be : **[31 Aug. 2021 (Evening)]**

(A) T (B) $\frac{4}{3}T$

(C) $\frac{3}{2}T$ (D) $\frac{3}{4}T$

66. For a body executing SHM :

(a) potential energy is always equal to its K.E.

(b) Average potential and kinetic energy over any given time interval are always equal

(c) Sum of the kinetic and potential energy at any point of time is constant

(d) Average K.E. in one time period is equal to average potential energy in one time period

Choose the most appropriate option from the options given below: **[31 Aug. 2021 (Evening)]**

(A) (c) and (d) (B) only (b)

(C) only (c) (D) (b) and (c)

67. A mass of 5 kg is connected to a spring. The potential energy curve of the simple harmonic motion executed by the system is shown in the figure. A simple pendulum of length 4 m has the same period of oscillation as the spring system. What is the value of acceleration due to gravity on the planet where these experiments are performed? **[01 Sep. 2021 (Evening)]**

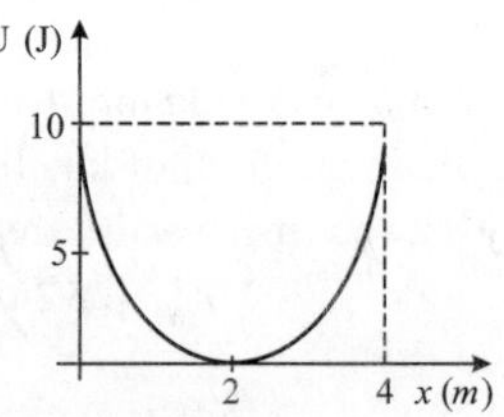

(A) 10 m/s^2 (B) 5 m/s^2

(C) 9.8 m/s^2 (D) 4 m/s^2

* * * * *

CHAPTER

14 WAVES

14.1 WAVE MOTION

1. A heavy ball of mass M is suspended from the ceiling of a car by a light string of mass m ($m << M$). When the car is at rest, the speed of transverse waves in the string is 60 ms^{-1}. When the car has acceleration a, the wave-speed increases to 60.5 ms^{-1}. The value of a, in terms of gravitational acceleration g, is closest to : **[9 Jan. 2019 (Morning)]**

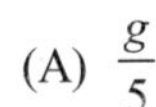

(A) $\frac{g}{5}$ (B) $\frac{g}{20}$

(C) $\frac{g}{10}$ (D) $\frac{g}{30}$

2. Equation of travelling wave on a stretched string of linear density 5 g/m is $y = 0.03 \sin(450t - 9x)$ where distance and time are measured is SI units. The tension in the string is : **[11 Jan. 2019 (Morning)]**

(A) 10N (B) 12.5N

(C) 7.5N (D) 5N

3. A travelling harmonic wave is represented by the equation $y(x, t) = 10^{-3} \sin(50t + 2x)$, where x and y are in meter and t is in seconds. Which of the following is a correct statement about the wave? The wave is propagating along the : **[12 Jan. 2019 (Morning)]**

(A) Negative x-axis with speed 25 ms^{-1}

(B) The wave is propagating along the positive x-axis with speed 25 ms^{-1}

(C) The wave is propagating along the positive x-axis with speed 100 ms^{-1}

(D) The wave is propagating along the negative x-axis with speed 100 ms^{-1}

4. The pressure wave, $P = 0.01 \sin[1000t - 3x]$ Nm^{-2}, corresponds to the sound produced by a vibrating blade on a day when atmospheric temperature is 0°C. On some other day, when temperature is T, the speed of sound produced by the same blade and at the same frequency is found to be 336 ms^{-1}. Approximate value of T is : **[9 April 2019 (Morning)]**

(A) 15°C (B) 12°C

(C) 4°C (D) 11°C

5. A progressive wave travelling along the positive x-direction is represented by $y(x, t) = A \sin(kx - \omega t + \phi)$. Its snapshot at $t = 0$ is given in the figure. For this wave, the phase ϕ is : **[12 April 2019 (Morning)]**

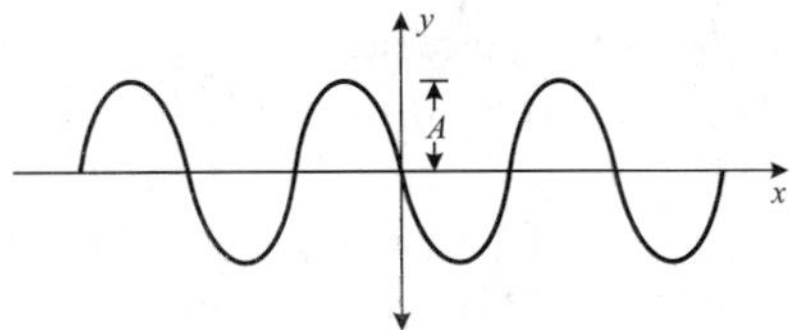

(A) 0 (B) $-\frac{\pi}{2}$

(C) π (D) $\frac{\pi}{2}$

6. A small speaker delivers 2 W of audio output. At what distance from the speaker will one detect 120 dB intensity sound? [Given reference intensity of sound as 10^{-12} W/m^2] : **[12 April 2019 (Evening)]**

(A) 10 cm (B) 30 cm

(C) 40 cm (D) 20 cm

7. Speed of a transverse wave on a straight wire (mass 6.0 g, length 60 cm and area of cross-section 1.0 mm^2) is 90 ms^{-1}. If the Young's modulus of wire is 16×10^{11} Nm^{-2}, the extension of wire over its natural length is : **[7 Jan. 2020 (Morning)]**

(A) 0.02 mm (B) 0.04 mm

(C) 0.03 mm (D) 0.01 mm

8. A transverse wave travels on a taut steel wire with a velocity of v when tension in it is 2.06×10^4 N. When the tension is changed to T, the velocity changed to $v/2$. The value of T is close to : **[8 Jan. 2020 (Evening)]**

(A) 10.2×10^2 N (B) 5.15×10^3 N

(C) 2.50×10^4 N (D) 30.5×10^4 N

9. A uniform thin rope of length 12 m and mass 6 kg hangs vertically from a rigid support and a block of mass 2 kg is attached to its free end. A transverse short wavetrain of wavelength 6 cm is produced at the lower end of the rope. What is the wavelength of the wavetrain (in cm) when it reaches the top of the rope ? **[3 Sept. 2020 (Morning)]**

(A) 9 (B) 12

(C) 6 (D) 3

10. For a transverse wave travelling along a straight line, the distance between two peaks (crests) is 5 m, while the distance between one crest and one trough is 1.5 m. The possible wavelengths (in m) of the waves are : **[4 Sept. 2020 (Morning)]**

(A) $1, 2, 3, \ldots$ (B) $\frac{1}{2}, \frac{1}{4}, \frac{1}{6}, \ldots$

(C) $1, 3, 5, \ldots$ (D) $\frac{1}{1}, \frac{1}{3}, \frac{1}{5}, \ldots$

11. Assume that the displacement(s) of air is proportional to the pressure difference (Δp) created by a sound wave. Displacement(s) further depends on the speed of sound (v), density of air (r) and the frequency (f). If $\Delta p \sim 10 Pa$, $v \sim 300$ m/s, $p \sim 1$ kg/m^3 and $f \sim 1000$ Hz, then s will be the order of (take multiplicative constant to be 1) : **[5 Sept. 2020 (Morning)]**

(A) 10 mm (B) $\frac{3}{100}$ mm

(C) 1 mm (D) $\frac{1}{10}$ mm

12. Which of the following equations represents a travelling wave ? **[24 Feb. 2021 (Evening)]**

(A) $y = Ae^{-x^2}(vt + \theta)$ (B) $y = Ae^{x} \cos(\omega t - \theta)$

(C) $y = A \sin x \cos \omega t$ (D) $y = A \sin(15x - 2t)$

13. The percentage increase in the speed of transverse waves produced in a stretched string if the tension is increased by 4%, will be ______%. **[25 Feb. 2021 (Evening)]**

14. The mass per unit length of a uniform wire is 0.135 g/cm. A transverse wave of the form $y = -0.21 \sin(x + 30t)$ is produced in it, where x is in meter and t is in second. Then, the expected value of tension in the wire is $x \times 10^{-2}$ N. Value of x is ______. (Round-off to the nearest integer). **[26 Feb. 2021 (Morning)]**

15. A sound wave of frequency 245 Hz travels with the speed of 300 ms^{-1} along the positive x-axis. Each point of the wave moves to and fro through a total distance of 6 cm. What will be the mathematical expression of this travelling wave ? **[17 March 2021 (Evening)]**

(A) $Y(x, t) = 0.03 [\sin 5.1 x - (0.2 \times 10^3)t]$

(B) $Y(x, t) = 0.06 [\sin 5.1 x - (1.5 \times 10^3)t]$

(C) $Y(x, t) = 0.06 [\sin 0.8 x - (0.5 \times 10^3)t]$

(D) $Y(x, t) = 0.03 [\sin 5.1 x - (1.5 \times 10^3)t]$

16. The amplitude of wave disturbance propagating in the positive x-direction is given by $y = \frac{1}{(1+x)^2}$ at time $t = 0$ and $y = \frac{1}{1+(x-2)^2}$ at $t = 1$ s, where x and y are in metres. The shape of wave doesn't change during the propagation. The velocity of the wave will be ______ m/s. **[20 July 2021 (Morning)]**

17. Two travelling waves produces a standing wave represented by equation. $y = 1.0$ mm $\cos(1.57 \text{ cm}^{-1})x \sin(78.5 \text{ s}^{-1})t$. The node closest to the origin in the region $x > 0$ will be at $x =$ ______ cm. **[26 Aug. 2021 (Morning)]**

* * * * *

CHAPTER

14 WAVES

14.2 STATIONARY WAVES AND BEATS

1. A standing wave is formed by the superposition of two waves travelling in opposite directions. The transverse displacement is given by $y(x, t) = 0.5 \sin\left(\frac{5\pi}{4}x\right) \cos(200\,\pi t)$. What is the speed of the travelling wave moving in the positive x direction ? (x and t are in meter and second, respectively) : **[2017 (Online)]**

(A) 160 m/s (B) 90 m/s
(C) 180 m/s (D) 120 m/s

2. Two wires W_1 and W_2 have the same radius r and respective densities ρ_1 and ρ_2 such that $\rho_2 = 4\rho_1$. They are joined together at the point O, as shown in the figure. The combination is used as a sonometer wire and kept under tension T. The point O is midway between the two bridges. When a stationary waves is set up in the composite wire, the joint is found to be a node. The ratio of the number of antinodes formed in W_1 to W_2 is : **[2017 (Online)]**

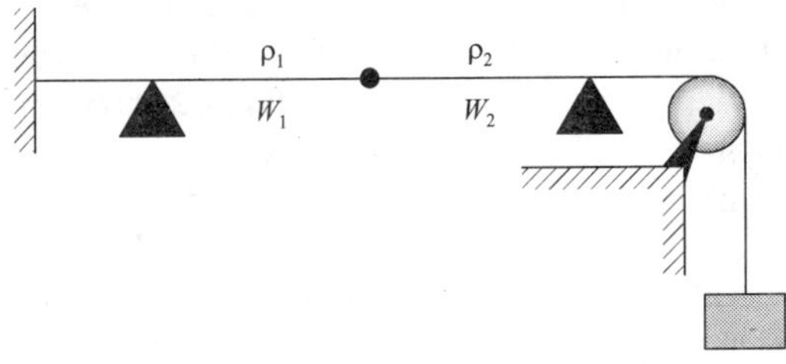

(A) 1 : 1 (B) 1 : 2
(C) 1 : 3 (D) 4 : 1

3. A granite rod of 60 cm length is clamped at its middle point and is set into longitudinal vibrations. The density of granite is 2.7×10^3 kg/m^3 and its Young's modulus is 9.27×10^{10} Pa. What will be the fundamental frequency of the longitudinal vibrations? **[2018]**

(A) 5 kHz (B) 2.5 kHz
(C) 10 kHz (D) 7.5 kHz

4. Two sitar strings, A and B, playing the note 'Dha' are slightly out of tune and produce beats and frequency 5 Hz. The tension of the string B is slightly increased and the beat frequency is found to decrease by 3 Hz. If the frequency of A is 425 Hz, the original frequency of B is : **[2018 (Online)]**

(A) 430 Hz (B) 428 Hz
(C) 422 Hz (D) 420 Hz

5. 5 beats/second are heard when a turning fork is sounded with a sonometer wire under tension, when the length of the sonometer wire is either 0.95 m or 1 m. The frequency of the fork will be : **[2018 (Online)]**

(A) 195 Hz (B) 251 Hz
(C) 150 Hz (D) 300 Hz

6. A tuning fork vibrates with frequency 256 Hz and gives one beat per second with the third normal mode of vibration of an open pipe. What is the length of the pipe ? (Speed of sound of air is 340 s^{-1}) : **[2018 (Online)]**

(A) 190 cm (B) 180 cm
(C) 220 cm (D) 200 cm

7. The end correction of a resonance column is 1 cm. If the shortest length resonating with the tuning fork is 10 cm, the next resonating length should be : **[2018 (Online)]**

(A) 32 cm (B) 40 cm
(C) 28 cm (D) 36 cm

8. A musician using an open flute of length 50 cm produces second harmonic sound waves. A person runs towards the musician from another end of a hall at a speed of 10 km/h. If the wave speed is 330 m/s, the frequency heard by the running person shall be close to : **[9 Jan. 2019 (Evening)]**

(A) 753 Hz (B) 500 Hz
(3) 333 Hz (D) 666 Hz

9. A string of length 1 m and mass 5 g is fixed at both ends. The tension in the string is 8.0 N. The string is set into vibration using an external vibrator of frequency 100 Hz. The separation between successive nodes on the string is close to : **[10 Jan. 2019 (Morning)]**

(A) 16.6 cm (B) 20.0 cm
(C) 10.0 cm (D) 33.3 cm

10. A closed organ pipe has a fundamental frequency of 1.5 kHz. The number of overtones that can be distinctly heard by a person with this organ pipe will be : (Assume that the highest frequency a person can hear is 20,000 Hz) : **[10 Jan. 2019 (Evening)]**

(A) 7 (B) 5
(C) 6 (D) 4

11. A resonance tube is old and has jagged end. It is still used in the laboratory to determine velocity of sound in air. A tuning fork of frequency 512 Hz produces first resonance when the tube is filled with water to a mark 11 cm below a reference mark, near the open end of the tube. The experiment is repeated with another fork of frequency 256 Hz which produces first resonance when water reaches a mark 27 cm below the reference mark. The velocity of sound in air, obtained in the experiment, is close to : **[12 Jan. 2019 (Evening)]**

(A) 328 ms^{-1} (B) 322 ms^{-1}
(C) 341 ms^{-1} (D) 335 ms^{-1}

12. A wire of length $2L$, is made by joining two wires A and B of same length but different radii r and $2r$ and made of the same material. It is vibrating at a frequency such that the joint of the two wires forms a node. If the number of antinodes in wire A is p and that in B is q then the ratio $p : q$ is : **[8 April 2019 (Morning)]**

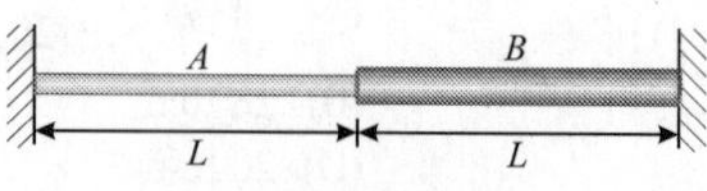

(A) 4 : 9 (B) 3 : 5
(C) 1 : 4 (D) 1 : 2

13. A string is clamped at both the ends and it is vibrating in its 4th harmonic. The equation of the stationary wave is $Y = 0.3 \sin(0.157x)\cos(200\pi t)$. The length of the string is : (All quantities are in SI units.) **[9 April 2019 (Morning)]**

(A) 20 m (B) 80 m
(C) 60 m (D) 40 m

14. A string 2.0 m long and fixed at its ends is driven by a 240 Hz vibrator. The string vibrates in its third harmonic mode. The speed of the wave and its fundamental frequency is : **[9 April 2019 (Evening)]**

(A) 320 m/s, 120 Hz (B) 180 m/s, 80 Hz
(C) 180 m/s, 120 Hz (D) 320 m/s, 80 Hz

15. The correct figure that shows, schematically, the wave pattern produced by superposition of two waves of frequencies 9 Hz and 11 Hz is : **[10 April 2019 (Evening)]**

(A)
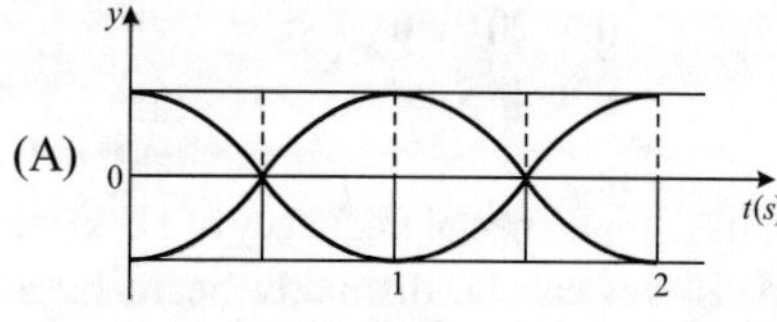

(B)
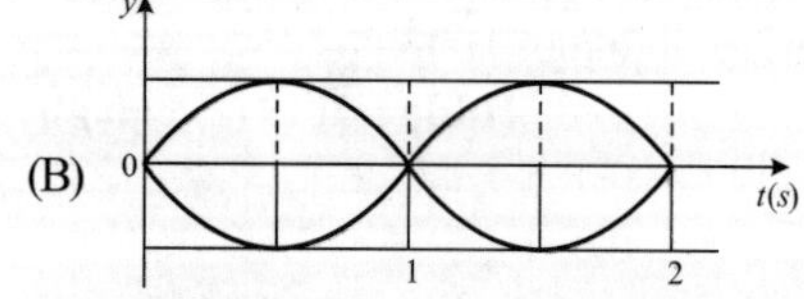

(C)
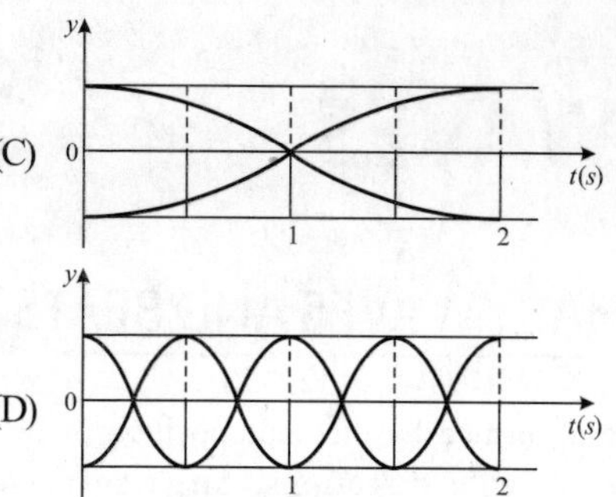

(D)

16. A tuning fork of frequency 480 Hz is used in an experiment for measuring speed of sound (v) in air by resonance tube method. Resonance is observed to occur at two successive lengths of the air column, $l_1 = 30$ cm and $l_2 = 70$ cm. Then v is equal to : **[12 April 2019 (Evening)]**

(A) 332 ms^{-1} (B) 379 ms^{-1}
(C) 384 ms^{-1} (D) 338 ms^{-1}

17. A one metre long (both ends open) organ pipe is kept in a gas that has double the density of air at STP. Assuming the speed of sound in air at STP is 300 m/s, the frequency difference between the fundamental and second harmonic of this pipe is ______ Hz. **[8 Jan. 2020 (Morning)]**

18. A wire of length L and mass per unit length 6.0×10^{-3} kgm^{-1} is put under tension of 540 N. Two consecutive frequencies that it resonates at are 420 Hz and 490 Hz. Then L in meters is : **[9 Jan. 2020 (Evening)]**

(A) 8.1 m (B) 5.1 m
(C) 1.1 m (D) 2.1 m

19. Two identical strings X and Z made of same material have tension T_X and T_Z in them. It their fundamental frequencies are 450 Hz and 300 Hz, respectively, then the ratio T_X/T_Z is : **[2 Sept. 2020 (Morning)]**

(A) 0.44 (B) 1.5
(C) 2.25 (D) 1.25

20. In a resonance tube experiment when the tube is filled with water up to height of 17.0 cm from bottom, it resonates with a given tuning fork. When the water level is raised the next resonance with the same tuning fork occurs at a height of 24.5 cm. If the velocity of sound in air is 330 m/s, the tuning fork frequency is : **[5 Sept. 2020 (Morning)]**

(A) 1100 Hz (B) 3300 Hz
(C) 2200 Hz (D) 550 Hz

21. A student is performing the experiment of resonance column. The diameter of the column tube is 6 cm. The frequency of the tuning fork is 504 Hz. Speed of the sound at the given temperature is 336 m/s. The zero of the metre scale coincides with the top end of the resonance column tube. The reading of the water level in the column when the first resonance occurs is : **[25 Feb. 2021 (Morning)]**

(A) 18.4 cm (B) 14.8 cm
(C) 16.6 cm (D) 13 cm

22. A tuning fork A of unknown frequency produces 5 beats/s with a fork of known frequency 340 Hz. When fork A is filed, the beat frequency decreases to 2 beats/s. What is the frequency of fork A ? **[26 Feb. 2021 (Evening)]**

(A) 342 Hz (B) 335 Hz
(C) 338 Hz (D) 345 Hz

23. A closed organ pipe of length L and an open organ pipe contain gases of densities ρ_1 and ρ_2 respectively. The compressibility of gases are equal in both the pipes. Both the pipes are vibrating in their first overtone with same frequency.

The length of the open pipe is $\frac{x}{3}L\sqrt{\frac{\rho_1}{\rho_2}}$ where x is ______.

(Round off to the Nearest Integer) **[16 March 2021 (Evening)]**

24. Two waves are simultaneously passing through a string and their equations are :

$$y_1 = A_1 \sin k(x - vt),\ y_2 = A_2 \sin k(x - vt + x_0)$$

Given amplitudes $A_1 = 12$ mm and $A_2 = 5$ mm, $x_0 = 3.5$ cm and wave number $k = 6.28\ \text{cm}^{-1}$. The amplitude of resulting, wave will be ______ mm. **[26 Aug. 2021 (Evening)]**

25. A tuning fork is vibrating at 250 Hz. The length of the shortest closed organ pipe that will resonate with the tuning fork will be ______ cm. (Take speed of sound in air as $340\ \text{ms}^{-1}$). **[27 Aug. 2021 (Evening)]**

26. A wire having a linear mass density 9.0×10^{-4} kg/m is stretched between two rigid supports with a tension of 900 N. the wire resonates at a frequency of 500 Hz. The next higher frequency at which the same wire resonates is 550 Hz. The length of the wire is ______ m : **[31 Aug. 2021 (Morning)]**

* * * * *

CHAPTER

14 WAVES

14.3 DOPPLER'S EFFECT

1. An observer is moving with half the speed of light towards a stationary microwave source emitting waves at frequency 10 GHz. What is the frequency of the microwave measured by the observer ? (Speed of light = 3×10^8 ms^{-1}) : **[2017]**

(A) 17.3 GHz (B) 15.3 GHz
(C) 10.1 GHz (D) 12.1 GHz

2. A train moves towards a stationary observer with speed 34 m/s. The train sounds a whistle and its frequency registered by the observer is f_1. If the speed of the train is reduced to 17 m/s, the frequency registered is f_2. If speed of sound is 340 m/s, then the ratio f_1/f_2 is : **[10 Jan. 2019 (Morning)]**

(A) 18/17 (B) 19/18
(C) 20/19 (D) 21/20

3. A person standing on an open ground hears the sound of a jet aeroplane, coming from north at an angle 60° with ground level. But he finds the aeroplane right vertically above his position. If v is the speed of sound, speed of the plane is : **[12 Jan. 2019 (Morning)]**

(A) $\frac{2v}{\sqrt{3}}$ (B) v
(C) $\frac{v}{2}$ (D) $\frac{\sqrt{3}}{2}v$

4. Two cars A and B are moving away from each other in opposite directions. Both the cars are moving with a speed of 20 ms^{-1} with respect to the ground. If an observer in car A detects a frequency 2000 Hz of the sound coming from car B, what is the natural frequency of the sound source in car B ? (speed of sound in air = 340 ms^{-1}) : **[9 April 2019 (Evening)]**

(A) 2250 Hz (B) 2060 Hz
(C) 2150 Hz (D) 2300 Hz

5. A stationary source emits sound waves of frequency 500 Hz. Two observers moving along a line passing through the source detect sound to be of frequencies 480 Hz and 530 Hz. Their respective speeds are, in ms^{-1}, (Given speed of sound = 300 m/s) : **[10 April 2019 (Morning)]**

(A) 16, 14 (B) 12, 18
(C) 12, 16 (D) 8, 18

6. A source of sound S is moving with a velocity of 50 m/s towards a stationary observer. The observer measures the frequency of the source as 1000 Hz. What will be the apparent frequency of the source when it is moving away from the observer after crossing him ? (Take velocity of sound in air is 350 m/s) **[10 April 2019 (Evening)]**

(A) 857 Hz (B) 807 Hz
(C) 750 Hz (D) 1143 Hz

7. A submarine (A) travelling at 18 km/hr is being chased along the line of its velocity by another submarine (B) travelling at 27 km/hr. B sends a sonar signal of 500 Hz to detect A and receives a reflected sound of frequency ν. The value of ν is close to : (Speed of sound in water = 1500 ms^{-1}) : **[12 April 2019 (Morning)]**

(A) 499 Hz (B) 502 Hz
(C) 507 Hz (D) 504 Hz

8. Two sources of sound S_1 and S_2 produce sound waves of same frequency 660 Hz. A listener is moving from source S_1 towards S_2 with a constant speed u m/s and he hears 10 beats/s. The velocity of sound is 330 m/s. Then, u equals : **[12 April 2019 (Evening)]**

(A) 2.5 m/s (B) 15.0 m/s
(C) 5.5 m/s (D) 10.0 m/s

9. A stationary observer receives sound from two identical tuning forks, one of which approaches and the other one recedes with the same speed (much less than the speed of sound). The observer hears 2 beats/sec. The oscillation frequency of each tuning fork is $\nu_0 = 1400$ Hz and the velocity of sound in air is 350 m/s. The speed of each tuning fork is close to : **[7 Jan. 2020 (Evening)]**

(A) $\frac{1}{8}$ m/s (B) $\frac{1}{2}$ m/s
(C) 1 m/s (D) $\frac{1}{4}$ m/s

10. The driver of a bus approaching a big wall notices that the frequency of his bus's horn changes from 420 Hz to 490 Hz, when he hears it after it gets reflected from the wall. Find the speed of the bus if speed of the sound is 330 ms^{-1} : **[4 Sept. 2020 (Evening)]**

(A) 91 kmh^{-1} (B) 71 kmh^{-1}
(C) 81 kmh^{-1} (D) 61 kmh^{-1}

11. A driver in a car, approaching a vertical wall notices that the frequency of his car horn, has changed from 440 Hz to 480 Hz, when it gets reflected from the wall. If the speed of sound in air is 345 m/s, then the speed of the car is :

[5 Sept. 2020 (Evening)]

(A) 36 km/hr (B) 24 km/hr
(C) 18 km/hr (D) 54 km/hr

12. A sound source S is moving along a straight track with speed v, and is emitting sound of frequency v_0 (see figure). An observer is standing at a finite distance, at the point O, from the track. The time variation of frequency heard by the observer is best represented by :
(t_0 represents the instant when the distance between the source and observer is minimum) **[6 Sept. 2020 (Morning)]**

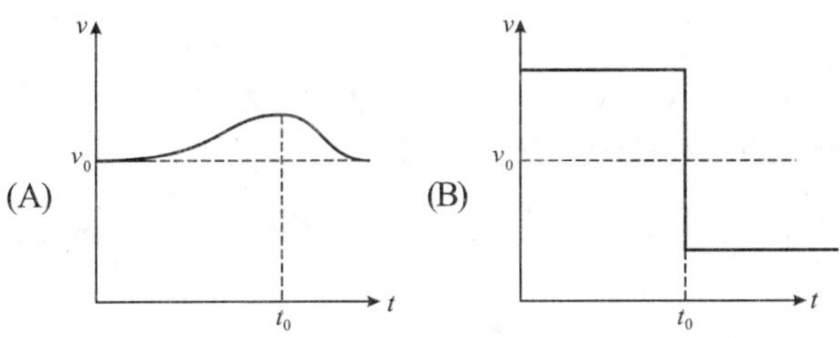

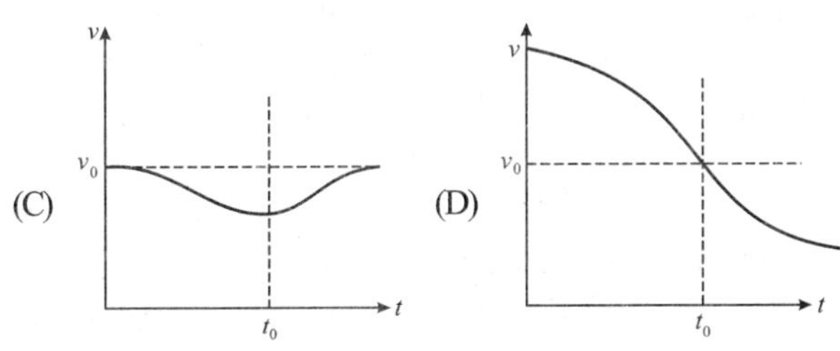

13. Two cars are approaching each other at an equal speed of 7.2 km/hr. When they see each other, both blow horns having frequency of 676 Hz. The beat frequency heard by each driver will be ______ Hz. [Velocity of sound in air is 340 m/s.]

[24 Feb. 2021 (Evening)]

14. A galaxy is moving away from the earth at a speed of 286 kms^{-1}. The shift in the wavelength of a red line at 630 nm is $x \times 10^{-10}$ m. The value of x, to the nearest integer, is ______.
[Take the value of speed of light c, as 3×10^8 ms^{-1}]

[18 March 2021 (Evening)]

15. The frequency of a car horn encountered a change from 400 Hz to 500 Hz, when the car approaches a vertical wall. If the speed of sound is 330 m/s. Then the speed of car is ______ km/h. **[20 July 2021 (Morning)]**

16. With what speed should a galaxy move outward, with respect to earth so that the sodium-D line at wavelength 5890 Å is observed at 5896 Å : **[20 July 2021 (Evening)]**

(A) 322 km/sec (B) 296 km/sec
(C) 306 km/sec (D) 336 km/sec

17. A source and a detector move away from each other in absence of wind with a speed of 20 m/s with respect to the ground. If the detector detects a frequency of 1800 Hz of the sound coming from the source, then the original frequency of source considering speed of sound in air 340 m/s will be ______Hz : **[26 Aug. 2021 (Morning)]**

18. Two cars X and Y are approaching each other with velocities 36 km/h and 72 km/h respectively. The frequency of a whistle sound as emitted by a passenger in car X, heard by the passenger in car Y is 1320 Hz. If the velocity of sound in air is 340 m/s, the actual frequency of the whistle sound produced is ______Hz.

[27 Aug. 2021 (Morning)]

* * * * *

CHAPTER

15 ELECTRIC CHARGES AND FIELDS

15.1 ELECTRIC FORCE AND ELECTRIC FIELD

1. A solid ball of radius R has a charge density ρ given by $\rho = \rho_0\left(1 - \frac{r}{R}\right)$ for $0 \le r \le R$. The electric field outside the ball is : **[2018 (Online)]**

(A) $\frac{\rho_0 R^3}{\epsilon_0 r^2}$ (B) $\frac{4\rho_0 R^3}{3\epsilon_0 r^2}$

(C) $\frac{3\rho_0 R^3}{4\epsilon_0 r^2}$ (D) $\frac{\rho_0 R^3}{12\epsilon_0 r^2}$

2. A body of mass M and charge q is connected to a spring of spring constant k. It is oscillating along x-direction about its equilibrium position, taken to be at $x = 0$, with an amplitude A. An electric field E is applied along the x-direction. Which of the following statements is correct ? **[2018 (Online)]**

(A) The total energy of the system is $\frac{1}{2}m\omega^2A^2 + \frac{1}{2}\frac{q^2E^2}{k}$

(B) The new equilibrium position is at a distance : $\frac{2qE}{k}$ from $x=0$

(C) The new equilibrium position is at a distance : $\frac{qE}{2k}$ from $x=0$

(D) The total energy of the system is $\frac{1}{2}m\omega^2A^2 - \frac{1}{2}\frac{q^2E^2}{k}$

3. Two identical conducting sphere A and B, carry equal charge. They are separated by a distance much larger than their diameter, and the force between them is F. A third identical conducting sphere, C, is uncharged. Sphere C is first touched to A, then to B, and then removed. As a result, the force between A and B would be equal to : **[2018 (Online)]**

(A) $\frac{3F}{4}$ (B) $\frac{F}{2}$

(C) F (D) $\frac{3F}{8}$

4. Three charges $+Q$, q, $+Q$ are placed respectively, at distance 0, $d/2$ and d from the origin, on the x-axis. If the net force experienced by $+Q$, placed at $x = 0$, is zero, then value of q is : **[9 Jan. 2019 (Morning)]**

(A) $+Q/2$ (B) $-Q/2$

(C) $-Q/4$ (D) $+Q/4$

5. For a uniformly charged ring of radius R, the electric field on its axis has the largest magnitude at a distance h from its centre. Then value of h is : **[9 Jan. 2019 (Morning)]**

(A) $\frac{R}{\sqrt{5}}$ (B) R

(C) $\frac{R}{\sqrt{2}}$ (D) $R\sqrt{2}$

6. Two point charges $q_1(\sqrt{10}\ \mu C)$ and $q_2(-25\ \mu C)$ are placed on the x-axis at $x = 1$ m and $x = 4$ m respectively. The electric field (in V/m) at a point $y = 3$ m on y-axis is,

$\left[\text{take} \frac{1}{4\pi\varepsilon_0} = 9\times10^9\ \text{Nm}^2\text{C}^{-2}\right]$: **[9 Jan. 2019 (Evening)]**

(A) $(-63\hat{i} + 27\hat{j})\times10^2$ (B) $(81\hat{i} - 81\hat{j})\times10^2$

(C) $(63\hat{i} - 27\hat{j})\times10^2$ (D) $(-81\hat{i} + 81\hat{j})\times10^2$

7. Charge is distributed within a sphere of radius R with a volume charge density $\rho(r) = \frac{A}{r^2}e^{-2r/a}$, where A and a are constants. If Q is the total charge of this charge distribution, the radius R is : **[9 Jan. 2019 (Evening)]**

(A) $\frac{a}{2}\log\left(1 - \frac{Q}{2\pi aA}\right)$ (B) $a\log\left(1 - \frac{Q}{2\pi aA}\right)$

(C) $a\log\left(\frac{1}{1 - \frac{Q}{2\pi aA}}\right)$ (D) $\frac{a}{2}\log\left(\frac{1}{1 - \frac{Q}{2\pi aA}}\right)$

8. The bob of a simple pendulum has mass 2 g and a charge of 5.0 μC. It is at rest in a uniform horizontal electric field of intensity 2000 V/m. At equilibrium, the angle that the pendulum makes with the vertical is : (take $g = 10$ m/s^2) : **[8 April 2019 (Morning)]**

(A) $\tan^{-1}(5.0)$ (B) $\tan^{-1}(2.0)$

(C) $\tan^{-1}(0.5)$ (D) $\tan^{-1}(0.2)$

9. Four point charges $-q$, $+q$, $+q$ and $-q$ are placed on y-axis at $y = -2d$, $y = -d$, $y = +d$ and $y = +2d$, respectively. The magnitude of the electric field E at a point on the x-axis at $x = D$, with $D >> d$, will behave as : **[9 April 2019 (Evening)]**

(A) $E \propto \dfrac{1}{D}$ (B) $E \propto \dfrac{1}{D^3}$

(C) $E \propto \dfrac{1}{D^2}$ (D) $E \propto \dfrac{1}{D^4}$

10. Let a total charge $2Q$ be distributed in a sphere of radius R, with the charge density given by $\rho(r) = kr$, where r is the distance from the centre. Two charges A and B, of $-Q$ each, are placed on diametrically opposite points, at equal distance a from the centre. If A and B do not experience any force, then : **[12 April 2019 (Evening)]**

(A) $a = \dfrac{3R}{2^{1/4}}$ (B) $a = R/\sqrt{3}$

(C) $a = 8^{-1/4} R$ (D) $a = 2^{-1/4} R$

11. Three charged particle A, B and C with charges $-4q$, $2q$ and $-2q$ are present on the circumference of a circle of radius d. The charged particles A, C and centre O of the circle formed an equilateral triangle as shown in figure. Electric field at O along x-direction is : **[8 Jan. 2020 (Morning)]**

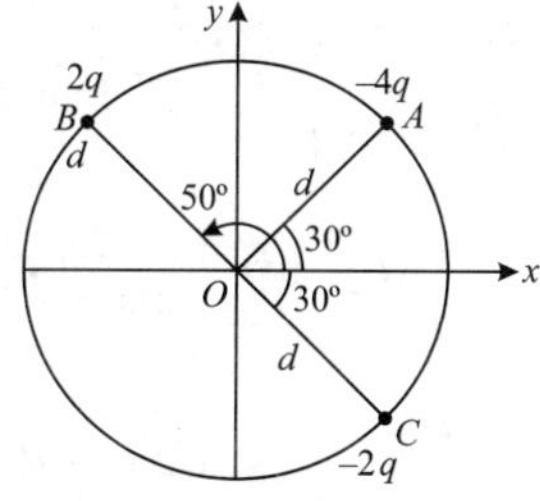

(A) $\dfrac{2\sqrt{3}q}{\pi\varepsilon_0 d^2}$ (B) $\dfrac{\sqrt{3}q}{4\pi\varepsilon_0 d^2}$

(C) $\dfrac{3\sqrt{3}q}{4\pi\varepsilon_0 d^2}$ (D) $\dfrac{\sqrt{3}q}{\pi\varepsilon_0 d^2}$

12. A particle of mass m and charge q is released from rest in a uniform electric field. If there is no other force on the particle, the dependence of its speed v on the distance x travelled by it is correctly given by (graphs are schematic and not drawn to scale) : **[8 Jan. 2020 (Evening)]**

(A) 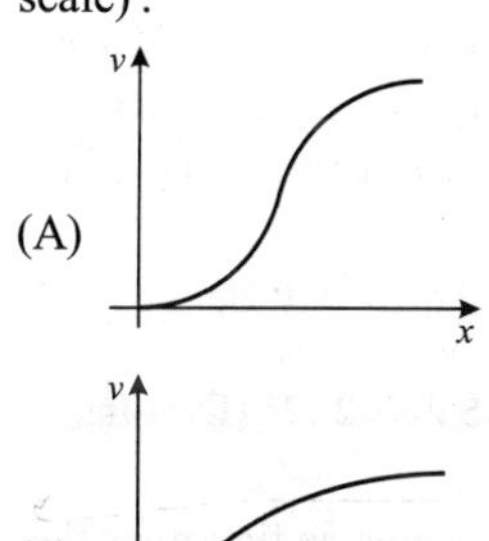

(B) 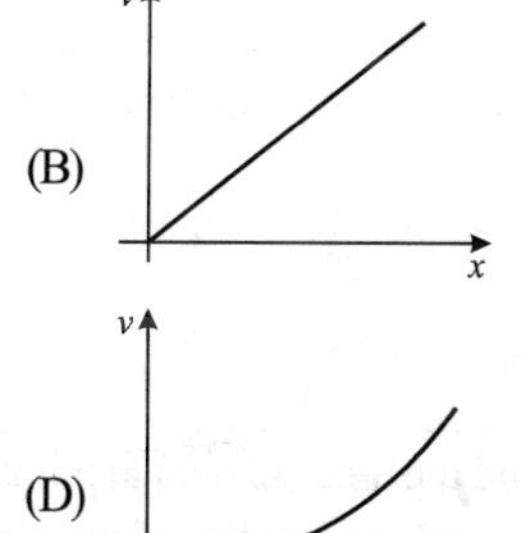

(C)

(D)

13. Consider a sphere of radius R which carries a uniform charge density ρ. If a sphere of radius $\dfrac{R}{2}$ is carved out of it, as shown, the ratio $\dfrac{|\vec{E}_A|}{|\vec{E}_B|}$ of magnitude of electric field $\vec{E}_A$ and $\vec{E}_B$, respectively, at points A and B due to the remaining portion is : **[9 Jan. 2020 (Morning)]**

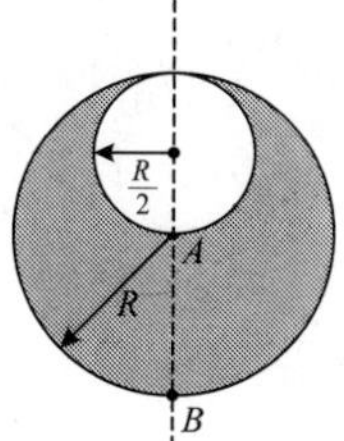

(A) $\dfrac{18}{54}$ (B) $\dfrac{21}{34}$

(C) $\dfrac{17}{54}$ (D) $\dfrac{18}{34}$

14. A charge Q is distributed over two concentric conducting thin spherical shells radii r and R $(R > r)$. If the surface charge densities on the two shells are equal, the electric potential at the common centre is : **[2 Sept. 2020 (Evening)]**

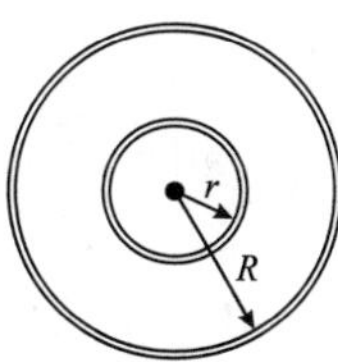

(A) $\dfrac{1}{4\pi\varepsilon_0}\dfrac{(R+2r)Q}{2(R^2+r^2)}$ (B) $\dfrac{1}{4\pi\varepsilon_0}\dfrac{(R+r)}{2(R^2+r^2)}Q$

(C) $\dfrac{1}{4\pi\varepsilon_0}\dfrac{(R+r)}{(R^2+r^2)}Q$ (D) $\dfrac{1}{4\pi\varepsilon_0}\dfrac{(2R+r)}{(R^2+r^2)}Q$

15. A small point mass carrying some positive charge on it, is released from the edge of a table. There is a uniform electric field in this region in the horizontal direction. Which of the following options then correctly describe the trajectory of the mass : (Curves are drawn schematically and are not to scale). **[2 Sept. 2020 (Evening)]**

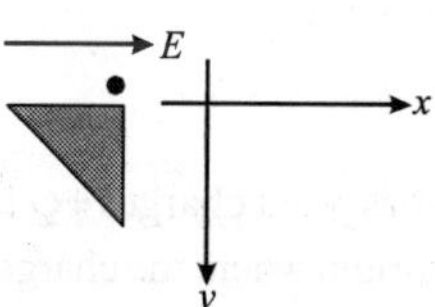

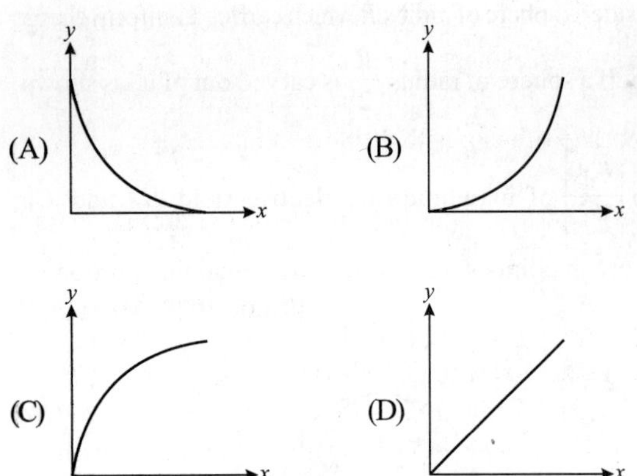

16. Charges Q_1 and Q_2 are at points A and B of a right angle triangle OAB (see figure). The resultant electric field at point O is perpendicular to the hypotenuse, then Q_1/Q_2 is proportional to : **[6 Sept. 2020 (Morning)]**

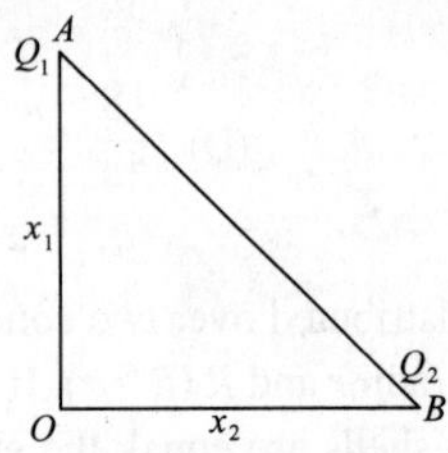

(A) $\frac{x_2^2}{x_1^2}$ (B) $\frac{x_1^3}{x_2^3}$

(C) $\frac{x_1}{x_2}$ (D) $\frac{x_2}{x_1}$

17. Consider the force F on a charge q due to a uniformly charged spherical shell of radius R carrying charge Q distributed uniformly over it. Which one of the following statements is true for F, if q is placed at distance r from the centre of the shell : **[6 Sept. 2020 (Evening)]**

(A) $F=\frac{1}{4\pi\varepsilon_0}\frac{Qq}{r^2}$ for $r>R$

(B) $\frac{1}{4\pi\varepsilon_0}\frac{qQ}{R^2}>F>0$ for $r<R$

(C) $F=\frac{1}{4\pi\varepsilon_0}\frac{Qq}{r^2}$ for all r

(D) $F=\frac{1}{4\pi\varepsilon_0}\frac{Qq}{R^2}$ for $r<R$

18. A cube of side a has point charges $+Q$ located at each of its vertices except at the origin where the charge is $-Q$. The electric field at the centre of cube is : **[24 Feb. 2021 (Morning)]**

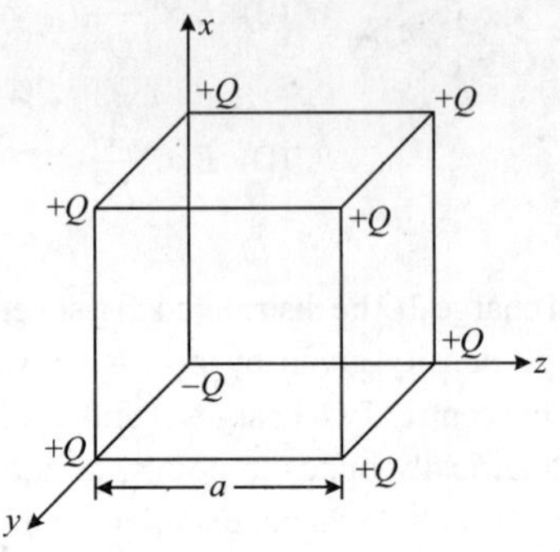

(A) $\frac{2Q}{3\sqrt{3}\pi\varepsilon_0 a^2}(\hat{x}+y+\hat{z})$ (B) $\frac{Q}{3\sqrt{3}\pi\varepsilon_0 a^2}(\hat{x}+y+\hat{z})$

(C) $\frac{-2Q}{3\sqrt{3}\pi\varepsilon_0 a^2}(\hat{x}+y+\hat{z})$ (D) $\frac{-Q}{3\sqrt{3}\pi\varepsilon_0 a^2}(\hat{x}+y+\hat{z})$

19. Two electrons each are fixed at a distance $2d$. A third charge proton placed at the midpoint is displaced slightly by a distance $x(x<<d)$ perpendicular to the line joining the two fixed charges. Proton will execute simple harmonic motion having angular frequency : (m = mass of charged particle) : **[24 Feb. 2021 (Evening)]**

(A) $\left(\frac{\pi\varepsilon_0 md^3}{2q^2}\right)^{1/2}$ (B) $\left(\frac{q^2}{2\pi\varepsilon_0 md^3}\right)^{1/2}$

(C) $\left(\frac{2q^2}{\pi\varepsilon_0 md^3}\right)^{1/2}$ (D) $\left(\frac{2\pi\varepsilon_0 md^3}{q^2}\right)^{1/2}$

20. Two identical conducting spheres with negligible volume have 2.1 nC and – 0.1 nC charges, respectively. They are brought into contact and then separated by a distance of 0.5 m. The electrostatic force acting between the spheres is ______ $\times 10^{-9}$ N. [Given : $4\pi\varepsilon_0=\frac{1}{9\times10^9}$ SI unit]. **[25 Feb. 2021 (Evening)]**

21. Two small spheres each of mass 10 mg are suspended from a point by threads 0.5 m long. They are equally charged and repel each other to a distance of 0.20 m. The charge on each of the sphere is $\frac{a}{21}\times10^{-8}$ C. The value of a will be ______. [Given $g=10$ ms^{-2}] **[25 Feb. 2021 (Evening)]**

22. Find the electric field at point P (as shown in figure) on the perpendicular bisector of a uniformly charged thin wire of length L carrying a charge Q. The distance of the point P from the centre of the rod is $a = \frac{\sqrt{3}}{2}L$: **[26 Feb. 2021 (Morning)]**

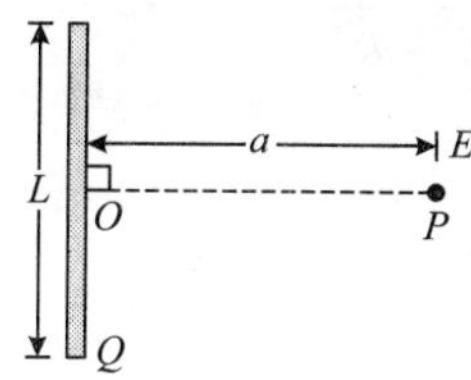

(A) $\frac{Q}{4\pi\varepsilon_0 L^2}$ (B) $\frac{\sqrt{3}Q}{4\pi\varepsilon_0 L^2}$

(C) $\frac{Q}{2\sqrt{3}\pi\varepsilon_0 L^2}$ (D) $\frac{Q}{3\pi\varepsilon_0 L^2}$

23. An inclined plane making an angle of 30° with the horizontal is placed in a uniform horizontal electric field $200\frac{N}{C}$ as shown in the figure. A body of mass 1 kg and charge 5 mC is allowed to slide down from rest at a height of 1 m. If the coefficient of friction is 0.2, find the time taken by the body to reach the bottom. **[26 Feb. 2021 (Evening)]**

[$g = 9.8$ m/s^2; $\sin 30° = \frac{1}{2}$; $\cos 30° = \frac{\sqrt{3}}{2}$]

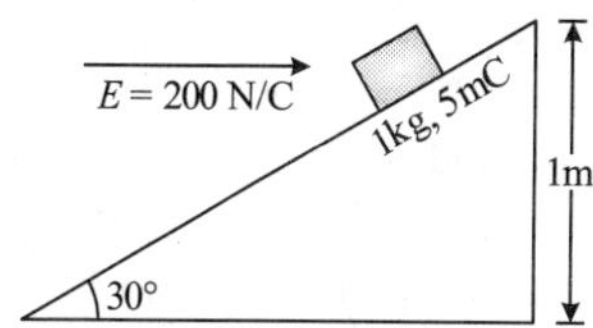

(A) 0.46 s (B) 2.3 s
(C) 0.92 s (D) 1.3 s

24. Find out the surface charge density at the intersection of point $x = 3$ m plane and x-axis, in the region of uniform line charge of 8 nC/m lying along the z-axis in free space.
[16 March 2021 (Evening)]

(A) 0.424 nC m^{-2} (B) 47.88 C/m
(C) 0.07 nC m^{-2} (D) 4.0 nC m^{-2}

25. An oil drop of radius 2 mm with a density 3g cm^{-3} is held stationary under a constant electric field 3.55×10^5 V m^{-1} in the Millikan's oil drop experiment. What is the number of excess electrons that the oil drop will possess ?
(consider $g = 9.81$ m/s^2) **[18 March 2021 (Morning)]**

(A) 48.8×10^{11} (B) 1.73×10^{10}
(C) 17.3×10^{10} (D) 1.73×10^{12}

26. An infinite number of point charges, each carrying 1 μC charge, are placed along the y-axis at $y = 1$ m, 2 m, 4 m, 8 m .... The total force on a 1 C point charge, placed at the origin, is $x \times 10^3$ N. The value of x, to the nearest integer, is______.

[Take $\frac{1}{4\pi\epsilon_0} = 9 \times 10^9$ Nm^2/C^2] **[18 March 2021 (Evening)]**

27. A certain charge Q is divided into two parts q and $(Q - q)$. How should the charges Q and q be divided so that q and $(Q - q)$. Placed at a certain distance apart experience maximum electrostatics repulsion : **[20 July 2021 (Morning)]**

$+q$ ———— r ———— $Q - q$

(A) $Q = 4q$ (B) $Q = 4q$

(C) $Q = \frac{q}{2}$ (D) $Q = 2q$

28. A body having specific charge 8 μC/g is resting on a frictionless plane at a distance 10 cm from the wall (as shown in the figure). It starts moving towards the wall when a uniform electric field of 100 V/m is applied horizontally towards the wall. If the collision of the body with the wall is perfectly elastic, the time period of the motion will be ______ s.
[20 July 2021 (Morning)]

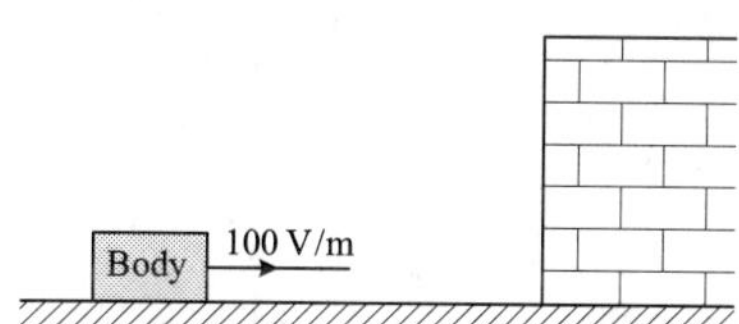

29. An electric dipole is placed on x-axis in proximity to a line charge of linear charge density 3.0×10^{-6} C/m. Line charge is placed on z-axis and positive and negative charge of dipole is at a distance of 10 mm and 12 mm from the origin respectively. If total force of 4 N is exerted on the dipole, find out the amount of positive or negative charge of the dipole :
[22 July 2021 (Evening)]

(A) 0.485 mC (B) 4.44 μC
(C) 815.1 nC (D) 8.8 μC

30. A particle of mass 1 mg and charge q is lying at the mid-point of two stationary particle kept at a distance 2 m when each is carrying same charge q. If the free charged particle is displaced from its equilibrium position through distance x ($x << 1$ m). The particle executes SHM. Its angular frequency of oscillation will be ______ $\times 10^5$ rad/s if $q^2 = 10$ C^2 :
[25 July 2021 (Morning)]

31. Two identical tennis balls each having mass 'm' and charge 'q' are suspended from a fixed point by threads of length 'l'.

What is the equilibrium separation when each thread makes a small angle 'θ' with the vertical ? **[27 July 2021 (Morning)]**

(A) $x=\left(\frac{q^2l^2}{2\pi\varepsilon_0 m^2 g}\right)^{1/3}$ (B) $x=\left(\frac{q^2l}{2\pi\varepsilon_0 mg}\right)^{1/3}$

(C) $x=\left(\frac{q^2l^2}{2\pi\varepsilon_0 m^2 g^2}\right)^{1/3}$ (D) $x=\left(\frac{q^2l}{2\pi\varepsilon_0 mg}\right)^{1/3}$

32. A solid metal sphere of radius R having charge q is enclosed inside the concentric spherical shell of inner radius a and outer radius b as shown in figure. The approximate variation of electric field $\vec{E}$ as a function of distance r from centre O is given by : **[26 Aug. 2021 (Morning)]**

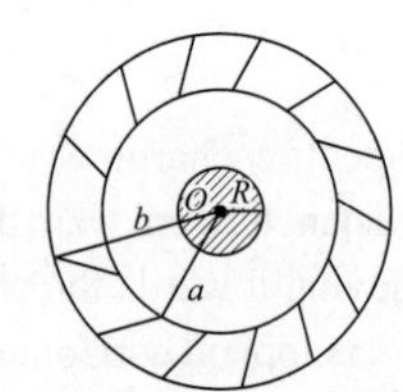

(A) 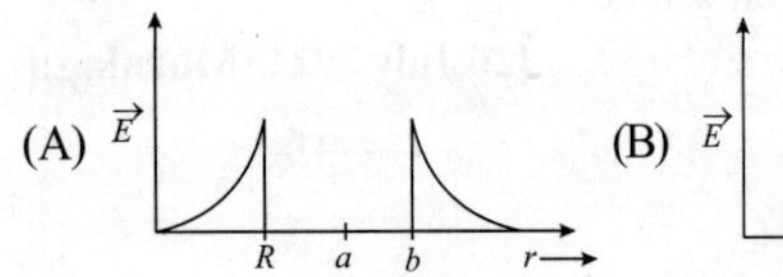

(B) 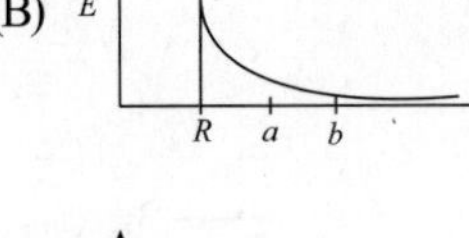

(C) 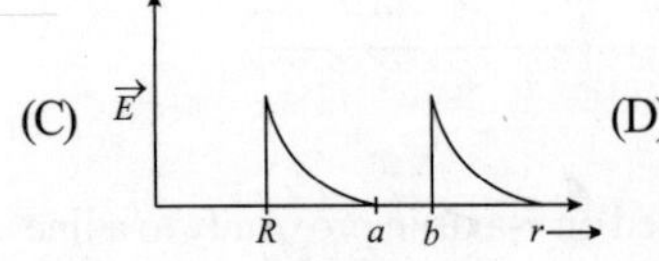

(D) 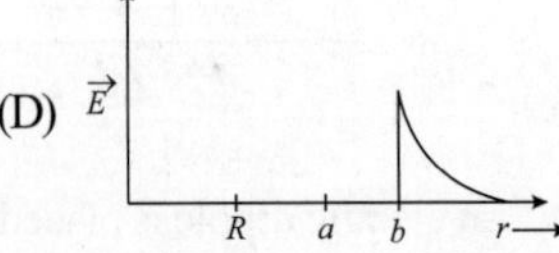

33. A uniformly charged disc of radius R having surface charge density σ is placed in the xy plane with its center at the origin. Find the electric field intensity along the z-axis at a distance Z from origin : **[27 Aug. 2021 (Morning)]**

(A) $E=\frac{\sigma}{2\varepsilon_0}\left(\frac{1}{Z^2+R^2}\right)+\frac{1}{Z^2}$

(B) $E=\frac{2\varepsilon_0}{2\sigma}\left(\frac{1}{(Z^2+R^2)^{1/2}}+Z\right)$

(C) $E=\frac{\sigma}{2\varepsilon_0}\left(1+\frac{Z}{(Z^2+R^2)^{1/2}}\right)$

(D) $E=\frac{\sigma}{2\varepsilon_0}\left(1-\frac{Z}{(Z^2+R^2)^{1/2}}\right)$

34. Figure shows a rod AB, which is bent in a 120° circular are of radius R. A charge $(-Q)$ is uniformly distributed over rod AB. What is the electric filed $\vec{E}$ at the centre of curvature O ? **[27 Aug. 2021 (Evening)]**

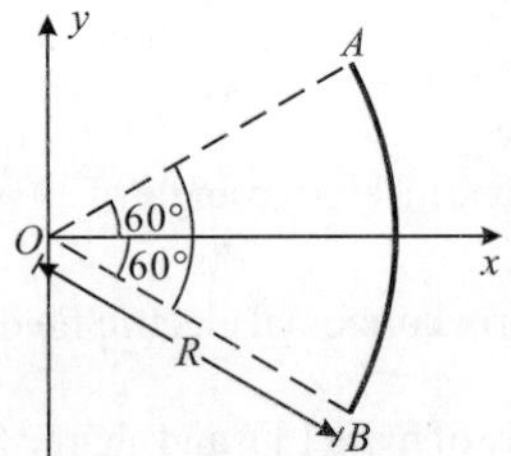

(A) $\frac{3\sqrt{3}Q}{8\pi^2\varepsilon_0 R^2}(\hat{i})$ (B) $\frac{3\sqrt{3}Q}{8\pi\varepsilon_0 R^2}(\hat{i})$

(C) $\frac{3\sqrt{3}Q}{8\pi^2\varepsilon_0 R^2}(-\hat{i})$ (D) $\frac{3\sqrt{3}Q}{16\pi^2\varepsilon_0 R^2}(\hat{i})$

35. Two particles A and B having charges 20 μC and –5 μC respectively are held fixed with a separation of 5 cm. At what position a third charged particle should be placed so that it does not experience a net electric force ? **[31 Aug. 2021 (Morning)]**

20μC –5μC
A 5 cm B

(A) At midpoint between two charges
(B) At 1.25 cm from a –5 μC between two charges
(C) At 5 cm from 20 μC on the left side of system
(D) At 5 cm from –5 μC on the right side

* * * * *

CHAPTER

15 ELECTRIC CHARGES AND FIELDS

15.2 ELECTRIC FLUX AND GAUSS'S LAW

1. Four closed surfaces and corresponding charge distributions are shown below :

Let the respective electric fluxes through the surfaces be Φ_1, Φ_2, Φ_3 and Φ_4. Then : **[2017 (Online)]**

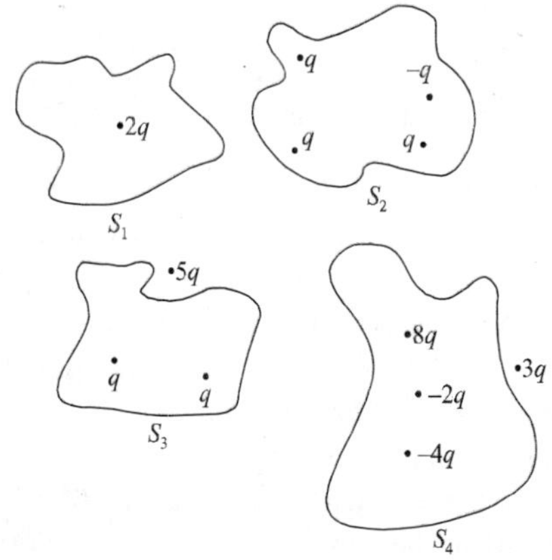

(A) $\Phi_1 < \Phi_2 = \Phi_3 > \Phi_4$ (B) $\Phi_1 > \Phi_2 > \Phi_3 > \Phi_4$
(C) $\Phi_1 = \Phi_2 = \Phi_3 = \Phi_4$ (D) $\Phi_1 > \Phi_3 ; \Phi_2 < \Phi_4$

2. A charge Q is placed at a distance $a/2$ above the centre of the square surface of edge a as shown in the figure. The electric flux through the square surface is : **[2018 (Online)]**

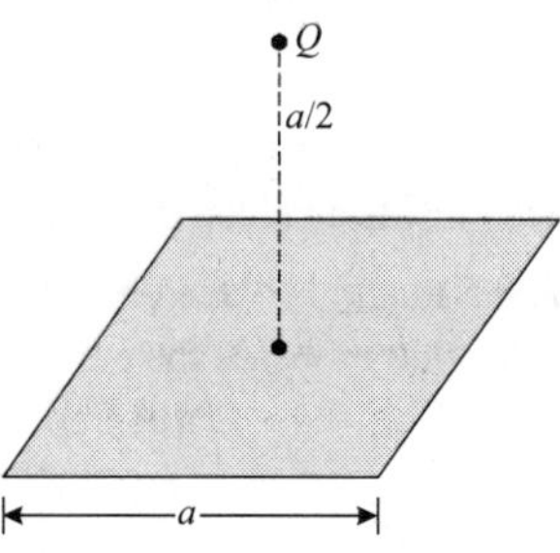

(A) $\dfrac{Q}{3\epsilon_0}$ (B) $\dfrac{Q}{6\epsilon_0}$

(C) $\dfrac{Q}{2\epsilon_0}$ (D) $\dfrac{Q}{\epsilon_0}$

3. In finding the electric field using Gauss Law the formula $|\vec{E}| = \dfrac{q_{enc}}{\varepsilon_0 |A|}$ is applicable. In the formula ε_0 is permittivity of free space, A is the area of Gaussian surface and q_{enc} is charge enclosed by the Gaussian surface. The equation can be used in which of the following situation ? **[8 Jan. 2020 (Morning)]**

(A) Only when the Gaussian surface is an equipotential surface
(B) Only when $|\vec{E}|$ = constant on the surface
(C) For any choice of Gaussian surface
(D) Only when the Gaussian surface is an equipotential surface and $|\vec{E}|$ is constant on the surface

4. An electric field $\vec{E} = 4x\hat{i} - (y^2+1)\hat{j}$ N/C passes through the box shown in figure. The flux of the electric field through surfaces $ABCD$ and $BCGF$ are marked as ϕ_I and ϕ_{II} respectively. The difference between $(\phi_I - \phi_{II})$ is (in Nm²/C) _______. **[9 Jan. 2020 (Evening)]**

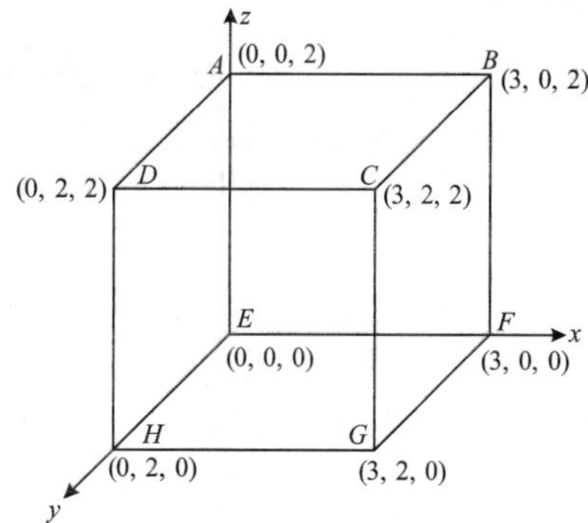

5. Two charged thin infinite plane sheets of uniform surface charge density σ_+ and σ_- where $|\sigma_+| > |\sigma_-|$ intersect at right angle. Which of the following best represents the electric field lines for this system : **[4 Sept. 2020 (Morning)]**

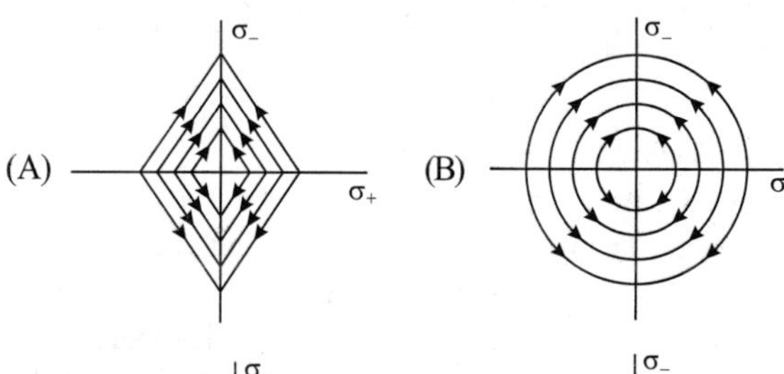

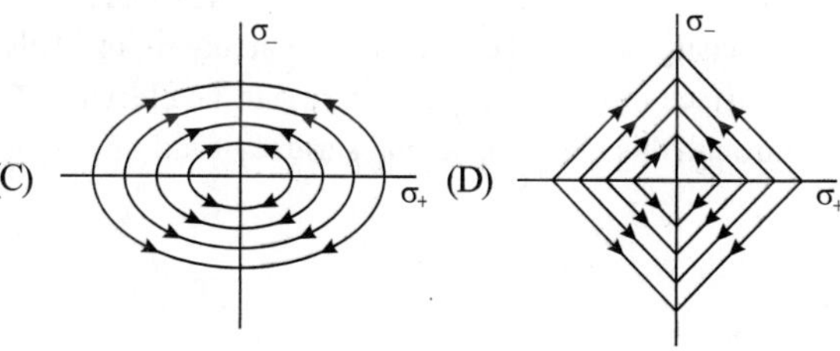

6. A point charge of + 12μC is at a distance 6 cm vertically above the centre of a square of side 12 cm as shown in figure. The magnitude of the electric flux through the square will be ______ $\times 10^3$ Nm^2/C **[24 Feb. 2021 (Evening)]**

7. The electric field in a region is given by

$$\vec{E} = \left(\frac{3}{5}E_0\hat{i} + \frac{4}{5}E_0\hat{j}\right)\frac{N}{C}$$

The ratio of flux of reported field through the rectangular surface of area 0.2 m^2 (parallel to y-z plane) to that of the surface of area 0.3 m^2 (parallel to x-z plane) is $a : b$, where a = ______.

[Here $\hat{i}$, $\hat{j}$ and $\hat{k}$ are unit vectors along x, y and z-axes respectively]. **[25 Feb. 2021 (Morning)]**

8. A charge q is placed at one corner of a cube as shown in figure. The flux of electrostatic field $\vec{E}$ through the shaded area is : **[25 Feb. 2021 (Evening)]**

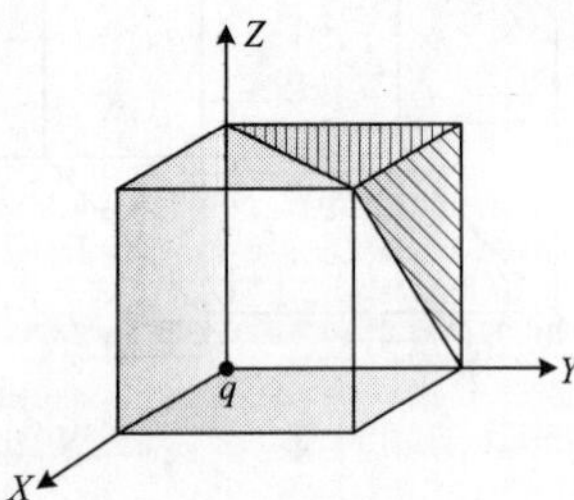

(A) $\frac{q}{4\varepsilon_0}$ (B) $\frac{q}{8\varepsilon_0}$

(C) $\frac{q}{48\varepsilon_0}$ (D) $\frac{q}{24\varepsilon_0}$

9. Given below are two statements :

Statement-I : An electric dipole is placed at the centre of a hollow sphere. The flux of electric field through the sphere is zero but the electric field is not zero anywhere in the sphere.

Statement-II : If R is the radius of a solid metallic sphere and Q be the total charge on it. The electric field at any point on the spherical surface of radius $r(< R)$ is zero but the electric flux passing through this closed spherical surface of radius r is not zero.

In the light of the above statements, choose the correct answer from the options given below : **[26 Feb. 2021 (Evening)]**

(A) Statement-I is true but Statement-II is false

(B) Both Statement-I and Statement-II are false

(C) Both Statement-I and Statement-II are true

(D) Statement-I is false but Statement-II is true

10. The electric field in a region is given by $\vec{E} = \frac{2}{5}E_0\hat{i} + \frac{3}{5}E_0\hat{j}$ with $E_0 = 4.0 \times 10^3\ \frac{N}{C}$. The flux of this field through a rectangular surface area 0.4 m^2 parallel to the $Y - Z$ plane is ______ Nm^2 C^{-1}. **[17 March 2021 (Evening)]**

11. The total charge enclosed in an incremental volume of 2×10^{-9} m^3 located at the origin is ______ nC, if electric flux density of its field is found as $D = e^{-x}\sin y\,\hat{i} - e^{-x}\cos y\,\hat{j} + 2z\,\hat{k}$ C/m^2. **[22 July 2021 (Evening)]**

12. Choose the incorrect statement : **[31 Aug. 2021 (Evening)]**

(a) The electric lines of force entering into a Gaussian surface provide negative flux

(b) A charge q is placed at the centre of a cube. The flux through all the faces will be the same

(c) In a uniform electric field net flux through a closed Gaussian surface containing no net charge, is zero.

(d) When electric field is parallel to a Gaussian surface, it provides a finite non-zero flux

Choose the most appropriate answer from the options given below :

(A) (b) and (d) only (B) (a) and (c) Only

(C) (d) Only (D) (c) and (d) Only

13. A cube is placed inside an electric field, $\vec{E} = 150y^2\,\hat{j}$. The side of the cube is 0.5 m and is placed in the field as shown in the given figure. The charge inside the cube is : **[01 Sep. 2021 (Evening)]**

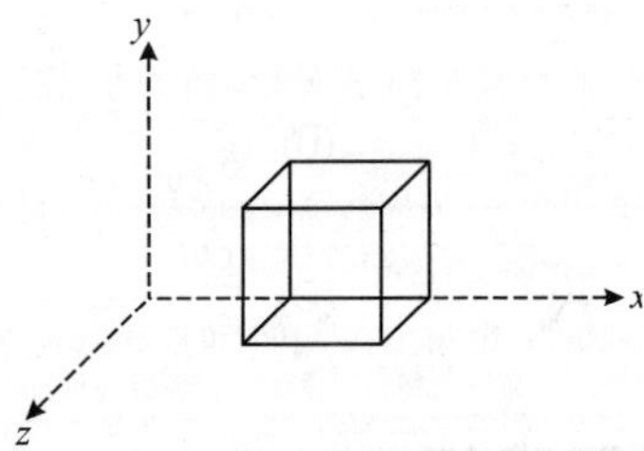

(A) 8.3×10^{-12} C (B) 8.3×10^{-11} C

(C) 3.8×10^{-12} C (D) 3.8×10^{-11} C

* * * * *

CHAPTER

16 ELECTRIC POTENTIAL AND CAPACITANCE

16.1 ELECTRIC POTENTIAL AND DIPOLE

1. There is a uniform electrostatic field in a region. The potential at various points on a small sphere centred at P, in the region, is found to vary between in the limits 589.0 V to 589.8 V. What is the potential at a point on the sphere whose radius vector makes an angle of 60° with the direction of the field ? **[2017 (Online)]**

(A) 589.5 V (B) 589.2 V

(C) 589.4 V (D) 589.6 V

2. Three concentric metal shells A, B and C of respective radii a, b and c $(a < b < c)$ have surface charge densities $+\sigma$, $-\sigma$ and $+\sigma$ respectively. The potential of shell B is : **[2018]**

(A) $\frac{\sigma}{\epsilon_0}\left[\frac{a^2-b^2}{a}+c\right]$ (B) $\frac{\sigma}{\epsilon_0}\left[\frac{a^2-b^2}{b}+c\right]$

(C) $\frac{\sigma}{\epsilon_0}\left[\frac{b^2-c^2}{b}+a\right]$ (D) $\frac{\sigma}{\epsilon_0}\left[\frac{b^2-c^2}{c}+a\right]$

3. A charge Q is distributed over three concentric spherical shells of radii a, b, c $(a < b < c)$ such that their surface charge densities are equal to one another. The total potential at a point at distance r from their common centre, where $r < a$, would be: **[10 Jan. 2019 (Morning)]**

(A) $\frac{Q}{4\pi\varepsilon_0(a+b+c)}$ (B) $\frac{Q(a+b+c)}{4\pi\varepsilon_0(a^2+b^2+c^2)}$

(C) $\frac{Q}{12\pi\varepsilon_0}\frac{ab+bc+ca}{abc}$ (D) $\frac{Q}{4\pi\varepsilon_0}\frac{(a^2+b^2+c^2)}{(a^3+b^3+c^3)}$

4. Two electric dipoles, A, B with respective dipole moments $\vec{d}_A = -4qa\hat{i}$ and $\vec{d}_B = -2qa\hat{i}$ placed on the x-axis with a separation R, as shown in the figure : **[10 Jan. 2019 (Morning)]**

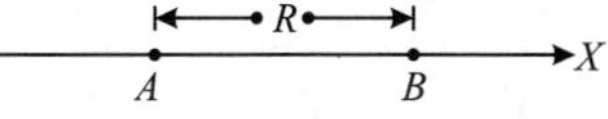

The distance from A at which both of them produce the same potential is :

(A) $\frac{\sqrt{2}R}{\sqrt{2}+1}$ (B) $\frac{R}{\sqrt{2}+1}$

(C) $\frac{\sqrt{2}\,R}{\sqrt{2}-1}$ (D) $\frac{R}{\sqrt{2}-1}$

5. Charges $-q$ and $+q$ located at A and B, respectively, constitute an electric dipole. Distance $AB = 2a$, O is the mid point of the dipole and OP is perpendicular to AB. A charge Q is placed at P where $OP = y$ and $y >> 2a$. The charge Q experiences and electrostatic force F. If Q is now moved along the equatorial line to P' such that $OP' = \left(\frac{y}{3}\right)$, the force on Q will be close to. $\left(\frac{y}{3} >> 2a\right)$: **[10 Jan. 2019 (Evening)]**

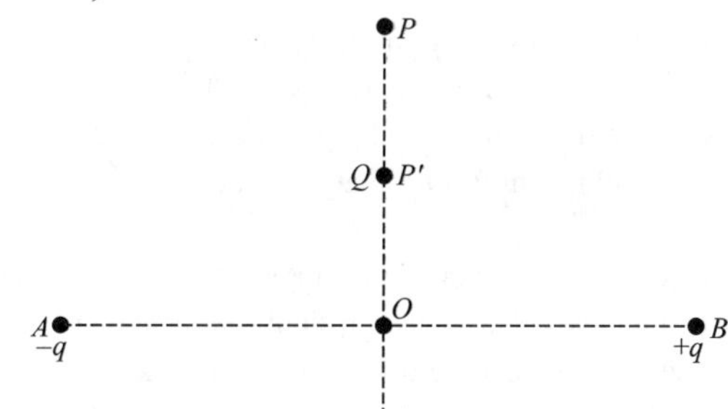

(A) $\frac{F}{3}$ (B) $3F$

(C) $9F$ (D) $27F$

3. Four equal point charges Q each are placed in the xy plane at (0, 2), (4, 2), (4, –2) and (0, –2). The work required to put a fifth charge Q at the origin of the coordinate system will be : **[10 Jan. 2019 (Evening)]**

(A) $\frac{Q^2}{2\sqrt{2}\pi\varepsilon_0}$ (B) $\frac{Q^2}{4\pi\varepsilon_0}\left(1+\frac{1}{\sqrt{5}}\right)$

(C) $\frac{Q^2}{4\pi\varepsilon_0}\left(1+\frac{1}{\sqrt{3}}\right)$ (D) $\frac{Q^2}{4\pi\varepsilon_0}$

7. The charges Q, $+q$ and $+q$ are placed at the vertices of a right-angle isosceles triangle as shown below. The net electrostatic energy of the configuration is zero, if the value of Q is: **[11 Jan. 2019 (Morning)]**

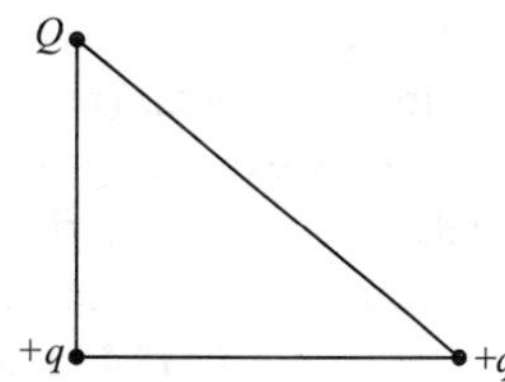

(A) $\frac{-\sqrt{2}q}{\sqrt{2}+1}$ (B) $-2q$

(C) $\frac{-q}{1+\sqrt{2}}$ (D) $+q$

8. The given graph shows variation (with distance r from centre) of : **[11 Jan. 2019 (Morning)]**

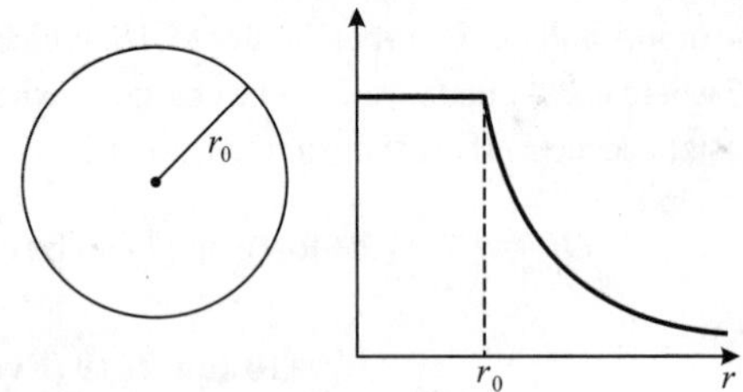

(A) Potential of a uniformly charged sphere
(B) Potential of a uniformly charged spherical shell
(C) Electric field of uniformly charged spherical shell
(D) Electric field of uniformly charged sphere

9. An electric field of 1000 V/m is applied to an electric dipole at angle of 45°. The value of electric dipole moment is 10^{-29} C.m. What is the potential energy of the electric dipole ? **[11 Jan. 2019 (Evening)]**

(A) -9×10^{-20} J (B) -7×10^{-27} J
(C) -10×10^{-29} J (D) -20×10^{-18} J

10. Determine the electric dipole moment of the system of three charges, placed on the vertices of an equilateral triangle, as shown in the figure : **[12 Jan. 2019 (Morning)]**

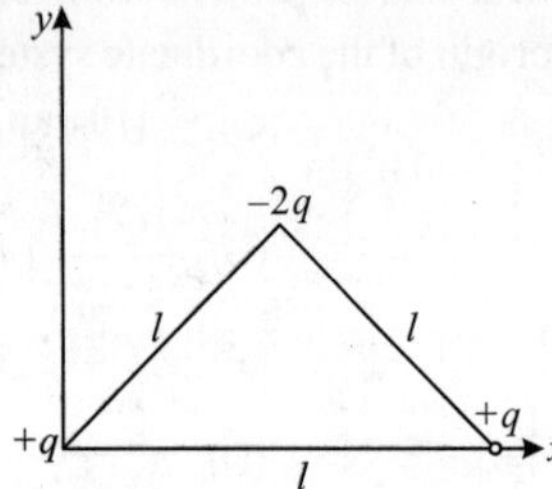

(A) $(ql)\frac{\hat{i}+\hat{j}}{\sqrt{2}}$ (B) $\sqrt{3}ql\frac{\hat{j}-\hat{i}}{\sqrt{2}}$

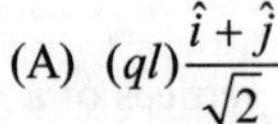

(C) $-\sqrt{3}ql\,\hat{j}$ (D) $2ql\,\hat{j}$

11. There is a uniform spherically symmetric surface charge density at a distance R_0 from the origin. The charge distribution is initially at rest and starts expanding because of mutual repulsion. The figure that represents best the speed $V(R(t))$ of the distribution as a function of its instantaneous radius $R(t)$ is : **[12 Jan. 2019 (Morning)]**

(A)
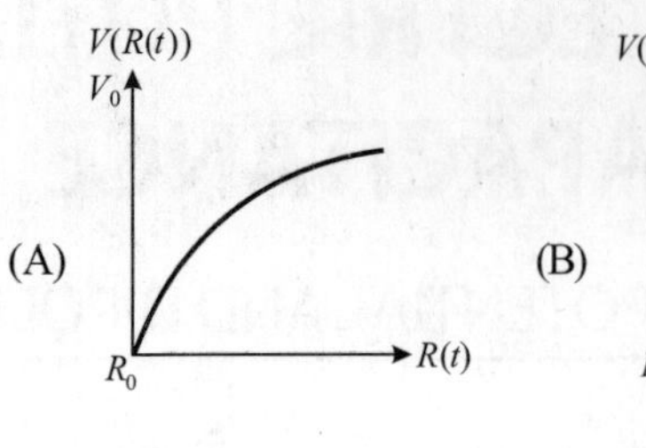

(B)
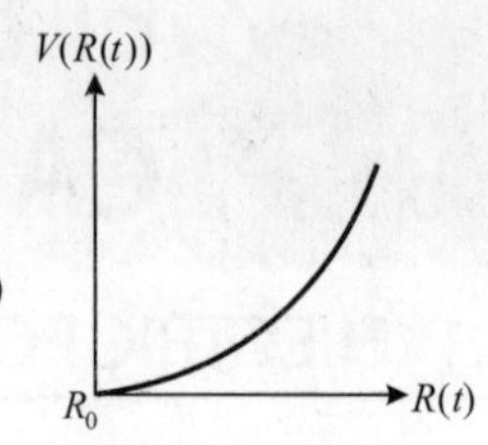

(C)
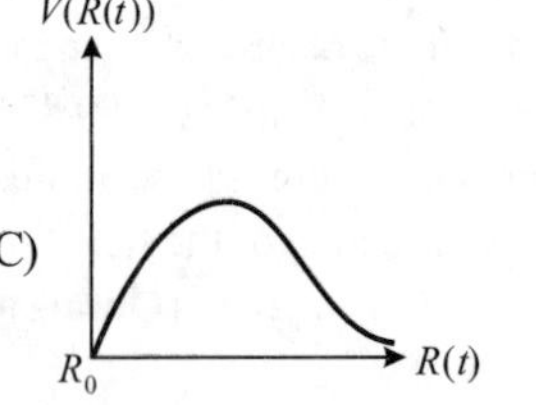

(D)
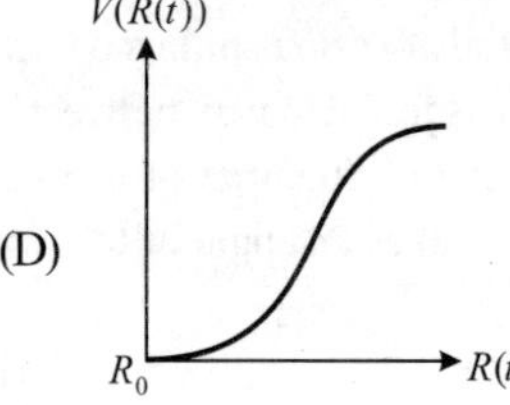

12. A solid conducting sphere, having a charge Q, is surrounded by an uncharged conducting hollow spherical shell. Let the potential difference between the surface of the solid sphere and that of the outer surface of the hollow shell be V. If the shell is now given a charge of $-4Q$, the new potential difference between the same two surfaces is : **[8 April 2019 (Morning)]**

(A) V (B) 2V
(C) -2V (D) 4V

13. An electric dipole is formed by two equal and opposite charges q with separation d. The charges have same mass m. It is kept in a uniform electric field E. If it is slightly rotated from its equilibrium orientation, then its angular frequency ω is : **[8 April 2019 (Evening)]**

(A) $\sqrt{\frac{qE}{2md}}$ (B) $2\sqrt{\frac{qE}{md}}$

(C) $\sqrt{\frac{2qE}{md}}$ (D) $\sqrt{\frac{qE}{md}}$

14. A positive point charge is released from rest at a distance r_0 from a positive line charge with uniform density. The speed (v) of the point charge, as a function of instantaneous distance r from line charge, is proportional to : **[8 April 2019 (Evening)]**

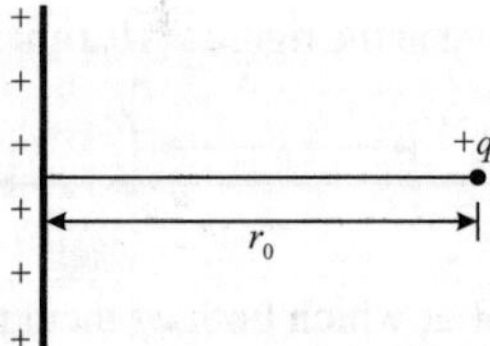

(A) $v \propto e^{+r/r_0}$ (B) $v \propto \ln\left(\frac{r}{r_0}\right)$

(C) $v \propto \left(\frac{r}{r_0}\right)$ (D) $v \propto \sqrt{\ln\left(\frac{r}{r_0}\right)}$

15. The electric field in a region is given by $\vec{E} = (Ax + B)\hat{i}$, where E is in NC^{-1} and x is in metres. The values of constants are $A = 20$ SI unit and $B = 10$ SI unit. If the potential at $x = 1$ is V_1 and that at $x = -5$ is V_2, then $V_1 - V_2$ is :

[8 April 2019 (Evening)]

(A) –48 V (B) –520 V

(C) 180 V (D) 320 V

16. A system of three charges are placed as shown in the figure.

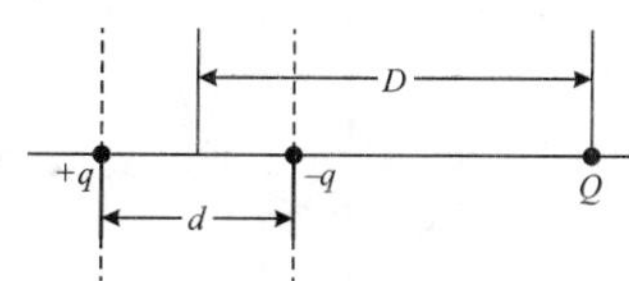

If $D >> d$, the potential energy of the system is best given by :

[9 April 2019 (Morning)]

(A) $\frac{1}{4\pi\varepsilon_0}\left[-\frac{q^2}{d} - \frac{qQd}{2D^2}\right]$ (B) $\frac{1}{4\pi\varepsilon_0}\left[+\frac{q^2}{d} + \frac{qQd}{D^2}\right]$

(C) $\frac{1}{4\pi\varepsilon_0}\left[-\frac{q^2}{d} + \frac{2qQd}{D^2}\right]$ (D) $\frac{1}{4\pi\varepsilon_0}\left[-\frac{q^2}{d} - \frac{qQd}{D^2}\right]$

17. A uniformly charged ring of radius $3a$ and total charge q is placed in xy-plane centred at origin. A point charge q is moving towards the ring along the z-axis and has speed u at $z = 4a$. The minimum value of u such that it crosses the origin is :

[10 April 2019 (Morning)]

(A) $\sqrt{\frac{2}{m}}\left(\frac{1}{15}\frac{q^2}{4\pi\varepsilon_0 a}\right)^{1/2}$ (B) $\sqrt{\frac{2}{m}}\left(\frac{2}{15}\frac{q^2}{4\pi\varepsilon_0 a}\right)^{1/2}$

(C) $\sqrt{\frac{2}{m}}\left(\frac{4}{15}\frac{q^2}{4\pi\varepsilon_0 a}\right)^{1/2}$ (D) $\sqrt{\frac{2}{m}}\left(\frac{1}{5}\frac{q^2}{4\pi\varepsilon_0 a}\right)^{1/2}$

18. In free space, a particle A of charge 1 μC is held fixed at a point P. Another particle B of the same charge and mass 4 μg is kept at a distance of 1 mm from P. If B is released, then its velocity at a distance of 9 mm from P is :

$\left[\text{Take } \frac{1}{4\pi\varepsilon_0} = 9 \times 10^9 \text{ Nm}^2\text{C}^{-2}\right]$ **[10 April 2019 (Evening)]**

(A) 2.0×10^3 m/s (B) 3.0×10^4 m/s

(C) 1.5×10^2 m/s (D) 1.0 m/s

19. Shown in the figure is a shell made of a conductor. It has inner radius a and outer radius b, and carries charge Q. At its centre is a dipole $\vec{p}$ as shown. In this case :

[12 April 2019 (Morning)]

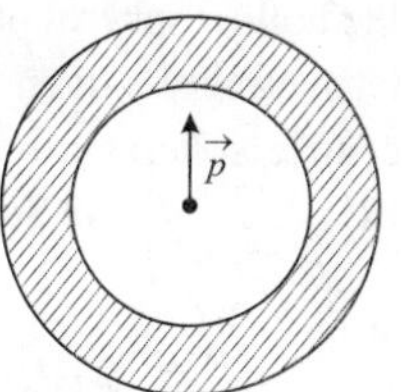

(A) Electric field outside the shell is the same as that of a point charge at the centre of the shell

(B) Surface charge density on the inner surface of the shell is zero everywhere

(C) Surface charge density on the inner surface is uniform and equal to $\frac{(Q/2)}{4\pi a^2}$

(D) Surface charge density on the outer surface depends on $|\vec{p}|$

20. A point dipole $\vec{p} = -p_0\hat{x}$ is kept at the origin. The potential and electric field due to this dipole on the y-axis at a distance d are, respectively : (Take $V = 0$ at infinity) :

[12 April 2019 (Morning)]

(A) $\frac{|\vec{p}|}{4\pi\varepsilon_0 d^2}, \frac{-\vec{p}}{4\pi\varepsilon_0 d^3}$ (B) $0, \frac{\vec{p}}{4\pi\varepsilon_0 d^3}$

(C) $\frac{|\vec{p}|}{4\pi\varepsilon_0 d^2}, \frac{\vec{p}}{4\pi\varepsilon_0 d^3}$ (D) $0, \frac{-\vec{p}}{4\pi\varepsilon_0 d^3}$

21. Consider two charged metallic spheres S_1 and S_2 of radii R_1 and R_2, respectively. The electric fields E_1 (on S_1) and E_2 (on S_2) on their surfaces are such that $E_1/E_2 = R_1/R_2$. Then the ratio V_1 (on S_1)/V_2 (on S_2) of the electrostatic potentials on each sphere is : **[8 Jan. 2020 (Evening)]**

(A) (R_2/R_1) (B) $(R_1/R_2)^3$

(C) R_1/R_2 (D) $(R_1/R_2)^2$

22. An electric dipole of moment $\vec{p} = (-\hat{i} - 3\hat{j} + 2\hat{k}) \times 10^{-29}$ C.m is at the origin (0, 0, 0). The electric field due to this dipole at $\vec{r} = +\hat{i} + 3\hat{j} + 5\hat{k}$ (note that $\vec{r} \cdot \vec{p} = 0$) is parallel to :

[9 Jan. 2020 (Morning)]

(A) $(-\hat{i} + 3\hat{j} - 2\hat{k})$ (B) $(+\hat{i} - 3\hat{j} - 2\hat{k})$

(C) $(+\hat{i} + 3\hat{j} - 2\hat{k})$ (D) $(-\hat{i} - 3\hat{j} + 2\hat{k})$

23. Two isolated conducting spheres S_1 and S_2 of radius $\frac{2}{3}R$ and $\frac{1}{3}R$ have 12 μC and – 3 μC charges, respectively, and are at a large distance from each other. They are now connected by a conducting wire. A long time after this is done the charges on S_1 and S_2 are respectively : **[3 Sept. 2020 (Morning)]**

(A) 6 μC and 3 μC (B) + 4.5 μC and – 4.5 μC

(C) 3 μC and 6 μC (D) 4.5 μC on both

24. Concentric metallic hollow spheres of radii R and $4R$ hold charges Q_1 and Q_2 respectively. Given that surface charge densities of the concentric spheres are equal, the potential difference $V(R) - V(4R)$ is : **[3 Sept. 2020 (Evening)]**

(A) $\frac{3Q_1}{16\pi\varepsilon_0 R}$ (B) $\frac{Q_2}{4\pi\varepsilon_0 R}$

(C) $\frac{3Q_1}{4\pi\varepsilon_0 R}$ (D) $\frac{3Q_2}{4\pi\varepsilon_0 R}$

25. Hydrogen ion and singly ionized helium atom are accelerated, from rest, through the same potential difference. The ratio of final speeds of hydrogen and helium ions is close to : **[3 Sept. 2020 (Evening)]**

(A) 5:7 (B) 1:2

(C) 10:7 (D) 2:1

26. A two point charges $4q$ and $-q$ are fixed on the x-axis at $x = -\frac{d}{2}$ and $x = \frac{d}{2}$, respectively. If a third point charge 'q' is taken from the origin to $x = d$ along the semicircle as shown in the figure, the energy of the charge will : **[4 Sept. 2020 (Morning)]**

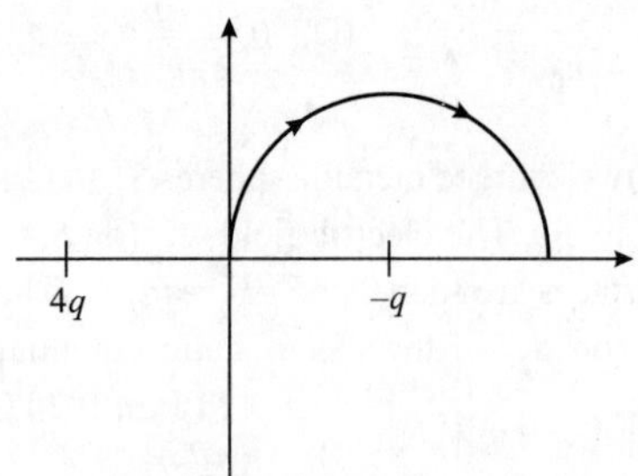

(A) Increase by $\frac{2q^2}{3\pi\varepsilon_0 d}$ (B) Increase by $\frac{3q^2}{4\pi\varepsilon_0 d}$

(C) Decrease by $\frac{4q^2}{3\pi\varepsilon_0 d}$ (D) Decrease by $\frac{q^2}{4\pi\varepsilon_0 d}$

27. A particle of charge q and mass m is subjected to an electric field $E = E_0(1 - ax^2)$ in the x-direction, where a and E_0 are constants. Initially the particle was at rest at $x = 0$. Other than the initial position the kinetic energy of the particle becomes zero when the distance of the particle from the origin is : **[4 Sept. 2020 (Evening)]**

(A) $\sqrt{\frac{2}{a}}$ (B) $\sqrt{\frac{1}{a}}$

(C) a (D) $\sqrt{\frac{3}{a}}$

28. A solid sphere of radius R carries a charge $(Q+q)$ distributed uniformly over its volume. A very small point like piece of it of mass m gets detached from the bottom of the sphere and falls down vertically under gravity. This piece carries charge q. If it acquires a speed v when it has fallen through a vertical height y (see figure), then : (assume the remaining portion to be spherical) : **[5 Sept. 2020 (Morning)]**

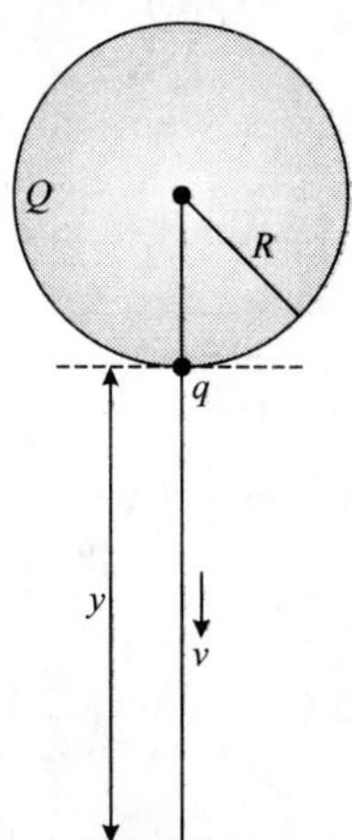

(A) $v^2 = 2y\left[\frac{qQ}{4\pi \in_0 R(R+y)m} + g\right]$

(B) $v^2 = y\left[\frac{qQ}{4\pi \in_0 R^2 ym} + g\right]$

(C) $v^2 = 2y\left[\frac{qQR}{4\pi \in_0 (R+y)^3 m} + g\right]$

(D) $v^2 = y\left[\frac{qQ}{4\pi \in_0 R(R+y)m} + g\right]$

29. Ten charges are placed on the circumference of a circle of radius R with constant angular separation between successive charges. Alternate charges 1, 3, 5, 7, 9 have charge $(+q)$ each, while 2, 4, 6, 8, 10 have charge $(-q)$ each. The potential V and the electric field E at the centre of the circle are respectively : (Take $V = 0$ at infinity) **[5 Sept. 2020 (Evening)]**

(A) $V = \frac{10q}{4\pi \in_0 R}; E = \frac{10q}{4\pi \in_0 R^2}$

(B) $V = 0, E = \frac{10q}{4\pi \in_0 R^2}$

(C) $V = 0, E = 0$

(D) $V = \frac{10q}{4\pi \in_0 R}; E = 0$

30. Two identical electric point dipoles have dipole moments $\vec{p}_1 = p\hat{i}$ and $\vec{p}_2 = -p\hat{i}$ and are held on the x-axis at distance 'a' from each other. When released, they move along the x-axis with the direction of their dipole moments remaining unchanged. If the mass of each dipole is 'm', their speed when they arc infinitely far apart is : **[6 Sept. 2020 (Evening)]**

(A) $\frac{p}{a}\sqrt{\frac{1}{\pi\varepsilon_0 ma}}$ (B) $\frac{p}{a}\sqrt{\frac{3}{2\pi\varepsilon_0 ma}}$

(C) $\frac{p}{a}\sqrt{\frac{1}{2\pi\varepsilon_0 ma}}$ (D) $\frac{p}{a}\sqrt{\frac{2}{\pi\varepsilon_0 ma}}$

31. 512 identical drops of mercury are charged to a potential of 2 V each. The drops are joined to form a single drop. The potential of this drop is ______ V. **[25 Feb. 2021 (Morning)]**

32. 27 similar drops of mercury are maintained at 10 V each. All these spherical drops combine into a single big drop. The potential energy of the bigger drops is ______ times that of a smaller drop. **[26 Feb. 2021 (Evening)]**

33. Two ideal electric dipoles A and B, having their dipole moment p_1 and p_2 respectively are placed on a plane with their centres at O as shown in the figure. At point C on the axis of dipole A, the resultant electric field is making an angle of 37° with the axis. The ratio of the dipole moment of A and B, $\frac{P_1}{P_2}$ is :

(Take $\sin 37° = \frac{3}{5}$) **[25 July 2021 (Evening)]**

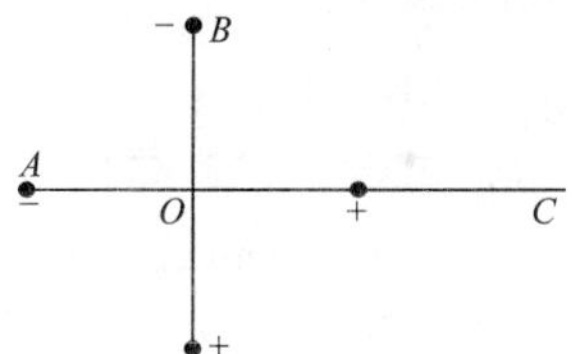

(A) $\frac{3}{2}$ (B) $\frac{2}{3}$

(C) $\frac{3}{8}$ (D) $\frac{4}{3}$

34. The two thin coaxial rings, each of radius a and having charges $+Q$ and $-Q$ respectively are separated by a distance of s. The potential difference between the centres of the two rings is : **[26 Aug. 2021 (Evening)]**

(A) $\frac{Q}{4\pi\varepsilon_0}\left[\frac{1}{a}-\frac{1}{\sqrt{s^2+a^2}}\right]$ (B) $\frac{Q}{4\pi\varepsilon_0}\left[\frac{1}{a}-\frac{1}{\sqrt{s^2+a^2}}\right]$

(C) $\frac{Q}{4\pi\varepsilon_0}\left[\frac{1}{a}+\frac{1}{\sqrt{s^2+a^2}}\right]$ (D) $\frac{Q}{2\pi\varepsilon_0}\left[\frac{1}{a}+\frac{1}{\sqrt{s^2+a^2}}\right]$

* * * * *

CHAPTER

16 ELECTRIC POTENTIAL AND CAPACITANCE

16.2 CAPACITANCE

1. A capacitance of 2 μF is required in an electrical circuit across a potential difference of 1.0 kV. A large number of 1 μF capacitors are available which can withstand a potential difference of not more than 300 V. The minimum number of capacitors required to achieve this is : **[2017]**

(A) 24 (B) 32
(C) 2 (D) 16

2. The energy stored in the electric field produced by a metal sphere is 4.5 J. If the sphere contains 4 μC charge, its radius will be : [Take : $\frac{1}{4\pi\varepsilon_0} = 9 \times 10^9$ N – m²/C²] **[2017 (Online)]**

(A) 20 mm (B) 32 mm
(C) 28 mm (D) 16 mm

3. A combination of parallel plate capacitors is maintained at a certain potential difference. When a 3 mm thick slab is introduced between all the plates, in order to maintain the same potential difference, the distance between the plates in increased by 2.4 mm. Find the dielectric constant of the slab : **[2017 (Online)]**

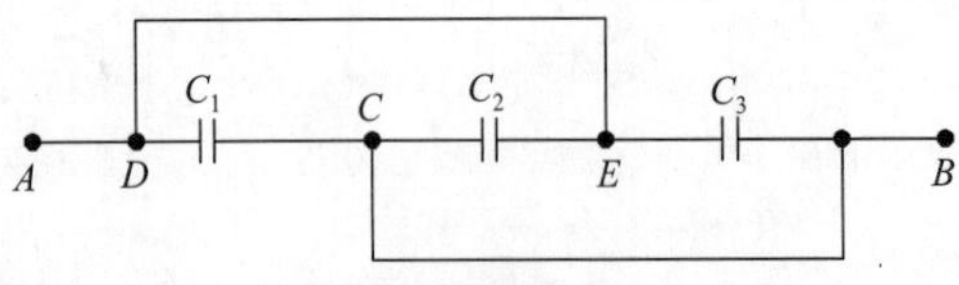

(A) 3 (B) 4
(C) 5 (D) 6

4. A parallel plate capacitor of capacitance 90 pF is connected to a battery of emf 20 V. If a dielectric material of dielectric constant $k = 5/3$ is inserted between the plates, the magnitude of the induced charge will be : **[2018]**

(A) 1.2 nC (B) 0.3 nC
(C) 2.4 nC (D) 0.9 nC

5. A parallel plate capacitor with area 200 cm² and separation between the plates 1.5 cm, is connected across a battery of emf V. If the force of attraction between the plates is 25×10^{-6} N, the value of V is approximately : **[2018 (Online)]**

$$\left(\epsilon_0 = 8.85 \times 10^{-12} \frac{C^2}{N.m^2}\right)$$

(A) 150 V (B) 100 V
(C) 250 V (D) 300 V

6. The equivalent capacitance between A and B in the circuit given below is : **[2018 (Online)]**

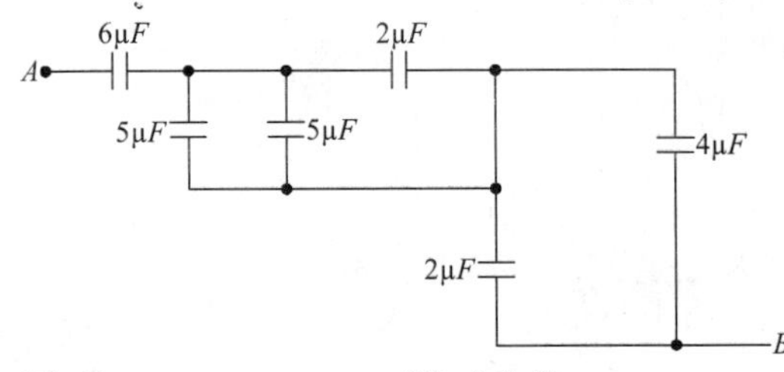

(A) 4.9 μF (B) 3.6 μF
(C) 5.4 μF (D) 2.4 μF

7. A capacitor $C_1 = 1$ μF is charged up to a voltage $V = 60$ V by connecting it to battery B through switch (1), Now C_1 is disconnected from battery and connected to a circuit consisting of two uncharged capacitors $C_2 = 3.0$ μF and $C_3 = 6.0$ μF through a switch (2) as shown in the figure. The sum of final charges on C_2 and C_3 is : **[2018 (Online)]**

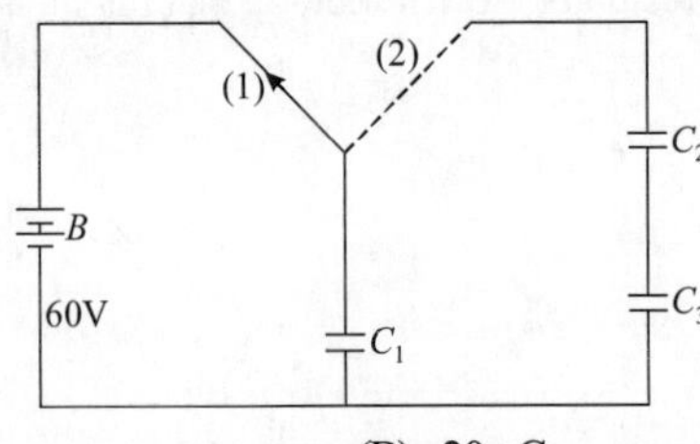

(A) 36 μC (B) 20 μC
(C) 54 μC (D) 40 μC

8. A parallel plate capacitor is made of two square plates of side 'a', separated by a distance d ($d << a$). The lower triangular portion is filled with a dielectric of dielectric constant K, as shown in the figure. Capacitance of this capacitor is : **[9 Jan. 2019 (Morning)]**

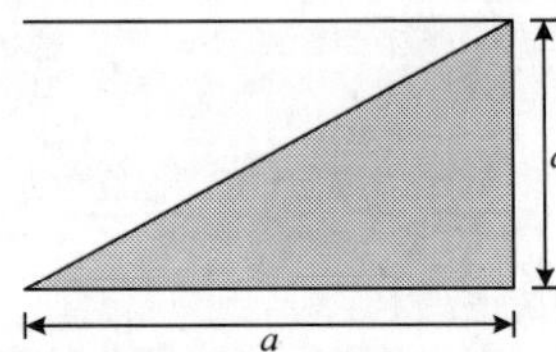

(A) $\frac{1}{2}\frac{k \in_0 a^2}{d}$ (B) $\frac{k \in_0 a^2}{d} \ln K$

(C) $\frac{k \in_0 a^2}{d(K-1)} \ln K$ (D) $\frac{k \in_0 a^2}{2d(K+1)}$

9. A parallel plate capacitor with square plates is filled with four dielectrics of dielectric constants K_1, K_2, K_3, K_4 arranged as shown in the figure. The effective dielectric constant K will be : **[9 Jan. 2019 (Evening)]**

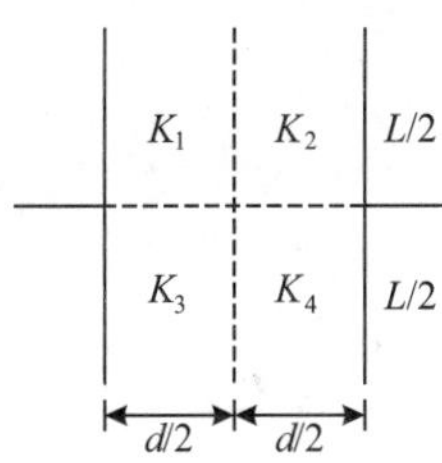

(A) $K = \dfrac{(K_1 + K_2)(K_3 + K_4)}{2(K_1 + K_2 + K_3 + K_4)}$

(B) $K = \dfrac{(K_1 + K_2)(K_3 + K_4)}{(K_1 + K_2 + K_3 + K_4)}$

(C) $K = \dfrac{(K_1 + K_4)(K_2 + K_3)}{2(K_1 + K_2 + K_3 + K_4)}$

(D) $K = \dfrac{(K_1 + K_3)(K_2 + K_4)}{K_1 + K_2 + K_3 + K_4}$

10. A parallel plate capacitor is of area 6 cm^2 and a separation 3 mm. The gap is filled with three dielectric materials of equal thickness (see figure) with dielectric constants $K_1 = 10$, $K_2 = 12$ and $K_3 = 14$. The dielectric constant of a material which when fully inserted in above capacitor, gives same capacitance would be : **[10 Jan. 2019 (Morning)]**

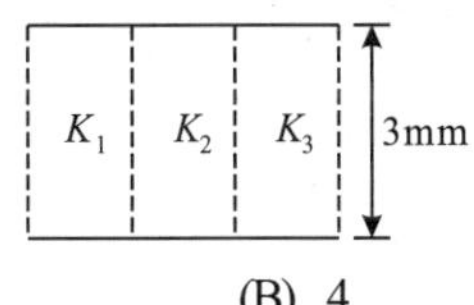

(A) 12 (B) 4
(C) 36 (D) 14

11. A parallel plate capacitor having capacitance 12 pF is charged by a battery to a potential difference of 10 V between its plates. The charging battery is now disconnected and a porcelain slab of dielectric constant 6.5 is slipped between the plates, the work done by the capacitor on the slab is : **[10 Jan. 2019 (Evening)]**

(A) 692 pJ (B) 60 pJ
(C) 508 pJ (D) 560 pJ

12. In the figure shown below, the charge on the left plate of the 10 μF capacitor is –30 μC. The charge on the right plate of the 6 μF capacitor is : **[11 Jan. 2019 (Morning)]**

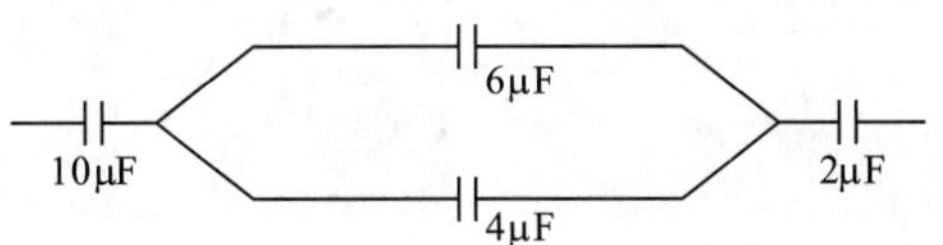

(A) –18 μC (B) –12 μC
(C) +12 μC (D) +18 μC

13. Seven capacitors, each of capacitance 2 μF, are to be connected in a configuration to obtain an effective capacitance of $\left(\dfrac{6}{13}\right)$ μF. Which of the combinations, shown in figures below, will achieve the desired value ? **[11 Jan. 2019 (Evening)]**

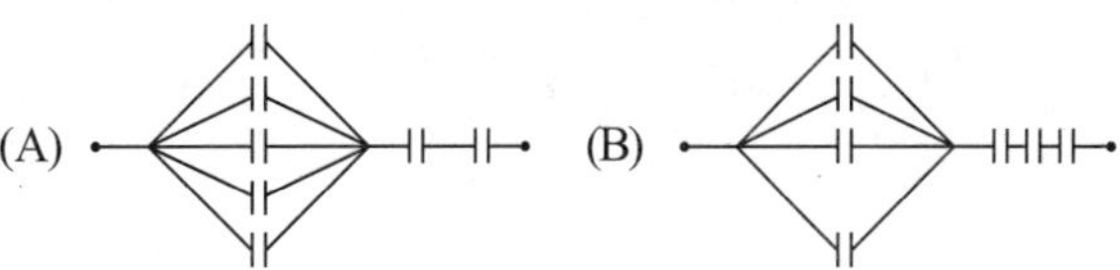

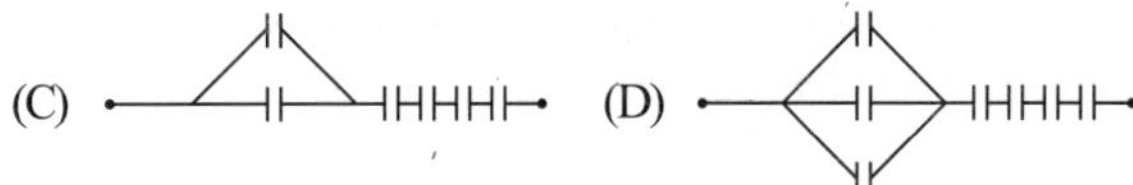

14. In the figure shown, after the switch 'S' is turned from position 'A' to position 'B', the energy dissipated in the circuit in terms of capacitance 'C' and total charge 'Q' is : **[12 Jan. 2019 (Morning)]**

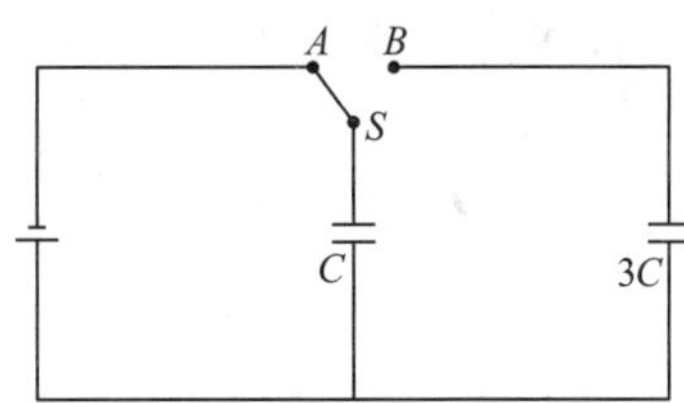

(A) $\dfrac{3}{8}\dfrac{Q^2}{C}$ (B) $\dfrac{3}{4}\dfrac{Q^2}{C}$

(C) $\dfrac{1}{8}\dfrac{Q^2}{C}$ (D) $\dfrac{5}{8}\dfrac{Q^2}{C}$

15. In the circuit shown, find C if the effective capacitance of the whole circuit is to be 0.5 μF. All values in the circuit are in μF. **[12 Jan. 2019 (Evening)]**

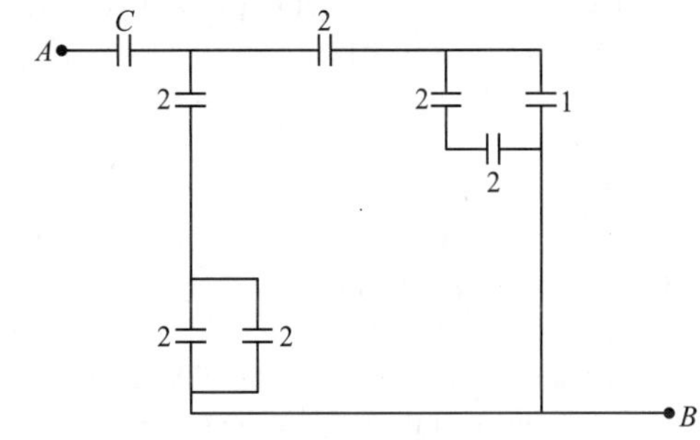

(A) $\dfrac{7}{10}\mu F$ (B) $\dfrac{7}{11}\mu F$

(C) $\dfrac{6}{5}\mu F$ (D) $4\ \mu F$

16. The charge on a capacitor plate in a circuit, as a function of time, is shown in the figure. What is the value of current at $t = 4$ s? **[12 Jan. 2019 (Evening)]**

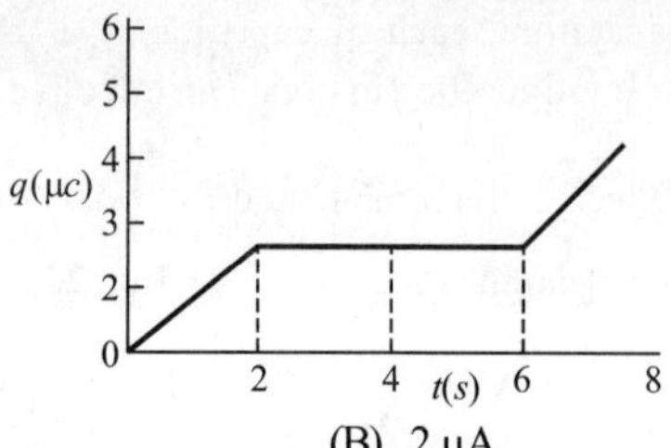

(A) 3 μA (B) 2 μA
(C) Zero (D) 1.5 μA

17. A parallel plate capacitor with plates of area 1 m^2 each, and at a separation of 0.1 m. If the electric field between the plates is 100 N/C, the magnitude of charge each plate is.

[Take $\varepsilon_0 = 8.85 \times 10^{-12} \dfrac{C^2}{N-m^2}$]: **[12 Jan. 2019 (Evening)]**

(A) 7.85×10^{-10} C (B) 6.85×10^{-10} C
(C) 9.85×10^{-10} C (D) 8.85×10^{-10} C

18. Voltage rating of a parallel plate capacitor is 500 V. Its dielectric can withstand a maximum electric field of 10^6 V/m. The plate area is 10^{-4} m^2. What is the dielectric constant of the capacitance is 15 pF ? [Given $\epsilon_0 = 8.86 \times 10^{-12}$ C^2/Nm^2] **[8 April 2019 (Morning)]**

(A) 3.8 (B) 4.5
(C) 6.2 (D) 8.5

19. A parallel plate capacitor has 1 μF capacitance. One of its two plates is given +2 μC charge and the other plate, +4 μC charge. The potential difference developed across the capacitor is : **[8 April 2019 (Evening)]**

(A) 5V (B) 2V
(C) 3V (D) 1V

20. A capacitor with capacitance 5 μF is charged to 5 μC. If the plates are pulled apart to reduce the capacitance to 2 μF, how much work is done ? **[9 April 2019 (Morning)]**

(A) 3.75×10^{-6} J (B) 2.55×10^{-6} J
(C) 2.16×10^{-6} J (D) 6.25×10^{-6} J

21. The parallel combination of two air filled parallel plate capacitors of capacitance C and nC is connected to a battery of voltage, V. When the capacitors are fully charged, the battery is removed and after that a dielectric material of dielectric constant K is placed between the two plates of the first capacitor. The new potential difference of the combined system is : **[9 April 2019 (Evening)]**

(A) $\dfrac{V}{K+n}$ (B) V

(C) $\dfrac{(n+1)V}{(K+n)}$ (D) $\dfrac{nV}{K+n}$

22. Figure shows charge (q) versus voltage (V) graph for series and parallel combination of two given capacitors. The capacitances are : **[10 April 2019 (Morning)]**

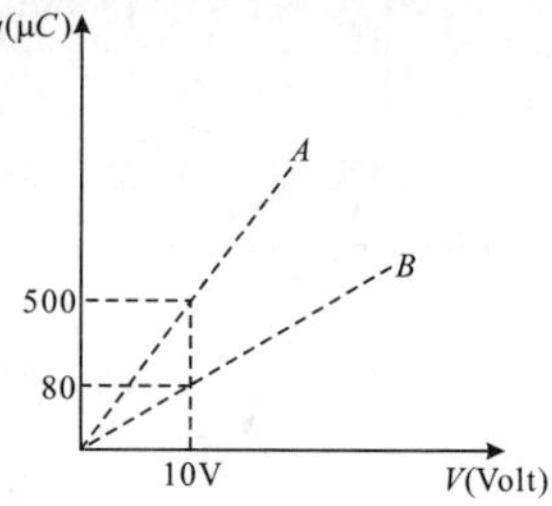

(A) 50 μF and 30 μF (B) 20 μF and 30 μF
(C) 60 μF and 40 μF (D) 40 μF and 10 μF

23. A simple pendulum of length L is placed between the plates of a parallel plate capacitor having electric field E, as shown in figure. Its bob has mass m and charge q. The time period of the pendulum is given by : **[10 April 2019 (Evening)]**

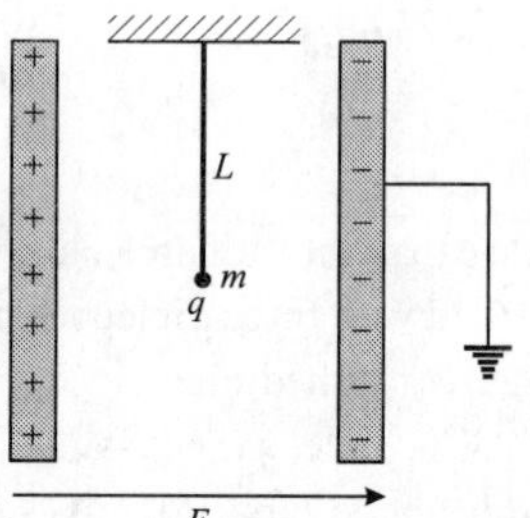

(A) $2\pi\sqrt{\dfrac{L}{\sqrt{g^2 + \left(\dfrac{qE}{m}\right)^2}}}$ (B) $2\pi\sqrt{\dfrac{L}{\left(g + \dfrac{qE}{m}\right)}}$

(C) $2\pi\sqrt{\dfrac{L}{\left(g - \dfrac{qE}{m}\right)}}$ (D) $2\pi\sqrt{\dfrac{L}{\sqrt{g^2 - \dfrac{q^2E^2}{m^2}}}}$

24. Two identical parallel plate capacitors, of capacitance C each, have plates of area A, separated by a distance d. The space between the plates of the two capacitors, is filled with three dielectrics, of equal thickness and dielectric constants K_1, K_2 and K_3. The first capacitor is filled as shown in figure-I, and the second one is filled as shown in figure-II. If these two modified capacitors are charged by the same potential V, the ratio of the energy stored in the two, would be (E_1 refers to capacitor (I) and E_2 to capacitor (II)) : **[12 April 2019 (Morning)]**

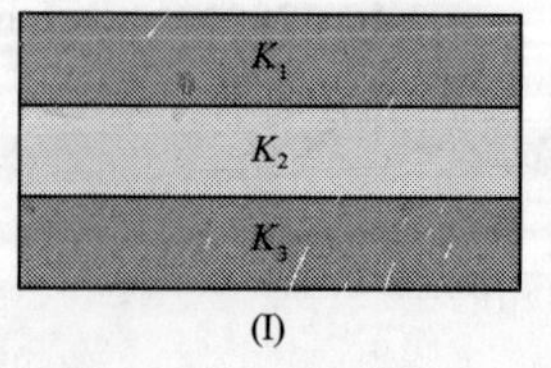

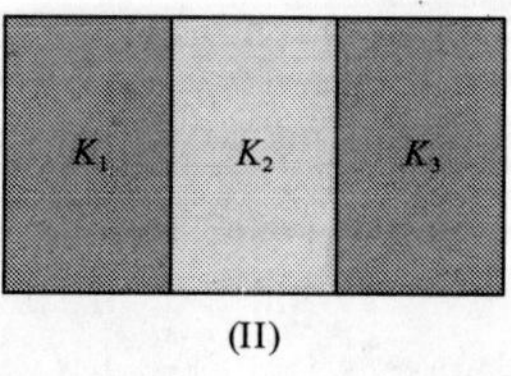

(A) $\dfrac{E_1}{E_2} = \dfrac{9K_1K_2K_3}{(K_1+K_2+K_3)(K_2K_3+K_3K_1+K_1K_2)}$

(B) $\dfrac{E_1}{E_2} = \dfrac{K_1K_2K_3}{(K_1+K_2+K_3)(K_2K_3+K_3K_1+K_1K_2)}$

(C) $\dfrac{E_1}{E_2} = \dfrac{(K_1+K_2+K_3)(K_2K_3+K_3K_1+K_1K_2)}{K_1K_2K_3}$

(D) $\dfrac{E_1}{E_2} = \dfrac{(K_1+K_2+K_3)(K_2K_3+K_3K_1+K_1K_2)}{9K_1K_2K_3}$

25. In the given circuit, the charge on 4 μF capacitor will be : **[12 April 2019 (Evening)]**

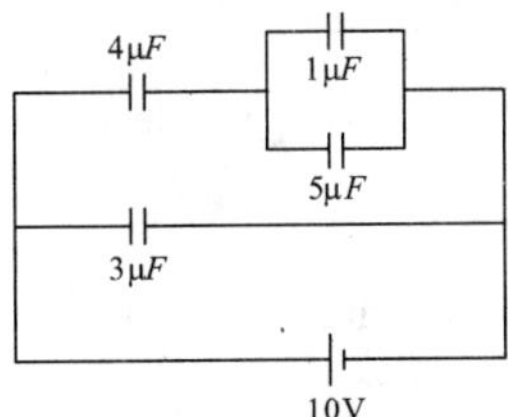

(A) 5.4 μC (B) 24 μC
(C) 13.4 μC (D) 9.6 μC

26. A parallel plate capacitor has plates of area A separated by distance 'd' between them. It is filled with a dielectric which has a dielectric constant that varies as $k(x) = K(1 + \alpha x)$ where 'x' is the distance measured from one of the plates. If $(\alpha d) << 1$, the total capacitance of the system is best given by the expression : **[7 Jan. 2020 (Morning)]**

(A) $\dfrac{AK\varepsilon_0}{d}\left(1+\dfrac{\alpha d}{2}\right)$ (B) $\dfrac{A\varepsilon_0 K}{d}\left(1+\left(\dfrac{\alpha d}{2}\right)^2\right)$

(C) $\dfrac{A\varepsilon_0 K}{d}\left(1+\dfrac{\alpha^2 d^2}{2}\right)$ (D) $\dfrac{AK\varepsilon_0}{d}(1+\alpha d)$

27. A 60 pF capacitor is fully charged by a 20 V supply. It is then disconnected from the supply and is connected to another uncharged 60 pF capacitor in parallel. The electrostatic energy that is lost in this process by the time the charge is redistributed between them is (in *nJ*) ______ . **[7 Jan. 2020 (Evening)]**

28. Effective capacitance of parallel combination of two capacitors C_1 and C_2 is 10 μF. When these capacitors are individually connected to a voltage source of 1 V, the energy stored in the capacitor C_2 is 4 times that of C_1. If these capacitors are connected in series, their effective capacitance will be : **[8 Jan. 2020 (Morning)]**

(A) 3.2 μF (B) 8.4 μF
(C) 1.6 μF (D) 4.2 μF

29. A capacitor is made of two square plates each of side 'a' making a very small angle α between them, as shown in figure. The capacitance will be close to : **[8 Jan. 2020 (Evening)]**

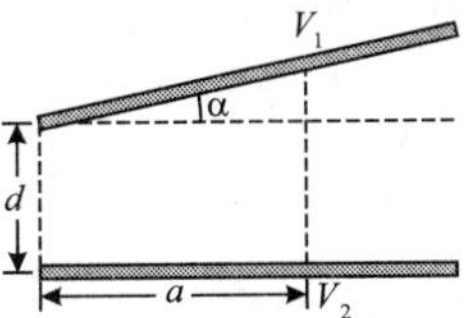

(A) $\dfrac{\varepsilon_0 a^2}{d}\left(1-\dfrac{3\alpha a}{2d}\right)$ (B) $\dfrac{\varepsilon_0 a^2}{d}\left(1-\dfrac{\alpha a}{4d}\right)$

(C) $\dfrac{\varepsilon_0 a^2}{d}\left(1+\dfrac{\alpha a}{d}\right)$ (D) $\dfrac{\varepsilon_0 a^2}{d}\left(1-\dfrac{\alpha a}{2d}\right)$

30. A 5 μF capacitor is charged fully by a 220 V supply. It is then disconnected from the supply and is connected in series to another uncharged 2.5 μF capacitor. If the energy change during the charge redistribution is $\dfrac{X}{100}$ J then value of X to the nearest integer is : **[2 Sept. 2020 (Morning)]**

31. A 10 μF capacitor is fully charged to a potential difference of 50 V. After removing the source voltage it is connected to an uncharged capacitor in parallel. Now the potential difference across them becomes 20 V. The capacitance of the second capacitor is : **[2 Sept. 2020 (Evening)]**

(A) 10 μF (B) 15 μF
(C) 20 μF (D) 30 μF

32. In the circuit shown in the figure, the total charge is 750 μC and the voltage across capacitor C_2 is 20 V. Then the charge on capacitor C_2 is : **[3 Sept. 2020 (Morning)]**

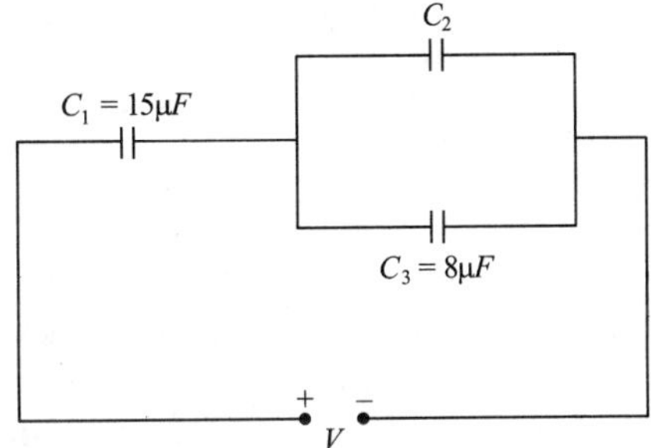

(A) 590 μC (B) 450 μC
(C) 650 μC (D) 160 μC

33. A capacitor C is fully charged with voltage V_0. After disconnecting the voltage source, it is connected in parallel with another uncharged capacitor of capacitance $\frac{C}{2}$. The energy loss in the process after the charge is distributed between the two capacitors is : **[4 Sept. 2020 (Evening)]**

(A) $\frac{1}{6}CV_0^2$ (B) $\frac{1}{2}CV_0^2$

(C) $\frac{1}{3}CV_0^2$ (D) $\frac{1}{4}CV_0^2$

34. Two capacitors of capacitances C and $2C$ are charged to potential differences V and $2V$, respectively. These are then connected in parallel in such a manner that the positive terminal of one is connected to the negative terminal of the other. The final energy of this configuration is : **[5 Sept. 2020 (Morning)]**

(A) $\frac{9}{2}CV^2$ (B) $\frac{25}{6}CV^2$

(C) Zero (D) $\frac{3}{2}CV^2$

35. A parallel plate capacitor has plate of length 'l', width 'w' and separation of plates is 'd'. It is connected to a battery of *emf* V. A dielectric slab of the same thickness 'd' and of dielectric constant $k = 4$ is being inserted between the plates of the capacitor. At what length of the slab inside plates, will be energy stored in the capacitor be two times the initial energy stored? **[5 Sept. 2020 (Evening)]**

(A) $l/4$ (B) $l/2$

(C) $l/3$ (D) $2l/3$

36. In the circuit shown, charge on the 5 μF capacitor is : **[5 Sept. 2020 (Evening)]**

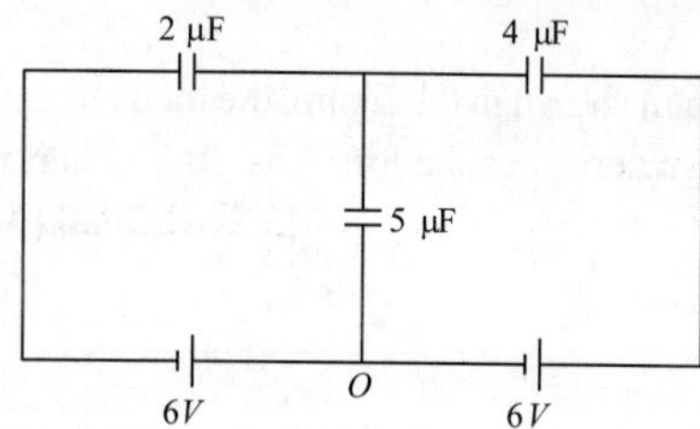

(A) 5.45 μC (B) 16.36 μC

(C) 10.90 μC (D) 18.00 μC

37. Two equal capacitors are first connected in series and then in parallel. The ratio of the equivalent capacities in the two cases will be : **[24 Feb. 2021 (Morning)]**

(A) 4 : 1 (B) 1 : 4

(C) 1 : 2 (D) 2 : 1

38. Consider the combination of 2 capacitors C_1 and C_2, with $C_2 > C_1$, when connected in parallel, the equivalent capacitance is $\frac{15}{4}$ times the equivalent capacitance of the same connected in series. Calculate the ratio of capacitors, $\frac{C_2}{C_1}$:

[26 Feb. 2021 (Morning)]

(A) $\frac{15}{4}$ (B) $\frac{29}{15}$

(C) $\frac{15}{11}$ (D) $\frac{111}{80}$

39. For changing the capacitance of a given parallel plate capacitor, a dielectric material of dielectric constant K is used, which has the same area as the plates of the capacitor. The thickness of the dielectric slab is $\frac{3}{4}d$, where 'd' is the separation between the plates of parallel plate capacitor. The new capacitance (C') in terms of original capacitance (C_0) is given by the following relation : **[16 March 2021 (Morning)]**

(A) $C' = \frac{3+K}{4K}C_0$ (B) $C' = \frac{4+K}{3}C_0$

(C) $C' = \frac{4K}{K+3}C_0$ (D) $C' = \frac{4}{3+K}C_0$

40. In a parallel plate capacitor set up, the plate area of capacitor is 2 m^2 and the plates are separated by 1 m. If the space between the plates are filled with a dielectric material of thickness 0.5 m and area 2 m^2 (see figure) the capacitance of the set-up will be ______ ε_0. (Dielectric constant of the material = 3.2) (Round off to the Nearest Integer) **[16 March 2021 (Evening)]**

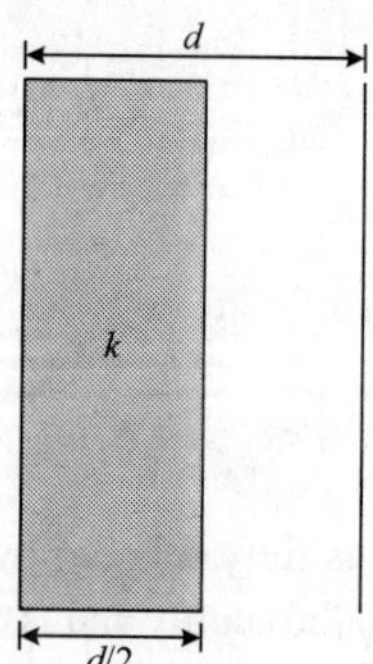

41. Four identical rectangular plates with length, $l = 2$ cm and breadth, $b = \frac{3}{2}$ cm are arranged as shown in figure. The equivalent capacitance between A and C is $\frac{x\varepsilon_0}{d}$. The value of x is _____. (Round off to the Nearest Integer) **[17 March 2021 (Morning)]**

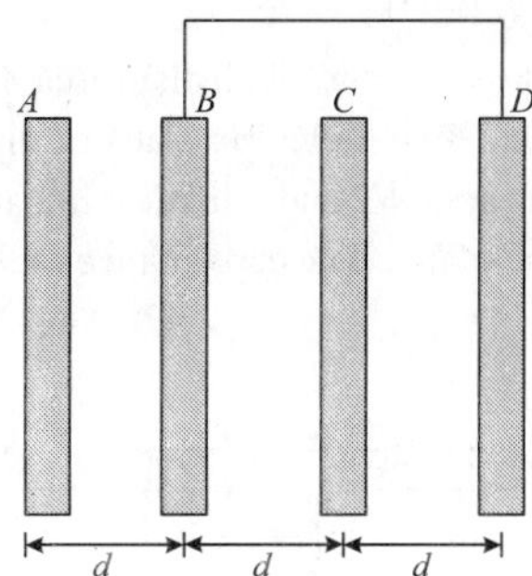

42. A parallel plate capacitor whose capacitance C is 14 pF is charged by a battery to a potential difference $V = 12$ V between its plates. The charging battery is now disconnected and a porcelain plate with $k = 7$ is inserted between the plates, then the plate would oscillate back and forth between the plates with a constant mechanical energy of ______ pJ.
(Assume no friction) **[17 March 2021 (Morning)]**

43. A 2 μF capacitor C_1 is first charged to a potential difference of 10 V using a battery. Then the battery is removed and the capacitor is connected to an uncharged capacitor C_2 of 8 μF. The charge in C_2 on equilibrium condition is______ μC. (Round off to the Nearest Integer) **[17 March 2021 (Evening)]**

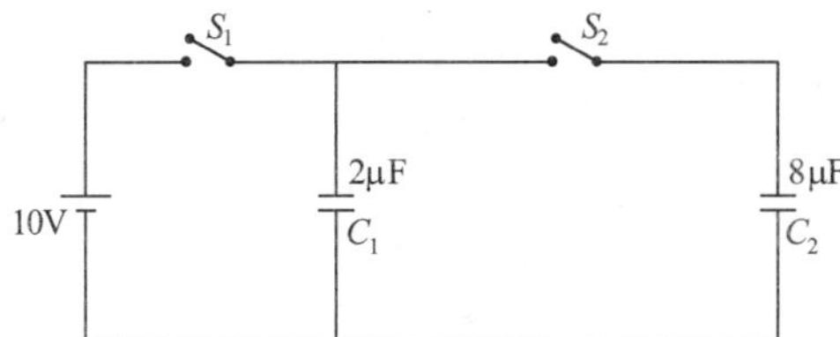

44. A parallel plate capacitor has plate area 100 m^2 and plate separation of 10 m. The space between the plates is filled up to a thickness 5 m with a material of dielectric constant of 10. The resultant capacitance of the system is 'x' pF.
The value of $\varepsilon_0 = 8.85 \times 10^{-12}$ F.m^{-1}.
The value of 'x' to the nearest integer is______.
[18 March 2021 (Morning)]

45. A parallel plate capacitor with plate area 'A' and distance of separation d is filled with a dielectric. What is the capacitance of the capacitor when permittivity of the dielectric varies as : **[25 July 2021 (Morning)]**

$$\varepsilon(x) = \varepsilon_0 + kx, \text{ for } \left(0 < x \le \frac{d}{2}\right)$$

$$\varepsilon(x) = \varepsilon_0 + k(d - x), \text{ for } \left(\frac{d}{2} \le x \le d\right)$$

Where x is the distance from one plate of capacitor.

(A) $\dfrac{kA}{2\ln\left(\dfrac{2\varepsilon_0 + kd}{2\varepsilon_0}\right)}$ (B) $\left(\varepsilon_0 + \dfrac{kd}{2}\right)^{2/kA}$

(C) $\dfrac{kA}{2}\ln\left(\dfrac{2\varepsilon_0}{2\varepsilon_0 - kd}\right)$ (D) 0

46. If q_f is the free charge on the capacitor plates and q_b is the bound charge on the dielectric slab of dielectric constant k placed between the capacitor plates, then bound charge q_b can be expressed as : **[25 July 2021 (Evening)]**

(A) $q_b = q_f\left(1 + \dfrac{1}{k}\right)$ (B) $q_b = q_f\left(1 - \dfrac{1}{\sqrt{k}}\right)$

(C) $q_b = q_f\left(1 + \dfrac{1}{\sqrt{k}}\right)$ (D) $q_b = q_f\left(1 - \dfrac{1}{k}\right)$

47. In the reported figure, a capacitor is formed by placing a compound dielectric between the plates of parallel plate capacitor. The expression for the capacity of the said capacitor will be : (Given area of plate $= A$) : **[27 July 2021 (Morning)]**

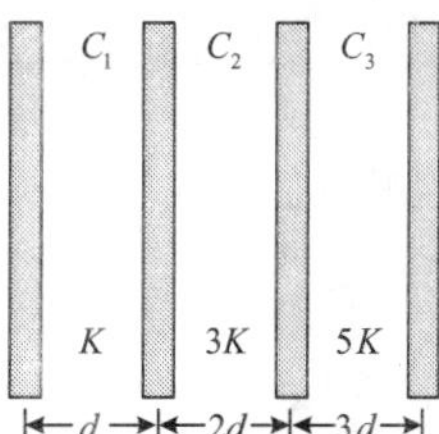

(A) $\dfrac{9}{6}\dfrac{K\varepsilon_0 A}{d}$ (B) $\dfrac{25}{6}\dfrac{K\varepsilon_0 A}{d}$

(C) $\dfrac{15}{6}\dfrac{K\varepsilon_0 A}{d}$ (D) $\dfrac{15}{34}\dfrac{K\varepsilon_0 A}{d}$

48. Two capacitors of capacities 2 C and C are joined in parallel and charged up to potential V. The battery is removed and the capacitor of capacity C is filled completely with a medium of dielectric constant K. The potential difference across the capacitors will now be : **[27 July 2021 (Morning)]**

(A) $\dfrac{V}{K}$ (B) $\dfrac{3V}{K+2}$

(C) $\dfrac{3V}{K}$ (D) $\dfrac{V}{K+2}$

49. A capacitor of capacitance $C = 1$ μF is suddenly connected to a battery of 100 volt through a resistance $R = 100\ \Omega$. The time taken for the capacitor to be charged to get 50 V is :
(Take $\ln 2 = 0.69$) : **[27 July 2021 (Morning)]**

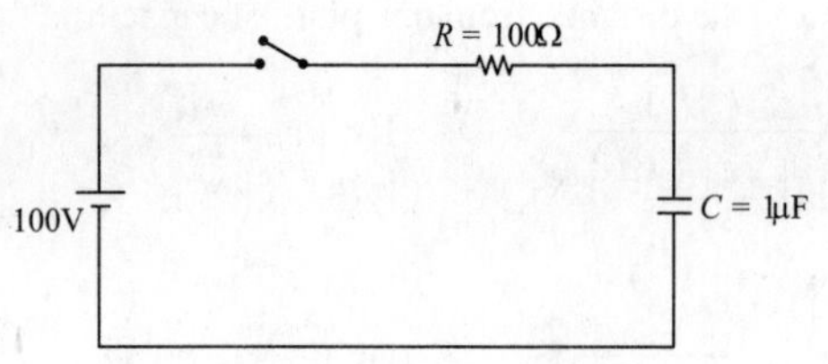

(A) 0.30×10^{-4} s (B) 0.69×10^{-4} s
(C) 1.44×10^{-4} s (D) 3.33×10^{-4} s

50. A simple pendulum of mass 'm', length 'l' and charge $+q$' suspended in the electric field produced by two conducting parallel plates as shown. The value of deflection of pendulum in equilibrium position will be : **[27 July 2021 (Evening)]**

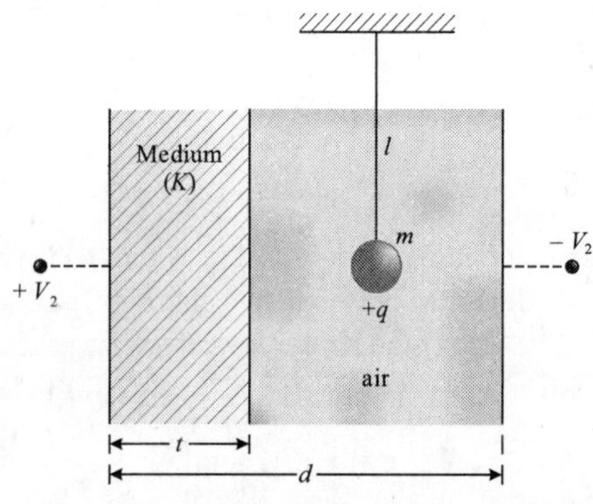

(A) $\tan^{-1}\left[\frac{q}{mg}+\frac{C_1(V_2-V_1)}{(C_1+C_2)(d-t)}\right]$

(B) $\tan^{-1}\left[\frac{q}{mg}+\frac{C_1(V_2+V_1)}{(C_1+C_2)(d-t)}\right]$

(C) $\tan^{-1}\left[\frac{q}{mg}+\frac{C_2(V_1+V_2)}{(C_1+C_2)(d-t)}\right]$

(D) $\tan^{-1}\left[\frac{q}{mg}+\frac{C_2(V_2-V_1)}{(C_1+C_2)(d-t)}\right]$

51. The material filled between the plates of a parallel plate capacitor has resistivity 200 Ωm. The value of capacitance of the capacitor is 2 pF. If a potential difference of 40 V is applied across the plates of the capacitor, then the value of leakage current following out of the capacitor is : (given the value of relative permittivity of material is 50)**[26 Aug. 2021 (Morning)]**

(A) 9.0 μA (B) 9.0 mA
(C) 0.9 mA (D) 0.9 μA

52. A parallel-plate capacitor with plate area A has separation d between the plates. Two dielectric slabs of dielectric constant K_1 and K_2 of same area $A/2$ and thickness $d/2$ are inserted in the space between the plates. The capacitance of the capacitor will be given by : **[26 Aug. 2021 (Evening)]**

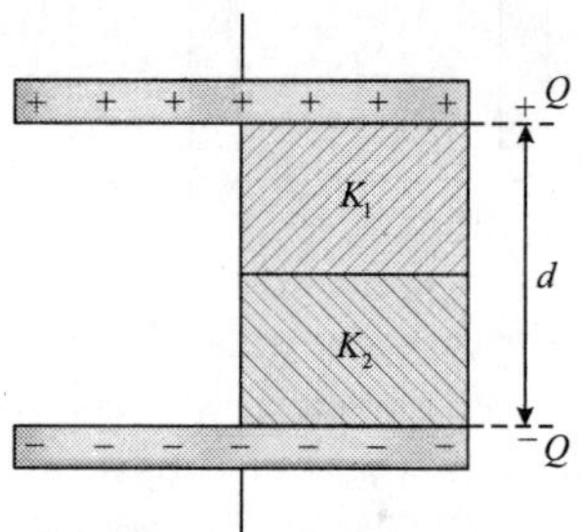

(A) $\frac{\varepsilon_0 A}{d}\left(\frac{1}{2}+\frac{2(K_1+K_2)}{K_1K_2}\right)$ (B) $\frac{\varepsilon_0 A}{d}\left(\frac{1}{2}+\frac{K_1+K_2}{K_1K_2}\right)$

(C) $\frac{\varepsilon_0 A}{d}\left(\frac{1}{2}+\frac{K_1K_2}{K_1+K_2}\right)$ (D) $\frac{\varepsilon_0 A}{d}\left(\frac{1}{2}+\frac{K_1K_2}{2(K_1+K_2)}\right)$

53. Three capacitors $C_1 = 2$ μF, $C_2 = 6$ μF and $C_3 = 12$ μF are connected as shown in figure. Find the ratio of the charges on capacitors C_1, C_2 and C_3 respectively : **[27 Aug. 2021 (Evening)]**

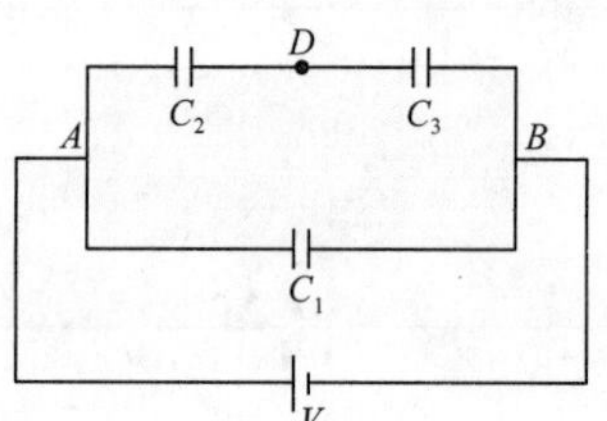

(A) 3 : 4 : 4 (B) 2 : 3 : 3
(C) 2 : 1 : 1 (D) 1 : 2 : 2

54. A parallel plate capacitor of capacitance 200 μF is connected to a battery of 200 V. A dielectric slab of dielectric constant 2 is now inserted into the space between plates of capacitor while the battery remain connected. The change in the electrostatic energy in the capacitor will be ______ ? **[31 Aug. 2021 (Evening)]**

* * * * *

CHAPTER

17 CURRENT ELECTRICITY

17.1 CURRENT ELECTRICITY

1. A uniform wire of length l and radius r has a resistance of 100 Ω. It is recast into a wire of radius $r/2$. The resistance of new wire will be : **[2017 (Online)]**

(A) 1600 Ω (B) 400 Ω

(C) 200 Ω (D) 100 Ω

2. A copper rod of cross-sectional area A carries a uniform current I through it. At temperature T, if the volume charge density of the rod is ρ, how long will the charges take to travel a distance d ? **[2018 (Online)]**

(A) $\frac{2\rho dA}{IT}$ (B) $\frac{2\rho dA}{I}$

(C) $\frac{\rho dA}{I}$ (D) $\frac{\rho dA}{IT}$

3. A heating element has a resistance of 100 Ω at room temperature. When it is connected to a supply of 220 V, a steady current of 2 A passes in it and temperature is 500°C more than room temperature. What is the temperature coefficient of resistance of the heating element ? **[2018 (Online)]**

(A) $1 \times 10^{-4}\,°C^{-1}$ (B) $5 \times 10^{-4}\,°C^{-1}$

(C) $2 \times 10^{-4}\,°C^{-1}$ (D) $0.5 \times 10^{-4}\,°C^{-1}$

4. A resistance is shown in the figure. Its value and tolerance are given respectively by : **[9 Jan. 2019 (Morning)]**

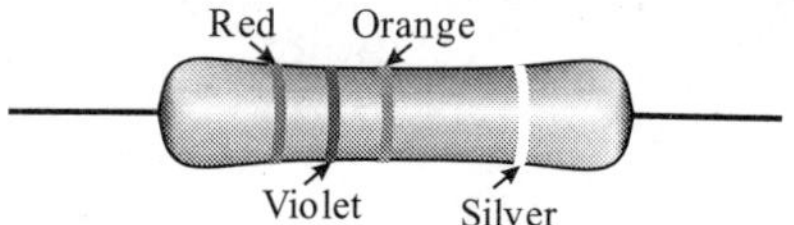

(A) 27 KΩ, 20% (B) 270 KΩ, 5%

(C) 270 KΩ, 10% (D) 27 KΩ, 10%

5. Drift speed of electrons, when 1.5 A of current flows in a copper wire of cross section 5 mm², is v. If the electron density in copper is 9×10^{28} /m³ the value of v in mm/s is close to (Take charge of electron to be $=1.6 \times 10^{-19}$ C) : **[9 Jan. 2019 (Morning)]**

(A) 0.2 (B) 3

(C) 2 (D) 0.02

6. A carbon resistance has a following colour code. What is the value of the resistance ? **[9 Jan. 2019 (Evening)]**

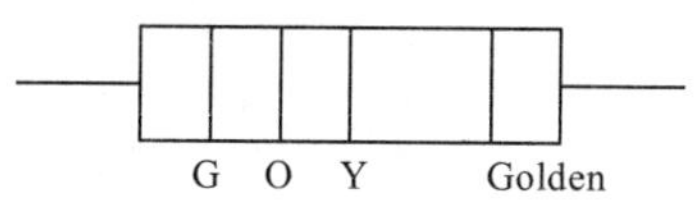

(A) 1.64 MΩ ± 5% (B) 530 kΩ ± 5%

(C) 64 kΩ ± 10% (D) 5.3 MΩ ± 5%

7. A 2 W carbon resistor is color coded with green, black, red and brown respectively. The maximum current which can be passed through this resistor is : **[10 Jan. 2019 (Morning)]**

(A) 63 mA (B) 0.4 mA

(C) 100 mA (D) 20 mA

8. A 200 Ω resistor has a certain color code. If one replaces the red color by green in the code, the new resistance will be : **[8 April 2019 (Morning)]**

(A) 100 Ω (B) 400 Ω

(C) 500 Ω (D) 300 Ω

9. A metal wire of resistance 3 Ω is elongated to make a uniform wire of double its previous length. This new wire is now bent and the ends joined to make a circle. If two points on this circle make an angle 60° at the centre, the equivalent resistance between these two points will be : **[9 April 2019 (Evening)]**

(A) $\frac{12}{5}\Omega$ (B) $\frac{5}{3}\Omega$

(C) $\frac{5}{2}\Omega$ (D) $\frac{7}{2}\Omega$

10. In a conductor, if the number of conduction electrons per unit volume is 8.5×10^{28} m^{-3} and mean free time is $25fs$ (femtosecond), it's approximate resistivity is : ($m_e = 9.1 \times 10^{-31}$ kg) : **[9 April 2019 (Evening)]**

(A) 10^{-5} Ωm (B) 10^{-6} Ωm

(C) 10^{-7} Ωm (D) 10^{-8} Ωm

11. A current of 5A passes through a copper conductor (resistivity $= 1.7 \times 10^{-8}$ Ωm) of radius of cross-section 5 mm. Find the mobility of the charges if their drift velocity is 1.1×10^{-3} m/s : **[10 April 2019 (Morning)]**

(A) 1.3 m²/Vs (B) 1.5 m²/Vs

(C) 1.8 m²/Vs (D) 1.0 m²/Vs

12. In an experiment, the resistance of a material is plotted as a function of temperature (in some range). As shown in the figure, it is a straight line. One may conclude that : **[10 April 2019 (Morning)]**

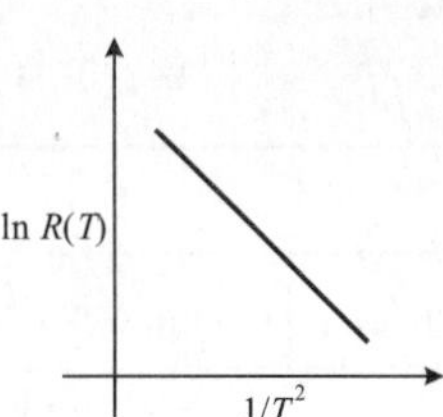

(A) $R(T)=\frac{R_0}{T^2}$ (B) $R(T)=R_0e^{-T^2/T_0^2}$

(C) $R(T)=R_0e^{-T_0^2/T^2}$ (D) $R(T)=R_0e^{T^2/T_0^2}$

13. Space between two concentric conducting spheres of radii a and $b(b>a)$ is filled with a medium of resistivity ρ. The resistance between the two spheres will be : **[10 April 2019 (Evening)]**

(A) $\frac{\rho}{4\pi}\left(\frac{1}{a}-\frac{1}{b}\right)$ (B) $\frac{\rho}{2\pi}\left(\frac{1}{a}-\frac{1}{b}\right)$

(C) $\frac{\rho}{2\pi}\left(\frac{1}{a}+\frac{1}{b}\right)$ (D) $\frac{\rho}{4\pi}\left(\frac{1}{a}+\frac{1}{b}\right)$

14. Consider four conducting materials copper, tungsten, mercury and aluminium with resistivity $\rho_C<\rho_T<\rho_M$ and ρ_A respectively. Then : **[2 Sept. 2020 (Morning)]**

(A) $\rho_A>\rho_T>\rho_C$ (B) $\rho_C>\rho_A>\rho_T$

(C) $\rho_A>\rho_M>\rho_C$ (D) $\rho_M>\rho_A>\rho_C$

15. A current through a wire depends on time as $i=\alpha_0 t+\beta t^2$ where α_0 = 20 A/s and β = 8 As^{-2}. Find the charge crossed through a section of the wire in 15 s: **[24 Feb. 2021 (Morning)]**

(A) 260 C (B) 2250 C

(C) 2100 C (D) 11250 C

16. A cylindrical wire of radius 0.5 mm and conductivity 5×10^7 S/m is subjected to an electric field of 10 mV/m. The expected value of current in the wire will be $x^3\pi\, mA$. The value of x is ______. **[24 Feb. 2021 (Evening)]**

17. A conducting wire of length 'l', area of cross-section A and electric resistivity ρ is connected between the terminals of a battery. A potential difference V is developed between its ends, causing an electric current. If the length of the wire of the same material is doubled and the area of cross-section is halved, the resultant current would be : **[16 March 2021 (Morning)]**

(A) $\frac{1}{4}\frac{VA}{\rho l}$ (B) $\frac{3}{4}\frac{VA}{\rho l}$

(C) $\frac{1}{4}\frac{\rho l}{VA}$ (D) $4\frac{VA}{\rho l}$

18. In the figure given, the electric current flowing through the 5 kΩ resistor is 'x' mA.

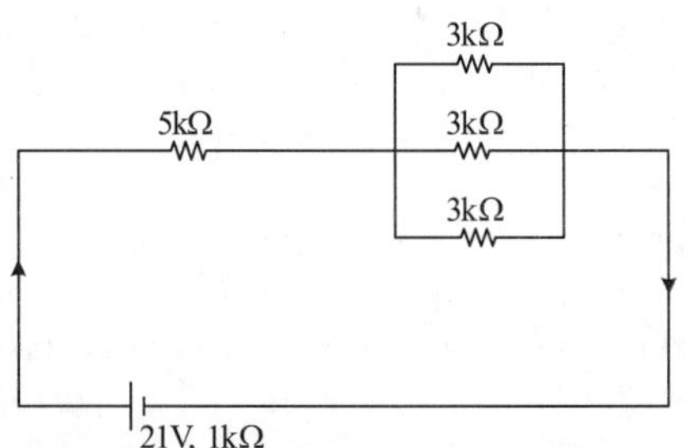

The value of x to the nearest integer is ______. **[16 March 2021 (Morning)]**

19. A current of 10 A exists in a wire of cross-sectional area of 5 mm^2 with a drift velocity of 2×10^{-3} ms^{-1}. The number of free electrons in each cubic meter of the wire is ______. **[17 March 2021 (Morning)]**

(A) 2×10^6 (B) 625×10^{25}

(C) 2×10^{25} (D) 1×10^{23}

20. In the experiment of Ohm's law, a potential difference of 5.0 V is applied across the end of a conductor of length 10.0 cm and diameter of 5.00 mm. The measured current in the conductor is 2.00 A. The maximum permissible percentage error in the resistivity of the conductor is : **[18 March 2021 (Morning)]**

(A) 3.9 (B) 8.4

(C) 7.5 (D) 3.0

21. A current of 5A is a passing through a non-linear magnesium wire of cross-section 0.04 m^2. At every point the direction of current density is at an angle of 60° with the unit vector of area of cross-section. The magnitude of electric field at every point of the conductor is :
(Resistivity of magnesium $\rho=44\times10^{-8}$ Ωm) : **[20 July 2021 (Morning)]**

(A) 11×10^{-7} V/m (B) 11×10^{-5} V/m

(C) 11×10^{-2} V/m (D) 11×10^{-3} V/m

22. In the given figure, a battery of emf E is connected across a conductor PQ of length 'l' and different area of cross-sections having radii r_1 and $r_2(r_2<r_1)$. Choose the correct option as one moves from P to Q: **[27 July 2021 (Morning)]**

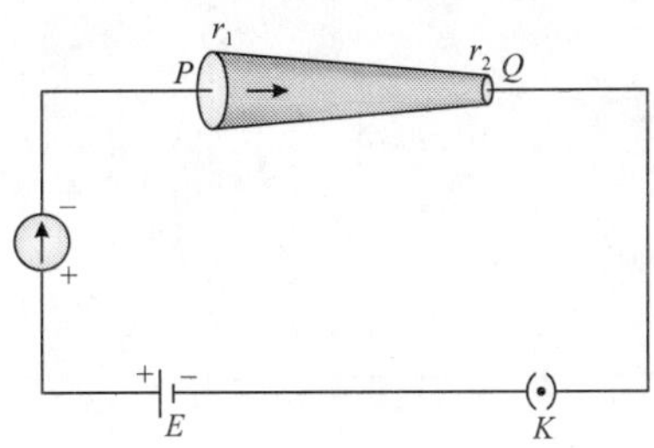

(A) Drift velocity of electron increases
(B) Electric field decreases
(C) All of these
(D) Electron current decreases

23. For the circuit shown, the value of current at time $t = 3.2$ s will be _______ A.

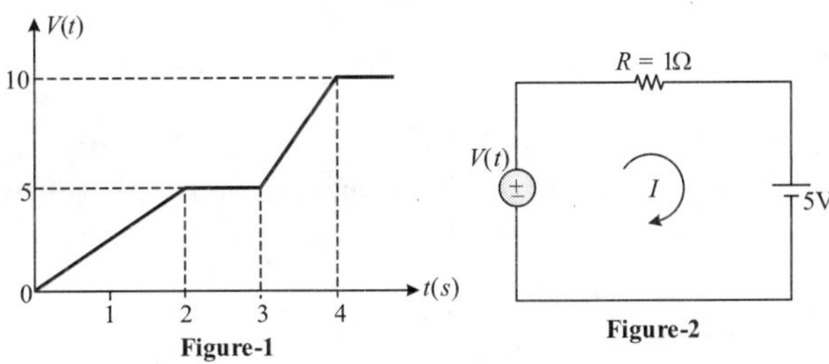

Figure-1 Figure-2

[Voltage distribution $V(t)$ is shown by figure-(1) and the circuit is shown in figure-(2)]. **[27 July 2021 (Evening)]**

24. What equal length of an iron wire a copper-nickel alloy wire, each of 2 mm diameter connected parallel to give an equivalent resistance of 3 Ω ? [Given resistivities of iron and copper-nickel alloy wire are 12 μΩcm and 51 μΩ cm respectively]

[26 Aug. 2021 (Morning)]

(A) 82 m (B) 97 m
(C) 110 m (D) 90 m

25. The colour coding on a carbon resistor is shown in the given figure. The resistance value of the given resistor is :

[27 Aug. 2021 (Evening)]

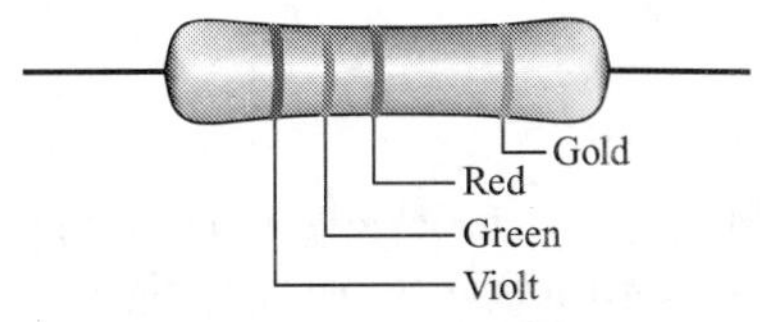

(A) $(7500 \pm 375)\Omega$ (B) $(5700 \pm 285)\Omega$
(C) $(7500 \pm 755)\Omega$ (D) $(5700 \pm 375)\Omega$

* * * * *

CHAPTER

17 CURRENT ELECTRICITY

17.2 CIRCUIT ANALYSIS

1. 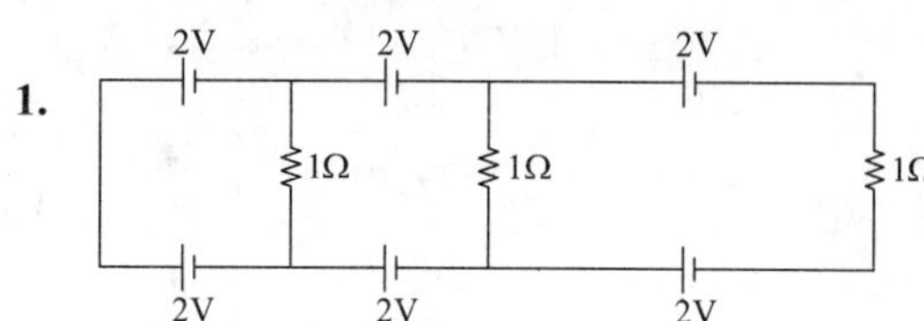

In the above circuit the current in each resistance is : **[2017]**

(A) 0.5 A (B) 0 A
(C) 1 A (D) 0.25 A

2. A 9 V battery with internal resistance of 0.5 Ω is connected across an infinite network as shown in the figure. All ammeters A_1, A_2, A_3 and voltmeter V are ideal : **[2017 (Online)]**

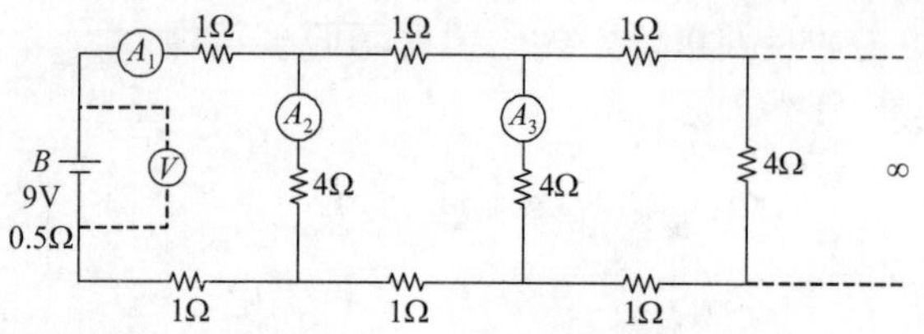

(A) Reading of A_1 is 2 A (B) Reading of A_1 is 18 A
(C) Reading of V is 9 V (D) Reading of V is 7 V

3. Two batteries with e.m.f. 12 V and 13 V are connected in parallel across a load resistor of 10 Ω. The internal resistance of the two batteries are 1 Ω and 2 Ω respectively. The voltage across the load lies between : **[2018]**

(A) 11.6 V and 11.7 V (B) 11.5 V and 11.6 V
(C) 11.4 and 11.5 V (D) 11.7 V and 11.8 V

4. In the given circuit all resistances are of volume R ohm each. The equivalent resistance between A and B is : **[2018 (Online)]**

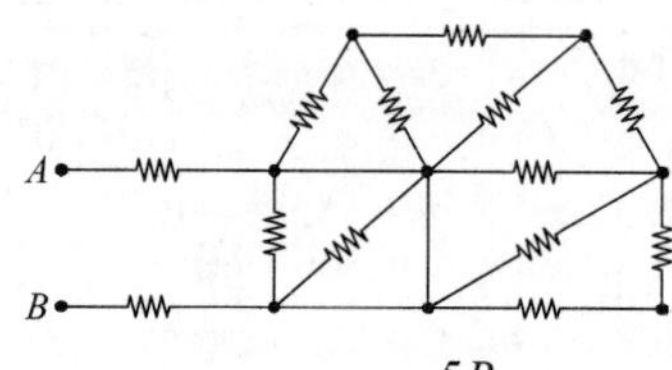

(A) $2R$ (B) $\frac{5R}{2}$
(C) $\frac{5R}{3}$ (D) $3R$

5. When the switch S, in the circuit shown, is closed, then the value of current i will be : **[9 Jan. 2019 (Morning)]**

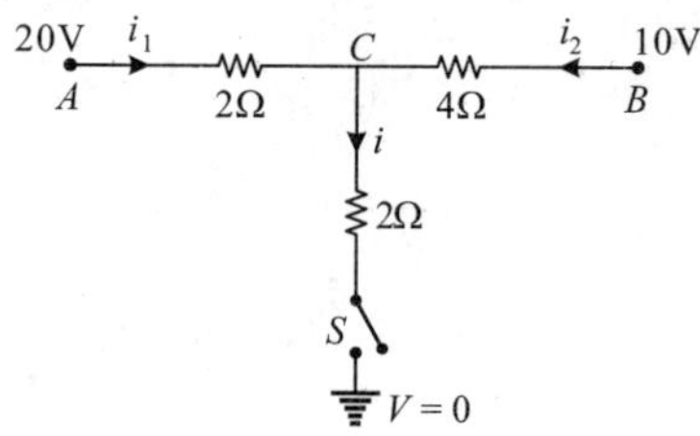

(A) 3 A (B) 5 A
(C) 4 A (D) 2 A

6. A uniform metallic wire has a resistance of 18 Ω and is bent into an equilateral triangle. Then, the resistance between any two vertices of the triangle is : **[10 Jan. 2019 (Morning)]**

(A) 8 Ω (B) 12 Ω
(C) 4 Ω (D) 2 Ω

7. In the given circuit the cells have zero internal resistance. The currents (in Amperes) passing through resistance R_1, and R_2 respectively, are : **[10 Jan. 2019 (Morning)]**

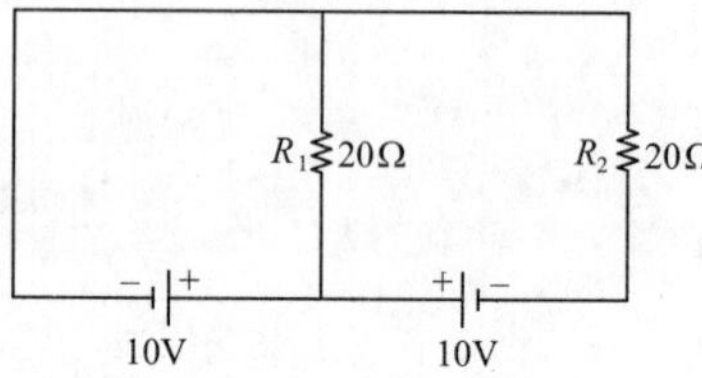

(A) 2, 2 (B) 0,1
(C) 1, 2 (D) 0.5, 0

8. In the circuit, the potential difference between A and B is: **[11 Jan. 2019 (Evening)]**

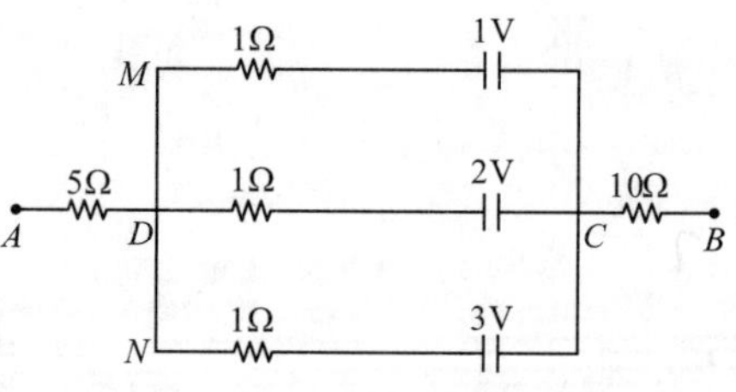

(A) 6 V (B) 1 V
(C) 3 V (D) 2 V

9. In the given circuit diagram, the currents, $I_1 = -0.3$ A, $I_4 = 0.8$ A and $I_5 = 0.4$ A, are flowing as shown. The currents

I_2, I_3 and I_6, respectively, are : **[12 Jan. 2019 (Evening)]**

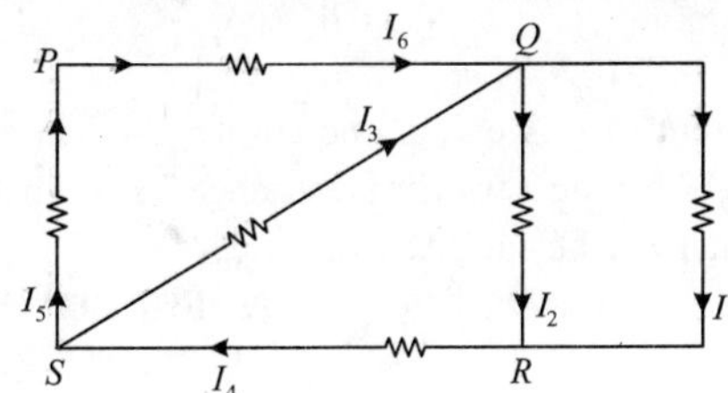

(A) 1.1 A, 0.4 A, 0.4 A (B) –0.4 A, 0.4 A, 1.1 A
(C) 0.4 A, 1.1 A, 0.4 A (D) 1.1 A, –0.4 A, 0.4 A

10. For the circuit shown, with $R_1 = 1.0\ \Omega$, $R_2 = 2.0\ \Omega$, $E_1 = 2$ V and $E_2 = E_3 = 4$ V, the potential difference between the points 'a' and 'b' is approximately (in V) : **[8 April 2019 (Morning)]**

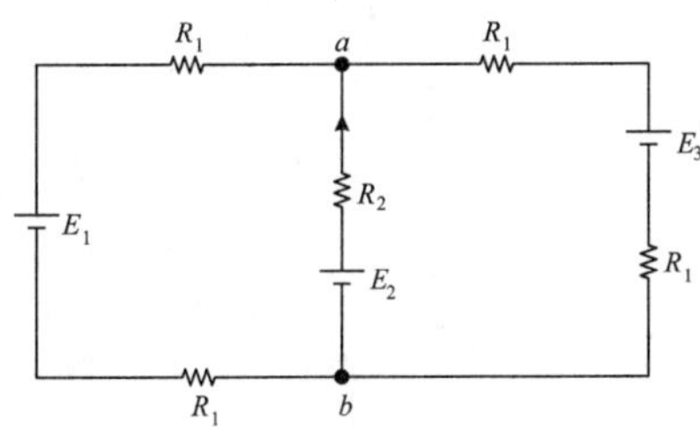

(A) 2.7 (B) 3.3
(C) 2.3 (D) 3.7

11. In the figure shown, what is the current (in Ampere) drawn from the battery ? You are given.
$R_1 = 15\ \Omega$, $R_2 = 10\ \Omega$, $R_3 = 20\ \Omega$, $R_4 = 5\ \Omega$, $R_5 = 25\ \Omega$, $R_6 = 30\ \Omega$, $E = 15$ V **[8 April 2019 (Evening)]**

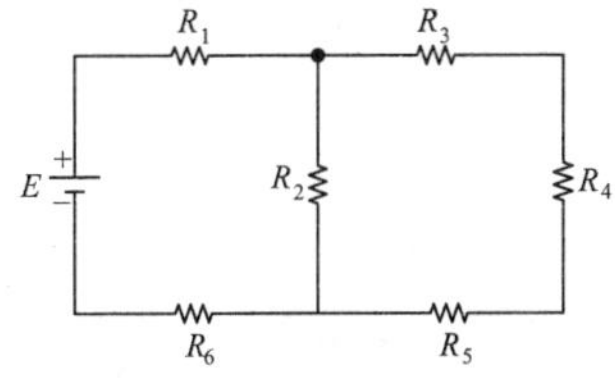

(A) 7/18 (B) 13/24
(C) 9/32 (D) 20/3

12. A wire of resistance R is bent to form a square $ABCD$ as shown in the figure. The effective resistance between E and C is : (E is mid-point of arm CD) : **[9 April 2019 (Morning)]**

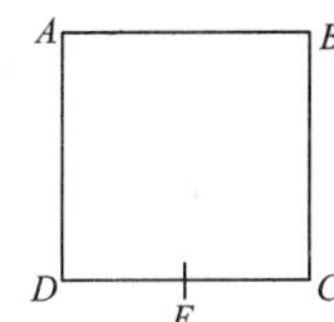

(A) R (B) $\frac{1}{16}R$

(C) $\frac{7}{64}R$ (D) $\frac{3}{4}R$

13. The current I_1(in A) flowing through 1 Ω resistor in the following circuit is : **[7 Jan. 2020 (Morning)]**

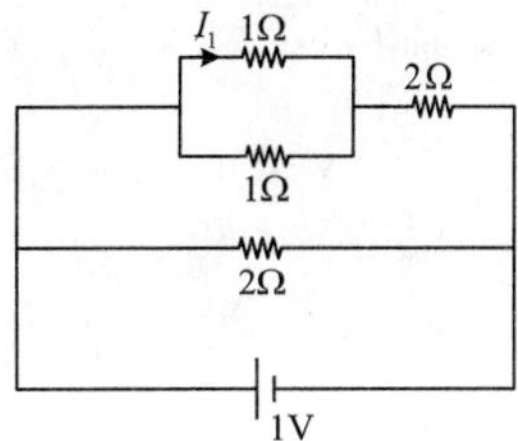

(A) 0.5 (B) 0.2
(C) 0.25 (D) 0.4

14. The series combination of two batteries, both of the same emf 10 V, but different internal resistance of 20 Ω and 5 Ω, is connected to the parallel combination of two resistors 30 Ω and RΩ. The voltage difference across the battery of internal resistance 20 Ω is zero, the value of R (in Ω) is ______.
[8 Jan. 2020 (Evening)]

15. In the given circuit diagram, a wire is joining points B and D. The current in this wire is : **[9 Jan. 2020 (Morning)]**

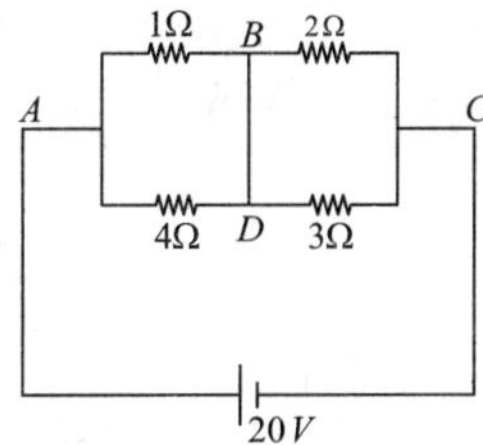

(A) 4 A (B) 2 A
(C) 0.4 A (D) Zero

16. The value of current i_1 flowing from A to C in the circuit diagram is : **[4 Sept. 2020 (Evening)]**

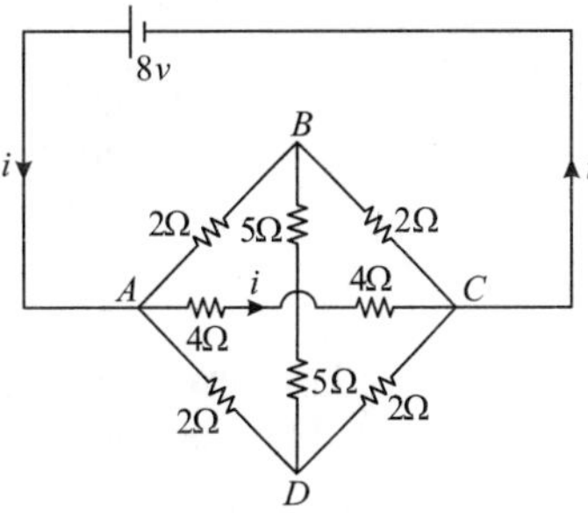

(A) 5 A (B) 2 A
(C) 4 A (D) 1 A

17. Four resistances 40 Ω, 60 Ω, 90 Ω and 110 Ω make the arms of a quadrilateral $ABCD$. Across AC is a battery of emf 40 V and internal resistance negligible. The potential difference across

BD is V is ______. **[4 Sept. 2020 (Evening)]**

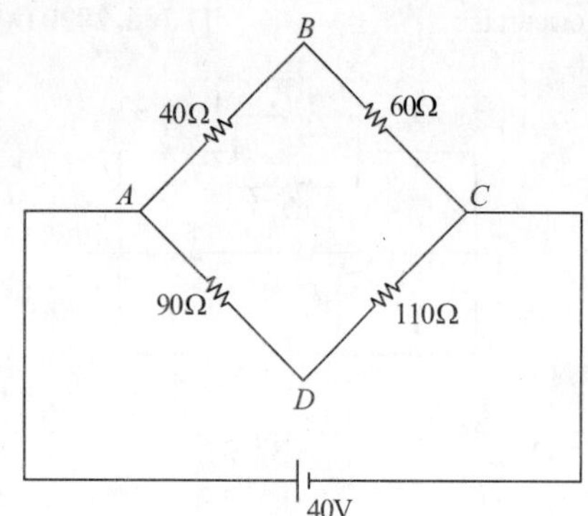

18. In the circuit, given in the figure currents in different branches and value of one resistor are shown. Then potential at point B with respect to the point A is :

[5 Sept. 2020 (Evening)]

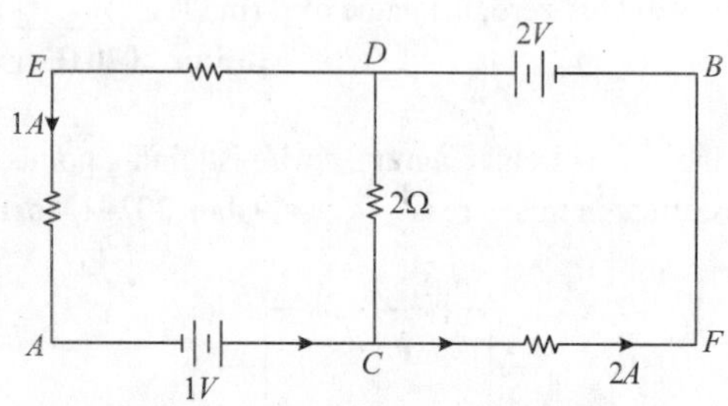

(A) +1 V (B) −1 V
(C) −2 V (D) +2 V

19. In the figure shown, the current in the 10 V battery is close to : **[6 Sept. 2020 (Evening)]**

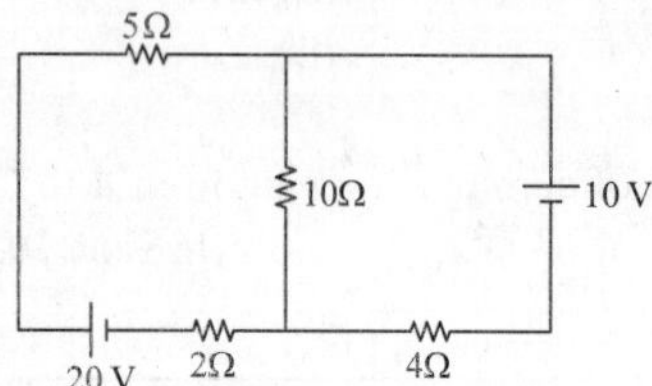

(A) 0.36 A from negative to positive terminal
(B) 0.71 A from positive to negative terminal
(C) 0.21 A from positive to negative terminal
(D) 0.42 A from positive to negative terminal

20. A cell E_1 of emf 6 V and internal resistance 2 Ω is connected with another cell E_2 of emf 4 V and internal resistance 8 Ω (as shown in the figure). The potential difference across points X and Y is : **[24 Feb. 2021 (Morning)]**

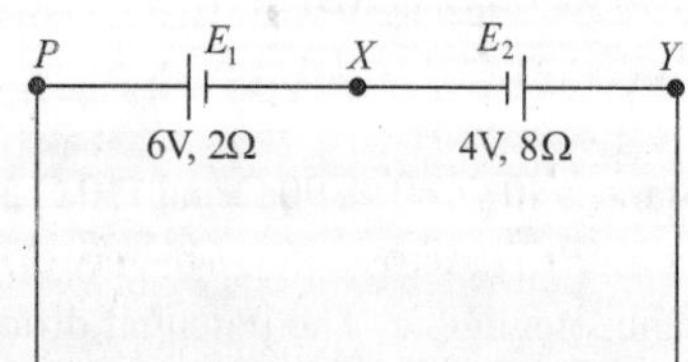

(A) 2.0 V (B) 10.0 V
(C) 5.6 V (D) 3.6 V

21. A current of 6 A enters one corner P of an equilateral triangle PQR having 3 wires of resistance 2 Ω each and leaves by the corner R. The currents i_1 in ampere is ______.

[25 Feb. 2021 (Evening)]

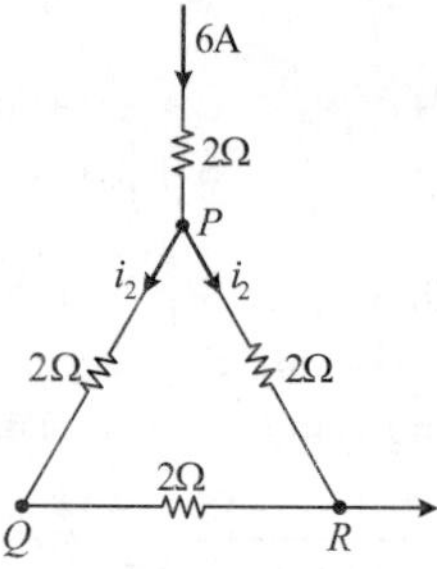

22. Five equal resistances are connected in a network as shown in figure. The net resistance between the points A and B is :

[26 Feb. 2021 (Morning)]

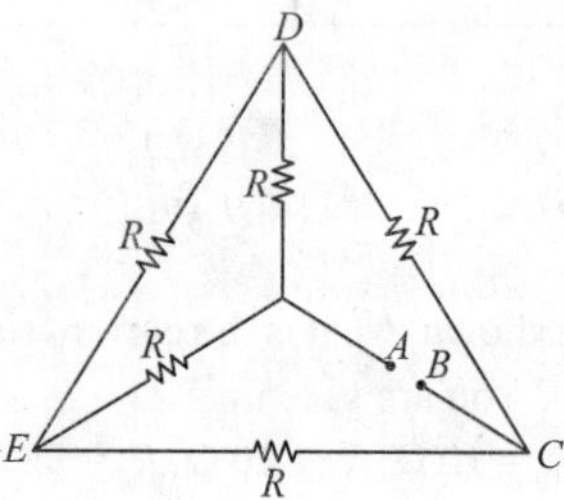

(A) $2R$ (B) $\frac{R}{2}$

(C) R (D) $\frac{3R}{2}$

23. In an electrical circuit, a battery is connected to pass 20 C of charge through it in a certain given time. The potential difference between two plates of the battery is maintained at 15 V. The workdone by the battery is ______ J.

[26 Feb. 2021 (Morning)]

24. The equivalent resistance of series combination of two resistors is 's'. When they are connected in parallel, the equivalent resistance is 'p'. If $s = np$, then the minimum value for n is ______. (Round off to the Nearest Integer)

[17 March 2021 (Morning)]

25. Two cells of emf $2E$ and E with internal resistance r_1 and r_2 respectively are connected in series to an external resistor R (see figure). The value of R, at which the potential difference across the terminals of the first cell becomes zero is :

[17 March 2021 (Evening)]

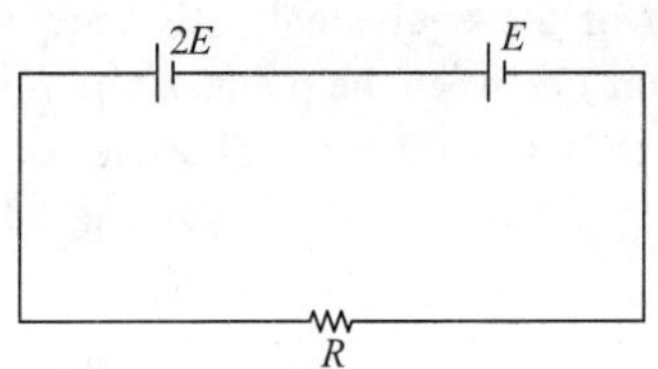

(A) $r_1 + r_2$ (B) $\frac{r_1}{2} - r_2$

(C) $\frac{r_1}{2} + r_2$ (D) $r_1 - r_2$

26. The four arms of a Wheatstone bridge have resistances as shown in the figure. A galvanometer of 15 Ω resistance is connected across *BD*. Calculate the current through the galvanometer when a potential difference of 10 V is maintained across *AC*. **[17 March 2021 (Evening)]**

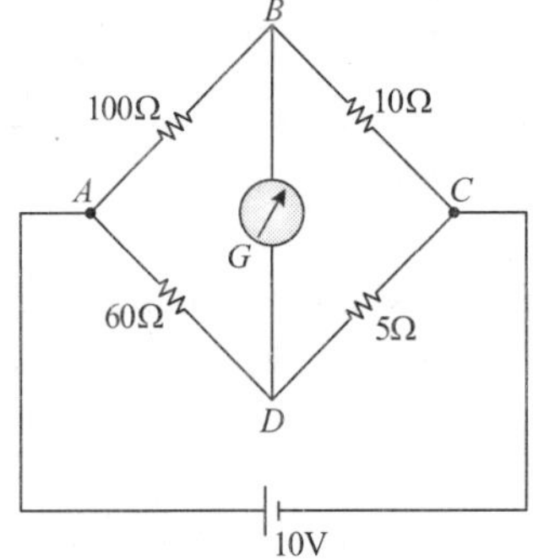

(A) 2.44 μA (B) 2.44 mA
(C) 4.87 mA (D) 4.87 μA

27. The voltage across the 10 Ω resistor in the given circuit is *x* volt.

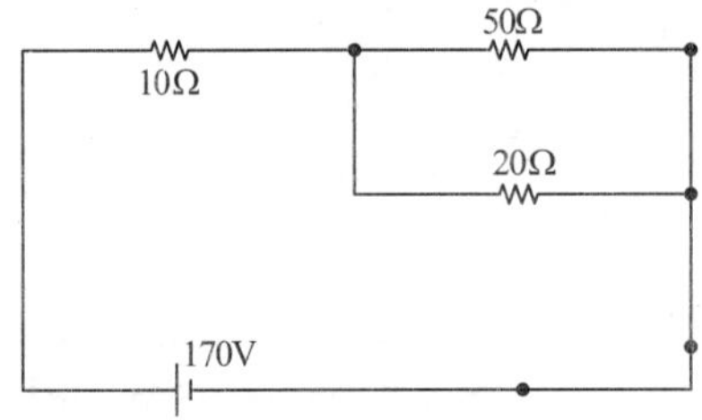

The value of '*x*' to the nearest integer is ______. **[18 March 2021 (Morning)]**

28. Two wires of same length and thickness having specific resistances 6 Ω cm and 3 Ω cm respectively are connected in parallel. The effective resistivity is ρΩ cm. The value of ρ, to the nearest integer, is______. **[18 March 2021 (Evening)]**

29. The value of current in the 6 Ω resistance is :

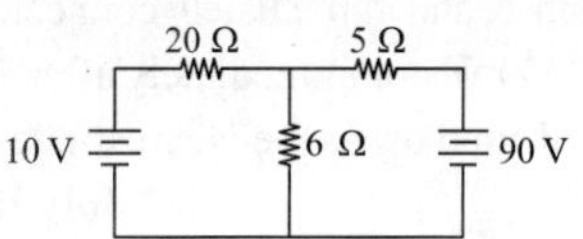

[20 July 2021 (Morning)]

(A) 4 A (B) 10 A
(C) 8 A (D) 6 A

30. In the given figure switches S_1 and S_2 are in open condition. The resistance across *ab* when the switches S_1 and S_2 are closed is ______Ω. **[20 July 2021 (Evening)]**

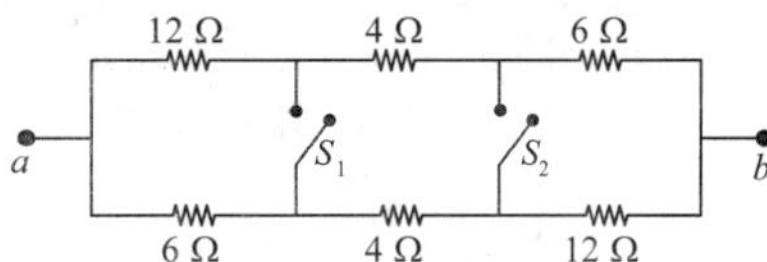

31. A copper (Cu) rod of length 25 cm and cross-sectional area 3 mm^2 is joined with a similar Aluminium (Al) rod as shown in figure. Find the resistance of the combination between the ends *A* and *B*. (Take Resistivity of Copper = 1.7×10^{-8} Ωm) : **[22 July 2021 (Evening)]**

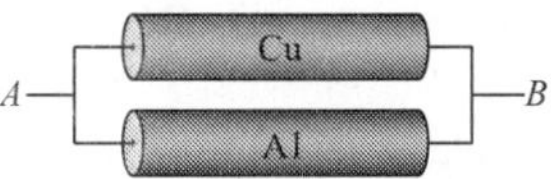

(A) 0.858 mΩ (B) 0.0858 mΩ
(C) 2.170 mΩ (D) 1.420 mΩ

32. In an electric circuit, a cell of certain emf provides a potential difference of 1.25 V across a load resistance of 5 Ω. However, it provides a potential difference of 1 V across a load resistance of 2 Ω. The emf of the cell is given by $\frac{x}{10}$ V. Then the value of *x* is ______. **[22 July 2021 (Evening)]**

33. An inductor of 10 mH is connected to a 20 V battery through a resistor of 10 Ω and a switch. After a long time, when maximum current is set up in the circuit, the current is switched off. The current in the circuit after 1 ms is $\frac{x}{100}$ *mA*. Then *x* is equal to ______. (Take $e^{-1} = 0.37$) **[25 July 2021 (Morning)]**

34. A 16 Ω wire is bend to from a square loop. A 9 V supply having internal resistance of 1 Ω is connected across one of its sides. The potential drop across the diagonals of the square loop is ______ $\times 10^{-1}$ V : **[25 July 2021 (Evening)]**

35. Consider an electrical circuit containing a two way switch *S*. Initially *S* is open and then T_1 is connected to T_2. As the current in $R = 6\ \Omega$ attains a maximum value of steady state level, T_1 is disconnected from T_2 and immediately connected to T_3. Potential drop across $r = 3\ \Omega$ resistor immediately after T_1 is connected to T_3 is ______V. (Round off to the Nearest integer).

[27 July 2021 (Morning)]

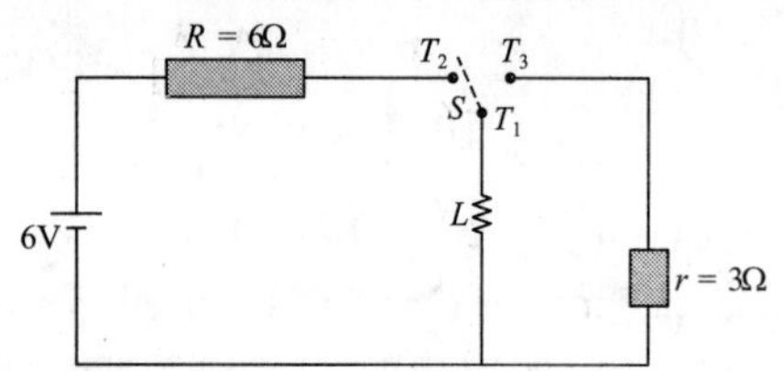

36. In the given figure, the emf of the cell is 2.2 V and if internal resistance is 0.6 Ω. Calculate the power dissipated in the whole circuit : **[26 Aug. 2021 (Morning)]**

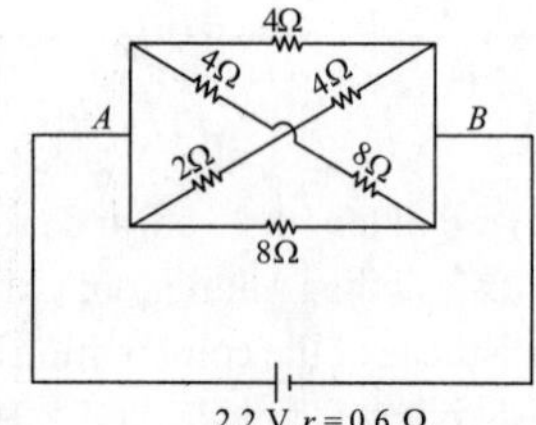

(A) 0.65 W (B) 4.4 W
(C) 1.32 W (D) 2.2 W

37. If you are provided a set of resistances 2 Ω, 4 Ω, and 8 Ω. Connect these resistances so as to obtain an equivalent resistance of $\frac{46}{3}\Omega$: **[26 Aug. 2021 (Evening)]**

(A) 2 Ω and 4 Ω are in parallel with 6 Ω and 8 Ω in series
(B) 6 Ω and 8 Ω are in parallel with 2 Ω and 4 Ω in series
(C) 2 Ω and 6 Ω are in parallel with 4 Ω and 8 Ω in series
(D) 4 Ω and 6 Ω are in parallel with 2 Ω and 8 Ω in series

38. Five identical cells each of internal resistance 1 Ω and emf 5 V are connected in series and in parallel with an external resistance R. For what value of R, current in series and parallel combination will remain the same ? **[27 Aug. 2021 (Morning)]**

(A) 10 Ω (B) 1 Ω
(C) 25 Ω (D) 5 Ω

39. First, a set of n equal resistors of 10 Ω each are connected in series to a battery of emf 20 V and internal resistance 10 Ω. A current I is observed to flow. Then, the n resistors are connected in parallel to the same battery. It is observed that the current is increased 20 times, then the value of n is ______.

[27 Aug. 2021 (Morning)]

40. The ratio of the equivalent resistance of the network (shown in figure) between the points a and b when switch is open and switch is closed is $x : 8$. The value of x is______.

[27 Aug. 2021 (Evening)]

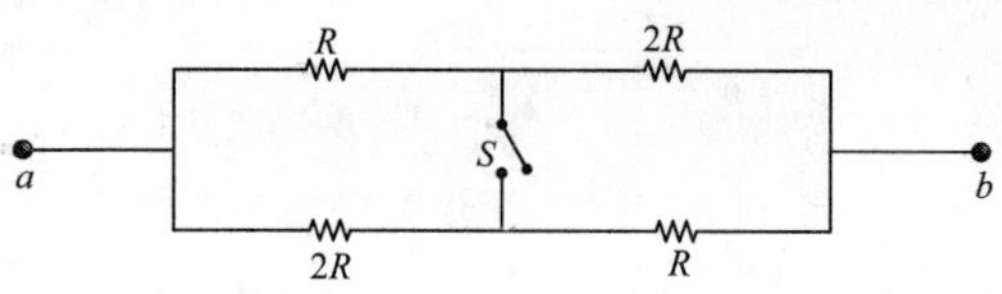

41. The voltage drop across 15 Ω resistance in the given figure will be ______V. **[31 Aug. 2021 (Morning)]**

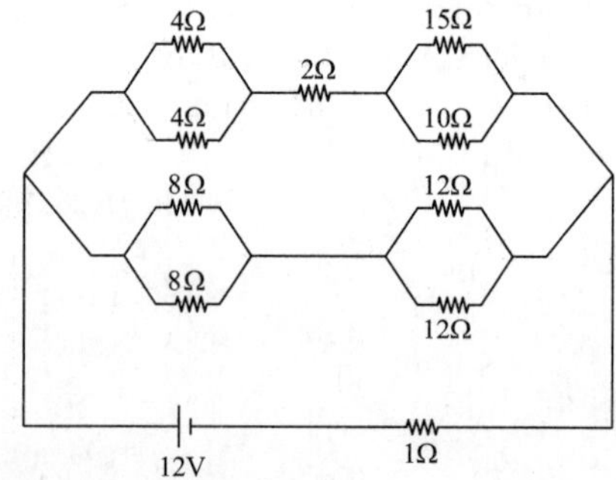

42. A square shaped wire with resistance of each side 3 Ω is bent to form a complete circle. The resistance between two diametrically opposite points of the circle in unit of Ω will be ______: **[31 Aug. 2021 (Morning)]**

43. The equivalent resistance of the given circuit between the terminals A and B is : **[31 Aug. 2021 (Evening)]**

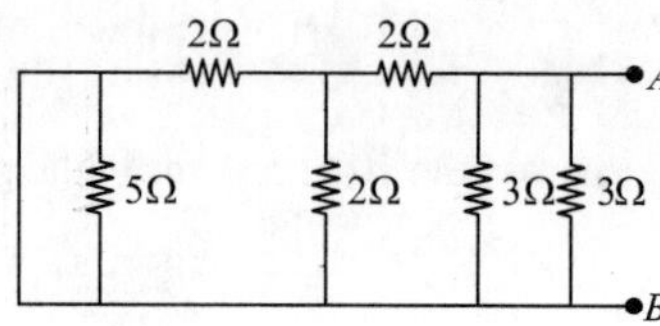

(A) 1 Ω (B) 3 Ω
(C) 0 Ω (D) $\frac{9}{2}\Omega$

44. Two resistors $R_1 = (4 \pm 0.4)\ \Omega$ and $R_2 = (4 \pm 0.4)\ \Omega$ are connected in parallel. The equivalent resistance of their parallel combination will be : **[01 Sep. 2021 (Evening)]**

(A) $(4 \pm 0.3)\ \Omega$ (B) $(2 \pm 0.4)\ \Omega$
(C) $(4 \pm 0.4)\ \Omega$ (D) $(2 \pm 0.3)\ \Omega$

* * * * *

CHAPTER

17 CURRENT ELECTRICITY

17.3 THERMAL EFFECTS OF CURRENT & RC CIRCUITS

1. In the given circuit diagram when the current reaches steady state in the circuit, the charge on the capacitor of capacitance C will be : **[2017]**

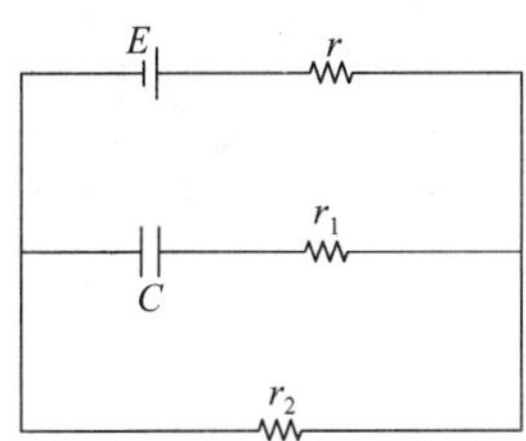

(A) $CE\frac{r_2}{(r+r_2)}$ (B) $CE\frac{r_1}{(r_1+r)}$

(C) CE (D) $CE\frac{r_1}{(r_2+r)}$

2. The figure shows three circuits I, II, and III which are connected to a 3 V battery. If the powers dissipated by the configurations I, II, and III are P_1, P_2 and P_3 respectively, then : **[2017 (Online)]**

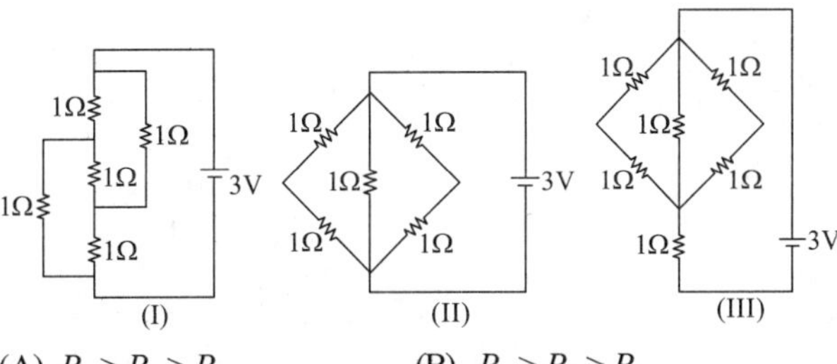

(A) $P_1 > P_2 > P_3$ (B) $P_1 > P_3 > P_2$
(C) $P_2 > P_1 > P_3$ (D) $P_3 > P_2 > P_1$

3. In the following circuit, the switch S is closed at $t = 0$. The charge on the capacitor C_1 as a function of time will be given by $\left(C_{eq} = \frac{C_2C_2}{C_1+C_2}\right)$. **[2018 (Online)]**

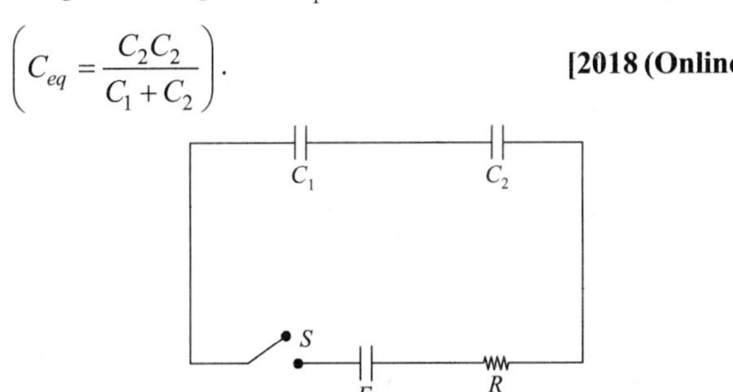

(A) $C_{eq}E[1-\exp(-t/RC_{eq})]$ (B) $C_1E[1-\exp(-tR/C_1)]$
(C) $C_2E[1-\exp(-t/RC_2)]$ (D) $C_{eq}E\exp(-t/RC_{eq})$

4. A constant voltage is applied between two ends of a metallic wire. If the length is halved and the radius of the wire is doubled, the rate of heat developed in the wire will be : **[2018 (Online)]**
(A) Increased 8 times (B) Doubled
(C) Halved (D) Unchanged

5. A current of 2 mA was passed through an unknown resistor which dissipated a power of 4.4 W. Dissipated power when an ideal power supply of 11 V is connected across it is :
[10 Jan. 2019 (Evening)]
(A) 11×10^{-5} W (B) 11×10^{-4} W
(C) 11×10^{5} W (D) 11×10^{-3} W

6. Two equal resistance when connected in series to a battery, consume electric power of 60 W. If these resistances are now connected in parallel combination to the same battery, the electric power consumed will be : **[11 Jan. 2019 (Morning)]**
(A) 60 W (B) 240 W
(C) 30 W (D) 120 W

7. Two electric bulbs, rated at (25 W, 220 V) and (100 W, 220 V), are connected in series across a 220 V voltage source. If the 25 W and 100 W bulbs draw powers P_1 and P_2 respectively, then :
[12 Jan. 2019 (Morning)]
(A) $P_1 = 9$ W, $P_2 = 16$ W (B) $P_1 = 4$ W, $P_2 = 16$ W
(C) $P_1 = 16$ W, $P_2 = 4$ W (D) P_1 16 W, $P_2 = 9$ W

8. A 20 Henry inductor coil is connected to a 10 ohm resistance in series as shown in figure. The time at which rate of dissipation of energy (joule's heat) across resistance is equal to the rate at which magnetic energy is stored in the inductor is :
[8 April 2019 (Morning)]

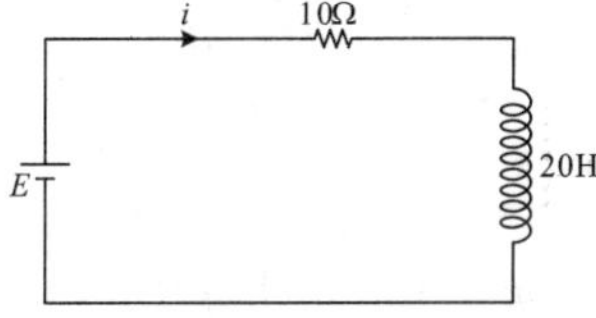

(A) $\frac{2}{\ln 2}$ (B) ln 2

(C) 2 ln 2 (D) $\frac{1}{2}\ln 2$

9. A cell of internal resistance r drives current through an external resistance R. The power delivered by the cell to the external resistance will be maximum when :

[8 April 2019 (Evening)]

(A) $R = 1000\,r$ (B) $R = 0.001\,r$

(C) $R = 2r$ (D) $R = r$

10. Determine the charge on the capacitor in the following circuit : **[9 April 2019 (Morning)]**

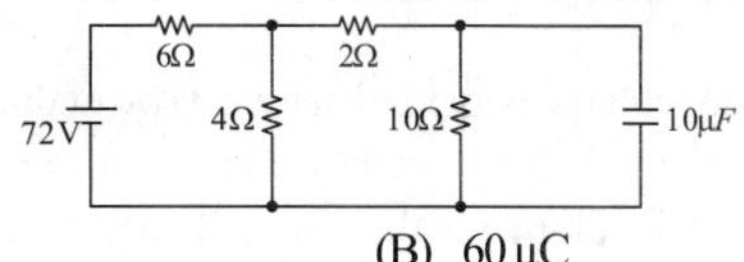

(A) 2 μC (B) 60 μC

(C) 200 μC (D) 10 μC

11. The resistive network shown below is connected to a D.C. source of 16 V. The power consumed by the network is 4 Watt. The value of R is : **[12 April 2019 (Morning)]**

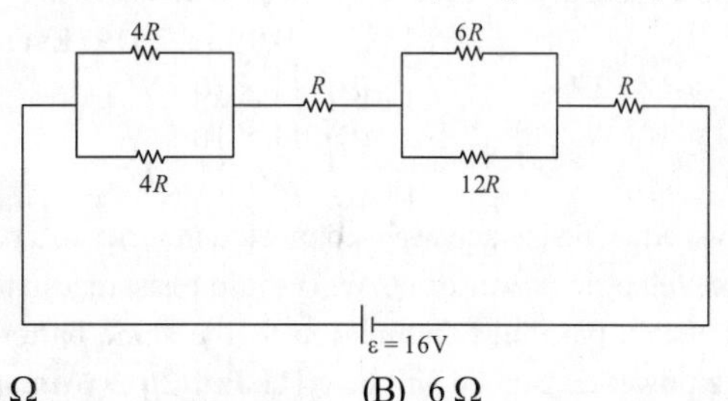

(A) 8 Ω (B) 6 Ω

(C) 1 Ω (D) 16 Ω

12. One kg of water, at 20ºC, is heated in an electric kettle whose heating element has a mean (temperature averaged) resistance of 20 Ω. The rms voltage in the mains is 200 V. Ignoring heat loss from the kettle, time taken for water to evaporate fully, is close to : [Specific heat of water = 4200 J/kgºC). Latent heat of water = 2260 kJ/kg] **[12 April 2019 (Evening)]**

(A) 3 minutes (B) 22 minutes

(C) 10 minutes (D) 16 minutes

13. In a building there are 15 bulbs of 45 W, 15 bulbs of 100 W, 15 small fans of 10 W and 2 heaters of 1 kW. The voltage of electric main is 220 V. The minimum fuse capacity (rated value) of the building will be : **[7 Jan. 2020 (Evening)]**

(A) 10 A (B) 25 A

(C) 15 A (D) 20 A

14. An ideal cell of emf 10 V is connected in circuit shown in figure. Each resistance is 2 Ω. The potential difference (in V) across the capacitor when it is fully charged is :

[2 Sept. 2020 (Evening)]

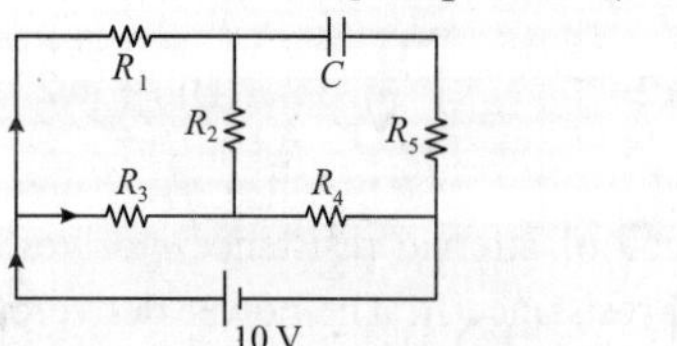

15. Model of a torch battery of length l to be made up of a thin cylindrical bar of radius a and a concentric thin cylindrical shell of radius b filled in between with an electrolyte of resistivity ρ (see figure). If the battery is connected to a resistance of value R, the maximum Joule heating in R will take place for :

[3 Sept. 2020 (Morning)]

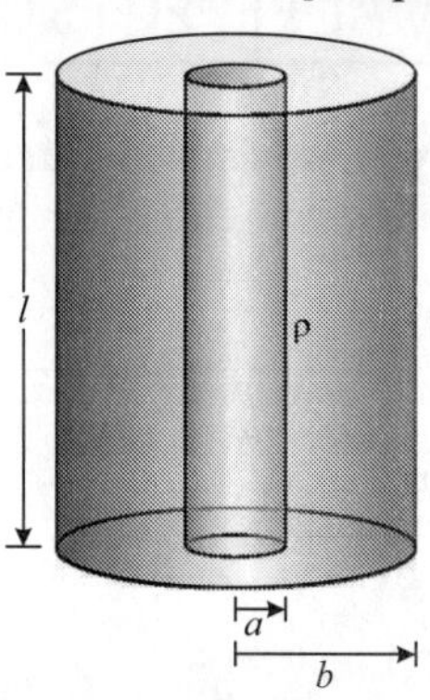

(A) $R = \dfrac{2\rho}{\pi l}\ln\left(\dfrac{b}{a}\right)$ (B) $R = \dfrac{\rho}{\pi l}\ln\left(\dfrac{b}{a}\right)$

(C) $R = \dfrac{\rho}{2\pi l}\left(\dfrac{b}{a}\right)$ (D) $R = \dfrac{\rho}{2\pi l}\ln\left(\dfrac{b}{a}\right)$

16. A battery of 3.0 V is connected to a resistor dissipating 0.5 W of power. If the terminal voltage of the battery is 2.5 V, the power dissipated within the internal resistance is :

[4 Sept. 2020 (Morning)]

(A) 0.50 W (B) 0.125 W

(C) 0.072 W (D) 0.10 W

17. An electrical power line, having a total resistance of 2 W, delivers 1 kW at 220 V. The efficiency of the transmission line is approximately : **[5 Sept. 2020 (Morning)]**

(A) 72% (B) 96%

(C) 91% (D) 85%

18. An RC circuit as shown in the figure is driven by a AC source generating a square wave. The output wave pattern monitored by CRO would look close to :

[16 March 2021 (Morning)]

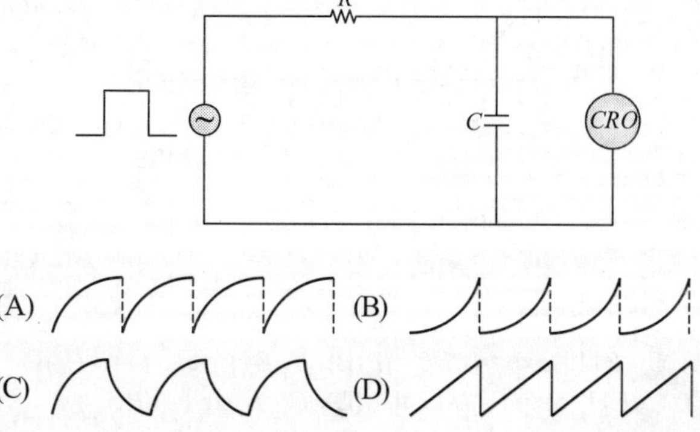

19. A resistor develops 500 J of thermal energy in 20 s when a current of 1.5 A is passed through it. If the current is increased from 1.5 A to 3 A, what will be the energy developed in 20 s. **[16 March 2021 (Evening)]**

(A) 1500 J (B) 1000 J
(C) 500 J (D) 2000 J

20. The energy dissipated by a resistor is 10 mJ in 1 s when an electric current of 2 mA flows through it. The resistance is ______ Ω. (Round off to the Nearest Integer) **[16 March 2021 (Evening)]**

21. The circuit shown in the figure consists of a charged capacitor of capacity 3 μF and a charge of 30 μC. At time $t = 0$, when the key is closed, the value of current flowing through the 5 MΩ resistor is 'x' μ-A. The value of x to the nearest integer is ______. **[18 March 2021 (Morning)]**

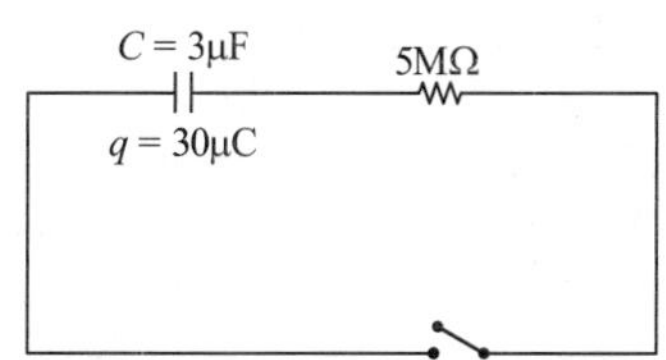

22. An electric bulb rated as 200 W at 100 V is used in a circuit having 200 V supply. The resistance R that must be put in series with the bulb so that the bulb delivers the same power is ______Ω. **[25 July 2021 (Morning)]**

23. An electric bulb of 500 W at 100 V is used in a circuit having a 200 V supply. Calculate the resistance R to be connected in series with the bulb so that the power delivered by the bulb is 500 W : **[26 Aug. 2021 (Evening)]**

(A) 10 Ω (B) 5 Ω
(C) 30 Ω (D) 20 Ω

24. Calculate the amount of charge on capacitor of 4 μF. The internal resistance of battery is 1 Ω : **[27 Aug. 2021 (Morning)]**

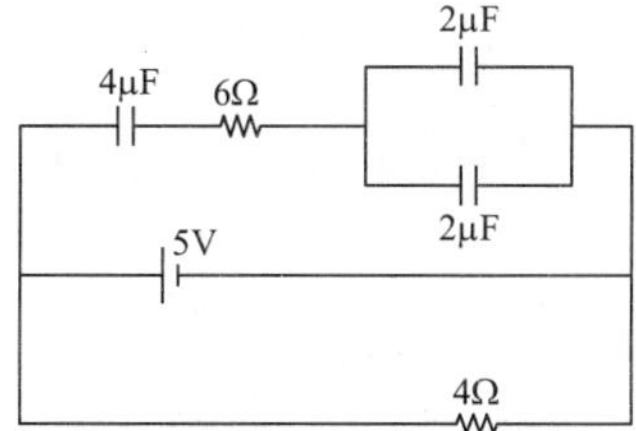

(A) Zero (B) 8 μC
(C) 16 μC (D) 4 μC

25. A capacitor of 50 μF is connected in a circuit as shown in figure. The charge on the upper plate of the capacitor is ______μC : **[31 Aug. 2021 (Morning)]**

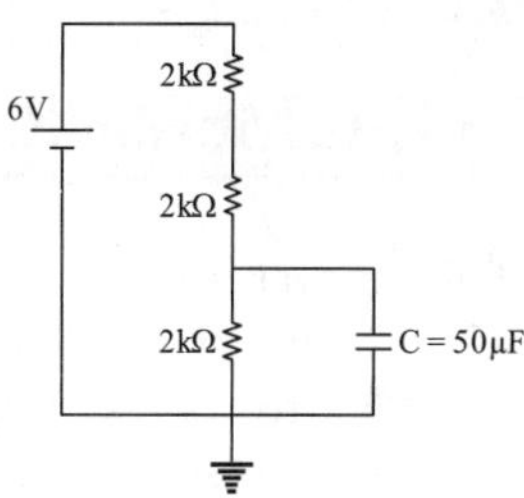

26. A resistor dissipated 192 J of energy in 1 s when a current of 4 A is passed through it. Now, when the current is doubled, the amount of thermal energy dissipated in 5 s is ______ J : **[31 Aug. 2021 (Evening)]**

27. A capacitor is connected to a 20 V battery through a resistance of 10 Ω. It is found that the potential difference across the capacitor rises to 2 V in 1 μs. The capacitance of the capacitor is ______ μF. Given $ln\left(\frac{10}{9}\right) = 0.105$.

[01 Sep. 2021 (Evening)]

(A) 1.85 (B) 9.52
(C) 0.95 (D) 0.105

28. For the given circuit the current i through the battery when the key is closed and the steady state has been reached is ______. **[01 Sep. 2021 (Evening)]**

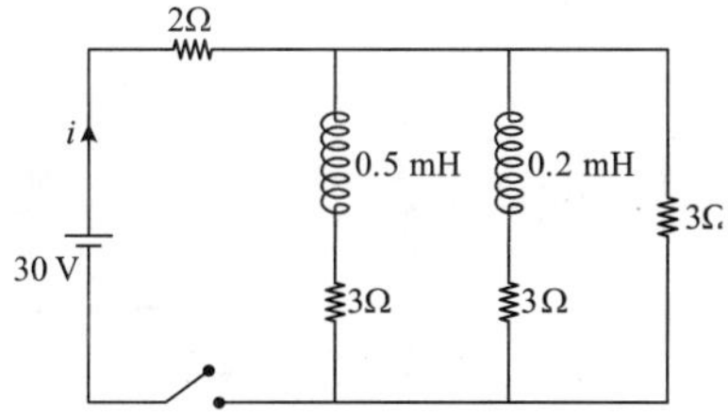

(A) 0 A (B) 25 A
(C) 10 A (D) 6 A

29. A uniform heating wire of resistance 36 Ω is connected across a potential difference of 240 V. The wire is then cut into half and a potential difference of 240 V is applied across each half separately. The ratio of power dissipation in first case to the total power dissipation in the second case would be 1 : x, where x is______. **[01 Sep. 2021 (Evening)]**

* * * * *

CHAPTER

17 CURRENT ELECTRICITY

17.4 ELECTRICAL MEASURING INSTRUMENTS

1. Which of the following statements is false ? **[2017]**

(A) A rheostat can be used as a potential divider

(B) Kirchhoff's second law represents energy conservation

(C) Wheatstone bridge is the most sensitive when all the four resistances are of the same order of magnitude

(D) In a balanced Wheatstone bridge if the cell and the galvanometer are exchanged, the null point is disturbed

2. When a current of 5 mA is passed through a galvanometer having a coil of resistance 15 Ω, it shows full scale deflection. The value of the resistance to be put in series with the galvanometer to convert it into to voltmeter of range 0 – 10 V is: **[2017]**

(A) $2.535 \times 10^3\,\Omega$ (B) $4.005 \times 10^3\,\Omega$

(C) $1.985 \times 10^3\,\Omega$ (D) $2.045 \times 10^3\,\Omega$

3. A potentiometer PQ is set up to compare two resistances as shown in the figure. The ammeter A in the circuit reads 1.0 A when two way key K_3 is open. The balance point is at a length l_1 cm from P when two way key K_3 is plugged in between 2 and 1, while the balance point is at a length l_2 cm from P when key K_3 is plugged in between 3 and 1. The ratio of two resistances $\frac{R_1}{R_2}$, is found to be : **[2017 (Online)]**

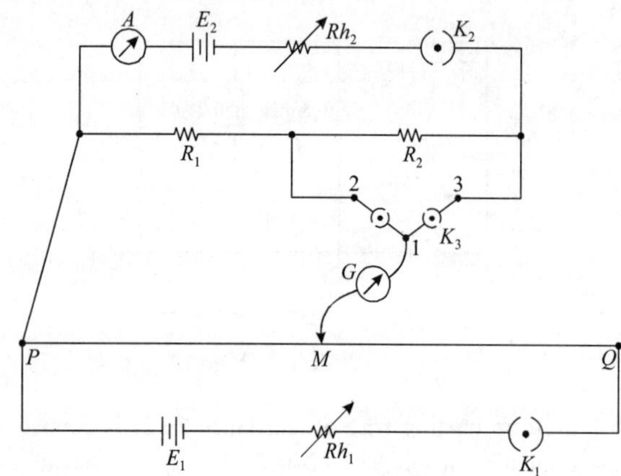

(A) $\frac{l_1}{l_1+l_2}$ (B) $\frac{l_2}{l_2-l_1}$

(C) $\frac{l_1}{l_1-l_2}$ (D) $\frac{l_1}{l_2-l_1}$

4. In a meter bridge experiment resistances are connected as shown in the figure. Initially resistance $P = 4\,\Omega$ and the neutral point N is at 60 cm from A. Now an unknown resistance R is connected in series to P and the new position of the neutral point is at 80 cm from A. The value of unknown resistance R is: **[2017 (Online)]**

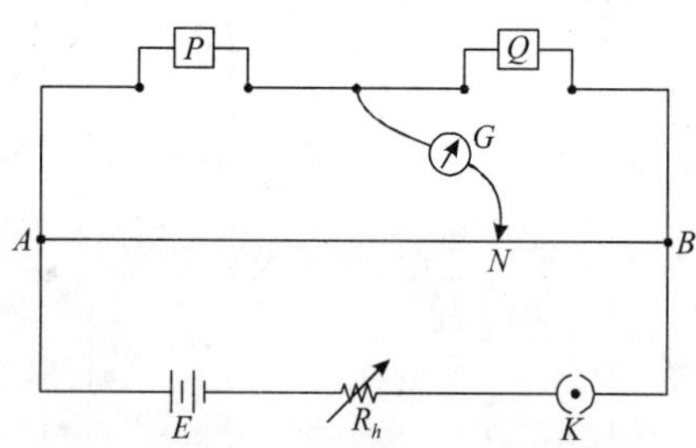

(A) $\frac{33}{5}\Omega$ (B) 6 Ω

(C) 7 Ω (D) $\frac{20}{3}\Omega$

5. In a potentiometer experiment, it is found that no current passes through the galvanometer when the terminals of the cell are connected across 52 cm of the potentiometer wire. If the cell is shunted by a resistance of 5 Ω, a balance is found when the cell is connected across 40 cm of the wire. Find the internal resistance of the cell : **[2018]**

(A) 1 Ω (B) 1.5 Ω

(C) 2 Ω (D) 2.5 Ω

6. On interchanging the resistances, the balance point of a meter bridge shifts to the left by 10 cm. The resistance of their series combination is 1 kΩ. How much was the resistance on the left slot before interchanging the resistances? **[2018]**

(A) 990 Ω (B) 505 Ω

(C) 550 Ω (D) 910 Ω

7. In a meter bridge, as shown in the figure, it is given that resistance $Y = 12.5\,\Omega$ and that the balance is obtained at a distance 39.5 cm from end A (by jockey J). After interchanging the resistances X and Y, a new balance point is found at a distance l_1 from end A. What are the values of X and l_1 ? **[2018 (Online)]**

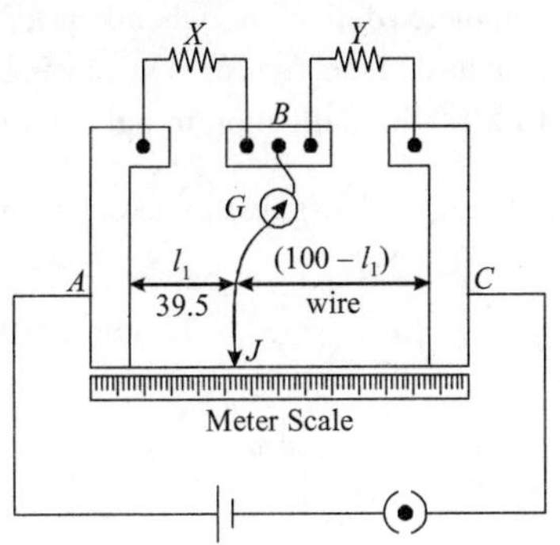

(A) 19.15 Ω and 39.5 cm (B) 8.16 Ω and 60.5 cm
(C) 19.15 Ω and 60.5 cm (D) 8.16 Ω and 39.5 cm

8. In a circuit for finding the resistance of a galvanometer by half deflection method, a 6 V battery and a high resistance of 11 kΩ are used. The figure of merit of the galvanometer 60 μA/division. In the absence of shunt resistance, the galvanometer produces a deflection of $\theta = 9$ divisions when current flows in the circuit. The value of the shunt resistance that can cause the deflection of $\theta/2$, is closest to : **[2018 (Online)]**

(A) 55 Ω (B) 110 Ω
(C) 220 Ω (D) 550 Ω

9. A galvanometer with its coil resistance 25 Ω requires a current of 1 mA for its full deflection. In order to construct an ammeter to read up to a current of 2 A, the approximate value of the shunt resistance should be : **[2018 (Online)]**

(A) 2.5×10^{-2} Ω (B) 1.25×10^{-3} Ω
(C) 2.5×10^{-3} Ω (D) 1.25×10^{-2} Ω

10. In the given circuit the internal resistance of the 18 V cell is negligible. If $R_1 = 400$ Ω, $R_3 = 100$ Ω and $R_4 = 500$ Ω and the reading of an ideal voltmeter across R_4 is 5 V, then the value R_2 will be : **[9 Jan. 2019 (Evening)]**

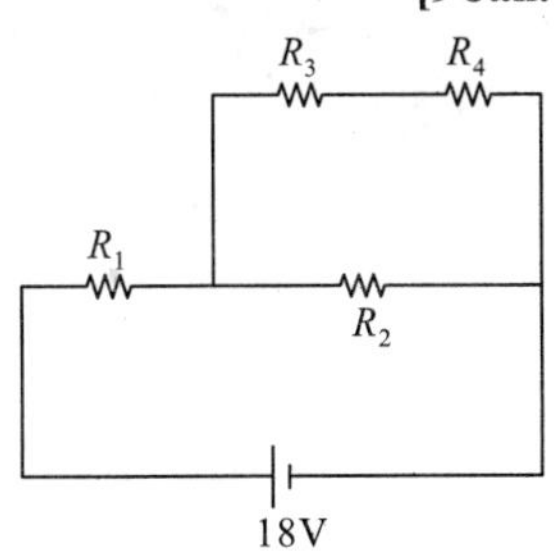

(A) 300 Ω (B) 230 Ω
(C) 450 Ω (D) 550 Ω

11. A potentiometer wire *AB* having length *L* and resistance 12*r* is joined to a cell *D* of emf ε and internal resistance *r*. A cell *C* having emf ε/2 and internal resistance 3*r* is connected. The length *AJ* at which the galvanometer as shown in figure shows no deflection is : **[10 Jan. 2019 (Morning)]**

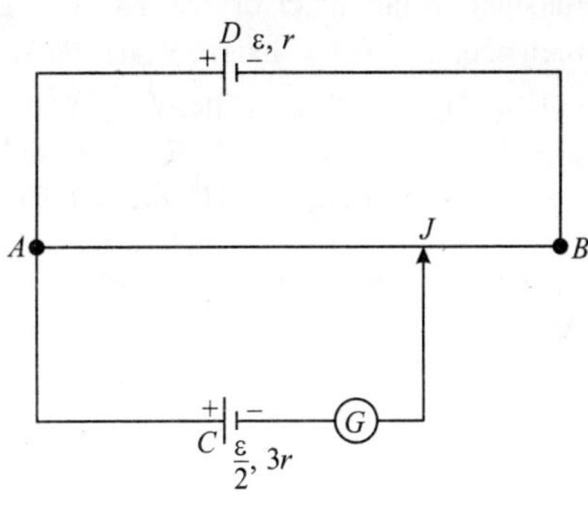

(A) $\frac{5}{12}L$ (B) $\frac{11}{24}L$
(C) $\frac{11}{12}L$ (D) $\frac{13}{24}L$

12. The actual value of resistance *R*, shown in the figure is 30 Ω. This is measured in an experiment as shown using the standard formula $R = \frac{V}{I}$, where *V* and *I* are the readings of the voltmeter and ammeter, respectively. If the measured value of *R* is 5% less, then the internal resistance of the voltmeter is : **[10 Jan. 2019 (Evening)]**

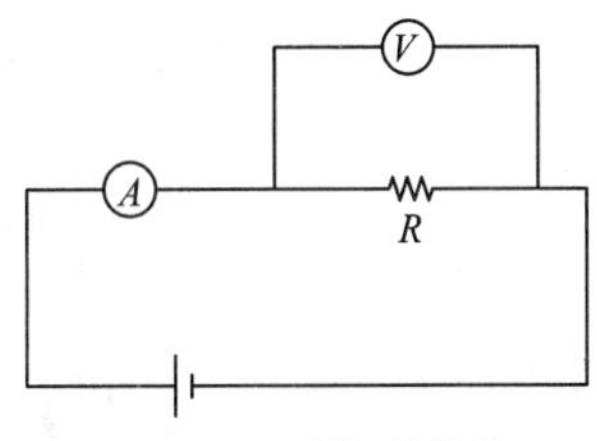

(A) 350 Ω (B) 570 Ω
(C) 35 Ω (D) 600 Ω

13. The Wheatstone bridge shown in Figure here, gets balanced when the carbon resistor used as R_1 has the colour code (Orange, Red, Brown). The resistors R_2 and R_4 are 80 Ω and 40 Ω, respectively.
Assuming that the colour code for the carbon resistors gives their accurate values, the colour code for the carbon resistor, used as R_3, would be : **[10 Jan. 2019 (Evening)]**

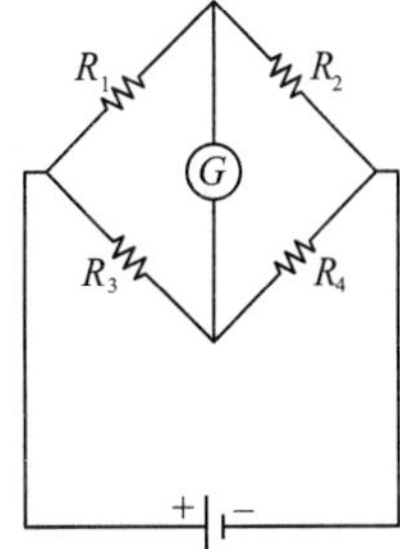

(A) Red, Green, Brown (B) Brown, Blue, Brown
(C) Grey, Black, Brown (D) Brown, Blue, Black

14. The resistance of the meter bridge AB in given figure is $4\ \Omega$. With a cell of emf $\varepsilon = 0.5$ V and rheostat resistance $R_h = 2\ \Omega$ the null point is obtained at some point J. When the cell is replaced by another one of emf $\varepsilon = \varepsilon_2$ the same null point J is found for $R_h = 6\ \Omega$. The emf ε_2 is : **[11 Jan. 2019 (Morning)]**

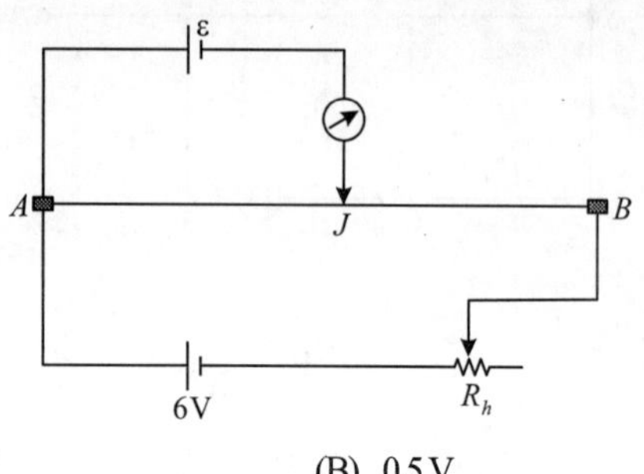

(A) 0.6 V (B) 0.5 V
(C) 0.3 V (D) 0.4 V

15. In a Wheatstone bridge (see figure). Resistances P and Q are approximately equal. When $R = 400\ \Omega$, the bridge is balanced. On interchanging P and Q, the value of R, for balance, is $405\ \Omega$. The value of X is close to : **[11 Jan. 2019 (Morning)]**

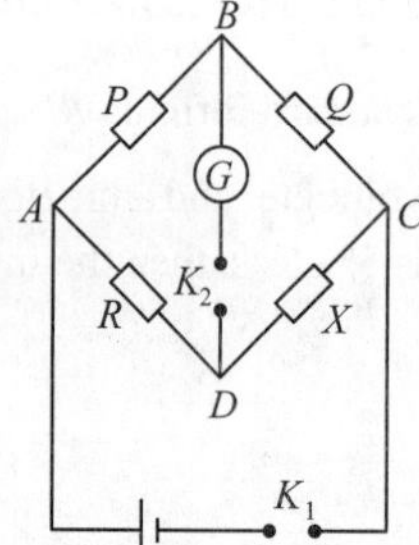

(A) 403.5 ohm
(B) 404.5 ohm
(C) 401.5 ohm
(D) 402.5 ohm

16. A galvanometer having a resistance of $20\ \Omega$ and 30 divisions on both sides has figure of merit 0.005 ampere/division. The resistance that should be connected in series such that it can be used as a voltmeter upto 15 volt, is : **[11 Jan. 2019 (Evening)]**

(A) $80\ \Omega$ (B) $120\ \Omega$
(C) $125\ \Omega$ (D) $100\ \Omega$

17. In the experimental set up of metre bridge shown in the figure, the null point is obtained at a distance of 40 cm from A. If a $10\ \Omega$ resistor is connected in series with R_1, the null point shifts by 10 cm. The resistance that should be connected in parallel with $(R_1 + 10)\ \Omega$ such that the null point shifts back to its initial position is : **[11 Jan. 2019 (Evening)]**

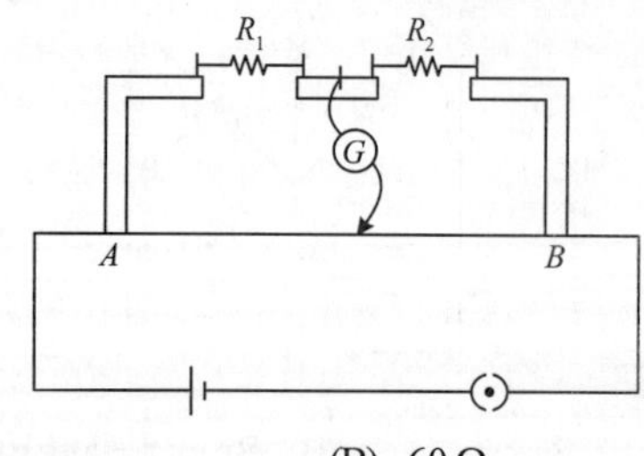

(A) $40\ \Omega$ (B) $60\ \Omega$
(C) $20\ \Omega$ (D) $30\ \Omega$

18. The galvanometer deflection, when key K_1 is closed but K_2 is open, equals θ_0 (see figure). On closing K_2 also and adjusting R_2 to $5\ \Omega$, the deflection in galvanometer becomes $\frac{\theta_0}{5}$. The resistance of the galvanometer is, then, given by [Neglect the internal resistance of battery] : **[12 Jan. 2019 (Morning)]**

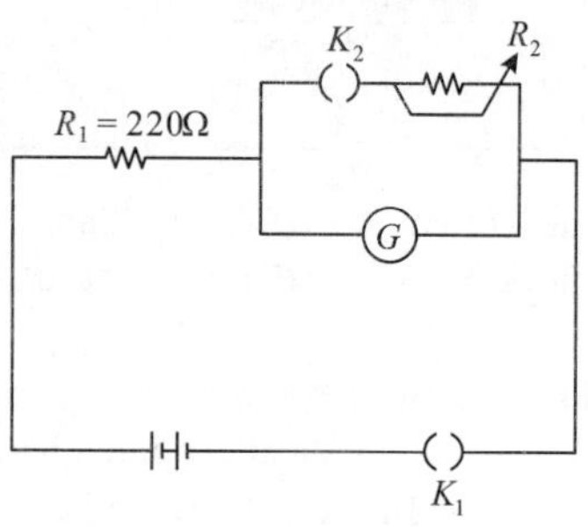

(A) $12\ \Omega$ (B) $25\ \Omega$
(C) $5\ \Omega$ (D) $22\ \Omega$

19. In a meter bridge, the wire of length 1 m has a non-uniform cross-section such that, the variation $\frac{dR}{dl}$ of its resistance R with length l is $\frac{dR}{dl} \propto \frac{1}{\sqrt{l}}$. Two equal resistances are connected as shown in the figure. The galvanometer has zero deflection when the jockey is at point P. What is the length AP ? **[12 Jan. 2019 (Morning)]**

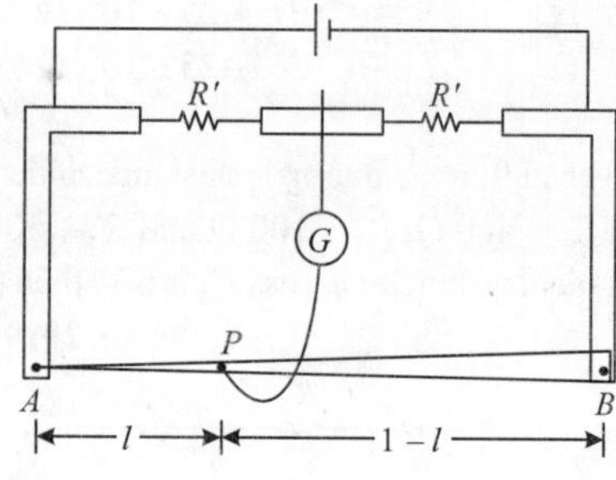

(A) 0.25 m (B) 0.3 m
(C) 0.35 m (D) 0.2 m

20. An ideal battery of 4 V and resistance R are connected in series in the primary circuit of a potentiometer of length 1 m and resistance $5\ \Omega$. The value of R, to give a potential difference of 5 mV across 10 cm of potentiometer wire, is : **[12 Jan. 2019 (Morning)]**

(A) $490\ \Omega$ (B) $480\ \Omega$
(C) $395\ \Omega$ (D) $495\ \Omega$

21. A galvanometer, whose resistance is 50 ohm, has 25 divisions in it. When a current of 4×10^{-4} A passes through it, its needle (pointer) deflects by one division. To use this galvanometer as a voltmeter of range 2.5 V, it should be connected to a resistance of : **[12 Jan. 2019 (Evening)]**

(A) 6250 ohm (B) 250 ohm
(C) 200 ohm (D) 6200 ohm

22. In the circuit shown, a four-wire potentiometer is made of a 400 cm long wire, which extends between A and B. The resistance per unit length of the potentiometer wire is $r = 0.01\ \Omega$/cm. If an ideal voltmeter is connected as shown with jockey J at 50 cm from end A, the expected reading of the voltmeter will be : **[8 April 2019 (Evening)]**

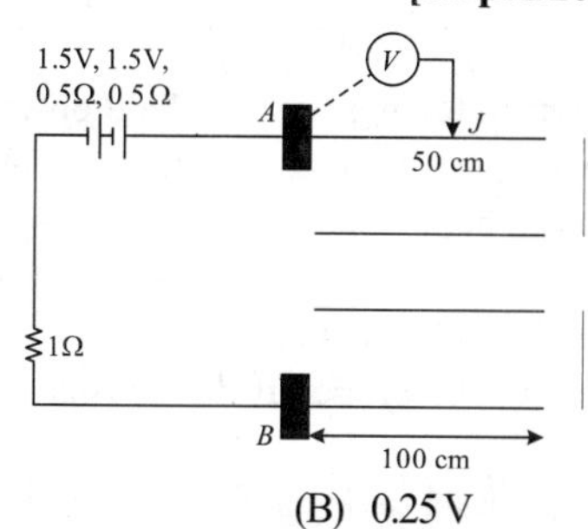

(A) 0.20 V (B) 0.25 V
(C) 0.75 V (D) 0.50 V

23. The resistance of a galvanometer is 50 ohm and the maximum current which can be passed through it is 0.002 A. What resistance must be connected to it in order to convert it into an ammeter of range 0 – 0.5 A? **[9 April 2019 (Evening)]**

(A) 0.2 ohm (B) 0.002 ohm
(C) 0.02 ohm (D) 0.5 ohm

24. In a meter bridge experiment, the circuit diagram and the corresponding observation table are shown in figure

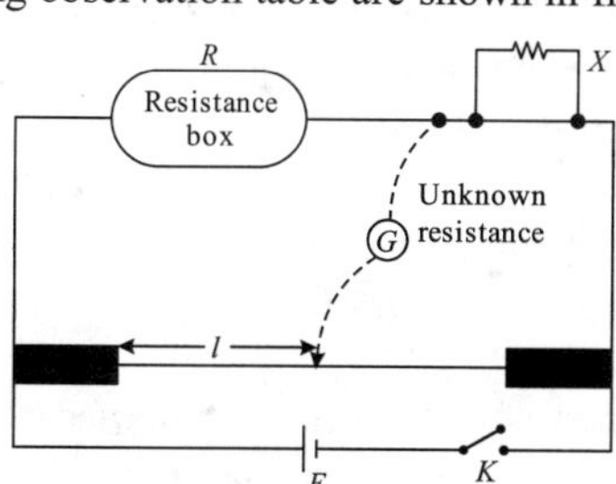

SI.No.	$R(\Omega)l$	l(cm)
1.	1000	60
2.	100	13
3.	10	1.5
4.	1	1.0

Which of the readings is inconsistent ? **[10 April 2019 (Morning)]**

(A) 4 (B) 1
(C) 2 (D) 3

25. In the given circuit, an ideal voltmeter connected across the 10 Ω resistance reads 2 V. The internal resistance r, of each cell is : **[10 April 2019 (Morning)]**

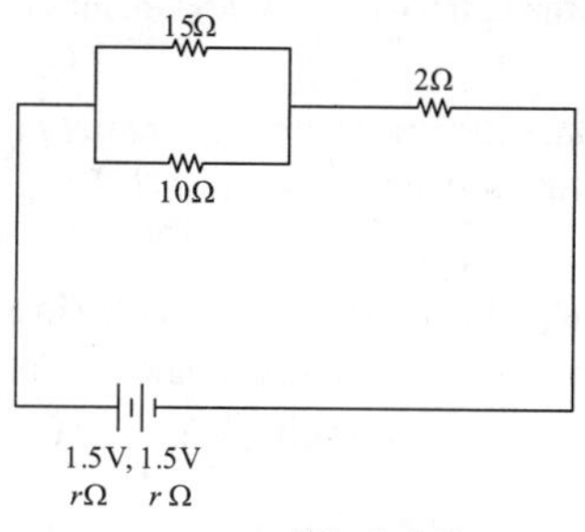

(A) 1 Ω (B) 1.5 Ω
(C) 0 Ω (D) 0.5 Ω

26. A galvanometer of resistance 100 Ω has 50 divisions on its scale and has sensitivity of 20 μA/division. It is to be converted to a voltmeter with three ranges, of 0 – 2 V, 0 – 10 V and 0 – 20 V. The appropriate circuit to do so is : **[12 April 2019 (Morning)]**

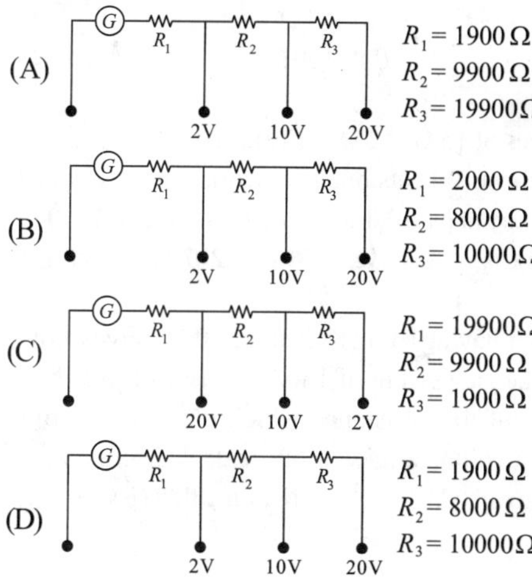

27. To verify Ohm's law, a student connects the voltmeter across the battery as, shown in the figure. The measured voltage is plotted as a function of the current, and the following graph is obtained. If V_0 is almost zero, identify the correct statement : **[12 April 2019 (Morning)]**

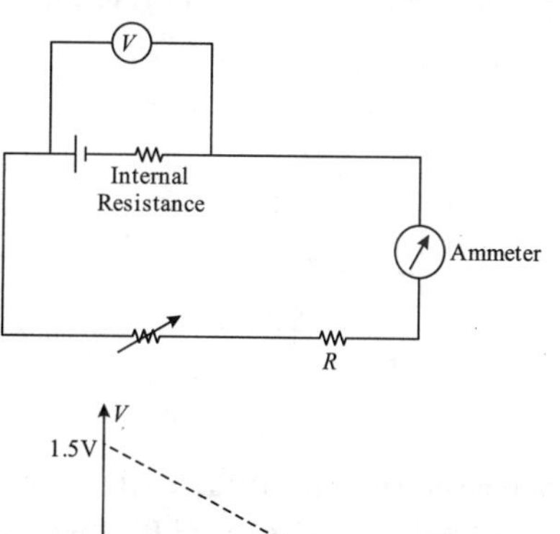

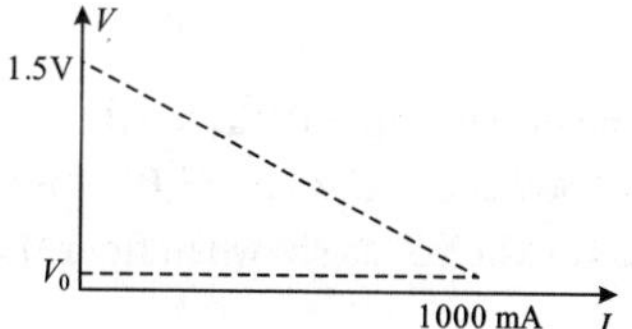

(A) The value of the resistance R is 1.5 Ω
(B) The emf of the battery is 1.5 V and the value of R is 1.5 Ω
(C) The emf of the battery is 1.5 V and its internal resistance is 1.5 Ω
(D) The potential difference across the battery is 1.5 V when it sends a current of 1000 mA

28. The balancing length for a cell is 560 cm in a potentiometer experiment. When an external resistance of 10 Ω is connected in parallel to the cell, the balancing length changes by 60 cm. If the internal resistance of the cell is $\frac{N}{10}\Omega$, where N is an integer then value of N is ______. **[7 Jan. 2020 (Evening)]**

29. The length of a potentiometer wire is 1200 cm and it carries a current of 60 mA. For a cell of emf 5 V and internal resistance of 20 Ω, the null point on it is found to be a 1000 cm. The resistance of whole wire is : **[8 Jan. 2020 (Morning)]**
(A) 120 Ω (B) 60 Ω
(C) 80 Ω (D) 100 Ω

30. Four resistances of 15 Ω, 12 Ω, 4 Ω and 10 Ω respectively in cyclic order to form. Wheatstone's network. The resistance that is to be connected in parallel with the resistance of 10 Ω to balance the network is ______ Ω. **[8 Jan. 2020 (Morning)]**

31. A galvanometer having a coil resistance 100 Ω gives a full scale deflection when a current of 1 mA is passed through it. What is the value of the resistance which can convert this galvanometer into a voltmeter giving full scale deflection for a potential difference of 10 V ? **[8 Jan. 2020 (Evening)]**
(A) 9.9 kΩ (B) 8.9 kΩ
(C) 7.9 kΩ (D) 10 kΩ

32. In a meter bridge experiment S is a standard resistance. R is a resistance wire. It Is found that balancing length is $l = 25$ cm. If R is replaced by a wire of half length and half diameter that of R of same material, then the balancing distance l' (in cm) will now be______. **[9 Jan. 2020 (Evening)]**

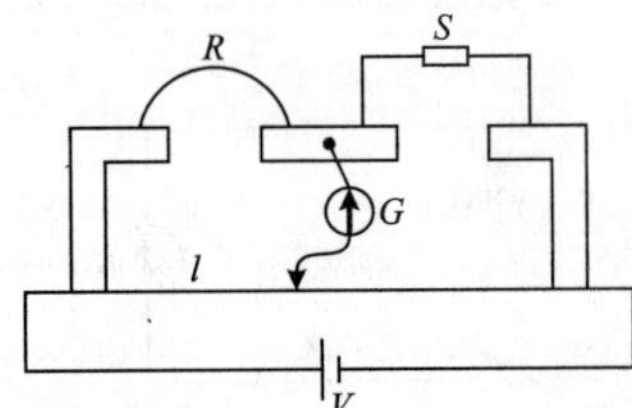

33. A potentiometer wire PQ of 1 m length is connected to a standard cell E_1. Another cell E_2 of emf 1.02 V is connected with a resistance 'r' and switch S (as shown in figure). With switch S open, the null position is obtained at a distance of 49 cm from Q. The potential gradient in the potentiometer wire is : **[2 Sept. 2020 (Evening)]**

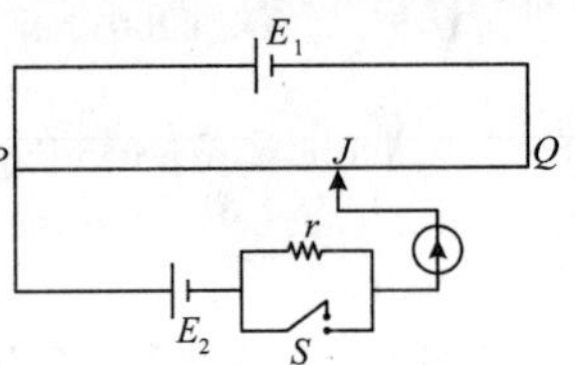

(A) 0.02 V/cm (B) 0.04 V/cm
(C) 0.01 V/cm (D) 0.03 V/cm

34. Which of the following will NOT be observed when a multimeter (operating in resistance measuring mode) probes connected across a component, are just reversed ? **[3 Sept. 2020 (Evening)]**
(A) Multimeter shows NO deflection in both cases *i.e.* before and after reversing the probes if the chosen component is capacitor
(B) Multimeter shows a deflection, accompanied by a splash of light out of connected component in one direction and NO deflection on reversing the probes if the chosen component is LED
(C) Multimeter shows NO deflection in both cases *i.e.* before and after reversing the probes if the chosen component is metal wire
(D) Multimeter shows an equal deflection in both cases *i.e.* before and after reversing the probes if the chosen component is resistor

35. Two resistors 400 Ω and 800 Ω are connected in series across a 6 V battery. The potential difference measured by a voltmeter of 10 $k\Omega$ across 400 Ω resistor is close to : **[3 Sept. 2020 (Evening)]**
(A) 2 V (B) 1.95 V
(C) 2.05 V (D) 1.8 V

36. A galvanometer of resistance G is converted into a voltmeter of range 0 – 1 V by connecting a resistance R_1 in series with it. The additional resistance that should be connected in series with R_1 to increase the range of the voltmeter to 0 – 2V will be : **[5 Sept. 2020 (Morning)]**
(A) R_1 (B) $R_1 + G$
(C) $R_1 - G$ (D) G

37. A galvanometer is used in laboratory for detecting the null point in electrical experiments. If, on passing a current of 6 mA it produces a deflection of 2°, its figure of merit is close to : **[5 Sept. 2020 (Evening)]**
(A) 3×10^{-3} A/div (B) 333° A/div
(C) 6×10^{-3} A/div (D) 666° A/div

38. A circuit to verify Ohm's law uses ammeter and voltmeter in series or parallel connected correctly to the resistor. In the circuit : **[6 Sept. 2020 (Evening)]**

(A) Ammeter is always connected in series and voltmeter in parallel.

(B) Both, ammeter and voltmeter must be connected in series.

(C) Both ammeter and voltmeter must be connected in parallel.

(D) Ammeter is always used in parallel and voltmeter is series.

39. In the given circuit of potentiometer, the potential difference E across AB (10 m length) is larger than E_1 and E_2 as well. For key K_1 (closed), the jockey is adjusted to touch the wire at point J_1 so that there is no deflection in the galvanometer. Now the first battery (E_1) is replaced by second battery (E_2) for working by making K_1 open and K_2 closed. The galvanometer gives then null deflection at J_2. The value of $\frac{E_1}{E_2}$ is $\frac{a}{b}$, where $a =$ ______. **[25 Feb. 2021 (Morning)]**

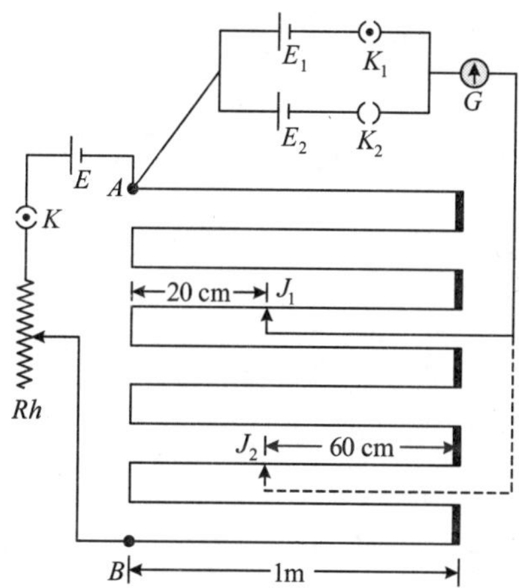

40. Consider a 72 cm long wire AB as shown in the figure. The galvanometer jockey is placed at P on AB at a distance x cm from A. The galvanometer shows zero deflection.

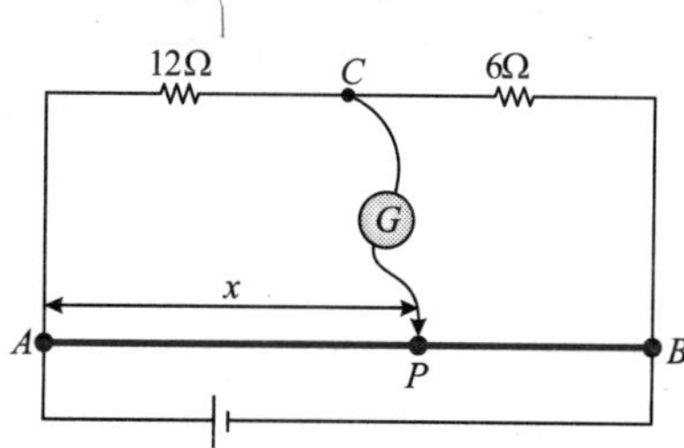

The value of x, to the nearest integer, is ? **[18 March 2021 (Evening)]**

41. In the given figure, there is a circuit of potentiometer of length $AB = 10$ m. The resistance per unit length of 0.1 Ω per cm. Across AB, a battery of emf E and internal resistance r is connected. The maximum value of emf measured by this potentiometer is : **[25 July 2021 (Morning)]**

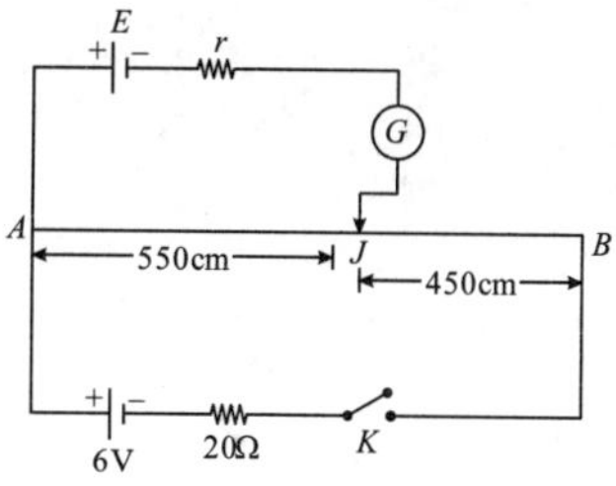

(A) 2..25 V (B) 6V

(C) 5V (D) 2.75 V

42. The given potentiometer has its wire of resistance 10 Ω When the sliding contact is in the middle of the potentiometer wire, the potential drop across 2 Ω resistor is : **[25 July 2021 (Evening)]**

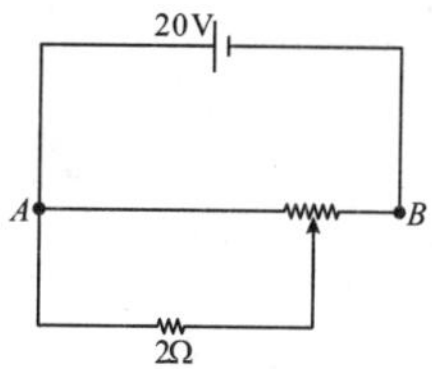

(A) $\frac{40}{9}$V (B) 5V

(C) $\frac{40}{11}$V (D) 10V

43. In the given potentiometer circuit arrangement, the balancing length AC is measured to be 250 cm. When the galvanometer connection is shifted from point (1) to point (2) in the given diagram, the balancing length becomes 400 cm. The ration of the emf of two cells, $\frac{\varepsilon_1}{\varepsilon_2}$ is : **[25 July 2021 (Evening)]**

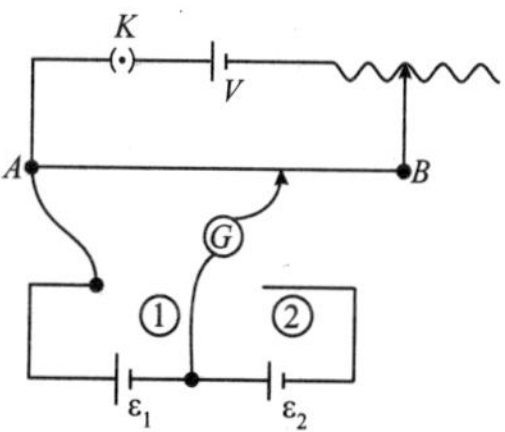

(A) $\frac{4}{3}$ (B) $\frac{5}{3}$

(C) $\frac{8}{5}$ (D) $\frac{3}{2}$

44. For full scale deflection of total 50 divisions, 50 mV voltage is required in galvanometer. The resistance of galvanometer if its current sensitivity is 2 div/mA will be :

[27 Aug. 2021 (Evening)]

(A) 1 Ω (B) 2 Ω
(C) 5 Ω (D) 4 Ω

45. Consider a galvanometer shunted with 5 Ω resistance and 2% of current passes through it. What is the resistance of the given galvanometer ? **[31 Aug. 2021 (Morning)]**

(A) 226 Ω (B) 245 Ω
(C) 344 Ω (D) 300 Ω

* * * * *

CHAPTER

18 MOVING CHARGES & MAGNETISM

18.1 MAGNETIC EFFECTS OF CURRENT

1. A current of 1 A is flowing on the sides of an equilateral triangles of side 4.5×10^{-2} m. The magnetic field at the centre of the triangle will be : **[2018 (Online)]**

(A) 4×10^{-5} Wb/m^2 (B) Zero
(C) 2×10^{-5} Wb/m^2 (D) 8×10^{-5} Wb/m^2

2. A Helmholtz coil has pair of loops, each with N turns and radius R. They are placed coaxially at distance R and the same current I flows through the loops in the same direction. The magnitude of magnetic field at P, midway between the centres A and C, is given by (Refer to figure) : **[2018 (Online)]**

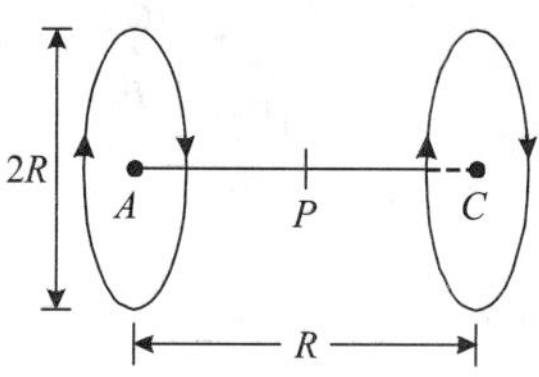

(A) $\frac{4N\mu_0 I}{5^{3/2}R}$ (B) $\frac{8N\mu_0 I}{5^{3/2}R}$

(C) $\frac{4N\mu_0 I}{5^{1/2}R}$ (D) $\frac{8N\mu_0 I}{5^{1/2}R}$

3. A current loop, having two circular arcs joined by two radial lines is shown in the figure. It carries a current of 10 A. The magnetic field at point O will be close to : **[9 Jan. 2019 (Morning)]**

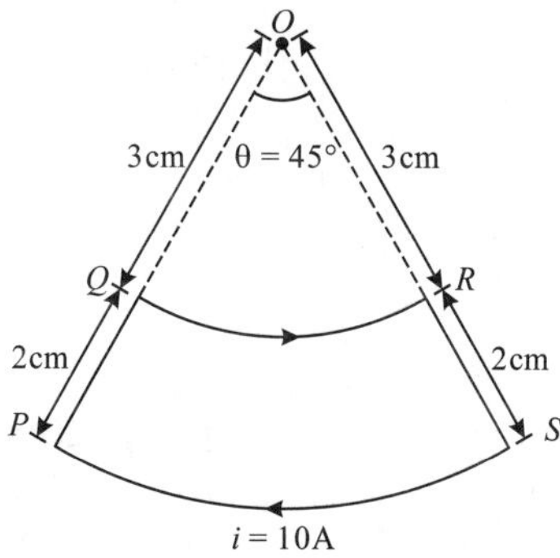

(A) 1.0×10^{-5} T (B) 1.5×10^{-5} T
(C) 1.0×10^{-7} T (D) 1.0×10^{-7} T

4. One of the two identical conducting wires of length L is bent in the form of a circular loop and the other one into a circular coil of N identical turns. If the same current is passed in both, the ratio of the magnetic field at the center of the loop (B_L) to that at the centre of the coil (B_C), then $\frac{B_L}{B_C}$ will be : **[9 Jan. 2019 (Evening)]**

(A) $\frac{1}{N}$ (B) N^2

(C) $\frac{1}{N^2}$ (D) N

5. As shown in the figure, two infinitely long, identical wires are bent by 90° and placed in such a way that the segments LP and QM are along the x-axis, while segments PS and QN are parallel to the y-axis. If $OP = OQ = 4$ cm, and the magnitude of the magnetic field at O is 10^{-4} T, and the two wires carry equal currents (see figure), the magnitude of the current in each wire and the direction of the magnetic field at O will be ($\mu_0 = 4\pi \times 10^{-7}$ NA^{-2}) : **[12 Jan. 2019 (Morning)]**

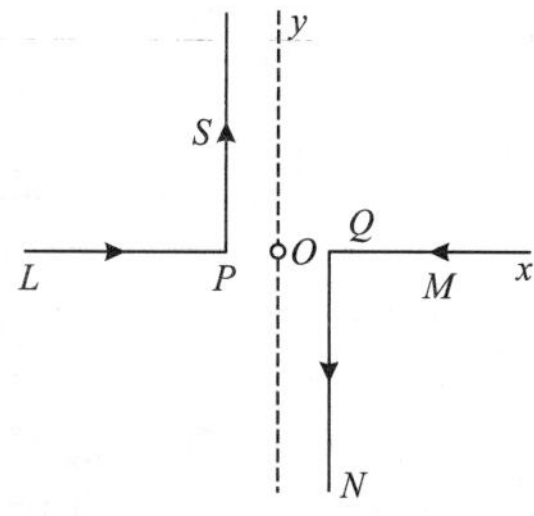

(A) 40 A, perpendicular into the page
(B) 40 A, perpendicular out of the page
(C) 20 A, perpendicular out of the page
(D) 20 A, perpendicular into the page

6. In SI units, the dimensions of $\sqrt{\frac{\epsilon_0}{\mu_0}}$ is : **[8 April 2019 (Morning)]**

(A) $A^{-1}TML^3$ (B) $A^2T^3M^{-1}L^{-2}$
(C) $AT^2M^{-1}L^{-1}$ (D) $AT^{-3}ML^{3/2}$

7. Two very long, straight, and insulated wires are kept at 90° angle from each other in xy-plane as shown in the figure. These wires carry currents of equal magnitude I, whose directions are shown in the figure. The net magnetic field at point P will be :

[8 April 2019 (Evening)]

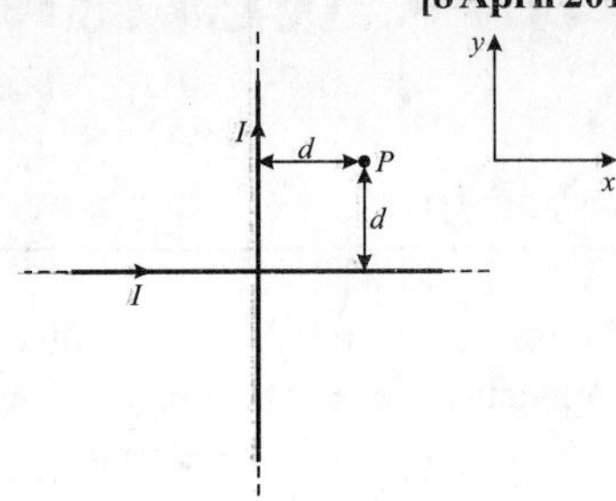

(A) Zero (B) $\frac{+\mu_0 I}{\pi d}(\hat{z})$

(C) $-\frac{\mu_0 I}{2\pi d}(\hat{x}+\hat{y})$ (D) $\frac{\mu_0 I}{2\pi d}(\hat{x}+\hat{y})$

8. The magnitude of the magnetic field at the center of an equilateral triangular loop of side 1 m which is carrying a current of 10 A is : [Take $\mu_0 = 4\pi \times 10^{-7}$ NA^{-2}] **[10 April 2019 (Evening)]**

(A) 18 μT (B) 3 μT

(C) 1 μT (D) 9 μT

9. Which of the following combinations has the dimension of electrical resistance (ε_0 is the permittivity of vacuum and μ_0 is the permeability of vacuum) ? **[12 April 2019 (Morning)]**

(A) $\sqrt{\frac{\varepsilon_0}{\mu_0}}$ (B) $\frac{\mu_0}{\varepsilon_0}$

(C) $\sqrt{\frac{\mu_0}{\varepsilon_0}}$ (D) $\frac{\varepsilon_0}{\mu_0}$

10. A thin ring of 10 cm radius carries a uniformly distributed charge. The ring rotates at a constant angular speed of 40π rad s^{-1} about its axis, perpendicular to its plane. If the magnetic field at its centre is 3.8×10^{-9} T, then the charge carried by the ring is close to ($\mu_0 = 4\pi \times 10^{-7}$ N/A^2) :

[12 April 2019 (Morning)]

(A) 2×10^{-6} C (B) 3×10^{-5} C

(C) 4×10^{-5} C (D) 7×10^{-6} C

11. Find the magnetic field at point P due to a straight line segment AB of length 6 cm carrying a current of 5A. (See figure) ($\mu_0 = 4\pi \times 10^{-7}$ N-A^{-2}) : **[12 April 2019 (Evening)]**

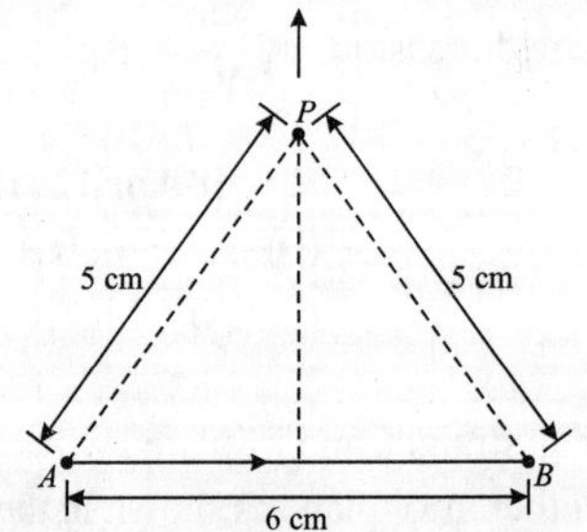

(A) 3.0×10^{-5} T (B) 2.5×10^{-5} T

(C) 2.0×10^{-5} T (D) 1.5×10^{-5} T

12. The dimension of $B^2/2\mu_0$, where B is magnetic field and μ_0 is the magnetic permeability of vacuum, is :

[7 Jan. 2020 (Evening)]

(A) $ML^{-1}T^{-2}$ (B) ML^2T^{-1}

(C) MLT^{-2} (D) ML^2T^{-2}

13. A very long wire $ABDMNDC$ is shown in figure carrying current I. AB and BC parts are straight, long and at right angle. At D wire forms a circular turn $DMND$ of radius R. AB, BC parts are tangential to circular turn at N and D. Magnetic field at the centre of circle is : **[8 Jan. 2020 (Evening)]**

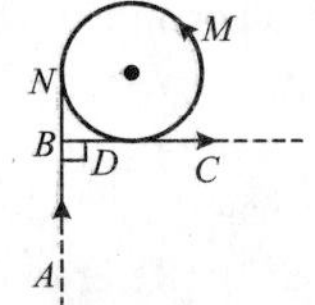

(A) $\frac{\mu_0 I}{2R}$

(B) $\frac{\mu_0 I}{2\pi R}(\pi+1)$

(C) $\frac{\mu_0 I}{2\pi R}\left(\pi+\frac{1}{\sqrt{2}}\right)$ (D) $\frac{\mu_0 I}{2\pi R}\left(\pi-\frac{1}{\sqrt{2}}\right)$

14. A long, straight wire of radius a carries a current distributed uniformly over its cross-section. The ratio of the magnetic fields due to the wire at distance $a/3$ and $2a$, respectively from the axis of the wire is : **[9 Jan. 2020 (Morning)]**

(A) $\frac{2}{3}$ (B) $\frac{3}{2}$

(C) $\frac{1}{2}$ (D) 2

15. Magnitude of magnetic field (in SI units) at the centre of a hexagonal shape coil of side 10 cm, 50 turns and carrying current I (Ampere) in units of $\mu_0 I/\pi$ is : **[3 Sept. 2020 (Morning)]**

(A) $250\sqrt{3}$ (B) $5\sqrt{3}$

(C) $500\sqrt{3}$ (D) $50\sqrt{3}$

16. A wire A, bent in the shape of an arc of a circle, carrying a current of $2A$ and having radius 2 cm and another wire B, also bent in the shape of arc of a circle, carrying a current of $3A$ and having radius of 4 cm, are placed as shown in the figure. The ratio of the magnetic fields due to the wires A and B at the common centre O is : **[4 Sept. 2020 (Morning)]**

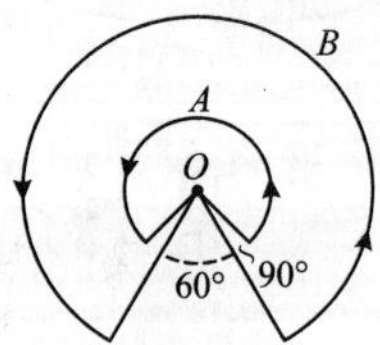

(A) 4:6 (B) 6:4

(C) 6:5 (D) 2:5

17. Magnetic fields at two points on the axis of a circular coil at a distance of 0.05 m and 0.2 m from the centre are in the ratio 8 : 1. The radius of coil is______. **[25 Feb. 2021 (Morning)]**

(A) 0.1 m (B) 0.2 m

(C) 1.0 m (D) 0.15 m

18. A charge Q is moving $\vec{dl}$ distance in the magnetic field $\vec{B}$. Find the value of work done by $\vec{B}$: **[16 March 2021 (Evening)]**

(A) 1 (B) Infinite

(C) Zero (D) –1

19. A solenoid of 1000 turns per metre has a core with relative permeability 500. Insulated windings of the solenoid carry an electric current of 5A. The magnetic flux density produced by the solenoid is : **[17 March 2021 (Morning)]**

(permeability of free space $= 4\pi \times 10^{-7}$ H/m)

(A) πT (B) $2 \times 10^{-3}\pi$T

(C) $\frac{\pi}{5}$T (D) $10^{-4}\pi$T

20. A hairpin like shape as shown in figure is made by bending a long current carrying wire. What is the magnitude of a magnetic field at point P which lies on the centre of the semicircle ? **[17 March 2021 (Evening)]**

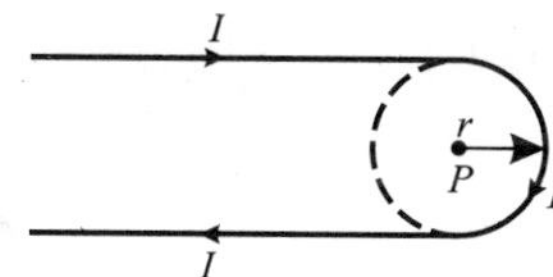

(A) $\frac{\mu_0 I}{4\pi r}(2-\pi)$ (B) $\frac{\mu_0 I}{4\pi r}(2+\pi)$

(C) $\frac{\mu_0 I}{2\pi r}(2+\pi)$ (D) $\frac{\mu_0 I}{2\pi r}(2-\pi)$

21. Figure-(A) and (B) show two long straight wires of circular cross-section (a and b with $a < b$), carrying current I which is uniformly distributed across the cross-section. The magnitude of magnetic field B varies with radius r and an be represented as : **[27 July 2021 (Evening)]**

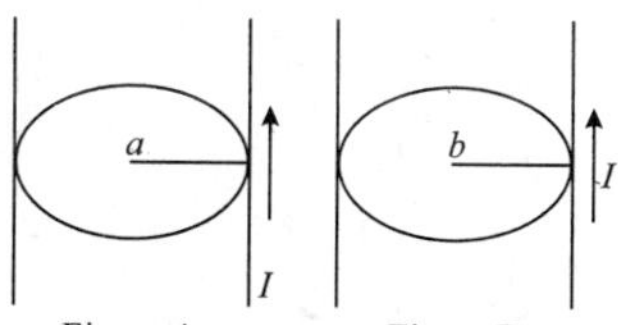

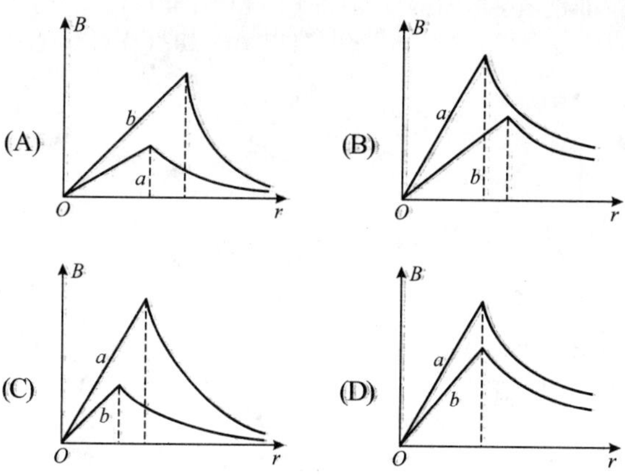

22. What will be the magnitude of electric field at point O as shown in figure ? Each side of the figure is l and perpendicular to each other ? **[27 July 2021 (Evening)]**

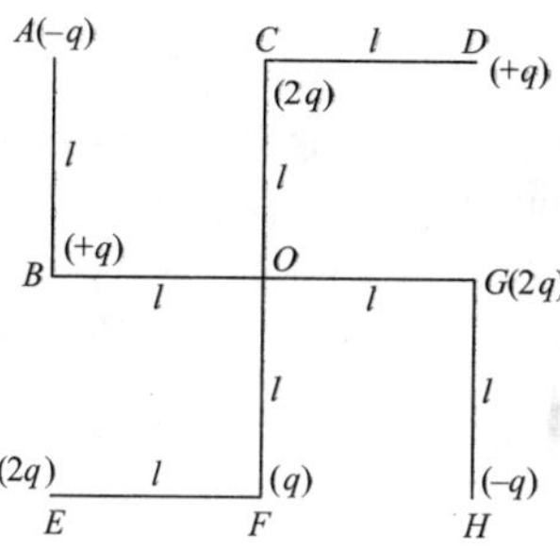

(A) $\frac{1}{4\pi\varepsilon_0}\frac{q}{(2l^2)}(2\sqrt{2}-1)$ (B) $\frac{1}{4\pi\varepsilon_0}\frac{q}{(2l^2)}(\sqrt{2})$

(C) $\frac{1}{4\pi\varepsilon_0}\frac{q}{l^2}$ (D) $\frac{1}{4\pi\varepsilon_0(2l)^2}$

23. The fractional change in the magnetic field intensity at distance 'r' from centre on the axis of current carrying coil of radius 'a' to the magnetic field intensity at centre of the same coil is : (Take $r << a$) **[26 Aug. 2021 (Morning)]**

(A) $\frac{2}{3}\frac{r^2}{a^2}$ (B) $\frac{3}{2}\frac{r^2}{a^2}$

(C) $\frac{2}{3}\frac{a^2}{r^2}$ (D) $\frac{3}{2}\frac{a^2}{r^2}$

24. A coaxial cable consists of an inner wire of radius a surrounded by an outer shell of inner and outer radii b and c respectively. The inner wire carries an electric current i_0, which is distributed uniformly across cross-sectional area. The outer shell carries an equal current in opposite direction and distributed uniformly. What will be the ratio of the magnetic field at a distance x from the axis when (i) $x < a$ and (ii) $a < x < b$?

[27 Aug. 2021 (Evening)]

(A) $\frac{a^2}{x^2}$ (B) $\frac{b^2 - a^2}{x^2}$

(C) $\frac{x^2}{b^2 - a^2}$ (D) $\frac{x^2}{a^2}$

25. A coil having N turns is wound tightly in the form of a spiral with inner and outer radii a and b respectively. Find the magnetic field at centre, when a current I passes through coil :

[31 Aug. 2021 (Morning)]

(A) $\frac{\mu_0 I}{a(a-b)}\left[\frac{1}{a} - \frac{1}{b}\right]$ (B) $\frac{\mu_0 IN}{2(b-a)}\log_e\left(\frac{b}{a}\right)$

(C) $\frac{\mu_0 I}{8}\left[\frac{a+b}{a-b}\right]$ (D) $\frac{\mu_0 I}{8}\left[\frac{a-b}{a-b}\right]$

26. A current of 1.5 A is flowing through a triangle, of side 9 cm each. The magnetic field at the centroid of the triangle is : (Assume that the current is flowing in the clockwise direction)

[31 Aug. 2021 (Evening)]

(A) $2\sqrt{3} \times 10^{-7}$ T, outside the plane of triangle

(B) 3×10^{-7} T, outside the plane of triangle

(C) $2\sqrt{3} \times 10^{-5}$ T, inside the plane of triangle

(D) 3×10^{-5} T, inside the plane of triangle

27. There are two infinitely long straight current carrying conductors and they are held at right angles to each other so that their common ends meet at the origin as shown in the figure given below. The ratio of current in both conductors is 1 : 1. The magnetic field at point P is ______.

[01 Sep. 2021 (Evening)]

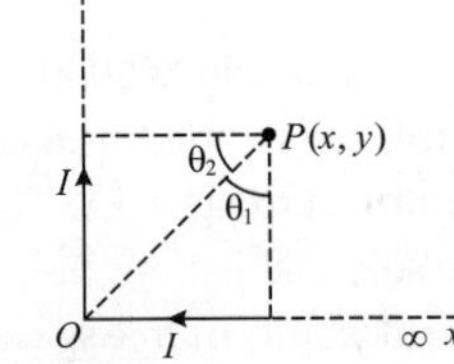

(A) $\frac{\mu_0 I}{4\pi xy}[\sqrt{x^2 + y^2} - (x+y)]$

(B) $\frac{\mu_0 Ixy}{4\pi}[\sqrt{x^2 + y^2} - (x+y)]$

(C) $\frac{\mu_0 Ixy}{4\pi}[\sqrt{x^2 + y^2} - (x+y)]$

(D) $\frac{\mu_0 I}{4\pi xy}[\sqrt{x^2 + y^2} + (x+y)]$

* * * * *

CHAPTER

18 MOVING CHARGES & MAGNETISM

18.2 MAGNETIC FORCE ON MOVING CHARGES

1. A uniform magnetic field B of 0.3 T is along the positive Z-direction. A rectangular loop ($abcd$) of sides 10 cm × 5 cm carries a current I of 12 A. Out of the following different orientations which one corresponds to stable equilibrium? **[2017 (Online)]**

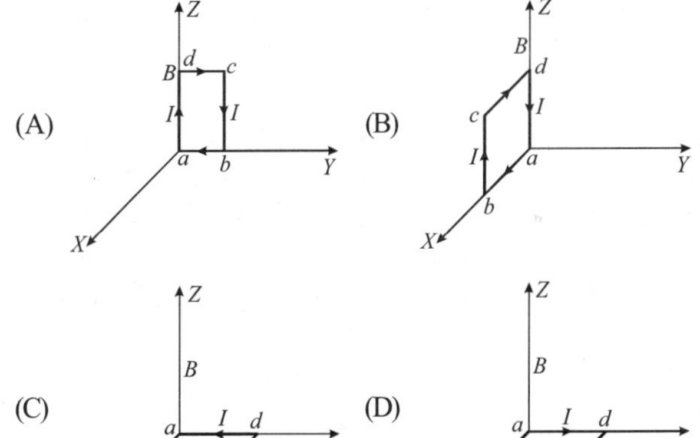

2. In a certain region static electric and magnetic fields exist. The magnetic field is given by $\vec{B} = B_0(\hat{i} + 2\hat{j} - 4\hat{k})$. If a test charge moving with a velocity $\vec{v} = v_0(3\hat{i} - \hat{j} + 2\hat{k})$ experiences no force in that region, then the electric field in the region, in SI units, is : **[2017 (Online)]**

(A) $\vec{E} = -v_0B_0(3\hat{i} - 2\hat{j} - 4\hat{k})$

(B) $\vec{E} = -v_0B_0(\hat{i} + \hat{j} + 7\hat{k})$

(C) $\vec{E} = v_0B_0(14\hat{j} + 7\hat{k})$

(D) $\vec{E} = -v_0B_0(14\hat{j} + 7\hat{k})$

3. A negative test charge is moving near a long straight wire. Carrying a current. The force acting on the test charge is parallel to the direction of the current. The motion of the charge is : **[2017 (Online)]**

(A) Away from the wire
(B) Towards the wire
(C) Parallel to the wire along the current
(D) Parallel to the wire opposite to the current

4. The dipole moment of a circular loop carrying a current I, is m and the magnetic field at the centre of the loop is B_1. When the dipole moment is doubled by keeping the current constant, the magnetic field at the centre of the loop is B_2. The ratio $\frac{B_1}{B_2}$ is : **[2018]**

(A) 2 (B) $\sqrt{3}$

(C) $\sqrt{2}$ (D) $\frac{1}{\sqrt{2}}$

5. An electron, a proton and an alpha particle having the same kinetic energy are moving in circular orbits of radii r_e, r_p, r_α respectively in a uniform magnetic field B. The relation between r_e, r_p, r_α is : **[2018]**

(A) $r_e > r_p = r_\alpha$ (B) $r_e < r_p = r_\alpha$

(C) $r_e < r_p < r_\alpha$ (D) $r_e < r_\alpha < r_p$

6. A charge q is spread uniformly over an insulated loop of radius r. If it is rotated with an angular velocity ω with respect to normal axis then the magnetic moment of the loop is : **[2018 (Online)]**

(A) $\frac{1}{2}q\omega r^2$ (B) $\frac{4}{3}q\omega r^2$

(C) $\frac{3}{2}q\omega r^2$ (D) $q\omega r^2$

7. An infinitely long current carrying wire and a small current carrying loop are in the plane of the paper as shown. The radius of the loop is 'a' and distance of its centre from the wire is d ($d >> a$). If the loop applies a force F on the wire then : **[9 Jan. 2019 (Morning)]**

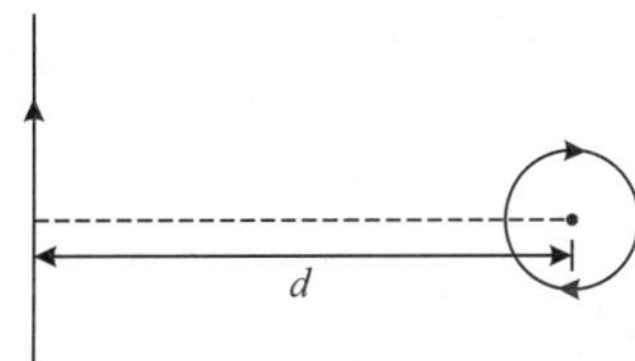

(A) $F \propto \left(\frac{a^2}{d^3}\right)$ (B) $F \propto \left(\frac{a}{d}\right)$

(C) $F \propto \left(\frac{a}{d}\right)^2$ (D) $F = 0$

8. A particle having the same charge as of electron moves in a circular path of radius 0.5 cm under the influence of a magnetic field of 0.5 T. If an electric field of 100 V/m makes it to move in a straight path, then the mass of the particle is (Given charge of electron $=1.6 \times 10^{-19}$ C) : **[9 Jan. 2019 (Evening)]**

(A) 2.0×10^{-24} kg (B) 1.6×10^{-19} kg
(C) 1.6×10^{-27} kg (D) 9.1×10^{-31} kg

9. An insulating thin rod of length l has a linear charge density $\rho(x) = \rho_0 \frac{x}{l}$ on it. The rod is rotated about an axis passing through the origin ($x = 0$) and perpendicular to the rod. If the rod makes n rotation per second, then the time averaged magnetic moment of the rod is : **[10 Jan. 2019 (Morning)]**

(A) $\frac{\pi}{4} n\rho l^3$ (B) $n\rho l^3$
(C) $\pi n\rho l^3$ (D) $\frac{\pi}{3} n\rho l^3$

10. In an experiment electrons are accelerated, from rest, by applying a voltage of 500 V. Calculate the radius of the path if a magnetic field 100 mT is then applied.
[Charge of the electron $= 1.6 \times 10^{-19}$ C
Mass of the electron $= 9.1 \times 10^{-31}$ kg] : **[11 Jan. 2019 (Morning)]**

(A) 7.5×10^{-4} m (B) 7.5×10^{-3} m
(C) 7.5 m (D) 7.5×10^{-2} m

11. The region between $y = 0$ and $y = d$ contains a magnetic field $\vec{B} = B\hat{z}$. A particle of mass m and charge q enters the region with a velocity $\vec{v} = v\hat{i}$. If $d = \frac{mv}{2qB}$, the acceleration of the charged particle at the point of its emergence at the other side is : **[11 Jan. 2019 (Evening)]**

(A) $\frac{qvB}{m}\left(\frac{\hat{i}+\hat{j}}{\sqrt{2}}\right)$ (B) $\frac{qvB}{m}\left(\frac{1}{2}\hat{i}-\frac{\sqrt{3}}{\sqrt{2}}\hat{j}\right)$
(C) $\frac{qvB}{m}\left(\frac{-\hat{j}+\hat{i}}{\sqrt{2}}\right)$ (D) $\frac{qvB}{m}\left(\frac{\sqrt{3}}{2}\hat{i}+\frac{1}{2}\hat{j}\right)$

12. A particle of mass m and charge q is in an electric and magnetic field given by

$$\vec{E} = 2\hat{i}+3\hat{j} \ ; \ \vec{B} = 4\hat{j}+6\hat{k}.$$

The charged particle is shifted from the origin to the point $P(x=1; y=1)$ along a straight path. The magnitude of the total work done is : **[11 Jan. 2019 (Evening)]**

(A) $(0.35)q$ (B) $(0.15)q$
(C) $(2.5)q$ (D) $5q$

13. A proton and an α-particle (with their masses in the ratio of 1 : 4 and charges in the ratio of 1 : 2) are accelerated from rest through a potential difference V. If a uniform magnetic field (B) is set up perpendicular to their velocities, the ratio of the radii $r_p : r_\alpha$ of the circular paths described by them will be : **[12 Jan. 2019 (Morning)]**

(A) $1:\sqrt{2}$ (B) 1 : 2
(C) 1 : 3 (D) $1:\sqrt{3}$

14. A circular coil having N turns and radius r carries a current I. It is held in the XZ plane in a magnetic field $B\hat{i}$. The torque on the coil due to the magnetic field is : **[8 April 2019 (Morning)]**

(A) $B\pi r^2 IN$ (B) $\frac{Br^2 I}{\pi N}$
(C) Zero (D) $\frac{B\pi r^2 I}{N}$

15. A moving coil galvanometer has resistance 50 Ω and it indicates full deflection at 4 mA current. A voltmeter is made using this galvanometer and a 5 kΩ resistance. The maximum voltage, that can be measured using this voltmeter, will be close to : **[9 April 2019 (Morning)]**

(A) 10V (B) 20V
(C) 40V (D) 15V

16. A rigid square loop of side 'a' and carrying current I_2 is lying on a horizontal surface near a long current I_1 carrying wire in the same plane as shown in figure. The net force on the loop due to wire will be : **[9 April 2019 (Morning)]**

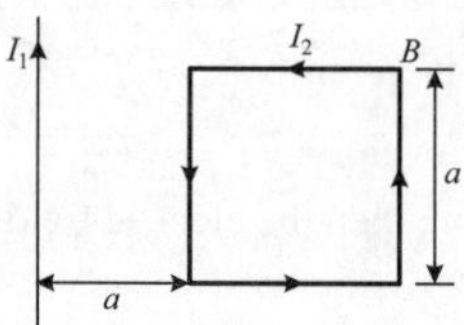

(A) Attractive and equal to $\frac{\mu_0 I_1 I_2}{3\pi}$

(B) Repulsive and equal to $\frac{\mu_0 I_1 I_2}{4\pi}$

(C) Repulsive and equal to $\frac{\mu_0 I_1 I_2}{2\pi}$

(D) Zero

17. A rectangular coil (Dimension 5 cm × 2.5 cm) with 100 turns, carrying a current of 3A in the clock-wise direction is kept centered at the origin and in the X-Z plane. A magnetic field of 1T is applied along X-axis. If the coil is tilted through 45° about Z-axis, then the torque on the coil is : **[9 April 2019 (Morning)]**

(A) 0.55 Nm (B) 0.27 Nm
(C) 0.38 Nm (D) 0.42 Nm

18. A moving coil galvanometer has a coil with 175 turns and area 1 cm^2. It uses a torsion band of torsion constant 10^{-6} N-m/rad. The coil is placed in a magnetic field B parallel to its plane. The coil deflects by 1° for a current of 1 mA. The value of B (in Tesla) is approximately : **[9 April 2019 (Evening)]**

(A) 10^{-3} (B) 10^{-1}
(C) 10^{-4} (D) 10^{-2}

19. A proton, an electron, and a Helium nucleus, have the same energy. They are in circular orbits in a plane due to magnetic field perpendicular to the plane. Let r_p, r_e and r_{He} be their respective radii, then : **[10 April 2019 (Morning)]**

(A) $r_e > r_p > r_{He}$ (B) $r_e < r_p < r_{He}$
(C) $r_e < r_p = r_{He}$ (D) $r_e > r_p = r_{He}$

20. Two wires A & B are carrying currents I_1 & I_2 as shown in the figure. The separation between them is d. A third wire C carrying a current I is to be kept parallel to them at a distance x from A such that the net force acting on it is zero. The possible values of x are : **[10 April 2019 (Morning)]**

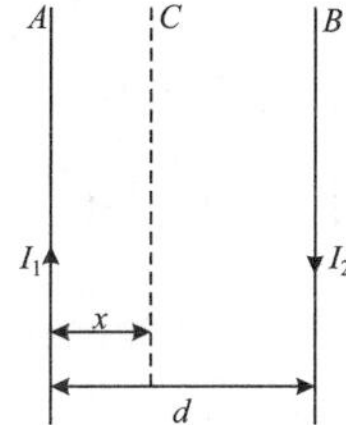

(A) $x = \left(\frac{I_1}{I_1 - I_2}\right)d$ and $x = \frac{I_2}{(I_1 + I_2)}d$

(B) $x = \pm\frac{I_1 d}{(I_1 - I_2)}$

(C) $x = \left(\frac{I_1}{I_1 + I_2}\right)d$ and $x = \frac{I_2}{(I_1 - I_2)}d$

(D) $x = \left(\frac{I_2}{I_1 + I_2}\right)d$ and $x = \left(\frac{I_2}{I_1 - I_2}\right)d$

21. A moving coil galvanometer allows a full scale current of 10^{-4} A. A series resistance of 2 $M\Omega$ is required to convert the above galvanometer into a voltmeter of range 0-5 V. Therefore the value of shunt resistance required to convert the above galvanometer into an ammeter of range 0-10 mA is : **[10 April 2019 (Morning)]**

(A) 200 Ω (B) 100 Ω
(C) 10 Ω (D) 500 Ω

22. A moving coil galvanometer, having a resistance G, produces full scale deflection when a current I_g flows through it. This galvanometer can be converted into (i) an ammeter of range 0 to I_0 ($I_0 > I_g$) by connecting a shunt resistance R_A to it and (ii) into a voltmeter of range 0 to $V(V = GI_0)$ by connecting a series resistance R_V to it. Then : **[12 April 2019 (Evening)]**

(A) $R_A R_V = G^2\left(\frac{I_g}{I_0 - I_g}\right)$ and $\frac{R_A}{R_V} = \left(\frac{I_0 - I_g}{I_g}\right)^2$

(B) $R_A R_V = G^2$ and $\frac{R_A}{R_V} = \left(\frac{I_g}{I_0 - I_g}\right)^2$

(C) $R_A R_V = G^2$ and $\frac{R_A}{R_V} = \frac{I_g}{(I_0 - I_g)}$

(D) $R_A R_V = G^2\left(\frac{I_0 - I_g}{I_g}\right)$ and $\frac{R_A}{R_V} = \left(\frac{I_g}{I_0 - I_g}\right)^2$

23. An electron, moving along the x-axis with an initial energy of 100 eV, enters a region of magnetic field $\vec{B} = (1.5 \times 10^{-3} T)\hat{k}$ at S (See figure). The field extends between $x = 0$ and $x = 2$ cm. The electron is detected at the point Q on a screen placed 8 cm away from the point S. The distance d between P and Q (on the screen) is : (electron's charge = 1.6×10^{-19} C, mass of electron = 9.1×10^{-31} kg) : **[12 April 2019 (Evening)]**

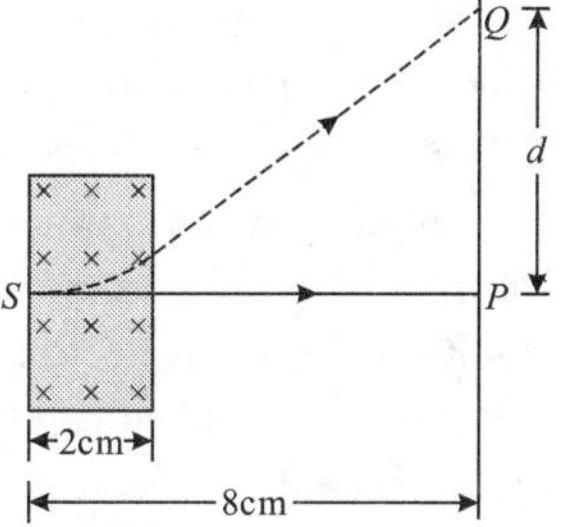

(A) 12.87 cm (B) 1.22 cm
(C) 11.65 cm (D) 2.25 cm

24. A particle of mass m and charge q has an initial velocity $\vec{v} = -v_0\hat{j}$. If an electric field $\vec{E} = E_0\hat{i}$ and magnetic field $\vec{B} = B_0\hat{i}$ act on the particle, its speed will double after a time : **[7 Jan. 2020 (Evening)]**

(A) $\frac{2mv_0}{qE_0}$ (B) $\frac{3mv_0}{qE_0}$

(C) $\frac{\sqrt{3}mv_0}{qE_0}$ (D) $\frac{\sqrt{2}mv_0}{qE_0}$

25. Protons with kinetic energy of 1 MeV moves from south to north. It gets an acceleration of 10^{12} m/s^2 by an applied magnetic field (west to east). The value of magnetic field : (Rest mass of proton is 1.6×10^{-27} kg) : **[8 Jan. 2020 (Morning)]**

(A) 71 mT (B) 7.1 mT
(C) 0.071 mT (D) 0.71 mT

26. A charged particle of mass 'm' and charge 'q' moving under the influence of uniform electric field $E\vec{i}$ and a uniform magnetic field $B\vec{k}$ follows a trajectory from point P to Q as shown in figure. The velocities at P and Q are respectively, $v\vec{i}$ and $-2v\vec{j}$. Then which of the following statements (a, b, c, d) are the correct ? (Trajectory shown is schematic and not to scale) :

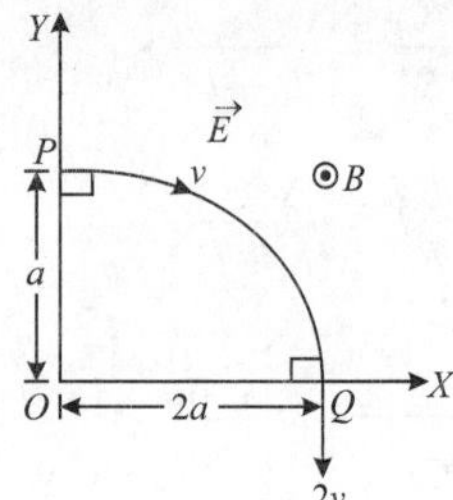

(a) $E=\frac{3}{4}\left(\frac{mv^2}{qa}\right)$

(b) Rate of work done by the electric field at P is $\frac{3}{4}\left(\frac{mv^3}{a}\right)$

(c) Rate of work done by both the fields at Q is zero

(d) The difference between the magnitude of angular momentum of the particle at P and Q is 2 mav.

[9 Jan. 2020 (Morning)]

(A) (a), (b), (c), (d) (B) (a), (b), (c)

(C) (b), (c), (d) (D) (a), (c), (d)

27. An electron gun is placed inside a long solenoid of radius R on its axis. The solenoid has n turns/length and carries a current I. The electron gun shoots an electron along the radius of the solenoid with speed v. If the electron does not hit the surface of the solenoid, maximum possible value of v is (all symbols have their standard meaning) : **[9 Jan. 2020 (Evening)]**

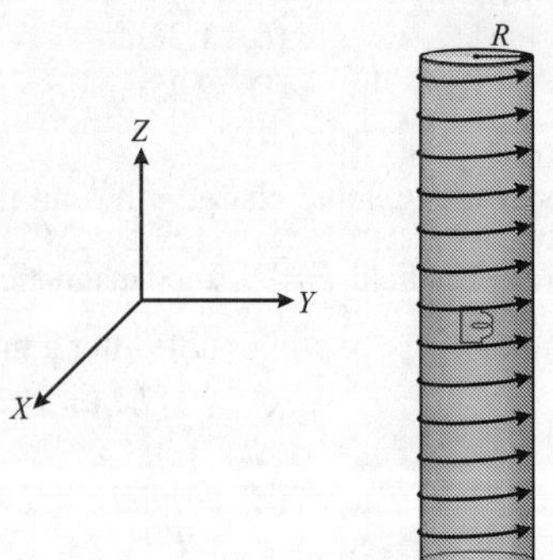

(A) $\frac{e\mu_0 nIR}{m}$ (B) $\frac{e\mu_0 nIR}{2m}$

(C) $\frac{2e\mu_0 nIR}{m}$ (D) $\frac{e\mu_0 nIR}{4m}$

28. A small circular loop of conducting wire has radius a and carries current I. It is placed in a uniform magnetic field B perpendicular to its plane such that when rotated slightly about its diameter and released, it starts performing simple harmonic motion of time period T. If the mass of the loop is m then : **[9 Jan. 2020 (Evening)]**

(A) $T=\sqrt{\frac{\pi m}{2IB}}$ (B) $T=\sqrt{\frac{2\pi m}{IB}}$

(C) $T=\sqrt{\frac{\pi m}{IB}}$ (D) $T=\sqrt{\frac{2m}{IB}}$

29. A beam of protons with speed 4×10^5 ms^{-1} enters a uniform magnetic field of 0.3 T at an angle of 60° to the magnetic field. The pitch of the resulting helical path of protons is close to : (Mass of the proton $=1.67\times10^{-27}$ kg, charge of the proton $=1.69\times10^{-19}$ C) **[2 Sept. 2020 (Morning)]**

(A) 12 cm (B) 4 cm

(C) 5 cm (D) 2 cm

30. A charged particle (mass m and charge q) moves along X-axis with velocity V_0. When it passes through the origin it enters a region having uniform electric field $\vec{E}=-E\hat{j}$ which extends upto $x=d$. Equation of path of electron in the region $x>d$ is : **[2 Sept. 2020 (Morning)]**

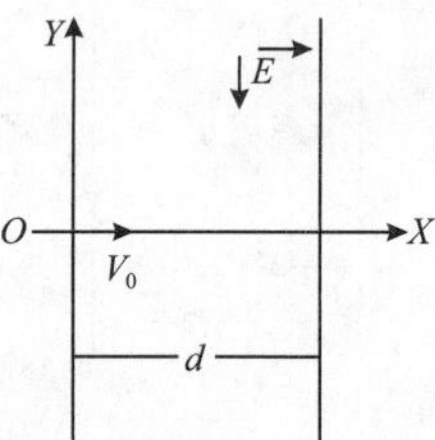

(A) $y=\frac{qEd}{mV_0^2}\left(\frac{d}{2}-x\right)$ (B) $y=\frac{qEd}{mV_0^2}(x-d)$

(C) $y=\frac{qEd}{mV_0^2}x$ (D) $y=\frac{qEd^2}{mV_0^2}x$

31. The figure shows a region of length 'l' with a uniform magnetic field of 0.3 T in it and a proton entering the region with velocity 4×10^5 ms^{-1} making an angle 60° with the field. If the proton completes 10 revolution by the time it cross the region shown, 'l' is close to (mass of proton $=1.67\times10^{-27}$ kg, charge of the proton $=1.6\times10^{-19}$ C) : **[2 Sept. 2020 (Evening)]**

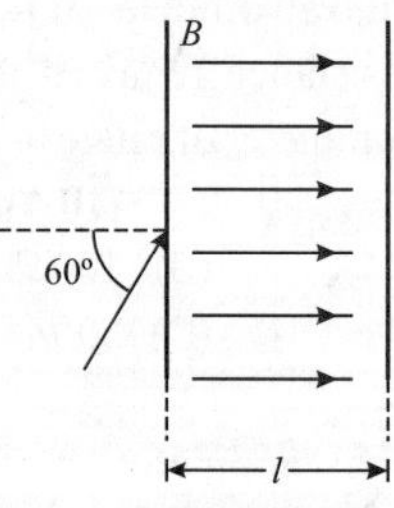

(A) 0.11 m (B) 0.22 m

(C) 0.44 m (D) 0.88 m

32. A wire carrying current I is bent in the shape $ABCDEFA$ as shown, where rectangle $ABCDA$ and $ADEFA$ are perpendicular to each other. If the sides of the rectangles are of lengths a and b, then the magnitude and direction of magnetic moment of the loop $ABCDEFA$ is : **[2 sept. 2020 Evening]**

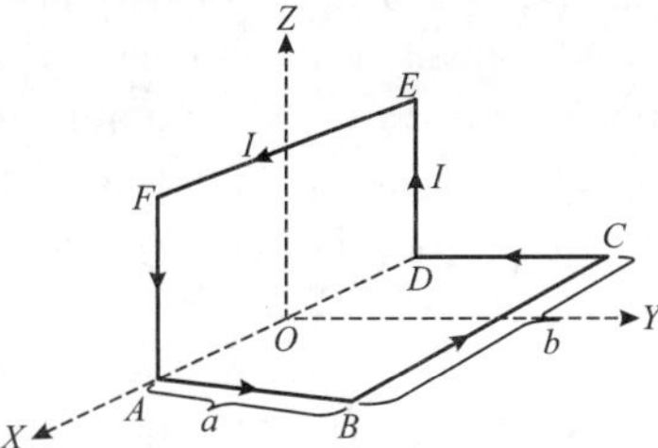

(A) $\sqrt{2}\,abI$, along $\left(\frac{\hat{j}}{\sqrt{2}}+\frac{\hat{k}}{\sqrt{2}}\right)$

(B) $\sqrt{2}\,abI$, along $\left(\frac{\hat{j}}{\sqrt{5}}+\frac{2\hat{k}}{\sqrt{5}}\right)$

(C) abI, along $\left(\frac{\hat{j}}{\sqrt{2}}+\frac{\hat{k}}{\sqrt{2}}\right)$

(D) abI, along $\left(\frac{\hat{j}}{\sqrt{5}}+\frac{2\hat{k}}{\sqrt{5}}\right)$

33. A charged particle carrying charge 1 μC is moving with velocity $(2\hat{i}+3\hat{j}+4\hat{k})\ \text{ms}^{-1}$. If an external magnetic field of $(5\hat{i}+3\hat{j}-6\hat{k})\times 10^{-3}$ T exists in the region where the particle is moving then the force on the particle is $\vec{F}\times 10^{-9}$ N. The vector $\vec{F}$ is : **[3 Sept. 2020 (Morning)]**

(A) $-0.30\,\hat{i}\ +0.32\,\hat{j}\ -0.09\,\hat{k}$

(B) $-300\,\hat{i}\ +320\,\hat{j}\ -90\,\hat{k}$

(C) $-30\,\hat{i}\ +32\,\hat{j}\ +9\,\hat{k}$

(D) $-3.0\,\hat{i}\ +3.2\,\hat{j}\ -0.9\,\hat{k}$

34. A galvanometer coil has 500 turns and each turn has an average area of 3×10^{-4} m^2. If a torque of 1.5 Nm is required to keep this coil parallel to magnetic field when a current of 0.5 A is flowing through it, the strength of the field (in T) is______: **[3 Sept. 2020 (Evening)]**

35. A circular coil has moment of inertia 0.8 kg m^2 around any diameter and is carrying current to produce a magnetic moment of 20 Am^2. The coil is kept initially in a vertical position and it can rotate freely around a horizontal diameter. When a uniform magnetic field of 4 T is applied along the vertical, it starts rotating around its horizontal diameter. The angular speed the coil acquires after rotating by 60° will be : **[4 Sept. 2020 (Evening)]**

(A) 10 rad s^{-1} (B) 20 π rad s^{-1}

(C) 10 π rad s^{-1} (D) 20 rad s^{-1}

36. A square loop of side $2a$, and carrying current I, is kept in XZ plane with its centre at origin. A long wire carrying the same current I is placed parallel to the z-axis and passing through the point $(0, b, 0)$, $(b >> a)$. The magnitude of the torque on the loop about z-axis is given by : **[5 Sept. 2020 (Morning)]**

(A) $\frac{2\mu_0 I^2 a^2}{\pi b}$ (B) $\frac{\mu_0 I^2 a^3}{2\pi b^2}$

(C) $\frac{\mu_0 I^2 a^2}{2\pi b}$ (D) $\frac{2\mu_0 I^2 a^3}{\pi b^2}$

37. An iron rod of volume 10^{-3} m^3 and relative permeability 1000 is placed as core in a solenoid with 10 turns/cm. If a current of 0.5 A is passed through the solenoid, then the magnetic moment of the rod will be : **[5 Sept. 2020 (Evening)]**

(A) 0.5×10^2 Am^2 (B) 50×10^2 Am^2

(C) 500×10^2 Am^2 (D) 5×10^2 Am^2

38. A particle of charge q and mass m is moving with a velocity $-v\hat{i}$ $(v\neq 0)$ towards a large screen placed in the $Y-Z$ plane at a distance d. If there is a magnetic field $\vec{B} = B_0\hat{k}$, the minimum value of v for which the particle will not hit the screen is : **[6 Sept. 2020 (Morning)]**

(A) $\frac{qdB_0}{2m}$ (B) $\frac{qdB_0}{m}$

(C) $\frac{2qdB_0}{m}$ (D) $\frac{qdB_0}{3m}$

39. An electron is moving along $+x$ direction with a velocity of 6×10^6 ms^{-1}. It enters a region of uniform electric field of 300 V/cm pointing along $+y$ direction. The magnitude and direction of the magnetic field set up in this region such that the electron keeps moving along the x direction will be : **[6 Sept. 2020 (Morning)]**

(A) 5×10^{-3} T, along $+z$ direction

(B) 3×10^{-4} T, along $-z$ direction

(C) 3×10^{-4} T, along $+z$ direction

(D) 5×10^{-3} T, along $-z$ direction

40. A particle moving in the xy plane experiences a velocity dependent force $\vec{F}=k(v_y\hat{i}+v_x\hat{j})$, where v_x and v_y are the x and y components of its velocity $\vec{v}$. If $\vec{a}$ is the acceleration of the particle, then which of the following statements is true for the particle : **[6 Sept. 2020 (Evening)]**

(A) Quantity $\vec{v}\cdot\vec{a}$ is constant in time

(B) Kinetic energy of particle is constant in time

(C) Quantity $\vec{v}\times\vec{a}$ is constant in time

(D) $\vec{F}$ arises due to a magnetic field

41. A charged particle going around in a circle can be considered to be a current loop. A particle of mass m carrying charge q is moving in a plane with speed v under the influence of magnetic field $\vec{B}$. The magnetic moment of this moving particle : **[6 Sept. 2020 (Evening)]**

(A) $-\frac{mv^2\vec{B}}{B^2}$ (B) $-\frac{mv^2\vec{B}}{2\pi B^2}$

(C) $-\frac{mv^2\vec{B}}{3B^2}$ (D) $-\frac{mv^2\vec{B}}{2B^2}$

42. A square loop of side $2a$ and carrying current I is kept in xz plane with its centre at origin. A long wire carrying the same current I is placed parallel to z-axis and passing through point (0, b, 0), ($b >> a$). The magnitude of torque on the loop about z-axis will be : **[6 Sept. 2020 (Evening)]**

(A) $\frac{2\mu_0 I^2 a^2 b}{\pi(a^2+b^2)}$ (B) $\frac{\mu_0 I^2 a^2 b}{2\pi(a^2+b^2)}$

(C) $\frac{\mu_0 I^2 a^2}{2\pi b}$ (D) $\frac{2\mu_0 I^2 a^2}{\pi b}$

43. A proton, a deuteron and an α particle are moving with same momentum in a uniform magnetic field. The ratio of magnetic forces acting on them is _____ and their speed is _____, in the ratio : **[25 Feb. 2021 (Morning)]**

(A) 1 : 2 : 4 and 1 : 1 : 2 (B) 2 : 1 : 1 and 4 : 2 : 1
(C) 4 : 2 : 1 and 2 : 1 : 1 (D) 1 : 2 : 4 and 2 : 1 : 1

44. A proton and an α-particle, having kinetic energies K_p and K_α, respectively, enter into a magnetic field at right angles. The ratio of the radii of trajectory of proton to that of α-particle is 2 : 1. The ratio of $K_p : K_\alpha$ is : **[18 March 2021 (Evening)]**

(A) 1:8 (B) 8:1
(C) 1:4 (D) 4:1

45. A deuteron and an alpha particle having equal kinetic energy enter perpendicularly into a magnetic filed. Let r_d and r_a be their respective radii of circular path. The value of $\frac{r_d}{r_a}$ is :

[20 July 2021 (Morning)]

(A) $\frac{1}{\sqrt{2}}$ (B) $\sqrt{2}$
(C) 2 (D) 1

46. Two ions having same mass have charge in the ratio 1 : 2. They are projected normally in a uniform magnetic field with their speeds in the ration 2 : 3. The ratio of the radii of their circular trajectories is : **[25 July 2021 (Evening)]**

(A) 1:4 (B) 2:3
(C) 4:3 (D) 3:1

47. If the maximum value of acceleration potential provided by a radio frequency oscillator is 12 kV. The number of revolution made by a proton in cyclotron to achieve one sixth of the speed of light is ______ .

[m_p = 1.67 × 10^{-27} kg, e = 1.6 × 10^{-19} C, Speed of light = 3 × 108 m/s]. **[26 Aug. 2021 (Evening)]**

48. Two ions of masses 4 amu and 16 amu have charges +2e and +3e respectively. These ions pass through the region of constant perpendicular magnetic field. The kinetic energy of both ions is same. Then : **[27 Aug. 2021 (Morning)]**

(A) No ion will be deflected
(B) Lighter ion will be deflected less than heavier ion
(C) Lighter ion will be deflected more than heavier ion
(D) Both ions will be deflected equally.

49. The magnetic field vector of an electromagnetic wave is given by $B = B_0\frac{\hat{i}+\hat{j}}{\sqrt{2}}$ cos($kz - \omega t$) where $\hat{i}, \hat{j}$ represents unit vector x and y-axis respectively. At t = 0 s, two electric changes q_1 of 4π coulomb and q_2 of 2π coulomb located at $\left(0,0,\frac{\pi}{k}\right)$ and $\left(0,0,\frac{3\pi}{k}\right)$, respectively, have the same velocity of 0.5 $c\hat{i}$, (where c is the velocity of light). The ratio of the force acting on charge q_1 to q_2 is : **[31 Aug. 2021 (Evening)]**

(A) 1: $\sqrt{2}$ (B) $\sqrt{2}$: 1
(C) 2$\sqrt{2}$:1 (D) 2:1

* * * * *

CHAPTER

19 MAGNETISM AND MATTER

19.1 CLASSICAL AND TERRESTRIAL MAGNETISM

1. A magnetic needle of magnetic moment 6.7×10^{-2} Am^2 and moment of inertia 7.5×10^{-6} kg m^2 is performing simple harmonic oscillations in a magnetic field of 0.01 T. Time taken for 10 complete oscillations is : **[2017]**

(A) 6.98 s (B) 8.76 s
(C) 6.65 s (D) 8.89 s

2. A Magnetic dipole in a constant magnetic field has : **[2017 (Online)]**

(A) Maximum potential energy when the torque is maximum
(B) Zero potential energy when the torque is minimum
(C) Zero potential energy when the torque is maximum
(D) Minimum potential energy when the torque is maximum

3. At some location on earth the horizontal component of earth's magnetic field is 18×10^{-6} T. At this location, magnetic needle of length 0.12 m and pole strength 1.8 Am is suspended from its mid-point using a thread, it makes 45° angle with horizontal in equilibrium. To keep this needle horizontal, the vertical force that should be applied at one of its ends is : **[10 Jan. 2019 (Evening)]**

(A) 3.6×10^{-5} N (B) 6.5×10^{-5} N
(C) 1.3×10^{-5} N (D) 1.8×10^{-5} N

4. A hoop and a solid cylinder of same mass and radius are made of a permanent magnetic material with their magnetic moment parallel to their respective axes. But the magnetic moment of hoop is twice of solid cylinder. They are placed in a uniform magnetic field in such a manner that their magnetic moments make a small angle with the field. If the oscillation periods of hoop and cylinder are T_h and T_c respectively, then: **[10 Jan. 2019 (Evening)]**

(A) $T_h = 0.5\,T_c$ (B) $T_h = 2\,T_c$
(C) $T_h = 1.5\,T_c$ (D) $T_h = T_c$

5. A square loop is carrying a steady current I and the magnitude of its magnetic dipole moment is m. If this square loop is changed to a circular loop and it carries the same current, the magnitude of the magnetic dipole moment of circular loop will be : **[10 April 2019 (Evening)]**

(A) $\frac{3m}{\pi}$ (B) $\frac{4m}{\pi}$
(C) $\frac{2m}{\pi}$ (D) $\frac{m}{\pi}$

6. A magnetic compass needle oscillates 30 times per minute at a place where the dip is 45°, and 40 times per minute where the dip is 30°. If B_1 and B_2 are respectively the total magnetic field due to the earth at the two places, then the ratio B_1/B_2 is best given by : **[12 April 2019 (Morning)]**

(A) 2.2 (B) 1.8
(C) 0.7 (D) 3.6

7. A small bar magnet placed with its axis at 30° with an external field of 0.06 T experiences a torque of 0.018 Nm. The minimum work required to rotate it from its stable to unstable equilibrium position is : **[4 Sept. 2020 (Morning)]**

(A) 9.2×10^{-3} J (B) 6.4×10^{-2} J
(C) 11.7×10^{-3} J (D) 7.2×10^{-2} J

8. A bar magnet of length 14 cm is placed in the magnetic meridian with its north pole pointing towards the geographic north pole. A neutral point is obtained at a distance of 18 cm from the center of the magnet. If $B_H = 0.4$ G, the magnetic moment of the magnet is (1 G = 10^{-4}T) **[16 March 2021 (Morning)]**

(A) 2.880×10^3 J T^{-1} (B) 2.880×10^2 J T^{-1}
(C) 2.880 J T^{-1} (D) 28.80 J T^{-1}

9. At an angle of 30° to the magnetic meridian, the apparent dip is 45°. Find the true dip : **[20 July 2021 (Evening)]**

(A) $\tan^{-1}\frac{\sqrt{3}}{2}$ (B) $\tan^{-1}\frac{1}{\sqrt{3}}$
(C) $\tan^{-1}\sqrt{3}$ (D) $\tan^{-1}\frac{2}{\sqrt{3}}$

10. Choose the correct option : **[22 July 2021 (Evening)]**

(A) True dip is always equal to apparent dip
(B) True dip is less than the apparent dip
(C) True dip is always greater than the apparent dip
(D) True dip is not mathematically related to apparent dip

* * * * *

CHAPTER 19

MAGNETISM AND MATTER

19.2 MAGNETIC PROPERTIES OF MATERIALS

1. The B-H curve for a ferromagnet is shown in the figure. The ferromagnet is placed inside a long solenoid with 1000 turns/cm. The current that should be passed in the solenoid to demagnetize the ferromagnet completely is :

[2018 (Online)]

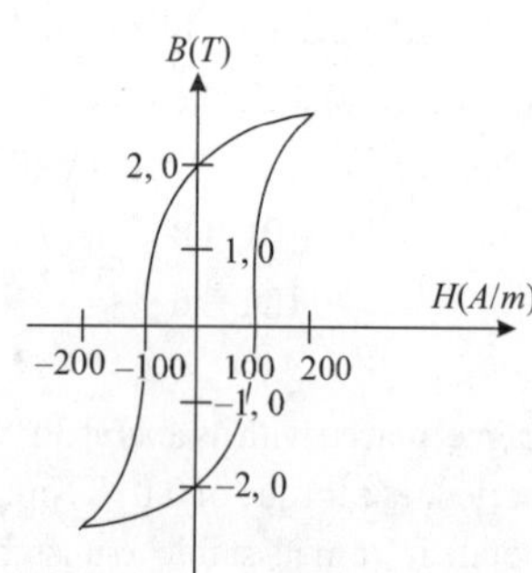

(A) 2 mA (B) 1 mA
(C) 40 μA (D) 20 μA

2. A bar magnet is demagnetized by inserting it inside a solenoid of length 0.2 m, 100 turns, and carrying a current of 5.2 A. The coercivity of the bar magnet is :

[9 Jan. 2019 (Morning)]

(A) 1200 A/m (B) 2600 A/m
(C) 520 A/m (D) 285 A/m

3. A paramagnetic substance in the form of a cube with sides 1 cm has a magnetic dipole moment of 20×10^{-6} J/T when a magnetic intensity of 60×10^3 A/m is applied. Its magnetic susceptibility is : **[11 Jan. 2019 (Evening)]**

(A) 2.3×10^{-2} (B) 3.3×10^{-2}
(C) 3.3×10^{-4} (D) 4.3×10^{-2}

4. A paramagnetic material has 10^{28} atoms/m^3. Its magnetic susceptibility at temperature 350 K is 2.8×10^{-4}. Its susceptibility at 300 K is : **[12 Jan. 2019 (Evening)]**

(A) 3.672×10^{-4} (B) 3.726×10^{-4}
(C) 3.267×10^{-4} (D) 2.672×10^{-4}

5. Two magnetic dipoles X and Y are placed at a separation d, with their axes perpendicular to each other. The dipole moment of Y is twice that of X. A particle of charge q is passing, through their midpoint P, at angle $\theta = 45°$ with the horizontal line, as shown in figure. What would be the magnitude of force on the particle at that instant ? (d is much larger than the dimensions of the dipole) : **[8 April 2019 (Evening)]**

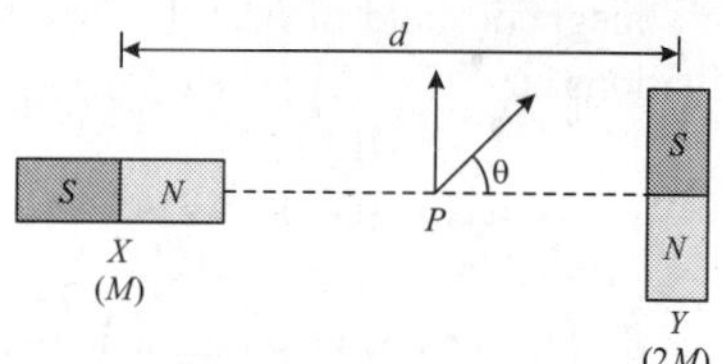

(A) $\sqrt{2}\left(\frac{\mu_0}{4\pi}\right)\frac{M}{(d/2)^3}\times qv$ (B) $\left(\frac{\mu_0}{4\pi}\right)\frac{2M}{(d/2)^3}\times qv$

(C) $\left(\frac{\mu_0}{4\pi}\right)\frac{M}{(d/2)^3}\times qv$ (D) 0

6. The figure gives experimentally measured B vs H variation in a ferromagnetic material. The retentivity, coercivity and saturation, respectively, of the material are :

[7 Jan. 2020 (Evening)]

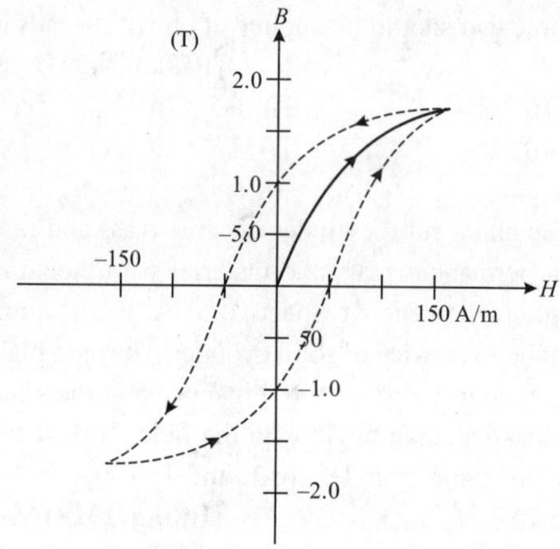

(A) 150 A/m, 1.0 T and 1.5 T (B) 1.0 T, 50 A/m and 1.5 T
(C) 1.5 T, 50 A/m and 1.0 T (D) 1.5 T, 50 A/m and 1.0 T

7. Magnetic materials used for making permanent magnets (P) and magnets in a transformer (T) have different properties of the following, which property best matches for the type of magnet required : **[2 Sept. 2020 (Morning)]**

(A) T : Large retentivity, small coercivity
(B) P : Small retentivity, large coercivity
(C) T : Large retentivity, large coercivity
(D) P : Large retentivity, large coercivity

8. A perfectly diamagnetic sphere has a small spherical cavity at its centre, which is filled with a paramagnetic substance. The whole system is placed in a uniform magnetic field $\vec{B}$. Then the field inside the paramagnetic substance is :

[3 Sept. 2020 (Evening)]

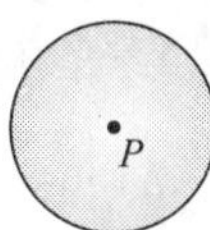

(A) Zero
(B) $\vec{B}$
(C) Much large than $|\vec{B}|$ but opposite to $\vec{B}$
(D) Much large than $|\vec{B}|$ and parallel to $\vec{B}$

9. A paramagnetic sample shows a net magnetisation of 6 A/m when it is placed in an external magnetic field of 0.4 T at a temperature of 4 K. When the sample is placed in an external magnetic field of 0.3 T at a temperature of 24 K, then the magnetisation will be : **[4 Sept. 2020 (Evening)]**

(A) 4 A/m (B) 0.75 A/m
(C) 2.25 A/m (D) 1 A/m

10. A soft ferromagnetic material is placed in an external magnetic field. The magnetic domains : **[24 Feb. 2021 (Evening)]**

(A) Increase in size but no change in orientation
(B) May increase or decrease in size and change its orientation
(C) Decrease in size and changes orientation
(D) Have no relation with external magnetic field

11. In a ferromagnetic material, below the curie temperature, a domain is defined as : **[25 Feb. 2021 (Evening)]**

(A) A macroscopic region with zero magnetization
(B) A macroscopic region with consecutive magnetic dipoles oriented in opposite direction
(C) A macroscopic region with randomly oriented magnetic dipoles
(D) A macroscopic region with saturation magnetization

12. Which of the following statements are correct?

(a) Electric monopoles do not exist whereas magnetic monopoles exist.
(b) Magnetic field lines due to a solenoid at its ends and outside cannot be completely straight and confined.
(c) Magnetic field lines are completely confined within a toroid.
(d) Magnetic field lines inside a bar magnet are not parallel.
(e) $\chi = -1$ is the condition for a perfect diamagnetic material, where χ is its magnetic susceptibility.

Choose the correct answer from the options given below :

[18 March 2021 (Evening)]

(A) (c) and (e) only (B) (b) and (d) only
(C) (a) and (b) only (D) (b) and (c) only

13. The magnetic susceptibility of a material of a rod is 499. Permeability in vacuum is $4\pi \times 10^{-7}$ H/m. Absolute permeability of the material of the rod is : **[20 July 2021 (Evening)]**

(A) $\pi \times 10^{-4}$ H/m (B) $3\pi \times 10^{-4}$ H/m
(C) $4\pi \times 10^{-4}$ H/m (D) $2\pi \times 10^{-4}$ H/m

14. Statement-I: The ferromagnetic property depends on temperature. At high temperature, ferromagnet becomes paramagnet.

Statement-II: At high temperature, the domain wall area of a ferromagnetic substance increases.

In the light of the above statements, choose the most appropriate answer from the option given below : **[22 July 2021 (Evening)]**

(A) Both Statement-I and Statement-II are true
(B) Statement-I is false but Statement-II is true
(C) Both Statement-I and Statement-II are false
(D) Statement-I is true but Statement-II is false

15. Two short magnetic dipoles m_1 and m_2 each having magnetic moment of 1 Am^2 are placed at point O and P respectively. The distance between OP is 1 meter. The torque experienced by the magnetic dipole m_2 due to the presence of m1 is ______ $\times 10^{-7}$ Nm : **[26 Aug. 2021 (Morning)]**

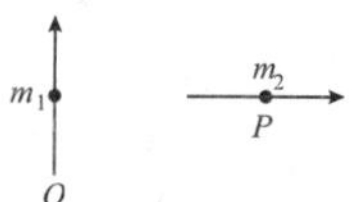

16. A coil in the shape of an equilateral triangle of side 10 cm lies in a vertical place between the pole pieces of permanent magnet producing a horizontal magnetic field 20 mT. The torque acting on the coil when a current of 0.2 A is passed through it and its plane becomes parallel to the magnetic field will be $\sqrt{x} \times 10^5$ Nm. The value of x is ______.

[26 Aug. 2021 (Evening)]

17. A uniform conducting wire of length $24a$, and resistance R is wound up as a current carrying coil in the shape of an equilateral triangle of side a and then in the form of a square of side a. The coil is connected to a voltage source V_0. The ratio of magnetic moment of the coils in case of equilateral triangle to that for square is $1 : y$ where y is ______.

[27 Aug. 2021 (Morning)]

18. A long solenoid with 1000 turns/m has a core material with relative permeability 500 and volume 10^3 cm^3. If the core material is replaced by another material having relative permeability of 750 with same volume maintain same current of 0.75 A in the solenoid. The fractional change in the magnetic moment of the core would be approximately $\left(\frac{x}{499}\right)$. Find the value of x :

[31 Aug. 2021 (Evening)]

19. Following plots show Magnetization (M) vs Magnetising field (H) and Magnetic susceptibility (χ) vs Temperature (T) graph : **[01 Sep. 2021 (Evening)]**

(a) 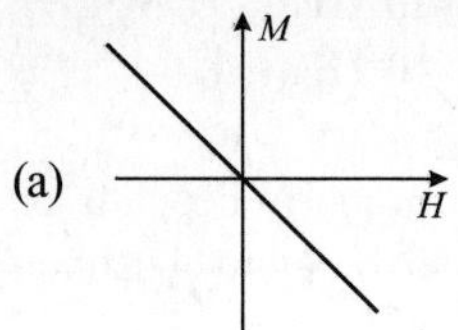

(b) 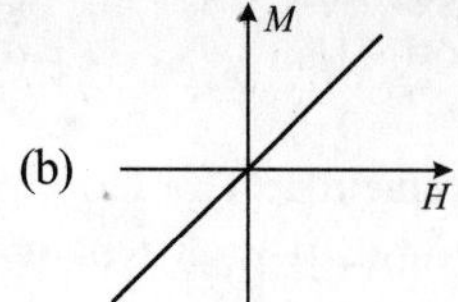

(c) 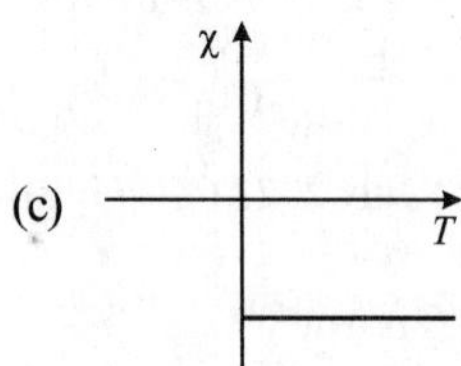

(d) 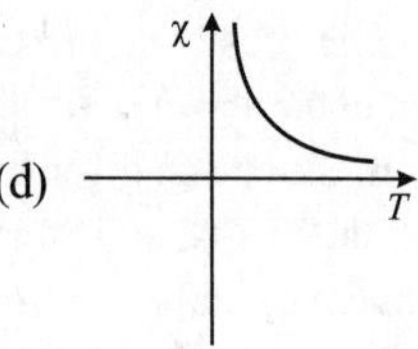

Which of the following combination will be represented by a diamagnetic material ?

(A) (b), (d) (B) (b), (c)
(C) (a), (d) (D) (a), (c)

* * * * *

CHAPTER

20 ELECTROMAGNETIC INDUCTION

20.1 ELECTROMAGNETIC INDUCTION

1. In a coil of resistance 100 Ω, a current is induced by changing the magnetic flux through it as shown in the figure. The magnitude of change in flux through the coil is : **[2017]**

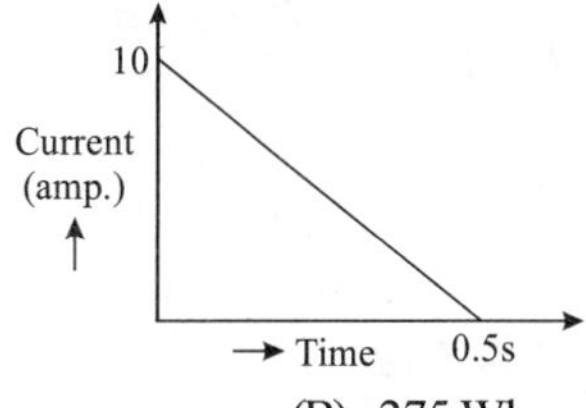

(A) 250 Wb (B) 275 Wb
(C) 200 Wb (D) 225 Wb

2. At the centre of a fixed large circular coil of radius R, a much smaller circular coil of radius r is placed. The two coils are concentric and are in the same plane. The larger coil carries a current I. The smaller coil is set to rotate with a constant angular velocity ω about an axis along their common diameter. Calculate the emf induced in the smaller coil after a time t of its start of rotation : **[2018 (Online)]**

(A) $\frac{\mu_0 I}{2R}\omega r^2 \sin \omega t$ (B) $\frac{\mu_0 I}{4R}\omega\pi r^2 \sin \omega t$

(C) $\frac{\mu_0 I}{2R}\omega\pi r^2 \sin \omega t$ (D) $\frac{\mu_0 I}{4R}\omega r^2 \sin \omega t$

3. A copper rod of mass m slides under gravity on two smooth parallel rails, with separation 1 and set at angle of θ with the horizontal. At the bottom, rails are joined by a resistance R. There is a uniform magnetic field B normal to the plane of the rails, as shown in the figure. The terminal speed of the copper rod is : **[2018 (Online)]**

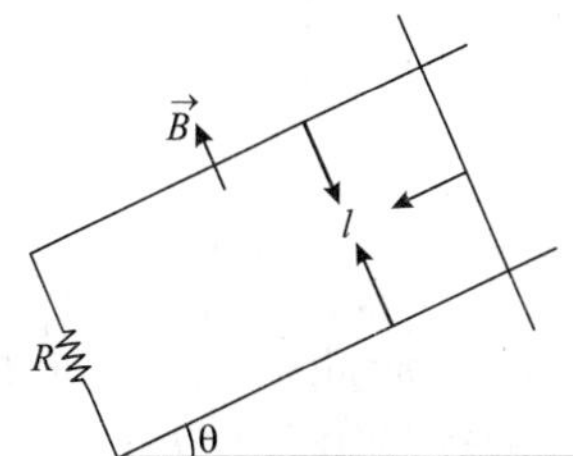

(A) $\frac{mgR\cos\theta}{B^2l^2}$ (B) $\frac{mgR\sin\theta}{B^2l^2}$

(C) $\frac{mgR\tan\theta}{B^2l^2}$ (D) $\frac{mgR\cot\theta}{B^2l^2}$

4. An ideal capacitor of capacitance 0.2 μF is charged to a potential difference of 10 V. The charging battery is then disconnected. The capacitor is then connected to an ideal inductor of self inductance 0.5 mH. The current at a time when the potential difference across the capacitor is 5 V, is : **[2018 (Online)]**

(A) 0.17 A (B) 0.15 A
(C) 0.34 A (D) 0.25 A

5. A coil of cross-sectional area A having n turns is placed in a uniform magnetic field B. When it is rotated with an angular velocity ω, the maximum e.m.f. induced in the coil will be : **[2018 (Online)]**

(A) $nBA\,\omega$ (B) $\frac{3}{2}\,nBA\,\omega$

(C) $3\,nBA\,\omega$ (D) $\frac{1}{2}\,nBA\,\omega$

6. A conducting circular loop made of a thin wire, has area 3.5×10^{-3} m^2 and resistance 10 Ω. It is placed perpendicular to a time dependent magnetic field $B(t) = (0.4\text{T})\sin(50\pi t)$. The field is uniform in space. Then the net charge flowing through the loop during $t = 0$ s and $t = 10$ ms is close to : **[9 Jan. 2019 (Morning)]**

(A) 14mC (B) 21 mC
(C) 6mC (D) 7mC

7. A solid metal cube of edge length 2 cm is moving in a positive y-direction at a constant speed of 6 m/s. There is a uniform magnetic field of 0.1 T in the positive z-direction. The potential difference between the two faces of the cube perpendicular to the x-axis, is : **[10 Jan. 2019 (Morning)]**

(A) 6mV (B) 1mV
(C) 12mV (D) 2mV

8. A magnet of total magnetic moment $10^{-2}\,\hat{i}$ A-m^2 is placed in a time varying magnetic field, $B\,\hat{i}$ (cost ωt) where $B = 1$ Tesla and ω = 0.125 rad/s. The work done for reversing the direction of the magnetic moment at $t = 1$ second, is : **[10 Jan. 2019 (Morning)]**

(A) 0.007 J (B) 0.014 J
(C) 0.01 J (D) 0.028 J

9. A 10 m long horizontal wire extends from North East to South West. It is falling with a speed of 5.0 ms^{-1}, at right angles to the horizontal component of the earth's magnetic field, of 0.3×10^{-4} Wb/m^2. The value of the induced emf in wire is :

[12 Jan. 2019 (Evening)]

(A) 2.5×10^{-3} V (B) 1.1×10^{-3} V
(C) 0.3×10^{-3} V (D) 1.5×10^{-3} V

10. A thin strip 10 cm long is on a *U* shaped wire of negligible resistance and it is connected to a spring of spring constant 0.5 Nm^{-1} (see figure). The assembly is kept in a uniform magnetic field of 0.1 T. If the strip is pulled from its equilibrium position and released, the number of oscillation it performs before its amplitude decreases by a factor of e is *N*. If the mass of the strip is 50 grams, its resistance 10 Ω and air drag negligible, *N* will be close to : **[8 April 2019 (Morning)]**

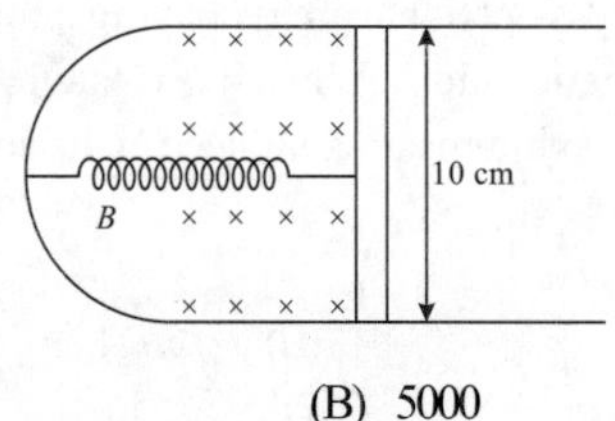

(A) 50000 (B) 5000
(C) 10000 (D) 1000

11. The figure shows a square loop *L* of side 5 cm which is connected to a network of resistances. The whole setup is moving towards right with a constant speed of 1 cms^{-1}. At some instant, a part of *L* is in a uniform magnetic field of 1*T*, perpendicular to the plane of the loop. If the resistance of *L* is 1.7 Ω, the current in the loop at that instant will be close to :

[12 April 2019 (Morning)]

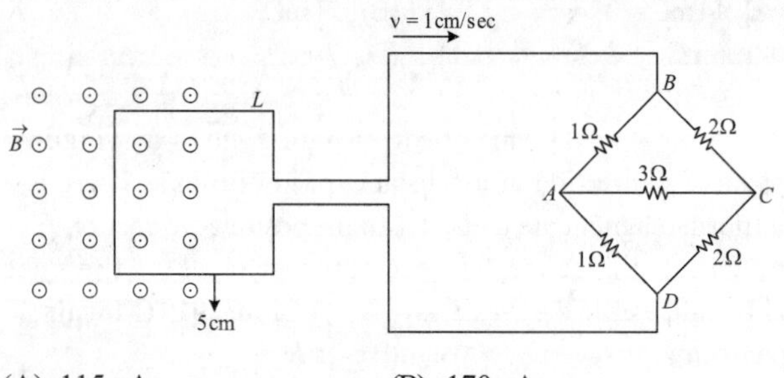

(A) 115 μA (B) 170 μA
(C) 60 μA (D) 150 μA

12. A long solenoid of radius *R* carries a time (*t*)-dependent current $I(t) = I_0t(1 - t)$. A ring of radius 2*R* is placed coaxially near its middle. During the time interval $0 \le t \le 1$, the induced current (I_R) and the induced EMF(V_R) in the ring change as :

[7 Jan. 2020 (Morning)]

(A) At $t = 0.5$ direction of I_R reverses and V_R is zero

(B) Direction of I_R remains unchanged and V_R is zero at $t = 0.25$

(C) Direction of I_R remains unchanged and V_R is maximum at $t = 0.5$

(D) At $t = 0.25$ direction of I_R reverses and V_R is maximum

13. Consider a circular coil of wire carrying constant current *I*, forming a magnetic dipole. The magnetic flux through an infinite plane that contains the circular coil and excluding the circular coil area is given by ϕ_i. The magnetic flux through the area of the circular coil area is given by ϕ_0. Which of the following option is correct ? **[7 Jan. 2020 (Morning)]**

(A) $\phi_i = -\phi_0$ (B) $\phi_i = \phi_0$
(C) $\phi_i < \phi_0$ (D) $\phi_i > \phi_0$

14. A loop *ABCDEFA* of straight edges has six corner points *A*(0,0,0), *B*(5,0,0), *C*(5,5,0), *D*(0, 5, 0), *E*(0, 5, 5) and *F*(0, 0, 5). The magnetic field in this region is $\vec{B} = (3\hat{i} + 4\hat{k})T$. The quantity of flux through the loop *ABCDEFA* (in *Wb*) is ______.

[7 Jan. 2020 (Morning)]

15. A planar loop of wire rotates in a uniform magnetic field. Initially, at $t = 0$, the plane of the loop is perpendicular to the magnetic field. If it rotates with a period of 10 s about an axis in its plane then the magnitude of induced emf will be maximum and minimum, respectively at : **[7 Jan. 2020 (Evening)]**

(A) 2.5 s and 7.5 s (B) 5.0 s and 7.5 s
(C) 5.0 s and 10.0 s (D) 2.5 s and 5.0 s

16. At time $t = 0$ magnetic field of 1000 Gauss is passing perpendicularly through the area defined by the closed loop shown in the figure. If the magnetic field reduces linearly to 500 Gauss, in the next 5s, then induced *EMF* in the loop is :

[8 Jan. 2020 (Morning)]

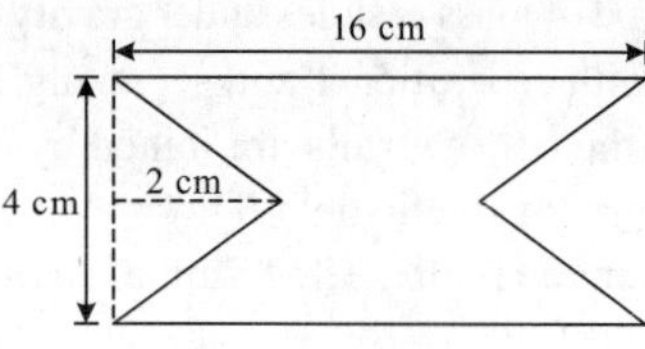

(A) 36 μ*V* (B) 48 μ*V*
(C) 56 μ*V* (D) 28 μ*V*

17. A circular coil of radius 10 cm is placed in a uniform magnetic field of 3.0×10^{-5} T with its plane perpendicular to the field initially. It is rotated at constant angular speed about an axis along the diameter of coil and perpendicular to magnetic field so that it undergoes half of rotation in 0.2s. The maximum value of *EMF* induced (in μV) in the coil will be close to the integer.

[2 Sept. 2020 (Morning)]

18. An elliptical loop having resistance *R*, of semi major axis *a*, and semi minor axis *b* is placed in a magnetic field as shown in the figure. If the loop is rotated about the x-axis with angular frequency ω, the average power loss in the loop due to Joule heating is : **[3 Sept. 2020 (Morning)]**

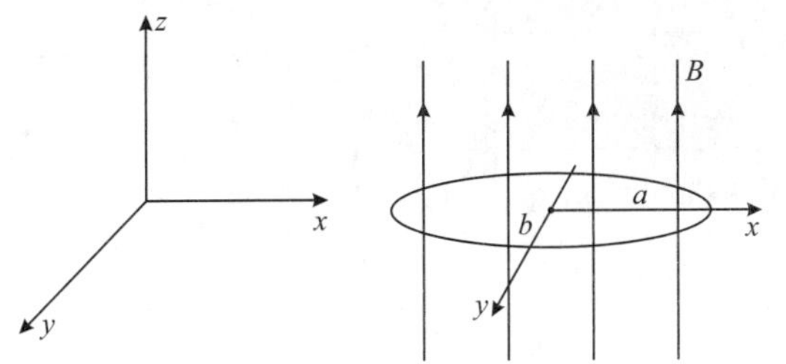

(A) $\frac{\pi^2 a^2 b^2 B^2 \omega^2}{2R}$ (B) Zero

(C) $\frac{\pi^2 a^2 b^2 B^2 \omega^2}{R}$ (D) $\frac{\pi abB\omega}{}$

19. A uniform magnetic field B exists in a direction perpendicular to the plane of a square loop made of a metal wire. The wire has a diameter of 4 mm and a total length of 30 cm. The magnetic field changes with time at a steady rate $dB/dt = 0.032\ \text{Ts}^{-1}$. The induced current in the loop is close to (Resistivity of the metal wire is 1.23×10^{-8} Wm) : **[3 Sept. 2020 (Evening)]**

(A) 0.61 A (B) 0.34 A

(C) 0.43 A (D) 0.53 A

20. A small bar magnet is moved through a coil at constant speed from one end to the other. Which of the following series of observations will be seen on the galvanometer G attached across the coil ? Three positions shown describe : (a) the magnet's entry (b) magnet is completely inside and (c) magnet's exit : **[4 Sept. 2020 (Morning)]**

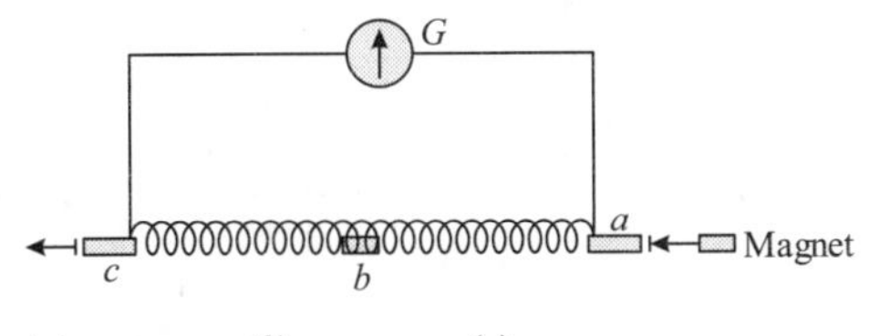

(A)
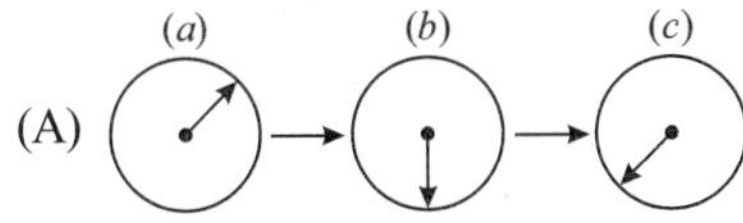

(B)
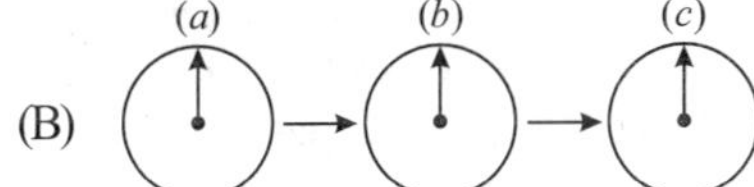

(C)
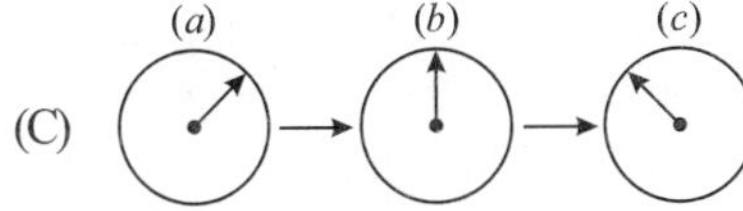

(D)
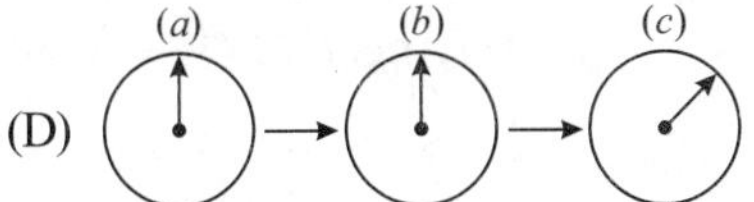

21. Two concentric circular coils, C_1 and C_2, are placed in the XY plane. C_1 has 500 turns, and a radius of 1 cm. C_2 has 200 turns and radius of 20 cm. C_2 carries a time dependent current $I(t) = (5t^2 - 2t + 3)$. A where t is in s. The emf induced in C_1 (in mV), at the instant $t = 1$s is $\frac{4}{x}$. The value of x is : **[5 Sept. 2020 (Morning)]**

22. An infinitely long straight wire carrying current I, one side opened rectangular loop and a conductor C with a sliding connector are located in the same plane, as shown in the figure. The connector has length l and resistance R. It slides to the right with a velocity v. The resistance of the conductor and the self ind... ce of the loop are negligible. The induced current ... the lo... ion of separation r, between the connector and the st... **[5 Sept. 2020 (Evening)]**

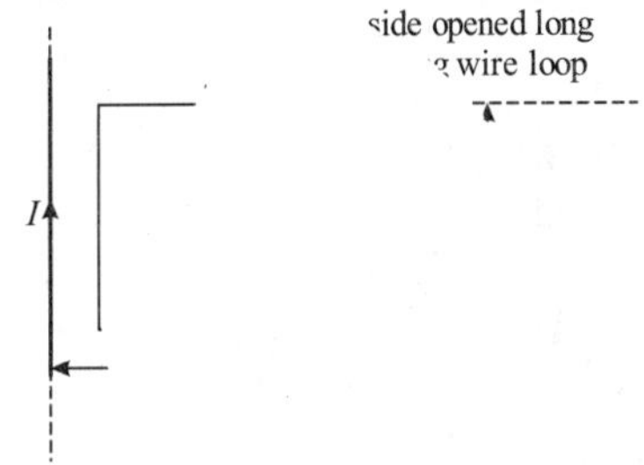

(A) $\frac{\mu_0}{\pi}\frac{Ivl}{Rr}$ (B) $\frac{}{2\pi\ Rr}$

(C) $\frac{2\mu_0}{\pi}\frac{Ivl}{Rr}$ (D) $\frac{\mu_0}{4\pi}\frac{Ivl}{Rr}$

23. An aeroplane, with its wings spread 10 m, is flying at a speed of 180 km/h in a horizontal direction. The total intensity of earth's field at that part is 2.5×10^{-4} Wb/m^2 and the angle of dip is 60°. The emf induced between the tips of the plane wings will be ______. **[26 Feb. 2021 (Evening)]**

(A) 62.50 mV (B) 54.125 mV

(C) 88.37 mV (D) 108.25 mV

24. A conducting bar of length L is free to slide on two parallel conducting rails as shown in the figure.

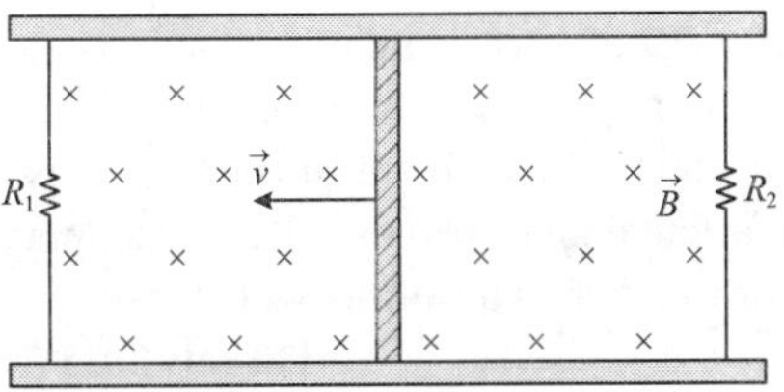

Two resistors R_1 and R_2 are connected across the ends of the rails. There is a uniform magnetic field $\vec{B}$ pointing into the page.

An external agent pulls the bar to the left at a constant speed v. The correct statement about the directions of induced currents I_1 and I_2 flowing through R_1 and R_2 respectively is :

[16 March 2021 (Morning)]

(A) Both I_1 and I_2 are in anticlockwise direction

(B) Both I_1 and I_2 are in clockwise direction

(C) I_1 is in clockwise direction and I_2 is in anticlockwise direction

(D) I_1 is in anticlockwise direction and I_2 is in clockwise direction

25. The magnetic field in a region is given by $\vec{B} = B_0\left(\frac{x}{a}\right)\hat{k}$. A square loop of side d is placed with its edges along the x and y axes. The loop is moved with a constant velocity $\vec{v} = v_0\hat{i}$. The emf induced in the loop is : **[16 March 2021 (Evening)]**

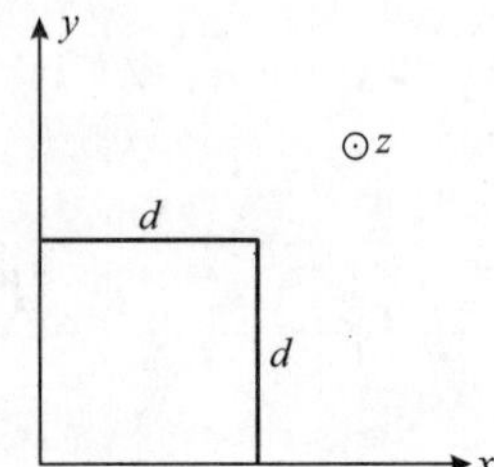

(A) $\frac{B_0 v_0^2 d}{2a}$ (B) $\frac{B_0 v_0 d}{2a}$

(C) $\frac{B_0 v_0 d^2}{a}$ (D) $\frac{B_0 v_0 d^2}{2a}$

26. A loop of flexible wire of irregular shape carrying current is placed in an external magnetic field. Identify the effect of the field on the wire. **[18 March 2021 (Morning)]**

(A) Loop assumes circular shape with its plane normal to the field.

(B) Loop assumes circular shape with its plane parallel to the field.

(C) Wire gets stretched to become straight.

(D) Shape of the loop remains unchanged.

27. The arm PQ of a rectangular conductor is moving from $x = 0$ to $x = 2b$ outwards and then inwards from $x = 2b$ to $x = 0$ as shown in the figure. A uniform magnetic field perpendicular to the plane is acting from $x = 0$ to $x = b$. Identify the graph showing the variation of different quantities with distance :

[20 July 2021 (Morning)]

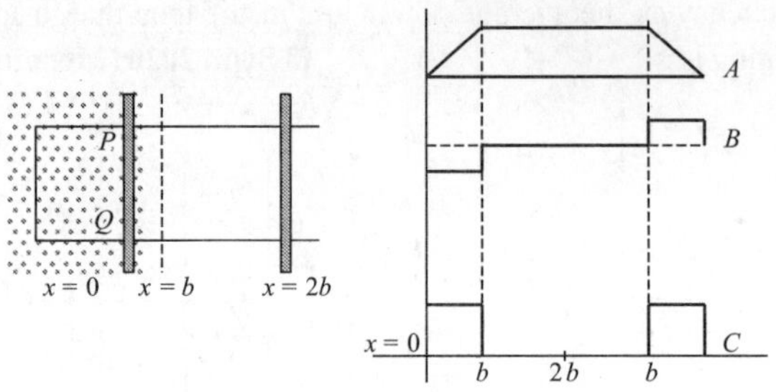

(A) A-Flux, B-power dissipated, C-EMF

(B) A-Flux, B-power dissipated, C-Flux

(C) A-Power dissipated, B-Flux, C-EMF

(D) A-Flux, B-EMF, C-Power dissipated

28. A circular conducting coil of radius 1 m is being heated by the change of magnetic field $\vec{B}$ passing perpendicular to the plane in which the coil is laid. The resistance of the coil is 2 μΩ. The magnetic field is slowly switched off such that its magnitude changes in time as $B = \frac{4}{\pi} \times 10^{-3}\left(1 - \frac{t}{100}\right)$ tesla. The energy dissipated by the coil before the magnetic field is switched off completely is $E =$ ______ mJ : **[25 July 2021 (Morning)]**

29. In the given figure the magnetic flux through the loop increases according to the relation $\phi_B(t) = 10t^2 + 20t$, where ϕ_B is in milliwebers and t is in seconds. The magnitude of current through $R = 2\Omega$ resistor at $t = 5$ s is ______ mA.

[27 July 2021 (Evening)]

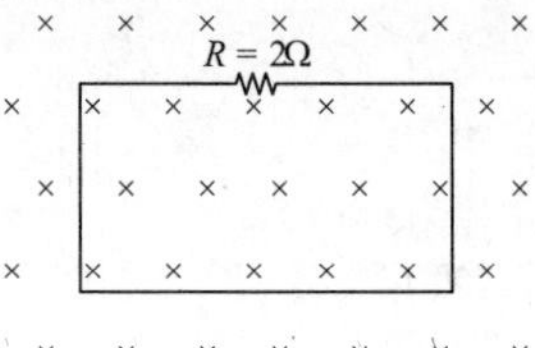

30. A circular coil of radius 8.0 cm and 20 turns is rotated about its vertical diameter with an angular speed of 50 rad s^{-1} in a uniform horizontal field of 3.0×10^{-2} T. The maximum emf induced the coil will be ______ $\times 10^{-2}$ volt (rounded off to the nearest integer). **[26 Aug. 2021 (Evening)]**

31. A bar magnet is passing through a conducting loop of radius R with velocity v. The radius of the bar magnet is such that it just passes through the loop. The induced e.m.f. in the loop can be represented by the approximate curve :

[27 Aug. 2021 (Morning)]

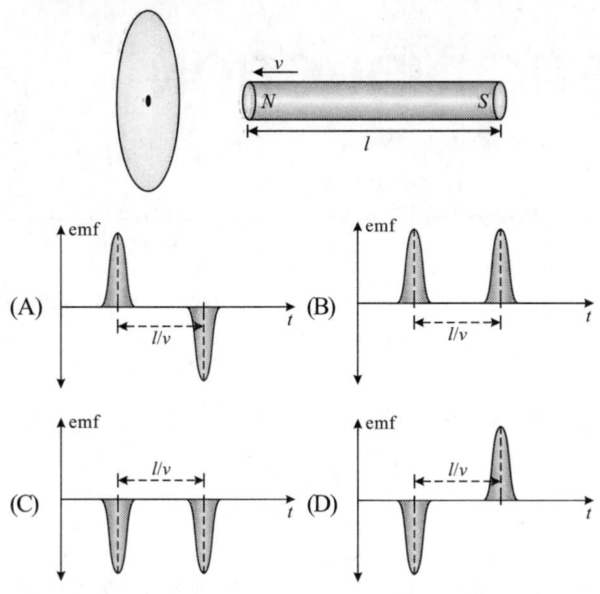

32. A constant magnetic field of 1 T is applied in the $x > 0$ region. A metallic circular ring of radius 1 m is moving with a constant velocity of 1 m/s along the x-axis. At $t = 0$ s, the centre O of the ring is at $x = -1$m. What will be the value of the induced emf in the ring at $t = 1$s ? (Assume the velocity of the ring does not change.) : **[27 Aug. 2021 (Evening)]**

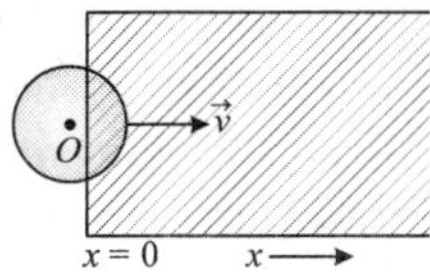

(A) 0V (B) 2V
(C) 2πV (D) 1V

33. A coil is placed in a magnetic field $\vec{B}$ as shown below. **[31 Aug. 2021 (Evening)]**

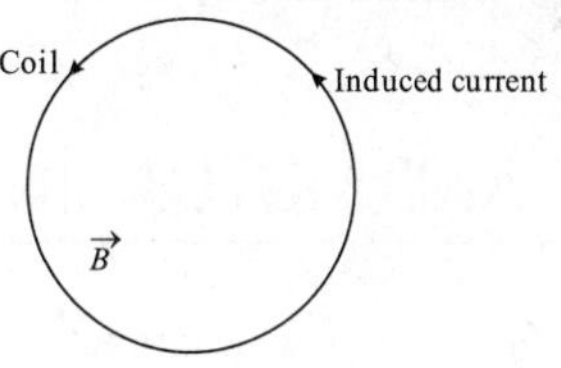

A current is induced in the coil because $\vec{B}$ is :
(A) parallel to the plane of coil and decreasing with time
(B) outward and increasing with time
(C) parallel to the plane of coil and increasing with time
(D) outward and decreasing with time

34. A square loop of side 20 cm and resistance 1 Ω is moved towards right a constant speed v_0. The right arm of the loop is in a uniform magnetic field of 5 T. The field is perpendicular to the plane of the plane of loop and is going into it. The loop is connected to a network of resistor each of value 4 Ω. What should be the value of v_0 so that a steady current of 2 mA flows in the loop ? **[01 Sep. 2021 (Evening)]**

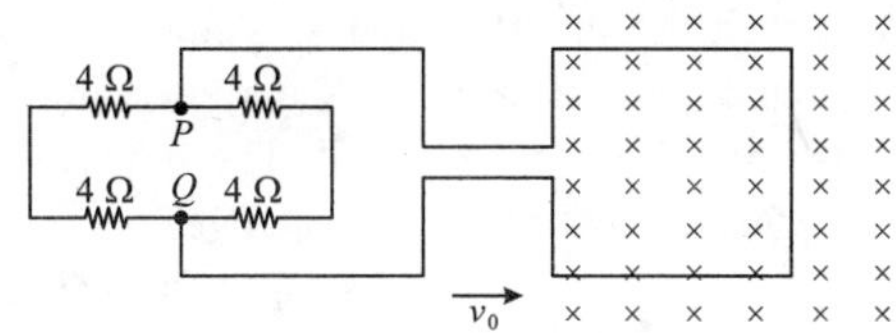

(A) 1 cm/s (B) 10^{-2} cm/s
(C) 10^2 m/s (D) 1 m/s

* * * * *

CHAPTER

20 ELECTROMAGNETIC INDUCTION

20.2 SELF AND MUTUAL INDUCTION

1. A power transmission line feeds input power at 2300 V to a step down transformer with its primary windings having 4000 turns. The output power is delivered at 230 V by the transformer. If the current in the primary of the transformer is 5 A and its efficiency is 90%, the output current would be :

[9 Jan. 2019 (Evening)]

(A) 25 A (B) 50 A
(C) 35 A (D) 45 A

2. The self induced emf of a coil is 25 volt. When the current in it is changed at uniform rate from 10 A to 25 A in 1 s, the change in the energy of the inductance is :

[10 Jan. 2019 (Evening)]

(A) 437.5 J (B) 637.5 J
(C) 740 J (D) 540 J

3. In the circuit shown, the switch S_1 is closed at time $t = 0$ and the switch S_2 is kept open. At some later time (t_0), the switch S_1 is opened and S_2 is closed. The behaviour of the current I as a function of time 't' is given by :

[11 Jan. 2019 (Morning)]

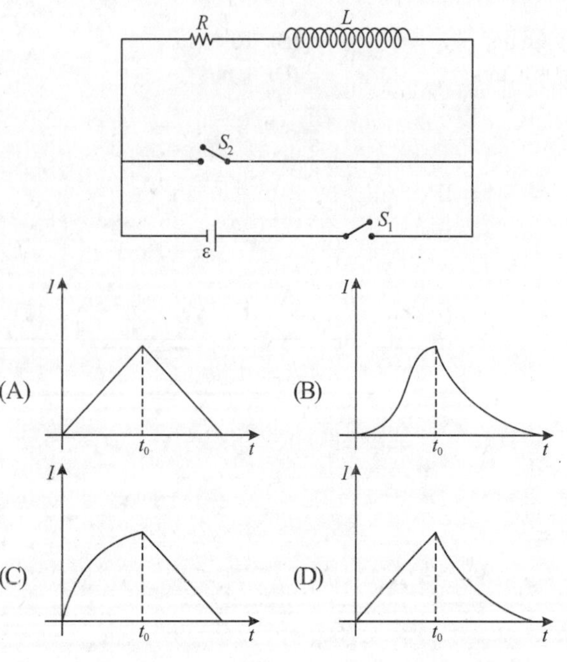

4. There are two long co-axial solenoids of same length l. The inner and outer coils have radii r_1 and r_2 and number of turns per unit length n_1 and n_2 respectively. The ratio of mutual inductance to the self-inductance of the inner-coil is :

[11 Jan. 2019 (Morning)]

(A) $\frac{n_2}{n_1} \cdot \frac{r_2^2}{r_1^2}$ (B) $\frac{n_2}{n_1} \cdot \frac{r_1}{r_2}$

(C) $\frac{n_1}{n_2}$ (D) $\frac{n_2}{n_1}$

5. A copper wire is wound on a wooden frame, whose shape is that of an equilateral triangle. If the linear dimension of each side of the frame is increased by a factor of 3, keeping the number of turns of the coil per unit length of the frame the same, then the self inductance of the coil : **[11 Jan. 2019 (Evening)]**

(A) Decreases by a factor of $9\sqrt{3}$
(B) Increases by a factor of 3
(C) Decreases by a factor of 9
(D) Increases by a factor of 27

6. In the figure shown, a circuit contains two identical resistors with resistance $R = 5\ \Omega$ and an inductance with $L = 2$ mH. An ideal battery of 15 V is connected in the circuit. What will be the current through the battery long after the switch is closed?

[12 Jan. 2019 (Morning)]

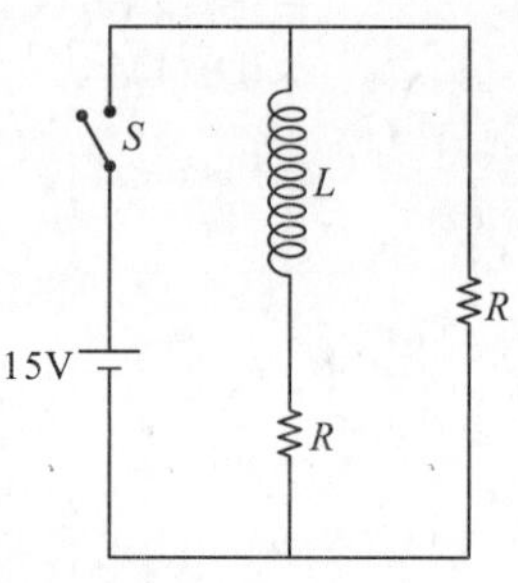

(A) 6 A (B) 7.5 A
(C) 5.5 A (D) 3 A

7. The total number of turns and cross-section area in a solenoid is fixed. However, its length L is varied by adjusting the separation between windings. The inductance of solenoid will be proportional to : **[9 April 2019 (Morning)]**

(A) $1/L^2$ (B) $1/L$
(C) L (D) L^2

8. Two coils 'P' and 'Q' are separated by some distance. When a current of 3 A flows through coil 'P', a magnetic flux of 10^{-3}Wb passes through 'Q'. No current is passed through 'Q'. When no current passes through 'P' and a current of $2A$ passes through 'Q', the flux through 'P' is : **[9 April 2019 (Evening)]**

(A) 6.67×10^{-3} Wb (B) 6.67×10^{-4} Wb
(C) 3.67×10^{-4} Wb (D) 3.67×10^{-3} Wb

9. A very long solenoid of radius R is carrying current $I(t) = kte^{-\alpha t}(k > 0)$, as a function of time $(t \geq 0)$. Counter clockwise current is taken to be positive. A circular conducting coil of radius $2R$ is placed in the equatorial plane of the solenoid and concentric with the solenoid. The current induced in the outer coil is correctly depicted, as a function of time, by : **[9 April 2019 (Evening)]**

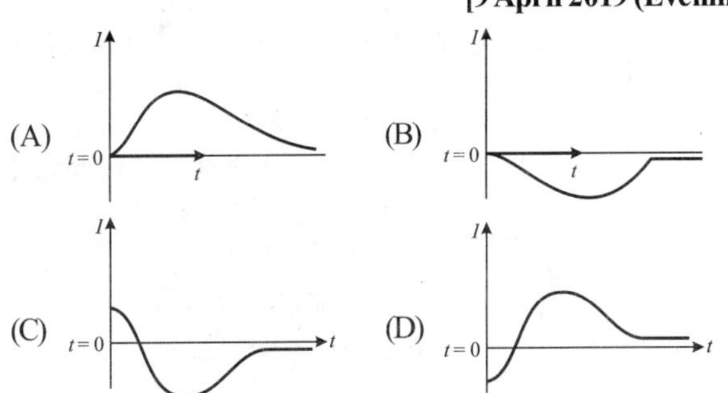

10. A coil of self inductance 10 mH and resistance 0.1 Ω is connected through a switch to a battery of internal resistance 0.9 Ω. After the switch is closed, the time taken for the current to attain 80% of the saturation value is : (Take ln 5 = 1.6) : **[10 April 2019 (Evening)]**

(A) 0.103 s (B) 0.016 s
(C) 0.002 s (D) 0.324 s

11. Consider the LR circuit shown in the figure. If the switch S is closed at $t = 0$ then the amount of charge that passes through the battery between $t = 0$ and $t = \frac{L}{R}$ is : **[12 April 2019 (Evening)]**

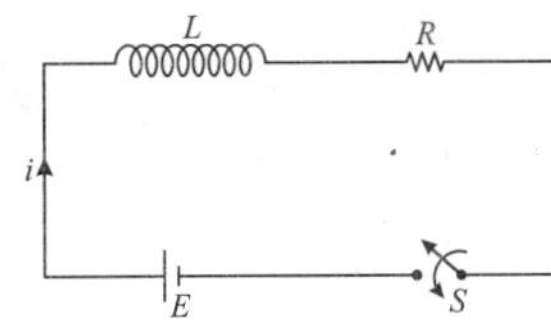

(A) $\frac{EL}{7.3R^2}$ (B) $\frac{EL}{2.7R^2}$
(C) $\frac{7.3EL}{R^2}$ (D) $\frac{2.7EL}{R^2}$

12. An emf of 20 V is applied at time $t = 0$ to a circuit containing in series 10 mH inductor and 5Ω resistor. The ratio of the currents at time $t = \infty$ and at $t = 40$ s is close to : (Take $e^2 = 7.389$) : **[7 Jan. 2020 (Evening)]**

(A) 1.06 (B) 1.15
(C) 1.46 (D) 0.84

13. A shown in the figure, a battery of emf ε is connected to an inductor L and resistance R in series. The switch is closed at $t = 0$. The total charge that flows from the battery, between $t = 0$ and $t = t_c$ (t_c is the time constant of the circuit) is : **[8 Jan. 2020 (Evening)]**

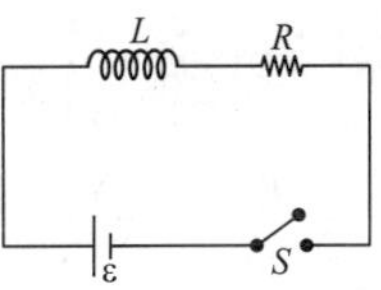

(A) $\frac{\varepsilon L}{R^2}\left(1 - \frac{1}{e}\right)$ (B) $\frac{\varepsilon R}{eL^2}$
(C) $\frac{\varepsilon L}{R^2}$ (D) $\frac{\varepsilon L}{eR^2}$

14. In a fluorescent lamp choke (a small transformer) 100 V of reverse voltage is produced when the choke current changes uniformly from 0.25 A to 0 in a duration of 0.025 ms. The self-inductance of the choke (in mH) is estimated to be ______ . **[9 Jan. 2020 (Morning)]**

15. A series L-R circuit is connected to a battery of *emf* V. If the circuit is switched on at $t = 0$, then the time at which the energy stored in the inductor reaches $(1/n)$ times of its maximum value, is : **[4 Sept. 2020 (Evening)]**

(A) $\frac{L}{R}\ln\left(\frac{\sqrt{n}-1}{\sqrt{n}}\right)$ (B) $\frac{L}{R}\ln\left(\frac{\sqrt{n}}{\sqrt{n}+1}\right)$

(C) $\frac{L}{R}\ln\left(\frac{\sqrt{n}}{\sqrt{n}-1}\right)$ (D) $\frac{L}{R}\ln\left(\frac{\sqrt{n}+1}{\sqrt{n}-1}\right)$

16. A part of a complete circuit is shown in the figure. At some instant, the value of current I is $1A$ and it is decreasing at a rate of $10^2 A\ s^{-1}$. The value of the potential difference $V_P - V_Q$, (in volts) at that instant, is : **[6 Sept. 2020 (Morning)]**

L = 50mH, I, 30 V, R = 2Ω, P, Q

17. A common transistor radio set requires 12 V (D.C.) for its operation. The D.C. source is constructed by using a transformer and a rectifier circuit, which are operated at 220 V (A.C.) on standard domestic A.C. supply. The number of turns of secondary coil are 24, then the number of turns of primary are ______. **[24 Feb. 2021 (Morning)]**

18. Figure shows a circuit that contains four identical resistors with resistance $R = 2.0\ \Omega$, two identical inductors with inductance $L = 2.0$ mH and an ideal battery with emf $E = 9$ V. The current 'i'

just after the switch 'S' is closed will be :

[24 Feb. 2021 (Evening)]

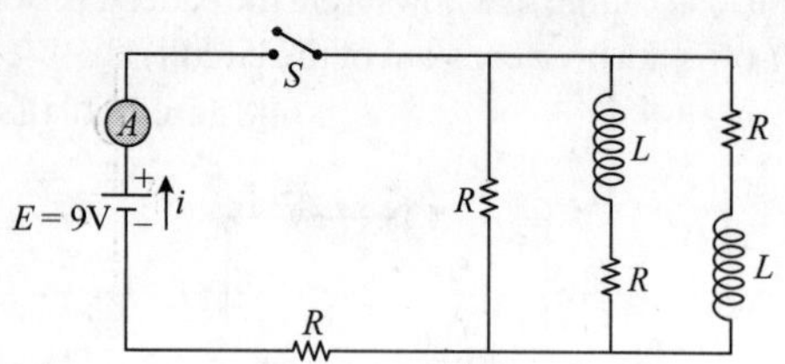

(A) 2.25 A (B) 3.0 A

(C) 3.37 A (D) 9 A

19. The current (i) at time $t = 0$ and $t = \infty$ respectively for the given circuit is : **[25 Feb. 2021 (Morning)]**

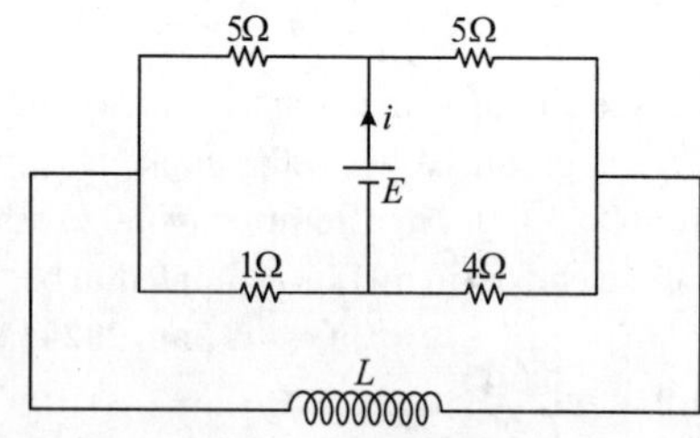

(A) $\frac{5E}{18}, \frac{10E}{33}$ (B) $\frac{10E}{33}, \frac{5E}{18}$

(C) $\frac{5E}{18}, \frac{18E}{55}$ (D) $\frac{18E}{55}, \frac{5E}{18}$

20. A coil of inductance 2 H having negligible resistance is connected to a source of supply whose voltage is given by $e = 3t$ volt. (where t is in second). If the voltage is applied when $t = 0$, then the energy stored in the coil after 4 s is ______ J. **[25 Feb. 2021 (Morning)]**

21. For the given circuit, comment on the type of transformer used : **[16 March, 2021 (Evening)]**

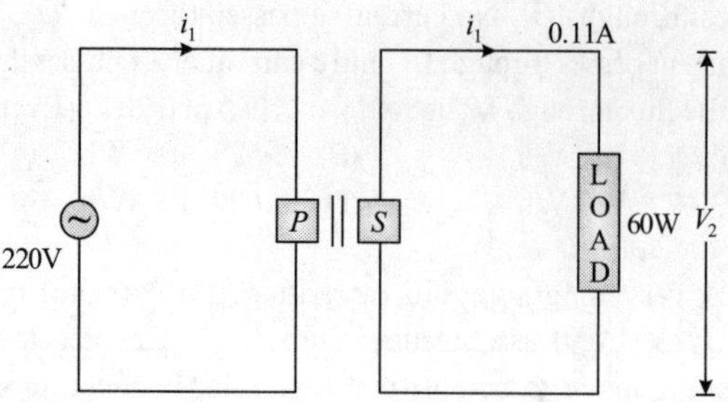

(A) Auxilliary transformer (B) Auto transformer

(C) Step-up transformer (D) Step down transformer

22. Four identical long solenoids A, B, C and D are connected to each other as shown in the figure. If the magnetic field at the center of A is 3 T, the field at the center of C would be : (Assume that the magnetic field is confined with in the volume of respective solenoid). **[18 March 2021 (Morning)]**

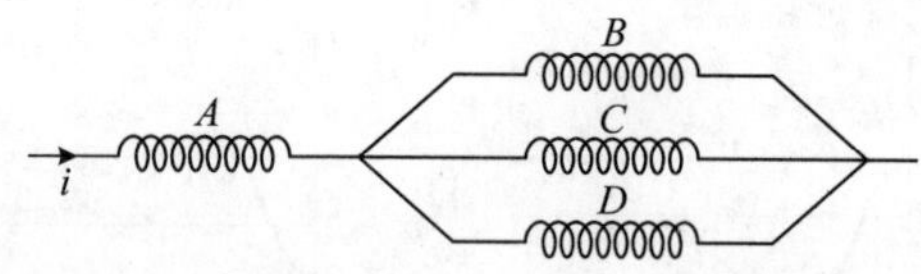

(A) 12 T (B) 6 T

(C) 9 T (D) 1 T

23. A small square loop of side a and one turn is placed inside a larger square loop of side b and one turn ($b >> a$). The two loops are coplanar with their centres coinciding. If a current I is passed in the square loop of side b, then the coefficient of mutual inductance between the two loops is: **[31 Aug. 2021 (Morning)]**

(A) $\frac{\mu_0}{4\pi} 8\sqrt{2} \frac{b^2}{a}$ (B) $\frac{\mu_0}{4\pi} \frac{8\sqrt{2}}{b}$

(C) $\frac{\mu_0}{4\pi} \frac{8\sqrt{2}}{a}$ (D) $\frac{\mu_0}{4\pi} 8\sqrt{2} \frac{a^2}{a}$

* * * * *

CHAPTER

21 ALTERNATING CURRENT

21.1 ALTERNATING CURRENT

1. A sinusoidal voltage of peak value 283 V and angular frequency 320/s is applied to a series *LCR* circuit. Given that $R = 5\ \Omega$, $L = 25$ mH and $C = 1000\ \mu F$. The total impedance, and phase difference between the voltage across the source and the current will respectively be : **[2017 (Online)]**

(A) $10\ \Omega$ and $\tan^{-1}\left(\frac{5}{3}\right)$ (B) $7\ \Omega$ and $45°$

(C) $10\ \Omega$ and $\tan^{-1}\left(\frac{8}{3}\right)$ (D) $7\ \Omega$ and $\tan^{-1}\left(\frac{5}{3}\right)$

2. A small circular loop of wire of radius *a* is located at the centre of a much larger circular wire loop of radius *b*. The two loops are in the same plane. The outer loop of radius *b* carries an alternating current $I = I_0 \cos(\omega t)$. The emf induced in the smaller inner loop is nearly: **[2017 (Online)]**

(A) $\frac{\pi\mu_0 I_0}{2}\cdot\frac{a^2}{b}\ \omega \sin(\omega t)$ (B) $\frac{\pi\mu_0 I_0}{2}\cdot\frac{a^2}{b}\ \omega \cos(\omega t)$

(C) $\pi\mu_0 I_0 \frac{a^2}{b}\ \omega \sin(\omega t)$ (D) $\frac{\pi\mu_0 I_0 b^2}{a}\ \omega \cos(\omega t)$

3. For an *RLC* circuit driven with voltage of amplitude v_m and frequency $\omega_0 = \frac{1}{\sqrt{LC}}$ the current exhibits resonance. The quality factor, Q is given by : **[2018]**

(A) $\frac{\omega_0 L}{R}$ (B) $\frac{\omega_0 R}{L}$

(C) $\frac{R}{(\omega_0 C)}$ (D) $\frac{CR}{\omega_0}$

4. In an AC circuit, the instantaneous e.m.f. and current are given by

$e = 100 \sin 30t$

$i = 20 \sin\left(30t - \frac{\pi}{4}\right)$

In one cycle of AC, the average power consumed by the circuit and the wattless current are, respectively : **[2018]**

(A) 50 W, 10 A (B) $\frac{1000}{\sqrt{2}}$ W, 10 A

(C) $\frac{50}{\sqrt{2}}$ W, 0 (D) 50 W, 0

5. A power transmission line feeds input power at 2300 V to a step down transformer with its primary windings having 4000 turns, given the output power at 230 V. If the current in the primary of the transformer is 5 A, and its efficiency is 90%, the output current would be : **[2018 (Online)]**

(A) 20 A (B) 40 A
(C) 45 A (D) 25 A

6. A series *AC* circuit containing an inductor (20 mH), a capacitor (120 μF) and a resistor (60 Ω) is driven by an *AC* source of 24 V/50 Hz. The energy dissipated in the circuit in 60 s is : **[9 Jan. 2019 (Evening)]**

(A) 2.26×10^3 J (B) 3.39×10^3 J
(C) 5.65×10^2 J (D) 5.17×10^2 J

7. 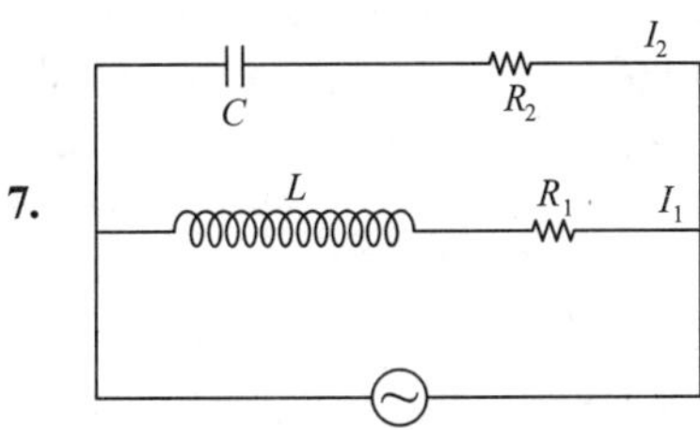

In the above circuit, $C = \frac{\sqrt{3}}{2}\mu F$, $R_2 = 20\ \Omega$, $L = \frac{\sqrt{3}}{10}$ *H* and $R_1 = 10\ \Omega$. Current in *L*-R_1 path is I_1 and in *C*-R_2 path it is I_2. The voltage of A.C source is given by $V = 200\sqrt{2}\sin(100t)$ volts. The phase difference between I_1 and I_2 is : **[12 Jan. 2019 (Evening)]**

(A) 30° (B) 0°
(C) 90° (D) 60°

8. An alternating voltage $v(t) = 220 \sin 100\ \pi t$ volt is applied to a purely resistance load of 50 Ω. The time taken for the current to rise from half of the peak value to the peak value is : **[8 April 2019 (Morning)]**

(A) 2.2 ms (B) 5 ms
(C) 3.3 ms (D) 7.2 ms

9. A circuit connected to an ac source of *emf* $e = e_0 \sin(100t)$ with *t* in seconds, gives a phase difference of $\pi/4$ between the emf *e* and current *i*. Which of the following circuits will exhibit this ? **[8 April 2019 (Evening)]**

(A) *RC* circuit with R = 1 kΩ and *C* = 1 μF

(B) RL circuit with $R = 1\ k\Omega$ and $L = 1$ mH
(C) RL circuit with $R = 1\ k\Omega$ and $L = 10$ mH
(D) RC circuit with $R = 1\ k\Omega$ and $C = 10\ \mu F$

10. A transformer consisting of 300 turns in the primary and 150 turns in the secondary gives output power of 2.2 kW. If the current in the secondary coil is 10 A, then the input voltage and current in the primary coil are : **[10 April 2019 (Morning)]**
(A) 220 V and 10 A (B) 440 V and 5 A
(C) 440 V and 20 A (D) 220 V and 20 A

11. A LCR circuit behaves like a damped harmonic oscillator. Comparing it with a physical spring-mass damped oscillator having damping constant 'b', the correct equivalence would be : **[7 Jan. 2020 (Morning)]**

(A) $L \leftrightarrow m, C \leftrightarrow \frac{1}{k}, R \leftrightarrow b$

(B) $L \leftrightarrow \frac{1}{b}, C \leftrightarrow \frac{1}{m}, R \leftrightarrow \frac{1}{k}$

(C) $L \leftrightarrow m, C \leftrightarrow k, R \leftrightarrow b$
(D) $L \leftrightarrow k, C \leftrightarrow b, R \leftrightarrow m$

12. In LC circuit the inductance $L = 40$ mH and capacitance $C = 100\ \mu F$. If a voltage $V(t) = 10 \sin (314\ t)$ is applied to the circuit, the current in the circuit is given as : **[9 Jan. 2020 (Evening)]**
(A) $0.52 \cos 314\ t$ (B) $0.52 \sin 314\ t$
(C) $10 \cos 314\ t$ (D) $5.2 \cos 314\ t$

13. An inductance coil has a reactance of 100 W. When an AC signal of frequency 1000 Hz is applied to the coil, the applied voltage leads the current by 45°. The self-inductance of the coil is : **[2 Sept. 2020 (Evening)]**
(A) 1.1×10^{-2} H (B) 1.1×10^{-1} H
(C) 5.5×10^{-5} H (D) 6.7×10^{-7} H

14. A 750 Hz, 20 V (r.m.s.) source is connected to a resistance of 100 Ω, an inductance of 0.1803 H and a capacitance of 10 μF all in series. The time in which the resistance (heat capacity 2 J/°C) will get heated by 10°C. (assume no loss of heat to the surroundings) is close to : **[3 Sept. 2020 (Morning)]**
(A) 418 s (B) 245 s
(C) 348 s (D) 365 s

15. In a series LR circuit, power of 400 W is dissipated from a source of 250 V, 50 Hz. The power factor of the circuit is 0.8. In order to bring the power factor to unity, a capacitor of value C is added in series to the L and R. Taking the value of C as $\left(\frac{n}{3\pi}\right)\mu F$, then value of n is : **[6 Sept. 2020 (Evening)]**

16. A resonance circuit having inductance and resistance 2×10^{-4} H and 6.28 Ω respectively oscillates at 10 MHz frequency. The value of quality factor of this resonator is ______. [$\pi = 3.14$] **[24 Feb. 2021 (Morning)]**

17. Match List-I with List-II : **[24 Feb. 2021 (Evening)]**

List-I	List-II
(a) Source of microwave frequency	(i) Radioactive decay of nucleus
(b) Source of infrared frequency	(ii) Magnetron
(c) Source of Gamma Rays	(iii) Inner shell electrons
(d) Source of X-rays	(iv) Vibration of atoms and molecules
	(v) LASER
	(vi) RC circuit

Choose the correct answer from the options given below :
(A) (a) → (vi), (b) → (v), (c) → (i), (d) → (iv)
(B) (a) → (ii), (b) → (iv), (c) → (i), (d) → (iii)
(C) (a) → (ii), (b) → (iv), (c) → (vi), (d) → (iii)
(D) (a) → (vi), (b) → (iv), (c) → (i), (d) → (v)

18. A series LCR circuit is designed to resonate at an angular frequency $\omega_0 = 10^5$ rad/s. The circuit draws 16 W power from 120 V source at resonance. The value of resistance 'R' in the circuit is ______ Ω. **[24 Feb. 2021 (Evening)]**

19. The angular frequency of alternating current in a LCR circuit is 100 rad/s. The components connected are shown in the figure. Find the value of inductance of the coil and capacity of condenser. **[25 Feb. 2021 (Morning)]**

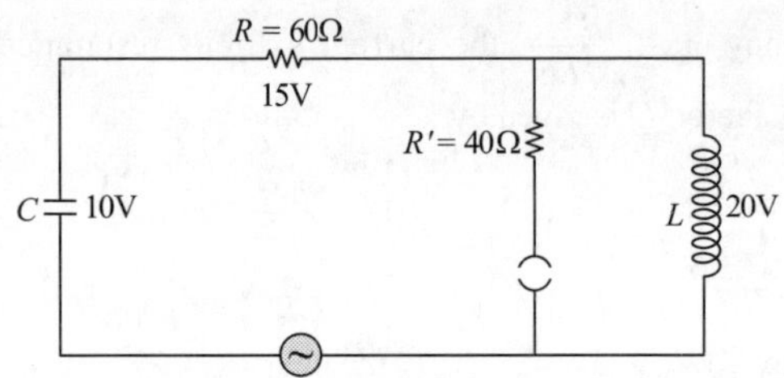

(A) 1.33 H and 250 μF (B) 1.33 H and 150 μF
(C) 0.8 H and 150 μF (D) 0.8 H and 250 μF

20. An LCR circuit contains resistance of 110 Ω and a supply of 220 V at 300 rad/s angular frequency. If only capacitance is removed from the circuit, current lags behind the voltage by 45°. If on the other hand, only inductor is removed the current leads by 45° with the applied voltage. The RMS current flowing in the circuit will be : **[25 Feb. 2021 (Evening)]**
(A) 2 A (B) 2.5 A
(C) 1.5 A (D) 1 A

21. Match List-I with List-II : **[25 Feb. 2021 (Evening)]**

List-I	List-II
(a) Rectifier	(i) Used either for stepping up or stepping down the a.c. voltage
(b) Stabilizer	(ii) Used to convert a.c. voltage into d.c. voltage
(c) Transformer	(iii) Used to remove any ripple in the rectified output voltage
(d) Filter	(iv) Used for constant output voltage even when the input voltage or load current change

(A) (a) $\to$ (ii), (b) $\to$ (iv), (c) $\to$ (i), (d) $\to$ (iii)
(B) (a) $\to$ (ii), (b) $\to$ (i), (c) $\to$ (iii), (d) $\to$ (iv)
(C) (a) $\to$ (ii), (b) $\to$ (i), (c) $\to$ (iv), (d) $\to$ (iii)
(D) (a) $\to$ (iii), (b) $\to$ (iv), (c) $\to$ (i), (d) $\to$ (ii)

22. An alternating current is given by the equation $i = i_1 \sin \omega t + i_2 \cos \omega t$. The rms current will be : **[26 Feb. 2021 (Morning)]**

(A) $\frac{1}{\sqrt{2}}(i_1 + i_2)^2$ (B) $\frac{1}{\sqrt{2}}(i_1 + i_2)$

(C) $\frac{1}{\sqrt{2}}(i_1^2 + i_2^2)^{1/2}$ (D) $\frac{1}{2}(i_1^2 + i_2^2)^{1/2}$

23. In a series LCR resonant circuit, the quality factor is measured as 100. If the inductance is increased by two fold and resistance is decreased by two fold, then the quality factor after this change will be ______. **[26 Feb. 2021 (Morning)]**

24. Find the peak current and resonant frequency of the following circuit (as shown in figure). **[26 Feb. 2021 (Evening)]**

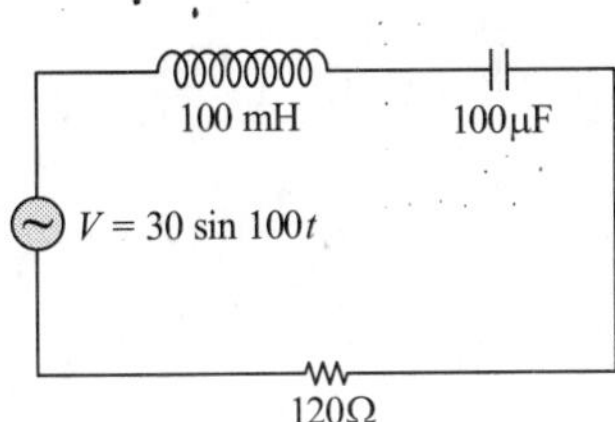

(A) 2 A and 50 Hz (B) 0.2 A and 100 Hz
(C) 2 A and 100 Hz (D) 0.2 A and 50 Hz

25. A sinusoidal voltage of peak value 250 V is applied to a series LCR circuit, in which $R = 8\ \Omega$, $L = 24$ mH and $C = 60\ \mu$F. The value of power dissipated at resonant condition is 'x' kW. The value of x to the nearest integer is ______. **[16 March 2021 (Morning)]**

26. An AC current is given by $I = I_1 \sin\omega t + I_2 \cos\omega t$. A hot wire ammeter will give a reading : **[17 March 2021 (Morning)]**

(A) $\sqrt{\frac{I_1^2 - I_2^2}{2}}$ (B) $\sqrt{\frac{I_1^2 + I_2^2}{2}}$

(C) $\frac{I_1 + I_2}{\sqrt{2}}$ (D) $\frac{I_1 + I_2}{2\sqrt{2}}$

27. Match List-I with List-II

List-I	List-II
(a) Phase difference between current and voltage in a purely resistive AC circuit	(i) $\frac{\pi}{2}$; current leads voltage
(b) Phase difference between current and voltage in a pure inductive AC circuit	(ii) zero
(c) Phase difference between current and voltage in a pure capacitive AC circuit	(iii) $\frac{\pi}{2}$; current lags voltage
(d) Phase difference between current and voltage in an LCR series circuit	(iv) $\tan^{-1}\left(\frac{X_C - X_L}{R}\right)$

Choose the most appropriate answer from the options given below : **[17 March 2021 (Evening)]**

(A) (a) $\to$ (i), (b) $\to$ (iii), (c) $\to$ (iv), (d) $\to$ (ii)
(B) (a) $\to$ (ii), (b) $\to$ (iv), (c) $\to$ (iii), (d) $\to$ (i)
(C) (a) $\to$ (ii), (b) $\to$ (iii), (c) $\to$ (iv), (d) $\to$ (i)
(D) (a) $\to$ (ii), (b) $\to$ (iii), (c) $\to$ (i), (d) $\to$ (iv)

28. What happens to the inductive reactance and the current in a purely inductive circuit if the frequency is halved ? **[17 March 2021 (Evening)]**

(A) Both, inductive reactance and current will be halved.
(B) Inductive reactance will be halved and current will be doubled.
(C) Inductive reactance will be doubled and current will be halved.
(D) Both, inducting reactance and current will be doubled.

29. In a series LCR resonance circuit, if we change the resistance only, from a lower to higher value : **[18 March 2021 (Morning)]**

(A) The bandwidth of resonance circuit will increase
(B) The resonance frequency will increase
(C) The quality factor will increase
(D) The quality factor and the resonance frequency will remain constant

30. An AC source rated 220 V, 50 Hz is connected to a resistor. The time taken by the current to change from its maximum to the rms value is : **[18 March 2021 (Morning)]**

(A) 2.5 ms (B) 25 ms

(C) 2.5 s (D) 0.25 ms

31. In a series *LCR* circuit, the inductive reactance (X_L) is 10 Ω and the capacitive reactance (X_C) is 4 Ω. The resistance (R) in the circuit is 6 Ω. The power factor of the circuit is : **[18 March 2021 (Evening)]**

(A) $\frac{1}{2}$ (B) $\frac{1}{2\sqrt{2}}$

(C) $\frac{1}{\sqrt{2}}$ (D) $\frac{\sqrt{3}}{2}$

32. The time taken for the magnetic energy to reach 25% of its maximum value, when a solenoid of resistance *R*, inductance *L* is connected to a battery, is : **[18 March 2021 (Evening)]**

(A) $\frac{L}{R}\ln 5$ (B) Infinite

(C) $\frac{L}{R}\ln 2$ (D) $\frac{L}{R}\ln 10$

33. In an *LCR* series circuit, an inductor 30 mH and a resistor 1 Ω are connected to an AC source of angular frequency 300 rad/s. The value of capacitance for which, the current leads the voltage by 45° is $\frac{1}{x} \times 10^{-3}$ F. Then the value of *x* is ______. **[20 July 2021 (Morning)]**

34. For a series *LCR* circuit with *R* = 100 Ω, *L* = 0.5 mH ad *C* = 0.1 pF connected across 220 V – 50 Hz AC supply, the phase angle between current and supplied voltage and the nature of the circuit is : **[20 July 2021 (Evening)]**

(A) 0°, resonance circuit

(B) ≈ 90°, predominantly capacitive circuit

(C) 0°, resistive circuit

(D) ≈ 90, predominantly inductive circuit

35. A series *LCR* circuit of *R* = 5 Ω, *L* = 20 mH and *C* = 0.5 μ*F* is connected across an *AC* supply of 250 *V*, having variable frequency. The power dissipated at resonance condition is ______ × 10² W. **[20 July 2021 (Evening)]**

36. In a circuit consisting of a capacitance and a generator with alternating emf $E_g = E_{g0} \sin \omega t$ V_C and I_C are the voltage and current. Correct phasor diagram for such circuit is : **[22 July 2021 (Evening)]**

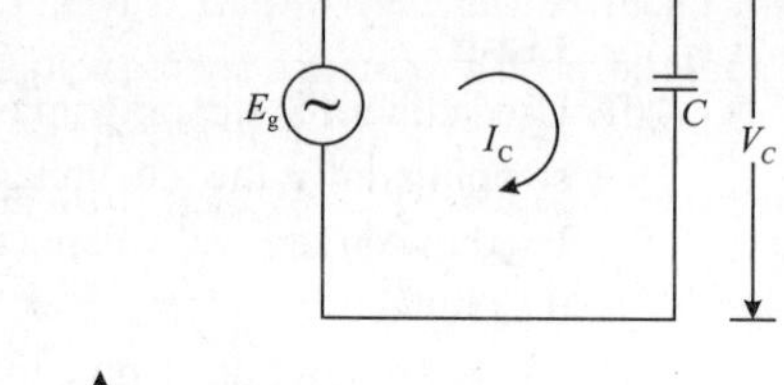

(A)
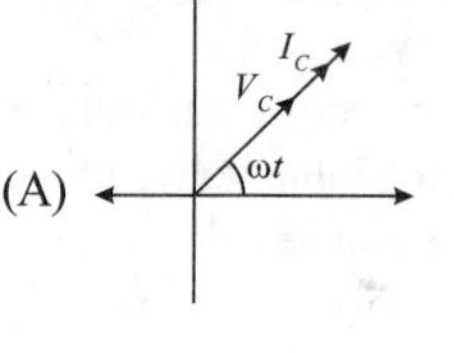

(B)
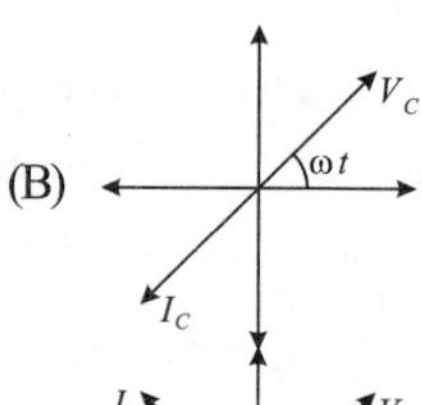

(C)
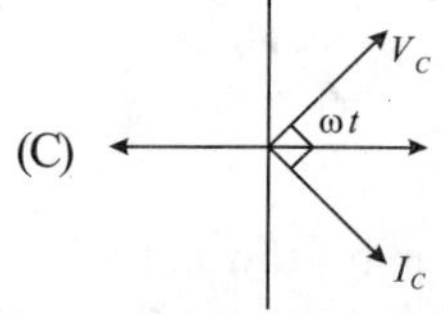

(D)
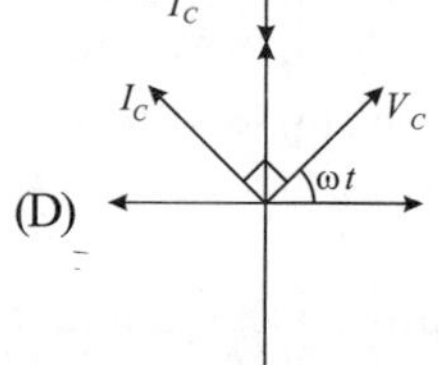

37. Match List-I with List-II : **[22 July 2021 (Evening)]**

List-I	List-II
(a) $\omega L > \frac{1}{\omega C}$	(i) Current is in phase with emf
(b) $\omega L = \frac{1}{\omega C}$	(ii) Current lags behind the applied emf
(c) $\omega L < \frac{1}{\omega C}$	(iii) Maximum current occurs
(d) Resonant frequency	(iv) Current leads the emf

Choose the correct answer from the option given below :

(A) (a)-(iv); (b)-(iii); (c)-(ii); (d)-(i)

(B) (a)-(ii); (b)-(i); (c)-(iv); (d)-(iii)

(C) (a)-(ii); (b)-(i); (c)-(iii); (d)-(iv)

(D) (a)-(iii); (b)-(i); (c)-(iv); (d)-(ii)

38. A 10 Ω resistance is connected across 220 V – 50 Hz AC supply. The time taken by the current to change from its maximum value to the rms value is : **[25 July 2021 (Evening)]**

(A) 3.0 ms (B) 1.5 ms

(C) 4.5 ms (D) 2.5 ms

39. Two circuits are shown in the figure (*a*) and (*b*). At a frequency of ______ rad/s the average power dissipated in one cycle will be same in both the circuits : **[25 July 2021 (Evening)]**

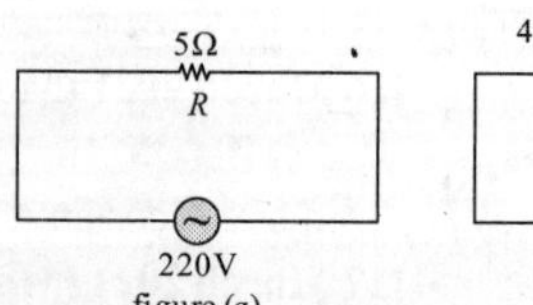

figure (*a*)

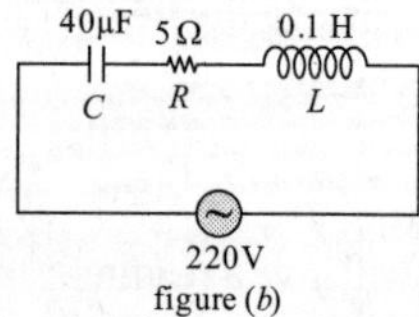

figure (*b*)

40. A 0.07 H inductor and 12 Ω resistor are connected in series to a 220 V, 50 Hz ac source. The approximate current in the circuit and phase angle between current and source voltage are respectively [Take π as $\frac{22}{7}$]: **[27 July 2021 (Morning)]**

(A) 8.8 A and $\tan^{-1}\left(\frac{6}{11}\right)$ (B) 0.88 A and $\tan^{-1}\left(\frac{11}{6}\right)$

(C) 8.8 A and $\tan^{-1}\left(\frac{11}{6}\right)$ (D) 88 A and $\tan^{-1}\left(\frac{11}{6}\right)$

41. A 100 Ω resistance, a 0.1 μF capacitor and an inductor are connected in series across a 250 V supply at variable frequency, Calculate the value of inductance of inductor at which resonance will occur, Given that the resonant frequency is 60 Hz : **[27 July 2021 (Evening)]**

(A) 7.03×10^{-5} H (B) 0.70 H
(C) 70.3 mH (D) 70.3 H

42. An inductor coil stores 64 J of magnetic field energy and dissipates energy at the rate of 640 W when a current of 8 A is passed through it. If this coil is joined across an ideal battery, find the time constant of the circuit in seconds : **[26 Aug. 2021 (Morning)]**

(A) 0.8 (B) 0.125
(C) 0.4 (D) 0.2

43. A series *LCR* circuit driven by 300 V at frequency of 50 Hz contains a resistance $R = 3$ kΩ, an inductor of inductive reactance $X_L = 250\,\pi\Omega$ and an unknown capacitor. The value of capacitance to maximize the average power should be : (take $\pi^2 = 10$) : **[26 Aug. 2021 (Morning)]**

(A) 4 μF (B) 400 μF
(C) 40 μF (D) 25 μF

44. In the given circuit the *AC* source has $\omega = 100$ rad s^{-1}. Considering the inductor and capacitor to be ideal, what will be the current *I* flowing through the circuit ? **[26 Aug. 2021 (Evening)]**

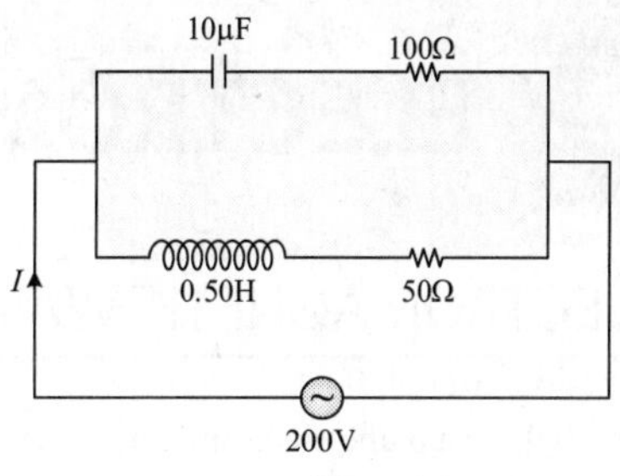

(A) 0.94 A (B) 5.9 A
(C) 4.24 A (D) 6 A

45. The alternating current is given by

$$i = \left\{\sqrt{42}\ \text{sir}\left(\frac{2\pi}{T}t\right) + 10\right\} A$$

The rms value of this current is ______ A. **[27 Aug. 2021 (Morning)]**

46. An AC circuit has an inductor and a resistor of resistance *R* in series, such that $X_L = 3R$. Now, a capacitor is added in series such that $X_C = 2R$. The ratio of new power factor with the old power factor of the circuit is $\sqrt{5} : x$. The value of *x* is______. **[27 Aug. 2021 (Evening)]**

47. In an ac circuit, an inductor, a capacitor and a resistor are connected in series with $X_L = R = X_C$. Impedance of this circuit is : **[31 Aug. 2021 (Morning)]**

(A) $R\sqrt{2}$ (B) R
(C) $2R^2$ (D) Zero

48. At very high frequencies, the effective impedance of the given circuit will be ______ Ω. **[31 Aug. 2021 (Evening)]**

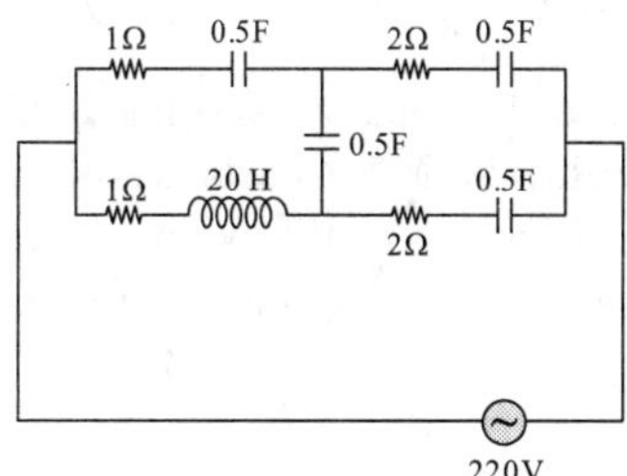

* * * * *

CHAPTER

22 ELECTROMAGNETIC WAVES

22.1 ELECTROMAGNETIC WAVE

1. Magnetic field in a plane electromagnetic wave is given by

$\vec{B} = B_0 \sin(kx + \omega t)\,\hat{j}T$

Expression for corresponding electric field will be :

Where c is speed of light. **[2017 (Online)]**

(A) $\vec{E} = B_0 c \sin(kx + \omega t)\,\hat{k}$ V/m

(B) $\vec{E} = \frac{B_0}{c} \sin(kx + \omega t)\,\hat{k}$ V/m

(C) $\vec{E} - B_0 c \sin(kx + \omega t)\,\hat{k}$ V/m

(D) $\vec{E} = B_0 c \sin(kx - \omega t)\,\hat{k}$ V/m

2. The electric field component of a monochromatic radiation is given by

$\vec{E} = 2E_0 \hat{i} \cos kz \cos \omega t$

Its magnetic field $\vec{B}$ is then given by : **[2017 (Online)]**

(A) $\frac{2E_0}{c}\hat{j} \sin kz \cos \omega t$ (B) $\frac{2E_0}{c}\hat{j} \cos kz \sin \omega t$

(C) $\frac{2E_0}{c}\hat{j} \sin kz \sin \omega t$ (D) $\frac{2E_0}{c}\hat{j} \cos kz \cos \omega t$

3. An EM wave from air enters a medium. The electric fields are $\vec{E_1} = E_{01}\,\hat{x} \cos\left[2\pi\nu\left(\frac{z}{c} - t\right)\right]$ in air and $\vec{E_2} = E_{02}\,\hat{x} \cos [k(2z - ct)]$ in medium, where the wave number k and frequency ν refer to their values in air. The medium is nonmagnetic. If ϵ_{r_1} and ϵ_{r_2} refer to relative permittivities of air and medium respectively, which of the following options is correct ? **[2018]**

(A) $\frac{\epsilon_{r_1}}{\epsilon_{r_2}} = 4$ (B) $\frac{\epsilon_{r_1}}{\epsilon_{r_2}} = 2$

(C) $\frac{\epsilon_{r_1}}{\epsilon_{r_2}} = \frac{1}{4}$ (D) $\frac{\epsilon_{r_1}}{\epsilon_{r_2}} = \frac{1}{2}$

4. A monochromatic beam of light has a frequency $\nu = \frac{3}{2\pi} \times 10^{12}$ Hz and is propagating along the direction $\frac{\hat{i} + \hat{j}}{\sqrt{2}}$.

It is polarized along the $\hat{k}$ direction. The acceptable form for the magnetic field is : **[2018 (Online)]**

(A) $k\frac{E_0}{C}\left(\frac{\hat{i} - \hat{j}}{\sqrt{2}}\right) \cos\left[10^4\left(\frac{\hat{i} - \hat{j}}{\sqrt{2}}\right)\cdot\vec{r} - (3\times10^{12})t\right]$

(B) $\frac{E_0}{C}\left(\frac{\hat{i} - \hat{j}}{\sqrt{2}}\right) \cos\left[10^4\left(\frac{\hat{i} + \hat{j}}{\sqrt{2}}\right)\cdot\vec{r} - (3\times10^{12})t\right]$

(C) $\frac{E_0}{C}\hat{k} \cos\left[10^4\left(\frac{\hat{i} + \hat{j}}{\sqrt{2}}\right)\cdot\vec{r} + (3\times10^{12})t\right]$

(D) $\frac{E_0}{C}\frac{(\hat{i} + \hat{j} + \hat{k})}{\sqrt{3}} \cos\left[10^4\left(\frac{\hat{i} + \hat{j}}{\sqrt{2}}\right)\cdot\vec{r} + (3\times10^{12})t\right]$

5. A plane electromagnetic wave of wavelength λ has an intensity I. It is propagating along the positive Y-direction. The allowed expressions for the electric and magnetic fields are given by : **[2018 (Online)]**

(A) $\vec{E} = \sqrt{\frac{I}{\epsilon_0 c}} \cos\left[\frac{2\pi}{\lambda}(y - ct)\right]\hat{i}$; $\vec{B} = \frac{1}{c}E\hat{k}$

(B) $\vec{E} = \sqrt{\frac{I}{\epsilon_0 c}} \cos\left[\frac{2\pi}{\lambda}(y - ct)\right]\hat{k}$; $\vec{B} = -\frac{1}{c}E\hat{k}$

(C) $\vec{E} = \sqrt{\frac{2I}{\epsilon_0 c}} \cos\left[\frac{2\pi}{\lambda}(y - ct)\right]\hat{k}$; $\vec{B} = +\frac{1}{c}E\hat{i}$

(D) $\vec{E} = \sqrt{\frac{2I}{\epsilon_0 c}} \cos\left[\frac{2\pi}{\lambda}(y + ct)\right]\hat{k}$; $\vec{B} = \frac{1}{c}E\hat{i}$

6. A plane polarized monochromatic *EM* wave is travelling a vacuum along z-direction such that at $t = t_1$ it is found that the electric field is zero at a spatial point z_1. The next zero that occurs in its neighbourhood is at z_2. The frequency of the electromagnetic wave is : **[2018 (Online)]**

(A) $\frac{3\times10^8}{|z_2 - z_1|}$ (B) $\frac{6\times10^8}{|z_2 - z_1|}$

(C) $\frac{1.5\times10^8}{|z_2 - z_1|}$ (D) $\frac{1}{t_1 + \frac{|z_2 - z_1|}{3\times10^8}}$

7. A plane electromagnetic wave of frequency 50 MHz travels in free space along the positive x-direction. At a particular point in space and time, $\vec{E} = 6.3\hat{j}$ V/m. The corresponding magnetic field $\vec{B}$, at that point will be : **[9 Jan. 2019 (Morning)]**

(A) $18.9 \times 10^{-8}\ \hat{k}T$ (B) $6.6 \times 10^{-8}\ \hat{k}T$
(C) $2.1 \times 10^{-8}\ \hat{k}T$ (D) $18.9 \times 10^{8}\ \hat{k}T$

8. The energy associated with electric field is (U_E) and with magnetic field is (U_B) for an electromagnetic wave in free space. Then : **[9 Jan. 2019 (Evening)]**

(A) $U_E = \frac{U_B}{2}$ (B) $U_E < U_B$
(C) $U_E = U_B$ (D) $U_E > U_B$

9. If the magnetic field of a plane electromagnetic wave is given by (The speed of light = 3×10^8/m/s)

$$B = 100 \times 10^{-6} \sin\left[2\pi \times 2 \times 10^{15}\left(t - \frac{x}{c}\right)\right]$$

then the maximum electric field associated with it is : **[10 Jan. 2019 (Morning)]**

(A) 4×10^4 N/C (B) 4.5×10^4 N/C
(C) 6×10^4 N/C (D) 3×10^4 N/C

10. The electric field of a plane polarized electromagnetic wave in free space at time $t = 0$ is given by an expression $\vec{E}(x, y) = 10\hat{j}$ $\cos[(6x + 8z)]$.

The magnetic field $\vec{B}(x, z, t)$ is given by (c is the velocity of light) : **[10 Jan. 2019 (Evening)]**

(A) $\frac{1}{c}(6\hat{k} + 8\hat{i})\cos[6x - 8z + 10ct]$

(B) $\frac{1}{c}(6\hat{k} - 8\hat{i})\cos[6x + 8z - 10ct]$

(C) $\frac{1}{c}(6\hat{k} + 8\hat{i})\cos[6x + 8z - 10ct]$

(D) $\frac{1}{c}(6\hat{k} - 8\hat{i})\cos[6x + 8z + 10ct]$

11. An electromagnetic wave of intensity 50 Wm^{-2} enters in a medium of refractive index 'n' without any loss. The ratio of the magnitudes of electric fields, and the ratio of the magnitudes of magnetic fields of the wave before and after entering into the medium are respectively, given by : **[11 Jan. 2019 (Morning)]**

(A) $\left(\frac{1}{\sqrt{n}}, \frac{1}{\sqrt{n}}\right)$ (B) $\left(\sqrt{n}, \frac{1}{\sqrt{n}}\right)$

(C) $(\sqrt{n}, \sqrt{n})$ (D) $\left(\frac{1}{\sqrt{n}}, \sqrt{n}\right)$

12. A 27 mW laser beam has a cross-sectional area of 10 mm^2. The magnitude of the maximum electric field in this electromagnetic wave is given by [Given permittivity of space $\epsilon_0 = 9 \times 10^{-12}$ SI units, Speed of light $c = 3 \times 10^8$ m/s]: **[11 Jan. 2019 (Evening)]**

(A) 1 kV/m (B) 2 kV/m
(C) 1.4 kV/m (D) 0.7 kV/m

13. A light wave is incident normally on a glass slab of refractive index 1.5. If 4% of light gets reflected and the amplitude of the electric field of the incident light is 30 V/m, then the amplitude of the electric field for the wave propagating in the glass medium will be : **[12 Jan. 2019 (Morning)]**

(A) 10 V/m (B) 24 V/m
(C) 30 V/m (D) 6 V/m

14. The mean intensity of radiation on the surface of the Sun is about 10^8 W/m^2. The rms value of the corresponding magnetic field is closest to : **[12 Jan. 2019 (Evening)]**

(A) 10^2 T (B) 10^{-4} T
(C) 1 T (D) 10^{-2} T

15. A plane electromagnetic wave travels in free space along the x-direction. The electric field component of the wave at a particular point of space and time is $E = 6$ V m^{-1} along y-direction. Its corresponding magnetic field component, B would be : **[8 April 2019 (Morning)]**

(A) 6×10^{-8} T along z-direction
(B) 6×10^{-8} T along x-direction
(C) 2×10^{-8} T along z-direction
(D) 2×10^{-8} T along y-direction

16. The magnetic field of an electromagnetic wave is given by.

$$\vec{B} = 1.6 \times 10^{-6}\cos(2 \times 10^7 z + 6 \times 10^{15} t)(2\hat{i} + \hat{j})\frac{Wb}{m^2}$$

The associated electric field will be : **[8 April 2019 (Evening)]**

(A) $\vec{E} = 4.8 \times 10^2 \cos(2 \times 10^7 z + 6 \times 10^{15} t)(\hat{i} - 2\hat{j})\frac{V}{m}$

(B) $\vec{E} = 4.8 \times 10^2 \cos(2 \times 10^7 z - 6 \times 10^{15} t)(2\hat{i} + \hat{j})\frac{V}{m}$

(C) $\vec{E} = 4.8 \times 10^2 \cos(2 \times 10^7 z - 6 \times 10^{15} t)(-2\hat{j} + \hat{i})\frac{V}{m}$

(D) $\vec{E} = 4.8 \times 10^2 \cos(2 \times 10^7 z + 6 \times 10^{15} t)(-\hat{i} + 2\hat{j})\frac{V}{m}$

17. The magnetic field of a plane electromagnetic wave is given by $\vec{B} = B_0\hat{i}[\cos(kz - \omega t)] + B_1\hat{j}\cos(kz + \omega t)$ where $B_0 = 3 \times 10^{-5}$ T and $B_1 = 2 \times 10^{-6}$ T. The rms value of the force experienced by a stationary charge $Q = 10^{-4}$ C at $z = 0$ is closest to : **[9 April 2019 (Morning)]**

(A) 0.9 N (B) 0.1 N
(C) 3×10^{-2} N (D) 0.6 N

18. The electric field of a plane electromagnetic wave is given by

$\vec{E} = E_0\hat{i}\cos(kz)\cos(\omega t)$

The corresponding magnetic field $\vec{B}$ is then given by : **[10 April 2019 (Morning)]**

(A) $\vec{B} = \frac{E_0}{C}\hat{j}\sin(kz)\cos(\omega t)$ (B) $\vec{B} = \frac{E_0}{C}\hat{j}\sin(kz)\sin(\omega t)$

(C) $\vec{B} = \frac{E_0}{C}\hat{k}\sin(kz)\cos(\omega t)$ (D) $\vec{B} = \frac{E_0}{C}\hat{j}\cos(kz)\sin(\omega t)$

19. An electromagnetic wave is represented by the electric field $\vec{E} = E_0\hat{n}\sin[\omega t + (6y - 8z)]$. Taking unit vectors in x, y and z directions to be $\hat{i}, \hat{j}, \hat{k}$, the direction of propagation $\hat{s}$, is: **[12 April 2019 (Morning)]**

(A) $\hat{s} = \frac{4\hat{j} - 3\hat{k}}{5}$ (B) $\hat{s} = \frac{3\hat{i} - 4\hat{j}}{5}$

(C) $\hat{s} = \left(\frac{-3\hat{j} + 4\hat{k}}{5}\right)$ (D) $\hat{s} = \frac{-4\hat{k} + 3\hat{j}}{5}$

20. A plane electromagnetic wave having a frequency $\nu = 23.9$ GHz propagates along the positive z-direction in free space. The peak value of the electric field is 60 V/m. Which among the following is the acceptable magnetic field component in the electromagnetic wave ? **[12 April 2019 (Evening)]**

(A) $\vec{B} = 2\times10^{7}\sin(0.5\times10^{3}z + 1.5\times10^{11}t)\hat{i}$

(B) $\vec{B} = 2\times10^{-7}\sin(1.5\times10^{2}x + 0.5\times10^{11}t)\hat{j}$

(C) $\vec{B} = 2\times10^{-7}\sin(0.5\times10^{3}z - 1.5\times10^{11}t)\hat{i}$

(D) $\vec{B} = 60\sin(0.5\times10^{3}x + 1.5\times10^{11}t)\hat{k}$

21. If the magnetic field in a plane electromagnetic wave is given by $\vec{B} = 3\times10^{-8}\sin(1.6\times10^{3}x + 48\times10^{10}t)\hat{j}\ T$, then what will be expression for electric field ? **[7 Jan. 2020 (Morning)]**

(A) $\vec{E} = (9\sin(1.6\times10^{3}x + 48\times10^{10}t)\hat{k}$ V/m)

(B) $\vec{E} = (3\times10^{-8}\sin(1.6\times10^{3}x + 48\times10^{10}t)\hat{i}$ V/m)

(C) $\vec{E} = (60\sin(1.6\times10^{3}x + 48\times10^{10}t)\hat{k}$ V/m)

(D) $\vec{E} = (3\times10^{-8}\sin(1.6\times10^{3}x + 48\times10^{10}t)\hat{j}$ V/m)

22. The electric field of a plane electromagnetic wave is given by

$$\vec{E} = E_0\frac{\hat{i} + \hat{j}}{\sqrt{2}}\cos(kz + \omega t)$$

At $t = 0$, a positively charged particle is at the point $(x, y, z) = \left(0, 0, \frac{\pi}{k}\right)$. If its instantaneous velocity at $(t = 0)$ is $\upsilon_0\hat{k}$, the force acting on it due to the wave is : **[7 Jan. 2020 (Evening)]**

(A) Zero (B) Parallel to $\frac{\hat{i} + \hat{j}}{\sqrt{2}}$

(C) Antiparallel to $\frac{\hat{i} + \hat{j}}{\sqrt{2}}$ (D) Parallel to $\hat{k}$

23. A plane electromagnetic wave of frequency 25 GHz is propagating in vacuum along the z-direction. At a particular point in space and time, the magnetic field is given by $\vec{B} = 5\times10^{-8}\hat{j}T$. The corresponding electric field $\vec{E}$ is (speed of light $c = 3\times10^{8}$ ms^{-1}) : **[8 Jan. 2020 (Evening)]**

(A) $1.66\times10^{-16}\hat{i}$ V/m (B) $15\hat{i}$ V/m

(C) $-1.66\times10^{-16}\hat{i}$ V/m (D) $-15\hat{i}$ V/m

24. The electric fields of two plane electromagnetic plane waves in vacuum are given by

$\vec{E}_1 = E_0\hat{j}\cos(\omega t - kx)$ and

$\vec{E}_2 = E_0\hat{k}\cos(\omega t - ky)$

At $t = 0$, a particle of charge q is at origin with a velocity $\vec{v} = 0.8c\hat{j}$ (c is the speed of light in vacuum). The instantaneous force experienced by the particle is : **[9 Jan. 2020 (Morning)]**

(A) $E_0q(-0.8\hat{i} + \hat{j} + \hat{k})$ (B) $E_0q(0.8\hat{i} - \hat{j} + 0.4\hat{k})$

(C) $E_0q(0.8\hat{i} + \hat{j} + 0.2\hat{k})$ (D) $E_0q(0.4\hat{i} - 3\hat{j} + 0.8\hat{k})$

25. A plane electromagnetic wave is propagating along the direction $\frac{\hat{i} + \hat{j}}{\sqrt{2}}$, with its polarization along the direction $\hat{k}$. The correct form of the magnetic field of the wave would be (here B_0 is an appropriate constant) : **[9 Jan. 2020 (Evening)]**

(A) $B_0\frac{\hat{i} - \hat{j}}{\sqrt{2}}\cos\left(\omega t - k\frac{\hat{i} + \hat{j}}{\sqrt{2}}\right)$

(B) $B_0\frac{\hat{i} + \hat{j}}{\sqrt{2}}\cos\left(\omega t - k\frac{\hat{i} + \hat{j}}{\sqrt{2}}\right)$

(C) $B_0\hat{k}\cos\left(\omega t - k\frac{\hat{i} + \hat{j}}{\sqrt{2}}\right)$

(D) $B_0\frac{\hat{j} - \hat{i}}{\sqrt{2}}\cos\left(\omega t + k\frac{\hat{i} + \hat{j}}{\sqrt{2}}\right)$

26. A plane electromagnetic wave, has frequency of 2.0×10^{10} Hz and its energy density is 1.02×10^{-8} J/m^3 in vacuum. The amplitude of the magnetic field of the wave is close to :

[Take: $\frac{1}{4\pi\varepsilon_0} = 9\times10^{9}\frac{Nm^2}{C^2}$ and speed of light $= 3\times10^{8}$ ms^{-1}]

[2 Sept. 2020 (Morning)]

(A) 180 nT (B) 160 nT

(C) 150 nT (D) 190 nT

27. In a plane electromagnetic wave, the directions of electric field and magnetic field are represented by $\hat{k}$ and $2\hat{i}-2\hat{j}$, respectively. What is the unit vector along direction of propagation of the wave : **[2 Sept. 2020 (Evening)]**

(A) $\frac{1}{\sqrt{2}}(\hat{i}+\hat{j})$ (B) $\frac{1}{\sqrt{5}}(\hat{i}+2\hat{j})$

(C) $\frac{1}{\sqrt{5}}(2\hat{i}+\hat{j})$ (D) $\frac{1}{\sqrt{2}}(\hat{j}+\hat{k})$

28. The magnetic field of a plane electromagnetic wave is given as $\vec{B}=3\times10^{-8}\sin[200\pi(y+ct)]\,\hat{i}\,T$. Where $c=3\times10^{8}\,ms^{-1}$ is the speed of light. The corresponding electric field is : **[3 Sept. 2020 (Morning)]**

(A) $\vec{E}=10^{-6}\sin[200\pi(y+ct)]\,\hat{k}$ V/m

(B) $\vec{E}=-9\sin[200\pi(y+ct)]\,\hat{k}$ V/m

(C) $\vec{E}=9\sin[200\pi(y+ct)]\,\hat{k}$ V/m

(D) $\vec{E}=3\times10^{-8}\sin[200\pi(y+ct)]\,\hat{k}$ V/m

29. The electric field of a plane electromagnetic wave propagating along the x direction in vacuum is $\vec{E}=E_0\hat{j}\cos(\omega t-kx)$. The magnetic field $\vec{B}$, at the moment $t=0$ is : **[3 Sept. 2020 (Evening)]**

(A) $\vec{B}=E_0\sqrt{\mu_0\in_0}\cos(kx)\,\hat{j}$

(B) $\vec{B}=\frac{E_0}{\sqrt{\mu_0\in_0}}\cos(kx)\,\hat{k}$

(C) $\vec{B}=E_0\sqrt{\mu_0\in_0}\cos(kx)\,\hat{k}$

(D) $\vec{B}=\frac{E_0}{\sqrt{\mu_0\in_0}}\cos(kx)\,\hat{j}$

30. Choose the correct option relating wavelengths of different parts of electromagnetic wave spectrum : **[4 Sept. 2020 (Morning)]**

(A) $\lambda_{x\text{-rays}}<\lambda_{\text{micro waves}}<\lambda_{\text{radio waves}}<\lambda_{\text{visible}}$

(B) $\lambda_{\text{visible}}>\lambda_{x\text{-rays}}>\lambda_{\text{radio waves}}>\lambda_{\text{micro waves}}$

(C) $\lambda_{\text{radio waves}}>\lambda_{\text{micro waves}}>\lambda_{\text{visible}}>\lambda_{x\text{-rays}}$

(D) $\lambda_{\text{visible}}<\lambda_{\text{micro waves}}<\lambda_{\text{radio waves}}<\lambda_{x\text{-rays}}$

31. The electric field of a plane electromagnetic wave is given by $\vec{E}=E_0(\hat{x}+\hat{y})\sin(kz-\omega t)$. Its magnetic field will be given by : **[4 Sept. 2020 (Evening)]**

(A) $\frac{E_0}{c}(\hat{x}-\hat{y})\cos(kz-\omega t)$ (B) $\frac{E_0}{c}(-\hat{x}+\hat{y})\sin(kz-\omega t)$

(C) $\frac{E_0}{c}(\hat{x}-\hat{y})\sin(kz-\omega t)$ (D) $\frac{E_0}{c}(\hat{x}+\hat{y})\sin(kz-\omega t)$

32. An electron is constrained to move along the y-axis with a speed of 0.1 c (c is the speed of light) in the presence of electromagnetic wave, whose electric field is $\vec{E}=30\hat{j}\sin(1.5\times10^{7}t-5\times10^{-2}x)$ V/m The maximum magnetic force experienced by the electron will be :
(Given $c=3\times10^{8}\,ms^{-1}$ and electron charge $=1.6\times10^{-19}$ C) : **[5 Sept. 2020 (Morning)]**

(A) 1.6×10^{-19} N (B) 4.8×10^{-19} N

(C) 3.2×10^{-18} N (D) 2.4×10^{-18} N

33. The correct match between the entries in column-I and column-II are : **[5 Sept. 2020 (Evening)]**

Column-I	Column-II
Radiation	**Wavelength**
(a) Microwave	(i) 100m
(b) Gamma rays	(ii) 10^{-15} m
(c) A.M. radio waves	(iii) 10^{-10} m
(d) X-rays	(iv) 10^{-3} m

(A) (a)-(ii), (b)-(i), (c)-(iv), (d)-(iii)
(B) (a)-(i), (b)-(iii), (c)-(iv), (d)-(ii)
(C) (a)-(iii), (b)-(ii), (c)-(i), (d)-(iv)
(D) (a)-(iv), (b)-(ii), (c)-(i), (d)-(iii)

34. An AC circuit has $R=100\,\Omega$, $C=2\,\mu F$ and $L=80$ mH, connected in series. The quality factor of the circuit is : **[6 Sept. 2020 (Morning)]**

(A) 0.5 (B) 2

(C) 20 (D) 400

35. Suppose that intensity of a laser is $\left(\frac{315}{\pi}\right)$ W/m^2. The rms electric field, in units of V/m associated with this source is close to the nearest integer is ($\in_0=8.86\times10^{-12}$ C^2 Nm^{-2}; $c=3\times10^{8}\,ms^{-1}$): **[6 Sept. 2020 (Morning)]**

36. For a plane electromagnetic wave, the magnetic field at a point x and time t is $\vec{B}(x,t)=[1.2\times10^{-7}\sin(0.5\times10^{3}x+1.5\times10^{11}t)]T$. The instantaneous electric field $\vec{E}$ corresponding to $\vec{B}$ is : (speed of light $c=3\times10^{8}\,ms^{-1}$) : **[6 Sept. 2020 (Evening)]**

(A) $\vec{E}(x,t)=[36\sin(0.5\times10^{3}x+1.5\times10^{11}t)\,\hat{k}]\frac{v}{m}$

(B) $\vec{E}(x,t)=[36\sin(0.5\times10^{3}x+1.5\times10^{11}t)\,\hat{j}]\frac{v}{m}$

(C) $\vec{E}(x,t)=[36\sin(1\times10^{3}x+0.5\times10^{11}t)\,\hat{j}]\frac{v}{m}$

(D) $\vec{E}(x,t)=[36\sin(1\times10^{3}x+1.5\times10^{11}t)\,\hat{j}]\frac{v}{m}$

37. An electromagnetic wave of frequency 5 GHz, is travelling in a medium whose relative electric permittivity and relative magnetic permeability both are 2. Its velocity in this medium is ______ $\times 10^7$ m/s. **[24 Feb. 2021 (Morning)]**

38. An electromagnetic wave of frequency 3 GHz enters a dielectric medium of relative electric permittivity 2.25 from vacuum. The wavelength of this wave in that medium will be ______ $\times 10^{-2}$ cm. **[24 Feb. 2021 (Evening)]**

39. The peak electric field produced by the radiation coming from the 8 W bulb at a distance of 10 m is $\frac{x}{10}\sqrt{\frac{\mu_0 c}{\pi}}\frac{V}{m}$. The bulb is a point source. The value of x is ______. **[25 Feb. 2021 (Evening)]**

40. A radiation is emitted by 1000 W bulb and it generates an electric field and magnetic field at P, placed at a distance of 2 m. The efficiency of the bulb is 1.25%. The value of peak electric field at P is $x \times 10^{-1}$ V/m. Value of x is ______. (Rounded-off to the nearest integer) **[26 Feb. 2021 (Morning)]**
[Take $\varepsilon_0 = 8.85 \times 10^{-12}\, C^2 N^{-1} m^{-2}$, $c = 3 \times 10^8\, ms^{-1}$]

41. A plane electromagnetic wave of frequency 500 MHz is travelling in vacuum along y-direction. At a particular point in space and time, $\vec{B} = 8.0 \times 10^{-8}\, \hat{z}$ T. The value of electric field at this point is : (speed of light $= 3 \times 10^8\, ms^{-1}$)
$\hat{x}, \hat{y}, \hat{z}$ are unit vectors along x, y and z direction. **[16 March 2021 (Morning)]**

(A) $-24\,\hat{x}$ V/m (B) $2.6\,\hat{x}$ V/m
(C) $24\,\hat{x}$ V/m (D) $-2.6\,\hat{y}$ V/m

42. For an electromagnetic wave travelling in free space, the relation between average energy densities due to electric (U_e) and magnetic (U_m) fields is : **[16 March 2021 (Morning)]**
(A) $U_e = U_m$ (B) $U_e > U_m$
(C) $U_e < U_m$ (D) $U_e \neq U_m$

43. The electric field intensity produced by the radiation coming from a 100 W bulb at a distance of 3m is E. The electric field intensity produced by the radiation coming from 60 W at the same distance is $\sqrt{\frac{x}{5}}E$. Where the value of x = ______. **[17 March 2021 (Evening)]**

44. Seawater at a frequency $f = 9 \times 10^2$ Hz, has permittivity $\varepsilon = 80\, \varepsilon_0$ and resistivity $\rho = 0.25\, \Omega$m. Imagine a parallel plate capacitor is immersed in seawater and is driven by an alternating voltage source $V(t) = V_0 \sin(2\pi f t)$. Then the conduction current density becomes 10^x times the displacement current density after time $t = \frac{1}{800}$ s. The value of x is ______.

(**Given :** $\frac{1}{4\pi\varepsilon_0} = 9 \times 10^9\, Nm^2 C^{-2}$) **[17 March 2021 (Evening)]**

45. A plane electromagnetic wave of frequency 100 MHz is travelling in vacuum along the x-direction. At a particular point in space and time, $\vec{B} = 2.0 \times 10^{-8}$ T $\hat{k}$. (Where, $\hat{k}$ is the unit vector along z-direction) What is $\vec{E}$ at this point ? **[18 March 2021 (Morning)]**

(A) $0.6\,\hat{j}$ V/m (B) $6.0\,\hat{k}$ V/m
(C) $6.0\,\hat{j}$ V/m (D) $0.6\,\hat{k}$ V/m

46. A plane electromagnetic wave propagating along y-direction can have the following pair of electric field ($\vec{E}$) and magnetic field ($\vec{B}$) components. **[18 March 2021 (Evening)]**
(A) E_y, B_y or E_z, B_z (B) E_y, B_x or E_x, B_y
(C) E_x, B_z or E_z, B_x (D) E_x, B_y or E_y, B_x

47. AC voltage $V(t) = 20 \sin \omega t$ of frequency 50 Hz is applied to a parallel plate capacitor. The separation between the plates is 2 mm and the area is 1 m^2. The amplitude of the oscillating displacement current for the applied AC voltage is ______.
(Take $\varepsilon_0 = 8.85 \times 10^{-12}$ F/m) **[20 July 2021 (Morning)]**
(A) 55.58 μA (B) 21.14 μA
(C) 27.79 μA (D) 83.37 μA

48. In an electromagnetic wave the electric field vector and magnetic field vector are given as $\vec{E} = E_0 \hat{i}$ and $\vec{B} = B_0 \hat{k}$ respectively. the direction of propagation of electromagnetic wave is along : **[20 July 2021 (Evening)]**

(A) $(-\hat{j})$ (B) $\hat{j}$
(C) $(-\hat{k})$ (D) $\hat{k}$

49. Intensity of sunlight is observed as 0.092 Wm^{-2} at a point in free space. What will be the peak value of magnetic field at that point ? ($\varepsilon_0 = 8.85 \times 10^{-12}\, C^2 N^{-1} m^{-2}$) **[22 July 2021 (Evening)]**
(A) 1.96×10^{-8} T (B) 5.88 T
(C) 2.77×10^{-8} T (D) 8.31 T

50. A linearly polarized electromagnetic wave in vacuum is $E = 3.1 \cos[(1.8)z - (5.4 \times 10^6)t]\, \hat{i}$ N/C is incident normally on a perfectly reflecting wall at $z = a$. Choose the correct option : **[25 July 2021 (Morning)]**

(A) The frequency of electromagnetic wave is 54×10^4 Hz

(B) The reflected wave will be

$3.1 \cos [(1.8)z + (5.4 \times 10^6)t]\ \hat{i}$ N/C

(C) The transmitted wave will be

$3.1 \cos [(1.8)z - (5.4 \times 10^6)t]\ \hat{i}$ N/C

(D) The wavelength is 5.4 m

51. The relative permittivity of distilled water is 81. The velocity of light in it will be : (Given $\mu_r = 1$) : **[27 July 2021 (Morning)]**

(A) 3.33×10^7 m/s (B) 2.33×10^7 m/s

(C) 5.33×10^7 m/s (D) 4.33×10^7 m/s

52. Light beam is described by $E = 800 \sin \omega\left(t - \frac{x}{c}\right)$. An electron is allowed to move normal to the propagation of light beam with a speed of 3×10^7. What is the maximum magnetic force exerted on the electron ? **[26 Aug. 2021 (Evening)]**

(A) 1.28×10^{-21} N (B) 1.28×10^{-18} N

(C) 12.8×10^{-18} N (D) 12.8×10^{-17} N

53. Electric field in a plane electromagnetic wave is given by $E = 50 \sin (500x - 10 \times 10^{10} t)$ V/m. The velocity of electromagnetic wave in this medium is : (Given C = speed of light in vacuum) : **[27 Aug. 2021 (Morning)]**

(A) $\frac{2}{3}C$ (B) C

(C) $\frac{C}{2}$ (D) $\frac{3}{2}C$

54. A plane electromagnetic wave with frequency of 30 MHz travels in free space. At particular point in space and time, electric field is 6 V/m. The magnetic field at this point will be $x \times 10^{-8}$ T. The value of x is______. **[27 Aug. 2021 (Evening)]**

55. The electric field in an electromagnetic wave is given by $E = (50\ NC^{-1}) \sin \omega(t - x/c)$. The energy contained in a cylinder of volume V is 5.5×10^{-12} J. The value of V is ______cm^3. (Given $\epsilon_0 = 8.8 \times 10^{-12}\ C^2 - N^{-1}\ m^{-2}$) **[31 Aug. 2021 (Morning)]**

56. Electric of a plane electromagnetic wave propagating through a non magnetic medium is given by

$$E = 20 \cos (2 \times 10^{10}\ t - 200\,x)\ \text{V/m}$$

The dielectric constant of the medium is equal to : (Take $\mu_r = 1$) **[01 Sep. 2021 (Evening)]**

(A) 3 (B) $\frac{1}{2}$

(C) 2 (D) 9

* * * * *

CHAPTER 23

RAY OPTICS AND OPTICAL INSTRUMENTS

23.1 GEOMETRICAL OPTICS I - REFLECTION OF LIGHT

1. A particle is oscillating on the X-axis with an amplitude 2 cm about the point $x_0 = 10$ cm with a frequency ω. A concave mirror of focal length 5 cm is placed at the origin (see figure) Identify the correct statements : **[2018 (Online)]**

(a) The image executes periodic motion

(b) The image executes non-periodic motion

(c) The turning points of the image are asymmetric w.r.t. the image of the point at $x = 10$ cm

(d) The distance between the turning points of the oscillation of the image is $\frac{100}{21}$ cm.

(A) (b), (d) (B) (b), (c)

(C) (a), (c), (d) (D) (a), (d)

2. Two plane mirrors are inclined to each other such that a ray of light incident on the first mirror (M_1) and parallel to the second mirror (M_2) is finally reflected from the second mirror (M_2) parallel to the first mirror (M_1). The angle between the two mirrors will be : **[9 Jan. 2019 (Evening)]**

(A) 90° (B) 45°

(C) 75° (D) 60°

3. A point source of light, S is placed at a distance L in front of the centre of plane mirror of width d which is hanging vertically on a wall. A man walks in front of the mirror along a line parallel to the mirror, at a distance $2L$ as shown below. The distance over which the man can see the image of the light source in the mirror is : **[12 Jan. 2019 (Morning)]**

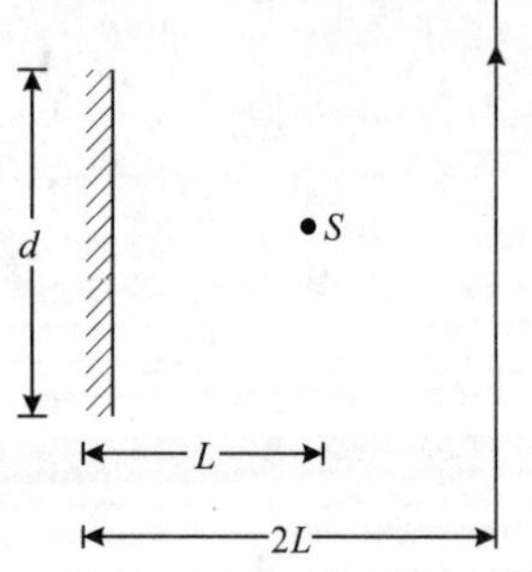

(A) $3d$ (B) $d/2$

(C) d (D) $2d$

4. A concave mirror for face viewing has focal length of 0.4 m. The distance at which you hold the mirror from your face in order to see your image upright with a magnification of 5 is : **[9 April 2019 (Morning)]**

(A) 1.60 m (B) 0.24 m

(C) 0.16 m (D) 0.32 m

5. A concave mirror has radius of curvature of 40 cm. It is at the bottom of a glass that has water filled up to 5 cm (see figure). If a small particle is floating on the surface of water, its image as seen, from directly above the glass, is at a distance d from the surface of water. The value of d is close to : (Refractive index of water = 1.33) : **[12 April 2019 (Morning)]**

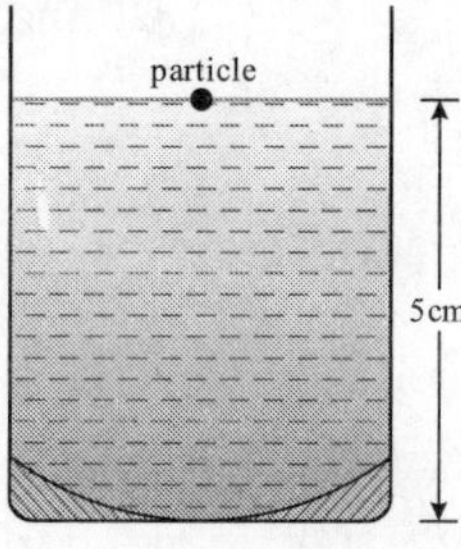

(A) 8.8 cm (B) 11.7 cm

(C) 6.7 cm (D) 13.4 cm

6. An object is gradually moving away from the focal point of a concave mirror along the axis of the mirror. The graphical representation of the magnitude of linear magnification (m) versus distance of the object from the mirror (x) is correctly given by : **[8 Jan. 2020 (Evening)]**

(Graphs are drawn schematically and are not to scale)

(A) 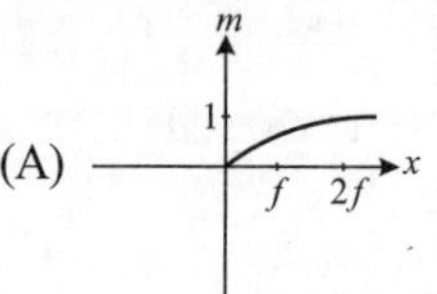

(B) 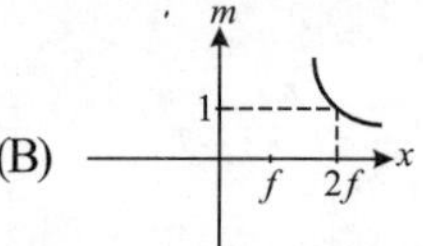

(C) 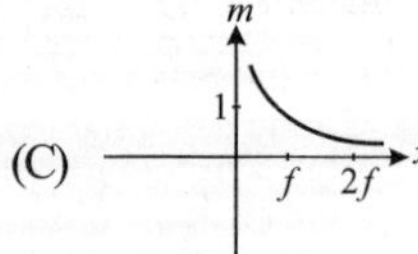

(D) 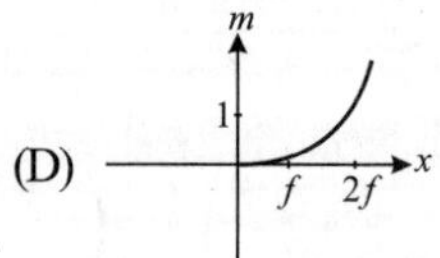

7. A spherical mirror is obtained as shown in the figure from a hollow glass sphere. If an object is positioned in front of the mirror, what will be the nature and magnification of the image of the object ? (Figure drawn as schematic and not to scale)

[2 Sept. 2020 (Morning)]

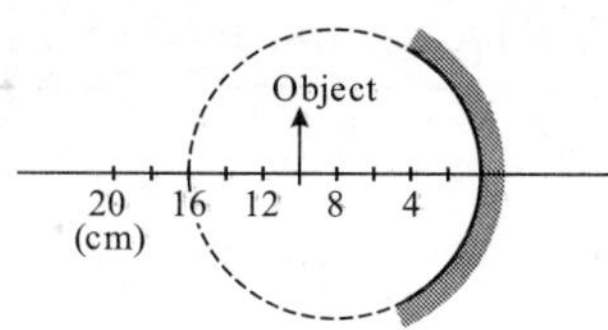

(A) Inverted, real and magnified
(B) Erect, virtual and magnified
(C) Erect, virtual and unmagnified
(D) Inverted, real and unmagnified

8. When an object is kept at a distance of 30 cm from a concave mirror, the image is formed at a distance of 10 cm from the mirror. If the object is moved with a speed of 9 cms^{-1}, the speed (in cms^{-1}) with which image moves at that instant is ______:

[3 Sept. 2020 (Evening)]

9. The focal length f is related to the radius of curvature r of the spherical convex mirror by : **[24 Feb. 2021 (Morning)]**

(A) $f=-r$ (B) $f=-\frac{1}{2}r$

(C) $f=+\frac{1}{2}r$ (D) $f=r$

10. A short straight object of height 100 cm lies before the central axis of a spherical mirror whose focal length has absolute value $|f|=40$ cm. The image of object produced by the mirror is of height 25 cm and has the same orientation of the object. One may conclude from the information : **[26 Feb. 2021 (Morning)]**

(A) Image is virtual, opposite side of convex mirror
(B) Image is virtual, opposite side of concave mirror
(C) Image is real, same side of convex mirror
(D) Image is real, same side of concave mirror

11. The incident ray, reflected ray and the outward drawn normal are denoted by the unit vectors $\vec{a}$, $\vec{b}$ and $\vec{c}$ respectively. Then choose the correct relation for these vectors.

[26 Feb. 2021 (Evening)]

(A) $\vec{b}=2\vec{a}+\vec{c}$ (B) $\vec{b}=\vec{a}-2(\vec{a}\cdot\vec{c})\vec{c}$

(C) $\vec{b}=\vec{a}-\vec{c}$ (D) $\vec{b}=\vec{a}+2\vec{c}$

12. A point source of light S, placed at a distance 60 cm infront of the centre of a plane mirror of width 50 cm, hangs vertically on a wall. A man walks infront of the mirror along a line parallel to the mirror at a distance 1.2 cm from it (see in the figure). The distance between the extreme points where he can see the image of the light source in the mirror is _________ cm.

[26 Feb. 2021 (Evening)]

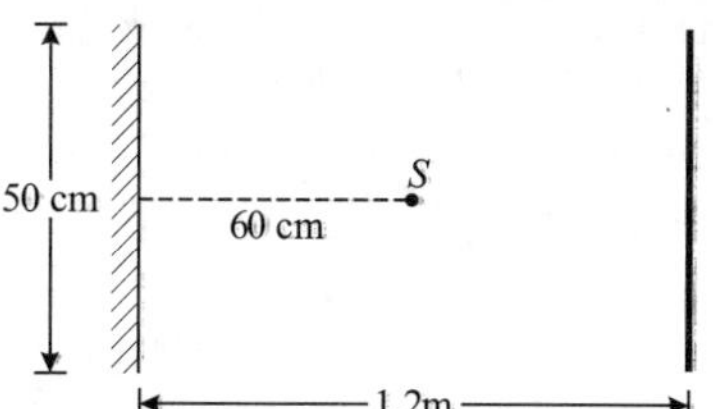

13. Car B overtakes another car A at relative speed of 40 ms^{-1}. How fast will the image of car B appear to move in the mirror of focal length 10 cm fitted in car A, when the car B is 1.9 m away from car A? **[26 Aug. 2021 (Morning)]**

(A) 0.2 ms^{-1} (B) 40 ms^{-1}
(C) 0.1 ms^{-1} (D) 4 ms^{-1}

14. An object is placed beyond the centre of curvature C of the given concave mirror. If the distance of the object is d_1 from C and the distance of the image formed is d_2 from C. The radius of curvature of this mirror is : **[27 Aug. 2021 (Morning)]**

(A) $\frac{d_1d_2}{d_1+d_2}$ (B) $\frac{2d_1d_2}{d_1+d_2}$

(C) $\frac{d_1d_2}{d_1-d_2}$ (D) $\frac{2d_1d_2}{d_1-d_2}$

15. Two plane mirror M_1 and M_2 are at right angle to each other shown. A point source P is placed at a and $2a'$ meter away from M_1 and M_2 respectively. The shortest distance between the images thus formed is : (Take $\sqrt{5}=2.3$)

[31 Aug. 2021 (Morning)]

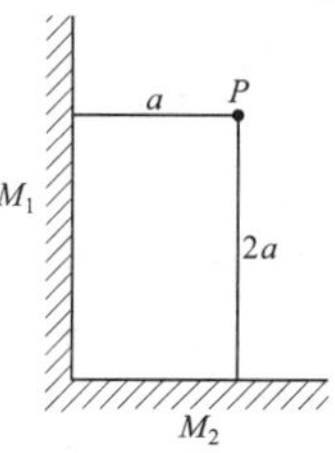

(A) 4.6 a (B) $2\sqrt{10}\,a$
(C) 3a (D) 2.3 a

* * * * *

CHAPTER

23 RAY OPTICS AND OPTICAL INSTRUMENTS

23.2 GEOMETRICAL OPTICS II - REFRACTION OF LIGHT

1. Let the refractive index of a denser medium with respect to a rarer medium be n_{12} and its critical angle be θ_C. At an angle of incidence A when light is travelling from denser medium to rarer medium, a part of the light is reflected and the rest is refracted and the angle between reflected and refracted rays is 90°. Angle A is given by : **[2017 (Online)]**

(A) $\frac{1}{\cos^{-1}(\sin\theta_C)}$ (B) $\frac{1}{\tan^{-1}(\sin\theta_C)}$

(C) $\cos^{-1}(\sin\theta_C)$ (D) $\tan^{-1}(\sin\theta_C)$

2. A ray of light is incident at an angle of 60° on one face of a prism of angle 30°. The emergent ray of light makes an angle of 30° with incident ray. The angle made by the emergent ray with second face of prism will be : **[2018 (Online)]**

(A) 30° (B) 90°

(C) 0° (D) 45°

3. The eye can be regarded as a single refracting surface. The radius of curvature of this surface is equal to that of cornea (7.8 mm). This surface separates two media of refractive indices 1 and 1.34. Calculate the distance from the refracting surface at which a parallel beam of light will come to focus : **[10 Jan. 2019 (Evening)]**

(A) 2 cm (B) 1 cm

(C) 3.1 cm (D) 4.0 cm

4. A monochromatic light is incident at a certain angle on an equilateral triangular prism and suffers minimum deviation. If the refractive index of the material of the prism is $\sqrt{3}$, then the angle of incidence is : **[11 Jan. 2019 (Evening)]**

(A) 30° (B) 45°

(C) 90° (D) 60°

5. In figure, the optical fiber is $l = 2$ m long and has a diameter of $d = 20$ μm. If a ray of light is incident on one end of the fiber at angle $\theta_1 = 40°$, the number of reflection it makes before emerging from the other end is close to :
(refractive index of fibre is 1.31 and sin 40° = 0.64) **[8 April 2019 (Morning)]**

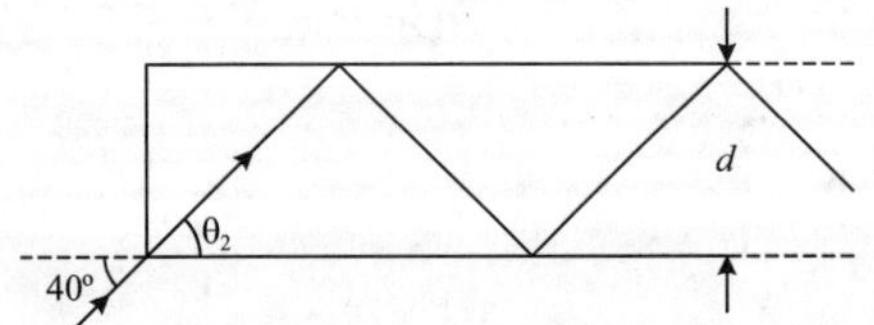

(A) 55000 (B) 57000

(C) 66000 (D) 45000

6. A ray of light AO in vacuum is incident on a glass slab at angle 60° and refracted at angle 30° along OB as shown in the figure. The optical path length of light ray from A to B is : **[10 April 2019 (Morning)]**

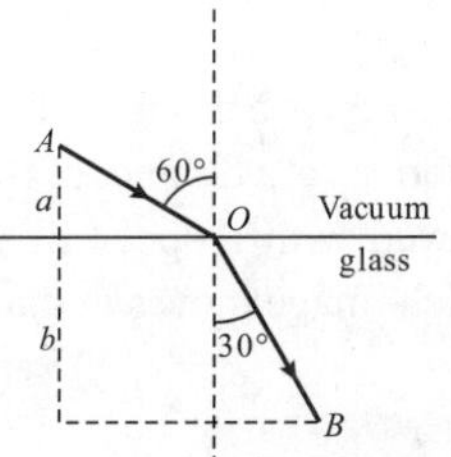

(A) $2a + 2b$ (B) $2a + \frac{2b}{3}$

(C) $\frac{2\sqrt{3}}{a} + 2b$ (D) $2a + \frac{2b}{\sqrt{3}}$

7. A transparent cube of side d, made of a material of refractive index μ_2, is immersed in a liquid of refractive index $\mu_1 (\mu_1 < \mu_2)$. A ray is incident on the face AB at an angle θ (shown in the figure). Total internal reflection takes place at point E on the face BC. The θ must satisfy : **[12 April 2019 (Evening)]**

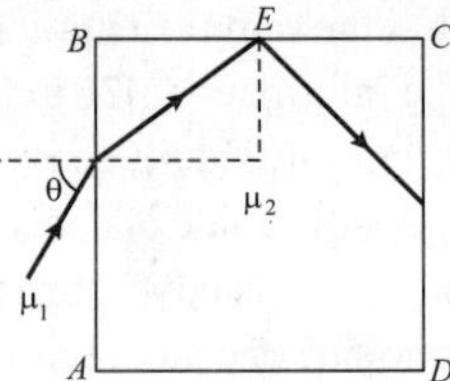

(A) $\theta < \sin^{-1}\frac{\mu_1}{\mu_2}$ (B) $\theta < \sin^{-1}\sqrt{\frac{\mu_2^2}{\mu_1^2} - 1}$

(C) $\theta > \sin^{-1}\frac{\mu_1}{\mu_2}$ (D) $\theta > \sin^{-1}\sqrt{\frac{\mu_2^2}{\mu_1^2} - 1}$

8. The critical angle of a medium for a specific wavelength, if the medium has relative permittivity 3 and relative permeability 4/3 for this wavelength, will be : **[8 Jan. 2020 (Morning)]**

(A) 60° (B) 15°

(C) 45° (D) 30°

9. A vessel of depth $2h$ is half filled with a liquid of refractive index $2\sqrt{2}$ and the upper half with another liquid of refractive index $\sqrt{2}$. The liquids are immiscible. The apparent depth of the inner surface of the bottom of vessel will be : **[9 Jan. 2020 (Morning)]**

(A) $\frac{h}{\sqrt{2}}$ (B) $\frac{3}{4}h\sqrt{2}$

(C) $\frac{h}{2(\sqrt{2}+1)}$ (D) $\frac{h}{3\sqrt{2}}$

10. There is a small source of light at some depth below the surface of water (refractive index = 4/3) in a tank of large cross sectional surface area. Neglecting any reflection from the bottom and absorption by water, percentage of light that emerges out of surface is (nearly) :

[Use the fact that surface area of a spherical cap of height h and radius of curvature r is $2\pi rh$] **[9 Jan. 2020 (Evening)]**

(A) 17% (B) 21%

(C) 34% (D) 50%

11. A light ray enters a solid glass sphere of refractive index $\mu = \sqrt{3}$ at an angle of incidence 60°. The ray is both reflected and refracted at the farther surface of the sphere. The angle (in degrees) between the reflected and refracted rays at this surface is : **[2 Sept. 2020 (Evening)]**

12. An observer can see through a small hole on the side of a jar (radius 15 cm) at a point at height of 15 cm from the bottom (see figure). The hole is at a height of 45 cm. When the jar is filled with a liquid up to a height of 30 cm the same observer can see the edge at the bottom of the jar. If the refractive index of the liquid $N/100$, where N is an integer, the value of N is______. **[3 Sept. 2020 (Morning)]**

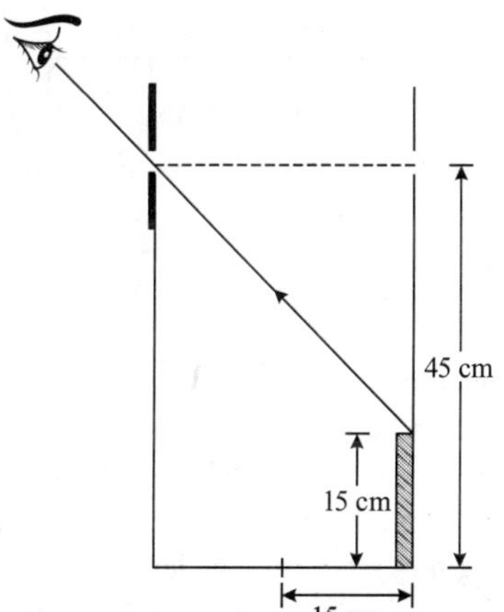

13. A prism of angle $A = 1°$ has a refractive index $\mu = 1.5$. A good estimate for the minimum angle of deviation (in degrees) is close to N/10. Value of N is ______. **[5 Sept. 2020 (Evening)]**

14. A double convex lens has power P and same radii of curvature R of both the surfaces. The radius of curvature of a surface of a plano-convex lens made of the same material with power $1.5\,P$ is : **[6 Sept. 2020 (Evening)]**

(A) $\frac{R}{2}$ (B) $2R$

(C) $\frac{3R}{2}$ (D) $\frac{R}{3}$

15. The refractive index of a converging lens is 1.4. What will be the focal length of this lens if it is placed in a medium of same refractive index ? (Assume the radii of curvature of the faces of lens are R_1 and R_2 respectively) **[16 March 2021 (Evening)]**

(A) 1 (B) Infinite

(C) $\frac{R_1R_2}{R_1 - R_2}$ (D) Zero

16. The thickness at the centre of a plano convex lens is 3 mm and the diameter is 6 cm. If the speed of light in the material of the lens is 2×10^8 ms^{-1}. The focal length of the lens is______. **[17 March 2021 (Morning)]**

(A) 0.30 cm (B) 15 cm

(C) 1.5 cm (D) 30 cm

17. The image of an object placed in air formed by a convex refracting surface is at a distance of 10 m behind the surface. The image is real and is at $\left(\frac{2}{3}\right)^{rd}$ of the distance of the object from the surface. The wavelength of light inside the surface is 2/3 times the wavelength in air. The radius of the curved surface is $\frac{x}{13}$ m. The value of 'x' is______. **[17 March 2021 (Evening)]**

18. Three rays of light, namely red (R), green (G) and blue (B) are incident on the face PQ of a right angled prism PQR as shown in figure. The refractive indices of the material of the prism for red, green and blue wavelength are 1.27, 1.42 and 1.49 respectively. The colour of the ray(s) emerging out of the face PR is : **[18 March 2021 (Evening)]**

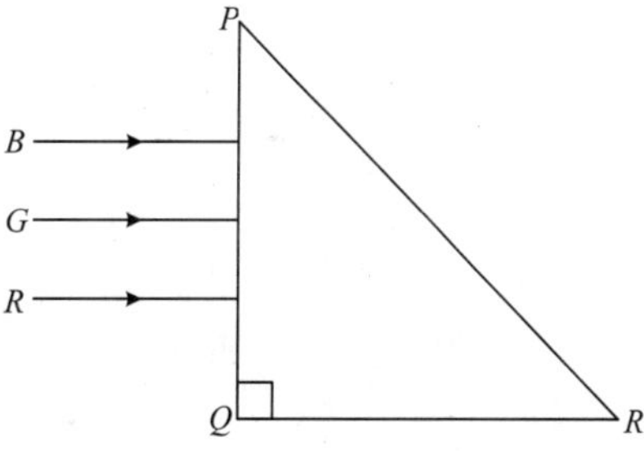

(A) Green (B) Red

(C) Blue and Green (D) Blue

19. Region I and II are separated by a spherical surface of radius 25 cm an object is kept in region I at a distance of 40 cm from the surface. The distance of the image from the surface : **[20 July 2021 (Morning)]**

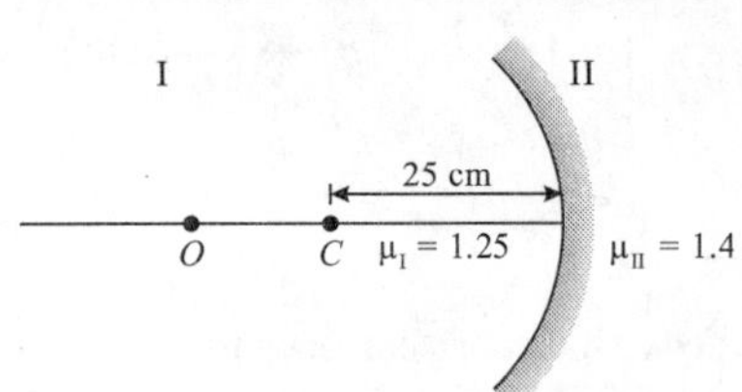

(A) 9.52 cm (B) 18.23 cm
(C) 55.44 cm (D) 37.58 cm

20. A ray of light passes from a denser medium to a rarer medium at an angle of incidence i. The reflected and refracted rays make an angle of 90° with each other. The angle of reflection and refraction are respectively r and r'. The critical angle is given by : **[22 July 2021 (Evening)]**

(A) $\sin^{-1}(\tan r)$ (B) $\sin^{-1}(\cot r)$
(C) $\tan^{-1}(\sin i)$ (D) $\sin^{-1}(\tan r')$

21. A ray of light passing through a prism ($\mu = \sqrt{3}$) suffers minimum deviation. It is found that the angle of incidence is double the angle of refraction within the prism. Then, the angle of prism is ______ (in degree). **[22 July 2021 (Evening)]**

22. A ray of laser of a wavelength 630 nm is incident at an angle of 30° at the diamond-air interface. It is going from diamond to air. The refractive index of diamond is 2.42 and that of air is 1. Choose the correct option : **[25 July 2021 (Morning)]**

(A) angle of refraction is 30° (B) angle of refraction is 24.41°
(C) refraction is not possible (D) angle of refraction is 53.4°

23. A ray of light entering from air into a denser medium of refractive index $\frac{4}{3}$, as shown in figure. The light ray suffers total internal reflection at the adjacent surface as shown. The maximum value of angle? Should be equal to : **[25 July 2021 (Evening)]**

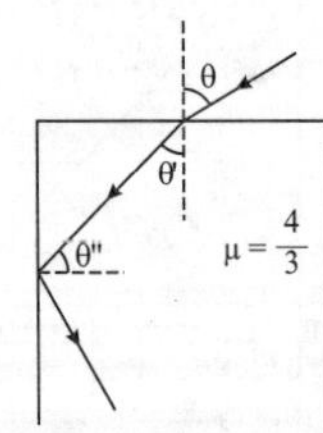

(A) $t^{1/2}$ (B) $t^{1/4}$
(C) $t^{3/2}$ (D) $t^{3/4}$

24. The expected graphical representation of the variation of angle of deviation 'δ' with angle of incidence 'i' in a prism is : **[27 July 2021 (Evening)]**

(A)
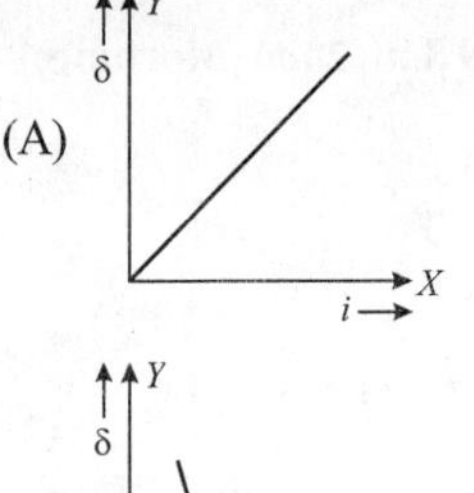

(B)
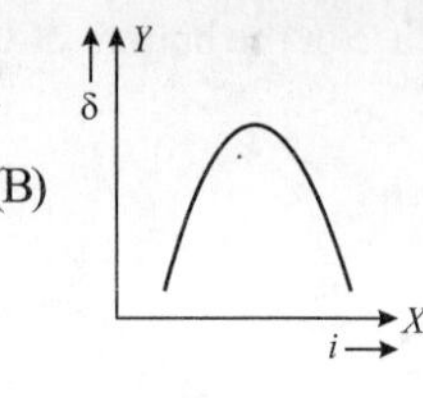

(C)
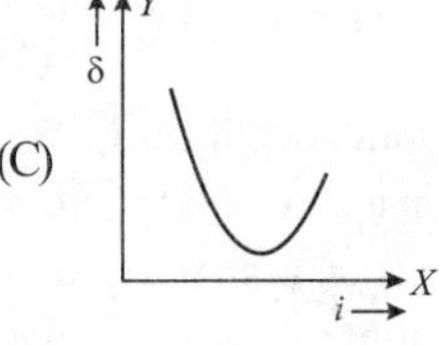

(D)
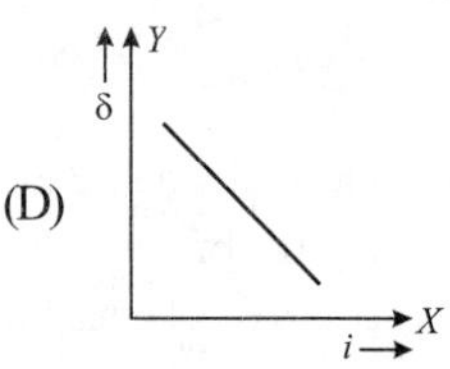

25. An object Is placed at a distance of 12 cm from a convex lens. A convex mirror of focal length 15 cm is placed on other side of lens at 8 cm as shown in the figure. Image of object coincides with the object. When the convex mirror is removed, a real and inverted image is formed at a position. The distance of the image from the object will be ______ (cm). **[26 Aug. 2021 (Evening)]**

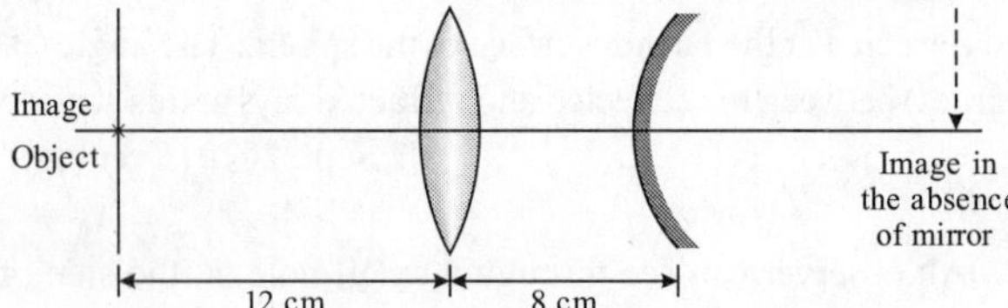

26. Find the distance of the image from object O, formed by the combination of lenses in the figure : **[27 Aug. 2021 (Morning)]**

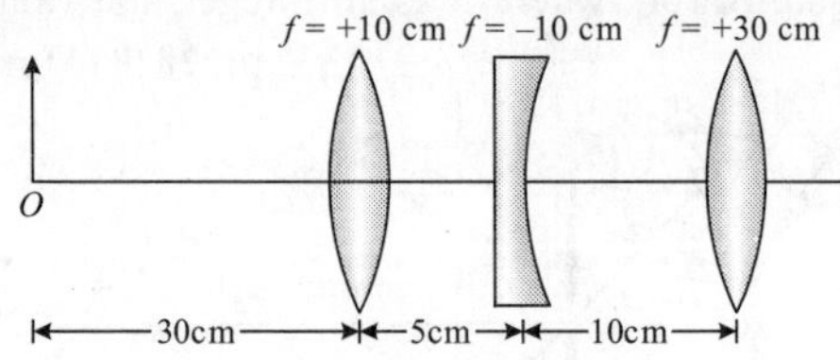

(A) 20 cm (B) 75 cm
(C) 10 cm (D) Infinity

27. Curved surfaces of a plano-convex lens of refractive index μ_1 and a plano-concave lens of refractive index μ_2 have equal radius of curvature as shown in figure. Find the ratio of radius of curvature to the focal length of the combined lenses : **[27 Aug. 2021 (Evening)]**

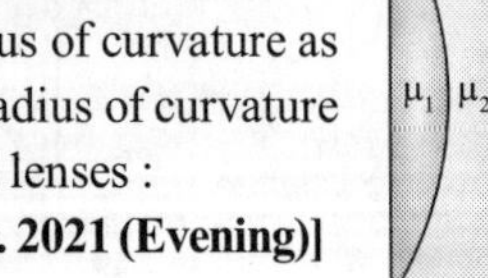

(A) $\frac{1}{\mu_2 - \mu_1}$ (B) $\frac{1}{\mu_1 - \mu_2}$

(C) $\mu_2 - \mu_1$ (D) $\mu_1 - \mu_2$

28. The temperature of an ideal gas in 3-dimensions is 300 K. The corresponding de-Broglie wavelength of the electron approximately at 300 K, is :
[m_e = mass of electron = 9×10^{-31} kg
h = Planck constant = 6.6×10^{-34} J s
k_B = Boltzmann constant = 1.38×10^{-23} JK^{-1}]

[01 Sep. 2021 (Evening)]

(A) 6.26 nm (B) 3.25 nm
(C) 8.46 nm (D) 2.26 nm

29. A glass tumbler having inner depth of 17.5 cm is kept on a table. A student starts pouring water (μ = 4/3) into it while looking at the surface of water from the above. When he feels that the tumbler is half filled, he stops pouring water. Up to what height, the tumbler is actually filled ?

[01 Sep. 2021 (Evening)]

(A) 7.5 cm (B) 10 cm
(C) 8.75 cm (D) 11.7 cm

* * * * *

CHAPTER 23

RAY OPTICS AND OPTICAL INSTRUMENTS

23.3 GEOMETRICAL OPTICS III - THIN LENSES

1. In an experiment a convex lens of focal length 15 cm is placed coaxially on an optical bench in front of a convex mirror at a distance of 5 cm from it. It is found that an object and its image coincide, if the object is placed at a distance of 20 cm from the lens. The focal length of the convex mirror is : **[2017 (Online)]**

(A) 27.5 cm (B) 20.0 cm
(C) 25.0 cm (D) 30.5 cm

2. A plano-convex lens becomes an optical system of 28 cm focal length when its plane surface is silvered and illuminated from left to right as shown in figure-A. If the same lens is instead silvered on the curved surface and illuminated from other side as in Figure-B, it acts like an optical system of focal length 10 cm. The refractive index of the material of lens if: **[2018 (Online)]**

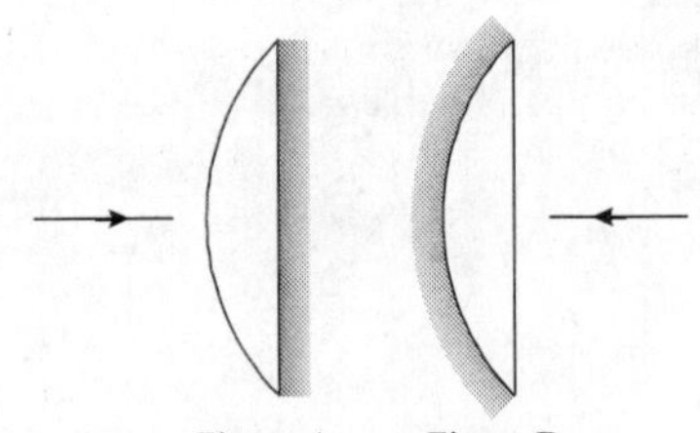

Figure-A Figure-B

(A) 1.50 (B) 1.55
(C) 1.75 (D) 1.51

3. A convex lens is put 10 cm from a light source and it makes a sharp image on a screen, kept 10 cm from the lens. Now a glass block (refractive index 1.5) of 1.5 cm thickness is placed in contact with the light source. To get the sharp image again, the screen is shifted by a distance d. Then d is : **[9 Jan. 2019 (Morning)]**

(A) 0.55 cm away from the lens
(B) 1.1 cm away from the lens
(C) 0.55 cm towards the lens
(D) 0

4. A plano-convex lens of refractive index μ_1 and focal length f_1 is kept in contact with another plano-concave lens of refractive index μ_2 and focal length f_2. If the radius of curvature of their spherical faces is R each and $f_1 = 2f_2$, then μ_1 and μ_2 are related as : **[10 Jan. 2019 (Morning)]**

(A) $\mu_1 + \mu_2 = 3$ (B) $2\mu_1 - \mu_2 = 1$
(C) $2\mu_2 - \mu_1 = 1$ (D) $3\mu_2 - 2\mu_1 = 1$

5. An object is at a distance of 20 m from a convex lens of focal length 0.3 m. The lens forms an image of the object. If the object moves away from the lens at a speed of 5 m/s, the speed and direction of the image will be : **[11 Jan. 2019 (Morning)]**

(A) 0.92×10^{-3} m/s away from the lens
(B) 2.26×10^{-3} m/s away from the lens
(C) 1.16×10^{-3} m/s towards the lens
(D) 3.22×10^{-3} m/s towards the lens

6. What is the position and nature of image formed by lens combination shown in figure? (f_1, f_2 are focal lengths) : **[12 Jan. 2019 (Morning)]**

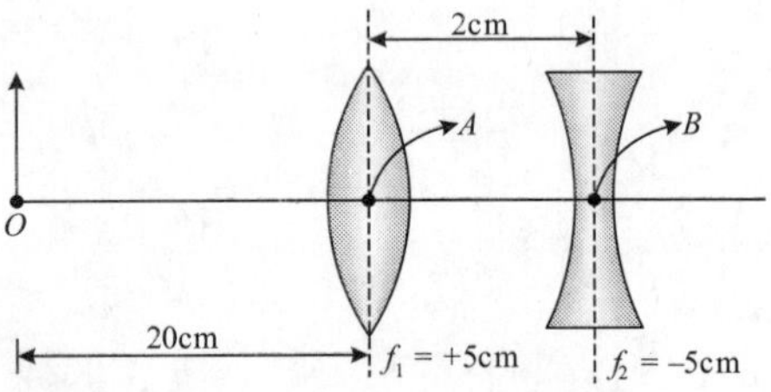

(A) 70 cm from point B at left, virtual
(B) 40 cm from point B at right, real
(C) $\frac{20}{3}$ cm from point B at right, real
(D) 70 cm from point B at right, real

7. Formation of real image using a biconvex lens is shown below.
If the whole set up is immersed in water without disturbing the object and the screen position, what will one observe on the screen ? **[12 Jan. 2019 (Evening)]**

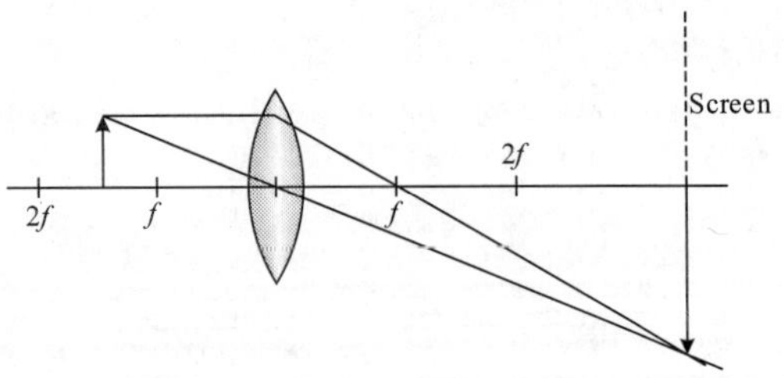

(A) Image disappears (B) No change
(C) Erect real image (D) Magnified image

8. A plano-convex lens (focal length f_2, refractive index μ_2, radius of curvature R) fits exactly into a plano-concave lens (focal length f_1, refractive index μ_1, radius of curvature R). Their plane surfaces are parallel to each other. Then, the focal length of the combination will be : **[12 Jan. 2019 (Evening)]**

(A) $f_1 - f_2$ (B) $f_1 + f_2$

(C) $\frac{R}{\mu_2 - \mu_1}$ (D) $\frac{2f_1f_2}{f_1 + f_2}$

9. An upright object is placed at a distance of 40 cm in front of a convergent lens of focal length 20 cm. A convergent mirror of focal length 10 cm is placed at a distance of 60 cm on the other side of the lens. The position and size of the final image will be : **[8 April 2019 (Morning)]**

(A) 40 cm from the convergent mirror, same size as the object

(B) 20 cm from the convergent mirror, same size as the object

(C) 20 cm from the convergent mirror, twice the size of the object

(D) 40 cm from the convergent lens, twice the size of the object

10. A convex lens (of focal length 20 cm) and a concave mirror, having their principal axes along the same lines, are kept 80 cm apart from each other. The concave mirror is to the right of the convex lens. When an object is kept at a distance of 30 cm to the left of the convex lens, its image remains at the same position even if the concave mirror is removed. The maximum distance of the object for which this concave mirror, by itself would produce a virtual image would be : **[8 April 2019 (Evening)]**

(A) 20 cm (B) 10 cm

(C) 25 cm (D) 30 cm

11. A thin convex lens L (refractive index = 1.5) is placed on a plane mirror M. When a pin is placed at A, such that $OA = 18$ cm, its real inverted image is formed at A itself, as shown in figure. When a liquid of refractive index μ_1 is put between the lens and the mirror, The pin has to be moved to A', such that $OA' = 27$ cm, to get its inverted real image at A' itself. The value of μ_1 will be: **[9 April 2019 (Evening)]**

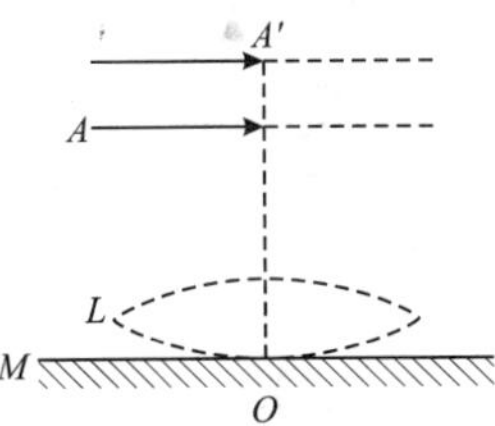

(A) $\sqrt{2}$ (B) $\frac{4}{3}$

(C) $\sqrt{3}$ (D) $\frac{3}{2}$

12. A convex lens of focal length 20 cm produces images of the same magnification 2 when an object is kept at two distances x_1 and $x_2 (x_1 > x_2)$ from the lens. The ratio of x_1 and x_2 is : **[9 April 2019 (Evening)]**

(A) 5 : 3 (B) 2 : 1

(C) 4 : 3 (D) 3 : 1

13. One plano-convex and one plano-concave lens of same radius of curvature 'R' but of different materials are joined side by side as shown in the figure. If the refractive index of the material of 1 is μ_1 and that of 2 is μ_2, Then the focal length of the combination is : **[10 April 2019 (Morning)]**

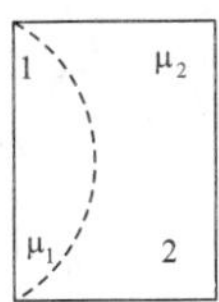

(A) $\frac{R}{2-(\mu_1 - \mu_2)}$ (B) $\frac{2R}{\mu_1 - \mu_2}$

(C) $\frac{R}{2(\mu_1 - \mu_2)}$ (D) $\frac{R}{\mu_1 - \mu_2}$

14. The graph shows how the magnification m produced by a thin lens varies with image distance v. What is the focal length of the lens used ? **[10 April 2019 (Evening)]**

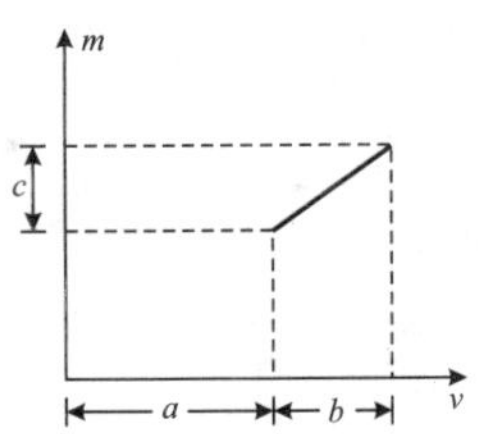

(A) $\frac{b^2c}{a}$ (B) $\frac{b^2}{ac}$

(C) $\frac{a}{c}$ (D) $\frac{b}{c}$

15. A thin lens made of glass (refractive index = 1.5) of focal length $f = 16$ cm is immersed in a liquid of refractive index 1.42. If its focal length in liquid is f_1, then the ratio f_1/f is closest to the integer : **[7 Jan. 2020 (Evening)]**

(A) 1 (B) 5

(C) 9 (D) 17

16. A point object in air is in front of the curved surface of a plano-convex lens. The radius of curvature of the curved surface is 30 cm and the refractive index of the lens material is 1.5, then the focal length of the lens (in cm) is ______. **[8 Jan. 2020 (Morning)]**

17. The distance between an object and a screen is 100 cm. A lens can produce real image of the object on the screen for two different positions between the screen and the object. The distance between these two positions is 40 cm. If the power of the lens is close to $\left(\frac{N}{100}\right)$ D where N is an integer, the value of N is ______ . **[4 Sept. 2020 (Evening)]**

18. For a concave lens of focal length f, the relation between object and image distance u and v, respectively, from its pole can best be represented by ($u = v$ is the reference line) : **[5 Sept. 2020 (Morning)]**

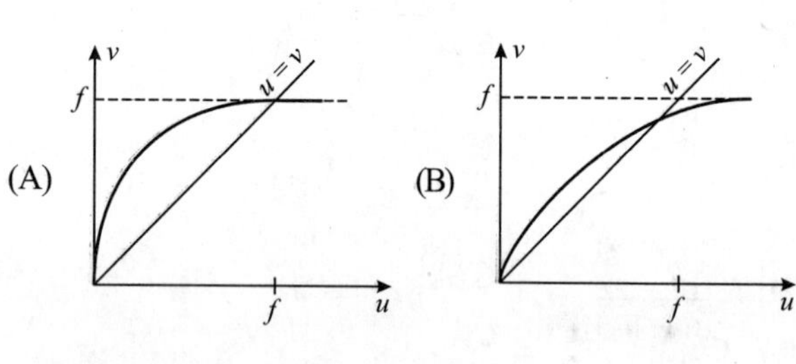

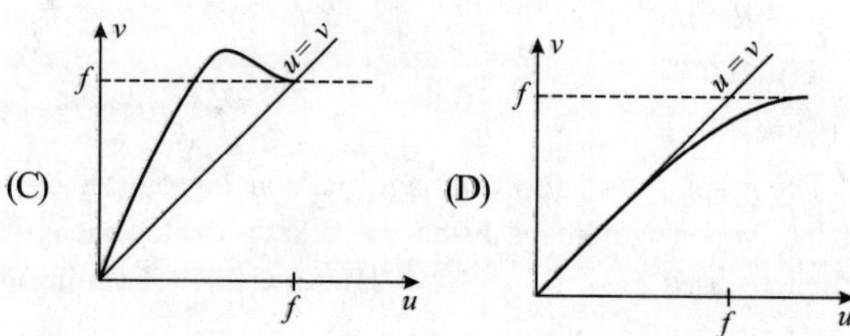

19. A point like object is placed at a distance of 1 m in front of a convex lens of focal length 0.5 m. A plane mirror is placed at a distance of 2 m behind the lens. The position and nature of the final image formed by the system is : **[6 Sept. 2020 (Morning)]**

(A) 1 m from the mirror, virtual
(B) 1 m from the mirror, real
(C) 2.6 m from the mirror, real
(D) 2.6 m from the mirror, virtual

20. An object is placed at the focus of concave lens having, focal length f. What is the magnification and distance of the image from the optical centre of the lens ? **[31 Aug. 2021 (Morning)]**

(A) 1/2, f/2 (B) 1/4, f/4
(C) 1, ∞ (D) very high, ∞

* * * * *

CHAPTER

23 RAY OPTICS AND OPTICAL INSTRUMENTS

23.4 DISPERSION OF LIGHT

1. A convergent doublet of separated lenses, corrected for spherical aberration, has resultant focal length of 10 cm. The separation between the two lenses is 2 cm. The focal lengths of the component lenses : **[2018 (Online)]**

(A) 18 cm, 20 cm (B) 10 cm, 12 cm
(C) 12 cm, 14 cm (D) 16 cm, 18 cm

2. The variation of refractive index of a crown glass thin prism with wavelength of the incident light is shown. Which of the following graphs is the correct one, if D_m is the angle of minimum deviation? **[11 Jan. 2019 (Morning)]**

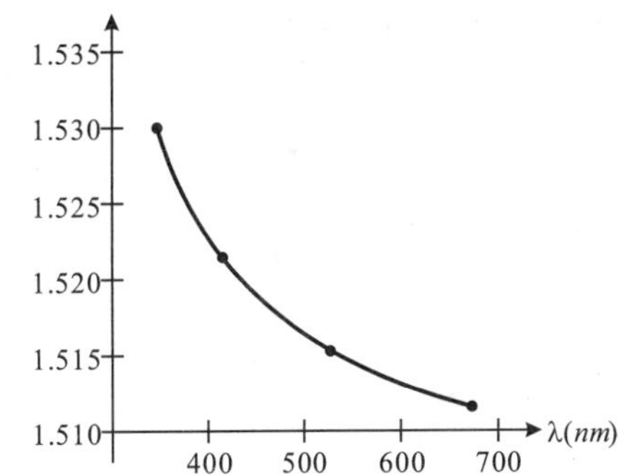

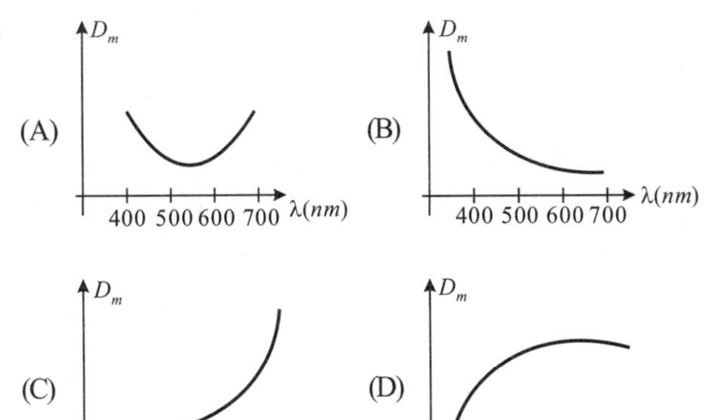

3. The angle of deviation through a prism is minimum when

(1) Incident ray and emergent ray are symmetric to the prism

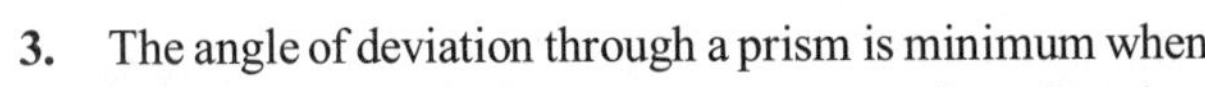

(2) The refracted ray inside the prism becomes parallel to its base

(3) Angle of incidence is equal to that of the angle of emergence

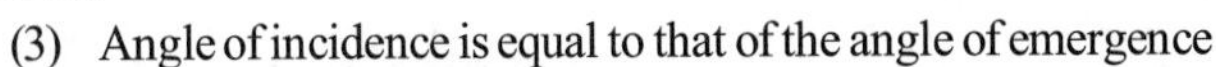

(4) When angle of emergence is double the angle of incidence

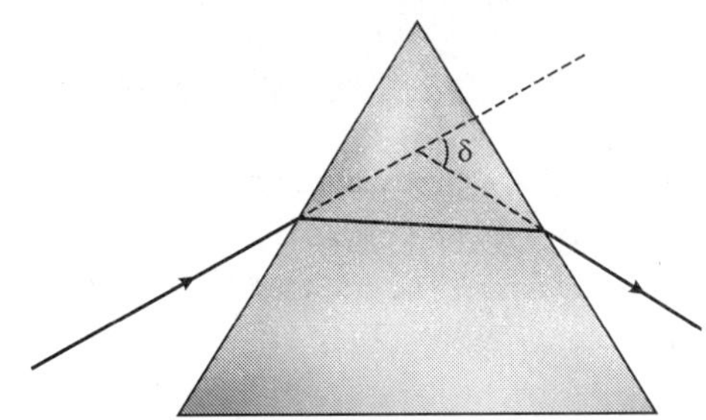

Choose the correct answer from the options given below : **[16 March 2021 (Morning)]**

(A) Statements (1), (2) and (3) are true
(B) Only statement (4) is true
(C) Only statements (1) and (2) are true
(D) Statements (2) and (3) are true

4. A deviation of 2° is produced in the yellow ray when prism of crown and flint glass are achromatically combined. Taking dispersive powers of crown and flint glass are 0.02 and 0.03 respectively and refractive index for yellow light for these glasses are 1.5 and 1.6 respectively. The refracting angles for crown glass prism will be ______ ° (in degree)
(Round off to the Nearest Integer) **[16 March 2021 (Evening)]**

5. A prism of refractive index μ and angle of prism A is placed in the position of minimum angle of deviation. If minimum angle of deviation is also A, then in terms of refractive index value of A is: **[25 July 2021 (Evening)]**

(A) $\sin^{-1}\left(\sqrt{\frac{\mu-1}{2}}\right)$ (B) $\sin^{-1}\left(\frac{\mu}{2}\right)$

(C) $\cos^{-1}\left(\frac{\mu}{2}\right)$ (D) $2\cos^{-1}\left(\frac{\mu}{2}\right)$

6. A prism of refractive index n_1 and another prism of refractive index n_2 are stuck together (as shown in the figure). n_1 and n_2 depend on λ, the wavelength of light, according to the relation

$$n_1 = 1.2 + \frac{10.8\times10^{-14}}{\lambda^2} \text{ and } n_2 = 1.45 + \frac{1.8\times10^{-14}}{\lambda^2}$$

The wavelength for which rays incident at any angle on the interface BC pass through without bending at that interface will be ______ nm. **[27 July 2021 (Morning)]**

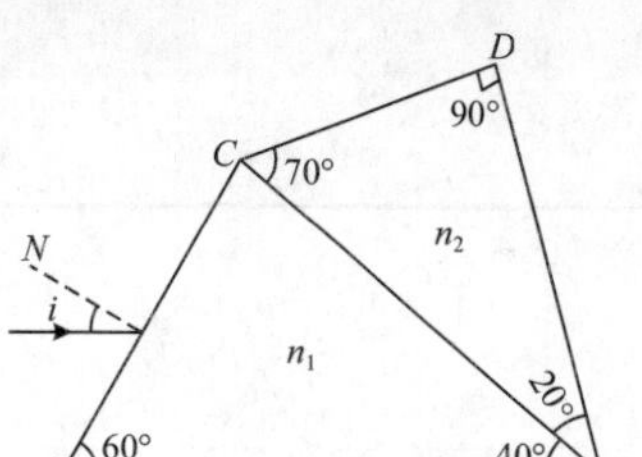

7. Cross-section view of a prism is the equilateral triangle ABC shown in the figure. The minimum deviation is observed using this prism when the angle of incidence is equal to the prism angle. The time taken by light to travel from P (mid point of BC) to A is ______ $\times 10^{-10}$s. (Given, speed of light in vacuum $= 3 \times 10^8$ m/s and $\cos 30° = \frac{\sqrt{3}}{2}$) **[31 Aug. 2021 (Evening)]**

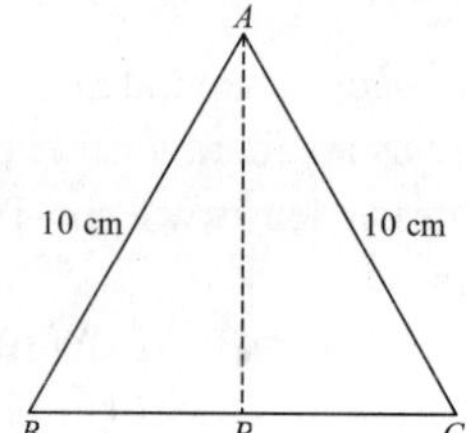

* * * * *

CHAPTER

23 RAY OPTICS AND OPTICAL INSTRUMENTS

23.5 OPTICAL INSTRUMENTS

1. Calculate the limit of resolution of a telescope objective having a diameter of 200 cm, if it has to detect light of wavelength 500 nm coming from a star : **[8 April 2019 (Evening)]**

(A) 305×10^{-9} radian (B) 152.5×10^{-9} radian

(C) 610×10^{-9} radian (D) 457.5×10^{-9} radian

2. Diameter of the objective lens of a telescope is 250 cm. For light of wavelength 600 nm coming from a distant object, the limit of resolution of the telescope is close to : **[9 April 2019 (Evening)]**

(A) 1.5×10^{-7} rad (B) 2.0×10^{-7} rad

(C) 3.0×10^{-7} rad (D) 4.5×10^{-7} rad

3. The value of numerical aperture of the objective lens of a microscope is 1.25. If light of wavelength 5000 Å is used, the minimum separation between two points, to be seen as distinct, will be : **[12 April 2019 (Morning)]**

(A) 0.24 μm (B) 0.48 μm

(C) 0.12 μm (D) 0.38 μm

4. If we need a magnification of 375 from a compound microscope of tube length 150 mm and an objective of focal length 5 mm, the focal length of the eye-piece, should be close to : **[7 Jan. 2020 (Morning)]**

(A) 22mm (B) 12mm

(C) 33mm (D) 2mm

5. The magnifying power of a telescope with tube 60 cm is 5. What is the focal length of its eye piece ? **[8 Jan. 2020 (Morning)]**

(A) 30 cm (B) 40 cm

(C) 20 cm (D) 10 cm

6. The aperture diameter of a telescope is 5m. The separation between the moon and the earth is 4×10^5 km. With light of wavelength of 5500 Å, the minimum separation between objects on the surface of moon, so that they are just resolved, is close to : **[9 Jan. 2020 (Morning)]**

(A) 20m (B) 600m

(C) 60m (D) 200m

7. In a compound microscope, the magnified virtual image is formed at a distance of 25 cm from the eye-piece. The focal length of its objective lens is 1 cm. If the magnification is 100 and the tube length of the microscope is 20 cm, then the focal length of the eye-piece lens (in cm) is ______ . **[4 Sept. 2020 (Morning)]**

8. A compound microscope consists of an objective lens of focal length 1 cm and an eye piece of focal length 5 cm with a separation of 10 cm. The distance between an object and the objective lens, at which the strain on the eye is minimum is $\frac{n}{40}$ cm. The value of n is : **[5 Sept. 2020 (Morning)]**

9. Given below are two statements : one is labeled as **Assertion A** and the other is labeled as **Reason R**

Assertion-A : For a simple microscope, the angular size of the object equals the angular size of the image

Reason-R : Magnification is achieved as the small object can be kept much closer to the eye than 25 cm and hence it subtends a large angle

In the light of the above statements, choose the most appropriate answer from the options given below : **[26 Feb. 2021 (Evening)]**

(A) **A** is true but **R** is false

(B) Both **A** and **R** are true and **R** is the correct explanation of **A**

(C) Both **A** and **R** are true but **R** is NOT the correct explanation of **A**

(D) **A** is false but **R** is true

10. Your friend is having eye sight problem. She is not able lo see clearly a distant uniform window mesh and it appears to her as non-uniform and distorted. The doctor diagnosed the problem as : **[18 March 2021 (Morning)]**

(A) Astigmatism

(B) Myopia with Astigmatism

(C) Presbyopia with Astigmatism

(D) Myopia and hypermetropia

11. The speed of electrons in a scanning electron microscope is 1×10^7 ms^{-1}. If the protons having the same speed are used instead of electrons, then the resolving power of scanning proton microscope will be changed by a factor of : **[18 March 2021 (Evening)]**

(A) 1837 (B) $\frac{1}{1837}$

(C) $\sqrt{1837}$ (D) $\frac{1}{\sqrt{1837}}$

12. An object viewed from a near point distance of 25 cm, using a microscopic lens with magnification '6', gives an unresolved image. A resolved image is observed at infinite distance with a total magnification double the earlier using an eyepiece along with the given lens and a tube of length 0.6 m, if the focal length of the eyepiece is equal to ______ cm.

[20 July 2021 (Morning)]

* * * * *

CHAPTER

24 WAVE OPTICS

24.1 INTERFERENCE OF LIGHT

1. In a Young's double slit experiment, slits are separated by 0.5 mm, and the screen is placed 150 cm away. A beam of light consisting of two wavelengths, 650 nm and 520 nm, is used to obtain interference fringes on the screen. The least distance from the common central maximum to the point where the bright fringes due to both the wavelengths coincide is : **[2017]**

(A) 9.75 mm (B) 15.6 mm
(C) 1.56 mm (D) 7.8 mm

2. Two coherent sources produce waves of different intensities which interfere. After interference, the ratio of the maximum intensity to the minimum intensity is 16. The intensity of the waves are in the ratio : **[9 Jan. 2019 (Morning)]**

(A) 4 : 1 (B) 25 : 9
(C) 16 : 9 (D) 5 : 3

3. In a Young's double slit experiment, the slits are placed 0.320 mm apart. Light of wavelength $\lambda = 500$ nm is incident on the slits. The total number of bright fringes that are observed in the angular range $-30° \le \theta \le 30°$ is : **[9 Jan. 2019 (Evening)]**

(A) 320 (B) 641
(C) 321 (D) 640

4. In a Young's double slit experiment with slit separation 0.1 mm, one observes a bright fringe at angle 1/40 rad by using light of wavelength λ_1. When the light of wavelength λ_2 is used a bright fringe is seen at the same angle in the same set up. Given that λ_1 and λ_2 are in visible range (380 nm to 740 nm), their values are : **[10 Jan. 2019 (Morning)]**

(A) 380 nm, 500 nm (B) 625 nm, 500 nm
(C) 380 nm, 525 nm (D) 400 nm, 500 nm

5. Consider a Young's double slit experiment as shown in figure. What should be the slit separation d in terms of wavelength λ such that the first minima occurs directly in front of the slit (S_1) ? **[10 Jan. 2019 (Evening)]**

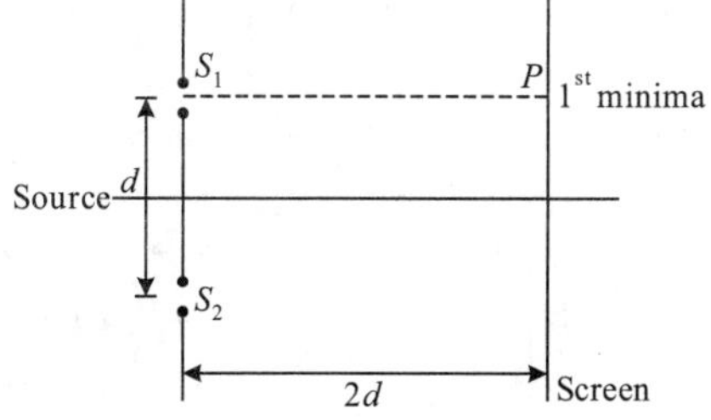

(A) $\frac{\lambda}{2(5-\sqrt{2})}$ (B) $\frac{\lambda}{(5-\sqrt{2})}$

(C) $\frac{\lambda}{(\sqrt{5}-2)}$ (D) $\frac{\lambda}{2(\sqrt{5}-2)}$

6. In a Young's double slit experiment, the path difference, at a certain point on the screen, between two interfering waves is 1/8th of wavelength. The ratio of the intensity at this point to that at the centre of a bright fringe is close to : **[11 Jan. 2019 (Morning)]**

(A) 0.94 (B) 0.74
(C) 0.85 (D) 0.80

7. In a double-slit experiment, green light (5303 Å) falls on a double slit having a separation of 19.44 μm and a width of 4.05 μm. The number of bright fringes between the first and the second diffraction minima is : **[11 Jan. 2019 (Evening)]**

(A) 09 (B) 10
(C) 04 (D) 05

8. In an interference experiment the ratio of amplitudes of coherent waves is $\frac{a_1}{a_2} = \frac{1}{3}$. The ratio of maximum and minimum intensities of fringes will be : **[8 April 2019 (Morning)]**

(A) 4 (B) 2
(C) 9 (D) 18

9. The figure shows a Young's double slit experimental setup. It is observed that when a thin transparent sheet of thickness t and refractive index μ is put in front of one of the slits, the central maximum gest shifted by a distance equal to n fringe widths. If the wavelength of light used is λ, t will be : **[9 April 2019 (Morning)]**

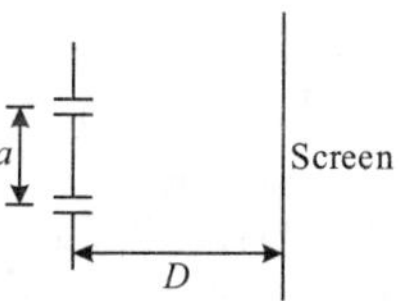

(A) $\frac{2D\lambda}{a(\mu-1)}$ (B) $\frac{D\lambda}{a(\mu-1)}$

(C) $\frac{2nD\lambda}{a(\mu-1)}$ (D) $\frac{nD\lambda}{a(\mu-1)}$

10. In a Young's double slit experiment, the ratio of the slit's width is 4 : 1. The ratio of the intensity of maxima to minima, close to the central fringe on the screen, will be : **[10 April 2019 (Evening)]**

(A) $(\sqrt{3}+1)^4 : 16$ (B) 9 : 1
(C) 4 : 1 (D) 25 : 9

11. In a double slit experiment, when a thin film of thickness t having refractive index μ is introduced in front of one of the slits, the maximum at the centre of the fringe pattern shifts by one fringe width. The value of t is (λ is the wavelength of the light used) : **[12 April 2019 (Morning)]**

(A) $\frac{\lambda}{2(\mu-1)}$ (B) $\frac{\lambda}{(2\mu-1)}$

(C) $\frac{2\lambda}{(\mu-1)}$ (D) $\frac{\lambda}{(\mu-1)}$

12. In a Young's double slit experiment, the separation between the slits is 0.15 mm. In the experiment, a source of light of wavelength 589 nm is used and the interference pattern is observed on a screen kept 1.5 m away. The separation between the successive bright fringes on the screen is : **[7 Jan. 2020 (Evening)]**

(A) 6.9 mm (B) 5.9 mm
(C) 4.9 mm (D) 3.9 mm

13. In a double slit experiment, at a certain point on the screen the path difference between the two interfering waves is $\frac{1}{8}$ th of a wavelength. The ratio of the intensity of light at that point to that at the centre of a bright fringe is : **[8 Jan. 2020 (Evening)]**

(A) 0.568 (B) 0.672
(C) 0.760 (D) 0.853

14. Three harmonic waves having equal frequency ν and same intensity I_0, have phase angles 0, $\frac{\pi}{4}$ and $-\frac{\pi}{4}$ respectively. When they are superimposed the intensity of the resultant wave is close to : **[9 Jan. 2020 (Morning)]**

(A) $5.8\, I_0$ (B) $0.2\, I_0$
(C) I_0 (D) $3\, I_0$

15. In a Young's double slit experiment 15 fringes are observed on a small portion of the screen when light of wavelength 500 nm is used. Ten fringes are observed on the same section of the screen when another light source of wavelength λ is used. Then the value of λ is (in nm) ______. **[9 Jan. 2020 (Evening)]**

16. Interference fringes are observed on a screen by illuminating two thin slits 1 mm apart with a light source (l = 632.8 nm). The distance between the screen and the slits is 100 cm. If a bright fringe is observed on a screen at a distance of 1.27 mm from the central bright fringe, then the path difference between the waves, which are reaching this point from the slits is close to : **[2 Sept. 2020 (Morning)]**

(A) 1.27 mm (B) 2 nm
(C) 2.87 nm (D) 2.05 mm

17. In a Young's double slit experiment, 16 fringes are observed in a certain segment of the screen when light of wavelength 700 nm is used. If the wavelength of light is changed to 400 nm, the number of fringes observed in the same segment of the screen would be : **[2 Sept. 2020 (Evening)]**

(A) 28 (B) 24
(C) 18 (D) 30

18. In a Young's double slit experiment, light of 500 nm is used to produce an interference pattern. When the distance between the slits is 0.05 mm, the angular width (in degree) of the fringes formed on the distance screen is close to : **[3 Sept. 2020 (Morning)]**

(A) 0.07° (B) 0.17°
(C) 1.7° (D) 0.57°

19. Two light waves having the same wavelength λ in vacuum are in phase initially. Then the first wave travels a path L_1 through a medium of refractive index n_1 while the second wave travels a path of length L_2 through a medium of refractive index n_2. After this the phase difference between the two waves is : **[3 Sept. 2020 (Evening)]**

(A) $\frac{2\pi}{\lambda}(n_1L_1 - n_2L_2)$ (B) $\frac{2\pi}{\lambda}\left(\frac{L_2}{n_1} - \frac{L_1}{n_2}\right)$

(C) $\frac{2\pi}{\lambda}\left(\frac{L_1}{n_1} - \frac{L_2}{n_2}\right)$ (D) $\frac{2\pi}{\lambda}(n_2L_1 - n_1L_2)$

20. Two coherent sources of sound, S_1 and S_2, produce sound waves of the same wavelength, λ = 1 m, in phase. S_1 and S_2 are placed 1.5 m apart (see figure) A listener, located at L, directly in front of S_2 finds that the intensity is at a minimum when he is 2 m away from S_2. The listener moves away from S_1, keeping his distance from S_2 fixed. The adjacent maximum of intensity is observed when the listener is at a distance d from S_1. Then, d is : **[5 Sept. 2020 (Evening)]**

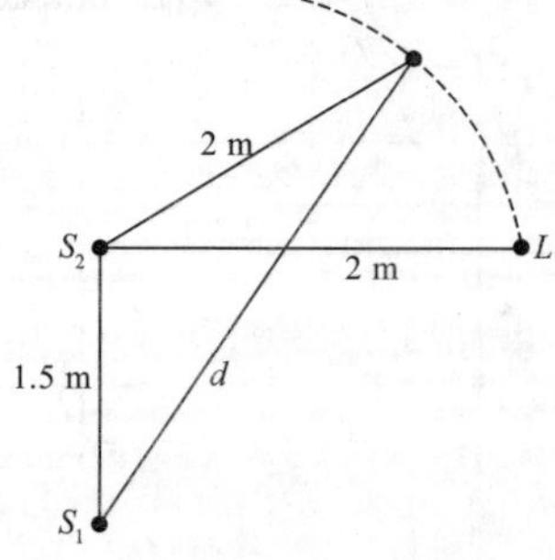

(A) 12m (B) 3m
(C) 5m (D) 2m

21. In the figure below, P and Q are two equally intense coherent sources emitting radiation of wavelength 20 m. The separation between P and Q is 5 m and the phase of P is ahead of that of Q by 90°. A, B and C are three distinct points of observation, each equidistant from the midpoint of PQ. The intensities of radiation at A, B, C will be in the ratio : **[6 Sept. 2020 (Morning)]**

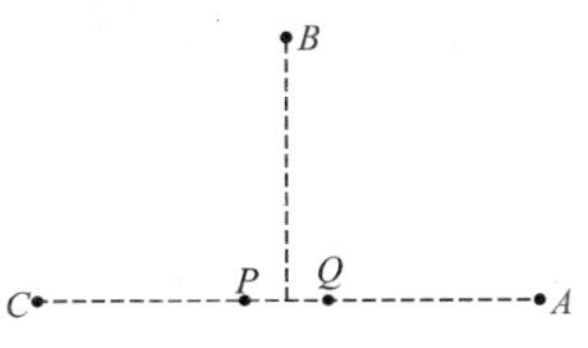

(A) 0 : 1 : 2 (B) 4 : 1 : 0
(C) 0 : 1 : 4 (D) 2 : 1 : 0

22. A Young's double-slit experiment is performed using monochromatic light of wavelength l. The intensity of light at a point on the screen, where the path difference is l, is K units. The intensity of light at a point where the path difference is $\frac{\lambda}{6}$ is given by $\frac{nK}{12}$, where n is an integer. The value of n is : **[6 Sept. 2020 (Evening)]**

23. In a Young's double slit experiment, the width of the one of the slit is three times the other slit. The amplitude of the light coming from a slit is proportional to the slit-width. Find the ratio of the maximum to the minimum intensity in the interference pattern : **[24 Feb. 2021 (Morning)]**

(A) 2 : 1 (B) 4 : 1
(C) 3 : 1 (D) 1 : 4

24. If the source of light used in a Young's double slit experiment is changed from red to violet : **[24 Feb. 2021 (Evening)]**

(A) The intensity of minima will increase
(B) The fringes will become brighter
(C) The central bright fringe will become a dark fringe
(D) Consecutive fringe lines will come closer

25. Two coherent light sources having intensity in the ratio $2x$ produce an interference pattern. The ratio $\frac{I_{max} - I_{min}}{I_{max} + I_{min}}$ will be : **[25 Feb. 2021 (Morning)]**

(A) $\frac{2\sqrt{2x}}{2x+1}$ (B) $\frac{\sqrt{2x}}{x+1}$
(C) $\frac{\sqrt{2x}}{2x+1}$ (D) $\frac{2\sqrt{2x}}{x+1}$

26. In a Young's double slit experiment two slits are separated by 2 mm and the screen is placed one meter away. When a light of wavelength 500 nm is used, the fringe separation will be : **[26 Feb. 2021 (Morning)]**

(A) 0.75 mm (B) 0.25 mm
(C) 1 mm (D) 0.50 mm

27. A fringe width of 6 mm was produced for two slits separated by 1 mm apart. The screen is placed 10 m away. The wavelength of light used is 'x' nm. The value of 'x' to the nearest integer is ______. **[16 March 2021 (Morning)]**

28. In the Young's double slit experiment, the distance between the slits varies in time as $d(t) = d_0 + a_0 \sin \omega t$; where d_0, ω and a_0 are constants. The difference between the largest fringe width and the smallest fringe width obtained over time is given as : **[25 July 2021 (Morning)]**

(A) $\frac{2\lambda D a_0}{(d_0^2 - a_0^2)}$ (B) $\frac{2\lambda D(d_0)}{(d_0^2 - a_0^2)}$
(C) $\frac{\lambda D}{d_0 - a_0}$ (D) $\frac{\lambda D}{d_0^2} a_0$

29. White light is passed through a double slit and interference is observed on a screen 1.5 m away. The separation between the slits is 0.3 mm. The first violet and red fringes are formed 2.0 mm and 3.5 mm away from the central white fringes. The difference in wavelength of red and violet light is______ nm : **[26 Aug. 2021 (Morning)]**

30. The light waves from two coherent sources have same intensity $I_1 = I_2 = I_0$. In interference pattern the intensity of light at minima is zero. What will be the intensity of light at maxima? **[27 Aug. 2021 (Evening)]**

(A) I_0 (B) $2I_0$
(C) $4I_0$ (D) $5I_0$

31. The width of one of the two slits in a Young's double slit experiment is three times the other slit. If the amplitude of the light coming from a slit is proportional to the slit-width, the ratio of maximum to minimum intensity in the interference pattern is $x : 4$ where x is______. **[01 Sep. 2021 (Evening)]**

* * * * *

CHAPTER

24 WAVE OPTICS

24.2 DIFFRACTION OF LIGHT

1. A single slit of width 0.1 mm is illuminated by a parallel beam of light of wavelength 6000 Å and diffraction bands are observed on a screen 0.5 m from the slit. The distance of the third dark band from the central bright band is : **[2017 (Online)]**

(A) 3mm (B) 9mm
(C) 4.5mm (D) 1.5mm

2. A single slit of width b is illuminated by a coherent monochromatic light of wavelength λ. If the second and fourth minima in the diffraction pattern at a distance 1 m from the slit are at 3 cm and 6 cm respectively from the central maximum, wheat is the width of the central maximum? (i.e. distance between first minimum on either side of the central maximum) : **[2017 (Online)]**

(A) 1.5 cm (B) 3.0 cm
(C) 4.5 cm (D) 6.0 cm

3. The angular width of the central maximum in a single slit diffraction pattern is 60°. The width of the slit is 1 μm. The slit is illuminated by monochromatic plane waves. If another slit of same width is made near it, Young's fringes can be observed on a screen placed at a distance 50 cm from the slits. If the observed fringe width is 1 cm, what is slit separation distance ? (i.e. distance between the centres of each slit) **[2018]**

(A) 25 μm (B) 50 μm
(C) 75 μm (D) 100 μm

4. Light of wavelength 550 nm falls normally on a slit of width 22.0×10^{-5} cm. The angular position of the second minima from the central maximum will be (in radians) : **[2018 (Online)]**

(A) $\frac{\pi}{8}$ (B) $\frac{\pi}{12}$
(C) $\frac{\pi}{4}$ (D) $\frac{\pi}{6}$

5. Visible light of wavelength 6000×10^{-8} cm falls normally on a single slit and produces a diffraction pattern. It is found that the second diffraction minimum is at 60° from the central maximum. If the first minimum is produced at θ_1, then θ_1 is close to : **[7 Jan. 2020 (Morning)]**

(A) 20° (B) 45°
(C) 30° (D) 25°

6. Orange light of wavelength 6000×10^{-10} m in illuminates a single slit of width 0.6×10^{-4} m. The maximum possible number of diffraction minima produced on both sides of the central maximum is ______. **[4 Sept. 2020 (Evening)]**

7. A beam of electrons of energy E scatters from a target having atomic spacing of 1 Å. The first maximum intensity occurs at $\theta = 60°$. Then E (in eV) is ______.
(Planck constant $h = 6.64 \times 10^{-34}$ Js, 1 eV $= 1.6 \times 10^{-19}$ J, electron mass $m = 9.1 \times 10^{-31}$ kg) : **[5 Sept. 2020 (Morning)]**

8. Consider the diffraction pattern obtained from the sunlight incident on a pinhole of diameter 0.1 μm. If the diameter of the pinhole is slightly increased, it will affect the diffraction pattern such that : **[25 Feb. 2021 (Evening)]**

(A) Its size decreases, but intensity increases
(B) Its size decreases, and intensity decreases
(C) Its size increases, and intensity increases
(D) Its size increases, but intensity decreases

* * * * *

CHAPTER

24 WAVE OPTICS

24.3 POLARIZATION OF LIGHT

1. Unpolarized light of intensity I passes through an ideal polarizer A. Another identical polarizer B is placed behind A. The intensity of light beyond B is found to be $I/2$. Now another identical polarizer C is placed between A and B. The intensity beyond B is now found to be $I/8$. The angle between polarizer A and C is : **[2018]**

(A) 0° (B) 30°

(C) 45° (D) 60°

2. A plane polarized light is incident on a polarizer with its pass axis making angle θ with x-axis, as shown in the figure. At four different values of θ, $\theta = 8^\circ, 38^\circ, 188^\circ$ and 218°, the observed intensities are same. What is the angle between the direction of polarization and x-axis : **[2018 (Online)]**

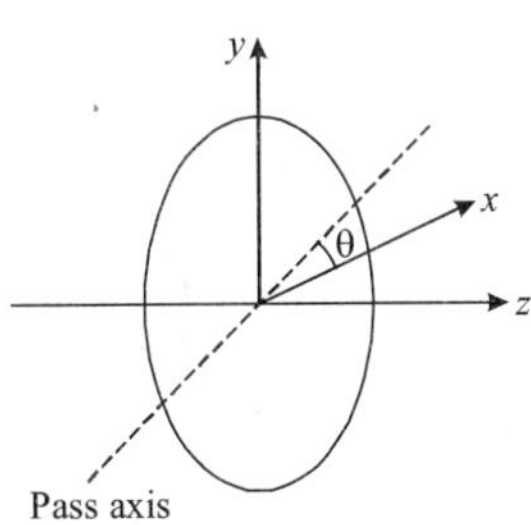

(A) 203° (B) 45°

(C) 98° (D) 128°

3. Unpolarized light of intensity I is incident on a system of two polarizers, A followed by B. The intensity of emergent light is $I/2$. If a third polarizer C is placed between A and B, the intensity of emergent light is reduced to $I/3$. The angle between the polarizers A and C is θ. Then : **[2018 (Online)]**

(A) $\cos\theta = \left(\frac{2}{3}\right)^{1/4}$ (B) $\cos\theta = \left(\frac{1}{3}\right)^{1/4}$

(C) $\cos\theta = \left(\frac{1}{3}\right)^{1/2}$ (D) $\cos\theta = \left(\frac{2}{3}\right)^{1/2}$

4. Consider a tank made of glass (refractive index 1.5) with a thick bottom. It is filled with a liquid of refractive index μ. A student finds that, irrespective of what the incident angle i (see figure) is for a beam of light entering the liquid, the light reflected from the liquid glass interface is never completely polarized. For this to happen, the minimum value of μ is : **[9 Jan. 2019 (Morning)]**

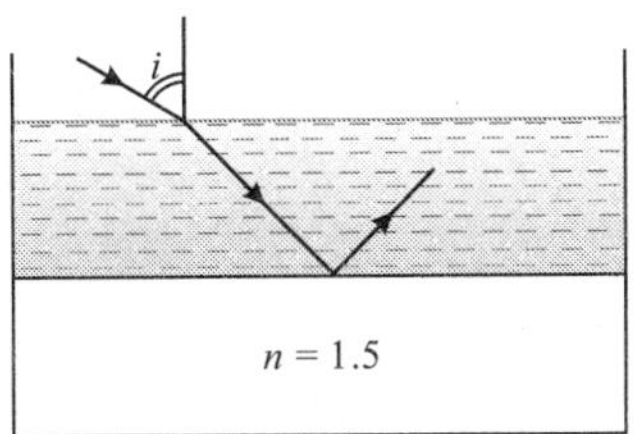

(A) $\frac{3}{\sqrt{5}}$ (B) $\frac{5}{\sqrt{3}}$

(C) $\sqrt{\frac{5}{3}}$ (D) $\frac{4}{3}$

5. A system of three polarizers P_1, P_2, P_3 is set up such that the pass axis of P_3 is crossed with respect to that of P_1. The pass axis of P_2 is inclined at 60° to the pass axis of P_3. When a beam of unpolarized light of intensity I_0 is incident on P_1, the intensity of light transmitted by the three polarizers is I. The ratio (I_0/I) equals (nearly) : **[12 April 2019 (Evening)]**

(A) 16.00 (B) 1.80

(C) 5.33 (D) 10.67

6. A polarizer-analyser set is adjusted such that the intensity of light coming out of the analyser is just 10% of the original intensity. Assuming that the polarizer-analyser set does not absorb any light, the angle by which the analyser need to be rotated further to reduce the output intensity to be zero, is : **[7 Jan. 2020 (Morning)]**

(A) 18.4° (B) 71.6°

(C) 90° (D) 45°

7. A beam of plane polarised light of large cross sectional area and uniform intensity of 3.3 Wm^{-2} falls normally on a polariser (cross sectional area 3×10^{-4} m^2) which rotates about its axis with an angular speed of 31.4 rad/s. The energy of light passing through the polariser per revolution, is close to : **[4 Sept. 2020 (Morning)]**

(A) 1.0×10^{-5} J (B) 5.0×10^{-4} J

(C) 1.0×10^{-4} J (D) 1.5×10^{-4} J

8. An unpolarized light beam is incident on the polarizer of a polarization experiment and the intensity of light beam emerging from the analyzer is measured as 100 Lumens. Now, if the analyzer is rotated around the horizontal axis (direction of light) by 30° in clockwise direction, the intensity of emerging light will be ______ Lumens. **[24 Feb. 2021 (Morning)]**

9. A source of light is placed in front of a screen. Intensity of light on the screen is I. Two Polaroids P_1 and P_2 are so placed in between the source of light and screen that the intensity of light on screen is $I/2$. P_2 should be rotated by an angle of ______ (degrees) so that the intensity of light on the screen becomes $3I/8$. **[26 Aug. 2021 (Evening)]**

* * * * *

CHAPTER

25 DUAL NATURE OF RADIATION, MATTER & X-RAYS

25.1 PHOTOELECTRIC EFFECT

1. The maximum velocity of the photoelectrons emitted from the surface is v when light of frequency n falls on a metal surface. If the incident frequency is increased to $3n$, the maximum velocity of the ejected photoelectrons will be : **[2017 (Online)]**

(A) Less than $\sqrt{3}v$ (B) v

(C) More than $\sqrt{3}v$ (D) Equal to $\sqrt{3}v$

2. A Laser light of wavelength 600 nm is used to weld retina detachment. If a Laser pulse of width 60 ms and power 0.5 kW is used the approximate number of photons in the pulse are : [Take Plank's constant $h = 6.62 \times 10^{-34}$ Js] : **[2017 (Online)]**

(A) 10^{20} (B) 10^{18}

(C) 10^{22} (D) 10^{19}

3. Surface of certain metal is first illuminated with light of wavelength $\lambda_1 = 350$ nm and then, by light of wavelength $\lambda_2 = 540$ nm. It is found that the maximum speed of the photo electrons in the two cases differ by a factor of 2. The work function of the metal (in eV) is close to :

(Energy of photon $= \dfrac{1240}{\lambda(\text{in } nm)}$ eV) **[9 Jan. 2019 (Morning)]**

(A) 1.8 (B) 1.4

(C) 2.5 (D) 5.6

4. The magnetic field associated with a light wave is given, at the origin, by

$B = B_0\,[\sin(3.14 \times 10^7)\,ct + \sin(6.28 \times 10^7)\,ct]$.

If this light falls on a silver plate having a work function of 4.7 eV, what will be the maximum kinetic energy of the photo electrons? ($c = 3 \times 10^8$ ms^{-1}, $h = 6.6 \times 10^{-34}$ J-s) : **[9 Jan. 2019 (Evening)]**

(A) 7.72 eV (B) 8.52 eV

(C) 12.5 eV (D) 6.82 eV

5. A metal plate of area 1×10^{-4} m^2 is illuminated by a radiation of intensity 16 mW/m^2. The work function of the metal is 5 eV. The energy of the incident photons is 10 eV and only 10% of it produces photo electrons. The number of emitted photo electrons per second and their maximum energy, respectively, will be. [1 eV $= 1.6 \times 10^{-19}$ J] : **[10 Jan. 2019 (Evening)]**

(A) 10^{10} and 5 eV (B) 10^{14} and 10 eV

(C) 10^{12} and 5 eV (D) 10^{11} and 5 eV

6. In a photoelectric experiment, the wavelength of the light incident on a metal is changed from 300 nm to 400 nm. The decrease in the stopping potential is close to :

$\left(\dfrac{hc}{e} = 1240\, nm - V\right)$ **[11 Jan. 2019 (Evening)]**

(A) 0.5 V (B) 1.0 V

(C) 2.0 V (D) 1.5 V

7. When a certain photosensitive surface is illuminated with monochromatic light of frequency ν, the stopping potential for the photo current is $-V_0/2$. When the surface is illuminated by monochromatic light of frequency ν/2, the stopping potential is $-V_0$. The threshold frequency for photoelectric emission is: **[12 Jan. 2019 (Evening)]**

(A) $\dfrac{3\nu}{2}$ (B) 2ν

(C) $\dfrac{4}{3}\nu$ (D) $\dfrac{5\nu}{3}$

8. The electric field of light wave is given as $\vec{E} = 10^{-3}\cos\left(\dfrac{2\pi x}{5\times10^{-7}} - 2\pi\times6\times10^{14}t\right)\hat{x}\dfrac{N}{C}$. This light falls on a metal plate of work function 2 eV. The stopping potential of the photoelectrons is.

Given, $E\,(\text{in } eV) = \dfrac{12375}{\lambda(\text{in Å})}$: **[9 April 2019 (Morning)]**

(A) 0.48 V (B) 2.0 V

(C) 2.48 V (D) 0.72 V

9. In a photoelectric effect experiment the threshold wavelength of the light is 380 nm. If the wavelength of incident light is 260 nm, the maximum kinetic energy of emitted electrons will be : Given $E\,(\text{in } eV) = \dfrac{1237}{\lambda(\text{in } nm)}$ **[10 April 2019 (Morning)]**

(A) 1.5 eV (B) 4.5 eV

(C) 15.1 eV (D) 3.0 eV

10. The stopping potential V_0 (in volt) as a function of frequency (ν) for a sodium emitter, is shown in the figure. The work function of sodium, from the data plotted in the figure, will be : (Given : Planck's constant $(h) = 6.63 \times 10^{-34}$ Js, electron charge $e = 1.6 \times 10^{-19}$ C) **[12 April 2019 (Morning)]**

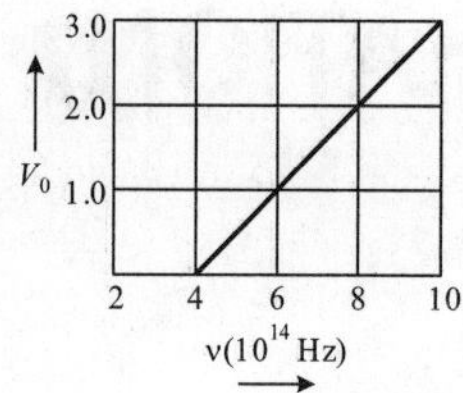

(A) 1.95 eV (B) 1.82 eV
(C) 1.66 eV (D) 2.12 eV

11. A beam of electromagnetic radiation of intensity 6.4×10^{-5} W/cm^2 is comprised of wavelength, $\lambda = 310$ nm. It falls normally on a metal (work function $\varphi = 2$ eV) of surface area of 1 cm^2. If one in 10^3 photons ejects an electron, total number of electrons ejected in 1 s is 10^x. (hc = 1240 $eVnm$, 1 $eV = 1.6 \times 10^{-19}$ J), then x is______. **[7 Jan. 2020 (Morning)]**

12. When photon of energy 4.0 eV strikes the surface of a metal A, the ejected photoelectrons have maximum kinetic energy $T_A eV$ end de-Broglie wavelength λ_A. The maximum kinetic energy of photoelectrons liberated from another metal B by photon of energy 4.50 eV is $T_B = (T_A - 1.5)$ eV. If the de-Broglie wavelength of these photoelectrons $\lambda_B = 2\lambda_A$, then the work function of metal B is : **[8 Jan. 2020 (Morning)]**

(A) 3 eV (B) 2 eV
(C) 4 eV (D) 1.5 eV

13. The dimension of stopping potential V_0 in photoelectric effect in units of Planck's constant 'h', speed of light 'c' and Gravitational constant 'G' and ampere A is :

[8 Jan. 2020 (Morning)]

(A) $h^2\ G^{3/2}\ c^{1/3}\ A^{-1}$ (B) $h^{-2/3}\ c^{-1/3}\ G^{4/3}\ A^{-1}$
(C) $h^{1/3}\ G^{2/3}\ c^{1/3}\ A^{-1}$ (D) $h^{2/3}\ c^{5/3}\ G^{1/3}\ A^{-1}$

14. Radiation, with wavelength 6561 Å falls on a metal surface to produce photoelectrons. The electrons are made to enter a uniform magnetic field of 3×10^{-4} T. If the radius of the largest circular path followed by the electrons is 10 mm, the work function of the metal is close to : **[9 Jan. 2020 (Morning)]**

(A) 1.8 eV (B) 1.1 eV
(C) 0.8 eV (D) 1.6 eV

15. When radiation of wavelength λ is used to illuminate a metallic surface, the stopping potential is V. When the same surface is illuminated with radiation of wavelength 3λ, the stopping potential is $V/4$. If the threshold wavelength for the metallic surface is $n\lambda$ then value of n will be :

[2 Sept. 2020 (Morning)]

16. When the wavelength of radiation falling on a metal is changed from 500 nm to 200 nm, the maximum kinetic energy of the photoelectrons becomes three times larger. The work function of the metal is close to : **[3 Sept. 2020 (Morning)]**

(A) 0.61 eV (B) 0.52 eV
(C) 0.81 eV (D) 1.02 eV

17. Given figure shows few data points in a photo electric effect experiment for a certain metal. The minimum energy for ejection of electron from its surface is : (Planck's constant $h = 6.62 \times 10^{-34}$ J.s) : **[4 Sept. 2020 (Morning)]**

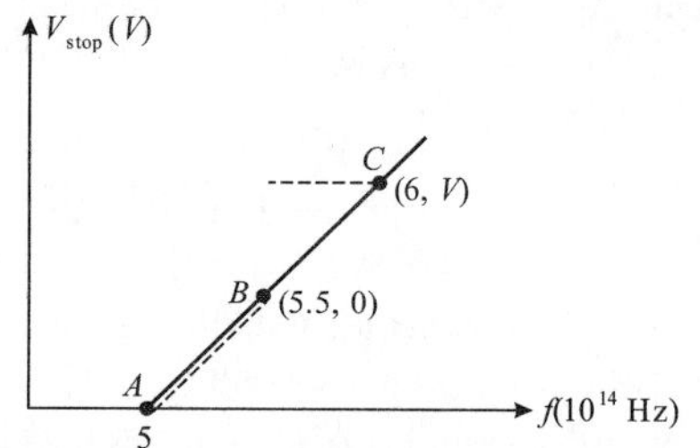

(A) 2.27 eV (B) 2.59 eV
(C) 1.93 eV (D) 2.10 eV

18. In a photoelectric effect experiment, the graph of stopping potential V versus reciprocal of wavelength obtained is shown in the figure. As the intensity of incident radiation is increased : **[4 Sept. 2020 (Evening)]**

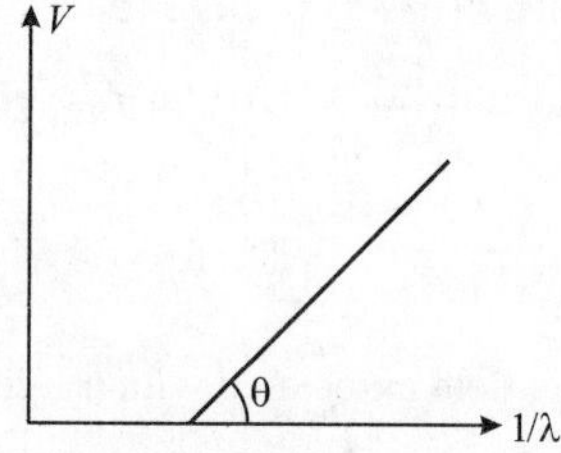

(A) Slope of the straight line get more steep
(B) Straight line shifts to left
(C) Graph does not change
(D) Straight line shifts to right

19. The surface of a metal is illuminated alternately with photons of energies $E_1 = 4$ eV and $E_2 = 2.5$ eV respectively. The ratio of maximum speeds of the photoelectrons emitted in the two cases is 2. The work function of the metal in (eV) is______.

[5 Sept. 2020 (Evening)]

20. The stopping potential for electrons emitted from a photosensitive surface illuminated by light of wavelength 491 nm is 0.710 V. When the incident wavelength is changed to a new value, the stopping potential is 1.43 V. The new wavelength is : (Consider hc = 1240 eV-nm) **[25 Feb. 2021 (Evening)]**

(A) 382 nm (B) 329 nm
(C) 309 nm (D) 400 nm

21. Two stream of photons, possessing energies equal to twice and ten times the work function of metal are incident on the metal surface successively. The value of ratio of maximum velocities of the photoelectrons emitted in the two respective cases is $x : y$. The value of x is ______. **[26 Feb. 2021 (Evening)]**

22. The stopping potential in the context of photoelectric effect depends on the following property of incident electromagnetic radiation : **[16 March 2021 (Morning)]**

(A) Phase (B) Intensity
(C) Amplitude (D) Frequency

23. If 2.5×10^{-6} N average force is exerted by a light wave on a non-reflecting surface of 30 cm^2 area during 40 minutes of time span, the energy flux of light just before it falls on the surface is ______ W/cm^2. (Round off to the Nearest Integer)
(Assume complete absorption and normal incidence conditions are there) **[17 March 2021 (Morning)]**

24. Two identical photocathodes receive the light of frequencies f_1 and f_2 respectively. If the velocities of the photo-electrons coming out are v_1 and v_2 respectively, then : **[17 March 2021 (Evening)]**

(A) $v_1^2 - v_2^2 = \frac{2h}{m}[f_1 - f_2]$

(B) $v_1^2 + v_2^2 = \frac{2h}{m}[f_1 + f_2]$

(C) $v_1 + v_2 = \left[\frac{2h}{m}(f_1 + f_2)\right]^{\frac{1}{2}}$

(D) $v_1 - v_2 = \left[\frac{2h}{m}(f_1 - f_2)\right]^{1/2}$

25. The radiation corresponding to $3 \rightarrow 2$ transition of a hydrogen atom falls on a gold surface to generate photoelectrons. These electrons are passed through a magnetic field of 5×10^{-4} T. Assume that the radius of the largest circular path followed by these electrons is 7 mm, the work function of the metal is : (Mass of electron $= 9.1 \times 10^{-31}$ kg) **[20 July 2021 (Morning)]**

(A) 0.82 eV (B) 0.16 eV
(C) 1.36 eV (D) 1.88 eV

26. A certain metallic surface is illuminated by monochromatic radiation of wavelength λ. The stopping potential for photoelectric current for this radiation is $3V_0$. If the same surface is illuminated with a radiation of wavelength 2λ, the stopping potential is V_0. The threshold wavelength of this surface for photoelectric effect is ______ λ. **[20 July 2021 (Evening)]**

27. When radiation of wavelength λ is incident on a metallic surface, the stopping potential of ejected photoelectrons is 4.8 V. If the same surface is illuminated by radiation of double the previous wavelength, then the stopping potential becomes 1.6 V. The threshold wavelength of the metal is : **[25 July 2021 (Evening)]**

(A) 8λ (B) 4λ
(C) 2λ (D) 6λ

28. A light beam of wavelength 500 nm is incident on a metal having work function of 1.25 eV, placed in a magnetic field of intensity B. The electrons emitted perpendicular to the magnetic field B, with maximum kinetic energy are bent into circular are of radius 30 cm. The value of B is ______ $\times 10^{-7}$ T.
(Given : $hc = 20 \times 10^{-26}$ J-m, mass of electron $= 9 \times 10^{-31}$ kg) **[25 July 2021 (Evening)]**

29. An electron and proton are separated by a large distance. The electron start approaching the proton with energy 3 eV. The proton captures the electron and form a hydrogen atom in second excited state. The resulting photon is incident on a photosensitive metal of threshold wavelength 4000 Å. What is the maximum kinetic energy of the emitted photoelectron ? **[27 July 2021 (Evening)]**

(A) No photoelectron would be emitted
(B) 1.41 eV
(C) 7.61 eV
(D) 3.3 eV

30. In a photoelectric experiment ultraviolet light of wavelength 280 nm used with lithium cathode having work function $\phi = 2.5$ eV. If the wavelength of incident light is switched to 400 nm, find out the change in the stopping potential. ($h = 6.63 \times 10^{-34}$ Js, c $= 3 \times 10^8$ ms^{-1}) **[26 Aug. 2021 (Morning)]**

(A) 0.6 V (B) 1.1 V
(C) 1.3 V (D) 1.9 V

31. The electric field in a plane electromagnetic wave is given by $\vec{E} = 200\cos\left[\left(\frac{0.5\times10^3}{m}\right)x - \left(1.5\times10^{11}\frac{rad}{s}\times t\right)\right]\hat{j}$ V/m. If this wave falls normally on a perfectly reflecting surface having an area of 100 cm^2. If the radiation pressure exerted by the $E.M.$ wave on the surface during a 10 minute exposure is $\frac{x}{10^9}\frac{N}{m^2}$.
Find the value of x: **[26 Aug. 2021 (Morning)]**

32. In a photoelectric experiment, increasing the intensity of incident light : **[27 Aug. 2021 (Morning)]**

(A) Increases the number of photons incident and the K.E. of the ejected electrons remains unchanged.

(B) Increases the frequency of photons incident and the K.E. of the ejected electrons remains unchanged.
(C) Increases the number of photons incident and also increases the K.E. of the ejected electrons.
(D) Increases the frequency of photons incident and increases the K.E. of the ejected electrons

33. A monochromatic neon lamp with wavelength of 670.5 illuminates a photo-sensitive material which has a stopping voltage of 0.48 V. What will be the stopping voltage if the source light is changed with another source of wavelength of 476.6 nm ? **[27 Aug. 2021 (Evening)]**
(A) 0.24 V (B) 1.5 V
(C) 1.25 V (D) 0.96 V

34. A free electron of 2.6 eV energy collides with a H^+ ion. This results in the formation of a hydrogen atoms in the first excites state and a photon is released. Find the frequency of the emitted photon. ($h = 6.6 \times 10^{-34}$ J S) **[31 Aug. 2021 (Evening)]**
(A) 1.45×1016 MHz (B) 1.45×10^9 MHz
(C) 9.0×10^{27} MHz (D) 0.19×10^{15} MHz

* * * * *

CHAPTER 25 DUAL NATURE OF RADIATION, MATTER & X-RAYS

25.2 MATTER WAVES & X-RAYS

1. A particle A of mass m and initial velocity v collides with a particle B of mass $m/2$ which is at rest. The collision is head on, and elastic. The ratio of the de-Broglie wavelengths λ_A to λ_B after the collision is : **[2017]**

(A) $\frac{\lambda_A}{\lambda_B} = \frac{2}{3}$ (B) $\frac{\lambda_A}{\lambda_B} = \frac{1}{2}$

(C) $\frac{\lambda_A}{\lambda_B} = \frac{1}{3}$ (D) $\frac{\lambda_A}{\lambda_B} = 2$

2. An electron beam is accelerated by a potential difference V to hit a metallic target to produce X-rays. It produces continuous as well as characteristic X-rays. If λ_{min} is the smallest possible wavelength of X-ray in the spectrum, the variation of log λ_{min} with log V is correctly represented in : **[2017]**

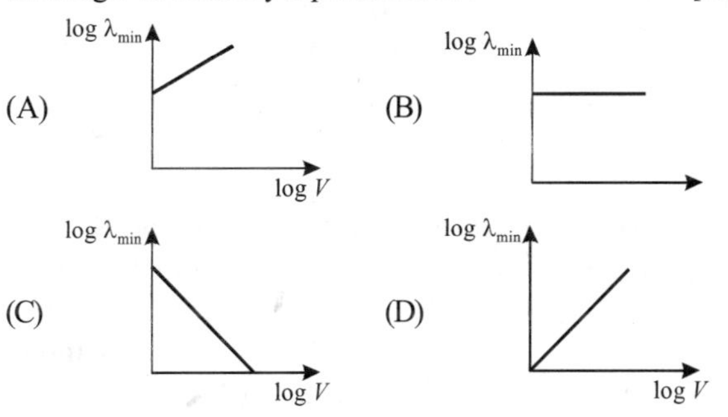

3. If the de Broglie wavelengths associated with a proton and an α-particle are equal, then the ratio of velocities of the proton and the α- particle will be : **[2018 (Online)]**

(A) 1 : 4 (B) 1 : 2

(C) 4 : 1 (D) 2 : 1

4. Two electrons are moving with non-relativistic speeds perpendicular to each other. If corresponding de Broglie wavelengths are λ_1 and λ_2, their de Broglie wavelength in the frame of reference attached to their centre of mass is : **[2018 (Online)]**

(A) $\lambda_{CM} = \lambda_1 = \lambda_2$ (B) $\frac{1}{\lambda_1} = \frac{1}{\lambda_1} + \frac{1}{\lambda_2}$

(C) $\lambda_{CM} = \frac{2\lambda_1\lambda_2}{\sqrt{\lambda_1^2 + \lambda_2^2}}$ (D) $\lambda_{CM} = \left(\frac{\lambda_1 + \lambda_2}{2}\right)$

5. The de-Broglie wavelength (λ_B) associated with the electron orbiting in the second excited state of hydrogen atom is related to that in the ground state (λ_G) by : **[2018 (Online)]**

(A) $\lambda_B = \lambda_G/3$ (B) $\lambda_B = \lambda_G/2$

(C) $\lambda_B = 2\lambda_G$ (D) $\lambda_B = 3\lambda_G$

6. In an electron microscope, the resolution that can be achieved is of the order of the wavelength of electrons used. To resolve a width of 7.5×10^{-12} m, the minimum electron energy required is close to : **[10 Jan. 2019 (Morning)]**

(A) 100 keV (B) 500 keV

(C) 25 keV (D) 1 keV

7. If the de-Broglie wavelength of an electron is equal to 10^{-3} times the wavelength of a photon of frequency 6×10^{14} Hz, then the speed of electron is equal to : **[11 Jan. 2019 (Morning)]**

(Speed of light = 3×10^8 m/s

Planck's constant = 6.63×10^{-34} J.s

Mass of electron = 9.1×10^{-31} kg)

(A) 1.45×10^6 m/s (B) 1.7×10^6 m/s

(C) 1.8×10^6 m/s (D) 1.1×10^6 m/s

8. A particle A of mass 'm' and charge 'q' is accelerated by a potential difference of 50 V. Another particle B of mass '4 m' and charge 'q' is accelerated by a potential difference of 2500 V.

The ratio of de-Broglie wavelengths $\frac{\lambda_A}{\lambda_B}$ is close to :

[12 Jan. 2019 (Morning)]

(A) 10.00 (B) 14.14

(C) 4.47 (D) 0.07

9. Two particles move at right angle to each other. Their de-Broglie wavelengths are λ_1 and λ_2 respectively. The particles suffer perfectly inelastic collision. The de-Broglie wavelength λ, of the final particle, is given by : **[8 April 2019 (Morning)]**

(A) $\lambda = \frac{\lambda_1 + \lambda_2}{2}$ (B) $\frac{2}{\lambda} = \frac{1}{\lambda_1} + \frac{1}{\lambda_2}$

(C) $\lambda = \sqrt{\lambda_1\lambda_2}$ (D) $\frac{1}{\lambda^2} = \frac{1}{\lambda_1^2} + \frac{1}{\lambda_2^2}$

10. A nucleus A, with a finite de-Broglie wavelength λ_A, undergoes spontaneous fission into two nuclei B and C of equal mass. B flies in the same direction as that of A, while C flies in the opposite direction with a velocity equal to half of that of B. The de-Broglie wavelengths λ_B and λ_C of B and C are respectively : **[8 April 2019 (Evening)]**

(A) $2\lambda_A, \lambda_A$ (B) $\lambda_A, 2\lambda_A$

(C) $\lambda_A, \frac{\lambda_A}{2}$ (D) $\frac{\lambda_A}{2}, \lambda_A$

11. A particle 'P' is formed due to a completely inelastic collision of particles 'x' and 'y' having de-Broglie wavelengths 'λ_x' and 'λ_y' respectively. If x and y were moving in opposite directions, then the de-Broglie wavelength of 'P' is : **[9 April 2019 (Evening)]**

(A) $\lambda_x + \lambda_y$ (B) $\frac{\lambda_x \lambda_y}{\lambda_x + \lambda_y}$

(C) $\frac{\lambda_x \lambda_y}{|\lambda_x - \lambda_y|}$ (D) $\lambda_x - \lambda_y$

12. 50 W/m^2 energy density of sunlight is normally incident on the surface of a solar panel. Some part of incident energy (25%) is reflected from the surface and the rest is absorbed. The force exerted on 1m^2 surface area will be close to ($c = 3 \times 10^8$ m/s) : **[9 April 2019 (Evening)]**

(A) 15×10^{-8} N (B) 35×10^{-8} N

(C) 10×10^{-8} N (D) 20×10^{-8} N

13. Light is incident normally on a completely absorbing surface with an energy flux of 25 Wcm^{-2}. If the surface has an area of 25 cm^{-2}, the momentum transferred to the surface in 40 min time duration will be : **[10 April 2019 (Evening)]**

(A) 5.0×10^{-3} Ns (B) 3.5×10^{-6} Ns

(C) 1.4×10^{-6} Ns (D) 6.3×10^{-4} Ns

14. A 2 mW laser operates at a wavelength of 500 nm. The number of photons that will be emitted per second is : [Given Planck's constant $h = 6.6 \times 10^{-34}$ Js, speed of light $c = 3.0 \times 10^8$ m/s] **[10 April 2019 (Evening)]**

(A) 2×10^{16} (B) 1.5×10^{16}

(C) 5×10^{15} (D) 1×10^{16}

15. An electron (of mass m) and a photon have the same energy E in the range of a few eV. The ratio of the de-Broglie wavelength associated with the electron and the wavelength of the photon is (c = speed of light in vacuum) : **[7 Jan. 2020 (Evening)]**

(A) $\left(\frac{E}{2m}\right)^{1/2}$ (B) $\frac{1}{c}\left(\frac{E}{2m}\right)^{1/2}$

(C) $c(2mE)^{1/2}$ (D) $\frac{1}{c}\left(\frac{2E}{m}\right)^{1/2}$

16. An electron (mass m) with initial velocity $\vec{v} = v_0\hat{i} + v_0\hat{j}$ is in an electric field $\vec{E} = -E_0\hat{k}$. If λ_0 is initial de-Broglie wavelength of electron, its de-Broglie wave length at time t is given by : **[8 Jan. 2020 (Evening)]**

(A) $\frac{\lambda_0\sqrt{2}}{\sqrt{1+\frac{e^2E^2t^2}{m^2v_0^2}}}$ (B) $\frac{\lambda_0}{\sqrt{2+\frac{e^2E^2t^2}{m^2v_0^2}}}$

(C) $\frac{\lambda_0}{\sqrt{1+\frac{e^2E^2t^2}{2m^2v_0^2}}}$ (D) $\frac{\lambda_0}{\sqrt{1+\frac{e^2E_0^2t^2}{m^2v_0^2}}}$

17. A particle moving with kinetic energy E has de Broglie wavelength λ. If energy ΔE is added to its energy, the wavelength become $\lambda/2$. Value of ΔE, is : **[9 Jan. 2020 (Morning)]**

(A) $2E$ (B) E

(C) $3E$ (D) $4E$

18. An electron of mass m and magnitude of charge $|e|$ initially at rest gets accelerated by a constant electric field E. The rate of change of de-Broglie wavelength of this electron at time t ignoring relativistic effects is : **[9 Jan. 2020 (Evening)]**

(A) $\frac{-h}{|e|Et^2}$ (B) $\frac{|e|Et}{h}$

(C) $-\frac{h}{|e|E\sqrt{t}}$ (D) $-\frac{h}{|e|Et}$

19. A particle is moving 5 times as fast as an electron. The ratio of the de-Broglie wavelength of the particle to that of the electron is 1.878×10^{-4}. The mass of the particle is close to : **[2 Sept. 2020 (Evening)]**

(A) 4.8×10^{-27} kg (B) 1.2×10^{-28} kg

(C) 9.1×10^{-31} kg (D) 9.7×10^{-28} kg

20. Two sources of light emit X-rays of wavelength 1 nm and visible light of wavelength 500 nm, respectively. Both the sources emit light of the same power 200 W. The ratio of the number density of photons of X-rays to the number density of photons of the visible light of the given wavelengths is : **[3 Sept. 2020 (Evening)]**

(A) $\frac{1}{500}$ (B) 500

(C) 250 (D) $\frac{1}{250}$

21. Particle A of mass $m_A = \frac{m}{2}$ moving along the x-axis with velocity v_0 collides elastically with another particle B at rest having mass $m_B = \frac{m}{3}$. If both particles move along the x-axis after the collision, the change $\Delta\lambda$ in de-Broglie wavelength of particle A, in terms of its de-Broglie wavelength (λ_0) before collision is : **[4 Sept. 2020 (Morning)]**

(A) $\Delta\lambda = 4\lambda_0$
(B) $\Delta\lambda = \frac{5}{2}\lambda_0$
(C) $\Delta\lambda = 2\lambda_0$
(D) $\Delta\lambda = \frac{3}{2}\lambda_0$

22. An electron, a doubly ionized helium ion (He^{++}) and a proton are having the same kinetic energy. The relation between their respective de-Broglie wavelengths λ_e, $\lambda_{He^{++}}$ and λ_p is : **[6 Sept. 2020 (Morning)]**

(A) $\lambda_e < \lambda_p < \lambda_{He^{++}}$
(B) $\lambda_e < \lambda_{He^{++}} = \lambda_p$
(C) $\lambda_e > \lambda_{He^{++}} > \lambda_p$
(D) $\lambda_e > \lambda_p > \lambda_{He^{++}}$

23. Assuming the nitrogen molecule is moving with r.m.s. velocity at 400 K, the de-Broglie wavelength of nitrogen molecule is close to :
(Given : Nitrogen molecule weight
4.64×10^{-26} kg, Boltzmann constant
1.38×10^{-23} J/K, Planck constant
6.63×10^{-34} J.s) **[6 Sept. 2020 (Evening)]**

(A) 0.34 Å
(B) 0.24 Å
(C) 0.20 Å
(D) 0.44 Å

24. Given below are two statements :
Statement-I : Two photons having equal linear momenta have equal wavelengths
Statement-II : If the wavelength of photon is decreased, then the momentum and energy of a photon will also decrease
In the light of the above statements, choose the correct answer from the options given below. **[24 Feb. 2021 (Morning)]**

(A) Statement-I is false but Statement-II is true
(B) Both Statement-I and Statement-II are true
(C) Statement-I is true but Statement-II is false
(D) Both Statement-I and Statement-II are false

25. An X-ray tube is operated at 1.24 million volt. The shortest wavelength of the produced photon will be : **[24 Feb. 2021 (Evening)]**

(A) 10^{-4} nm
(B) 10^{-2} nm
(C) 10^{-3} nm
(D) 10^{-1} nm

26. The de-Broglie wavelength of a proton and α-particle are equal. The ratio of their velocities is : **[24 Feb. 2021 (Evening)]**

(A) 4 : 3
(B) 1 : 4
(C) 4 : 2
(D) 4 : 1

27. An α particle and a proton are accelerated from rest by a potential difference of 200 V. After this, their de-Broglie wavelengths are λ_α and λ_p respectively. $\frac{\lambda_p}{\lambda_\alpha}$ is : **[25 Feb. 2021 (Morning)]**

(A) 7.8
(B) 8
(C) 2.8
(D) 3.8

28. An electron of mass m_e and a proton of mass $m_p = 1836\, m_e$ are moving with the same speed. The ratio of their de-Broglie wavelength $\frac{\lambda_{electron}}{\lambda_{proton}}$ will be : **[25 Feb. 2021 (Evening)]**

(A) $\frac{1}{1836}$
(B) 918
(C) 1836
(D) 1

29. The wavelength of an X-ray beam is 10 Å. The mass of a fictitious particle having the same energy as that of the X-ray photons is $\frac{x}{3}h$ kg. The value of x is ______.
(h = Planck's constant). **[25 Feb. 2021 (Evening)]**

30. Given below are two statements : one is labelled as **Assertion A** and the other is labelled as **Reason R.**
Assertion-A : An electron microscope can achieve better resolving power than an optical microscope
Reason-R : The de-Broglie's wavelength of the electrons emitted from an electron gun is much less than wavelength of visible light
In the light of the above statements, choose the correct answer from the options given below : **[26 Feb. 2021 (Morning)]**

(A) **A** is true but **R** is false
(B) **A** is false but **R** is true
(C) Both **A** and **R** are true but **R** is NOT the correct explanation of **A**
(D) Both **A** and **R** are true and **R** is the correct explanation of **A**

31. The recoil speed of a hydrogen atom after it emits a photon in going from $n = 5$ state to $n = 1$ state will be : **[26 Feb. 2021 (Evening)]**

(A) 3.25 m/s
(B) 4.34 m/s
(C) 4.17 m/s
(D) 2.19 m/s

32. The de-Broglie wavelength associated with an electron and a proton were calculated by accelerating them through same potential of 100 V. What should nearly be the ratio of their wavelengths ? ($m_p = 1.00727$ u, $m_e = 0.00055$ u) **[16 March 2021 (Evening)]**

(A) 1860 : 1
(B) $(1860)^2$: 1
(C) 41.4 : 1
(D) 43 : 1

33. Red light differs from blue light as they have : **[16 March 2021 (Evening)]**

(A) Different frequencies and different wavelengths
(B) Different frequencies and same wavelengths
(C) Same frequencies and same wavelengths
(D) Same frequencies and different wavelengths

34. An electron of mass m and a photon have same energy E. The ratio of wavelength of electron to that of photon is : (c being the velocity of light) **[17 March 2021 (Morning)]**

(A) $\frac{1}{c}\left(\frac{2m}{E}\right)^{1/2}$ (B) $\frac{1}{c}\left(\frac{E}{2m}\right)^{1/2}$

(C) $\left(\frac{E}{2m}\right)^{1/2}$ (D) $c\,(2mE)^{1/2}$

35. A particle is travelling 4 times as fast as an electron. Assuming the ratio of de-Broglie wavelength of a particle to that of electron is 2 : 1, the mass of the particle is : **[18 March 2021 (Morning)]**

(A) $\frac{1}{16}$ times the mass of e^- (B) 8 times the mass of e^-

(C) 16 times the mass of e^- (D) $\frac{1}{8}$ times the mass of e^-

36. An electron having de-Broglie wavelength λ is incident on a target in a X-ray tube. Cut-of wavelength of emitted X-ray is : **[20 July 2021 (Evening)]**

(A) $\frac{2m^2c^2\lambda^2}{h^2}$ (B) $\frac{hc}{mc}$

(C) 0 (D) $\frac{2mc\lambda^2}{h}$

37. An electron of mass m_e and a proton of mass m_p are accelerated through the same potential difference. The ratio of the de-Broglie wavelength associated with the electron to that with the proton is : **[22 July 2021 (Evening)]**

(A) $\frac{m_p}{m_e}$ (B) 1

(C) $\frac{m_e}{m_p}$ (D) $\sqrt{\frac{m_p}{m_e}}$

38. A particle of mass $4M$ at rest disintegrates into two particles of mass M and $3M$ respectively having non zero velocities. The ratio of de-Broglie wavelength of particle of mass M to that of mass $3M$ will be : **[25 July 2021 (Morning)]**

(A) $1 : \sqrt{3}$ (B) 3 : 1

(C) 1 : 1 (D) 1 : 3

39. What should be the order of arrangement of de-Broglie wavelength of electron (λ_e), an α-particle (λ_α) and proton (λ_p) given that all have the same kinetic energy ? **[25 July 2021 (Morning)]**

(A) $\lambda_e = \lambda_p > \lambda_\alpha$ (B) $\lambda_e < \lambda_p < \lambda_\alpha$

(C) $\lambda_e = \lambda_p = \lambda_\alpha$ (D) $\lambda_e > \lambda_p > \lambda_\alpha$

40. An electron moving with speed v and a photon moving with speed c, have same de-Broglie wavelength. The ratio of kinetic energy of electron to that of photon is : **[25 July 2021 (Evening)]**

(A) $\frac{v}{3c}$ (B) $\frac{3c}{v}$

(C) $\frac{v}{2c}$ (D) $\frac{2c}{v}$

41. A particle of mass 9.1×10^{-31} kg travels in a medium with a speed of 10^6 m/s and a photon of a radiation of linear momentum 10^{-27} kg m/s travels in vacuum. The wavelength of photon is ______ times the wavelength of the particle. **[27 July 2021 (Morning)]**

42. The K_α X-ray of molybdenum has wavelength 0.071 nm. If the energy of a molybdenum atom with a K electron knocked out is 27.5 keV, the energy of this atom when an L electron is knocked out will be____keV. (Round off to the nearest integer) ($h = 4.14 \times 10^{-15}$ eVs, $c = 3 \times 108$ ms^{-1}). **[27 July 2021 (Evening)]**

43. The de-Broglie wavelength of a particle having kinetic energy E is λ. How much extra energy must be given to this particle so that the de-Broglie wavelength reduces to 75% of the initial value ? **[26 Aug. 2021 (Evening)]**

(A) E (B) $\frac{7}{9}E$

(C) $\frac{16}{9}E$ (D) $\frac{7}{9}E$

44. A monochromatic neon lamp with wavelength of 670.5 illuminates a photo-sensitive material which has a stopping voltage of 0.48 V. What will be the stopping voltage if the source light is changed with another source of wavelength of 476.6 nm ? **[27 Aug. 2021 (Evening)]**

(A) 0.24 V (B) 1.5 V

(C) 1.25 V (D) 0.96 V

45. A moving proton and electron have the same de-Broglie wavelength. If K and P denote the kinetic energy and momentum respectively. Then choose the correct option : **[31 Aug. 2021 (Morning)]**

(A) $K_p < K_e$ and $P_p = P_e$ (B) $K_p > K_e$ and $P_p = P_e$

(C) $K_p = K_e$ and $P_p = P_e$ (D) $K_p < K_e$ and $P_p < P_e$

* * * * *

CHAPTER

26 ATOMS

26.1 ATOMIC STRUCTURE

1. Some energy levels of a molecules are shown in the figure. The ratio of the wavelengths $r = \lambda_1/\lambda_2$, is given by: **[2017]**

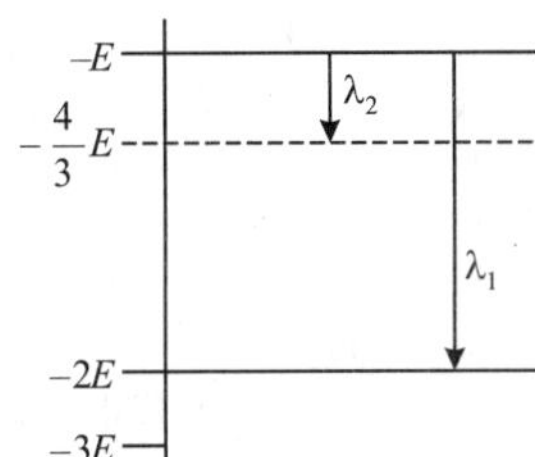

(A) $r = \frac{3}{4}$ (B) $r = \frac{1}{3}$

(C) $r = \frac{4}{3}$ (D) $r = \frac{2}{3}$

2. According to Bohr's theory, the time averaged magnetic field at the centre (i.e. nucleus) of a hydrogen atom due to the motion of electrons in the nth orbit is proportional to :
(n = principal quantum number) **[2017 (Online)]**

(A) n^{-4} (B) n^{-5}

(C) n^{-3} (D) n^{-2}

3. The acceleration of an electron in the first orbit of the hydrogen atom ($z = 1$) is : **[2017 (Online)]**

(A) $\frac{h^2}{\pi^2 m^2 r^3}$ (B) $\frac{h^2}{8\pi^2 m^2 r^3}$

(C) $\frac{h^2}{4\pi^2 m^2 r^3}$ (D) $\frac{h^2}{4\pi m^2 r^3}$

4. If the series limit frequency of the Lyman series is v_1, then the series limit frequency of the P-fund series at : **[2018]**

(A) $25\, v_L$ (B) $16\, v_L$

(C) $v_L/16$ (D) $v_L/25$

5. An electron from various excited states of hydrogen atom emit radiation to come to the ground state. Let λ_n, λ_g be the de Broglie wavelength of the electron in the nth state and the ground state respectively. Let Λ_A be the wavelength of the emitted photon in the transition from the nth state to the ground state. For large n, (A, B are constants) **[2018]**

(A) $\Lambda_n \approx A + \frac{B}{\lambda_n^2}$ (B) $\Lambda_n \approx A + B\lambda_n$

(C) $\Lambda_n^2 \approx A + B\lambda_n^2$ (D) $\Lambda_n^2 \approx \lambda$

6. Muon (μ^{-1}) is negatively charged ($|q| = |e|$) with a mass $m_\mu = 200\, m_e$, where m_e is the mass of the electron and e is the electronic charge. If μ^{-1} is bound to a proton to form a hydrogen like atom, identify the correct statements : **[2018 (Online)]**

(a) Radius of the muonic orbit is 200 times smaller than that of the electron

(b) the speed of the μ^{-1} in the nth orbit is 1/200 times that of the electron in the nth orbit

(c) The ionization energy of muonic atom is 200 times more than that of an hydrogen atom

(d) The momentum of the muon in the nth orbit is 200 times more than that of the electron

(A) (a), (b), (d) (B) (b), (d)

(C) (c), (d) (D) (a), (c), (d)

7. The energy required to remove the electron from a singly ionized Helium atom is 2.2 times the energy required to remove an electron from Helium atom. The total energy required to ionize the Helium atom completely is : **[2018 (Online)]**

(A) 20 eV (B) 79 eV

(C) 109 eV (D) 34 eV

8. A hydrogen atom, initially in the ground state is excited by absorbing a photon of wavelength 980 Å. The radius of the atom in the excited state, in terms of Bohr radius a_0, will be :
(h_c = 12500 eV – Å) **[11 Jan. 2019 (Morning)]**

(A) $9\, a_0$ (B) $25\, a_0$

(C) $4\, a_0$ (D) $16\, a_0$

9. In a hydrogen like atom, when an electron jumps from the M-shell to the L-shell, the wavelength of emitted radiation is λ. If an electron jumps from N-shell to the L-shell, the wavelength of emitted radiation will be : **[11 Jan. 2019 (Evening)]**

(A) $\frac{27}{20}\lambda$ (B) $\frac{16}{25}\lambda$

(C) $\frac{20}{27}\lambda$ (D) $\frac{25}{16}\lambda$

10. A particle of mass m moves in a circular orbit in a central potential field $U(r) = 1/2kr^2$. If Bohr's quantization conditions are applied, radii of possible orbitals and energy levels vary with quantum number n as : **[12 Jan. 2019 (Morning)]**

(A) $r_n \propto n^2, E_n \propto \frac{1}{n^2}$ (B) $r_n \propto \sqrt{n}, E_n \propto \frac{1}{n}$

(C) $r_n \propto n, E_n \propto n$ (D) $r_n \propto \sqrt{n}\ , E_n \propto n$

11. In a Frank-Hertz experiment, an electron of energy 5.6 eV passes through mercury vapour and emerges with an energy 0.7 eV. The minimum wavelength of photons emitted by mercury atoms is close to : **[12 Jan. 2019 (Evening)]**

(A) 2020 nm (B) 220 nm

(C) 250 nm (D) 1700 nm

12. Radiation coming from transitions $n = 2$ to $n = 1$ of hydrogen atoms fall on He^+ ions in $n = 1$ and $n = 2$ states. The possible transition of helium ions as they absorb energy from the radiation is : **[8 April 2019 (Morning)]**

(A) $n = 1 \rightarrow n = 4$ (B) $n = 2 \rightarrow n = 4$

(C) $n = 2 \rightarrow n = 5$ (D) $n = 2 \rightarrow n = 3$

13. Taking the wavelength of first Balmer line in hydrogen spectrum ($n = 3$ to $n = 2$) as 660 nm, the wavelength of the 2nd Balmer line ($n = 4$ to $n = 2$) will be : **[9 April 2019 (Morning)]**

(A) 889.2 nm (B) 642.7 nm

(C) 488.9 nm (D) 388.9 nm

14. A He^+ ion is in its first excited state. Its ionization energy is: **[9 April 2019 (Evening)]**

(A) 6.04 eV (B) 13.60 eV

(C) 54.40 eV (D) 48.36 eV

15. In Li^{++}, electron in first Bohr orbit is excited to a level by a radiation of wavelength λ. When the ion gets deexcited to the ground state in all possible ways (including intermediate emissions), a total of six spectral lines are observed. What is the value of λ ? (Given : $h = 6.63 \times 10^{-34}$ Js ; $c = 3 \times 10^8$ ms^{-1}) **[10 April 2019 (Evening)]**

(A) 9.4 nm (B) 12.3 nm

(C) 10.8 nm (D) 11.4 nm

16. An excited He^+ ion emits two photons in succession, with wavelengths 108.5 nm and 30.4 nm, in making a transition to ground state. The quantum number n, corresponding to its initial excited state is (for photon of wavelength λ, energy $E = \frac{1240\text{eV}}{\lambda(\text{in } nm)}$): **[12 April 2019 (Morning)]**

(A) $n = 5$ (B) $n = 4$

(C) $n = 6$ (D) $n = 7$

17. The electron in a hydrogen atom first jumps from the third excited state to the second excited state and subsequently to the first excited state. The ratio of the respective wavelengths, λ_1/λ_2, of the photons emitted in this process is : **[12 April 2019 (Evening)]**

(A) 9/7 (B) 7/5

(C) 27/5 (D) 20/7

18. Consider an electron in a hydrogen atom, revolving in its second excited state (having radius 4.65 Å). The de-Broglie wavelength of this electron is : **[12 April 2019 (Evening)]**

(A) 12.9 Å (B) 3.5 Å

(C) 9.7 Å (D) 6.6 Å

19. The time period of revolution of electron in its ground state orbit in a hydrogen atom is 1.6×10^{-16} s. The frequency of revolution of the electron in its first excited state (in s^{-1}) is : **[7 Jan. 2020 (Morning)]**

(A) 6.2×10^{15} (B) 5.6×10^{12}

(C) 7.8×10^{14} (D) 1.6×10^{14}

20. The graph which depicts the results of Rutherford gold foil experiment with α-particles is :

θ : Scattering angle

Y : Number of scattered α-particles detected (Plots are schematic and not to scale) : **[8 Jan. 2020 (Morning)]**

(A) 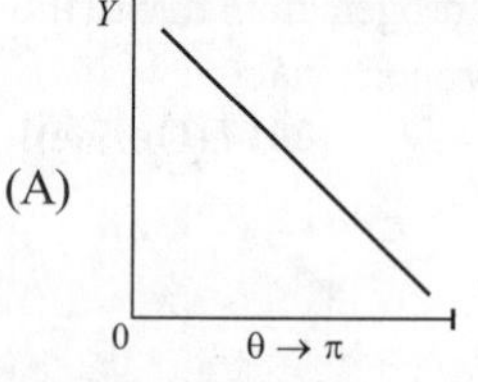

(B) 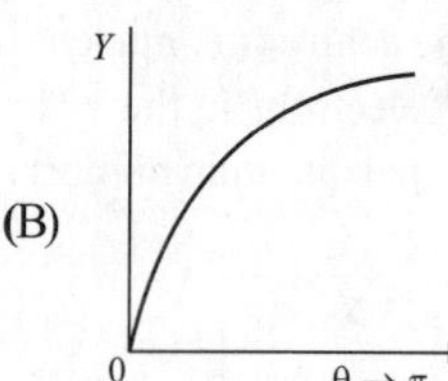

(C) 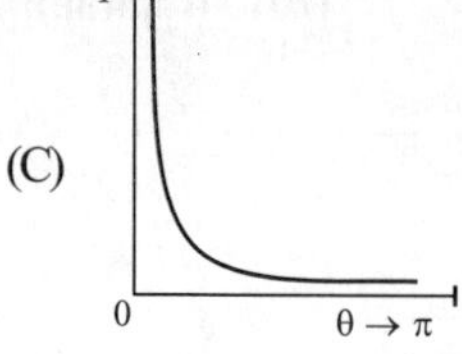

(D) 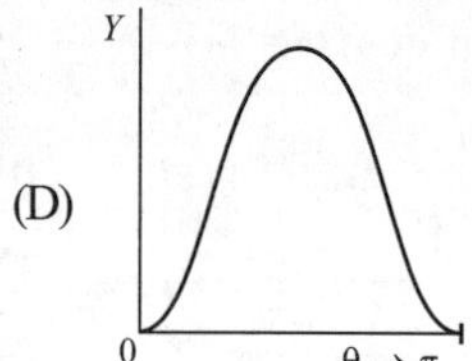

21. The first member of the Balmer series of hydrogen atom has a wavelength of 6561 Å. The wavelength of the second member of the Balmer series (in *nm*) is ______. **[8 Jan. 2020 (Evening)]**

22. The energy required to ionise a hydrogen like ion in its ground state is 9 Rydbergs. What is the wavelength of the radiation emitted when the electron in this ion jumps from the second excited state to the ground state ? **[9 Jan. 2020 (Evening)]**

(A) 35.8 nm (B) 24.2 nm

(C) 8.6 nm (D) 11.4 nm

23. In a hydrogen atom the electron makes a transition from $(n+1)^{th}$ level to the nth level. If $n >> l$, the frequency of radiation emitted is proportional to : **[2 Sept. 2020 (Evening)]**

(A) $\frac{1}{n^4}$ (B) $\frac{1}{n^3}$

(C) $\frac{1}{n^2}$ (D) $\frac{1}{n}$

24. In the line spectra of hydrogen atom, difference between the largest and the shortest wavelengths of the Lyman series is 304 Å. The corresponding difference for the Paschen series in Å is ______. **[4 Sept. 2020 (Morning)]**

25. A particle of mass 200 MeV/c^2 collides with a hydrogen atom at rest. Soon after the collision the particle comes to rest, and the atom recoils and goes to its first excited state. The initial kinetic energy of the particle (in eV) is $\frac{N}{4}$. The value of N is :
(Given the mass of the hydrogen atom to be 1 GeV/c^2) : **[5 Sept. 2020 (Morning)]**

26. In the given figure, the energy levels of hydrogen atom have been shown along with some transitions marked A, B, C, D and E. The transitions A, B and C respectively represent : **[24 Feb. 2021 (Morning)]**

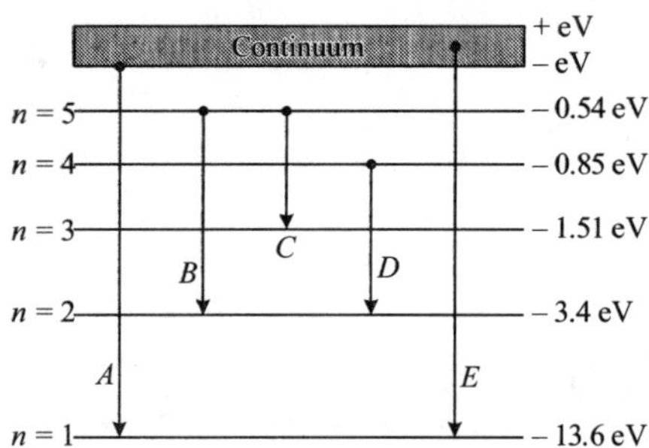

(A) The series limit of Lyman series, third member of Balmer series and second member of Paschen series
(B) The ionization potential of hydrogen, second member of Balmer series and third member of Paschen series
(C) The first member of the Lyman series, third member of Balmer series and second member of Paschen series
(D) The series limit of Lyman series, second member of Balmer series and second member of Paschen series

27. According to Bohr atom model, in which of the following transitions will the frequency be maximum ? **[24 Feb. 2021 (Evening)]**

(A) $n=2$ to $n=1$ (B) $n=4$ to $n=3$
(C) $n=3$ to $n=2$ (D) $n=5$ to $n=4$

28. The wavelength of the photon emitted by a hydrogen atom when an electron makes a transition from $n=2$ to $n=1$ state is : **[25 Feb. 2021 (Evening)]**

(A) 121.8 nm (B) 194.8 nm
(C) 913.3 nm (D) 490.7 nm

29. If λ_1 and λ_2 are the wavelength of the third member of Lyman and first member of the Paschen series respectively, then the value of $\lambda_1 : \lambda_2$ is : **[26 Feb. 2021 (Morning)]**

(A) 1 : 9 (B) 7 : 108
(C) 1 : 3 (D) 7 : 135

30. The first three spectral lines of H-atom in the Balmer series are given $\lambda_1, \lambda_2, \lambda_3$ considering the Bohr atomic model, the wave lengths of first and third spectral lines $\left(\frac{\lambda_1}{\lambda_3}\right)$ are related by a factor of approximately 'x' × 10^{-1}. The value of x, to the nearest integer, is ______. **[16 March 2021 (Morning)]**

31. If an electron is moving in the nth orbit of the hydrogen atom, then its velocity (v_n) for the nth orbit is given as : **[17 March 2021 (Morning)]**

(A) $v_n \propto n$ (B) $v_n \propto \frac{1}{n}$

(C) $v_n \propto n^2$ (D) $v_n \propto \frac{1}{n^2}$

32. Which level of the single ionized carbon has the same energy as the ground state energy of hydrogen atom? **[17 March 2021 (Morning)]**

(A) 1 (B) 6
(C) 4 (D) 8

33. The atomic hydrogen emits a line spectrum consisting of various series. Which series of hydrogen atomic spectra is lying in the visible region ? **[17 March 2021 (Evening)]**

(A) Brackett series (B) Paschen series
(C) Lyman series (D) Balmer series

34. Imagine that the electron in a hydrogen atom is replaced by a muon (μ). The mass of muon particle is 207 times that of an electron and charge is equal to the charge of an electron. The ionization potential of this hydrogen atom will be : **[18 March 2021 (Morning)]**

(A) 13.6 eV (B) 2815.2 eV
(C) 331.2 eV (D) 27.2 eV

35. Consider a situation in which reverse biased current of a particular P-N junction increases when it is exposed to a light of wavelength ≤ 621 nm. During this process, enhancement in

carrier concentration takes place due to generation of hole-electron pairs. The value of band gap is nearly :
(Take $hc/e = 1224$ nm) **[22 July 2021 (Evening)]**
(A) 4 eV (B) 1 eV
(C) 2 eV (D) 0.5 eV

36. In Bohr's atomic model, the electron is assumed to revolve in a circular orbit of radius 0.5 Å. If the speed of electron is 2.2×10^6 m/s, then the current associated with the electron will be ______ $\times 10^{-2}$ mA. [Take π as $\frac{22}{7}$] **[27 July 2021 (Morning)]**

37. Consider the following statements :
[27 July 2021 (Evening)]
(1) Atoms of each element emit characteristics spectrum
(2) According to Bohr's postulate, an electron in a hydrogen atom, revolves in a certain stationary orbit
(3) The density of nuclear matter depends on the size of the nucleus
(4) Neutrons always decay into protons
(5) Radioactivity is an indication of the instability of nuclei.

Choose the correct answer from the options given below :
(A) 1, 2 and 5 only (B) 1, 3 and 5 only
(C) 2 and 4 only (D) 1, 2, 3, 4 and 5

38. The difference in the number of waves when yellow light propagates through air and vacuum columns of the same thickness is one. The thickness of the column is ______ mm. [Refractive index of air = 1.0003, wavelength of yellow light in vaccum = 6000 Å]. **[27 July 2021 (Evening)]**

39. A particular hydrogen like ion radiation of frequency 2.92×10^{15} Hz when it makes transition from $n = 3$ to $n = 1$. The frequency in Hz of radiation emitted in transition from $n = 2$ to $n = 1$ will be : **[26 Aug. 2021 (Morning)]**
(A) 6.57×10^{15} (B) 2.46×10^{15}
(C) 4.38×10^{15} (D) 0.44×10^{15}

40. X different wavelength may be observed in the spectrum from a hydrogen sample if the atoms are exited to states with principle quantum number $n = 6$? The value of X is ______.
[27 Aug. 2021 (Evening)]

* * * * *

CHAPTER

27 NUCLEI

27.1 RADIOACTIVITY

1. A radioactive nucleus A with a half life T, decays into a nucleus B. At $t = 0$, there is no nucleus B. At sometime t, the ratio of the number of B to that of A is 0.3. Then, t is given by : **[2017]**

(A) $t = T\log(1.3)$ (B) $t = \frac{T}{\log(1.3)}$

(C) $t = T\frac{\log(2)}{\log(1.3)}$ (D) $t = T\frac{\log(1.3)}{\log(2)}$

2. A solution containing active cobalt $^{60}_{27}Co$ having activity of 0.8 μCi and decay constant λ is injected in an animal's body. If 1 cm³ of blood is drawn from the animal's body after 10 hrs of injection, the activity found was 300 decays per minute. What is the volume of blood that is flowing in the body ?
($1\text{Ci} = 3.7 \times 10^{10}$ decay per second and at $t = 10$ hrs $e^{-\lambda t} = 0.84$) **[2018 (Online)]**

(A) 6 litres (B) 7 litres
(C) 4 litres (D) 5 litres

3. At some instant, a radioactive sample S_1 having an activity 5 μCi has twice the number of nuclei as another sample S_2 which has an activity of 10 μCi. The half lives of S_1 and S_2 are : **[2018 (Online)]**

(A) 10 years and 20 years, respectively
(B) 5 years and 20 years, respectively
(C) 20 years and 10 years, respectively
(D) 20 years and 5 years, respectively

4. A sample of radioactive material A, that has an activity of 10 mCi (1 Ci = 3.7×10^{10} decays/s), has twice the number of nuclei as another sample of a different radioactive material B which has an activity of 20 mCi. The correct choices for half-lives of A and B would then be respectively : **[9 Jan. 2019 (Morning)]**

(A) 20 days and 5 days (B) 20 days and 10 days
(C) 5 days and 10 days (D) 10 days and 40 days

5. At a given instant, say $t = 0$, two radioactive substances A and B have equal activities. The ratio R_B/R_A of their activities after time t itself decays with time t as e^{-3t}. If the half-life of A is ln2, the half-life of B is : **[9 Jan. 2019 (Evening)]**

(A) $\frac{\ln 2}{2}$ (B) $2\ln 2$

(C) $\frac{\ln 2}{4}$ (D) $4\ln 2$

6. Using a nuclear counter the count rate of emitted particles from a radioactive source is measured. At $t = 0$ it was 1600 counts per second and $t = 8$ seconds it was 100 counts per second. The count rate observed, as counts per second, at $t = 6$ seconds is close to : **[10 Jan. 2019 (Morning)]**

(A) 150 (B) 360
(C) 200 (D) 400

7. In a radioactive decay chain, the initial nucleus is $^{232}_{90}Th$. At the end there are 6 α-particles and 4 β-particles which are emitted. If the end nucleus, If $^{A}_{Z}X$, A and Z are given by : **[12 Jan. 2019 (Evening)]**

(A) $A = 208; Z = 80$ (B) $A = 202; Z = 80$
(C) $A = 200; Z = 81$ (D) $A = 208; Z = 82$

8. Two radioactive materials A and B have decay constants 10λ and λ, respectively. It initially they have the same number of nuclei, then the ratio of the number of nuclei of A to that of B will be $1/e$ after a time : **[10 April 2019 (Morning)]**

(A) $\frac{11}{10\lambda}$ (B) $\frac{1}{9\lambda}$

(C) $\frac{1}{10\lambda}$ (D) $\frac{1}{11\lambda}$

9. Two radioactive substances A and B have decay constants 5λ and λ respectively. At $t = 0$, a sample has the same number of the two nuclei. The time taken for the ratio of the number of nuclei to become $\left(\frac{1}{e}\right)^2$ will be : **[10 April 2019 (Evening)]**

(A) $1/4\lambda$ (B) $1/\lambda$
(C) $1/2\lambda$ (D) $2/\lambda$

10. Half lives of two radioactive nuclei A and B are 10 minutes and 20 minutes, respectively. If, initially a sample has equal number of nuclei, then after 60 minutes, the ratio of decayed numbers of nuclei A and B will be : **[12 April 2019 (Evening)]**

(A) 9 : 8 (B) 1 : 8
(C) 8 : 1 (D) 3 : 8

11. The activity of a radioactive sample falls from 700 s^{-1} to 500 s^{-1} in 30 minutes. Its half life is close to :

[7 Jan. 2020 (Evening)]

(A) 66 min (B) 52 min
(C) 72 min (D) 62 min

12. In a radioactive material, fraction of active material remaining after time t is 9/16. The fraction that was remaining after $t/2$ is

[3 Sept. 2020 (Morning)]

(A) $\frac{3}{4}$ (B) $\frac{7}{8}$
(C) $\frac{4}{5}$ (D) $\frac{3}{5}$

13. Activities of three radioactive substances A, B and C are represented by the curves A, B and C, in the figure. Then their half-lives $T_{1/2}(A) : T_{1/2}(B) : T_{1/2}(C)$ are in the ratio :

[5 Sept. 2020 (Morning)]

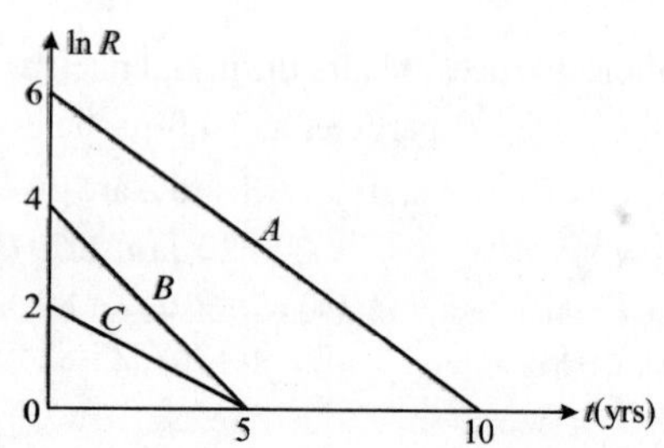

(A) 3 : 2 : 1 (B) 4 : 3 : 1
(C) 2 : 1 : 3 (D) 2 : 1 : 1

14. A radioactive nucleus decays by two different processes. The half life for the first process is 10 s and that for the second is 100 s. The effective half life of the nucleus is close to :

[5 Sept. 2020 (Evening)]

(A) 9 s (B) 55 s
(C) 6 s (D) 12 s

15. Two radioactive substances X and Y originally have N_1 and N_2 nuclei respectively. Half life of X is half of the half life of Y. After three lives of Y, number of nuclei of both are equal. The ratio $\frac{N_1}{N_2}$ will be equal to : **[25 Feb. 2021 (Morning)]**

(A) $\frac{8}{1}$ (B) $\frac{3}{1}$
(C) $\frac{1}{3}$ (D) $\frac{1}{8}$

16. A radioactive sample is undergoing α decay. At any time t_1, its activity is A and another time t_2, the activity is $A/5$. What is the average life time for the sample ?

[26 Feb. 2021 (Evening)]

(A) $\frac{t_2 - t_1}{\ln 5}$ (B) $\frac{\ln(t_2 + t_1)}{2}$
(C) $\frac{\ln 5}{t_2 - t_1}$ (D) $\frac{t_1 - t_2}{\ln 5}$

17. The half-life of Au^{198} is 2.7 days. The activity of 1.50 mg of Au^{198} if its atomic weight is 198 g mol^{-1} is, ($N_A = 6 \times 10^{23}$/mol)

[16 March 2021 (Evening)]

(A) 240 Ci (B) 357 Ci
(C) 535 Ci (D) 252 Ci

18. Calculate the time interval between 33% decay and 67% decay if half-life of a substance is 20 minutes :

[16 March 2021 (Evening)]

(A) 60 minutes (B) 20 minutes
(C) 40 minutes (D) 13 minutes

19. A radioactive sample disintegrates via two independent decay processes having half lives $T^{(1)}_{1/2}$ and $T^{(2)}_{1/2}$ respectively. The effective half-life $T_{1/2}$ of the nuclei is :

[18 March 2021 (Morning)]

(A) $T_{1/2} = \frac{T^{(1)}_{1/2} + T^{(2)}_{1/2}}{T^{(1)}_{1/2} - T^{(2)}_{1/2}}$ (B) $T_{1/2} = T^{(1)}_{1/2} + T^{(2)}_{1/2}$

(C) $T_{1/2} = \frac{T^{(1)}_{1/2} T^{(2)}_{1/2}}{T^{(1)}_{1/2} + T^{(2)}_{1/2}}$ (D) None of the above

20. A radioactive material decays by simultaneous emissions of two particles with half lives of 1400 years and 700 years respectively. What will be the time after which one third of the material remains ? (Take ln 3 = 1.1) : **[20 July 2021 (Morning)]**

(A) 700 years (B) 340 years
(C) 740 years (D) 1110 years

21. For a certain radioactive process the graph between lnR (logarithm of activity) and t(in second) is obtained as shown in the figure. Then the value of half-life for the unknown radioactive material is approximately :

[20 July 2021 (Evening)]

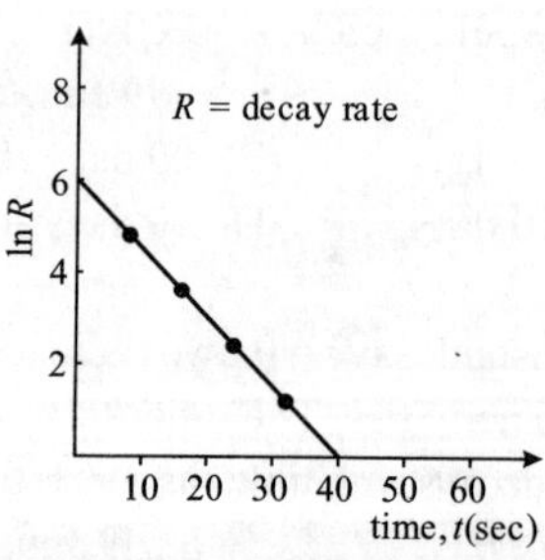

(A) 4.62 s (B) 2.62 s
(C) 9.15 s (D) 6.93 s

22. A radioactive substance decays to $\left(\frac{1}{16}\right)^{th}$ of its initial activity in 80 days. The half life of the radioactive substance expressed in days is ______. **[20 July 2021 (Evening)]**

23. Some nuclei of a radioactive material are undergoing radioactive decay. The time gap between the instances when a quarter of the nuclei have decayed and when half of the nuclei have decayed is given as: (where λ is the decay constant) **[25 July 2021 (Morning)]**

(A) 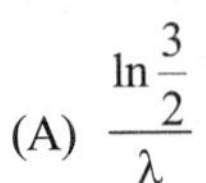 $\frac{\ln\frac{3}{2}}{\lambda}$ (B) $\frac{1}{2}\frac{\ln 2}{\lambda}$

(C) $\frac{\ln 2}{\lambda}$ (D) $\frac{2\ln 2}{\lambda}$

24. The half-life of ^{198}Au is 3 days. If atomic weight of ^{198}Au is 198 g/mol then the activity of 2 mg ^{198}Au is [in dps] **[25 July 2021 (Morning)]**

(A) 2.67×10^{12} (B) 6.06×10^{18}

(C) 32.36×10^{12} (D) 16.18×10^{12}

25. The nuclear activity of a radioactive element becomes $\left(\frac{1}{8}\right)^{th}$ of its initial value in 30 years. The half-life of radioactive element is ______ years : **[25 July 2021 (Evening)]**

26. If f' denotes the ratio of the number of nuclei decayed (N_d) to the number of nuclei at $t = 0(N_0)$ then for a collection of radioactive nuclei, the rate of change of f with respect to time is given as : [x is the radioactive decay constant] : **[27 July 2021 (Morning)]**

(A) $\lambda e^{-\lambda t}$ (B) $\lambda(1-e^{-\lambda t})$

(C) $(-\lambda e^{-\lambda t})$ (D) $-\lambda(1-e^{-\lambda t})$

27. A radioactive sample has an average life of 30 ms and is decaying. A capacitor of capacitance 200 μF is first charged and later connected with resistor 'R'. If the ratio of charge on capacitor to the activity of radioactive sample is fixed with respect to time then the value of 'R' should be ______Ω. **[27 July 2021 (Morning)]**

28. At time $t = 0$, material is composed of two radioactive atoms A and B, where $N_A(0) = 2N_B(0)$. The decay constant of both kind of radioactive atoms is λ. However, A disintegrates to B and B disintegrates to C. Which of the following figures represents the evolution of $N_B(t)/N_B(0)$ with respect to time t?

$\begin{bmatrix} N_A(0) = \text{No. of } A \text{ atoms at } t=0 \\ N_B(0) = \text{No. of } B \text{ atoms at } t=0 \end{bmatrix}$ **[26 Aug. 2021 (Evening)]**

(A) 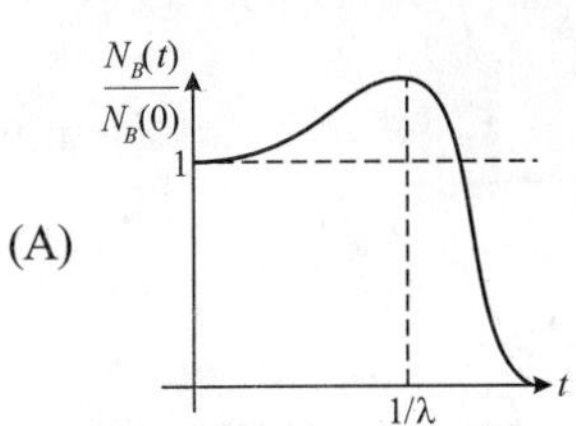

(B) 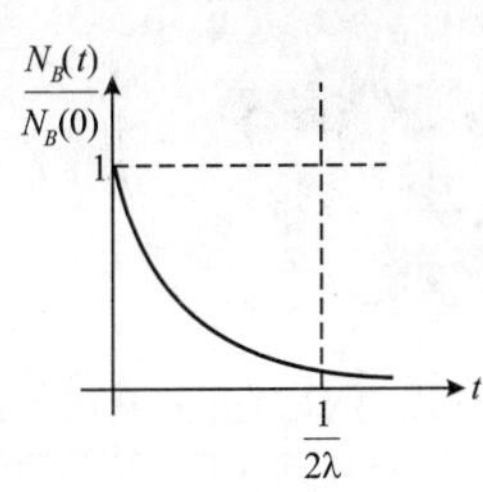

(C) 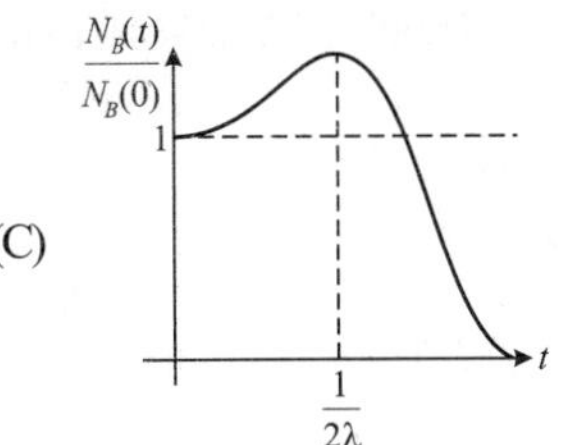

(D) 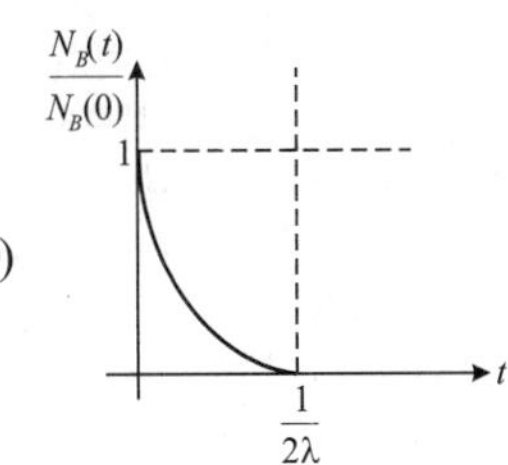

29. There are 10^{10} radioactive nuclei in a given radioactive element. Its half–life time is 1 minute. How many nuclei will remain after 30 seconds ? ($\sqrt{2} = 1.414$) : **[27 Aug. 2021 (Morning)]**

(A) 4×10^{10} (B) 7×10^{9}

(C) 10^{5} (D) 2×10^{10}

30. A sample of a radioactive nucleus A disintegrates to another radioactive nucleus B, which in turn disintegrates to some other stable nucleus C. Plot of a graph showing the variation of number of atoms of nucleus B vesus time is : (Assume that at $t = 0$, there are no B atoms in the sample) **[31 Aug. 2021 (Morning)]**

(A) 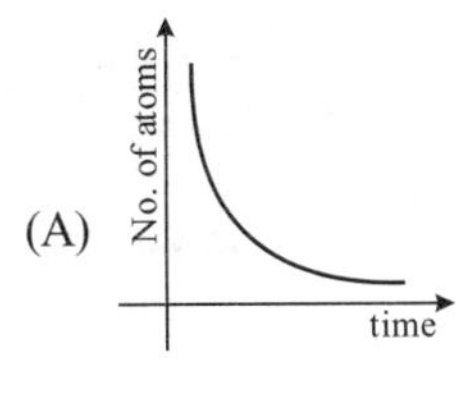

(B) 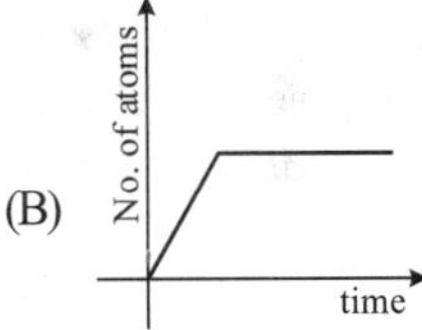

(C) 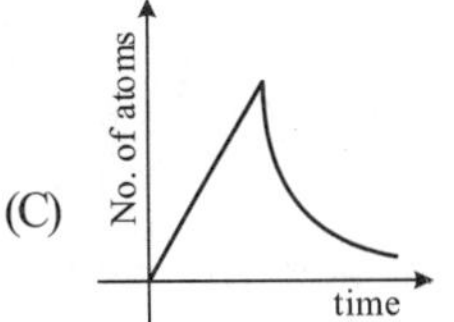

(D) 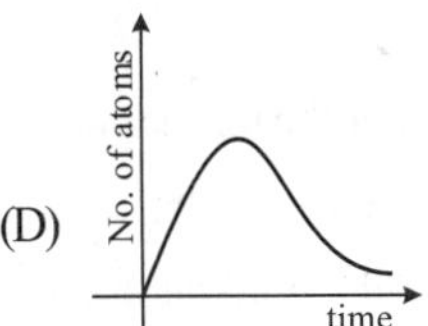

31. The half life period of a radioactive element x is same as the mean life time of another radioactive element y. Initially they have the same number of atoms. Then : **[01 Sep. 2021 (Evening)]**

(A) x and y have same decay rate initially and later on different decay rate

(B) x and y decay at the same rate always

(C) x – will decay faster than y

(D) y – will decay faster than x

* * * * *

CHAPTER

27 NUCLEI

27.2 NUCLEAR PHYSICS

1. Two deuterons undergo nuclear fusion to form a Helium nucleus. Energy released in this process is : (given binding energy per nucleon for deuteron = 1.1 MeV and for helium = 7.0 MeV): **[2017 (Online)]**

(A) 30.2 MeV (B) 32.4 MeV
(C) 23.6 MeV (D) 25.8 MeV

2. Imagine that a reactor converts all given mass into energy and that it operates at a power level of 10^9 watt. The mass of the fuel consumed per hour in the reactor will be : (velocity of light, c is 3×10^8 m/s) : **[2017 (Online)]**

(A) 0.96 gm (B) 0.8 gm
(C) 4×10^{-2} gm (D) 6.6×10^{-5} gm

3. Both the nucleus and the atom of some element are in their respectively first excited states. They get de-excited by emitting photons of wavelengths λ_N, λ_A respectively. The ratio $\frac{\lambda_N}{\lambda_A}$ is closet to : **[2018 (Online)]**

(A) 10^{-6} (B) 10
(C) 10^{-1} (D) 10^{-10}

4. An unstable heavy nucleus at rest breaks into two nuclei which move away with velocities in the ratio of 8 : 27. The ratio of the radii of the nuclei (assumed to be spherical) is : **[2018 (Online)]**

(A) 8 : 27 (B) 2 : 3
(C) 3 : 2 (D) 4 : 9

5. Consider the nuclear fission

$Ne^{20} \rightarrow 2He^4 + C^{12}$

Given that the binding energy/nucleon of Ne^{20}, He^4 and C^{12} are, respectively, 8.03 MeV, 7.07 MeV and 7.86 MeV, identify the correct statement : **[10 Jan. 2019 (Evening)]**

(A) 8.3 MeV energy will be released
(B) Energy of 12.4 MeV will be supplied
(C) Energy of 11.9 MeV has to be supplied
(D) Energy of 3.6 MeV will be released

6. The ratio of mass densities of nuclei of ^{40}Ca and ^{16}O is close to : **[8 April 2019 (Evening)]**

(A) 1 (B) 2
(C) 0.1 (D) 5

7. In a reactor, 2 kg of ${}_{92}U^{235}$ fuel is fully used up in 30 days. The energy released per fission is 200 MeV. Given that the Avogadro number, $N = 6.023 \times 10^{26}$ per kilo mole and 1 eV = 1.6×10^{-19} J. The power output of the reactor is close to : **[2 Sept. 2020 (Morning)]**

(A) 125 MW (B) 60 MW
(C) 35 MW (D) 54 MW

8. The radius R of a nucleus of mass number A can be estimated by the formula $R = (1.3 \times 10^{-15})A^{1/3}$ m. It follows that the mass density of a nucleus is of the order of : ($M_{prot.} \cong M_{neut.} \simeq 1.6 \times 10^{-27}$ kg) : **[3 Sept. 2020 (Evening)]**

(A) 10^{24} kg m^{-3} (B) 10^3 kg m^{-3}
(C) 10^{17} kg m^{-3} (D) 10^{10} kg m^{-3}

9. Find the binding energy per nucleon for ${}^{120}_{50}Sn$. Mass of proton $m_p = 1.00783$ U, mass of neutron $m_n = 1.00867$ U and mass of tin nucleus $m_{Sn} = 119.902199$ U. (Take 1 U = 931 MeV) : **[4 Sept. 2020 (Evening)]**

(A) 8.5 MeV (B) 7.5 MeV
(C) 8.0 MeV (D) 9.0 MeV

10. You are given that Mass of ${}^7_3Li = 7.0160\,u$, Mass of ${}^4_2He = 4.0026\,u$ and Mass of ${}^1_1H = 1.0079\,u$. When 20 g of ${}^7_3Li$ is converted into ${}^4_2He$ by proton capture, the energy liberated, (in kWh), is : [Mass of nucleon = 1 GeV/c^2] **[6 Sept. 2020 (Morning)]**

(A) 8×10^6 (B) 1.33×10^6
(C) 6.82×10^5 (D) 4.5×10^5

11. Given the masses of various atomic particles $m_p = 1.0072$ u, $m_n = 1.0087$ u, $m_e = 0.000548$ u, $m_{\bar{v}} = 0$, $m_d = 2.0141$ u, where p is proton, n is neutron, e is electron, $\bar{v}$ is antineutrino and d is deuteron. Which of the following process is allowed by momentum and energy conservation : **[6 Sept. 2020 (Evening)]**

(A) $n + p \rightarrow d + g$ (B) $e^+ + e^- \rightarrow g$
(C) $n + n \rightarrow$ deuterium atom (electron bound to the nucleus)
(D) $p \rightarrow n + e^+ + \bar{v}$

12. The decay of a proton to neutron is : **[18 March 2021 (Evening)]**

(A) not possible as proton mass is less than the neutron mass
(B) possible only inside the nucleus

(C) not possible but neutron to proton conversion is possible
(D) always possible as it is associated only with β^+ decay

13. A nucleus of mass M emits γ-rays photon of frequency 'ν'. The loss of internal energy by the nucleus is :
[Take 'c' as the speed of electromagnetic wave] :
[20 July 2021 (Morning)]

(A) 0
(B) $h\nu\left[1+\frac{h\nu}{2Mc^2}\right]$
(C) $h\nu$
(D) $h\nu\left[1-\frac{h\nu}{2Mc^2}\right]$

14. A nucleus with mass number 184 initially at rest emits an α-particle. If the Q value of the reaction is 5.5 MeV, calculate the kinetic energy of the α-particle : **[22 July 2021 (Evening)]**
(A) 5.5 MeV
(B) 5.0 MeV
(C) 5.38 MeV
(D) 0.12 MeV

15. From the given data, the amount of energy required to break the nucleus of aluminium $^{27}_{13}$Al is ______ $x \times 10^{-3}$ J.
Mass of neutron = 1.00866 u Mass of proton = 1.00726 u Mass of Aluminium nucleus = 27.18846 u (Assume 1 u corresponds to x J of energy) (Round off to the nearest integer)
[25 July 2021 (Evening)]

* * * * *

CHAPTER

28 SEMICONDUCTORS ELECTRONICS

28.1 SEMICONDUCTOR

1. What is the conductivity of a semiconductor sample having electron concentration of 5×10^{18} m^{-3}, hole concentration of 5×10^{19} m^{-3}, electron mobility of 2.0 m^2 V^{-1} s^{-1} and hole mobility of 0.01 m^2 V^{-1} s^{-1}? **[2017 (Online)]**
(Take charge of electron as 1.6×10^{-19} C) :
(A) 1.68 $(\Omega - m)^{-1}$ (B) 1.83 $(\Omega - m)^{-1}$
(C) 0.59 $(\Omega - m)^{-1}$ (D) 1.20 $(\Omega - m)^{-1}$

2. Mobility of electrons in a semiconductor is defined as the ratio of their drift velocity to the applied electric field. If, for an *n*-type semiconductor, the density of electrons is 10^{19} m^{-3} and their mobility is 1.6 m^2/(V.s) then the resistivity of the semiconductor (since it is an *n*-type semiconductor contribution of holes is ignored) is close to : **[9 Jan. 2019 (Morning)]**
(A) 2 Ωm (B) 0.4 Ωm
(C) 4 Ωm (D) 0.2 Ωm

3. For extrinsic semiconductors; when doping level is increased; **[25 Feb. 2021 (Evening)]**
(A) Fermi-level of *p*-type semiconductors will go downward and Fermi-level of *n*-type semiconductor will go upward
(B) Fermi-level of *p* and *n*-type semiconductors will not be affected
(C) Fermi-level of *p*-type semiconductor will go upward and Fermi-level of *n*-type semiconductors will go downward
(D) Fermi-level of both *p*-type and *n*-type semiconductors will go upward for $T > T_F K$ and downward for $T < T_F K$, where T_F is Fermi temperature

4. In a semiconductor, the number density of intrinsic charge carries at 27°C is 1.5×10^{16}/m^3. If the semiconductor is doped with impurity atom, the hole density increases to 4.5×10^{22}/m^3. The electron density in the doped semiconductor is ______ 10^9/m^3 : **[25 July 2021 (Evening)]**

5. A transistor is connected in common emitter circuit configuration, the collector supply voltage is 10 V and the voltage drop across a resistor of 1000 Ω in the collector circuit is 0.6 V. If the current gain factor (β) is 24, then the base current is ______ μA. (Round off to the nearest integer). **[27 July 2021 (Morning)]**

6. Statement-I : By doping semiconductor with pentavalent material, the density increases.
Statement-II : The *n*-type semiconductor has net negative charge.
In the light of the above statements, choose the most appropriate answer from the options given below : **[26 Aug. 2021 (Morning)]**
(A) Both Statement I and Statement II are false
(B) Both Statement I and Statement II are true
(C) Statement I is false but Statement II is true
(D) Statement I is true but Statement II is false

* * * * *

CHAPTER

28 SEMICONDUCTORS ELECTRONICS

28.2 PN JUNCTION AND DIODES

1. The V-I characteristic of a diode is shown in the figure. The ratio of forward to reverse bias resistance is : **[2017 (Online)]**

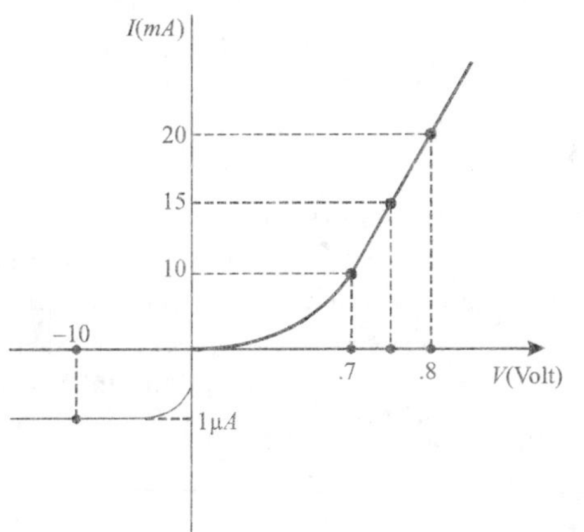

(A) 10 (B) 10^{-6}
(C) 10^{6} (D) 100

2. The reading of the ammeter for a silicon diode in the given circuit is : **[2018]**

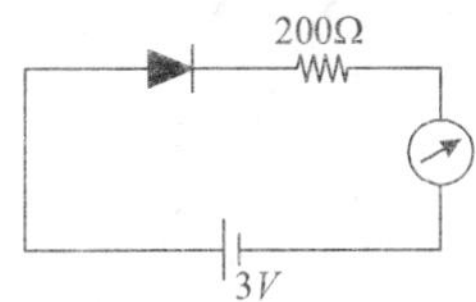

(A) 0 (B) 15 mA
(C) 11.5mA (D) 13.5 mA

3. In the given circuit, the current through zener diode is : **[2018 (Online)]**

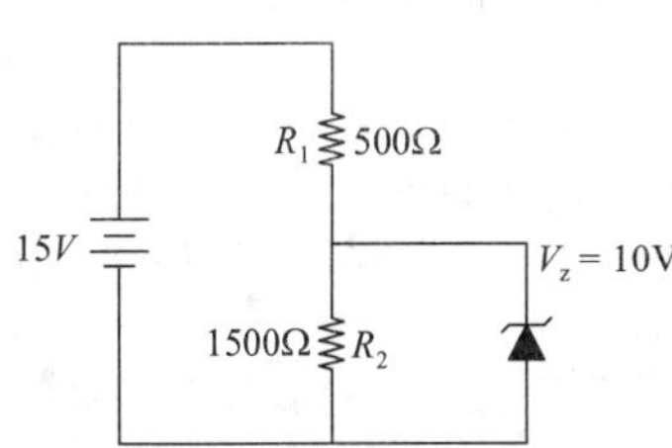

(A) 2.5 mA (B) 3.3 mA
(C) 5.5mA (D) 6.7 mA

4. *Ge* and *Si* diodes start conducting at 0.3 V and 0.7 V respectively. In the following figure if *Ge* diode connection are reversed, the value of V_0 changes by : (assume that the *Ge* diode has large breakdown voltage) : **[9 Jan. 2019 (Evening)]**

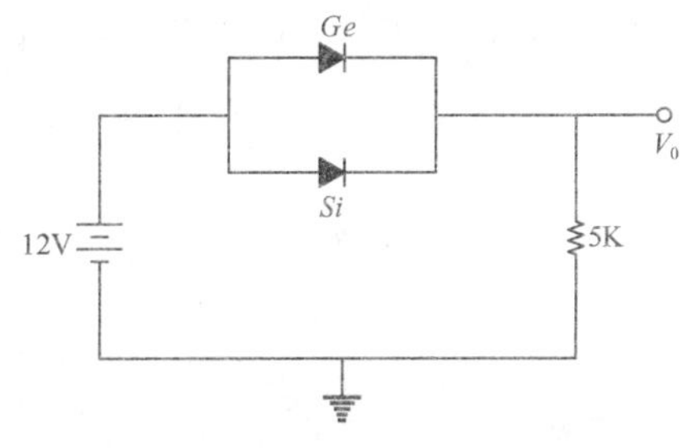

(A) 0.6 V (B) 0.8 V
(C) 0.4 V (D) 0.2 V

5. For the circuit shown below, the current through the Zener diode is : **[10 Jan. 2019 (Evening)]**

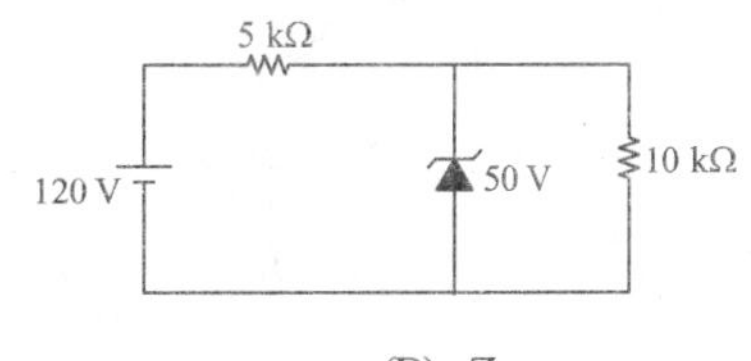

(A) 5 mA (B) Zero
(C) 14 mA (D) 9 mA

6. In the given circuit the current through Zener Diode is close to : **[11 Jan. 2019 (Morning)]**

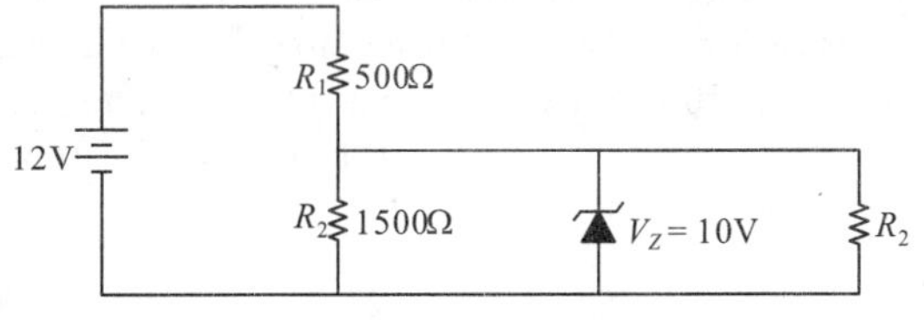

(A) 6.0 mA (B) 4.0 mA
(C) 6.7 mA (D) 0.0 mA

7. The circuit shown below contains two ideal diodes, each with a forward resistance of 50 Ω. If the battery voltage is 6 V, the current through the 100 Ω resistance (in Amperes) is : **[11 Jan. 2019 (Evening)]**

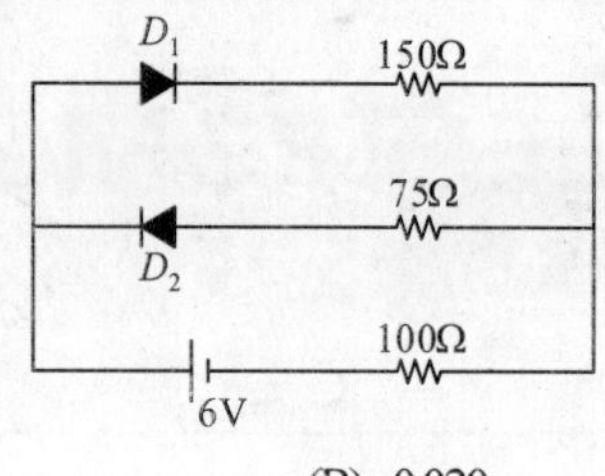

(A) 0.027 (B) 0.020
(C) 0.030 (D) 0.036

8. The reverse breakdown voltage of a Zener diode is 5.6 V in the given circuit. The current I_Z through the Zener is :
[8 April 2019 (Morning)]

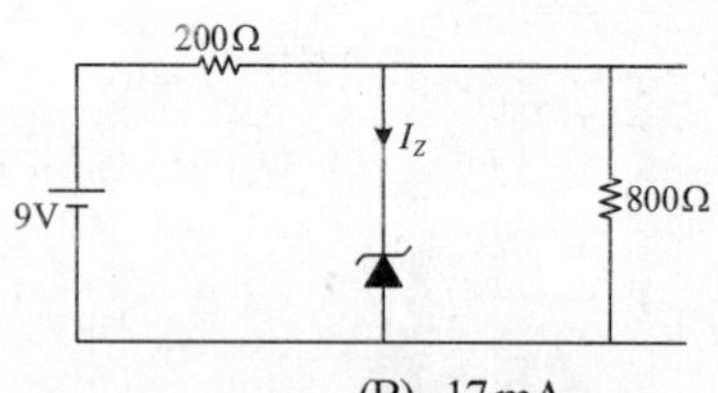

(A) 7 mA (B) 17 mA
(C) 10 mA (D) 15 mA

9. The figure represents a voltage regulator circuit using a Zener diode. The breakdown voltage of the Zener diode is 6 V and the load resistance is $R_L = 4\ k\Omega$. The series resistance of the circuit is $R_i = 1\ k\Omega$. If the battery voltage V_B varies from 8 V to 16 V, what are the minimum and maximum values of the current through Zener diode ? **[10 April 2019 (Evening)]**

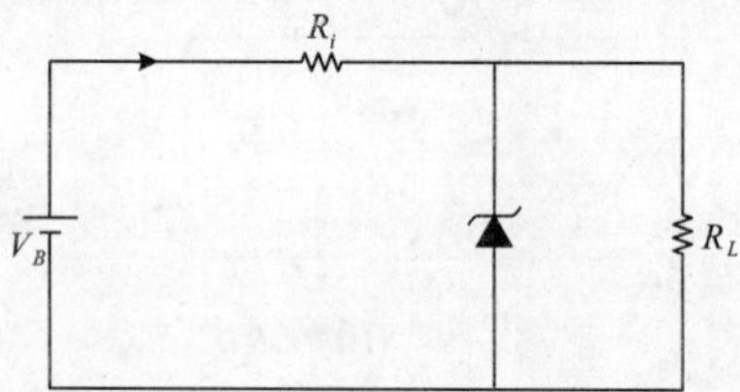

(A) 0.5 mA ; 6 mA (B) 0.5 mA ; 8.5 mA
(C) 1.5 mA ; 8.5 mA (D) 1 mA ; 8.5 mA

10. Figure shown a *DC* voltage regulator circuit, with a Zener diode of breakdown voltage = 6V. If the unregulated input voltage varies between 10V to 16V, then what is the maximum Zener current ? **[12 April 2019 (Evening)]**

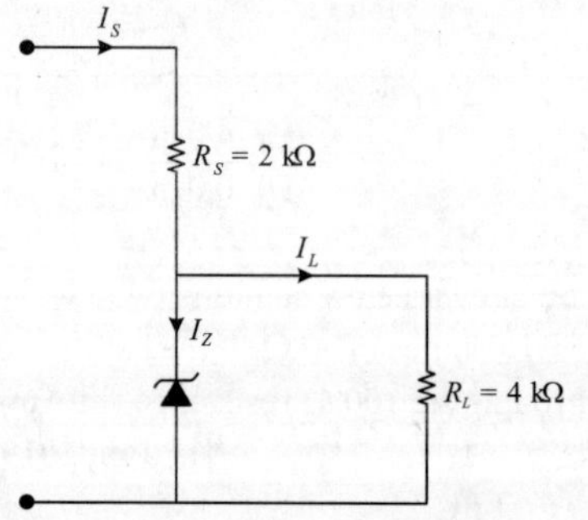

(A) 2.5 mA (B) 3.5 mA
(C) 7.5 mA (D) 1.5 mA

11. In the figure, potential difference between *A* and *B* is :
[7 Jan. 2020 (Evening)]

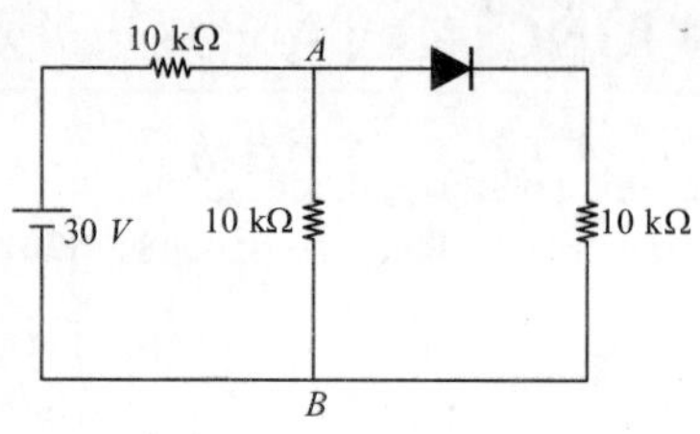

(A) 5V (B) 10V
(C) Zero (D) 15V

12. Both the diodes used in the circuit shown are assumed to be ideal and have negligible resistance when these are forward biased. Built in potential in each diode is 0.7 V. For the input voltages shown in the figure, the voltage (in Volts) at point *A* is _____ . **[9 Jan. 2020 (Morning)]**

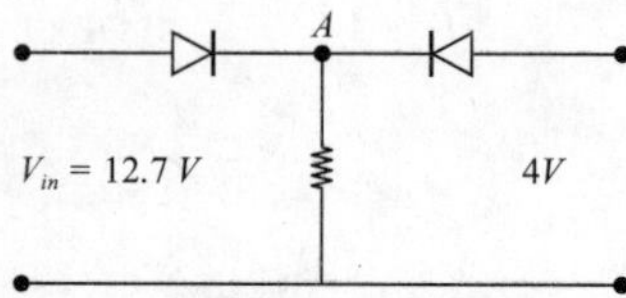

13. The current *i* in the network is : **[9 Jan. 2020 (Evening)]**

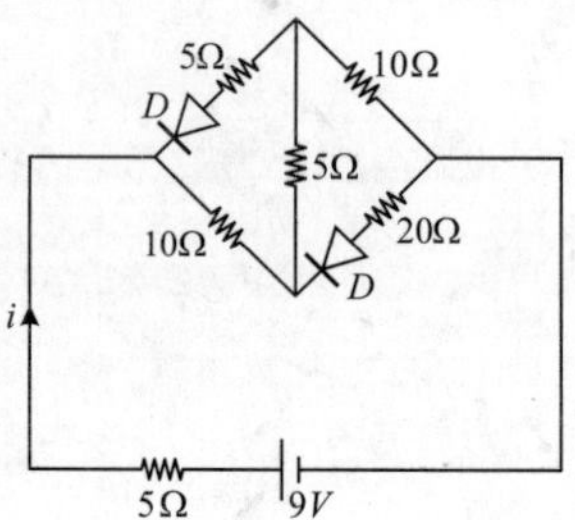

(A) 0 A (B) 0.6 A
(C) 0.3 A (D) 0.2 A

14. Two identical capacitors *A* and *B*, charged to the same potential 5*V* are connected in two different circuits as shown below at time $t = 0$. If the charge on capacitors *A* and *B* at time $t = CR$ is Q_A and Q_B respectively, then (Here *e* is the base of natural logarithm) **[9 Jan. 2020 (Evening)]**

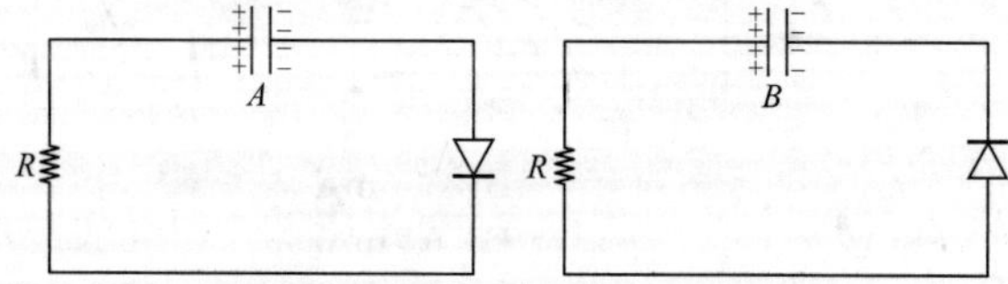

(A) $Q_A = VC, Q_B = \frac{VC}{e}$ (B) $Q_A = \frac{CV}{2}, Q_B = \frac{VC}{e}$

(C) $Q_A = VC, Q_B = CV$ (D) $Q_A = \frac{VC}{e}, Q_B = \frac{CV}{2}$

15. The circuit shown below is working as a 8*V* dc regulated voltage source. When 12*V* is used as input, the power dissipated (in *mW*) in each diode is; (considering both zener diodes are identical) ______. **[9 Jan. 2020 (Evening)]**

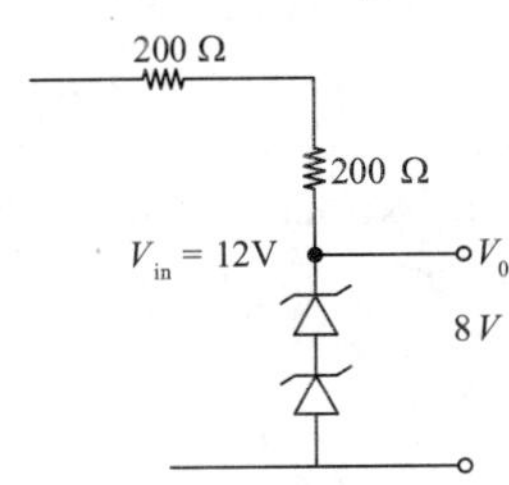

16. When a diode is forward biased, it has a voltage drop of 0.5 V. The safe limit of current through the diode is 10 mA. If a battery of emf 1.5 V is used in the circuit, the value of minimum resistance to be connected in series with the diode so that the current does not exceed the safe limit is : **[3 Sept. 2020 (Morning)]**

(A) 100 Ω (B) 50 Ω
(C) 300 Ω (D) 200 Ω

17. If a semiconductor photodiode can detect a photon with a maximum wavelength of 400 nm, then its band gap energy is : **[3 Sept. 2020 (Evening)]**

Planck's constant $h = 6.63 \times 10^{-34}$ Js
Speed of light $c = 3 \times 10^8$ m/s
(A) 2.0 eV (B) 1.5 eV
(C) 3.1 eV (D) 1.1 eV

18. Take the breakdown voltage of the zener diode used in the given circuit as 6 V. For the input voltage shown in figure below, the time variation of the output voltage is : (Graphs drawn are schematic and not to scale) : **[4 Sept. 2020 (Morning)]**

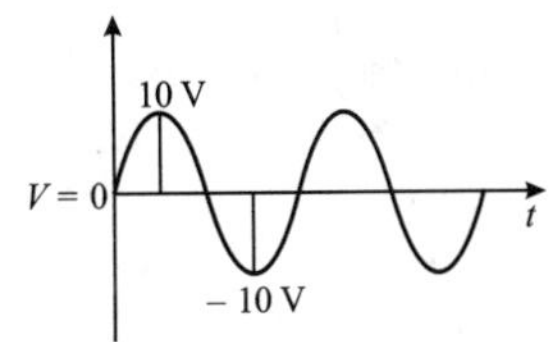

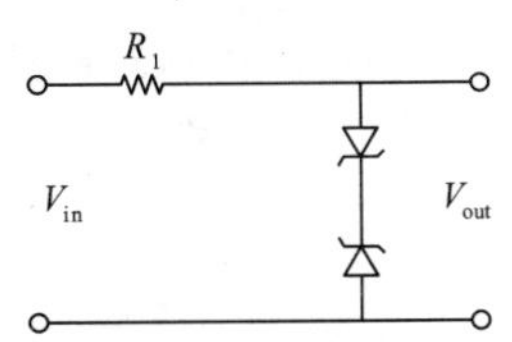

(A)
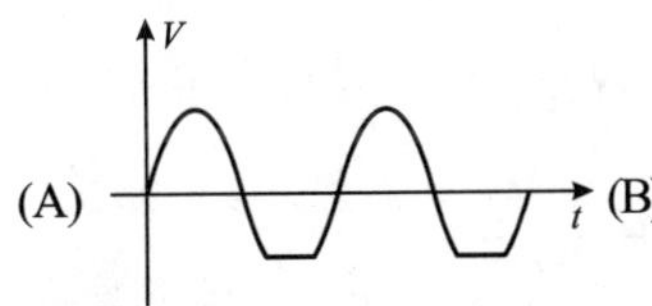

(B)
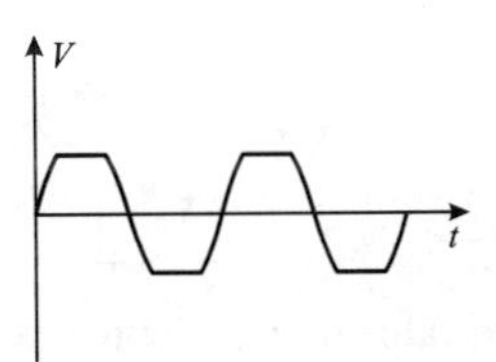

(C) (D)
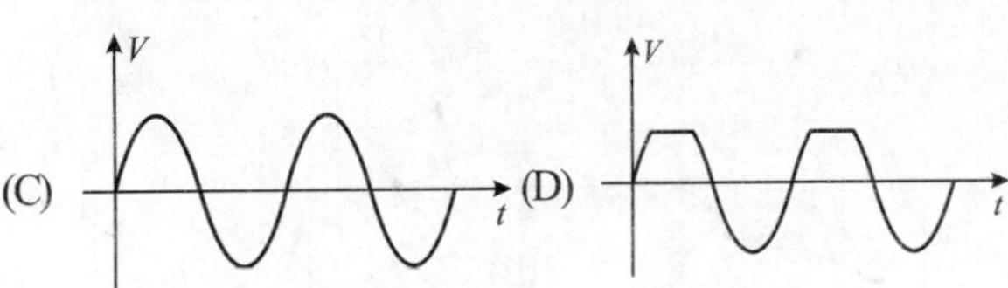

19. With increasing biasing voltage of a photodiode, the photocurrent magnitude : **[5 Sept. 2020 (Morning)]**

(A) Increases initially and saturates finally
(B) Increases initially and after attaining certain value, it decreases
(C) Increases linearly
(D) Remains constant

20. Two Zener diodes (*A* and *B*) having breakdown voltages of 6V and 4V respectively, are connected as shown in the circuit below. The output voltage V_O variation with input voltage linearly increasing with time, is given by : ($V_{input} = 0\ V$ at $t = 0$) (figures are qualitative) : **[5 Sept. 2020 (Evening)]**

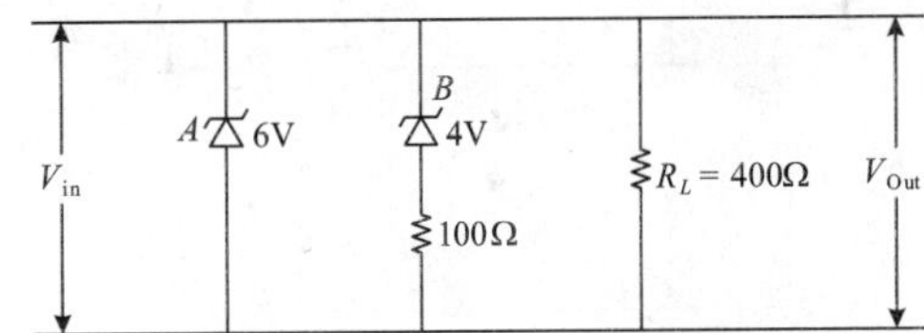

(A) (B)
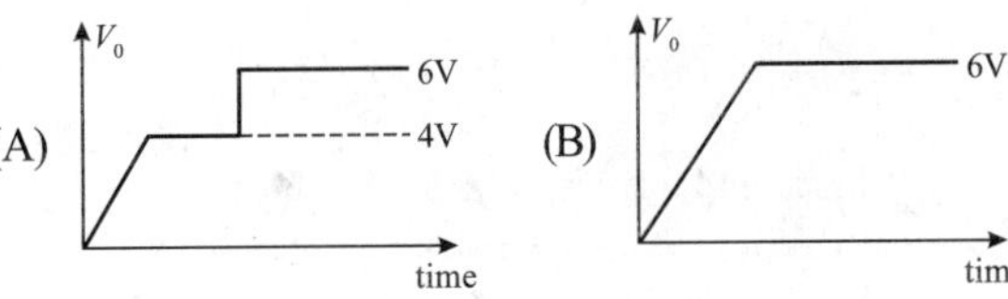

(C) (D)
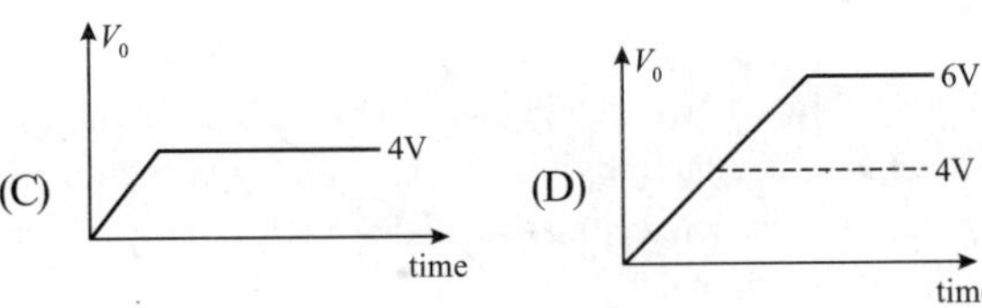

21. For the given input voltage waveform $V_{in}(t)$, the output voltage waveform $V_D(t)$, across the capacitor is correctly depicted by : **[6 Sept. 2020 (Morning)]**

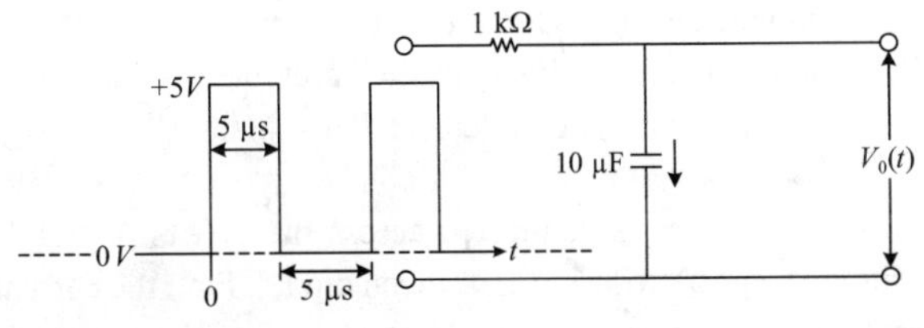

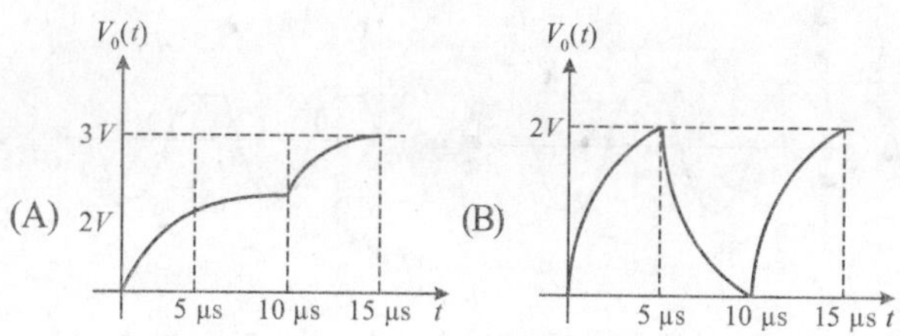

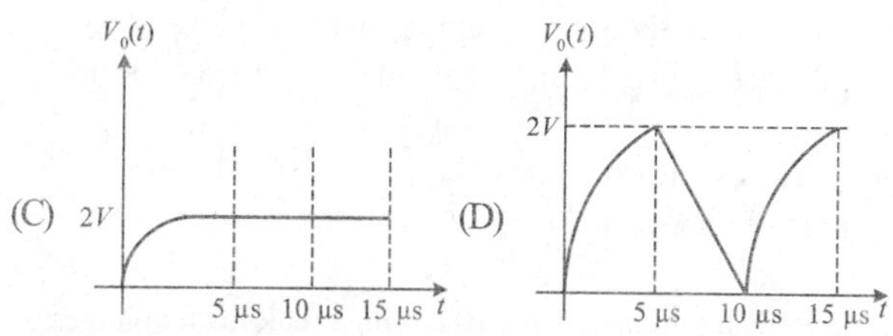

22. In connection with the circuit drawn below, the value of current flowing through 2 kΩ resistor is _____ $\times 10^{-4}$ A. **[24 Feb. 2021 (Morning)]**

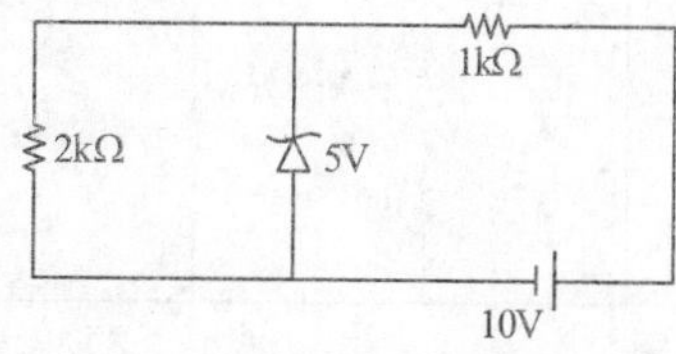

23. Zener breakdown occurs in a *p-n* junction having *p* and *n* both : **[24 Feb. 2021 (Evening)]**

(A) Heavily doped and have wide depletion layer
(B) Heavily doped and have narrow depletion layer
(C) Lightly doped and have wide depletion layer
(D) Lightly doped and have narrow depletion layer

24. Given below are two statements : **[24 Feb. 2021 (Evening)]**

Statement-I : PN-Junction diodes can be used to function as transistor, simply by connecting two diodes, back to back, which acts as the base terminal.

Statement-II : In the study of transistor, the amplification factor β indicates ratio of the collector current to the base current

In the light of the above statements, choose the correct answer from the options given below :

(A) Statement-I is true but Statement-II is false
(B) Statement-I is false but Statement-II is true
(C) Both Statement-I and Statement-II are true
(D) Both Statement-I and Statement-II are false

25. A 5 V battery is connected across the points *X* and *Y*. Assume D_1 and D_2 to be normal silicon diodes. Find the current supplied by the battery if the +*ve* terminal of the battery is connected to point *X* : **[25 Feb. 2021 (Morning)]**

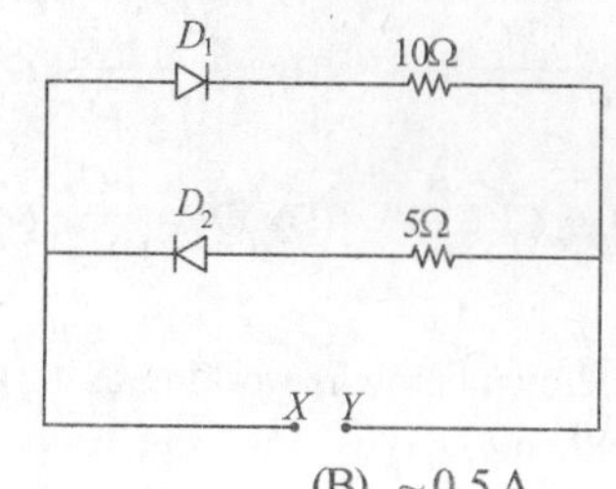

(A) ~0.86 A (B) ~0.5 A
(C) ~0.43 A (D) ~1.5 A

26. LED is constructed from *Ga-As-P* semiconducting material. The energy gap of this LED is 1.9 eV. Calculate the wavelength of light emitted and its colour. **[26 Feb. 2021 (Morning)]**
[$h = 6.63 \times 10^{-34}$ Js and $c = 3 \times 10^8$ ms^{-1}] :

(A) 1046 nm and blue colour (B) 1046 nm and red colour
(C) 654 nm and red colour (D) 654 nm and orange colour

27. The circuit contains two diodes each with a forward resistance of 50 Ω and with infinite reverse resistance. If the battery voltage is 6 V, the current through the 120 Ω resistance is _____ *mA*. **[26 Feb. 2021 (Morning)]**

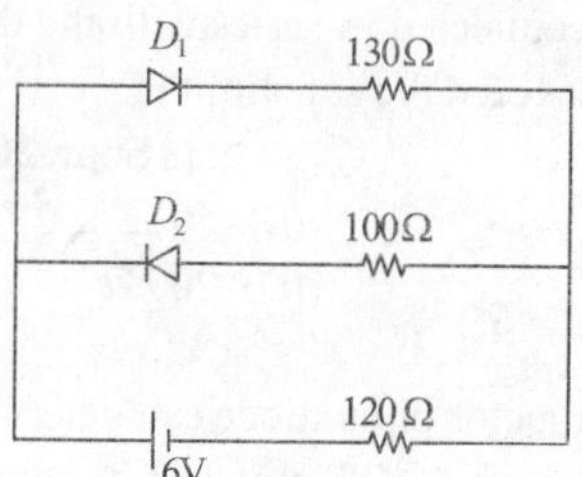

28. The zener diode has a V_z = 30 V. The current passing through the diode for the following circuit is _______ *mA*. **[26 Feb. 2021 (Evening)]**

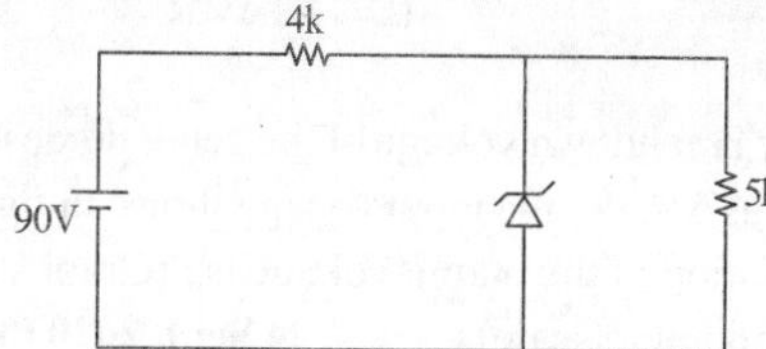

29. The value of power dissipated across the zener diode (V_z = 15 V) connected in the circuit as shown in the figure is $x \times 10^{-1}$ watt.

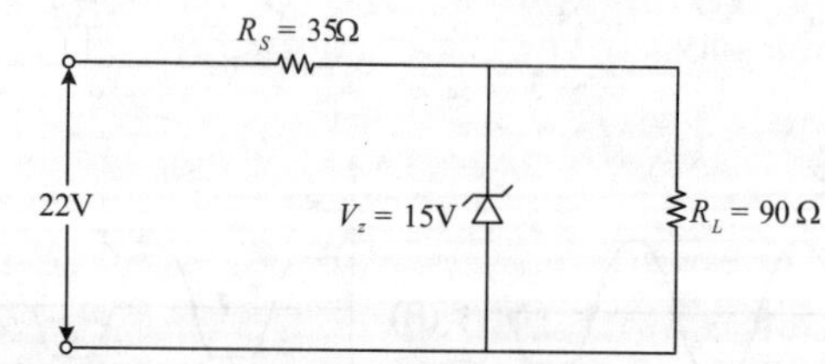

The value of *x*, to the nearest integer, is _______. **[16 March 2021 (Morning)]**

30. For the circuit shown below, calculate the value of I_z :

[20 July 2021 (Morning)]

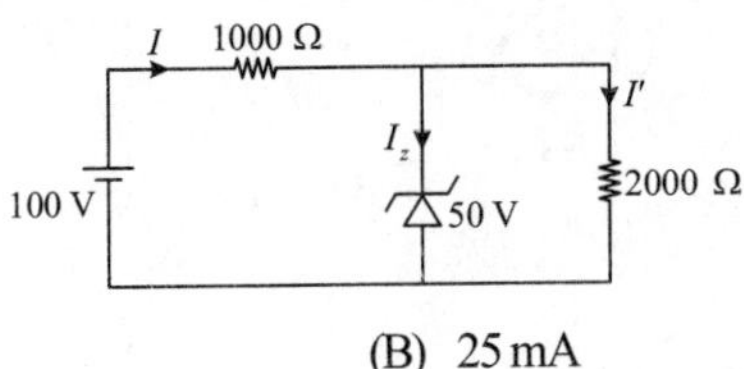

(A) 0.15 A
(B) 25 mA
(C) 0.05 A
(D) 0.1 A

31. For the forward biased diode characteristics shown in the figure, the dynamic resistance at $I_D = 3$ mÅ will be ______ Ω.

[20 July 2021 (Evening)]

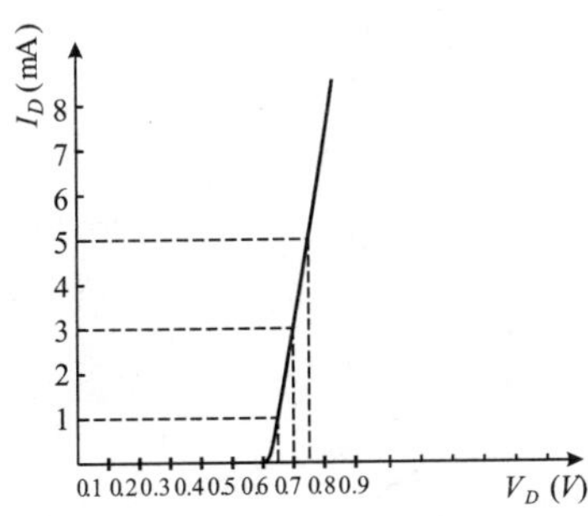

32. A zener diode having zener voltage 8 V and power dissipation rating of 0.5 W is connected across a potential divider arranged with maximum potential drop across zener diode is as shown in the diagram. The value of protective resistance R_P is ______ Ω.

[20 July 2021 (Evening)]

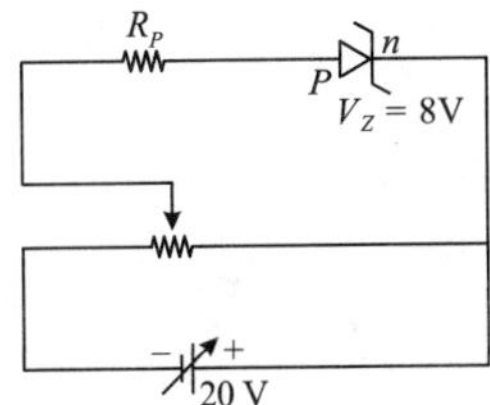

33. In a given circuit diagram, a 5 V zener diode along with a series resistance is connected across a 50 V power supply. The minimum value of the resistance required, if the maximum zener current is 90 mA will be ______ Ω. **[22 July 2021 (Evening)]**

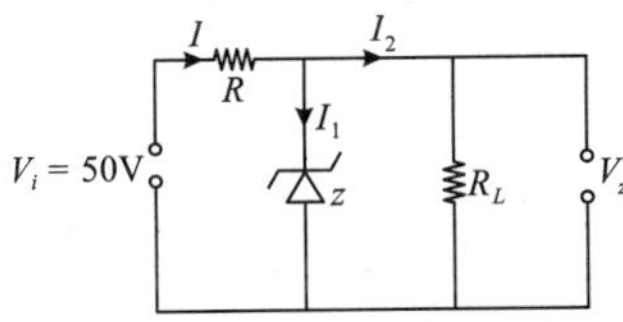

34. For the given circuit, the power across zener diode is ______ *mW*.

[26 Aug. 2021 (Evening)]

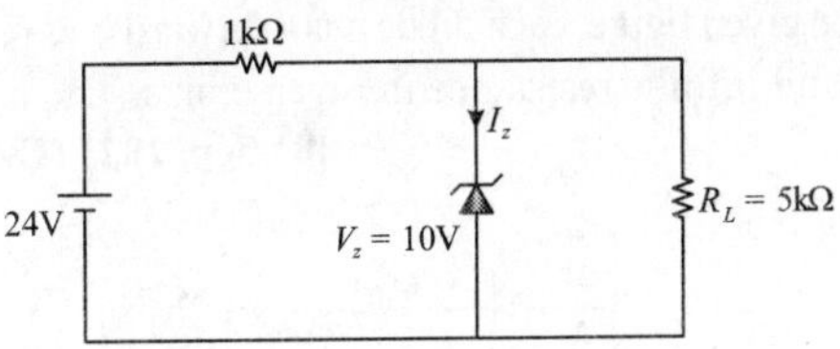

35. A zener diode of power rating 2 W is to be used as a voltage regulator. If the zener diode has a breakdown of 10 V and it has to regulate voltage fluctuated between 6 V and 14 V, the value of R_S for safe operation should be ______ Ω.

[27 Aug. 2021 (Evening)]

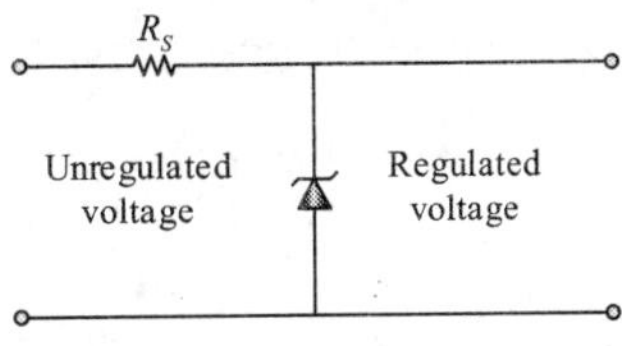

36. Choose the correct waveform that can represent the voltage across R of the following circuit, assuming the diode is ideal one :

[31 Aug. 2021 (Morning)]

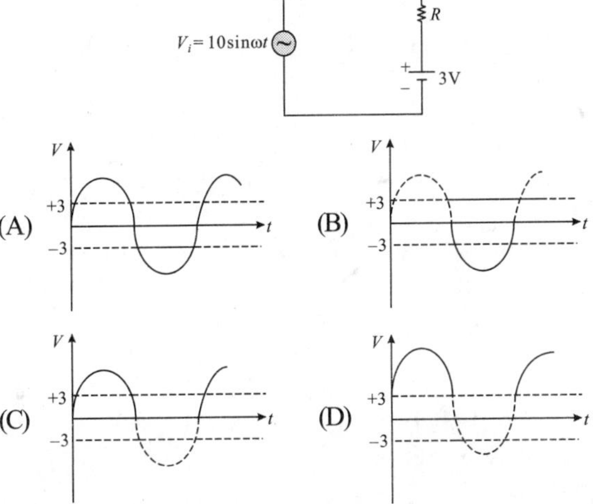

37. Statement-I : To get a steady dc output from the pulsating voltage received from a full wave rectifier we can connect a capacitor across the output parallel to the load R_L.

Statement-II : To get a steady dc output from the pulsating voltage received from a full wave rectifier we can connect an inductor in series with R_L.

In the light of the above statements, choose the most appropriate answer from the options given below :

[31 Aug. 2021 (Evening)]

(A) Both Statement-I and Statement-II are false
(B) Statement-I is false but Statement-II is true
(C) Both Statement-I and Statement-II are true
(D) Statement-I is true but Statement-II is false

38. In the given figure, each diode has a forward bias resistance of 30 Ω and infinite resistance in reverse bias. The current I_1 will be : **[01 Sep. 2021 (Evening)]**

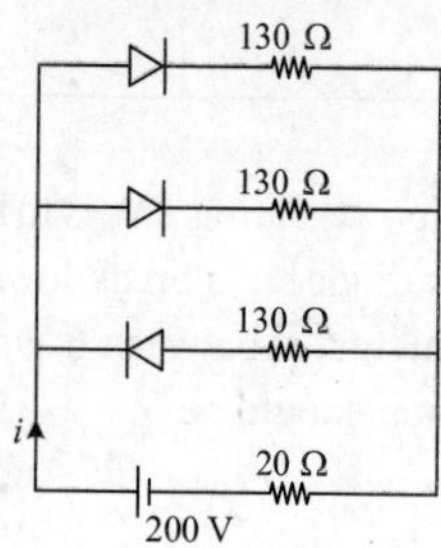

(A) 2 A
(B) 3.75 A
(C) 2.73 A
(D) 2.35 A

* * * * *

CHAPTER

28 SEMICONDUCTORS ELECTRONICS

28.3 TRANSISTORS

1. The current gain of a common emitter amplifier is 69. If the emitter current is 7.0 mA, collector current is : **[2017 (Online)]**

(A) 9.6 mA (B) 6.9 mA
(C) 0.69 mA (D) 69 mA

2. In a common emitter amplifier circuit using an *n-p-n* transistor, the phase difference between the input and the output voltages will be : **[2017 (Online)]**

(A) 135° (B) 180°
(C) 45° (D) 90°

3. In a common emitter configuration with suitable bias, it is given than R_L is the load resistance and R_{BE} is small signal dynamic resistance (input side). Then, voltage gain, current gain and power gain are given, respectively, by :
(β is current gain, I_B, I_C, I_E are respectively base, collector and emitter currents :) **[2018 (Online)]**

(A) $\beta\frac{R_L}{R_{BE}}, \frac{\Delta I_E}{\Delta I_B}, \beta^2\frac{R_L}{R_{BE}}$ (B) $\beta^2\frac{R_L}{R_{BE}}, \frac{\Delta I_C}{\Delta I_B}, \beta\frac{R_L}{R_{BE}}$

(C) $\beta^2\frac{R_L}{R_{BE}}, \frac{\Delta I_C}{\Delta I_E}, \beta^2\frac{R_L}{R_{BE}}$ (D) $\beta\frac{R_L}{R_{BE}}, \frac{\Delta I_C}{\Delta I_B}, \beta^2\frac{R_L}{R_{BE}}$

4. In the figure, given that V_{BB} supply can vary from 0 to 5.0 V, V_{CC} = 5 V, β_{dc} = 200, R_B = 100 kΩ, R_C = 1 kΩ and V_{BE} = 1.0 V, The minimum base current and the input voltage at which the transistor will go to saturation, will be, respectively : **[12 Jan. 2019 (Evening)]**

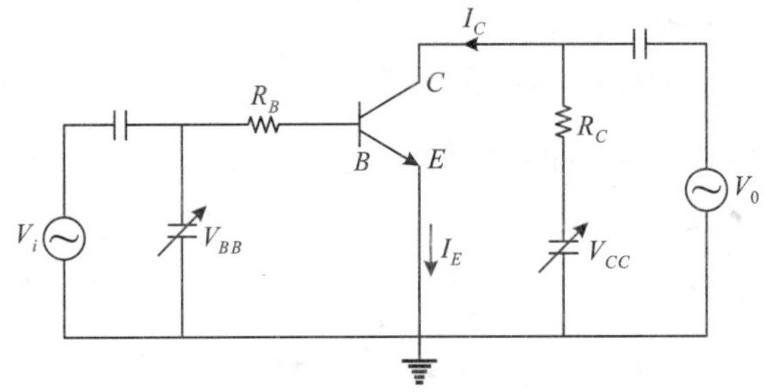

(A) 20 μA and 3.5 V (B) 25 μA and 3.5 V
(C) 25 μA and 2.5 V (D) 20 μA and 2.8 V

5. A common emitter amplifier circuit, built using an npn transistor, is shown in the figure. Its *dc* current gain is 250, R_C = 1 kΩ and V_{CC} = 10 V. What is the minimum base current for V_{CE} to reach saturation ? **[8 April 2019 (Evening)]**

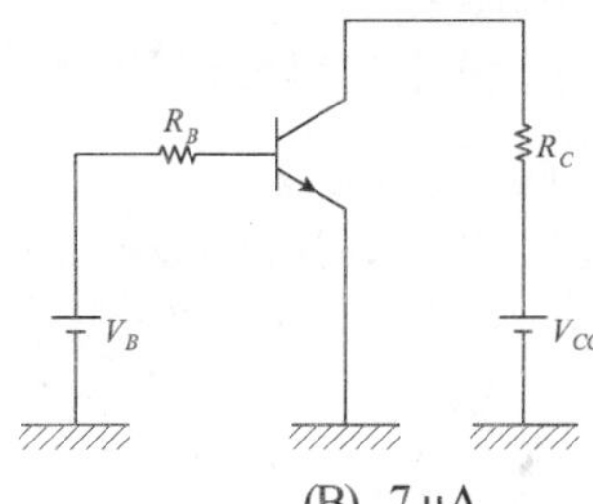

(A) 100 μA (B) 7 μA
(C) 40 μA (D) 10 μA

6. An *NPN* transistor is used in common emitter configuration as an amplifier with 1 kΩ load resistance. Signal voltage of 10 mV is applied across the base-emitter. This produces a 3 mA change in the collector current and 15μ*A* change in the base current of the amplifier. The input resistance and voltage gain are : **[9 April 2019 (Morning)]**

(A) 0.33 kΩ, 1.5 (B) 0.67 kΩ, 200
(C) 0.33 kΩ, 300 (D) 0.67 kΩ, 300

7. An *npn* transistor operates as a common emitter amplifier, with a power gain of 60 dB. The input circuit resistance is 100 Ω and the output load resistance is 10 kΩ. The common emitter current gain β is : **[10 April 2019 (Morning)]**

(A) 60 (B) 10^4
(C) 6×10^2 (D) 10^2

8. The transfer characteristic curve of a transistor, having input and output resistance 100 Ω and 100 kΩ respectively, is shown in the figure. The Voltage and Power gain, are respectively : **[12 April 2019 (Morning)]**

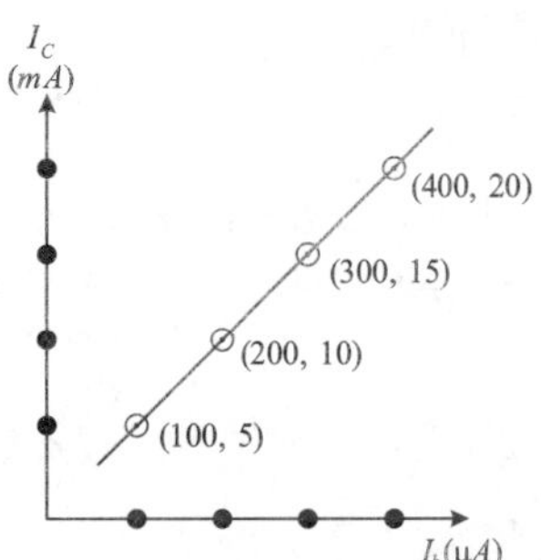

(A) $5 \times 10^4, 5 \times 10^5$ (B) $5 \times 10^4, 5 \times 10^6$
(C) $5 \times 10^4, 2.5 \times 10^6$ (D) $2.5 \times 10^4, 2.5 \times 10^6$

9. Boolean relation at the output stage-Y for the following circuit is : **[8 Jan. 2020 (Morning)]**

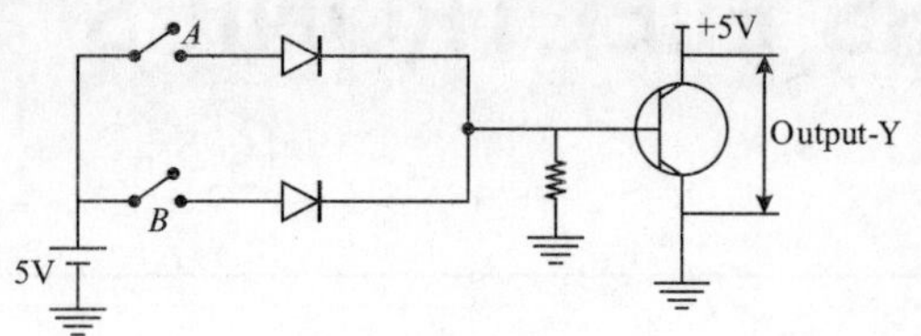

(A) $A+B$ (B) $\overline{A}+\overline{B}$

(C) $\overline{A}\cdot\overline{B}$ (D) $A\cdot B$

10. The output characteristics of a transistor is shown in the figure. When VCE is 10 V and $1C = 4.0\ \mu A$, then value of back is : **[6 Sept. 2020 (Evening)]**

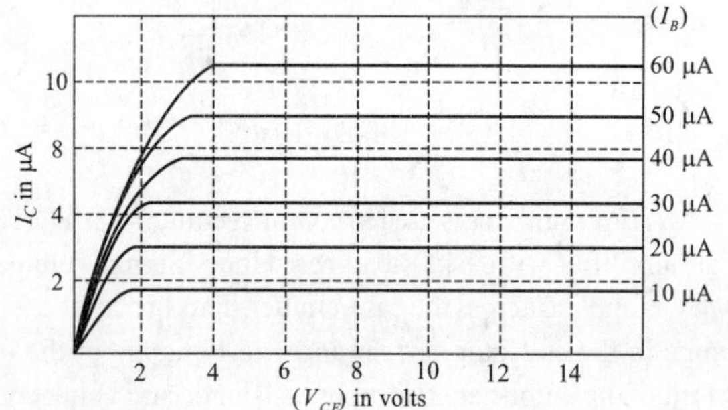

11. If an emitter current is changed by 4 mA, the collector current changes by 3.5 mA. the value of β will be : **[24 Feb. 2021 (Morning)]**

(A) 7 (B) 3.5

(C) 0.875 (D) 0.5

12. The same size images are formed by a convex lens when the object is placed at 20 cm or at 10 cm from the lens. The focal length of convex lens is ______ cm. **[25 Feb. 2021 (Morning)]**

13. An npn transistor operates as a common emitter amplifier with a power gain of 10^6. The input circuit resistance is 100 Ω and the output load resistance is 10 KΩ. The common emitter current gain 'β' will be ______. (Round off to the Nearest Integer) **[18 March 2021 (Morning)]**

14. The correct relation between α (ratio of collector current to emitter current) and β (ratio of collector current to base current) of a transistor is : **[18 March**

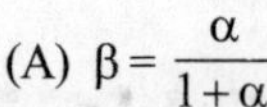

(A) $\beta = \dfrac{\alpha}{1+\alpha}$ (B) $\alpha = \dfrac{\beta}{1-\alpha}$

(C) $\beta = \dfrac{1}{1-\alpha}$ (D) $\alpha = \dfrac{\beta}{1+\beta}$

15. The typical output characteristics curve for a transistor working in the common-emitter configuration is shown in the figure.

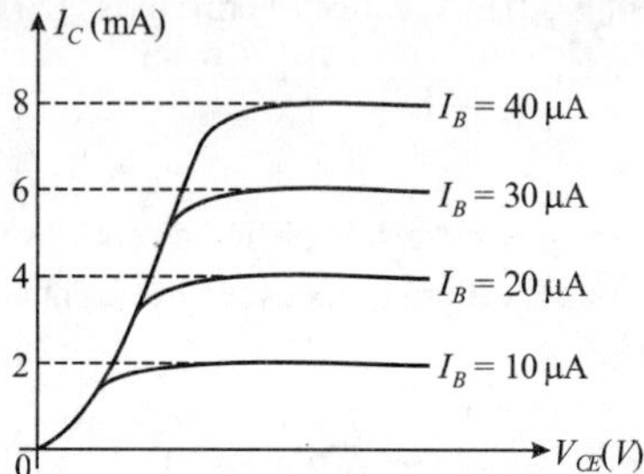

Calculate the estimated current gain from the above figure. **[18 March 2021 (Evening)]**

16. For a transistor in CE mode to be used as an amplifier, it must be operated in : **[27 Aug. 2021 (Morning)]**

(A) Cut-off region only (B) The active region only

(C) Saturation region only (D) Both cut-off and Saturation

17. A circuit is arranged as shown in figure. The output voltage V_0 is equal to ______ V. **[27 Aug. 2021 (Morning)]**

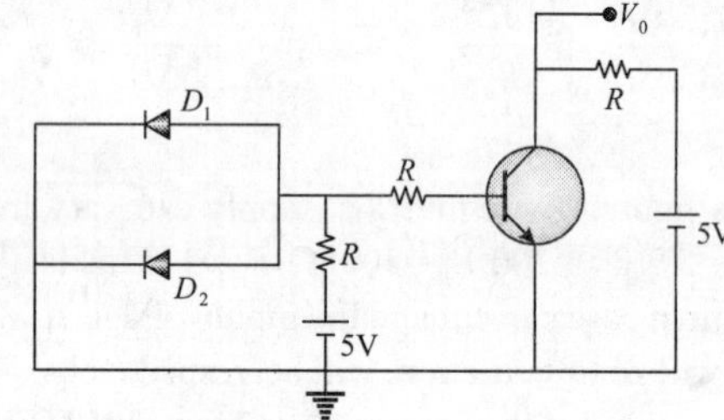

18. For a transistor α and β are given as $\alpha = \dfrac{I_C}{I_E}$ and $\beta = \dfrac{I_C}{I_B}$. Then the correct relation between α and β will be : **[27 Aug. 2021 (Evening)]**

(A) $\alpha = \dfrac{1-\beta}{\beta}$ (B) $\alpha\beta = 1$

(C) $\beta = \dfrac{\alpha}{1-\alpha}$ (D) $\alpha = \dfrac{\beta}{1-\beta}$

* * * * *

CHAPTER 28 SEMICONDUCTORS ELECTRONICS

28.4 LOGIC GATES

1. Truth table for the given circuit will be : **[2018 (Online)]**

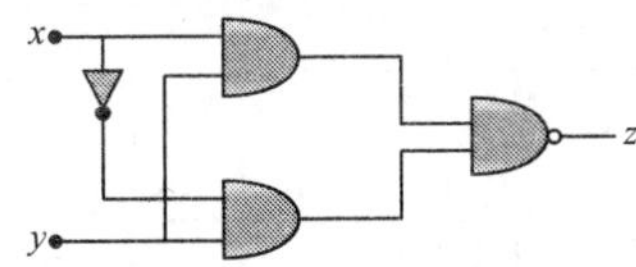

(A)

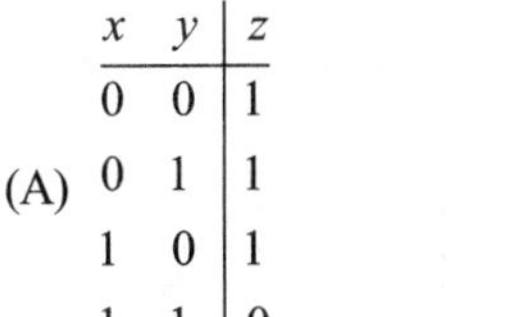

x	y	z
0	0	1
0	1	1
1	0	1
1	1	0

(B)

x	y	z
0	0	0
0	1	0
1	0	0
1	1	1

(C)

x	y	z
0	0	1
0	1	1
1	0	1
1	1	1

(D)

x	y	z
0	0	0
0	1	1
1	0	1
1	1	1

2. To get output '1' at R, for the given logic gate circuit the input values must be : **[10 Jan. 2019 (Morning)]**

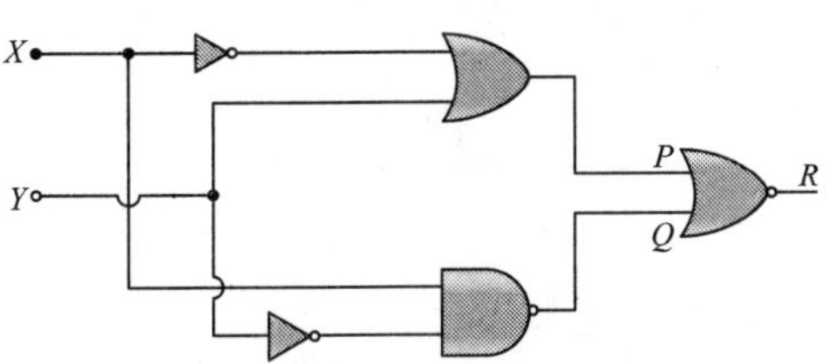

(A) $X=0, Y=1$ (B) $X=1, Y=1$
(C) $X=0, Y=0$ (D) $X=1, Y=0$

3. The output of the given logic circuit is : **[12 Jan. 2019 (Morning)]**

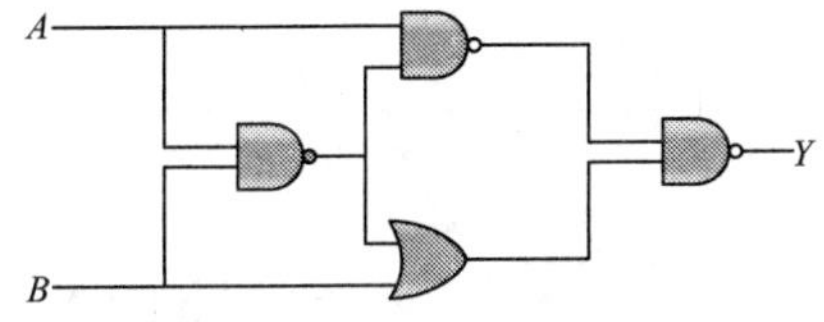

(A) $\overline{A}B$ (B) $A\overline{B}$

(C) $AB+\overline{AB}$ (D) $A\overline{B}+\overline{A}B$

4. The logic gate equivalent to the given logic circuit is : **[9 April 2019 (Evening)]**

(A) OR (B) AND
(C) NOR (D) NAND

5. The truth table for the circuit given in the figure is: **[12 April 2019 (Morning)]**

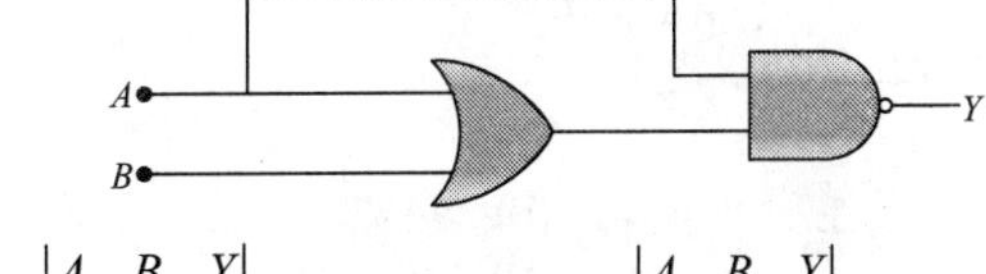

(A)

A	B	Y
0	0	1
0	1	1
1	0	0
1	1	0

(B)

A	B	Y
0	0	0
0	1	0
1	0	1
1	1	1

(C)

A	B	Y
0	0	1
0	1	0
1	0	0
1	1	0

(D)

A	B	Y
0	0	1
0	1	1
1	0	1
1	1	1

6. Which of the following gives a reversible operation ? **[7 Jan. 2020 (Morning)]**

(A) 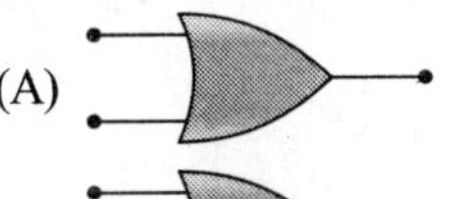 (B) 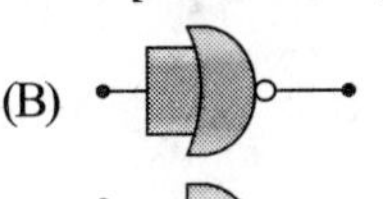

(C) 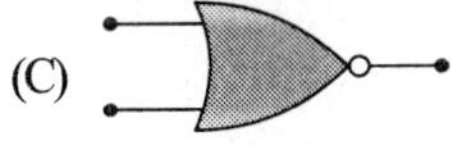 (D)

7. In the given circuit, value of Y is : **[8 Jan. 2020 (Evening)]**

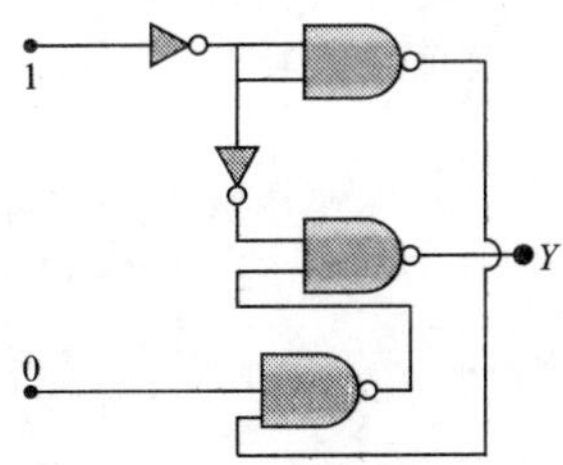

(A) Will not execute (B) 0
(C) Toggles between 0 and 1 (D) 1

8. In the following digital circuit, what will be the output at 'Z', when the input (A, B) are (1, 0), (0, 0), (1, 1), (0, 1) :

[2 Sept. 2020 (Evening)]

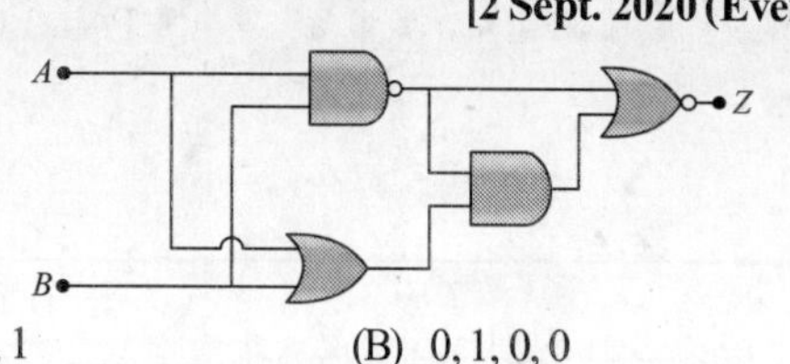

(A) 1, 0, 1, 1 (B) 0, 1, 0, 0
(C) 0, 0, 1, 0 (D) 1, 1, 0, 1

9. Identify the operation performed by the circuit given below :

[4 Sept. 2020 (Evening)]

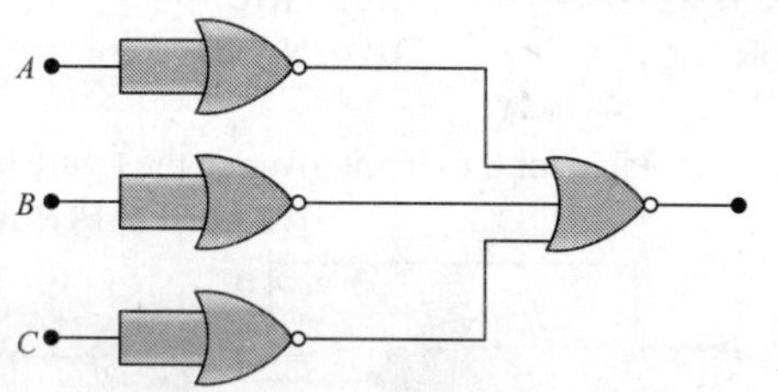

(A) AND (B) NAND
(C) OR (D) NOT

10. Identify the correct output signal Y in the given combination of gates (as shown) for the given inputs A and B :

[6 Sept. 2020 (Morning)]

(B)

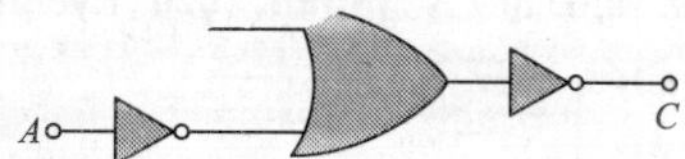

The logic circuit shown above is equivalent to :

[24 Feb. 2021 (Evening)]

(A) 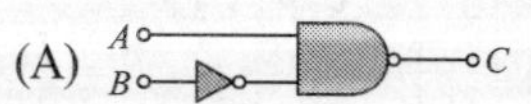 (B) 

(C) 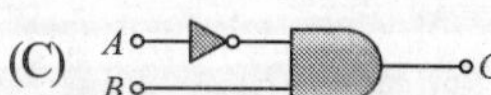 (D) 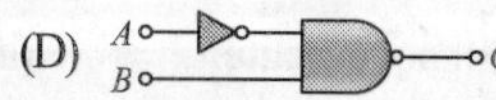

12. The truth table for the following logic circuit is :

[25 Feb. 2021 (Evening)]

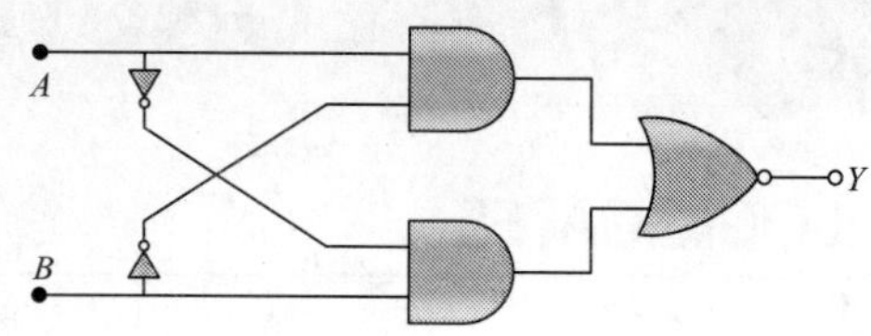

(A)

A	B	Y
0	0	1
0	1	0
1	0	1
1	1	0

(B)

A	B	Y
0	0	0
0	1	1
1	0	0
1	1	1

(C)

A	B	Y
0	0	1
0	1	0
1	0	0
1	1	1

(D)

A	B	Y
0	0	0
0	1	1
1	0	1
1	1	0

13. Draw the output signal Y in the given combination of gates.

[26 Feb. 2021 (Evening)]

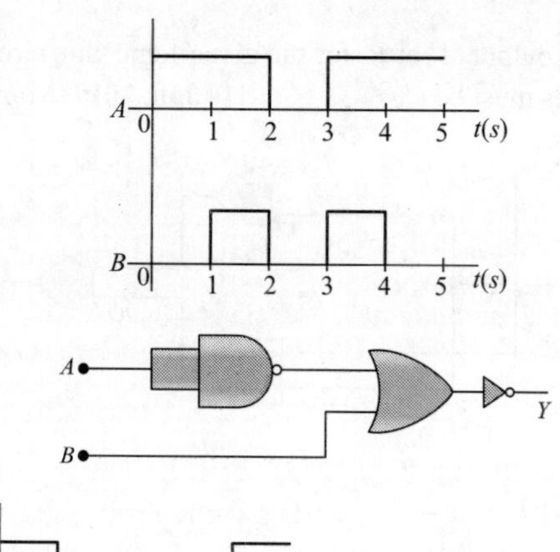

(A)

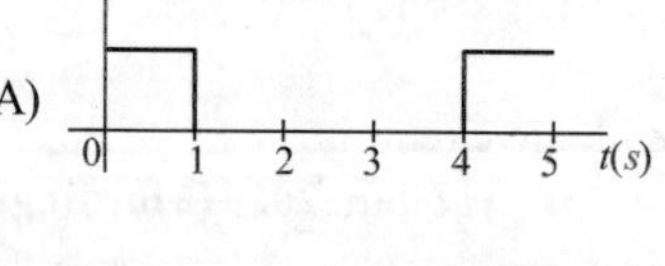

(B)

(C)

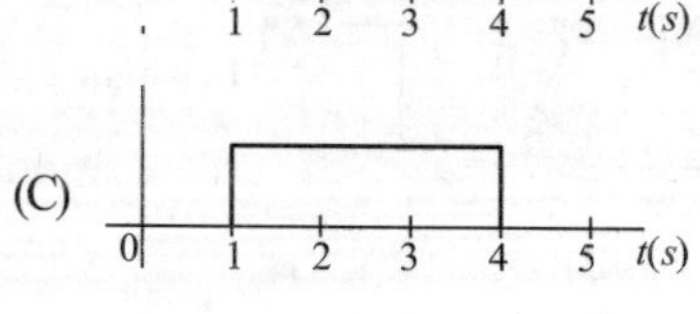

(D)

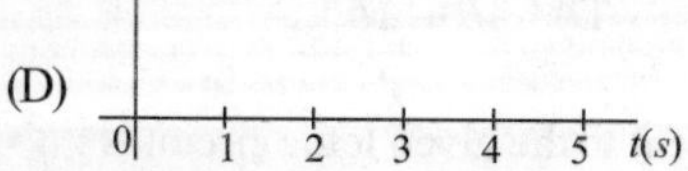

14. In the logic circuit shown in the figure, if input A and B are 0 to 1 respectively, the output at Y would be 'x'. The value of x is ______. **[16 March 2021 (Morning)]**

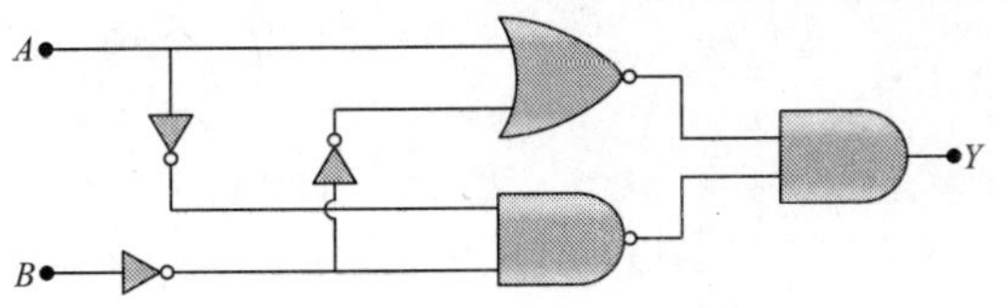

15. The following logic gate is equivalent to : **[16 March 2021 (Evening)]**

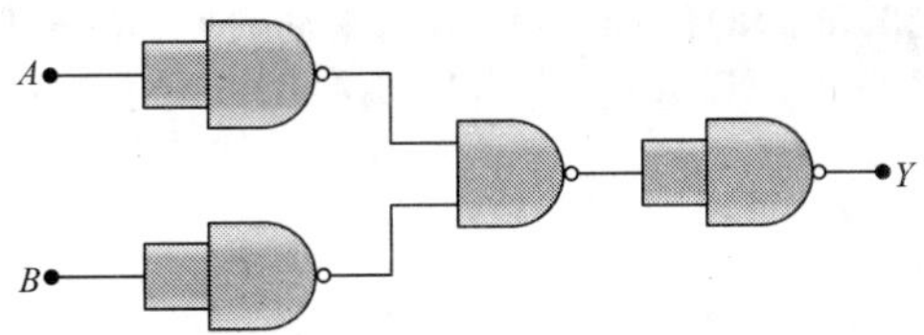

(A) NOR Gate (B) OR Gate
(C) AND Gate (D) NAND Gate

16. The output of the given combination gates represents : **[17 March 2021 (Morning)]**

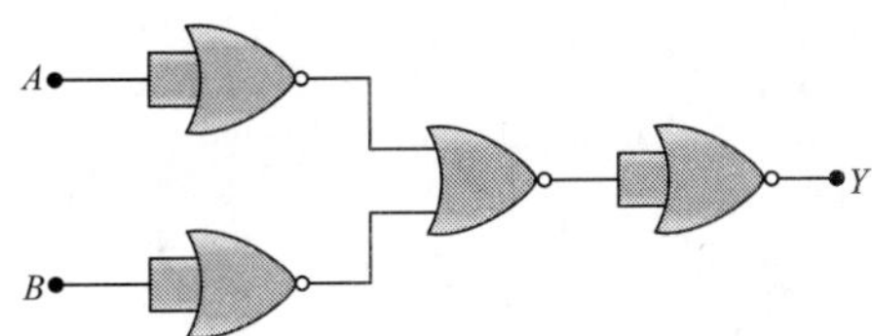

(A) XOR Gate (B) NAND Gate
(C) AND Gate (D) NOR Gate

17. Which one of the following will be the output of the given circuit ? **[17 March 2021 (Evening)]**

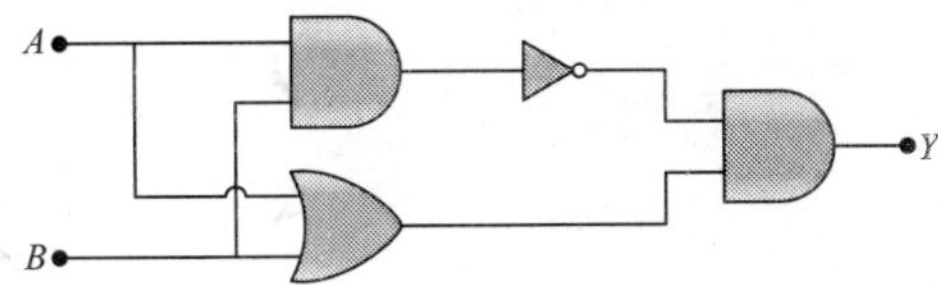

(A) NOR Gate (B) NAND Gate
(C) AND Gate (D) XOR Gate

18. Identify the logic operation carried out : **[25 July 2021 (Morning)]**

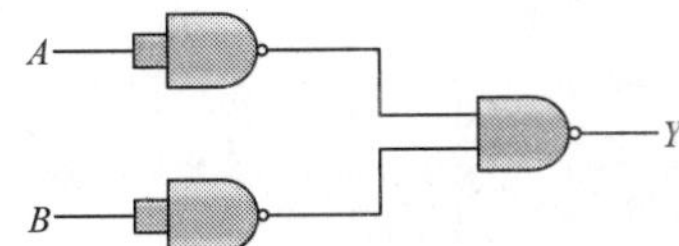

(A) *OR* (B) *NOR*
(C) *NAND* (D) *AND*

19. Find the truth table for the function Y for inputs A and B represented in the following figure : **[27 July 2021 (Evening)]**

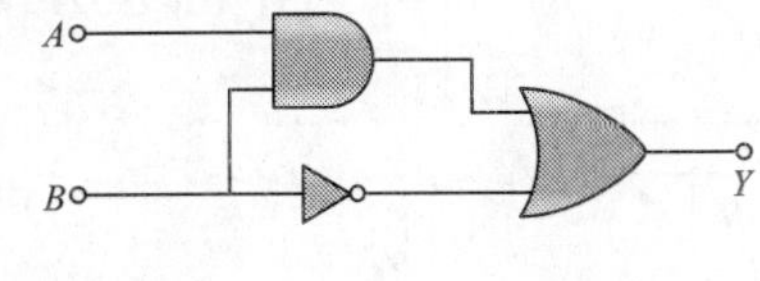

(A)

A	B	Y
0	0	0
0	1	1
1	0	0
1	1	0

(B)

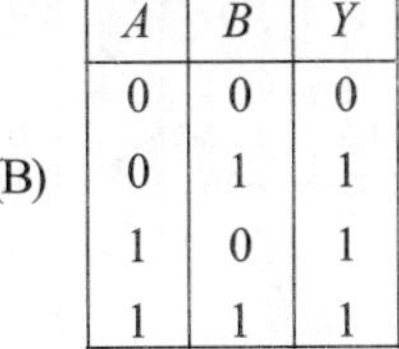

A	B	Y
0	0	0
0	1	1
1	0	1
1	1	1

(C)

A	B	Y
0	0	1
0	1	0
1	0	1
1	1	1

(D)

A	B	Y
0	0	0
0	1	0
1	0	0
1	1	1

20. Identify the logic operation carried out by the given circuit : **[26 Aug. 2021 (Morning)]**

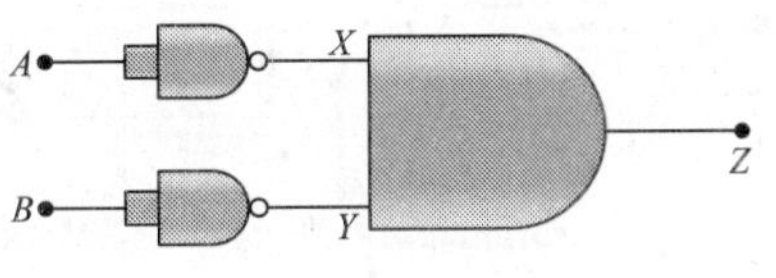

(A) *NOR* (B) *OR*
(C) *NAND* (D) *AND*

21. Four NOR gates are connected as shown in figure. The truth table for the given figure is : **[26 Aug. 2021 (Evening)]**

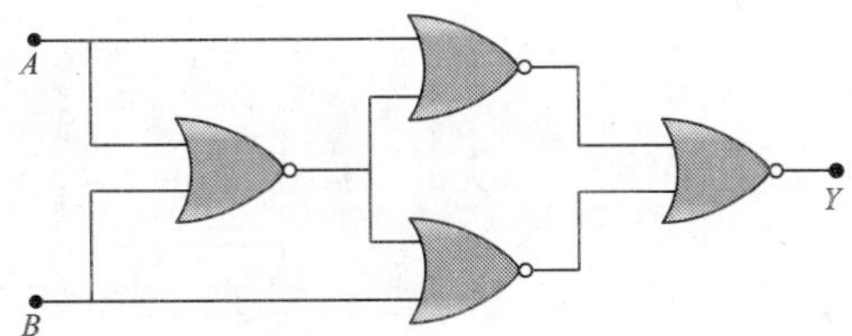

(A)

A	B	Y
0	0	1
0	1	0
1	0	0
1	1	1

(B)

A	B	Y
0	0	1
0	1	0
1	0	1
1	1	0

(C)

A	B	Y
0	0	0
0	1	1
1	0	0
1	1	1

(D)

A	B	Y
0	0	0
0	1	1
1	0	1
1	1	0

22. In the following logic circuit the sequence of the inputs A, B are (0, 0), (0, 1), (1, 0) and (1, 1). The output Y for this sequence will be : **[31 Aug. 2021 (Morning)]**

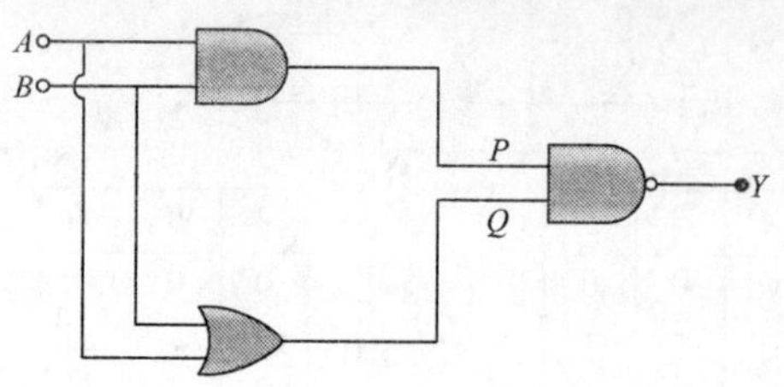

(A) 0, 0, 1, 1
(B) 1, 0, 1, 0
(C) 0, 1, 0, 1
(D) 1, 1, 1, 0

23. If V_A and V_B are the input voltages (either 5 V or 0 V) and V_0 is the output voltage then the two gates represented in the following circuits (A) and (B) are : **[31 Aug. 2021 (Evening)]**

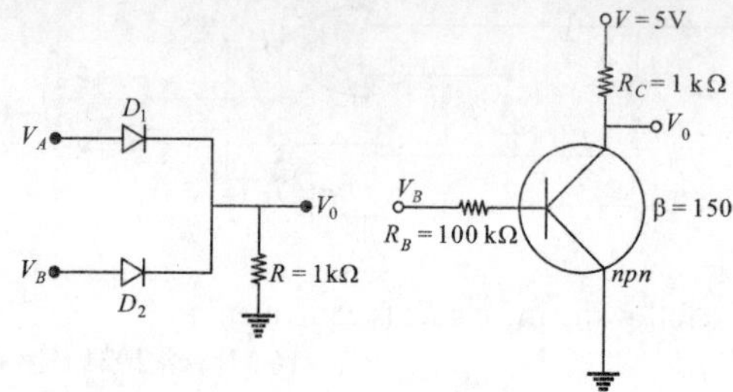

(A) AND and NOT Gate
(B) OR and NOT Gate
(C) AND and OR Gate
(D) NAND and NOR Gate

* * * * *

COMMUNICATION SYSTEMS

29.1 COMMUNICATION SYSTEMS

1. A signal is to be transmitted through a wave of wavelength λ, using a linear antenna. The length l of the antenna and effective power radiated P_{eff} will be given respectively as : (K is a constant of proportionality) : **[2017 (Online)]**

(A) $\lambda, P_{eff} = K\left(\frac{1}{\lambda}\right)^2$ (B) $\frac{\lambda}{8}, P_{eff} = K\left(\frac{1}{\lambda}\right)$

(C) $\frac{\lambda}{16}, P_{eff} = K\left(\frac{1}{\lambda}\right)^3$ (D) $\frac{5}{\lambda}, P_{eff} = K\left(\frac{1}{\lambda}\right)^{\frac{1}{2}}$

2. A *TV* transmission tower has a height of 140 m and the height of the receiving antenna is 40 m. What is the maximum distance upto which signals can be broadcasted from this tower in *LOS* (Line of Sight) mode ? (Given : radius of earth = 6.4×10^6 m) : **[10 Jan. 2019 (Morning)]**

(A) 80 km (B) 48 km
(C) 40 km (D) 65 km

3. To double the covering range of a *TV* transmission tower, its height should be multiplied by : **[12 Jan. 2019 (Evening)]**

(A) $\frac{1}{\sqrt{2}}$ (B) 4
(C) $\sqrt{2}$ (D) 2

4. In a line of sight radio communication, a distance of about 50 km is kept between the transmitting and receiving antennas. If the height of the receiving antenna is 70 m, then the minimum height of the transmitting antenna should be.
(Radius of the Earth = 6.4×10^6 m) : **[8 April 2019 (Evening)]**

(A) 40 m (B) 51 m
(C) 32 m (D) 20 m

5. Given below in the left column are different modes of communication using the kinds of waves given the right column: **[10 April 2019 (Morning)]**

A.	Optical Fibre communication	P.	Ultrasound
B.	Radar	Q.	Infrared Light
C.	Sonar	R.	Microwaves
D.	Mobile Phones	S.	Radio Waves

(A) A-S, B-Q, C-R, D-P (B) A-R, B-P, C-S, D-Q
(C) A-Q, B-S, C-R, D-P (D) A-Q, B-S, C-P, D-R

6. A signal of 0.1 kW is transmitted in a cable. The attenuation of cable is – 5 dB per km and cable length is 20 km. The power received at receiver is 10^{-x} W. The value of x is ______.

[Given in dB = $10\log_{10}\left(\frac{P_0}{P_i}\right)$]. **[24 Feb. 2021 (Evening)]**

7. A transmitting station releases waves of wavelength 960 m. A capacitor of 2.56 μF is used in the resonant circuit. The self inductance of coil necessary for resonance is ______ $\times 10^{-8}$ H. **[25 Feb. 2021 (Morning)]**

8. If the highest frequency modulating a carrier is 5 kHz, then the number of AM broadcast stations accommodated in a 90 kHz bandwidth are ______. **[26 Feb. 2021 (Evening)]**

9. A 25 m long antenna is mounted on an antenna tower. The height of the antenna tower is 75 m. The wavelength (in meter) of the signal transmitted by this antenna would be : **[16 March 2021 (Morning)]**

(A) 300 (B) 400
(C) 200 (D) 100

10. Two identical antennas mounted on identical towers are separated from each other by a distance of 45 km. What should nearly be the minimum height of receiving antenna to receive the signals in line of sight ? (Assume radius of earth is 6400 km) **[16 March 2021 (Evening)]**

(A) 19.77 m (B) 39.55 m
(C) 79.1 m (D) 158.2 m

11. For VHF signal broadcasting, ______ km^2 of maximum service area will be covered by an antenna tower of height 30 m, if the receiving antenna is placed at ground. Let radius of the earth be 6400 km. (Round off to the Nearest Integer)
(Take π as 3.14) **[17 March 2021 (Morning)]**

12. A carrier signal $C(t) = 25 \sin(2.512 \times 10^{10} t)$ is amplitude modulated by a message signal $m(t) = 5 \sin(1.57 \times 10^8 t)$ and transmitted through an antenna. What will be the bandwidth of the modulated signal ? **[17 March 2021 (Evening)]**

(A) 8 GHz (B) 2.01 GHz
(C) 1987.5 MHz (D) 50 MHz

13. Match List-I with List-II :

List-I

(a) 10 km height over earth's surface
(b) 70 km height over earth's surface
(c) 180 km height over earth's surface
(d) 270 km height over earth's surface

List-II

(i) Thermosphere
(ii) Mesosphere
(iii) Stratosphere
(iv) Troposphere **[18 March 2021 (Morning)]**

(A) (a) → (iv), (b) → (iii), (c) → (ii), (d) → (i)
(B) (a) → (i), (b) → (iv), (c) → (iii), (d) → (ii)
(C) (a) → (iii), (b) → (ii), (c) → (i), (d) → (iv)
(D) (a) → (ii), (b) → (i), (c) → (iv), (d) → (iii)

14. A *TV* transmission tower antenna is at a height of 20 m. Suppose that the receiving antenna is at
(i) ground level
(ii) a height of 5 m
The increase in antenna range in case (ii) relative to case (i) is n%. Calculate the value of n, to the nearest integer. **[18 March 2021 (Evening)]**

15. What should be the height of transmitting antenna and the population covered if the television telecast is to cover a radius of 150 km ? The average population density around the tower is 2000/ km^2 and the value of $R_e = 6.5 \times 10^6$ m : **[22 July 2021 (Evening)]**

(A) Height = 1731 m, Population Covered = 1413×10^5
(B) Height = 1600 m, Population Covered = 2×10^5
(C) Height = 1800 m, Population Covered = 1413×10^8
(D) Height = 1241 m, Population Covered = 7×10^5

16. In amplitude Modulation, the message signal is given as

$$V_m(t) = 10 \sin(2\pi \times 10^5 t) \text{ volts}$$

and Carrier signal

$$V_C(t) = 20 \sin(2\pi \times 10^7 t) \text{ volts and}$$

The modulated signal now contains the message signal with lower side band and upper side band frequency, therefore the bandwidth of modulated signal is α kHz. The value of is α is : **[25 July 2021 (Morning)]**

(A) 50 kHz (B) 200 kHz
(C) 100 kHz (D) 0

17. The amplitude of upper and lower side bands of A.M. wave where a carrier signal with frequency 11.21 MHz, peak voltage 15 V is amplitude modulated by a 7.7 kHz sine wave of 5 V amplitude are $\frac{a}{10}$ V and $\frac{b}{10}$ V respectively. Then the value of $\frac{a}{b}$ is ______. **[27 July 2021 (Morning)]**

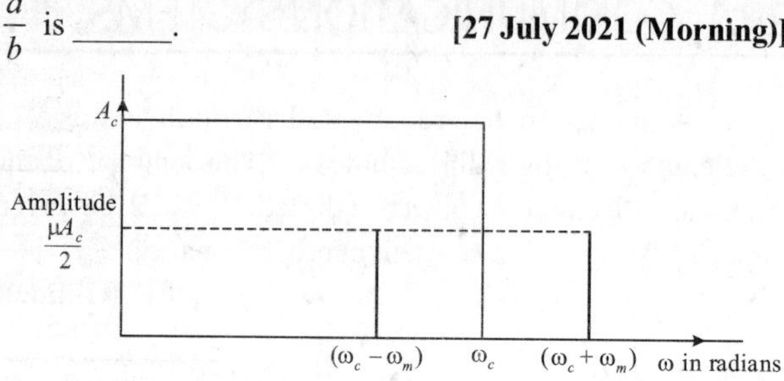

18. A transmitting antenna at top of a tower has a height of 50 m and the height of receiving antenna is 80 m. What is the range of communication for Line of Sight (LoS) mode ?
[use radius of earth = 6400 km] **[26 Aug. 2021 (Evening)]**

(A) 80.2 km (B) 57.28 km
(C) 144.1 km (D) 45.5 km

19. A transmitting antenna has a height of 320 m and that of receiving antenna is 2000 m. The maximum distance between them for satisfactory communication in line of sight mode is d. The value of d is______ km. **[27 Aug. 2021 (Morning)]**

20. An antenna is mounted on a 400 m tall building. What will be the wavelength of signal that can be radiated effectively by the transmission tower upto a range of 44 km ? **[27 Aug. 2021 (Evening)]**

(A) 605 m (B) 302 m
(C) 37.8 m (D) 75.6 m

21. If the sum of the heights of transmitting and receiving antennas in the line of sight of communication is fixed at 160 m, then the maximum range of *LOS* communication is ______ km (Take radius of Earth = 6400 km) **[31 Aug. 2021 (Morning)]**

22. A bandwidth of 6 MHz is available for *A.M.* transmission. If the maximum audio signal frequency used for modulating the carrier wave is not to exceed 6 kHz. The number of stations that can be broadcasted within this band simultaneously without interfering with each other will be______. **[31 Aug. 2021 (Evening)]**

* * * * *

CHAPTER

29 COMMUNICATION SYSTEMS

29.2 MODULATION AND DEMODULATION

1. In amplitude modulations, sinusoidal carrier frequency used is denoted by ω_c and the signal frequency is denoted by ω_m. The bandwidth ($\Delta\omega_m$) of the signal is such that $\Delta\omega_m < \omega_c$. Which of the following frequencies is not contained in the modulated wave ? **[2017]**

(A) $\omega_m + \omega_c$ (B) $\omega_c - \omega_m$

(C) ω_m (D) ω_c

2. A signal of frequency 20 kHz and peak voltage of 5 Volt is used to modulate a carrier wave of frequency 1.2 MHz and peak voltage 25 Volts. Choose the correct statement :

(A) Modulation index = 5, side frequency bands are at 1400 kHz and 1000 kHz

(B) Modulation index = 5, side frequency bands are at 21.2 kHz and 18.8 kHz

(C) Modulation index = 0.8, side frequency bands are at 1180 kHz and 1220 kHz

(D) Modulation index = 0.2, side frequency bands are at 1220 kHz and 1180 kHz **[2017 (Online)]**

3. A carrier wave of peak voltage 14 V is used for transmitting a message signal. The peak voltage of modulating signal given to achieve a modulation index of 80% will be : **[2018]**

(A) 11.2 V (B) 7 V

(C) 22.4 V (D) 28 V

4. A telephonic communication service is working at carrier frequency of 10 GHz. Only 10% of it is utilized for transmission. How many telephonic channels can be transmitted simultaneously if each channel requires a bandwidth of 5 kHz ? **[2018]**

(A) 2×10^3 (B) 2×10^4

(C) 2×10^5 (D) 2×10^6

5. The carrier frequency of a transmitter is provided by a tank circuit of a coil of inductance 49 μH and a capacitance of 2.5 nF. It is modulated by an audio signal of 12 kHz. The frequency range occupied by the side bands is : **[2018 (Online)]**

(A) 18 kHz – 30 kHz (B) 63 kHz – 75 kHz

(C) 442 kHz – 466 kHz (D) 13482 kHz – 13494 kHz

6. The number of amplitude modulated broadcast stations that can be accommodated in a 300 kHz band width for the highest modulating frequency 15 kHz will be : **[2018 (Online)]**

(A) 20 (B) 10

(C) 8 (D) 15

7. In a communication system operating at wavelength 800 nm, only one percent of source frequency is available as signal bandwidth. The number of channels accommodated for transmitting *TV* signals of band width 6 MHz are (Take velocity of light $c = 3 \times 10^8$ m/s, $h = 6.6 \times 10^{-34}$ J-s) : **[9 Jan. 2019 (Evening)]**

(A) 3.75×10^6 (B) 4.87×10^5

(C) 3.86×10^6 (D) 6.25×10^5

8. The modulation frequency of an *AM* radio station is 250 kHz, which is 10% of the carrier wave. If another *AM* station approaches you for license what broadcast frequency will you allot ? **[10 Jan. 2019 (Evening)]**

(A) 2750 kHz (B) 2000 kHz

(C) 2250 kHz (D) 2900 kHz

9. An amplitude modulated signal is given by $V(t) = 10[1 + 0.3 \cos(2.2 \times 10^4)] \sin(5.5 \times 10^5 t)$. Here t is in seconds. The sideband frequencies (in kHz) are, [Given $\pi = 22/7$] : **[11 Jan. 2019 (Morning)]**

(A) 1785 and 1715 (B) 892.5 and 857.5

(C) 89.25 and 85.75 (D) 178.5 smf 171.5

10. An amplitude modulated signal is plotted below : **[11 Jan. 2019 (Evening)]**

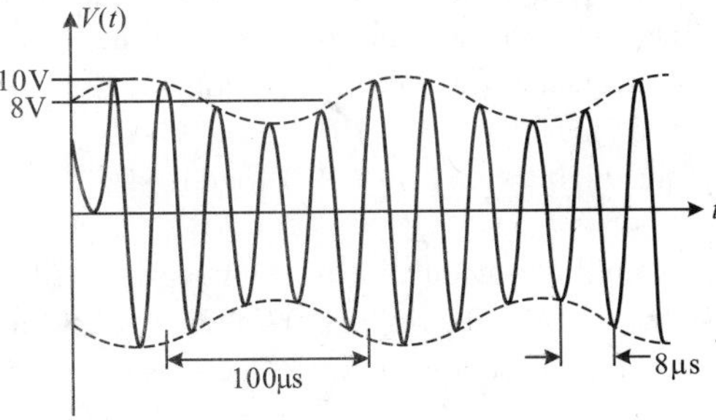

Which one of the following best describes the above signal ?

(A) $(9 + \sin(2.5\pi \times 10^5 t)) \sin(2\pi \times 10^4 t)V$

(B) $(9 + \sin(4\pi \times 10^4 t)) \sin(5\pi \times 10^5 t)V$

(C) $(1 + 9\sin(2\pi \times 10^4 t)) \sin(2.5\pi \times 10^5 t)V$

(D) $(9 + \sin(2\pi \times 10^4 t)) \sin(2.5\pi \times 10^5 t)V$

11. A 100 V carrier wave is made to vary between 160 V and 40 V by a modulating signal. What is the modulation index? **[12 Jan. 2019 (Morning)]**

(A) 0.6 (B) 0.5

(C) 0.3 (D) 0.4

12. The wavelength of the carrier waves in a modern optical fiber communication network is close to : **[8 April 2019 (Morning)]**

(A) 600 nm (B) 900 nm

(C) 2400 nm (D) 1500 nm

13. A signal $A \cos \omega t$ is transmitted using $v_0 \sin \omega_0 t$ as carrier wave. The correct amplitude modulated (AM) signal is : **[9 April 2019 (Morning)]**

(A) $v_0 \sin \omega_0 t + A \cos \omega t$

(B) $v_0 \sin \omega_0 t + \frac{A}{2}\sin(\omega_0 - \omega)t + \frac{A}{2}\sin(\omega_0 + \omega)t$

(C) $(v_0 + A) \cos \omega t \sin \omega_0 t$

(D) $v_0 \sin [\omega_0 (1 + 0.01 A \sin \omega t) t]$

14. The physical sizes of the transmitter and receiver antenna in a communication system are : **[9 April 2019 (Evening)]**

(A) Proportional to carrier frequency

(B) Inversely proportional to modulation frequency

(C) Inversely proportional to carrier frequency

(D) Independent of both carrier and modulation frequency

15. A message signal of frequency 100 MHz and peak voltage 100 V is used to execute amplitude modulation on a carrier wave of frequency 300 GHz and peak voltage 400 V. The modulation index and difference between the two side band frequencies are : **[10 April 2019 (Morning)]**

(A) 4; 1×10^8 Hz (B) 0.25; 1×10^8 Hz

(C) 4; 2×10^8 Hz (D) 0.25; 2×10^8 Hz

16. In an amplitude modulator circuit, the carrier wave is given by, $C(t) = 4 \sin (20000\ \pi t)$ while modulating signal is given by, $m(t) = 2 \sin (2000\ \pi t)$. The values of modulation index and lower side band frequency are : **[12 April 2019 (Evening)]**

(A) 0.5 and 9 kHz (B) 0.5 and 10 kHz

(C) 0.3 and 9 kHz (D) 0.4 and 10 kHz

17. An amplitude modulated wave is represented by the expression $v_m = 5(1 + 0.6 \cos 6280t) \sin (2\pi \times 10^4 t)$ volts. The minimum and maximum amplitudes of the amplitude modulated wave are, respectively : **[2 Sept. 2020 (Morning)]**

(A) 5 V, 8 V (B) $\frac{3}{2}$ V, 5 V

(C) $\frac{5}{2}$ V, 8 V (D) 3 V, 5 V

18. An audio signal $v_m = 20 \sin 2\pi (1500\, t)$ amplitude modulates a carrier $v_c = 80 \sin 2\pi\ (100{,}000t)$. The value of percent modulation is ______. **[24 Feb. 2021 (Morning)]**

19. Given below are two statements :

Statement-I : A speech signal of 2 kHz is used to modulate a carrier signal of 1 MHz. The bandwidth requirement for the signal is 4 kHz

Statement-II : The side band frequencies are 1002 kHz and 988 kHz

In the light of the above statements, choose the correct answer from the options given below : **[25 Feb. 2021 (Morning)]**

(A) Both Statement-I and Statement-II are true

(B) Statement-I is true but Statement-II is false

(C) Both Statement-I and Statement-II are false

(D) Statement-I is false but Statement-II is true

20. If a message signal of frequency 'f_m' is amplitude modulated with a carrier signal of frequency 'f_c' and radiated through an antenna, the wavelength of the corresponding signal in air is : **[25 Feb. 2021 (Evening)]**

(A) $\frac{c}{f_m}$ (B) $\frac{c}{f_c + f_m}$

(C) $\frac{c}{f_c - f_m}$ (D) $\frac{c}{f_c}$

21. The maximum and minimum amplitude of an amplitude modulated wave is 16 V and 8 V respectively. The modulation index for tis amplitude modulated wave is $x \times 10^{-2}$. The value of x is ________. **[26 Feb. 2021 (Morning)]**

22. A carrier wave $V_C(t) = 160 \sin (2\pi \times 10^6 t)$ volts is made to vary between $V_{max} = 200$ V and $V_{min} = 120$ V by a message signal $V_m(t) = A_m \sin (2\pi \times 10^3 t)$ volt. The peak voltage A_m of the modulating signal is ______. **[20 July 2021 (Morning)]**

23. A message signal of frequency 20 kHz and peak voltage of 20 volt is used to modulate a carrier wave of frequency 1 MHz and peak voltage of 20 volt. The modulation index will be ______. **[25 July 2021 (Evening)]**

24. The maximum amplitude for an amplitude modulated wave is found to be 12 V while the minimum amplitude is found to be 3 V. The modulation index is $0.6\, x$ where x is _____ **[27 July 2021 (Evening)]**

25. An amplitude modulated wave is represented by $C_m(t) = 10\ (1 + 0.2 \cos 12560\ t) \sin (111 \times 10^4\, t)$ volts. The modulating frequency in kHz will be_____: **[26 Aug. 2021 (Morning)]**

26. A carrier wave with amplitude of 250 V is amplitude modulated by a sinusoidal base band signal of amplitude 150 V. The ratio of minimum amplitude to maximum amplitude for the amplitude modulated wave is 50 : x, then value of x is_____. **[01 Sep. 2021 (Evening)]**

* * * * *

CHAPTER

30 ERRORS AND EXPERIMENTS

30.1 ERRORS AND EXPERIMENTS

1. The following observations were taken for determining surface tension T of water by capillary method :
Diameter of capillary, $D = 1.25 \times 10^{-2}$ m, rise of water, $h = 1.45 \times 10^{-2}$ m. Using $g = 9.80$ m/s^2 and the simplified relation $T = \frac{rhg}{2} \times 10^3$ N/m, the possible error in surface tension is closest to : **[2017]**

(A) 2.4% (B) 10%
(C) 0.15% (D) 1.5%

2. A physical quantity P is described by the relation $P = a^{1/2} b^2 c^3 d^{-4}$. If the relative errors in the measurement of a, b, c, and d respectively, are 2%, 1%, 3% and 5%, then the relative error in P will be : **[2017 (Online)]**

(A) 8% (B) 12%
(C) 32% (D) 25%

3. The percentage errors in quantities P, Q, R and S are 0.5%, 1%, 3% and 1.5% respectively in the measurement of a physical quantity $A = \frac{P^3 Q^2}{\sqrt{RS}}$.

The maximum percentage error in the value of A will be : **[2018 (Online)]**

(A) 8.5% (B) 6.0%
(C) 7.5% (D) 6.5%

4. In a screw gauge, 5 complete rotations of the screw cause it to move a linear distance of 0.25 cm. There are 100 circular scale divisions. The thickness of a wire measured by this screw gauge gives a reading of 4 main scale divisions and 30 circular scale divisions. Assuming negligible zero error, the thickness of the wire is : **[2018 (Online)]**

(A) 0.0430 cm (B) 0.3150 cm
(C) 0.4300 cm (D) 0.2150 cm

5. The relative error in the determination of the surface area of a sphere is α. Then the relative error in the determination of its volume is : **[2018 (Online)]**

(A) $\frac{2}{3}\alpha$ (B) $\frac{1}{3}\alpha$

(C) $\frac{3}{2}\alpha$ (D) α

6. The relative uncertainty in the period of a satellite orbiting around the earth is 10^{-2}. If the relative uncertainty in the radius of the orbit is negligible, the relative uncertainty in the mass of the earth is : **[2018 (Online)]**

(A) 3×10^{-2} (B) 10^{-2}
(C) 2×10^{-2} (D) 6×10^{-2}

7. The density of a material in the shape of a cube is determined by measuring three sides of the cube and its mass. If the relative errors in measuring the mass and length are respectively 1.5% and 1%, the maximum error in determining the density is : **[2018 (Online)]**

(A) 2.5% (B) 3.5%
(C) 4.5% (D) 6%

8. A copper wire is stretched to make it 0.5% longer. The percentage change in its electrical resistance, if its volume remains unchanged is : **[9 Jan. 2019 (Morning)]**

(A) 2.5% (B) 0.5%
(C) 1.0% (D) 2.0%

9. The pitch and the number of divisions, on the circular scale, for a given screw gauge are 0.5 mm and 100 respectively. When the screw gauge is fully tightened without any object, the zero of its circular scale lies 3 divisions below the reference line. The readings of the main scale and the circular scale, for a thin sheet, are 5.5 mm and 48 respectively, the thickness of this sheet is: **[9 Jan. 2019 (Evening)]**

(A) 5.755 m (B) 5.725 mm
(C) 5.740 m (D) 5.950 mm

10. The diameter and height of a cylinder are measured by a meter scale to be 12.6 ± 0.1 cm and 34.2 ± 0.1 cm, respectively. What will be the value of its volume in appropriate significant figures ? **[10 Jan. 2019 (Evening)]**

(A) 4260 ± 80 cm^3 (B) 4300 ± 80 cm^3
(C) 4264.4 ± 81.0 cm^3 (D) 4264 ± 81 cm^3

11. The least count of the main scale of a screw gauge is 1 mm. The minimum number of divisions on its circular scale required to measure 5 μm diameter of wire is : **[12 Jan. 2019 (Morning)]**

(A) 50 (B) 100
(C) 200 (D) 500

12. In a simple pendulum experiment for determination of acceleration due to gravity (g), time taken for 20 oscillations is measured by using a watch of 1 second least count. The mean value of time taken comes out to be 30 s. The length of pendulum is measured by using a meter scale of least count 1 mm and the value obtained is 55.0 cm. The percentage error in the determination of g is close to : **[8 April 2019 (Evening)]**

(A) 0.7% (B) 0.2%
(C) 3.5% (D) 6.8%

13. In the density measurement of a cube, the mass and edge length are measured as (10.00 ± 0.10) kg and (0.10 ± 0.01) m, respectively. The error in the measurement of density is : **[9 April 2019 (Morning)]**

(A) 0.10 kg/m^3 (B) 0.31 kg/m^3
(C) 0.07 kg/m^3 (D) 0.01 kg/m^3

14. The area of a square is 5.29 cm^2. The area of 7 such squares taking into account the significant figures is : **[9 April 2019 (Evening)]**

(A) 37 cm^2 (B) 37.0 cm^2
(C) 37.03 cm^2 (D) 37.030 cm^2

15. A simple pendulum is being used to determine the value of gravitational acceleration g at a certain place. The length of the pendulum is 25.0 cm and a stop watch with 1s resolution measures the time taken for 40 oscillations to be 50 s. The accuracy in g is : **[8 Jan. 2020 (Evening)]**

(A) 3.40% (B) 5.40%
(C) 4.40% (D) 2.40%

16. If the screw on a screw-gauge is given six rotations, it moves by 3 mm on the main scale. If there are 50 divisions on the circular scale the least count of the screw gauge is : **[9 Jan. 2020 (Morning)]**

(A) 0.001 mm (B) 0.001 cm
(C) 0.02 mm (D) 0.01 cm

17. For the four sets of three measured physical quantities as given below. Which of the following options is correct ?

(i) $A_1 = 24.36, B_1 = 0.0724, C_1 = 256.2$
(ii) $A_2 = 24.44, B_2 = 16.082, C_2 = 240.2$
(iii) $A_3 = 25.2, B_3 = 19.2812, C_3 = 236.183$
(iv) $A_4 = 25, B_4 = 236.191, C_4 = 19.5$ **[9 Jan. 2020 (Evening)]**

(A) $A_4 + B_4 + C_4 < A_1 + B_1 + C_1 < A_3 + B_3 + C_3 < A_2 + B_2 + C_2$
(B) $A_1 + B_1 + C_1 < A_3 + B_3 + C_3 < A_2 + B_2 + C_2 < A_4 + B_4 + C_4$
(C) $A_1 + B_1 + C_1 = A_2 + B_2 + C_2 = A_3 + B_3 + C_3 = A_4 + B_4 + C_4$
(D) $A_4 + B_4 + C_4 < A_1 + B_1 + C_1 = A_2 + B_2 + C_2 = A_3 + B_3 + C_3$

18. The least count of the main scale of a vernier callipers is 1 mm. Its vernier scale is divided into 10 divisions and coincide with 9 divisions of the main scale. When jaws are touching each other, the 7th division of vernier scale coincides with a division of main scale and the zero of vernier scale is lying right side of the zero of main scale. When this vernier is used to measure length of a cylinder, the zero of the vernier scale is between 3.1 cm and 3.2 cm and 4th *VSD* coincides with a main scale division. The length of the cylinder is : (*VSD* is vernier scale division) : **[2 Sept. 2020 (Morning)]**

(A) 3.21 cm (B) 2.99 cm
(C) 3.2 cm (D) 3.07 cm

19. Using screw gauge of pitch 0.1 cm and 50 divisions on its circular scale, the thickness of an object is measured. It should correctly be recorded as : **[3 Sept. 2020 (Morning)]**

(A) 2.123 cm (B) 2.125 cm
(C) 2.121 cm (D) 2.124 cm

20. A physical quantity z depends on four observables a, b, c and d, as $z = \frac{a^2 b^{\frac{2}{3}}}{\sqrt{c}\, d^3}$. The percentage of error in the measurement of a, b, c and d 2%, 1.5%, 4% and 2.5% respectively. The percentage of error in z is : **[5 Sept. 2020 (Morning)]**

(A) 12.25% (B) 14.5%
(C) 16.5% (D) 13.5%

21. A screw gauge has 50 divisions on its circular scale. The circular scale is 4 units ahead of the pitch scale marking, prior to use. Upon one complete rotation of the circular scale, a displacement of 0.5 mm is noticed on the pitch scale. The nature of zero error involved, and the least count of the screw gauge, are respectively : **[6 Sept. 2020 (Morning)]**

(A) Negative, 2 μm (B) Positive, 10 μm
(C) Positive, 0.1 μm (D) Positive, 0.1 mm

22. The density of a solid metal sphere is determined by measuring its mass and its diameter. The maximum error in the density of the sphere is $\left(\frac{x}{100}\right)$%. If the relative errors in measuring the mass and the diameter are 6.0% and 1.5% respectively, the value of x is : **[6 Sept. 2020 (Morning)]**

23. A student measuring the diameter of a pencil of circular cross-section with the help of a vernier scale records the following four readings 5.50 mm, 5.55 mm, 5.45 mm; 5.65 mm. The average of these four readings is 5.5375 mm and the standard deviation of the data is 0.07395 mm. The average diameter of the pencil should therefore be recorded as : **[6 Sept. 2020 (Evening)]**

(A) (5.5375 ± 0.0739) mm (B) (5.538 ± 0.074) mm
(C) (5.54 ± 0.07) mm (D) (5.5375 ± 0.0740) mm

24. The pitch of the screw gauge is 1 mm and there are 100 divisions on the circular scale. When nothing is put in between the jaws, the zero of the circular scale lies 8 division below the reference line. When a wire is placed between the jaws, the first linear scale division in clearly visible while 72nd division on circular scale coincides with the reference line. The radius of the wire is : **[25 Feb. 2021 (Morning)]**

(A) 1.64 mm (B) 0.90 mm
(C) 1.80 mm (D) 0.82 mm

25. A wire of 1 Ω has a length of 1 m. It is stretched till its length increases by 25%. The percentage change in resistance to the nearest integer is : **[26 Feb. 2021 (Evening)]**

(A) 12.5% (B) 25%
(C) 76% (D) 56%

26. The resistance $R = \frac{V}{I}$, where $V = (50 \pm 2)V$ and $I = (20 \pm 0.2)$A. The percentage error in R is 'x' %. The value of 'x' to the nearest integer is ______. **[16 March 2021 (Morning)]**

27. The vernier scale used for measurement has a positive zero error of 0.2 mm. If while taking a measurement it was noted that '0' on the vernier scale lies between 8.5 cm and 8.6 cm, vernier coincidence is 6, then the correct value of measurement is______ cm. (least count = 0.01 cm) **[17 March 2021 (Morning)]**

(A) 8.36 cm (B) 8.54 cm
(C) 8.58 cm (D) 8.56 cm

28. The time period of a simple pendulum is given by $T = 2\pi\sqrt{\frac{l}{g}}$. The measured value of the length of pendulum is 10 cm known to a 1 mm accuracy. The time for 200 oscillations of the pendulum is found to be 100 second using a clock of 1s resolution. The percentage accuracy in the determination of 'g' using this pendulum is 'x'. The value of 'x' to the nearest integer is : **[18 March 2021 (Morning)]**

(A) 2% (B) 3%
(C) 5% (D) 4%

29. The radius of a sphere is measured to be (7.50 ± 0.85) cm. Suppose the percentage error in its volume is x. The value of x, to the nearest x, is______. **[18 March 2021 (Evening)]**

30. Three students S_1, S_2 and S_3 perform an experiment for determining the acceleration due to gravity (g) using a simple pendulum. They use different lengths of pendulum and record time for different number of oscillations. The observations are shown in the table. **[22 July 2021 (Evening)]**

Students No.	Length of Pendulum (cm)	No. of oscillations (n)	Total time for n oscillations	Time period (s)
1	64.0	8	128.0	16.0
2	64.0	4	64.0	16.0
3	20.0	4	36.0	9.0

(Least count of length = 0.1 cm least count for time = 0.1 s)

If E_1, E_2 and E_3 are the percentage errors in 'g' for students 1, 2 and 3 respectively, then the minimum percentage error is obtained by student no. ______ .

31. Student A and student B used two screw gauges of equal pitch and 100 equal circular divisions to measure the radius of a given wire. The actual value of the radius of the wire is 0.322 cm. The absolute value of the difference between the final circular scale readings observed by the students A and B is ______. [Figures shows position of reference 'O' when jaws of screw gauge are closed] Given pitch = 0.1 cm : **[25 July 2021 (Morning)]**

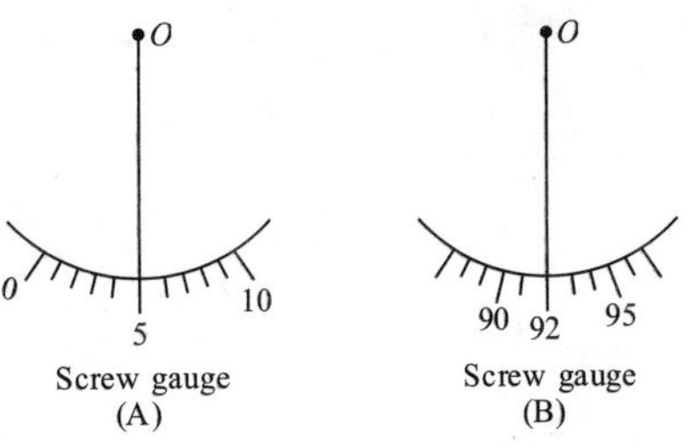

32. The value of aluminum susceptibility is 2.2×10^{-5}. The percentage increase in the magnetic field if space within a current carrying toroid is filled with aluminum is $\frac{x}{10^4}$. Then the value of x is ______. **[25 July 2021 (Morning)]**

33. Assertion-A : If in five complete rotations of the circular scale, the distance travelled on main scale of the screw gauge is 5 mm and there are 50 total divisions on circular scale, then least count is 0.001 cm

Reason-R : $\text{Least Count} = \frac{\text{Pitch}}{\text{Total division on circular scale}}$

In the light of the above statements, choose the most appropriate answer from the options given below : **[27 July 2021 (Morning)]**

(A) Both **A** and **R** are correct and R is the correct explanation of **A**
(B) **A** is not correct but **R** is correct
(C) **A** is correct but **R** is not correct
(D) Both **A** and **R** are correct and **R** is not the correct explanation of **A**

34. A physical quantity 'y' is represented by the formula $y = m^2 r^{-4} g^x l^{-3/2}$. If the percentage errors found in y, m, r, l and g are 18, 1, 0.5, 4 and p respectively, then find the value of x and p: **[27 July 2021 (Evening)]**

(A) 4 and ± 3 (B) 8 and ± 2

(C) $\frac{16}{3}$ and $\pm\frac{3}{2}$ (D) 5 and ± 2

35. The acceleration due to gravity is found upto an accuracy of 4% on a planet. The energy supplied to a simple pendulum of known mass m to undertake oscillations of time period T is being estimated. If time period is measured to an accuracy of 3%, the accuracy to which E is known as ______ %.

[26 Aug. 2021 (Evening)]

* * * * *

ANSWER KEY

Ch-1.1 Measurements and Dimensional Analysis

1. (A) **2.** (B) **3.** (D) **4.** (D) **5.** (B)
6. (B) **7.** (C) **8.** (B) **9.** (C) **10.** (C)
11. (A) **12.** (B) **13.** (D) **14.** (C) **15.** (B)
16. (B) **17.** (A) **18.** (C) **19.** (D) **20.** (D)
21. (D) **22.** (25) **23.** (D) **24.** (A) **25.** (A)
26. (B) **27.** (D) **28.** (D) **29.** (B) **30.** (A)
31. (C) **32.** (C) **33.** (A) **34.** (D) **35.** (B)
36. (D) **37.** (52)

2.1 Motion in One Dimension

1. (A) **2.** (C) **3.** (B) **4.** (B) **5.** (C)
6. (A) **7.** (D) **8.** (B) **9.** (A) **10.** (D)
11. (C) **12.** (C) **13.** (A) **14.** (D) **15.** (B)
16. (8) **17.** (3) **18.** (20) **19.** (A) **20.** (C)
21. (D) **22.** (C) **23.** (A) **24.** (B) **25.** (B)
26. (A) **27.** (12) **28.** (C) **29.** (B) **30.** (B)
31. (C) **32.** (A) **33.** (C) **34.** (B) **35.** (A)
36. (C) **37.** (50) **38.** (12) **39.** (D) **40.** (1)

3.1 Motion in Two Dimensions

1. (A) **2.** (B) **3.** (C) **4.** (D) **5.** (B)
6. (D) **7.** (C) **8.** (C) **9.** (D) **10.** (B)
11. (A) **12.** (C) **13.** (B) **14.** (580) **15.** (A)
16. (C) **17.** (C) **18.** (B) **19.** (D) **20.** (180)
21. (C) **22.** (82) **23.** (None) **24.** (5) **25.** (5)
26. (B) **27.** (C) **28.** (C) **29.** (3) **30.** (B)
31. (B) **32.** (A) **33.** (B) **34.** (30) **35.** (B)
36. (C) **37.** (B) **38.** (A) **39.** (A) **40.** (D)
41. (30) **42.** (B) **43.** (B) **44.** (B)

3.2 Relative Motion

1. (C) **2.** (D) **3.** (D) **4.** (B) **5.** (D)
6. (120) **7.** (B) **8.** (C)

4.1 Newton's Laws of Motion

1. (B) **2.** (A) **3.** (A) **4.** (90°) **5.** (B)
6. (D) **7.** (B) **8.** (492) **9.** (500) **10.** (B)
11. (30) **12.** (10) **13.** (A) **14.** (C) **15.** (D)
16. (D) **17.** (D) **18.** (D) **19.** (D)

4.2 Friction

1. (B) **2.** (B) **3.** (D) **4.** (A) **5.** (B)
6. (A) **7.** (B) **8.** (D) **9.** (25) **10.** (25)
11. (10) **12.** (B) **13.** (21) **14.** (5) **15.** (C)
16. (3) **17.** (15) **18.** (3)

5.1 Work Energy Power

1. (C) **2.** (A) **3.** (C) **4.** (B) **5.** (A)
6. (C) **7.** (B) **8.** (B) **9.** (A) **10.** (A)
11. (A) **12.** (B) **13.** (B) **14.** (C) **15.** (C)
16. (10) **17.** (C) **18.** (B) **19.** (3) **20.** (150)
21. (C) **22.** (346) **23.** (B) **24.** (18) **25.** (C)
26. (1) **27.** (B) **28.** (A) **29.** (3) **30.** (B)
31. (10) **32.** (6) **33.** (B) **34.** (C) **35.** (D)
36. (450) **37.** (A) **38.** (C) **39.** (40) **40.** (2)
41. (1) **42.** (A) **43.** (A) **44.** (16)

5.2 Circular Motion

1. (D) **2.** (D) **3.** (B) **4.** (D) **5.** (D)
6. (B) **7.** (D) **8.** (B) **9.** (A) **10.** (A)
11. (5) **12.** (A) **13.** (D) **14.** (D) **15.** (728)
16. (D) **17.** (200) **18.** (2) **19.** (02) **20.** (D)
21. (B)

6.1 System of Particles

1. (A) **2.** (A) **3.** (B) **4.** (A) **5.** (A)
6. (D) **7.** (A) **8.** (C) **9.** (C) **10.** (A)
11. (A) **12.** (A) **13.** (D) **14.** (C) **15.** (B)
16. (B) **17.** (C) **18.** (C) **19.** (D) **20.** (B)
21. (A) **22.** (C) **23.** (B) **24.** (D) **25.** (1)
26. (C) **27.** (D) **28.** (B) **29.** (D) **30.** (C)
31. (D) **32.** (10) **33.** (23) **34.** (A) **35.** (C)
36. (D) **37.** (120) **38.** (D) **39.** (3.00) **40.** (1)
41. (A) **42.** (2) **43.** (1) **44.** (B) **45.** (C)
46. (30) **47.** (C) **48.** (D) **49.** (4) **50.** (20)
51. (A) **52.** (C) **53.** (2) **54.** (A) **55.** (400)
56. (D)

6.2 Rigid Body Dynamics

1. (D) **2.** (D) **3.** (B) **4.** (B) **5.** (B)
6. (C) **7.** (C) **8.** (D) **9.** (A) **10.** (C)
11. (A) **12.** (B) **13.** (C) **14.** (D) **15.** (B)
16. (A) **17.** (A) **18.** (C) **19.** (C) **20.** (C)
21. (D) **22.** (D) **23.** (B) **24.** (B) **25.** (A)
26. (None) **27.** (B) **28.** (B) **29.** (D) **30.** (A)
31. (C) **32.** (C) **33.** (D) **34.** (A) **35.** (D)
36. (C) **37.** (D) **38.** (B) **39.** (B) **40.** (D)
41. (D) **42.** (D) **43.** (D) **44.** (C) **45.** (B)
46. (A) **47.** (75) **48.** (B) **49.** (A) **50.** (A)

51. (15) **52.** (B) **53.** (D) **54.** (D) **55.** (D)
56. (C) **57.** (B) **58.** (9) **59.** (B) **60.** (25)
61. (11) **62.** (20) **63.** (B) **64.** (D) **65.** (C)
66. (195) **67.** (20) **68.** (A) **69.** (B) **70.** (D)
71. (C) **72.** (8) **73.** (None) **74.** (C) **75.** (D)
76. (C) **77.** (C) **78.** (20) **79.** (20) **80.** (3)
81. (A) **82.** (D) **83.** (4) **84.** (C) **85.** (C)
86. (C) **87.** (D) **88.** (C) **89.** (4) **90.** (2)
91. (C) **92.** (3) **93.** (A) **94.** (2) **95.** (C)
96. (B) **97.** (A) **98.** (25) **99.** (10) **100.** (4)
101. (A) **102.** (D) **103.** (9) **104.** (52) **105.** (D)
106. (A) **107.** (C) **108.** (C) **109.** (B) **110.** (6)

7.1 Gravitational Field

1. (C) **2.** (B) **3.** (A) **4.** (D) **5.** (B)
6. (B) **7.** (C) **8.** (B) **9.** (B) **10.** (D)
11. (D) **12.** (D) **13.** (A) **14.** (B) **15.** (A)
16. (D) **17.** (D) **18.** (D) **19.** (B) **20.** (C)
21. (4) **22.** (3) **23.** (A) **24.** (A) **25.** (D)
26. (D) **27.** (4) **28.** (C) **29.** (D) **30.** (C)
31. (2) **32.** (C) **33.** (A) **34.** (C)

7.2 Satellite Motion

1. (C) **2.** (C) **3.** (C) **4.** (B) **5.** (D)
6. (A) **7.** (C) **8.** (C) **9.** (D) **10.** (B)
11. (B) **12.** (B) **13.** (16) **14.** (A) **15.** (A)
16. (C) **17.** (C) **18.** (C) **19.** (A) **20.** (B)
21. (B) **22.** (D) **23.** (C) **24.** (C) **25.** (C)
26. (10) **27.** (B) **28.** (A) **29.** (64) **30.** (C)
31. (B) **32.** (C) **33.** (B) **34.** (3)

8.1 Elasticity

1. (B) **2.** (C) **3.** (D) **4.** (B) **5.** (A)
6. (None) **7.** (None) **8.** (B) **9.** (4) **10.** (D)
11. (35) **12.** (B) **13.** (A) **14.** (A) **15.** (2)
16. (A) **17.** (D) **18.** (D) **19.** (D) **20.** (B)
21. (32) **22.** (A) **23.** (C) **24.** (C) **25.** (5)
26. (D) **27.** (A) **28.** (20) **29.** (40) **30.** (B)
31. (500) **32.** (D) **33.** (5) **34.** (B)

9.1 Fluid Statics

1. (C) **2.** (C) **3.** (B) **4.** (D) **5.** (A)
6. (C) **7.** (C) **8.** (B) **9.** (25600) **10.** (A)
11. (A) **12.** (B) **13.** (D)

9.2 Fluid Dynamics

1. (B) **2.** (D) **3.** (B) **4.** (C) **5.** (B)
6. (D) **7.** (C) **8.** (B) **9.** (3) **10.** (C)
11. (6)

9.3 Viscosity

1. (D) **2.** (C) **3.** (D) **4.** (B) 5. (D)

9.4 Surface Tension

1. (D) **2.** (C) **3.** (C) **4.** (B) **5.** (C)
6. (A) **7.** (101) **8.** (A) **9.** (B) **10.** (C)
11. (2)

10.1 Thermal Expansion

1. (C) **2.** (C) **3.** (B) **4.** (C) **5.** (B)
6. (B) **7.** (60) **8.** (A) **9.** (B) **10.** (20)
11. (D) **12.** (A) **13.** (D) **14.** (D) **15.** (C)
16. (8)

10.2 Calorimetry

1. (D) **2.** (D) **3.** (D) **4.** (B) **5.** (A)
6. (B) **7.** (C) **8.** (C) **9.** (40) **10.** (50)
11. (B) **12.** (A) **13.** (B) **14.** (C) **15.** (C)
16. (C) **17.** (D)

10.3 Heat Transfer

1. (D) **2.** (A) **3.** (D) **4.** (A) **5.** (B)
6. (A) **7.** (D) **8.** (A) **9.** (A) **10.** (A)
11. (57) **12.** (C) **13.** (2) **14.** (D)

11.1 Laws of Thermodynamics

1. (B) **2.** (D) **3.** (A) **4.** (C) **5.** (A)
6. (B) **7.** (B) **8.** (A) **9.** (C) **10.** (D)
11. (B) **12.** (A) **13.** (C) **14.** (A) **15.** (C)
16. (D) **17.** (B) **18.** (A) **19.** (A) **20.** (B)
21. (600) **22.** (B) **23.** (C) **24.** (A) **25.** (D)
26. (C) **27.** (B) **28.** (B) **29.** (D)
30. (1818 to 1819) **31.** (B) **32.** (4) **33.** (C)
34. (B) **35.** (D) **36.** (8791) **37.** (266) **38.** (A)
39. (150) **40.** (A) **41.** (C) **42.** (D) **43.** (D)
44. (19) **45.** (D) **46.** (A) **47.** (D) **48.** (B)
49. (3600) **50.** (50) **51.** (B) **52.** (208) **53.** (26)
54. (B) **55.** (25) **56.** (60) **57.** (A) **58.** (–113)
59. (A) **60.** (D) **61.** (B) **62.** (A) **63.** (B)
64. (100) **65.** (17258) **66.** (C) **67.** (D) **68.** (A)
69. (D) **70.** (C) **71.** (D) **72.** (C) **73.** (A)
74. (C) **75.** (500) **76.** (C) **77.** (480) **78.** (25)

12.1 Kinetic Theory of Gases

1. (A) **2.** (B) **3.** (D) **4.** (A) **5.** (C)
6. (D) **7.** (D) **8.** (C) **9.** (D) **10.** (B)

11. (A) 12. (None) 13. (C) 14. (C) 15. (B)

16. (D) 17. (C) 18. (C) 19. (C) 20. (D)

21. (C) 22. (D) 23. (C) 24. (A) 25. (41)

26. (5) 27. (C) 28. (D) 29. (400) 30. (C)

31. (C) 32. (D) 33. (B) 34. (B) 35. (A)

36. (D) 37. (C) 38. (D) 39. (D) 40. (B)

41. (B) 42. (A) 43. (25) 44. (D) 45. (A)

46. (C) 47. (D) 48. (C) 49. (B) 50. (B)

51. (A) 52. (500)

13.1 Simple Harmonic Motion

1. (B) 2. (C) 3. (B) 4. (D) 5. (B)

6. (B) 7. (C) 8. (D) 9. (B) 10. (D)

11. (B) 12. (C) 13. (D) 14. (None) 15. (C)

16. (A) 17. (C) 18. (A) 19. (D) 20. (D)

21. (A) 22. (D) 23. (B) 24. (D) 25. (D)

26. (A) 27. (B) 28. (A) 29. (C) 30. (C)

31. (C) 32. (D) 33. (D) 34. (C) 35. (A)

36. (B) 37. (A) 38. (C) 39. (3) 40. (7)

41. (D) 42. (A) 43. (C) 44. (2) 45. (None)

46. (D) 47. (6) 48. (D) 49. (C, D) 50. (1)

51. (D) 52. (B) 53. (A) 54. (10) 55. (A)

56. (A) 57. (8) 58. (C) 59. (2) 60. (A)

61. (2) 62. (D) 63. (1) 64. (8) 65. (B)

66. (A) 67. (D)

14.1 Wave Motion

1. (A) 2. (B) 3. (A) 4. (C) 5. (C)

6. (C) 7. (C) 8. (B) 9. (B) 10. (D)

11. (B) 12. (D) 13. (2) 14. (1215) 15. (D)

16. (2) 17. (1)

14.2 Stationary Waves and Beats

1. (A) 2. (B) 3. (A) 4. (D) 5. (A)

6. (D) 7. (A) 8. (D) 9. (B) 10. (A)

11. (A) 12. (D) 13. (B) 14. (D) 15. (D)

16. (C) 17. (106.06) 18. (D) 19. (C) 20. (C)

21. (B) 22. (B) 23. (4) 24. (7) 25. (34)

26. (10)

14.3 Doppler's Effect

1. (A) 2. (B) 3. (C) 4. (A) 5. (B)

6. (C) 7. (B) 8. (A) 9. (D) 10. (A)

11. (D) 12. (D) 13. (8) 14. (6) 15. (132)

16. (C) 17. (2025) 18. (1210)

15.1 Electric Force and Electric Field

1. (D) 2. (A) 3. (D) 4. (C) 5. (C)

6. (C) 7. (D) 8. (C) 9. (D) 10. (C)

11. (D) 12. (C) 13. (D) 14. (C) 15. (D)

16. (C) 17. (A) 18. (C) 19. (B) 20. (36)

21. (20) 22. (C) 23. (D) 24. (A) 25. (B)

26. (12) 27. (D) 28. (1) 29. (B) 30. (6000)

31. (B) 32. (C) 33. (D) 34. (A) 35. (D)

15.2 Electric Flux and Gauss Law

1. (C) 2. (B) 3. (D) 4. (–48.00) 5. (A)

6. (226) 7. (1) 8. (D) 9. (A) 10. (640)

11. (4) 12. (C) 13. (B)

16.1 Electric Potential and Dipole

1. (B) 2. (B) 3. (B) 4. (C) 5. (D)

6. (B) 7. (A) 8. (B) 9. (B) 10. (C)

11. (A) 12. (A) 13. (C) 14. (D) 15. (C)

16. (D) 17. (B) 18. (A) 19. (A) 20. (D)

21. (D) 22. (C) 23. (A) 24. (A) 25. (D)

26. (C) 27. (D) 28. (A) 29. (C) 30. (C)

31. (128) 32. (243) 33. (B) 34. (B)

16.2 Capacitance

1. (B) 2. (D) 3. (C) 4. (A) 5. (C)

6. (D) 7. (D) 8. (C) 9. (None) 10. (A)

11. (C) 12. (D) 13. (D) 14. (A) 15. (B)

16. (C) 17. (D) 18. (D) 19. (D) 20. (A)

21. (C) 22. (D) 23. (A) 24. (A) 25. (B)

26. (A) 27. (6) 28 (C) 29. (D) 30. (–4.03)

31. (B) 32. (A) 33. (A) 34. (D) 35. (C)

36. (B) 37. (B) 38. (None) 39. (C) 40. (C)

41. (2) 42. (864) 43. (16) 44. (161) 45. (A)

46. (D) 47. (D) 48. (B) 49. (B) 50. (C)

51. (C) 52. (C) 53. (D) 54. (4)

17.1 Current Electricity

1. (A) 2. (C) 3. (C) 4. (D) 5. (D)

6. (B) 7. (D) 8. (C) 9. (B) 10. (D)

11. (D) 12. (C) 13. (B) 14. (D) 15. (D)

16. (5) 17. (A) 18. (3) 19. (B) 20. (A)

21. (B) 22. (A) 23. (01) 24. (B) 25. (A)

17.2 Circuit Analysis

1. (B) 2. (A) 3. (B) 4. (A) 5. (B)

6. (C) 7 (D) 8. (D) 9. (A) 10. (B)

11. (C) 12. (C) 13. (B) 14. (30) 15. (B)

16. (D) 17. (2) 18. (A) 19. (C) 20. (C)

21. (2) 22. (C) 23. (300) 24. (4) 25. (B)

26. (C) 27. (70) 28. (4) 29. (B) 30. (10)
31. (A) 32. (15) 33. (74) 34. (45) 35. (3)
36. (D) 37. (A) 38. (B) 39. (20) 40. (9)
41. (1) 42. (3) 43. (A) 44. (D)

17.3 Thermal Effects of Current & RC Circuits

1. (A) 2. (C) 3. (A) 4. (A) 5. (A)
6. (B) 7. (C) 8. (C) 9. (D) 10. (C)
11. (A) 12. (B) 13. (D) 14. (8) 15. (D)
16. (D) 17. (B) 18. (C) 19. (D) 20. (2500)
21. (2) 22. (50) 23. (D) 24. (B) 25. (100)
26. (3840) 27. (C) 28. (C) 29. (4)

17.4 Electrical Measuring Instruments

1. (D) 2. (C) 3. (D) 4. (D) 5. (B)
6. (C) 7. (B) 8. (B) 9. (D) 10. (A)
11. (D) 12. (B) 13. (B) 14. (C) 15. (D)
16. (A) 17. (B) 18. (D) 19. (A) 20. (C)
21. (C) 22. (B) 23. (A) 24. (A) 25. (D)
26. (D) 27. (C) 28. (12) 29. (D) 30. (10)
31. (A) 32. (40) 33. (A) 34. (A) 35. (B)
36. (B) 37. (A) 38. (A) 39. (1) 40. (48)
41. (C) 42. (A) 43. (B) 44. (B) 45. (B)

18.1 Magnetic Effect of Current

1. (A) 2. (B) 3. (A) 4. (C) 5. (D)
6. (B) 7. (A) 8. (A) 9. (C) 10. (B)
11. (D) 12. (A) 13. (C) 14. (A) 15. (C)
16. (C) 17. (A) 18. (C) 19. (A) 20. (B)
21. (B) 22. (A) 23. (B) 24. (D) 25. (B)
26. (D) 27. (D)

18.2 Magnetic Force on Moving Charges

1. (C) 2. (D) 3. (B) 4. (C) 5. (B)
6. (A) 7. (C) 8. (A) 9. (A) 10. (A)
11. (None) 12. (D) 13. (A) 14. (A) 15. (B)
16. (B) 17. (B) 18. (A) 19. (C) 20. (B)
21. (None) 22. (B) 23. (A) 24. (C) 25. (D)
26. (B) 27. (B) 28. (B) 29. (B) 30. (A)
31. (C) 32. (A) 33. (C) 34. (20) 35. (A)
36. (A) 37. (D) 38. (B) 39. (A) 40. (C)
41. (D) 42. (A) 43. (B) 44. (D) 45. (B)
46. (C) 47. (543) 48. (C) 49. (D)

19.1 Classical and Terrestrial Magnetism

1. (C) 2. (C) 3. (B) 4. (D) 5. (B)
6. (C) 7. (D) 8. (C) 9. (A) 10. (B)

19.2 Magnetic Properties of Materials

1. (B) 2. (B) 3. (C) 4. (C) 5. (D)
6. (B) 7. (D) 8. (A) 9. (B) 10. (B)
11. (D) 12. (A) 13. (D) 14. (D) 15. (1)
16. (3) 17. (3) 18. (250) 19. (D)

20.1 Electromagnetic Induction

1. (A) 2. (C) 3. (B) 4. (A) 5. (A)
6. (A) 7. (C) 8. (B) 9. (D) 10. (B)
11. (B) 12. (A) 13. (A) 14. (175) 15. (D)
16. (C) 17. (15) 18. (A) 19. (A) 20. (C)
21. (5) 22. (B) 23. (D) 24. (C) 25. (C)
26. (A) 27. (D) 28. (80) 29. (60) 30. (60)
31. (D) 32. (B) 33. (D) 34. (A)

20.2 Self and Mutual Induction

1. (D) 2. (A) 3. (B) 4. (D) 5. (B)
6. (A) 7. (B) 8. (B) 9. (B) 10. (B)
11. (B) 12. (A) 13. (D) 14. (10) 15. (C)
16. (33) 17. (440) 18. (A) 19. (A) 20. (144)
21. (C) 22. (D) 23. (D)

21.1 Alternating Current

1. (B) 2. (A) 3. (A) 4. (B) 5. (C)
6. (D) 7. (None) 8. (C) 9. (D) 10. (B)
11. (A) 12. (A) 13. (A) 14. (C) 15. (400)
16. (2000) 17. (B) 18. (900) 19. (D) 20. (A)
21. (A) 22. (C) 23. (283) 24. (D) 25. (4)
26. (B) 27. (D) 28. (B) 29. (A) 30. (A)
31. (C) 32. (C) 33. (3) 34. (B) 35. (125)
36. (D) 37. (B) 38. (D) 39. (500) 40. (C)
41. (D) 42. (D) 43. (A) 44. (None) 45. (11)
46. (1) 47. (B) 48. (2)

22.1 Electromagnetic Wave

1. (A) 2. (C) 3. (C) 4. (B) 5. (C)
6. (C) 7. (C) 8. (C) 9. (D) 10. (B)
11. (B) 12. (C) 13. (B) 14. (B) 15. (C)
16. (D) 17. (D) 18. (B) 19. (C) 20. (C)
21. (A) 22. (C) 23. (B) 24. (C) 25. (A)
26. (B) 27. (A) 28. (B) 29. (C) 30. (C)
31. (B) 32. (B) 33. (D) 34. (B) 35. (194)
36. (B) 37. (15) 38. (667) 39. (2) 40. (137)
41. (A) 42. (A) 43. (3) 44. (6) 45. (C)
46. (C) 47. (C) 48. (A) 49. (C) 50. (B)
51. (A) 52. (C) 53. (A) 54. (2) 55. (500)
56. (D)

23.1 Geometrical Optics I - Reflection of Light

1. (C) 2. (D) 3. (A) 4. (D) 5. (A)
6. (B) 7. (D) 8. (1) 9. (A) 10. (B)
11. (B) 12. (150) 13. (C) 14. (D) 15. (A)

23.2 Geometrical Optics II - Refraction of Light

1. (D) 2. (C) 3. (C) 4. (D) 5. (B)
6. (A) 7. (B) 8. (D) 9. (B) 10. (A)
11. (90) 12. (158) 13. (5.00) 14. (D) 15. (B)
16. (D) 17. (30) 18. (B) 19. (D) 20. (A)
21. (60) 22. (C) 23. (C) 24. (C) 25. (50)
26. (B) 27. (D) 28. (A) 29. (B)

23.3 Geometrical Optics III - Thin Lenses

1. (A) 2. (B) 3. (A) 4. (B) 5. (C)
6. (D) 7. (A) 8. (C) 9. (B) 10. (B)
11. (B) 12. (D) 13. (D) 14. (D) 15. (C)
16. (60) 17. (476) 18. (D) 19. (C) 20. (A)

23.4 Dispersion of Light

1. (A) 2. (B) 3. (A) 4. (12) 5. (D)
6. (600) 7. (5)

23.5 Optical Instruments

1. (A) 2. (C) 3. (A) 4. (A) 5. (D)
6. (C) 7. (5) 8. (50) 9. (C) 10. (B)
11. (A) 12. (25)

24.1 Interference of Light

1. (D) 2. (B) 3. (B) 4. (B) 5. (D)
6. (C) 7. (D) 8. (A) 9. (None) 10. (B)
11. (D) 12. (B) 13. (D) 14. (A) 15. (750)
16. (A) 17. (A) 18. (D) 19. (A) 20. (B)
21. (D) 22. (9) 23. (B) 24. (D) 25. (A)
26. (A) 27. (600) 28. (A) 29. (300) 30. (C)
31. (16)

24.2 Diffraction of Light

1. (B) 2. (B) 3. (B) 4. (D) 5. (D)
6. (198) 7. (50) 8. (A)

24.3 Polarization of Light

1. (C) 2. (A) 3. (A) 4. (A) 5. (D)
6. (A) 7. (C) 8. (75) 9. (30)

25.1 Photoelectric Effect

1. (C) 2. (A) 3. (A) 4. (A) 5. (D)
6. (B) 7. (None) 8. (A) 9. (A) 10. (C)
11. (11) 12. (C) 13. (Bonus) 14. (B) 15. (9)
16. (A) 17. (A) 18. (C) 19. (2) 20. (A)
21. (1) 22. (D) 23. (25) 24. (A) 25. (A)
26. (4) 27. (B) 28. (125) 29. (B) 30. (C)
31. (1) 32. (A) 33. (C) 34. (B)

25.2 Matter Waves and X-Ray

1. (D) 2. (C) 3. (C) 4. (C) 5. (D)
6. (C) 7. (A) 8. (B) 9. (D) 10. (D)
11. (C) 12. (D) 13. (A) 14. (B) 15. (B)
16. (C) 17. (C) 18. (A) 19. (D) 20. (A)
21. (A) 22. (D) 23. (B) 24. (C) 25. (C)
26. (D) 27. (C) 28. (C) 29. (10) 30. (D)
31. (C) 32. (D) 33. (A) 34. (B) 35. (D)
36. (D) 37. (D) 38. (C) 39. (D) 40. (C)
41. (910) 42. (10) 43. (B) 44. (C) 45. (A)

26.1 Atomic Structure

1. (B) 2. (B) 3. (C) 4. (D) 5. (A)
6. (D) 7. (B) 8. (D) 9. (C) 10. (D)
11. (C) 12. (B) 13. (C) 14. (B) 15. (C)
16. (A) 17. (D) 18. (C) 19. (C) 20. (C)
21. (486) 22. (D) 23. (B) 24. (10553) 25. (51)
26. (A) 27. (A) 28. (A) 29. (D) 30. (15)
31. (B) 32. (B) 33. (D) 34. (B) 35. (C)
36. (112) 37. (A) 38. (2) 39. (B) 40. (15)

27.1 Radioactivity

1. (D) 2. (D) 3. (D) 4. (A) 5. (C)
6. (C) 7. (D) 8. (B) 9. (C) 10. (A)
11. (D) 12. (A) 13. (C) 14. (A) 15. (A)
16. (A) 17. (B) 18. (B) 19. (C) 20. (C)
21. (A) 22. (3) 23. (A) 24. (D) 25. (10)
26. (A) 27. (150) 28. (A) 29. (B) 30. (D)
31. (D)

27.2 Nuclear Physics

1. (C) 2. (C) 3. (A) 4. (C) 5. (C)
6. (A) 7. (B) 8. (C) 9. (A) 10. (B)
11. (A) 12. (B) 13. (B) 14. (C) 15. (27)

28.1 Semiconductor

1. (A) 2. (B) 3. (A) 4. (5) 5. (25)
6. (D)

28.2 PN Junction and Diodes

1. (B)	**2.** (C)	**3.** (B)	**4.** (C)	**5.** (D)
6. (D)	**7.** (B)	**8.** (C)	**9.** (B)	**10.** (B)
11. (B)	**12.** (12)	**13.** (C)	**14.** (A)	**15.** (40)
16. (A)	**17.** (C)	**18.** (B)	**19.** (A)	**20.** (D)
21. (A)	**22.** (25)	**23.** (B)	**24.** (B)	**25.** (C)
26. (C)	**27.** (20)	**28.** (9)	**29.** (5)	**30.** (B)
31. (25)	**32.** (192)	**33.** (500)	**34.** (120)	**35.** (20)
36. (D)	**37.** (C)	**38.** (A)		

28.3 Transistors

1. (B)	**2.** (B)	**3.** (D)	**4.** (B)	**5.** (C)
6. (D)	**7.** (D)	**8.** (C)	**9.** (C)	**10.** (150)
11. (A)	**12.** (15)	**13.** (100)	**14.** (D)	**15.** (200)
16. (B)	**17.** (5)	**18.** (C)		

28.4 Logic Gates

1. (C)	**2.** (D)	**3.** (B)	**4.** (A)	**5.** (A)
6. (B)	**7.** (B)	**8.** (C)	**9.** (A)	**10.** (None)
11. (D)	**12.** (C)	**13.** (A)	**14.** (0)	**15.** (A)
16. (B)	**17.** (D)	**18.** (A)	**19.** (C)	**20.** (A)
21. (A)	**22.** (D)	**23.** (B)		

29.1 Communication Systems

1. (A)	**2.** (D)	**3.** (B)	**4.** (C)	**5.** (D)
6. (8)	**7.** (10)	**8.** (9)	**9.** (D)	**10.** (B)
11. (1206)	**12.** (D)	**13.** (A)	**14.** (50)	**15.** (A)
16. (B)	**17.** (1)	**18.** (B)	**19.** (224)	**20.** (A)
21. (64)	**22.** (500)			

29.2 Modulation and Demodulation

1. (C)	**2.** (D)	**3.** (A)	**4.** (C)	**5.** (C)
6. (B)	**7.** (D)	**8.** (B)	**9.** (C)	**10.** (D)
11. (A)	**12.** (D)	**13.** (B)	**14.** (C)	**15.** (D)
16. (A)	**17.** (None)	**18.** (25)	**19.** (A)	**20.** (D)
21. (33)	**22.** (40)	**23.** (1)	**24.** (1)	**25.** (2)
26. (200)				

30.1 Errors and Experiments

1. (D)	**2.** (C)	**3.** (D)	**4.** (D)	**5.** (C)
6. (C)	**7.** (C)	**8.** (C)	**9.** (B)	**10.** (A)
11. (C)	**12.** (D)	**13.** (None)	**14.** (C)	**15.** (C)
16. (B)	**17.** (Bonus)	**18.** (D)	**19.** (D)	**20.** (B)
21. (B)	**22.** (1050)	**23.** (C)	**24.** (D)	**25.** (D)
26. (5)	**27.** (B)	**28.** (B)	**29.** (34)	**30.** (1)
31. (13)	**32.** (22)	**33.** (B)	**34.** (C)	**35.** (14)

* * * * *

SOLUTIONS

1.1 Measurements and Dimensional Analysis

Sol. 1 Dimensions of mass in terms of *T*, *C* and *h* are given as

$$D[\text{M}] = D(T)^a . D(C)^b . D(h)^c$$

$$[\text{M}] = [\text{T}]^a[\text{LT}^{-1}]^b[\text{ML}^2\text{T}^{-1}]^c$$

Comparing LHS & RHS dimensions, we get

$$c = 1$$

$$a - b - c = 0 \Rightarrow a = -1$$

$$b + 2c = 0 \Rightarrow b = -2c = -2$$

$$\Rightarrow \quad D[\text{M}] = [\text{T}^{-1}\text{C}^{-2}\text{h}^1]$$

Ans. (A)

Sol. 2 Let the plank length (l_p) in terms of *G*, *h* and *c* is given as

$$l_p = KG^a h^b c^c \qquad \ldots(1)$$

$$\Rightarrow \quad D(l_p) = D(G^a) \cdot D(h^b) \cdot D(c^c)$$

$$\Rightarrow \quad [\text{L}] = [\text{M}^{-1}\text{L}^3\text{T}^{-2}]^a [\text{ML}^2\text{T}^{-1}]^b [\text{LT}^{-1}]^c$$

Comparing dimensions of L, M & T on LHS & RHS gives

$$1 = 3a + 2b + c$$

$$0 = -a + b \Rightarrow a = b$$

$$0 = -2a - b - c \Rightarrow c = -3a$$

$$\Rightarrow \quad 1 = 5a - 3a \Rightarrow a = 1/2$$

$$\Rightarrow \quad b = 1/2$$

$$c = -3/2$$

From equation-(1) $l_p = KG^{1/2}h^{1/2}c^{-3/2} = K\sqrt{\frac{Gh}{c^3}}$

For all given options we take $K = 1$ **Ans. (B)**

Sol. 3 We use gravitational force as

$$F = \frac{GM^2}{R^2}$$

$$\Rightarrow \quad [G] = [\text{M}^{-1}\text{L}^3\text{T}^{-2}]$$

Also as $\quad E = h\nu$

$$\Rightarrow \quad [h] = [\text{ML}^2\text{T}^{-1}]$$

and $\quad [c] = [\text{LT}^{-1}]$

Now we use

$$t \propto G^x h^y C^z$$

$$[\text{T}] = [\text{M}^{-1}\text{L}^3\text{T}^{-2}]^x [\text{ML}^2\text{T}^{-1}]^y [\text{LT}^{-1}]^z$$

$$[\text{M}^0\text{L}^0\text{T}^1] = [\text{M}^{-x+y}\,\text{L}^{3x+2y+z}\,\text{T}^{-2x-y-z}]$$

Comparing the powers of M, L, T in LHS & RHS, we have

$$-x + y = 0$$

$$\Rightarrow \quad x = y$$

and $\quad 3x + 2y + z = 0$

$$\Rightarrow \quad 5x + z = 0 \qquad \ldots(1)$$

and $\quad -2x - y - z = 1$

$$\Rightarrow \quad 3x + z = -1 \qquad \ldots(2)$$

On solving (1) & (2) $x = y = \frac{1}{2}, z = -\frac{5}{2}$

$$\Rightarrow \quad t \propto \sqrt{\frac{Gh}{c^5}}$$

Ans. (D)

Sol. 4 Density of material is

$$\rho = \frac{128\text{kg}}{\text{m}^3} = \frac{128 \times (50 \times 20\text{g})}{(25 \times 4\text{cm})}$$

$$\Rightarrow \quad \rho = \frac{128}{64}(20) \text{ units}$$

$$\Rightarrow \quad \rho = 40 \text{ units}$$

Ans. (D)

Sol. 5 Force is given as

$$F = \alpha\beta e^{\left(\frac{-x^2}{\alpha KT}\right)}$$

Here the exponential power must be dimensionless, so we have

$$D\left(\frac{x^2}{\alpha KT}\right) = \text{M}^0\text{L}^0\text{T}^0$$

$$\Rightarrow \quad \frac{L^2}{[\alpha]ML^2T^{-2}} = \text{M}^0\text{L}^0\text{T}^0$$

$$\Rightarrow \quad [\alpha] = \text{M}^{-1}\text{T}^2$$

And we use

$$[F] = [\alpha]\,[\beta]$$

$$\text{ML}^1\text{T}^{-2} = \text{M}^{-1}\text{T}^2[\beta]$$

$$\Rightarrow \quad [\beta] = \text{M}^2\text{L}^1\text{T}^{-4}$$

Ans. (B)

Sol. 6 We use young modulus $= \frac{\text{stress}}{\text{strain}}$

$$\Rightarrow \quad \frac{F}{S} = Y \cdot \frac{\Delta l}{l}$$

$$\Rightarrow \quad [\text{Y}] = \frac{[\text{F}]}{[\text{S}]}$$

If dimensions of *Y* in unit of *V*, *A* and *F* are *x*, *y*, *z* then we use

$$[\text{Y}] = \frac{[\text{F}]}{[\text{S}]}$$

$$\Rightarrow \quad [\text{V}]^x [\text{A}]^y [\text{F}]^z = \frac{[\text{MLT}^{-2}]}{[\text{L}^2]}$$

$$[\text{LT}^{-1}]^x [\text{LT}^{-2}]^y [\text{MLT}^{-2}]^z = [\text{ML}^{-1}\,\text{T}^{-2}]$$

Comparing dimensions in LHS & RHS, we get

$$z = 1$$

$$x + y + z = -1$$

$$\Rightarrow \quad x + y = -2$$

$$-x - 2y - 2z = -2$$

$$\Rightarrow \quad x + 2y = 0$$

$$\Rightarrow \quad y = 2 \text{ and } x = -4$$

$$\Rightarrow \quad [Y] = V^{-4}A^2F^1$$ **Ans. (B)**

Sol. 7 $\left[\frac{l}{r}\right] = T$

and $[CV] = AT$

$$\Rightarrow \quad \left[\frac{l}{rCV}\right] = \frac{T}{AT} = A^{-1}$$ **Ans. (C)**

Sol. 8 $p = kS^aI^bh^c$

Where k is dimensionless constant

$$MLT^{-1} = (MT^{-2})^a (ML^2)^b (ML^2T^{-1})^c$$

$$a + b + c = 1$$

$$2b + 2c = 1$$

$$-2a - c = -1$$

Solving, we get

$$a = \frac{1}{2}$$

$$b = \frac{1}{2}$$

$$c = 0$$

$$[P] = S^{1/2}I^{1/2}h^0$$ **Ans. (B)**

Sol. 9 Given that $X = 5YZ^2$

$$\Rightarrow \quad Y = \frac{X}{5Z^2}$$

Dimensionally we use

$$[Y] = \frac{[X]}{[Z^2]}$$

$$\Rightarrow \quad [Y] = \frac{A^2 \cdot M^{-1}L^{-2} \cdot T^4}{(MA^{-1}T^{-2})^2}$$

$$\Rightarrow \quad [Y] = M^{-3} \cdot L^{-2} \cdot T^8 \cdot A^4$$ **Ans. (C)**

Sol. 10 Dimension of h, c and G are

$$[h] = M^1L^2T^{-1}$$

$$[c] = L^1T^{-1}$$

$$[G] = M^{-1}L^3T^{-2}$$

So dimensions of f can be calculated as

$$[f] = \sqrt{\frac{M^1L^2T^{-1} \times L^5T^{-5}}{M^{-1}L^3T^{-2}}} = M^1L^2T^{-2}$$ **Ans. (C)**

Sol. 11 Dimensions of Young's modulus are considered as x y z then we have

$$[Y] = F^xA^yV^z$$

Comparing dimensions in term of MLT, we have

$$M^1L^{-1}T^{-2} = [MLT^{-2}]^x[L^2]^y[LT^{-1}]^z$$

$$M^1L^1T^{-2} = [M]^x[L]^{x+2y+z}[T]^{-2x-z}$$

Comparing LHS and RHS dimensions, we get

$$x = 1 \quad \ldots(1)$$

$$x + 2y + z = -1 \quad \ldots(2)$$

$$-2x - z = -2 \quad \ldots(3)$$

Solving (1), (2) and (3), we get

$$x = 1$$

$$y = -1$$

$$z = 0$$

$$[Y] = FA^{-1}V^0$$ **Ans. (A)**

Sol. 12 If dimensions of fundamental quantities are taken as x, y and z, we have

$$D[E] = [P]^x[A]^y[T]^z$$

$$\Rightarrow \quad ML^2T^{-2} = [MLT^{-1}]^x[L^2]^y[T]^z$$

$$\Rightarrow \quad ML^2T^{-2} = M^xL^{x+2y}T^{-x+z}$$

Comparing dimensions of LHS and RHS, we have

$$x = 1$$

and $x + 2y = 2$

and $1 + 2y = 2$

$$\Rightarrow \quad y = \frac{1}{2}$$

$$\Rightarrow \quad -x + z = -2$$

$$\Rightarrow \quad -1 + z = -2$$

$$\Rightarrow \quad z = -1$$

$$D[E] = [PA^{1/2}T^{-1}]$$ **Ans. (B)**

Sol. 13 By steady state conduction, heat flow rate is given as

$$\frac{d\theta}{dt} = \frac{kA(T_2 - T_1)}{l}$$

$$\Rightarrow \quad D[k] = \frac{D[d\theta/dt][L]}{[L^2][\theta]} = \frac{[ML^2T^{-3}][L]}{[L^2][\theta]}$$

$$\Rightarrow \quad D[k] = [MLT^{-3}\theta^{-1}]$$ **Ans. (D)**

Sol. 14 Dimensions of x is given in terms of MLT as

$$[x] = \frac{[ML^2][MLT^{-2}][LT^{-1}]^2}{[ML^2T^{-2}][L^4]}$$

$$[x]=[ML^{-1}T^{-2}]=\frac{[ML^2T^{-2}]}{[L^3]}=\frac{\text{Energy}}{\text{volume}}$$

Thus dimensional formula of x matches with energy density.

Ans. (C)

Sol. 15 Dimensions of x, y and z are

$$x=\frac{1}{\sqrt{\mu_0\varepsilon_0}}=\text{speed of light}$$

$$\Rightarrow \quad [x]=[LT^{-1}]$$

$$y=\frac{E}{B}=\text{speed of light}$$

$$\Rightarrow \quad [y]=[L^1T^{-1}]$$

$$z=\frac{l}{RC}=\frac{\text{length}}{\text{time}}=\text{speed}$$

$$\Rightarrow \quad [z]=[L^1T^{-1}]$$

So, x, y, z all have the same dimensions. **Ans. (B)**

Sol. 16 The dimensions of any value in exponent must be dimensionless so in the given expression $\frac{x^2}{\alpha kT}$ is dimensionless, this gives

$$\Rightarrow \quad D[\alpha]=\frac{[x^2]}{[kT]}=\frac{L^2}{ML^2T^{-2}}=M^{-1}T^2$$

and $\quad D[W]=[\alpha][\beta]^2$

$$\Rightarrow \quad D[\beta]=\sqrt{\frac{ML^2T^{-2}}{M^{-1}T^2}}=M^1L^1T^{-2}$$

Ans. (B)

Sol. 17 By direct dimensions of the given physical quantities, we can match the values in given columns. **Ans. (A)**

Sol. 18 By the generic expressions of force and energy, we have

$$F=\frac{1}{4\pi\epsilon_0}\frac{e^2}{r^2} \text{ and } E=\frac{hc}{\lambda}$$

$$\left[\frac{e^2}{4\pi\varepsilon_0}\times\frac{1}{hc}\right]=\frac{Fr^2}{E\lambda}$$

$$\Rightarrow \quad D\left(\frac{Fr^2}{E\lambda}\right)=[M^0L^0T^0]$$

Ans. (C)

Sol. 19 The dimensions of any value in exponent must be dimensionless so in the given expression kT which has dimensions of energy is such that $\frac{\beta x^2}{kT}$ is dimensionless, so we use

$$[\beta][L^2]=[ML^2T^{-2}]$$

$$\Rightarrow \quad D[\beta]=[MT^{-2}]$$

Here $\alpha^2\beta$ has dimensions of work, so we use

$$[\alpha^2][MT^{-2}]=[ML^2T^{-2}]$$

$$\Rightarrow \quad D[\alpha]=[M^0LT^0]$$

Ans. (D)

Sol. 20 The given expression can be rewritten as

$$\lambda=\frac{C}{V}=\frac{Q/V}{V}=\frac{Q}{V^2}$$

Voltage, can be expressed as

$$V=\frac{\text{work}}{Q}$$

$$\Rightarrow \quad \lambda=\frac{Q^3}{(\text{work})^2}=\frac{(\text{It})^3}{(F.s)^2}$$

$$\Rightarrow \quad D[\lambda]=\frac{[I^3T^3]}{[ML^2T^{-2}]^2}=[M^{-2}L^{-4}I^3T^7]$$

Ans. (D)

Sol. 21 If one vernier scale division length is a' then we use

$$(n-1)a=n(a')$$

$$\Rightarrow \quad a'=\frac{(n-1)a}{n}$$

Least count of the vernier callipers is given as

$$LC=1\text{ MSD}-1\text{ VSD}$$

$$\Rightarrow \quad LC=(a-a')\text{ cm}$$

$$\Rightarrow \quad LC=a-\frac{(n-1)a}{n}$$

$$\Rightarrow \quad LC=\frac{na-na+a}{n}=\frac{a}{n}\text{ cm}$$

$$\Rightarrow \quad LC=\left(\frac{10a}{n}\right)\text{mm}$$

Ans. (D)

Sol. 22 Volume of 100 drops is equal to the volume of solution layer and if the acid layer of thickness t is there in the solution, we have

$$4t=100\times\frac{4}{3}\pi r^3\times 0.01$$

$$\Rightarrow \quad 4t=100\times\frac{4\pi}{3}\times\frac{3}{40\pi}\times10^{-9}\times0.01$$

$$\Rightarrow \quad x=25$$

Ans. (25)

Sol. 23 Dimensions of entropy, Boltzman's constant, gas constant, mechanical equivalent of heat are given as

$$[S]=\left[\frac{Q}{\Delta T}\right]=\frac{ML^2T^{-2}}{K}$$

$$[K]=[S]=\frac{ML^2T^{-2}}{K}$$

$$[R]=\left[\frac{\text{Energy}}{\text{nT}}\right]=\frac{ML^2T^{-2}}{\text{mol K}}$$

$$[J]=M^0L^0T^0$$

$$[\mu KR]=[J\beta^2]$$

$$[\beta]=ML^2T^{-2}K^{-1}$$

$$[\alpha^2]=\left[\frac{S}{\beta}\right]=\frac{ML^2T^{-2}}{K\times ML^2T^{-2}K^{-1}};\ \alpha=M^0L^0T^0$$

Ans. (D)

Sol. 24 Mass can be written in terms of t, v and l with dimensions x, y and z respectively and given as

$$M\propto t^x v^y l^z$$

$$M^0L^0T^0=t^x[LT^{-1}]^y[ML^2T^{-1}]^z$$

$$M^1L^0T^0=t^{x-y-z}L^{y+2z}M^z$$

On comparing powers at LHS and RHS, we get

$$z=1 \quad \ldots(1)$$

$$x-y-z=0 \quad \ldots(2)$$

$$y+2z=0 \quad \ldots(3)$$

$$\Rightarrow \quad y+2\times 1=0$$

$$\Rightarrow \quad y=-2$$

$$\Rightarrow \quad x-(-2)-1=0$$

$$\Rightarrow \quad x=-1$$

$$\Rightarrow \quad M\propto t^{-1}v^{-2}l^1$$

$$\Rightarrow \quad [M]\propto[t^{-1}v^{-2}l]$$

Ans. (A)

Sol. 25 By homogeneity of given equation, the dimensions of constants are taken as

$$A=MLT^{-2},\ B=L^{-1},\ D=T^{-1}$$

Thus dimension formula of $\frac{AD}{B}$ can be written as

$$\left[\frac{AD}{B}\right]=\frac{MLT^{-2}T^{-1}}{L^{-1}}=ML^2T^{-3}$$

Ans. (A)

Sol. 26 (a) Capacitance

$$C=\frac{q}{V}=\frac{q}{\left(\frac{W}{q}\right)}=\frac{q^2}{W}=\frac{(it)^2}{f\cdot x}=\frac{[AT]^2}{[M^1L^2T^{-2}]}$$

$$C=[M^{-1}L^{-2}T^4A^2]$$

(b) $$F=\frac{kq^2}{r^2}=\frac{1}{4\pi\varepsilon_0}\frac{q^2}{r^2}$$

$$\varepsilon_0=\frac{q^2}{4\pi Fr^2}=\frac{[AT]^2}{[M^1L^1T^{-2}][L^2]}$$

$$\varepsilon_0=[M^{-1}L^{-3}T^4A^2]$$

(c) $$C=\frac{1}{\sqrt{\mu_0\varepsilon_0}}\Rightarrow C^2=\frac{1}{\mu_0\varepsilon_0}$$

$$\mu_0=\frac{1}{[C^2\varepsilon_0]}=\frac{1}{[L^2T^{-2}][M^{-1}L^{-3}T^4A^2]}$$

$$\mu_0=[M^1L^1T^{-2}A^{-2}]$$

(d) $$F=qE$$

$$E=\frac{F}{q}=\frac{M^1L^1T^{-2}}{A^1T^1}$$

$$E=[M^1L^1T^{-3}A^{-1}]$$

So correct option is (B) **Ans. (B)**

Sol. 27 Lease Count of screw Gauge is given as

$$LC=\frac{0.5}{50}=0.01\text{ mm}$$

Zero error of screw Gauge $=5\times LC=5\times 0.01=0.05$ mm

Reading of screw gauge given as

R = main scale reading + circular scale reading × $L.C.$ – Zero error

$$\Rightarrow \quad R=5+20\times 0.01-0.05=5.15\text{ mm}$$

Ans. (D)

Sol. 28 Dimensions of all physical quantities in the expression of P are substituted as

$$P=EL^2M^{-2}G^{-2}$$

$$\Rightarrow \quad P=[ML^2T^{-2}][ML^2T^{-1}]2\,[M^{-5}][M^{-1}L^3T^{-2}]^{-2}$$

$$\Rightarrow \quad P=[M^0L^0T^0]$$

Ans. (D)

Sol. 29 By dimensional analysis lists can be directly matched with appropriate options. **Ans. (B)**

Sol. 30 Permeability of free space has dimensions and all other option are dimensionless quantities. **Ans. (A)**

Sol. 31 The dimensional expression of electric field and magnetising field can be written as

$$D\left[\frac{E}{H}\right]=\frac{\text{newton/coulomb}}{\text{ampere/m}}$$

$$\Rightarrow \quad D\left[\frac{E}{H}\right]=\text{Joule/(second-Ampere}^2\text{)}$$

Ans. (C)

Sol. 32 By dimensional analysis list I can be matched directly with List II. **Ans. (C)**

Sol. 33 Dimension of density if taken as a, b and c then we have

$$d \propto F^a L^b T^c$$

$$\Rightarrow \quad [ML^{-3}] = [MLT^{-2}]^a [L^b] [T^c]$$

$$\Rightarrow \quad [ML^{-3}] = [M^a L^{a+b} T^{-2a+c}]$$

Comparing dimensions of LHS and RHS, we have

$$a = 1 \quad \ldots(1)$$

$$a + b = -3 \quad \ldots(2)$$

$$-2a + c = 0 \quad \ldots(3)$$

Solving equations (1), (2) and (3), we have

$$c = 2$$

$$b = -4$$

$$\Rightarrow \quad [d] = [FL^{-4}T^2]$$ **Ans. (A)**

Sol. 34 List I can directly be matched with the dimensions of the physical quantities listed in List II. **Ans. (D)**

Sol. 35 Option-(A) is the expression of displacement current. option-(C) is the expression of capillary height and option-(D) is the work calculation in rotational motion. Expression of option -(B) is that of volume flow rate whereas in its LHS volume is given so this is dimensionally wrong. **Ans. (B)**

Sol. 36 If the dimensions are taken as a, b and c respectively, we use

$$[M] = [V]^a [T]^b [F]^c$$

$$\Rightarrow \quad [M] = [LT^{-1}]^a [T]^b [MLT^{-2}]^c$$

$$\Rightarrow \quad [M] = [M]^c [L]^{a+c} [T]^{-a+b-2c}$$

Comparing dimensions at LHS and RHS gives

$$c = 1, \; a + c = 0 \text{ and } b - a - 2c = 0$$

Solving these equations, we get

$$a = -1 \text{ and } b = -1 + 2 = 1$$

$$\Rightarrow \quad [M] = [V]^{-1} [T]^1 [F]^1$$ **Ans. (D)**

Sol. 37 For 9 main scale divisions are coinciding with 10 vernier scale divisions, we use

$$9MSD = 10\,VSD$$

$$\Rightarrow \quad 1\,VSD = 0.9\,\text{mm}$$

Least count of vernier callipers can be calculated as

$$LC = \frac{MSD - VSD}{MSD} = \frac{1 - 0.9}{1} = 0.1\,\text{mm}$$

Reading = Main scale reading + $LC \times$ vernier scale reading

$$\text{Reading} = 10 + 0.1 \times 8$$

$$\text{Reading} = 10.8$$

Considering zero error into account, final diameter of spherical bob is given as

$$d = 10.8\,\text{mm} - 0.04\,\text{cm}$$

$$\Rightarrow \quad d = 10.8\,\text{mm} - 0.4\,\text{mm}$$

$$\Rightarrow \quad d = 10.4\,\text{mm}$$

Eadius of the bob is calculated as

$$r = 5.2\,\text{mm}$$

$$\Rightarrow \quad r = 0.52\,\text{cm} = 52 \times 10^{-2}\,\text{cm}$$ **Ans. (52)**

2.1 Motion in One Dimension

Sol. 1

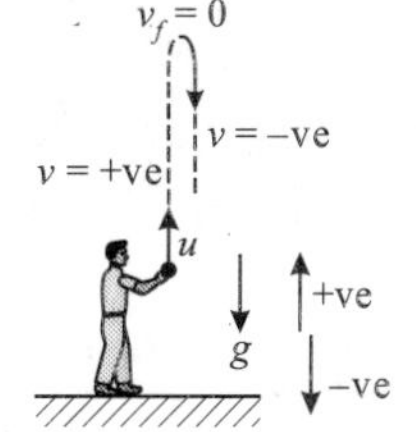

Acceleration $a = \dfrac{dv}{dt} = -g$ (constant)

Option (D) can be discarded as at one time instant velocity in two directions is NOT possible and options (B) can also be discarded as after topmost point direction (sign) of velocity is not changing in this case and option (C) can be discarded as velocity increases from beginning in this case. **Ans. (A)**

Sol. 2 v +ve, $v_f = 0$, x; a −ve, speed decreases

Acceleration of object is taken as

$$a = -v\frac{dv}{dx}$$

slope of v-x curve $\dfrac{dv}{dx} = -\dfrac{a}{v}$

with time initially v decreases so

$$\left|\frac{dv}{dx}\right| = \frac{a}{v} \text{ increases}$$

This is only shown in option (C). **Ans. (C)**

Sol. 3

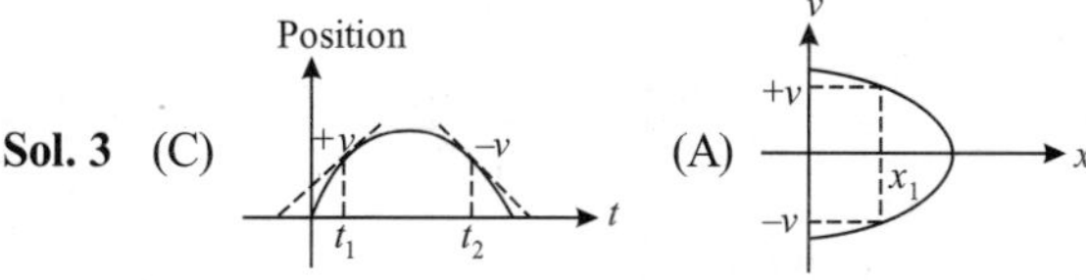

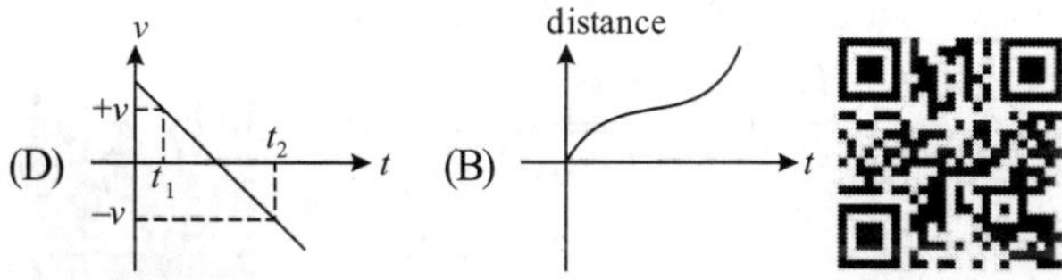

Actual distance-time curve for the given situation is shown above which speed is first increasing then decreasing as in the option (B) curve. **Ans. (B)**

Sol. 4 In first case $u_1 = 40$ kph

Stopping distance $s_1 = 40\,\text{m}$

$\Rightarrow$ We use $s_1 = \dfrac{u_1^2}{2a}$

$$\Rightarrow \quad a = \frac{u_1^2}{2s_1} = \frac{1600}{2\times 40} = 20 \text{ units}$$

In second case $u_2 = 80$ km/h

Stopping distance $s_2 = \dfrac{u_2^2}{2a} = \dfrac{6400}{2\times 20} = 160\,\text{m}$ **Ans. (B)**

Sol. 5 Distance travelled by car in 15s is

$$S_{car} = \frac{1}{2}\left(\frac{45}{15}\right)(15)^2 = \frac{1}{2}\times 45\times 15 = \frac{675}{2}\,\text{m}$$

Distance travelled by scooter in 15s is

$$S_{scooter} = vt = 30\times 15 = 450\,\text{m}$$

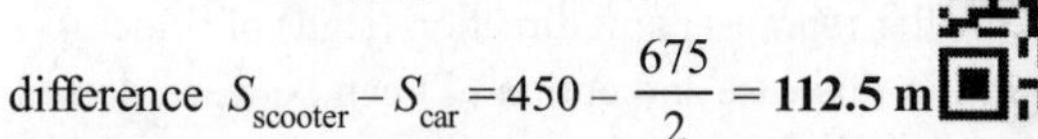

difference $S_{scooter} - S_{car} = 450 - \dfrac{675}{2} = \mathbf{112.5\ m}$

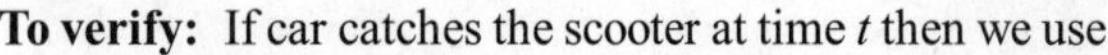

To verify: If car catches the scooter at time t then we use

$$S_{car\ t} = S_{scooter\ t} \qquad [t > 15\text{s}]$$

$$\frac{675}{2} + 45(t-15) = 30\,t$$

$$\Rightarrow \quad 15t = \frac{675}{2} \Rightarrow t = \mathbf{22.5\ s}$$ **Ans. (C)**

Sol. 6 The time taken by car to reach point P is

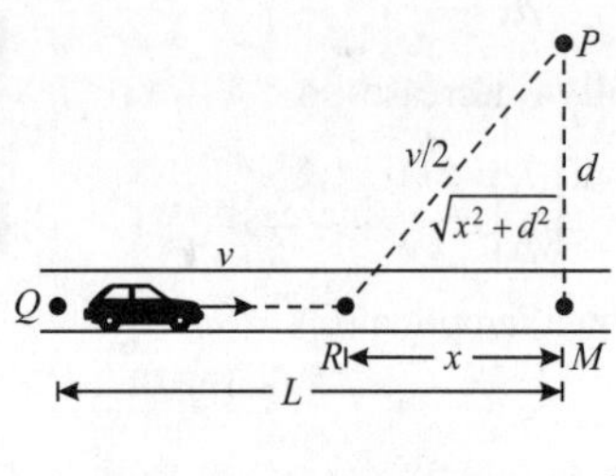

$$t = \frac{L-x}{v} + \frac{\sqrt{x^2+d^2}}{v/2}$$

t is minimum when $\dfrac{dt}{dx} = 0$

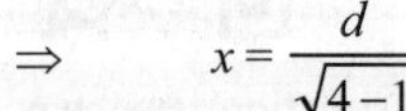

$$\Rightarrow \quad \frac{dt}{dx} = \frac{1}{v}\left[-1 + 2\frac{1}{2\sqrt{x^2+d^2}}\cdot 2x\right] = 0$$

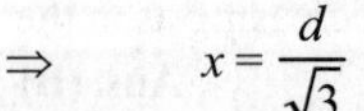

$$\Rightarrow \quad x = \frac{d}{\sqrt{4-1}}$$

$$\Rightarrow \quad x = \frac{d}{\sqrt{3}}$$ **Ans. (A)**

Sol. 7 For A & B let time taken by A is t_0

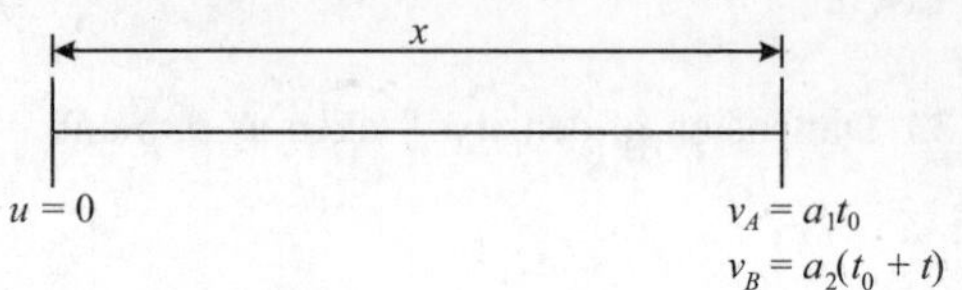

Given that

$$v_A - v_B = v = (a_1 - a_2)t_0 - a_2 t \qquad \ldots(1)$$

$$\Rightarrow \quad x_B = x_A = \frac{1}{2}a_1 t_0^2 = \frac{1}{2}a_2(t_0+t)^2$$

$$\Rightarrow \quad \sqrt{a_1}\,t_0 = \sqrt{a_2}\,(t_0+t)$$

$$\Rightarrow \quad (\sqrt{a_2} - \sqrt{a_2})t_0 = \sqrt{a_2}\,t \qquad \ldots(2)$$

Putting t_0 in equation, we have

$$\Rightarrow \quad v = (a_1 - a_2)\left(\frac{\sqrt{a_2}\,t}{\sqrt{a_1}-\sqrt{a_2}}\right) - a_2 t$$

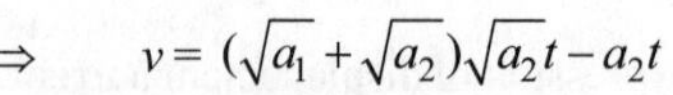

$$\Rightarrow \quad v = (\sqrt{a_1}+\sqrt{a_2})\sqrt{a_2}\,t - a_2 t$$

$$\Rightarrow \quad v = \sqrt{a_1 a_2}\,t + a_2 t - a_2 t$$

$$\Rightarrow \quad v = \sqrt{a_1 a_2}\,t$$ **Ans. (D)**

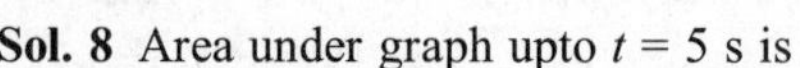

Sol. 8 Area under graph upto $t = 5$ s is

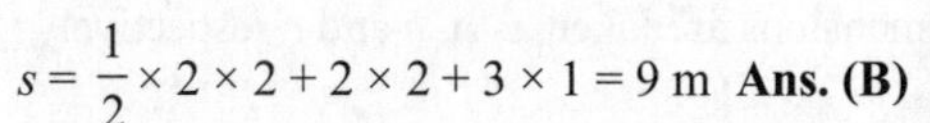

$$s = \frac{1}{2}\times 2\times 2 + 2\times 2 + 3\times 1 = 9\ \text{m}$$ **Ans. (B)**

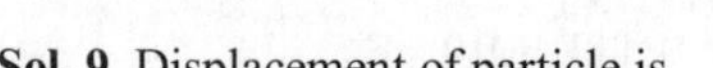

Sol. 9 Displacement of particle is

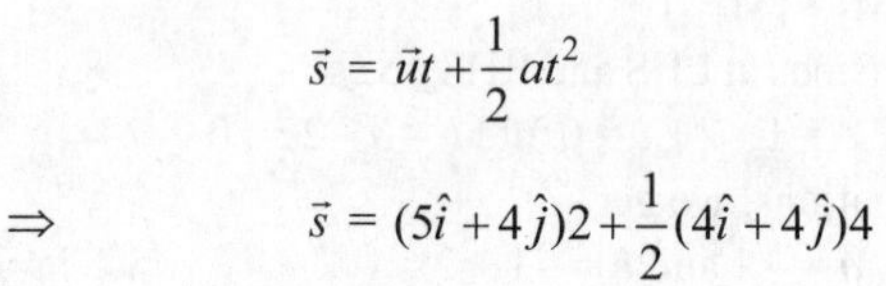

$$\vec{s} = \vec{u}t + \frac{1}{2}at^2$$

$$\Rightarrow \quad \vec{s} = (5\hat{i}+4\hat{j})2 + \frac{1}{2}(4\hat{i}+4\hat{j})4$$

$$\Rightarrow \quad \vec{s} = 10\hat{i} + 8\hat{j} + 8\hat{i} + 8\hat{j}$$

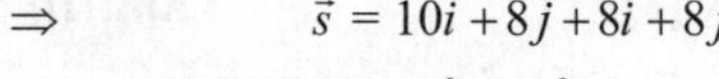

$$\Rightarrow \quad \vec{r}_f - \vec{r}_i = 18\hat{i} + 16\hat{j}$$

$$\Rightarrow \quad \vec{r}_f = 20\hat{i} + 20\hat{j} \qquad (\text{As } \vec{r}_i = 2\hat{i} + 4\hat{j})$$

$$\Rightarrow \quad |\vec{r}_f| = 20\sqrt{2}$$ **Ans. (A)**

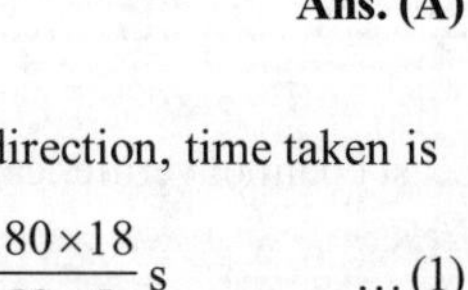

Sol. 10 When trains travel in same direction, time taken is

$$t_1 = \frac{180\times 18}{(80-30)\times 5} = \frac{180\times 18}{50\times 5}\,\text{s} \qquad \ldots(1)$$

When trains travel in opposite direction, time taken is

$$t_2 = \frac{180\times 18}{(80+30)\times 5} = \frac{180\times 18}{110\times 5}\,\text{s} \qquad \ldots(2)$$

From (1) and (2), we have

$$\frac{t_1}{t_2} = \frac{11}{5}$$ **Ans. (D)**

Sol. 11 Given initial velocity $u = 0$ and acceleration is constant

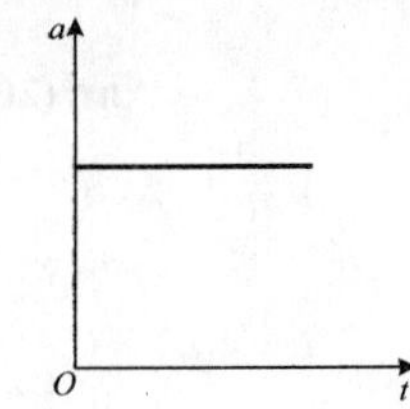

At time t, velocity is given as

$$v = 0 + at$$

$$\Rightarrow \quad v = at$$

and displacement is given as

$$x = 0(t) + \frac{1}{2}at^2$$

$$\Rightarrow \quad x = \frac{1}{2}at^2$$

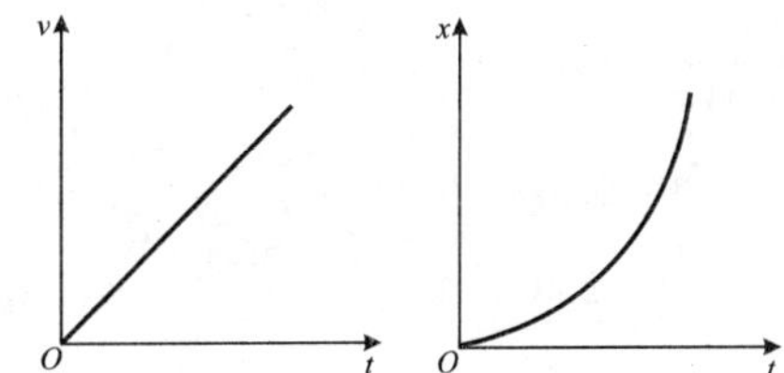

Graph (1) ; (2) and (4) are correct. **Ans. (C)**

Sol. 12 Given that

$$|\vec{A}_1| = 3$$

$$|\vec{A}_2| = 5$$

$$|\vec{A}_1 + \vec{A}_2| = 5$$

$$\Rightarrow \quad |\vec{A}_1 + \vec{A}_2| = \sqrt{|\vec{A}_1|^2 + |\vec{A}_2|^2 + 2|\vec{A}_1||\vec{A}_2|\cos\theta}$$

$$\Rightarrow \quad 5 = \sqrt{9 + 25 + 2 \times 3 \times 5\cos\theta}$$

$$\Rightarrow \quad \cos\theta = -\frac{9}{2 \times 3 \times 5} = -\frac{3}{10}$$

$$\Rightarrow \quad (2\vec{A}_1 + 3\vec{A}_2)\cdot(3\vec{A}_1 - 2\vec{A}_2)$$

$$= 6|\vec{A}_1|^2 + 9\vec{A}_1\cdot\vec{A}_2 - 4\vec{A}_1\vec{A}_2 - 6|\vec{A}_2|^2$$

$$= 54 + 5 \times 3 \times 5\left(-\frac{3}{10}\right) - 6 \times 25$$

$$= 54 - 150 - \frac{45}{2} = -118.5$$

Ans. (C)

Sol. 13 Momentum

$$p = mv \qquad \ldots(1)$$

For motion under gravity, we use

$$h = \frac{u^2 - v^2}{2g} \qquad \ldots(2)$$

$$h = \frac{u^2 - p^2/m^2}{2g}$$

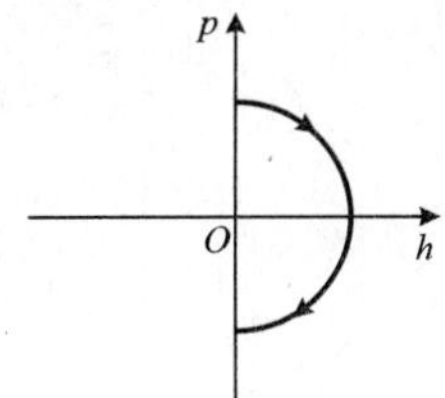

Ans. (A)

Sol. 14 Given that

$$x = at + bt^2 - ct^3$$

$$\Rightarrow \quad v = \frac{dx}{dt} = a + 2bt - 3ct^2$$

$$\Rightarrow \quad a = \frac{dv}{dt} = 2b - 6ct = 0$$

$$\Rightarrow \text{At} \quad t = \frac{b}{3c}, a = 0$$

$$\Rightarrow \quad v_{\left(at\ t=\frac{b}{3c}\right)} = a + 2b\left(\frac{b}{3c}\right) - 3c\left(\frac{b}{3c}\right)$$

$$= a + \frac{b^2}{3c}$$

Ans. (D)

Sol. 15 Given that

$$v = b\sqrt{x}$$

$$\Rightarrow \quad \frac{dv}{dt} = \frac{b}{2\sqrt{x}}\frac{dx}{dt}$$

$$\Rightarrow \quad a = \frac{bv}{2\sqrt{x}}$$

$$\Rightarrow \quad a = \frac{b(b\sqrt{x})}{2\sqrt{x}} = \frac{b^2}{2}$$

For constant acceleration velocity after time $t = \tau$ is given as

$$v = 0 + at$$

$$\Rightarrow \quad v = \frac{b^2}{2}\tau$$

Ans. (B)

Sol. 16 Total time taken by ball to travel 81 m is t s and for complete 100 m is $\left(t + \frac{1}{2}\right)$ s, by speed equation, we use

$$\Rightarrow \quad 81 = \frac{1}{2} \times g \times t^2$$

$$\Rightarrow \quad t = 9\sqrt{\frac{2}{g}}$$

$$\Rightarrow \quad 100 = \frac{1}{2} \times g \times \left(t + \frac{1}{2}\right)^2$$

$$\Rightarrow \quad t + \frac{1}{2} = 10\sqrt{\frac{2}{g}}$$

$$\Rightarrow \quad 9\sqrt{\frac{2}{g}} + \frac{1}{2} = 10\sqrt{\frac{2}{g}}$$

$$\Rightarrow \quad \frac{1}{2} = \sqrt{\frac{2}{g}}$$

$$\Rightarrow \quad a = 8 \text{ m/s}^2$$ **Ans. (8)**

Sol. 17 From the given expression, we have

$$x = \sqrt{at^2 + 2bt + c}$$

Velocity of particle is given as

$$v = \frac{dx}{dt} = \frac{1}{2\sqrt{at^2 + 2bt + c}} \times (2at + 2b)$$

$$\Rightarrow \quad v = \frac{at + b}{x}$$

$$\Rightarrow \quad vx = at + b$$

Differentiating above expression gives

$$vdx + xdv = adt$$

$$\Rightarrow \quad x\frac{dv}{dx} + v = a \times \frac{dt}{dx}$$

Multiplying both side by v, we get

$$\left(v\frac{dv}{dx}\right)x + v^2 = a$$

We use acceleration $f = v\frac{dv}{dx}$ so we have

$$fx = a - v^2$$

$$\Rightarrow \quad fx = a - \left(\frac{at + b}{x}\right)^2$$

$$\Rightarrow \quad fx = \frac{ax^2 - (at + b)^2}{x^2}$$

$$\Rightarrow \quad fx = \frac{a(at^2 + 2bt + c) - (at + b)^2}{x^2}$$

$$\Rightarrow \quad fx = \frac{ac - b^2}{x^2}$$

$$\Rightarrow \quad f = \frac{ac - b^2}{x^3}$$

$$\Rightarrow \quad f \propto \frac{1}{x^3}$$

$$\Rightarrow \quad n = 3$$ **Ans. (3.00)**

Sol. 18 Distance = Area under curve

$$s = \frac{1}{2} \times 5 \times 8 = 20 \text{ m}$$ **Ans. (20)**

Sol. 19

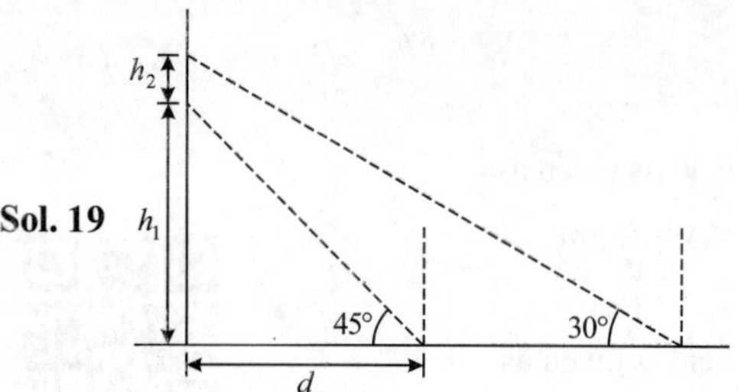

From above figure, we use

$$\frac{h_1}{d} = \tan 45^\circ$$

$$\Rightarrow \quad h_1 = d \quad \ldots(1)$$

and $$\frac{h_1 + h_2}{d + 2.464\,d} = \tan 30^\circ$$

$$\Rightarrow \quad (h_1 + h_2) \times \sqrt{3} = 3.46d$$

$$\Rightarrow \quad (h_1 + h_2) = \frac{3.46\,d}{\sqrt{3}}$$

$$\Rightarrow \quad d + h_2 = \frac{3.46\,d}{\sqrt{3}}$$

$$\Rightarrow \quad h_2 = d$$ **Ans. (A)**

Sol. 20

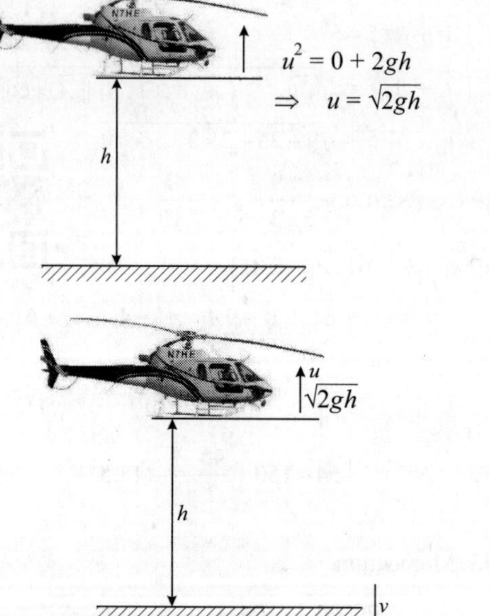

For food packet we use

$$-h = \sqrt{2gh}t - \frac{1}{2}gt^2$$

$$\Rightarrow \quad t^2 - 2\sqrt{\frac{2h}{g}}t - \frac{2h}{g} = 0$$

$$\Rightarrow \quad t = 2\sqrt{\frac{2h}{g}} \pm \sqrt{\frac{8h}{g} + \frac{2h}{g}}$$

$$\Rightarrow \quad t = \sqrt{\frac{2h}{g}} + 2\sqrt{\frac{h}{g}}$$

$$\Rightarrow \quad t = (2+\sqrt{2})\sqrt{\frac{h}{g}} = 3.4\sqrt{\frac{h}{g}}$$

Ans. (C)

Sol. 21 Total distance covered can be given by area under the *v-t* curve. In given figure

$$OS = 4 + \frac{1}{3} = \frac{13}{3}$$

$$SD = 2 - \frac{1}{3} = \frac{5}{3}$$

Area of *OABS* is

$$A_1 = \frac{1}{2}\left[\frac{13}{3} + 1\right] \times 4 = \frac{32}{3}\text{ m}$$

Area of *SCD* is

$$A_2 = \frac{1}{2} \times \frac{5}{3} \times 2 = \frac{5}{3}\text{ m}$$

Total distance covered

$$s = |A_1| + |A_2|$$

$$\Rightarrow \quad s = \frac{32}{3} + \frac{5}{3}\text{ m}$$

$$\Rightarrow \quad s = \frac{37}{3}\text{ m}$$

Ans. (D)

Sol. 22 Slope of velocity-time graph gives acceleration so acceleration time graph can be drawn as given below from the given graph. AM part of curve has negative slope and MB part has positive slope and it is constant also so acceleration time curve can be drawn as

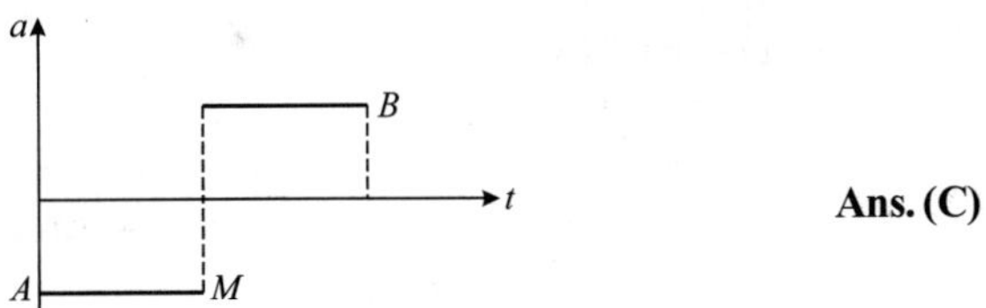

Ans. (C)

Sol. 23 Same as Question asked in 2012 Online. **Ans. (A)**

Sol. 24 Speed of first particle after falling 5m is given as

$$v = \sqrt{2gh} = 10\text{ m/s}$$

Time to fall 5m is

$$t = \sqrt{\frac{2h}{g}} = 1\text{ s}$$

Time required for first particle to meet second one can be given as

$$t' = \frac{20}{10} = 2\text{ s}$$

Time taken by I[st] particle to reach ground = 1 + 2 = 3 s
Thus height of building is given as the distance fallen by first stone in 3s, calculated as

$$H = \frac{1}{2}g(3)^2 = 45\text{ m}$$

Ans. (B)

Sol. 25 As in this situation scooter is accelerating and retarding in same journey so we should draw *v-t* curve of the situation as

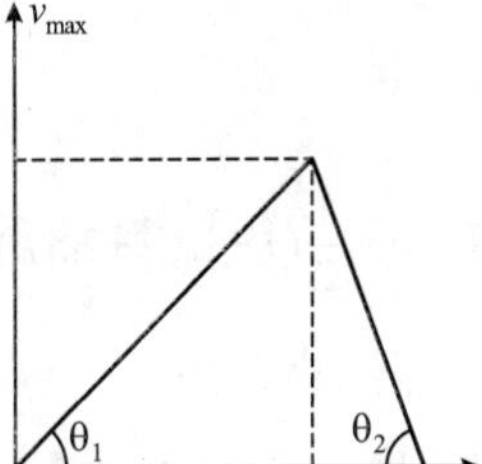

In above curve, we use

$$\tan\theta_1 = a_1 = \frac{v_{max}}{t_1}$$

and

$$\tan\theta_2 = a_2 = \frac{v_{max}}{t_2}$$

From above two equations, we have

$$\frac{t_1}{t_2} = \frac{a_2}{a_1}$$

Ans. (B)

Sol. 26 For motion of bicycle from $0 \le x \le 200$, velocity is given as

$$v = \frac{1}{5}x + 10$$

Thus acceleration is calculated as

$$a = \frac{vdv}{dx} = \left(\frac{x}{5} + 10\right)\left(\frac{1}{5}\right)$$

$$\Rightarrow \quad a = \frac{x}{25} + 2$$

At $x = 0$, $a = 2\text{ m/s}^2$ and at $x = 200$, $a = 10\text{ m/s}^2$ so upto $x = 200$ m, acceleration vs displacement curve will be a straight line and then onward $a = 0$. Hence the curve should be drawan like the one shown in figure below.

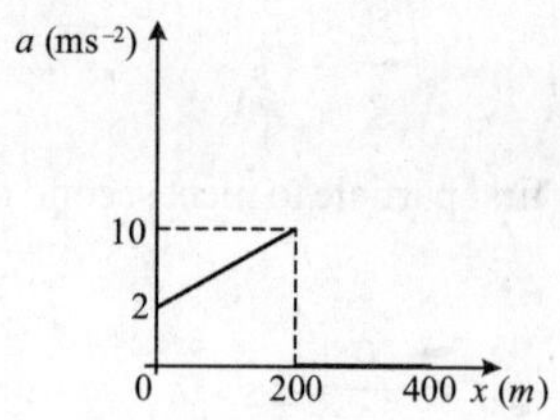

None of the given option matches with this. Most approriate shape of curve matches with option (A), or none would be correct. **Ans. (A)**

Sol. 27 Acceleration of the body is given as

$$\vec{a}=\frac{\vec{F}}{m}=\frac{2\hat{i}+3\hat{j}+5\hat{k}}{2}=\hat{i}+1.5\hat{j}+2.5\hat{k}$$

After time t, the displacement is given as

$$\vec{s}=\vec{u}t+\frac{1}{2}\vec{a}t^2$$

$$\Rightarrow \quad \vec{s}=0+\frac{1}{2}(\hat{i}+1.5\hat{j}+2.5\hat{k})(16)$$

$$\Rightarrow \quad \vec{s}=8\hat{i}+12\hat{j}+20\hat{k}$$

$$\Rightarrow \quad b=12$$ **Ans. (12)**

Sol. 28 If car accelerates for time t_1 and retards for time t_2 and the maximum speed of car in this journey is v_0, the velocity time curve of the car can be drawn as shown in figure below

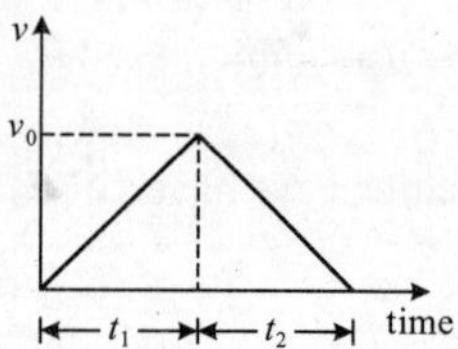

From graph we can use

$$v_0=\alpha t_1=\beta t_2$$

Total time elapsed is given as

$$t_1+t_2=t$$

$$\Rightarrow \quad v_0\left(\frac{1}{\alpha}+\frac{1}{\beta}\right)=t$$

$$\Rightarrow \quad v_0=\frac{\alpha\beta t}{\alpha+\beta}$$

Total distance travelled can be calculated by the area under velocity time curve, given as

$$s=\frac{1}{2}\times t\times v_0=\frac{1}{2}\times t\times\frac{\alpha\beta t}{\alpha+\beta}=\frac{\alpha\beta t^2}{2(\alpha+\beta)}$$ **Ans. (C)**

Sol. 29 The displacement of the particle can be calculated as

$$v=v_0+gt+Ft^2$$

$$\Rightarrow \quad \frac{ds}{dt}=v_0+gt+Ft^2$$

$$\Rightarrow \quad \int ds=\int_0^1(v_0+gt+Ft^2)\,dt$$

$$\Rightarrow \quad s=\left[v_0t+\frac{gt^2}{2}+\frac{Ft^3}{3}\right]_0^1$$

$$\Rightarrow \quad s=v_0+\frac{g}{2}+\frac{F}{3}$$ **Ans. (B)**

Sol. 30 For constant acceleration velocity time graph is straight line with positive slope and *x-t* graph should be an opening upward parabola so option (B) is appropriate among all given options. **Ans. (B)**

Sol. 31 From the given graph, velocity can be expressed as displacement function as

$$v=-\left(\frac{v_0}{x_0}\right)x+v_0$$

Acceleration of particle can be given as

$$a=\frac{vdv}{dx}$$

$$\Rightarrow \quad a=\left[-\left(\frac{v_0}{x_0}\right)x+v_0\right]\left[-\frac{v_0}{x_0}\right]$$

$$\Rightarrow \quad a=\left(\frac{v_0}{x_0}\right)^2x-\frac{v_0^2}{x_0}$$

Thus appropriate graph matches with option (C). **Ans. (C)**

Sol. 32 If a drop comes from the tab after every t seconds then distance travelled by 1st drop in 4 s is given as

$$s_1=\frac{1}{2}at_1^2=78.4\text{ m}$$

Distance travelled by succeeding drop in $(4-t)$ s

$$s_2=\frac{1}{2}a(4-t)^2$$

$$s_1-s_2=34.3$$

$$78.4-4.9(4-t)^2=34.3$$

$$(4-t)^2=9$$

$$4-t=3$$

$$t=1\text{ s}$$ **Ans. (A)**

Sol. 33 Differentiating the given time equation, we have

$$1=2mxv+nv$$

$$1=v(2mx+n)$$

Again differentiating we have

$$\frac{dv}{dt}\times(2mx+n)+2mv^2=0$$

$$\Rightarrow \quad a=-2mv^3$$ **Ans. (C)**

Sol. 34 For the given expression of velocity, we use

$$dx = vdt$$

$$\Rightarrow \quad x = \int_1^2 \alpha t dt + \int_1^2 \beta t^2 dt$$

$$\Rightarrow \quad x = \left[\frac{\alpha t^2}{2}\right]_1^2 + \left[\frac{\beta t^3}{3}\right]_1^2$$

$$\Rightarrow \quad x = \frac{3}{2}\alpha + \frac{7}{3}\beta$$

Ans. (B)

Sol. 35 For stone moving upward at 10 m/s, we use

$$75 = -10t + \frac{1}{2}gt^2$$

$$\Rightarrow \quad 75 = -10t + 5t^2$$

$$\Rightarrow \quad t^2 = 2t - 15 = 0$$

$$\Rightarrow \quad t = 5 \text{ s}$$

Height of balloon at time t is given as

$$H = vt + 75$$

$$\Rightarrow \quad H = 10 \times 5 + 75 = 125 \text{ m}$$

Ans. (A)

Sol. 36 Throw velocity of the ball to reach a height h is given as

$$v = \sqrt{2gh}$$

If t is the time required to reach a height $h/3$ then we use

$$\frac{h}{3} = \sqrt{2gh}t - \frac{1}{2}gt^2$$

$$\Rightarrow \quad gt^2 - 2\sqrt{2gh}t + \frac{2h}{3} = 0$$

$$\Rightarrow \quad t = \frac{2\sqrt{2gh} \pm \sqrt{8gh - 4g \times \frac{2h}{3}}}{2g} = \frac{2\sqrt{2gh} \pm \sqrt{\frac{16gh}{3}}}{2g}$$

$$\Rightarrow \quad t = \frac{2\sqrt{2gh} \pm 4\sqrt{\frac{gh}{3}}}{2g}$$

$$\Rightarrow \quad \frac{t_1}{t_2} = \frac{2\sqrt{2gh} - 4\sqrt{\frac{gh}{3}}}{2\sqrt{2gh} + 4\sqrt{\frac{gh}{3}}}$$

$$\Rightarrow \quad \frac{t_1}{t_2} = \frac{2\sqrt{2} - \frac{4}{\sqrt{3}}}{2\sqrt{2} + \frac{4}{\sqrt{3}}} = \frac{\sqrt{3} - \sqrt{2}}{\sqrt{3} + \sqrt{2}}$$

Ans. (C)

Sol. 37 $\quad S_1 = S_2$

$$35t + \frac{1}{2}(-g)t^2 = 35(t-3) + \frac{1}{2}(-g)(t-3)^2$$

$$35t - \frac{1}{2}gt^2 = 35t - 35 \times 3 - \frac{1}{2}g(t^2 - 6t + 9)$$

$$35 \times 3 + 45 = 30t$$

$$t = \frac{150}{30} = 5$$

$$\text{height } h = 35 \times 5 - \frac{1}{2} \times 10 \times 5^2$$

$$h = 175 - 125 = 50 \text{ m}$$

Ans. (50)

Sol. 38 Acceleration of body is given as

$$a = \frac{vdv}{dx}$$

$$\Rightarrow \quad v = \sqrt{5000 + 24x} \times \frac{1}{2\sqrt{5000 + 24x}} \times 24$$

$$\Rightarrow \quad a = 12 \text{ m/s}^2$$

Ans. (12)

Sol. 39 First drop strikes the floor at $t = 2$ s when third drop begins to fall. The distance from the nozzle is given as

$$h = \frac{1}{2}g(2)^2$$

2nd drop starts at $t = 1$ s so its depth below the nozzle is given as

$$h' = \frac{1}{2}g(1)^2$$

$$\Rightarrow \quad \frac{h}{h'} = \frac{4}{1}$$

$$\Rightarrow \quad h' = \frac{h}{4} = \frac{9.8}{4}$$

Height of 2nd drop above the floor is given as

$$H = h - h' = 9.8 - \frac{9.8}{4} = \frac{3}{4} \times 9.8 = 7.35 \text{ m}$$

Ans. (D)

Sol. 40 Acceleration of particle is given as

$$a = v\frac{dv}{dx}$$

From the graph we have the function of v^2 variation with x, given as

$$v^2 = 2x + 20$$

$$\Rightarrow \quad a = v\frac{dv}{dx} = 1 \text{ m/s}^2$$

Ans. (1)

3.1 Motion in Two Dimensions

Sol. 1 If we consider

$$\vec{C} = a\hat{i} + b\hat{j}$$

$$\vec{A}.\vec{C} = a + b$$

$$\vec{B}.\vec{C} = 2a - b$$

$$\vec{A}.\vec{B} = 2 - 1 = 1$$

$$a + b = 1 \quad \ldots(1)$$

$$\Rightarrow \quad 2a - b = 1 \quad \ldots(2)$$

Solving-(1) and (2) gives

$$a = \frac{1}{3};\ b = \frac{2}{3}$$

$$\Rightarrow \quad |\vec{C}| = \sqrt{a^2 + b^2}$$

$$= \sqrt{\left(\frac{1}{3}\right)^2\left(\frac{2}{3}\right)^2} = \sqrt{\frac{5}{9}}$$

Ans. (A)

Sol. 2 From the given velocity function, we use

$$\frac{dx}{dt} = ky \quad \text{and} \quad \frac{dy}{dt} = kx$$

$$\Rightarrow \quad \frac{dy}{dx} = \frac{\frac{dy}{dt}}{\frac{dx}{dt}} = \frac{x}{y}$$

$$\Rightarrow \quad ydy = xdx$$

Integrating both side, we get

$$y^2 = x^2 + C$$

Ans. (B)

Sol. 3 Velocity components of particle are

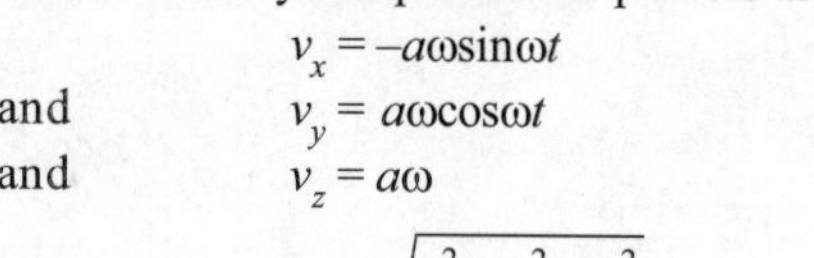

$$v_x = -a\omega\sin\omega t$$

and $$v_y = a\omega\cos\omega t$$

and $$v_z = a\omega$$

$$\Rightarrow \quad v = \sqrt{v_x^2 + v_y^2 + v_z^2}$$

$$\Rightarrow \quad v = \sqrt{2}a\omega$$

Ans. (C)

Sol. 4 Range of projectile is

$$R = \frac{u^2 \sin 2\theta}{g}$$

Area covered on ground is

$$A = \pi R^2$$

$$\Rightarrow \quad A \propto R^2$$

$$\Rightarrow \quad A \propto u^4$$

$$\Rightarrow \quad \frac{A_1}{A_2} = \frac{u_1^4}{u_2^4} = \left[\frac{1}{2}\right]^4 = \frac{1}{16}$$

Ans. (D)

Sol. 5 We have position vectors of points G and H are

$$\vec{r}_g = \frac{a}{2}\hat{i} + \frac{a}{2}\hat{k}$$

$$\vec{r}_H = \frac{a}{2}\hat{j} + \frac{a}{2}\hat{k}$$

$$\vec{r}_H - \vec{r}_g = \frac{a}{2}(\hat{j} - \hat{i})$$

Ans. (B)

Sol. 6 Resultant of $2F$ and $3F$ is given as

$$4F^2 + 9F^2 + 12F^2 \cos\theta = R^2 \quad \ldots(1)$$

When second force becomes $6F$ then resultant becomes $2R$, so we use

$$4F^2 + 36F^2 + 24F^2 \cos\theta = 4R^2 \quad \ldots(2)$$

From equation-(1) and (2), we have

$$4F^2 + 36F^2 + 24F^2 \cos\theta$$

$$= 4(13F^2 + 12F^2\cos\theta)$$

$$= 52F^2 + 48F^2\cos\theta$$

$$\Rightarrow \quad \cos\theta = -\frac{12F^2}{24F^2} = -\frac{1}{2}$$

$$\Rightarrow \quad \theta = 120°$$

Ans. (D)

Sol. 7 If θ is angle between $\vec{A}$ and $\vec{B}$, and $|\vec{A}| = |\vec{B}| = a$ then we have

$$|\vec{A} + \vec{B}| = 2a \cos\theta/2 \quad \ldots(1)$$

$$|\vec{A} - \vec{B}| = 2a\cos\frac{(\pi - \theta)}{2} = 2a \sin\theta/2 \quad \ldots(2)$$

$$\Rightarrow \quad n\left(2a\cos\frac{\theta}{2}\right) = 2a\frac{\sin\theta}{2}$$

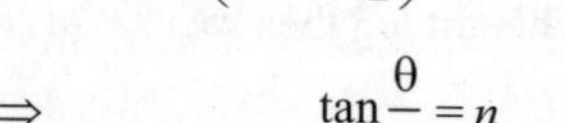

$$\Rightarrow \quad \tan\frac{\theta}{2} = n$$

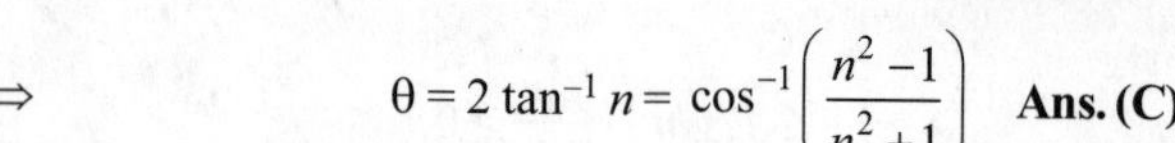

$$\Rightarrow \quad \theta = 2\tan^{-1} n = \cos^{-1}\left(\frac{n^2 - 1}{n^2 + 1}\right)$$

Ans. (C)

Sol. 8 Projection velocity components are

$$u_x = 10\cos 60° = 5 \text{ m/s}$$

$$u_y = 10\cos 30° = 5\sqrt{3} \text{ m/s}$$

velocity after $t = 1$s is given as

$$v_x = 5 \text{ m/s}$$

$$v_y = (5\sqrt{3} - 10) \text{ m/s} (-\text{ve} \Rightarrow \text{downward motion})$$

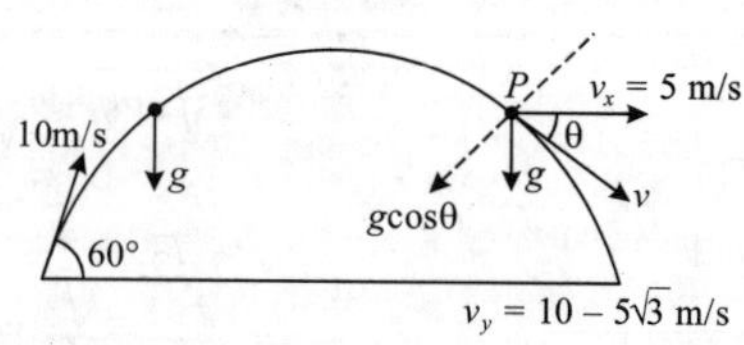

At point P, normal acceleration of body is

$$a_n = \frac{v^2}{R}$$

$$\Rightarrow \quad R = \frac{v_x^2 + v_y^2}{a_n} = \frac{25 + 100 + 75 - 100\sqrt{3}}{10\cos\theta}$$

Where $\tan\theta = \left|\frac{v_y}{v_x}\right| = \frac{10-5\sqrt{3}}{5} = 2-\sqrt{3}$

$\Rightarrow \quad \theta = 15^\circ$

$\Rightarrow \quad R = \frac{100(2-\sqrt{3})}{10\cos 15^\circ} = 2.8\text{ m}$ **Ans. (C)**

Sol. 9 Given $\vec{r} = 15t^2\hat{i} + (4-20t^2)\hat{j}$

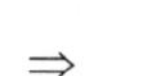

$\Rightarrow \quad \vec{v} = \frac{d\vec{r}}{dt} = 30t\hat{i} + (-40t)\hat{j}$

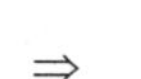

$\Rightarrow \quad \vec{a} = \frac{d\vec{v}}{dt} = 30\hat{i} - 40\hat{j}$

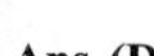

$\Rightarrow \quad |\vec{a}| = 50\text{ m/s}^2$ **Ans. (D)**

Sol. 10 Time of flight is

$$t = \frac{2\times 2\times \sin 15^\circ}{g\cos 30^\circ}$$

Range on inclined plane is

$$R = 2\cos 15^\circ \times t - \frac{1}{2}g\sin 30^\circ t^2$$

Solving, we get

$R \simeq 20\text{ cm}$ **Ans. (B)**

Sol. 11 Equation of trajectory is given as

$$y = 2x - 9x^2 \quad \ldots(1)$$

Comparing with general equation of projectile

$$y = x\tan\theta - \frac{g}{2u^2\cos^2\theta}\cdot x^2 \quad \ldots(2)$$

We get $\tan\theta = 2$

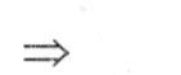

$\Rightarrow \quad \cos\theta = \frac{1}{\sqrt{5}}$

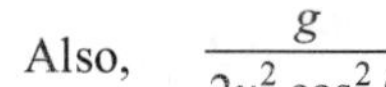

Also, $\frac{g}{2u^2\cos^2\theta} = 9$

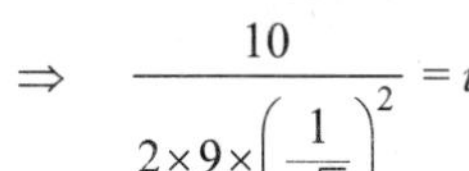

$\Rightarrow \quad \frac{10}{2\times 9\times\left(\frac{1}{\sqrt{5}}\right)^2} = u^2$

$\Rightarrow \quad u^2 = \frac{25}{9}$

$\Rightarrow \quad u = \frac{5}{3}\text{ m/s}$ **Ans. (A)**

Sol. 12 Range will be same for time t_1 & t_2, so angles of projection will be 'θ' & '90° – θ'

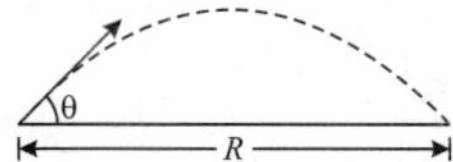

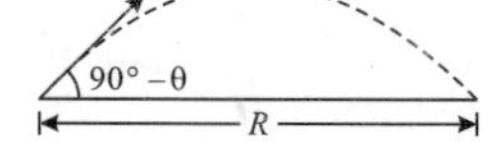

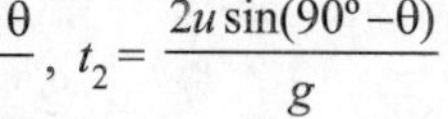

$$t_1 = \frac{2u\sin\theta}{g},\ t_2 = \frac{2u\sin(90^\circ-\theta)}{g}$$

and $R = \frac{u^2\sin 2\theta}{g}$

$$t_1 t_2 = \frac{4u^2\sin\theta\cos\theta}{g^2} = \frac{2}{g}\left[\frac{2u^2\sin\theta\cos\theta}{g}\right]$$

$$= \frac{2R}{g}$$ **Ans. (C)**

Sol. 13

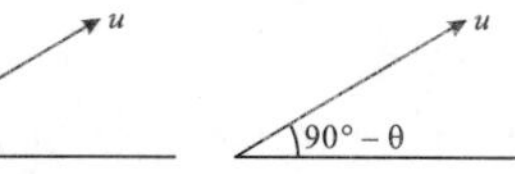

For same range angle of projection will be θ & 90 – θ, so we use

$$R = \frac{u^2 2\sin\theta\cos\theta}{g}$$

and time of flights are

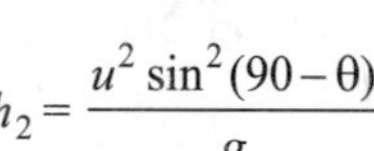

$$h_1 = \frac{u^2\sin^2\theta}{g}$$

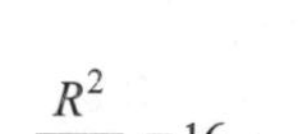

$$h_2 = \frac{u^2\sin^2(90-\theta)}{g}$$

$\Rightarrow \quad \frac{R^2}{h_1h_2} = 16$ **Ans. (B)**

Sol. 14 Position along x-direction is given as

$$x = 10 + 8t - 3t^2$$

$\Rightarrow \quad v_{x_1} = 8 - 6t$

$\Rightarrow \quad (v_{x_1})_{t=1} = 2\hat{i}$

Position along y-direction is given as

$$y = 5 - 8t^3$$

$\Rightarrow \quad v_{y_2} = -24t^2$

$\Rightarrow \quad (v_{y_2})_{t=1} = -24\hat{j}$

Now given that

$$\sqrt{v} = \sqrt{v_{x_1}^2 + v_{y_2}^2}$$

$$\sqrt{v} = \sqrt{(24)^2 + (2)^2} = \sqrt{580}$$

$\Rightarrow \quad v = 580\text{ m/s}$

Sol. 15 Given that

$$\vec{r}(t) = \cos\omega t\,\hat{i}$$

Velocity vector is given as

$$\vec{v} = \frac{\overrightarrow{dr}}{dt} = -\omega$$

Acceleration vector is given as

$$\vec{a} = \frac{d\vec{v}}{dt} = -\omega^2 \cos \omega t\, \hat{i} - \omega^2 \sin \omega t\, \hat{j} = -\omega^2 \vec{r}$$

$\Rightarrow \quad \vec{v} \cdot \vec{r} = 0$ **Ans. (A)**

Sol. 16 For uniform acceleration motion in both x and y direction, we use

$$x = u_x t + \frac{1}{2} a_x t^2 \quad \ldots(1)$$

and

$$y = u_y t + \frac{1}{2} a_y t^2 \quad \ldots(2)$$

$\Rightarrow \quad 32 = 0 \times t + \frac{1}{2}(4)(t)^2$ [Given that $u_y = 0$]

$\Rightarrow \quad t^2 = 16$

$\Rightarrow \quad t = 4$ s

Along x-direction, we use from equation-(1)

$$x = 3 \times 4 + \frac{1}{2} \times 6 \times 4^2$$

$\Rightarrow \quad x = 12 + 48 = 60$ m **Ans. (C)**

Sol. 17 Given that

$$\vec{u} = 5\hat{j} \text{ m/s}$$

$$\vec{a} = 10\hat{i} + 4\hat{j} \text{ m/s}$$

The coordinate $(20, y_0)$ after in time t are given as

Along x-direction

$$s_x = u_x t + \frac{1}{2} a_x t^2$$

$\Rightarrow \quad 20 - 0 = 0 + \frac{1}{2} \times 10 \times t^2$

$\Rightarrow \quad t = 2$ s

Along y-direction

$$s_y = u_y \times t + \frac{1}{2} a_y t^2$$

$\Rightarrow \quad y_0 = 5 \times 2 + \frac{1}{2} 4 \times 2^2 = 18$ m **Ans. (C)**

Sol. 18 As we ... v that for a closed polygon

$\overrightarrow{OA} + \overrightarrow{OB} + \ldots \overrightarrow{OD} + \overrightarrow{OE} + \overrightarrow{OF} + \overrightarrow{OG} + \overrightarrow{OH} = \vec{0}$

By triangle law of vec... ition, in the given octagon, we can use

$\overrightarrow{AB} = \overrightarrow{AO} + \ldots$

$\overrightarrow{AC} = \overrightarrow{AO} + \ldots$

$\overrightarrow{AD} = \ldots$

$\overrightarrow{AE} = \ldots$

$\overrightarrow{AF} = \overrightarrow{AO} \ldots$

$\overrightarrow{AG} = \overrightarrow{AO} + \overrightarrow{OG}$

$\overrightarrow{AH} = \overrightarrow{AO} + \overrightarrow{OH}$

$\Rightarrow \overrightarrow{AB} + \overrightarrow{AC} + \overrightarrow{AD} + \overrightarrow{AE} + \overrightarrow{AF} + \overrightarrow{AG} + \overrightarrow{AH}$

$= (7\,\overrightarrow{AO}) + \overrightarrow{OB} + \overrightarrow{OC} + \overrightarrow{OD} + \overrightarrow{OE} + \overrightarrow{OF} + \overrightarrow{OG} + \overrightarrow{OH}$

$= (7\,\overrightarrow{AO}) + \overrightarrow{AO} = 8\overrightarrow{AO}$

$= 8(2\hat{i} + 3\hat{j} - 4\hat{k}) = 16\hat{i} + 24\hat{j} - 32\hat{k}$ **Ans. (B)**

Sol. 19 Figure below shows the situation explained in the question

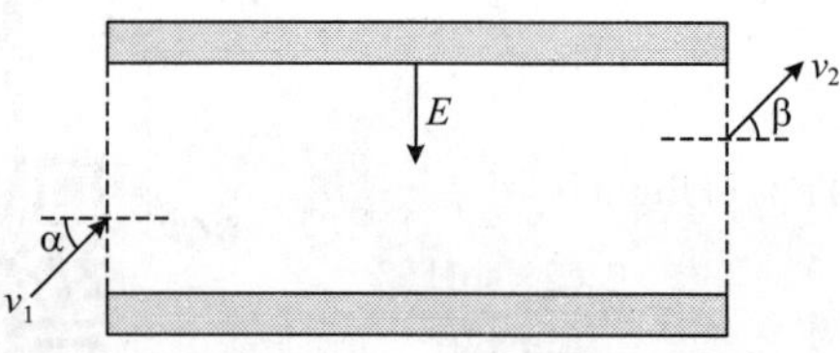

As acceleratoin is perpendicular to plates so velocity along the plate will remain constant

$\Rightarrow \quad v_1 \cos \alpha = v_2 \cos \beta$

$\Rightarrow \quad \frac{K_1}{K_2} = \frac{v_1^2}{v_2^2} = \frac{\cos^2 \beta}{\cos^2 \alpha}$ **Ans. (D)**

Sol. 20 For the given vector equation, we can write

$$-PQ \sin \theta = PQ \sin \theta$$

$\Rightarrow \quad \theta = 180°$ **Ans. (180)**

Sol. 21 From the given trajectory equation

$$y = \alpha x - \beta x^2$$

By comparing with the standard trajectory equation

$$y = x \tan \theta - \frac{1}{2} \frac{gx^2}{u^2 \cos^2 \theta}$$

We have

$$\tan \theta = \alpha$$

$\Rightarrow \quad \theta = \tan^{-1} \alpha$

and

$$\beta = \frac{1}{2} \frac{g}{u^2 \cos^2 \theta}$$

$\Rightarrow \quad u^2 = \frac{g}{2\beta \cos^2 \theta}$

Thus maximum height during flight can be given as

$$H = \frac{u^2 \sin^2 \theta}{2g} = \frac{g}{2\beta \cos^2 \theta} \frac{\sin^2 \theta}{2g}$$

$\Rightarrow \quad H = \frac{\tan^2 \theta}{4\beta} = \frac{\alpha^2}{4\beta}$ **Ans. (C)**

Sol. 22 Then angles between rods and vertical are shownin figure below.

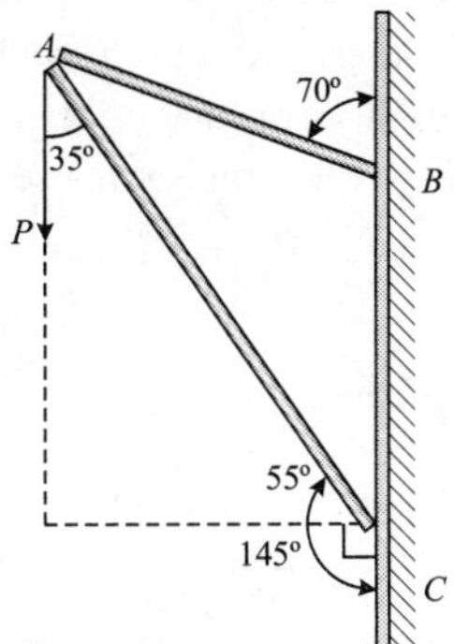

The component along AC of applied force is given as

$$P_c = 100 \cos 35° \text{ N}$$

$$\Rightarrow \quad P_c = 100 \times 0.819 \text{ N}$$

$$\Rightarrow \quad P_c = 81.9 \text{ N} \approx 82 \text{ N}$$ **Ans. (82)**

Sol. 23 Velocity of mosquito after 2 s is given as

$$\vec{v}_{t=2} = 2\hat{i} + 6\hat{j} + 9\hat{k}$$

Angle made by direction of motion of mosquito from the coordinate axes are given as

$$\cos^{-1}\frac{2}{11} \text{ (from } x\text{-axis)} = \tan^{-1}\frac{\sqrt{117}}{2}$$

$$\cos^{-1}\frac{6}{11} \text{ (from } y\text{-axis)} = \tan^{-1}\frac{\sqrt{85}}{6}$$

$$\cos^{-1}\frac{9}{11} \text{ (from } z\text{-axis)} = \tan^{-1}\frac{\sqrt{40}}{9}$$

None of the given option is correct. **Ans. (None)**

Sol. 24 To reach a point directly opposite on other bank of river, swimmer has to swimg such that the resultant velocity of swimmer should be directly toward the point on other bank of river where he has to reach. Below diagram shows the resultant velocity.

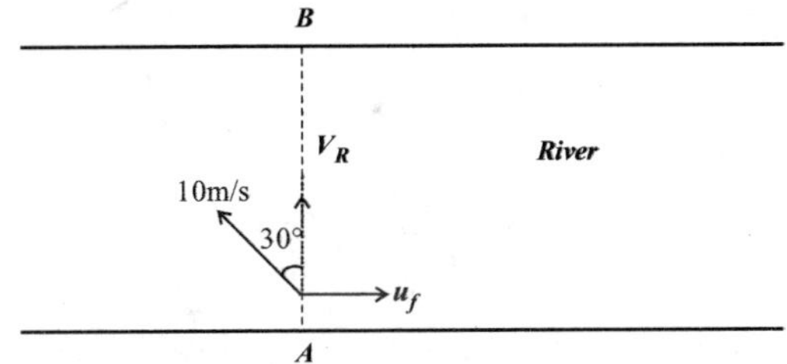

From above figure line AB is perpendicular to rivier flow when

$$10 \sin 30° = x$$

$$\Rightarrow \quad x = 5 \text{ m/s}$$ **Ans. (5)**

Sol. 25 Velocity vectors of projectile at points A and B are given as

$$\vec{u} = u \cos 45\,\hat{i} + u \sin 45\,\hat{j}$$

$$\vec{v} = v \cos 45\,\hat{i} - v \sin 45\,\hat{j}$$

Change in momentum can be calculated as

$$|\overrightarrow{\Delta P}| = m(\vec{v} - \vec{u})$$

$$\Rightarrow \quad \Delta P = 2mu \sin 45°$$

$$\Rightarrow \quad \Delta P = 2 \times 5 \times 10^{-3} \times 5\sqrt{2} \times \frac{1}{\sqrt{2}} \text{ N-s}$$

$$\Rightarrow \quad \Delta P = 50 \times 10^{-3} = 5 \times 10^{-2} \text{N-s}$$ **Ans. (5)**

Sol. 26 From the given relation in question, we use

$$\vec{A} \cdot \vec{B} = |\vec{A} \times \vec{B}|$$

$$\Rightarrow \quad AB \cos \theta = AB \sin \theta$$

$$\Rightarrow \quad \theta = 45°$$

$$\Rightarrow \quad |\vec{A} - \vec{B}| = \sqrt{A^2 + B^2 - 2AB \cos 45°}$$

$$\Rightarrow \quad |\vec{A} - \vec{B}| = \sqrt{A^2 + B^2 - \sqrt{2}AB}$$ **Ans. (B)**

Sol. 27 As per the given condition in question, we have

$$|\vec{P} + \vec{Q}| = n|\vec{P} - \vec{Q}|$$

$$\Rightarrow |\vec{P}|^2 + |\vec{Q}|^2 + 2|\vec{P}||\vec{Q}|\cos\theta = n^2(|\vec{P}|^2 + |\vec{Q}|^2 - 2|\vec{P}||\vec{Q}|\cos\theta)$$

$$\Rightarrow \quad 2 + 2\cos\theta = n^2(2 - 2\cos\theta)$$

$$\Rightarrow \quad \theta = \cos^{-1}\left(\frac{n^2 - 1}{n^2 + 1}\right)$$ **Ans. (C)**

Sol. 28 Projection of vector $\vec{A}$ on $\vec{B}$ is given as

$$p = |\vec{A}|\cos\theta\,\hat{B}$$

$$p = \frac{\vec{A} \cdot \vec{B}}{|\vec{B}|}\hat{B} = \hat{i} + \hat{j}$$ **Ans. (C)**

Sol. 29 Direction vector of motion of particle $\vec{P}$ can be given as

$$\vec{P} = K(\vec{A} \times \vec{B})$$

$$\begin{bmatrix} i & j & k \\ 1 & 1 & 0 \\ 0 & 1 & 1 \end{bmatrix} = K(\hat{i} - \hat{j} + \hat{k})$$

Direction vector of motion of particle $\vec{Q}$ can be given as

$$\vec{Q} = F(\vec{A} \times \vec{C})$$

$$\begin{bmatrix} \hat{i} & \hat{j} & \hat{k} \\ 1 & 1 & 0 \\ -1 & 1 & 0 \end{bmatrix} = 2F(\hat{k})$$

Angle between the direction of motion of P and Q can be calculated as

$$\cos\theta = \frac{\vec{P} \cdot \vec{Q}}{|\vec{P}||\vec{Q}|} = \frac{1}{\sqrt{3}}$$

Sol. 30 All the options in List-I can be directly related with the vector triangles given in List-II using triangle law of vector addition. **Ans. (B)**

Sol. 31 It is given that

$$|\vec{X}-\vec{Y}| = n|\vec{X}+\vec{Y}|$$

$$\Rightarrow \ |\vec{X}|^2+|\vec{Y}|^2-2|\vec{X}||\vec{Y}|\cos\theta = n^2[|\vec{X}|^2+|\vec{Y}|^2+2|\vec{X}||\vec{Y}|\cos\theta]$$

As magnitude of X is equal to that of Y, we use

$$2|\vec{X}|^2-2|\vec{X}|^2\cos\theta = 2n^2|\vec{X}|^2+2n^2|\vec{X}|^2\cos\theta$$

$$\Rightarrow \quad 1-\cos\theta = n^2+n^2\cos\theta$$

$$\Rightarrow \quad \cos\theta = \frac{1-n^2}{1+n^2}$$

$$\Rightarrow \quad \theta = \cos^{-1}\frac{1-n^2}{1+n^2}$$ **Ans. (B)**

Sol. 32 Acceleration vector of body is given as

$$\vec{a} = 8\hat{i}+2\hat{j}$$

As acceleration is constant, displacement is given as

$$\vec{s} = \vec{u}t+\frac{1}{2}\vec{a}t^2$$

$$\Rightarrow \quad \vec{s} = \frac{1}{2}(8\hat{i}+2\hat{j})\times 100$$

$$\Rightarrow \quad \vec{s} = 400\hat{i}+100\hat{j}$$ **Ans. (A)**

Sol. 33 By law of vector addition we have

$$\overrightarrow{OA}+\overrightarrow{AB} = \overrightarrow{OB} \quad \ldots(1)$$

$$\overrightarrow{OB}+\overrightarrow{BC} = \overrightarrow{OC} \quad \ldots(2)$$

$$\overrightarrow{OC}+\overrightarrow{CD} = \overrightarrow{OD} \quad \ldots(3)$$

$$\Rightarrow \quad \overrightarrow{AB} = \overrightarrow{OB}-\overrightarrow{OA}$$

$$\Rightarrow \quad \overrightarrow{AC} = \overrightarrow{OC}-\overrightarrow{OA}$$

$$\Rightarrow \quad \overrightarrow{AD} = \overrightarrow{OD}-\overrightarrow{OA} = -\overrightarrow{OA}-\overrightarrow{OA}$$

$$\Rightarrow \quad \overrightarrow{AB}+\overrightarrow{AC}+\overrightarrow{AD} = \overrightarrow{OB}+\overrightarrow{OC}-4\overrightarrow{OA}$$

$$\Rightarrow \quad \overrightarrow{AB}+\overrightarrow{AC}+\overrightarrow{AD} = \overrightarrow{OB}+\overrightarrow{OC}+4\overrightarrow{AO} \quad \ldots(4)$$

Hence Assertion-A is correct. Now from equations-(1), (2) & (3), we get

$$\overrightarrow{AB}+\overrightarrow{BC}+\overrightarrow{CD} = \overrightarrow{OB}+\overrightarrow{OC}+\overrightarrow{OD}-\overrightarrow{OA}-\overrightarrow{OB}-\overrightarrow{OC}$$

$$\overrightarrow{AB}+\overrightarrow{BC}+\overrightarrow{CD} = -\overrightarrow{OA}-\overrightarrow{OA}$$

$$\overrightarrow{AB}+\overrightarrow{BC}+\overrightarrow{CD} = -2\overrightarrow{OA}$$

$$\overrightarrow{AB}+\overrightarrow{BC}+\overrightarrow{CD} = 2\overrightarrow{AO} \quad \ldots(5)$$

With above analysis, we can state that both A and R are correct but R is not the correct explanation of A.

Sol. 34 Velocity of man with respect to ground can be given as

$$\vec{V}_M = \vec{V}_{MR}+\vec{V}_R$$

Vector addition of the velocities is shown below in which V_M should be along line AB. Given that velocity of man with respect to river and that of river is same so we use

$$V_{MR} = V_R$$

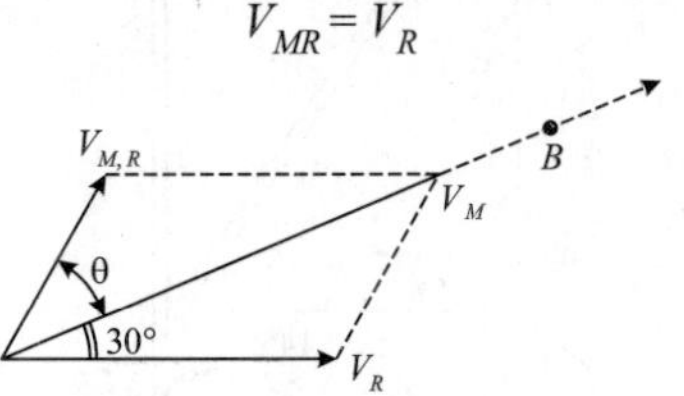

As the velocities of man and that of river are same the resultant velocity of man with respect to ground vector will be the angle bisector of the two vectors in the diagram shown so we have $\theta = 30°$. **Ans. (30)**

Sol. 35 From the given figure, the given vectors in terms of unit vectors are given as

$$\overrightarrow{OA} = R[\cos30°\,\hat{i}+\sin30°\,\hat{j}\,]$$

and $$\overrightarrow{OB} = R[\cos60°\,\hat{i}+(-\sin60°)\,\hat{j}\,]$$

and $$\overrightarrow{OC} = R[-\cos45°\,\hat{i}+\sin45°\,\hat{j}\,]$$

$$\Rightarrow \overrightarrow{OA}+\overrightarrow{OB}-\overrightarrow{OC} = R\left[\frac{1}{2}+\frac{\sqrt{3}}{2}+\frac{1}{\sqrt{2}}\right]\hat{i}+R\left[\frac{1}{2}-\frac{\sqrt{3}}{2}-\frac{1}{\sqrt{2}}\right]\hat{j}$$

Angle of above vector with x-axis is given as

$$\tan\alpha = \frac{\frac{1}{2}-\frac{\sqrt{3}}{2}-\frac{1}{\sqrt{2}}}{\frac{1}{2}+\frac{\sqrt{3}}{2}+\frac{1}{\sqrt{2}}} = \frac{1-\sqrt{3}-\sqrt{2}}{1+\sqrt{3}+\sqrt{2}}$$

$$\Rightarrow \quad \alpha = \tan^{-1}\left[\frac{1-\sqrt{3}-\sqrt{2}}{1+\sqrt{3}+\sqrt{2}}\right]$$ **Ans. (B)**

Sol. 36 Figure below shows the vector subtraction of the two given vectors

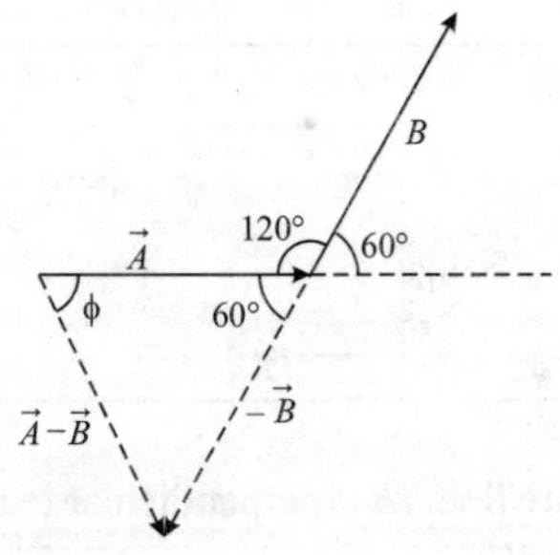

In above triangle, we have

$$\text{tna}\,\phi = \frac{B\sin60°}{A-B\cos60°}$$

$$\Rightarrow \quad \phi = \tan^{-1}\left(\frac{B\sin60°}{A-B\cos60°}\right) = \tan^{-1}\left(\frac{\sqrt{3}B}{2A-B}\right)$$ **Ans. (C)**

Sol. 37 Due to horizontally flying plane the horizontal component of velocity of bomb & fighter jet will be same so, bomb will remains just below the jet always during its projetile motion with respect to ground and with respect to an observer sitting in the plane the bomb appear to fall vertically in a straight line. **Ans. (B)**

Sol. 38 Resultant can be calculated by resolving all vectors into the components alsong x and y directions as

$$(\vec{R}) = \hat{i}\,(10\cos 30° + 20\cos 60° - 15\cos 30° - 15\cos 45° + 20\cos 45°) + \hat{j}\,(10\sin 30° + 20\sin 60° + 15\sin 30° - 15\sin 45° - 20\cos 45°)$$

$$(\vec{R}) = 9.25\hat{i} + 5\hat{j}$$ **Ans. (A)**

Sol. 39 Maximum height of football during projectile is given as

$$H = \frac{u^2\sin^2\theta}{2g} = \frac{(25)^2 \times \left(\frac{1}{2}\right)}{2\times 10} = \frac{128}{8}m$$

Time to reach maximum height is given as

$$t = \frac{T}{2}$$

$$\Rightarrow \quad t = \frac{u\sin\theta}{g} = \frac{25\left(\frac{1}{\sqrt{2}}\right)}{10}$$

$$\Rightarrow \quad td = \frac{25}{10\sqrt{2}} = \frac{5}{2\sqrt{2}}$$ **Ans. (A)**

Sol. 40 Situation described in question is shown in figure below

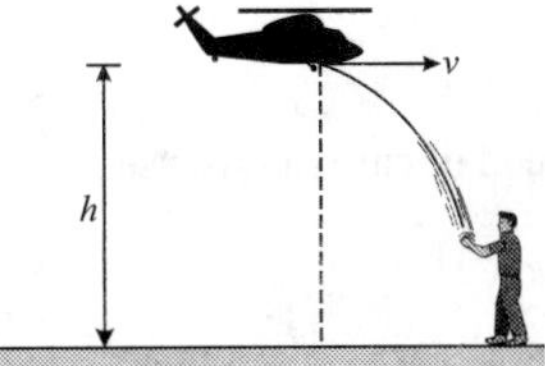

The food packet starts with the same speed of helicopter horizontally when dropped so horizontal range of the food packet is given as

$$R = v \times \sqrt{\frac{2h}{g}}$$

Horizontal distance, travelled by helicopter in this period will be same as that of R as horizontal speed of helicopter and food packet is same. Thus, vertical distance between helicopter and person is h. Thus from the point food packet is dropped, the distance between man and helicopter is given as

$$l = \sqrt{\frac{2v^2h}{g} + h^2}$$ **Ans. (D)**

Sol. 41 Figure below shows the car ascending above the inclined plane and the pendulum inside it.

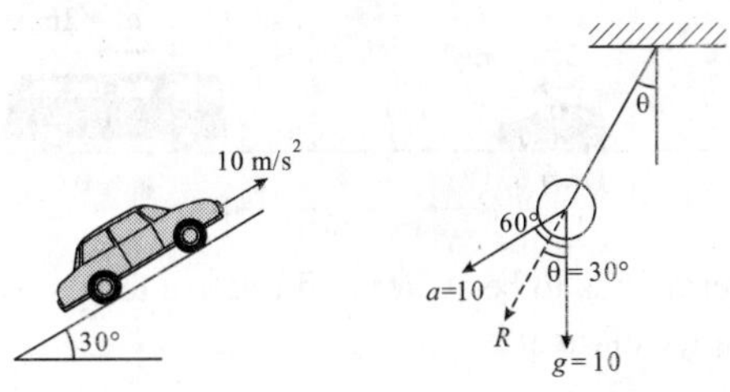

As shown in free body diagram of the pendulum, it experiences a pseudo force in direction opposite to the motion of car. We can clearly see that the resultant of this pseudo force and the weight of bob will be acting along the line at 30° from vertical so the string will be aligned along this line. **Ans. (30)**

Sol. 42 Three force vectors which represents the three sides of a triangle in same order then sum of these three must be zero thus statement-I is true. Statement-II is also stating the same thing in slightly changed language so this is also true. **Ans. (B)**

Sol. 43 For forces, $\vec{F}_1$ and $\vec{F}_2$ at θ_1

$$F_{net\,1} = \sqrt{P^2+Q^2+P^2+Q^2+2(P^2+Q^2)\cos\theta_1}$$

$$F_{net\,2} = \sqrt{P^2+Q^2+P^2+Q^2+2(P^2+Q^2)\cos\theta_2}$$

If $F_{net\,1} = \sqrt{3(P^2+Q^2)} = \sqrt{2(P^2+Q^2)+2(P^2+Q^2)\cos\theta_1}$

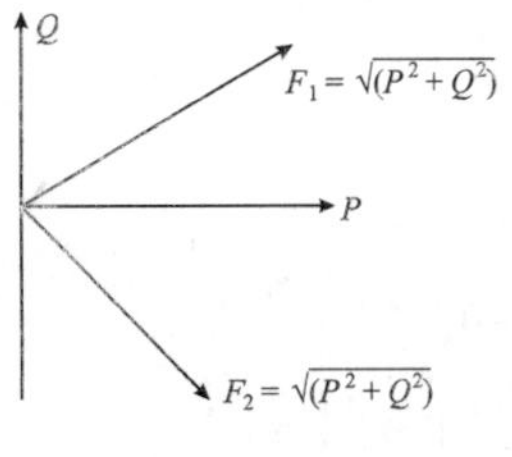

$$\Rightarrow \quad \cos\theta_1 = \frac{P^2+Q^2}{2(P^2+Q^2)}$$

$$\Rightarrow \quad \theta_1 = 60°$$

$$\Rightarrow \quad F_{net\,2} = \sqrt{2(P^2+Q^2)}$$

$$\Rightarrow \quad F_{net\,2} = \sqrt{2(P^2+Q^2)+2(P^2+Q^2)\cos\theta_2}$$

$$\Rightarrow \quad \cos\theta_2 = 0$$

$$\Rightarrow \quad \theta_2 = 90°$$ **Ans. (B)**

Sol. 44 As the given angles are complimentary angles, as

$$42° + 48° = 90°$$

Thus range for the two projectiles will be same so $R_1 = R_2$ and as we have $\theta_1 < \theta_2$ we can conclude that $H_1 < H_2$ **Ans. (B)**

3.2 Relative Motion

Sol. 1

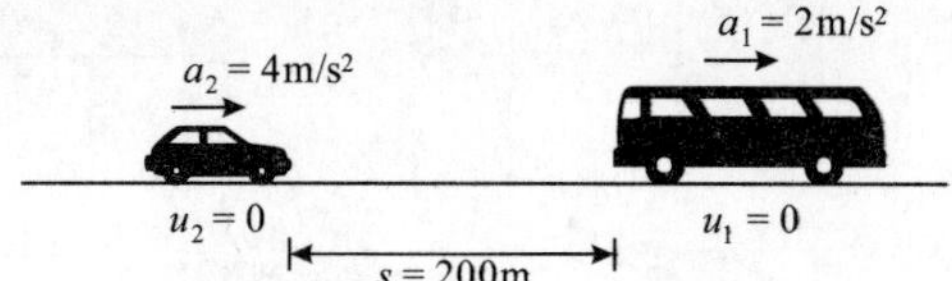

Considering bus to be at rest and relative to bus, speed and acceleration of car are

$$u_{CB} = 0 = u_2 - 4,$$

$$a_{CB} = a_2 - a_1 = 4 - 2 = 2m/s^2$$

time to bus by car (rel. to bus) a distance of 200 m is

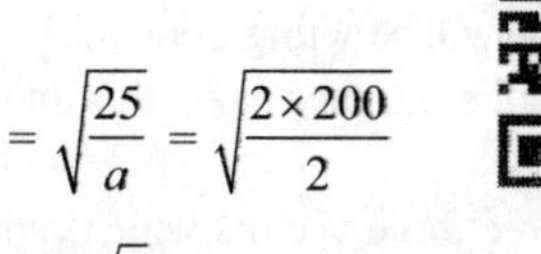

$$t = \sqrt{\frac{25}{a}} = \sqrt{\frac{2 \times 200}{2}}$$

$$= 10\sqrt{2}\text{ s}$$

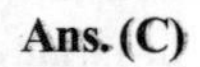

Ans. (C)

Sol. 2 If we take the position of ship 'A' as origin then positions and velocities of both ships can be given as :

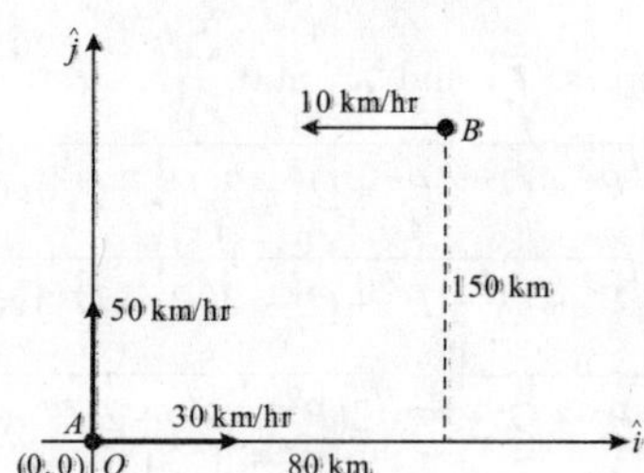

$$\vec{v}_A = (30\hat{i} + 50\hat{j})\text{ km/hr}$$

$$\vec{v}_B = -10\hat{i}\text{ km/hr}$$

$$\vec{r}_A = 0\hat{i} + 0\hat{j}$$

$$\vec{r}_B = (80\hat{i} + 150\hat{j})\text{ km}$$

Time after which distance between them will be minimum

$$t = -\frac{\vec{r}_{BA} \cdot \vec{v}_{BA}}{|\vec{v}_{BA}|^2};$$

where $\vec{r}_{BA} = (80\hat{i} + 150\hat{j})$ km

$$\vec{v}_{BA} = -10\hat{i} - (30\hat{i} + 50\hat{j})$$

$$(-40\hat{i} - 50\hat{j})\text{ km/hr}$$

$$\Rightarrow \quad t = -\frac{(80\hat{i} + 150\hat{j}) \cdot (-40\hat{i} - 50\hat{j})}{|-40\hat{i} - 50\hat{j}|^2}$$

$$= \frac{3200 + 7500}{4100}\text{ hr} = \frac{10700}{4100}\text{ hr} = 2.6\text{ hrs}$$ **Ans. (D)**

Sol. 3

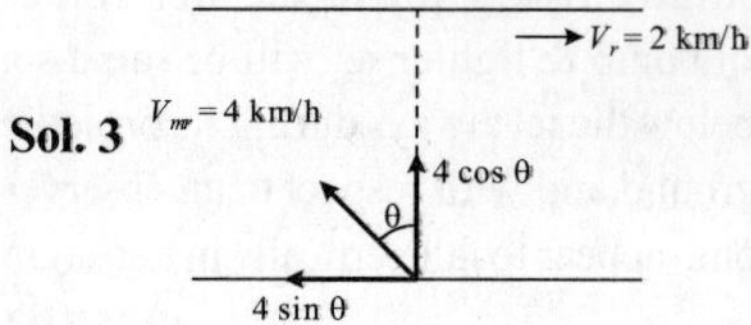

For swimmer to cross the river straight

$$\Rightarrow \quad 4 \sin\theta = 2$$

$$\Rightarrow \quad \sin\theta = \frac{1}{2}$$

$$\Rightarrow \quad \theta = 30°$$

So, angle with direction of river flow

$$= 90° + \theta = 120°$$ **Ans. (D)**

Sol. 4 Figure below shows the given situation

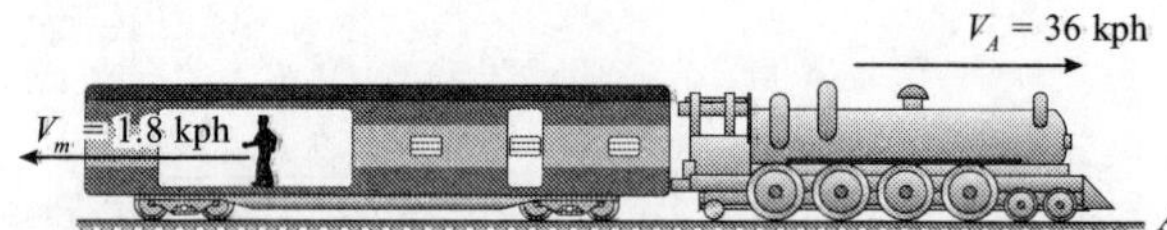

Speed of man w.r. to train B is

$$\vec{V}_{mB} = \vec{V}_m + \vec{V}_A - \vec{V}_B$$

$$= 1.8 - 36 - 72$$

$$= 106.2\text{ km/hr}$$

$$= 29.5\text{ m/s}$$ **Ans. (B)**

Sol. 5 When speed of car is v, we use

$$\overrightarrow{V_{RC}} = \overrightarrow{V_R} - \overrightarrow{V_C}$$

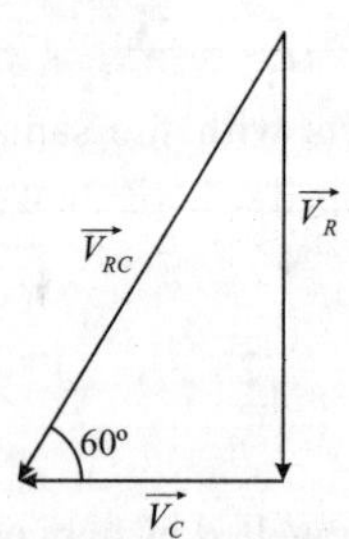

$$\tan 60° = \frac{V_R}{V} = \sqrt{3}$$

$$\Rightarrow \quad V_R = \sqrt{3}v \quad \ldots(1)$$

When speed of car changes to $V'_C = V(1 + \beta)$, we use

$$\overrightarrow{V_{RC'}} = \overrightarrow{V_R} - \overrightarrow{V_{C'}}$$

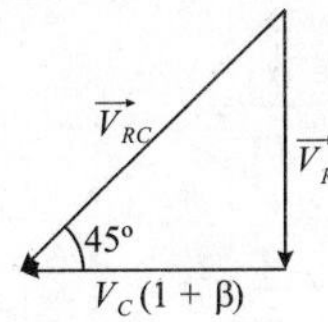

$$\Rightarrow \qquad \tan 45^\circ = \frac{V_R}{V(1+\beta)} = 1$$

$$\Rightarrow \qquad V_R = v(1+\beta) \qquad \ldots(2)$$

From (1) and (2)

$$1+\beta = \sqrt{3}$$

$$\Rightarrow \qquad \beta = 1.732 - 1 = 0.732$$

Ans. (D)

Sol. 6 To reach just opposite point on the other bank from where he started, velocity component of swimmer along the river must be equal to rivir velocity, so we use

$$12\sin\theta = v_r$$

where θ is the angle from the line joining starting point and opposite point on the other bank of river.

$$\Rightarrow \qquad \sin\theta = \frac{1}{2}$$

$$\Rightarrow \qquad \theta = 30^\circ$$

Thus angle made by the swimmer velocity with the river will be given as

$$\alpha = 30^\circ + 90^\circ = 120^\circ$$

Ans. (120)

Sol. 7 If we consider $\hat{i}$ and $\hat{j}$ as unit vectors along north and east directions, the net displacement of butterfly can be given as

$$\vec{D} = vt$$

$$\Rightarrow \qquad \vec{D} = \{(4\hat{i} + 4\hat{j} + (-\hat{j}))\} \times 3\text{s} = (4\hat{i} + 3\hat{j}) \times 3\text{s}$$

$$\Rightarrow \qquad |\vec{D}| = 15\,\text{m}$$

Ans. (B)

Sol. 8 If the length of escalator is L, then from the situation and information given in the question we have

Speed of man relative to escalator is calculated as

$$v_m = \frac{L}{t_1}$$

Speed of escalator motion is calculated as

$$v_e = \frac{L}{t_2}$$

Speed of man relative to ground when escalator is moving can be given as

$$v_{mG} = \frac{L}{t_1} + \frac{L}{t_2}$$

Thus time taken by the man to walk on the moving escalator is calculated as

$$t = \frac{L}{\frac{L}{t_1} + \frac{L}{t_2}} = \frac{t_1 t_2}{t_1 + t_2}$$

Ans. (C)

4.1 Newton's Laws of Motion

Sol. 1

Equilibrium, we use

$$\tan 45^\circ = \frac{100}{F}$$

$$\Rightarrow \qquad F = 100\,\text{N}$$

Ans. (B)

Sol. 2 By Newton's second law, we use

$$\frac{dp}{dt} = F = kt$$

$$\Rightarrow \qquad \int_p^{3p} dp = \int_O^T kt\,dt$$

$$\Rightarrow \qquad 2p = \frac{kT^2}{2}$$

$$\Rightarrow \qquad T = 2\sqrt{\frac{p}{k}}$$

Ans. (A)

Sol. 3 FBD of the mass is shown in figure

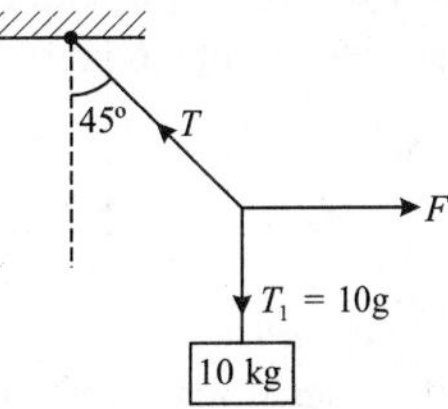

For equilibrium of system, we use

$$T\sin 45^\circ = F \qquad \ldots(1)$$

and

$$T\cos 45^\circ = 10g \qquad \ldots(2)$$

$$\Rightarrow \qquad F = 10g$$

$$\Rightarrow \qquad F = 100\,\text{N}$$

Ans. (A)

Sol. 4 Figure-(1) shows the initial given condition

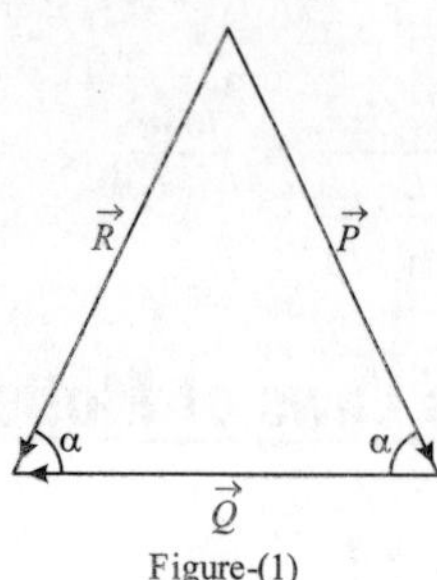

Figure-(1)

After $\vec{P}$ is made twice then the situation is shown in figure-(2) below

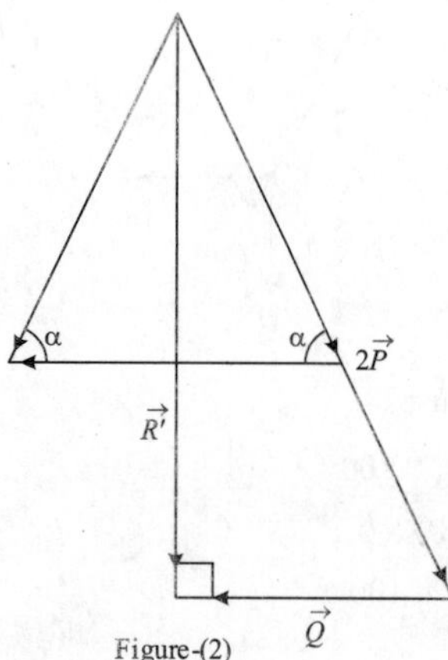

Figure-(2)

Thus angle here is 90°. **Ans. (90°)**

Sol. 5 White going up acceleration of ball is given as

$$\frac{vdv}{dy} = a = -(kv^2 + g)$$

$$\Rightarrow \quad \int_u^0 \frac{vdv}{kv^2+g} = \int_0^H -dy$$

$$\Rightarrow \quad \frac{1}{2k}\ln\left(\frac{ku^2+g}{g}\right) = H$$ **Ans. (B)**

Sol. 6 Force on planet due to increase in mass an be taken as Opposing force on spaceship due to increase in mass is given as

$$F = v\frac{dM_{(t)}}{dt} = -bv^3 [\text{As } F_{ext} = 0]$$

Thus acceleration of spaceship is given as

$$a = \frac{F}{m_{(t)}} = -\frac{bv^3}{m_{(t)}}$$ **Ans. (D)**

Sol. 7 The damping force on particle is givn as

$$F = -\alpha x^2 = ma$$

$$\Rightarrow \quad a = \frac{-\alpha x^2}{m}$$

$$\Rightarrow \quad \frac{vdv}{dx} = -\frac{\alpha}{m}x^2$$

For the motion of particle from starting instant to coming to rest can be integrated as

$$\Rightarrow \quad \int_{v_0}^0 v\,dv = \int_0^x -\frac{\alpha}{m}x^2 dx$$

$$\Rightarrow \quad \left(\frac{v^2}{2}\right)_{v_0}^0 = -\frac{\alpha}{m}\left(\frac{x^3}{3}\right)_0^x$$

$$\Rightarrow \quad \frac{-v_0^2}{2} = -\frac{\alpha}{m}\frac{x^3}{3}$$

$$\Rightarrow \quad x = \left(\frac{3mv_0^2}{2\alpha}\right)^{1/3}$$

Option (B) out of all options given is most appropriate here.

Ans. (B)

Sol. 8 Figure below shows the situation explained in the question. When lift is at rest, we use

$$N = mg$$

$$\Rightarrow \quad 60 \times 10 = 600 \text{ N}$$

When lift is moving with downward acceleration. In the frame of lift pseudo force will act on man in upward direction, so in frame of life, as man is at rest, we have

$$N' = M(g-a) = 60\,(10 - 1.8) = 492 \text{ N}$$

Ans. (492)

Sol. 9 The applied force is given as

$$\vec{F} = 20\hat{i} + 10\hat{j}$$

Thus acceleration of box can be given as

$$\vec{a} = \frac{\vec{F}}{m} = \frac{20\hat{i}+10\hat{j}}{2} = 10\hat{i} + 5\hat{j}$$

Displacement of box can now be given as

$$\vec{s} = \frac{1}{2}\vec{a}t^2 = \frac{1}{2}(10\hat{i}+5\hat{j}) \times (10)^2$$

$$= 50(10\hat{i}+5\hat{j})m$$

Displacement along x-axis here is 500 m **Ans. (500)**

Sol. 10 As friction provides the necessary centripetal force for circular turning, we have

$$\mu_s N = \frac{mv^2}{R}$$

If F_L is the negative lift acting on the car in downward direction, then for its equilibrium in vertical direction, we use

$$N = \frac{mv^2}{\mu_s R} = mg + F_L$$

$$\Rightarrow \quad F_L = \frac{mv^2}{\mu_s R} - mg \qquad \textbf{Ans. (B)}$$

Sol. 11 The free body diagram of the man can be drawn as

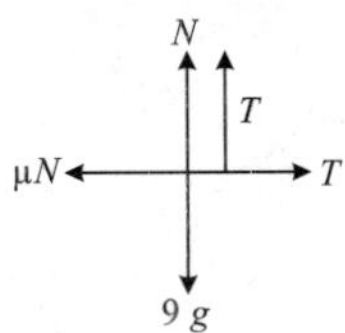

For equilibrium of man, we use

$$N + T = 90$$

$$\Rightarrow \quad T = \mu N = 0.5\,(90 - T)$$

$$\Rightarrow \quad 1.5\,T = 45$$

$$\Rightarrow \quad T = 30\,\text{N}$$

Tension in string is the force exerted by man on rope.

Ans. (30)

Sol. 12 The acceleration of bullet due to resistance force of wood can be calculated as

$$v^2 = u^2 + 2as$$

$$\Rightarrow \quad 0 = (10)^2 + 2\,(-a)\left(\frac{1}{2}\right)$$

$$\Rightarrow \quad a = 100\,\text{m/s}^2$$

Thus retarding force on bullet is given as

$$F = ma = (0.1)(100) = 10\ \text{N} \qquad \textbf{Ans. (10)}$$

Sol. 13 As system is going down at acceleration 0.2 m/s^2, equation of motion of system can be written as

$$70\,g - N = 70 \times 0.2$$

$$\Rightarrow \quad N = 700 - 14 = 686\,\text{N} \qquad \textbf{Ans. (A)}$$

Sol. 14 Acceleration of ball is given as

$$a = \frac{F}{m}$$

$$a = \frac{F_0}{m}\left[1 - \left(\frac{t-T}{T}\right)^2\right]$$

Velocity of ball as a function of time can be calculated as

$$\frac{dv}{dt} = \frac{F_0}{m}\left[1 - \left(\frac{t-T}{T}\right)^2\right]$$

$$\Rightarrow \quad \int_0^v dv = \frac{F_0}{m}\int_0^{2T}\left[1 - \left(\frac{t-T}{T}\right)^2\right]dt$$

$$\Rightarrow \quad v = \frac{F_0}{m}\left[t - \frac{1}{3T^2}(t-T)^3\right]_0^{2T}$$

$$\Rightarrow \quad v = \frac{F_0}{m}\left\{\left[2T - \frac{1}{3T^2}(2T-T)^3\right] - \left[0 - \frac{(-T)^3}{3T^2}\right]\right\}$$

$$\Rightarrow \quad v = \frac{4F_0T}{3M} \qquad \textbf{Ans. (C)}$$

Sol. 15 As rocket accelerates upward so thurst due to gases ejection on rocket is given as

$$f_T = mg + ma$$

$$\Rightarrow \quad \frac{dm}{dt}v_r = mg + ma$$

$$\Rightarrow \quad \frac{dm}{dt}(500) = 1000\,(10 + 20)$$

$$\Rightarrow \quad \frac{dm}{dt} = 60\ \text{kg/s} \qquad \textbf{Ans. (D)}$$

Sol. 16 Tension in the string of an Atwood's machine is given as

$$T = \frac{2m_1m_2}{(m_1 + m_2)}g$$

$$\Rightarrow \quad T = \frac{2 \times 3 \times 5 \times 10}{8} = \frac{75}{2}$$

Stress in wire is given as

$$\text{Stress} = \frac{T}{A}$$

$$\Rightarrow \quad \frac{24}{\pi} \times 10^{-2} = \frac{75}{2(\pi r^2)}$$

$$\Rightarrow \quad r^2 = \frac{75}{48} \times 100$$

$$\Rightarrow \quad r^2 = \left(\frac{25}{2}\right)^2$$

$$\Rightarrow \quad r = 12.5\ \text{cm} \qquad \textbf{Ans. (D)}$$

Sol. 17 Figure below shows the tension in strings and acceleration of blocks considered for equation of motion

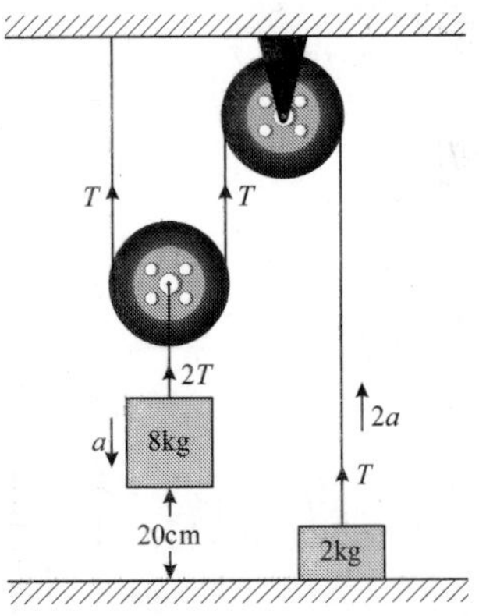

Equation of motion of the two blocks is written as

$$8g - 2T = 8a$$

and $\quad T-2g=4a$

$\Rightarrow \quad 4g=16a$

$\Rightarrow \quad a=\frac{5}{2}\text{m/s}^2$

By speed equation, displacement of 8 kg block is written as

$$s=ut+\frac{1}{2}at^2$$

$\Rightarrow \quad 0.2=0+\frac{1}{2}\times\frac{5}{2}\times t^2$

$\Rightarrow \quad t=0.4\text{ s}$ **Ans. (D)**

Sol. 18 As the object is moving at constant velocity so net force on object must be zero. In this case while moving up the force would be positive and balances its weight component along the incline and while moving down force must be negative(acting upward) and balacing the weight component along the incline. Thus most appropriate choice here is option-(D) among all options. **Ans. (D)**

Sol. 19 Figure below shows the free body diagram representing forces acting on the wedge and block

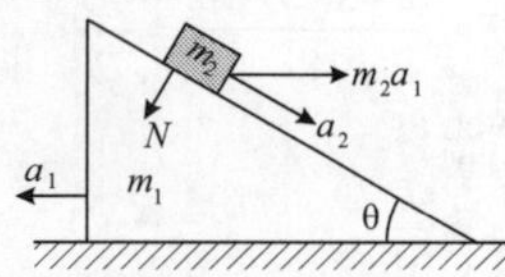

Equaton of motion of wedge is written as

$$N\cos 60^\circ = Ma_1 = 16a_1$$

$\Rightarrow \quad N=32a_1$

For block along the direction normal to incline, including pseudo force, we balance forces as

$$N=8g\cos 30^\circ - 8a_1\sin 30^\circ$$

$\Rightarrow \quad 32a_1=4\sqrt{3}-4a_1$

$\Rightarrow \quad a_1=\frac{\sqrt{3}}{9}g$

Along the incline for the block we use

$$8g\sin 30^\circ + 8a_1\cos 30^\circ = ma_2 = 8a_2$$

$\Rightarrow \quad a_2=\frac{g}{2}+\frac{\sqrt{3}}{9}g\cdot\frac{\sqrt{3}}{2}=\frac{2g}{3}$ **Ans. (D)**

4.2 Friction

Sol. 1

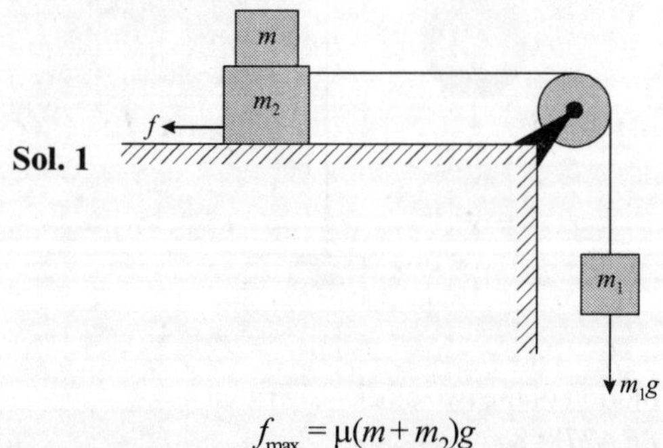

$$f_{\max}=\mu(m+m_2)g$$

To stop the motion, we use

$$m_1g \le \mu(m+m_2)g$$

$\Rightarrow \quad 5 \le 0.15\,(m+10)$

$\Rightarrow \quad \frac{100}{3} \le m+10 \Rightarrow m \ge 23.33\text{ kg}$ **Ans. (B)**

Sol. 2

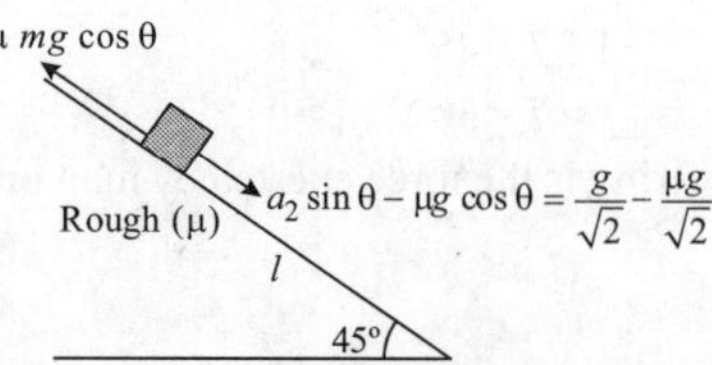

Time taken to slide is

$$t_1=\sqrt{\frac{2l}{a_1}}=\sqrt{\frac{2\sqrt{2}l}{g}}$$

Given that $\quad t_2=nt_1$

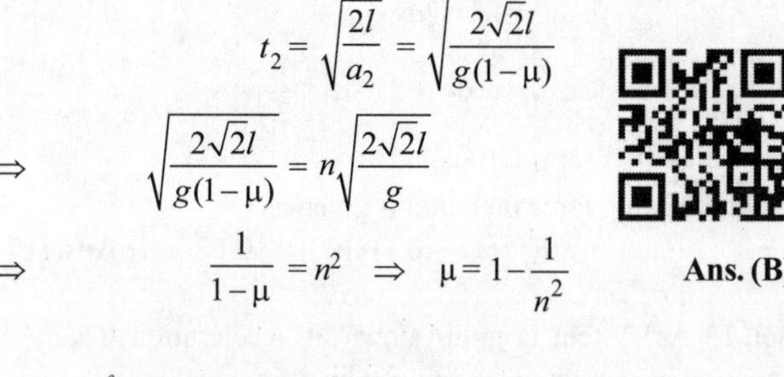

Time taken to slide is

$$t_2=\sqrt{\frac{2l}{a_2}}=\sqrt{\frac{2\sqrt{2}l}{g(1-\mu)}}$$

$\Rightarrow \quad \sqrt{\frac{2\sqrt{2}l}{g(1-\mu)}}=n\sqrt{\frac{2\sqrt{2}l}{g}}$

$\Rightarrow \quad \frac{1}{1-\mu}=n^2 \Rightarrow \mu=1-\frac{1}{n^2}$ **Ans. (B)**

Sol. 3

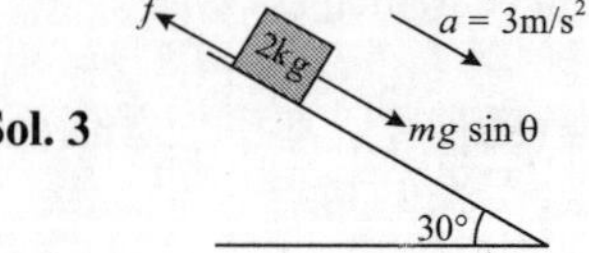

Equation motion of block

$$mg\sin\theta - f = ma$$

$\Rightarrow \quad 2\times 10\times\frac{1}{2}-f=2\times 3$

$\Rightarrow \quad f=10-6=4\text{ N}$

Equation of motion of block

$$F-mg\sin\theta-f=ma$$

$\Rightarrow \quad F-10-4=6$

$\Rightarrow \quad F=20\text{ N}$ **Ans. (D)**

Sol. 4 To keep the block at rest *FBD* is shown in figure. At equilibrium of block we use

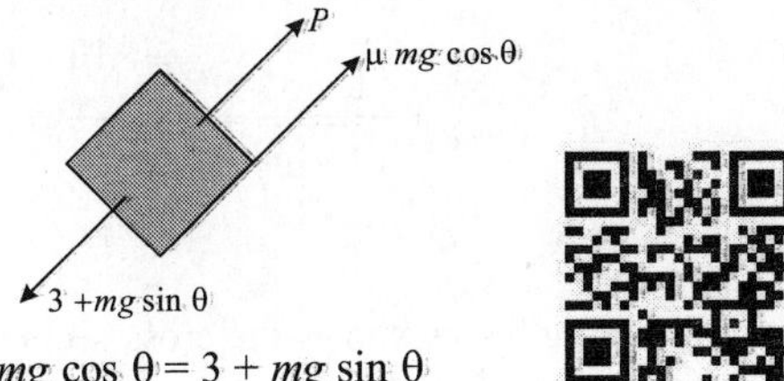

$$P + \mu mg \cos\theta = 3 + mg \sin\theta$$

$$\Rightarrow \qquad P = 3 + 10 \times 10 \times \frac{1}{\sqrt{2}} - 0.6 \times 10 \times 10 \times \frac{1}{\sqrt{2}}$$

$$\Rightarrow \qquad P = 3 + 20\sqrt{2} = 31.28\text{ N}$$ **Ans. (A)**

Sol. 5 For limiting equilibrium down the plane, we use

$$2 + mg\sin 30 = \mu mg \cos 30^\circ \qquad \ldots(1)$$

for limiting equilibrium up the plane, we use

$$10 = mg \sin 30 + \mu mg \cos 30^\circ \qquad \ldots(2)$$

from (1) and (2)

$$\Rightarrow \qquad 6 = \mu mg \cos 30^\circ$$

and $$4 = mg \sin 30^\circ$$

$$\Rightarrow \qquad \frac{3}{2} = \mu \times \sqrt{3}$$

$$\Rightarrow \qquad \mu = \frac{\sqrt{3}}{2}$$ **Ans. (B)**

Sol. 6

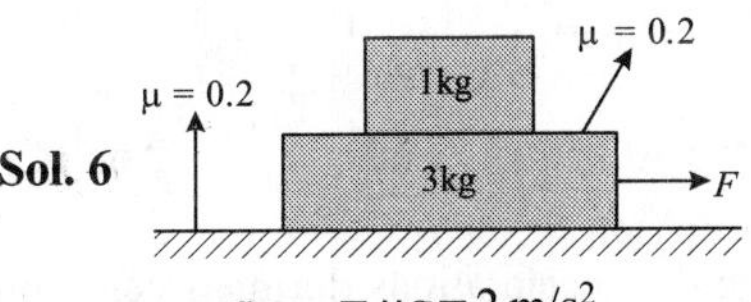

$$a_{Amax} = \mu g = 2\text{ m/s}^2$$

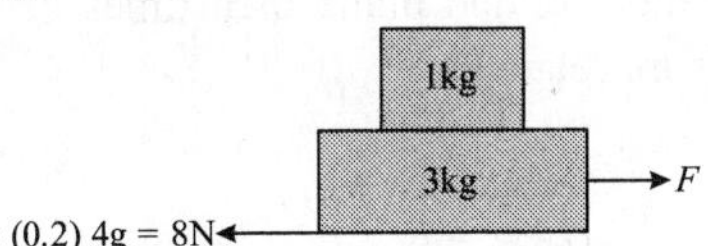

Equation of motion of blocks for combined motion is

$$F - 8 = 4 \times 2$$

$$F = 16\text{N}$$ **Ans. (A)**

Sol. 7 FBD of the two cases are shown below

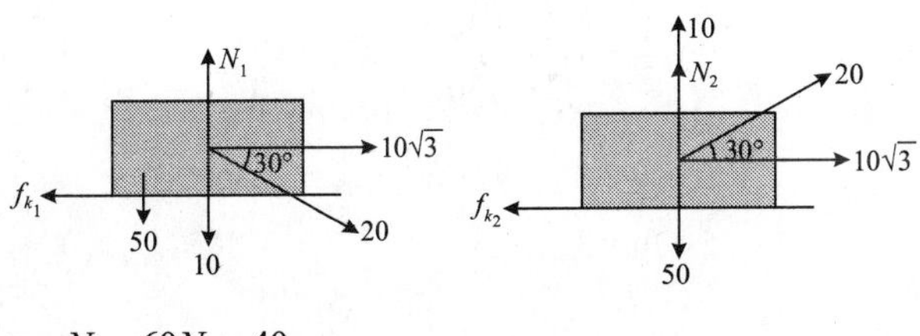

$$N_1 = 60 N_2 = 40$$

$$a_1 = \frac{10\sqrt{3} - 0.2 \times 60}{5}$$

$$a_2 = \frac{10\sqrt{3} - 0.2 \times 40}{5}$$

$$a_1 - a_2 = 0.8$$ **Ans. (B)**

Sol. 8 At the point of slipping as shown in figure, we use

$$\tan\theta = \mu = \frac{3}{4}$$

$$\Rightarrow \qquad \cos\theta = \frac{4}{5}$$

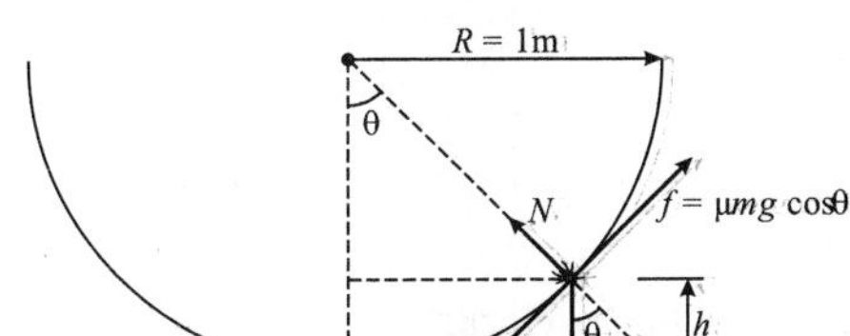

Value of h is given as

$$h = R(1 - \cos\theta) = R\left(1 - \frac{4}{5}\right) = \frac{R}{5}$$ **Ans. (D)**

Sol. 9 Figure below shows the free body diagram of the block as described in the question

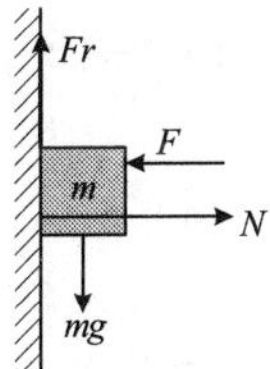

As block is at rest, in this case, we use

$$fr - mg = 0 \qquad \ldots(1)$$

and $$F - N = 0 \qquad \ldots(2)$$

$$\Rightarrow \qquad fr = \mu N = \mu F \qquad \ldots(3)$$

From equations-(1) and (3), we use

$$\mu F = mg$$

$$\Rightarrow \qquad F = \frac{0.5 \times 10}{0.2} = 25\text{ N}$$ **Ans. (25)**

Sol. 10 Figure below shows the situation explained in the question

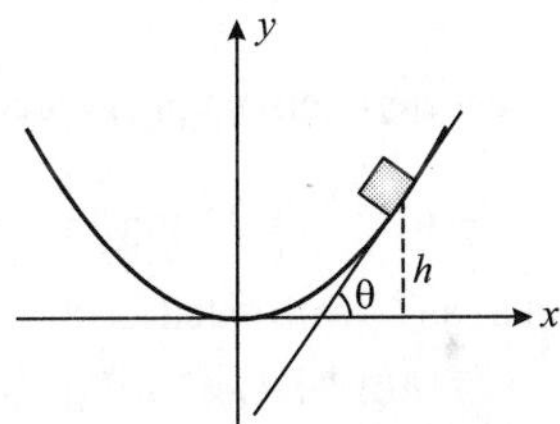

In this situation, at maximum height, block will experience the maximum friction. Thus at this height the slope angle of tangent must be angle of repose, so we use

$$\tan\theta = \frac{dy}{dx} = \frac{2x}{4} = \frac{x}{2} = 0.5$$

$$\Rightarrow \quad x = 1$$

$$\Rightarrow \quad y = \frac{x^2}{4} = 0.25\,\text{m} = 25\,\text{cm}$$ **Ans. (25)**

Sol. 11 Figure below shows the free body diagram of the block with force components shown on it

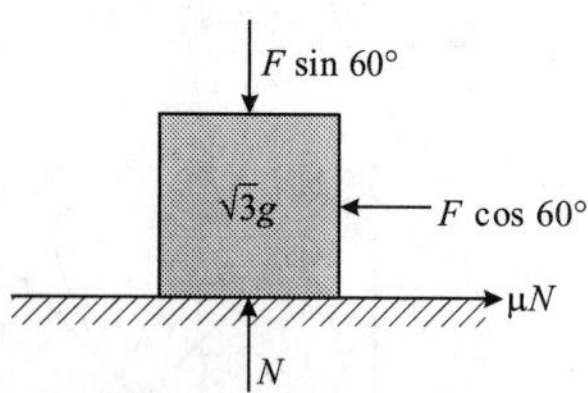

For limiting situation of block, we use

$$F\cos 60° = \mu N$$

$$\Rightarrow \quad \frac{F}{2} = \frac{1}{3\sqrt{3}}N \qquad \ldots(1)$$

$$\text{and} \quad N = \sin 60° + \sqrt{3}g \qquad \ldots(2)$$

From equations-(1) & (2), we use

$$\frac{F}{2} = \frac{1}{3\sqrt{3}}\left(\frac{F\sqrt{3}}{2} + \sqrt{3}g\right)$$

$$\Rightarrow \quad F = g = 10\,\text{N} = x$$ **Ans. (10)**

Sol. 12

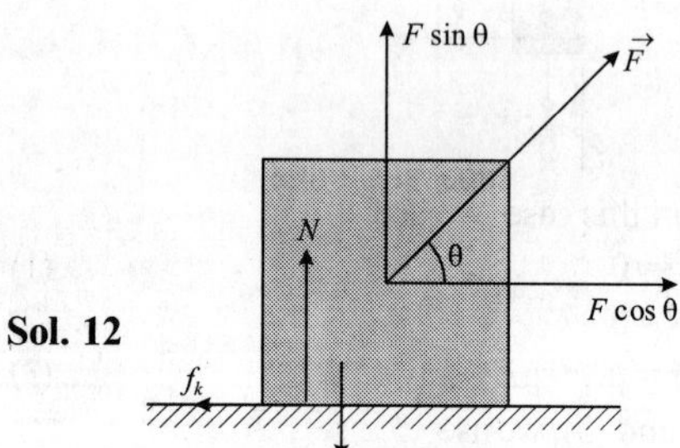

$$N = mg - F\sin\theta$$

$$\Rightarrow \quad F\cos\theta - \mu_k N = ma$$

$$\Rightarrow \quad F\cos\theta - \mu_k(mg - F\sin\theta) = ma$$

$$\Rightarrow \quad a = \frac{F}{m}\cos\theta - \mu_k\left(g - \frac{F}{m}\sin\theta\right)$$ **Ans. (B)**

Sol. 13 Maximum acceleration of upper block can be given as

$$a_{max} = \mu g = \frac{3}{7} \times 9.8\,\text{m/s}^2$$

The maximum force can be calculated as

$$F = (M+m)\,a_{max} = 5\,a_{max}$$

$$\Rightarrow \quad F = 21\ \text{Newton}$$ **Ans. (21)**

Sol. 14 The minimum force required to slide a body placed on a horizontal surface with friction coefficient μ, is given as

$$F_{min} = \frac{\mu mg}{\sqrt{1+\mu^2}} = \frac{\frac{1}{\sqrt{3}} \times 10}{\frac{2}{\sqrt{3}}} = 5\,\text{N}$$ **Ans. (5)**

Sol. 15 Figure below shows the cross section of the vehicle turing on a banked road.

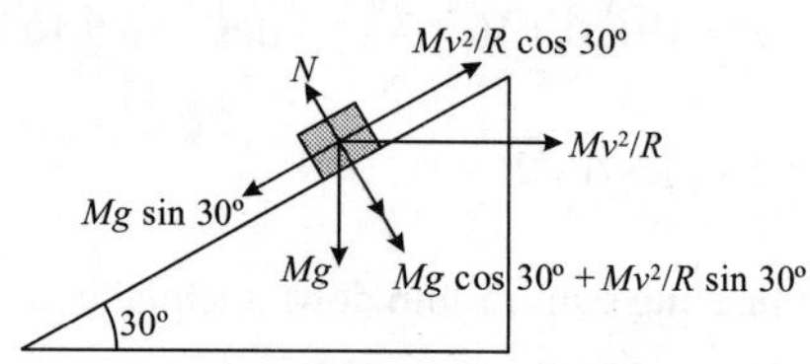

Along the direction normal to inclined plane, forces can be balanced, which gives

$$N = mg\cos 30° + \frac{mv^2}{R}\sin 30°$$

$$\Rightarrow \quad N - mg\cos 30° = \frac{mv^2}{R}\sin 30° \qquad \ldots(1)$$

Along inclined plane, force equation is written as

$$mg\sin 30° + \mu_s N = \frac{mv^2}{R}\cos 30° \qquad \ldots(2)$$

From equations-(1) & (2), we have

$$\frac{N - mg\cos 30°}{mg\sin 30° + \mu_s N} = \tan 30°$$

$$\Rightarrow \quad N = 10.2 \times 10^3\,\text{kg m/s}^2$$ **Ans. (C)**

Sol. 16 If a_A and a_D are the accelerations during ascent and descent of the inclined plane then times of ascending and descending are related as

$$\frac{t_A^2}{t_D^2} = \frac{a_C}{a_A}$$

$$\Rightarrow \quad \frac{t_A^2}{t_D^2} = \frac{g\sin\theta - \mu g\cos\theta}{g\sin\theta + \mu g\cos\theta}$$

$$\Rightarrow \quad \frac{t_A}{t_D} = \sqrt{\frac{g\sin\theta - \mu g\cos\theta}{g\sin\theta + \mu g\cos\theta}}$$

$$\Rightarrow \quad \frac{1}{2} = \sqrt{\frac{1-\sqrt{3}\mu}{1+\sqrt{3}\mu}}$$

$$\Rightarrow \quad 1 + \sqrt{3}\mu = 4 - 4\sqrt{3}$$

$$\Rightarrow \quad 5\sqrt{3}\mu = 3$$

$$\Rightarrow \quad \mu = \frac{\sqrt{3}}{5}$$ **Ans. (3)**

Sol. 17 Maximum horizontal force F_{max} at which the two blocks will move together will be such that friction between the blocks f_s between 2kg & 1 kg will be maximum. Common acceleration of the blocks is given as

$$a_{max}=\frac{F}{3}$$

The acceleration of 1 kg block is only due to the static friction, which is given as

$$f_s=0.5\times 10=5\,\text{N}$$

Thus maximum acceleration of 1 kg block due to static friction is given as

$$a_{1\,kg}=f_s/m=5\,\text{m/s}^2$$

For the two blocks to move at common acceleration, we have

$$F=(1+2)\,a_{max}=3\times 5=15\,\text{N}$$ **Ans. (15)**

Sol. 18 Acceleration of body on a smooth incline is given as

$$a=g\sin 30^\circ$$

Displacement covered by the body is given by

$$s=ut+\frac{1}{2}at^2$$

$$\Rightarrow \quad s=\frac{1}{2}\frac{g}{2}T^2=\frac{g}{4}T^2 \quad \ldots(1)$$

On rough incline, acceleration is given as

$$a=g\sin 30^\circ-\mu g\cos 30^\circ$$

Displacement covered by the body is given by

$$s=ut+\frac{1}{2}at^2$$

$$\Rightarrow \quad s=\frac{1}{4}g(1-\sqrt{3}\mu)(\alpha T)^2 \quad \ldots(2)$$

Using equations-(1) and (2), we have

$$\frac{1}{4}gT^2=\frac{1}{4}g(1-\sqrt{3}\mu)\,\alpha^2T^2$$

$$\Rightarrow \quad 1-\sqrt{3}g=\frac{1}{a^2}$$

$$\Rightarrow \quad g=\left(\frac{\alpha^2-1}{\alpha^2}\right)\frac{1}{\sqrt{3}}$$

$$\Rightarrow \quad x=3$$ **Ans. (3)**

5.1 Work Energy Power

Sol. 1 Work done by force = Gain in KE $=\frac{1}{2}mv^2$

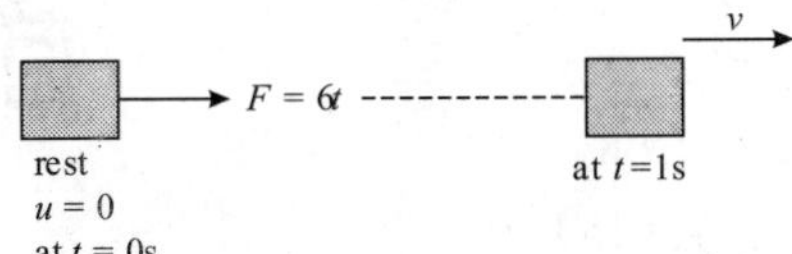

$$\text{Acceleration of particle } a=\frac{F}{m}=\frac{6t}{1}=6t\,\text{m/s}^2$$

$$\Rightarrow \quad \frac{dv}{dt}=6t$$

$$\Rightarrow \quad \int_0^v dv=\int_0^1 6t\,dt$$

$$\Rightarrow \quad v=6\left(\frac{t^2}{2}\right)_0^1=6\times\frac{1}{2}=3\,\text{m/s}$$

$$=\frac{1}{2}\times 1\times(3)^2=\frac{9}{2}=4.5\,\text{J}$$ **Ans. (C)**

Sol. 2 At $t=0$, $v=10$ m/s

and at $t=10$s, $\frac{1}{2}mv_f^2=\frac{1}{8}m(10)^2$

$$\Rightarrow \quad v_f^2=\frac{100}{4}=25 \quad \Rightarrow v_f=5\,\text{m/s}$$

$$F=-kv^2$$

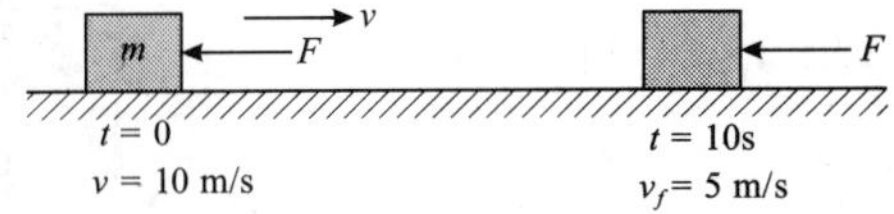

$$\Rightarrow \quad \text{Acceleration } a=\frac{F}{m}=-\frac{k}{m}v^2$$

$$\Rightarrow \quad \frac{dv}{dt}=-\frac{k}{m}v^2$$

$$\Rightarrow \quad \int_{10}^{5}\frac{dv}{v^2}=\int_0^{10}\frac{k}{m}dt$$

$$\Rightarrow \quad \left[+\frac{1}{v}\right]_{10}^{5}=+\frac{k}{m}[t]_0^{10}$$

$$\Rightarrow \quad \frac{1}{5}-\frac{1}{10}=\frac{k}{m}\times 10$$

$$\Rightarrow \quad k=10^{-3}\left[\frac{1}{5}-\frac{1}{10}\right]=10^{-4}\,\text{kg/m}$$ **Ans. (A)**

Sol. 3 Force on particle in the given potential field is given as

$$\vec{F}=-\frac{dU}{dr}\hat{r}=\frac{k}{r^3}\hat{r}$$

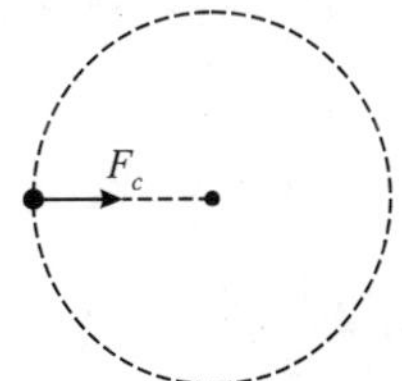

For stable circular motion we use

$$\frac{k}{r^3} = \frac{mv^2}{r}$$

$$\Rightarrow \quad mv^2 = \frac{k}{r^2}$$

$\Rightarrow$ The kinetic energy of particle is

$$\text{KE} = \frac{1}{2}\text{mv}^2 = \frac{k}{2r^2}$$

Total energy of particle TE = KE + PE

$$= \frac{k}{2r^2} - \frac{k}{2r^2} = 0$$

Ans. (C)

Sol. 4 If particle moves at speed v in circular orbit then

$$\frac{16}{r} + r^3 = \frac{mv^2}{r} \Rightarrow mv^2 = \left(\frac{16}{r} + r^3\right) \times r = 16 + r^4$$

Kinetic energy of particle $\frac{1}{2}mv^2 = \frac{1}{2}(16 + r^4)$

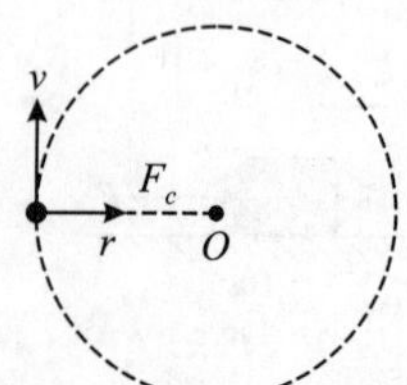

Ratio of kinetic energy of the two particles is given as

$$\frac{k_1}{k_2} = \frac{\frac{1}{2}(16+(1)^4)}{\frac{1}{2}(16+(4)^4)} = \frac{16+1}{16+256} = \frac{17}{272} = 0.06$$

Ans. (B)

Sol. 5 Velocity of body is

$$v = a\sqrt{s}$$

$$\Rightarrow \quad \frac{ds}{dt} = a\sqrt{s}$$

$$\Rightarrow \quad \int_0^s \frac{ds}{\sqrt{s}} = \int_0^t a\,dt$$

$$\Rightarrow \quad 2\sqrt{s} = at$$

$$\Rightarrow \quad s = \frac{1}{4}a^2t^2$$

$$\Rightarrow \quad v = \frac{ds}{dt} = \frac{1}{2}a^2t$$

By work-energy theorem we have work done by all forces on body is

$$W = \Delta K = \frac{1}{2}mv^2 - 0$$

$$\Rightarrow \quad W = \frac{1}{2}m\left(\frac{1}{2}a^2t\right)^2 = \frac{1}{8}ma^4t^2$$

Ans. (A)

Alternative :

$$W = F \cdot s \quad \left[\text{As } F = m\left(\frac{1}{2}a^2\right) \rightarrow \text{constant}\right]$$

$$\Rightarrow \quad W = \frac{1}{2}ma^2\left(\frac{1}{4}a^2t^2\right) = \frac{1}{8}ma^4t^2$$

Ans. (A)

Sol. 6 Maximum speed is at mean position (equilibrium) where

$$x = \frac{F}{k}$$

Using work energy theorem, we have

$$F(x) - \frac{1}{2}kx^2 = \frac{1}{2}mv^2 - 0$$

$$\Rightarrow \quad F\left(\frac{F}{k}\right) - \frac{1}{2}k\left(\frac{F}{k}\right)^2 = \frac{1}{2}mv^2$$

$$\Rightarrow \quad v_{\max} = \frac{F}{\sqrt{mk}}$$

Ans. (C)

Sol. 7 Given that

$$x = 3t^2 + 5$$

$$\Rightarrow \quad v = \frac{dx}{dt} = 6t + 0$$

at $t = 0$; $v = 0$

and at $t = 5$ sec ; $v = 30$ m/s

By work energy theorem, we use

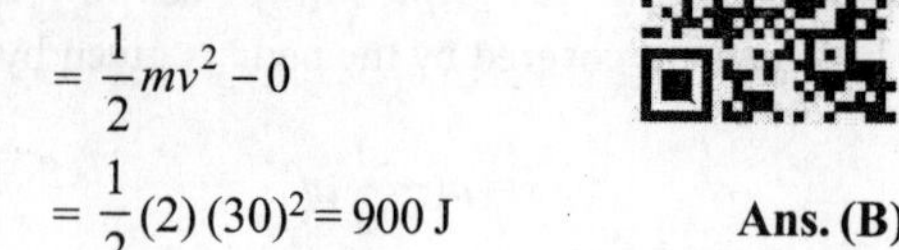

$$\text{Work done} = \frac{1}{2}mv^2 - 0$$

$$= \frac{1}{2}(2)(30)^2 = 900 \text{ J}$$

Ans. (B)

Sol. 8 For block we use

$$N - mg = \frac{mg}{2}$$

$$\Rightarrow \quad N = \frac{3mg}{2}$$

Work done by normal reaction is

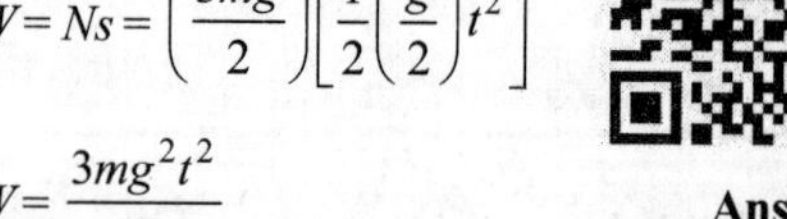

$$W = Ns = \left(\frac{3mg}{2}\right)\left[\frac{1}{2}\left(\frac{g}{2}\right)t^2\right]$$

$$\Rightarrow \quad W = \frac{3mg^2t^2}{8}$$

Ans. (B)

Sol. 9 Work done

$$W = \vec{F} \cdot \vec{d} = (3\hat{i} - 12\hat{j}) \cdot (4\hat{i})$$

$$\Rightarrow \quad W = 12 \text{ J}$$

By work energy theorem, we use

$$K_i + W_{\text{net}} = K_f$$

$$\Rightarrow \quad 3 + 12 = K_f$$

$$\Rightarrow \quad K_f = 15 \text{ J}$$

Ans. (A)

Sol. 10 Velocity of 1 kg block just before it collides with 3kg block

$$= \sqrt{2gh} = \sqrt{2000}\text{ m/s}$$

Applying momentum conversation just before and just after collision

$$1 \times \sqrt{2000} = 4v$$

$$\Rightarrow \quad v = \frac{\sqrt{2000}}{4}\text{ m/s}$$

Initial compression of spring

$$1.25 \times 10^6 x_0 = 30$$

$$\Rightarrow \quad x_0 \simeq 0$$

Applying work energy theorem, we have

$$W_g + W_{sp} = \Delta KE$$

$$\frac{1}{2}(m_1 + m_2)v^2 + (m_1 + m_2)gx + \frac{1}{2}kx^2 = 0$$

$$\frac{1}{2} \times 4 \times v^2 + 40 \times x - \frac{1}{2} \times 1.25 \times 10^6 x^2 = 0$$

Solving, we get $x \simeq 4$ cm **Ans. (A)**

Sol. 11 According to work energy theorem

Work done by force on the particle = Change in KE

Work done = Area under F-x graph

$$\Rightarrow \quad W = \int F.dx$$

$$\Rightarrow \quad W = 2\times 2 + \frac{(2+3)\times 1}{2}$$

$$\Rightarrow \quad W = KE_{\text{final}} - KE_{\text{initial}} = 6.5$$

As $KE_{\text{initial}} = 0$

$$\Rightarrow \quad KE_{\text{final}} = 6.5\text{ J}$$ **Ans. (A)**

Sol. 12 Mass of the hanging part $= \dfrac{M}{n}$

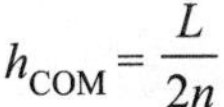

$$h_{\text{COM}} = \frac{L}{2n}$$

Work done by gravity in lifting the cable is

$$W = mgh_{\text{COM}} = \left(\frac{M}{n}\right)g\left(\frac{L}{2n}\right) = \frac{MgL}{2n^2}$$ **Ans. (B)**

Sol. 13 Considering gravity, acceleration of ball is

$$\frac{dv}{dt} = -(g + \gamma v^2)$$

$$\Rightarrow \quad \frac{g}{\gamma}\left(\frac{dv}{\frac{g}{\gamma} + v^2}\right) = -gdt$$

Integrating in limits from

$t = 0$ to $t = t$ and $v = V_0$ to $v = 0$

$$\int_{V_0}^{0} \frac{g}{\gamma}\left(\frac{dv}{\frac{g}{\gamma} + v^2}\right) = -\int_0^t gdt$$

$$\Rightarrow \quad -gt = -\sqrt{\frac{g}{\gamma}}\tan^{-1}\left(\frac{V_0}{\sqrt{\frac{g}{\gamma}}}\right)$$

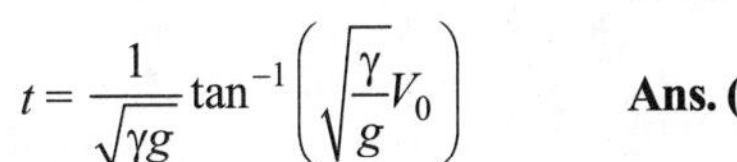

$$\Rightarrow \quad t = \frac{1}{\sqrt{\gamma g}}\tan^{-1}\left(\sqrt{\frac{\gamma}{g}}V_0\right)$$ **Ans. (B)**

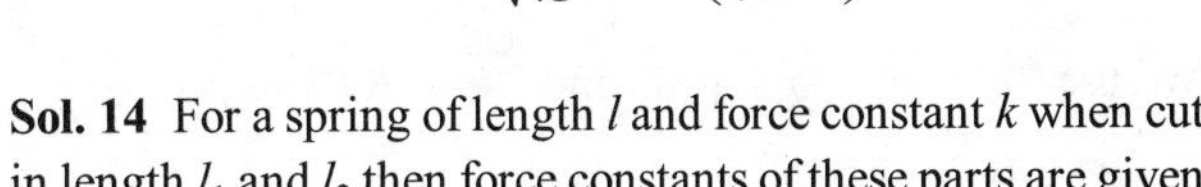

Sol. 14 For a spring of length l and force constant k when cut in length l_1 and l_2 then force constants of these parts are given as

$$k_1 = \frac{kl}{l_1} \quad \ldots(1)$$

$$k_2 = \frac{kl}{l_2}$$

$$\Rightarrow \quad \frac{k_1}{k_2} = \frac{l_2}{l_1}$$

$$\Rightarrow \quad \frac{k_1}{k_2} = \frac{l_2}{nl_2} = \frac{1}{n}$$ **Ans (C)**

Sol. 15 Upward force by motor on elevator is

$$F = 2000\,g + f_r = 20000 + 4000$$

$$F = 24000\text{ N}$$

At full load with speed v, power of motor is

$$\Rightarrow \quad 60 \times 746 = (24000)\,v$$

$$\Rightarrow \quad v = \frac{60 \times 746}{24000} = 1.865$$

$$\Rightarrow \quad v \approx 1.9\text{ m/s}$$ **Ans. (C)**

Sol. 16 Using energy conservation between point A & P, we have

$$mgh_1 = mgh_2 + KE$$

$$mg \times 2 = mg \times 1 + KE$$

$$\Rightarrow \quad KE = mg \times 1 = 10\text{ J}$$ **Ans. (10)**

Sol. 17 Free body diagram of elevator is

T

6800N + 9200N + 6000N

As elevator moving with constant speed, so we use

$$T = 6800 + 9200 + 6000$$

$$\Rightarrow \quad T = 22000\text{ N}$$

Max power is given as

$$P_{max} = T.v = 22000 \times 3$$

$$\Rightarrow \quad P_{max} = 66000\text{ W}$$

Ans. (C)

Sol. 18 Work done by force is given as

$$W = \int_{\vec{r}_i}^{\vec{r}_g} \vec{F} \cdot \vec{dr}$$

We use $\vec{F} = -x\hat{i} + y\hat{i}$ and $\vec{dr} = dx\hat{i} + dy\hat{j}$

$$\Rightarrow \quad W = \int_1^0 -xdx + \int_0^1 ydy$$

$$\Rightarrow \quad W = \left[\frac{-x^2}{2}\right]_1^0 + \left[\frac{y^2}{2}\right]_0^1$$

$$\Rightarrow \quad W = -\left(\frac{0^2}{2} - \frac{1^2}{2}\right) + \left(\frac{1^2}{2} - \frac{0^2}{2}\right)$$

$$\Rightarrow \quad W = 1\text{ J}$$

Ans. (B)

Sol. 19 Using work-energy theorem, we use

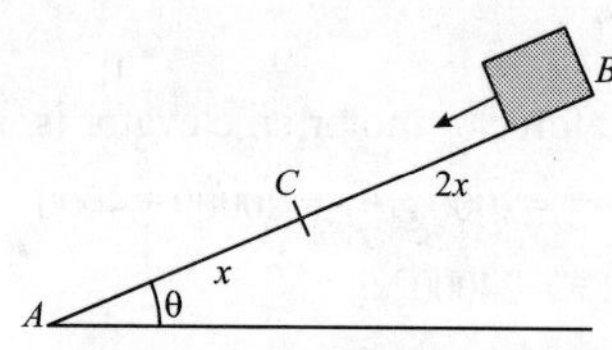

$$mg\sin\theta\,(x + 2x) - \mu mg\cos\theta\, x = 0$$

$$\Rightarrow \quad 3\sin\theta - \mu\cos\theta = 0$$

$$\Rightarrow \quad \mu = 3\tan\theta$$

$$\Rightarrow \quad k = 3$$

Ans. (3)

Sol. 20 By work energy theorem, we use

$$Fl = mgh$$

$$\Rightarrow \quad F \times 0.2 = 0.15 \times 10 \times 20$$

$$\Rightarrow \quad F = \frac{0.15 \times 200}{0.2} = 150\text{ N}$$

Ans. (150)

Sol. 21 If power is constant, we use

$$K = Pt$$

$$\Rightarrow \quad \frac{1}{2}mv^2 = Pt$$

$$\Rightarrow \quad \frac{ds}{dt} = v = \sqrt{\frac{2Pt}{m}}$$

$$\Rightarrow \quad \int_0^s ds = \int_0^t \sqrt{\frac{2Pt}{m}}\,dt$$

$$\Rightarrow \quad s = \sqrt{\frac{2P}{m}}\left(\frac{2}{3}t^{3/2}\right)$$

$$\Rightarrow \quad s \propto t^{3/2}$$

Ans. (C)

Sol. 22 When block goes up its acceleration is

$$a_1 = -g(\sin\theta + \mu\cos\theta)$$

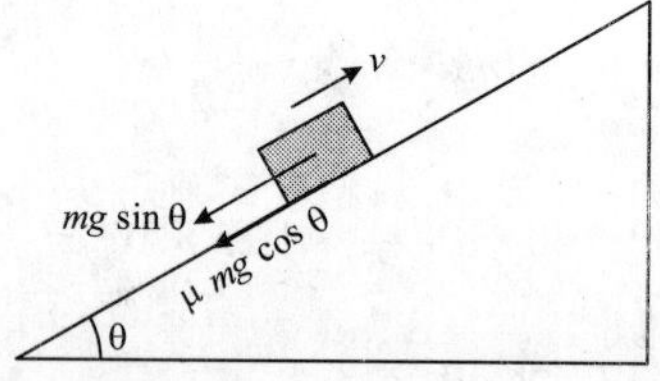

distance it goes up the plane is given as

$$l = \frac{v_0^2}{2a_1}$$

When it comes down its acceleration is given as

$$a_2 = g(\sin\theta - \mu g\cos\theta)$$

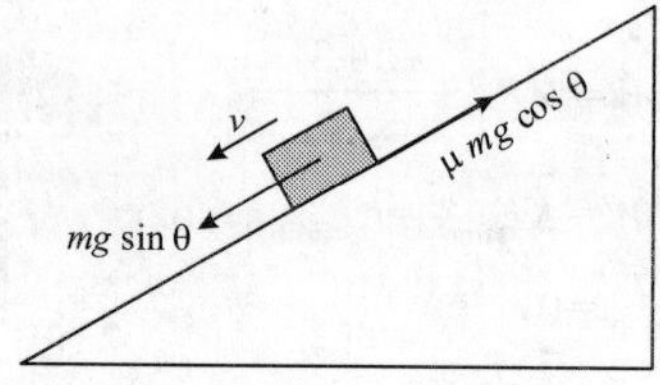

Velocity at bottom point is given as

$$\left(\frac{v_2}{2}\right)^2 = 2a_2 l$$

$$\Rightarrow \quad \frac{v_0^2}{4} = 2g(\sin\theta - \mu\cos\theta) \cdot \frac{v_0^2}{2g(\sin\theta + \mu\cos\theta)}$$

$$\Rightarrow \quad \sin\theta + \mu\cos\theta = 4(\sin\theta - \mu\cos\theta)$$

At $\theta = 30°$

$$\frac{1}{2} + \frac{\mu\sqrt{3}}{2} = 4\left(\frac{1}{2} - \frac{\mu\sqrt{3}}{2}\right)$$

$$\Rightarrow \quad 1 + \mu\sqrt{3} = 4 - 4\sqrt{3}\mu$$

$$\Rightarrow \quad 5\sqrt{3}\mu = 3$$

$$\Rightarrow \quad \mu = \frac{\sqrt{3}}{5} = 0.3464 = \frac{I}{1000}$$

$$\Rightarrow \quad I = 346$$

Ans. (346)

Sol. 23 Given that applied force with position is

$$F = 200\text{ N for } x \le 15\text{ m}$$

$$\left. F = 200 - \frac{100}{15}(x-15) \right\}$$

for $15 \le x < 30$

$$\Rightarrow \quad F = 300 - \frac{100}{15}x$$

Total work done by force is

$$W = 200 \times 15 + \int_{15}^{30} \left(30 - \frac{100}{15}x\right) dx$$

$$\Rightarrow \quad W = 300 + 300 \times 15 - \frac{100}{15} \cdot \left[\frac{(30)^2}{2} - \frac{(15)^2}{2}\right]$$

$$\Rightarrow \quad W = 7500 - \frac{100}{30}[900 - 225]$$

$$\Rightarrow \quad W = 7500 - 2250 = 5250\text{ J}$$ **Ans. (B)**

Sol. 24 As power of engine in constant we use

$$P = mva = mv\frac{dv}{dt}$$

$$\Rightarrow \quad \int_0^v v\,dv = \frac{P}{m}\int_0^t dt$$

$$\Rightarrow \quad v = \sqrt{\frac{2Pt}{m}} = \frac{dx}{dt}$$

$$\Rightarrow \quad \int_0^s dx = \sqrt{\frac{2P}{m}}\int_0^t t^{1/2} dt$$

$$\Rightarrow \quad s = \sqrt{\frac{2P}{m}} \cdot \frac{t^{3/2}}{(3/2)} = \frac{2}{3}\sqrt{\frac{2P}{m}}t^{3/2}$$

$$\Rightarrow \quad s = \frac{2}{3}\sqrt{\frac{2\times 1}{2}} \times (9)^{3/2}$$

$$\Rightarrow \quad s = \frac{2}{3} \times 27 = 18\text{ m}$$ **Ans. (18.00)**

Sol. 25 The field force between particles is given as

$$F = \left|\frac{dU}{dr}\right| = -(A(-6r^{-7})) + B(-12r^{-13})$$

$$\Rightarrow \quad F = \frac{6A}{r^7} - \frac{12B}{r^{13}}$$

At equilibrium, we have $F = 0$, so we use

$$0 = \frac{6A}{r^7} - \frac{12B}{r^{13}}$$

$$\Rightarrow \quad \frac{6A}{12B} = \frac{1}{r^6}$$

$$\Rightarrow \quad r = \left(\frac{2B}{A}\right)^{1/6}$$

At equilibrium position, potential energy is given as

$$U_{\left(r = \left(\frac{2B}{A}\right)^{1/6}\right)} = -\frac{A}{2B/A} + \frac{B}{4B^2/A^2}$$

$$\Rightarrow \quad U = \frac{-A^2}{2B} + \frac{A^2}{4B} = \frac{-A^2}{4B}$$ **Ans. (C)**

Sol. 26 For equilibrium of molecules, the net field force acting must be zero, so we use

$$\frac{dU}{dr} = 0$$

$$\Rightarrow \quad \frac{-10\alpha}{r^{11}} + \frac{5\beta}{r^6} = 0$$

$$\Rightarrow \quad \frac{5\beta}{r^6} = \frac{10\alpha}{r^{11}}$$

$$\Rightarrow \quad r = \left(\frac{2\alpha}{\beta}\right)^{\frac{1}{5}}$$

$$\Rightarrow \quad a = 1$$ **Ans. (1)**

Sol. 27 Initial kinetic energy of ball is given as

$$K_i = \frac{1}{2}mu^2 = \frac{1}{2} \times 0.5 \times 20 \times 20 = 100\text{ J}$$

After deflection it moves with 5% of K_i, thus final kinetic energy of ball is given as

$$K_f = \frac{5}{100} \times K_i$$

$$\Rightarrow \quad K_f = \frac{5}{100} \times 100$$

$$\Rightarrow \quad K_f = 5\text{ J}$$

Now, let the final speed be v m/s, then

$$K_f = 5 = \frac{1}{2}mv^2$$

$$\Rightarrow \quad v^2 = 20$$

$$\Rightarrow \quad v = \sqrt{20} = 4.47\text{ m/s}$$ **Ans. (B)**

Sol. 28 As C collides elastically, it comes to rest and A moves with speed v. Then spring compression will be maximum when A and B moves with equal speeds i.e. $v/2$. At this instant maximum compression can be given by conservation of energy, calculated as

$$\frac{1}{2}(2m)\left(\frac{v}{2}\right)^2 = \frac{1}{2}kx^2$$

$$\Rightarrow \quad x = \sqrt{\frac{m}{2k}}v$$ **Ans. (A)**

Sol. 29 Centripetal force provided by potential field is given as

$$F=\frac{-dU}{dr}=-4U_0r^3=\frac{mv^2}{r}$$

$\Rightarrow \quad mv^2=4U_0r^4$

$\Rightarrow \quad v\propto r^2$

By using Bohr's quantization postulate, we have

$$mvr=\frac{nh}{2\pi}$$

$\Rightarrow \quad r^3\propto n$

$\Rightarrow \quad r\propto n^{1/3}$

$\Rightarrow \quad \alpha=3$ **Ans. (3)**

Sol. 30 For the machine delivering constant power to a box, the force exerted on the box and its speed are related as

$$Fv=C$$

$\Rightarrow \quad m\left(\frac{dv}{dt}\right)v=C$

$\Rightarrow \quad \frac{v^2}{2}\propto t$

$\Rightarrow \quad v\propto t^{1/2}$

$\Rightarrow \quad \frac{dx}{dt}\propto t^{1/2}$

$\Rightarrow \quad x\propto t^{3/2}$ **Ans. (B)**

Sol. 31 By work energy theorem from point A and B, we have

$$mgh=\frac{1}{2}mv^2$$

$\Rightarrow \quad 10\times10\times5=\frac{1}{2}(10)v^2$

$\Rightarrow \quad v=10\text{ m/s}$ **Ans. (10)**

Sol. 32 If the spring gets compressed by a distance y, by work energy theorem we use

$$\frac{1}{2}mv^2=\frac{1}{2}ky^2$$

$\Rightarrow \quad y=\sqrt{\frac{m}{k}}\cdot v=\sqrt{\frac{4}{100}}\times10=2\text{m}$

Thus final length of spring is given as

$$x=8-2=6\text{ m}$$ **Ans. (6)**

Sol. 33 Kinetic energy of a moving body is related to its liniear momentum as

$$K=\frac{P^2}{2m}$$

$\Rightarrow \quad P\propto\sqrt{K}$

$\Rightarrow \quad \frac{P_2}{P_1}=\sqrt{\frac{K_2}{K_1}}=\sqrt{\frac{4K}{K}}=2$

Percentage change in momentum is given as

$$\frac{P_2-P_1}{P_1}\%=\left(\frac{P_2}{P_1}-1\right)\times100=(2-1)\times100=100$$

$\Rightarrow \quad \frac{\Delta P}{P_1}\%=100\%$ **Ans. (B)**

Sol. 34 As power supplied is constant to the body, its energy after time t is given as Pt.

$$Pt=\frac{1}{2}mv^2$$

$$v\propto\sqrt{t}$$

$\Rightarrow \quad \frac{ds}{dt}=C\sqrt{t}$

Distance covered by the body in t seconds can be calculated as

$$\int_0^L ds=C\int_0^t t^{1/2}dt$$

$\Rightarrow \quad L=\frac{2Ct^{3/2}}{3}$

$\Rightarrow \quad L\propto t^{3/2}$ **Ans. (C)**

Sol. 35 The work done in lowering the suitcase by the porter can be calculated as

$$W_{\text{person}}=mgh\cos180^\circ=-784\times80\times10^{-2}=-627.2\text{ J}$$

Ans. (D)

Sol. 36 Work done by the force is given as

$$W=\int F\cdot dy$$

$\Rightarrow \quad W=\int_0^{10}(5y+20)dy=\left[\frac{5y^2}{2}+20y\right]_0^{10}$

$\Rightarrow \quad \frac{5\times100}{2}+200=450\text{ J}$ **Ans. (450)**

Sol. 37 Energy supplied by engine in time t is given as

$$E=Pt=\frac{1}{2}mv^2$$

$$v=\sqrt{\frac{2pt}{m}}$$

$\Rightarrow \quad \frac{ds}{dt}=\sqrt{\frac{2p}{m}}\sqrt{t}$

$\Rightarrow \quad \int_0^s ds=\sqrt{\frac{2p}{m}}\int_0^t t^{1/2}dt$

$$\Rightarrow \quad s = \frac{2\sqrt{\frac{2p}{m}}t^{3/2}}{3}$$

$$\Rightarrow \quad t^{3/2} = \frac{3s}{2\sqrt{\frac{2p}{m}}}$$

$$\Rightarrow \quad s = \left(\frac{8P}{9m}\right)^{1/3} t^{3/2}$$ **Ans. (A)**

Sol. 38 By conservation of energy for the particle we use

$KE + U = \text{Total energy} = \text{constant}$

At $x = x_2$ from graph we can see that its potential energy is zero thus its kinetic energy is maximum. **Ans. (C)**

Sol. 39 The initial and final situation of chain are shown in the diagram below

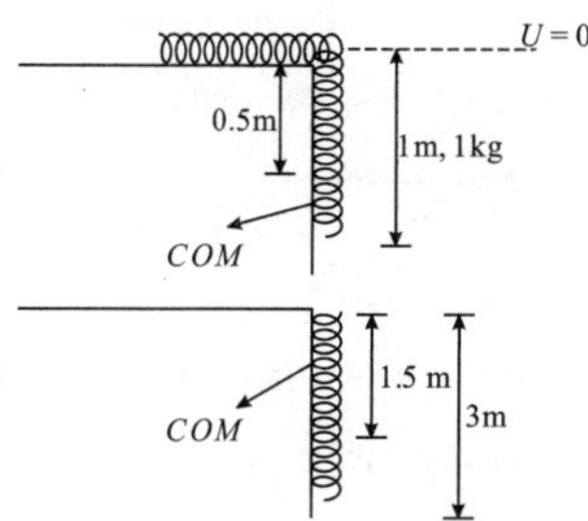

Using conservation of energy, we have

$$U_i + K_i = U_f + K_f$$

$$\Rightarrow \quad -1 \times 10 \times \left(\frac{1}{2}\right) + 0 = -3 \times 10 \times 1.5 + K_f$$

$$\Rightarrow \quad K_f = 45 - 5$$

$$\Rightarrow \quad K_f = 40\text{ J}$$ **Ans. (40)**

Sol. 40 As work done given is same in the two cases, we use

$$W_1 = W_2$$

$$\Rightarrow \quad F_1 s \cos 45° = F_2 s \cos 60°$$

$$\Rightarrow \quad \frac{F_1}{F_2} = \frac{1}{\sqrt{2}}$$

$$\Rightarrow \quad x = 2$$ **Ans. (2)**

Sol. 41 Figure below shows the situation described in question for the splitting process

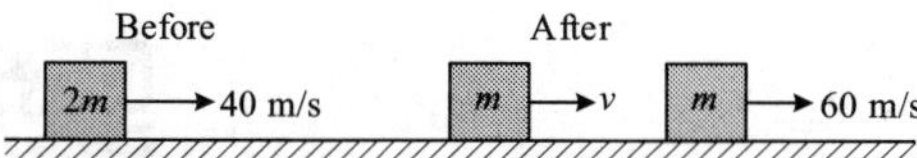

By conservation of linear momentum, we have

$$2m \times 40 = 60 \times m + mv$$

$$\Rightarrow \quad v = 20\text{ m/s}$$

Change in kinetic energy is given as

$$\Delta K = K_f - K_i = \frac{1}{2} m \times (20)^2 + \frac{1}{2} m \times (60)^2 - \frac{1}{2} 2\text{m} \times (40)^2$$

$$\Rightarrow \quad \Delta K = \frac{1}{2} m (800)$$

$$K_i = \frac{1}{2} 2\text{m} \times (40)^2 = \frac{1}{2} m \times 3200$$

Fractional change in kinetic energy is given as

$$f = \frac{\Delta K}{K_i} = \frac{1}{4}$$ **Ans. (1)**

Sol. 42 Figure below describes the situation explained in the question.

By conservation of linear momentum, we have

$$m \times 40 = \frac{2}{3} mv + \frac{m}{3} \times 60$$

$$\Rightarrow \quad 120 = 2v + 60$$

$$\Rightarrow \quad v = 30\text{ m/s}$$

Now $\quad K_i = \frac{1}{2} \times m \times 40^2 = 800\text{ m}$

$$K_f = \frac{1}{2} \times \frac{2m}{3} (30)^2 + \frac{1}{2} \times \frac{m}{3} \times (60)^2$$

$$\Rightarrow \quad K_T = \frac{1}{6}\{m(1800 + 3600)\} = \frac{m(5400)}{6} = 900\text{ m}$$

$$\Rightarrow \quad \frac{\Delta KE}{K_i} = \frac{900m - 800}{800m} = \frac{1}{8}$$ **Ans. (A)**

Sol. 43 By work energy theorem, we use

$$0 + mgh + W_f = \frac{1}{2} mv^2$$

$$\Rightarrow \quad W_f = -mgh + \frac{1}{2} m (0.8)^2 gh$$

$$\Rightarrow \quad W_f = -mgh + 0.32\,mgh$$

$$\Rightarrow \quad W_f = -mgh\,(0.68) = -6.8\,mgh$$ **Ans. (A)**

Sol. 44 By work energy theorem, we can write

$$\frac{1}{2} mv^2 + W_{\text{friction}} + W_{\text{spring}} = 0$$

$$\Rightarrow \quad -\frac{90}{100}\left(\frac{1}{2} mv^2\right) + W_{\text{spring}} = -\frac{1}{2} mv^2$$

$$\Rightarrow \quad W_{\text{spring}} = -\frac{10}{100} \times \frac{1}{2} mv^2$$

$$\Rightarrow \quad -\frac{1}{2} kx^2 = -\frac{1}{20} mv^2$$

$$\Rightarrow \quad k = \frac{40000 \times (20)^2}{10 \times (1)^2} = 16 \times 10^5\text{N/m}$$ **Ans. (16)**

5.2 Circular Motion

Sol. 1

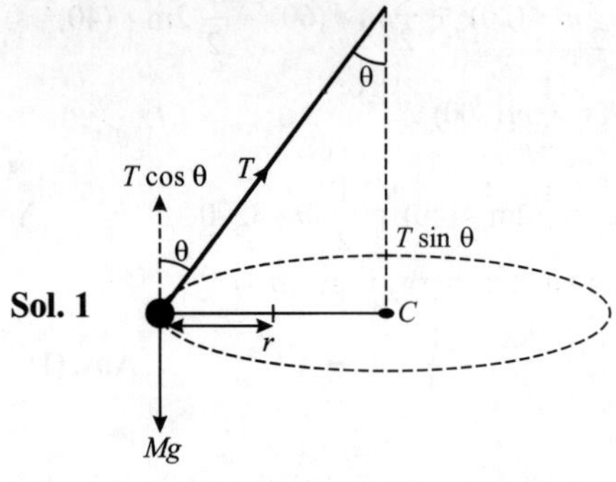

For stable circular motion, we use

$$T\sin\theta = \frac{mv^2}{r} \quad \ldots(1)$$

$$T\cos\theta = mg \quad \ldots(2)$$

$$(1)/(2) \Rightarrow \quad \tan\theta = \frac{v^2}{rg}$$

$$\Rightarrow \quad v = \sqrt{rg\tan\theta}$$

$$\Rightarrow \quad v = \sqrt{0.4\times10\times1} = 2\text{ m/s}$$ **Ans. (D)**

Sol. 2

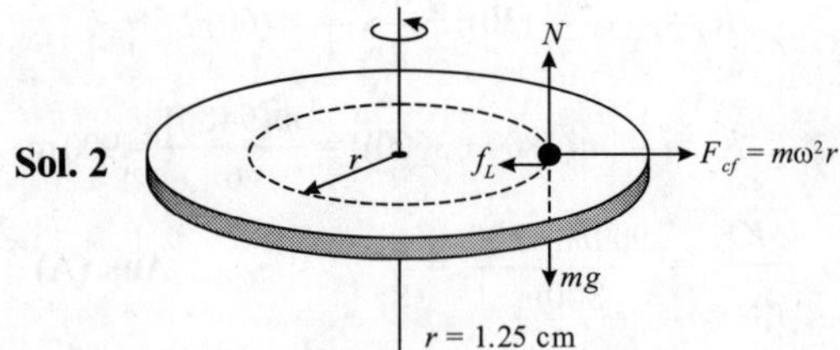

Angular speed $\omega = 2\pi n = 2\pi\times3.5 = 7\pi$ rad/s

in frame of disc $\quad F_{cf} = m\omega^2 r$

If coin is at rest on disc

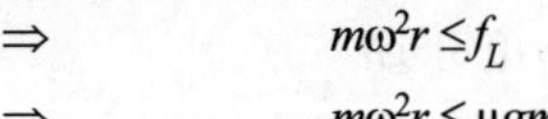

$$\Rightarrow \quad m\omega^2 r \le f_L$$

$$\Rightarrow \quad m\omega^2 r \le \mu gm$$

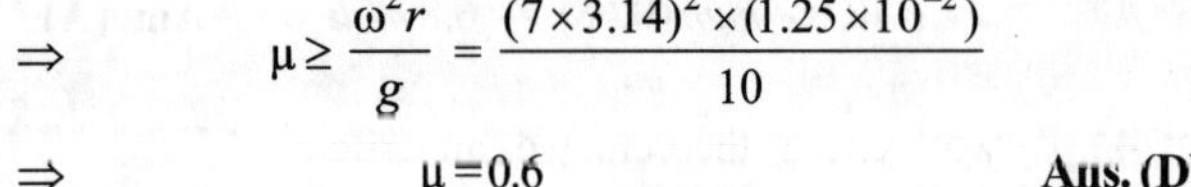

$$\Rightarrow \quad \mu \ge \frac{\omega^2 r}{g} = \frac{(7\times3.14)^2\times(1.25\times10^{-2})}{10}$$

$$\Rightarrow \quad \mu = 0.6$$ **Ans. (D)**

Sol. 3

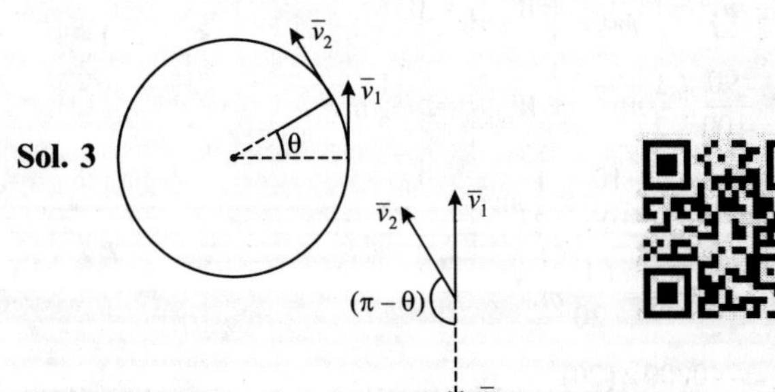

For above circular motion, change in velocity is

$$|\Delta\bar{v}| = \sqrt{v_1^2 + v_2^2 + 2v_1v_2\cos(\pi-\theta)}$$

$$\Rightarrow \quad |\Delta\bar{v}| = 2v\sin\frac{\theta}{2} \qquad \text{since } [|\bar{v}_1| = |\bar{v}_2|]$$

$$\Rightarrow \quad |\Delta\bar{v}| = (2\times10)\times\sin(30°)$$

$$\Rightarrow \quad |\Delta\bar{v}| = 10\text{ m/s}$$ **Ans. (B)**

Sol. 4 At $t = \dfrac{\pi}{2\omega}$ angle transversed by particles is

$$\theta = \omega t = \omega\frac{\pi}{2\omega} = \frac{\pi}{2}$$

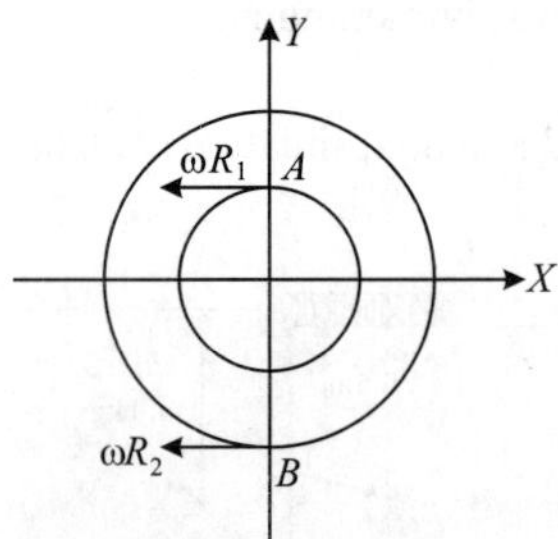

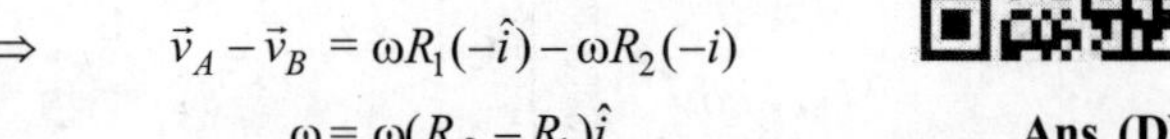

$$\Rightarrow \quad \vec{v}_A - \vec{v}_B = \omega R_1(-\hat{i}) - \omega R_2(-i)$$

$$\omega = \omega(R_2 - R_1)\hat{i}$$ **Ans. (D)**

Sol. 5 FBD of bead is shown in figure below

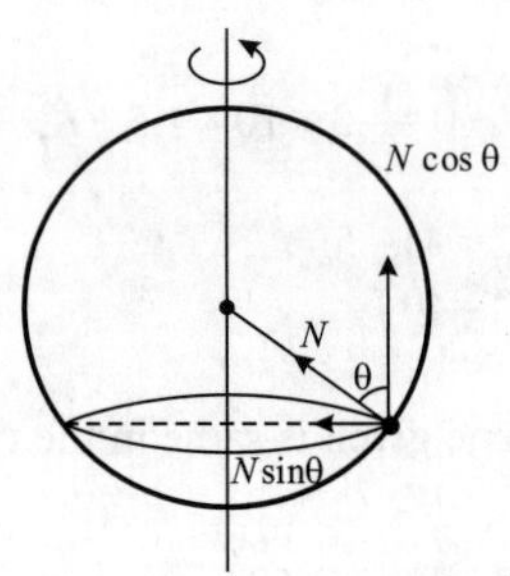

For equilibrium of bead in frame of wire, we have

$$N\sin\theta = m\frac{r}{2}\omega^2 \quad \ldots(1)$$

$$N\cos\theta = mg \quad \ldots(2)$$

$$\Rightarrow \quad \tan\theta = \frac{r\omega^2}{2g}$$

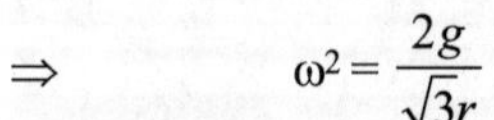

$$\Rightarrow \quad \frac{r}{2\frac{\sqrt{3}r}{2}} = \frac{r\omega^2}{2g}$$

$$\Rightarrow \quad \omega^2 = \frac{2g}{\sqrt{3}r}$$ **Ans. (D)**

Sol. 6 FBD of particle is shown in figure

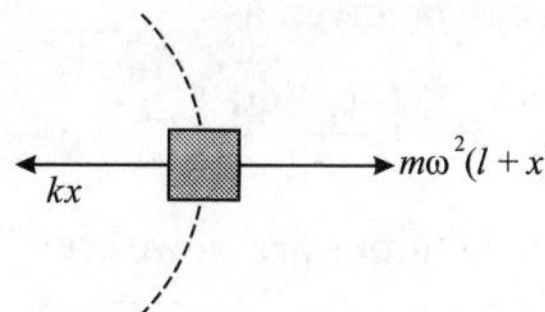

For uniform circular motion, we use

$$kx = ml\omega^2 (l + x)$$

$$\Rightarrow \qquad x = \frac{ml\omega^2}{k - m\omega^2}$$ **Ans. (B)**

Sol. 7 Free body diagram of mass in frame of disc is shown in figure

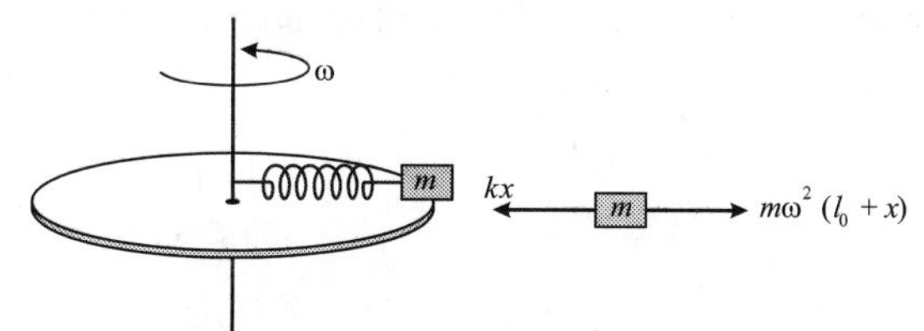

For equilibrium of mass m disc frame, we use

$$kx = m\omega^2(l_0 + x)$$

$$x = \frac{m\omega^2 l_0}{k - m\omega^2} \approx \frac{m\omega^2 l_0}{k}$$

Relative change in spring length is

$$\frac{x}{l_0} = \frac{m\omega^2}{k}$$ **Ans. (D)**

Sol. 8 Figure below shows the forces acting on bead

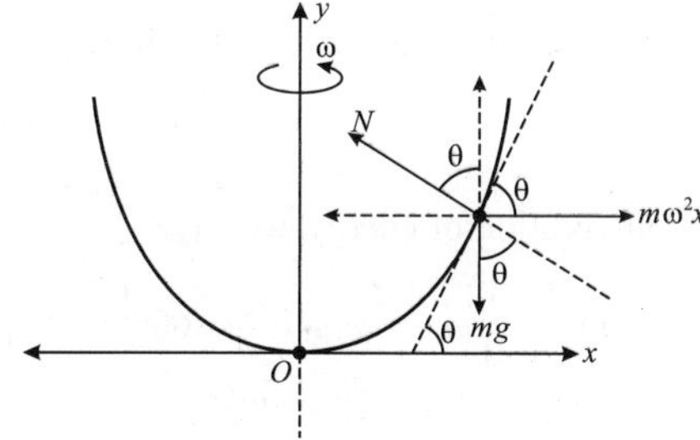

For equilibrium of bead in frame of wire, we use

$$m\omega^2 x \cos\theta = mg \sin\theta$$

$$\Rightarrow \qquad \omega^2 x = g \tan\theta$$

Slope of parabola at point P is $\frac{dy}{dx} = \tan\theta$

$$\Rightarrow \qquad \frac{dy}{dx} = \frac{\omega^2 x}{g} = 8Cx \qquad [\text{As } y = 4Cx^2]$$

$$\Rightarrow \qquad \omega^2 = 8Cg$$

$$\Rightarrow \qquad \omega = \sqrt{8Cg}$$ **Ans. (B)**

Sol. 9 As given is theory explanation shape of curved surface is given as

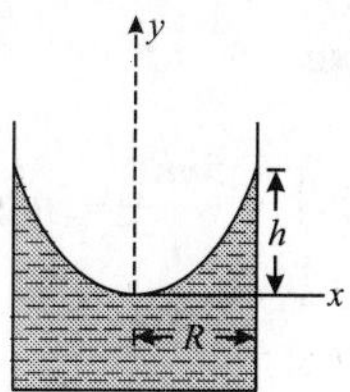

$$y = \frac{\omega^2 x^2}{2g}$$

$\Rightarrow$ at $x = R$, we have

$$h = \frac{\omega^2 R^2}{2g} = \frac{\omega^2 (5)^2}{2g} = \frac{25\omega^2}{2g}$$ **Ans. (A)**

Sol. 10 Angular speed of clock's second's hand is given as

$$\omega = \frac{2\pi}{T} = \frac{2\pi}{60} = 0.105 \text{ rad/sec}$$

Acceleration of tip of hand is

$$a = \omega^2 R$$

$$\Rightarrow \qquad a = (0.105)^2 (0.1) \text{ m/s}^2$$

$$\Rightarrow \qquad a = 0.0011 \text{ m/s}^2$$

$$\Rightarrow \qquad a = 1.1 \times 10^{-3} \text{ m/s}^2$$ **Ans. (A)**

Sol. 11 In the vertical circular motion if the speed of bob at bottom most position be v_1 and at the topmost position be v_2 and during motion, we know that maximum tension occurs at bottom most position and minimum at the topmost position. By conservation of energy for bob, we have

$$\frac{1}{2}mv_1^2 = \frac{1}{2}mv_2^2 + mg(2l)$$

$$\Rightarrow \qquad v_1^2 = v_2^2 + 4gl \qquad \ldots(1)$$

At the bottommost position of bob, we have

$$T_{max} - mg = \frac{mv_1^2}{l}$$

$$\Rightarrow \qquad T_{max} = mg + \frac{mv_1^2}{l} \qquad \ldots(2)$$

At the topmost position of bob, we have

$$T_{min} + mg = \frac{mv_2^2}{l}$$

$$\Rightarrow \qquad T_{min} = \frac{mv_2^2}{l} - mg \qquad \ldots(3)$$

From equations-(2) and (3), we have

$$\frac{T_{max}}{T_{min}} = \frac{5}{1}$$

$$\Rightarrow \quad \frac{mg + \dfrac{mv_1^2}{l}}{\dfrac{mv_2^2}{l} - mg} = \frac{5}{1}$$

$$\Rightarrow \quad mg + \frac{m}{l}[v_2^2 + 4gl] = \frac{5mv_2^2}{l} - 5mg$$

$$\Rightarrow \quad 10\,mg = \frac{4mv_2^2}{l}$$

$$\Rightarrow \quad v_2^2 = \frac{10\times10\times1}{4} = 25$$

$$\Rightarrow \quad v_2 = 5\text{ m/s}$$

Ans. (5)

Sol. 12 Given central force is expressed as

$$F \propto \frac{1}{R^3}$$

$$\Rightarrow \quad \frac{K}{R^3} = m\omega^2 R$$

$$\Rightarrow \quad \omega^2 = \frac{K}{m} \times \frac{1}{R^4}$$

If T is the time period of revolution, we use

$$\left(\frac{2\pi}{T}\right)^2 = \frac{K}{m} \times \frac{1}{R^4}$$

$$\Rightarrow \quad T^2 \propto R^4$$

$$\Rightarrow \quad T \propto R^2$$

Ans. (A)

Sol. 13 Normal force in this case provides the necessary centripetal force for uniform circular motion to block, so we use

$$N = m\omega^2 R$$

$$\Rightarrow \quad N = m\left[\frac{4\pi^2}{T^2}\right]R$$

$$\Rightarrow \quad N = 0.2 \times \left[\frac{4(3.14)^2}{(40)^2}\right] \times 0.2 = 9.859 \times 10^{-4}\text{ N}$$

Ans. (D)

Sol. 14 The maximum speed of safe turning without skidding is given as

$$v_{max} = \sqrt{\mu Rg} = \sqrt{(0.2)\times2\times9.8}$$

$$\Rightarrow \quad v_{max} = 1.97\text{ m/s} = 7.1\text{ km/hr}$$

The given peed is lower than v_{max} so it can take safe turn.

For safe turning speed on banked road, we use

$$v_{max} = \sqrt{Rg\left[\frac{\tan\theta + \mu}{1 - \mu\tan\theta}\right]}$$

$$\Rightarrow \quad v_{max} = \sqrt{2\times9.8\left[\frac{1+0.2}{1-0.2}\right]} = 5.42\text{ m/s} = 19.5\text{ km/hr}$$

The given speed is lower than v_{max} so it can take safe turn.

Ans. (D)

Sol. 15 The angle of revolution can be calculated by average angular speed and can be given as

$$\theta = \left(\frac{\omega_1 + \omega_2}{2}\right)t$$

If total number of revolutions are N, we use

$$2\pi N = 2\pi\left(\frac{900 + 2460}{60\times2}\right)\times 26$$

$$\Rightarrow \quad N = 728$$

Ans. (728)

Sol. 16 In the given potential field, force on particle is calculated as

$$F = -\frac{dU}{dr} = -\frac{C}{r^2}$$

For circular motion of particle under this force, we have

$$|F| = \frac{mv^2}{r}$$

$$\Rightarrow \quad \frac{C}{r^2} = \frac{mv^2}{r}$$

$$\Rightarrow \quad v^2 \propto \frac{1}{r}$$

Ans. (D)

Sol. 17 Initial and final angular speeds in the process are given as

$$\omega_0 = 600\text{ rpm} = 10\text{ rps}$$

$$\omega = 1800\text{ rpm} = 30\text{ rps}$$

Angular acceleration of the body can be calculated as

$$\alpha = \frac{30-10}{10} = 2\text{rps}^2$$

Number of revolutions body turned in the given time duration can be calculated as

$$\theta = \omega_0 t + \frac{1}{2}\alpha t^2 = 10\times10 + \frac{1}{2}\times2\times(100) = 200\text{ rev}$$

Ans. (200)

Sol. 18 By conservation of energy, we have

$$\frac{1}{2}mu^2 = \frac{1}{2}mv^2 + mgl(1 - \cos60°)$$

$$\Rightarrow \quad u^2 = v^2 + 2gl(1 - \cos60°)$$

$$\Rightarrow \quad 9 = v^2 + 20\times1/2\times1/2$$

$$\Rightarrow \quad 9 = v^2 + 5$$

$$\Rightarrow \quad v = 2\text{ m/s}$$

Ans. (2)

Sol. 19 Situation shown in figure below describes the state of block when it leaves the surface

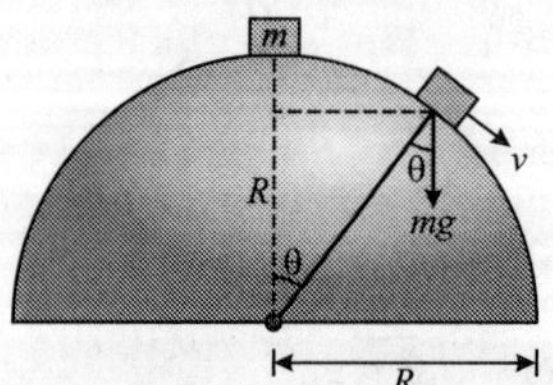

Using work energy theorem gives

$$Mg(R - R\cos\theta) = 1/2\, mv^2$$

$$\Rightarrow \quad v = \sqrt{2gR(1-\cos\theta)}$$

To loose contact, normal reaction at that point should be zero, so we use

$$\frac{mv^2}{R} = mg\cos\theta$$

$$\Rightarrow \quad 2mg(1-\cos\theta) = mg\cos\theta$$

$$\Rightarrow \quad 2 - 2\cos\theta = \cos\theta$$

$$\Rightarrow \quad \cos\theta = 2/3 = \frac{h}{R} = \frac{h}{3}$$

$$\Rightarrow \quad h = 2\text{ m}$$ **Ans. (02.00)**

Sol. 20 Figure below shows the situation described in the question. For stable horizontal circular motion, forces on bob must be balanced in the rotating frame of reference. In this case from the given data we have $\theta = 45°$ so we use

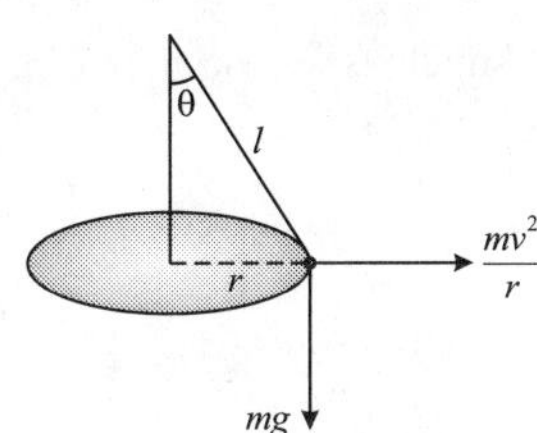

$$\tan\theta = \frac{\frac{mv^2}{r}}{mg} = \frac{v^2}{Rg} = 1$$

$$\Rightarrow \quad v^2 = rg$$

$$\Rightarrow \quad v = \sqrt{rg}$$ **Ans. (D)**

Sol. 21 Length of Arc in terms of radius is given as

$$l = R\theta$$

$$4.4 \times 9.46 \times 10^{15} = R\theta$$

Where $\theta = 4\text{ s} = 4 \times 4.843 \times 10^{-6} = 1.94 \times 10^5$ rad

$$\Rightarrow \quad 4.4 \times 9.46 \times 10^{15} = R \times 1.94 \times 10^{-5}$$

$$\Rightarrow \quad R = 2.1455 \times 10^{21}\text{ m}$$

$$\Rightarrow \quad v = 8\text{ AU} = 8 \times 1.5 \times 10^{11}\text{ m/s}$$

$$\Rightarrow \quad v = 12 \times 10^{11}\text{ m/s}$$

Distance covered in 4 revolution is given as

$$s = 4 \times 2\pi R \text{ metre}$$

Time taken in covering 4 revolution is given as

$$t = \frac{\text{distance}}{\text{speed}} = \frac{4 \times 2\pi R}{12 \times 10^{11}};$$

$$\Rightarrow \quad t = \frac{8 \times 3.14 \times 2.1455 \times 10^{21}}{12 \times 10^{11}}$$

$$\Rightarrow \quad t = 4.5 \times 10^{10}\text{ sec}$$ **Ans. (B)**

6.1 System of Particles

Sol. 1

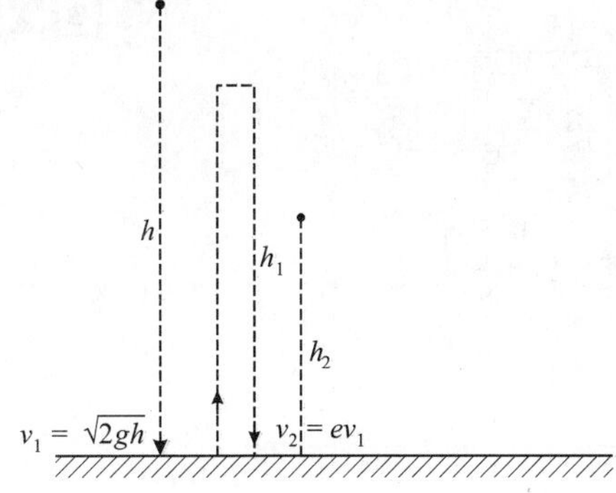

Given that

$$\frac{1}{2}mv_2^2 = \frac{1}{2}\left(\frac{1}{2}mv_1^2\right)$$

$$\Rightarrow \quad v_2 = \frac{1}{\sqrt{2}}v_1$$

We consider coefficient of restitution for this collision to be

$$e = \frac{1}{\sqrt{2}}$$

height attained by object after first impact

$$h_1 = \frac{v_2^2}{2g} = e^2\left(\frac{v_1^2}{2g}\right) = e^2 h$$

after second impact

$$h_2 = e^2 h_1 = e^4 h$$

at $t \to \infty$ total distance covered is

$$s = h + 2h_1 + 2h_2 + \ldots\infty$$

$$\Rightarrow \quad s = h(1 + 2e^2 + 2e^4 + \ldots\infty)$$

$$\Rightarrow \quad s = h\left[1 + 2e^2\left(\frac{1}{1-e^2}\right)\right] = h\left(\frac{1+e^2}{1-e^2}\right)$$

$$\Rightarrow \quad s = h\left(\frac{1+\frac{1}{2}}{1-\frac{1}{2}}\right) = 3h$$ **Ans. (A)**

Sol. 2 By conservation of linear momentum, we have along x-direction $\to mv\sin 30° - mv\sin 45° = 2mv_1\cos\theta$

$$\Rightarrow \quad \frac{mv}{2} - \frac{mv}{\sqrt{2}} = 2mv_1\cos\theta \qquad \ldots(1)$$

along y-direction $\to$

$$mv\cos 30° + mv\cos 45° = 2mv_1\sin\theta$$

$$\Rightarrow \quad \frac{\sqrt{3}}{2}mv + \frac{mv}{\sqrt{2}} = 2mv_1\sin\theta \qquad \ldots(2)$$

Here $\frac{(2)}{(1)}$ gives

$$\Rightarrow \tan\theta = \frac{\frac{\sqrt{3}}{2}+\frac{1}{\sqrt{2}}}{\frac{1}{2}-\frac{1}{\sqrt{2}}} = \frac{\sqrt{3}+\sqrt{2}}{1-\sqrt{2}}$$ **Ans. (A)**

Sol. 3 m (v_0) → m (rest) ⟶ m (v_1), m (v_2)

Given that

$$\frac{1}{2}mv_1^2 + \frac{1}{2}mv_2^2 = \frac{3}{2}\left(\frac{1}{2}mv_0^2\right)$$

$$\Rightarrow \quad v_1^2 + v_2^2 = \frac{3}{2}v_0^2 \quad \ldots(1)$$

By conservation of linear momentum, we have

$$mv_0 = mv_1 + mv_2$$

$$\Rightarrow \quad v_1 + v_2 = b_0 \quad \ldots(2)$$

$$(2)^2 \Rightarrow \quad v_1^2 + v_2^2 + 2v_1v_2 = v_0^2$$

$$(2)^2 - (1) \Rightarrow \quad 2v_1v_2 = -\frac{v_0^2}{2}$$

The relative velocity of separation is

$$(v_2 - v_1)^2 = v_1^2 + v_2^2 - 2v_1v_2$$

$$\Rightarrow \quad (v_2 - v_1)^2 = \frac{3}{2}v_0^2 + \frac{v_0^2}{2} = 2v_0^2$$

$$\Rightarrow \quad v_2 - v_1 = \sqrt{2}v_0$$ **Ans. (B)**

Sol. 4

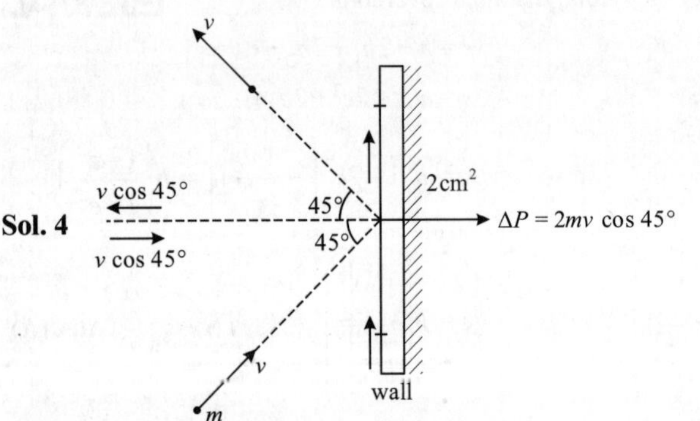

Change in momentum of H-molecule in one collision

Force on wall = change in momentum/sec due to collision on wall.

$$= 2mv\cos 45° \times 10^{23}\ \text{N}$$

$$\text{Pressure on wall} = \frac{F}{A}$$

$$\Rightarrow \quad P = \frac{2mv\left(\frac{1}{\sqrt{2}}\right)\times 10^{23}}{2\times 10^{-4}}$$

$$\Rightarrow \quad P = \frac{2\times 3.32\times 10^{-27}\times 10^{3}\times\frac{1}{\sqrt{3}}\times 10^{23}}{2\times 10^{-4}}$$

$$= 2.35\times 10^{3}\ \text{N/m}^2$$ **Ans. (A)**

Sol. 5 n (m) (u) → D/C $(2m)/(12m)$ ⟶ n (m) (v_1), D/C $(2m)/(12m)$ (v_2)

For elastic collision, final speed of neutron is given as

$$v_1 = \frac{m_1 - m_2}{m_1 + m_2}\cdot u_1 + \frac{2m_2}{m_1 + m_2}u_2$$

$$\Rightarrow \quad v_1 = -\frac{1}{3}u$$

Frictional loss in KE of neutron

$$P_d = \frac{\frac{1}{2}mu^2 - \frac{1}{2}m\left(\frac{u}{3}\right)^2}{\frac{1}{2}mu^2} = \frac{8}{9} = 0.89$$

For collision with C-nucleus we use $m_2 = 12m$ so final speed of neutron is given as

$$v_1' = -\frac{11}{13}u$$

fractional loss in KE of neutron

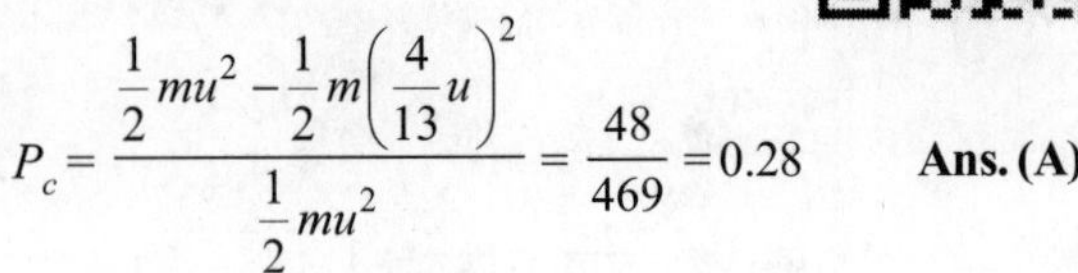

$$P_c = \frac{\frac{1}{2}mu^2 - \frac{1}{2}m\left(\frac{4}{13}u\right)^2}{\frac{1}{2}mu^2} = \frac{48}{469} = 0.28$$ **Ans. (A)**

Sol. 6

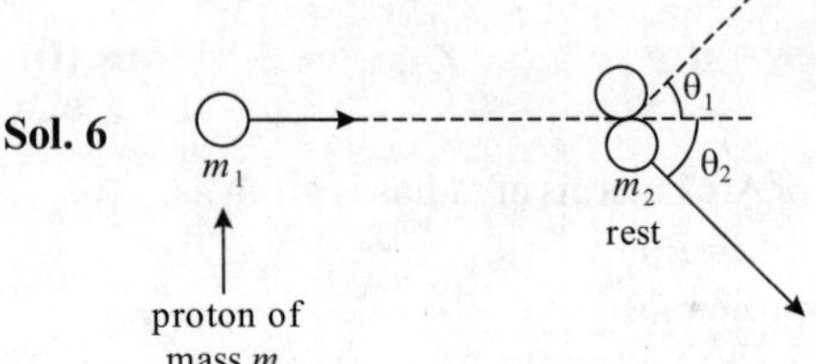

We have discussed that for an elastic collision if $m_1 = m_2$

Then divergence angle is

$$\theta_1 + \theta_2 = 90°$$

here $m_2 = m$ for $\theta_1 + \theta_2 = 90°$ **Ans. (D)**

Sol. 7 $k_i = \frac{1}{2}mv_0^2$

From linear momentum conservation

$$mv_0 = (2m + M)v_f$$

$$\Rightarrow \quad v_f = \frac{mv_0}{2m + M}$$

Given that

$$\frac{k_i}{k_f} = 6$$

$$\Rightarrow \quad \frac{\frac{1}{2}mv_0^2}{\frac{1}{2}(2m+M)\left(\frac{mv_0}{2m+M}\right)^2} = 6$$

$$\Rightarrow \quad \frac{2m+M}{m} = 6$$

$$\Rightarrow \quad \frac{M}{m} = 4$$

Ans. (A)

Sol. 8

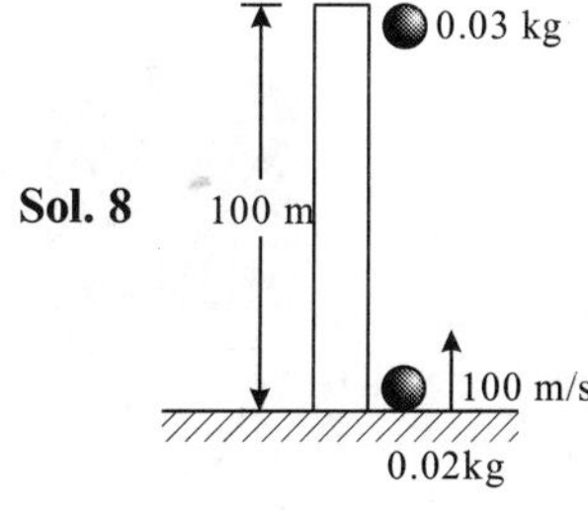

Time taken for the particles to collide,

$$t = \frac{d}{V_{rel}} = \frac{100}{100} = 1 \text{ sec}$$

In 1s distance fallen by wood is

$$h_1 = \frac{1}{2}gt^2 = \frac{1}{2} \times 10 \times (1)^2 = 5 \text{ m}$$

Speed of wood just before collision = gt = 10 m/s & speed of bullet just before collision v-gt

$$= 100 - 10 = 90 \text{ m/s}$$

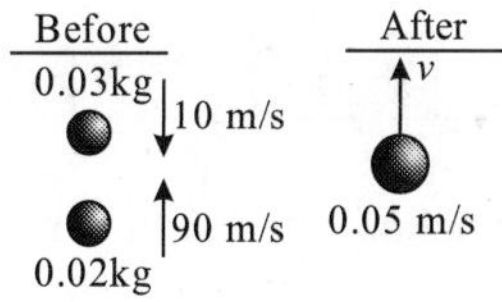

Now, conservation of linear momentum just before and after the collision

$$-(0.03)(1v) + (0.02)(9v) = (0.05)v$$

$$\Rightarrow \quad 150 = 5v$$

$$\Rightarrow \quad v = 30 \text{ m/s}$$

Maximum height reached by body $h = \dfrac{v^2}{2g}$

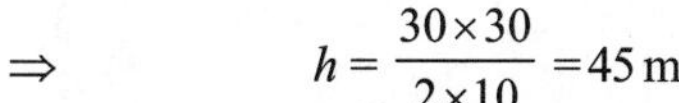

$$\Rightarrow \quad h = \frac{30 \times 30}{2 \times 10} = 45 \text{ m}$$

$\Rightarrow$ Height above tower = 45 – 5 = 40 m **Ans. (C)**

Sol. 9

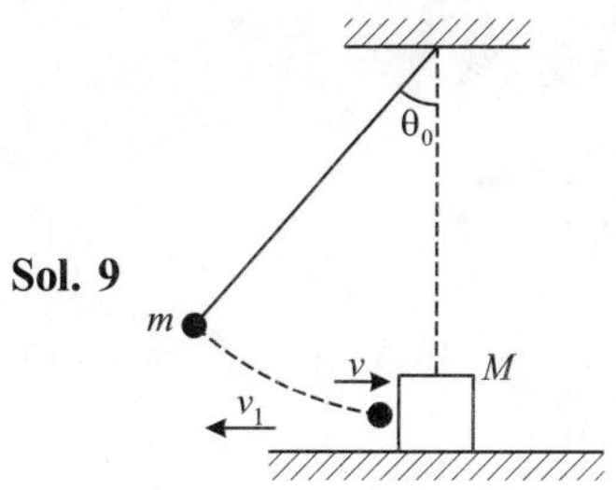

If m strikes M at speed v and rebound at v_1, we have

$$v = \sqrt{2gl(1-\cos\theta_0)}$$

and

$$v_1 = \sqrt{2gl(1-\cos\theta_1)}$$

By momentum conservation, we use

$$m\sqrt{2gl(1-\cos\theta_0)} = MV_m - m\sqrt{2gl(1-\cos\theta)}$$

$$\Rightarrow \quad m\sqrt{2gl}\{\sqrt{1-\cos\theta_0} + \sqrt{1-\cos\theta_1}\} = MV_m \quad \ldots(1)$$

For elastic collision, we use

$$e = 1 = \frac{V_m + \sqrt{2gl(1-\cos\theta_1)}}{\sqrt{2gl(1-\cos\theta_0)}}$$

$$\Rightarrow \quad \sqrt{2gl}(\sqrt{1-\cos\theta_0} - \sqrt{1-\cos\theta_1}) = V_m \quad \ldots(2)$$

(1)/(2) gives

$$\frac{(\sqrt{1-\cos\theta_0} + \sqrt{1-\cos\theta_1})}{(\sqrt{1-\cos\theta_0} - \sqrt{1-\cos\theta_1})} = \frac{M}{m}$$

$$\Rightarrow \quad \frac{m-M}{m+M} = \frac{\sqrt{1-\cos\theta_1}}{\sqrt{1-\cos\theta_0}} = \frac{\sin\left(\frac{\theta_1}{2}\right)}{\sin\left(\frac{\theta_0}{2}\right)}$$

$$\Rightarrow \quad \frac{M}{m} = \frac{\theta_0 - \theta_1}{\theta_0 + \theta_1}$$

$$\Rightarrow \quad M = m\left(\frac{\theta_0 - \theta_1}{\theta_0 + \theta_1}\right)$$

Ans. (C)

Sol. 10

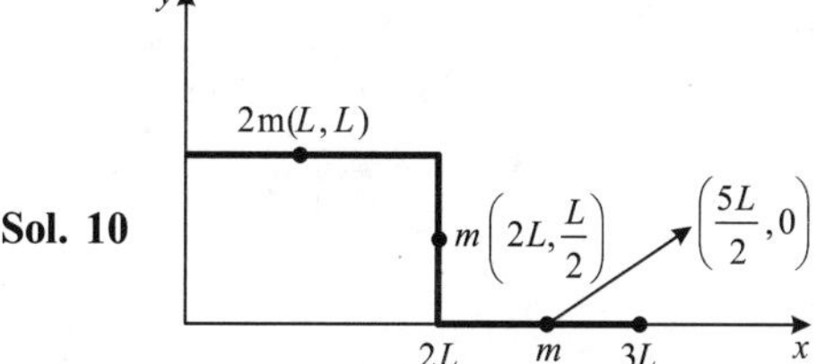

Centre of mass of three segments of bar are shown in figure. So centre of mass coordinates (X_{cm}, Y_{cm}) are given as

$$X_{cm} = \frac{2mL + 2mL + \frac{5mL}{2}}{4m} = \frac{13}{8}L$$

$$Y_{cm} = \frac{2m \times L + m \times \left(\frac{L}{2}\right) + m \times 0}{4m} = \frac{5L}{8}$$

Ans. (A)

Sol. 11 α N V_1 V_2

Using momentum conservation

$$mv_0 = Mv_2 - mv_1 \qquad \ldots(1)$$

Given that

$$\frac{1}{2}mv_1^2 = 0.36\times\frac{1}{2}mv_0^2$$

$$\Rightarrow \qquad v_1 = 0.6v_0$$

$$\Rightarrow \qquad \frac{1}{2}Mv_2^2 = 0.64\times\frac{1}{2}mv_0^2$$

$$\Rightarrow \qquad v_2 = \sqrt{\frac{m}{M}}\times 0.8v_0$$

From equation-(1)

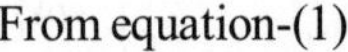

$$mv_0 = \sqrt{mM}\times 0.8v_0 - m\times 0.6v_0$$

$$\Rightarrow \qquad 1.6\,\text{m} = 0.8\sqrt{mM}$$

$$\Rightarrow \qquad 4\,\text{m}^2 = mM$$

$$\Rightarrow \qquad M = 4\,\text{m}$$

Ans. (A)

Sol. 12

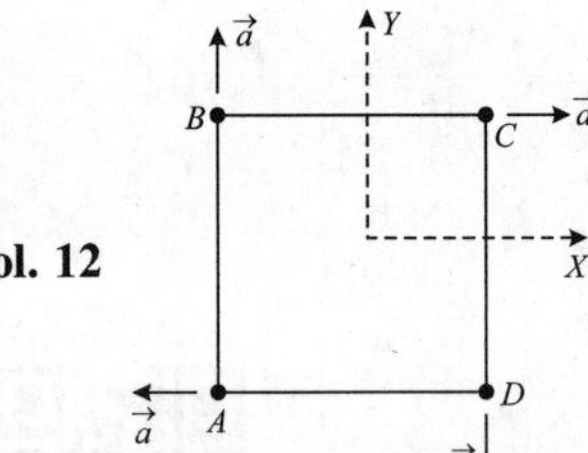

The acceleration vector of four particles are given as

$$\vec{a}_A = -a\hat{i}$$

$$\vec{a}_B = a\hat{j}$$

$$\vec{a}_C = a\hat{i}$$

$$\vec{a}_D = -a\hat{j}$$

Thus acceleration of centre of mass of this system is

$$\vec{a}_{cm} = \frac{m_a\vec{a}_a + m_b\vec{a}_b + m_c\vec{a}_c + m_d\vec{a}_d}{m_a + m_b + m_c + m_d}$$

$$\Rightarrow \qquad \vec{a}_{cm} = \frac{-ma\hat{i} + 2ma\hat{j} + 3ma\hat{i} - 4ma\hat{j}}{10\text{m}}$$

$$= \frac{2ma\hat{i} - 2ma\hat{j}}{10\text{m}}$$

$$= \frac{a}{5}\hat{i} - \frac{a}{5}\hat{j}$$

$$= \frac{a}{5}(\hat{i} - \hat{j})$$

Ans. (A)

Sol. 13 In figure below O is COM of complete sheet and C is COM of the shaded portion which is cut off

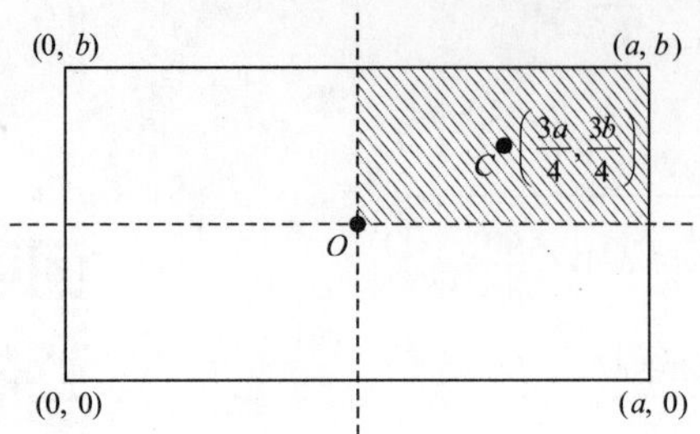

x-coordinate of COM of remaing sheet will be

$$x = \frac{M\dfrac{a}{2} - \dfrac{M}{4}\times\dfrac{3a}{4}}{M - \dfrac{M}{4}}$$

$$\Rightarrow \qquad x = \frac{\dfrac{a}{2} - \dfrac{3a}{16}}{\dfrac{3}{4}} = \frac{\dfrac{5a}{16}}{\dfrac{3}{4}} = \frac{5a}{12}$$

y-coordinate of COM of remaining sheet will be

$$y = \frac{M\dfrac{b}{2} - \dfrac{M}{4}\times\dfrac{3b}{4}}{M - \dfrac{M}{4}} = \frac{5b}{12}$$

Ans. (D)

Sol. 14 Using linear momentum conservation, we have

$$m_1v_1\hat{i} + m_2v_2\hat{i} = m_1v_3\hat{i} + m_2v_4\hat{i}$$

$$m_1v_1 + 0.5m_1v_2 = m_1(0.5v_1) + 0.5m_1v_4$$

$$0.5m_1v_1 = 0.5m_1(v_4 - v_2)$$

$$v_1 = v_4 - v_2$$

Ans. (C)

Sol. 15 v_0 2kg m Rest $\Rightarrow$ $v_0/4$ v 2kg m

Using conservation of linear momentum, we have

$$2v_0 = 2\left(\frac{v_0}{4}\right) + mv$$

$$\Rightarrow \qquad 2v_0 = \frac{v_0}{2} + mv$$

$$\Rightarrow \qquad \frac{3v_0}{2} = mv \qquad \ldots(1)$$

As collision is elastic, we use $e = 1$, so we have

$$v_{separation} = v_{approch}$$

$$\Rightarrow \qquad v - \frac{v_0}{4} = v_0$$

$$\Rightarrow \qquad \frac{5v_0}{4} = v \qquad \ldots(2)$$

From (1) and (2), we use

$$\Rightarrow \quad \frac{3v_0}{2} = m\left(\frac{5v_0}{4}\right)$$

$$\Rightarrow \quad m = \frac{6}{5} = 1.2\text{ kg} \qquad \textbf{Ans. (B)}$$

Sol. 16

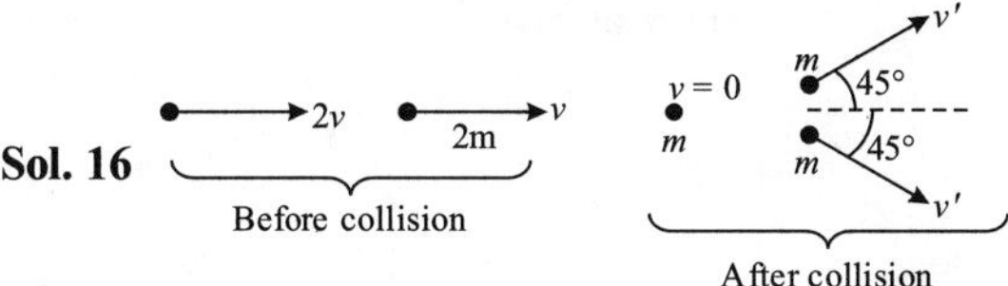

Using linear momentum conservation, we have

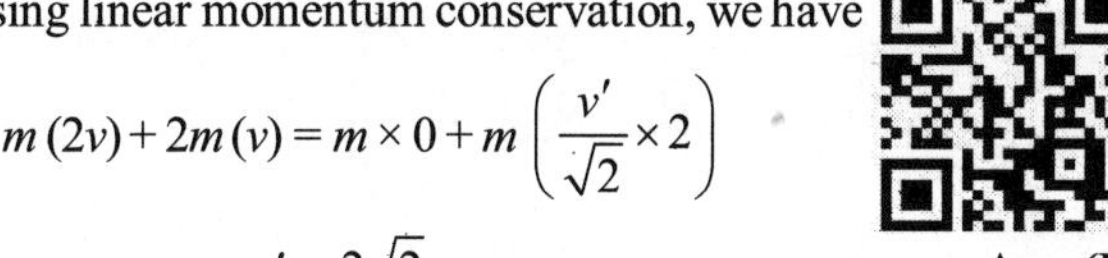

$$m(2v) + 2m(v) = m \times 0 + m\left(\frac{v'}{\sqrt{2}} \times 2\right)$$

$$\Rightarrow \quad v' = 2\sqrt{2}\, v \qquad \textbf{Ans. (B)}$$

Sol. 17

Initial State: m, v, $M = 4m$, rest. Final State: v_c, h.

Using linear momentum conservation, we have

$$mv = (m + M)v_c$$

$$\Rightarrow \quad v_c = \frac{v}{5}$$

Using work energy theorem, we have

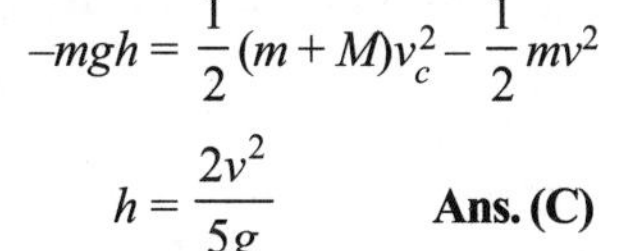

$$-mgh = \frac{1}{2}(m + M)v_c^2 - \frac{1}{2}mv^2$$

$$\Rightarrow \quad h = \frac{2v^2}{5g} \qquad \textbf{Ans. (C)}$$

Sol. 18 By conservation of momentum, we use

in X-direction $M(10)\cos 30° + 2M(5)\cos 45° = 2Mv_1 \cos 30° + Mv_2 \cos 45°$

$$5\sqrt{3} + 5\sqrt{2} = \sqrt{3}\, v_1 = \frac{v_2}{\sqrt{2}} \qquad \ldots(1)$$

in Y-direction $10M \sin 30° - 2M(5)\sin 45° = Mv_2 \sin 45° - 2Mv_1 \sin 30°$

$$5 - 5\sqrt{2} = \frac{v_2}{\sqrt{2}} - v_1 \qquad \ldots(2)$$

(1) (2) $5 \times 1.732 + 10 \times 1.414 - 5 = (\sqrt{3} + 1)v_1$

$$\Rightarrow \quad v_1 = \frac{8.66 + 14.14 - 5}{2.732}$$

$$\Rightarrow \quad v_1 = 6.51\text{ m/s}$$

From (2) we use

$$v_2 = (5 - 7.07 + 6.51) \times \sqrt{2}$$

$$\Rightarrow \quad v_2 = 4.44 \times 1.414$$

$$\Rightarrow \quad v_2 = 6.28\text{ m/s} \qquad \textbf{Ans. (C)}$$

Sol. 19

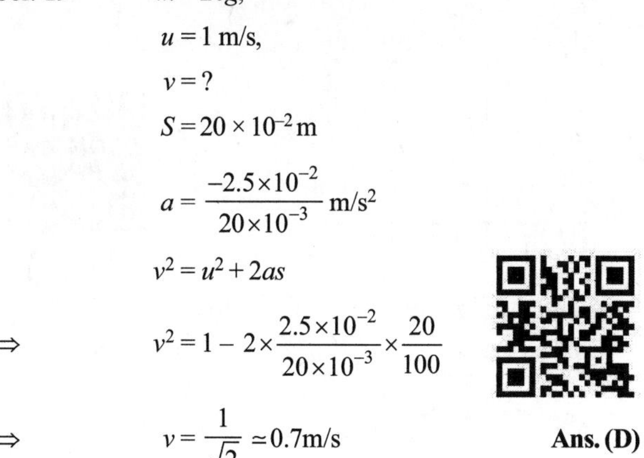

$m = 20g,$

$u = 1\text{ m/s},$

$v = ?$

$S = 20 \times 10^{-2}\text{ m}$

$$a = \frac{-2.5 \times 10^{-2}}{20 \times 10^{-3}}\text{ m/s}^2$$

$$v^2 = u^2 + 2as$$

$$\Rightarrow \quad v^2 = 1 - 2 \times \frac{2.5 \times 10^{-2}}{20 \times 10^{-3}} \times \frac{20}{100}$$

$$\Rightarrow \quad v = \frac{1}{\sqrt{2}} \simeq 0.7\text{m/s} \qquad \textbf{Ans. (D)}$$

Sol. 20 By conservation of angular momentum we use

$$Mv_0L = Mv_1(L - l)$$

$$\Rightarrow \quad v_1 = v_0\left(\frac{L}{L - l}\right) \qquad \ldots(1)$$

Gain in KE of system in

$$\Delta k = \frac{1}{2}M(v_1^2 - v_0^2)$$

$$= \frac{1}{2}Mv_0^2\left[\left(\frac{L}{L - l}\right)^2 - 1\right]$$

$$= \frac{1}{2}Mv_0^2\left[\left(1 - \frac{l}{L}\right)^{-2} - 1\right]$$

$$= Mv_0^2 \cdot \frac{l}{L}$$

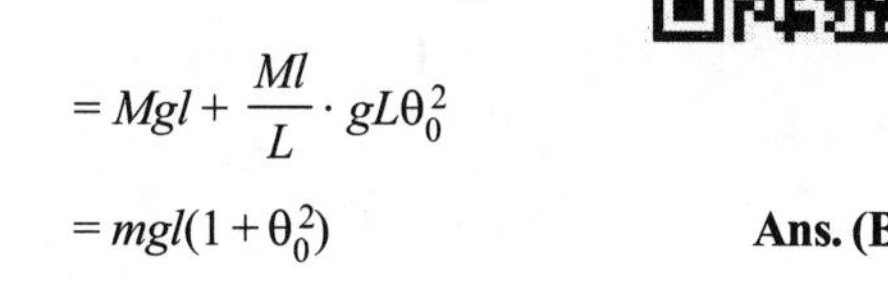

Total W/D by person is $W = Mgl + \dfrac{Mv_0^2 l}{L}$

$$= Mgl + \frac{Ml}{L} \cdot gL\theta_0^2$$

$$= mgl(1 + \theta_0^2) \qquad \textbf{Ans. (B)}$$

Sol. 21

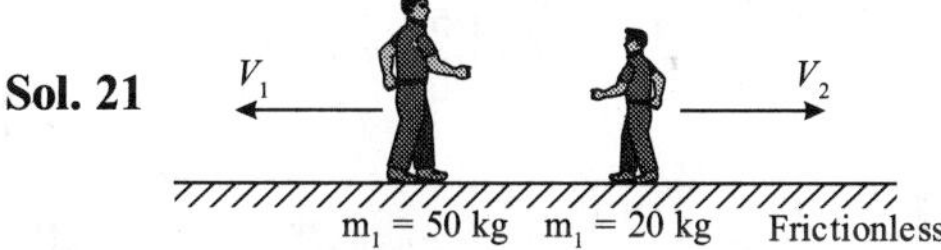

By conservation of momentum, we use

$$m_1v_1 = m_2v_2$$

$$50v_1 = 20v_2$$

and $\quad v_1 + v_2 = 0.7$

$$\Rightarrow \quad v_1 + 2.5v_1 = 0.7$$

$$\Rightarrow \quad v_1 = \frac{0.7}{3.5} = 0.2\text{ m/a} \qquad \textbf{Ans. (A)}$$

Sol. 22

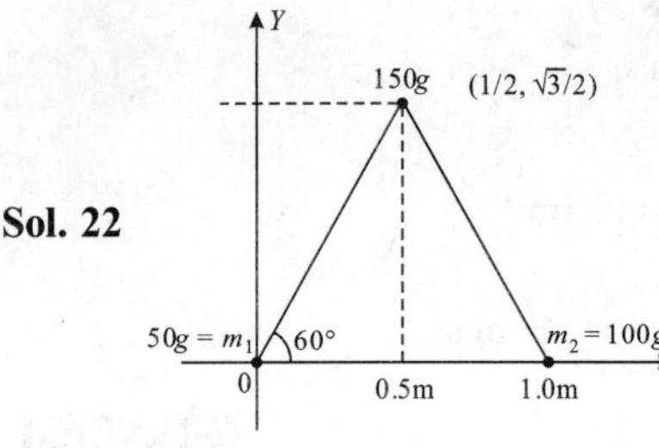

Co-ordinates of COM are

$$x_c = \frac{m_1x_1 + m_2x_2 + m_3x_3}{m_1 + m_2 + m_3} = \frac{50\times 0 + 100\times 1 + 150\times 0.5}{300}$$

$$= \frac{175}{300} = \frac{7}{12}\text{ m}$$

$$y_c = \frac{m_1y_1 + m_2y_2 + m_3y_3}{m_1 + m_2 + m_3} = \frac{50\times 0\times 100\times 0 + 150\times \sqrt{3}/2}{300}$$

$$= \frac{\sqrt{3}}{4}\text{ m}$$ **Ans. (C)**

Sol. 23 Using coordinate axis in given situation, considering 1 kg mass at origin we use

$$x_{cm} = \frac{1(0) + 1.5(3) + 2.5(0)}{5} = 0.9\text{ cm}$$

$$y_{cm} = \frac{1(0) + 1.5(0) + 2.5(4)}{5} = 2\text{ cm}$$ **Ans. (B)**

Sol. 24 In figure we consider masses m_1 and m_2 of two sections shown of 3 kg and 1 kg at centre coordinates (0.5, 1.5) and (1.5, 2.5)

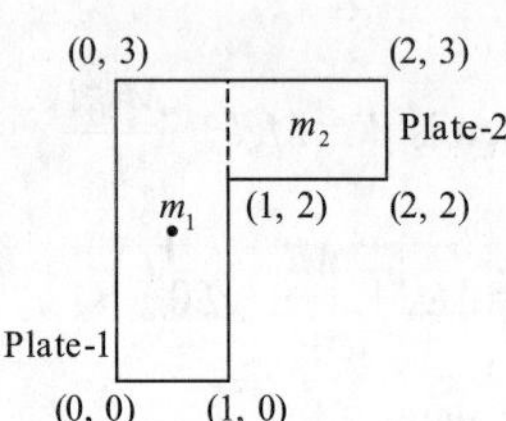

Thus coordinates of centre of mass are given as

$$x_{cm} = \frac{m_1x_1 + m_2x_2}{m_1 + m_2} = \frac{3\times 0.5 + 1\times 1.5}{4} = 0.75$$

$$y_{cm} = \frac{m_1y_1 + m_2y_2}{m_1 + m_2} = \frac{3\times 1.5 + 1\times 2.5}{4} = 1.75$$ **Ans. (D)**

Sol. 25 Using conservation of momentum, we have

$$(0.1)(3\hat{i}) + (0.1)(5\hat{j}) = (0.1)(4)(\hat{i} + \hat{j}) + (0.1)\vec{v}$$

$$\Rightarrow \quad \vec{v} = -\hat{i} + \hat{j}$$

$$\Rightarrow \quad |\vec{v}| = \sqrt{2}$$

Kinetic energy of B is

$$K = \frac{1}{2}mv^2 = \frac{1}{2}(0.1)(2) = \frac{1}{10}$$

$$\Rightarrow \quad x = 1$$ **Ans. (1)**

Sol. 26 Let $m_1 \rightarrow$ mass of body

$m_2 \rightarrow$ mass of cavity

Using mass moment concept of centre of mass, we have

$$m_1R_1 = m_2R_2$$

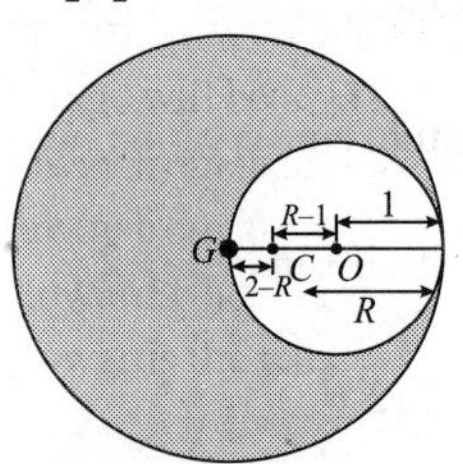

$$\Rightarrow \left(\frac{4}{3}\pi R^3\rho - \frac{4}{3}\pi(1)^3\rho\right)(2 - R) = \frac{4}{3}\pi(1)^3\rho \times (R - 1)$$

$$\Rightarrow (R^3 - 1)(2 - R) = R - 1$$

$$\Rightarrow (R^2 + R + 1)(2 - R) = 1$$ **Ans. (C)**

Sol. 27 Particles will collide in air after time given as

$$t_0 = \frac{h}{\sqrt{2gh}} = \sqrt{\frac{h}{2g}}$$

At the instant of collision velocities of the two particle are

$$v_A = gt_0 = \sqrt{\frac{gh}{2}}$$

$$v_B = u_B - gt_0 = \sqrt{\frac{gh}{2}}$$

$$\Rightarrow \quad v_A = v_B = \sqrt{\frac{gh}{2}}$$

Height at which collision takes place is

$$h_1 = h - \frac{1}{2}gt_0^2 = \frac{3h}{4}$$

As after collision, combined mass comes to rest, time taken to reach ground is given as

$$t_1 = \sqrt{\frac{2\times(3h/4)}{g}} = \sqrt{\frac{3h}{2g}}$$ **Ans. (D)**

Sol. 28 Using momentum conservation, we have

$$mu\,\hat{i} + mu\left(\frac{\hat{i} + \hat{j}}{2}\right) = (m + m)\vec{v}$$

$$\Rightarrow \quad \vec{v} = \frac{3}{4}u\hat{i} + \frac{u}{4}\hat{j}$$

$$\Rightarrow \quad v = \frac{u}{4}\sqrt{10}$$

After collision final kinetic energy is given as

$$K_i = \frac{1}{2}2\text{m}\left(\frac{u}{4}\sqrt{10}\right)^2 = \frac{5}{8}\text{mu}^2$$

Initial kinetic energy before collision is

$$K_i = \frac{1}{2}\text{mu}^2 + \frac{1}{2}\text{m}\left(\frac{u}{\sqrt{2}}\right)^2 = \frac{6}{8}\text{mu}^2$$

Loss in kinetic energy is given as

$$\Delta K = K_i - K_f = \frac{1}{8}\text{mu}^2$$ **Ans. (B)**

Sol. 29 Distance of centre of mass of rod from one end is given as

$$x_{cm} = \frac{\int xdm}{\int dm} = \frac{\int(\rho dx)x}{\int dm}$$

$$\Rightarrow \quad x_{cm} = \frac{\int_0^L\left(a+\frac{bx^2}{L^2}\right)xdx}{\int_0^L\left(a+\frac{bx^2}{L^2}\right)dx}$$

$$\Rightarrow \quad x_{cm} = \frac{\frac{aL^2}{2}+\frac{b}{L^2}\cdot\frac{L^4}{4}}{aL+\frac{b}{L^2}\cdot\frac{L^3}{3}}$$

$$\Rightarrow \quad x_{cm} = \frac{\left(\frac{4a+2b}{8}\right)L}{\frac{(3a+b)}{3}} = \frac{3}{4}\frac{(2a+b)L}{(3a+b)}$$ **Ans. (D)**

Sol. 30 Velocity of particle at highest point is given as

$$V_x = u\cos 60° = \frac{u}{2}$$

Using momentum conservation at the highest point, final velocity of combined mass is given as

$$\frac{mu}{2} + mu = 2mv'$$

$$\Rightarrow \quad v' = \frac{3v}{4}$$

Range of combined particle after collision is given as

$$R = \frac{3v}{4}\sqrt{\frac{2H}{g}}$$

$$\Rightarrow \quad R = \frac{3v}{4}\sqrt{\frac{2}{g}\left(\frac{u^2\sin^2 60°}{2g}\right)}$$

$$\Rightarrow \quad R = \frac{3}{4}\cdot\frac{\sqrt{3}}{2}\cdot\frac{u^2}{g}$$

$$\Rightarrow \quad R = \frac{3\sqrt{3}u^2}{8g}$$ **Ans. (C)**

Sol. 31 By momentum conservation principle, we have

$$m(u\hat{i}) + 3m(0) = mv\hat{j} + 3m\vec{v}_1$$

$$\Rightarrow \quad mu\hat{i} - mv\hat{j} = 3m\vec{v}_1$$

$$\Rightarrow \quad \vec{v}_1 = \frac{u\hat{i} - v\hat{j}}{3}$$

$$\Rightarrow \quad |\vec{v}_1| = \frac{\sqrt{u^2+v^2}}{3}$$

$$\Rightarrow \quad v_1^2 = \frac{u^2+v^2}{9} \quad \ldots(1)$$

As collision is perfectly elastic, we have

$$KE_i = KE_j$$

$$\Rightarrow \quad \frac{1}{2}mu^2 + \frac{1}{2}3m(0)^2 = \frac{1}{2}mv^2 + \frac{1}{2}3mv_1^2$$

$$\Rightarrow \quad u^2 = v^2 + 3v_1^2$$

$$\Rightarrow \quad u^2 = v^2 + 3\frac{(u^2+v^2)}{9}$$

$$\Rightarrow \quad 3u^2 = 3v^2 + u^2 + v^2$$

$$\Rightarrow \quad 2u^2 = 4v^2$$

$$\Rightarrow \quad v = \frac{u}{\sqrt{2}}$$ **Ans. (D)**

Sol. 32

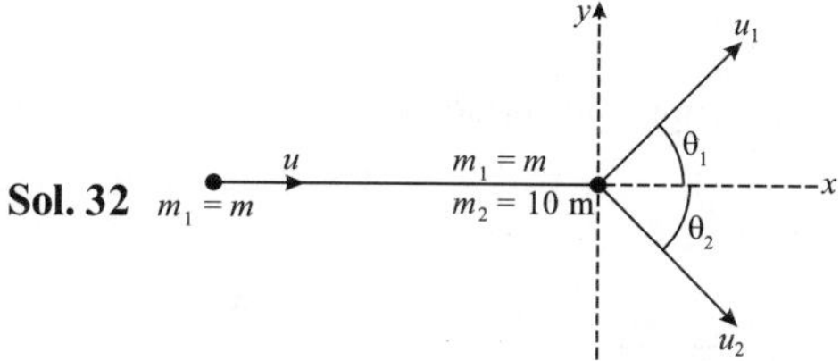

By momentum conservation along y direction, we have

$$m_1u_1\sin\theta_1 = m_2u_2\sin\theta_2$$

$$\Rightarrow \quad mu_1\sin\theta_1 = 10mu_2\sin\theta_2 \quad \ldots(1)$$

Given that

$$kE_{m_{1(\text{final})}} = \frac{1}{2}kE_{m_{1(\text{initial})}}$$

$$\Rightarrow \quad \frac{1}{2}mu_1^2 = \frac{1}{2}\times\frac{1}{2}mu^2$$

$$\Rightarrow \quad u_1 = \frac{u}{\sqrt{2}} \quad \ldots(2)$$

As collision is elastic, we use

$$\frac{1}{2}mu^2 = \frac{1}{2}mu_1^2 + \frac{1}{2}(10\,m)u_2^2$$

$$\Rightarrow \quad \frac{1}{2}mu^2 = \frac{1}{2}\times\frac{1}{2}mu^2 + \frac{1}{2}\times 10m\cdot u_2^2$$

$$\Rightarrow \quad \frac{1}{4}mu^2 = \frac{1}{2}\times 10\times mu_2^2$$

$$\Rightarrow \quad u_2 = \frac{u}{\sqrt{20}} \qquad \ldots(3)$$

From (1), (2) and (3), we have

$$\frac{u}{\sqrt{2}}\sin\theta_1 = 10\cdot\frac{u}{\sqrt{20}}\sin\theta_2$$

$$\Rightarrow \quad \sin\theta_1 = \sqrt{10}\sin\theta_2$$

$$\Rightarrow \quad n = 10$$

Ans. (10)

Sol. 33

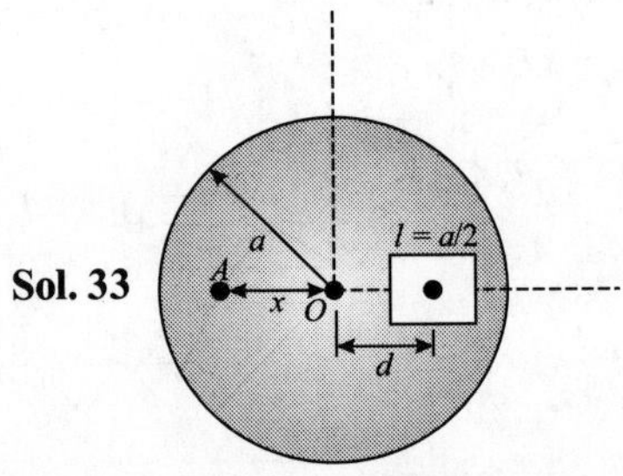

If centre of mass of remaining portion is at point A by mass moment concept, we have

$$m_1 x = m_2\left(\frac{a}{2}\right)$$

Hence m_1 is the mass of complete disc and m_2 is the removed mass from square hole.

As masses are proportional to surface are, we use

$$\left(\pi a^2 - \frac{a^2}{4}\right)x = \frac{a^2}{4}\left(\frac{a}{2}\right)$$

$$\Rightarrow \quad x = \frac{a}{2(4\pi - 1)}$$

$$\Rightarrow \quad X = 2(4\pi - 1)$$

$$\Rightarrow \quad X = 8\pi - 2 = 23.12$$

Thus nearest integral value of X is 23. **Ans. (23)**

Sol. 34

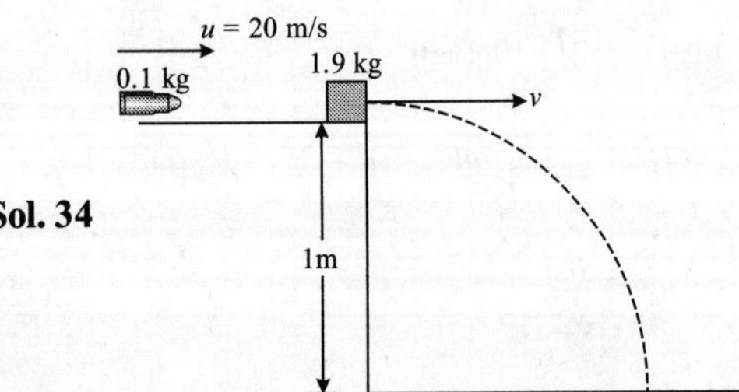

By linear momentum conservation we use

$$0.1\times 20 = 2v$$

$$\Rightarrow \quad v = 1\text{ m/s}$$

Final kinetic energy of combined system is given as

$$K_f = mgh + \frac{1}{2}mv^2$$

$$\Rightarrow \quad K_f = 2\times 10\times 1 + \frac{1}{2}\times 2(1)^2$$

$$\Rightarrow \quad K_f = 20 + 1 = 21\text{ J}$$

Ans. (A)

Sol. 35 At a height h speed is given as

$$v = \sqrt{2g(H-h)}$$

The graph is parabolic and most appropriate option is (C) which can be analyzed by slope variation. **Ans. (C)**

Sol. 36 By conservation of momentum when last block starts moving

$$mv = 16\,mv'$$

$$\Rightarrow \quad v' = \frac{v}{16}$$

Initial energy $E_1 = \frac{1}{2}mv^2$

Final energy $E_2 = \frac{1}{2}(16m)\left(\frac{v}{16}\right)^2 = \frac{1}{32}mv^2$

Energy loss $\Delta E = E_1 - E_2 = \frac{15}{32}mv^2$

Percentage energy loss is given as

$$\%\text{ loss} = \frac{\frac{15}{32}mv^2}{\frac{1}{2}mv^2}\times 100 = \frac{15}{16}\%$$

Ans. (D)

Sol. 37 Figure below shows the oblique collision in the given situation

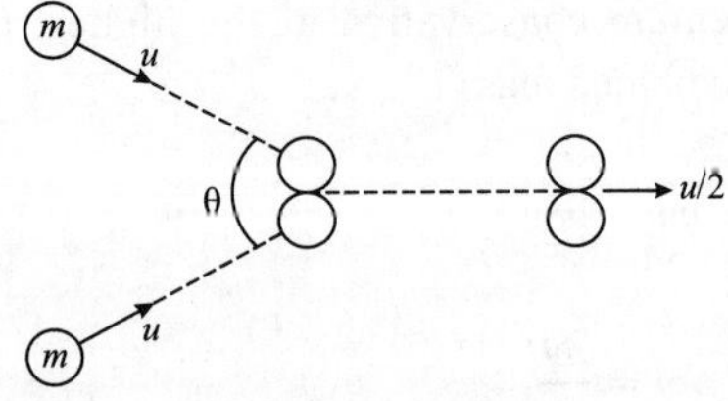

By conservation of momentum, we use

$$2mu\cos\frac{\theta}{2} = 2m\left(\frac{u}{2}\right)$$

$$\Rightarrow \quad \cos\frac{\theta}{2} = \frac{1}{2}$$

$$\Rightarrow \quad \frac{\theta}{2} = 60^\circ$$

$$\Rightarrow \quad \theta = 120^\circ$$

Ans. (120)

Sol. 38 Using conservation of linear momentum before and after collision, we have

$$m_1\vec{u} = m_1\vec{v}_1 + m_2\vec{v}_2$$

$$\Rightarrow \quad 2m_2(\sqrt{3}\hat{i}+\hat{j})+0 = 2m_2(\hat{i}+\sqrt{3}\hat{j})+m_2\vec{v}_2$$

$$\Rightarrow \quad \vec{v}_2 = 2(\sqrt{3}\hat{i}+\hat{j})-2(\hat{i}+\sqrt{3}\hat{j})$$

$$\Rightarrow \quad \vec{v}_2 = 2(\sqrt{3}\hat{i}-\hat{j})+2(\hat{i}-\sqrt{3}\hat{j})$$

$$\Rightarrow \quad \vec{v}_2 = 2(\sqrt{3}\hat{i}-\hat{j})(\hat{i}-\hat{j})$$

Angle between $\vec{v}_1$ and $\vec{v}_2$ is given as

$$\cos\theta = \frac{\vec{v}_1\cdot\vec{v}_2}{|\vec{v}||\vec{v}_2|} = \frac{2(\sqrt{3}-1)(1-\sqrt{3})}{2\times 2\sqrt{2}(\sqrt{3}-1)}$$

$$\Rightarrow \quad \cos\theta = \frac{1-\sqrt{3}}{2\sqrt{2}}$$

$$\Rightarrow \quad \theta = 105^\circ$$

Ans. (D)

Sol. 39 Centre of mass of hemisphere from its base centre is given as

$$x = \frac{3R}{8} = 3\text{ cm}$$

$$x = 3$$

Ans. (3)

Sol. 40 The situation described in the question is shown in figure below

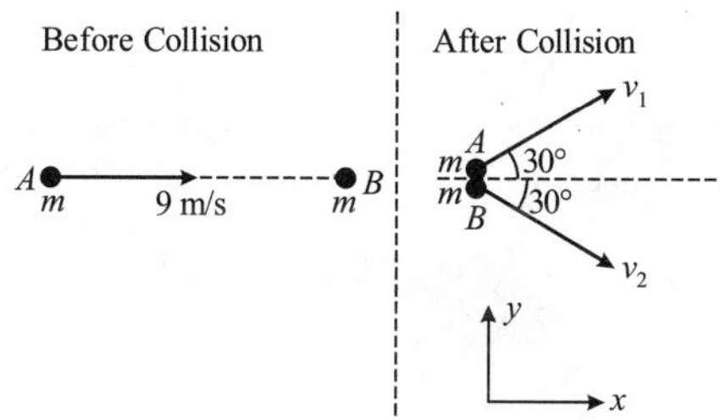

From conservation of momentum along the direction perpendicular to initial line of motion, we have

$$mv_2\sin 30^\circ = mv_1\sin 30^\circ$$

$$\Rightarrow \quad v_2 = v_1$$

$$\Rightarrow \quad \frac{v_1}{v_2} = 1$$

Ans. (1)

Sol. 41 If σ be the uniform mass density of disc then center of mass of the given object can be calculated as

$$x_{COM} = \frac{(\sigma\pi a^2)a - \sigma\pi\left(\frac{a^2}{4}\right)\times\frac{3a}{2}}{\sigma\pi a^2 - \frac{\sigma\pi a^2}{4}}$$

$$\Rightarrow \quad x_{COM} = \frac{a-\frac{3a}{8}}{1-\frac{1}{4}} = \frac{5a}{6}$$

Ans. (A)

Sol. 42 The kinetic energy of a particle is related to its momentum as

$$K = \frac{P^2}{2m}$$

As momentum are equal for both solids, so we have

$$K \propto \frac{1}{m}$$

$$\Rightarrow \quad \frac{K_A}{K_B} = \frac{m_B}{m_A} = \frac{2}{1}$$

Ans. (2)

Sol. 43 The kinetic energy of a particle is related to its momentum as

$$K = \frac{P^2}{2m}$$

As kinetic energies of both are same, we use

$$\frac{p_1^2}{2\times 4} = \frac{p_2^2}{2\times 16}$$

$$\Rightarrow \quad \frac{p_1}{p_2} = \frac{1}{2}$$

Ans. (1)

Sol. 44 For elastic collision, we use coefficient of restitution value $e = 1$ & second body was initially at rest, so final velocity can be given as

$$v_2 = \frac{2m_1u_1}{m_1+m_2} = \frac{2u(M)}{M+m}$$

As $M >> m$, we have

$$v_2 = 2u$$

Ans. (B)

Sol. 45 The acceleration of center of mass of the given system can be written as

$$a_{cm} = \frac{m_1a_1+m_2a_2}{m_1+m_2} = \frac{F}{2M}$$

$$\Rightarrow \quad \frac{F}{2M} = \frac{Ma+Ma_B}{2M}$$

$$\Rightarrow \quad a_B = \frac{F-Ma}{M}$$

Ans. (C)

Sol. 46 From conservation of momentum along X-axis, we have

$$\vec{P}_i = \vec{P}_f$$

$$\Rightarrow \quad 10\times 10\sqrt{3} = 200\cos\theta$$

$$\Rightarrow \quad \cos\theta = \frac{\sqrt{3}}{2}$$

$$\Rightarrow \quad \theta = 30^\circ$$

Ans. (30)

Sol. 47 If v_1 is the speed of combined system after collision, then by energy conservation, after collision, we use

$$(M+m)gh = \frac{1}{2}(M+m)v_1^2$$

$$\Rightarrow \quad v_1 = \sqrt{2gh}$$

Using momentum conservation, just before and just after collision, we have

$$mv = (M+m)v_1$$

$$\Rightarrow \quad v = \left(\frac{M+m}{m}\right)$$

$$\Rightarrow \quad v_1 = \frac{6}{10\times 10^{-3}} \times \sqrt{2\times 9.8\times 9.8\times 10^{-2}}$$

$$\Rightarrow \quad v_1 \approx 831.55 \text{ m/s}$$ **Ans. (C)**

Sol. 48 Speed of ball after falling through a distance h is given as

$$v_0 = \sqrt{2gh}$$

Bouncing speed is calculated as

$$v = e\sqrt{2gh} = \sqrt{2gh'}$$

$$\Rightarrow \quad e = 0.9$$

Total distance travelled by ball until it comes to rest after infinte bounces is calculated as

$$s = h + 2e^2h + 2e^4 h + \ldots$$

Total time of flight of ball after infinite bounces is calculated as

$$t = \sqrt{\frac{2h}{g}} + 2e\sqrt{\frac{2h}{g}} + 2e^2\sqrt{\frac{2h}{g}} + \ldots$$

Thus average speed of ball is given as

$$v_{av} = \frac{s}{t} = 2.5 \text{ m/s}$$ **Ans. (D)**

Sol. 49 The center of mass of the quarter disc is derived in Physics Galaxy Vol 1, chapter 4 so we use the direct result as

$\left(\frac{4a}{3\pi}, \frac{4a}{3\pi}\right)$. Thus $x = 4$. **Ans. (4)**

Sol. 50 If the velocity of second piece is taken as v then by conservation of linear momentum, we have

$$10(10\sqrt{3})\hat{i} = (10)(10\hat{j}) + 10\vec{v}$$

$$\Rightarrow \quad \vec{v} = 10\sqrt{3}\hat{i} - 10\hat{j}$$

$$\Rightarrow \quad |\vec{v}| = \sqrt{300+100} = \sqrt{400} = 20 \text{ m/s}$$ **Ans. (20)**

Sol. 51 Figure below shows the situation described in the question

By conservation of linear momentum, we have

$$m_1v_1 = -m_1v + m_2v$$

$$\Rightarrow \quad v_1 = -v + \frac{m_2}{m_1}v$$

$$\Rightarrow \quad \frac{(v_1+v)}{v} = \frac{m_2}{m_1} \qquad \ldots(1)$$

For elastic collision, we use coefficient of restitution as

$$e = \frac{2v}{v_1} = 1$$

$$\Rightarrow \quad v = \frac{v_1}{2}$$

From equation-(1), we have

$$\frac{v_1 + v_1/2}{v_1/2} = \frac{m_2}{m_1}$$

$$\Rightarrow \quad 3 = \frac{m_2}{m_1}$$ **Ans. (A)**

Sol. 52 Figure below shows firing of a bullet from gun as described in the question

By conservation of linear momentum, we use

$$0 = m_Gv_G + M_Bv_b$$

$$\Rightarrow \quad 0 = 4v_G + \frac{4}{1000}v_b$$

$$\Rightarrow \quad v_b = -1000\,v_G \qquad \ldots(1)$$

$$v_{bG} = v_b - v_G$$

$$\Rightarrow \quad 50 = v_B - v_G \qquad \ldots(2)$$

$$\Rightarrow \quad 50 = -1001\,v_G$$

$$v_G \approx 0.05 \text{ m/s}$$

Impulse $= m_Gv_G = 4\times 0.05 = 0.2$ kg m/s **Ans. (C)**

Sol. 53 The center of mass of a half ring is located at the bisector of the half ring at a distance y_{cm} from its center, given as

$$y_{cm} = \frac{2R}{\pi}$$

$$\Rightarrow \quad x = 2$$ **Ans. (2)**

Sol. 54 By conservation of linear momentum, we have

$$9m = mv_A + 2mv_B$$

$\Rightarrow$ $$9 = v_A + 2v_B \quad ...(1)$$

$\Rightarrow$ $$9 = v_B - v_A \quad ...(2)$$

Solving equations-(1) and (2) gives

$$v_B = 6 \text{ m/s}$$

For collision between B & C, we have

$$2mv_B = 4mv_C$$

$\Rightarrow$ $$v_C = 3 \text{ m/s}$$ **Ans. (A)**

Sol. 55 For pendulum to complete the vertical circle, final velocity of system (bullet + mass) is given as

$$v_f = \sqrt{5gR}$$

Using conservation of linear momentum before and after collision is given as

$$M\sqrt{5gR} = m(v + 100)$$

$\Rightarrow$ $$v = \frac{M}{m}\sqrt{5gr} - 100$$

$\Rightarrow$ $$v = 500 - 100 = 400 \text{ m/s} = 400 \text{ m/s}$$

Ans. (400)

Sol. 56 Figure below shows the situation described in the question.

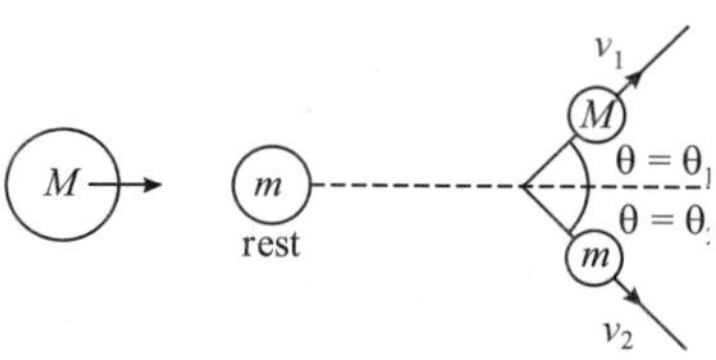

Using linear momentum conservationalong y-direction, we have

$$mv_2 \sin\theta = Mv_1 \sin\theta \quad ...(1)$$

$\Rightarrow$ $$mv_2 = Mv_1$$

Along x-direction, we have

$$Mv_0 = mv_2\cos\theta + Mv_1\cos\theta \quad ...(2)$$

From equations-(1) and (2), we have

$$Mv_0 = mv_2\cos\theta + mv_2\cos\theta$$

$\Delta\Rightarrow$ $$mv_0 = 2mv_2\cos\theta \quad ...(3)$$

As collision is elastic, we use $e = 1$

$$\frac{v_2 - v_1\cos 2\theta}{v_0\cos\theta} = 1$$

$\Rightarrow$ $$v_2 - v_1\cos 2\theta = v_0 \quad ...(4)$$

$\Rightarrow$ $$v_2 - \frac{mv_2}{M}\cos 2\theta = v_0\cos\theta$$

From equation-(3), we have

$$v_2\left[1 - \frac{m}{M}\cos 2\theta\right] = v_0\cos\theta$$

$\Rightarrow$ $$\frac{mv_0}{2m\cos\theta}\left[1 - \frac{m}{M}\cos 2\theta\right] = v_0\cos\theta$$

$\Rightarrow$ $$M - m\cos 2\theta = 2m\cos 2\theta$$

$\Rightarrow$ $$M = 2m\cos^2\theta + m(2\cos^2\theta - 1)$$

$\Rightarrow$ $$\frac{M}{m} = 4\cos^2\theta - 1$$

$\Rightarrow$ $$\left(\frac{M}{m}\right)_{max} = 4 - 1 = 3$$ **Ans. (D)**

6.2 Rigid Body Dynamics

Sol. 1 MI of the given cylinder about AOR shown is

$$I = \frac{1}{12}Ml^2 + \frac{1}{4}MR^2$$

$\Rightarrow$ $$I = \frac{1}{12}Ml^2 + \frac{1}{4}M\left(\frac{M}{\rho\pi l}\right)$$

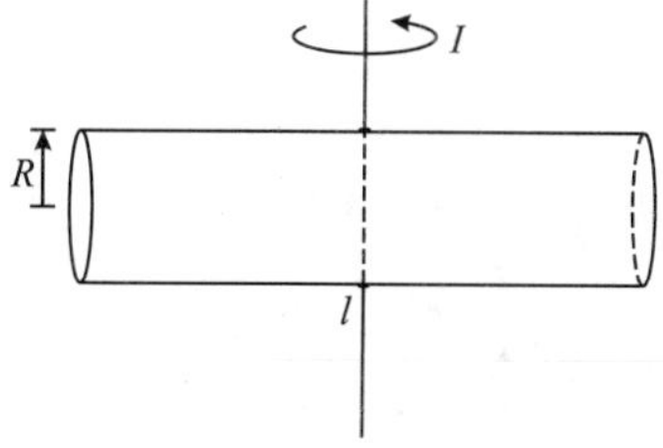

for I to be minimum. We use $\frac{dI}{dl} = 0$

$\Rightarrow$ $$\frac{dI}{dl} = \frac{1}{6}Ml + \frac{M^2}{4\rho\pi}\left(-\frac{1}{l^2}\right) = 0$$

$\Rightarrow$ $$l^3 = \frac{6M}{4\rho\pi} = \frac{3}{2}\frac{\rho\pi R^2 l}{\rho\pi}$$

$\Rightarrow$ $$\frac{l^2}{R^2} = \frac{3}{2} \Rightarrow \frac{l}{R} = \sqrt{\frac{3}{2}}$$ **Ans. (D)**

Sol. 2 Torque of rod's weight about point O is

$$\tau = Mg \times \frac{l}{2}\sin\theta$$

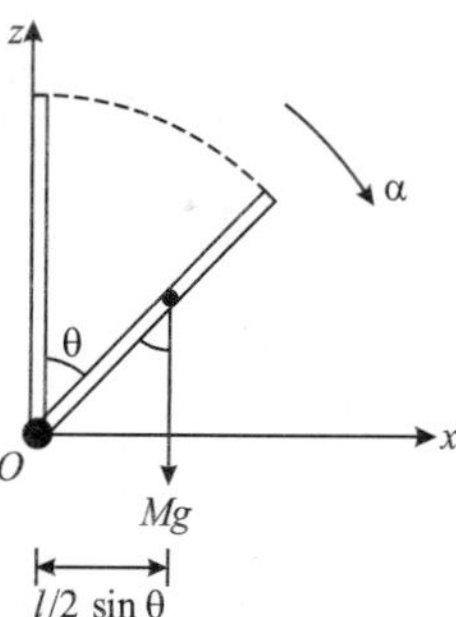

If α be the angular acceleration of rod, we use

$$Mg\frac{l}{2}\sin\theta = \left(\frac{Ml^2}{3}\right)\alpha$$

$\Rightarrow$ $$\alpha = \frac{3g}{2l}\sin\theta$$ **Ans. (D)**

Sol. 3 MI of whole disc is

$$I_0 = \frac{1}{2}MR^2$$

Mass of small disc is

$$m = \frac{M}{\pi R^2}\pi\left(\frac{R}{4}\right)^2 = \frac{M}{16}$$

MI of small disc about O is

$$I_{\text{hole}} = I_c + md^2$$

$$I_{\text{hole}} = \frac{1}{2}m\left(\frac{R}{4}\right)^2 + m\left(\frac{3R}{4}\right)^2$$

$$\Rightarrow \quad I_{\text{hole}} = \frac{1}{2}\left(\frac{M}{16}\right)\left(\frac{R^2}{16}\right) + \left(\frac{M}{16}\right)\left(\frac{9R^2}{16}\right)$$

$$\Rightarrow \quad I_{\text{hole}} = \frac{19}{512}MR^2$$

MI of given remaining portion is

$$I = I_0 - I_{\text{hole}}$$

$$\Rightarrow \quad I = \left(\frac{1}{2} - \frac{19}{512}\right)MR^2 = \frac{237}{512}MR^2$$ **Ans. (B)**

Sol. 4 By mass distribution property for same shape of objects of different dimensions MI expression remain same.

For lamina ABC $\quad I_0 = \gamma ML^2$

$$\Rightarrow \quad I_{DEF} = \gamma ml^2; \text{ where } m_{DEF} = \frac{1}{4}M \text{ and } l = \frac{l}{2}$$

$$\Rightarrow \quad I_{DEF} = \gamma\left(\frac{M}{4}\right)\left(\frac{L}{2}\right)^2 = \frac{\gamma}{16}ML^2 = \frac{I_0}{16}$$

MI of remaining part of lamina after removing DEF is

$$I = I_0 - I_{DEF} = \frac{15I_0}{16}$$ **Ans. (B)**

Sol. 5 In the given situation we have

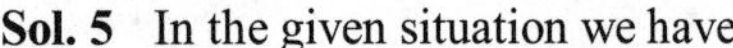

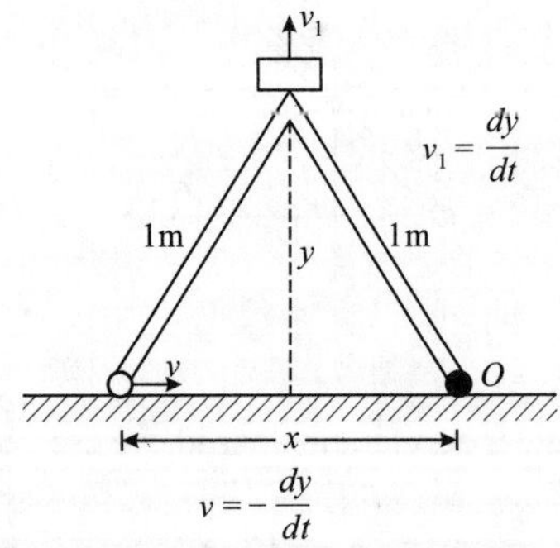

$$\left(\frac{x}{2}\right)^2 + y^2 = 1$$

Differentiate w.r.t. to time

$$\frac{x}{2}\frac{dx}{dt} + 2y\frac{dy}{dt} = 0$$

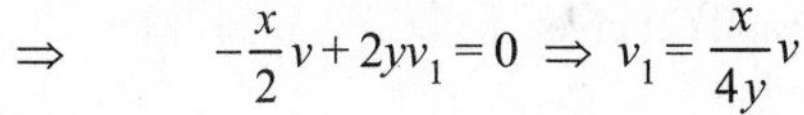

$$\Rightarrow \quad -\frac{x}{2}v + 2yv_1 = 0 \Rightarrow v_1 = \frac{x}{4y}v$$ [v = constant]

As with time $x \downarrow$ and $y \uparrow \Rightarrow v_1$ decreases **Ans. (B)**

Sol. 6 For equilibrium of beam balance we use

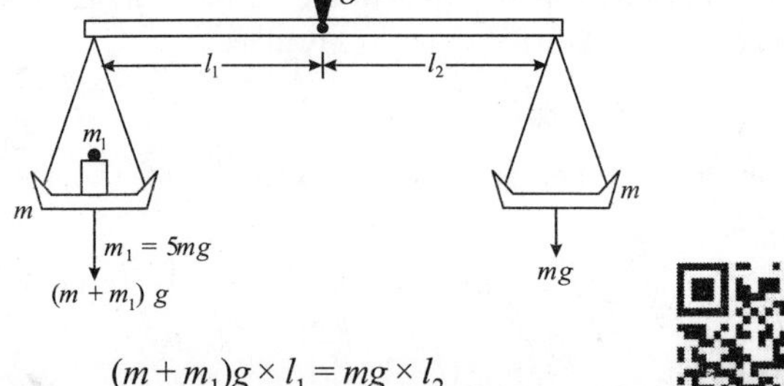

$$(m + m_1)g \times l_1 = mg \times l_2$$

$$\Rightarrow \quad l_1 < l_2$$ **Ans. (C)**

Sol. 7 The central force on particle is given as

$$F = \frac{k}{R^n} = m\omega^2 R \quad \text{(for circular motion)}$$

$$\Rightarrow \quad \omega = \sqrt{\frac{k}{mR^{n+1}}}$$

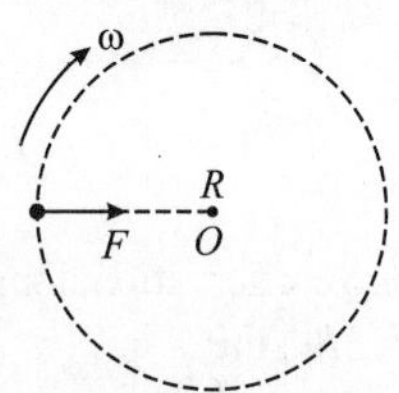

Time period of revolution is

$$T = \frac{2\pi}{\omega} = 2\pi\sqrt{\frac{mR^{n+1}}{k}}$$

$$\Rightarrow \quad T \propto R^{\frac{n+1}{2}}$$ **Ans. (C)**

Sol. 8 MI of this system about centre O of central disc is

$$I_0 = \frac{1}{2}MR^2 + 6\left(\frac{1}{2}MR^2 + M(2R)^2\right)$$

$$I_0 = \frac{7}{2}MR^2 + 24MR^2 = \frac{55}{2}MR^2$$

Now using parallel axes theorem for the system, its MI about P is given

$$I_P = I_0 + 7M(3R)^2 = \left(\frac{55}{2} + 63\right)MR^2$$

$$\Rightarrow \quad I_P = \frac{181}{2}MR^2$$ **Ans. (D)**

Sol. 9 Mass per unit area of disc $= \dfrac{9M}{\pi R^2}$

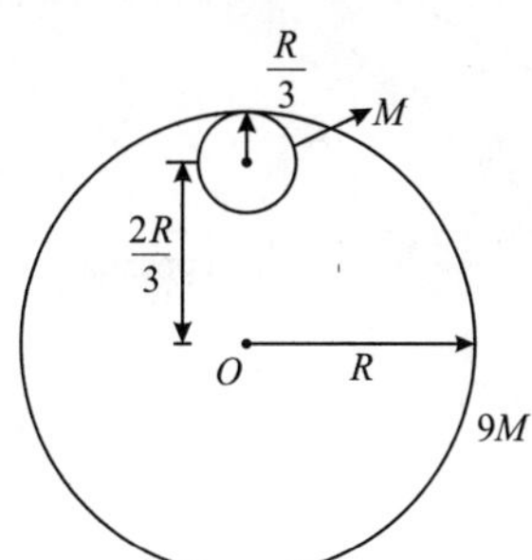

Mass of small disc is

$$m = \frac{9M}{\pi R^2}\times\pi\left(\frac{R}{3}\right)^2 = M$$

MI of remaining disc is

$$I_0 = (I_{9M} - I_M)_{\text{about } O}$$

$$I_0 = \frac{1}{2}(9M)R^2 - \left[\frac{1}{2}M\left(\frac{R}{3}\right)^2 + M\left(\frac{2R}{3}\right)^2\right]$$

$$\Rightarrow \qquad I_0 = \frac{9}{2}MR^2 - \frac{1}{2}MR^2 = 4MR^2$$ **Ans. (A)**

Sol. 10 Moment of force (Torque) about point A will be maximum when line AB is $\perp$ direction of $\vec{F}$.

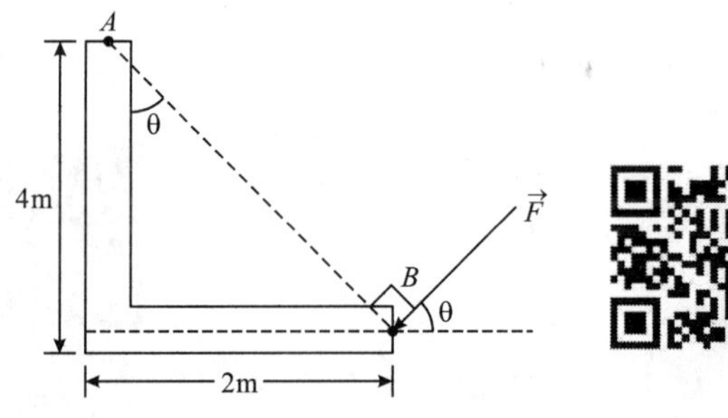

For this we use $\tan\theta = \dfrac{2}{4} = \dfrac{1}{2}$ **Ans. (C)**

Sol. 11 To keep the rod horizontal net force and torque on it must be zero.

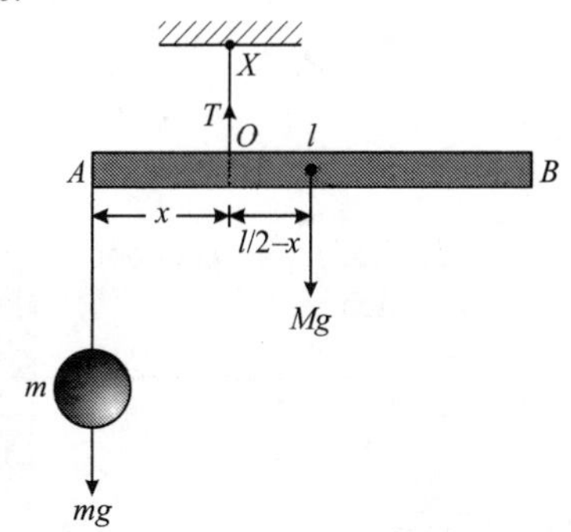

$$mg\,x = Mg\left(\frac{l}{2} - x\right)$$

Where $mgx \rightarrow$ Anticlockwise torque about O

and $Mg\left(\dfrac{l}{2} - x\right) \rightarrow$ Clockwise torque about O

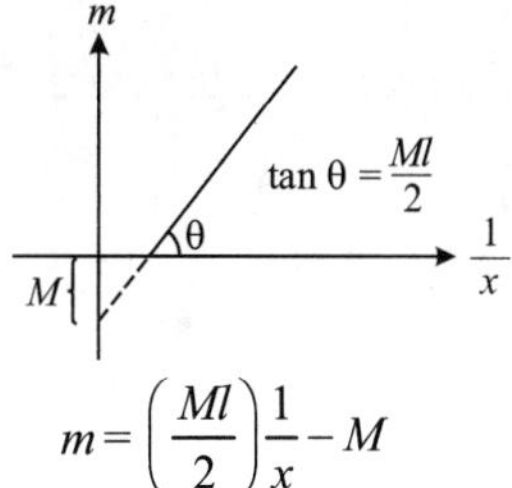

$$m = \left(\frac{Ml}{2}\right)\frac{1}{x} - M$$

where $\dfrac{Ml}{2} \rightarrow$ Slope

and $M \rightarrow$ Intercept on Y axis **Ans. (A)**

Sol. 12 After masses stick to rod, final COM of rod remains at mid point of rod as mass moments of m and $2m$ about O are equal & on opposite sides.

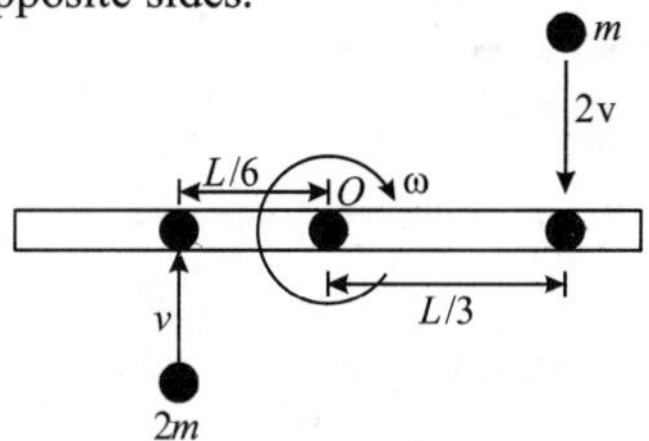

As no external torque is acting on system.

$\Rightarrow$ Total angular momentum will remain conserved about O.

$$L_i = L_f$$

$$2m(v)\left(\frac{L}{6}\right) + m(2v)\left(\frac{L}{3}\right) = I\omega$$

$$\Rightarrow \qquad \frac{mvL}{3} + \frac{2mvL}{3} = \left[\frac{1}{12}(8m)L^2 + m\left(\frac{L}{3}\right)^2 + 2m\left(\frac{L}{6}\right)^2\right]\omega$$

$$mvL = mL^2\omega\left[\frac{2}{3} + \frac{1}{9} + \frac{1}{18}\right] = \frac{15}{18}mL^2\omega$$

$$\Rightarrow \qquad mvL = \frac{5}{6}mL^2\omega$$

$$\Rightarrow \qquad \omega = \frac{6v}{5L}$$ **Ans. (B)**

Sol. 13 Ratio of I_z and $I_{z'}$ is

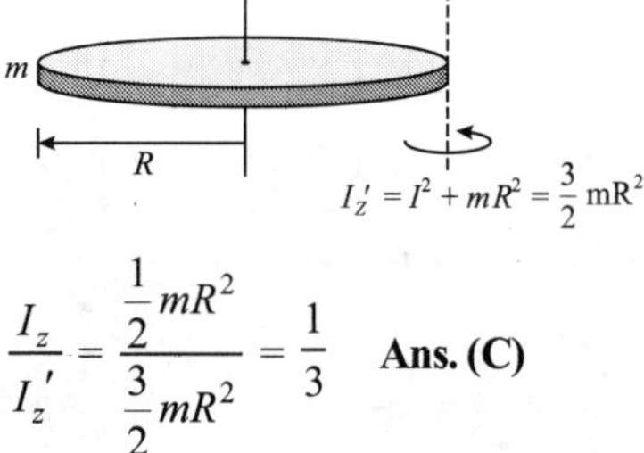

$$\frac{I_z}{I_z'} = \frac{\frac{1}{2}mR^2}{\frac{3}{2}mR^2} = \frac{1}{3}$$ **Ans. (C)**

Sol. 14 The COM of rod has fallen by

$$h = \frac{l}{2}\sin\alpha$$

By work-energy theorem, we have

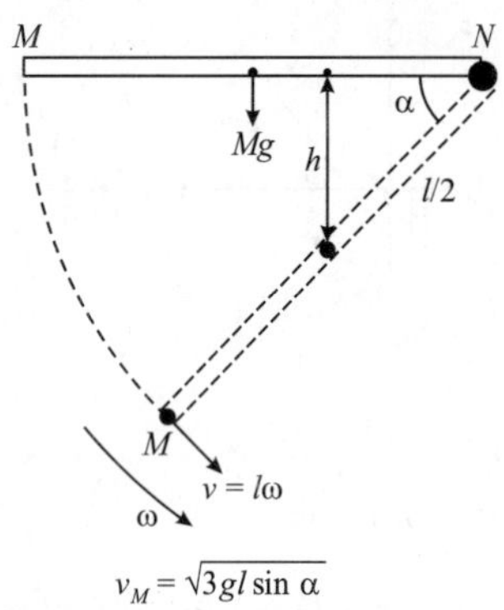

$$mg\left(\frac{l}{2}\sin\alpha\right) = \frac{1}{2}I\omega^2$$

$$\Rightarrow \quad mg\frac{l}{2}\sin\alpha = \frac{1}{2}\left(\frac{ml^2}{3}\right)\omega^2$$

$$\Rightarrow \quad \omega = \sqrt{\frac{3g\sin\alpha}{l}}$$

Ans. (D)

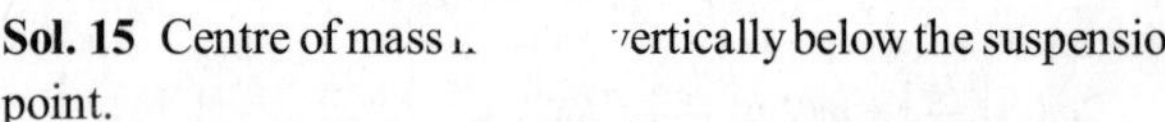

Sol. 15 Centre of mass ... vertically below the suspension point.

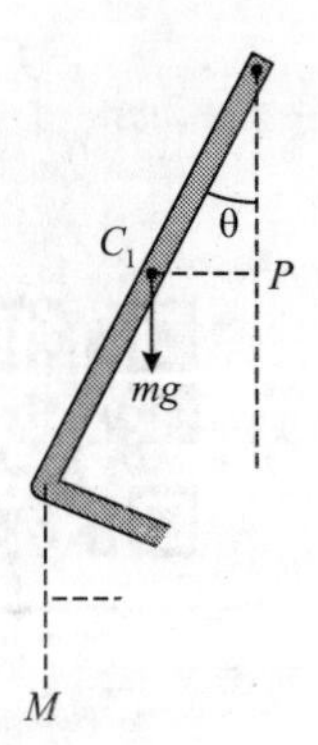

Balancing ... point, we have

... $\cos\theta - L\sin\theta\Big)$

$$\Rightarrow \quad \frac{3}{2}mgL\sin\theta \ \ldots \ \cos\theta$$

$$\Rightarrow \quad \tan\theta = \frac{1}{3}$$

Ans. (B)

Sol. 16

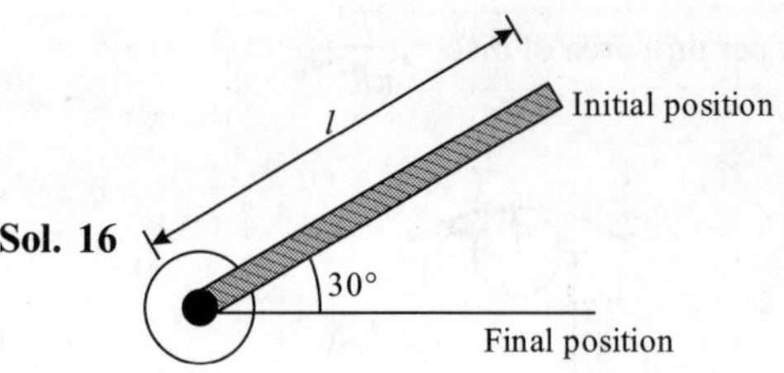

$$mg\left(\frac{l}{2}\right)\sin 30^\circ = \frac{1}{2}I\omega^2$$

$$\Rightarrow \quad \frac{1}{2}\left(\frac{ml^2}{3}\right)\omega^2 = \frac{mgl}{4}$$

$$\Rightarrow \quad \omega = \sqrt{\frac{3g}{2l}} = \sqrt{\frac{3\times 10}{2\times 0.5}}$$

$$\Rightarrow \quad \omega = \sqrt{30}\ \text{rad/s}$$

Ans. (A)

Sol. 17

Consider a strip of radius x & thickness dx,

Torque due to friction on this strip

$$\Rightarrow \quad \tau = \int d\tau = \int_0^R \frac{\mu F \cdot 2\pi x dx}{\pi R^2} \times x$$

$$\Rightarrow \quad \tau = \frac{2\mu F}{R^2}\cdot\frac{R^3}{3}$$

$$\Rightarrow \quad \tau = \frac{2\mu FR}{3}$$

Ans. (A)

Sol. 18

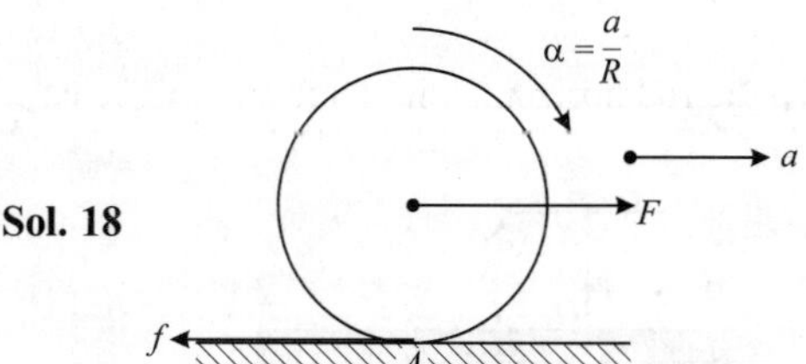

Equation of motion of cylinder about bottom point of contact A is written as

$$FR = \frac{3}{2}MR^2\alpha$$

$$\Rightarrow \quad \alpha = \frac{2F}{3MR}$$

Ans. (C)

Sol. 19

Applying torque equation about point P, we have

$$2M_0g(2l) - 5M_0 gl = I\alpha$$

Where $I = 2M_0(2l)^2 + 5M_0$

$$l^2 = 13M_0l^2d$$

$$\Rightarrow \quad \alpha = -\frac{M_0gl}{13M_0l^2}$$

$$\Rightarrow \quad \alpha = -\frac{g}{13l}$$

$$\Rightarrow \quad \alpha = \frac{g}{13l} \text{ anticlockwise}$$

Ans. (C)

Sol. 20 For Balls (considered solid spheres)

Using parallel axis theorem, the moment of inertia of each balls is

$$I_{\text{ball}} = \frac{2}{5}MR^2 + M(2R)^2$$

$$\Rightarrow \quad I_{\text{ball}} = \frac{22}{5}MR^2$$

For two balls it is $\frac{44}{5}MR^2$

Moment of inertia of rod is

$$I_{\text{rod}} = \frac{ML^2}{12} = \frac{M(2R)^2}{12} = \frac{MR^2}{3}$$

$$I_{\text{system}} = I_{\text{Ball}} + I_{\text{rod}}$$

$$\Rightarrow \quad I_{\text{system}} = \frac{44}{5}MR^2 + \frac{MR^2}{3}$$

$$\Rightarrow \quad I_{\text{system}} = \frac{137}{15}MR^2$$

Ans. (C)

Sol. 21 Suppose M is mass and a is side of larger triangle, then $\frac{M}{4}$ and $\frac{a}{2}$ will be mass and side length of smaller triangle. By symmetry expression of moment of inertia of a triangle remain same as

$$I = \gamma ML^2$$

So we use

$$\frac{I_{\text{removed}}}{I_{\text{original}}} = \frac{\frac{M}{4}\cdot\left(\frac{a}{2}\right)^2}{M\,(a)^2}$$

$$I_{\text{removed}} = \frac{I_0}{16}$$

Thus moment of inertia of remaining triangle is

$$I = I_0 - \frac{I_0}{16} = \frac{15I_0}{16}$$

Ans. (D)

Sol. 22 Torque for F_2 force

$$\overrightarrow{F_2} = \frac{F}{2}(-\hat{i}) + \frac{F\sqrt{3}}{2}(-\hat{j})$$

$$\vec{r}_2 = 0\hat{i} + 6\hat{j}$$

$$\vec{\tau}_{F_2} = \vec{r}_2 \times \overrightarrow{F_2} = 3F\hat{k}$$

Torque for F_1 force

$$\overrightarrow{F_2} = F\hat{k}$$

$$\vec{r}_1 = 2\hat{i} + 3\hat{j}$$

$$\vec{\tau}_{F_1} = \vec{r}_1 \times \overrightarrow{F_1} = 3F\hat{i} + 2F(-\hat{j})$$

$$\vec{\tau}_{\text{net}} = \vec{\tau}_{F_1} + \vec{\tau}_{F_2}$$

$$\Rightarrow \quad \vec{\tau}_{\text{net}} = 3F\hat{i} + 2F(-\hat{j}) + 3F(\hat{k})$$

Ans. (D)

Sol. 23

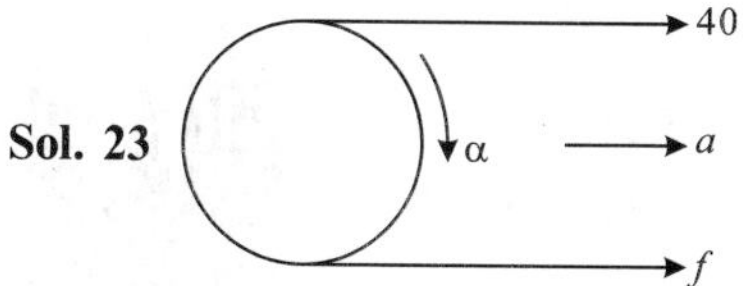

For pure rolling

$$\alpha = \frac{a}{R}$$

Equation of motion of cylinder is

$$40 + f = m(R\alpha) \quad \ldots(1)$$

$$40 \times R - f \times R = mR^2\alpha$$

$$\Rightarrow \quad 40 - f = mR\alpha \quad \ldots(2)$$

From (1) and (2)

$$\Rightarrow \quad \alpha = \frac{40}{mR} = \frac{40}{5\times 0.5}$$

$$= 16 \text{ rad/s}^2$$

Ans. (B)

Sol. 24 Torque on particle is given as

$$\tau = Fr\sin\theta$$

$$\Rightarrow \quad 2.5 = 1 \times 5\sin\theta$$

$$\Rightarrow \quad \sin\theta = 0.5 = \frac{1}{2}$$

$$\Rightarrow \quad \theta = \frac{\pi}{6}$$

Ans. (B)

Sol. 25 Moment of inertia of system is given as

$$I=\frac{MR^2}{2}+2\left(\frac{MR^2}{4}+MR^2\right)$$

$$\Rightarrow \quad I=\frac{MR^2}{2}+\frac{MR^2}{2}+2MR^2$$

$$\Rightarrow \quad I=3MR^2$$

Ans. (A)

Sol. 26 Moment of inertia of hollow cylinder is

$$I=\frac{1}{2}M(R_1^2+R_2^2)=\frac{1}{2}M(10^2+20^2)=250\,M$$

Moment of inertia of a solid cylinder is

$$I=\frac{1}{2}MR^2=250\,M$$

$$\Rightarrow \quad R=\sqrt{5000}=22.36\text{ cm}$$

Ans. (None)

Sol. 27 Using Work Energy Theorem from A to B, we have

$$\frac{1}{2}mv_A^2+mgh=\frac{1}{2}mv_B^2$$

$$\Rightarrow \quad 2gh=v_B^2-v_A^2$$

$$\Rightarrow \quad 2\times 10\times 10=v_B^2-5^2$$

$$\Rightarrow \quad v_B=15\text{ m/s}$$

Angular momentum about O is given as

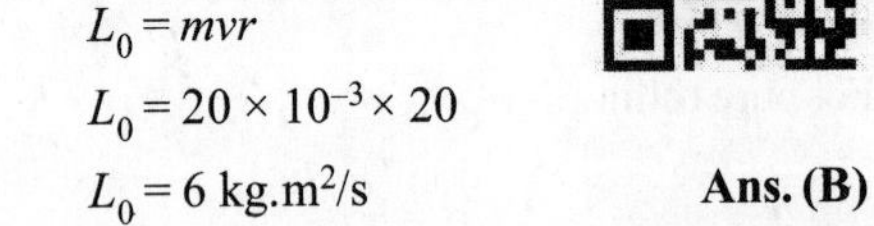

$$L_0=mvr$$

$$\Rightarrow \quad L_0=20\times 10^{-3}\times 20$$

$$\Rightarrow \quad L_0=6\text{ kg.m}^2\text{/s}$$

Ans. (B)

Sol. 28 By parallel axes theorem, we use

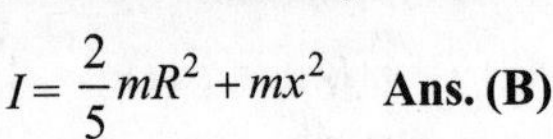

$$I=\frac{2}{5}mR^2+mx^2$$ **Ans. (B)**

Sol. 29 By parallel axes theorem, we have

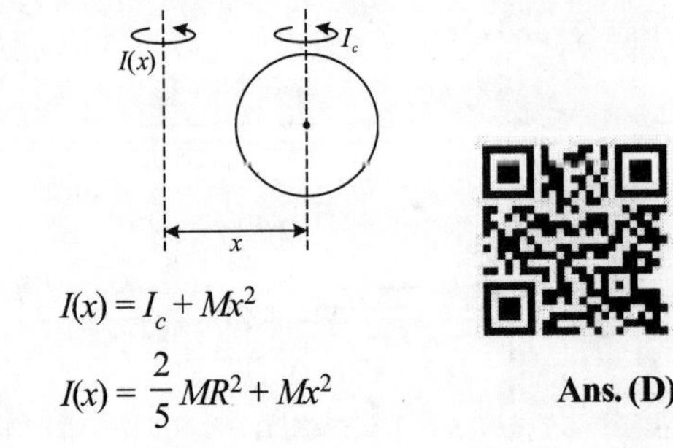

$$I(x)=I_c+Mx^2$$

$$\Rightarrow \quad I(x)=\frac{2}{5}MR^2+Mx^2$$

Ans. (D)

Sol. 30 As the box is in contact with the edge for $\Delta t=0.01$ s. So angular impulse on box is

$$\Delta L=\tau\cdot\Delta t=mg\cdot\frac{l}{2}\times\Delta t=I\omega$$

Gain in angular speed of box

$$\omega=\frac{mgl\Delta t}{2I}=\frac{mgl\Delta t}{2\left(\frac{ml^2}{3}\right)}=\frac{3g\Delta t}{2l}$$

$$\Rightarrow \quad \omega=\frac{3\times 10\times 0.01}{2\times 0.3}=0.5\text{ rad/s}$$

Time taken in reaching ground is

$$t=\sqrt{\frac{2h}{g}}=\sqrt{\frac{2\times 5}{10}}=1\text{s}$$

In 1s angle rotated by box is

$$\theta=\omega t=0.5\times 1=0.5\text{ rad}$$

Ans. (A)

Sol. 31 Angular impulse = change in angular momentum

$$\tau\Delta t=\Delta L$$

$$\Rightarrow \quad mg\frac{l}{2}\times .01=\frac{ml^2}{3}\omega$$

$$\Rightarrow \quad \omega=\frac{3g\times 0.01}{2l}$$

$$\omega=\frac{3\times 10\times .01}{2\times 0.3}$$

$$\omega=\frac{1}{2}=0.5\text{ rad/s}$$

time taken by rod to hit the ground

$$t=\sqrt{\frac{2h}{g}}=\sqrt{\frac{2\times 5}{10}}=1\text{ s}$$

in this time angle rotate by rod

$$\theta=\omega t=0.5\times 1=0.5\text{ rad}$$

Ans. (C)

Sol. 32 If moment of inertia of rolling object is taken as I and speed of centre of mass at bottom of incline is v, we use

$$\frac{1}{2}\left(m+\frac{I}{R^2}\right)v^2=mgh$$

If radius of gyration is k, then, we use $I=mk^2$, so we have

$$h=\frac{\left(1+\frac{k^2}{R^2}\right)v^2}{2g} \quad \ldots(1)$$

For given object we use

$$\frac{k_{\text{ring}}}{R_{\text{ring}}}=1$$

$$\frac{k_{\text{solid cylinder}}}{R_{\text{solid cylinder}}}=\frac{1}{\sqrt{2}}$$

$$\Rightarrow \quad \frac{k_{\text{solid sphere}}}{R_{\text{solid sphere}}}=\sqrt{\frac{2}{5}}$$

From equation-(1) we have

$$h_1 : h_2 : h_3 :: (1+1) : \left(1+\frac{1}{2}\right) : \left(1+\frac{2}{5}\right) :: 20 : 15 : 14$$

Therefor **most appropriate option is (B)** which in not in proper sequence. **Ans. (C)**

Sol. 33 Kinetic energy of disc is given as

$$KE = \frac{1}{2}I\omega^2 = k\theta^2$$

$$\Rightarrow \quad \omega^2 = \frac{2k\theta^2}{I}$$

$$\Rightarrow \quad \omega = \sqrt{\frac{2k}{I}}\,\theta \quad \dots(1)$$

Differentiating-(1) wrt time, we get angular acceleration

$$\alpha = \frac{d\omega}{dt} = \sqrt{\frac{2k}{I}}\left(\frac{d\theta}{dt}\right)$$

$$\Rightarrow \quad \alpha = \sqrt{\frac{2k}{I}} \cdot \sqrt{\frac{2k}{I}}\,\theta$$

$$\Rightarrow \quad \alpha = \frac{2k}{I}\theta$$

Ans. (D)

Sol. 34 Kinetic energy of rotating body is

$$KE = \frac{1}{2}I\omega^2$$

$$\Rightarrow \quad 1200 = \frac{1}{2}1.5 \times \omega^2$$

$$\Rightarrow \quad \omega^2 = \frac{1200 \times 2}{1.5} = 1600$$

$$\Rightarrow \quad \omega = 40 \text{ rad/s}^2$$

If body rotates for time t at angular acceleration $\alpha = 20$ rad/s^2, we use

$$\omega = \omega_0 + \alpha t$$

$$40 = 0 + 20t$$

$$t = 2 \text{ s}$$

Ans. (A)

Sol. 35 Applying angular momentum conservation, about axis of rotation, we have

$$L_i = L_f$$

$$\frac{ML^2}{12}\omega_0 = \left(\frac{ML^2}{12} + m\left(\frac{L}{2}\right)^2 \times 2\right)\omega$$

$$\Rightarrow \quad \omega = \frac{M\omega_0}{M + 6m}$$

Ans. (D)

Sol. 36 We consider an elemental strip of radius r and width dr as shown in figure mass of elemental strip is

$$dm = \sigma(r) \cdot 2\pi r dr$$

Moment of inertia of elemental strip is

$$dI = dmr^2$$

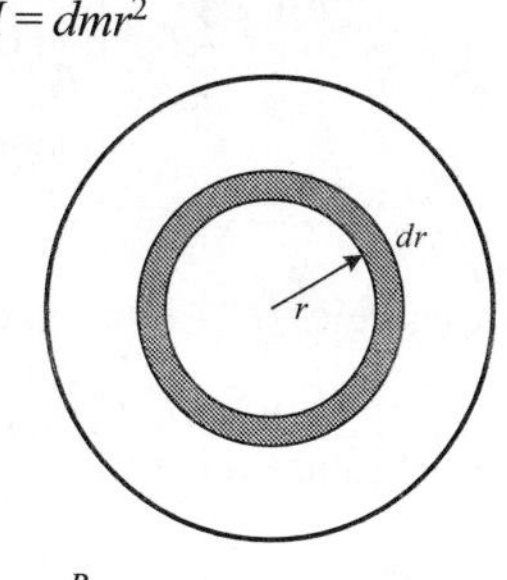

$$\Rightarrow \quad I_{\text{Disc}} = \int_0^R (dm)r^2$$

$$\Rightarrow \quad I_{\text{Disc}} = \int_0^R (\sigma 2\pi r dr)r^2$$

$$\Rightarrow \quad I_{\text{Disc}} = \int_0^R (kr^2\, 2\pi r dr)r^2$$

$$\Rightarrow \quad I_{\text{Disc}} = 2\pi k\int_0^R r^5 dr = \frac{2\pi kR^6}{6} = \frac{\pi kR^6}{3} \quad \dots(1)$$

Total mass of disc is

$$M = \int dm = \int_0^R 2\pi r dr\ (kr^2)$$

$$\Rightarrow \quad M = 2\pi k\int_0^R r^3 dr$$

$$\Rightarrow \quad M = 2\pi k \left.\frac{r^4}{4}\right|_0^R = \frac{\pi kR^4}{2} \quad \dots(2)$$

From equation-(1) and (2), we have

$$I_{\text{Disc}} = \frac{\pi kR^6}{3} = \left(\frac{\pi kR^4}{2}\right) \times \frac{2}{3}R^2$$

$$\Rightarrow \quad I_{\text{Disc}} = M \times \frac{2}{3}R^2$$

$$\Rightarrow \quad I_{\text{Disc}} = \frac{2}{3}MR^2$$

Ans. (C)

Sol. 37 Initial KE of system is

$$E_i = \frac{1}{2}I_1 \times \omega_1^2 + \frac{1}{2}\left(\frac{I_1}{2}\right)\left(\frac{\omega_1^2}{4}\right)$$

$$\Rightarrow \quad E_i = \frac{I_1\omega_1^2}{2}\left(\frac{9}{8}\right) = \frac{9}{16}I_1\omega_1^2$$

By conservation of angular momentum, we have

$$I_1\omega_1 + \frac{I_1\omega_1}{4} = \frac{3I_1}{2}\omega$$

$$\Rightarrow \quad \frac{5}{4}I_1\omega_1 = \frac{3I_1}{2}\omega$$

$$\Rightarrow \quad \omega = \frac{5}{6}\omega_1$$

Final KE of system is

$$E_f = \frac{1}{2} I_f \omega^2 = \frac{1}{2}\left(\frac{3}{2} I_1\right)\left(\frac{5}{6}\omega_1\right)^2$$

$$\Rightarrow \quad E_f = \frac{25}{48} I_1 \omega_1^2$$

Difference in kinetic energies $(E_f - E_i)$ is given as

$$E_f - E_i = I_1\omega_1^2\left(\frac{25}{48} - \frac{9}{16}\right) = \frac{-2}{48} I_1\omega_1^2$$

$$\Rightarrow \quad E_f - E_i = \frac{-I_1\omega_1^2}{24}$$

Ans. (D)

Sol. 38 Acceleration of particle

$$\overrightarrow{a_x} = \frac{d^2x}{dt^2} = -a\omega_1^2 \cos\omega_1 t\,\hat{i}$$

$$\overrightarrow{a_y} = \frac{d^2y}{dt^2} = -b\omega_2^2 \sin\omega_2 t\,\hat{i}$$

$$\vec{a}_{t=0} = -a\omega_1^2\,\hat{i}$$

Position vector of particle at $t = 0 \rightarrow \vec{r} = (x_0 + a)\,\hat{i} + y_0\,\hat{j}$

Torque on particle is $\vec{\tau} = \vec{r} \times \vec{F} = \vec{r} \times m\,\vec{a}_{t=0}$

$$= [(x_0 + a)\,\hat{i} + y_0\,\hat{j}] \times m[-a\omega_1^2\,\hat{i}]$$

$$= my_0\,a\omega_1^2\,\hat{k}$$

Ans. (B)

Sol. 39 Angular acceleration of coin is

$$\alpha = \frac{\Delta\omega}{\Delta t} = \frac{25 \times 2\pi}{5} = 10\pi \text{ rad/s}^2$$

Torque on coin is

$$\tau = \left(\frac{5}{4} MR^2\right)\alpha$$

$$\Rightarrow \quad \tau = \frac{5}{4} \times 5 \times 10^{-3} \times (10^{-2})^2 \times 10\pi$$

$$\Rightarrow \quad \tau = 1.9625 \times 10^{-5} \text{ Nm} \simeq 2.0 \times 10^{-5} \text{ Nm}$$ **Ans. (B)**

Sol. 40 Velocity at $t = 2$s is

$$\vec{v}_{(t)} - \frac{\overrightarrow{dr}}{dt} - 2\,\hat{i} - 6t\,\hat{j}\,\big|_{t=2s}\,(2\,\hat{i} - 12\,\hat{j})$$

Position vector at $t = 2$s is

$$\vec{r}_{(t)} = (4\,\hat{i} - 12\,\hat{j})$$

Angular momentum at $t = 2$s is

$$\vec{L} = m(\vec{r} \times \vec{v})$$

$$= 2(4\,\hat{i} - 12\,\hat{j}) \times (2\,\hat{i} - 12\,\hat{j})$$

$$= 2[-48\hat{k} + 24\hat{k}]$$

$$= -48\hat{k}$$

Ans. (D)

Sol. 41 For disc $m_1 = \frac{7M}{8}$, $R_1 = 2R$

MI of disc $I_1 = \frac{1}{2} m_1 R_1^2 = \frac{1}{2}\left(\frac{7M}{8}\right)(2R)^2 = \frac{7}{4} MR^2$ ...(1)

For new solid sphere $m_2 = \frac{M}{8}$,

$R_2 = \frac{R}{2}$ [we use $\frac{1}{8}\left(\frac{4}{3}\pi R^3\right) = \frac{4}{3}\pi R_2^3 \Rightarrow R_2 = \frac{R}{2}$]

MI of new sphere $I_2 = \frac{2}{5} m_2 R_2^2 = \frac{2}{5}\left(\frac{M}{8}\right)\left(\frac{R}{2}\right)^2$

$$= \frac{1}{80} MR^2 \quad \text{...(2)}$$

From (1) and (2) $\frac{I_1}{I_2} = \frac{\frac{7}{4}MR^2}{\frac{1}{80}MR^2} = 140$

Ans. (D)

Sol. 42 Area of elemental strip $= ds = 2\pi x\, dx$

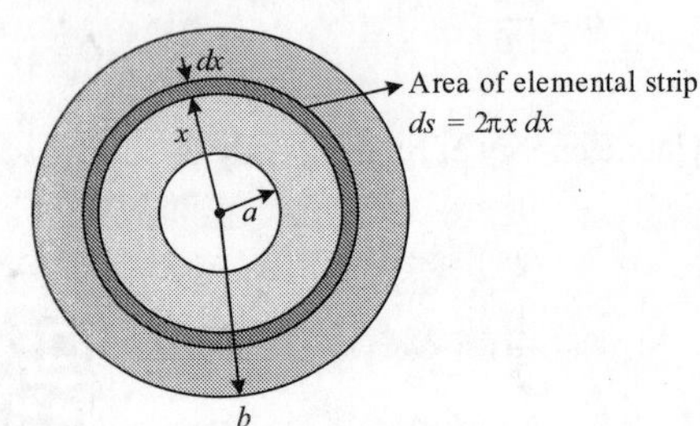

Mass of elemental strip $dm = \frac{\sigma_0 ds}{x} = \frac{\sigma_0}{x} \cdot 2\pi x\,dx = 2\pi\sigma_0\,dx$

Total mass of disc $m = \int dm = 2\pi\sigma_0 = \int_a^b dx = 2\pi\sigma_0\,(b - a)$

MI of elemental strip $dI = dmx^2$

$\Rightarrow \quad dI = 2\pi\sigma_0 x^2 dx$

$$\Rightarrow \quad I = \int dI = \int_a^b 2\pi\sigma_0 x^2\,dx = \frac{2\pi\sigma_0}{3} \cdot (b^3 - a^3)$$

Radius of gyration of disc is

$$k = \sqrt{\frac{I}{m}} = \sqrt{\frac{(b^3 - a^3)}{3(b-a)}} = \sqrt{\frac{a^2 + b^2 + ab}{3}}$$

Ans. (D)

Sol. 43

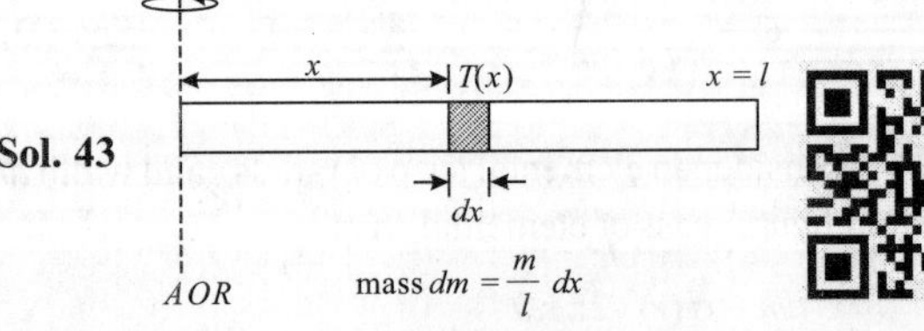

At a distance x from AOR, tension in rod will be due to the centrifugal force on part of rod outside it in frame of rod.
Centrifugal force on dm is

$$dF = dm\omega^2 x$$

Total centrifugal force on outer part of rod i.e. from x to l is

$$F = T_{(x)} = \int dF = \int_x^l \frac{m}{l}\omega^2 x dx = \frac{m\omega^2}{2l}(l^2 - x^2)$$

This is a parabolic function **Ans. (D)**

Sol. 44 Using work energy theorem, we use

$$mgh = \frac{1}{2}mv^2 + \frac{1}{2}\times\left(\frac{1}{2}mr^2\right)\times\frac{v^2}{r^2} = \frac{3}{4}mv^2$$

$$\Rightarrow \quad v = \sqrt{\frac{4}{3}gh}$$

Angular speed is given as

$$\omega = \frac{v}{r} = \frac{1}{r}\sqrt{\frac{4gh}{3}}$$ **Ans. (C)**

Sol. 45 Moment of inertia of given system is taken as

$$m\frac{l^2}{12} + m\frac{l^2}{16} = mk^2$$

$$\Rightarrow \quad \frac{7l^2}{48} = k^2$$

$$\Rightarrow \quad k = \sqrt{\frac{7}{48}}\cdot l$$ **Ans. (B)**

Sol. 46 We consider an elemental ring in disc as shown
Moment of inertia of this ring is given as

$$dI = dmr^2$$

$$\Rightarrow \quad dI = \sigma 2\pi r\, dr\, r^2$$

$$\Rightarrow \quad dI = 2\pi(A + Br)\cdot r^3\cdot dr$$

Total moment of inertia of disc is

$$I = \int dI = 2\pi\int_0^a (Ar^3 + Br^4)dr$$

$$\Rightarrow \quad I = 2\pi a^4\left(\frac{A}{4} + \frac{aB}{5}\right)$$ **Ans. (A)**

Sol. 47 Forces are shown in figure below

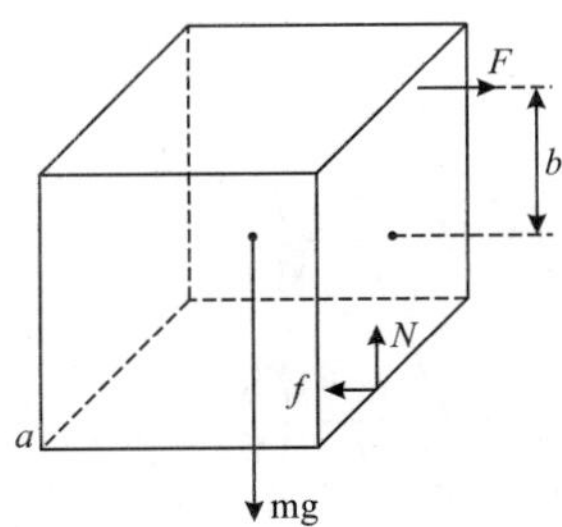

To move the block

$$F = \mu mg \quad \ldots(1)$$

For bock not to topple, we use

$$F\left(b + \frac{a}{2}\right) = mg\,\frac{a}{2} \quad \ldots(2)$$

$$\Rightarrow \quad \mu mg\left(b + \frac{a}{2}\right) = mg\times\frac{a}{2}$$

$$\Rightarrow \quad \left(b + \frac{a}{2}\right)\mu = \frac{a}{2}$$

$$\Rightarrow \quad 0.8b + 0.4a = a$$

$$\Rightarrow \quad 0.8b = 0.6a$$

$$\Rightarrow \quad \frac{b}{a} = \frac{3}{4}$$ **Ans. (75)**

Sol. 48 Figure below shows the given situation

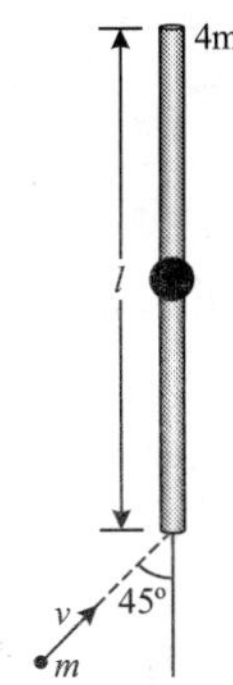

Using by conservation of angular momentum about the hinge, we have

$$m\left(\frac{v}{\sqrt{2}}\right)\left(\frac{l}{2}\right) = \left[\frac{4ml^2}{12} + \frac{ml^2}{4}\right]\omega$$

$$\Rightarrow \quad \omega = \frac{3\sqrt{2}}{7}\left(\frac{v}{l}\right)$$ **Ans. (B)**

Sol. 49 Kinetic energy of rolling sphere is given as

$$KE = \frac{1}{2}mv^2 + \frac{1}{2}I\omega^2$$

$$\Rightarrow \quad KE = \frac{1}{2}mv^2\left(1 + \frac{K^2}{R^2}\right)$$

$$\Rightarrow \quad KE = 8.75\times 10^{-4}\,\text{J}$$ **Ans. (A)**

Sol. 50 Using parallel axis theorem, moment of inertia of system is given as

$$I_0 = 3\times\left[\frac{2}{5}M\left(\frac{d}{2}\right)^2 + M\left(\frac{d}{\sqrt{3}}\right)^2\right] = \frac{13}{10}\text{Md}^2 \quad \ldots(1)$$

Moment of inertia about axis passing through A is given as

$$I_A = I_0 + 3M\left(\frac{d}{\sqrt{3}}\right)^2$$

$$\Rightarrow \quad I_A = \frac{13}{10}\text{Md}^2 + \text{Md}^2$$

$$\Rightarrow \quad I_A = \frac{23}{10}\text{Md}^2 \quad \ldots(2)$$

From (1) and (2), we have

$$\frac{I_0}{I_A} = \frac{13}{23}$$

Ans. (A)

Sol. 51 Using work energy theorem, we have

$$\Rightarrow \quad mg\left(\frac{l}{2}\sin 30^\circ\right) = \frac{1}{2}I\omega^2$$

$$\Rightarrow \quad mg \times \frac{1}{2} \times \frac{1}{2} = \frac{1}{2}\left(\frac{m(1)^2}{3}\right)\omega^2$$

$$\Rightarrow \quad \omega^2 = \frac{3g}{2}$$

$$\Rightarrow \quad \omega = \sqrt{15}\text{ rad/s}$$

$$\Rightarrow \quad n = 15$$

Ans. (15.00)

Sol. 52 Using work energy theorem, we have

$$(m_1 - m_2)gh = \frac{1}{2}(m_1 + m_2)v^2 + \frac{1}{2}I\omega^2$$

$$\Rightarrow (m_1 - m_2)gh = \frac{1}{2}(m_1 + m_2)(\omega R)^2 + \frac{1}{2}I\omega^2$$

$$\Rightarrow (m_1 - m_2)gh = \frac{\omega^2}{2}[(m_1 + m_2)R^2 + I]$$

$$\Rightarrow \quad \omega = \sqrt{\frac{2(m_1 - m_2)gh}{(m_1 + m_2)R^2 + I}}$$

Ans. (B)

Sol. 53 Figure below shows the forces acting on rod

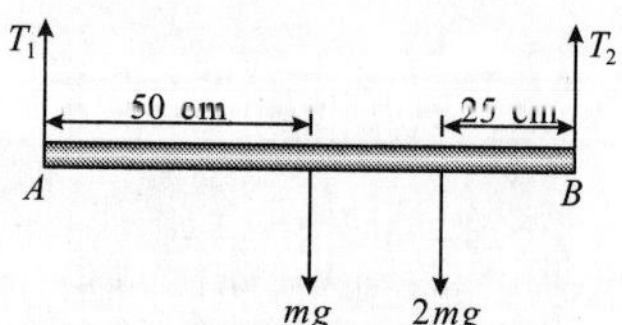

For equilibrium of rod, we use torque about point B

$$\tau_b = 0$$

$$2mg \times 25 + mg \times 50 - T_1 \times 100 = 0$$

$$\Rightarrow \quad 100\,T_1 = 100mg$$

$$\Rightarrow \quad T_1 = mg$$

Ans. (D)

Sol. 54 For maximum torque F must be applied normal to radius as shown in figure

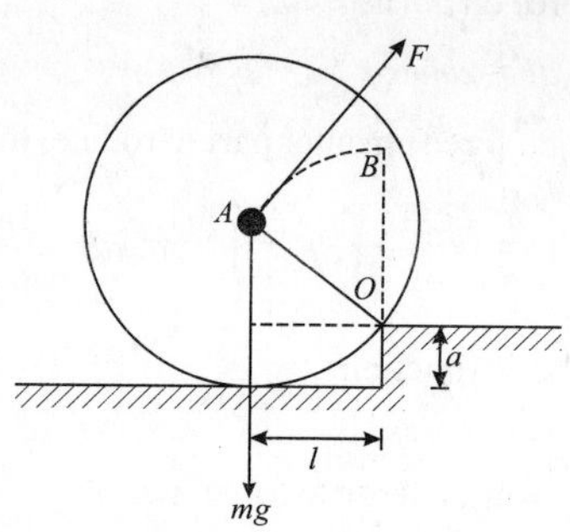

To pull over the edge force should be able to lift cylinder centre from A to B so at point A torque of F must be more than that of gravity about O, so we use

$$FR \geq Mgl$$

$$\Rightarrow \quad F \geq \frac{Mg}{R}\sqrt{R^2 - (R-a)^2}$$

$$\Rightarrow \quad F \geq Mg\sqrt{1 - \left(\frac{R-a}{R}\right)^2}$$

Ans. (D)

Sol. 55 By angular momentum conservation, we use

$$L_1 + L_2 = L_{\text{final}}$$

$$\Rightarrow \quad I_1\omega_1 + I_2\omega_2 = (I_1 + I_2)\omega_f$$

$$\Rightarrow \quad 0.1 \times 10 + 0.2 \times 5 = (0.1 + 0.2) \times \omega_f$$

$$\Rightarrow \quad \omega_f = \frac{20}{3}$$

Kinetic energy of combined disc system can be given as

$$\Rightarrow \quad K = \frac{1}{2}(I_1 + I_2)\omega_f^2$$

$$\Rightarrow \quad K = \frac{1}{2}(0.1 + 0.2).\left(\frac{20}{3}\right)^2$$

$$\Rightarrow \quad K = \frac{0.3}{2} \times \frac{400}{9} = \frac{120}{18} = \frac{20}{3}\text{J}$$

Ans. (D)

Sol. 56 For constant mass, volume is also constant, so we use

$$\pi R^2 L = \text{Constant}$$

$$\Rightarrow \quad R^2 L = C \quad \ldots(1)$$

Moment of inertia as given as

$$I = M\left(\frac{R^2}{4} + \frac{L^2}{12}\right)$$

$$\Rightarrow \quad I = M\left(\frac{C}{4L} + \frac{L^2}{12}\right)$$

For I to be minimum, we use

$$\frac{dI}{dL} = M\left[-\frac{C}{4L^2}+\frac{L}{6}\right]=0$$

$$\Rightarrow \quad \frac{C}{4L^2} = \frac{L}{6}$$

$$\Rightarrow \quad L=\left[\frac{3C}{2}\right]^{1/3}$$

From equation-(1), we use

$$R^2 = \frac{C}{L} = \frac{C}{\left[\frac{3}{2}C\right]^{1/3}} = \frac{C^{2/3}}{\left(\frac{3}{2}\right)^{1/3}}$$

$$\Rightarrow \quad R = \frac{C^{1/3}}{\left(\frac{3}{2}\right)^{1/6}}$$

$$\Rightarrow \quad \frac{L}{R} = \frac{\left(\frac{3}{2}\right)^{1/3}\cdot(C)^{1/3}}{C^{1/3}}\times\left(\frac{3}{2}\right)^{1/6}$$

$$\Rightarrow \quad \frac{L}{R} = \left(\frac{3}{2}\right)^{1/2} = \sqrt{\frac{3}{2}}$$

Ans. (C)

Sol. 57 By angular momentum conservation, we have

$$mvl = \frac{Ml^2}{3}\omega \times ml^2\omega$$

$$\Rightarrow \quad \omega = \frac{mvl}{\left(ml^2+\frac{Ml^2}{3}\right)}$$

$$\Rightarrow \quad \omega = \frac{1\times 6\times 1}{\frac{2}{3}+1} = \frac{18}{5} \text{ rad/s}$$

By energy conservation, we have

$$\frac{1}{2}\left(M\frac{l^2}{3}\right)\omega^2 + \frac{1}{2}(ml^2)\omega^2 = mgl(1-\cos\theta) + \frac{Mgl}{2}(1-\cos\theta)$$

$$\Rightarrow \quad \frac{1}{2}\left(\frac{2\times(1)^2}{3}\right)\left(\frac{18}{5}\right)^2 + \frac{1}{2}(1\times(1)^2)\left(\frac{18}{5}\right)^2 = (1+1)\times 10\times 1(1-\cos\theta)$$

$$\Rightarrow \quad \frac{5}{6}\times\left(\frac{18}{5}\right)^2 = 20(1-\cos\theta)$$

$$\Rightarrow \quad 1-\cos\theta = \frac{18}{5}\times\frac{3}{20}$$

$$\Rightarrow \quad \cos\theta = 1-\frac{27}{50}$$

$$\Rightarrow \quad \theta \simeq 63°$$

Ans. (B)

Sol. 58 By conservation of angular momentum, we use

$$\left(80R^2+\frac{200R^2}{2}\right)\omega_i = \left(0+\frac{200R^2}{2}\right)\omega_f$$

$$\Rightarrow \quad 180\,\omega_i = 100\,\omega_f$$

$$\Rightarrow \quad \omega_f = 1.8\,\omega_i = 1.8\times 5 = 9 \text{ rev/min}$$

Ans. (9)

Sol. 59 Horizontal and vertical forces given in system are shown in below figure are

$$F_H = m\omega^2\frac{l}{2}\sin\theta$$

$$F_V = mg$$

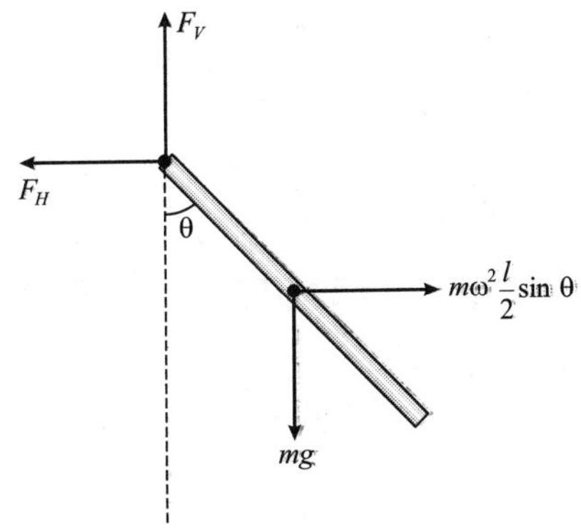

As stated in question, we use

$$mg\times\frac{l}{2}\sin\theta - m\omega^2\frac{l}{2}\sin\theta\cdot\frac{l}{2}\cos\theta = \frac{ml^2}{12}\omega^2\sin\theta\cos\theta$$

$$\Rightarrow \quad \cos\theta = \frac{3g}{2\omega^2 l}$$

Ans. (B)

Sol. 60 Moment of inertia about axis EX is given as

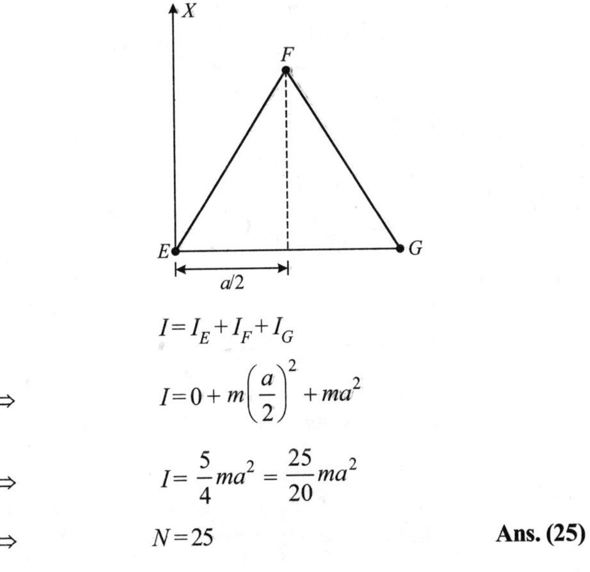

$$I = I_E + I_F + I_G$$

$$\Rightarrow \quad I = 0 + m\left(\frac{a}{2}\right)^2 + ma^2$$

$$\Rightarrow \quad I = \frac{5}{4}ma^2 = \frac{25}{20}ma^2$$

$$\Rightarrow \quad N = 25$$

Ans. (25)

Sol. 61 Moment of inertia of an equilateral triangle about its centroid is

$$I_0 = \frac{ma^2}{12}$$

ΔADE is also an equilateral triangle of side $\frac{a}{2}$ so its moment of inertia about its centroid is given as

$$I_{ADE} = \frac{m_1\left(\frac{a}{2}\right)^2}{12} = \frac{m_1 a^2}{48} = \frac{ma^2}{192} \quad \left(m_1 = \frac{m}{4}\right)$$

MI of system about *G* after removing ΔADE is given as

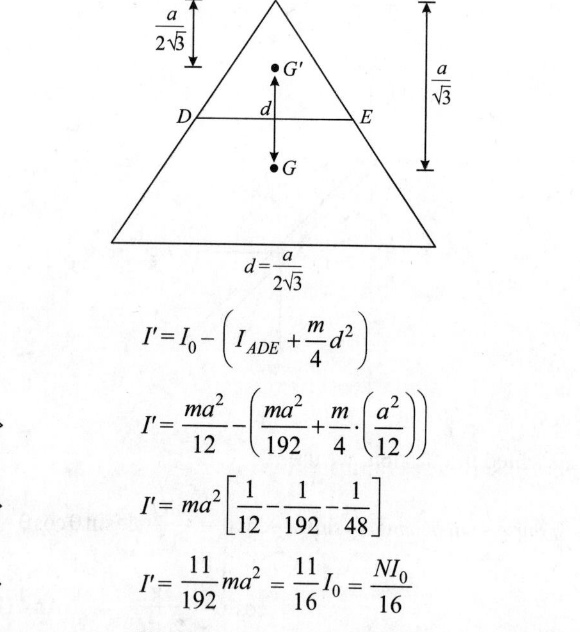

$$I' = I_0 - \left(I_{ADE} + \frac{m}{4}d^2\right)$$

$$\Rightarrow \quad I' = \frac{ma^2}{12} - \left(\frac{ma^2}{192} + \frac{m}{4}\cdot\left(\frac{a^2}{12}\right)\right)$$

$$\Rightarrow \quad I' = ma^2\left[\frac{1}{12} - \frac{1}{192} - \frac{1}{48}\right]$$

$$\Rightarrow \quad I' = \frac{11}{192}ma^2 = \frac{11}{16}I_0 = \frac{NI_0}{16}$$

Thus $\quad N = 11$ **Ans. (11)**

Sol. 62 By angular momentum conservation, we use

$$\frac{1}{2}MR^2\omega_1 = \left[\frac{1}{2}MR^2 + \frac{1}{2}M\left(\frac{R}{2}\right)^2\right]\omega_2$$

$$\Rightarrow \quad \omega_2 = \frac{4\omega_1}{5}$$

Initial kinetic energy

$$K_1 = \frac{1}{2}\left(\frac{1}{2}MR^2\right)\omega_1^2$$

Final kinetic energy

$$K_2 = \frac{1}{2}\left(\frac{5}{8}MR^2\right)\left(\frac{4\omega_1}{5}\right)^2$$

$$\Rightarrow \quad K_2 = \frac{1}{2}\left(\frac{2}{5}MR^2\right)\omega_1^2$$

Kinetic energy loss

$$\Delta K = K_1 - K_2 = \frac{1}{2}\left(\frac{1}{10}MR^2\right)\omega^2$$

Percentage loss *P* is given as

$$P = \frac{\Delta K}{K_1}\times 100 = \frac{\frac{1}{10}MR^2}{\frac{1}{2}MR^2}\times 100$$

$$\Rightarrow \quad P = 20\%$$ **Ans. (20)**

Sol. 63 Moment of inertia of the two discs are given as

$$I_1 = \frac{1}{2}MR^2 = \frac{1}{2}(\rho\pi R^2)R^2$$

and $$I_2 = \frac{1}{2}M'(\alpha R)^2 = \frac{1}{2}(\rho\pi\alpha^2R^2)\alpha^2R^2$$

Where ρ is surface mass density of discs

Given that $$\frac{I_1}{I_2} = \frac{1}{\alpha^4} = \frac{1}{16}$$

$$\Rightarrow \quad \alpha = 2$$ **Ans. (B)**

Sol. 64 Moment of inertia about axis passing through *O* is

$$I_0 = \frac{M}{12}(l^2 + b^2) = \frac{M}{12}((80)^2 + (60)^2)$$

MI about axis passing through *O'* is

$$I_0' = I_0 + Md^2$$

$$\Rightarrow \quad I_{0'} = I_0 + M(50)^2$$

$$\Rightarrow \quad I_{0'} = \frac{M}{12}((80)^2 + (60)^2) + M(50)^2$$

$$\Rightarrow \quad \frac{I_0}{I_{0'}} = \frac{\frac{M}{12}((80)^2 + (60)^2)}{\frac{M}{12}((80)^2 + (60)^2) + M(50)^2} = \frac{1}{4}$$

Ans. (D)

Sol. 65 On coupling of another wheel we use conservation of angular momentum as

$$I\omega = (I + 3I)\omega_1$$

$$\Rightarrow \quad \omega_1 = \frac{\omega}{4}$$

Initial kinetic energy of wheel

$$K_1 = \frac{1}{2}I\omega^2$$

Final kinetic energy of coupled wheels

$$K_2 = \frac{1}{2}(4I)\left(\frac{\omega}{4}\right)^2 = \frac{1}{8}I\omega^2$$

Fractional loss in kinetic energy

$$f = \frac{K_1 - K_2}{K_1} = \frac{\frac{1}{2} - \frac{1}{8}}{\frac{1}{2}} = \frac{3}{4}$$ **Ans. (C)**

Sol. 66 Torque of force is given as

$$\vec{\tau} = \vec{r} \times \vec{F}$$

Here $\vec{r} = \vec{r}_2 - \vec{r}_1 = (3\hat{i} + \hat{j} - 2\hat{k})$

$$\Rightarrow \quad \vec{\tau} = (3\hat{i} + \hat{j} - 2\hat{k}) \times (\hat{i} + 2\hat{j} + 3\hat{k})$$

$$\Rightarrow \quad \vec{\tau} = (7\hat{i} - 11\hat{j} + 5\hat{k})$$

$$\Rightarrow \quad |\vec{\tau}| = \sqrt{49+121+25} = \sqrt{195} \text{ N-m}$$ **Ans. (195)**

Sol. 67 By conservation of angular momentum, we use

$$L_i = L_f$$

$$mvl = \left(ml^2 + \frac{ML^2}{3}\right)\omega$$

$$\Rightarrow \quad 0.1 \times 80 \times 1 = \left(0.1 \times (1)^2 + \frac{0.9 \times (1)^2}{3}\right)\omega^2$$

$$\Rightarrow \quad \omega^2 = \frac{8}{0.4} = 20 \text{ rad/s}$$ **Ans. (20)**

Sol. 68 Moment of inertia of a hollow cone is $\frac{1}{2}MR^2$ as already derived in theory. **Ans. (A)**

Sol. 69 Moment of inertia of all point masses fixed on square about given axis of rotation is given as

$$I = 0 + 2\left(m\left(\frac{l}{\sqrt{2}}\right)^2\right) + m(\sqrt{2}\,l)^2$$

$$\Rightarrow \quad I = ml^2 + 2ml^2 = 3ml^2$$

Angular momentum of system is given as

$$L = I\omega = 3ml^2\omega$$ **Ans. (B)**

Sol. 70

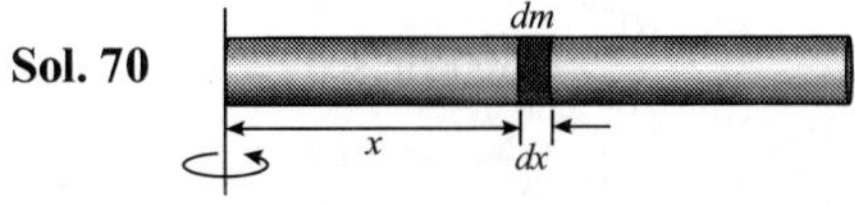

We consider an element as shown in figure. The moment of inertia of element is given as

$$I = \int dm x^2 = \int \lambda dx \cdot x^2$$

$$\Rightarrow \quad I = \int_0^L x^2 \lambda_0 \left(1 + \frac{x}{L}\right) dx$$

$$\Rightarrow \quad I = \lambda_0 \int_0^L \left(x^2 + \frac{x^3}{L}\right) dx$$

$$\Rightarrow \quad I = \lambda\left[\frac{L^3}{3} + \frac{L^3}{4}\right]$$

$$\Rightarrow \quad I = \frac{7L^3\lambda_0}{12} \quad \ldots(1)$$

Mass of rod is given as

$$M = \int_0^L \lambda dx = \int_0^L \lambda_0 \left(1 + \frac{x}{L}\right) dx$$

$$\Rightarrow \quad M = \lambda_0\left(L + \frac{L}{2}\right) = \frac{3\lambda_0 L}{2} \quad \ldots(2)$$

From (1) & (2), we use

$$I = \frac{7}{12} L^3 \left(\frac{2M}{3L}\right)$$

$$\Rightarrow \quad I = \frac{7}{18} ML^2$$ **Ans. (D)**

Sol. 71 The given moment of inertias are standard values written below

For a ring about diameter $I_1 = \frac{MR^2}{2}$

For a Disc about central axis $I_2 = \frac{MR^2}{2}$

For a Solid cylinder about central axis $I_3 = \frac{MR^2}{2}$

For a Solid sphere about central axis $I_4 = \frac{2}{5} MR^2$

$$\Rightarrow \quad I_1 = I_2 = I_3 > I_4$$ **Ans. (C)**

Sol. 72 The situation described in question is shown in figure below

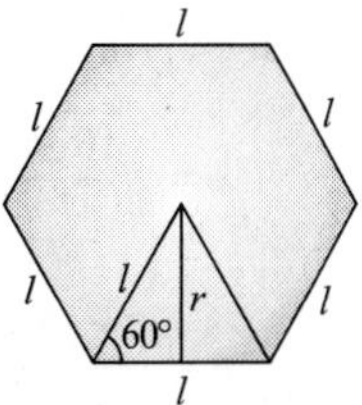

If m is the mass of one side of hexagon and r is the distance of side center from the center of hexagon, we have

$$6l = 2.4 \text{ m}$$

$$\Rightarrow \quad l = 0.4 \text{ m}$$

and $\sin 60° = \frac{r}{l}$

$$\Rightarrow \quad r = l \sin 60° = \frac{l\sqrt{3}}{2}$$

Moment of inertia of the system is given as

$$I = \left[\frac{ml^2}{12} + mr^2\right]6$$

$$\Rightarrow \quad I = \left[\frac{ml^2}{12} + m\left(\frac{l\sqrt{3}}{2}\right)^2\right]6$$

$$\Rightarrow \quad I = 5\,ml^2 = 5 \times 1 \times 0.16 = 0.8 \text{ kg-m}^2$$ **Ans. (8)**

Sol. 73 When the sphere starts up the plane, its angular speed changes which can be calculated by using angular momentum conservation as shown in figure below

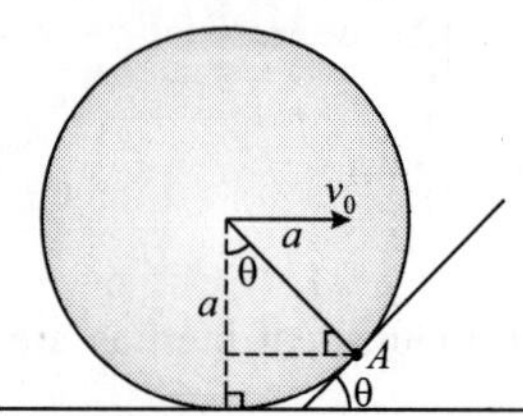

As initial speed of sphere on horizontal plane is v_0 and sphere starts moving up the incline plane with speed v, we use angular momentum conservation about point *A*, gives

$$mv_0\, a\cos\theta + \frac{2}{5}ma^2\omega = mva + \frac{2}{5}ma^2\omega'$$

$$\Rightarrow \quad mv_0 a\left[\frac{2}{5} + \cos\theta\right] = \frac{7}{5}mva$$

$$\Rightarrow \quad v = \frac{5}{7} = v_0\left[\frac{2}{5} + \cos\theta\right]$$

If sphere ascends up the plan to a height *h*, by conservation of energy, we use

$$\frac{1}{2}mv^2 + \frac{1}{2}I\omega^2 = \frac{7}{10}mv^2 = mgh$$

$$\Rightarrow \quad v = \sqrt{\frac{10gh}{7}} = v_0\left[\frac{2}{5} + \cos\theta\right]$$

$$\Rightarrow \quad h = \frac{7v_0^2(2 + 5\cos\theta)^2}{250g} = l\sin\theta$$

Thus none of the given options are correct. **Ans. (None)**

Sol. 74 The situation described in question is shown in figure below

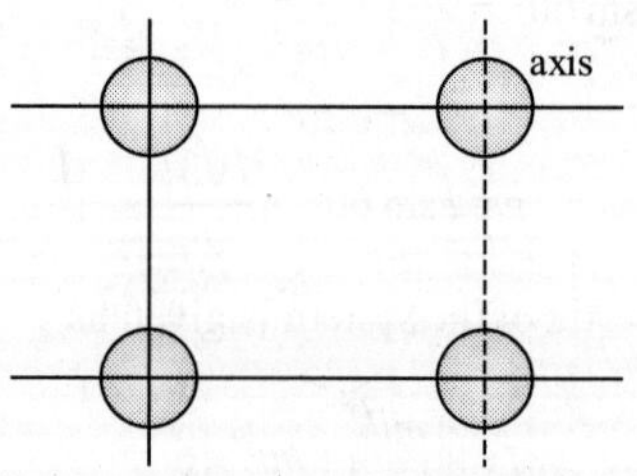

The moment of inertia of the system is given by using parallel axes theorem as

$$I = 2\times\left(\frac{2}{5}ma^2\right) + 2\times\left(\frac{2}{5}ma^2 + mb^2\right)$$

$$\Rightarrow \quad I = \frac{8}{5}ma^2 + 2mb^2$$ **Ans. (C)**

Sol. 75 Figure below shows the situation described in the question

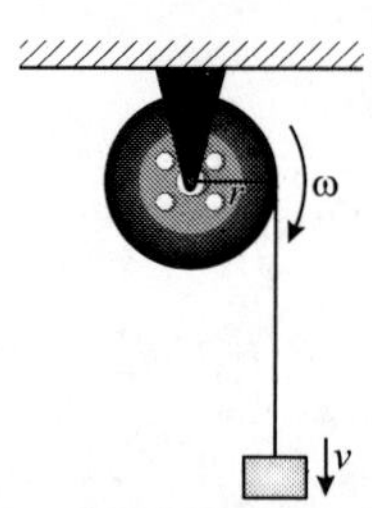

By conservation of energy after falling through a distance *h*, we have

$$mgh = \frac{1}{2}I\omega^2 + \frac{1}{2}mv^2$$

As $v = \omega r$, we have

$$mgh = \frac{1}{2}I\omega^2 + \frac{1}{2}m\omega^2 r^2$$

$$\Rightarrow \quad \frac{2mgh}{(I + mr^2)} = \omega^2$$ **Ans. (D)**

Sol. 76 By angular momentum conservation in elliptical orbit of comet, we use

$$mv_1 r_1 = mv_2 r_2$$

$$\Rightarrow \quad v_1 = \frac{48\times10^{14}}{1.6\times10^{12}} = 3000 \text{ m/sec}$$

$$\Rightarrow \quad v_1 = 3\times10^3 \text{ m/sec}$$ **Ans. (C)**

Sol. 77 Moment of inertia of the masses about the axis of rotation AA' as shown in figure is calculated as

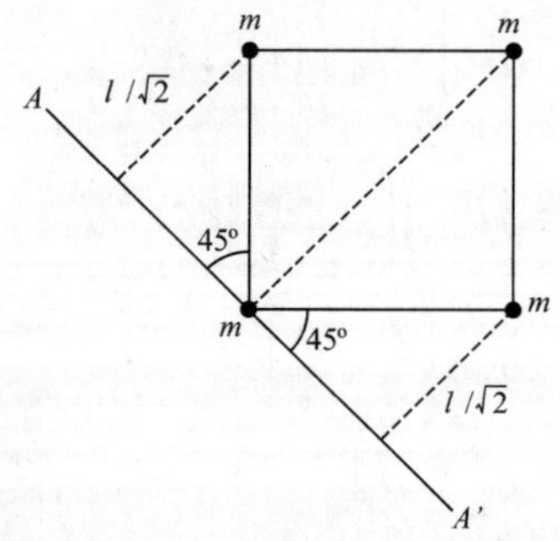

Moment of Inertia of the masses about axis AA' is given as

$$I = m(0)^2 + m(l\sqrt{2})^2 + m\left(\frac{l}{\sqrt{2}}\right)^2 + m\left(\frac{l}{\sqrt{2}}\right)^2 = 3\,ml^2$$

Ans. (C)

Sol. 78 The angular acceleration of the disc can be calculated as

$$\alpha = \frac{\tau}{I} = \frac{F.R.}{mR^2/2} = \frac{2F}{mR}$$

$$\Rightarrow \quad \alpha = \frac{2\times 200}{20\times(0.2)} = 10\text{ rad/s}^2$$

Angular speed after revolution by an angle $\Delta\theta$ is given as

$$\omega^2 = \omega_0^2 + 2\alpha\Delta\theta$$

$$\Rightarrow \quad (50)^2 = 0^2 + 2(10)\,\Delta\theta$$

$$\Rightarrow \quad \Delta\theta = \frac{2500}{20}$$

$$\Rightarrow \quad \Delta\theta = 125\text{ rad}$$

No. of revolution $= \frac{125}{2\pi} = \frac{125}{6.28} \approx 20$ revolution **Ans. (20)**

Sol. 79 Torque is calculated as

$$\vec{\tau} = \vec{r}\times\vec{F}$$

Position vector of the point of appliation of force is given as

$$\vec{r} = (2\hat{i}) - (2\hat{i} + 3\hat{j} + 4\hat{k}) = -3\hat{j} - 4\hat{k}$$

Torque calculation is done now as

$$\vec{\tau} = \vec{r}\times\vec{F} = \begin{vmatrix} \hat{i} & \hat{j} & \hat{k} \\ 0 & -3 & -4 \\ 4 & 3 & 4 \end{vmatrix}$$

$$\Rightarrow \quad \vec{\tau} = \hat{i}\,(-12+12) - \hat{j}\,(0+16) + \hat{k}\,(0+12)$$

$$\Rightarrow \quad \vec{\tau} = -16\hat{i} + 12\hat{k}$$

$$\Rightarrow \quad |\vec{\tau}| = \sqrt{16^2 + 12^2} = 20$$ **Ans. (20)**

Sol. 80 Acceleration of disc in pure rolling on incline plance can be given as

$$a = \frac{g\sin\theta}{1+\frac{I}{mR^2}} = \frac{g\sin\theta}{1+\frac{1}{2}} = \frac{2}{3}\,g\sin\theta$$

$$\Rightarrow \quad b = 3$$ **Ans. (3)**

Sol. 81 Position vectors of point P with respect to O is given as

$$\vec{r}_1 = 5\hat{i} + 5\sqrt{3}\hat{j}$$

And position vector of point P with respect to Q is given as

$$\vec{r}_2 = -5\hat{i} + 5\sqrt{3}\hat{j}$$

Torque about O is calculated as

$$\vec{\tau}_O = \vec{r}_1\times\vec{F} = (-15-20\sqrt{3})\,\hat{k} = (15+20\sqrt{3})(-\hat{k})$$

Torque about Q is calculated as

$$\vec{\tau}_Q = \vec{r}_2\times\vec{F} = (-15+20\sqrt{3})\,\hat{k} = (15-20\sqrt{3})(-\hat{k})$$ **Ans. (A)**

Sol. 82 The angular momentum of a moving particle is given as

$$\vec{L} = m(\vec{r}\times\vec{v})$$

Now for point A, always the direction of angular momentum will be along the axis of circle by right hand thumb rule and it also remain constant in magnitude as mvr. With respect to point B, the magnitude remain constant but direction will continuously changing as it is always in normal direction to line joining the mass M and point B. **Ans. (D)**

Sol. 83 The acceleration of the rolling body on the inclined plane without slipping is given as

$$a = \frac{g\sin\theta}{1+\frac{k^2}{R^2}}$$

Thus time of motion can be calculated as

$$t = \sqrt{\frac{2s}{a}} = \sqrt{\frac{2s}{g\sin\theta}\left(1+\frac{k^2}{R^2}\right)}$$

For least time, radius of gyration k should be least and out of given options for solid sphere it is minimum. **Ans. (4)**

Sol. 84 Acceleration of a body in pure rolling on inclined plane is given as

$$a = \frac{g\sin\theta}{1+\frac{I}{mR^2}} = \frac{5}{7}\times\frac{10}{2} = \frac{25}{7}$$

Time taken by sphere to come back to starting point can be calculated as

$$t = \frac{2v_0}{a} = \frac{2\times 1\times 7}{25} = 0.56\text{ s}$$ **Ans. (C)**

Sol. 85 Using conservation of angular momentum, we have

$$(Mr^2)\omega = (Mr^2 + 2mr^2)\omega'$$

$$\Rightarrow \quad \omega' = \frac{M\omega}{M+2m}$$ **Ans. (C)**

Sol. 86 As the wire is bent into a semicircle of radius r, we have

$$\pi r = L$$

$$\Rightarrow \quad r = \frac{L}{\pi}$$

The moment of inertia of the wire can be given as

$$I = Mr^2 = \frac{ML^2}{\pi^2}$$ **Ans. (C)**

Sol. 87 In Physics Galaxy vol. 1 chapter 6, we have derived that the areal velocity of a planet around the Sun is given as

$$\frac{dA}{dt} = \frac{L}{2m}$$ **Ans. (D)**

Sol. 88 As per situation described in question the cylinder is placed on the rough inclined plane. It is not given that it is sliding or moving so first we consider it in equilibrium. Forces acting on it are shown in figure below.

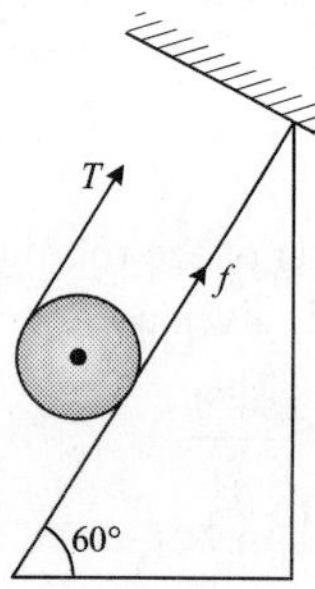

$$T + f = mg\sin\theta \quad \ldots(1)$$

$$TR - fR = 0 \quad \ldots(2)$$

Solving we get

$$T = f = \frac{mg\sin\theta}{2}$$

At $\theta = 60°$ and $\mu = 0.4$, we have

$$\mu mg\cos 60° < \frac{mg\sin 60°}{2}$$

Thus cylinder will not remain in equilibrium and will start moving down and friction will be maintained at kinetic friction. As separately kinetic friction coefficient is not specified, the friction value is given as

$$f = \mu N$$

$$\Rightarrow \quad f = \mu mg\cos 60° = \frac{mg}{5}$$ **Ans. (C)**

Sol. 89 Figure below shows the situation described in the question for collision of a particle with the rod.

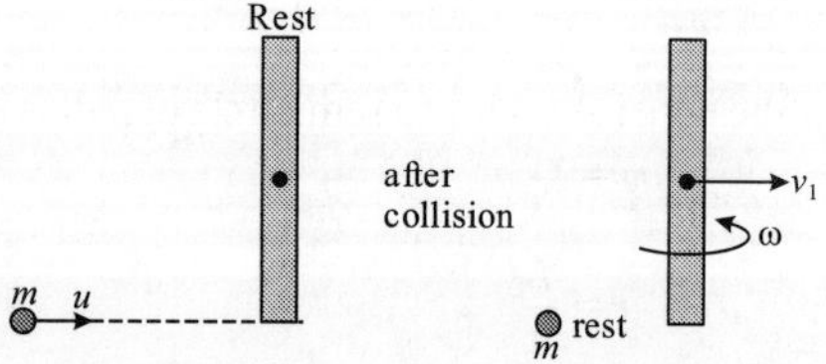

Conservation of angular momentum about centre of mass of rod

$$mu\left(\frac{L}{2}\right) = \frac{ML^2}{12}(\omega) \quad \ldots(1)$$

By conservation of linear momentum, we have

$$mu = Mv_1 \quad \ldots(2)$$

For elastic collision, we use coefficient of restitution as

$$1 = \frac{v_1 + \omega\frac{L}{2}}{u} \quad \ldots(3)$$

From equations-(1), (2) and (3), we have

$$u = \frac{m}{M}u + \frac{6mu}{2M}$$

$$1 = \frac{4m}{M}$$

$$\Rightarrow \quad \frac{m}{M} = \frac{1}{4}$$ **Ans. (4)**

Sol. 90 When the disc slides along the inclined plane, its acceleration is given as

$$a_1 = g\sin\theta$$

Thus displacement of disc can be written as

$$L = ut_1 + \frac{1}{2}a_1t_1^2 = \frac{1}{2}g\sin\theta\cdot t_1^2 \quad \ldots(1)$$

When disc will be in pure rolling and coming down the inclined plane, its acceleration can be given as

$$a_2 = \frac{g\sin\theta}{1 + k^2/R^2} = \frac{g\sin\theta}{1 + 1/2} = \frac{2}{3}g\sin\theta$$

Now displacement of disc can be written as

$$L = ut_2 + \frac{1}{2}a_2t_2^2 = \frac{1}{2}\cdot\frac{2}{3}g\sin\theta\cdot t_2^2 \quad \ldots(2)$$

From equations-(1) & (2), we have

$$\frac{t_2}{t_1} = \sqrt{\frac{3}{2}}$$ **Ans. (2)**

Sol. 91 In the situation described in question, we have

$$\frac{1}{2}I\omega^2 = \frac{1}{2}\times\frac{1}{2}mv^2$$

For pure rolling without slipping, we use

$$v = R\omega$$

$$\Rightarrow \quad \frac{1}{2}I\omega^2 = \frac{1}{4}mR^2\omega^2$$

$$\Rightarrow \quad I = \frac{1}{2}mR^2$$

So the body is a solid cylinder. **Ans. (C)**

Sol. 92 Considering bodies of same mass, given that

$$K.E_{ring} = K.E_{solid\ cylinder}$$

$$\frac{1}{2}M_R v_R^2 + \frac{1}{2}I_R\omega_R^2 = \frac{1}{2}M_C v_C^2 + \frac{1}{2}I_C\omega_C^2$$

Moment of inertia of ring and that of solid cylinder are given as

$I_R = M_R R^2$ and $I_C = 1/2\ M_C R^2$

Considering same mass, we have

$$2M_R V_R^2 = \frac{3}{2}M_C V_C^2$$

$$\Rightarrow \quad \frac{V_R}{V_C} = \frac{\sqrt{3}}{2}$$ **Ans. (3)**

Sol. 93 By conservation of energy, we have

$$mgh = \frac{1}{2}I_{cm}\omega^2 + \frac{1}{2}mv^2$$

$$\Rightarrow \quad v = \sqrt{\frac{2gh}{1+\frac{I_{cm}}{mR^2}}}$$

From above expression, it can be seen with increase in moment of inertia, speed decreases. **Ans. (A)**

Sol. 94 Situation described in question is shown in figure below. The speed of the particle on rim at the level of center is calculated by the vector sum of the two velocities in the particle is calculated below

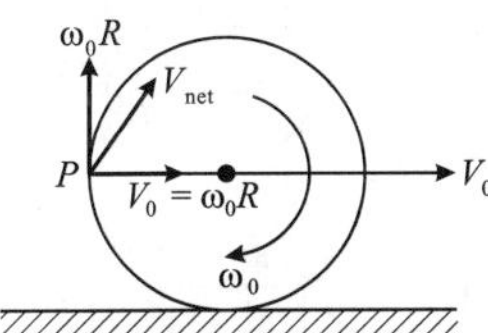

$$(V_P)_{net} = \sqrt{V_0^2 + (\omega_0 R)^2} = \sqrt{V_0^2 + V_0^2} = \sqrt{2}V_0$$ **Ans. (2)**

Sol. 95 Statement in Assertion ***A*** is false as moment of inertia about all given axes are different and are related by perpendicular axes theorem. Statement in Reason ***R*** is true as rotational motion for a rigid body is defined for a body which has fixed mass and shape and is called a rigid body. **Ans. (C)**

Sol. 96 Figure below shows the situation described in question.

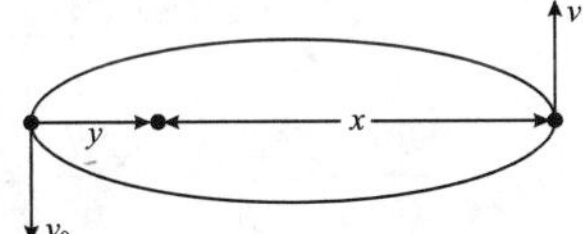

About the Sunc, by angular momentum conservation for motion of the planet, we use

$$mv_0 x_2 = mvx_1$$

$$\Rightarrow \quad v = \frac{v_0 x_2}{x_1}$$ **Ans. (B)**

Sol. 97 Ratio of impulse

$$\frac{J_1}{J_2} = \frac{\Delta P_1}{\Delta P_2} = \frac{m(u+u)}{m\left(\frac{u}{\sqrt{2}}+\frac{u}{\sqrt{2}}\right)} = \sqrt{2}:1$$

Ans. (A)

Sol. 98 From linear momentum conservation for the collision explained in the question, we have

$$2\times 4 + 0 = 2\times 1 + m_2 v_2$$

As collision is elastic ($e = 1$), we use

$$v_2 - v_1 = e(u_1 - u_2)$$

$$\Rightarrow \quad v_2 - 1 = 1(4-0)$$

$$\Rightarrow \quad v_2 = 5$$

$$\Rightarrow \quad 8 = 2 + m_2 + 5$$

$$\Rightarrow \quad m_2 = 6/5$$

The speed of centre of mass of the two bodies is calculated as

$$v_{cm} = \frac{m_1 v_1 + m_2 v_2}{m_1 + v_2} = \frac{2\times 4+0}{2+\frac{6}{5}} = 2.5\text{ m/s}$$

$$\Rightarrow \quad x = 25$$ **Ans. (25)**

Sol. 99 The velocity vector of the particle at an instant t is given as

$$\vec{v} = \frac{d\vec{r}}{dt} = 20\alpha t\,\hat{i} + 5\beta\,\hat{j}$$

The angular momentum of the particle about origin is given as

$$\vec{L} = m(\vec{r}\times\vec{v})$$

$$\Rightarrow \quad \vec{L} = m(10\alpha t^2\,\hat{i} + 5\beta(t-5)\,\hat{j})\times(20\alpha t\,\hat{i} + 5\beta\,\hat{j})$$

From above expression, at $t = 0$, $\vec{L} = 0$, thus at any time t, $L = 0$ occurs when

$$\vec{L} = m(50\alpha\beta t\,\hat{k} - 100\,\alpha\beta(t-5))\,\hat{k}$$

$$\Rightarrow \quad 0 = 50\text{ m}\,\alpha\beta\,[t - 2(t-5)]\,\hat{k}$$

$$\Rightarrow \quad t - 2t + 10 = 0$$

$$\Rightarrow \quad t = 10\text{ s}$$ **Ans. (10)**

Sol. 100 Angular speed at 600 rpm is calculated as

$$\omega = \frac{600\times 2\pi}{60} = 20\pi\text{ rad/s}$$

For constant acceleration, we use

$$\omega_f = \omega_i + \alpha t$$

$$\Rightarrow \quad 0 = 20\pi - \alpha(10)$$

$$\Rightarrow \quad \alpha = 2\pi \text{ rad/s}^2$$

Torque required at this angular accelration is given as

$$\tau = I \times \alpha$$

$$\Rightarrow \quad \tau = \frac{mR^2}{2} \times 2\pi$$

$$\Rightarrow \quad \tau = \frac{10 \times 0.04}{2} \times 2\pi = 4 \times 10^{-4}\pi$$

$$\Rightarrow \quad x = 4$$

Ans. (4)

Sol. 101 If surface mass density of discs is considered to be σ, then their masses can be written as

$$M_1 = \pi R^2\sigma,$$

$$M_2 = \pi r^2\sigma$$

Moment of inertia of the two discs is given as

$$I_1 = \frac{1}{2}M_1R^2 = \frac{\pi\sigma R^4}{2}$$

and

$$I_2 = \frac{1}{4}M_2r^2 = \frac{\pi\sigma r^4}{4}$$

$$\Rightarrow \quad \frac{I_1}{I_2} = \frac{2R^4}{r^4}$$

Ans. (A)

Sol. 102 Moment of inertia of the rod in respective options of List-I are given below -

(a) [rod of mass m, length L, axis through centre] $ML^2/12$

(b) [rod of mass $2m$, length L, axis through end] $\frac{(2M)L^2}{3} = \frac{2ML^2}{3}$

(c) [rod of mass m, length $2L$, axis through centre] $\frac{M(2L)^2}{12} = \frac{ML^2}{3}$

(d) [rod of mass $2m$, length $2L$, axis through end] $\frac{(2M)(2L)^2}{3} = \frac{8ML^2}{3}$

Ans. (D)

Sol. 103 As tangential velocities of the two wheels are same due to chain, we use

$$3R \times \omega_P = R \times \omega_Q$$

$$\Rightarrow \quad \frac{\omega_Q}{\omega_P} = 3$$

As kinetic energy of the two are same, we use

$$\frac{1}{2}I_P\omega_P^2 = \frac{1}{2}I_Q\omega_Q^2;$$

$$\Rightarrow \quad \frac{I_P}{I_Q} = \left(\frac{\omega_Q}{\omega_P}\right)^2 = 9$$

Ans. (9)

Sol. 104 Moment of inertia of racket about the given axis of rotation

$$I_A = \left[\frac{Mr^2}{2} + M\left(7r - \frac{r}{2}\right)^2\right] + \left[\frac{M(6r)^2}{12} + \left(M\left(3r - \frac{r}{2}\right)^2\right)\right]$$

$$\Rightarrow \quad I_A = \left\{\frac{Mr^2}{2} + M\frac{(169)r^2}{4}\right\} + \left\{M3r^2 + \frac{M25r^2}{4}\right\}$$

$$\Rightarrow \quad I_A = \frac{171Mr^2}{4} + \frac{37Mr^2}{4} = \frac{208}{4}Mr^2 = 52Mr^2$$

Ans. (52)

Sol. 105 Moment of inertia of the cylinder can be calculated by using parallel axes theorem as

$$I = \frac{Mr^2}{2} + Ml^2 = \frac{\rho(\pi r^2)lr^2}{2} + (\rho\pi r^2 l)\frac{l^2}{4}$$

$$\Rightarrow \quad I = \rho\pi r^2 l\left(\frac{r^2}{2} + \frac{l^2}{4}\right)$$

$$\Rightarrow \quad \rho = \frac{I}{\pi r^2 l\left(\frac{r^2}{2} + \frac{l^2}{4}\right)}$$

$$\Rightarrow \quad \rho = \frac{2.7}{\pi(0.2)^2(0.8)\left(\frac{(0.2)^2}{2} + \frac{(0.8)^2}{4}\right)}$$

$$\Rightarrow \quad \rho = 1.49 \times 10^2 \text{ kg/m}^3$$

Ans. (D)

Sol. 106 By parallel axes theorem, we have

$$I_Z = I_{cm} + M\left(\frac{l}{\sqrt{2}}\right)^2$$

$$I_Z = \frac{Ml^2}{6} + \frac{Ml^2}{2} = \frac{4Ml^2}{6} = \frac{2Ml^2}{3}$$

Ans. (A)

Sol. 107 Using angular momentum conservation before and after contact of the discs, we use

$$I_1\omega_1 - I_2\omega_2 = (I_1 + I_2)\omega$$

$$\Rightarrow \quad \omega = \frac{I_1\omega_1 - I_2\omega_2}{I_1 + I_2}$$

Loss of energy can be calculated as

$$\Delta E = \frac{1}{2}I_1\omega_1^2 + \frac{1}{2}I_2\omega_2^2 - \frac{1}{2}(I_1 + I_2)\omega^2$$

$$\Rightarrow \quad \Delta E = \frac{1}{2}I_1\omega_1^2 + \frac{1}{2}I_2\omega_2^2 - \frac{1}{2}(I_1 + I_2)\left(\frac{I_1\omega_1 - I_2\omega_2}{I_1 + I_2}\right)^2$$

$$\Rightarrow \quad \Delta E = \frac{1}{2}\frac{I_1I_2}{(I_1 + I_2)}(\omega_1 - \omega_2)^2$$

$$\Rightarrow \quad \Delta E = \frac{I_1I_2(\omega_1 - \omega_2)^2}{2(I_1 + I_2)}$$ **Ans. (C)**

Sol. 108 As it is uniform circular motion, angular momentum is given as

$$\vec{L} = \vec{r} \times \vec{p} = \vec{r} \times m\vec{v} = mvr$$

Thus it will remain same in magnitude and direction about its center. **Ans. (C)**

Sol. 109 Figure below shows the situation described in the question.

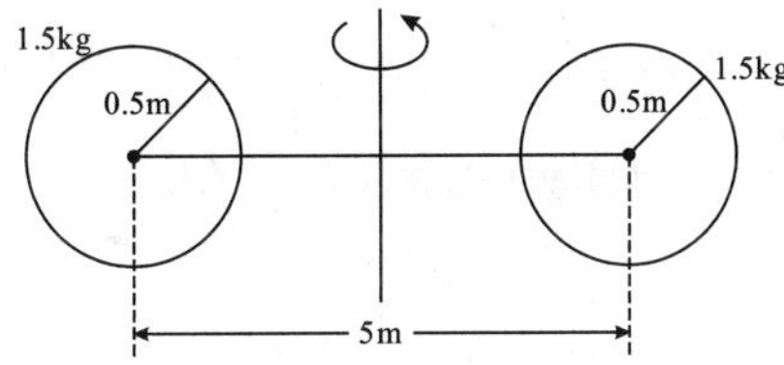

The moment of inertia of the above system can be calculated by parallel axes theorem as

$$I = \left(\frac{2}{5}MR^2 + Mx^2\right) \times 2$$

$$\Rightarrow \quad I = \left[\frac{2}{5} \times 1.5 \times (0.5)^2 + 1.5 \times (2.5)^2\right] \times 2$$

$$\Rightarrow \quad I = [0.15 + 9.375] \times 2 = 19.05 \text{ kgm}^2$$ **Ans. (B)**

Sol. 110 By work energy theorem, we have

$$mgl = \frac{1}{2}\frac{ml^2}{3}\omega^2$$

$$\Rightarrow \quad \omega = \sqrt{\frac{6g}{l}}$$

$$\Rightarrow \quad v = l\omega = \sqrt{3.6g} = 6 \text{ m/s}$$ **Ans. (6)**

7.1 Gravitational Field

Sol. 1 Weight of a person of mass m on earth surface is given as

$$W = mg_s$$

If earth is rotating at angular speed ω, value of g on earth's equator is

$$g_{eq} = g_s - \omega^2 R_e$$

For weight to be $\frac{3}{4}W$, we use

$$g_{eq} = \frac{3}{4}g_s = g_s - \omega^2 R_e$$

$$\Rightarrow \quad \omega^2 R_e = \frac{g_s}{4}$$

$$\Rightarrow \quad \omega = \sqrt{\frac{g_s}{4R_e}} = \sqrt{\frac{10}{4 \times 6400 \times 1000}}$$

$$\Rightarrow \quad \omega = \frac{1}{16 \times 10^2} = 0.625 \times 10^{-3} \text{ rad/s}$$ **Ans. (C)**

Sol. 2

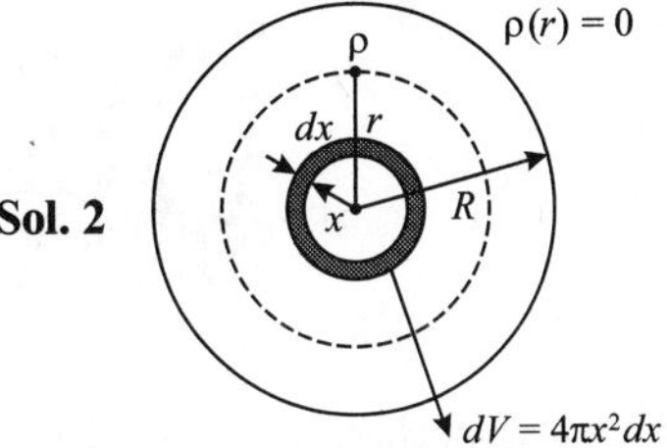

The mass of elemental shell of radius x and thickness dx is given as

$$dm = \rho(x) \times 4\rho x^2 dx$$

Total mass contained within the inner sphere of radius r is given as

$$m = \int dm = \int_0^r \frac{k}{x} \cdot 4\pi x^2 dx$$

$$\Rightarrow \quad m = 4\pi k \int_0^r x\, dx = 2\pi k r^2$$

acceleration of a test particle at point P is given as

$$a = -\frac{Gm}{r^2} = -\frac{G}{r^2} \cdot 2\pi k r^2 = -2\pi kG$$

–ve sign only denotes inward direction.

$|a| \to$ constant

independent of r for $r < R$ **Ans. (B)**

Sol. 3

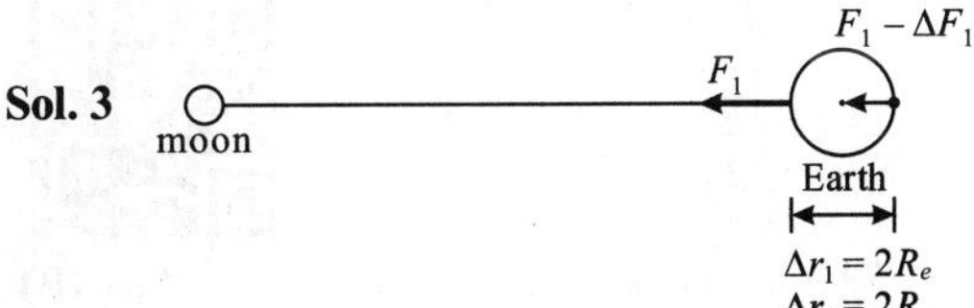

Similarly for Sun and earth this difference of force is given as Force between earth and moon is

$$F_1 = \frac{GM_eM_m}{r_1^2}$$

$$\Rightarrow \qquad \Delta F_1 = -\frac{2GM_eM_m}{r_1^3}\cdot\Delta r_1 \qquad \ldots(1)$$

and

$$\Delta F_2 = -\frac{2GM_eM_S}{r_2^3}\cdot\Delta r_2 \qquad \ldots(2)$$

From (a) (2) $\Rightarrow$ $\dfrac{\Delta F_1}{\Delta F_2} = \dfrac{M_m}{M_S}\times\left(\dfrac{r_2}{r_1}\right)^3$

$$\Rightarrow \qquad \frac{\Delta F_1}{\Delta F_2} = \left(\frac{8\times10^{22}}{2\times10^{30}}\right)\times\left(\frac{150\times10^6}{0.4\times10^6}\right)^3 = 2.1$$

Ans. (A)

Sol. 4

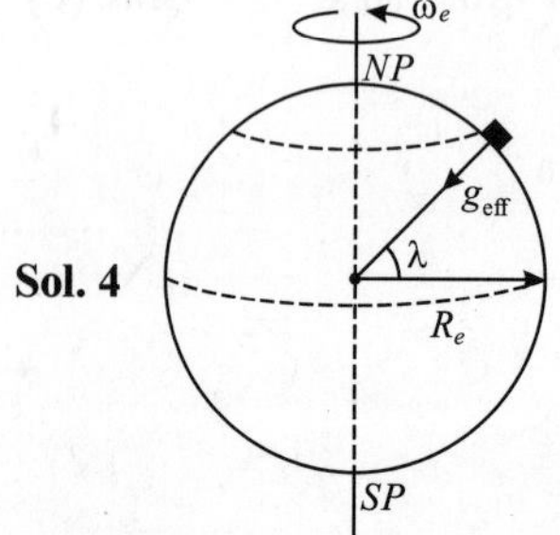

As effective gravity at a point on earth surface at latitude λ is given as

$$g_{eff} = g_s - \omega_e^2R_e\cos^2\lambda \qquad \ldots(1)$$

at poles $\qquad \lambda = \pm 90^\circ \Rightarrow g_{eff} = g_s$

at all points on earth other than poles

$$g_{eff} < g_s$$

If ω_e increases $\Rightarrow$ g_{eff} everywhere decreases except poles

weight $= mg_{eff}$

Ans. (D)

Sol. 5 Acceleration due to gravity is

$$g - \frac{GM}{R^2}$$

$$\frac{g_p}{g_e} = \frac{M_e}{M_e}\left(\frac{R_e}{R_p}\right)^2 = 3\left(\frac{1}{3}\right)^2 = \frac{1}{3}$$

For a simple pendulum, we have

$$T \propto \frac{1}{\sqrt{g}}$$

$$\Rightarrow \qquad \frac{T_p}{T_e} = \sqrt{\frac{g_e}{g_p}} = \sqrt{3}$$

$$\Rightarrow \qquad T_p = 2\sqrt{3}\text{s}$$

Ans. (B)

Sol. 6

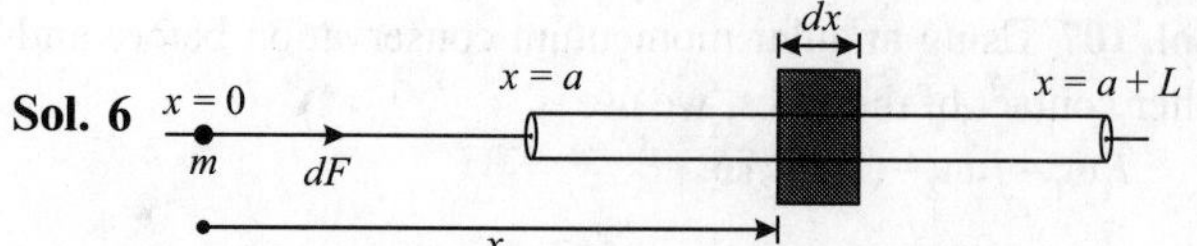

Mass of element of width dx considered is

$$dm = (A + Bx^2)dx$$

Force on this element due to m is given as

$$dF = \frac{Gm\,dm}{x^2} = \frac{Gm}{x^2}(A + Bx^2)dx$$

Net force on rod is

$$F = \int_a^{a+L}\frac{Gm}{x^2}(A + Bx^2)dx$$

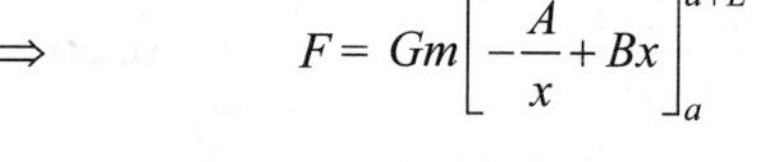

$$\Rightarrow \qquad F = Gm\left[-\frac{A}{x} + Bx\right]_a^{a+L}$$

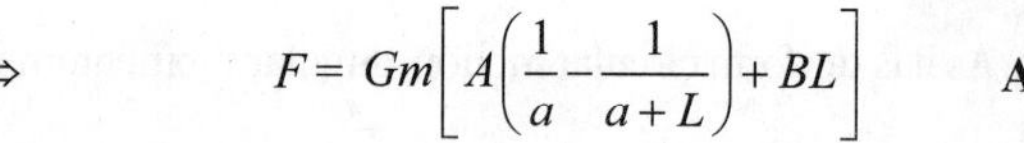

$$\Rightarrow \qquad F = Gm\left[A\left(\frac{1}{a} - \frac{1}{a+L}\right) + BL\right]$$

Ans. (B)

Sol. 7

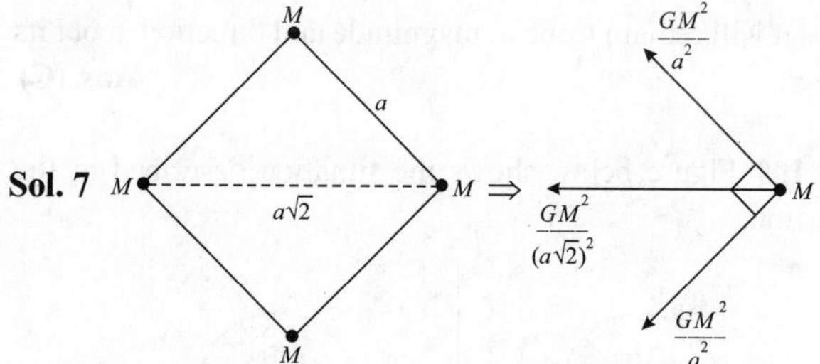

Net force on particle towards centre of circle is

$$F_C = \frac{GM^2}{2a^2} + \frac{GM^2}{a^2}\sqrt{2}$$

$$\Rightarrow \qquad F_C = \frac{GM^2}{a^2}\left(\frac{1}{2} + \sqrt{2}\right)$$

This force will act as centripetal force for circular motion in a circle of radius $\dfrac{a}{\sqrt{2}}$, given as

$$F_C = \frac{mv^2}{r}$$

$$\Rightarrow \qquad \frac{mv^2}{\frac{a}{\sqrt{2}}} = \frac{GM^2}{a^2}\left(\frac{1}{2} + \sqrt{2}\right)$$

$$\Rightarrow \qquad v^2 = \frac{GM}{a}\left(\frac{1}{2\sqrt{2}} + 1\right)$$

$$\Rightarrow \qquad v^2 = \frac{GM}{a}(1.35)$$

$$\Rightarrow \qquad v = 1.16\sqrt{\frac{GM}{a}}$$

Ans. (C)

Sol. 8

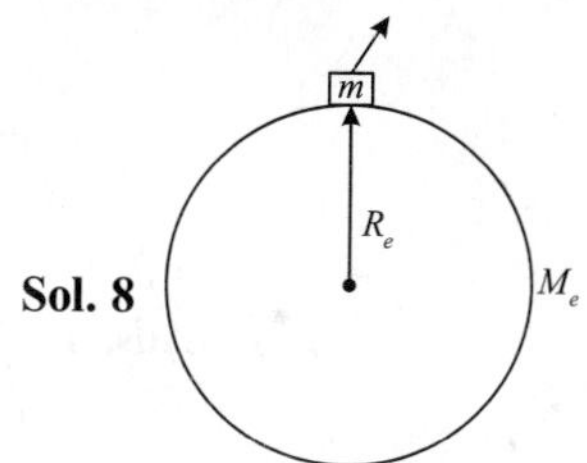

Energy required (minimum) for rocket to escape is

$$E = \frac{GM_e m}{R_e}$$

For moon, we have

$$R_m = \frac{R_e}{4} \quad \Rightarrow \quad V_m = \frac{V_e}{64}$$

$$\Rightarrow \quad M_m = \frac{M_e}{64}$$

Minimum energy required for rocket to escape from moon is

$$E_{moon} = \frac{GM_m m}{R_m} \qquad (\because M_m = \frac{M_e}{64}, R_m = \frac{R_e}{4})$$

$$E_{moon} = \frac{E}{16}$$ **Ans. (B)**

Sol. 9

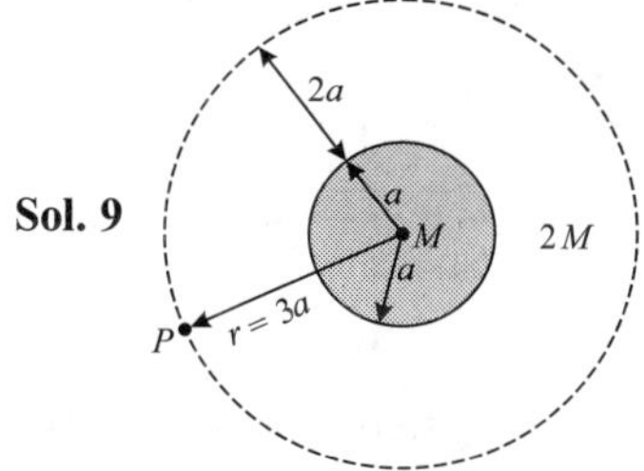

At point P gravitational field is due to the whole spherically symmetric mass $3M$ which will act as concentrated at centre for this point, so

$$g_P = \frac{G(3M)}{r^2} = \frac{G(3M)}{(3a)^2} = \frac{GM}{3a^2}$$ **Ans. (B)**

Sol. 10 Acceleration due to gravity at height h above earth surface is given as

$$g_h = \frac{GM_e}{(R_e+h)^2}$$

Given that $\frac{GM_e}{(R_e+h)^2} = \frac{1}{2}\frac{GM_e}{R_e^2}$

$$\Rightarrow \quad \sqrt{2}R_e = R_e + h$$

$$\Rightarrow \quad h = (\sqrt{2}-1)R_e$$

$$= 0.414 \times 6.4 \times 10^6 \text{ m}$$

$$= 2.6 \times 10^6 \text{ m}$$ **Ans. (D)**

Sol. 11 Weight at equator is given as

$$W_{eq} = W_{pole} - m\omega^2 R_e$$

$$\Rightarrow \quad W_{eq} = 196 - \frac{196}{10} \times \left(\frac{2\pi}{86400}\right)^2 \times 6.4 \times 10^5$$

$$= 195.34 \text{ N}$$ **Ans. (D)**

Sol. 12 Gravitational field on the surface of a solid sphere is given as

$$g_s = \frac{GM}{R^2}$$

From the graphs, we have

$$\frac{GM_1}{(1)^2} = 2 \text{ and } \frac{GM_2}{(2)^2} = 3$$

$$\Rightarrow \quad \frac{M_1}{M_2} = \frac{1}{6}$$ **Ans. (D)**

Sol. 13 Value of g at a height h above earth surface is

$$g_{out} = \frac{GM}{(R+h)^2}$$

Value of g at a depth h below the earth surface is

$$g_{in} = \frac{GM(R-h)}{R^3}$$

Given that

$$g_{out} = g_{in}$$

$$\Rightarrow \quad \frac{GM}{(R+h)^2} = \frac{GM(R-h)}{R^3}$$

$$\Rightarrow \quad (R-h)(R+h)^2 = R^3$$

Simplifying above equation gives

$$h^2 + Rh - R^2 = 0$$

$$\Rightarrow \quad h = \frac{-R + \sqrt{R^2 + 4R^2}}{2}$$

$$\Rightarrow \quad h = \frac{(\sqrt{5}-1)R}{2}$$ **Ans. (A)**

Sol. 14 Mass contained in a region of radius r is

$$m = \int_0^r \rho(r) \cdot 4\pi r^2 dr$$

$$m = \int_0^r \rho_0\left(1 - \frac{r^2}{R^2}\right) 4\pi r^2 dr$$

$$\Rightarrow \quad m = 4\pi\rho_0\left[\frac{r^3}{3} - \frac{r^5}{5R^2}\right]$$

Gravitational field due to this expression of EF, we use

$$g=\frac{Gm}{r^2}$$

$$\Rightarrow \quad g=\frac{G}{r^2}\left(4\pi\rho_0\left(\frac{r^3}{3}-\frac{r^5}{5R^2}\right)\right)$$

$$\Rightarrow \quad g=4\pi G\rho_0\left(\frac{r}{3}-\frac{r^3}{5R^2}\right)$$

g is maximum when

$$\frac{dg}{dr}=0$$

$$\Rightarrow \quad \frac{dg}{dr}=4\pi G\rho_0\left(\frac{1}{3}-\frac{3r^2}{5R^2}\right)=0$$

$$\Rightarrow \quad r=\sqrt{\frac{5}{9}}R$$

Ans. (B)

Sol. 15 Potential on X-axis is given as

$$V=-\int_{\infty}^{x}E_g\cdot dx=-\int_{\infty}^{x}\frac{Ax}{(x^2+a^2)^{3/2}}dx$$

$$\Rightarrow \quad V=\left[\frac{A}{(x^2+a^2)^{1/2}}\right]_{\infty}^{x}=\frac{A}{\sqrt{x^2+a^2}}$$

Ans. (A)

Sol. 16 Gravity at a height $h=\frac{R}{2}$ above earth surface is given as

$$g_{h\ \text{up}}=\frac{GM}{\left(R+\frac{R}{2}\right)^2}$$

gravity at a depth d below earth surface is given as

$$g_{d\ \text{in}}=\frac{GM(R-d)}{R^3}$$

Given that

$$g_{h\ \text{up}}=g_{d\ \text{in}}$$

$$\Rightarrow \quad \frac{1}{\left(\frac{3R}{2}\right)^2}=\frac{R-d}{R^3}$$

$$\Rightarrow \quad \frac{4R}{9}=R-d$$

$$\Rightarrow \quad \frac{d}{R}=\frac{5}{9}$$

Ans. (D)

Sol. 17 Gravity on equator of earth is given as

$$g_e=g_s-\omega^2R$$

Gravity at a height h above earth surface is

$$g_h=g_s\left(1-\frac{2h}{R}\right) \quad \text{[For } h<<R\text{]}$$

Given that $\quad g_e=g_h$

$$\Rightarrow \quad g_s-\omega^2R=g_s-\frac{2g_sh}{R}$$

$$\Rightarrow \quad h=\frac{\omega^2R^2}{2g}$$

Ans. (D)

Sol. 18 Weight of body on pole is given to be

$$mg=49\text{ N}$$

At equator due to rotation, the effective value of g is given as

$$g_e=g-R\omega^2$$

Thus weight at equator is given as

$$W_e=mg_e=m(g-R\omega^2)$$

$$W_P>W_e$$

As $W_P=49$ N so the only possible option is $W_e=48.83$ N

Ans. (D)

Sol. 19 If mass of complete sphere is m', the mass of removed part will be $m'/8$. The force on A due to complete sphere is calculated as

$$F_1=mg_O=\frac{Gmm'}{9R^2}$$

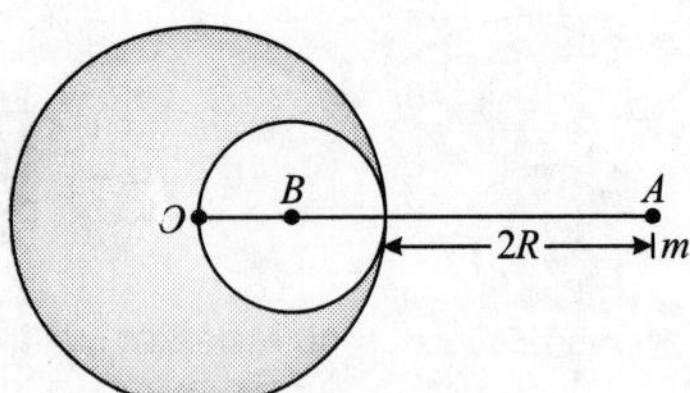

The net force on A due to the remaining sphere after removal of the small sphere from it is given as

$$F_2=m\left[\frac{G.m'}{(3R)^2}-\frac{G.m'/8}{(5R/2)^2}\right]$$

$$\Rightarrow \quad F_2=\frac{Gm'}{9R^2}-\frac{Gm'\times 4}{8\times 25}=\left(\frac{1}{9}-\frac{1}{50}\right)\frac{Gm'}{R^2}$$

$$\Rightarrow \quad F_2=\frac{41}{50\times 9}\cdot\frac{Gm'}{R^2}$$

$$\Rightarrow \quad \frac{F_1}{F_2}=\frac{1}{9}\times\frac{50\times 9}{41}=\frac{50}{41}$$

Ans. (B)

Sol. 20 Gravitational field of ring at the center of sphere is given as

$$g_Y=-\frac{Gmx}{(R^2+x^2)^{3/2}}$$

Force between sphere & ring can be given as

$$F=\frac{GmM(\sqrt{8}R)}{(R^2+8R^2)^{3/2}}=\frac{GmM}{R^2}\times\frac{\sqrt{8}}{27}$$

Ans. (C)

Sol. 21 Gravitational field at point A is given as

$$g_A = \frac{GM(r)}{R^3}$$

Gravitational field at point C is given as

$$g_C = \frac{GM}{\left(R+\frac{R}{2}\right)^2}$$

It is given that field at A is equal to that at point C, so we use

$$g_A = g_C$$

$$\Rightarrow \quad \frac{r}{R^3} = \frac{1}{\frac{9}{4}R^2}$$

$$\Rightarrow \quad OA = r = \frac{4R}{9}$$

The distance AB is given as

$$AB = R - r = \frac{5R}{9}$$

$$\Rightarrow \quad OA : AB = 4 : 5$$

Ans. (4)

Sol. 22 As already derived in Physics Galaxy book volume III, Self energy of a solid sphere of mass M and radius R which is the amount of energy required to break it up into its constituents and separate upto infinity is given as

$$U = \frac{3}{5}\frac{GM^2}{R}$$

$$\Rightarrow \quad x = 3$$

Ans. (3)

Sol. 23 At any point between Earth and Mars, gravitational field can be given as

$\vec{g}$ (at any point) = $\vec{g}_{\text{Earth}} + \vec{g}_{\text{mars}}$.

As we move away from earth, at some point between value of g decrease to zero where $\vec{g}_E + \vec{g}_M = 0$ then again increases upto the surface of Mars. **Ans. (A)**

Sol. 24 In the given system of binary star both the planets will move by the mutual gravitational interaction, so they will have same angular velocity and hence same time period. **Ans. (A)**

Sol. 25 By conservation of energy, the speed of body at a height y is given by

$$\frac{1}{2}mv^2 - \frac{GMm}{R+y} = 0$$

$$\Rightarrow \quad v = \sqrt{\frac{2GM}{R+y}}$$

$$\Rightarrow \quad \frac{dy}{dt} = \sqrt{\frac{2Gm}{R+y}}$$

If t is the time take for body to heach a height h above the surface of earth, we have

$$\Rightarrow \quad \int_0^h \sqrt{R+y}\,dy = \sqrt{2Gm}\int_0^t dt$$

$$\Rightarrow \quad \left[\frac{2(R+y)^{3/2}}{3}\right]_0^h = \sqrt{2Gm}\,t$$

$$\Rightarrow \quad \frac{2}{3}[(R+h)^{3/2} - R^{3/2}] = \sqrt{2Gm}\,t$$

$$\Rightarrow \quad t = \frac{\frac{2}{3}[(R+h)^{3/2} - R^{3/2}]}{\sqrt{2Gm}}$$

$$\Rightarrow \quad \frac{1}{3}\sqrt{\frac{2R_e}{g}}\left[\left(1+\frac{h}{R_e}\right)^{3/2} - 1\right]$$

Ans. (D)

Sol. 26 Given that mass of planet is twice that of Earth so we use

$$2M_E = M_P$$

If ρ is the density of planet and the Earth, we have

$$2\rho \times \frac{4}{3}R_E^3 = \rho \times \frac{4}{3}\pi R_P^3$$

$$\Rightarrow \quad R_P = 2^{1/3}R_E$$

Acceleration due to gravity on surface of planet is given as

$$g_P = \frac{GM_P}{R_P^2}$$

$$\Rightarrow \quad g_P = \frac{G2M_E}{(2^{1/3}R_E)^2} = \frac{G2M_E}{2^{2/3}R_E^2}$$

$$\Rightarrow \quad g_P = 2^{1/3}g_e$$

Object weight on planet = $2^{1/3}$ of the object weight on earth

$$\Rightarrow \quad W_P = 2^{1/3}\,\text{W}$$

Ans. (D)

Sol. 27 Let at a distance x from planet A, the net gravitational field becomes zero then x is given as

$$\frac{GM}{x^2} = \frac{G \times 9M}{(8R-x)^2}$$

$$(3x)^2 = (8R-x)^2$$

$$x = 2R$$

Thus, particle should be projected so that it cross a minimum distance of $2R$ as after this point other planet gravity will pull it, so we have

$$\frac{1}{2}mv^2 - \frac{GMm}{R} - \frac{G(9M)m}{7R} = -\frac{GMm}{2R} - \frac{G(9M)m}{6R}$$

$$\Rightarrow \quad \frac{1}{2}v^2 = \frac{2}{7}\frac{GM}{R}$$

$$\Rightarrow \quad v = \sqrt{\frac{4}{7}\frac{GM}{R}}$$

Ans. (4)

Sol. 28 Resulting gravitational attraction provides the necessary centripetal force for uniform circular motion, so we use

$$\frac{Gm^2}{4R^2} = m\omega^2 R$$

$$\Rightarrow \quad \omega = \sqrt{\frac{Gm}{4R^3}}$$

$$\Rightarrow \quad \omega = \frac{1}{2}\sqrt{\frac{G}{R^3}}$$ **Ans. (C)**

Sol. 29 For revolution of moon in circular motion, we use

$$m\omega^2 R = \frac{GmM}{R^2}$$

$$\Rightarrow \quad \omega^2 = \frac{GM}{R^3}$$

$$\Rightarrow \quad T = \frac{2\pi}{\omega} = \frac{2\pi}{\sqrt{\frac{GM}{R^3}}}$$

$$\Rightarrow \quad \frac{GM}{R^3} = \frac{4\pi^2}{T^2}$$

$$\Rightarrow \quad M = \frac{4\pi^2 R^3}{GT^2} = \frac{6\times10^{11}\times(9\times10^6)^3}{(7\times3600+30\times60)^2}$$

$$\Rightarrow \quad M = 6\times10^{23}\text{ kg}$$ **Ans. (D)**

Sol. 30 Gravitational field inside a spherical shell is zero everywhere inside so option (a) and (c) are correct. Gravitational potential due to uniform spherical shell is same everywhere inside so option (d) is also correct. **Ans. (C)**

Sol. 31 Gravitational potential energy of the system is given as

$$U = \frac{Gm^2}{a} + \frac{G(M-m)^2}{a} + \frac{2Gm(M-m)}{a} + \frac{2Gm(M-m)}{\sqrt{2}a}$$

The energy becomes maximum when $\frac{dU}{dm} = 0$

$$\Rightarrow \quad \frac{dU}{dm} = \frac{G}{a}(2m - 2(M-m) + (M-2m)\,2 +$$

$$\sqrt{2}(M-2m)\,(4m - 4m - 2\sqrt{2}m) +$$

$$(-2M + 2M + \sqrt{2}M) = 0$$

$$\Rightarrow \sqrt{2}M = 2\sqrt{2}m$$

$$\Rightarrow \quad \frac{M}{m} = 2$$ **Ans. (2)**

Sol. 32 Inside the given spherical shell, the gravitational potential can be written as

$$V_P = -\frac{GM}{R} - \frac{Gm}{r}$$

$$\Rightarrow \quad V_P = -\frac{G(100)}{50} - \frac{G(50)}{25}$$

$$\Rightarrow \quad V_P = -4G$$ **Ans. (C)**

Sol. 33 Value of g at a height r above the Earth's surface and at the same depth below the surface is given in the ratio calculated below as

$$\frac{g_h}{g_d} = \frac{\dfrac{GM}{(R_E+r)^2}}{\dfrac{GM}{R_E^3}(R_E-r)} = \frac{R_E^3}{(R_E-r)^2(R_E-r)}$$

$$\Rightarrow \quad \frac{g_d}{g_h} = \frac{(R_E+r)^2(R_E-r)}{R_E^3}$$

$$\Rightarrow \quad \frac{g_d}{g_h} = \frac{R_E^3 - R_E r^2 + rR_E^2 - r^3}{R_E^3}$$

$$\Rightarrow \quad \frac{g_d}{g_h} = 1 - \frac{r^2}{R_E^2} + \frac{r}{R_E} - \left(\frac{r}{R_E}\right)^3$$ **Ans. (A)**

Sol. 34 Figure below shows the forces acting on each particle due to others. Net force on one particle is given as

$$F_{net} = F_1 + 2F_2\cos 45^\circ$$

$$\Rightarrow \quad F_{net} = \frac{GM^2}{(2R)^2} + 2\times\frac{GM^2}{(\sqrt{2}R)^2}\times\frac{1}{\sqrt{2}}$$

$$\Rightarrow \quad F_{net} = \frac{GM^2}{4R^2} + \frac{GM^2}{\sqrt{2}R^2} = \frac{(4+\sqrt{2})GM^2}{4\sqrt{2}R^2}$$

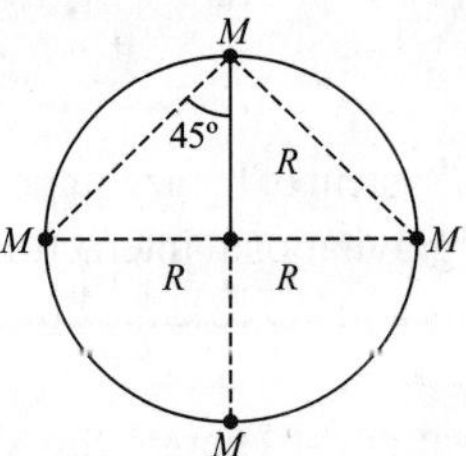

For ciruclar motion this force will provide the necessary centripetal force so we have

$$F_C = F_{net}$$

$$\frac{Mv^2}{R} = \frac{(4+\sqrt{2})GM^2}{4\sqrt{2}R^2}$$

$$\Rightarrow \quad v = \sqrt{\frac{(4+\sqrt{2})}{4\sqrt{2}}\frac{GM}{R}}$$ **Ans. (C)**

7.2 Satellite Motion

Sol. 1 Initial State

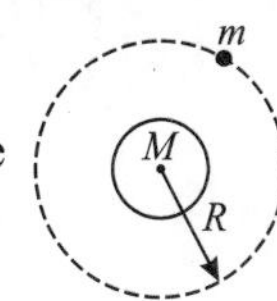

Total energy of body is

$$E_i = -\sqrt{\frac{GMm}{2R}}$$

Final State

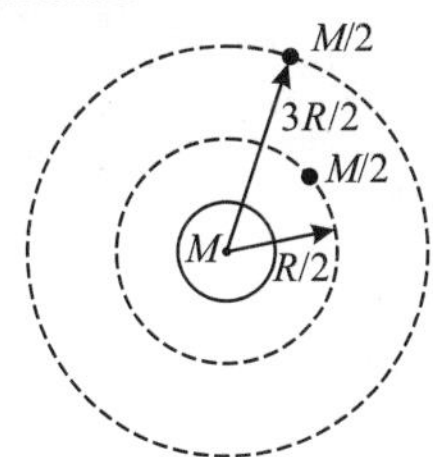

$$E_f = -\frac{GM\left(\frac{m}{2}\right)}{2\left(\frac{R}{2}\right)} - \frac{GM\left(\frac{m}{2}\right)}{2\left(3\frac{R}{2}\right)}$$

$$\Rightarrow \quad E_f = -\frac{2}{3}\frac{GMm}{R}$$

Difference in final and initial energies is

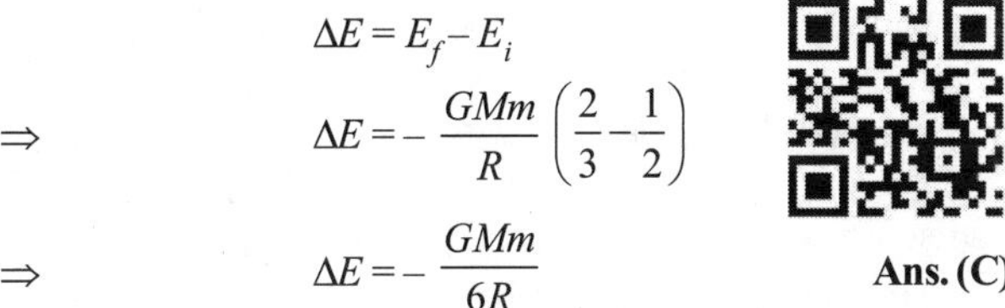

$$\Delta E = E_f - E_i$$

$$\Rightarrow \quad \Delta E = -\frac{GMm}{R}\left(\frac{2}{3} - \frac{1}{2}\right)$$

$$\Rightarrow \quad \Delta E = -\frac{GMm}{6R}$$

Ans. (C)

Sol. 2 By Kepler's II law areal velocity of a planet is constant even in elliptical path and given as

$$\frac{dA}{dt} = \frac{L}{2m}$$ **Ans. (C)**

Sol. 3 $U_{surface} + E_1 = U_h$

KE of satellite is zero at earth surface & at height h, we use

$$-\frac{GM_e m}{R_e} + E_1 = -\frac{GM_e m}{(R_e + h)}$$

$$E_1 = GM_e m\left(\frac{1}{R_e} - \frac{1}{R_e + h}\right)$$

$$E_1 = \frac{GM_e m}{(R_e + h)} \times \frac{h}{R_e}$$

Gravitational attraction at height h will be

$$F_G = mg = \frac{mv^2}{(R_e + h)}$$

$$\Rightarrow \quad \frac{GM_e m}{(R_e + h)^2} = \frac{mv^2}{(R_e + h)}$$

$$\Rightarrow \quad mv^2 = \frac{GM_e m}{(R_e + h)}$$

$$E_2 = \frac{1}{2}mv^2 = \frac{GM_e m}{2(R_e + h)}$$

Given that $\quad E_1 = E_2$

$$\Rightarrow \quad \frac{h}{R_e} = \frac{1}{2}$$

$$\Rightarrow \quad h = \frac{R_e}{2} = 3200\text{ km}$$

Ans. (C)

Sol. 4 At height r from center of earth orbital velocity is given as

$$v = \sqrt{\frac{GM}{r}}$$

By energy conservation, we have

$$KE \text{ of '}m\text{'} + \left(-\frac{GMm}{r}\right) = 0 + 0$$

$$\Rightarrow \quad KE \text{ of '}m\text{'} = \frac{GMm}{r} = \left(\sqrt{\frac{GM}{r}}\right)^2 m = mv^2$$

Ans. (B)

Sol. 5 If M is mass of star m is mass of meteorite then by energy conservation between O & ∞, we use

$$-\frac{GMm}{r} + \frac{-GMm}{r} + \frac{1}{2}mv^2 = 0 + 0$$

$$\Rightarrow \quad v = \sqrt{\frac{4GM}{r}} = 2.8 \times 10^5\text{ m/s}$$ **Ans. (D)**

Sol. 6 Orbital speed of satellite is

$$v_0 = \sqrt{g(R+h)} \approx \sqrt{gR}$$

Escape velocity is

$$v_e = \sqrt{2g(R+h)} \approx \sqrt{2gR}$$

Increment in speed required is

$$\Delta v = v_e - v_0 = (\sqrt{2} - 1)\sqrt{gR}$$

Ans. (A)

Sol. 7 As after impact the resulting velocity of the combined body is at some angle to the radial direction so resulting motion will be an elliptical orbit. **Ans. (C)**

Sol. 8 Orbital speed of satellite in orbit of radius P_c is given

$$v = \sqrt{\frac{GM}{r}}$$

Kinetic energies of the two satellites are

$$T_A = \frac{1}{2}m_A v_A^2$$

and $\quad T_B = \frac{1}{2} m_B v_B^2$

$$\Rightarrow \quad \frac{T_A}{T_B} = \frac{m \times \frac{GM}{R}}{2m \times \frac{GM}{2R}}$$

$$\Rightarrow \quad \frac{T_A}{T_B} = 1 \qquad \textbf{Ans (C)}$$

Sol. 9 Mass contained in the elemental shell is

$$dm = \rho \cdot 4\pi x^2 dx$$

$$M = \int dm = \int_0^R \frac{k}{x^2} \cdot 4\pi x^2\, dx = 4\pi kR$$

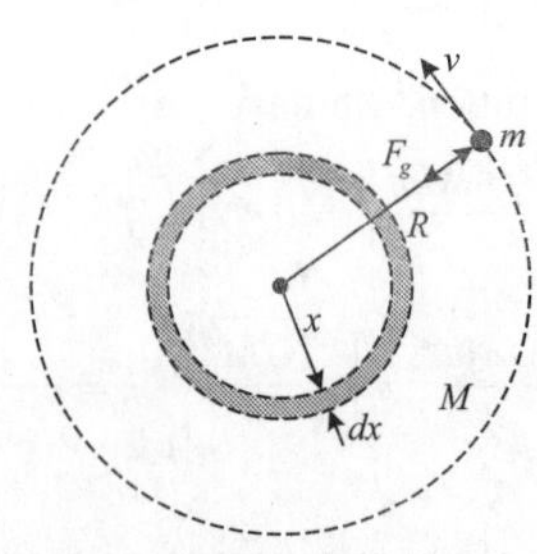

Gravitational force on mass in orbit is

$$F_g = \frac{GMm}{R^2} = \frac{G(4\pi kR)m}{R^2}$$

$$\Rightarrow \quad F_g = \frac{4\pi kGm}{R}$$

For circular motion, we use

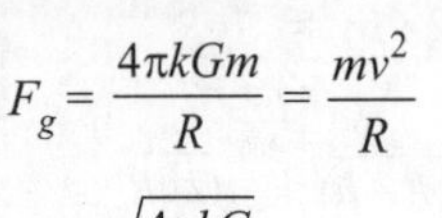

$$F_g = \frac{4\pi kGm}{R} = \frac{mv^2}{R}$$

$$\Rightarrow \quad v = \sqrt{4\pi kG}$$

Time period $\quad T = \frac{2\pi R}{\sqrt{4\pi kG}} \quad \Rightarrow \quad \frac{T}{R} =$ constant **Ans. (D)**

Sol. 10 Orbital speed of spaceship is

$$v = \sqrt{\frac{Gm}{r}}$$

Where $r = R_P + h = 2.02 \times 10^6$ m

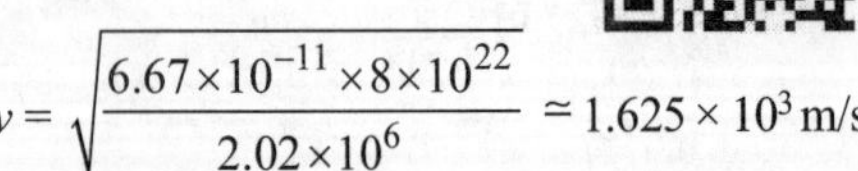

$$v = \sqrt{\frac{6.67 \times 10^{-11} \times 8 \times 10^{22}}{2.02 \times 10^6}} \simeq 1.625 \times 10^3 \text{ m/s}$$

Time period of revolutions of spaceship is

$$T = \frac{2\pi r}{v}$$

Number of revolutions in 24 hrs are

$$n = \frac{24 \times 3600}{T} = \frac{24 \times 3600 \times 1.625 \times 10^3}{2 \times 3.14 \times 2.02 \times 10^6}$$

$$\simeq 11 \qquad \textbf{Ans. (B)}$$

Sol. 11 Gravity on surface of a planet is given as

$$g_P = \frac{GM_P}{R_P^2}$$

$$\frac{g_e}{g_P} = \frac{9}{4} \frac{M_e / R_e^2}{M_P / R_P^2} \quad \left(\text{here } M_P = \frac{1}{9} M_e\right)$$

$$\Rightarrow \quad \frac{9}{4} = \frac{9R_P^2}{R_e^2} \Rightarrow R_P = \frac{R_e}{2} = \frac{R}{2} \qquad \textbf{Ans. (B)}$$

Sol. 12 Using energy conservation, we have

$$K_i + U_i = K_f + U_f$$

$$\frac{1}{2}mu^2 + \left(-\frac{GMm}{R}\right) = \frac{1}{2}mv^2 - \frac{GMm}{2R}$$

$$v = \sqrt{u^2 - \frac{GM}{R}} \qquad \ldots(1)$$

At height R rocket must carry all the momentum radial direction and along tangent to orbit it should carry the momentum such that satellite has left only with orbital speed $\sqrt{\frac{GM}{2R}}$ for circular motion. Thus by momentum conservation, we use

Along orbital direction (along tangent)

$$\frac{m}{10} v_{R0} = \frac{9m}{10} \sqrt{\frac{GM}{2R}}$$

$$v_{R_0} = 9\sqrt{\frac{GM}{2R}} \qquad \ldots(2)$$

Along radial direction

$$\frac{m}{10} v_{RR} = mv$$

$$\Rightarrow \quad \frac{m}{10} v_{RR} = m\sqrt{u^2 - \frac{GM}{R}} \qquad \ldots(3)$$

$$\Rightarrow \quad v_{RR} = 10\sqrt{u^2 - \frac{GM}{R}}$$

Now kinetic energy of rocket is given as

$$K = \frac{1}{2} m(v_{R0}^2 + v_{RR}^2)$$

$$\Rightarrow \quad K = \frac{m}{20}\left(81\frac{GM}{2R} + 100u^2 - 100\frac{GM}{R}\right)$$

$$\Rightarrow \quad K=\frac{m}{20}\left(100u^2-\frac{119GM}{2R}\right)$$

$$\Rightarrow \quad K=5m\left(u^2-\frac{119GM}{200R}\right)$$ **Ans. (B)**

Sol. 13 Using conservation of energy, we have

$$P_1+K_1=P_2+K_2$$

$$\Rightarrow \quad -\frac{GM_em}{10R}+\frac{1}{2}mv_0^2=-\frac{GM_em}{R}+\frac{1}{2}mv^2$$

$$\Rightarrow \quad +\frac{9}{10}\times\frac{GM_em}{R}+\frac{1}{2}mv_0^2=\frac{1}{2}mv^2$$

$$\Rightarrow \quad \frac{9}{10}\times\frac{1}{2}M\times v_e^2+\frac{1}{2}mv_0^2=\frac{1}{2}mv^2$$

$$\Rightarrow \quad v^2=\frac{9}{10}v_e^2+v_0^2$$

$$\Rightarrow \quad v^2=\frac{9}{10}\times(11.2)^2+(12)^2$$

$$\Rightarrow \quad v^2=112.896+144$$

$$\Rightarrow \quad v=16.027\text{ km/s}$$

$$\Rightarrow \quad v=16\text{ km/s}$$ **Ans. (16)**

Sol. 14 The orbital speed of body is given as

$$v_0=\sqrt{\frac{GM}{R}}$$

After collision, if combined mass moves with speed v_1, we use by conservation of momentum

$$mv_0+\frac{m}{2}\frac{v_0}{2}=\left(\frac{3m}{2}\right)v_1$$

$$\Rightarrow \quad v_1=\frac{5v_0}{6}$$

As this speed after collision is less than escape speed $\sqrt{2}v_0$, so the combined body cannot from escape gravitational field. So its motion will be elliptical around the planet. **Ans. (A)**

Sol. 15 Escape velocity from a planet is given as

$$v_e=\sqrt{\frac{2GM}{R}}$$

For planets A and B, we use

$$v_A=\sqrt{\frac{2GM}{R}}$$

and $$v_B=\sqrt{\frac{2G(M/2)}{R/2}}=\sqrt{\frac{2GM}{R}}$$

$$\Rightarrow \quad \frac{v_A}{v_B}=1=\frac{n}{4}\Rightarrow n=4$$ **Ans. (A)**

Sol. 16 Mass of galaxy inside radius R is given as

$$M=\int_0^R \rho_{(r)}\cdot 4\pi r^2dr$$

$$\Rightarrow \quad M=\int_0^R \frac{K}{r}\cdot 4\pi r^2dr$$

$$\Rightarrow \quad M=4\pi K\left[\frac{R^2}{2}\right]=2\pi KR^2$$

For orbital motion of star of mass m, we use

$$\frac{GMm}{R^2}=\frac{mv^2}{R}$$

$$\Rightarrow \quad v=\sqrt{\frac{GM}{R}}=\sqrt{2\pi KGR}$$

Time period of revolution is given as

$$T=\frac{2\pi R}{v}=\frac{2\pi R}{\sqrt{2\pi KGR}}$$

$$\Rightarrow \quad T=\sqrt{\frac{2\pi R}{KG}}$$

$$\Rightarrow \quad T\propto\sqrt{R}$$

$$\Rightarrow \quad T^2\propto R$$ **Ans. (C)**

Sol. 17 Figure below shows the trajectory of satellite after firing rockets

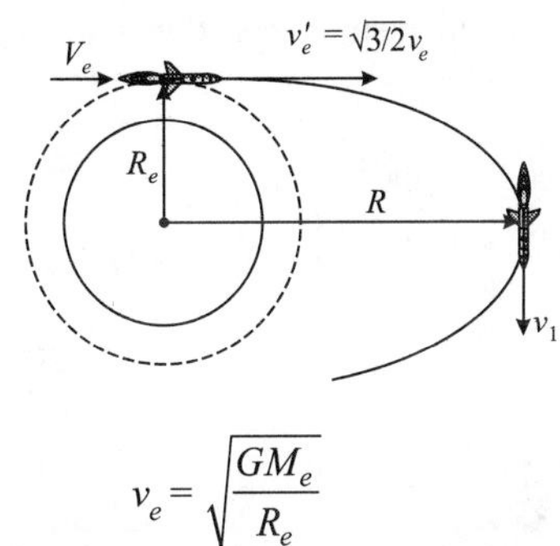

$$v_e=\sqrt{\frac{GM_e}{R_e}}$$

By conservation of angular momentum, we have

$$v_e'\cdot R_e=v_1R$$

$$\Rightarrow \quad \sqrt{\frac{3}{2}\frac{GM_e}{R_e}}\cdot R_e=v_1R$$

$$\Rightarrow \quad v_1=\sqrt{\frac{3}{2}\frac{GM_em}{R_e}}\cdot\frac{R_e}{R}$$

By conservation of energy we use

$$-\frac{GM_em}{R_e}+\frac{3}{4}\frac{GM_em}{R_e}=-\frac{GM_em}{R}+\frac{1}{2}mv_1^2$$

$$\Rightarrow \quad -\frac{GM_em}{4R_e}=-\frac{GM_em}{R}+\frac{1}{2}m\left(\frac{3}{2}\frac{GM_em}{R_e}\right)\cdot\frac{R_e^2}{R^2}$$

$$\Rightarrow \quad \frac{3}{4}\frac{R_e}{R^2}-\frac{1}{R}+\frac{1}{4R_e}=0$$

$$\Rightarrow \quad \frac{1}{R}=\frac{1\pm\sqrt{1-4\left(\frac{3}{4}\right)\left(\frac{1}{4}\right)}}{2\times\frac{3}{4}R_e}$$

$$\Rightarrow \quad \frac{1}{R}=\frac{1\pm\frac{1}{2}}{\frac{3}{2}R_e}$$

$$\Rightarrow \quad R=3R_e$$

Here R_e is not possible as $R > R_e$ **Ans. (B)**

Sol. 18 Orbital speed very close to planet surface is given as

$$v_0=\sqrt{\frac{GM}{R}}$$

escape velocity from the planet surface is given as

$$v_e=\sqrt{\frac{2GM}{R}}$$

$$\Rightarrow \quad \frac{v_0}{v_e}=\frac{1}{\sqrt{2}}$$ **Ans. (C)**

Sol. 19 By angular momentum conservation, we use

$$v_{max}r_{min}=v_{min}r_{max}$$

$$\Rightarrow \quad \frac{r_{min}}{r_{max}}=\frac{v_{min}}{v_{max}}=\frac{1}{6}$$ **Ans. (A)**

Sol. 20

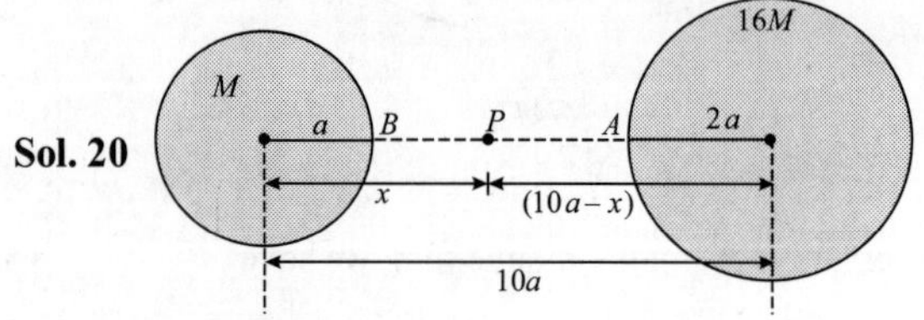

Null point P can be obtained as

$$\frac{GM}{x^2}=\frac{G(16M)}{(10a-x)^2}$$

$$\Rightarrow \quad \frac{1}{x}=\frac{4}{(10a-x)}$$

$$\Rightarrow \quad 4x=10a-x$$

$$\Rightarrow \quad x=2a \qquad \ldots(1)$$

For body fired from point A at speed v, it should cross null point P in order to reach point B so by energy conservation, we use

$$-\frac{GMm}{8a}-\frac{G(16M)m}{2a}+\frac{1}{2}mv^2$$

$$=-\frac{GMm}{2a}-\frac{G(16M)m}{8a}$$

$$\Rightarrow \quad v^2=2GM\left[\frac{1}{8a}+\frac{16}{2a}-\frac{1}{2a}-\frac{16}{8a}\right]$$

$$\Rightarrow \quad v^2=2GM\left[\frac{1+64-4-16}{8a}\right]$$

$$\Rightarrow \quad v^2=2GM\left[\frac{45}{8a}\right]$$

$$\Rightarrow \quad v=\sqrt{\frac{90GM}{8a}}$$

$$\Rightarrow \quad v=\frac{3}{2}\sqrt{\frac{5GM}{a}}$$ **Ans. (B)**

Sol. 21 The situation described in question is shown in figure below

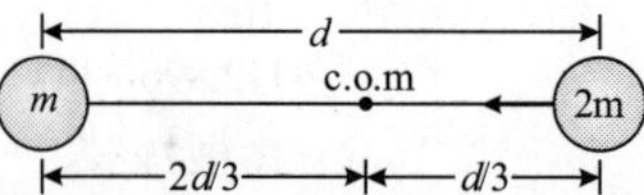

The force acting between the masses will provide the necessary centripetal force for circular motion which is given as

$$F=\frac{G(2m)m}{d^2}=(2m)\omega^2(d/3)$$

$$\Rightarrow \quad \frac{Gm}{d^2}=\omega^2\frac{d}{3}$$

$$\Rightarrow \quad \omega=\sqrt{\frac{3Gm}{d^3}}$$

Thus time period of revolution is given as

$$T=\frac{2\pi}{\omega}=2\pi\sqrt{\frac{d^3}{3Gm}}$$ **Ans. (B)**

Sol. 22 The ratio of time period of revolution is taken as

$$\frac{T_1}{T_2}=\frac{1}{8}$$

$$\Rightarrow \quad \frac{2\pi/\omega_1}{2\pi/\omega_2}=\frac{1}{8}$$

$$\Rightarrow \quad \frac{\omega_1}{\omega_2}=\frac{8}{1}$$ **Ans. (D)**

Sol. 23 Figure below shows the situation described in the question. The net gravitational force on one particle provides the necessary centripetal force for circular motion

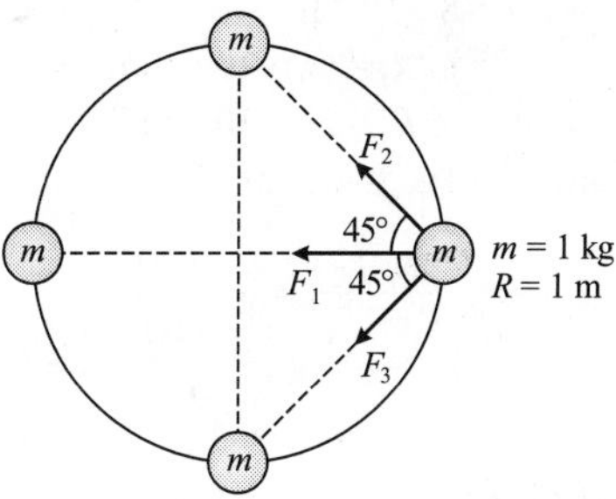

The forces shown in figure are calculated as

$$F_1 = \frac{Gmm}{(2R)^2} = \frac{Gm^2}{4R^2}$$

$$F_2 = \frac{Gmm}{(\sqrt{2}R)^2} = \frac{Gm^2}{2R^2}$$

$$F_3 = \frac{Gmm}{(\sqrt{2}R)^2} = \frac{Gm^2}{2R^2}$$

Resultant of these forces on mass can be given as

$$F_{net} = F_1 + F_2 \cos 45° + F_3 \cos 45°$$

$$\Rightarrow \quad F_{net} = \frac{Gm^2}{4R^2} + \frac{Gm^2}{2R^2}\frac{1}{\sqrt{2}} + \frac{Gm^2}{2R^2}\frac{1}{\sqrt{2}}$$

$$\Rightarrow \quad F_{net} = \frac{Gm^2}{R^2}\left(\frac{1}{4} + \frac{1}{2\sqrt{2}} + \frac{1}{2\sqrt{2}}\right)$$

$$\Rightarrow \quad F_{net} = \frac{Gm^2}{R^2}\left(\frac{1}{4} + \frac{1}{\sqrt{2}}\right) = \frac{Gm^2}{4R^2}(1+2\sqrt{2})$$

$$\Rightarrow \quad F_{net} = \frac{Gm^2}{4R^2}(1+2\sqrt{2}) = \frac{mv^2}{R}$$

$$\Rightarrow \quad v = \frac{\sqrt{G(1+2\sqrt{2})}}{2}$$ **Ans. (C)**

Sol. 24 Escape velocity on the surface of a planet is given as

$$v_e = \sqrt{\frac{2GM}{R}}$$

For the given two planets, as escape velocities are equal, we use

$$\frac{M_1}{R_1} = \frac{M_2}{R_2}$$

$$\Rightarrow \quad M_1R_2 = M_2R_1$$

Hence reason **R** is not correct. **Ans. (C)**

Sol. 25 The time period of revolution of satelllite is given as

$$T = 2\pi\sqrt{\frac{r^3}{GM}}$$

For planets A and B, we use

$$T_A = 2\pi\sqrt{\frac{(6400+600)\times 10^3}{GM}}$$

$$= 2\pi \times 10^9\sqrt{\frac{7^3}{GM}}$$

and $$T_B = 2\pi \times 10^9\sqrt{\frac{8^3}{GM}}$$

$$\Rightarrow \quad T_B - T_A = \frac{2\pi 10^9}{\sqrt{GM}}[8\sqrt{8} - 7\sqrt{7}]$$

$$= 314 \times 4.107 = 1.289 \times 10^3 \text{ s}$$ **Ans. (C)**

Sol. 26 By conservation of energy from surface of earth to the point of maximum height is given as

$$\frac{-GMm}{11R} = \frac{-GMm}{R} + \frac{1}{2}mv^2$$

$$\Rightarrow \quad v = \sqrt{\frac{20GM}{11R}}$$ **Ans. (10)**

Sol. 27 According to Keplar's second law, areal velocity of the position vector of planet remain constant. **Ans. (B)**

Sol. 28 Situation described in question is shown in figure below

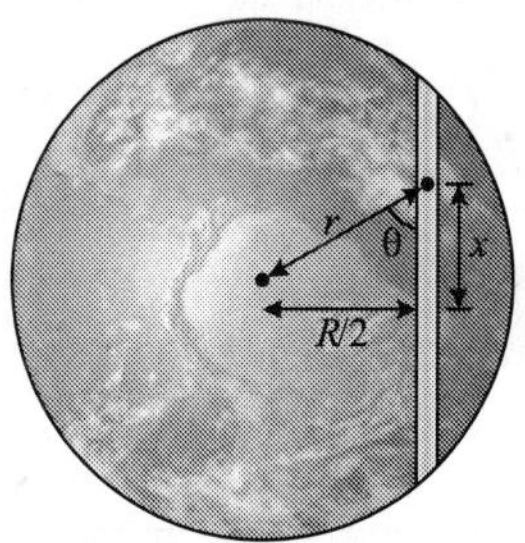

When particle is at a distance x from the center of tunnel, the force acting along the tunnel is given as

$$F = -\left(\frac{GMmr}{R^3}\right)\cos\theta = -\frac{gm}{R}x$$

$$\Rightarrow \quad a = -\frac{g}{R}x$$

Comparing above acceleration with the acceleration of SHM, it gives

$$\omega^2 = \frac{g}{R}$$

$$\Rightarrow \quad T = 2\pi\sqrt{\frac{R}{g}}$$ **Ans. (A)**

Sol. 29 Escape velocity is calculated as

$$v_e = \sqrt{\frac{2Gm}{R}} \qquad \ldots(1)$$

If escape velocity is to be increased to 10 times, we use

$$10v_e = \sqrt{\frac{2GM}{R'}} \quad \ldots(2)$$

From equations-(1) and (2), we have

$$10 = \sqrt{\frac{R}{R'}}$$

$$\Rightarrow \quad R' = \frac{R}{100} = \frac{6400}{100} = 64\text{ km}$$ **Ans. (64)**

Sol. 30 By Kepler's third law, we have

$$T \propto R^{3/2}$$

$$\Rightarrow \quad \frac{24}{T} = \left(\frac{12R}{3R}\right)^{3/2}$$

$$\Rightarrow \quad T = 3\text{ hr}$$ **Ans. (C)**

Sol. 31 From Keplar's third law we have

$$T^2 \propto R^3$$

$$\Rightarrow \quad \left(\frac{T'}{T}\right)^2 = \left(\frac{9R}{R}\right)^3$$

$$\Rightarrow \quad T'^2 = T^2 \times 9^3$$

$$\Rightarrow \quad T' = 27\,T$$ **Ans. (B)**

Sol. 32 Time period of a satellite in circular orbit around earth is related to its orbit radius as

$$T \propto R^{3/2}$$

$$\Rightarrow \quad \frac{\Delta T}{T} = \frac{3}{2} \times \frac{\Delta R}{R} = \frac{3}{2} \times 0.02 \times 100 = 3\%$$ **Ans. (C)**

Sol. 33 Situation described in question is shown in figure below

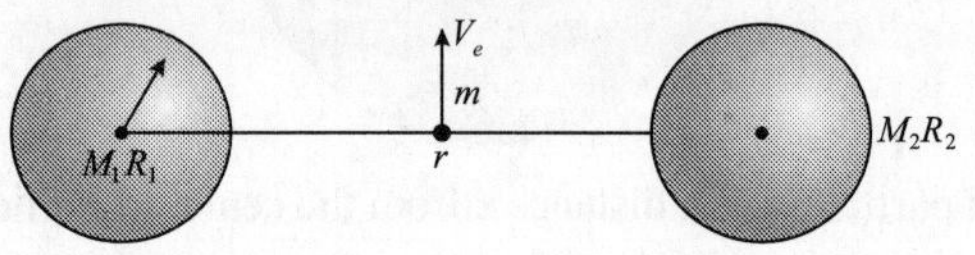

For mass m to escape final interaction energy of this mass to become zero, so we use

$$-\left(\frac{GM}{r/2} + \frac{GM_2}{r/2}\right)m + \frac{1}{2}mv_e^2 = 0$$

$$\Rightarrow \quad v_e = \sqrt{\frac{4G(M_1 + M_2)}{r}}$$ **Ans. (B)**

Sol. 34 By Keplar's third law, we use

$$T^2 \propto R^3$$

$$\Rightarrow \quad \left(\frac{R_2}{R_1}\right)^3 = \left(\frac{T_2}{T_1}\right)^2$$

$$\Rightarrow \quad \frac{R_2}{R_1} = \left(\frac{1}{8}\right)^{2/3} = 4$$

$$\Rightarrow \quad R = 8 \times 10^3\text{ km}$$

$$v_1 = \omega_1 R_1 = 4\pi \times 10^3\text{ km/h}$$

$$v_2 = \omega_2 R_2 = 2\pi \times 10^3\text{ km/h}$$

Relative angular speed is given as

$$\omega = \frac{V_1 - V_2}{R_2 - R_1} = \frac{2\pi \times 10^3}{6 \times 10^3} = \frac{\pi}{3}\text{ rad/hour}$$

$$\Rightarrow \quad x = 3$$ **Ans. (3)**

8.1 Elasticity

Sol. 1

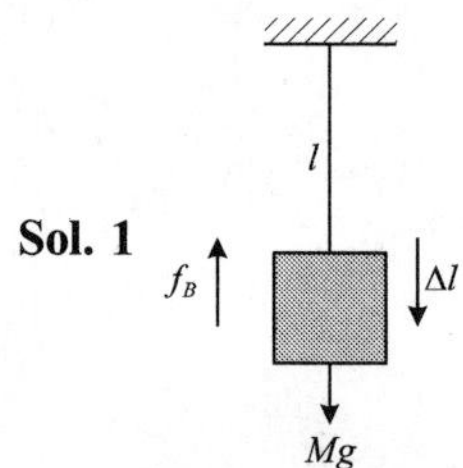

Young's modulus of wire is given as

$$Y = \frac{\text{Stress}}{\text{Strain}} = \frac{Mg/s}{\Delta l/l}$$

$$\Rightarrow \quad \Delta l \;\alpha\; MG \quad \ldots(1)$$

If load is immersed in liquid then the new extension in wire can be given as

$$\Delta l' \;\alpha\; Mg_{\text{eff}} = Mg\left(1 - \frac{\rho_l}{\rho_s}\right) = Mg\left(1 - \frac{2}{8}\right) = \frac{3}{4}Mg \quad \ldots(2)$$

$$(1)/(2) \Rightarrow \quad \frac{\Delta l}{\Delta l'} = \frac{4}{3}$$

$$\Rightarrow \quad \Delta l' = \frac{3}{4} \times \Delta l = \frac{3}{4} \times 4\text{ mm} = 3\text{ mm}$$ **Ans (B)**

Sol. 2 Due to stretch cord stores elastic PE and on releasing, this PE is transformed into KE of stone.

Energy density of cord

$$= \frac{1}{2} \times \text{stress} \times \text{strain}$$

$$= \frac{1}{2} \times Y \times (\text{strain})^2 = \frac{1}{2}Y\left(\frac{\Delta l}{l}\right)^2$$

PE in cord $\quad E_C = \frac{1}{2}Y\left(\frac{\Delta l}{l}\right)^2 \times A \times l$

By energy conservation

$$E_C = \frac{1}{2}mv^2$$

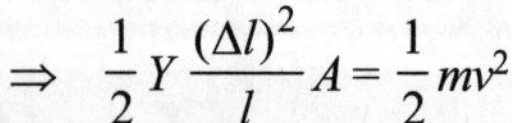

$$\Rightarrow \quad \frac{1}{2}Y\frac{(\Delta l)^2}{l}A = \frac{1}{2}mv^2$$

$$\Rightarrow \quad Y = \frac{mv^2 l}{(\Delta l^2)A} = \frac{0.02\times(20)^2\times 42\times 10^{-2}}{(20\times 10^{-2})^2\times\pi\times(3\times 10^{-3})^2}$$

$$\Rightarrow \quad Y \cong 3\times 10^6\ \text{N/m}^2$$

Ans. (C)

Sol. 3 Tensile stresss in wire will be

$$\text{Stress} = \frac{\text{Tensile force}}{\text{Cross section Area}}$$

$$= \frac{mg}{\pi R^2} = \frac{4\times 3.1\pi}{\pi\times 4\times 10^{-6}}\ \text{Nm}^{-2}$$

$$= 3.1\times 10^6\ \text{Nm}^{-2}$$

Ans. (D)

Sol. 4 Given that

$$\frac{Y_A}{Y_B} = \frac{7}{4};\ L_A = 2\,\text{m};\ L_B = 1.5\,\text{m};\ r_A = R;\ r_B = 2\,\text{mm}$$

Using $$Y = \frac{\text{Sress}}{\text{Strain}} = \frac{F/\pi r^2}{x/L}$$

$$\Rightarrow \quad \frac{F}{x} = \frac{Y\pi r^2}{L} \quad \text{Same for } A \text{ and } B$$

$$\Rightarrow \quad \frac{Y_A\pi r_A^2}{L_A} = \frac{Y_B\pi r_B^2}{L_B}$$

$$\Rightarrow \quad \frac{7}{4}\times\frac{R^2}{2} = \frac{(2\times 10^{-3})^2}{1.5}$$

$$\Rightarrow \quad R \simeq 1.74\times 10^{-3}\,\text{m} \simeq 1.7\,\text{mm}$$

Ans. (B)

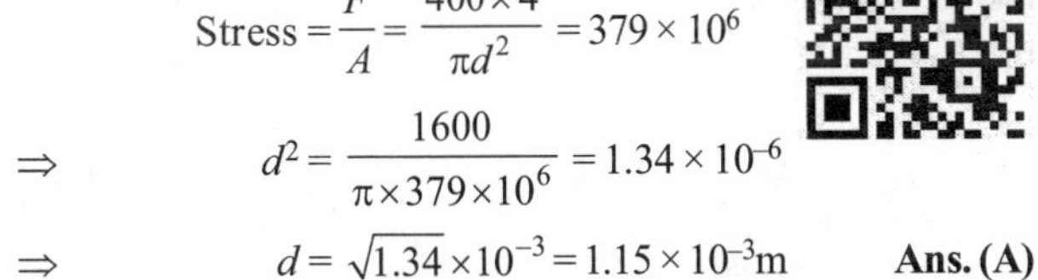

Sol. 5 Stress in rod is

$$\text{Stress} = \frac{F}{A} = \frac{400\times 4}{\pi d^2} = 379\times 10^6$$

$$\Rightarrow \quad d^2 = \frac{1600}{\pi\times 379\times 10^6} = 1.34\times 10^{-6}$$

$$\Rightarrow \quad d = \sqrt{1.34}\times 10^{-3} = 1.15\times 10^{-3}\text{m}$$

Ans. (A)

Sol. 6

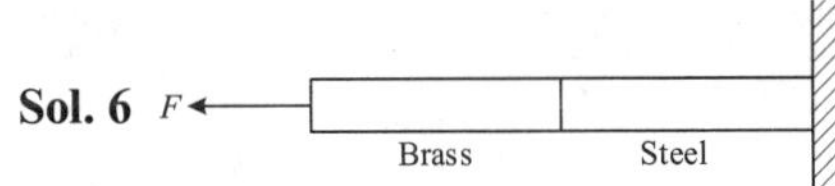

Equivalent force constaints for brass and steel wires are

$$k_1 = \frac{Y_1 A_1}{l_1} = \frac{120\times 10^9\times A}{1}$$

$$k_2 = \frac{Y_2 A_2}{l_2} = \frac{60\times 10^9\times A}{1}$$

For series combination, combined force constent is

$$k_{eq} = \frac{k_1 k_2}{k_1\times k_2} = \frac{120\times 60}{180}\times 10^9\times A$$

$$\Rightarrow \quad k_{eq} = 40\times 10^9\times \text{A}$$

For applied force F if net elongation is x, we have

$$F = k_{eq}(x)$$

$$\Rightarrow \quad F = (40\times 10^9)A\cdot(0.2\times 10^{-3})$$

$$\Rightarrow \quad \text{Stress} = \frac{F}{A} = 8\times 10^6\ \text{N/m}^2$$

None of the given option is correct **Ans. (None)**

Sol. 7 Thermal contraction is given as

$$\Delta l = l\alpha\Delta T$$

If Δl is elengation in wire due to hanging mass, we use Young's Modulus

$$Y = \frac{\text{Sress}}{\text{Strain}} = \frac{Mg/A}{\Delta l/l} = \frac{Mg}{A\alpha\Delta T}$$

$$\Rightarrow \quad M = \frac{YA\alpha\Delta T}{g}$$

$$= \frac{10^4\times\pi(10^{-3})^2\times 10^{-5}\times 20}{10} = 2\pi = 6.28\,\text{kg}$$

None of the given options is correct. **Ans. (None)**

Sol. 8 The expension in rod due to heating is

$$\Delta L = \alpha LT \qquad \text{(Coefficient of linear expension)}$$

Compressional stress in rod is $= \dfrac{F}{\pi r^2}$

The compression due to F, exactly balances the thermal expension so its length remain unchanged.

Using young's modulus, we have

$$Y = \frac{\text{Sress}}{\text{Strain}} = \frac{F/\pi r^2}{\Delta L/L} = \frac{F}{\pi r^2 dT} \Rightarrow \alpha = \frac{F}{\pi r^2 YT}$$

Coefficient of volume expension

$$\gamma = 3\alpha = \frac{3F}{\pi r^2 YT}$$

Ans. (B)

Sol. 9 Maximum tension in wire is given as

$$T = m\omega^2 l$$

Breaking stress is wire is given as

$$S = \frac{T}{A} = \frac{m\omega^2 l}{A}$$

$$\Rightarrow \quad \omega^2 = \frac{4.8\times 10^7\times(10^{-2}\times 10^{-4})}{10\times 0.3} = 16$$

$$\Rightarrow \quad \omega = 4$$

Ans. (4)

Sol. 10 Elastic energy stored per unit volume is given as

$$u = \frac{1}{2} \times \text{stress} \times \text{strain}$$

$$\Rightarrow \quad u = \frac{(\text{stress})^2}{2Y}$$

Given that

$$\frac{u_1}{u_2} = \frac{1}{4}$$

$$\Rightarrow \quad 4u_1 = u_2$$

$$4\left(\frac{1}{Y}\right)\left[\frac{mg}{\pi d_1^2/4}\right]^2 = \left(\frac{1}{2Y}\right)\left[\frac{mg}{\pi d_2^2/4}\right]^2$$

$$\Rightarrow \quad 4 = \left(\frac{d_1}{d_2}\right)^4$$

$$\Rightarrow \quad \frac{d_1}{d_2} = \sqrt{2} : 1$$ **Ans. (D)**

Sol. 11 Strain in wire is 4.9×10^{-4}

Stress in wire is given as

$$\text{Stress} = Y \times \text{Strain}$$
$$= 9 \times 10^{10} \times 4.9 \times 10^{-4}\,\text{N/m}^2$$

Lowest vibration frequency of clamped wire is

$$n_f = \frac{1}{2l}\sqrt{\frac{T}{\mu}} = \frac{1}{2l}\sqrt{\frac{T}{\rho A}} = \frac{1}{2l}\sqrt{\frac{\text{Stress}}{\rho}}$$

$$\Rightarrow \quad n_f = \frac{1}{2\times 1} \times \sqrt{\frac{9\times 10^{10} \times 4.9\times 10^{-4}}{9\times 10^{3}}}$$

$$\Rightarrow \quad n_f = \frac{1}{2} \times \sqrt{4.9\times 10^{3}}$$

$$\Rightarrow \quad n_f = \frac{7}{2} \times 10 = 35\,\text{Hz}$$ **Ans. (35)**

Sol. 12 Bulk modulus of material is given as

$$B = \frac{\Delta P}{\left|\frac{\Delta V}{V}\right|}$$

$$\Rightarrow \quad \left|\frac{\Delta V}{V}\right| = \frac{\Delta P}{B} = \frac{4\times 10^9}{8\times 10^{10}} = \frac{1}{20}$$

$$\Rightarrow \quad \frac{\Delta l}{l} = \frac{1}{3}\left|\frac{DV}{V}\right| = \frac{1}{3} \times \frac{1}{20} = \frac{1}{60}$$

Percentage change in length of cube is given as

$$\frac{\Delta l}{l} \times 100 = \frac{1}{60} \times 100 = \frac{5}{3} = 1.67\%$$ **Ans. (B)**

Sol. 13 An elastic wire can be considered as a spring of force constant k given as

$$k = \frac{YA}{l}$$

Thus time period of oscillations of object can be given as

$$T = 2\pi\sqrt{\frac{m}{k}}$$

So frequency of oscillations is given as

$$f = \frac{1}{2\pi}\sqrt{\frac{k}{m}} = \frac{1}{2\pi}\sqrt{\frac{YA}{ml}}$$ **Ans. (A)**

Sol. 14 Modulus of elasticity are related as

$$Y = 3K(1 - 2\sigma)$$

$$\Rightarrow \quad \sigma = \frac{1}{2}\left(1 - \frac{Y}{3K}\right) \qquad \ldots(2)$$

and

$$Y = 2\eta(1+\sigma)$$

$$\Rightarrow \quad \sigma = \frac{Y}{2\eta} - 1 \qquad \ldots(2)$$

From equation-(1) and (2), we have

$$\frac{1}{2}\left(1 - \frac{Y}{3k}\right) = \frac{Y}{2\eta} - 1$$

$$\Rightarrow \quad \frac{Y}{3K} = \frac{3\eta - Y}{\eta}$$

$$\Rightarrow \quad K = \frac{\eta Y}{9\eta - 3Y}$$ **Ans. (A)**

Sol. 15 The force on the wire can be related to the elongation in wire as

$$F = Y.A.\frac{\Delta l}{l}$$

$$\Rightarrow \quad \Delta l = \frac{F}{Y.A.} \cdot l = \frac{F.l}{Y.\pi r^2}$$

$$\Rightarrow \quad \Delta l \propto \frac{l}{r^2}$$

$$\Rightarrow \quad \frac{\Delta l_2}{\Delta l_1} = \left(\frac{l_2}{l_1}\right)\left(\frac{r_1}{r_2}\right)^2 = (2)\left(\frac{1}{2}\right)^2 = \frac{1}{2}$$

$$\Rightarrow \quad \Delta l_2 = \frac{\Delta l_1}{2} = \frac{0.04}{2} = 0.02\,\text{m}$$

$$\Rightarrow \quad \Delta l_2 = 2\,\text{cm}$$ **Ans. (2)**

Sol. 16 Density of the material can be written as

$$\rho = \frac{M}{V}$$

$$\Rightarrow \quad \frac{d\rho}{\rho} = -\frac{dV}{V} \quad \text{(As mass is constant)}$$

Bulk modulus of material is given as

$$K = -\frac{P}{\frac{dV}{V}}$$

$$\Rightarrow \quad -\frac{dV}{V} = \frac{P}{K}$$

$$\Rightarrow \quad \frac{d\rho}{\rho} = \frac{P}{k}$$

$$\Rightarrow \quad d\rho = \frac{\rho P}{K}$$

Ans. (A)

Sol. 17 For the elastic wire we can use tension in wire is proportional to its elongation, so we use

$$T = k(\Delta l)$$

If l_0 is the natural original length of wire then we use

$$T_1 = k(l_1 - l_0)$$

and $$T_2 = k(l_2 - l_0)$$

$$\Rightarrow \quad \frac{T_1}{T_2} = \frac{l_1 - l_0}{l_2 - l_0}$$

$$\Rightarrow \quad l_0 = \frac{T_2 l_1 - T_1 l_2}{T_2 - T_1}$$

Ans. (D)

Sol. 18 Young's Modulus of wire can be given as

$$Y = \frac{\text{Stress}}{\text{Strain}} = \frac{FL}{Al} = \frac{mg \cdot L}{\pi R^2 \cdot l}$$

Fractional error in calculation of Young's modulus is given as

$$\frac{\Delta Y}{Y} = \frac{\Delta m}{m} + \frac{\Delta L}{L} + 2 \cdot \frac{\Delta R}{R} + \frac{\Delta l}{l}$$

Percentage error is calculated as

$$\frac{\Delta Y}{Y} \times 100 = 100\left[\frac{1}{1000} + \frac{1}{1000} + 2\left(\frac{0.001}{0.2}\right) + \frac{0.001}{0.5}\right]$$

$$\Rightarrow \quad \frac{\Delta Y}{Y} \times 100 = \frac{1}{10} + \frac{1}{10} + 1 + \frac{1}{5} = \frac{14}{10} = 1.4\%$$

Ans. (D)

Sol. 19 Pressure at a depth h is given as

$$P = h\rho g$$

Bulk modulus of the medium which is the requierd ratio asked in question, is given as

$$\beta = \frac{p}{\frac{\Delta V}{V}} = \frac{2\times10^3 \times 10^3 \times 9.8}{1.36\times10^{-2}}$$

$$\Rightarrow \quad \beta = 1.44 \times 10^9 \text{ N/m}^2$$

Ans. (D)

Sol. 20 Fringe width in YDSE setup is given as

$$\beta = \frac{\lambda D}{d} = \frac{5890\times10^{-10}\times0.5}{0.5\times10^{-3}} = 589 \times 10^{-6}\text{ m}$$

Distance between first and third bright fringe can be calculated as

$$2\beta = 2\times589 \times 10^{-6}\text{ m} = 1178 \times 10^{-6}\text{ m}$$

Ans. (B)

Sol. 21 For wire A stress can be given as

$$\frac{F}{\pi r^2} = Y\left(\frac{2mm}{aL}\right) \qquad \ldots(1)$$

For wire B stress is given as

$$\frac{E}{\pi.16r^2} = Y\left(\frac{4mm}{bL}\right) \qquad \ldots(2)$$

From equations-(1) & (2), we get

$$16 = \frac{2b}{4a}$$

$$\Rightarrow \quad \frac{a}{b} = \frac{1}{32}$$

Ans (32)

Sol. 22 For adiabatic process, we use

$$PV^{\gamma} = \text{constant}$$

Differentiating the above equation, gives

$$\frac{dP}{dV} = -\frac{\gamma P}{V}$$

$$\Rightarrow \quad \frac{dP}{P} = -\frac{\gamma dV}{V}$$

Ans. (A)

Sol. 23 Let initial [illegible]ngth of wire be L_0 and cross sectional area A, then stress in wi[illegible] as

$$\frac{T}{[illegible]}$$

So for the two valu[illegible]

$$\frac{T_1}{A}$$

and $$\frac{T_2}{A}$$

$$\Rightarrow \quad \frac{T_1}{T_2} = \frac{L_1 - L_0}{L_2 - L_0} \qquad T_1L_0 = T_2L_0;\ L_0 = \frac{L_2T_1 - L_1T_2}{T_1 - T_2}$$

Ans. (C)

Sol. 24 Let initial l[illegible] l_0 and area A, the stress in wire at the two states [illegible]ven as

$$\frac{T_1}{A} = \frac{[illegible]\ l_0)}{l_0}$$

and $$\frac{T_2}{A} = \frac{Y(l_2 - l_0)}{l_0}$$

$$\Rightarrow \quad \frac{T_1}{T_2} = \frac{l_1 - l_0}{l_2 - l_0}$$

$$\Rightarrow \quad T_1 l_2 - T_1 l_0 = T_2 l_0 - T_2 l_0$$

$$\Rightarrow \quad l_0 = \frac{l_1 T_2 - l_2 T_1}{T_2 - T_1}$$ **Ans. (C)**

Sol. 25 Elastic energy stored in the track is calculated as

$$U = \frac{1}{2} \times \text{stress} \times \text{strain} \times \text{volume}$$

$$U = \frac{1}{2} \times \text{stress} \times \text{strain} \times \text{volume} \times Al$$

Energy stored per unit length in track is given as

$$\Rightarrow \quad \frac{U}{l} = \frac{1}{2} \times \text{stress} \times \text{strain} \times \text{volume} \times A$$

$$\Rightarrow \quad \frac{U}{l} = \frac{1}{2} \times Y \times (\text{strain})^2 \times A$$

$$\Rightarrow \quad \frac{U}{l} = \frac{1}{2} \times Y\left(\frac{\Delta l}{l}\right)^2 \times A = \frac{1}{2} Y \left(\frac{l\alpha\Delta t}{l}\right)^2 \times A = \frac{1}{2} Y\alpha^2 \Delta t^2 A$$

$$\Rightarrow \quad \frac{U}{l} = \frac{1}{2} \times 10^{11} \times 10^{-10} \times 10 \times 10 \times 10^{-2} = 5 \text{ J/m}$$ **Ans. (5)**

Sol. 26 As the two wires are joined end to end (series combination) so for series combination of two wires or springs, equivalent spring constant is given as

$$K_{eff} = \frac{K_1 K_2}{K_1 + K_2}$$

$$\Rightarrow \quad \frac{YA}{2l} = \frac{\frac{Y_1 A}{l} \times \frac{Y_2 A}{l}}{\frac{Y_1 A}{l} + \frac{Y_2 A}{l}}$$

$$\Rightarrow \quad Y = \frac{2Y_1 Y_2}{Y_1 + Y_2}$$ **Ans. (D)**

Sol. 27 Fringe width in YDSE is given as

$$\beta = \frac{\lambda D}{d}$$

As $\lambda_{\text{blue}} < \lambda_{\text{orange}} \Rightarrow \beta_{\text{blue}} < \beta_{\text{orange}}$ **Ans. (A)**

[illegible] When the stone is projected, it gains kinetic energy by ela[illegible]rgy of catapult, so we use

$$\frac{1}{2}kx^2 = \frac{1}{2}mv^2$$

$$\Rightarrow \quad [illegible]^2 = mv^2$$

$$\Rightarrow \quad \frac{0.5 \times 10^9 \times 10^{-6}}{0.1} \times 0.04 \times 0.04 = 20 \times 10^{-3} v^2$$

$$\Rightarrow \quad v = 20 \text{ m/s}$$ **Ans. (20)**

Sol. 29 Maximum tension supported by the two wires can be written as

$$F_{1\max} = \text{stress} \times \text{area} = 1.25 \times 8 \times 10^2 = 1000 \text{ N}$$

$$F_{2\max} = 1.25 \times 8 \times 10^2 = 500 \text{ N}$$

To calculate the maximum weight which can be placed in pan, it should increase the tension in wire W_2 to the maximum value so we use

$$F_{2\max} = (10 + M)g = 500 \text{ N}$$

$$\Rightarrow \quad M = 40 \text{ kg}$$ **Ans. (40)**

Sol. 30 Elongation in rod due to its self weight can be given as

$$\Delta L = \frac{Wl}{2YA}$$

$$\Rightarrow \quad \Delta L = \frac{10 \times 0.2}{2 \times 2 \times 10^{11} \times 100 \times 10^{-4}} = 5 \times 10^{-10} \text{ m}$$

Ans. (B)

Sol. 31 Volume strain in rubber ball is calculated as

$$\frac{-dV}{V} = \frac{0.5}{100} = 0.005 = 0.005$$

Bulk modulus of the ball can be written as

$$B = \frac{-dP}{\frac{dV}{V}} = \frac{\rho g h}{\frac{-dV}{V}}$$

$$\Rightarrow \quad 9.8 \times 10^8 = \frac{10^3 \times 9.8 \times h}{5 \times 10^{-3}}$$

$$\Rightarrow \quad h = 500 \text{ m}$$ **Ans. (500)**

Sol. 32 In the situation described in question, we can consider total area under stress is four times that of each columns so longitudinal strain can be given as

$$\frac{\Delta L}{L} = \frac{mg}{4AY}$$

$$\Rightarrow \quad \frac{\Delta L}{L} = \frac{50 \times 10^3 \times 9.8}{4\pi(100^2 - 50^2) \times 10^{-4} \times 2 \times 10^{11}}$$

$$\Rightarrow \quad \frac{\Delta L}{L} = \frac{5 \times 10^5}{4 \times \pi \times 150 \times 50 \times 10^{-4} \times 2 \times 10^{11}}$$

$$\Rightarrow \quad \frac{\Delta L}{L} = \frac{1}{12\pi} \times 10^{-5} = 2.60 \times 10^{-7}$$ **Ans. (D)**

Sol. 33 The distance between nth bright fridge on both side of central maxima is 2.4 cm. So this would be the fringe width on interference pattern. Distance between fourth bright fringe on the two sides of central bright fringe will be 8 time sthe fringe width, given as

$$\Delta y = \frac{D\lambda}{d} \times 8$$

The wavelength of light can be obtained as

$$\Rightarrow \quad \lambda = \frac{d\Delta y}{8D} = \frac{0.3\times10^{-3}\times2.4\times10^{-2}}{8\times1.5}$$

$$\Rightarrow \quad \lambda = \frac{0.12\times10^{-5}}{2} = 0.06\times10^{-5}$$

$$\Rightarrow \quad \lambda = 6\times10^{-7}\,\text{m}$$

Thus frequency of light is given as

$$f = \frac{c}{6\times10^{-7}} = 0.5\times10^{+15}$$

$$\Rightarrow \quad f = 5\times10^{14}\,\text{Hz}$$

Ans. (5)

Sol. 34 From the given expression of Young's modulus, we calculate the fractional error in measurement of Y as

$$Y = \frac{MgL^3}{4bd^3\delta}$$

$$\Rightarrow \quad \frac{dY}{Y} = \frac{dM}{M} + 3\frac{dL}{L} + 3\frac{dd}{d} + \frac{db}{b} + \frac{d\delta}{\delta}$$

$$\Rightarrow \quad \frac{dY}{Y} = 0.0005 + 3\times0.001 + 3\times0.0025 + 0.0025 + 0.002$$

$$\Rightarrow \quad \frac{dY}{Y} = 0.0155$$

Ans. (B)

9.1 Fluid Statics

Sol. 1 In initial state we use for floatation of block of volume v_B

$$V_B\rho_B g = \frac{4}{5}V_B\rho_w g$$

$$\Rightarrow \quad \rho_B = \frac{4}{5}\rho_w \qquad \ldots(1)$$

Later after pouring oil over the water, we use

$$V_B\rho_B g = \frac{V_B}{2}\rho_w g + \frac{V_B}{2}\rho_0 g$$

$$\Rightarrow \quad \frac{4}{5}\rho_w = \frac{1}{2}\rho_w + \frac{1}{2}\rho_0$$

$$\Rightarrow \quad \frac{3}{10}\rho_w = \frac{1}{2}\rho_0$$

$$\Rightarrow \quad \frac{P_0}{\rho_w} = \frac{6}{10} = 0.6$$

Ans. (C)

Sol. 2 At a depth d_1, we have

$$P_{atm} + d_1\rho g = P_1 = 5.05\times10^6 \qquad \ldots(1)$$

At a depth d_2, we have

$$\text{P}_{atm} + d_2\rho g = P_2 = 8.08\times10^6 \qquad \ldots(2)$$

$$(2) - (1) \Rightarrow (d_2 - d_1)\rho g = 3.03\times10^6$$

$$\Rightarrow \quad d_2 - d_1 = \frac{3.03\times10^6}{10^3\times10} = 3.03\times10^2 = 303\,\text{m}$$

Ans. (C)

Sol. 3

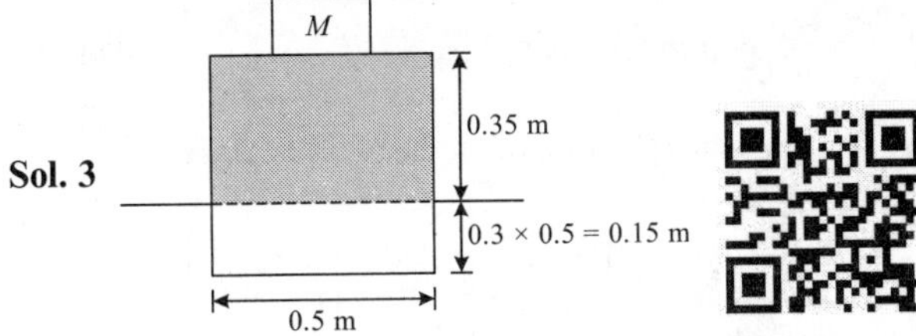

To fully submerge first submerge the complete block weight of mass M is to be balanced by buoyancy due to the 70% part of cubical block outside initially

$$\Rightarrow \quad Mg = (0.5\times0.5\times0.35)\times\rho_w\cdot g$$

$$\Rightarrow \quad M = (0.25\times0.35)\times10^3 = 87.5\,\text{kg}$$

Ans. (B)

Sol. 4 For minimum density of liquid, sphere will be floating while completely submerged so the buoyant force must be balancing the weight. Mass of sphere is given as

$$m = \int_0^R \rho(4\pi r^2 dr)$$

$$\Rightarrow \quad m = \rho_0\int_0^R\left(1 - \frac{r^2}{R^2}\right)4\pi r^2 dr$$

$$\Rightarrow \quad m = 4\pi\rho_0\left[\frac{R^3}{3} - \frac{R^3}{5}\right] = \frac{8}{15}\pi\rho_0 R^3$$

For floating, we use

$$mg = V\rho_l g$$

$$\Rightarrow \quad \frac{8}{15}4\pi\rho_0 R^3 = \frac{4}{3}\pi R^3\rho_l g$$

$$\Rightarrow \quad \rho_l = \frac{2}{5}\rho_0$$

Ans. (D)

Sol. 5 Average pressure on upper part side wall of liquid is $\frac{\rho gh}{2}$, so force on wall MN is given as

$$F_1 = \frac{\rho gh}{2}\times$$

Average pressure on lov ·ll of liquid is

$\left(\rho gh + \frac{2\rho gh}{2}\right)$, so force or 's

$$F_2 = \Big(\rho g$$

$$\Rightarrow \quad \frac{F_1}{F_2} = \frac{1}{4}$$

Ans. (A)

Sol. 6 For bubble, we use

$$F_B - mg = ma$$

$$\Rightarrow \quad V\rho_\omega g - mg = ma$$

$$\Rightarrow \quad m = \frac{V\rho_\omega g}{g+a} = \frac{\frac{4}{3}\times 3.14\times(1)^3\times 1\times 980}{980+9.8}$$

$$\Rightarrow \quad m \simeq 4.15 \text{ gm}$$

Ans. (C)

Sol. 7 Initial energy of liquid in the two vessels is

$$U_1 = (\rho S x_1)g\left(\frac{x_1}{2}\right) + (\rho S x_2)g\left(\frac{x_2}{2}\right)$$

If after equilibrium, final level becomes x_0, we use

$$U_2 = 2(\rho S x_0)g\left(\frac{x_0}{2}\right)$$

As volume of liquid is constant we use

$$x_0 = \frac{x_1+x_2}{2}$$

Change in energy is given as

$$\Delta U = U_1 - U_2 = \rho S g\left[\left(\frac{x_1+x_2}{4}\right) - x_0^2\right]$$

$$\Rightarrow \quad \Delta U = \rho S g\left[\frac{x_1^2}{2}+\frac{x_2^2}{2}-\left(\frac{x_1+x_2}{2}\right)^2\right]$$

$$\Rightarrow \quad \Delta U = \frac{1}{4}\rho S g(x_1 - x_2)^2$$

Ans. (C)

Sol. 8 For floating of shell, we use

$$\frac{4}{3}\pi(R^3 - r^3)\rho_s g = \frac{4}{3}\pi R^3 \rho_w g$$

$$\Rightarrow \quad 1-\left(\frac{r}{R}\right)^3 = \frac{8}{27}$$

$$\Rightarrow \quad \frac{r}{R} = \left(\frac{19}{27}\right)^{1/3} = \frac{19^{1/3}}{3}$$

$$\Rightarrow \quad r = 0.88 \simeq \frac{8}{9}R$$

Ans. (B)

Sol. 9 For a hydraulic press in the given situation, we can use pressure on both sides are equal so we have

$$\frac{100\times g}{A_2} = \frac{mg}{A_1} \quad \ldots(1)$$

Let m mass can lift M_0 in second case then, we use

$$\frac{M_0 g}{16A_2} = \frac{mg}{A_1/16} \quad \ldots(2)$$

From equation (1) and (2) we get

$$\frac{M_0}{16\times 100} = 16$$

$$\Rightarrow \quad M_0 = 25600 \text{ kg}$$

Ans. (25600)

Sol. 10 At a depth d, pressure on submarine is given as

$$P_1 = \rho g d + P_0 = 3\times 10^5 \text{ Pa}$$

$$\Rightarrow \quad \rho g d = 2\times 10^5 \text{ Pa}$$

When the depth is doubled, pressure becomes

$$P_2 = 2\rho g d + P_0$$

$$\Rightarrow \quad P_2 = 4\times 10^5 + 10^5$$

$$\Rightarrow \quad P_2 = 5\times 10^5 \text{ Pa}$$

Percentage increase in pressure is given as

$$\%\Delta = \frac{P_2 - P_1}{P_1}\times 100$$

$$\Rightarrow \quad \%\Delta = \frac{5\times 10^5 - 3\times 10^5}{3\times 10^5}\times 100 = \frac{200}{3}\%$$

Ans. (A)

Sol. 11 Terminal speed of raindrop is given as

$$v_T = \frac{2R^2(f_w - f_a)g}{9\eta}$$

$$\Rightarrow \quad v_T = \frac{2\times(0.002)^2\times(1000-1.2)\times 10}{9\times 1.8\times 10^{-5}}$$

$$\Rightarrow \quad v_T \approx 4.94$$

Ans. (A)

Sol. 12 Figure below shows the situation described in the quesion.

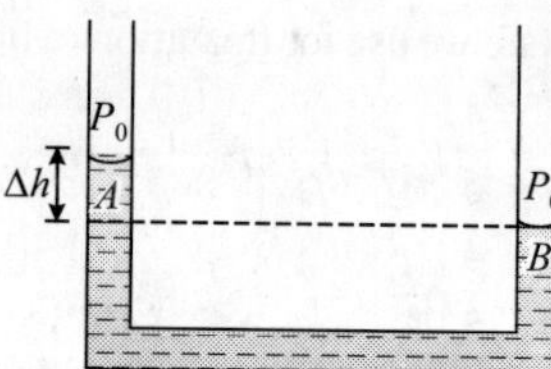

The pressure at a point just below the meniscus is $2T/R$ lower than that of atmospheric pressure so at the bottom of the tube, from left and right arm of tube, pressure is calculated from the two arms as

$$P_0 - \frac{2T}{R_1} + h_1\rho g = P_0 - \frac{2T}{R_2} + h_2\rho g$$

$$\Rightarrow \quad (h_1 - h_2)\rho g = 2T\left[\frac{1}{\frac{5}{2}\times 10^{-3}} - \frac{1}{\frac{8}{2}\times 10^{-3}}\right];$$

$$\Rightarrow \quad h_1 - h_2 = \frac{2\times 7.3\times 10^{-2}}{1000\times 10\times 10^{-3}}\times\frac{3}{20} = 2.19 \text{ mm}$$

Ans. (B)

Sol. 13 For an oil drop in experiment no electric force acts on uncharged drop so viscous force balances its weight, given as

Viscous force = weight

$\Rightarrow$ Viscous force $= mg = \rho \times \left(\frac{4}{3}\pi r^3\right) \times g$

$\Rightarrow$ Viscous force $= 3.9 \times 10^{-10}$ N **Ans. (D)**

9.2 Fluid Dynamics

Sol. 1 By equation of continuity, we use

In flow volume = Out flow volume

$$\Rightarrow \quad \frac{0.74}{60} = (\pi \times 4 \times 10^{-4}) \times \sqrt{2gh}$$

$$\Rightarrow \quad \sqrt{2gh} = \frac{74 \times 100}{240\pi}$$

$$\Rightarrow \quad 2gh = \frac{740 \times 740}{24 \times 24 \times 10}$$

$$\Rightarrow \quad h = \frac{74 \times 74}{2 \times 24 \times 24}$$

$$\Rightarrow \quad h \approx 4.8 \text{ m}$$ **Ans. (B)**

Sol. 2

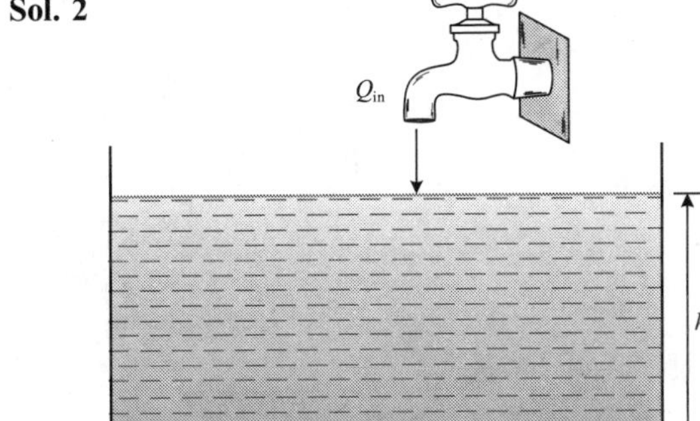

As height of water column is constant, we use water inflow rate (Q_{in}) = water outflow rate (Q_{out})

$$\Rightarrow \quad Q_{in} = 10^{-4} \text{ m}^3/\text{s}$$

$$\Rightarrow \quad Q_{out} = Av = 10^{-4} \times \sqrt{2gh}$$

$$\Rightarrow \quad 10^{-4} = 10^{-4}\sqrt{20 \times h}$$

$$\Rightarrow \quad h = \frac{1}{20} \text{ m} = 0.05 \text{ m} = 5 \text{ cm}$$ **Ans. (D)**

Sol. 3 Momentum per second carried by liquid per second is $\rho A v^2$ where area of hose pipe is $A = \pi a^2$

Net force due to reflected liquid $= 2 \times \left[\frac{1}{4}\rho A v^2\right]$

Net force due to stopped liquid $= \frac{1}{4}\rho A v^2$

Total force $F_{net} = \frac{3}{4}\rho A v^2$

Net pressure $= p = \frac{F_{net}}{A} = \frac{3}{4}\rho v^2$ **Ans. (B)**

Sol. 4

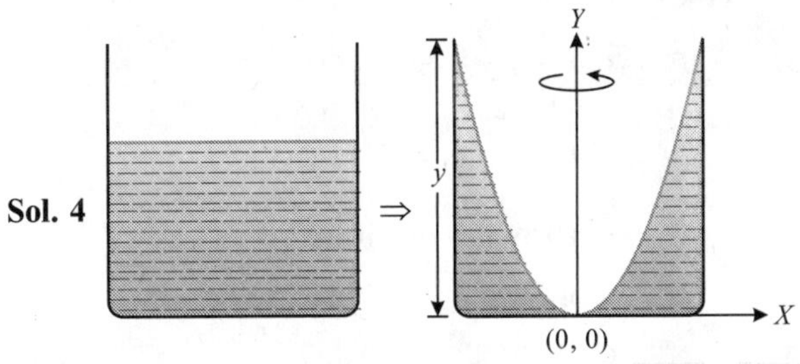

For the figure shown above equation of the liquid surface profile in a rotating container is given as

$$y = \frac{\omega^2 x^2}{2g} = \frac{(2 \times 2\pi)^2 \times (0.05)^2}{20} \simeq 2 \text{ cm}$$ **Ans. (C)**

Sol. 5

$A_1 = 10^{-4}\text{m}^2$
$v_1 = 1$ m/s
$h = 0.15$ m
g
A_2
v_2

By equation of continuity we use

$$A_1 v_1 = A_2 v_2 \qquad \ldots (1)$$

By acceleration due to gravity, we have

$$v_2^2 = v_1^2 + 2gh$$

$$v_2 = \sqrt{(1)^2 + 2 \times 10 \times 0.15} = \sqrt{4} = 2 \text{ m/s}$$

From equation-(1)

$$A_2 = \frac{A_1 v_1}{v_2} = \frac{10^{-4} \times 1}{2} = 5 \times 10^{-5} \text{ m}^2$$ **Ans. (B)**

Sol. 6 By continuity equation, we use

$$A_1 v_1 = A_2 v_2$$

$$\Rightarrow \quad \frac{v_{min}}{v_{max}} = \frac{A_{min}}{A_{max}}$$

$$\Rightarrow \quad \frac{v_{min}}{v_{max}} = \left(\frac{4.8}{6.4}\right)^2$$

$$\Rightarrow \quad \frac{v_{min}}{v_{max}} = \frac{9}{16}$$ **Ans. (D)**

Sol. 7 Rate of flow of water through the tube is given as

$$Q = A_A v_A = A_B v_B$$

$$\Rightarrow \quad (40)v_A = (20)v_B$$

$$\Rightarrow \quad v_B = 2v_A \qquad \ldots(1)$$

Applying Bernoulli's theorem at A and B, we have

$$P_A + \frac{1}{2}\rho v_A^2 = P_B + \frac{1}{2}\rho v_B^2$$

$$\Rightarrow \quad P_A - P_B = \frac{1}{2}\rho\,(v_B^2 - v_A^2)$$

$$\Rightarrow \quad 700 = \frac{1}{2} \times 1000\,(4v_A^2 - v_A^2)$$

$$\Rightarrow \quad v_A = 0.68 \text{ m/s} = 68 \text{ cm/s}$$

$$\Rightarrow \quad Q = A_A v_A$$

$$\Rightarrow \quad Q = (40)(68) = 2720 \text{ cm}^3/\text{s}$$

Ans. (C)

Sol. 8 Using Bernoulli's Equation at the two points, we have

$$P + \frac{1}{2}\rho v^2 = \frac{P}{2} + \frac{1}{2}\rho V^2$$

$$\Rightarrow \quad \frac{2P}{2\rho} + \frac{1}{2}\frac{\rho v^2}{\rho} \times 2 = V^2$$

$$\Rightarrow \quad V = \sqrt{\frac{P}{\rho} + v^2}$$

Ans. (B)

Sol. 9 In the given situation described in question, we use Bernoulli's equation just inside and outside the opening, it gives

$$\left(P_0 + \frac{mg}{A}\right) + \rho g H + \frac{1}{2}\rho v_1^2 = P_0 + 0 + \frac{1}{2}\rho v^2 \qquad \ldots(1)$$

As cross sectional area of vessel is very large, we can neglect v_1 which gives

$$v = \sqrt{2gH + \frac{2mg}{A\rho}}$$

$$\Rightarrow \quad v = \sqrt{8 + 1.2} = 3.033 \text{ m/s} \sim 3 \text{m/s}$$

Ans. (3)

Sol. 10 Force exerted by ejection of water on the container can be given by rate of momentum ejection per unit time, calculated as

$$F = \left(\frac{dm}{dt}\right)v = (\rho a v)v$$

$$\Rightarrow \quad F = \rho a v^2$$

The velocity of efflux of water is given as

$$v = \sqrt{2gh}$$

For vessel not to slide, we use

$$\mu m g \geq \rho a (2gh)$$

$$\Rightarrow \quad \mu \rho (Ah) g \geq \rho a 2gh$$

$$\Rightarrow \quad \mu_{\min} = \frac{2a}{A}$$

Ans. (C)

Sol. 11 R is maximum $dR/dh = 0$, which gives

$$h' = h = H/2 = 12/2 = 6 \text{ m}$$

Ans. (6)

9.3 Viscosity

Sol. 1

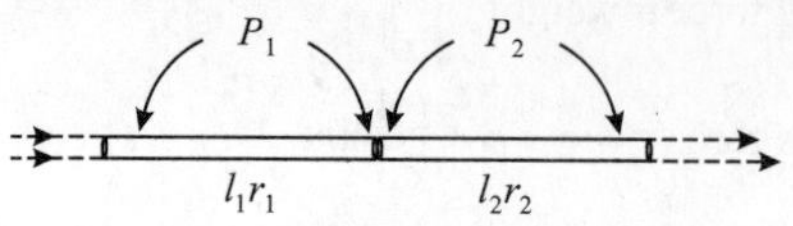

For viscous fluid flowing in a capillary tube we use

flow rate $\quad \dfrac{dQ}{dt} = \dfrac{\pi r^4}{8\eta l}\Delta P$

As tubes in this case are in series, so we use

$$\left.\frac{dQ}{dt}\right|_1 = \left.\frac{dQ}{dt}\right|_2$$

$$\frac{\pi r_1^4}{8\eta l_1}P_1 = \frac{\pi r_2^4}{8\eta l_2}.P_2$$

$$\Rightarrow \quad \frac{r_1^4}{l_1}P_1 = \frac{r_2^4}{l_1/4}(4P_1)$$

$$\Rightarrow \quad r_1^4 = 16 r_2^4$$

$$\Rightarrow \quad r_2 = \left(\frac{r_1^4}{16}\right)^{1/4} = \frac{r_1}{2}$$

Ans. (D)

Sol. 2 Volume flow rate of water is

$$Q = vA = v(\pi r^2)$$

$$\Rightarrow \quad v = \frac{Q}{\pi r} = \frac{100 \times 10^{-3}}{60} \times \frac{1}{\pi \times 25 \times 10^{-4}}$$

$$\Rightarrow \quad v = \frac{2}{3\pi} \text{ m/s}$$

Reynolds Number is given as

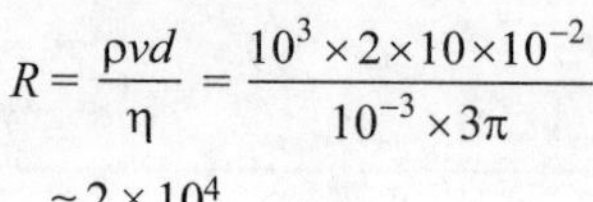

$$R = \frac{\rho v d}{\eta} = \frac{10^3 \times 2 \times 10 \times 10^{-2}}{10^{-3} \times 3\pi}$$

$$\simeq 2 \times 10^4$$

Thus order R is 10^4

Ans. (C)

Sol. 3 We have

$$V_T = \frac{2}{9}\left(\frac{r^2 g}{\eta}\right)(\rho_0 - \rho_l)$$

$$\Rightarrow \quad v_T \propto r^2$$

As mass of sphere will remain same before and after breaking, we have

$$\Rightarrow \quad \rho\left(\frac{4}{3}\pi R^3\right) = \rho\left(\frac{4}{3}\pi R^2\right) \times 27$$

$$\Rightarrow \quad r = \frac{R}{3}$$

$$\Rightarrow \quad \frac{v_1}{v_2} = \frac{R^2}{r^2} = 9$$

An.s (D)

Sol. 4 After falling through a height h as the velocity of ball is equal to terminal velocity, we use

$$\sqrt{2gh} = \frac{2}{9}\frac{r^2 g}{\eta}(\rho_l - \rho)$$

$$\Rightarrow \quad h = \frac{2}{81}\frac{r^4 g\,(\rho_l - \rho)^2}{\eta^2}$$

$$\Rightarrow \quad h \propto r^4$$

Ans. (B)

Sol. 5 The nature of flow whether streamline or unsteady or turbulent is determined by Reynolds Number, calculated as

$$R_e = \frac{\rho v D}{\eta}$$

As a standard data we can use for the value of Raynolds number less than 1000 flow remains steady and above 1000 upto 2000 it remain unsteady and above 2000 it becomes turbulent. Calculating the Raynolds number for the given situtions, we have

$$R_i = 10^3 \times \frac{0.18\times 10^{-3}}{\pi\times(0.5\times 10^{-2})^2\times 60} \times \frac{1\times 10^{-2}}{10^{-3}} = 382.16$$

$$R_f = 10^3 \times \frac{0.48\times 10^{-3}}{\pi\times(0.5\times 10^{-2})^2\times 60} \times \frac{1\times 10^{-2}}{10^{-3}} = 1019.09$$

Thus flow changes from steady to unsteady. **Ans. (D)**

9.4 Surface Tension

Sol. 1 Pressure inside a soap bubble of radius r is given as

$$P_{in} = P_{atm} + \frac{4T}{r} \qquad \ldots(1)$$

If volume of soap bubble increases at constant rate, we have

$$V \propto t$$

$$\Rightarrow \quad r \propto t^{1/3}$$

$$\Rightarrow \quad r = kt^{1/3}$$

From equation-(1), we have

$$P_{in} = P_{atm} + \frac{4T}{kt^{1/3}}$$

Thus graph between P and $\frac{1}{t^{1/3}}$ is a straight line with positive slope and positive intercept. **Ans. (D)**

Sol. 2 Height of liquid rise in capillary tube

$$h = \frac{2T\cos\theta_C}{\rho r g}$$

$$\Rightarrow \quad h \propto \frac{1}{r}$$

When radius becomes double height become half

$$\Rightarrow \quad h' = \frac{h}{2}$$

Now, $\quad M = \pi r^2 h \times \rho$

and $\quad M' = \pi(2r)^2(h/2)\times\rho = 2M$ **Ans. (C)**

Sol. 3 Capillary height is given as

$$h = \frac{2T\cos\theta}{r\rho g}$$

For H_g, we use $h = \dfrac{2T_{Hg}\cos\theta_{Hg}}{r_1\rho_{Hg}g}$

and for water, we use

$$h = \frac{2T_w\cos\theta_w}{r_2\rho_w g}$$

$$\Rightarrow \quad \frac{T_{Hg}\cos\theta_{Hg}}{r_1\rho_{Hg}} = \frac{T_w\cos\theta_w}{r_2\rho_w}$$

$$\Rightarrow \quad \frac{r_1}{r_2} = \left|\frac{T_{Hg}\cos\theta_H}{T_w\cos\theta_w}\times\frac{\rho_w}{\rho_{Hg}}\right|$$

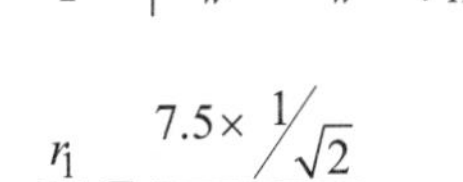

$$\Rightarrow \quad \frac{r_1}{r_2} = \frac{7.5\times 1/\sqrt{2}}{13.6}$$

$$\Rightarrow \quad \frac{r_1}{r_2} = \frac{5}{9\sqrt{2}} = \frac{5}{12.72} \simeq \frac{2}{5}$$

Ans. (C)

Sol. 4 For floatation of droplet, net upward force on droplet balances its weight. Net upward force on droplet is

$$F_{up} = F_{Buoyancy} + F_{surface\ tension}$$

$$\Rightarrow \quad F_{up} = \left(\frac{2}{3}\pi R^3\right)\rho g + T(2\pi R)$$

For equilibrium of droplet, we use

$$\left(\frac{2}{3}\pi R^3\right)\rho g + T(2\pi R) = d\left(\frac{4}{3}\pi R^3\right)g$$

$$\Rightarrow \quad T(2\pi R) = \left(\frac{2}{3}\pi R^3\right)g\,[2d - \rho]$$

$$\Rightarrow \quad R = \sqrt{\frac{3T}{(2d-\rho)g}}$$

Ans. (B)

Force of surface tension can be upward (as given in question) so in this case liquid does not wet the droplet.

Sol. 5 Figure below shows that angle of contact is $\theta = 30^\circ$

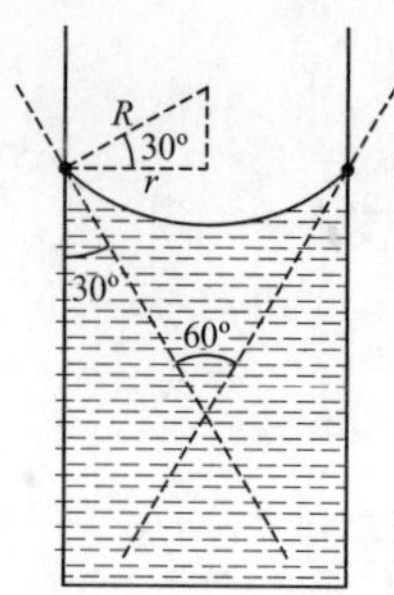

Capillary height of liquid in tube is given as

$$h = \frac{2T}{\rho g r \cos\theta}$$

$$\Rightarrow \quad h = \frac{2\times 0.05}{667\times 10\times\left(\frac{0.15\times 10^{-3}\times 2}{\sqrt{3}}\right)}$$

$$\Rightarrow \quad h = 0.087\text{ m}$$ **Ans. (C)**

Sol. 6 By excess pressure formula, we use

$$0.01 = \frac{4T}{R_1}$$

and $$0.02 = \frac{4T}{R_2}$$

$$\Rightarrow \quad \frac{R_1}{R_2} = 2$$

$$\Rightarrow \quad R_1 = 2R_2$$

Ratio of volumes is given as

$$\frac{V_1}{V_2} = \frac{R_1^3}{R_2^3} = 8$$ **Ans. (A)**

Sol. 7 Capillary rise tube is given as

$$h = \frac{2T\cos\theta}{r\rho g}$$

$$\Rightarrow \quad T = \frac{\rho g r h}{2\cos\theta}$$

$$\Rightarrow \quad T = \frac{(900)(10)(15\times 10^{-5})(15\times 10^{-2})}{2}$$

$$\Rightarrow \quad = 101.25\times 10^{-3}\text{ N/m}$$

$$\Rightarrow \quad = 101.25\, m\text{N/m}$$ **Ans. (101)**

Sol. 8 Excess pressure at common surface of the two bubbles is given as

$$P_{ex} = 4T\left(\frac{1}{a} - \frac{1}{b}\right) = \frac{4T}{r}$$

$$\Rightarrow \quad \frac{1}{r} = \frac{1}{a} - \frac{1}{b}$$

$$\Rightarrow \quad r = \frac{ab}{b-a}$$ **Ans. (A)**

Sol. 9 When the drops coalesce, volume of liquid remain conserved, given as

$$2\times\frac{4}{3}\pi R^3 = \frac{4}{3}\pi r^3$$

$$\Rightarrow \quad \frac{R}{r} = \left(\frac{1}{2}\right)^{1/3} \quad \ldots(1)$$

Ratio of before and after combining, ratio of surface energy of the liquid is given as

$$\frac{U_{\text{before}}}{U_{\text{after}}} = \frac{2\times S\times 4\pi R^2}{S\times 4\pi r^2} = 2\times\left(\frac{R}{r}\right)^2 = \frac{2^{1/3}}{1}$$

Ans. (B)

Sol. 10 When the two bubbles combine, the total surface energy remain conserved as temperature of system is not increasing, thus for surface tension σ we use

$$\sigma A_1 + \sigma A_2 = \sigma A$$

$$\Rightarrow \quad \sigma[2\times 4\pi r_1^2] + \sigma[2\times 4\pi r_2^2] = \sigma[2\times 4\pi r^2]$$

$$\Rightarrow \quad r_1^2 + r_2^2 = r^2$$

$$\Rightarrow \quad r = \sqrt{r_1^2 + r_2^2}$$ **Ans. (C)**

Sol. 11 If R is the radius of equivalent soap bubble then pressure inside the 3 cm bubble is given as

$$P_0 + \frac{4T}{6} + \frac{4T}{3} = P_0 + \frac{4T}{R}$$

$$\Rightarrow \quad \frac{1}{6} + \frac{1}{3} = \frac{1}{R}$$

$$\Rightarrow \quad R = 2\text{ cm}$$ **Ans. (2)**

10.1 Thermal Expansion

Sol. 1 For a cube of volume V, if on heating its volume increases by ΔV then we use for ΔT rise in temperature

$$\Delta V = V\gamma\Delta T = V(3\alpha)\Delta T \quad \ldots(1)$$

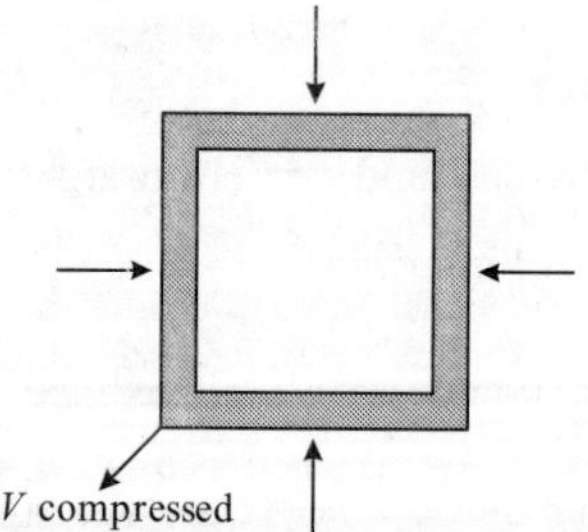

On applying pressure to cube if its volume is compressed by ΔV then we have

Bulk modulus

$$K = \frac{P}{\left|\frac{\Delta V}{V}\right|} = \frac{P}{3\alpha\Delta T}$$

$$\Rightarrow \quad \Delta T = \frac{P}{3\alpha K}$$

for same size ΔV increased (due to thermal expansion).

Ans. (C)

Sol. 2

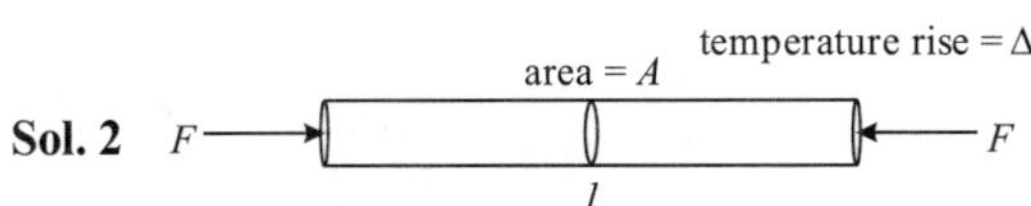

As length of rod → constant

⇒ Thermal expansion due to rise in temperature is compensated by elastic compression due to compressive force F.

Here we use

$$Y = \frac{\text{Stress}}{\text{Strain}} = \frac{F/A}{\Delta l/l}$$

Thermal expansion

$$\Delta l = \alpha l \Delta T$$

$$\Rightarrow \quad Y = \frac{F/A}{\alpha\Delta T}$$

$$\Rightarrow \quad F = YA\alpha\Delta T$$

Ans. (C)

Sol. 3

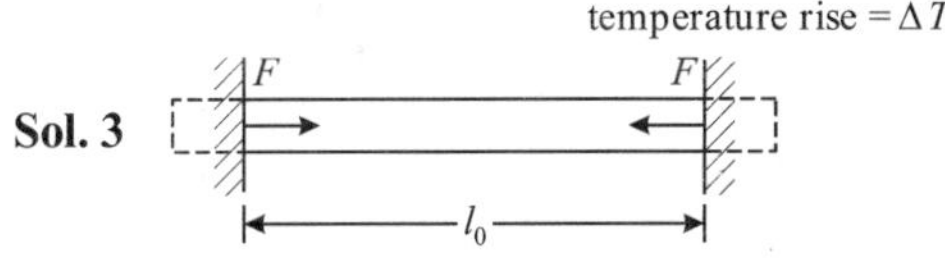

Due to the external force at ends, then thermal expansion in rail due to rise in temperature in compensated by elastic compression due to external force.

$$\Rightarrow \quad Y = \frac{\text{Stress}}{\text{Strain}} = \frac{F/A}{\Delta l/l_0}$$

Thermal expansion

$$\Delta l = \alpha l_0 \Delta T$$

$$\Rightarrow \quad Y = \frac{F/A}{\alpha\Delta T}$$

$$\Rightarrow \quad F = YA\alpha\Delta T$$

$$\Rightarrow \quad F = 2 \times 10^{11} \times 40 \times 10^{-4} \times 1.2 \times 10^{-5} \times 10$$

$$\Rightarrow \quad F = 9.6 \times 10^{4}\,\text{N}$$

$$\Rightarrow \quad F \simeq 1 \times 10^{5}\,\text{N}$$

Ans. (B)

Sol. 4 Young's modulus

$$Y = \frac{\text{Stress}}{\text{Strain}}$$

$$\Rightarrow \quad Y = \frac{F/A}{(\Delta l/l)}$$

$$\Rightarrow \quad Y = \frac{F}{A(\alpha\Delta T)}$$

Ans. (C)

Sol. 5 Given that

$$\Delta l_1 = \Delta l_2$$

$$\Rightarrow \quad l\alpha_1\Delta T_1 = l\alpha_2\Delta T_2$$

$$\Rightarrow \quad \frac{\alpha_1}{\alpha_2} = \frac{\Delta T_1}{\Delta T_2}$$

$$\Rightarrow \quad \frac{4}{3} = \frac{T-30}{180-30}$$

$$\Rightarrow \quad T = 230°\text{C}$$

Ans. (B)

Sol. 6 For the given temperature and scale, we use

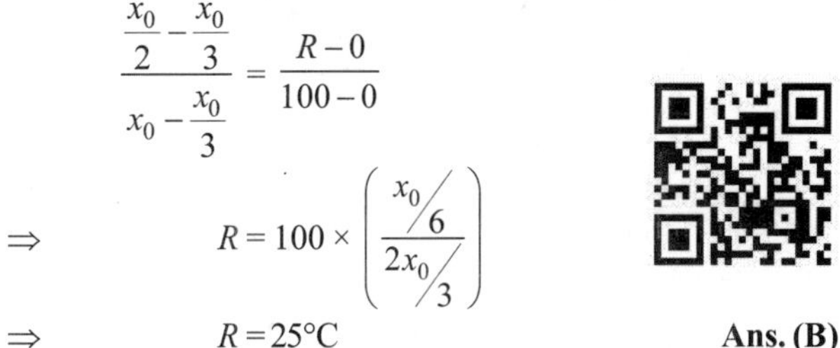

$$\frac{\frac{x_0}{2} - \frac{x_0}{3}}{x_0 - \frac{x_0}{3}} = \frac{R-0}{100-0}$$

$$\Rightarrow \quad R = 100 \times \left(\frac{x_0/6}{2x_0/3}\right)$$

$$\Rightarrow \quad R = 25°\text{C}$$

Ans. (B)

Sol. 7 Coefficient of volume expansion is given as

$$\gamma = \alpha_x + \alpha_y + \alpha_z$$

$$\Rightarrow \quad \gamma = 5 \times 10^{-5} + 5 \times 10^{-6} + 5 \times 10^{-6}$$

$$\Rightarrow \quad \gamma = (50 + 5 + 5) \times 10^{-6}$$

$$\Rightarrow \quad \gamma = 60 \times 10^{-6}$$

$$\Rightarrow \quad C = 60$$

Ans. (60)

Sol. 8 Initially we use

$$\rho_0 S(80)\,g = mg \qquad \ldots(1)$$

later $$\rho S(79)\,g = mg \qquad \ldots(2)$$

$$\Rightarrow \quad \frac{\rho}{\rho_0} = \frac{80}{79} = 1.012$$

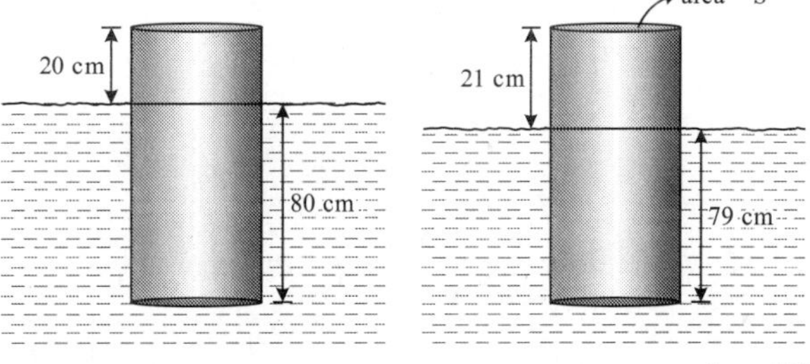

Ans. (A)

Sol. 9 It is given that by thermal expansion

$$\frac{\Delta l}{l} = \frac{0.02}{100} = \propto \Delta T$$

$$\Rightarrow \quad \propto = \frac{2\times10^{-4}}{\Delta T}$$

Thus volume expansion coefficient is

$$\gamma = 3\alpha = \frac{6\times10^{-4}}{\Delta T}$$

Density variation is given as

$$\Delta\rho = -\rho\gamma\Delta T$$

$$\Rightarrow \quad \frac{\Delta\rho}{\rho} = -\gamma\Delta T = -6\times10^{-4} = -0.06\%$$ **Ans. (B)**

Sol. 10 The unfitted volume of beaker is given as

$$\Delta V = (V_0 - V_m)$$

After increasing temperature by ΔT, we use

$$\Delta V' = (V_0' - V_m')$$

As $\Delta V' = \Delta V$, we use

$$V_0 - V_m = V_0(1+\gamma_b\Delta T) - V_m(1+\gamma_M\Delta T)$$

$$\Rightarrow \quad V_0\gamma_b = V_m\gamma_m$$

$$\Rightarrow \quad V_m = \frac{V_0\gamma_b}{\gamma_m} = \frac{(500)(6\times10^{-6})}{(1.5\times10^{-4})}$$

$$\Rightarrow \quad V_m = 20\text{ cm}^3$$ **Ans. (20)**

Sol. 11 At T°C, Total length of combination s

$$L = L_1 + L_2$$

L_1, α_1 L_2, α_2

At $T+\Delta T$, Total length becomes

$$L'_{eq} = L'_1 + L'_2$$

$(L_1+L_2), \alpha_{avg}$

There we us $L'_1 = L_1(1+\alpha_1\Delta T)$

and $L'_2 = L_2(1+\alpha_2\Delta T)$

$$\Rightarrow \quad L'_{eq} = (L_1+L_2)(1+\alpha_{avg}\Delta T)$$

$$\Rightarrow \quad (L_1+L_2)(1+\alpha_{avg}\Delta T) = L_1 + L_2 + L_1\alpha_1\Delta T + L_2\alpha_2\Delta T$$

$$\Rightarrow \quad (L_1+L_2)\alpha_{avg} = L_1\alpha_1 + L_2\alpha_2$$

$$\Rightarrow \quad \alpha_{avg} = \frac{L_1\alpha_1 + L_2\alpha_2}{L_1+L_2}$$ **Ans. (D)**

Sol. 12 Volume increment can be directly given as

$$\Delta V = V\gamma\Delta T$$

$$\Rightarrow \quad \Delta V = 3a^3\alpha\Delta T$$ **Ans. (A)**

Sol. 13 As rod is lying freely on smooth surface, it expands as no force is developed in rod so no stress will be there and in above case R is not the correct explanation of A. **Ans. (D)**

Sol. 14 As $\alpha_A > \alpha_B$ so on placing in the cold bath lenth of A will decrease more than length of B due to which system will bend toward left. **Ans. (D)**

Sol. 15 Resistance as a function of temperature is given as

$$R' = R(1+\alpha\Delta t)$$

$$\Rightarrow \quad 20 = 16(1+\alpha.85)$$

$$\Rightarrow \quad \alpha = \frac{\frac{20}{16}-1}{85} = \frac{1}{4\times85} = 0.003°\text{C}^{-1}$$ **Ans. (C)**

Sol. 16 Thermal force acting on the rod can be written as

$$F = Ay\alpha\Delta T$$

$$\Rightarrow \quad F = (10\times10^{-4})\times(2\times10^{11})(10^{-5})(400) = 8\times10^{-5}\text{ N}$$

$$\Rightarrow \quad x = 8$$ **Ans. (8)**

10.2 Calorimetry

Sol. 1

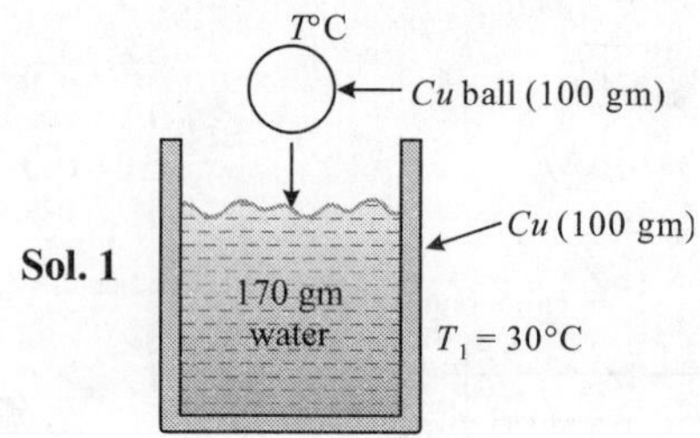

As $T_f = 75°C$

So rise in temperature of calorimeter + water

$$\Delta T = 75° - 30°$$

$$\Rightarrow \quad \Delta T = 45°C$$

As final temperature of mixture is 75°C

$\Rightarrow$ Heat lost by Cu-ball = Heat gained by Cu-calorimeter + water

$$\Rightarrow \quad 100\times0.1\times(T-75) = 100\times0.1\times45 + 170\times1\times45$$

$$\Rightarrow \quad 10\text{T} - 750 = 450 + 7650$$

$$\Rightarrow \quad 10\text{T} = 8100 + 750 = 8850$$

$$\Rightarrow \quad T = \frac{8850}{10} = 885°C$$ **Ans. (D)**

Sol. 2

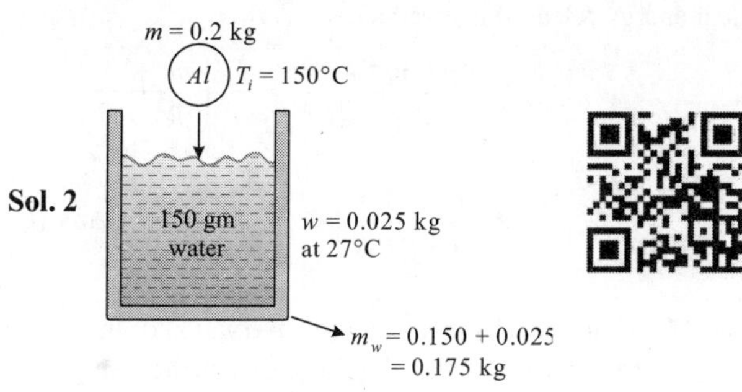

Equilibrium temperature attained = 40°C

Heat lost by Al - sphere = heat gained by calorimeter + water

$$0.2 \times s_M \times 110 = 0.175 \times 4200 \times 13$$

$$\Rightarrow \quad s_{Al} = \frac{0.175 \times 4200 \times 13}{0.2 \times 110} = 434.31 \text{ J/kg°C}$$

Ans. (D)

Sol. 3 By energy conservation, we use

$$192 \times s \times (100 - 21.5)$$
$$= 128 \times 394 \times (21.5 - 8.4)$$
$$+ 240 \times 4200 \times (21.5 - 8.4)$$

$$\Rightarrow \quad s = 916 \text{ J/kg-K}$$

Ans. (D)

Sol. 4 Let amount of ice is m gm. According to principal of calorimeter heat taken by ice = heat given by water

$$\Rightarrow \quad 20 \times 2.1 \times m + (m - 20) \times 334$$
$$= 50 \times 4.2 \times 40$$
$$376\text{m} = 8400 + 6680$$
$$m = 40.1\text{g}$$ **Ans. (B)**

Sol. 5 By conservation of energy, we use

$$100 \times s_A \times [100 - 90] = 50 \times s_B \times (90 - 75)$$

$$\Rightarrow \quad 2s_A = 1.5 s_B$$

$$\Rightarrow \quad s_A = \frac{3}{4} s_B$$

And $\quad 100 \times s_A \times [100 - T] = 50 \times s_B (T - 50)$

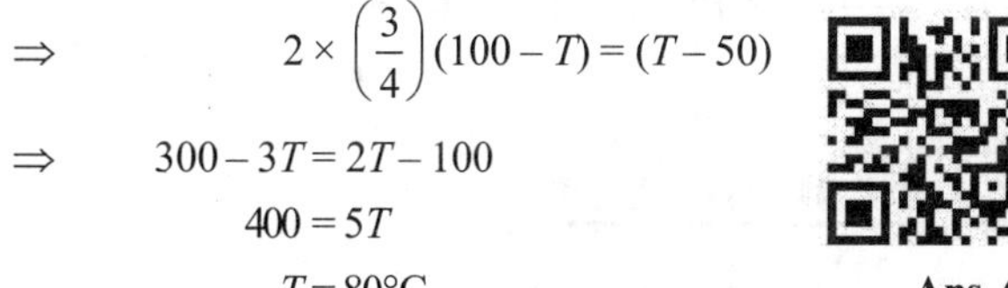

$$\Rightarrow \quad 2 \times \left(\frac{3}{4}\right)(100 - T) = (T - 50)$$

$$\Rightarrow \quad 300 - 3T = 2T - 100$$
$$400 = 5T$$
$$T = 80°\text{C}$$

Ans. (A)

Sol. 6 By conservation of energy, we use

$$0.1 \times 400 \times (500 - T) = 0.5 \times 4200 \times (T - 30) + 800\,(T - 30)$$

$$\Rightarrow \quad 40(500 - T) = (T - 30)(2100 + 800)$$

$$\Rightarrow \quad 20000 - 40T = 2900\,T - 30 \times 2900$$

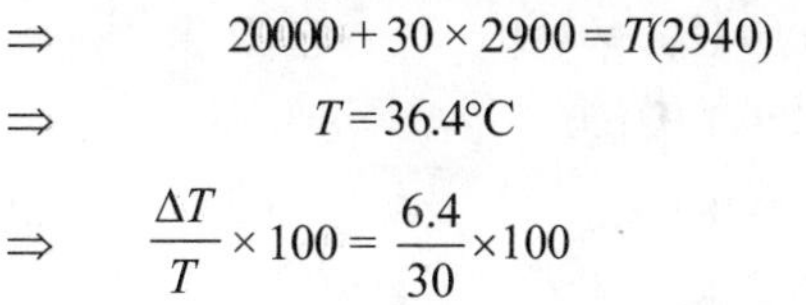

$$\Rightarrow \quad 20000 + 30 \times 2900 = T(2940)$$

$$\Rightarrow \quad T = 36.4°\text{C}$$

$$\Rightarrow \quad \frac{\Delta T}{T} \times 100 = \frac{6.4}{30} \times 100$$
$$= 21.3\%$$ **Ans. (B)**

Sol. 7

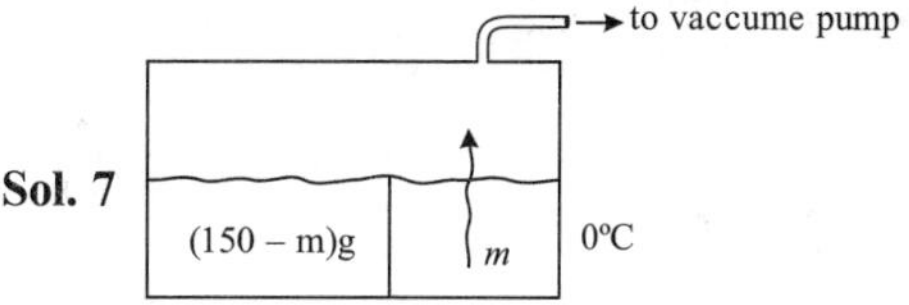

If mass m gets evaporated

$\Rightarrow$ heat required is

$$\Delta Q_{\text{evap}} = mL_v$$

This heat is supplied by remaining water in vessel which will transform into ice

$\Rightarrow$ heat supplied is $\Delta Q_{\text{freeze}} = (150 - m)\,L_f$

$$\Rightarrow \quad \Delta Q_{\text{evap}} = \Delta Q_{\text{freeze}}$$
$$mL_v = (150 - m)\,L_f$$
$$m = \frac{150 L_f}{L_v + L_f} = \frac{150 \times 3.36 \times 10^5}{(21 + 3.36) \times 10^5}$$
$$m \simeq 20\text{g}$$

Ans. (C)

Sol. 8 Heat lost by water = Heat gained by ice

$$\Rightarrow \quad M_2 S_w \Delta T_w = M_1 S_i \Delta T_i + M_1 L_f$$

$$\Rightarrow \quad M_2 \times 1 \times 50 = M_1 \times 0.5 \times 10 + M_1 L_f$$

$$\Rightarrow \quad 50\,M_2 = 5M_1 + M_1 L_f$$

$$\Rightarrow \quad L_f = \frac{50 M_2}{M_1} - 5$$

Ans. (C)

Sol. 9 For exchange of heat in mixture, we use

Heat Supplied = Heat Gained

$$M \times 540 + M \times 1 \times 60 = 200 \times 80 + 200 \times 1 \times 4$$

$$\Rightarrow \quad M = 40 \text{ gm}$$

Ans. (40)

Sol. 10 According to table using heat exchange between water amounts, we use

Case-I : $1(T_1 - 60) + 2\,(T_2 - 60) = 0$

$$\Rightarrow \quad 1T_1 + 2T_2 = (1 + 2)60°$$
$$T_1 + 2T_2 = 180 \quad \ldots(1)$$

Case-II : $2(T_1 - 60) + 1\,(T_2 - 30) = 0$

$$\Rightarrow \quad T_2 + 2T_3 = (1 + 2)30°$$
$$T_2 + 2T_3 = 90 \quad \ldots(2)$$

Case-III : $2(T_1-60)+1\,(T_3-60)=0$

$$\Rightarrow \quad 2T_1+1T_3=(1+2)\,60°$$

$$2T_1+T_3=180 \qquad \ldots(3)$$

Adding (1) + (2) + (3) gives

$$3(T_1+T_2+T_3)=450$$

$$\Rightarrow \quad T_1+T_2+T_3=150° \qquad \ldots(4)$$

Case-IV : $(T_1-\theta)+(T_2-\theta)+(T_3-\theta)=0$

$$\Rightarrow \quad T_1+T_2+T_3=(1+1+1)\,\theta$$

$$\Rightarrow \quad 150°=3\theta$$

$$\Rightarrow \quad \theta=50°\text{C}$$

Ans. (50)

Sol. 11 Given mixture can be taken as

Mixture = 200 g water at 25°C + m gram steam at 100°C

Heat supplied by steam = $m\times 540+m\times 1\times 69=609m$

Heat absorbed by water = $200\times 1\times 6=1200$ cal

by conservation of energy, we use

$$1200=609\,m$$

$$\Rightarrow \quad m=\frac{1200}{609}\simeq 2\text{ g}$$

Ans. (B)

Sol. 12 Here the water will supply heat to ice for melting

$$m_w s_w \Delta\theta=m_{ice}L_{ice}$$

$$m_{ice}=\frac{0.2\times 4200\times 25}{3.4\times 10^5}$$

$$\Rightarrow \quad m_{ice}=0.0617\text{ kg}$$

$$\Rightarrow \quad m_{ice}=61.7\text{ gm}$$

Remaining ice will remain un-melted so correct answer is (A).

Ans. (A)

Sol. 13 Given that half of KE is converted to heat, we use

$$\frac{1}{2}\times\frac{1}{2}mv^2=ms\Delta T$$

$$\Rightarrow \quad \Delta T=\frac{(210)^2}{4\times 0.03\times 4200}\simeq 87.5°\text{C}$$

Ans. (B)

Sol. 14 If n drops combine to form a bigger drop, the radius of bigger drop is given as

$$n\times\frac{4}{3}\pi r^3=\frac{4}{3}\pi R^3$$

$$\Rightarrow \quad n^{1/3}r=R$$

Total change in surface energy of the liquid is given by product of surface tension and change in area, given as

$$U=(n(4\pi r^2)-4\pi R^2)T$$

$$\Rightarrow \quad U=4\pi T(nr^2-R^2)$$

Heat energy released is given as

$$U=\frac{4\pi T(nr^2-R^2)}{J\times\frac{4}{3}\pi R^3}=\frac{3T}{J}\left(\frac{nr^2}{R^3}-\frac{1}{R}\right)$$

$$\Rightarrow \quad U=\frac{3T}{J}\left(\frac{1}{r}-\frac{1}{R}\right)$$

Ans. (C)

Sol. 15 When A & B are mixed, by conservation of energy we consider heat supplied by one liquid is equal to the heat gained by other liquid

$$m_1S_1(16-10)=m_2S_2(20-16)$$

$$\Rightarrow \quad 6m_1S_1=4m_2S_2 \qquad \ldots(1)$$

When B & C are mixed, we use

$$m_2S_2(26-20)=m_3S_3(30-26)$$

$$\Rightarrow \quad 6m_2S_2=4m_3S_3 \qquad \ldots(2)$$

From equation (1) and (2), we have

$$9m_1S_1=4m_3S_3$$

When A & C are mixed, we use

$$m_1S_1(T-10)=m_3S_3(30-T)$$

$$\Rightarrow \quad T=\frac{30m_3S_3+10m_1S_1}{m_1S_1+m_3S_3}$$

$$\Rightarrow \quad T=\frac{10(3m_3S_3+m_1S_1)}{m_3S_3+m_1S_1}$$

$$\Rightarrow \quad T=\frac{10\left(3m_3S_3+\frac{4}{9}m_3S_3\right)}{m_3S_3+\frac{4}{9}m_3S_3}$$

$$\Rightarrow \quad T=\frac{10\times 31}{13}=23.84°\text{C}$$

Ans. (C)

Sol. 16 By conservation of energy we have

$$mgh=ms\Delta\theta$$

$$\Rightarrow \quad g\times 63=\Delta\theta\times 4.2\times 10^3$$

$$\Rightarrow \quad \Delta\theta=0.147°\text{C}$$

Ans. (C)

Sol. 17 Heat supplied from resistance can be written as

$$i^2Rt=ms\Delta T+mL$$

$$\Rightarrow \quad (0.5)^2\times 4\times 10^3\times t=(10^3\times 1\times 10^{-4})\times 2400\times 10 + (10^3\times 1\times 10^{-4})\times 330\times 10^3$$

$$\Rightarrow \quad t=24\times 10^3\times 10^{-4}+33\times 10^4\times 10^{-4}=2.4+33=35.4\text{ s}$$

Ans. (D)

10.3 Heat Transfer

Sol. 1 Equivalent circuit of the given system of rods is shown in figure below.

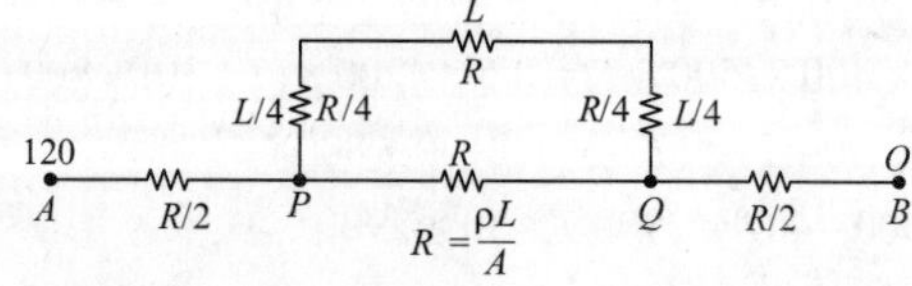

Thermal current is given as

$$I_{th} = \frac{\Delta T}{R_{eq.}} = \frac{(120)5}{8R} = \frac{120\times 5}{8R}$$

Temperature difference of P and Q is given as

$$\Delta T_{PQ} = I_{th} R_{PQ} = \left(\frac{120\times 5}{8R}\right)\times\frac{3}{5}R = \frac{360}{8} = 45°\text{C}$$ **Ans. (D)**

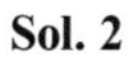

Sol. 2

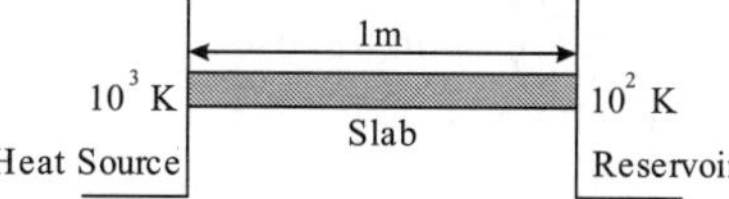

Rate of heat conduction through slab in steady state is given as

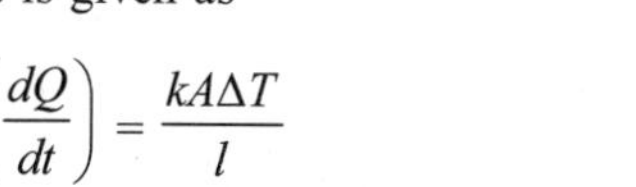

$$\left(\frac{dQ}{dt}\right) = \frac{kA\Delta T}{l}$$

Energy flux (w/m^2) through slab is given as

$$\frac{1}{A}\left(\frac{dQ}{dt}\right) = \frac{(0.1)(900)}{1} = 90\text{W/m}^2$$ **Ans. (A)**

Sol. 3

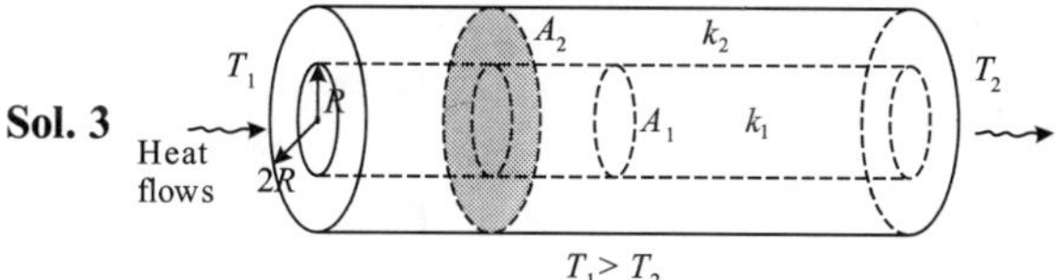

In parallel combination, equivalent thermal conductivity is given as

$$K_{eq} = \frac{k_1 A_1 + k_2 A_2}{A_1 + A_2}$$

$$K_{eq} = \frac{k_1(\pi R^2) + k_2(\pi(2R)^2 - \pi R^2)}{4\pi R^2}$$

$$K_{eq} = \frac{k_1 + 3k_2}{4}$$ **Ans. (D)**

Sol. 4 By Newton's law of cooling

$$\frac{dT}{dt} = -\frac{4eA\sigma T_0^3}{ms}\cdot(\Delta T)$$

$$\Rightarrow \quad \left|\frac{dT}{dt}\right| \propto \frac{1}{ms}$$

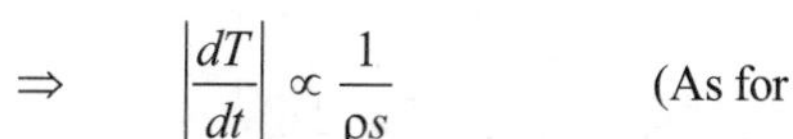

$$\Rightarrow \quad \left|\frac{dT}{dt}\right| \propto \frac{1}{\rho s}$$ (As for equal volume $m \propto \rho$)

Here $(\rho s)_A = 8\times 10^2 \times 2000 = 16 \times 15^5$ SI units

$(\rho s)_B = 10^3 \times 4000 = 4 \times 10^6$ SI units

$$(\rho s)_B > (\rho s)_A \Rightarrow \left|\frac{dT}{dt}\right|_B < \left|\frac{dT}{dt}\right|_A$$

Here T vs t curve of A is steeper than that of B. **Ans. (A)**

Sol. 5 If interface temperature is T, we use in steady state

$$\frac{3kA(\theta_2 - T)}{d} = \frac{kA(T-\theta_1)}{3d}$$

$$\Rightarrow \quad 3\theta_2 - 3T = \frac{T-\theta_1}{3}$$

$$\Rightarrow \quad 10T = 9\theta_2 + \theta_1$$

$$\Rightarrow \quad T = \frac{9\theta_2 + \theta_1}{10}$$ **Ans. (B)**

Sol. 6 Using average form of Newton's law of cooling it gives

$$\frac{50-40}{300} = k\left(\frac{50+40}{2} - 20\right) \quad \ldots(1)$$

and

$$\frac{40-T}{300} = k\left(\frac{40+T}{2} - 20\right) \quad \ldots(2)$$

Dividing (1) and (2), it gives

$$\frac{10}{40-T} = \frac{25}{T/2}$$

$$\Rightarrow \quad 5T = 100 - 25T$$

$$\Rightarrow \quad 30T = 1000$$

$$\Rightarrow \quad T = \frac{100}{3}\,°\text{C}$$ **Ans. (A)**

Sol. 7 Solar constant has units of Joule/s-m^2 thus its dimensional formula is given as

$$\left[\frac{\text{Joule}}{\text{sec}-m^2}\right] \equiv \frac{[ML^2T^{-2}]}{[T][L^2]} = MT^{-3}$$ **Ans. (D)**

Sol. 8

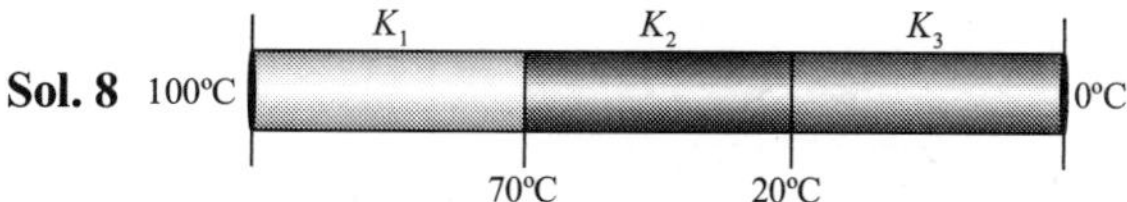

We consider rods have same length l and area of cross-section A so for steady state conductor, we use

$$\left(\frac{\Delta Q}{\Delta t}\right)_{AB} = \left(\frac{\Delta Q}{\Delta t}\right)_{BC} = \left(\frac{\Delta Q}{\Delta t}\right)_{CD}$$

$$\frac{(100-70)K_1 A}{l} = \frac{(70-20)K_2 A}{l} = \frac{(20-0)K_3 A}{l}$$

$\Rightarrow \quad 30K_1 = 50K_2 = 20K_3$

$\Rightarrow \quad 3K_1 = 2K_3$

$\Rightarrow \quad \frac{K_1}{K_3} = \frac{2}{3}$

and $\quad \frac{K_2}{K_3} = \frac{2}{5} = 2:5$ **Ans. (A)**

Sol. 9 Heat flow rate through the two slabswill be same through both

$$\Rightarrow \quad \frac{\theta_1 - \theta}{R_1} = \frac{\theta - \theta_2}{R_2}$$

$$R_2\theta_1 - R_2\theta = R_1\theta - R_1\theta_2$$

$$\theta = \frac{R_2\theta_1 + R_1\theta_2}{R_1 + R_2}$$ **Ans. (A)**

Sol. 10 For the two wires connected in series, thermal resistance can be calculated by series combination as

$$R_{\text{eff}} = \frac{l}{K_1 A} + \frac{l}{K_2 A} = \frac{2l}{K_{eq} A}$$

$$\Rightarrow \quad K_{\text{eq}} = \frac{2K_1K_2}{K_1 + K_2}$$ **Ans. (A)**

Sol. 11 Average rate of cooling of body is written as

$$\frac{\Delta T}{t} = K\left(\frac{T_1 + T_2}{2} - T_0\right)$$

$$\Rightarrow \quad \frac{75-65}{5} = K\left(\frac{75+65}{2} - 25\right) \quad \ldots(1)$$

$$\Rightarrow \quad \frac{65-T}{5} = K\left(\frac{T+65}{2} - 25\right) \quad \ldots(2)$$

From equations-(1) and (2), we have

$$\frac{65-T}{75-65} = \frac{\frac{T+65}{2} - 25}{\frac{75+65}{2} - 25}$$

$$\Rightarrow \quad \frac{65-T}{10} = \frac{T+15}{90}$$

$$\Rightarrow \quad 90 \times 65 - 90T = 10T + 10 \times 15$$

$$\Rightarrow \quad 100T = 90 \times 65 - 15 \times 10$$

$$\Rightarrow \quad T = 57°C$$ **Ans. (57)**

Sol. 12 By average form of Newton's law of cooling, we have

$$\frac{61-59}{4} = K\left[\frac{61+59}{2} - 30\right] \quad \ldots(1)$$

$$\frac{51-49}{t} = K\left[\frac{51+49}{2} - 30\right] \quad \ldots(2)$$

Dividing equation-(1) and (2), wehave

$$\frac{t}{4} = \frac{30}{20}$$

$$\Rightarrow \quad t = \frac{4 \times 30}{20} = 6 \text{ min}$$ **Ans. (C)**

Sol. 13 Replacing the rods with thermal resistance, equivalent circuit can be shown in figure below

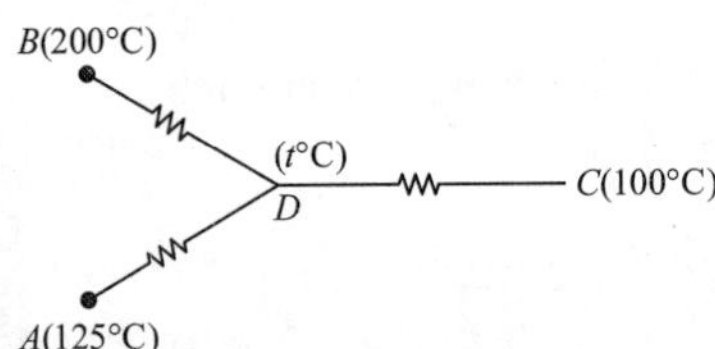

By Kirchoff's law total heat current at a junction must be zero, so we use

$$Q_{BD} + Q_{AD} = Q_{DC}$$

$$\Rightarrow \quad \frac{KA(200-t)}{\frac{l}{2}} + \frac{KA(125-t)}{l} = \frac{(t-100)KA}{\frac{l}{2}}$$

$$\Rightarrow \quad 400 - 2t + 125 - t = 2t - 200$$

$$\Rightarrow \quad 725 = 5t$$

$$\Rightarrow \quad t = 145°$$

$$\Rightarrow \quad Q_{AD} = \frac{KA(125-145)}{l} = \left|-\frac{20}{\frac{l}{KA}}\right| = \left|-\frac{20}{10}W\right| = 2W$$

Ans. (2)

Sol. 14 Figure below shows the situation described in the question.

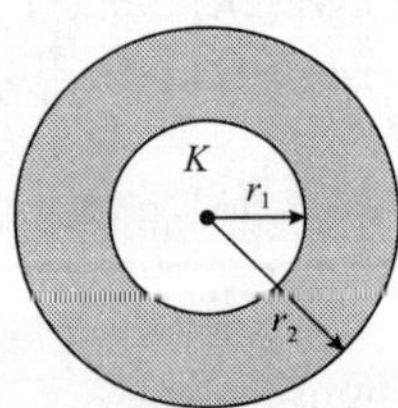

For this situation the equivalent thermal resistance of the shell is given as

$$\text{Equivalent resistance } R_{th} = \frac{r_2 - r_1}{K4\pi r_2 r_1}$$

So heat current through the shell material in radial direction is calculated as

$$Q = \frac{\theta_2 - \theta_1}{R_{th}} = \frac{4\pi K r_1 r_2(\theta_2 - \theta_1)}{(r_2 - r_1)}$$ **Ans. (D)**

11.1 Laws of Thermodynamics

Sol. 1 Work done by the gas in the cycle is

$$W = P_0V_0 \text{ (area enclosed)}$$

Heat supplied in cycle can be given as

$$\Delta Q_{supply} = Q_{AB} + Q_{BC} = nC_V(T_B - T_A) + nC_P(T_C - T_B)$$

Here we use $C_V = 1.5R$ and $C_P = C_V + R = 2.5R$

$$\Rightarrow \quad \Delta Q_{supply} = 1.5(nRT_B - nRT_A) + 2.5(nRT_C - nRT_B)$$

By Gas law, we use

$$\Delta Q_{supply} = 1.5(2P_0V_0 - P_0V_0) + 2.5(4P_0V_0 - 2P_0V_0)$$

$$\Rightarrow \quad \Delta Q_{supply} = 1.5P_0V_0 + 5P_0V_0$$

$$\Rightarrow \quad \Delta Q_{supply} = 6.5P_0V_0$$

Engine efficiency

$$\eta = \frac{W}{\Delta Q_{supply}}$$

$$= \frac{P_0V_0}{(13/2)P_0V_0} = \frac{2}{13} = 0.15 \quad \textbf{Ans. (B)}$$

Sol. 2 By First law of thermodynamics, we have for ACB

$$\Delta Q_{ACB} = \Delta U_{ACB} + \Delta W_{ACB}$$

$$\Rightarrow \quad 60 = \Delta U_{ACB} + 30 \text{ J}$$

$$\Rightarrow \quad \Delta U_{ACB} = 30 \text{ J} = \Delta U_{ADB}$$

For process ADB

$$\Delta Q_{ADB} = \Delta U_{ADB} + \Delta W_{ADB}$$

$$\Delta Q_{ADB} = 30 \text{ J} + 10 \text{ J} = 40 \text{ J}$$

Ans. (D)

Sol. 3 For closed vessel, heat supplied is given as

$$\Delta Q = nC_v\Delta T$$

$$\Rightarrow \quad \Delta Q = \frac{15}{28} \times \frac{5R}{2} \times (4T - T) \quad [T = 300 \text{ K}]$$

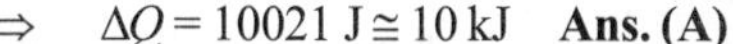

$$\Rightarrow \quad \Delta Q = 10021 \text{ J} \cong 10 \text{ kJ} \quad \textbf{Ans. (A)}$$

Sol. 4 Given system of carnot engines can be drawn as

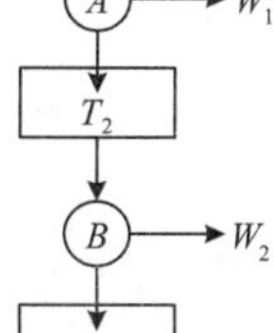

As we use $\quad W_1 = W_2$

$$\Rightarrow \quad \Delta Q_1 = \Delta Q_2$$

$$\Rightarrow \quad T_2 - T_1 = T_3 - T_2$$

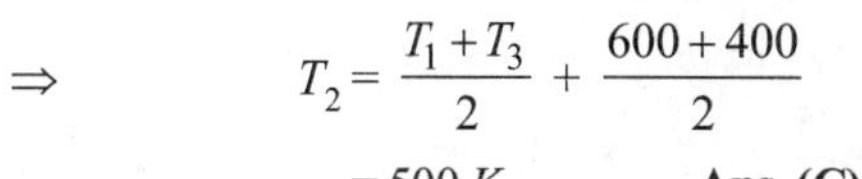

$$\Rightarrow \quad T_2 = \frac{T_1 + T_3}{2} + \frac{600 + 400}{2}$$

$$= 500\,K$$

Ans. (C)

Sol. 5 As engines are equally efficient, we use

$$\eta = 1 - \frac{T_2}{T_1} = 1 - \frac{T_3}{T_2} = 1 - \frac{T_4}{T_3}$$

$$\Rightarrow \quad \frac{T_2}{T_1} = \frac{T_3}{T_4} = \frac{T_4}{T_3}$$

$$\Rightarrow \quad T_2 = \sqrt{T_1T_3}$$

and $\quad T_3 = \sqrt{T_2T_4}$

$$\Rightarrow \quad T_2^{3/4} = \sqrt{T_1^{1/2}T_4^{1/4}}$$

$$\Rightarrow \quad T_2 = T_1^{2/3}T_4^{1/3}$$

and $\quad T_3 = T_1^{1/3}T_4^{2/3}$ **Ans. (A)**

Sol. 6 Work done by gas is

$$W = P\Delta V = nR\Delta T$$

$$\Rightarrow \quad W = \frac{1}{2} \times 8.31 \times 70$$

$$\Rightarrow \quad W = 291 \text{ J}$$

Ans. (B)

Sol. 7 For adiabatic process we use $TV^{\gamma-1}$ = constant and for a diatomic gas we use

$$\gamma - 1 = \frac{7}{5} - 1 = \frac{2}{5}$$

$$\Rightarrow \quad x = \frac{2}{5} \quad \textbf{Ans. (B)}$$

Sol. 8 Given that $VT = K$

$$\Rightarrow \quad V\left(\frac{PV}{nR}\right) = K$$

$$\Rightarrow \quad PV^2 = K$$

For a polytropic process, molar specific heat of gas is given as

$$C = \frac{R}{1-n} + C_V \qquad (n = 2)$$

$$\Rightarrow \quad C = \frac{R}{1-2} + \frac{3R}{2} = \frac{R}{2}$$

Heat supplied to gas is

$$\Delta Q = nC\,\Delta T$$

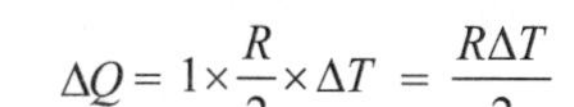

$$\Rightarrow \quad \Delta Q = 1 \times \frac{R}{2} \times \Delta T = \frac{R\Delta T}{2}$$

Ans. (A)

Sol. 9 Since P–V indicator diagram is given, so work done by gas is area enclosed by the cyclic diagram

$$\Rightarrow \quad \text{Work done by gas} = \frac{1}{2} \times 4 \times 5 \text{ J}$$

$$\Rightarrow \quad W = 10 \text{ J} \quad \textbf{Ans. (C)}$$

Sol. 10 Process $a \rightarrow$ constant $\Rightarrow$ Isobaric

Process $b \rightarrow$ Isothermal
Process $c \rightarrow$ Adiabatic
Process $d \rightarrow$ Isochoric

Adiabatic curve is steeper them isothermal curve

Order as specified in que $\rightarrow$ d, a, b, c **Ans. (D)**

Sol. 11

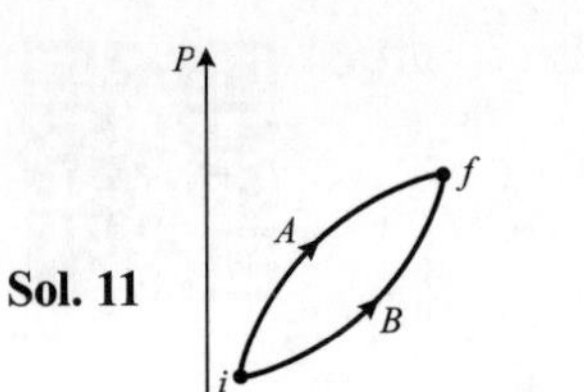

Initial and final states for both the processes are same so we use

$$\Delta U_A = \Delta U_B$$

Work done during process A is greater than in process B so using First Law of thermodynamics, we have

$$\Delta Q = \Delta U + W$$

$$\Rightarrow \quad \Delta Q_A > \Delta Q_B \quad \textbf{Ans. (B)}$$

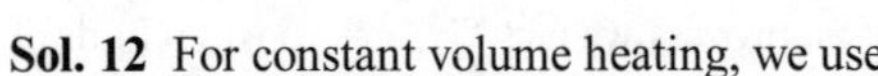

Sol. 12 For constant volume heating, we use

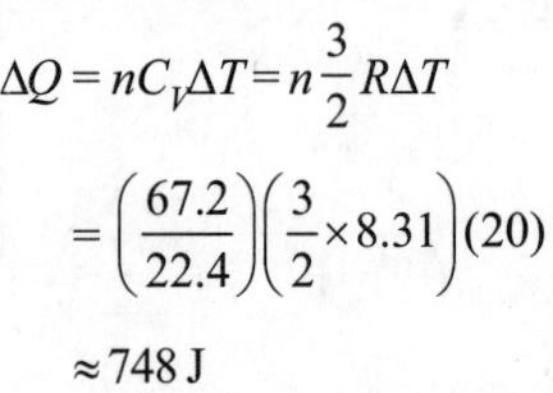

$$\Delta Q = nC_V \Delta T = n\frac{3}{2}R\Delta T$$

$$= \left(\frac{67.2}{22.4}\right)\left(\frac{3}{2} \times 8.31\right)(20)$$

$$\approx 748 \text{ J}$$ **Ans. (A)**

Sol. 13 For isobaric process work done by gas is given as

Heat supplied is $\quad W = nR\Delta T$

$$\Delta Q = (C_v + nR)\Delta T$$

$$\Rightarrow \quad \frac{W}{\Delta Q} = \frac{nR}{C_v + nR}$$ **Ans. (C)**

Sol. 14 Heat required for constant volume and constant pressure to raise gas temperature by ΔT is

$$Q = nC_v \Delta T$$

$$Q' = nC_p \Delta T$$

$$\Rightarrow \quad \frac{Q'}{Q} = \frac{C_p}{C_v}$$

For diatomic gas we have

$$\frac{C_p}{C_v} = \gamma = \frac{7}{5}$$

$$\Rightarrow \quad Q' = \frac{7}{5}Q$$ **Ans. (A)**

Sol. 15 Temperature of gas ($n = 1$) is given as

$$PV = RT \Rightarrow T = \frac{PV}{R} = \frac{P_0 V}{R}\left(1 - \frac{1}{2}\left(\frac{V_0}{V}\right)^2\right)$$

At $V = V_0$, temperature is

$$T_1 = \frac{P_0 V_0}{R}\left[1 - \frac{1}{2}\right] = \frac{P_0 V_0}{2R}$$

At $V = 2V_0$, temperature is

$$T_2 = \frac{2P_0 V_0}{R}\left[1 - \frac{1}{2}\left(\frac{1}{4}\right)\right] = \frac{7P_0 V_0}{4R}$$

Temperature change

$$\Delta T = T_2 - T_1 = \frac{7P_0 V_0}{4R} - \frac{P_0 V_0}{2R} = \frac{5P_0 V_0}{4R}$$ **Ans. (C)**

Sol. 16 Work done along path abc is

$$W_{abc} = W_{ab}$$

As $\quad W_{bc} = 0 \quad$ (being an isochoric process)

By First law of thermodynamics, we use

$$W_{ab} = \Delta Q_{ab} - \Delta U_{ab}$$

In a cyclic process net

$$\Delta U_{cycle} = 0$$

$$\Delta U_{ab} + \Delta U_{bc} + \Delta U_{ca} = 0$$

$$\Rightarrow \quad \Delta U_{ab} + 60 - 180 = 0 \qquad [\Delta U_{bc} = \Delta Q_{bc}]$$

$$\Rightarrow \quad \Delta U_{ab} = 120 \text{ J}$$

$$= 250 - 120 = 130 \text{ J}$$ **Ans. (D)**

Sol. 17 Efficiency of Carnot engine

$$\eta = 1 - \frac{T_{sink}}{T_{source}}$$

$$\Rightarrow \quad \frac{1}{6} = 1 - \frac{T_{sink}}{T_{source}}$$

$$\Rightarrow \quad \frac{T_{sink}}{T_{source}} = \frac{5}{6} \quad \ldots(1)$$

Efficiency gets doubled when temperature of sink decreases by 62°C so we use

$$\frac{2}{6} = 1 - \frac{T_{sink} - 62}{T_{source}}$$

$$\Rightarrow \quad \frac{62}{T_{source}} = \frac{1}{6} \quad \ldots(2)$$

$$\Rightarrow \quad T_{source} = 372 \text{ K} = 99^\circ\text{C}$$

From equation-(1), we have

$$T_{sink} = \frac{5}{6} \times 372 = 310 \text{ K} = 37^\circ\text{C}$$ **Ans. (B)**

Sol. 18 For a diatomic gas, we have

$$C_p = \frac{7}{2}R$$

As gas undergoes isobaric process, heat absorbed by gas is

$$\Rightarrow \quad \Delta Q = nC_P\Delta T$$

In isobaric process, work done is

Also, $\Delta W = nR\Delta T = 10\text{J (given)}$

$$\Rightarrow \quad \Delta Q = n\left(\frac{7}{2}R\Delta T\right) = \frac{7}{2}(nR\Delta T)$$

$$= \frac{7}{2}\times 10 = 35\text{ J}$$

Ans. (A)

Sol. 19 Work done in adiabatic process is given as

$$W = \frac{nR(T_1 - T_2)}{\gamma - 1} = \frac{P_1V_1 - P_2V_2}{0.4}$$

$$\Rightarrow \quad W = \frac{100\times 1 - \frac{100}{4.6555}\times 3}{0.4} = 88.90$$

Thus closest answer is option (A) **Ans. (A)**

Sol. 20 For equivalent r for mixture we use

$$\frac{n_1 + n_2}{r_{eq} - 1} = \frac{n_1}{r_1 - 1} + \frac{n_2}{r_2 - 1}$$

$$\Rightarrow \quad \frac{5}{r_{eq} - 1} = \frac{2}{2/3} + \frac{3}{1/3}$$

$$\Rightarrow \quad r_{eq} = \frac{17}{12} = 1.416$$

$$\simeq 1.42$$

Ans. (B)

Sol. 21 For Carnot Cycle, we use

$$\frac{Q_1}{Q_2} = \frac{T_1}{T_2}$$

$$\Rightarrow \quad \frac{Q_2 + 1200}{Q_2} = \frac{900}{300} \qquad [\text{As } Q_1 = Q_2 + W]$$

$$\Rightarrow \quad Q_2 + 1200 = 3Q$$

$$\Rightarrow \quad Q_2 = 600\text{ J}$$

Ans. (600)

Sol. 22 Electric field due to each charged sheet is given as

$$E = \frac{\sigma}{2\varepsilon_0}$$

Which in direction normal to planes as shown in figure

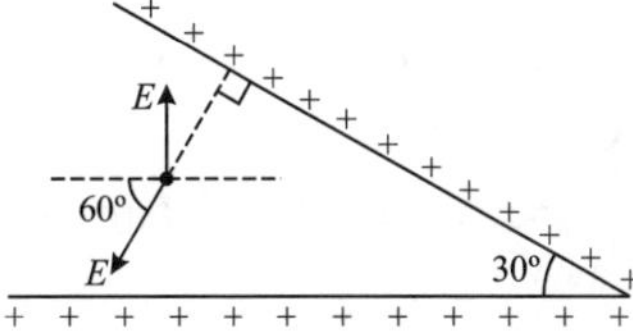

Net electric field between plates is given as

$$\vec{E}_R = E\cos 60°\,(-\hat{x}) + (E - E\sin 60°)\,(\hat{y})$$

$$\Rightarrow \quad \vec{E}_R = \frac{\sigma}{2\varepsilon_0}\left[-\frac{\hat{x}}{2} + \left(1 - \frac{\sqrt{3}}{2}\right)\hat{y}\right]$$

Ans. (B)

Sol. 23 For the engines, we use

$$\frac{Q}{Q_0} = \frac{T_1}{T}$$

and $$\frac{Q_0}{Q_2} = \frac{T_1}{T}$$

As work is equal by both engines, we use

$$\frac{Q_2 + W}{Q_2} = \frac{T_1}{T}$$

and $$\frac{Q_2}{Q_2 - W} = \frac{T}{T_2}$$

Solving we get

$$T = \frac{T_1 + T_2}{2}$$

Ans. (C)

Sol. 24 Mean collision time is taken as

$$t \propto \frac{V}{\sqrt{T}} \qquad \ldots(1)$$

And for adiabatic process, we use

$$TV^{\gamma - 1} = \text{constant} \qquad \ldots(2)$$

$$\Rightarrow \quad t \propto V^{\frac{\gamma+1}{2}}$$

Ans. (A)

Sol. 25 Given processes are

$x \to y$ Isobaric, $y \to z \Rightarrow$ Isochoric, $z \to x \Rightarrow$ Isothermal

So by slope analysis correct option is (D) **Ans. (D)**

Sol. 26 Efficiency for refrigerator cycle is

$$\eta = \frac{W}{Q_1} = \frac{W}{W + Q_2}$$

$$\frac{1}{10} = \frac{10}{10 + Q_2}$$

$$\Rightarrow \quad Q_2 = 90\text{ J}$$

Heat absorbed from the reservoir at lower temperature is 90 J

Ans. (C)

Sol. 27 C_P/C_V for mixture is given as

$$\gamma_{\text{mixture}} = \frac{n_1C_{P_1} + n_2C_{P_2}}{n_1C_{V_1} + n_2C_{V_2}}$$

$$\Rightarrow \quad \gamma_{\text{mixture}} = \frac{n\times\left(\frac{5R}{2}\right) + 2n\left(\frac{7R}{2}\right)}{n\times\frac{3R}{2} + 2n\left(\frac{5R}{2}\right)}$$

$$\Rightarrow \quad \gamma_{\text{mixture}} = \frac{19}{13}$$

Ans. (B)

Sol. 28 For a diatomic gas molecule there are 5 degrees of freedom if vibration mode is absent and if vibration mode is there then degrees of freedom will be 7 so specific heats in the two cases are $C_V^A = \frac{5}{2}R$ and $C_V^B = \frac{7}{2}R \Rightarrow \frac{C_V^A}{C_V^B} = \frac{5}{7}$.

Ans. (B)

Sol. 29 In the given *PV* diagram process 2 to 3 is isobaric & process 3 to 1 is isochoric. This is shown only in option (D).

Ans. (D)

Sol. 30 For an adiabatic process, we use

$$PV^{\gamma} = \text{constant}$$

and $\quad TV^{\gamma-1} = C$

For the given situation, we use

$$300 \times V^{\frac{7}{5}-1} = T_2\left(\frac{V}{16}\right)^{\frac{7}{5}-1}$$

$$\Rightarrow \quad T_2 = 300 \times 2^{4\times\frac{2}{5}}$$

For isobaric process, we use

$$\frac{V_2}{T_2} = \frac{V_3}{T_3}$$

Using $V_3 = 2V_2$, we get

$$\frac{1}{2} = \frac{T_2}{T_3} \Rightarrow T_3 = 2T_2$$

$$\Rightarrow \quad T_3 = 2 \times 300 \times 2^{\frac{8}{5}} = 1818.85$$

Ans. (1818 to 1819)

Sol. 31 Total internal energy of mixture is given as

$$U = \frac{f_1}{2}n_1RT + \frac{f_2}{2}n_2RT$$

$$\Rightarrow \quad U = \frac{5}{2} \times 3RT + \frac{3}{2} \times 5RT = 15RT$$

Ans. (B)

Sol. 32 As volume compresses to *V*/10, we use

$$TV^{\gamma-1} = \text{Constant} \qquad [r = 7/5 \text{ for air}]$$

$$\Rightarrow \quad T_1V_1^{2/5} = T_2\left(\frac{V}{10}\right)^{2/5}$$

$$\Rightarrow \quad T_2 = T(10)^{2/5}$$

Change in internal energy in adiabatic process is

$$\Delta U = \frac{R(T_2 - T_1)}{\gamma - 1}$$

$$\Rightarrow \quad \Delta U = \frac{5 \times 8.31 \times 293 \times (10^{2/5} - 1)}{\frac{7}{5} - 1}$$

$$\cong 4 \times 10^3 \text{ J}$$

$$\Rightarrow \quad x = 4$$

Ans. (4)

Sol. 33 Efficiency of heat engine is given as

$$\eta = \frac{\text{Work done}}{\text{Heat supplied}}$$

$$\Rightarrow \quad \eta = \frac{1915 - 40 + 125 - Q}{1915 + 125} = 0.5$$

$$\Rightarrow \quad \frac{2000 - Q}{2040} = 0.5$$

$$\Rightarrow \quad 2000 - Q = 1020$$

$$\Rightarrow \quad Q = 980 \text{ J}$$

Ans. (C)

Sol. 34 As all sudden process are considered irreversible and in these processes no time is available for heat exchange so considered adiabatic.

Ans. (B)

Sol. 35 For triangular molecule degrees of freedom are given translational *dof* = 3

Rotational, *dof* = 3

Thus internal energy of gas is given as

$$U = \frac{f}{2}nRT = 3RT$$

Ans. (D)

Sol. 36 For minimum work, we use Carnot cycle so, coefficient of performance is given as

$$COP = \frac{Q_{in}}{W} = \frac{Q_{in}}{Q_{out} - Q_{in}} = \frac{273}{300 - 273} = \frac{273}{27}$$

$$\Rightarrow \quad W = \frac{27}{273} \times Q_{in} = \frac{27}{273} \times 100 \times 80 \text{ cal}$$

Heat rejected by refrigerator is given as

$$Q_{out} = W + Q_1$$

$$\Rightarrow \quad Q_{out} = \frac{27}{273} \times 100 \times 80 + 100 \times 80 \text{ cal}$$

$$\Rightarrow \quad Q_{out} = 8791.2 \text{ cal}$$

Ans. (8791)

Sol. 37 For mixture of gases, equilibrium temperature is given as

$$T_{eq} = \frac{n_1T_1 + n_2T_2}{n_1 + n_2} = \frac{0.1 \times 200 + 0.05 \times 400}{0.1 + 0.05}$$

$$\Rightarrow \quad T_{eq} = \frac{20 + 20}{0.15} = \frac{40}{0.15} = \frac{800}{3}$$

$$\Rightarrow \quad T_{eq} = 266.66 \text{ K}$$

Ans. (266)

Sol. 38 (I) In Adiabatic process no heat exchange take place so option (B) is correct match

(II) In Isothermal process no change internal energy so option (D) is correct match

(III) In Isochoric process V = constant. So $\Delta\omega = 0$

(IV) In isobaric process as P = constant option (C) is correct match.

Ans. (A)

Sol. 39 By gas law we use

$$PV = nRT$$

At constant pressure

$$P\Delta V = nR\Delta T$$

$$\Rightarrow \quad \Delta T = \left(\frac{P}{nR}\right)\Delta V$$

At constant temperature

$$P\Delta V + V\Delta P = 0$$

$$\Rightarrow \quad \Delta P = -\frac{P\Delta V}{V}$$

Value of C is given as

$$C = \left|\frac{\Delta T}{\Delta P}\right| = \frac{V}{nR} = \frac{T}{P} = \frac{300}{2}$$

$$\Rightarrow \quad C = 150\text{ K/atm}$$ **Ans. (150)**

Sol. 40 From the given indicter diagram, we can see for all processes as ΔT is same

$$\Delta U = nC_v\Delta T = \text{same}$$

$AB \rightarrow$ volume is increasing $\Rightarrow W > 0$

$AD \rightarrow$ volume is decreasing $\Rightarrow W < 0$

$AC \rightarrow$ volume is constant $\Rightarrow W = 0$ **Ans. (A)**

Sol. 41 Using gas law, we use

$$n = \frac{PV}{RT}$$

Mean kinetic energy of a molecule is given as

$$\frac{3}{2}kT = 4\times 10^{-14}$$

$$\Rightarrow \quad RT = \frac{8}{3}\times 10^{-14}\times N_{AV} \qquad \left[\text{As } k = \frac{R}{N_{AV}}\right]$$

Total molecules in gas are given as

$$N = \frac{PV}{RT}\times N_{AV}$$

$$\Rightarrow \quad N = \frac{2\times 13.6\times 980\times 4}{\frac{8}{3}\times 10^{-14}} = 3.99\times 10^{18}$$ **Ans. (C)**

Sol. 42 In adiabatic process we use

$$PV^{\gamma} = \text{constant}$$

$$\Rightarrow \quad P\left(\frac{m}{\rho}\right)^{\gamma} = \text{constant}$$

$$\Rightarrow \quad P \propto \rho^{\gamma}$$

$$\Rightarrow \quad \frac{P_f}{P_i} = \left(\frac{\rho_f}{\rho_i}\right)^{\gamma} = (32)^{7/5} = 2^7 = 128$$ **Ans. (D)**

Sol. 43 Internal energy of gas for f degrees of freedom is given as

$$U = \frac{f}{2}nRT = \frac{5}{2}RT$$

Ratio of specific heats is given as

$$\gamma = \frac{C_P}{C_V}$$

$$\Rightarrow \quad \gamma = \frac{f+2}{f} = \frac{5+2}{5} = \frac{7}{5}$$ **Ans. (D)**

Sol. 44 Work done in cycle $ABCDA$ is given as

$$W_{ABCDA} = 2P_0V_0$$

Total heat input in cycle is given as

$$Q_{in} = Q_{AB} + Q_{BC}$$

Here, we use in process AB

$$Q_{AB} = nC(T_B - T_A)$$

$$\Rightarrow \quad Q_{AB} = \frac{n3R}{2}(T_B - T_A)$$

$$\Rightarrow \quad Q_{AB} = \frac{3}{2}(P_BV_B - P_AV_A)$$

$$\Rightarrow \quad Q_{AB} = \frac{3}{2}(3P_BV_0 = P_0V_0) = 3P_0V_0$$

For process BC, we use

$$Q_{BC} = nC_P(T_C - T_B)$$

$$\Rightarrow \quad Q_{BC} = \frac{n5R}{2}(T_C - T_B)$$

$$\Rightarrow \quad Q_{BC} = \frac{5}{2}(P_CV_C - P_BV_B)$$

$$\Rightarrow \quad Q_{BC} = \frac{5}{2}(6P_0V_0 - 3P_0V_0) = \frac{15}{2}P_0V_0$$

Efficiency of cycle is given as

$$\eta = \frac{W}{Q_{in}}\times 100$$

$$\Rightarrow \quad \eta = \frac{2P_0V_0}{3P_0V_0 + \frac{15}{2}P_0V_0}\times 100$$

$$\Rightarrow \quad \eta = \frac{400}{21} = 19.04 \approx 19$$ **Ans. (19)**

Sol. 45 All the given processes in list-I are by definition related to the parameters given in list II. **Ans. (D)**

Sol. 46 Work done in the process AB is given as

$$W_{AB} = nRT\ln\left(\frac{v_2}{v_1}\right) = nRT\ln\left(\frac{2V}{V}\right) = nRT\ln 2$$

Work done in process BC and CA are given as

$$W_{BC} = P\Delta V = nR\Delta T = nR\left(\frac{T}{2} - T\right) = -nR\frac{T}{2}$$

$$W_{CA} = 0$$

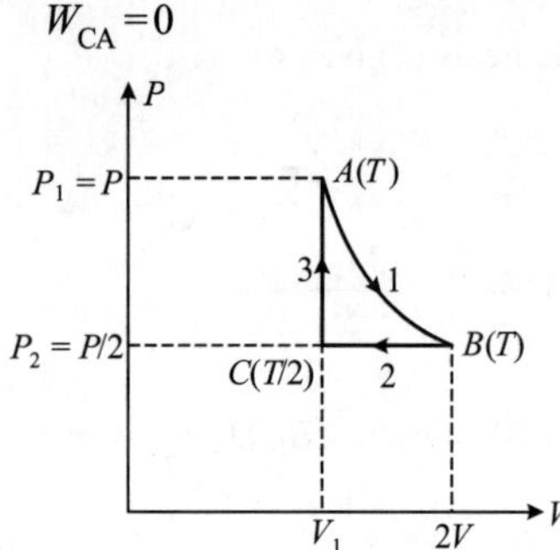

Net work done in the cycle is given by sum of all

$$W_{net} = nRT\left(\ln 2 - \frac{1}{2}\right)$$ **Ans. (A)**

Sol. 47 Work done in process AB is given as

$$W_{AB} = P_1V_1 \ln\left[\frac{2V_1}{V_1}\right] = P_1V_1 \ln(2)$$

Work done in process BC is given as

$$W_{BC} = 0$$

Work done in process CA is given as

$$W_{CA} = \frac{P_1V_1 - \frac{P_1}{4} \times 2V_1}{1-\gamma}$$

$$\Rightarrow \quad W_{CA} = \frac{P_1V_1\left[1 - \frac{1}{2}\right]}{1-\gamma} = \frac{P_1V_1}{2(1-\gamma)}$$

Net work done is calculated as

$$W_{net} = W_{AB} + W_{BC} + W_{CA}$$

$$\Rightarrow \quad W_{net} = P_1V_1 \ln(2) + 0 + \frac{P_1V_1}{2(1-\gamma)}$$

$$\Rightarrow \quad W_{net} = RT\left[\ln(2) - \frac{1}{2(\gamma-1)}\right]$$ **Ans. (D)**

Sol. 48 For an isobaric process, we use

$$dU = nC_v dT$$

$$dQ = nC_p dT$$

$$dW = PdV = nRdT$$

$$\Rightarrow \quad dU : dQ : dW = C_v : C_p : R$$

$$\Rightarrow \quad dU : dQ : dW = \frac{5R}{2} : \frac{7R}{2} : R = 5 : 7 : 2$$ **Ans. (B)**

Sol. 49 When container is suddendly stopped and due to this the change in temperature is ΔT, we have

$$\frac{1}{2}mv^2 = nC_v\Delta T$$

$$\Rightarrow \quad \frac{1}{2}m \times (30)^2 = \frac{m}{M} \times \frac{3R}{2} \times \Delta T$$

$$\Rightarrow \quad \Delta T = \frac{3600}{3R}$$ **Ans. (3600)**

Sol. 50 For the given thermodynamic process, we have

$$\Delta T = 300 - 100 = 200°C$$

As $P = kV^3$ we use by gas law $PV = nRT$ we have

$$kV^4 = nRT$$

$$\Rightarrow \quad 4kV^3\, dV = nRdT$$

$$\Rightarrow \quad PdV = nRdT/4$$

Work done in the process is given as

$$W = \int PdV = \int \frac{nRdT}{4} = \frac{nR}{4}\Delta T$$

$$\Rightarrow \quad W = \frac{200}{4} \times nR = 50nR$$ **Ans. (50)**

Sol. 51 For the thermodynamic process $PV^{½} = Constant$, we use by gas law

$$\frac{nRT}{V} V^{½} = C$$

$$\Rightarrow \quad T = C_1 V^{½}$$

$$\Rightarrow \quad \frac{T_2}{T_1} = \left(\frac{V_2}{V_1}\right)^{1/2} = \left(\frac{2V_1}{V_1}\right)^{1/2}$$

$$\Rightarrow \quad \frac{T_2}{T_1} = \sqrt{2}$$ **Ans. (B)**

Sol. 52 Efficiency of given heat engine is

$$\eta = \frac{1}{4} = 1 - \frac{T_2}{T_1}$$

$$\Rightarrow \quad \frac{T_2}{T_1} = \frac{3}{4}$$

When temperature of sink is reduced by 52 K, efficiency is doubled so we have

$$\frac{T_2 - 52}{T_1} = \frac{1}{2}$$

$$\Rightarrow \quad \frac{0.75T_1 - 52}{T_1} = \frac{1}{2}$$

$$\Rightarrow \quad T_1 = 208\text{ K}$$ **Ans. (208)**

Sol. 53 If equilibrium pressure of mixture is P atm. then internal energy of mixture is given as

$$U_1 + U_2 = U_{mixutre}$$

$$\Rightarrow \quad \frac{f}{2}P_1V_1 + \frac{f}{2}P_2V_2 = \frac{f}{2}P(V_1 + V_2)$$

$$\Rightarrow \quad \frac{f}{2}(2)(4.5) + \frac{f}{2}(3)(5.5) = \frac{f}{2}P(4.5 + 5.5)$$

$$\Rightarrow \quad P = 2.55 = x \times 10^{-1}\text{ atm}$$

$$\Rightarrow \quad x = 25.5 \approx 26 \text{ (Nearest integer)}$$

Ans. (26)

Sol. 54 The thermodynamics process is given as

$$U = 3PV + 4$$

If the gas has f degrees of freedom then we can write

$$\frac{nf}{2}RT = 3PV + 4$$

$$\Rightarrow \quad \frac{f}{2}PV = 3PV + 4$$

$$\Rightarrow \quad f = 6 + \frac{8}{PV}$$

Since degree of freedom is more than 6 therefore gas is polyatomic. **Ans. (B)**

Sol. 55 By first law of thermodynamics, we have

$$Q = \Delta U + W$$

$$\Rightarrow \quad Q = \Delta U + \frac{Q}{5}$$

$$\Rightarrow \quad \Delta U = \frac{4Q}{5}$$

$$\Rightarrow \quad nC_v\Delta T = \frac{4}{5}nC\Delta T$$

$$\Rightarrow \quad \frac{5}{4}C_V = C$$

$$\Rightarrow \quad C = \frac{5}{4}\left(\frac{f}{2}\right)R = \frac{5}{4}\left(\frac{5}{2}\right)R$$

$$\Rightarrow \quad C = \frac{25}{8}R$$

$$\Rightarrow \quad x = 25$$ **Ans. (25)**

Sol. 56 Work done in the process is given as

$$W = \int PdV \qquad \ldots(1)$$

$$\Rightarrow \quad W = \int \frac{nRT}{V}dV$$

$$\Rightarrow \quad W = \int \frac{nRT}{KT^{2/3}} \cdot dV$$

From the process equation, we have

$$dV = \frac{2}{3}KT^{-1/3}dT$$

$$\Rightarrow \quad W = \int_{T_1}^{T_2} \frac{nRT}{KT^{2/3}}\frac{2}{3}K\frac{1}{T^{1/3}}dT$$

$$\Rightarrow \quad W = \frac{2}{3}nR \times (T_2 - T_1)$$

Given that $T_2 - T_1 = 90$ K, so we have

$$\Rightarrow \quad W = \frac{2}{3}nR \times 90 = 60\,nR$$

Considering 1 mole of gas, $W = 60\,R$. **Ans. (60)**

Sol. 57 Heat and work are considered as path functions for thermodynamic processes and related by first law of thermodynamics as

$$\Delta Q = \Delta U + \Delta W$$

Since work done by gas depends on type of process which is determined by path and ΔU depends just on initial and final states, so ΔQ, the heat supplied, also depend on path of process. **Ans. (A)**

Sol. 58 Efficiency of heat engine is given as

$$\eta = 0.60 = 1 = \frac{T_L}{T_H}$$

$$\Rightarrow \quad \frac{T_L}{T_H} = 0.4$$

$$\Rightarrow \quad T_L = 0.4 \times 400 = 160\text{ K} = -113°\text{C}$$ **Ans. (–113)**

Sol. 59 As each vibrational mode has 2 degrees of freedom so number of vibrational degrees of freedom will be 48 and total degrees of freedom are given as

$$f = 3 + 3 + 48 = 54$$

Thus ratio of specific heats is given as

$$\gamma = 1 + \frac{2}{f} = \frac{28}{27} = 1.03$$ **Ans. (A)**

Sol. 60 The efficiency of carnot engine is given as

$$\eta = \frac{T_2}{T_1} = \frac{Q_2}{Q_1} = \frac{Q_1 - W}{Q_1}.$$

$$\Rightarrow \quad \frac{400}{800} = 1 - \frac{W}{Q_1}$$

$$\Rightarrow \quad \frac{W}{Q_1} = 1 - \frac{1}{2} = \frac{1}{2}$$

$$\Rightarrow \quad Q_1 = 2W = 2400\,\text{J}$$ **Ans. (D)**

Sol. 61 Option (a) is wrong as as in adiabatic process $V \neq$ constant and 0ption (b) is wrong as in isothermal process $P \neq$ constant.

In option (c) the slope of adiabatic curve on V–T diagram is negative and increasing which matches with its process equation hence correct and similarly in option (d) the slope of adiabatic curve on P–T diagram is positive and increasing which also matches with its process equation hence correct.

Ans. (B)

Sol. 62 The adiabatic process is from C to D can be calculated as

$$W_{CD} = \frac{P_2V_2 - P_1V_1}{1-\gamma} = \frac{P_DV_D - P_CV_C}{1-\gamma}$$

$$\Rightarrow \quad W_{CD} = \frac{200(3) - (100)(4)}{1 - 1.4} = -500\,\text{J}$$ **Ans. (A)**

Sol. 63 As work is zero, for isochoric process, we use

$$Q = \Delta U$$

$$\Rightarrow \quad Q = nC_v\Delta T = 4 \times \frac{5R}{2} \times 50 = 500\,R$$ **Ans. (B)**

Sol. 64 In the given cyclic process the heat absorbed by the system can be calculated by area enclosed by the cyclic curve, given as

$$\Delta Q = \pi ab$$

$$\Rightarrow \quad \Delta Q = \left[\frac{40-20}{2} \times 10^3\right] \times \left[\frac{40-20}{2} \times 10^{-3}\right]$$

$$\Rightarrow \quad \Delta Q = 100\,\pi\ \text{Joule}$$ **Ans. (100)**

Sol. 65 Work in the isothermal process AB shown in the indicator diagram is given as

$$W = nRT \ln \frac{V_2}{V_1}$$

$$\Rightarrow \quad W = 1 \times 8.3 \times 300 \ln 2 = 1725.8\,\text{J}$$ **Ans. (17258)**

Sol. 66 For adiabatic process, the process equation can be written as

$$\frac{T_1}{T_2} = \left(\frac{V_2}{V_1}\right)^{\gamma-1}$$

$$\Rightarrow \quad \frac{T_1}{T_2} = \left(\frac{V_2}{V_1}\right)^{5/3-1} = \left(\frac{Al_2}{Al_1}\right)^{2/3} = \left(\frac{l_2}{l_1}\right)^{2/3}$$ **Ans. (C)**

Sol. 67 Heat supplied can be written as

$$Q = nC\Delta T$$

Rate of heat supplied can be written as

$$\frac{dQ}{dt} = nC\frac{dT}{dt}$$

Rate of heat supply is same for both gases, so we use

$$n_1C_1\left(\frac{dT}{dt}\right)_1 = n_2C_2\left(\frac{dT}{dt}\right)_2$$

$$\Rightarrow \quad \frac{C_1}{C_2} = \frac{\left(\frac{dT}{dt}\right)_2}{\left(\frac{dT}{dt}\right)_1} = \frac{\frac{90}{6}}{\frac{120}{3}} = \frac{90 \times 3}{120 \times 6} = \frac{3}{8}$$ **Ans. (D)**

Sol. 68 Considering Carnot cycle in the given heat engine as temperature based efficiency calculation is to be done. Thus efficiency of a carnot engine is given as

$$\eta = \left(1 - \frac{T_2}{T_1}\right)$$

$$\Rightarrow \quad \frac{T_2}{T_1} = 1 - \eta = 1 - \frac{1}{6} \qquad \ldots(1)$$

$$\Rightarrow \quad \frac{T_2 - 62}{T_1} = 1 - \frac{1}{3} \qquad \ldots(2)$$

From equations-(1) & (2), we use

$$\frac{T_2}{T_2 - 62} = \frac{5}{6} \times \frac{3}{2} = \frac{5}{4}$$

$$\Rightarrow \quad T_2 = 5 \times 62$$

From equation-(1), we have

$$T_1 = \frac{T_2}{1-\eta} = \frac{5 \times 62}{1 - \frac{1}{6}} = 5 \times 62 \times \frac{6}{5} = 372\text{K}$$

$$\Rightarrow \quad T_1 = 372 - 273 = 99°\text{C}$$ **Ans. (A)**

Sol. 69 From the given curve in question, we have

$$T_A = T_B$$

and $$T_C = T_D$$

Work done in process BC is given as

$$W_{BC} = \frac{nR(T_B - T_C)}{\gamma - 1}$$

Work done in process AD is given as

$$W_{AD} = \frac{nR(T_A - T_D)}{\gamma - 1}$$

$$\Rightarrow \quad W_{AD} = W_{BC} \qquad \textbf{Ans. (D)}$$

Sol. 70 For four vibrational modes total 8 vibrational degrees of freedom are added so total degress of freedom of gas molecules are

$$f = 6 + 8 = 14$$

Work done in adiabatic process is given as

$$W = \frac{P_1V_1 - P_2V_2}{\gamma - 1} = \frac{nR(T_1 - T_2)}{\gamma - 1}$$

$$\Rightarrow \quad W = \frac{1 \times 8.314 \times (-10)}{\left(\frac{8}{7} - 1\right)} \quad \left(as\ \gamma = 1 + \frac{2}{f} = \frac{8}{7}\right)$$

$$\Rightarrow \quad W = -582\text{ J} \qquad \textbf{Ans. (C)}$$

Sol. 71 Figure below shows the two carnot engines working in series as described in question

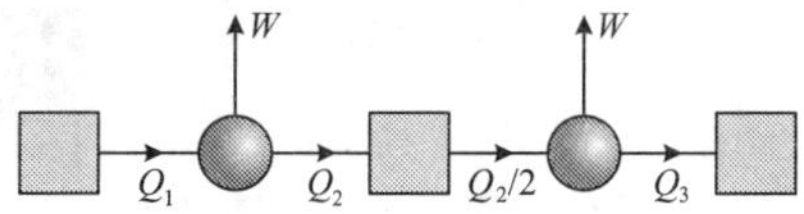

Given that work done in the two engines are equal so we use

$$W = Q_1 - Q_2$$

$$W = \frac{Q_2}{2} - Q_3$$

$$\Rightarrow \quad Q_1 - Q_2 = \frac{Q_2}{2} - Q_3$$

$$\Rightarrow \quad Q_1 + Q_3 = \frac{3}{2}Q_2$$

$$\Rightarrow \quad T_1 + \frac{T_3}{2} = \frac{3}{2}T$$

$$\Rightarrow \quad T = \frac{2}{3}T_1 + \frac{1}{3}T_3 \qquad \textbf{Ans. (D)}$$

Sol. 72 Rate of energy supply is given as

$$\frac{6000}{60}\text{ J/s} = 100\text{ J/s}$$

By first law of thermodynamics, for a time t, we have

$$\Delta U = Q - W$$

$$2500 = 100t - 90t$$

$$\Rightarrow \quad t = 250\text{ s} \qquad \textbf{Ans. (C)}$$

Sol. 73 For a refrigerator if Q_1 is the amount of heat rejected per second to surrounding and Q_2 is the heat extracted per second from the cooling chamber and W is the work done per second by compressor then we use

$$W + Q_2 = Q_1$$

$$\Rightarrow \quad W = Q_1 - Q_2$$

$$\Rightarrow \quad \frac{Q_2}{W} = \frac{Q_2}{Q_1 - Q_2}$$

$$\Rightarrow \quad \frac{Q_2}{W} = \frac{T_2}{T_1 - T_2}$$

$$\Rightarrow \quad \frac{Q_2}{35} = \frac{263}{298 - 263}$$

$$\Rightarrow \quad Q_2 = 263\text{ J/s} \qquad \textbf{Ans. (A)}$$

Sol. 74 Coefficient of volume expansion is given as

$$\gamma = \frac{1}{V}\frac{dV}{dt}$$

$$\Rightarrow \quad PT^3 = \text{constant}$$

$$\Rightarrow \quad \frac{1}{V}\frac{dV}{dt} = \text{Constant}$$

$$\Rightarrow \quad T^4V^{-1} = C$$

$$\Rightarrow \quad V^{-1}(4T^3dT) + T^4\left(\frac{-1}{V^2}dV\right) = 0$$

$$\Rightarrow \quad \frac{dV}{dT} = \frac{4T^3/V}{T^4/V^2} = \frac{4V}{T}$$

$$\Rightarrow \quad \gamma = \frac{4}{T} \qquad \textbf{Ans. (C)}$$

Sol. 75 Efficiency of a heat engine is given as

$$\eta = \frac{W}{Q_H} = 1 - \frac{T_L}{T_H}$$

$$\Rightarrow \quad \frac{60}{300} = 1 - \frac{400}{T_H}$$

$$\Rightarrow \quad T_H = 500\text{ K} \qquad \textbf{Ans. (500)}$$

Sol. 76 Efficiency of the heat engine is given as

$$\eta = 1 - \frac{T_L}{T_H}$$

$$\Rightarrow \quad \frac{1}{4} = 1 - \frac{T_L}{T_H}$$

$$\Rightarrow \quad \frac{T_L}{T_H} = \frac{3}{4}$$

When temperature of sink is recuced by 58°C, its efficiency becomes double so we have

$$2\eta = \frac{1}{2} = 1 - \frac{T_L - 58}{T_H}$$

$$\Rightarrow \quad \frac{T_L - 58}{T_H} = \frac{1}{2}$$

$$\Rightarrow \quad \frac{T_L}{T_L - 58} = \frac{3/4}{1/2}$$

$$\Rightarrow \quad 2T_L = 3T_L - 174$$

$$\Rightarrow \quad T_L = 174\,°C$$

Ans. (C)

Sol. 77 As volume is compressed adiabatically, we use

$$P_i V_i^\gamma = P_f V_f^\gamma$$

$$\Rightarrow \quad 200 \times 10^3 \left(\frac{1200\times10^{-6}}{300\times10^{-6}}\right)^\gamma = P_f$$

$$\Rightarrow \quad P_f = 200 \times 10^3 \times 4^{3/2}$$

$$\Rightarrow \quad P_f = 200 \times 10^3 \times 8 = 1600 \times 10^3 = 16 \times 10^5\, Pa$$

Work done in adiabatic process is calculated as

$$W = \frac{P_i V_i - P_f V_f}{\gamma - 1}$$

$$\Rightarrow \quad W = \frac{2\times10^5\times1200\times10^{-6} - 16\times10^5\times300\times10^{-6}}{1.5-1}$$

$$\Rightarrow \quad W = \frac{240-480}{0.5} = -240\times2 = -480\text{ J}$$

Ans. (480)

Sol. 78 Ratio of change in internal energy to the work done can be given as

$$\frac{\Delta U}{W} = \frac{\frac{5}{2}nR\Delta T}{nR\Delta T} = \frac{5}{2} = \frac{25}{10}$$

$$\Rightarrow \quad x = 25$$

Ans. (25)

12.1 Kinetic Theory of Gases

Sol. 1 If C_p and C_v are molar specific heats (J/mol-k) then we use for all gases

$$C_p - C_v = R = 8.314\text{ J/mol-k}$$

for C_p and C_v taken as specific heats (J/kg-k) we use

for H_2 gas $\quad C_p - C_v = \dfrac{R}{M_{H_2}} = \dfrac{R}{2\times10^{-3}} = a$

for N_2 gas $\quad C_p - C_v = \dfrac{R}{28\times10^{-3}} = b$

$$\Rightarrow \quad a = 14b$$

Ans. (A)

Sol. 2 From Gas law we use no of molecules in room are given as

$$N = n.A_v = \left(\frac{PV}{RT}\right)\cdot Av$$

Initial molecules in room

$$n_i = \frac{PVA_v}{RT_i} \quad (\text{where } T_i = 273 + 17 = 290\text{ k})$$

Final molecules in room

$$n_f = \frac{PVA_v}{RT_f} \quad (\text{where } T_f = 273 + 27 = 300\text{ k})$$

$$\Rightarrow \quad \Delta n = n_f - n_i = \frac{PVA_v}{R}\left[\frac{1}{T_f} - \frac{1}{T_i}\right]$$

$$\Rightarrow \quad \Delta n = \frac{10^5\times30\times6.023\times10^{23}}{8.314}\left[\frac{1}{300} - \frac{1}{290}\right]$$

$$\Rightarrow \quad \Delta n = -2.5\times10^{25}$$

Ans. (D)

Sol. 3 For a gas molecule having f degrees of freedom, we use

$$C_v = \frac{f}{2}R \quad \text{and} \quad C_p = \frac{f+2}{2}R$$

for $f = 5$ we have

$$C_v = \frac{5}{2}R \text{ and } C_p = \frac{7}{2}R$$

$$\Rightarrow \quad \gamma = \frac{C_p}{C_v} = \frac{7}{5}$$

Ans. (D)

Sol. 4 The total KE of a gas having n moles is given as

$$U = \frac{f}{2}nRT$$

[where $f \rightarrow$ degrees of freedom]

for diatomic gas $f = 5$

for monoatomic gas $f = 3$

Initially N moles of diatomic gas

$$\Rightarrow \quad U_i = \frac{f}{2}NRT = \frac{5}{2}NRT \quad \ldots(1)$$

Initially there are $(N-n)$ moles of diatomic gas and $2n$ moles of monoatomic gas

$$\Rightarrow \quad U_f = \frac{5}{2}(N\text{-}n)RT + \frac{3}{2}(2n)RT$$

$$\Rightarrow \quad U_f = \frac{5}{2}NRT + 3nRT - \frac{5}{2}nRT$$

$$\Rightarrow \quad U_f = \frac{5}{2}NRT + \frac{1}{2}nRT \quad \ldots(2)$$

from (1) and (2) change in total KE of gas

$$\Delta U = U_f - U_i = \frac{1}{2}nRT$$

Ans. (A)

Sol. 5 By Gas law

$$P = \frac{nRT}{V}$$

$\Rightarrow$ T = constant from 1 to 2

$$P_1 > P_2$$

Ans. (C)

Sol. 6 By Gas law, we use

$$\frac{V_{rms}(He)}{V_{rms}(Ar)} = \sqrt{\frac{M_{Ar}}{M_{He}}}$$

$$= \sqrt{\frac{40}{4}} \cong 3.16$$

Ans. (D)

Sol. 7 For a monoatomic gas, energy of thermal motion

$$U = \frac{3}{2}kT \times N$$

$\Rightarrow$ $U = \frac{3}{2}nRT = \frac{3}{2}PV$ [Volume $V = \frac{2}{8} = \frac{1}{4}\,m^3$]

$\Rightarrow$ $U = \frac{3}{2} \times 4 \times 10^4 \times \frac{1}{4}$

$\Rightarrow$ $U = 1.5 \times 10^4$ J

Order is = 10^4 J **Ans. (D)**

Sol. 8 Total internal energy of a gaseous mixture is given as

$$U = \frac{f_1}{2}n_1RT + \frac{f_2}{2}n_2RT$$

$\Rightarrow$ $U = \frac{5}{2}(3RT) + \frac{3}{2} \times (5RT)$

$\Rightarrow$ $U = 15RT$

Ans. (C)

Sol. 9 Energy $= \frac{f}{2}nRT = \frac{f}{2}PV$

$\Rightarrow$ $E = \frac{f}{2}(3\times10^6)(2)$

$\Rightarrow$ $E = f \times 3 \times 10^6$

Considering gas is monatomic for ideal case, we use

$\Rightarrow$ $f = 3$

$\Rightarrow$ $E = 9 \times 10^6$ J **Ans. (D)**

Sol. 10

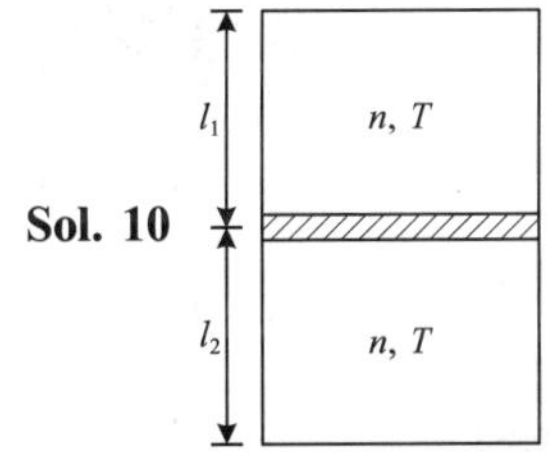

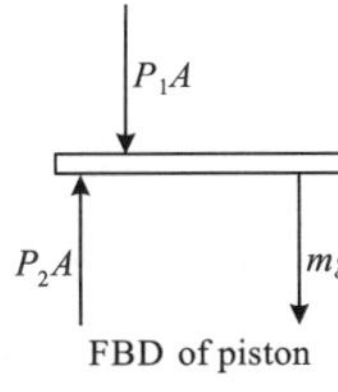

At equilibrium, we use

$$P_2A = P_1A + mg$$

$\Rightarrow$ $\frac{nRT.A}{Al_2} = \frac{nRT.A}{Al_1} + mg$

$\Rightarrow$ $nRT\left(\frac{1}{l_2} - \frac{1}{l_1}\right) = mg$

$\Rightarrow$ $m = \frac{nRT}{g}\left(\frac{l_1 - l_2}{l_1 \cdot l_2}\right)$

Ans (B)

Sol. 11 Mean time between two successive collisions is given as

$$t_m = \frac{\text{average separation}}{\text{mean velocity}}$$

Average separation between molecules is

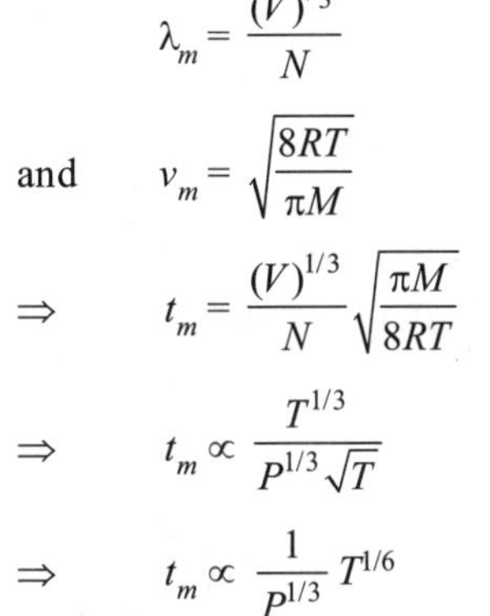

$$\lambda_m = \frac{(V)^{1/3}}{N}$$

and $v_m = \sqrt{\frac{8RT}{\pi M}}$

$\Rightarrow$ $t_m = \frac{(V)^{1/3}}{N}\sqrt{\frac{\pi M}{8RT}}$

$\Rightarrow$ $t_m \propto \frac{T^{1/3}}{P^{1/3}\sqrt{T}}$

$\Rightarrow$ $t_m \propto \frac{1}{P^{1/3}}T^{1/6}$

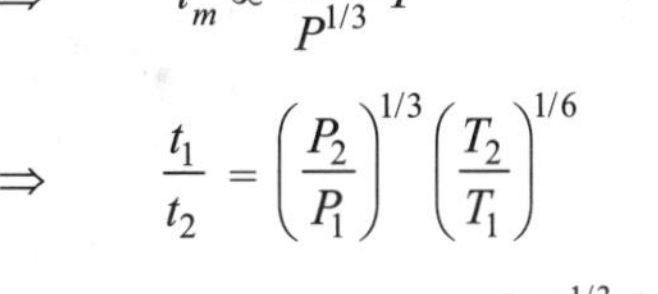

$\Rightarrow$ $\frac{t_1}{t_2} = \left(\frac{P_2}{P_1}\right)^{1/3}\left(\frac{T_2}{T_1}\right)^{1/6}$

$\Rightarrow$ $t_2 = 6 \times 10^{-8} \times \left(\frac{1}{2}\right)^{1/3}\left(\frac{3}{5}\right)^{1/6}$

$\Rightarrow$ $t_2 = 6 \times 10^{-8}/1.26 \times 1.09 = 4.37 \times 10^{-8}$ s **Ans. (A)**

Sol. 12

Magnitude of change in momentum per collision = $2mv$

$$\text{Pressure} = \frac{\text{Force}}{\text{Area}} = \frac{N(2mv)}{1}$$

$$p = \frac{10^{22} \times 2\times10^{-26} \times 10^4}{1}$$

$\Rightarrow$ $p = 2\,N/m^2$

Ans. (None)

Sol. 13 RMS velocity of hydrogen molecules is given as

$$v_{rms} = \sqrt{\frac{3RT}{M}}$$

If RMS velocity is equal to escape velocity then we use

$$\sqrt{\frac{3RT}{M}} = \sqrt{2gRe}$$

$$\Rightarrow \quad T = \frac{2gR_eM_e}{3R} = \frac{2\times10\times6.4\times10^6\times2\times10^{-3}}{3\times8.314}$$

$$\Rightarrow \quad T \simeq 10\times10^3 = 10^4 \; 6.4\times10^6$$

$$\Rightarrow \quad T = \frac{4\times10\times6.4\times10^6}{3\times1.38\times6.02\times10^3}$$

$$\Rightarrow \quad T = 10\times10^3 = 10^4\text{K}$$ **Ans. (C)**

Sol. 14 According to law of equipartion of energy, we use

$$\frac{1}{2}m(v_{rms}^2) = 3\times\frac{1}{2}K_bT$$

$$\Rightarrow \quad T = \frac{m\bar{v}_{rms}^2}{3k}$$ **Ans. (C)**

Sol. 15 RMS speed is given as

$$v_{\text{rms}} = \sqrt{\frac{3RT}{M}}$$

$$\Rightarrow \quad v_{\text{rms}} \propto \sqrt{T}$$

For the given data, we use

$$\frac{v}{200} = \sqrt{\frac{500}{400}}$$

$$\Rightarrow \quad \frac{v}{200} = \frac{\sqrt{5}}{2}$$

$$\Rightarrow \quad v = 100\sqrt{5}\text{ m/s}$$ **Ans. (B)**

Sol. 16 For A, we use

$$R = C_p - C_v = 7$$

$$C_v = \frac{fR}{2} = 22$$

$$\Rightarrow \quad f = \frac{44}{7} = 6.3$$

$f \simeq 6$ → 5(Rotation + Translational), 1(Vibration)

For B, we use

$$R = C_p - C_v = 9$$

$$C_v = \frac{fR}{2} = 21$$

$$\Rightarrow \quad f = \frac{42}{9}$$

$f \simeq 5$ → 5(Rotation + Translational), 0(Vibration) **Ans. (D)**

Sol. 17 Mean free path of O_2 gas

$$\lambda = \frac{RT}{\sqrt{2}\pi d^2N_AP} = \frac{V}{\sqrt{2}\pi d^2N_A}$$

Average collision rate is given as

$$r_c = \frac{v_{av}}{\lambda} = \frac{\sqrt{8/3\pi}\times v_{ms}}{\lambda}$$

$$\Rightarrow \quad r_c = \frac{\sqrt{8/3\pi}\times200\text{ m/s}}{25\times10^{-3}}\times\sqrt{2}\pi(2\times0.3\times10^{-9})\times6.023\times10^{23}$$

$$\Rightarrow \quad r_c \simeq 0.18\times10^{10}\text{ sec}^{-1} \sim 10^{10}\text{ sec}^{-1}$$ **Ans. (C)**

Sol. 18 C_V of a mixture of gases is given as

$$C_{V_{eq}} = \frac{n_1C_{V_1} + n_2C_{V_2}}{n_1+n_2}$$

For He, $C_{V_1} = \frac{3R}{2}$ and For H_2, $C_{V_2} = \frac{5R}{2}$

$$\Rightarrow \quad C_{V_{eq}} = \frac{2\times 3R/2 + 3\times 5R/2}{2+3}$$

$$\Rightarrow \quad C_{V_{eq}} = \frac{21}{10}R = \frac{21}{10}\times8.314 = 17.46\text{ J/mol K}$$ **Ans. (C)**

Sol. 19

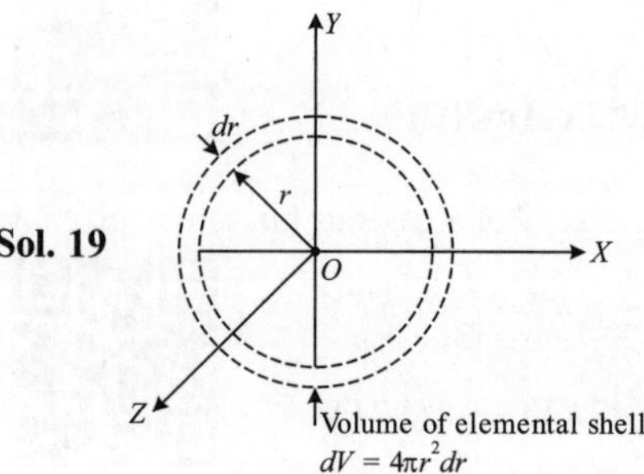

Number of molecule in the elemental shell is

$$dN = n(r)\cdot dV$$

$$\Rightarrow \quad dN = n_0e^{-\alpha r^4}\cdot 4\pi r^2dr$$

Total molecules

$$N = \int dN = 4\pi n_0\int_0^\infty e^{-\alpha r^4}\cdot r^2dr$$

Let $\quad t = \alpha r^4 \Rightarrow dt = 4\alpha r^3dr$

$$\Rightarrow \quad r^2dr = \frac{dt}{4\alpha r} \quad \Rightarrow \quad r = \frac{t^{1/4}}{\alpha^{1/4}}$$

$$\Rightarrow \quad N = 4\pi n_0\int_0^\infty e^{-t}\cdot\frac{dt\,\alpha^{1/4}}{4\alpha t^{1/4}}$$

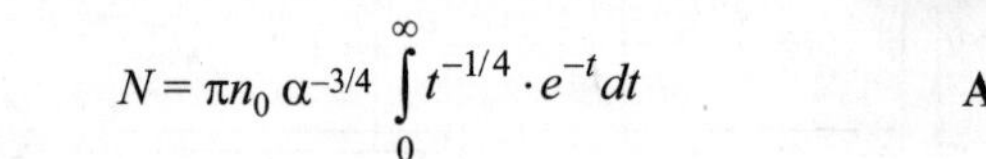

$$\Rightarrow \quad N = \pi n_0\,\alpha^{-3/4}\int_0^\infty t^{-1/4}\cdot e^{-t}dt$$ **Ans. (C)**

Sol. 20 Mean free time $\propto \frac{1}{\text{Mean speed}}$

$$\Rightarrow \quad t \propto \frac{1}{\sqrt{T}}$$ **Ans. (D)**

Sol. 21 Mean free path is given as

$$\lambda = \frac{1}{\sqrt{2}\pi n_v d^2}$$

Mean time between successive collision is given as

$$\tau = \frac{\lambda}{v_m} = \frac{1}{\sqrt{2}\pi n_v d^2 v_m} = \frac{1}{\sqrt{2}\pi n_v d^2}\sqrt{\frac{\pi M}{8RT}}$$

For two gases, we use

$$\frac{\tau_1}{\tau_2} = \sqrt{\frac{M_1}{M_2}\frac{d_2^2}{d_1^2}}$$

$$\Rightarrow \quad \frac{\tau_1}{\tau_2} = \sqrt{\frac{40}{140}}\frac{(0.1)^2}{(0.07)^2} = 1.09$$

Thus most appropriate answer among given option is (C).

Ans. (C)

Sol. 22 The mean free path of molecules of an ideal gas is given as

$$\lambda = \frac{V}{\sqrt{2}\pi d^2 N}$$

Due to rise in temperature speed of molecules increases so mean collision time between molecules decreases and mean free path is independent of temperature. **Ans. (D)**

Sol. 23 Given that at constant pressure

$$\Delta Q = nC_P\Delta T$$

$$\Rightarrow \quad 160 = n \times C_P(50) \quad \ldots(1)$$

At constant volume, we use

$$240 = nC_V(100) \quad \ldots(2)$$

from (1) & (2)

$$\gamma = \frac{C_P}{C_V} = \frac{160/50}{240/100} = \frac{4}{3}$$

As gas has $\gamma = \frac{4}{3}$

$$\Rightarrow \quad \gamma = \frac{f+2}{f}$$

$$\Rightarrow \quad f = 6$$

Ans. (C)

Sol. 24 For monoatomic molecule $\rightarrow f = 3$

$$\Rightarrow \quad \gamma = \frac{f+2}{f} = \frac{5}{3}$$

For Diatomic rigid molecules $\rightarrow f = 5$

$$\Rightarrow \quad \gamma = \frac{7}{5}$$

For Diatomic non rigid molecules $\rightarrow f = 7$

$$\Rightarrow \quad \gamma = \frac{9}{7}$$

For Triatomic rigid molecules $\rightarrow f = 6$

$$\Rightarrow \quad \gamma = \frac{8}{6} = \frac{4}{3}$$

Ans. (A)

Sol. 25 RMS speed of gas molecule is given as

$$V_{rms} = \sqrt{\frac{3RT}{M}}$$

Given that

$$V_{N_2} = V_{H_2}$$

$$\Rightarrow \quad \sqrt{\frac{3RT_{N_2}}{M_{N_2}}} = \sqrt{\frac{3RT_{H_2}}{M_{H_2}}}$$

$$\Rightarrow \quad \frac{573}{28} = \frac{T_{H_2}}{2}$$

$$\Rightarrow \quad T_{H_2} = 40.928\text{ K}$$

Ans. (41)

Sol. 26 By gas law we use

$$P_1V_1 = nR(250) \quad \ldots(1)$$

If $\frac{n}{4}$ mole molecules gets dissociated then final number of moles will become $n + \frac{n}{4} = \frac{5n}{4}$ so we use again by gas law

$$P_2(2V_1) = \frac{5n}{4}R(2000) \quad \ldots(2)$$

From (1) and (2), we have

$$\frac{P_1}{2P_2} = \frac{4\times 250}{5\times 2000}$$

$$\Rightarrow \quad \frac{P_1}{P_2} = \frac{1}{5} \text{ or } \frac{P_2}{P_1} = 5$$

Ans. (5)

Sol. 27 As mean speed of molecules is $v_m \propto \sqrt{T}$ mean time between collisions is given as

$$t_m = \frac{\lambda}{v_m}$$

$$\Rightarrow \quad t_m \propto \frac{1}{\sqrt{T}}$$

Ans. (C)

Sol. 28 From the assumption of Kinetic Theory, the molecules of gas collide with the container walls and continuously suffer momentum change which causes the force on the wall and hence pressure. **Ans. (D)**

Sol. 29 Root mean square speed of molecules of the gas is given as

$$v_{rms} = \sqrt{\frac{3RT}{M}}$$

$$\Rightarrow \quad v_{rms} \propto \sqrt{T}$$

$$\Rightarrow \quad \frac{(v_{rms})_2}{(v_{rms})_1} = \sqrt{\frac{T_2}{T_1}} = \sqrt{\frac{400}{300}} = \frac{2}{\sqrt{3}}$$

$$\Rightarrow \quad (v_{rms})_2 = \frac{2}{\sqrt{3}}(v_{rms})_1 = \frac{2}{\sqrt{3}} \times 200$$

$$\Rightarrow \quad (v_{rms})_2 = \frac{400}{\sqrt{3}} \text{ m/s}$$

Ans. (400)

Sol. 30 In a diatomic molecule, translational degree of freedom are 3 and rotational degree of freedom are 2 so statement II is false. Statement I is by definition true. **Ans. (C)**

Sol. 31 Gas law for the mixture of gases is written as

$$PV = (n_1 + n_2 + n_3)RT$$

$$\Rightarrow \quad PV = \left[\frac{16}{32} + \frac{28}{28} + \frac{44}{44}\right]RT$$

$$\Rightarrow \quad PV = \left[\frac{1}{2} + 1 + 1\right]RT$$

$$\Rightarrow \quad P = \frac{5}{2}\frac{RT}{V}$$

Ans. (C)

Sol. 32 Mean free path at temperature T for a gas is given as

$$\lambda = \frac{RT}{\sqrt{2}\pi d^2 N_A P}$$

$$\Rightarrow \quad \lambda = 102 \text{ nm}$$

Ans. (D)

Sol. 33 If the final temperature of the gaseous mixture is T, by conservation of internal energy we use

$$\frac{f_1}{2} n_1 R\,\Delta T + \frac{f_2}{2} n_2 R \Delta T = 0$$

$$\Rightarrow \quad \frac{f_1}{2} n_1 R\,(T_1 - T) + \frac{f_2}{2} n_2 R\,(T_2 - T) = 0$$

$$\Rightarrow \quad T = \frac{f_1 n_1 R T_1 + f_2 n_2 R T_2}{f_1 n_1 R + f_2 n_2 R} = \frac{f_1 n_1 T_1 + f_2 n_2 T_2}{f_1 n_1 + f_2 n_2}$$

Ans. (B)

Sol. 34 For two vibrational modes of gas molecule it has 4 vibrational degress of freedom and assuming non linear molecule, total degress of freedom can be given as

$$f = 4 + 3 + 3 = 10$$

$$\Rightarrow \quad \beta = \frac{C_p}{C_v} = 1 + \frac{2}{f} = \frac{12}{10} = 1.2$$

Ans. (B)

Sol. 35 According to law of equipartition of energy, the energy associated with each degree of freedom per molecule is $\frac{1}{2}k_B T$. **Ans. (A)**

Sol. 36 Conceptually it can be stated that at molecular level both the gas molecules (identical) get mixed and total number of possible states available for molecular motion will increase so final entropy of the system will be $S_1 + S_2$. **Ans. (D)**

Sol. 37 Root mean square velocity of gas molecules is given as

$$v_{rms} = \sqrt{\frac{3RT}{M}}$$

Average speed of gas molecules is given as

$$v_{avg} = \sqrt{\frac{8}{\pi}\frac{RT}{M}}$$

$$\Rightarrow \quad \frac{v_{rms}}{v_{avg}} = \sqrt{\frac{3\pi}{8}}$$

Ans. (C)

Sol. 38 Root mean square speed of gas molecules is given as

$$v_{rms} = \sqrt{\frac{3RT}{m}}$$

as it is inversely proportional to the square root of mass of molecules, option (D) is correct. **Ans. (D)**

Sol. 39 For a gas, its specific heats at constant volume and at constant pressure are given in terms of degrees of freedom as

$$C_V = \frac{fR}{2}$$

and

$$C_P = \left(\frac{f}{2} + 1\right)R$$

$$\Rightarrow \quad \gamma = \frac{C_P}{C_V} = 1 + \frac{2}{f}$$

$$\Rightarrow \quad f = \frac{2}{\gamma - 1}$$

Ans. (D)

Sol. 40 For an ideal gas, by gas law, we have

$$PV = nRT$$

$$\Rightarrow \quad PV \propto T$$

Ans. (B)

Sol. 41 By law of equipartition of energy the average kinetic energy for a molecule having 3 degrees of freedom (monoatomic gas) is given as

$$E = \frac{f}{2}k_B T = \frac{3}{2}k_B T$$

Ans. (B)

Sol. 42 For an ideal gas the difference of specific heats is R and as gas behaviour deviates from ideal the difference also deviates away from R. At very high temperature real gases also behave like ideal gases hence in this case state P can be considered at higher temperature. **Ans. (A)**

Sol. 43 Mean free path $= \frac{1}{\sqrt{2}d^2n}$

$$\frac{x_1}{x_2} = \left(\frac{d_2}{d_1}\right)^2$$

$$\left(\frac{1}{2}\right)^2 = 0.25 = 25 \times 10^{-2}$$ **Ans. (25)**

Sol. 44 Kinetic energy of a gas molecule at absolute temperature T is given as

$$K = \frac{3}{2}kT$$

By gas law, we have

$$PV = \frac{NRT}{N_A}$$

$$\Rightarrow \quad N = \frac{PV}{\left(\frac{2}{3}\right)k} = \frac{2\times10^{-3}}{2/3\times10^{-9}} = 1.5 \times 10^{11}$$ **Ans. (D)**

Sol. 45 RMS speed of gas molecules is given as

$$v_{rms} = \sqrt{\frac{3RT}{M}} \propto \frac{1}{\sqrt{M}}$$

As $M_{H_2} < M_{O_2} < M_{CO_2}$, we have

$$V_{H_2} > V_{O_2} > V_{CO_2}$$ **Ans. (A)**

Sol. 46 For gas A using gas law, we have

$$P_A V = n_A RT$$

$$\Rightarrow \quad P_A = \frac{(1)RT}{V}$$

For gas B, using gas law, we have

$$P_B V = n_B RT$$

$$\Rightarrow \quad P_B = \frac{(2)RT}{V}$$

The pressure in mixure of gases is given as

$$P = P_A + P_B$$

$$\Rightarrow \quad P = \frac{3RT}{V} = \frac{3\times8.314(400)}{4\times10^{-3}}$$

$$\Rightarrow \quad P = 24.9 \times 10^5 \text{ Pa}$$ **Ans. (C)**

Sol. 47 In floating condition of balloon, we use

$$V\rho_{air}g = mg$$

$$V\rho_{air} = 185$$

When pressure and temperature changes, if new load which the balloon can carry is m' then we use

$$V\rho_{air}g = m'g$$

$$\Rightarrow \quad \rho_{air}\left(\frac{P'}{P}\times\frac{T}{T'}\right)V = m'$$

$$\Rightarrow \quad 185\left[\frac{45}{76}\times\frac{300}{266}\right] = m'$$

$$\Rightarrow \quad m' = 123.54 \text{ kg}$$ **Ans. (D)**

Sol. 48 Root mean square speed of molecules are related to molecular moss as

$$V_{rms} \propto \frac{1}{\sqrt{M}}$$

$$\Rightarrow \quad \frac{(V_{rms})O_2}{(V_{rms})H_2} = \sqrt{\frac{m_{H_2}}{m_{O_2}}} = \sqrt{\frac{2}{32}} = \frac{1}{4}$$

$$(V_{RMS})_{H_2} = 4(V_{RMS})_{O_2} = 4 \times 160 = 640 \text{ m/s}$$ **Ans. (C)**

Sol. 49 Pressure variation with volume is given as

$$\frac{dp}{dV} = -aP$$

$$\Rightarrow \quad \int_{P_0}^{P}\frac{dP}{P} = -\int_0^V a\,dV$$

$$\Rightarrow \quad ln\frac{P}{P_0} = -aV$$

$$\Rightarrow \quad P = P_0 e^{-aV}$$

By gas law for one mole of gas, we use

$$PV = RT$$

$$\Rightarrow \quad P + V\frac{dP}{dV} = RT$$

$$\Rightarrow \quad P + V\frac{dP}{dV} = R\frac{dT}{dV}$$

For maximum Temperature, we use

$$\frac{dT}{dV} = 0$$

$$\Rightarrow \quad P - (aP)V = 0$$

$$\Rightarrow \quad V = \frac{1}{a}$$

$$\Rightarrow \quad T = \frac{PV}{R} = \frac{P_0e^{-1}}{Ra} = \frac{P_0}{eRa}$$ **Ans. (B)**

Sol. 50 If m_1 and m_2 are the masses of oxygen and hydrogen in the misture then number of moles of O_2 are given as

$$n_1 = \frac{m_1}{32}$$

Number of moles of H_2 in mixture are given as

$$n_2 = \frac{m_2}{2}$$

By ideal gas equation, we have

$$PV = \left[\frac{m_1}{32} + \frac{m_2}{2}\right]\frac{25}{3} \times 300$$

$$\Rightarrow \quad 400 \times 10^3 \times 500 \times 10^{-6} = \left[\frac{m_1}{32} + \frac{m_2}{2}\right] 25 \times 100$$

$$\Rightarrow \quad 2 \times 10^2 = \left[\frac{m_1}{m_2} + \frac{1}{32} + \frac{1}{2}\right] m_2 \times 25 \times 100$$

$$\Rightarrow \quad \frac{2}{25} = \left[\frac{m_1}{m_2} \times \frac{1}{32} + \frac{1}{2}\right] m_2 \quad \ldots(1)$$

As the mass of mixture is 0.76 g, we use

$$m_1 + m_2 = 0.76$$

$$\Rightarrow \quad m_2\left[1 + \frac{m_1}{m_2}\right] = \frac{76}{100} \quad \ldots(2)$$

From equations-(1) and (2), we have

$$\frac{\left[\frac{m_1}{m_2} \times \frac{1}{32} + \frac{1}{2}\right]}{\left[\frac{m_1}{m_2} + 1\right]} = \frac{2}{25} \times \frac{100}{76}$$

Here if we consider $\frac{m_1}{m_2} = x$, we have

$$\frac{x+16}{32(x+1)} = \frac{8}{76}$$

$$\Rightarrow \quad 76x + 76 \times 16 = 8 \times 32x + 8 \times 32$$

$$\Rightarrow \quad 76x + 1216 = 256x + 256$$

$$\Rightarrow \quad 960 = 180x$$

$$\Rightarrow \quad x = \frac{96}{18} = \frac{16}{3}$$

Ans. (B)

Sol. 51 Kinetic energy of each particle of either of electron or protons gases is given as

$$K = \frac{3}{2}kT$$

$$\Rightarrow \quad \frac{P^2}{2m} = \frac{3}{2}kT$$

$$\Rightarrow \quad P = \sqrt{3kTm}$$

Using Heisenberg uncertainty principle, we have

$$\Delta x.\,\Delta P = \frac{h}{4\pi}$$

$$\Rightarrow \quad \Delta x = \frac{h}{4\pi\Delta P} = \frac{h}{4\pi\sqrt{3kTm}}$$

$$\Rightarrow \quad \Delta x \propto \frac{1}{\sqrt{m}}$$

$$\Rightarrow \quad \frac{\Delta x_e}{\Delta x_p} = \sqrt{\frac{m_p}{m_e}}$$

Ans. (A)

Sol. 52 As stated in question that translation kinetic energy of N_2 molecule is equal to the kinetic energy of electron, we use

$$\frac{3}{2}\text{kT} = \text{eV}$$

$$\Rightarrow \quad \frac{3}{2} \times 1.38 \times 10^{-3}\,\text{T} = 1.6 \times 10^{-19} \times 0.1$$

$$\Rightarrow \quad T = 773\,\text{K}$$

$$\Rightarrow \quad T = 773 - 273 = 500^\circ\text{C}$$

Ans. (500)

13.1 Simple Harmonic Motion

Sol. 1 For a particle starting SHM from mean position, equation of motion is given as

$$y = A\sin\omega t$$

$$\Rightarrow \quad \text{particle's velocity } v = \frac{dy}{dt} = A\omega\cos\omega t$$

$$\Rightarrow \quad \text{KE} = \frac{1}{2}mv^2 = \frac{1}{2}m\omega^2A^2\cos^2\omega t$$

KE_{max} at $t = 0$

In option (B), KE oscillates at twice the frequency of SHM.

Ans. (B)

Sol. 2 Case-1:

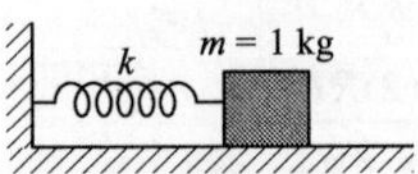

Oscillation frequency of block

$$f = \frac{1}{2\pi}\sqrt{\frac{k}{m}} = 1\,\text{Hz}$$

$$\Rightarrow \quad k = 4\pi^2\,\text{N/m}$$

Case-2:

In parallel effective force constant of spring

$$k_{eff} = k_1 + k_2 = 2k = 8\pi^2 \text{ N/m}$$

$$\text{Oscillation frequency } f = \frac{1}{2\pi}\sqrt{\frac{2k}{M}}$$

$$= \frac{1}{2\pi}\sqrt{\frac{8\pi^2}{8}} = \frac{1}{2}\text{Hz}$$ **Ans. (C)**

Sol. 3 Time period of oscillation of a pendulum is given as

$$T = 2\pi\sqrt{\frac{l}{g}}$$

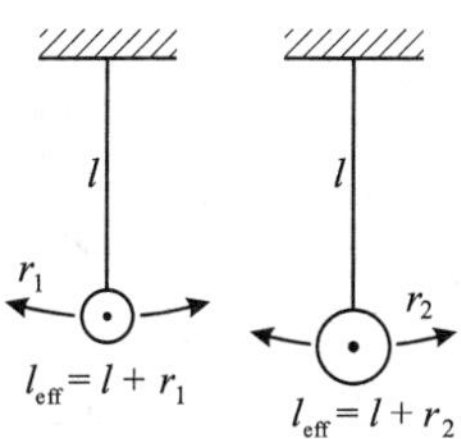

For uniform mass distribution COM is taken at centre of bob.

$$\ln T = \ln\frac{2\pi}{\sqrt{g}} + \ln\sqrt{l}$$

Relative change in period

$$\frac{dT}{T} = \frac{1}{2}\frac{dl}{l} \qquad \text{here } dl = |r_2 - r_1|$$

$$\Rightarrow \quad 5 \times 10^{-4} = \frac{1}{2}\frac{|r_2 - r_1|}{1\text{m}}$$

$$\Rightarrow \quad |r_2 - r_1| = 2 \times 5 \times 10^{-4} = 10^{-3}\text{ m} = 1 \text{ mm}$$ **Ans.(B)**

Sol. 4 For damped oscillations, the oscillation amplitude is given as

$$A = A_0 e^{\frac{-bt}{2m}} \qquad \ldots(1)$$

$b \rightarrow$ damping coefficient

$m \rightarrow$ mass of block

As we know mechanical energy of oscillation $E \propto A^2$

E will be half when $A \rightarrow \frac{A}{\sqrt{2}}$

from equation-(1), if at $t = t$, $A_0 \rightarrow \frac{A}{\sqrt{2}}$

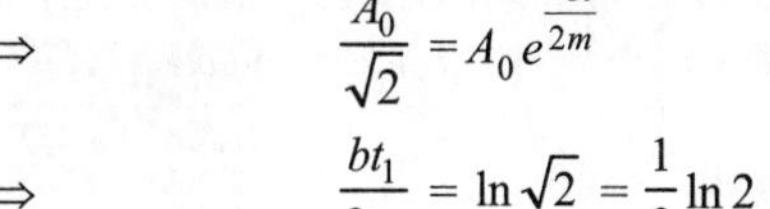

$$\Rightarrow \quad \frac{A_0}{\sqrt{2}} = A_0 e^{\frac{-bt}{2m}}$$

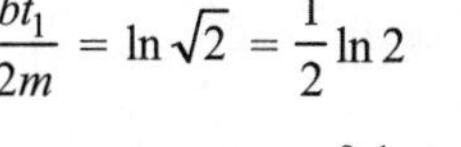

$$\Rightarrow \quad \frac{bt_1}{2m} = \ln\sqrt{2} = \frac{1}{2}\ln 2$$

$$\Rightarrow \quad t_1 = \frac{m}{b}\ln(2) = \frac{0.1}{0.01} \times 0.693$$

$$\Rightarrow \quad t_1 = 6.93 \text{ s}$$ **Ans. (D)**

Sol. 5 If in SHM amplitude is A and angular frequency is ω, then we have

max. velocity $\quad v_{max} = A\omega$

max. acceleration $\quad a_{max} = \omega^2 A$

Given that $\quad \frac{a_{max}}{v_{max}} = \omega = 10 \text{ s}^{-1}$

General equation of SHM can be taken as

$$y = A\sin(\omega t + \alpha)$$

at $\quad t = 0, y = 5\text{m}, \alpha = \frac{\pi}{4}$

$$5 = A\sin\left(\frac{\pi}{4}\right)$$

$$\Rightarrow \quad 5 = A\left(\frac{1}{\sqrt{2}}\right)$$

$$\Rightarrow \quad A = 5\sqrt{2}\text{ m}$$

$$\Rightarrow \quad a_{max} = \omega^2 A = (10)^2 \times 5\sqrt{2}$$

$$= 500\sqrt{2}\text{ m/s}^2$$ **Ans. (B)**

Sol. 6 Mass of one atom of silver is

$$m = \frac{M}{Av} = \frac{108 \times 10^{-3}}{6.02 \times 10^{23}}$$

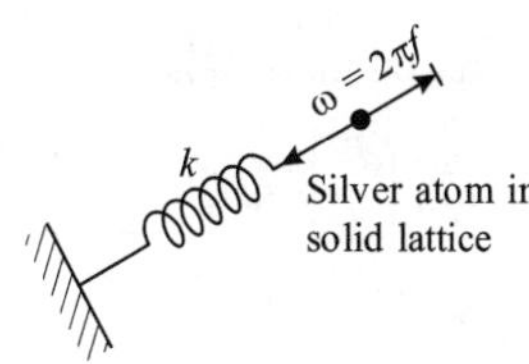

Spring represent the bonds connecting the silver atom with other atoms.

For SHM angular frequency of atom is given as

$$\omega = \sqrt{\frac{k}{m}}$$

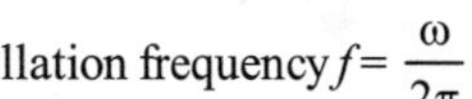

$$\Rightarrow \quad \text{Oscillation frequency } f = \frac{\omega}{2\pi}$$

$$\Rightarrow \quad 10^{12} = \frac{1}{2 \times 3.14}\sqrt{\frac{k \times 6.02 \times 10^{23}}{108 \times 10^{-3}}}$$

$$\Rightarrow \quad k = 7.1 \text{ N/m}$$ **Ans. (B)**

Sol. 7 for $a = b$, general equation of Lissajous's figure is given as

$$\frac{x^2}{A^2}+\frac{y^2}{B^2}-\frac{2xy}{AB}\cos\delta = \sin^2\delta$$

(B) $\frac{x^2}{A^2}+\frac{y^2}{B^2}-\frac{2xy}{A^2}(0) = 1 \Rightarrow x^2+y^2 = A^2 \rightarrow$ circle (✗)

(C) $\frac{x^2}{A^2}+\frac{y^2}{B^2}-\frac{2xy}{AB}(0) = 1 \Rightarrow \frac{x^2}{A^2}+\frac{y^2}{B^2} = 1 \rightarrow$ ellipse (✓)

(D) $\frac{x^2}{A^2}+\frac{y^2}{B^2}-\frac{2xy}{AB}(1) = 0 \rightarrow$ Not a parabola. (✗) **Ans. (C)**

Sol. 8 If particle starts SHM from its mean position (say) then equation of SHM is

$$y = A\sin\omega t$$

With given conditions, we have

$$a = A\sin\omega t_0 \quad \ldots(1)$$

$$b = A\sin 2\omega t_0 \quad \ldots(2)$$

$$c = A\sin 3\omega t_0 \quad \ldots(3)$$

$(1)+(3) \Rightarrow \quad a + c = 2A\sin 2\omega t_0 \cos\omega t_0$

$$\Rightarrow \quad \cos\omega t_0 = \frac{a+c}{2b}$$

$$\Rightarrow \quad \omega t_0 = \cos^{-1}\left(\frac{a+c}{2b}\right)$$

$$\omega = \frac{1}{t_0}\cos^{-1}\left(\frac{a+c}{2b}\right)$$

$$f = \frac{\omega}{2\pi} = \frac{1}{2\pi t_0}\cos^{-1}\left(\frac{a+c}{2b}\right)$$ **Ans. (D)**

Sol. 9 Restoring force on mass M is

$$F = \left|\frac{dV}{dx}\right| = k(x - X) = 0 \text{ (a mean position)}$$

$$\Rightarrow \quad x = X$$

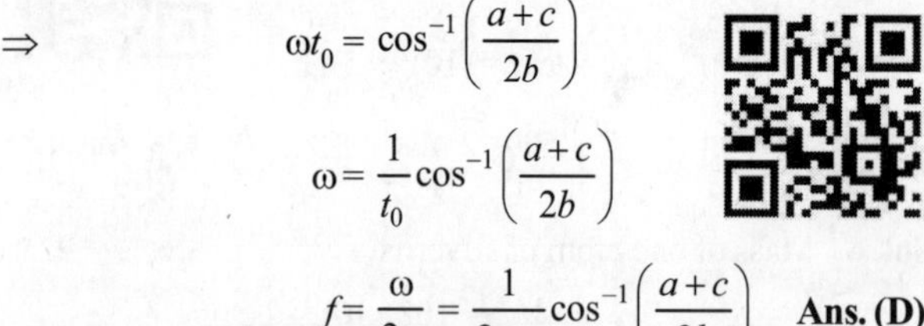

After 13th collision velocity of combined mass $M + 13m$ is given as

$$v = \frac{mu}{M+13m} = \frac{5\times 1}{10+65} = \frac{5}{75} = \frac{1}{15}$$

at mean position

When block reaches extreme position, its KE at mean position is transformed into PE at extreme position.

$\Rightarrow$ By conservation of energy

$$\frac{1}{2}(M+13m)v^2 = \frac{1}{2}k(X - A - x)^2$$

$$\Rightarrow \quad \frac{1}{2}(10+65)\left(\frac{1}{15}\right)^2 = \frac{1}{2}\times 1\times A^2$$

$$\Rightarrow \quad A^2 = \frac{75}{15\times 15} = \frac{1}{3} \Rightarrow A = \frac{1}{\sqrt{3}}$$ **Ans. (B)**

Sol. 10 Angular frequency of SHM of rod is

$$\omega = \sqrt{\frac{k}{I}}$$

$$\Rightarrow \quad \omega = \sqrt{\frac{3k}{ml^2}} \qquad \left(I = \frac{ml^2}{9}+\frac{m}{2}\cdot\frac{4l^2}{9} = \frac{ml^2}{3}\right)$$

$$\dot{\theta}_{max} = \omega\theta_0 \qquad \text{[maximum angular speed]}$$

tension in rod is given as

$$\Rightarrow \quad T = m\dot{\theta}^2 r_1$$

$$\Rightarrow \quad T = m\dot{\theta}^2\frac{l}{3}$$

$$\Rightarrow \quad T = m\omega^2\theta_0^2\frac{l}{3}$$

$$\Rightarrow \quad T = m\frac{3k}{ml^2}\theta_0^2\frac{l}{3}$$

$$\Rightarrow \quad T = \frac{k\theta_0^2}{l}$$ **Ans. (D)**

Sol. 11 Angular frequency of torsional oscillations is given by

$$\omega = \sqrt{\frac{k}{I}}$$

$$\omega_1 = \sqrt{\frac{k}{\frac{M(2L)^2}{12}}} \qquad \text{[Oscillations of rod]}$$

$$\omega_2 = \sqrt{\frac{k}{\frac{M(2L)^2}{12}+2m\left(\frac{L}{2}\right)^2}} \qquad \text{[Oscillations of rod with masses]}$$

$$\omega_2 = 0.8\omega_1$$

$\Rightarrow$ $$\frac{m}{M} = 0.375$$ **Ans. (B)**

Sol. 12 Potential energy $(U) = \frac{1}{2}kx^2$

Kinetic energy $(K) = \frac{1}{2}kA^2 - \frac{1}{2}kx^2$

Given that, $U = k$

$\Rightarrow$ $$\frac{1}{2}kx^2 = \frac{1}{2}kA^2 - \frac{1}{2}kx^2$$

$\Rightarrow$ $$x = \pm\frac{A}{\sqrt{2}}$$ **Ans. (C)**

Sol. 13 Velocity and acceleration of particle in SHM's

$$v = \omega\sqrt{A^2 - x^2} \quad \ldots(1)$$

$$a = -\omega^2 x \quad \ldots(2)$$

Given that

$$|v| = |a| \quad \ldots(3)$$

$$\omega\sqrt{A^2 - x^2} = \omega^2 x$$

$\Rightarrow$ $A^2 - x^2 = \omega^2 x^2$

$\Rightarrow$ $5^2 - 4^2 = \omega^2(4^2)$

$\Rightarrow$ $3 = \omega \times 4$

$\Rightarrow$ $\omega = 3/4$

$$T = 2\pi/\omega = \frac{8\pi}{3}$$ **Ans. (D)**

Sol. 14

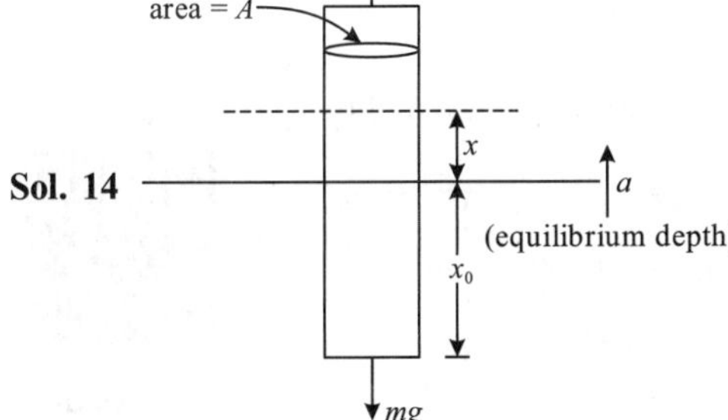

At equilibrium

$$mg = \rho_\omega A x_0 g$$

When further pushed down by x, we use

$$\rho_\omega A(x + x_0)g - mg = ma$$

$\Rightarrow$ $$\rho_\omega Axg = ma$$

$\Rightarrow$ $$a = \frac{\rho_\omega Ag}{m}x$$

Comparing with

$a = \omega^2 x$ we get

$$\omega = \sqrt{\frac{\rho_\omega Ag}{m}}$$

$\Rightarrow$ $$\omega = \sqrt{\frac{10^3 \times \pi(2.5)^2 \times 10^{-4} \times 10}{310 \times 10^{-6} \times 10^3}}$$

$\Rightarrow$ $$\omega = \sqrt{63.30} = 7.95 \text{ rad/s}$$ **Ans. (None)**

Sol. 15 Kinetic energy of particle in SHM is

$$K = \frac{1}{2}m\omega^2A^2\cos^2\omega t \quad \ldots(1)$$

Potential energy of particle in SHM is

$$U = \frac{1}{2}m\omega^2A^2\sin^2\omega t \quad \ldots(2)$$

$\Rightarrow$ $$\frac{K}{U} = \cot^2\omega t = \cot^2\frac{\pi}{90}(210) = \frac{1}{3}$$ **Ans. (C)**

Sol. 16 Angular frequency of simple pendulum is given as

$$\omega = \sqrt{\frac{g_{eff}}{l}}$$

$\Rightarrow$ $$\frac{\Delta\omega}{\omega} = \frac{1}{2}\frac{\Delta g_{eff}}{g_{eff}}$$

$\Rightarrow$ $$\Delta\omega = \frac{1}{2}\frac{\Delta g_{eff}}{g_{eff}} \times \omega \quad \ldots(1)$$

Due to vertical oscillations of support change in g_{eff} is given as

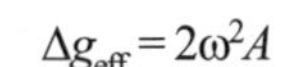

$$\Delta g_{eff} = 2\omega^2 A$$

So from equation-(1), we have

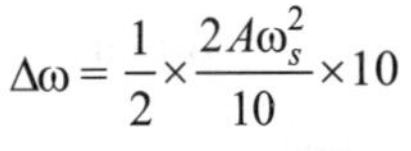

$$\Delta\omega = \frac{1}{2} \times \frac{2A\omega_s^2}{10} \times 10$$

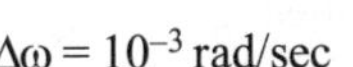

$\Rightarrow$ $$\Delta\omega = 10^{-3} \text{ rad/sec}$$ **Ans. (A)**

Sol. 17 Maximum kinetic energy at lowest point B is given as

$$K = mgl(1 - \cos\theta) \quad \ldots(1)$$

Where θ = angular amplitude of pendulum

For the given conditions, we use from equation-(1)

$$K_1 = mgl(1 - \cos\theta)$$

and $$K_2 = mgl(2l)(1 - \cos\theta)$$

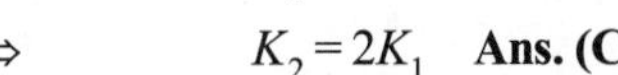

$\Rightarrow$ $K_2 = 2K_1$ **Ans. (C)**

Sol. 18

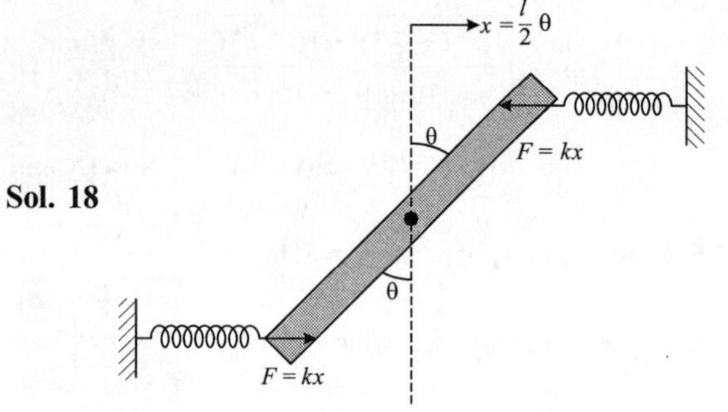

Restoring torque on rod due to its angular displacement is

$$\tau = -2kx + \frac{l}{2}$$

$$\Rightarrow \quad \tau = -\left(\frac{kl^2}{2}\right)\theta$$

Angular acceleration of rod is

$$\alpha = \frac{\tau}{I} = -\frac{\frac{kl^2}{2}\theta}{(Ml^2/12)}$$

$$\Rightarrow \quad \alpha = -\frac{6k}{M}\cdot\theta$$

Comparing with

$\alpha = -\omega^2\theta$, we get

$$\omega = \sqrt{\frac{6k}{M}}$$

$$\Rightarrow \quad f = \frac{\omega}{2\pi} = \frac{1}{2\pi}\sqrt{\frac{6k}{M}}$$ **Ans. (A)**

Sol. 19 SHM equation given as

$$y = 5\left[\sin(3\pi t) + \sqrt{3}\cos(3\pi t)\right]$$

Above equation cab be rewritten as

$$y = 10\sin\left(3\pi t + \frac{\pi}{3}\right)$$

$\Rightarrow$ Amplitude = 10 cm

$$\Rightarrow \quad T = \frac{2\pi}{\omega} = \frac{2\pi}{3\pi} = \frac{2}{3}\text{ s}$$ **Ans. (D)**

Sol. 20 For a damped oscillator with time amplitude decreases exponentially

$$\Rightarrow \quad A_t = A_0 e^{-\gamma t} \quad \ldots(1)$$

Given that after 10 oscilations (2 sec)

$$A = \frac{A_0}{2}$$

$\Rightarrow$ From (1) $\frac{A_0}{2} = A_0 e^{-2\gamma}$

$\Rightarrow$ $\ln(2) = 2\gamma \Rightarrow \gamma = \frac{\ln(2)}{2}$

$$\frac{A_0}{1000} = A_0 e^{-\gamma t}$$

$$\ln(1000) = \frac{\ln 2}{2}\cdot 2$$

$$\Rightarrow \quad t = \frac{6\ln(10)}{\ln 2} = 19.93\text{ s} \simeq 20\text{ s}$$ **Ans. (B)**

Sol. 21 For a simple pendulum

$$T = 2\pi\sqrt{\frac{L}{g_{eff}}}$$

Situation 1 : When pendulum is in air, we use $g_{\text{eff}} = g$

Situation 2 : When pendulum is in liquid, we use

$$g_{\text{eff}} = g\left(1 - \frac{\rho_{\text{liquid}}}{\rho_{\text{body}}}\right) = g\left(1 - \frac{1}{16}\right) = \frac{15g}{16}$$

$$\Rightarrow \quad \frac{T'}{T} = \frac{2\pi\sqrt{\frac{L}{15g/16}}}{2\pi\sqrt{\frac{L}{g}}}$$

$$\Rightarrow \quad T' = \frac{4T}{\sqrt{15}}$$ **Ans. (A)**

Sol. 22 By law of conservation of energy, we use

$$\frac{1}{2}kx^2 = (m_1 s_1 + m_2 s_2)\Delta T$$

$$\Rightarrow \quad \Delta T = \frac{16\times10^{-2}}{4384} = 3.65\times10^{-5}$$ **Ans. (D)**

Sol. 23 For damped oscillation, amplitude is given as

$$A = A_0 e^{-0.1t} = \frac{A_0}{2}$$

$$\Rightarrow \quad \ln 2 = 0.1t$$

$$\Rightarrow \quad t = 10\ln 2 = 6.93$$

$$\Rightarrow \quad \approx 7\text{ sec}$$ **Ans. (B)**

Sol. 24 (i) At $t = \frac{3T}{4}$ displacement is zero where velocity is maximum and acceleration force is zero.

(ii) At $t = T$, displacement is maximum so acceleration is maximum.

(iii) At $t = \frac{T}{4}$, displacement is zero so velocity is maximum.

(iv) at $t = \frac{T}{2}$ it shows displacement is maximum where kinetic energy = 0 and potential energy is maximum.

Hence (i), (ii) and (iii) are correct. **Ans. (D)**

Sol. 25 Angular frequency of SHM of block is given as

Initial $\omega_1 = \sqrt{\frac{K}{m}}$

Initial $\omega_2 = \sqrt{\frac{K}{m/2}}$

At mean position mass breaks off at maximum speed so we use

$$\omega_1 A_1 = \omega_2 A_2$$

$\Rightarrow$ $A_2 = \frac{\omega_1 A_1}{\omega_2} = \frac{A_1}{\sqrt{2}}$ **Ans. (D)**

Sol. 26 The ring acts like a compound pendulum. In its oscillations in plane, time period is given as

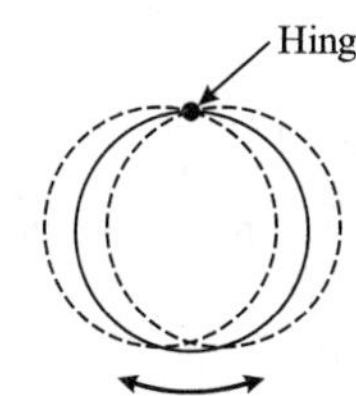

$$T_1 = 2\pi\sqrt{\frac{2MR^2}{MgR}} = 2\pi\sqrt{\frac{2R}{g}}$$

In its oscillations back and forth. Perpendicular to vertical plane, its time period is given as

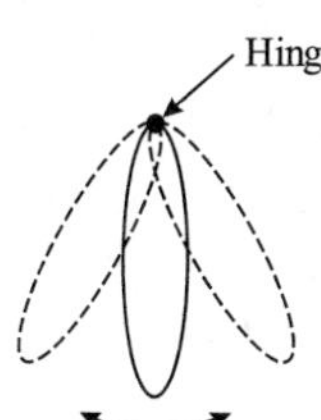

$$T_2 = 2\pi\sqrt{\frac{\frac{3}{2}MR^2}{MgR}} = 2\pi\sqrt{\frac{3M}{2g}}$$

$\Rightarrow$ $\frac{T_1}{T_2} = \frac{\sqrt{2}}{\sqrt{3/2}} = \frac{2}{\sqrt{3}}$ **Ans. (A)**

Sol. 27 Given motion equation is

$$y = y_0 \sin^2 \omega t = \frac{y_0}{2}(1 - \cos 2\omega t)$$

$\Rightarrow$ $y - \frac{y_0}{2} = -\frac{y_0}{2}\cos 2\omega t$

As y is measured from lower end of unstretched spring here $\frac{y_0}{2}$ is the amplitude of SHM. Thus we use

$$\frac{y_0}{2} = \frac{mg}{k}$$

SHM frequency is given as

$$2\omega = \sqrt{\frac{k}{m}} = \sqrt{\frac{2g}{y_0}}$$

$\Rightarrow$ $\omega = \sqrt{\frac{g}{2y_0}}$ **Ans. (B)**

Sol. 28 When mass m is placed on M, momentum of system remains conserved and after placing the mass m angular frequency of oscillations and amplitude changes. Here we use

$$MA\omega = (m+M)A'\omega'$$

$\Rightarrow$ $MA\sqrt{\frac{k}{M}} = (m+M)A'\sqrt{\frac{k}{m+M}}$

$\Rightarrow$ $A' = A\sqrt{\frac{M}{M+m}}$ **Ans. (A)**

Sol. 29 For a particle executing SHM, the equation of motion and velocity of particle is given as

$$x = A\sin(\omega t + \phi)$$

$$v = \omega A\cos(\omega t + \phi)$$

$\Rightarrow$ $\frac{v^2}{\omega^2 A^2} + \frac{x^2}{A^2} = 1$

Above is the equation of ellipse between v and x plotted on a coordinate system. **Ans. (C)**

Sol. 30 In the given situation when block is displaced, the two springs will deform by same distance and exert restoring force on block in same direction thus we can consider the two springs to be connected in parallel so we use

$$K_{eq} = K_1 + K_2 = K + K = 2K$$

Thus time period of oscillations of body is given as

$$T = 2\pi\sqrt{\frac{M}{K_{eq}}} = 2\pi\sqrt{\frac{M}{2K}}$$

$$f = \frac{1}{T} = \frac{1}{2\pi}\sqrt{\frac{2K}{M}}$$ **Ans. (C)**

Sol. 31 The time period of small oscillations of a simple pendulum is given as

$$T = 2\pi\sqrt{\frac{L}{g}}$$

$\Rightarrow$ $g = \frac{4\pi^2 l}{T^2}$

$\Rightarrow$ $\frac{\Delta g}{g} = \frac{\Delta l}{l} + \frac{2\Delta T}{T}$

$\Rightarrow$ $\frac{\Delta g}{g} = \frac{1\times10^{-3}}{1} + 2\times\frac{0.01}{1.95} = 0.0113$ or 1.13%

Ans. (C)

Sol. 32 For the simple pendulum, we have

$$T=2\pi\sqrt{\frac{l}{g}}$$

$$\Rightarrow \quad 2=2\pi\sqrt{\frac{2}{g}}$$

$$\Rightarrow \quad g=2\pi^2$$

Ans. (D)

Sol. 33 At $t=0$ if the initial displacement of particle is $A/2$, and it is moving along negative x-direction, the initial phase can be calculated by projection of a particle in uniform circular motion on its diameter as explained in the chapter of SHM in Physics Galaxy Book Vol 2. This is given as

$$\frac{\pi}{2}+\frac{\pi}{3}=\frac{5\pi}{6}$$

Ans. (D)

Sol. 34 In the given situation when block is displaced, the two springs will deform by same distance and exert restoring force on block in same direction thus we can consider the two springs to be connected in parallel so we use

$$k_{eq}=k_1+k_2$$

$$\Rightarrow \quad k_{eq}=4k$$

$$\Rightarrow \quad T=2\pi\sqrt{\frac{m}{k_{eq}}}$$

Ans. (C)

Sol. 35 At M the acceleration of particle will be maximum as it is one extreme position. The time period of oscillation in this SHM is given as

$$T=\frac{0.1}{30^\circ}\times 360^\circ=1.2\text{ s}$$

Angular frequency of SHM in this case can be given as

$$\omega=\frac{2\pi}{T}=\frac{5\pi}{3}\text{ rad/s}$$

At M, the restroing force on particle per unit mass is given as

$$F=m\omega^2A$$

$$\Rightarrow \quad \frac{F}{m}-\omega^2A$$

Ans. (A)

Sol. 36 For series combination of springs, we use

$$\frac{1}{k_{eq}}=\frac{1}{k_1}+\frac{1}{k_2}=\frac{1}{k}+\frac{1}{k}$$

$$\Rightarrow \quad k_{eq}=\frac{k}{2}$$

Thus time period of oscillation can be given as

$$T=2\pi\sqrt{\frac{M}{k_{eq}}}$$

$$\Rightarrow \quad T'=2\pi\sqrt{\frac{M}{k}}\times\sqrt{2}$$

$$\Rightarrow \quad T'=\sqrt{2}T$$

Ans. (B)

Sol. 37 This question is same as the one asked on 24 Feb, 2021 evening shift. Refer solution of this question. **Ans. (A)**

Sol. 38 Second pendulum has a time period of 2 s thusstatement-I is false but with this it is clear that from one extreme to other it takes only half the time period i.e. 1 s thus statement-II is true. **Ans. (C)**

Sol. 39 The speed of particle in SHM as a function of displacement is given as

$$v=\omega\sqrt{A^2-y^2}$$

Maximum speed occurs at mean position i.e. at y=0, thus we have

$$v_{max}=A\omega$$

So half of maximum speed occurs at a position, given as

$$\frac{A\omega}{2}=\omega\sqrt{A^2-y^2}$$

$$\Rightarrow \quad \frac{A^2}{4}=A^2-y^2$$

$$\Rightarrow \quad y^2=\frac{3A^2}{4}$$

$$\Rightarrow \quad y=\frac{\sqrt{3}}{2}A$$

Ans. (3)

Sol. 40 In the given oscillation $\frac{5}{8}^{th}$ of oscillation can also be written as $\left(\frac{1}{2}+\frac{1}{8}\right)^{th}$ of the oscillation, which can be shown on projection diagram of uniform circular motion as given below.

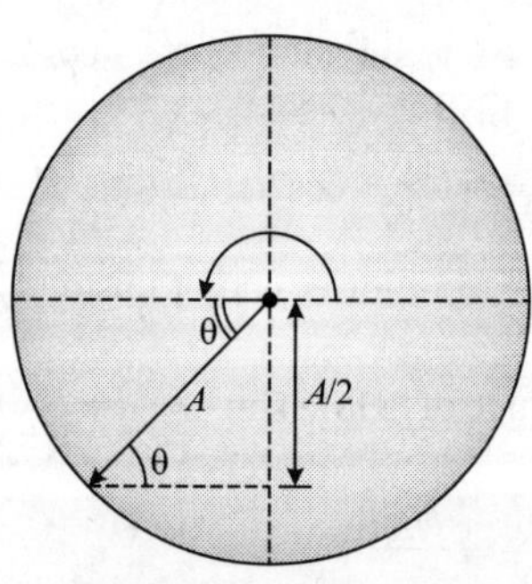

From this diagram if t is the time take for this part of oscillation, we use

$$\pi+\theta=\omega t$$

$$\Rightarrow \quad \frac{7\pi}{6}=\left(\frac{2\pi}{T}\right)t$$

$$\Rightarrow \quad t=\frac{7T}{12}$$

Ans. (7)

Sol. 41 When lift is stationary, we use

$$T=2\pi\sqrt{\frac{L}{g}}$$

When lift is moving upwards, effective gravity is taken as

$$g_{\text{eff}}=g+\frac{g}{2}=\frac{3g}{2}$$

New time period now is given as

$$T'=2\pi\sqrt{\frac{L}{g_{eff}}}=2\pi\sqrt{\frac{2L}{3g}}$$

$$\Rightarrow \quad T'=\sqrt{\frac{2}{3}}\,T$$

Ans. (D)

Sol. 42 By radioactive decay law we have

$$A=A_0e^{-\gamma t}=A_0e^{-bt/2\text{m}}$$

$$\Rightarrow \quad \frac{A_0}{2}=A_0e^{-bt/2\text{m}}$$

$$\Rightarrow \quad \frac{bt}{2m}=\ln 2$$

$$\Rightarrow \quad t=\frac{2m}{b}\ln 2=\frac{2\times 500\times 0.693}{20}=34.65\text{ s}$$

Ans. (A)

Sol. 43 For SHM, at a point where kinetic and potential energy are equal, we have

$$KE=PE$$

$$\Rightarrow \quad \frac{1}{2}m\omega^2(A^2-x^2)=\frac{1}{2}m\omega^2x^2$$

$$\Rightarrow \quad A^2-x^2=x^2$$

$$\Rightarrow \quad 2x^2=A^2$$

$$\Rightarrow \quad x=\pm\frac{A}{\sqrt{2}}$$

Ans. (C)

Sol. 44 Time period of oscillations of the two SHM cases in figure-1 and figure-2 are given as

$$T_a=2\pi\sqrt{\frac{M}{K}}$$

and

$$T_b=2\pi\sqrt{\frac{M}{K/2}}$$

$$\Rightarrow \quad \frac{T_b}{T_a}=\sqrt{2}=\sqrt{x}$$

$$\Rightarrow \quad x=2$$

Ans. (2)

Sol. 45 Amplitude of damped oscillations is given as a function of time as

$$A=A_0e^{-bt/2m}$$

$$\Rightarrow \quad \ln 2=\frac{b}{2m}\times 120$$

$$\Rightarrow \quad \frac{0.693\times 2\times 1}{120}=b=1.16\times 10^{-2}\text{kg/sec}$$

Ans. (None)

Sol. 46 As the maximum velocities are equal, we use

$$A_1\omega_1=A_2\omega_2$$

$$\Rightarrow \quad A_1\sqrt{\frac{k_1}{m}}=A_2\sqrt{\frac{k_2}{m}}$$

$$\Rightarrow \quad \frac{A_1}{A_2}=\sqrt{\frac{k_2}{k_1}}$$

Ans. (D)

Sol. 47 Half amplitude displacement corresponds to a phase change of $\pi/6$ rad which is one twelfth of the complete oscillation so the time taken can be given as

$$t=\frac{T}{12}=\frac{2}{12}=\frac{1}{6}$$

$$\Rightarrow \quad a=6$$

Ans. (6)

Sol. 48 As objects on earth surface starts to float, we use

$$mg=m\omega^2R$$

$$\Rightarrow \quad \omega=\sqrt{\frac{g}{R}} \qquad \ldots(1)$$

If duration of the day (time period of rotation of earth) is T, we use

$$T=\frac{2\pi}{\omega} \qquad \ldots(2)$$

$$\Rightarrow \quad T=2\pi\sqrt{\frac{R}{g}}=2\pi\sqrt{\frac{6400\times 10^3}{10}}$$

$$\Rightarrow \quad \frac{T}{60}=83.775\text{ min}\sim 84\text{ min}$$

Ans. (D)

Sol. 49 The angular frequency of SHM having period π/ω is given as

$$\omega'=\frac{2\pi}{T}=2\omega$$

The functions representing oscillations with angular frequency 2ω are given in option (C) and (D). Option (D) is having its mean position at origin whereas in option (C) it is displaced fro origin but both are the correct answer to this question. **Ans. (C, D)**

Sol. 50 Figure below shows the situation described in the question.

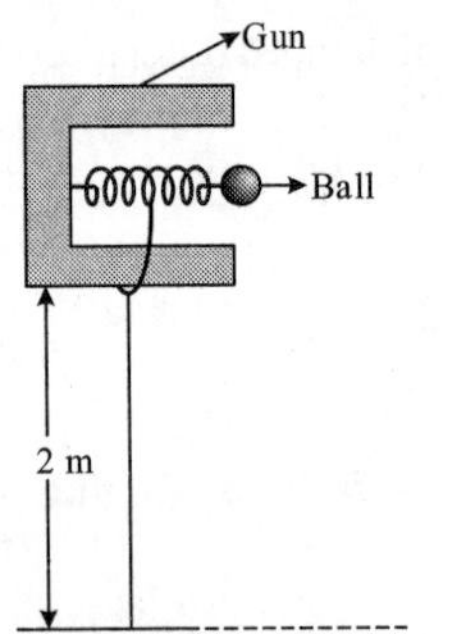

By conservation of energy, we can calculate the kinetic energy of the projected ball as

$$\frac{1}{2}kx^2 = \frac{1}{2}mv^2$$

$$\Rightarrow \quad v = x\sqrt{\frac{x}{m}}$$

$$\Rightarrow \quad v = 0.05 \times \sqrt{\frac{100}{0.1}} = 0.5\sqrt{10}\text{ m/s}$$

Time of flight of ball can be given as

$$T = \sqrt{\frac{2H}{g}} = \sqrt{\frac{2\times 2}{10}} = \frac{2}{\sqrt{10}}\text{ s}$$

Range of ball on ground can be calculated as

$$d = ut$$

$$\Rightarrow \quad d = 0.5\sqrt{10} \times \left(\frac{2}{\sqrt{10}}\right) = 1\text{ m} \qquad \textbf{Ans. (1)}$$

Sol. 51 Velocity of particle during SHM at a displacement x from its equilibrium position is given as

$$v = \omega\sqrt{A^2 - x^2}$$

$$\Rightarrow \quad v_1 = \omega\sqrt{A^2 - x_1^2}$$

$$\Rightarrow \quad v_2 = \omega\sqrt{A^2 - x_2^2}$$

$$\Rightarrow \quad \left(\frac{v_1}{\omega}\right)^2 - \left(\frac{v_2}{\omega}\right)^2 = x_2^2 - x_1^2$$

$$\Rightarrow \quad \omega^2 = \frac{v_1^2 - v_2^2}{x_2^2 - x_1^2}$$

$$\Rightarrow \quad \omega = \sqrt{\frac{v_1^2 - v_2^2}{x_2^2 - x_1^2}}$$

Thus time period of SHM is given as

$$T = \frac{2\pi}{\omega} = 2\pi\sqrt{\frac{x_2^2 - x_1^2}{v_1^2 - v_2^2}} \qquad \textbf{Ans. (D)}$$

Sol. 52 When length is reduced to one sixteenth of initial value, the modified time period can be given as

$$T = 2\pi\sqrt{\frac{l}{g}}$$

$$\Rightarrow \quad T' = 2\pi\sqrt{\frac{l}{16g}} = \frac{T}{4} \qquad \textbf{Ans. (B)}$$

Sol. 53 From the given equation of motion, we use

$$x(t) = A\sin\omega t + B\cos\omega t \qquad \ldots(1)$$

and $$v(t) = A\omega\cos\omega t - B\omega\sin\omega t \qquad \ldots(2)$$

$$\Rightarrow \quad x(0) = B \text{ and } v(0) = A\omega$$

In Equation-(1) we can substitute

$$A = C\sin\phi$$

and $$B = C\cos\phi$$

After substitution, it gives

$$x(t) = C\cos(\omega t - \phi)$$

Where $C = \sqrt{A^2 + B^2} \Rightarrow C = \sqrt{\frac{v(0)^2}{\omega^2} + x(0)^2}$

and $\phi = \tan^{-1}\left(\frac{A}{B}\right) \Rightarrow \phi = \tan^{-1}\left(\frac{v(0)}{x(0)\omega}\right)$ **Ans. (A)**

Sol. 54 For free oscillations of the blocks placed on smooth surface, we analyse the motion by considering reduced mass of one body with respect to another, given as

$$\mu = \frac{m_1 m_2}{m_1 + m_2}$$

$$\Rightarrow \quad \mu = \frac{200 \times 800}{200 + 800} = 160\text{ g} = 0.16\text{ kg}$$

As springs are connected in series, effective force constant is given as

$$K_{eq} = \frac{k_1 k_2}{k_1 + k_2} = \frac{4k \times k}{4k + k} = \frac{4}{5}k = \frac{4}{5} \times 20 = 16\text{ N/m}$$

Angular frequency of oscillations can now be given as

$$\omega = \sqrt{\frac{K_{eq}}{\mu}} = \sqrt{\frac{16}{0.16}} = 10\text{ rad/s} \qquad \textbf{Ans. (10)}$$

Sol. 55 At midway point, the speed of particle in SHM is given as

$$v_{A/2} = \omega\sqrt{A^2 - x^2}$$

$$\Rightarrow \quad v_{A/2} = \omega\sqrt{A^2 - \left(\frac{A}{2}\right)^2} = \omega\left(\frac{\sqrt{3}}{2}A\right) = \frac{\sqrt{3}}{2}V_{max}$$

$$\Rightarrow \quad KE = \frac{1}{2}m\left(\frac{\sqrt{3}}{2}V_{max}\right)^2$$

$$\Rightarrow \quad TE = \frac{1}{2}m(V_{max})^2$$

$$\Rightarrow \quad \frac{KE}{TE} = \frac{3}{4}$$

Ans. (A)

Sol. 56 Speed of particle at a displacement x is given as

$$v = \omega\sqrt{a^2 - x^2}$$

Given that

$$\frac{3}{4}KE_{max} = KE_x$$

$$\Rightarrow \quad \frac{3}{4}\left(\frac{1}{2}m\omega^2 A^2\right) = \frac{1}{2}m\omega^2(a^2 - x^2)$$

$$\Rightarrow \quad \frac{3}{4}a^2 = a^2 - x^2$$

$$\Rightarrow \quad x^2 = \frac{a^2}{4}$$

$$\Rightarrow \quad x = \pm\frac{a}{2}$$

Ans. (A)

Sol. 57 Time period of oscillations of the magnetic needle in magnetic field is given as

$$T = 2\pi\sqrt{\frac{I}{MB}}$$

$$\Rightarrow \quad \frac{1}{2} = 2\pi\sqrt{\frac{5\times10^{-6}}{9.85\times10^{-2}\times B}}$$

$$\Rightarrow \quad \frac{1}{2} = 2\pi\sqrt{\frac{5\times10^{-6}}{\pi^2\times B\times10^{-2}}}$$

$$\Rightarrow \quad B = 8\times10^{-3}\,\mathrm{T} = 8\,mT$$

Ans. (8)

Sol. 58 At $t = \frac{T}{4}$ particle is at extreme position where potential energy of SHM is maximum, given as

$$U_{max} = \frac{1}{2}m\omega^2A^2,$$

$$\Rightarrow \quad \frac{1}{2}\times0.5\times\left(\frac{2\pi}{0.2}\right)^2\times(0.05)^2 = 0.6186 \approx 0.62$$

Ans. (C)

Sol. 59 From the given SHM equation, we use

$$2 = A\sin(0+\phi)$$

$$\Rightarrow \quad \sin\phi = \frac{2}{A}$$

$$\Rightarrow \quad \cos\phi\sqrt{1-\frac{4}{A^2}} = \sqrt{\frac{A^2-4}{A^2}} \qquad \ldots(1)$$

Velocity of particle in SHM for given equation is written as

$$v = A\omega\cos(\omega t+\phi)$$

$$\Rightarrow \quad 2\omega = A\omega\cos(0+\phi)$$

$$\Rightarrow \quad 2 = A\cos\phi$$

$$\Rightarrow \quad A = \frac{2}{\cos\phi} \qquad \ldots(2)$$

From equations-(1) & (2), we have

$$A = \frac{2}{\sqrt{\frac{A^2-4}{A^2}}}$$

$$\Rightarrow \quad A = \frac{2A}{\sqrt{A^2-4}}$$

$$\Rightarrow \quad \sqrt{A^2-4} = 2$$

$$\Rightarrow \quad A = \sqrt{8} = \pm\,2\sqrt{2}$$

Ans. (2)

Sol. 60 Time period of simple pendulum is given as

$$T = 2\pi\sqrt{\frac{l}{g}}$$

$$\Rightarrow \quad T \propto l^{1/2}$$

$$\Rightarrow \quad \frac{\Delta T}{T} = \frac{1}{2}\frac{\Delta l}{l}$$

$$\Rightarrow \quad \frac{\Delta T}{24\times3600} = \frac{1}{2}\times\frac{0.1}{100}$$

$$\Rightarrow \quad \Delta T = \frac{1}{2}\times\frac{0.1}{100}\times24\times3600$$

$$\Rightarrow \quad \Delta T = 43.2\,\mathrm{s}$$

Ans. (A)

Sol. 61 Amplitude of firs SHM is given as

$$A_1 = 5$$

Second SHM equation can be re-written as

$$x_2 = 5\sqrt{2}\,[\sin2\pi t + \cos2\pi t]$$

$$\Rightarrow \quad x_2 = 10\left[\sin\left(2\pi T + \frac{\pi}{4}\right)\right]$$

Thus amplitude of second SHM is given as

$$A_2 = 10$$

$$\Rightarrow \quad A_2 = 2A_1$$

Ans. (2)

Sol. 62 For the given SHM of the particle, its equation is written as

$$x = x_0 \sin \omega t$$

Potential energy of particle in SHM is expressed as

$$U(x) = \frac{1}{2}kx^2 = \frac{1}{2}kx_0^2 \sin^2 \omega t$$

$\Rightarrow$ $U(x) = C \sin^2 \omega t$ **Ans. (D)**

Sol. 63 Amplitude of first wave is 10 and for second wave, amplitude can be calculated by rewriting the equation as

$$y_2 = 5[\sin 3\pi t + \sqrt{3} \cos \pi t]$$

$$\Rightarrow \quad y_2 = 10 \sin\left(3\pi t + \frac{\pi}{3}\right)$$

$\Rightarrow$ $\frac{A_2}{A_1} = \frac{10}{10} = 1$ **Ans. (1)**

Sol. 64 When kinetic and potential energies are equal in SHM at a position y, we have

$$K = U$$

$$\Rightarrow \quad \frac{1}{2}k(A^2 - y^2) = \frac{1}{2}ky^2$$

$$\Rightarrow \quad y^2 = \frac{A^2}{2}$$

$$\Rightarrow \quad y = \pm \frac{A}{\sqrt{2}}$$

SHM equation is taken as

$$y = A\sin\omega t$$

$$\Rightarrow \quad \frac{A}{\sqrt{2}} = A \sin\omega t$$

$$\Rightarrow \quad \sin\frac{\pi}{4} = \sin\omega t$$

At $\omega t = \frac{\pi}{4}$ we have $t = \frac{T}{8}$ which gives $x = 8$. **Ans. (8)**

Sol. 65 Time period of a simple pendulum is given as

$$T - 2\pi\sqrt{\frac{l}{g}}$$

Inside liquid, due to buoyancy force the effective gravity decreases, which is given as

$$g_{\text{eff}} = g - \frac{g}{4} = \frac{3g}{4}$$

Thus new time period is given as

$$T' = 2\pi\sqrt{\frac{4l/3}{3g/4}} = 2\pi\sqrt{\frac{16l}{9g}}$$

$\Rightarrow$ $T' = \frac{4}{3}T$ **Ans. (B)**

Sol. 66 In a body exeucting SHM, kinetic and potential energy of particles continuously changes such that total energy remain constant so option (a) and (b) are not correct and option (c0 and (d) are correct. **Ans. (A)**

Sol. 67 From figure, amplitude of oscillation is 2 m. The energy of oscillations can be given as

$$\frac{1}{2}m\omega^2 A^2 = 10$$

$$\Rightarrow \quad \frac{1}{2}m\omega^2 (2)^2 = 10$$

$$\Rightarrow \quad m\omega^2 = 5$$

$$\Rightarrow \quad 5(\omega^2) = 5$$

$$\Rightarrow \quad \omega^2 = 1$$

$$\Rightarrow \quad \omega = 1$$

Time period of oscillation can be calculated as

$$T = \frac{2\pi}{\omega} = 2\pi$$

$$\Rightarrow \quad 2\pi = \sqrt{\frac{l}{g_p}} = 2\pi$$

$$\Rightarrow \quad 2\pi = \sqrt{\frac{4}{g_p}} = 2\pi$$

$\Rightarrow$ $g_p = 4 \text{ m/s}^2$ **Ans. (D)**

14.1 Wave Motion

Sol. 1 When car is at rest, wave speed is given as

$$60 = \sqrt{\frac{Mg}{\mu}}$$

When car is accelerating, we use

$$60.5 = \sqrt{\frac{M(g^2 + a^2)^{1/2}}{\mu}}$$

$$\Rightarrow \quad \frac{60.5}{60} = \sqrt{\sqrt{\frac{g^2 + a^2}{g^2}}}$$

$$\Rightarrow \quad \left(1 + \frac{0.5}{60}\right)^4 = \frac{g^2 + a^2}{g^2} = 1 + \frac{2}{60}$$

$$\Rightarrow \quad g^2 + a^2 = g^2 + g^2 \times \frac{2}{60}$$

$$\Rightarrow \quad a = g\sqrt{\frac{2}{60}} = \frac{g}{\sqrt{30}} = \frac{g}{5.47}$$

$\Rightarrow$ $a \simeq \frac{g}{5}$ **Ans. (A)**

Sol. 2 Given equation is

$$y = 0.03 \sin(450t - 9x)$$

Comparing with

$$y = A \sin(\omega t - kx), \text{ we get}$$

$$v = \frac{\omega}{k} = \frac{450}{9} = 50 \text{ m/s}$$

Velocity of transverse waves on a string gives

$$v = \sqrt{\frac{T}{\mu}}$$

$\Rightarrow$ $$\frac{T}{\mu} = 2500$$

$\Rightarrow$ $$T = 2500 \times 5 \times 10^{-3}$$

$\Rightarrow$ $$T = 12.5 \text{ N}$$ **Ans. (B)**

Sol. 3 Above wave is moving along $-ve$ x-axis with speed given as

$$v = \frac{\omega}{K}$$

$\Rightarrow$ $$v = \frac{50}{2} = 25 \text{ m/s}$$ **Ans. (A)**

Sol. 4 Speed of wave from wave equation is written as

$$v = -\frac{(\text{coeffecient of } t)}{(\text{coeffecient of } x)}$$

$$v = -\frac{1000}{(-3)} = \frac{1000}{3} \text{ m/s}$$

As speed of wave $v \propto \sqrt{T}$, we have

$$= \frac{1000/3}{336} = \sqrt{\frac{273}{T}}$$

$\Rightarrow$ $$T = 277.41 \text{ K}$$

$\Rightarrow$ $$T = 4.41°\text{C}$$ **Ans. (C)**

Sol. 5

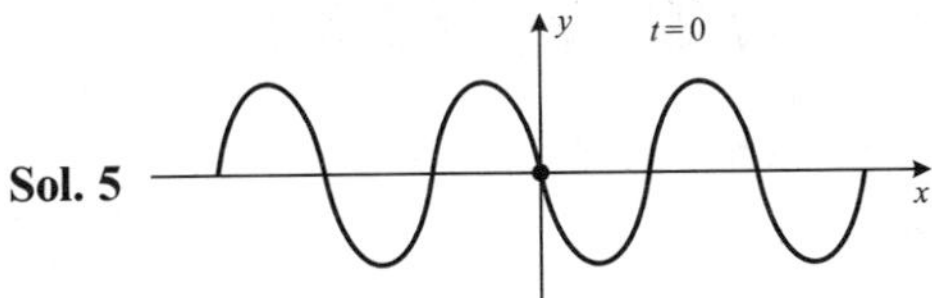

For the given wave equation

$$y = A \sin(kx - wt + \phi)$$

at $x = 0, t = 0, y = 0$ and slope is negative

$\Rightarrow$ $$\frac{\partial y}{\partial x} = Ak \cos(kx - wt + \phi) = -\text{ve}$$

$\Rightarrow$ $\phi = x$ (From the given options) **Ans. (C)**

Sol. 6 Loudness of sound is given by

$$L(dB) = 10 \log \frac{I}{I_0}$$

$\Rightarrow$ $$120 = 10 \log\left(\frac{I}{I_0}\right)$$

$\Rightarrow$ $$I = 1 \text{W/m}^2$$

For isotropic point sources, intensity at a distance r is given as

$$I = \frac{P}{4\pi r^2} = \frac{2}{4\pi r^2}$$

$\Rightarrow$ $$r = \sqrt{\frac{2}{4\pi}} = \sqrt{\frac{1}{2\pi}} m$$ (Taking $I = 1$ W/m²)

$\Rightarrow$ $$r = 0.399 \text{ m} \approx 40 \text{ cm}$$ **Ans. (C)**

Sol. 7 Velocity of waves on wire is given as

$$v = \sqrt{\frac{T}{\mu}}$$

$\Rightarrow$ $$90 = \sqrt{\frac{\frac{YA}{l}\Delta l}{\frac{m}{l}}} = \sqrt{\frac{16 \times 10^{11} \times 10^{-6} \times \Delta l}{6 \times 10^{-3}}}$$

$\Rightarrow$ $$\Delta l = \frac{8100 \times 3}{8} \times 10^{-8} \cong 3 \times 10^{-5} \text{ m}$$

$$= 0.03 \text{ mm}$$ **Ans. (C)**

Sol. 8 Velocity of transverse wave in a string is $v \propto \sqrt{T}$.

When velocity is changed to $v/2$ then tension changes to $T/4$, so new tension is

$$T' = \frac{2.06 \times 10^4}{4}$$

$\Rightarrow$ $$T' = 5.15 \times 10^3 \text{ N}$$ **Ans. (B)**

Sol. 9 Tension in repeat bottom is

$$T_1 = 2g = 20 \text{ N}$$

tension in rope at top is

$$T_2 = 8g = 80 \text{ N}$$

Wave speed in rope is given as

$$v = \sqrt{\frac{T}{\mu}}$$

As frequency of pulse is same at bottom and top, we use

$$\frac{v_1}{\lambda_1} = \frac{v_2}{\lambda_2}$$

$$\Rightarrow \quad \lambda_2 = \frac{v_2}{v_1} \times \lambda_1 = \sqrt{\frac{T_2}{T_1}} \times \lambda_1$$

$$\Rightarrow \quad \lambda_2 = \sqrt{\frac{80}{20}} \times 6 = 12\text{ cm}$$ **Ans. (B)**

Sol. 10 Given that

$$5\text{m} = N_2\lambda$$

and $$1.5\text{m} = (2N_1 + 1)\lambda/2$$

$$\Rightarrow \quad \frac{5}{1.5} = \frac{2N_2}{2N_1 + 1}$$

$$\Rightarrow \quad 10N_1 + 5 = 3N_2$$

Thus for $N_1 = 1 \Rightarrow N_2 = 5 \Rightarrow \lambda = 1$

and $N_1 = 4 \Rightarrow N_2 = 15 \Rightarrow \lambda = 1/3$

and $N_1 = 7 \Rightarrow N_2 = 25 \Rightarrow \lambda = 1/5$ **Ans. (D)**

Sol. 11 Pressure amplitude is given as

$$\Delta P_0 = \frac{2\pi}{\lambda} BA$$

$$\Rightarrow \quad \Delta P_0 = \frac{2\pi}{\lambda} \times \rho v^2 \times A = \frac{w}{v} \times \rho v^2 \times A$$

$$\Rightarrow \quad A = \frac{\Delta P_0}{\rho v \omega} = \frac{10}{1 \times 300 \times 2 \times 3.14 \times 1000}$$

$$\Rightarrow \quad A = \frac{10}{300 \times 6.28} \times 10^{-3}\text{ m}$$

Among all given options, (B) is correct. **Ans. (B)**

Sol. 12 A travelling wave is represented by a function given as

$$y = F(x, t)$$

For travelling wave y should be a function of x and t and the function must be in form of $(x \pm vt)$ where $\pm$ represents the direction of propagation of wave. Out of all given option, option (D) is appropriate here. **Ans. (D)**

Sol. 13 The speed of transverse waves on a stretched string is given as

$$V = \sqrt{\frac{T}{\mu}}$$

$$\Rightarrow \quad \frac{\Delta V}{V} = \frac{1}{2}\frac{\Delta T}{T}$$ **Ans. (2)**

Sol. 14 From the given wave equation

$$y = -0.21 \sin(x + 30t)$$

The wave velocity is given as

$$v = \frac{\omega}{k} = \frac{30}{1} = 30\text{ m/s}$$

The speed of transverse waves on a stretched string is given as

$$v = \sqrt{\frac{T}{\mu}}$$

$$\Rightarrow \quad T = v^2\mu = (30)^2(0.0135) = 12.15$$

$$\Rightarrow \quad T = x \times 10^{-2}\text{ N}$$

$$\Rightarrow \quad x = 1215$$ **Ans. (1215)**

Sol. 15 The angular frequency of the wave is given as

$$\omega = 2\pi f = 1.5 \times 10^3\text{ rad/s}$$

Amplitude of the wave is calculated as

$$A = \frac{6}{2} = 3\text{ cm} = 0.03\text{ m}$$

This matches with option (D) only. **Ans. (D)**

Sol. 16 The general equation of a wave disturbance propagating with a speed v is given as

$$y = \frac{1}{1 + (x - vt)^2}$$

By comparing it with given equation at $t = 1$s, we have

$$v = 2\text{ m/s}$$ **Ans. (2)**

Sol. 17 At node of standing wave, displacement of particle remain zero always so we have

$$1\cos(1.57x)\sin(178.5)t = 0$$

$$\Rightarrow \quad \cos(1.57x) = 0$$

$$\Rightarrow \quad 1.57\times = \frac{\pi}{2}$$

$$\Rightarrow \quad x = 1\text{ cm}$$ **Ans. (1)**

14.2 Stationary Waves and Beats

Sol. 1 General equation for stationary wave is given as-

$$y(x, t) = 2A \sin kx \cos \omega t \quad \ldots(1)$$

Comparing equation-(1) with the given equation we have

$$k = \frac{5\pi}{4} \text{ and } \omega = 200\pi$$

$\Rightarrow$ Wave speed is given as

$$v = \frac{\omega}{k} = \frac{200\pi}{5\pi/4} = \frac{200 \times 4}{5}$$

$$\Rightarrow \quad v = 160\text{ m/s}$$ **Ans. (A)**

Sol. 2 If wire W_1 oscillates in p loops then its frequency is

$$\nu_1 = \frac{p}{2l}\sqrt{\frac{T}{\mu_1}} = \frac{p}{2l}\sqrt{\frac{T}{\rho_1 A}}$$

If wire W_2 oscillates in q loops then its frequency is

$$\nu_2 = \frac{q}{2l}\sqrt{\frac{T}{\mu_2}} = \frac{q}{2l}\sqrt{\frac{T}{\rho_2 A}}$$

p and q are no. of Antinodes in wires W_1 and W_2.
For stationary wave in composite wire, we have

$$\nu_1 = \nu_2$$

$$\Rightarrow \quad \frac{p}{2l}\sqrt{\frac{T}{\rho_1 A}} = \frac{q}{2l}\sqrt{\frac{T}{\rho_2 A}}$$

$$\Rightarrow \quad \frac{p}{q} = \frac{1}{2}$$

Ans. (B)

Sol. 3

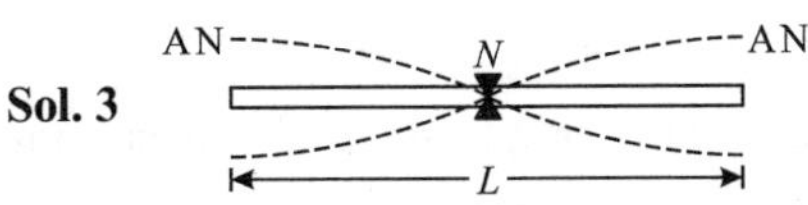

Fundamental frequency oscillations

here $\quad L = \lambda/2$

$\Rightarrow \quad \lambda = 2L$

Frequency of oscillation

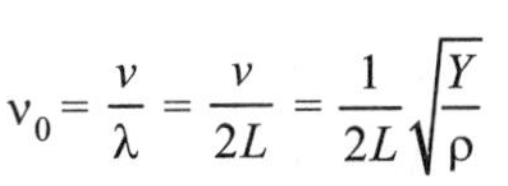

$$\nu_0 = \frac{v}{\lambda} = \frac{v}{2L} = \frac{1}{2L}\sqrt{\frac{Y}{\rho}}$$

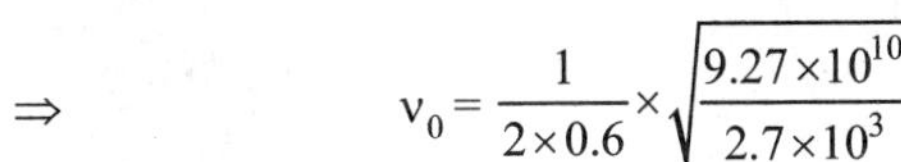

$$\Rightarrow \quad \nu_0 = \frac{1}{2\times 0.6}\times\sqrt{\frac{9.27\times 10^{10}}{2.7\times 10^{3}}}$$

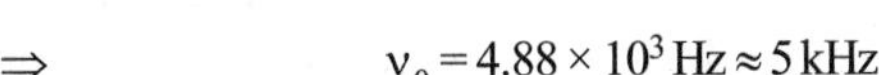

$$\Rightarrow \quad \nu_0 = 4.88\times 10^3\ \text{Hz} \approx 5\ \text{kHz}$$

Ans. (A)

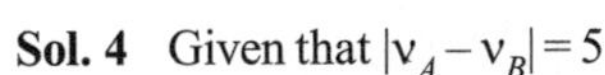

Sol. 4 Given that $|\nu_A - \nu_B| = 5$

either $\nu_B = 420$ Hz or 430 Hz As $\nu_A = 425$ Hz

As tension in B is increased

$\Rightarrow$ ν_B increases and new beat frequency with A becomes 3 Hz.

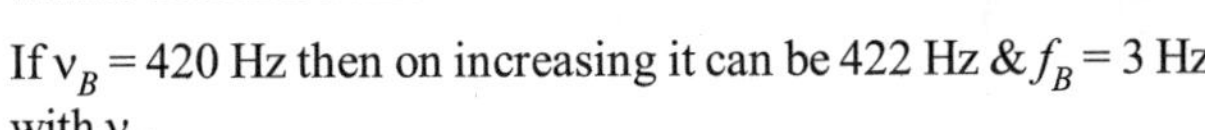

If $\nu_B = 420$ Hz then on increasing it can be 422 Hz & $f_B = 3$ Hz with ν_A.

If $\nu_B = 430$ Hz then on increasing always $f_B > 5$ Hz with ν_A.

Ans. (D)

Sol. 5 For a sonometer wire oscillation frequency is given as

$$n_s = \frac{v}{2l}$$

at $l = 0.95$ m, $\quad n_{s_1} = \dfrac{v}{2\times 0.95}$

and at $l = 1$ m, $\quad n_{s_2} = \dfrac{v}{2\times 1}$

$\Rightarrow \quad n_{s_1} > n_{s_2}$

As tuning fork produces 5 beats/s with both n_{s_1} and n_{s_2}

$\Rightarrow \quad n_{s_1} > n_f > n_{s_2}$

$$\Rightarrow \quad n_f - 5 = n_{s_2} = \frac{v}{1.9} \qquad \ldots(1)$$

$$\text{and} \quad n_f - 5 = n_{s_1} = \frac{v}{2} \qquad \ldots(2)$$

(2)/(1) gives $\quad \dfrac{n_f + 5}{n_f - 5} = \dfrac{20}{19}$

$\Rightarrow \quad 19n_f + 95 = 20n_f - 100$

$\Rightarrow \quad n_f = 195$ Hz

Ans. (A)

Sol. 6 Frequency of 3rd Harmonic of open pipe is

$$n_3 = \frac{3v}{2l}$$

As beat frequency is 1 beat/s

$$\Rightarrow \quad \frac{3v}{2l} = 256 \pm 1 = 255\ \text{Hz or } 257\ \text{Hz}$$

Case-I: $\quad \dfrac{3v}{2l} = 255$

$$\Rightarrow \quad l = \frac{3v}{2\times 255} = \frac{3\times 340}{2\times 255}$$

$$= 2\ \text{m} = 200\ \text{cm}$$

Case-II: $\quad \dfrac{3v}{2l} = 257$

$$\Rightarrow \quad l = \frac{3v}{2\times 257} = \frac{3\times 340}{2\times 257}$$

$\Rightarrow \quad l \simeq 1.97$ m $= 197$ cm

Ans. (D)

Sol. 7 First resonance length is 10 cm

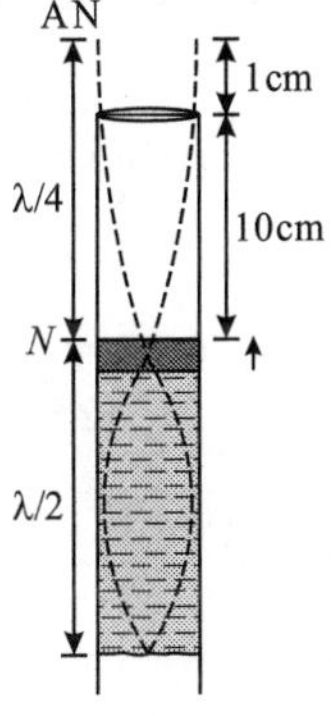

$$\Rightarrow \quad \frac{\lambda}{4} = 10 + 1 = 11\ \text{cm}$$

Second resonance occurs when length of column is approx $3\lambda/4$.

$$\frac{3\lambda}{4} = 3 \times 11 = 33 \text{ cm}$$

Resonance column length now is

$$\frac{3\lambda}{4} - e = 33 - 1 = 32 \text{ cm} \quad \textbf{Ans. (A)}$$

Sol. 8 Frequency of the sound produced by flute is given as

$$f = 2\left(\frac{v}{2l}\right) = \frac{2 \times 330}{2 \times 0.5} = 660\text{Hz}$$

Velocity of observer,

$$v_0 = 10 \times \frac{5}{18} = \frac{25}{9} \text{ m/s}$$

Frequency detected by observer is

$$f_{ap} = \left[\frac{v + v_0}{v}\right] f$$

$$\Rightarrow \quad f' = \left[\frac{\frac{25}{9} + 330}{330}\right] 660$$

$$\Rightarrow \quad f' = 335.56 \times 2 = 671.12 \text{ Hz}$$

Ans. (D)

Sol. 9 Velocity of transverse wave on string

$$v = \sqrt{\frac{T}{\mu}} = \sqrt{\frac{8}{5} \times 1000} = 40 \text{ m/s}$$

Now, wavelength of wave

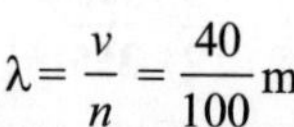

$$\lambda = \frac{v}{n} = \frac{40}{100} \text{m}$$

Separation between successive nodes given as

$$\frac{\lambda}{2} = \frac{20}{100} \text{ m} = 20 \text{ cm}$$

Ans. (B)

Sol. 10 For closed organ pipe, resonant frequencies are odd multiple of fundamental frequency. For maximum, we use

$(2n + 1) f_0 \leq 20,000$

(f_0 is fundamental frequency = 1.5 kHz)

$$\Rightarrow \quad n = 6$$

So total number of overtones that can be heard are 7 (0 to 6)

Ans. (A)

Sol. 11 If antinode is produced at a distance x above the reference mark, with frequency 512 Hz, we have

$$512 = \frac{v}{4(11 + x)} \quad \ldots(1)$$

with 256 Hz frequency we use

$$256 = \frac{v}{4(27 + x)} \quad \ldots(2)$$

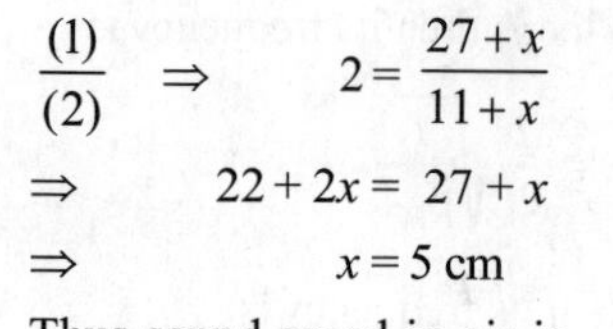

$$\frac{(1)}{(2)} \Rightarrow \quad 2 = \frac{27 + x}{11 + x}$$

$$\Rightarrow \quad 22 + 2x = 27 + x$$

$$\Rightarrow \quad x = 5 \text{ cm}$$

Thus sound speed in air is

$$v = 512 \times 4(11 + x)$$

$$\Rightarrow \quad v = 512 \times 4 \times 0.16$$

$$\Rightarrow \quad v = 327.68 \text{ m/s}$$

Ans. (A)

Sol. 12

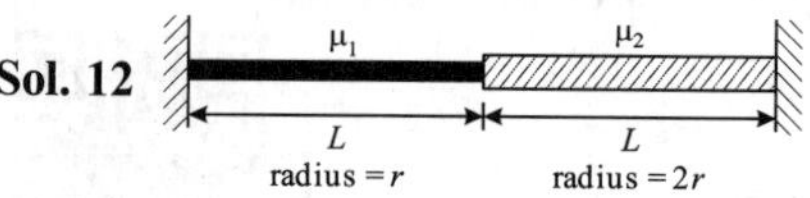

Let mass per unit length of wires are μ_1 and μ_2 respectively

As materials are same, so density ρ is same, we use

$$\mu_1 = \frac{\rho \pi r^2 L}{L} = \mu$$

and $$\mu_2 = \frac{\rho 4 \pi r^2 L}{L} = 4\mu$$

Tension in both are same, taken as T, let speed of wave in wires are v_1 and v_2, we have

$$v_1 = \sqrt{\frac{T}{\mu}} = v;$$

$$v_2 = \sqrt{\frac{T}{4\mu}} = \frac{v}{2}$$

So fundamental frequencies in both wires are given as

$$f_{01} = \frac{v_1}{2L} = \frac{v}{2L}$$

and $$f_{02} = \frac{v_2}{2L} = \frac{v}{4L}$$

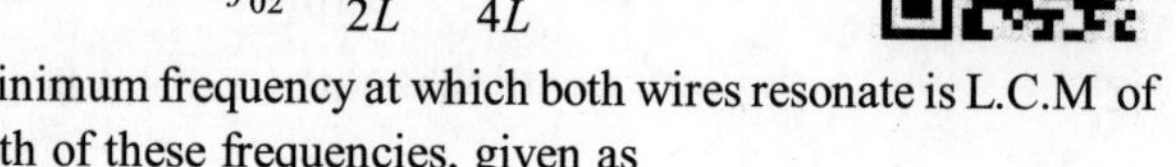

Minimum frequency at which both wires resonate is L.C.M of both of these frequencies, given as

$$f_0 = \frac{v}{2L}$$

Hence number of loops in wires are 1 and 2 respectively as shown below

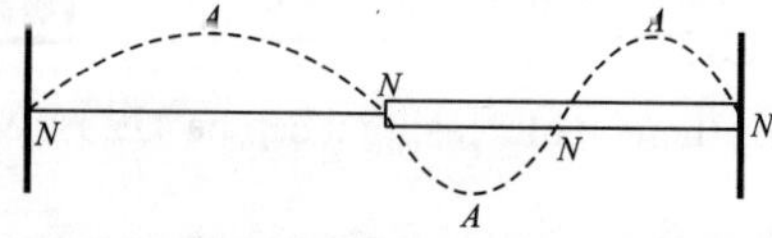

So, ratio of no. of antinodes is 1 : 2 **Ans. (D)**

Sol. 13 4th harmonic oscillations in clamped wire are shown below

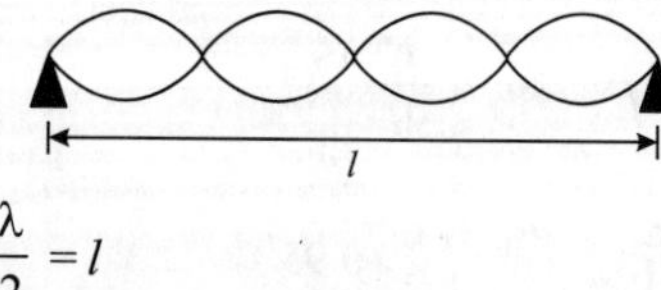

$$\Rightarrow \quad 4\frac{\lambda}{2} = l$$

$$\Rightarrow \quad 2\lambda = l \Rightarrow \lambda = \frac{l}{2}$$

From given equation, we use

$$\frac{2\pi}{\lambda} = 0.157$$

$$\lambda = 40$$

$$l = 2\lambda$$

$$\Rightarrow \quad l = 80\,\text{m}$$

Ans. (B)

Sol. 14 As string vibrates in 3rd harmonic, we use

$$3\left(\frac{v}{2l}\right) = 240$$

$$\Rightarrow \quad 3\frac{v}{2\times 2} = 240$$

$$\Rightarrow \quad v = 320\,\text{m/s}$$

Fundamental frequency of string is given as

$$n_0 = \frac{v}{2l} = \frac{320}{2\times 2} = 80\,\text{Hz}$$

Ans. (D)

Sol. 15 Beat frequency of the given waves is

$$f_{beat} = 11 - 9 = 2\,\text{Hz}$$

Thus in one second there are two maxima and minima occur. This is corresponding to option (D). **Ans. (D)**

Sol. 16 In resonance tube experiment, wave speed is given as

$$v = 2f(l_2 - l_1)$$

$$\Rightarrow \quad v = 2\times 480\times(70-30)\times 10^{-2}$$

$$\Rightarrow \quad v = 960\times 40\times 10^{-2}$$

$$\Rightarrow \quad v = 38400\times 10^{-2}\,\text{m/s}$$

$$\Rightarrow \quad v = 384\,\text{m/s}$$

Ans. (C)

Sol. 17 Speed of sound in air is given as

$$v_s = \sqrt{\frac{\gamma P}{\rho}}$$

$$\Rightarrow \quad \frac{v_{gas}}{v_{air}} = \sqrt{\frac{\rho_{air}}{\rho_{gas}}}$$

$$\Rightarrow \quad \frac{v_{gas}}{300} = \frac{1}{\sqrt{2}}$$

$$\Rightarrow \quad v_{gas} = \frac{300}{\sqrt{2}}\,\text{m/s}$$

$$\Rightarrow \quad v_{gas} = 150\sqrt{2}\,\text{m/s}$$

For open organ pipe, we use

$$n_2 - n_1 = \frac{v_{gas}}{2l} = \frac{150\sqrt{2}}{2(1)} = 75\sqrt{2}\ \text{Hz}$$

$$\Rightarrow \quad \Delta n = 106.06\,\text{Hz}$$

Ans. (106.06)

Sol. 18 For n^{th} harmonic, frequency is given as

$$\frac{nv}{2l} = 420$$

For next $(n+1)^{th}$ harmonic, we use

$$\frac{(n+1)v}{2l} = 490$$

$$\Rightarrow \quad \frac{v}{2l} = 70$$

$$\Rightarrow \quad l = \frac{v}{140} = \frac{1}{140}\sqrt{\frac{540}{6\times 10^{-3}}} = \frac{1}{140}\sqrt{90\times 10^3}$$

$$\Rightarrow \quad l = \frac{300}{140} = 2.142\,\text{m}$$

Ans. (D)

Sol. 19 Fundamental frequency of oscillation for a clamped string is given as

$$f = \frac{l}{2l}\sqrt{\frac{T}{\mu}}$$

For identical string l and m will be same

$$\Rightarrow \quad f \propto \sqrt{T}$$

$$\Rightarrow \quad \frac{450}{300} = \sqrt{\frac{T_x}{T_y}}$$

$$\Rightarrow \quad \frac{T_x}{T_y} = \frac{9}{4} = 2.25$$

Ans. (C)

Sol. 20 For two successive resonance lengths in tube, we use

$$\Rightarrow \quad \lambda = 2(l_2 - l_1)$$

$$\Rightarrow \quad \lambda = 2\times(24.5-17)$$

$$\Rightarrow \quad \lambda = 2\times 7.5 = 15\,\text{cm}$$

As $v = f\lambda$, we use

$$\Rightarrow \quad 330 = f\times 15\times 10^{-2}$$

$$\Rightarrow \quad f = \frac{330}{15}\times 100\ \text{Hz}$$

$$\Rightarrow \quad f = \frac{110\times 100}{5}\ \text{Hz}$$

$$\Rightarrow \quad f = 2200\,\text{Hz}$$

Ans. (C)

Sol. 21 As the diameter of the column is given so it implies we have to consider the end correction $e = 0.3d$ in the tube length so, first resonance length can be given as

$$l + e = \frac{\lambda}{4} = \frac{V}{4f}$$

$$\Rightarrow \quad l = 16.66 - 0.3\times 6 = 14.866\,\text{cm}$$

$$\Rightarrow \quad l = 14.8\,\text{cm}$$

Ans. (B)

Sol. 22 Initially beat frequency is given as 5 Hz so initial frequency of for A is either 345 Hz or 335 Hz and after filing the frequency of fork A increases. It is also given that beat frequency decreases so frequency of A must be 335 Hz and as its final frequency is increasing due to which beat frequency also decrease to 2 beats/s. **Ans. (B)**

Sol. 23 As both pipes are oscillating at same frequency, if length of open pipe of L', we use

$$f_c = f_0$$

$$\Rightarrow \quad \frac{3V_C}{4L} = \frac{2V_0}{2L'}$$

$$\Rightarrow \quad \frac{3V_C}{4L} = \frac{V_0}{L'}$$

$$\Rightarrow \quad L' = \frac{4L}{3}\frac{V_0}{V_C} = \frac{4L}{3}\sqrt{\frac{B.\rho_1}{\rho_2.B}} \quad (B \text{ is bulk modulus})$$

$$\Rightarrow \quad L' = \frac{4L}{3}\sqrt{\frac{\rho_1}{\rho_2}}$$

$$\Rightarrow \quad x = 4$$ **Ans. (4)**

Sol. 24 The phase difference between the two waves at the is given as

$$\Delta\phi = kx_0 = \frac{2\pi}{\lambda}x_0 = 2\pi \times 3.5 = 7\pi \text{ rad}$$

Amplitude of resulting wave can be given as

$$\Rightarrow \quad A_R = \sqrt{A_1^2 + A_2^2 + 2A_1A_2\cos(\Delta\phi)}$$

$$\Rightarrow \quad A_R = \sqrt{12^2 + 5^2 + 2\times 12\times 5\times \cos(7\pi)}$$

$$\Rightarrow \quad A_R = \sqrt{(12-5)^2} = 7$$ **Ans. (7)**

Sol. 25 Closed organ pipe resonate at frequencies given as

$$f = (2n+1)\frac{v}{4l}$$

Minimum length, at which resonance occurs can be calculated for

$$n = 0$$

$$\Rightarrow \quad f_0 = \frac{v}{4l}$$

$$\Rightarrow \quad l = \frac{v}{4f_0}$$

$$\Rightarrow \quad l = \frac{340}{4\times 250} = 34 \text{ cm}$$ **Ans. (34)**

Sol. 26 Resonating frequency of the wire stretched between the rigid supports is given as

$$f_n = \frac{nv}{2l} = 500$$

Next higher frequency is given as

$$f_{n+1} = \frac{(n+1)v}{2l} = 500$$

$$\Rightarrow \quad \frac{n+1}{n} = \frac{11}{10}$$

$$\Rightarrow \quad n = 10$$

$$\Rightarrow \quad l = \frac{nv}{2f_n}\left(v = \sqrt{\frac{T}{\mu}}\right)$$

$$\Rightarrow \quad l = \frac{10}{2\times 500}\times\sqrt{\frac{900}{9\times 10^{-4}}} = 10 \text{ m}$$ **Ans. (10)**

14.3 Doppler's Effect

Sol. 1

$v_0 = C/2$ → Observer C ← Stationary source n_0

The apparent frequency observed by observer is given by relativistic relation as

$$n_{ap} = n_0\sqrt{\frac{C-V_0}{C+V_0}} = n_0\sqrt{\frac{C+C/2}{C-C/2}} \quad [\because v_0 = -C/2]$$

$$\Rightarrow \quad n_{ap} = n_0\left[\frac{3C/2}{C/2}\right]^{1/2} = 10\times 10^9\times\sqrt{3}$$

$$\Rightarrow \quad n_{ap} = 17.32 \text{ GHz}$$ **Ans. (A)**

Sol. 2 Apparent frequency of sound observed is

$$f_{ap} = f_0\left[\frac{c_2 \pm v_0}{c_2 \mp v_s}\right]$$

when $v_s = 34$ m/s

$$\Rightarrow \quad f_1 = f_0\left[\frac{340}{340-34}\right]$$

when $v_s = 17$ m/s

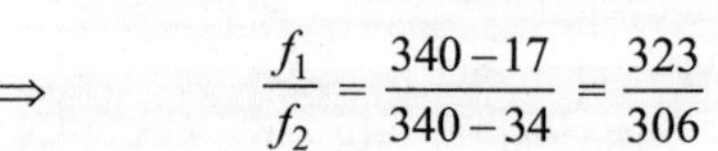

$$f_2 = f_0\left[\frac{340}{340-17}\right]$$

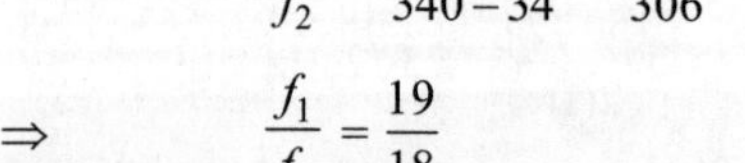

$$\Rightarrow \quad \frac{f_1}{f_2} = \frac{340-17}{340-34} = \frac{323}{306}$$

$$\Rightarrow \quad \frac{f_1}{f_2} = \frac{19}{18}$$ **Ans. (B)**

Sol. 3 Sound appears to be coming from location B and plane in at A, so we use

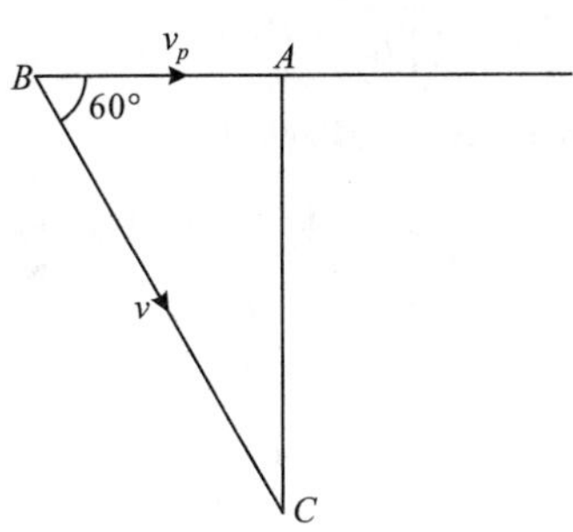

$$AB = v_p \times t \qquad (v_p \rightarrow \text{velocity of plane})$$

$$BC = vt$$

From figure

$$\cos 60° = \frac{AB}{BC}$$

$$\Rightarrow \qquad \frac{1}{2} = \frac{v_p \times t}{vt}$$

$$\Rightarrow \qquad v_p = \frac{v}{2}$$

Ans. (C)

Sol. 4 

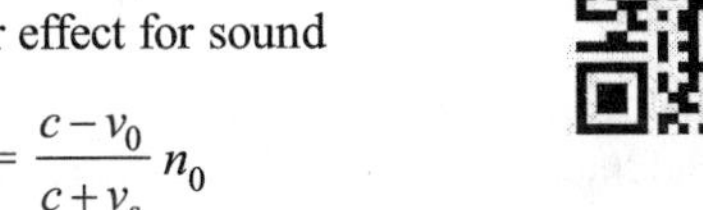

Applying Doppler effect for sound

$$x = \frac{c - v_0}{c + v_s} n_0$$

(v_0 & v_s is taken ⊕ when approaching each other)

$$\Rightarrow \qquad 2000 = \frac{340 - 20}{340 + 20} f_0$$

$$\Rightarrow \qquad f_0 = 2250 \text{ Hz}$$

Ans. (A)

Sol. 5 Apparent frequency for the recording observer is given as

$$f_1 = 480 = \frac{300 - v_1}{300} \times 500$$

$$\Rightarrow \qquad \frac{1440}{5} = 300 - v_1$$

$$\Rightarrow \qquad v_1 = \frac{60}{5} = 12 \text{ m/s}$$

Apparent frequency for the approaching observer is given as

$$530 = \frac{300 + v_2}{300} \times 500$$

$$\Rightarrow \qquad 1590 = 1500 + 5v_2$$

$$\Rightarrow \qquad 5v_2 = 90$$

$$\Rightarrow \qquad v_2 = 18 \text{ m/s}$$

Ans. (B)

Sol. 6 When source is approaching as shown below, we use

50 m/s
O ← S

$$f_{\text{app}} = \left(\frac{c - 0}{c - 50}\right) f_{\text{source}} \qquad (c \rightarrow \text{speed of second})$$

$$\Rightarrow \qquad 1000 = \left(\frac{350}{300}\right) f_{\text{source}}$$

$$\Rightarrow \qquad f_{\text{source}} = \frac{1000 \times 300}{350} \text{ Hz}$$

When source is recording as shown below, we use

50 m/s
← S O

$$f'_{\text{app}} = \left(\frac{c}{c + 50}\right) \cdot f_{\text{source}}$$

$$\Rightarrow \qquad f'_{\text{app}} = \frac{350}{400} \times 1000 \times \frac{300}{350}$$

$$\Rightarrow \qquad f'_{\text{app}} = 750 \text{ Hz}$$

Ans. (C)

Sol. 7

27 km/hr 7.5 m/s — B, n_0 = 500 Hz; C_w; 18 km/hr 5 m/s — A, n_{ap_1}

Frequency of signal detected/received by A

$$n_{ap_1} = n_0\left(\frac{C_w - v_{OA}}{C_w - v_{SB}}\right) = 500\left(\frac{1500 - 5}{1500 - 7.5}\right)$$

$$\Rightarrow \quad n_{ap_1} = \frac{500 \times 1495}{1492.5} \text{ Hz}$$

Now n_{ap_1} will be reflected by A as source and frequency detected by B for this reflected sound is

$$n_{ap_2} = n_{ap_1}\left(\frac{C_w + v_{OB}}{C_w + v_{SA}}\right) = \frac{500 \times 1495}{1492.5} \times \left(\frac{1500 + 7.5}{1500 + 5}\right)$$

$$\Rightarrow \quad n_{ap_2} \simeq 502 \text{ Hz}$$

Ans. (B)

Sol. 8 n_0 = 660 Hz $\qquad$ n_0 = 660 Hz

S_1 → n_{ap_1} → O (observer) → u ← n_{ap_2} ← S_2

Apparent frequency received from S_2 is

$$n_{ap_2} = n_0\left[\frac{C + 4}{C}\right]$$

Apparent frequency received from S_1 is

$$n_{ap_1} = n_0\left[\frac{C - 4}{C}\right]$$

Beat frequency heard by observer

$$f_B = n_{ap_2} - n_{ap_1} = n_0\left[\frac{C+4}{C}\right] - n_0\left[\frac{C-4}{C}\right] = n_0\left(\frac{24}{C}\right)$$

$\Rightarrow$ $10 = 660 \times \dfrac{24}{330}$ $\Rightarrow$ $4 = \dfrac{10}{4} = 2.5$ m/s **Ans. (A)**

Sol. 9  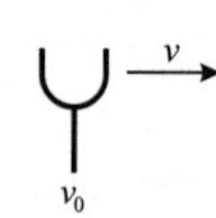

Frequency from fork A

$$\nu_A = \left(\frac{c}{c-v}\right)\nu_0$$

Frequency from fork B

$$\nu_B = \left(\frac{c}{c+v}\right)\nu_0$$

Beat frequency

$\Rightarrow$ $f_B = \nu_1 - \nu_2$

$\Rightarrow$ $f_B = c\nu_0\left(\dfrac{1}{c-v} - \dfrac{1}{c+v}\right)$

$\Rightarrow$ $f_B = c\nu_0\left(\dfrac{c+v-c+v}{c^2-v^2}\right) = \dfrac{2c\nu_0^2 v}{c^2-v^2}$

$\Rightarrow$ $f_B \approx \dfrac{2c\nu_0 v}{c^2}$

$\Rightarrow$ $f_B = \dfrac{2\nu_0 v}{c}$

$\Rightarrow$ $2 = \dfrac{2\times1400\times v}{350}$

$\Rightarrow$ $v = \dfrac{1}{4}$ m/s **Ans. (D)**

Sol. 10 Apparent frequency of bus as received by wall is

$$f_w = 420 \times \left(\frac{330}{330 - v_B}\right)$$

Apparent frequency as received by driver is

$$f_D = 420 \times \left(\frac{330+v_0}{330}\right)\left(\frac{330}{330-v_B}\right)$$

$\Rightarrow$ $490 = 420 \times \left(\dfrac{330+v_B}{330-v_B}\right)$

$\Rightarrow$ $\dfrac{7}{6} = \dfrac{330+v_B}{330-v_B}$

$\Rightarrow$ $v_B = \dfrac{330}{13}$ m/s

$\Rightarrow$ $v_B = \dfrac{330}{13}\times\dfrac{18}{5}$

≈ 91 km/hr **Ans. (A)**

Sol. 11 Frequency received by wall is

$$h_\omega = h_0\left[\frac{C}{C-V_C}\right]$$

reflected frequency received by driver is

$$h_d = h_\omega\left[\frac{C+V_C}{C}\right] = n_0\left(\frac{C+V_C}{C-V_C}\right)$$

$\Rightarrow$ $480 = 440\dfrac{(C+V_C)}{(C-V_C)}$

$\Rightarrow$ $12C - 12V_C = 11C + 11V_C$

$\Rightarrow$ $23V_C = C$

$\Rightarrow$ $V_C = \dfrac{C}{23} = \dfrac{345}{23} = 15$ m/s

$\Rightarrow$ $V_C = 54$ kph **Ans. (D)**

Sol. 12 Observer is located at finite distance from track as shown in figure.

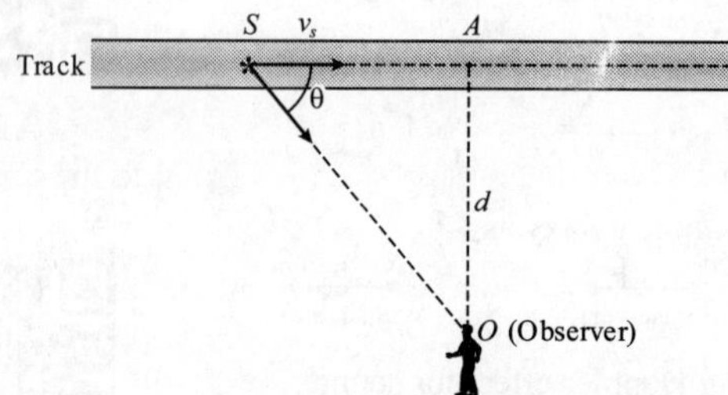

When source crosses point A the apparent frequency which was higher then ν_0 will drop below ν_0 as it started receding the observer. There correct option can be (B) or (D) but (B) is possible only when observer is located on track so in this case (D) is correct answer. This can also be mathematically verified with expression of apparent frequency, given as

$$f_{\text{app}} = \nu_0\left[\frac{v_{\text{sound}}}{v_{\text{sound}} - v_s\cos\theta}\right]$$ **Ans. (D)**

Sol. 13 Frequency of sound received by first car, which comes from second car is given as

$$f_1 = f_0\left(\frac{340+2}{340-2}\right) = 676\times\left(\frac{342}{338}\right) = 684 \text{ Hz}$$

Frequency of sound from its own horn remains the same at 676 Hz so beat frequency is given as

$$f_2 = f_0 = 676 \text{ Hz}$$

$\Rightarrow$ $f_1 - f_2 = 684 - 676 = 8$ beats/s **Ans. (8)**

Sol. 14 By Doppler's effect in light we use

$$\frac{\Delta\lambda}{\lambda}c = v$$

$\Rightarrow$ $\Delta\lambda = \dfrac{v}{c}\times\lambda$

$$\Rightarrow \quad \Delta\lambda = \frac{286}{3\times10^5}\times 630\times10^{-9} = 6\times10^{-10}$$

$$\Rightarrow \quad x = 6$$ **Ans. (6)**

Sol. 15 Frequency received by wall as observer is given as

$$f' = \left(\frac{v_s}{v_s - v}\right) f_0$$

Reflected frequency from wall as received by man is given as

$$f'' = \left(\frac{v_s + v}{v_s}\right) f'$$

$$\Rightarrow \quad f'' = \left(\frac{v_s + v}{v_s}\right)\left(\frac{v_s}{v_s - v}\right) f_0$$

$$\Rightarrow \quad f'' = \left(\frac{v_s + v}{v_s - v}\right) f_0$$

$$\Rightarrow \quad 500 = \left(\frac{330 + v}{330 - v}\right) 400$$

$$\Rightarrow \quad v = \frac{330}{9}\times\frac{18}{5} = 132\ \text{km/hr}$$ **Ans. (132)**

Sol. 16 The wavelength change $\Delta\lambda$ is related to the receding velocity of the galaxy as

$$\frac{v_{rel}}{c} = \frac{\Delta\lambda}{\lambda}$$

$$\Rightarrow \quad V_{\text{rel}} = \frac{6}{5890}\times 3\times10^8 \approx 306\ \text{km/s}$$ **Ans. (C)**

Sol. 17 The apparent frequency observed by detector in the situation described in question is expressed as

$$f' = f\left(\frac{v - v_0}{v + v_s}\right)$$

$$\Rightarrow \quad 1800 = f\left[\frac{340 - 20}{324 + 20}\right]$$

$$\Rightarrow \quad f = \frac{1800\times360}{320}$$

$$\Rightarrow \quad f = 2050\ \text{Hz}$$ **Ans. (2025)**

Sol. 18 The apparent frequency received by Y is given as

$$f_{app} = f_0\left(\frac{v + v_0}{v - v_s}\right)$$

$$1320 = f_0\left(\frac{340 + 20}{340 - 10}\right)$$

$$f_0 = 1320\times\frac{33}{36} = 1210\ \text{Hz}$$ **Ans. (1210)**

15.1 Electric Force and Electric Field

Sol. 1

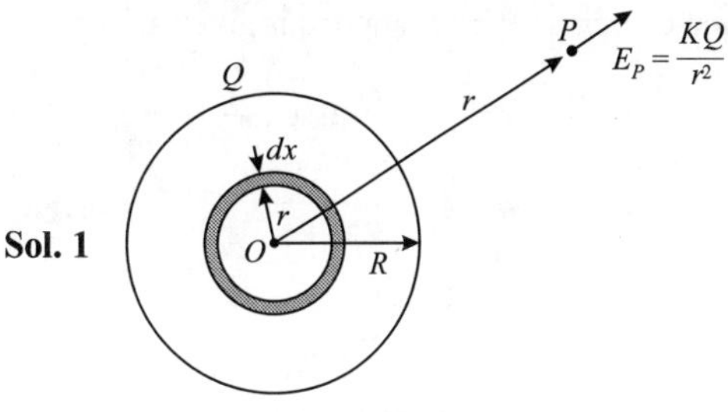

To calculate total charge on ball we consider on elemental shell inside the ball as shown of volume given as

$$dV = 4\pi r^2 dr$$

(For spherically symmetric charge distribution)

charge in the elemental shell thickness is

$$dq = \rho dV = \rho_0 = \left(1 - \frac{r}{R}\right)\cdot 4\pi r^2 dr$$

Total ball's charge is given as

$$Q = \int dq = \int_0^R 4\pi\rho_0\left(1 - \frac{r}{R}\right)\cdot r^2 dr$$

$$\Rightarrow \quad Q = 4\pi\rho_0 \int_0^R \left(r^2 - \frac{r^3}{R}\right) dr$$

$$\Rightarrow \quad Q = 4\pi\rho_0\left[\frac{r^3}{3} - \frac{r^4}{4R}\right]_0^R$$

$$\Rightarrow \quad Q = 4\pi\rho_0\left[\frac{R^3}{3} - \frac{R^3}{4}\right]$$

$$\Rightarrow \quad Q = 4\pi\rho_0\left(\frac{R^3}{12}\right) = \frac{1}{3}\pi\rho_0 R^3$$

Electric field at an outer point of ball is given as

$$E_P = \frac{KQ}{r^2} = \frac{1}{4\pi\epsilon_0}\cdot\frac{\left(\frac{1}{3}\pi\rho_0 R^3\right)}{r^2} = \frac{\rho_0 R^3}{12\epsilon_0 r^2}$$ **Ans. (D)**

Sol. 2

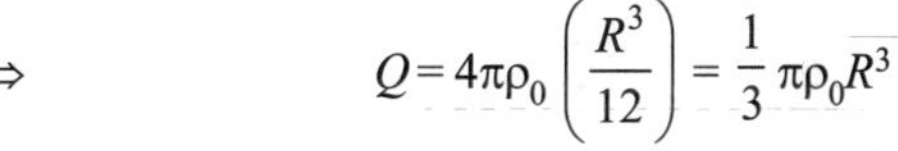

Oscillation energy of system $E_T = \frac{1}{2} m\omega^2 A^2$

PE of oscillations at equilibrium position = 0

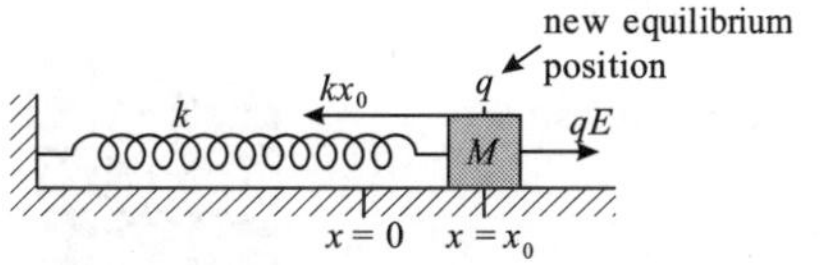

At equilibrium position $kx_0 = qE$

$$\Rightarrow \quad x_0 = \frac{qE}{k}$$

PE of oscillation at equilibrium position $= \frac{1}{2}kx_0^2 = \frac{1}{2}\frac{q^2E^2}{k}$

Total oscillation energy after EF is applied is given as

$$E_T = \frac{1}{2}m\omega^2A^2 + U_{\text{equilibrium}}$$

$$E_T = \frac{1}{2}m\omega^2A^2 + \frac{1}{2}\frac{q^2E^2}{k}$$ **Ans. (A)**

Sol. 3

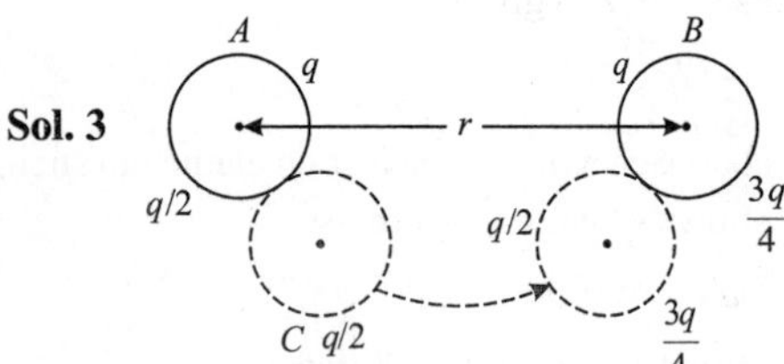

When C is touched to A, charge q is divided equally among A and C

$\Rightarrow$ final charge of A is $q_A = \frac{q}{2}$

Now C is touched to B, charge $\left(q + \frac{q}{2} = \frac{3q}{2}\right)$ is divided equally among B and C

$\Rightarrow$ Final charge on B is $q_B = \frac{3q}{4}$

So final force between A and B is

$$F' = \frac{K\left(\frac{a}{2}\right)\left(\frac{3q}{4}\right)}{r^2} = \frac{3}{8}\frac{Kq^2}{r^2} = \frac{3}{8}F$$ **Ans. (D)**

Sol. 4

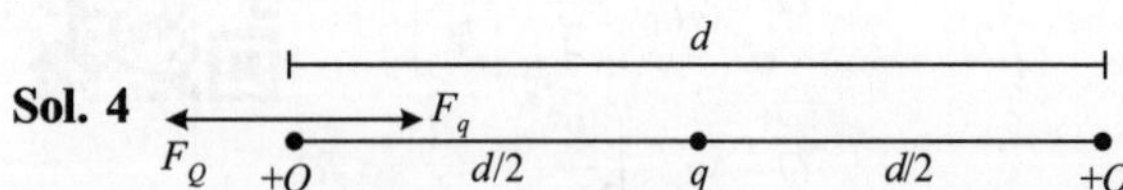

For equilibrium of charge Q, we use

$$\overrightarrow{F_Q} + \overrightarrow{F_q} = 0$$

$$\overrightarrow{F_Q} = -\overrightarrow{F_q}$$

$$\frac{kQQ}{d^2} = -\frac{kQq}{(d/2)^2}$$

$$\Rightarrow \quad q = -\frac{Q}{4}$$ **Ans. (C)**

Sol. 5 Electric field on axis of ring at a distance h is given as

$$E = \frac{kQx}{(h^2 + R^2)^{3/2}}$$

For maximum electric field, we use

$$\frac{dE}{dx} = 0$$

Solving we get

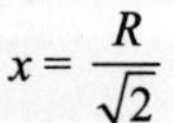

$$x = \frac{R}{\sqrt{2}}$$ **Ans. (C)**

Sol. 6

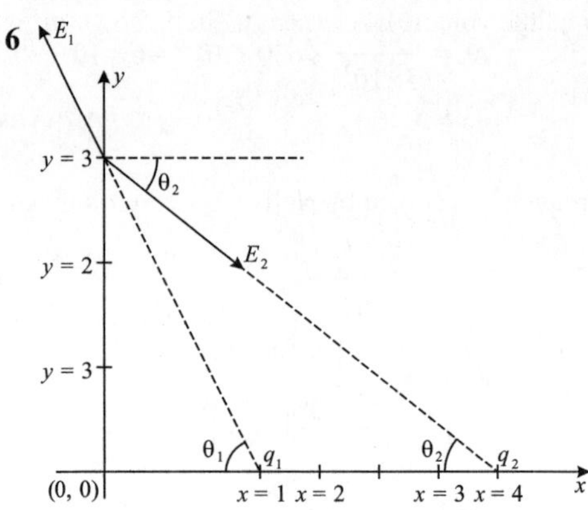

Let $\vec{E}_1$ & $\vec{E}_2$ are the values of electric field due to q_1 & q_2 respectively are magnitude of

$$E_2 = \frac{1}{4\pi\epsilon_0}\frac{q_2}{r^2}$$

$$E_2 = \frac{9\times10^9\times(25)\times10^{-6}}{(4^2+3^2)}\text{ V/m}$$

$$E_2 = 9\times10^3\text{ V/m}$$

$$\Rightarrow \quad \vec{E}_2 = 9\times10^3(\cos\theta_2\hat{i} - \sin\theta_2\hat{j})\text{ V/m}$$

Where $\tan\theta_2 = \frac{3}{4}$

$$\Rightarrow \quad \vec{E}_2 = 9\times10^3\left(\frac{4}{5}\hat{i} - \frac{3}{5}\hat{j}\right) = (72\hat{i} - 54\hat{j})\times10^2\text{ V/m}$$

Magnitude of

$$E_1 = \frac{1}{4\pi\epsilon_0}\frac{\sqrt{10}\times10^{-6}}{(1^2+3^2)}$$

$$\Rightarrow \quad E_1 = (9\times10^9)\times\sqrt{10}\times10^{-7}\text{ V/m}$$

$$\Rightarrow \quad E_1 = 9\sqrt{10}\times10^2\text{ V/m}$$

$$\Rightarrow \quad \vec{E}_1 = 9\sqrt{10}\times10^2\left[\cos\theta_1(-\hat{i}) + \sin\theta_1\hat{j}\right]\text{V/m}$$

Where $\tan\theta_1 = 3$

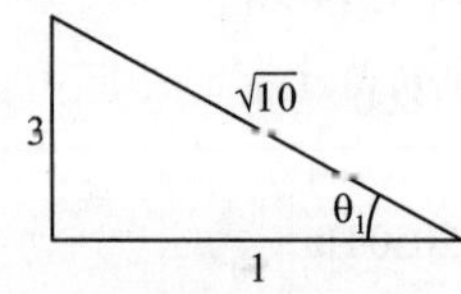

$$\Rightarrow \quad E_1 = 9\times\sqrt{10}\times10^2\left[\frac{1}{\sqrt{10}}(-\hat{i}) + \frac{3}{\sqrt{10}}\hat{j}\right]\text{V/m}$$

$$\Rightarrow \quad E_1 = 9\times10^2\left[-\hat{i} + 3\hat{j}\right] = \left[-9\hat{i} + 27\hat{j}\right]10^2\text{ V/m}$$

$$\Rightarrow \quad \vec{E}_{\text{net}} = \vec{E}_1 + \vec{E}_2 = (63\hat{i} - 27\hat{j})\times10^2\text{ V/m}$$ **Ans. (C)**

Sol. 7 Total charge in spherical volume is given as

$$Q = \int\rho dV$$

Where dV is the volume of elemental shall shown in figure

$$\Rightarrow \quad Q = \int_0^R \frac{A}{r^2} e^{-2r/a}(4\pi r^2 dr)$$

$$\Rightarrow \quad Q = 4\pi A \int_0^R e^{-2r/a} dr$$

$$\Rightarrow \quad Q = 4\pi A \left(\frac{e^{-2r/a}}{-\frac{2}{a}} \right)_0^R$$

$$\Rightarrow \quad Q = 4\pi A \left(-\frac{a}{2} \right)(e^{-2R/a} - 1)$$

$$\Rightarrow \quad Q = 2\pi a A(1 - e^{-2R/a})$$

$$\Rightarrow \quad R = \frac{a}{2} \log \left(\frac{1}{1 - \frac{Q}{2\pi a A}} \right)$$

$dV = 4\pi r^2 dr$

Ans. (D)

Sol. 8

For equilibrium of bob, we use

$$\tan\theta = \frac{qE}{mg} = \frac{5\times10^{-6}\times2000}{2\times10^{-3}\times10}$$

$$\Rightarrow \quad \tan\theta = \frac{1}{2}$$

$$\Rightarrow \quad \theta = \tan^{-1}(0.5)$$

Ans. (C)

Sol. 9

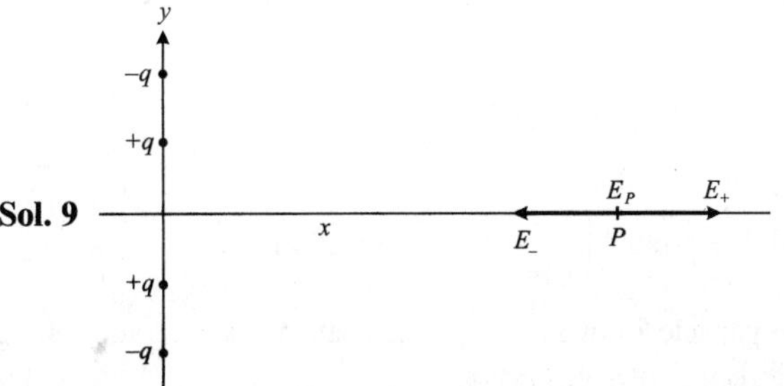

Electric field due to $+q$ charges at point P is

$$E_+ = \frac{K(2q)x}{(x^2+d^2)^{3/2}} = \frac{2kqD}{(D^2+d^2)^{3/2}}$$

Electric field due to $-q$ charges at point P is

$$E_- = \frac{K(2q)x}{(x^2+4d^2)^{3/2}} = \frac{2kqD}{(D^2+4d^2)^{3/2}}$$

Net Electric field at point P is given as

$$E_P = E_+ - E_- = 2kqD\left[\frac{1}{(D^2+d^2)^{3/2}} - \frac{1}{(D^2+4d^2)^{3/2}}\right]$$

For $D >> d$ we use $\Rightarrow E_P = E_P = \frac{2kqD}{D^3}$

$$\left[\left(1+\frac{d^2}{D^2}\right)^{-3/2} - \left(1+\frac{4d^2}{D^2}\right)^{-3/2}\right]$$

$$\Rightarrow \quad E_P = \frac{2kq}{D^2}\left[\left(1-\frac{3d^2}{2D^2}\right) - \left(1 - \frac{3}{2}\cdot\frac{4d^2}{D^2}\right)\right]$$

$$\Rightarrow \quad E_P = \frac{2kq}{D^2}\left[1 - \frac{3d^2}{2D^2} - 1 + \frac{6d^2}{D^2}\right] = \frac{9kqd^2}{D^4}$$

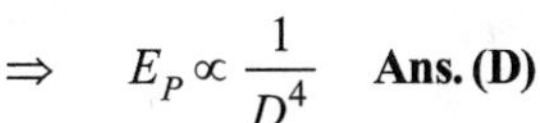

$$\Rightarrow \quad E_P \propto \frac{1}{D^4}$$ **Ans. (D)**

Sol. 10

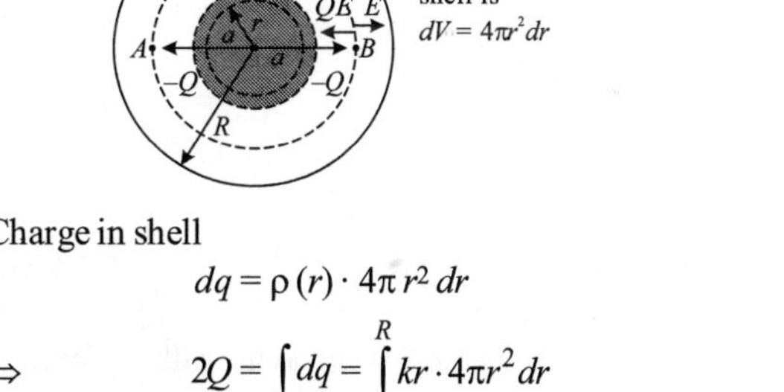

Charge in shell

$$dq = \rho(r)\cdot 4\pi r^2 dr$$

$$\Rightarrow \quad 2Q = \int dq = \int_0^R kr\cdot 4\pi r^2 dr$$

$$\Rightarrow \quad 2Q = 4\pi \frac{R^4}{4} = \pi k R^4$$

$$\Rightarrow \quad k = \frac{2Q}{\pi R^4} \quad \ldots(1)$$

Electric field at a distance a from centre is

$$E_a = \left(\frac{1}{4\pi\epsilon_0}\right)\frac{q_a}{a^2}$$

Where $$q_a = \int_0^a kr\cdot 4\pi r^2 dr$$

$$\Rightarrow \quad q_a = 4\pi \frac{kq^4}{4}$$

$$\Rightarrow \quad E_a = \frac{1}{4\pi\epsilon_0}\cdot\frac{\pi k a^4}{a^2} = \frac{1}{4\pi\epsilon_0}\cdot \pi k a^2$$

Repulsive force between A and B is given as

$$F = \frac{1}{4\pi\epsilon_0}\cdot\frac{Q^2}{(2a)^2}$$

As net force on A and B is zero, we use

$$\Rightarrow \quad F = QE$$

$$\frac{1}{4\pi\epsilon_0}\cdot\frac{Q^2}{4a^2} = Q\cdot\frac{1}{4\pi\epsilon_0}\cdot \pi k a^2$$

$$\Rightarrow \quad \frac{Q}{4a^2} = \pi\left(\frac{2Q}{\pi R^4}\right)\cdot a^2 \Rightarrow a^4 = \frac{1}{8}R^4$$

$$\Rightarrow \quad a = 8^{-1/4}\cdot R$$ **Ans. (C)**

Sol. 11 Summing up all components of electric field along X-direction, we get

$$E_x = \frac{K(4q)}{R^2}\cos 30° + \frac{K(2q)}{R^2}\cos 30° + \frac{K(2q)}{R^2}\cos 30°$$

$$\Rightarrow \quad E_x = (2\sqrt{3}+\sqrt{3}+\sqrt{3})\,\text{kg}/R^2 = \frac{\sqrt{3}q}{\pi\varepsilon_0 d^2}$$ **Ans. (D)**

Sol. 12 Speed after travelling a distance x is given as

$$v^2 = u^2 + 2ax$$

$$\Rightarrow \quad v^2 = 0 + 2\left(\frac{qE}{m}\right)x$$

$$\Rightarrow \quad v^2 = \frac{2qE}{m}x$$

$$\Rightarrow \quad v \propto \sqrt{x}$$

Thus option (C) is correct. **Ans. (C)**

Sol. 13 If we fill the empty space of cavity with $+\rho$ and $-\rho$ charge density in this region, the electric fields at points A and B are

$$E_A = 0 + \frac{k\rho.\frac{4}{3}\pi\left(\frac{R}{2}\right)^3}{\left(\frac{R}{2}\right)^2} = k\rho\,\frac{4}{3}\pi\left(\frac{R}{2}\right) \qquad \ldots(1)$$

$$\text{and } E_B = \frac{k\rho\cdot\frac{4}{3}\pi R^3}{R^2} - \frac{k\rho\cdot\frac{4}{3}\pi\left(\frac{R}{2}\right)^3}{\left(\frac{3R}{2}\right)^2}$$

$$\Rightarrow \quad E_B = k\rho\frac{4}{3}\pi R - k\rho\frac{4}{3}\pi\frac{R}{18} = k\rho\cdot\frac{4}{3}\pi\left(\frac{17R}{18}\right) \qquad \ldots(2)$$

Using equation-(1) and (2), we have

$$\frac{E_A}{E_B} = \frac{9}{17} = \frac{18}{34}$$ **Ans. (D)**

Sol. 14 Let the charges on inner and outer spheres are q_1 and q_2.

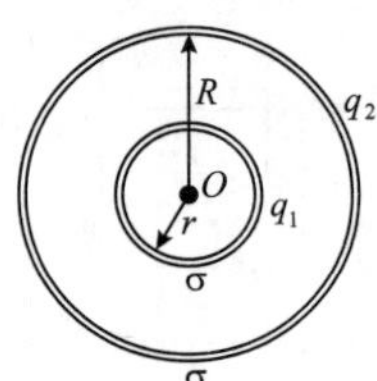

As charge density is same on both shells, we use

$$\sigma = \frac{q_1}{4\pi r^2} = \frac{q_2}{4\pi R^2}$$

$$\Rightarrow \quad \frac{q_1}{q_2} = \frac{r^2}{R^2}$$

$$q_1 + q_2 = Q$$

$$\Rightarrow \quad \frac{q_2 r^2}{R^2} + q_2 = Q$$

$$\Rightarrow \quad q_2 = \frac{qR^2}{(r^2+R^2)}$$

$$\Rightarrow \quad q_1 = \frac{r^2}{R^2}\cdot\frac{qR^2}{(R^2+r^2)} = \frac{qr^2}{(R^2+r^2)}$$

Potential at common centre is given as

$$V_C = \frac{kQ_1}{r} + \frac{kQ_2}{R}$$

$$\Rightarrow \quad V_C = K\left[\frac{Qr^2}{r(R^2+r^2)} + \frac{QR^2}{R(R^2+r^2)}\right]$$

$$\Rightarrow \quad V_C = \frac{KQ(r+R)}{(R^2+r^2)}$$

$$\Rightarrow \quad V_C = \frac{1}{4\pi\epsilon_0}\,\frac{(R+r)}{(R^2+r^2)}Q$$ **Ans. (C)**

Sol. 15 As initial speed of particle is zero, particle starts moving in the direction of acceleration which is given as

$$a = \sqrt{g^2 + \left(\frac{qE}{m}\right)^2}$$

at angle $\theta = \tan^{-1}\left(\frac{qE}{mg}\right)$ from Y direction.

Thus particle follows a straight line path and acceleration along the direction of acceleration. **Ans. (D)**

Sol. 16 Electric field at O due to Q_1 and Q_2 are

$$\vec{E}_1 = -\frac{kQ_1}{x_1^2}\hat{i}$$

and $$\vec{E}_2 = -\frac{kQ_2}{x_2^2}\hat{i}$$

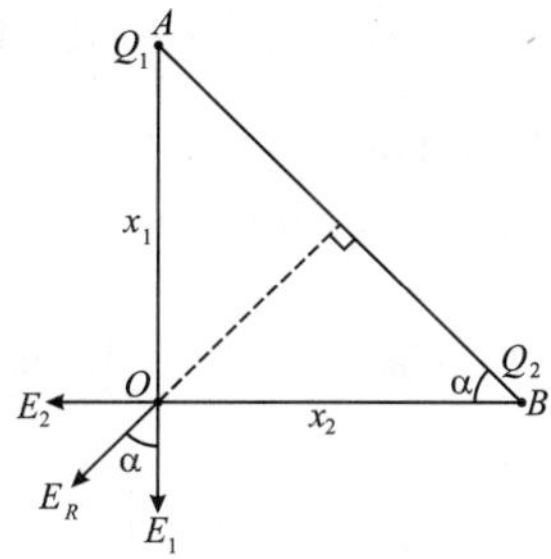

For $\overrightarrow{E_R}$ to be normal to line AB, we use

$$\tan\alpha = \frac{E_2}{E_1} = \frac{x_1}{x_2}$$

$$\Rightarrow \quad \frac{Q_2}{Q_1} = \frac{x_2}{x_1}$$ **Ans. (C)**

Sol. 17 For a uniformly charged body electric field is always in radial direction for all outside points so force on charge q (outside) is given as

$$F = \frac{1}{4\pi\epsilon_0}\cdot\frac{qQ}{r^2}$$

and at all interior points as

$$E = 0,\ F = 0$$ **Ans. (A)**

Sol. 18 If we replace $-Q$ charge at origin by a combination of one $+Q$ and one $-2Q$ charge then the situation remains the same. Now due to $+Q$ charge at every corner of cube the electric field at center of cube is zero and the net electric field at center is because of only $-2Q$ charge kept at origin, which is given as

$$\vec{E} = \frac{kq\vec{r}}{r^3} = \frac{1(-2Q)\frac{a}{2}(\hat{x}+\hat{y}+\hat{z})}{4\pi\varepsilon_0\left(\frac{a}{2}\sqrt{3}\right)^3}$$

$$\Rightarrow \quad \vec{E} = \frac{-2Q(\hat{x}+\hat{y}+\hat{z})}{3\sqrt{3}\pi a^2\varepsilon_0}$$ **Ans. (C)**

Sol. 19 The given situation described in question is shown in figure below

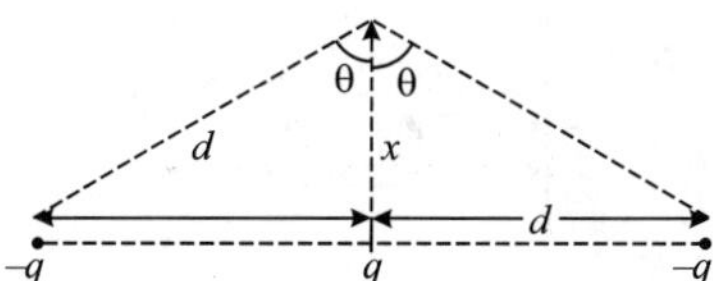

For the small displacement of charge at center, restoring force on it is given as

$$F_R = -[2F\cos\theta]$$

$$\Rightarrow \quad F_R = -2\cdot\frac{1}{4\pi\varepsilon_0}\cdot\frac{q^2}{(\sqrt{d^2+x^2})^2}\cdot\frac{x}{\sqrt{d^2+x^2}}$$

$$\Rightarrow \quad F_R = -\frac{q^2}{2\pi\varepsilon_0}\frac{x}{(d^2+x^2)^{3/2}}$$

For $x << d$, we have

$$F_R = -\frac{q^2}{2\pi\varepsilon_0 d^3}x$$

The acceleration of proton is given as

$$\Rightarrow \quad a = -\frac{q^2}{2\pi\varepsilon_0 .md^3}x$$

Comparing with equation of SHM ($a = -\omega^2 x$)

$$\Rightarrow \quad \omega = \sqrt{\frac{q^2}{2\pi\varepsilon_0 md^3}}$$ **Ans. (B)**

Sol. 20 The final charges after conduction on each sphere will be given as

$$q = \frac{(2.1-0.1)}{2}nC = 1nC$$

The force between the spheres can now be given as

$$F = \frac{9\times10^9\times10^{-18}}{(0.5)^2} = 36\times10^{-9}$$ **Ans. (36)**

Sol. 21 The situation described in question is shown in figure below

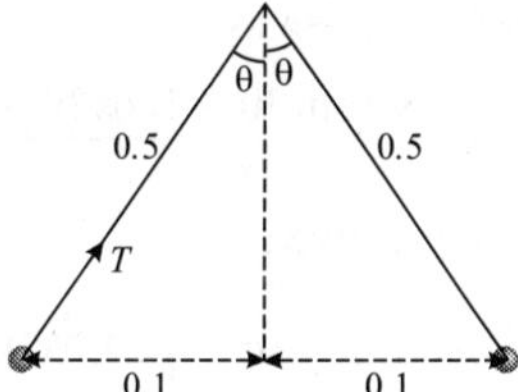

For equilibrium of sphere, we use

$$T\cos\theta = mg = 10\times10^{-6}\times10 = 10^{-4}$$

$$T\sin\theta = \frac{9\times10^9\times q^2}{0.04} = F$$

$$\Rightarrow \quad \tan\theta = \frac{0.1}{\sqrt{0.24}} = \frac{F}{mg}$$

$$\Rightarrow \quad q = \frac{2\sqrt{10}}{3\sqrt{\sqrt{24}}}\times10^{-8}$$

$$\Rightarrow \quad 0.95\times10^{-8}=\frac{a}{21}\times10^{-8}$$

$$\Rightarrow \quad a=20$$ **Ans. (20)**

Sol. 22 The electric field due to the rod at point P is given as

$$E=\frac{k\lambda}{a}(\sin\theta_1+\sin\theta_2)$$

$$\Rightarrow \quad E=\frac{1}{4\pi\varepsilon_0}\times\frac{Q}{L}\times\frac{1}{\left(\frac{\sqrt{3}L}{2}\right)}\times(2\sin\theta)$$

Where angle θ is given as

$$\tan\theta=\frac{L/2}{\frac{\sqrt{3}L}{2}}=\frac{1}{\sqrt{3}}$$

$$\Rightarrow \quad \sin\theta=\frac{1}{2}$$

$$\Rightarrow \quad E=\frac{1}{4\pi\varepsilon_0}\times\frac{2Q}{\sqrt{3}L^2}\times\left(2\times\frac{1}{2}\right)=\frac{Q}{2\sqrt{3}\pi\varepsilon_0L^2}$$

Ans. (C)

Sol. 23 Free body diagram of the body is shown in figure below

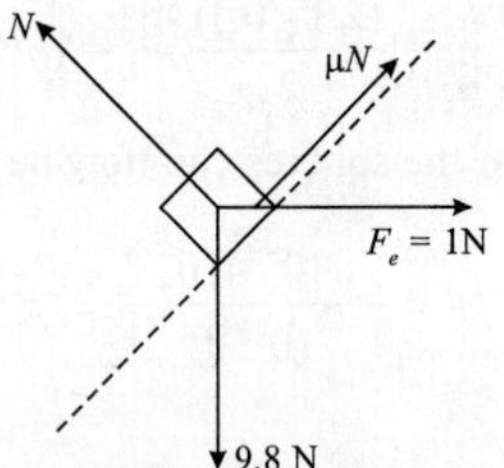

The normal reaction on body is given as

$$N=9.8\cos30^\circ+1\sin30^\circ\approx9\text{ N}$$

Acceleration of body is given as

$$a=\frac{9.8\sin30^\circ-1\cos30^\circ-\mu N}{1}=2.233\text{ m/s}^2$$

Displacement of body is given as

$$s=ut+\frac{1}{2}at^2=\frac{1}{2}(2.233)t^2\sin30^\circ$$

$$\Rightarrow \quad t\approx1.3\text{ sec}$$ **Ans. (D)**

Sol. 24 Inside the plane as electric field should be zero so for σ induced surface charge on plane, we have

$$\frac{2K\lambda}{r}=\frac{\sigma}{\varepsilon_0}$$

$$\Rightarrow \quad \sigma=\frac{2\times9\times10^9\times8\times10^{-9}\times8.85\times10^{-12}}{3}$$

$$\Rightarrow \quad \sigma=0.424\times10^{-9}\,\frac{C}{m^2}$$ **Ans. (A)**

Sol. 25 For equilibrium of oil drop we have

$$qE=Mg$$

$$\Rightarrow \quad neE=\rho\left(\frac{4}{3}\pi r^3\right)\times g$$

$$\Rightarrow \quad n\times1.6\times10^{-19}\times3.55\times10^5=3\times10^3\times\frac{4}{3}\times\pi\times(2\times10^{-3})^3\times9.81$$

$$\Rightarrow \quad n=173\times10^{(3-9-5+19)}$$

$$\Rightarrow \quad n=1.73\times10^{10}$$ **Ans. (B)**

Sol. 26 Force on 1 C charge can be calculated by Coulomb's law as

$$F=k(1\text{C})(1\mu\text{C})\left[1+\frac{1}{2^2}+\frac{1}{4^2}+\frac{1}{8^2}+\dots\right]$$

$$\Rightarrow \quad F=9\times10^3\left[\frac{1}{1-\frac{1}{4}}\right]=12\times10^3\text{N}$$ **Ans. (12)**

Sol. 27 The force between the charges is given as

$$F=\frac{Kq_1q_2}{r^2}=\frac{K(q)(Q-q)}{r^2}$$

The force will be maximum when

$$\frac{dF}{dq}=0$$

$$\Rightarrow \quad Q-2q=0$$

$$\Rightarrow \quad Q=2q$$ **Ans. (D)**

Sol. 28 Acceleration of the body due to electric field is given as

$$a=\frac{qE}{m}=\frac{8\times10^{-6}}{10^{-3}}\times100=0.8\text{ m/s}^2$$

As electric field is switched on, ball first strikes to wall and returns back thus time of one oscillation can be calculated as

$$s=ut_1+\frac{1}{2}at_1^2$$

$$0.1=\frac{1}{2}\times0.8t_1^2;\ t_1=\frac{1}{2}\text{s}$$

Thus time period $T=2\times\frac{1}{2}=1\text{ s}$ **Ans. (1)**

Sol. 29 Situation described in question is shown in figure below

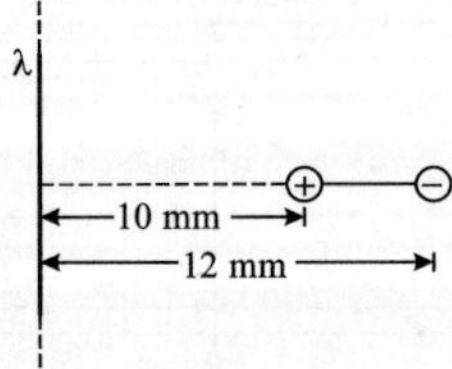

Let each charge in dipole is of magnitude Q, net force on dipole can be given as

$$F_{net} = \frac{2k\lambda Q}{r_1} + \frac{2k\lambda(-Q)}{r_2}$$

$$\Rightarrow \quad 4N = 2k\lambda Q\left[\frac{1}{r_2} - \frac{1}{r_2}\right]$$

$$\Rightarrow \quad 4N = 2 \times 9 \times 10^9 \times 3 \times 10^{-6}\, Q\left[\frac{1}{10} - \frac{1}{12}\right] \times 10^3$$

$$\Rightarrow \quad 4N = 54 \times 10^6\, Q \times \frac{1}{60}$$

$$\Rightarrow \quad Q = 4.44 \times 10^{-6}\,\text{C} = 4.44\,\mu\text{C}$$ **Ans. (B)**

Sol. 30 The situation described in question is shown in figure

q ●———●q x ●———● q
2m

The restoring force on particle executing SHM is given as

$$F_{net} = \frac{Kq^2}{(1-x)^2} - \frac{Kq^2}{(1+x)^2} = Kq^2\left[\frac{4x}{(1^2 - x^2)^2}\right]$$

$$\Rightarrow \quad a = \frac{4Kq^2x}{m}$$

$$\Rightarrow \quad \omega = 2\sqrt{\frac{Kq^2}{m}} = 2 \times \sqrt{\frac{9 \times 10^9 \times 10}{10^{-6}}} = 6000$$

Ans. (6000)

Sol. 31 Situation described in question is shown in figure below

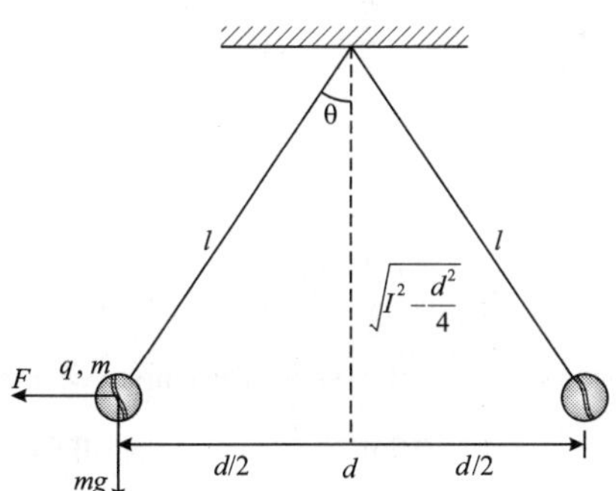

Force acting on charges is given as

$$F = \frac{kq^2}{d^2}$$

At equilibrium, we use

$$\tan\theta = \frac{\frac{kq^2}{d^2}}{mg} \approx \sin\theta$$

$$\Rightarrow \quad \frac{kq^2}{mgd^2} = \frac{d}{2l}$$

$$\Rightarrow \quad d^3 = \frac{2kq^2l}{mg}$$

$$\Rightarrow \quad d = \left(\frac{2kq^2l}{mg}\right)^{1/3}$$

$$\Rightarrow \quad d = \left(\frac{2q^2l}{4\pi\varepsilon_0 mg}\right)^{1/3} = \left(\frac{q^2l}{2\pi\varepsilon_0 mg}\right)^{1/3}$$ **Ans. (B)**

Sol. 32 Inside the metal shell electric field is zero and as charge is at the central sphere, electric lines of forces are radial and electric field will be inversely proportional to the square of distance from center of sphere so proper variation of electric field as a function of r is shown in figure below.

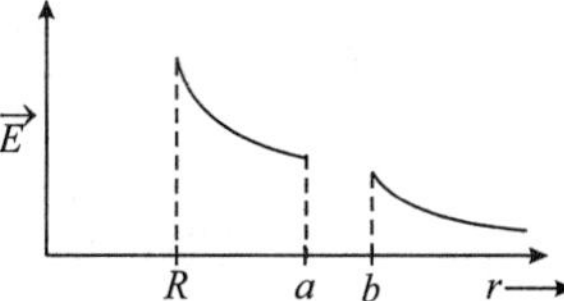

Out of given options, none of the option is correct but option (C) is most appropriate here. **Ans. (C)**

Sol. 33 This expression of electric field due to a uniformly surface charged disc is derived and explained in Physics Galaxy Vol 3A in detail. Final expression is written as

$$E = \frac{\sigma}{2\varepsilon_0}\left[1 - \frac{Z}{\sqrt{R^2 + Z^2}}\right]$$ **Ans. (D)**

Sol. 34 Electric field due to a uniformly charged circular arc at its centre of curvature is given as

$$E = \frac{2k\lambda}{R}\sin\frac{\theta}{2}\hat{i}$$

$$\Rightarrow \quad E = \frac{2k}{R}\left(\frac{Q}{2R\frac{\pi}{3}}\right)\sin 60°\hat{i}$$

$$\Rightarrow \quad E = \frac{3kQ}{\pi R^2}\frac{\sqrt{3}}{2}\hat{i} = \frac{3\sqrt{3}}{2\pi}\frac{kQ}{R^2}\hat{i}$$ **Ans. (A)**

Sol. 35 Third charge should be placed near -5μC charge at x distance as shown in figure below for which it does not experience any electric force.

+20μC ●———● −5μC ● Q
5 cm, x

The electric force on this charge due to the two charges are balanced, so we use

$$\frac{K(20)Q}{(5+x)^2} = \frac{K(5)Q}{X^2}$$

$$\Rightarrow \quad 4x^2 = (5+x)^2$$

$$\Rightarrow \quad 4x^2 = 25 + x^2 + 10x$$

$$\Rightarrow \quad 3x^2 - 10x - 25 = 0$$

$$\Rightarrow \quad x = 5$$ **Ans. (D)**

15.2 Electric Flux and Gauss's Law

Sol. 1 Using Gauss's law we have electric flux coming out from any closed surface is given as

$$\phi_s = \frac{q_{\text{enclosed by S}}}{\epsilon_0}$$

flux through surfaces S_1, S_2, S_3 and S_4 are

$$\phi_1 = \frac{2q}{\epsilon_0};$$

$$\phi_2 = \frac{3q-q}{\epsilon_0} = \frac{2q}{\epsilon_0};$$

$$\phi_3 = \frac{2q}{\epsilon_0};$$

$$\phi_4 = \frac{8q-2q-4q}{\epsilon_0} = \frac{2q}{\epsilon_0}$$

$\Rightarrow$ $$\phi_1 = \phi_2 = \phi_3 = \phi_4$$ **Ans. (C)**

Sol. 2

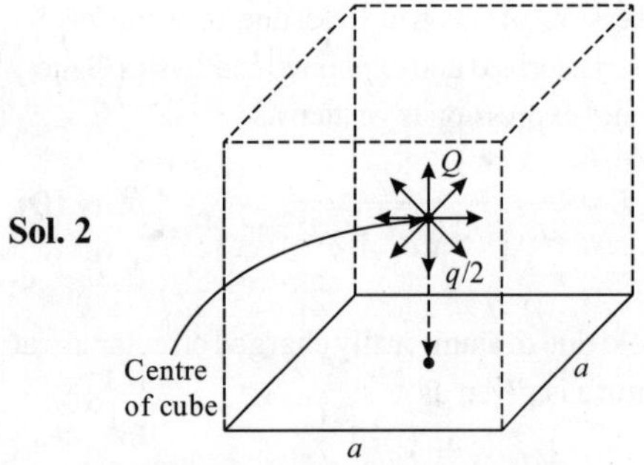

As all six surfaces (squares of side a) of imaginary cube considered are symmetrically located w.r. to charge Q at centre. Total electric flux coming out from cubical surface is

$$\phi_0 = \frac{Q}{\epsilon_0}$$

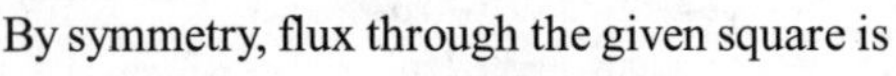

By symmetry, flux through the given square is

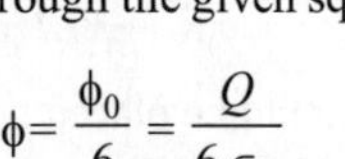

$$\phi = \frac{\phi_0}{6} = \frac{Q}{6\epsilon_0}$$ **Ans. (B)**

Sol. 3 By Gauss's law we use

$$\oint \vec{E}\cdot\vec{dA} = \frac{q_{enc}}{\varepsilon_0}$$

If $\vec{E}$ is normal to surface and uniform over the surface then we can use

$$\oint \vec{E}\cdot\vec{dA} = EA$$

$\Rightarrow$ $$E = \frac{q_{enc}}{\varepsilon_0 A}$$

Thus option-(D) is correct. **Ans. (D)**

Sol. 4 Electric flux through face $ABCD$ is

$$\phi_I = 0 \quad \text{as} \quad E_z = 0$$

Electric flux through face $BCGF$ is

$$\phi_{II} = E_x \cdot S = (4\times 3)\times(2\times 2) = 48\ \text{Nm}^2/\text{C}$$

$\Rightarrow$ $$\phi_I - \phi_{II} = 0 - 48 = -48\ \text{Nm}^2/\text{C}$$ **Ans. (– 48)**

Sol. 5

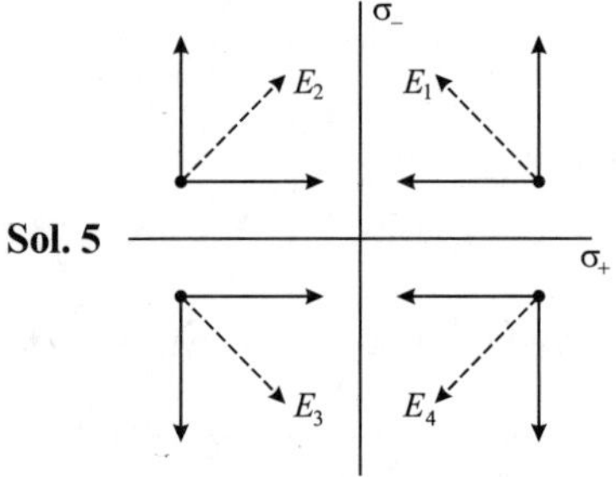

In figure above E_1, E_2, E_3 and E_4 represent the resultants electric field in region due to the two sheets. As $|\sigma_+| > |\sigma_-|$ option (A) is more appropriate. **Ans. (A)**

Sol. 6 By symmetry, the magnitude of electric flux through the square can be given as one sixth of the total flux originating from the charge as the square can be considered as one of the six symmetric faces of a cube at the center of which the charge is considered to be kept.

$$\phi = \frac{1}{6}\left(\frac{q}{\varepsilon_0}\right) = \frac{12\times10^{-6}}{6\times8.85\times10^{-12}}$$

$$= 225.98\times10^3\ \frac{Nm^2}{s} = 226\times10^3\ \frac{Nm^2}{C}$$ **Ans. (226)**

Sol. 7 The electric field vector is given as

$$\vec{E} = \left(\frac{3E_0}{5}\hat{i} + \frac{4E_0}{5}\hat{j}\right)\frac{N}{C}$$

The area vectors of the given surfaces can be written as

$$\vec{A_1} = 0.2\text{m}^2\,\hat{i} \qquad \text{(parallel to } y\text{-}z \text{ plane)}$$

and $$\vec{A_2} = 0.3\text{m}^2\,\hat{j} \qquad \text{(parallel to } x\text{-}z \text{ plane)}$$

The electric flux through these areas can be given as

$$\phi_a = \left[\frac{3E_0}{5}\hat{i} + \frac{4E_0}{5}\hat{j}\right]\cdot[0.2\hat{i}] = \frac{3\times0.2}{5}E_0$$

and $$\phi_b = \left[\frac{3E_0}{5}\hat{i} + \frac{4E_0}{5}\hat{j}\right]\cdot[0.3\hat{j}] = \frac{4\times0.3}{5}E_0$$

The ratio as asked is given as

$$\frac{\phi_a}{\phi_b} = \frac{0.6}{1.2} = \frac{1}{2} = \frac{a}{b}$$

$\Rightarrow$ $$a : b = 1 : 2$$

$\Rightarrow$ $$a = 1$$ **Ans. (1)**

Sol. 8 Total flux through the cube can be given by symmetry of charge placed at one vertex, which is given as

$$\phi = \frac{q}{8\epsilon_0}$$

The flux through the cube side surfaces *ABEH*, *ADGH*, *ABCD* will be zero and for the remaining three surfaces flux will be equal which can be given as

$$\phi(EFGH) = \phi(DCFG) = \phi(EBCF) = \frac{1}{3}\left(\frac{q}{8\epsilon_0}\right) = \frac{q}{24\epsilon_0}$$

Ans. (D)

Sol. 9 Figure below shows the situation described in statement-I, in which it can be observed that total charge enclosed in the hollow sphere is zero but due to electric dipole the electric field is not zero anywhere in surrounding of dipole thus statement-I is true.

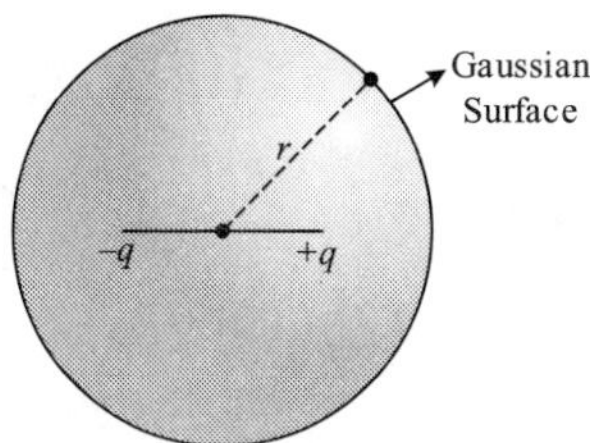

As per situation given in Statement-II, in case of a metal sphere always electric field inside is zero so electric flux through any surface considered inside the sphere is always zero thus statement-II is false. **Ans. (A)**

Sol. 10 Flux through the surface is calculated by using only *x* component of electric field as area is parallel to YZ plane. Thus required flux is given as

$$\phi = E_x A$$

$$\Rightarrow \frac{2}{5} \times 4 \times 10^3 \times 0.4 = 640$$ **Ans. (640)**

Sol. 11 Using vector form of Gauss's law, we have

$$\vec{\nabla} \cdot \vec{E} = \frac{\rho}{\varepsilon_0}$$

$$\Rightarrow \frac{\delta E}{\delta x} + \frac{\delta E}{\delta y} + \frac{\delta E}{\delta z} = \frac{\rho}{\varepsilon_0}$$

$$\Rightarrow \rho = 2\varepsilon_0$$

$$\Rightarrow Q = 2 \times 10^{-9} \times \rho = 4\varepsilon_0 \text{ nC}$$ **Ans. (4)**

Sol. 12 For a given surface ingoing flux is taken negative because field and surface area vectors are in opposite direction hence option (a) is correct. Due to symmetry of cube, flux through all faces due to a point charge is same hence option (b) is correct. In case of uniform field, number of lines entering into a closed surface are equal to the number of lines coming out hence option (c) is correct. When electric field is parallel to a surface, flux through the surface must be zero hence option (d) is incorrect. **Ans. (C)**

Sol. 13 Using gauss law, we have

$$\oint \vec{E} \cdot \vec{ds} = \frac{q_{in}}{\varepsilon_0}$$

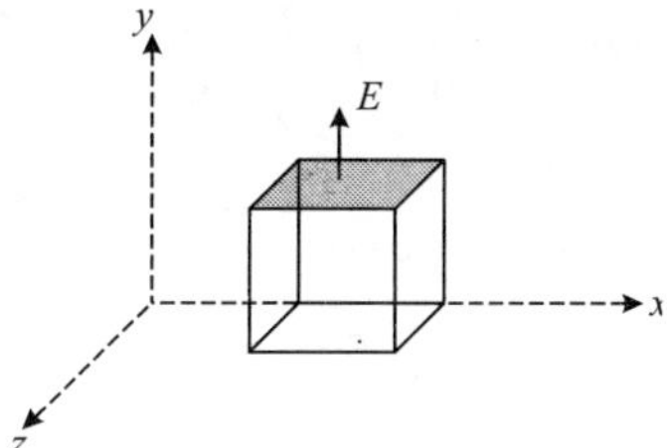

The electric flux coming out of the cube is from the top face as shown in above figure, which can be calculated as

$$\varepsilon_0 (150)(0.5)^2 \times (0.5)^2 = q_{in}$$

$$\Rightarrow q_{in} = \frac{8.85 \times 10^{-12} \times 150}{16}$$

$$\Rightarrow q_{in} = 8.3 \times 10^{-11} \text{ C}$$ **Ans. (B)**

16.1 Electric Potential and Dipole

Sol. 1 Here $V_P > V_M > V_B$

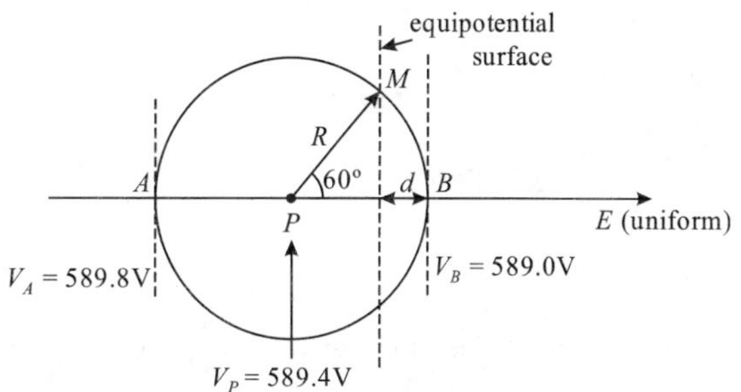

There exist only option (B) which is 589.2 V satisfying the given condition. **Ans. (B)**

Alternative : In given figure shown we use here

$$d = R(1 - \cos 60^\circ) = \frac{R}{2}$$

We use from the given values, we have

$$V_A - V_B = E(2R) = 0.8$$

$$\Rightarrow ER = 0.4$$

$$V_M - V_B = Ed = \frac{ER}{2}$$

$$\Rightarrow V_M = \frac{ER}{2} + V_B$$

$$\Rightarrow V_M = 0.2 + 589.0 = 589.2 \text{ V}$$ **Ans. (B)**

Sol. 2 Charges on shells A, B and C are

$$q_A = +\sigma \times 4\pi a^2$$
$$q_B = -\sigma \times 4\pi b^2$$
$$q_C = -\sigma \times 4\pi e^2$$

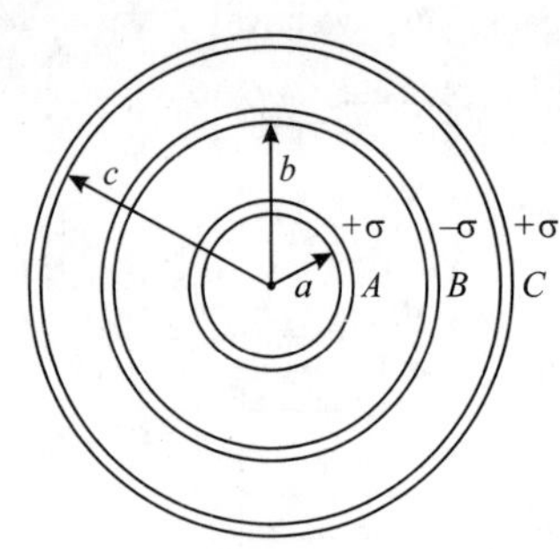

Potential of shell B due to all the shell charges is given as

$$V_B = V_{q_A} + V_{q_B} + V_{q_C}$$

$$\Rightarrow \quad V_B = \frac{Kq_A}{b} + \frac{Kq_B}{b} + \frac{Kq_C}{C}$$

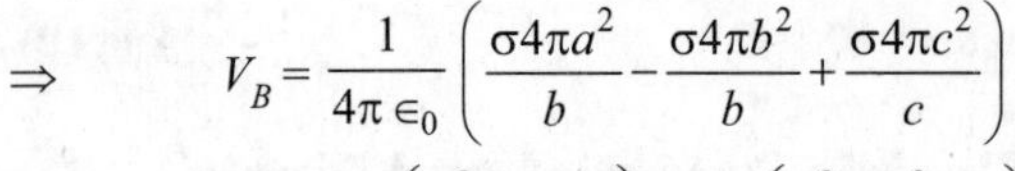

$$\Rightarrow \quad V_B = \frac{1}{4\pi \epsilon_0}\left(\frac{\sigma 4\pi a^2}{b} - \frac{\sigma 4\pi b^2}{b} + \frac{\sigma 4\pi c^2}{c}\right)$$

$$\Rightarrow \quad V_B = \frac{\sigma}{\epsilon_0}\left(\frac{a^2}{b} - b + c\right) = \frac{\sigma}{\epsilon_0}\left(\frac{a^2 - b^2}{b} + c\right)$$

Ans. (B)

Sol. 3

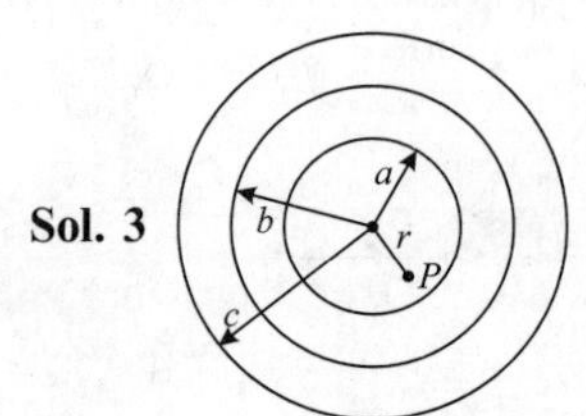

Potential at point P, due to the three shells is given as

$$V = \frac{kQ_a}{a} + \frac{kQ_b}{b} + \frac{kQ_c}{c}$$

As $Q_a : Q_b : Q_c :: a^2 : b^2 : c^2$ [since $\sigma_a = \sigma_b = \sigma_c$]

$$\Rightarrow \quad Q_a = \left[\frac{a^2}{a^2 + b^2 + c^2}\right]Q$$

$$\Rightarrow \quad Q_b = \left[\frac{b^2}{a^2 + b^2 + c^2}\right]Q$$

$$\Rightarrow \quad Q_c = \left[\frac{c^2}{a^2 + b^2 + c^2}\right]Q$$

$$\Rightarrow \quad V = \frac{Q}{4\pi \epsilon_0}\left[\frac{(a+b+c)}{a^2 + b^2 + c^2}\right]$$

Ans. (B)

Sol. 4 At a distance x on right of dipole B, we use

$$V = \frac{4qa}{(R+x)^2} = \frac{2qa}{(x^2)}$$

$$\Rightarrow \quad \sqrt{2}x = R + x$$

$$\Rightarrow \quad x = \frac{R}{\sqrt{2} - 1}$$

Distance of this point from A is

$$x_A = \frac{R}{\sqrt{2} - 1} + R = \frac{\sqrt{2}R}{\sqrt{2} - 1}$$

Ans. (C)

Sol. 5 Electric field at equatorial points of dipole is given as

$$E = \frac{Kp}{r^3}$$

At point P, force on Q is

$$F = \frac{Kp}{r^3}Q \qquad \ldots(1)$$

At point P', force on Q is

$$F' = \frac{KpQ}{(r/3)^3} = 27F$$

Ans. (D)

Sol. 6 Figure below shows the given charge configuration

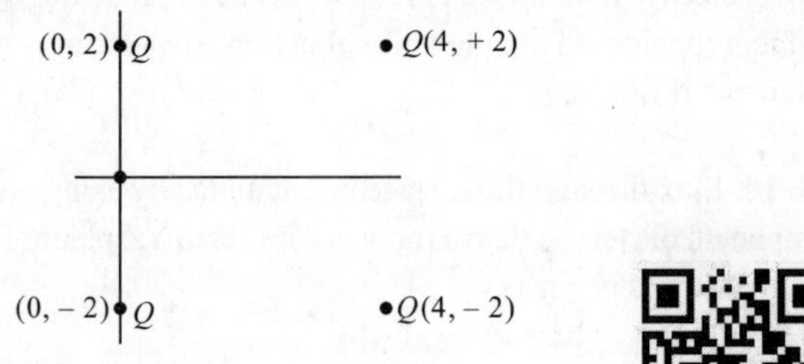

Potential at origin is given as

$$V_0 = \frac{KQ}{2} + \frac{KQ}{2} + \frac{KQ}{\sqrt{20}} + \frac{KQ}{\sqrt{20}}$$

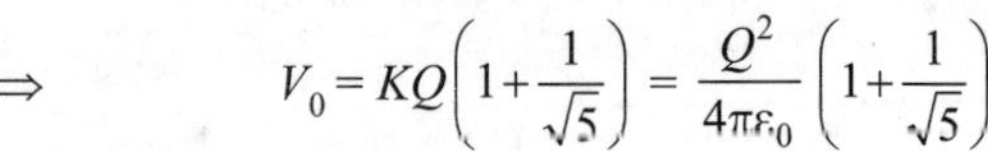

$$\Rightarrow \quad V_0 = KQ\left(1 + \frac{1}{\sqrt{5}}\right) = \frac{Q^2}{4\pi\varepsilon_0}\left(1 + \frac{1}{\sqrt{5}}\right)$$

Work required to put a fifth charge Q at origin is given as

$$W = QV = \frac{Q^2}{4\pi\varepsilon_0}\left(1 + \frac{1}{\sqrt{5}}\right)$$

Ans. (B)

Sol. 7 Total electrostatic energy of system is given as

$$U = \left[\frac{q^2}{a} + \frac{Qq}{a} + \frac{Qq}{a\sqrt{2}}\right] = 0$$

$$\Rightarrow \quad q = -Q\left[1 + \frac{1}{\sqrt{2}}\right]$$

$$\Rightarrow \qquad Q = \frac{-q\sqrt{2}}{\sqrt{2}+1}$$

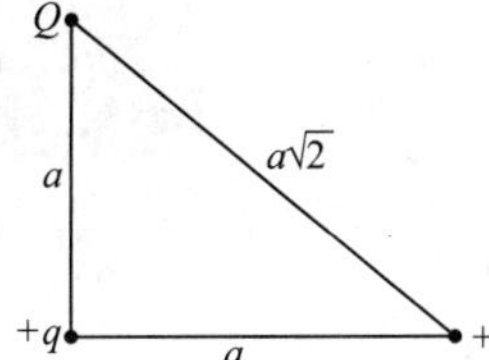
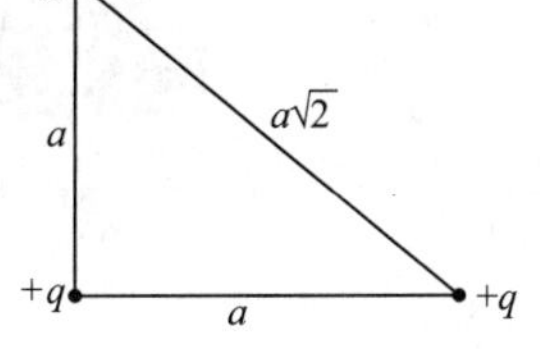

Ans. (A)

Sol. 8 Graph in figure shows that from centre to surface of sphere the physical quantity is constant than at outer points it decreases with distance so it must be electric potential due to a uniformly charged conducting shell. **Ans. (B)**

Sol. 9 Potential energy of a dipole in electric field given as

$$U = -\vec{P}\cdot\vec{E}$$

$$\Rightarrow \qquad U = -PE\cos\theta$$

$$\Rightarrow \qquad U = -(10^{-29})(10^{3})\cos 45^\circ$$

$$\Rightarrow \qquad U = -0.707\times 10^{-26}\text{ J}$$

$$\Rightarrow \qquad U = -7\times 10^{-27}\text{ J}$$

Ans. (B)

Sol. 10

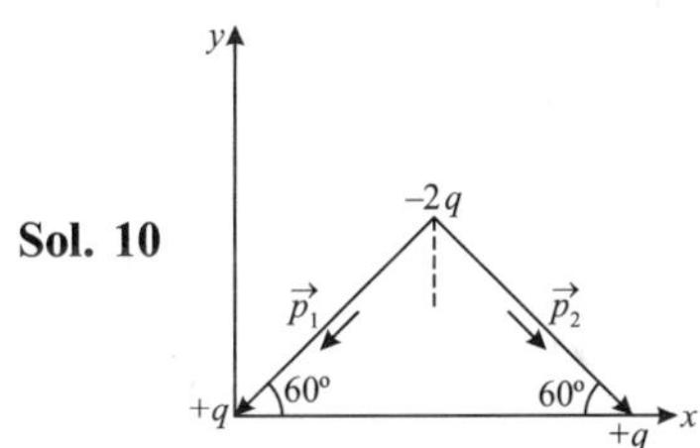

Given system is equivalent to two dipoles as shown above where dipole moments are given as

$$\vec{p}_1 = ql(-\cos 60^\circ\hat{i} - \sin 60^\circ\hat{j}) = \frac{ql}{2}(\hat{i}+\sqrt{3}\hat{j})$$

$$\vec{p}_2 = ql(\cos 60^\circ\hat{i} - \sin 60^\circ\hat{j}) = \frac{ql}{2}(\hat{i}-\sqrt{3}\hat{j})$$

Net dipole moment of system is

$$\vec{p}_{\text{net}} = \vec{p}_1 + \vec{p}_2 = -ql\sqrt{3}\hat{j}$$

Ans. (C)

Sol. 11 At any instant 't' total energy of charge distribution is constant thus by energy conservation, we use

$$\frac{1}{2}mv^2 + \frac{KQ^2}{2R} = 0 + \frac{KQ^2}{2R_0}$$

$$\Rightarrow \qquad \frac{1}{2}mV^2 = \frac{KQ^2}{2R_0} - \frac{KQ^2}{2R}$$

$$\Rightarrow \qquad v = \sqrt{\frac{2}{m}\frac{KQ^2}{2}\cdot\left(\frac{1}{R_0}-\frac{1}{R}\right)}$$

$$\Rightarrow \qquad v = \sqrt{\frac{KQ^2}{m}\left(\frac{1}{R_0}-\frac{1}{R}\right)}$$

From above expression we can see that with increase in R, v increases and $\frac{dv}{dR}$ decreases. **Ans. (A)**

Sol. 12 As given in the first case, outer shall is given charge $+Q$

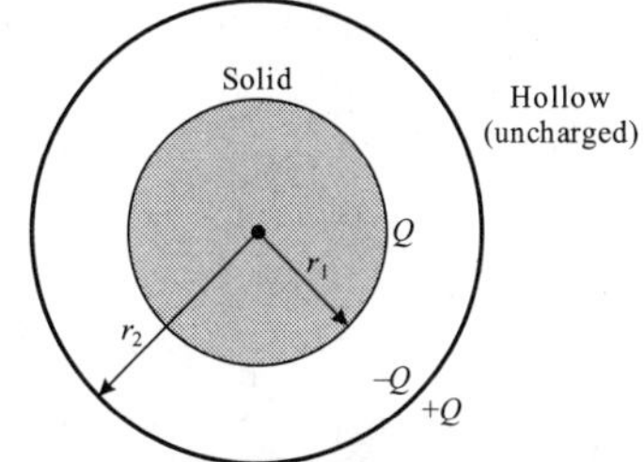

The potential difference of inner and outer spheres is given as

$$V_{\text{in}} - V_{\text{out}} = \left(\frac{kQ}{r_1}\right) - \left(\frac{kQ}{r_2}\right)$$

$$\Rightarrow \qquad V_{\text{in}} - V_{\text{out}} = kQ\left(\frac{1}{r_1}-\frac{1}{r_2}\right) = V$$

In the second case, outer shall is given charge $-4Q$

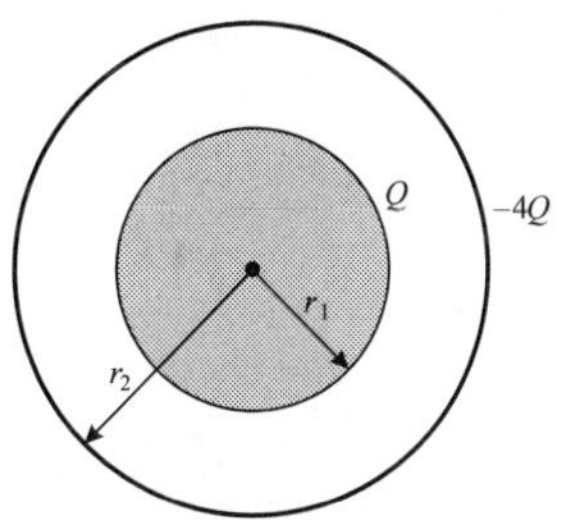

New potential difference is given as

$$V_{\text{in}} - V_{\text{out}} = \left(\frac{kQ}{r_1} - \frac{4kQ}{r_2}\right) - \left(\frac{kQ}{r_2} - \frac{4kQ}{r_2}\right)$$

$$= \frac{kQ}{r_1} - \frac{kQ}{r_2}$$

$$= kQ\left(\frac{1}{r_1}-\frac{1}{r_2}\right) = V$$

Ans. (A)

Sol. 13

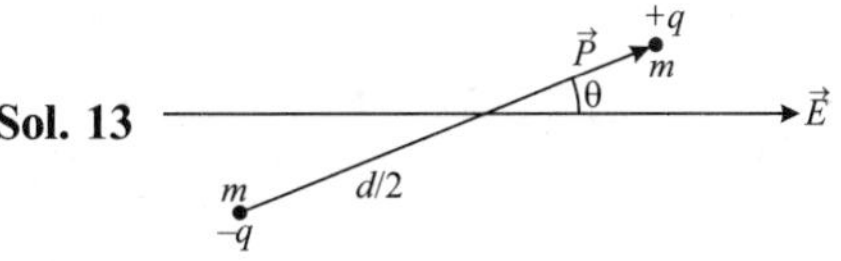

Moment of inertia of charges

$$I = 2 \times m\left(\frac{d}{2}\right)^2 = \frac{md^2}{2}$$

Torque on dipole is

$$\tau = PE \sin\theta = (qd) E \sin\theta = I\alpha$$

$\Rightarrow \qquad \alpha = -\dfrac{qEd}{I}\cdot\theta$ [As for small θ, $\sin\theta \simeq \theta$]

– sign denotes θ to be in opposite direction to α

$$\Rightarrow \qquad \alpha = -\frac{2qE}{md}\cdot\theta$$

Comparing with $\alpha = -\omega^2\theta$ we get $\omega = \sqrt{\dfrac{2qE}{md}}$ **Ans. (C)**

Sol. 14

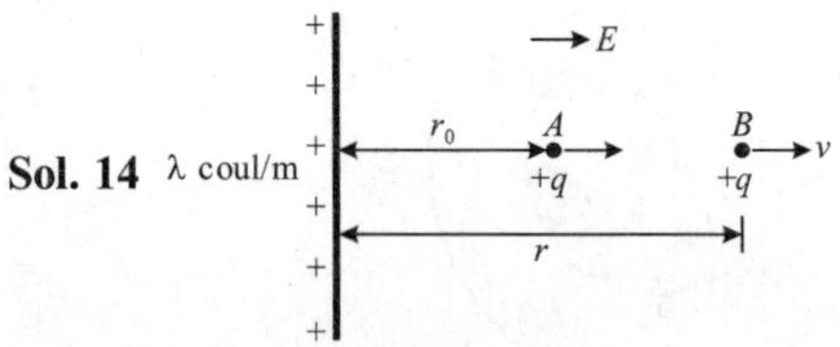

By work energy theorem, we use

$$\frac{1}{2}mv^2 = -q(V_A - V_B) \qquad \ldots(1)$$

$$\Rightarrow \qquad V_A - V_B = \int_{r_0}^{r} E\cdot dr = \int_{r_0}^{r}\frac{\lambda}{2\pi\epsilon_0 r}dr = \frac{\lambda}{2\pi\epsilon_0}[\ln r]_{r_0}^{r}$$

$$\Rightarrow \qquad V_A - V_B = \frac{\lambda}{2\pi\epsilon_0}\ln\left(\frac{r}{r_0}\right)$$

From equation-(1), we use

$$\frac{1}{2}mv^2 = q\left[\frac{\lambda}{2\pi\epsilon_0}\ln\left(\frac{r}{r_0}\right)\right]$$

$\Rightarrow \qquad v \propto \sqrt{\ln\left(\dfrac{r}{r_0}\right)}$ **Ans. (D)**

Sol. 15 Given electric field is

$$\vec{E} = (20x + 10)\hat{i}$$

Potential difference is given as

$$V_1 - V_2 = -\int_{-5}^{1}(20x + 10)dx$$

$$\Rightarrow \qquad V_1 - V_2 = -(10x^2 + 10x)_{-5}^{1}$$

$$\Rightarrow \qquad V_1 - V_2 = 10(25 - 5 - 1 - 1)$$

$\Rightarrow \qquad V_1 - V_2 = 180\text{ V}$ **Ans. (C)**

Sol. 16

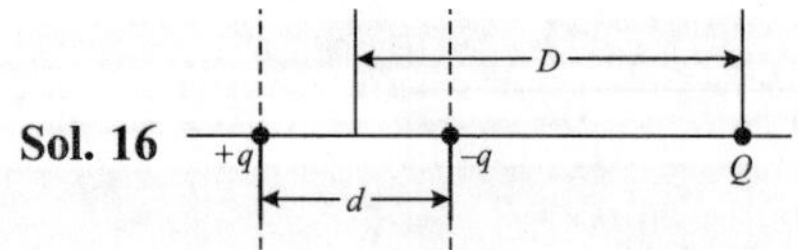

Total potential energy of this system is given as

$$U_{total} = U_{self\ of\ dipole} + U_{interaction}$$

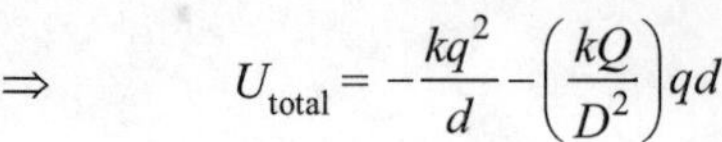

$$\Rightarrow \qquad U_{total} = -\frac{kq^2}{d} - \left(\frac{kQ}{D^2}\right)qd$$

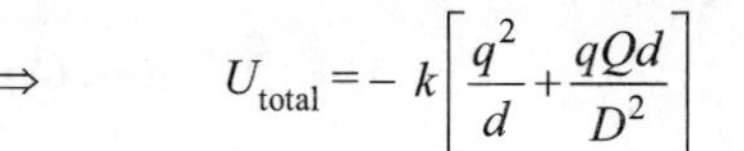

$\Rightarrow \qquad U_{total} = -k\left[\dfrac{q^2}{d} + \dfrac{qQd}{D^2}\right]$ **Ans. (D)**

Sol. 17 By conservation of energy, we have

$$U_i + K_i = U_f + K_f$$

$$\Rightarrow \qquad \frac{Kq^2}{\sqrt{16a^2 + 9a^2}} + \frac{1}{2}mv^2 = \frac{Kq^2}{3a}$$

$$\Rightarrow \qquad \frac{1}{2}mv^2 = \frac{Kq^2}{a}\left(\frac{1}{3} - \frac{1}{5}\right) = \frac{2Kq^2}{15a}$$

$\Rightarrow \qquad v = \sqrt{\dfrac{4Kq^2}{15ma}}$ **Ans. (B)**

Sol. 18 By energy conservation, we have

$$\frac{Kq_1q_2}{r_1} - \frac{Kq_1q_2}{r_2} = \frac{1}{2}mv^2$$

$$\Rightarrow \qquad \frac{(9\times10^9)\times10^{-12}}{10^{-3}} - \frac{(9\times10^9)\times10^{-12}}{9\times10^{-3}} = \frac{1}{2}\times(4\times10^{-6})v^2$$

$$\Rightarrow \qquad v^2 = 4\times10^6$$

$\Rightarrow \qquad v = 2\times10^3\text{ m/s}$ **Ans. (A)**

Sol. 19 Due to inside dipole charge induced on inner surface will be non uniform as shown below

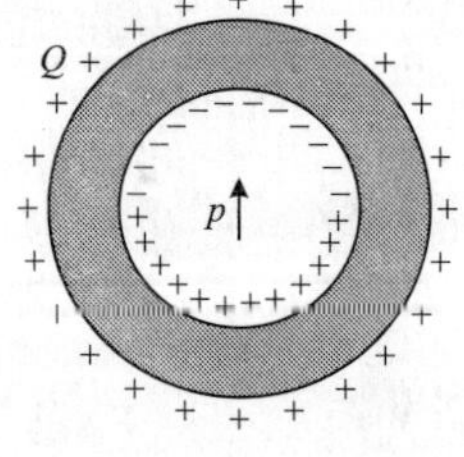

In this case as total charge of dipole is 0, so charge induced on outside surface is 0 due to it. Due to non uniform electric field of dipole, the charge induced on inner surface is non zero and non uniform so, for any observer outside the shell, the resultant electric field is due to Q which is uniformly distributed on outer surface only and it is given as

$E = \dfrac{KQ}{r^2} = \dfrac{1}{4\pi\epsilon_0}\cdot\dfrac{Q}{r^2}$ **Ans. (A)**

Sol. 20 At point M in figure as both charges of dipole are at same distance potential at M is

$$V = 0$$

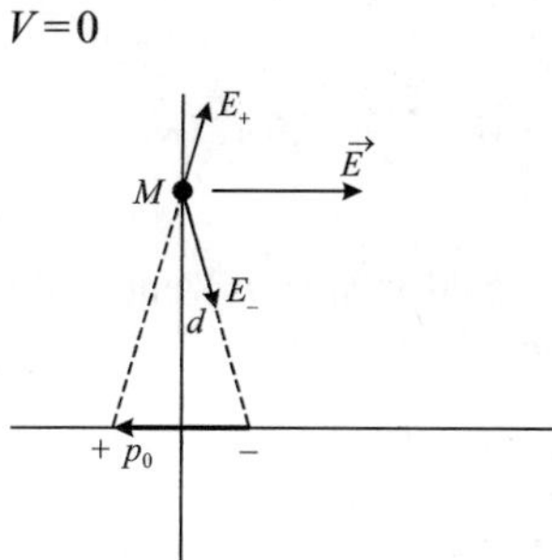

As M is an equatorial point of dipole electric field at M is given as

$$E = \frac{Kp}{d^3}$$

Vectorially, it is written as

$$\vec{E} = -\frac{1}{4\pi \epsilon_0} \frac{p_0}{d^3}$$ **Ans. (D)**

Sol. 21 Electric fields on surfaces of spheres are

$$E_1 = \frac{KQ_1}{R_1^2}$$

$$E_2 = \frac{KQ_2}{R_2^2}$$

Given that ratio of electric fields is

$$\frac{E_1}{E_2} = \frac{R_1}{R_2}$$

$$\Rightarrow \quad \frac{\frac{KQ_1}{R_1^2}}{\frac{KQ_2}{R_2^2}} = \frac{R_1}{R_2}$$

$$\Rightarrow \quad \frac{Q_1}{Q_2} = \frac{R_1^3}{R_2^3}$$

$$\Rightarrow \quad \frac{V_1}{V_2} = \frac{KQ_1/R_1}{KQ_2/R_2} = \frac{R_1^2}{R_2^2}$$ **Ans. (D)**

Sol. 22 As $\vec{r}$ and $\vec{p}$ are perpendicular to each other so the point is on the equatorial plane and electric field at this point will be in direction opposite to the dipole moment so we use

$$\hat{E} = -\hat{p}$$

Thus $\vec{E}$ is parallel to $-\vec{p}$ **Ans. (C)**

Sol. 23 As total charge remain conserved, if final charges are q_1' and q_2', we use

$$q_1 + q_2 = q_1' + q_2' = 12\ \mu C - 3\ \mu C = 9\ \mu C$$

As spheres are connected their potential must be same

$$\Rightarrow \quad V_1 = V_2$$

$$\Rightarrow \quad \frac{Kq_1'}{\left(\frac{2R}{3}\right)} = \frac{Kq_2'}{\left(\frac{R}{3}\right)}$$

$$\Rightarrow \quad q_1' = 2q_2'$$

$$\Rightarrow \quad 2q_2' + q_2' = 9\ \mu C$$

$$\Rightarrow \quad q_2' = 3\ \mu C$$

and $q_1' = 6\ \mu C$ **Ans. (A)**

Sol. 24 Potential of inner sphere is given as

$$V_i = \frac{KQ_1}{R} + \frac{KQ_2}{4R} \quad \ldots(1)$$

and that of outer sphere is given as

$$V_O = \frac{KQ_1}{4R} + \frac{KQ_2}{4R} \quad \ldots(2)$$

As surface density of charges are equal, we use

$$\frac{Q_1}{4\pi R_1^2} = \frac{Q_2}{4\pi R_2^2}$$

$$\Rightarrow \quad \frac{Q_1}{R} = \frac{Q_2}{4R}$$

$$\Rightarrow \quad Q_2 = 4Q_1$$

From (1) & (2) we use

$$V_i - V_O = \frac{3}{4}\frac{KQ_1}{R} = \frac{3Q_1}{16\pi \epsilon_0 R}$$ **Ans. (A)**

Sol. 25 For a charge, kinetic energy attained due to electric potential difference is given as

$$q\Delta V = \frac{1}{2}mv^2$$

$$\Rightarrow \quad v = \sqrt{\frac{2q\Delta V}{m}}$$

$$\Rightarrow \quad \frac{v_1}{v_2} = \sqrt{\frac{q \times m_0}{q \times m_1}} = 2$$ **Ans. (D)**

Sol. 26 Due to $-\frac{q}{2}$ there is no charge in energy of q so energy change is only due to $4q$, which is given as

$$\Delta U = 4kq^2\left[\frac{1}{\frac{3d}{2}} - \frac{1}{\frac{d}{2}}\right]$$

$$\Rightarrow \quad \Delta U = \frac{8kq^2}{d}\left[\frac{1}{3} - 1\right] = -\frac{16kq^2}{3d}$$

$$\Rightarrow \quad \Delta U = -\frac{4q^2}{3\pi \epsilon_0 d} \text{ (Decreased)}$$ **Ans. (C)**

Sol. 27 Particle given kinetic energy due to work done by electric field so kinetic energy becomes zero again when

$$W=\int_0^{x_1} qEdx = \int_0^{x_1} qE_0(1-ax^2)dx = 0$$

$$\Rightarrow \quad qE_0\left[x_1-\frac{ax_1^3}{3}\right]=0$$

$$\Rightarrow \quad x_1=\sqrt{\frac{3}{a}}$$ **Ans. (D)**

Sol. 28 By work energy theorem for mass m, we use

$$\frac{kQq}{R}+mgy=\frac{kQq}{R+y}+\frac{1}{2}mv^2$$

$$\Rightarrow \quad v^2=2gy+\frac{2kQqy}{mR(R+y)}$$ **Ans. (A)**

Sol. 29 By symmetry of charges $E_C=0$ and sum of total charges is zero, so at centre $V_C=0$. **Ans. (C)**

Sol. 30 Initial interaction energy of dipoles is given as

$$U=-\vec{p_1}\cdot\vec{E_2}$$

$$\Rightarrow \quad U=-p_1\left(\frac{2Kp_2}{a^3}\right)\cos 180^\circ$$

$$=\frac{2Kp_1p_2}{a^3}=\frac{1}{2\pi\epsilon_0}\cdot\frac{p^2}{a^3}$$

At infinite separation interaction energy becomes zero so by conservation of energy, we have

$$\frac{p^2}{2\pi\epsilon_0 a^3}=2\times\frac{1}{2}mv^2$$

$$\Rightarrow \quad v=\sqrt{\frac{p^2}{2\pi\epsilon_0 ma^3}}=\frac{p}{a}\sqrt{\frac{1}{2\pi\epsilon_0 ma}}$$ **Ans. (C)**

Sol. 31 As 512 drops are joined to form a single drop, the final charge on this bigger drop is given as

$$Q=512q$$

As total liquid volume is same and if R is the final radius of the bigger drop, we have

$$512\times\frac{4}{3}\pi r^3=\frac{4}{3}\pi R^3$$

$$\Rightarrow \quad 2^9r^3=R^3$$

$$\Rightarrow \quad R=8r$$

The final potential of the bigger drop thus formed is given as

$$V=\frac{KQ}{R}=\frac{K(512q)}{8r}$$

Given that potential of each small drop is 2 V so we have

$$2=\frac{Kq}{R}$$

$$\Rightarrow \quad V=\frac{512}{8}\times 2=128\text{ V}$$ **Ans. (128)**

Sol. 32 As total liquid volume is same and if R is the final radius of the bigger drop, we have

$$27\times\frac{4}{3}\pi r^3=\frac{4}{3}\pi R^3$$

$$\Rightarrow \quad 3^3r^3=R^3$$

$$\Rightarrow \quad R=3r$$

Potential energy of smaller drop is given as

$$U_1=\frac{Kq^2}{2r}$$

Potential energy of bigger drop is given as

$$U=\frac{KQ^2}{2R}$$

$$U=\frac{K(27q)^2}{2R}=\frac{K(27)(27)q^2}{2(3r)}$$

$$U=\frac{(27)(27)}{3}\left(\frac{Kq^2}{2r}\right)=243\,U_1$$ **Ans. (243)**

Sol. 33 The electric field due to $\vec{P_1}$ at an axial point, is given as

$$E_a=\frac{2KP_1}{r^3}$$

$$\Rightarrow \quad E\cos 37^\circ=\frac{2KP_1}{r^3} \quad \ldots(1)$$

Electric field due to $\vec{P_2}$ at equatorial point which is the axis point of first dipole is given as

$$E_e=\frac{KP_2}{r^3}$$

$$\Rightarrow \quad E\sin 37^\circ=\frac{KP_2}{r^3} \quad \ldots(3)$$

$$\Rightarrow \quad \frac{\frac{2KP_1}{r^3}}{\frac{KP_2}{r^3}}=\frac{E\cos 37^\circ}{E\sin 37^\circ}$$

$$\Rightarrow \quad \frac{2P_1}{P_2}=\frac{4}{3}$$

$$\Rightarrow \quad \frac{P_1}{P_2}=\frac{2}{3}$$ **Ans. (B)**

Sol. 34 Potentials at the center of the two rings

$$V_1 = \frac{KQ}{a} + \frac{-KQ}{\sqrt{s^2+a^2}}$$

and $$V_2 = \frac{KQ}{\sqrt{s^2+a^2}} + \frac{-KQ}{a}$$

$$\Rightarrow \quad V_1 - V_2 = \frac{Q}{2\pi\varepsilon_0}\left[\frac{1}{a} - \frac{1}{\sqrt{s^2+a^2}}\right]$$ **Ans. (B)**

16.2 Capacitance

Sol. 1 Supply voltage = 1000 V

Potential is equally divided in identical capacitors in series so for max 300V p.d. we need 4 capacitors (each 1μF) in series across supply.

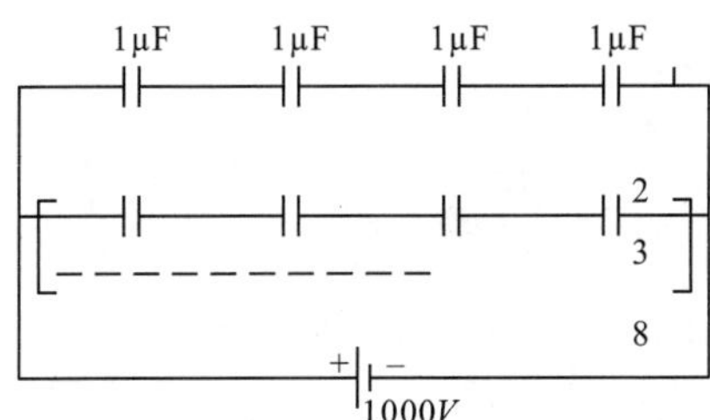

$$C_{eq} = \frac{C}{4} = 0.25\ \mu\text{F}$$

To make 2μF capacitance we have to connect 8 such branches (4 in series) in parallel

$$\Rightarrow \quad C_F = 0.25 \times 8 = 2\mu\text{F}$$

Thus total capacitors required = 4 × 8 = 32 **Ans. (B)**

Sol. 2 The capacitance of an isolated metal sphere is

$$C = 4\pi\epsilon_0 R$$

If q is the charge on sphere than energy stored in its EF is (Self energy of charged sphere)

$$U = \frac{q^2}{2C} = 4.5$$

$$\Rightarrow \quad (4 \times 10^{-6}) = 4.5 \times 2 \times 4\pi\epsilon_0 R$$

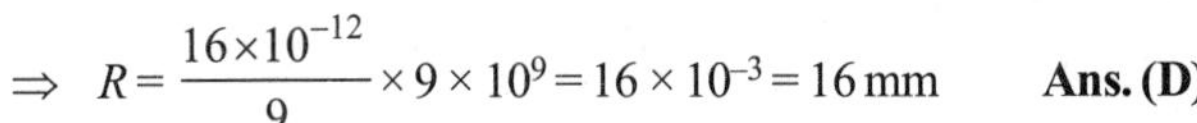

$$\Rightarrow \quad R = \frac{16 \times 10^{-12}}{9} \times 9 \times 10^9 = 16 \times 10^{-3} = 16\text{ mm}$$ **Ans. (D)**

Sol. 3 For a given capacitor (parallel plate)

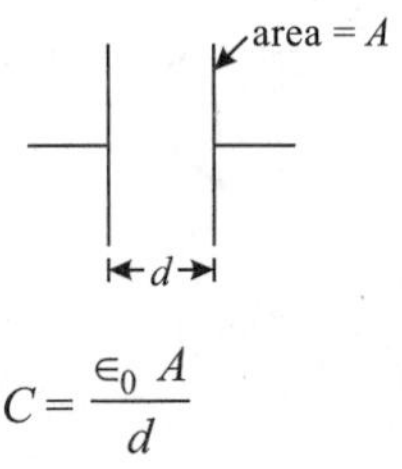

$$C = \frac{\epsilon_0 A}{d}$$

For a given charge to capacitor the potential difference across plates is

$$V = \frac{q}{C} \quad \Rightarrow \quad \text{As } C \uparrow \Rightarrow V \downarrow$$

To maintain the potential difference across capacitors, C must remain constant if d is increased to $d + 2.4$ ⇒ now capacitance with slab is given as

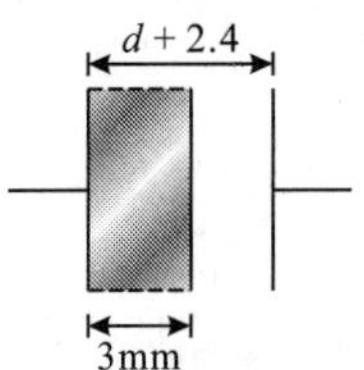

$$C' = \frac{\epsilon_0 A}{d' - t + \frac{t}{k}} \qquad t \to \text{thickness of slab}$$

To maintain the potential difference we use

$$C' = C$$

$$\frac{\epsilon_0 A}{d' - t + \frac{t}{k}} = \frac{\epsilon_0 A}{d}$$

$$\Rightarrow \quad d + 2.4 - 3 + \frac{3}{k} = d$$

$$\Rightarrow \quad \frac{3}{k} = 0.6 \quad \Rightarrow \quad k = 5$$ **Ans. (C)**

Sol. 4

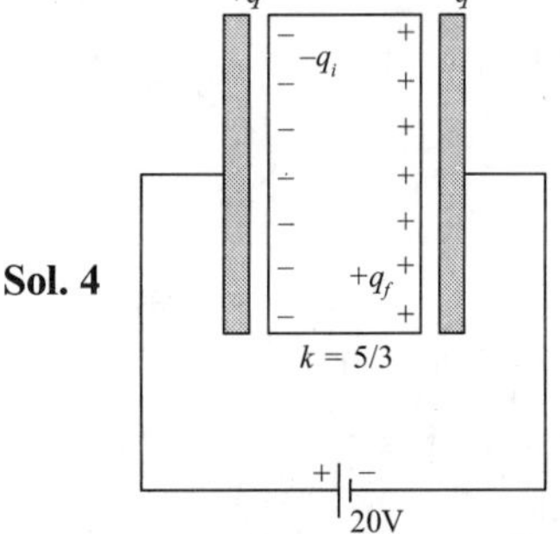

$$C = 90\text{pF} \rightarrow C' = \frac{5}{3} \times 90 = 150\text{ pF}$$

Capacitance with dielectric

In steady state charge on capacitor plates is

$$q = C'V$$

$$q = 3000\text{ pC}$$

Free charge on plates

Induced charges on the surface of dielectric is given as

$$q_i = q\left(1 - \frac{1}{k}\right) = 3000\left(1 - \frac{1}{5/3}\right)$$

$$\Rightarrow \quad q_i = 3000 \times \frac{2}{5} = 1200\text{ pC} = 1.2\text{ nC}$$ **Ans. (A)**

Sol. 5 If charge on a capacitor plates is q then force of attraction between plates is given as

$$F=\frac{q^2}{2\in_0 A}=\frac{C^2V^2}{2\in_0 A} \quad (\text{As } q=CV)$$

and $C=\frac{\in_0 A}{d} \Rightarrow F=\frac{\in_0 AV^2}{2d^2}$

$$\Rightarrow \quad V=\sqrt{\frac{2Fd^2}{\in_0 A}}$$

$$=\sqrt{\frac{2\times25\times10^{-6}\times(1.5\times10^{-2})^2}{8.85\times10^{-12}\times200\times10^{-4}}}\approx 250\,\text{V} \quad \textbf{Ans. (C)}$$

Sol. 6 Simplifying the above circuit, it can be redrawn as

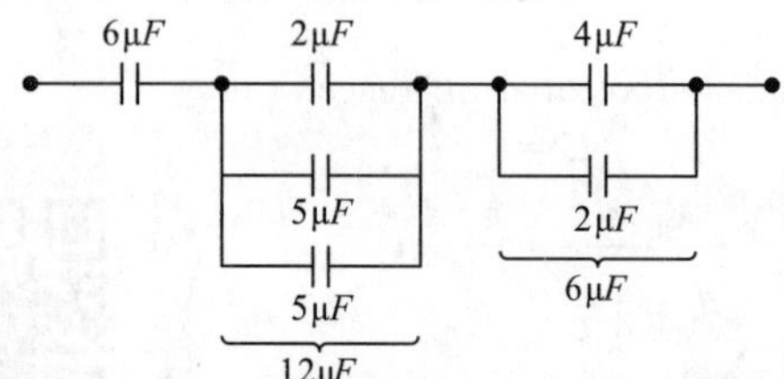

Here equivalent capacitance across AB is given as

$$\frac{1}{C_{AB}}=\frac{1}{6}+\frac{1}{12}+\frac{1}{6}$$

$$\frac{1}{C_{AB}}=\frac{2+1+2}{12}=\frac{5}{12}$$

$$\Rightarrow \quad C_{AB}=\frac{12}{5}=2.4\,\mu\text{F} \quad \textbf{Ans. (D)}$$

Sol. 7 When C_1 is charged by battery then its charge will be

$$q_{C_1}=C_1V=1\times60=60\,\mu\text{C}$$

Combination of C_2 and C_3 in series is

$$C'=\frac{C_2C_3}{C_2+C_3}=\frac{3\times6}{3+6}=2\mu\text{F}$$

After switch is shifted to position (2), the circuit will be

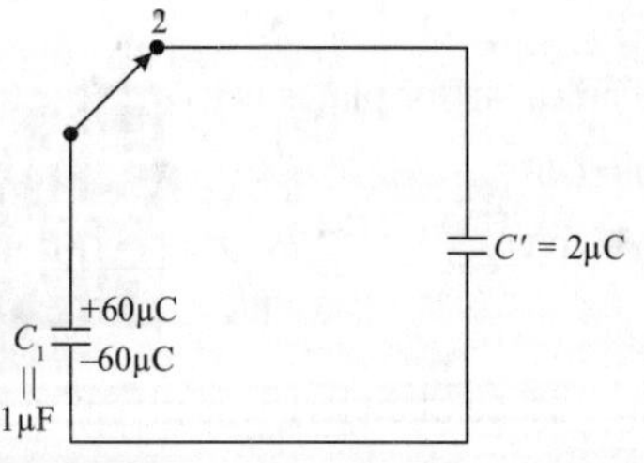

Final voltage across C_1 and C' is

$$V_f=\frac{q}{C_1+C'}=\frac{60}{1+2}=20\text{V}$$

$\Rightarrow$ Final charge on C' is

$$q_C=C'V_f=2\times20=40\,\mu\text{C} \quad \textbf{Ans. (D)}$$

Sol. 8

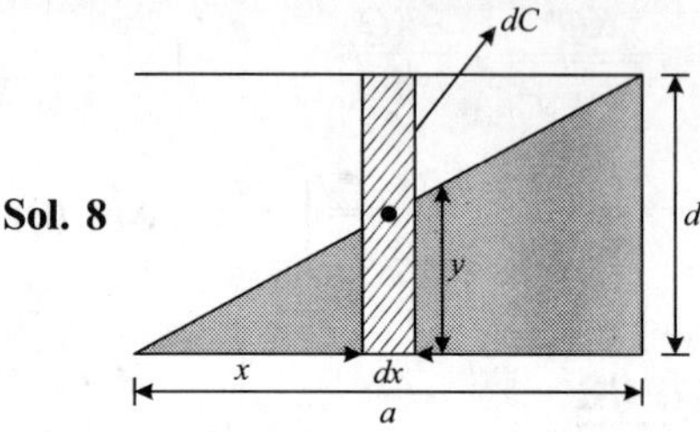

Here we consider elemental capacitor dC as shown in figure for which, we use

$$\frac{y}{x}=\frac{d}{a}$$

$$y=\frac{d}{a}x$$

$$dy=\frac{d}{a}(dx)$$

Capacitance of elemental capacitor is taken as

$$\frac{1}{dC}=\frac{y}{k\in_0\cdot adx}+\frac{(d-y)}{\in_0 adx}$$

$$\Rightarrow \quad \frac{1}{dC}=\frac{1}{\in_0 adx}\left(\frac{y}{k}+d-y\right)$$

All elemental capacitors are considered in parallel so equivalent capacitance is

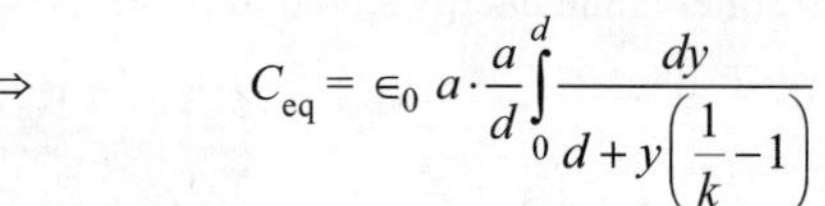

$$C_{eq}=\int dC=\int\frac{\in_0 adx}{\frac{y}{k}+d-y}$$

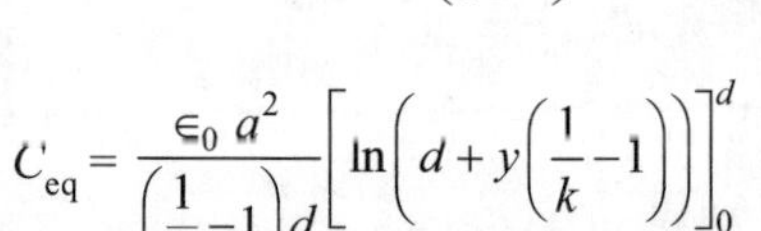

$$\Rightarrow \quad C_{eq}=\in_0 a\cdot\frac{a}{d}\int_0^d\frac{dy}{d+y\left(\frac{1}{k}-1\right)}$$

$$\Rightarrow \quad C_{eq}=\frac{\in_0 a^2}{\left(\frac{1}{k}-1\right)d}\left[\ln\left(d+y\left(\frac{1}{k}-1\right)\right)\right]_0^d$$

$$\Rightarrow \quad C_{eq}=\frac{k\in_0 a^2}{(1-k)d}\ln\left(\frac{d+d\left(\frac{1}{k}-1\right)}{d}\right)$$

$$\Rightarrow \quad C_{eq}=\frac{k\in_0 a^2}{(1-k)d}\ln\left(\frac{1}{k}\right)=\frac{k\in_0 a^2\ln k}{(k-1)d} \quad \textbf{Ans. (C)}$$

Sol. 9 Equivalent circuit of given system is shown below

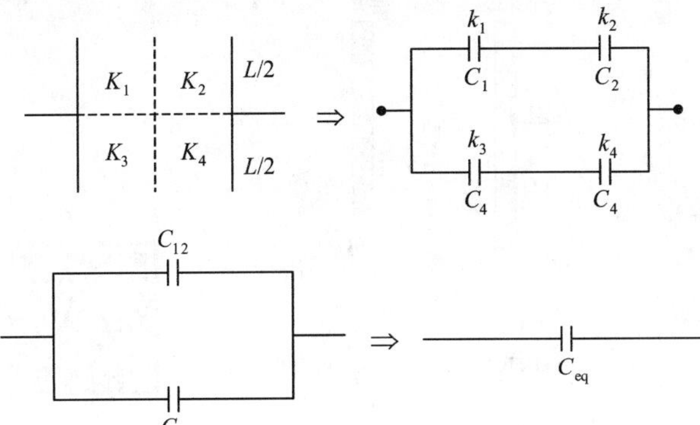

Thus equivalent capacitance is given as

$$C_{12} = \frac{C_1C_2}{C_1+C_2} = \frac{\frac{k_1 \in_0 \frac{L}{2}\times L}{d/2} \cdot \frac{k_2\left[\in_0 \frac{L}{2}\times L\right]}{d/2}}{(k_1+k_2)\left[\frac{\in_0 \cdot \frac{L}{2}\times L}{d/2}\right]}$$

$$C_{12} = \frac{k_1k_2}{k_1+k_2}\frac{\in_0 L^2}{d}$$

In the same way we get,

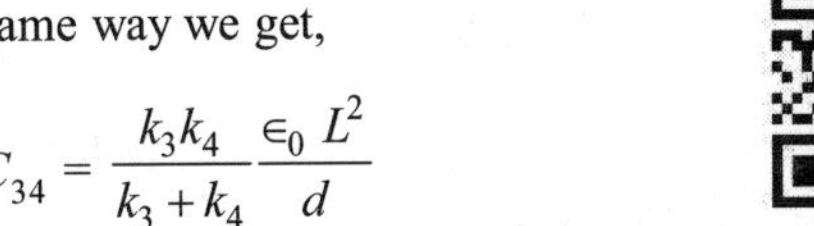

$$C_{34} = \frac{k_3k_4}{k_3+k_4}\frac{\in_0 L^2}{d}$$

$$\Rightarrow \quad C_{eq} = C_{12} + C_{34} = \left[\frac{k_1k_2}{k_1+k_2} + \frac{k_3k_4}{k_3+k_4}\right]\frac{\in_0 L^2}{d} \quad \ldots(1)$$

Comparing with

$$C_{eq} = \frac{k \in_0 L^2}{d} \quad \ldots(2)$$

$$\Rightarrow \quad k_{eq} = \frac{k_1k_2(k_3+k_4)+k_3k_4(k_1+k_2)}{(k_1+k_2)(k_3+k_4)}$$

This does not match with any of the options, so none of the given options is correct **Ans. (None)**

Sol. 10 Let dielectric constant of material used be K

$$\Rightarrow \quad \frac{10 \in_0 A/3}{d} + \frac{12 \in_0 A/3}{d} + \frac{14 \in_0 A/3}{d} = \frac{K \in_0 A}{d}$$

$\Rightarrow K = 12$ **Ans. (A)**

Sol. 11 Initial energy of capacitor

$$U_i = \frac{1}{2}CV^2 = \frac{1}{2} \times 12 \times 10^{-12} \times (10)^2$$

$$\Rightarrow \quad U_i = 600 \text{ pJ}$$

Since battery is disconnected so charge remain same at 120pc. Final energy of capacitor will be

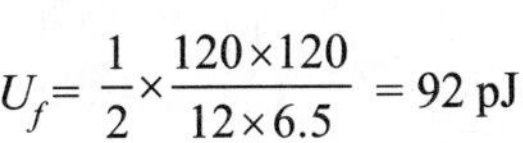

$$U_f = \frac{q^2}{2C}$$

$$\Rightarrow \quad U_f = \frac{1}{2}\times\frac{120\times120}{12\times6.5} = 92 \text{ pJ}$$

Work done by capacitor is

$$W = U_i - U_f = 600 - 92 = 508 \text{ pJ}$$ **Ans. (C)**

Sol. 12

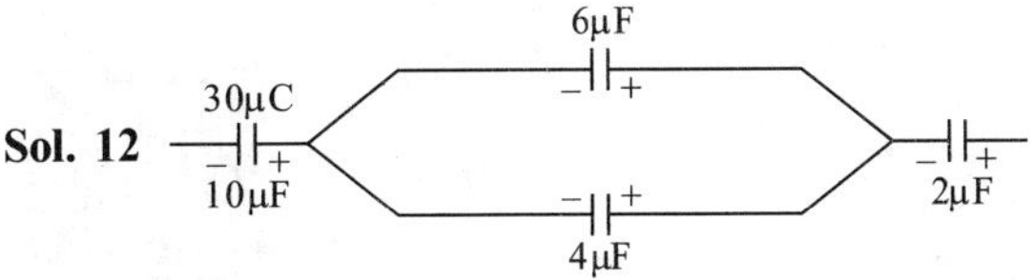

6μF & 4μF are in parallel & total charge on this combination is 30 μC

$\Rightarrow$ Charge on 6μF capacitor

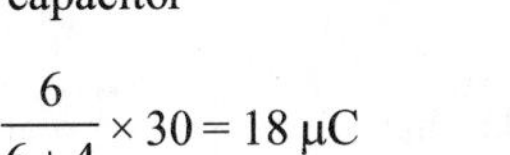

$$= \frac{6}{6+4}\times 30 = 18 \text{ μC}$$

Thus charge is on right plate of 6 μF capacitor is +18 μC. **Ans. (D)**

Sol. 13 Equivalent capacitance is given as

$$C_{eq} = \frac{6}{13}\mu F = \frac{3C}{13} \qquad [\text{As } C = 2\mu F \text{ each}]$$

Therefore three capacitors must be in parallel with 4 in series so, we have

$$\frac{1}{C_{eq}} = \frac{1}{3C} + \frac{1}{C} + \frac{1}{C} + \frac{1}{C} + \frac{1}{C}$$

$$C_{eq} = \frac{3C}{13} = \frac{6}{13}\mu F$$ **Ans. (D)**

Sol. 14 Initial energy in capacitor is

$$U_i = \frac{Q^2}{2C}$$

After turning the switch to position *B*, find energy in circuit is

$$U_f = \frac{Q^2}{8C}$$

Heat dissipated is given as

$$H = U_i - U_f = \frac{3Q^2}{8C}$$ **Ans. (A)**

Sol. 15

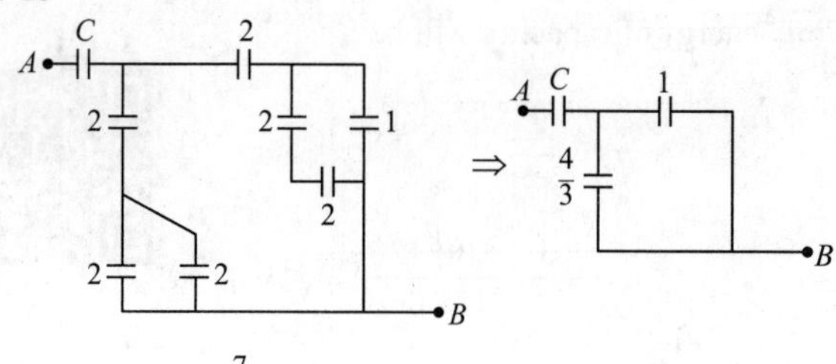

$$\Rightarrow \quad A \;—\; C \;—\; \frac{7}{3} \;—\; B$$

Equivalent capacitance across A and B is given as

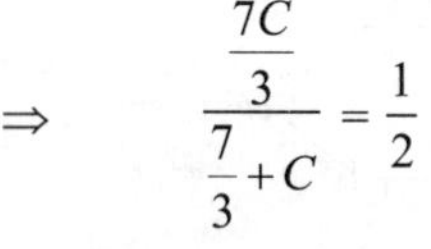

$$\Rightarrow \quad \frac{\frac{7C}{3}}{\frac{7}{3}+C} = \frac{1}{2}$$

$$\Rightarrow \quad 14C = 7 + 3C$$

$$\Rightarrow \quad C = \frac{7}{11}\ \mu F$$

Ans (B)

Sol. 16 At $t = 4$s, charge on capacitor plates is constant so current in circuit will be zero.

Ans. (C)

Sol. 17 Electric field between plates of a capacitor is given as

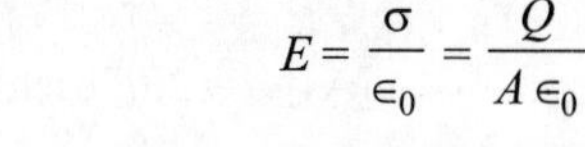

$$E = \frac{\sigma}{\epsilon_0} = \frac{Q}{A\epsilon_0}$$

$$\Rightarrow \quad Q = AE\epsilon_0$$

$$\Rightarrow \quad Q = (1)(100)(8.85 \times 10^{-12})$$

$$\Rightarrow \quad Q = 8.85 \times 10^{-10}\text{C}$$

Ans. (D)

Sol. 18 Maximum electric field between plates

$$E_m = 10^6\ \text{V/m}$$

If d is the plate separation, we use

$$E_m = \frac{V_m}{d}$$

$$\Rightarrow \quad d = \frac{V_m}{E_m} = \frac{500}{10^6} = 5 \times 10^{-4}\ \text{m}$$

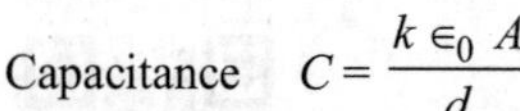

Capacitance $C = \dfrac{k\epsilon_0 A}{d}$

$$\Rightarrow \quad k = \frac{Cd}{\epsilon_0 A} = \frac{15\times10^{-12}\times5\times10^{-4}}{8.85\times10^{-12}\times10^{-4}}$$

$$= \frac{15\times5}{8.85} = 8.465 \sim 8.5$$

Ans. (D)

Sol. 19

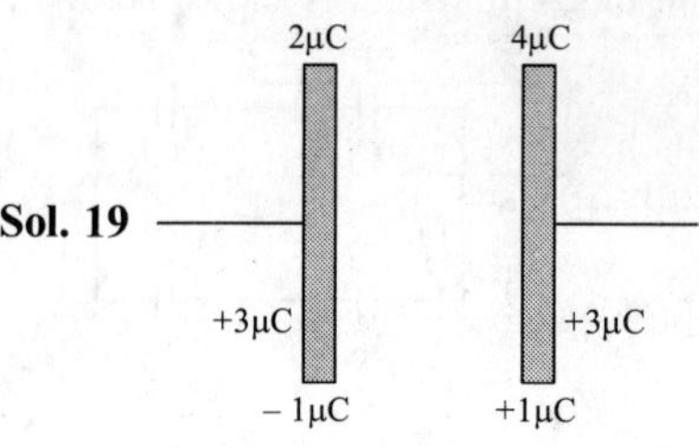

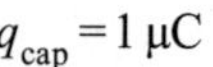

$$q_{\text{cap}} = 1\ \mu\text{C}$$

⇒ Potential difference across plates

$$V = \frac{q}{C} = \frac{1\mu C}{1\mu F} = 1\ \text{V}$$

Ans. (D)

Sol. 20 Work done in pulling plates apart is given as

$$W = U_f - U_i$$

$$\Rightarrow \quad W = \frac{q^2}{2C_f} - \frac{q^2}{2C_i}$$

$$\Rightarrow \quad W = \frac{(5\times10^{-6})^2}{2}\left(\frac{1}{2\times10^{-6}} - \frac{1}{5\times10^{-6}}\right)$$

$$\Rightarrow \quad W = \frac{15}{4} \times 10^{-6}$$

$$\Rightarrow \quad W = 3.75 \times 10^{-6}\ \text{J}$$

Ans. (A)

Sol. 21

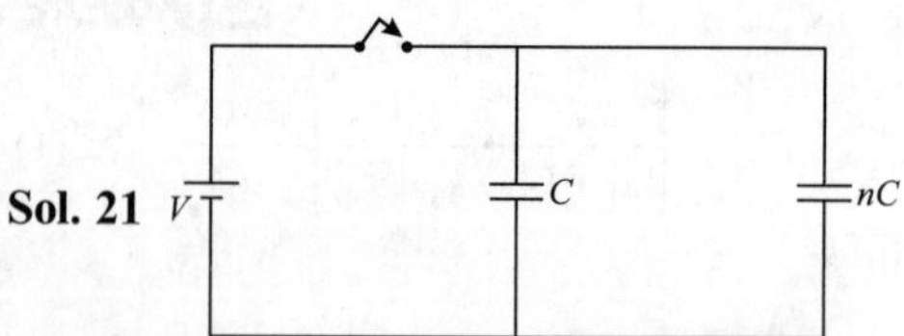

After fully charging, battery is disconnected from circuit so charge on system remain constant. after disconnecting battery system now is shown below

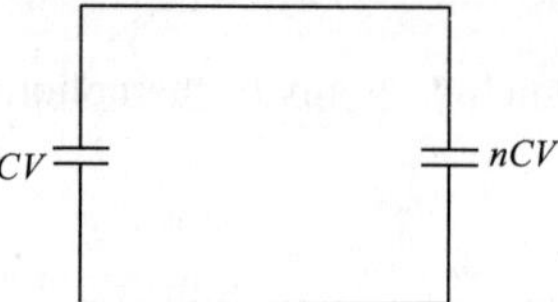

Total charge of the system

$$= CV + nCV$$

$$= (n+1)CV$$

After the insertion of dielectric of constant K, if final potential difference becomes V_f, we use

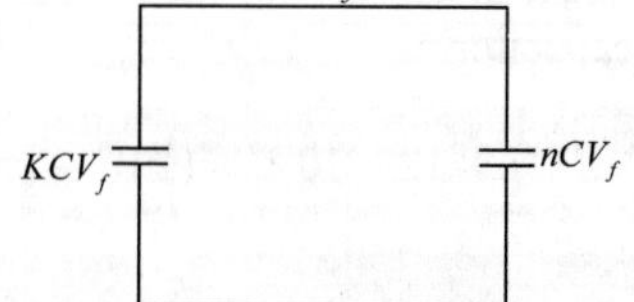

$$V_f = \frac{\text{total charge}}{\text{total capacitance}}$$

$$\Rightarrow \quad V_f = \frac{(n+1)CV}{KC+nC} = \frac{(n+1)V}{K+n} \qquad \textbf{Ans. (C)}$$

Sol. 22 As $q = CV$ so the slope of graph will give capacitance. Slope will be more in parallel combination. Hence capacitance in parallel should be 50 μF & in series combination should be 8 μF. From the given option only in option 40 μF & 10 μF it is obtained as

$$C_{\text{parallel}} = 40 + 10 = 50\mu\text{F}$$

$$C_{\text{series}} = \frac{40 \times 10}{40+10} = 8\,\mu\text{F} \qquad \textbf{Ans. (D)}$$

Sol. 23 In electric field effective gravity for pendulum is given as

$$g_{\text{eff}} = \sqrt{g^2 + \left(\frac{qE}{m}\right)^2}$$

Time period of pendulum is taken as

$$T = 2\pi\sqrt{\frac{l}{g_{\text{eff}}}}$$

$$\Rightarrow \quad T = 2\pi\sqrt{\frac{l}{\sqrt{g^2 + \left(\frac{qE}{m}\right)^2}}} \qquad \textbf{Ans. (A)}$$

Sol. 24 In first case we consider the three capacitors of each dielectric are in series.

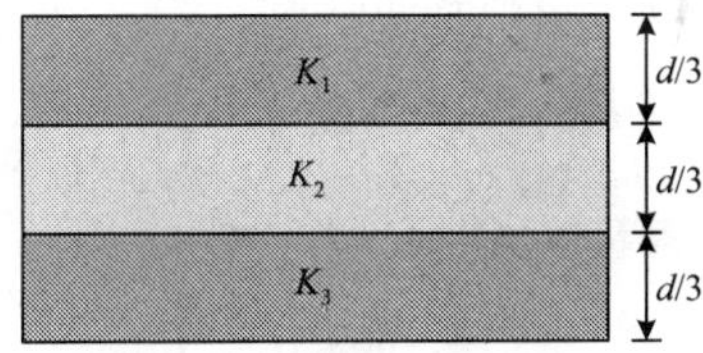

Capacitance for each dielectric is given as

$$C_1 = \frac{3\varepsilon_0 AK_1}{d}$$

$$C_2 = \frac{3\varepsilon_0 AK_2}{d}$$

$$C_3 = \frac{3\varepsilon_0 AK_3}{d}$$

Combination capacitance is taken as

$$\frac{1}{C_{\text{eq}}} = \frac{1}{C_1} + \frac{1}{C_2} + \frac{1}{C_3}$$

$$\Rightarrow \quad C_{\text{eq}} = \frac{3\varepsilon_0 AK_1K_2K_3}{d(K_1K_2 + K_2K_3 + K_3K_1)} \qquad \ldots(1)$$

In second case we consider capacitance of each dielectric are in parallel

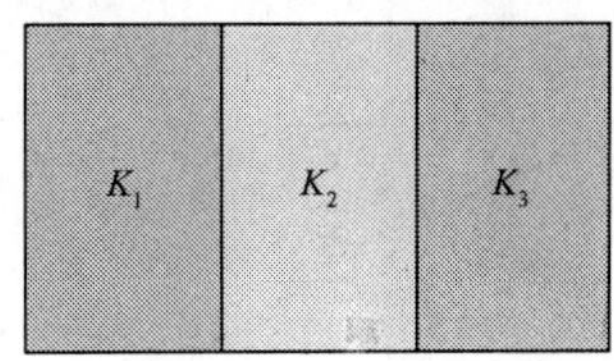

Capacitance of each dielectric is given as

$$C_1 = \frac{\varepsilon_0 K_1 A}{3d}$$

$$C_2 = \frac{\varepsilon_0 K_2 A}{3d}$$

$$C_3 = \frac{\varepsilon_0 K_3 A}{3d}$$

Combination capacitance is taken as

$$C'_{\text{eq}} = C_1 + C_2 + C_3$$

$$C'_{\text{eq}} = \frac{\varepsilon_0 A}{3d}(K_1 + K_2 + K_3) \qquad \ldots(2)$$

Ratio of energy stored in two cases is

$$\frac{E_1}{E_2} = \frac{\frac{1}{2}C_{eq}.V^2}{\frac{1}{2}C'_{eq}V^2}$$

$$\Rightarrow \quad \frac{E_1}{E_2} = \frac{9K_1K_2K_3}{(K_1+K_2+K_3)(K_1K_2+K_2K_3+K_3K_1)} \qquad \textbf{Ans. (A)}$$

Sol. 25

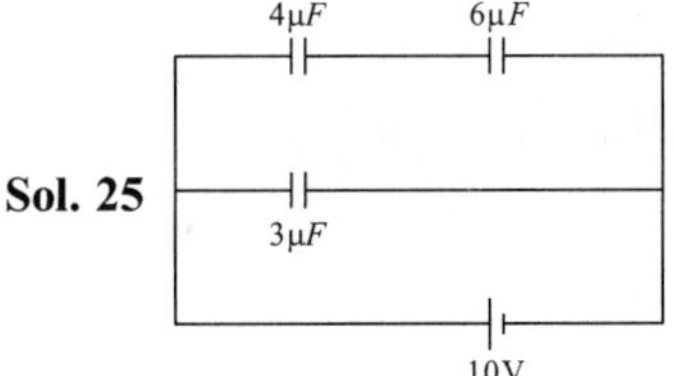

In above circuit we consider 4μF and 6 μF capacitors in series

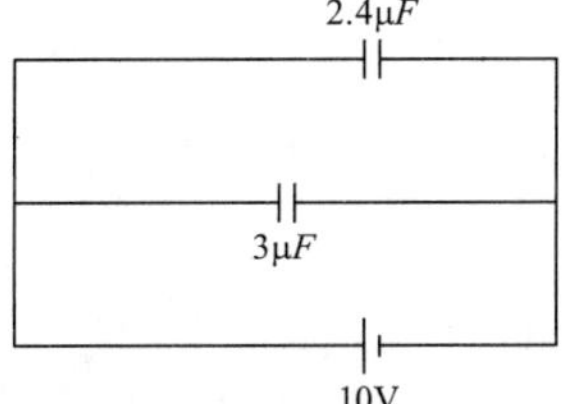

Charge supplied by battery to 2.4 μF capacitor is

$$q = 2.4 \times 10 = 24\,\mu C$$

In series both capacitors will have same charge of 24 μC.

Ans. (B)

Sol. 26 Here we use,

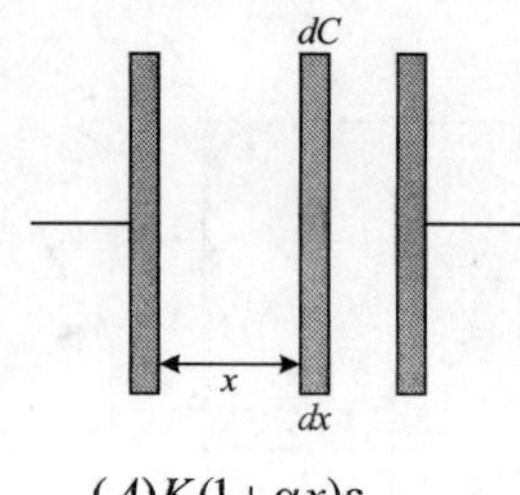

$$dC = \frac{(A)K(1+\alpha x)\varepsilon_0}{dx}$$

Now all such elemental capacitors are is series so we use,

$$\frac{1}{C} = \int \frac{1}{dC} = \int_0^d \frac{dx}{AK\varepsilon_0(1+\alpha x)}$$

$$\Rightarrow \quad \frac{1}{C} = \int_0^d \frac{1}{AK\varepsilon_0}\cdot(1-\alpha x)dx$$

$$\Rightarrow \quad \frac{1}{C} = \frac{1}{AK\varepsilon_0}\left[d - \frac{\alpha d^2}{2}\right] = \frac{d}{AK\varepsilon_0}\left(1-\frac{\alpha d}{2}\right)$$

$$\Rightarrow \quad \frac{1}{C} = \frac{d}{AK\varepsilon_0}\left(1-\frac{\alpha d}{2}\right)$$

$$\Rightarrow \quad C = \frac{AK\varepsilon_0}{d}\left(1+\frac{\alpha d}{2}\right)$$ **Ans. (A)**

Sol. 27 After second connection charge is equally distributed in both capacitors

+Q C ⇒ C Q/2 Q/2 C

$$Q = CV$$

Energy loss is given as

$$\Delta Q_L = \frac{Q^2}{2C} - \left[\frac{(Q/2)^2}{2C}\times 2\right] = \frac{Q^2}{4C}$$

$$\Rightarrow \quad \Delta Q_L = \frac{1}{4}CV^2$$

$$\Rightarrow \quad \Delta Q_L = \frac{1}{4}\times 60\times 10^{-12}\times 4\times 10^2$$

$$\Rightarrow \quad \Delta Q_L = 6\text{ nJ}$$ **Ans. (6)**

Sol. 28 In parallel combination, we use

$$C_1 + C_2 = 10 \qquad \ldots(1)$$

For energy stored, it is given that

$$\frac{1}{2}C_2V^2 = 4\times\frac{1}{2}C_1V^2$$

$$\Rightarrow \quad C_2 = 4C_1 \qquad \ldots(2)$$

$$\Rightarrow \quad C_1 = 2$$

$$\& \quad C_2 = 8$$

For series combination, we use

$$C_{eq} = \frac{C_1C_2}{C_1+C_2} = 1.6\ \mu\text{F}$$ **Ans. (C)**

Sol. 29 Assume small element dx at a distance x from left end

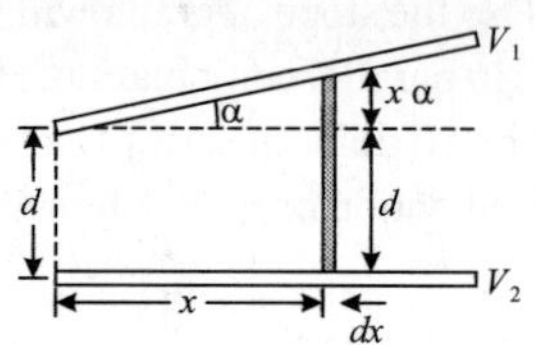

Capacitance of this element is given as

$$dC = \frac{\varepsilon_0(adx)}{d+x\alpha}$$

Thus total capacitance is given as

$$C = \int_0^a \frac{\varepsilon_0 adx}{d+x\alpha}$$

$$\Rightarrow \quad C = \frac{\varepsilon_0 a}{\alpha}\ln\left(\frac{1+\alpha a}{d}\right)\Big|_0^a \quad \left(\text{We use } \ln(1+x)\approx x - \frac{x^2}{2}\right)$$

$$\Rightarrow \quad C = \frac{\varepsilon_0 a^2}{d}\left(1-\frac{\alpha a}{2d}\right)$$ **Ans. (D)**

Sol. 30 Initial energy stored in capacitor is

$$U_i = \frac{1}{2}\times 5\times 10^{-6}(220)^2$$

Final common potential of capacitors is given as

$$V = \frac{220\times 5 + 0\times 2.5}{5+2.5} = 220\times\frac{2}{3}$$

Final energy stored in capacitors is

$$U_f = \frac{1}{2}(5+2.5)\times 10^{-6}\left(220\times\frac{2}{3}\right)^2$$

$$\Delta U = U_f - U_i$$

$$\Rightarrow \quad \Delta U = -403.33\times 10^{-4}$$

$$\Rightarrow \quad -403.33\times 10^{-4} = \frac{X}{100}$$

$$X = -4.03$$ **Ans. (–4.03)**

Sol. 31 Initial charges on capacitors are

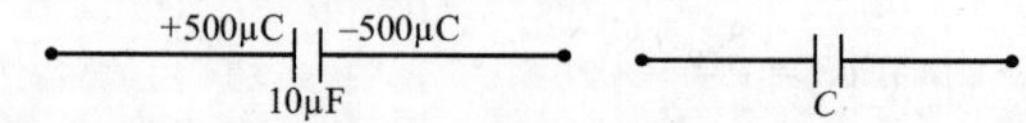

After connection

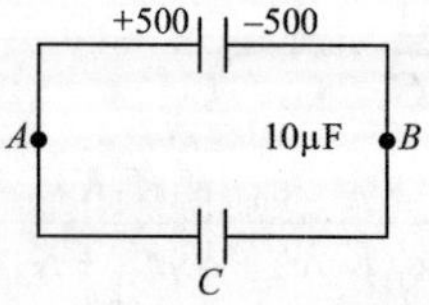

Potential difference across A and B is

$$V_{AB} = \frac{q_{\text{total}}}{q_{\text{total}}} = \frac{500}{C+10} = 20$$

$$\Rightarrow \quad 20C + 200 = 500$$

$$\Rightarrow \quad 20C = 300$$

$$\Rightarrow \quad C = 15\ \mu\text{F}$$ **Ans. (B)**

Sol. 32 If total charge is 750 μC then charge on C_2 is given as

$$q_{C_2} = 750 - q_{C_3} = 750 - 8 \times 20 = 590\ \mu\text{C}$$

Ans. (A)

Sol. 33 Initial energy in capacitor is

$$U_1 = \frac{1}{2}CV_0^2$$

After connection with $\frac{C}{2}$ final potential of system is given as

$$V_f = \frac{CV_0}{C + \frac{C}{2}} = \frac{2V_0}{3}$$

Final energy in system is

$$U_2 = \frac{1}{2}\left(\frac{3C}{2}\right)\left(\frac{2V_0}{3}\right)^2 = \frac{1}{2}CV_0^2$$

Energy loss is given as

$$\Delta U_L = U_1 - U_2 = \left(\frac{1}{2} - \frac{1}{3}\right)CV_0^2 = \frac{1}{6}CV_0^2$$

Ans. (A)

Sol. 34 Initial charges on capacitors are given as

$$q_1 = CV$$

$$q_2 = 2C \times 2V = 4CV$$

When capacitors are connected with opposite polarity, find change on combination is

$$q_f = 4CV - CV = 3CV$$

$+CV$ $-CV$

$-4CV$ $+4CV$

Thus final energy in combination is

$$U_f = \frac{q_f^2}{2(3C)} = \frac{(3CV)^2}{6C} = \frac{3}{2}CV^2$$ **Ans. (D)**

Sol. 35

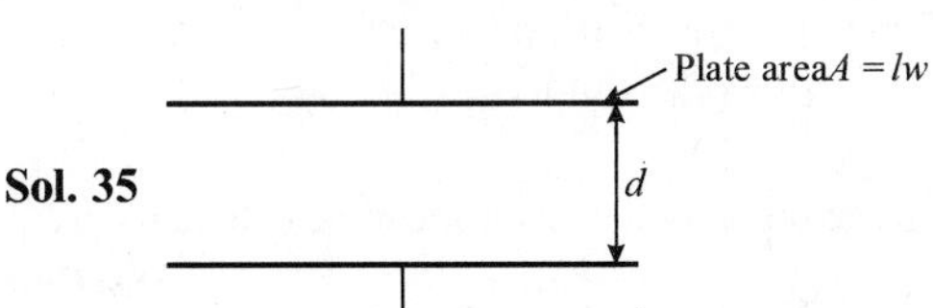

Capacitance of empty capacitor shown above is given as

$$C = \frac{\epsilon_0\, lw}{d}$$

$k = 4$ $(l-x)$ d x

Capacitance of capacitor with dielectric shown above is given as

$$C_f = \frac{\epsilon_0\,(4)xw}{d} + \frac{\epsilon_0\,(l-x)w}{d}$$

For given applied potential difference final energy be two times if capacitance is two times, so we use

$$C_f = 2C$$

$$\frac{4\epsilon_0\, xw}{d} + \frac{3\epsilon_0\, lw}{d} - \frac{\epsilon_0\, xw}{d} = 2\frac{\epsilon_0\, lw}{d}$$

$$\Rightarrow \quad \frac{3\epsilon_0\, xw}{d} = \frac{\epsilon_0\, lw}{d}$$

$$\Rightarrow \quad x = \frac{l}{3}$$ **Ans. (C)**

Sol. 36 Figure below shows the potential distribution on different points of circuit by

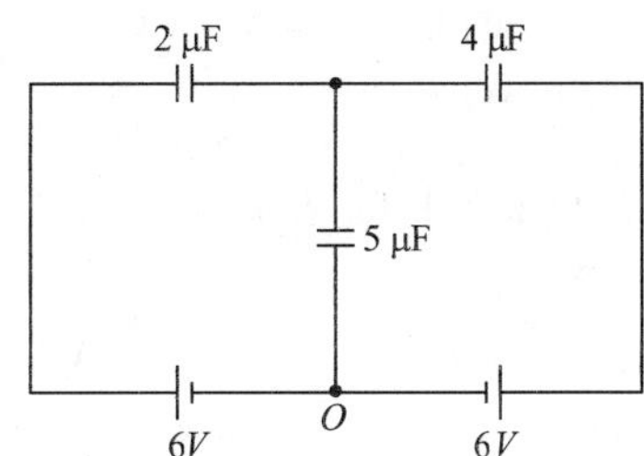

Taking potential of point O at 0

Writing equation for x, we have

$$q_1 + q_2 + q_3 = 0$$

$$2(x-6) + 4(x-6) + 5(x) = 0$$

$$\Rightarrow \quad x = \frac{36}{11}$$

$$\Rightarrow \quad q_3 = \frac{36(5)}{11} = \frac{180}{11}$$

$$q_3 = 16.36\ \mu\text{C}$$ **Ans. (B)**

Sol. 37 For series combination of capacitances, we have

$$\frac{1}{C_{eq_1}} = \frac{1}{C} + \frac{1}{C}$$

$$\Rightarrow \quad C_{eq_1} = \frac{C}{2}$$

For parallel combination, we have

$$C_{eq2} = C + C$$

$$\Rightarrow \quad C_{eq2} = 2C$$

$$\Rightarrow \quad \frac{C_{eq1}}{C_{eq2}} = \frac{(C/2)}{2C} = \frac{1}{4} = 1:4$$

Ans. (B)

Sol. 38 When capacitors are connected in parallel, we have

$$C_{eq} = C_1 + C_2$$

When capacitors are connected in series, we have

$$C'_{eq} = \frac{C_1C_2}{C_1 + C_2}$$

Given that $\quad C_1 + C_2 = \frac{15}{4}\left(\frac{C_1C_2}{C_1 + C_2}\right)$

$$\Rightarrow \quad 4(C_1 + C_2)^2 = 15C_1C_2$$

$$\Rightarrow \quad 4C_1^2 + 4C_2^2 - 7C_1C_2 = 0$$

Dividing all terms of the equation by C_1^2, we have

$$4\left(\frac{C_2}{C_1}\right)^2 - \frac{7C_2}{C_1} + 4 = 0$$

Considering $\frac{C_2}{C_1} = x$, we get

$$4x^2 - 7x + 4 = 0$$

For this quadratic equation, $b^2 - 4ac = 49 - 64 < 0$ thus no solution exits to this equation. **Ans. (None)**

Sol. 39 The capacitance before insertion of slab is given as

$$C_0 = \frac{\epsilon_0 A}{d}$$

After insertion of slab capacitor looks like shown in figure below which can be considered as combination of two capacitors in series, one with dielectric and one with air

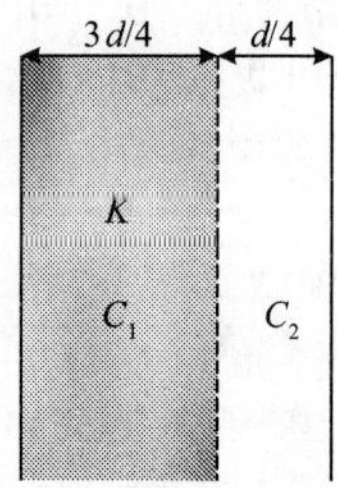

The equivalent capacitance of this system C' is given as

$$\frac{1}{C'} = \frac{1}{C_1} + \frac{1}{C_2}$$

$$\Rightarrow \quad \frac{1}{C'} = \frac{(3d/4)}{\epsilon_0 KA} + \frac{d/4}{\epsilon_0 A}$$

$$\Rightarrow \quad \frac{1}{C'} = \frac{d}{4\epsilon_0 A}\left(\frac{3+K}{K}\right)$$

$$\Rightarrow \quad C' = \frac{4KC_0}{(3+K)}$$

Ans. (C)

Sol. 40 Capacitance of given capacitor partially filled with dielectric is given as

$$C = \frac{\varepsilon_0 A}{\frac{d}{2K} + \frac{d}{2}} = \frac{2\varepsilon_0 A}{\frac{d}{K} + d}$$

$$\Rightarrow \quad C = \frac{2\times 2\varepsilon_0}{\frac{1}{3.2} + 1} = \frac{4\times 3.2}{4.2}\varepsilon_0 = 3.04\,\varepsilon_0$$

Ans. (C)

Sol. 41 The equivalent combination of capacitor from the plates setup can be drawn as shown in figure below

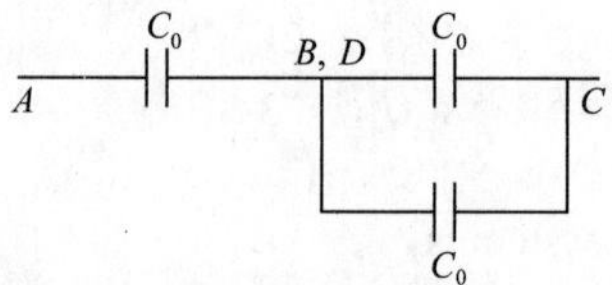

Thus equivalent capacitance is calculated as

$$C_{eq} = \frac{2C_0}{3} = \frac{2}{3}\frac{\epsilon_0 A}{d}$$

$$\Rightarrow \quad C_{eq} = \frac{2\epsilon_0}{3d} \times \left(2\times\frac{3}{2}\right) = 2$$

Ans. (2)

Sol. 42 The initial energy stored in the capacitor is given as

$$U_i = \frac{1}{2} \times 14 \times 12 \times 12\,\text{pJ} = 1008\,\text{pJ}$$

Final energy stored in capacitor after insertion of plate is given as

$$U_f = \frac{1008}{7}\,\text{pJ} = 144\,\text{pJ}$$

The mechanical energy of oscillation can be considered as the difference in energies as it remains there in the plate as kinetic energy at the mean position and as potential energy at extreme position, calculated as

$$\Delta U = 1008 - 144 = 864\,\text{pJ}$$

Ans. (864)

Sol. 43 When first capacitor is charged to 10 V then it will be charged to $C_1V = 20\,\mu\text{C}$ charge which will remain constant once battery is disconnected. When it is connected to other capacitor then same charge is distributed between the two and if final potential difference becomes V', we have

$$20 = (C_1 + C_2)\,V$$

$$\Rightarrow \quad V = 2\,\text{V}$$

Thus final charge on second capacitor can be calculated as

$$Q_2 = C_2V = 16\,\mu\text{C}$$

Ans. (16)

Sol. 44 The capacitance of the dielectric filled part of capacitor is calculated as

$$C_1 = \frac{k \in_0 A}{d}$$

$$\Rightarrow \quad C_1 = \frac{10 \in_0 (100)}{5} = 200 \in_0$$

For the air filled part of capacitor, capacitance is given as

$$C_2 = \frac{\in_0 (100)}{5} = 20 \in_0$$

As C_1 & C_2 are in series so, for equivalent capacitance, we use

$$C_{eq} = \frac{C_1 C_2}{C_1 + C_2} = \frac{4000 \in_0}{220}$$

$$\Rightarrow \quad C_{eq} = 160.9 \times 10^{-12} = 161 \text{ pF}$$ **Ans. (161)**

Sol. 45 To calculate the capacitance, we consider an elemental capacitor of plate separation dx as shown in figure. All such elemental capacitors will be considered in series so the capacitance of given system is calculated as

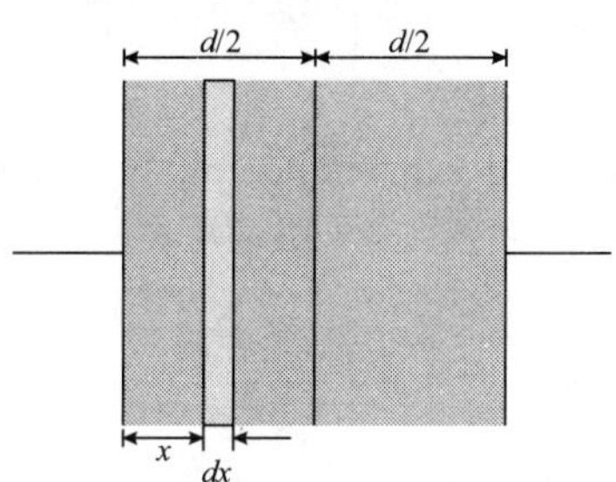

$$\frac{1}{C_{eq}} = \int \frac{1}{dC} = \int \frac{dx}{A\varepsilon}$$

$$\Rightarrow \quad \frac{1}{C_{eq}} = \int_0^{d/2} \frac{dx}{A(\varepsilon_0 + kx)} + \int_{d/2}^{d} \frac{dx}{A[\varepsilon_0 + k(d-x)]}$$

$$\Rightarrow \quad \frac{1}{C_{eq}} = \frac{1}{A}\left\{\frac{1}{k}[\ln(\varepsilon_0 + kx)]_0^{d/2} + \frac{1}{2}[\ln(\varepsilon_0 + k(d-x))]_{d/2}^{d}\right\}$$

$$\Rightarrow \quad \frac{1}{C_{eq}} = \frac{1}{Ak} \times 2\ln\frac{\varepsilon_0 + \frac{kd}{2}}{\varepsilon_0}$$

$$\Rightarrow \quad C_{eq} = \frac{kA}{2ln\left(\frac{2\varepsilon_0 + kx}{2\varepsilon_0}\right)}$$ **Ans. (A)**

Sol. 46 Bound charge on capacitor plate can be directly given by the expression

$$q_b = q_f\left(1 - \frac{1}{K}\right)$$ **Ans. (D)**

Sol. 47 The given figure can be considered as three capacitors in series combination as shown in figure below

●—||—||—||—●

The capacitances of the three capacitors can be given as

$$C_1 = \frac{A\varepsilon_0 K}{d}; C_2 = \frac{3KA\varepsilon_0}{2d}; C_3 = \frac{5KA\varepsilon_0}{3d}$$

Equivalent capacitance of the given combination is calculated as

$$\frac{1}{C_{eq}} = \frac{1}{C_1} + \frac{1}{C_2} + \frac{1}{C_3}$$

$$\Rightarrow \quad \frac{1}{C_{eq}} = \frac{d}{A\varepsilon_0 K}\left(1 + \frac{2}{3} + \frac{3}{5}\right) = \frac{34}{15}\frac{d}{A\varepsilon_0 K}$$

$$\Rightarrow \quad C_{eq} = \frac{15}{34}\frac{A\varepsilon_0 K}{d}$$ **Ans. (D)**

Sol. 48 After removing the battery, the combination is shown as

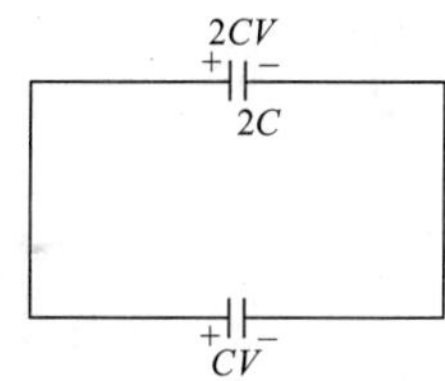

If after insertion of dielectric in capacitor of capacitance C, the voltage across the combination becomes V_1 then by conservation of charge, we have

$$2CV + CV = 2C(V_1 + Ck\,V_1)$$

$$\Rightarrow \quad V_1 = \frac{3CV}{2C + CK}$$

$$\Rightarrow \quad V_1 = \frac{3V}{2 + K}$$ **Ans. (B)**

Sol. 49 During charging capacitor charge is given as

$$V = V_0 e^{-\frac{t}{RC}}$$

$$\Rightarrow \quad 50 = 100e^{-\frac{t}{RC}}$$

$$\Rightarrow \quad t = RC \ln 2$$

$$\Rightarrow \quad t = 100 \times 10^{-6} \ln 2$$

$$\Rightarrow \quad t = 10^{-4} \times 0.693 \text{ sec} = 69.3 \,\mu\text{ sec}$$ **Ans. (B)**

Sol. 50 Considering one part of capacitor with dielectric as C_2 and remaining with pendulum as C_1, the equivalent circuit is drawn as shown below

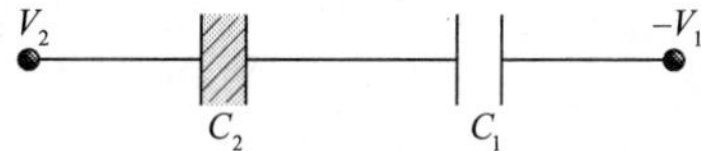

The potential difference across the capacitors is given as

$$V = V_2 - (-V_1) = V_1 + V_2$$

Charge on each capacitor can be given as

$$q = \frac{C_1C_2}{C_1+C_2}(V_1+V_2)$$

Potential difference across C_1 can be calculated as

$$V_1 = \frac{Q}{C_1} = \frac{C_2(V_1+V_2)}{(C_1+C_2)}$$

The electric field between plates of C_1 is given as

$$E = \frac{V_1}{(d-t)}$$

$$\Rightarrow \qquad E = \frac{C_2(V_1+V_2)}{(C_1+C_2)(d-t)}$$

For equilibrium of pendulum bob, its free body diagram is shown in figure below

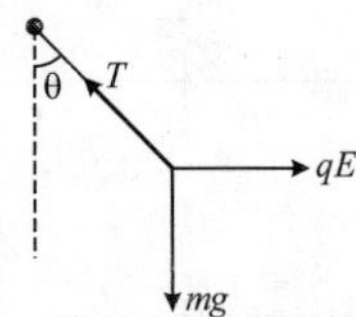

Force on pendulum bob due to electric field is given as

$$F = Eq = \frac{qC_2}{(C_1+C_2)}\frac{(V_1+V_2)}{(d-t)}$$

The equilibrium angle of string is given as

$$\theta = \tan^{-1}\left(\frac{qE}{mg}\right)$$

$$\Rightarrow \qquad \theta = \tan^{-1}\left[\frac{q}{mg}\times\frac{C_2(V_1+V_2)}{(C_1+C_2)(d-t)}\right]$$ **Ans. (C)**

Sol. 51 For a capacitor with leaky dielectric we use

$$RC = \rho K \epsilon_0$$

Resistance of the leaky dielectric is given as

$$R = \frac{\rho \epsilon_0 K}{C}$$

Leakage current in circuit is calculated as

$$I = \frac{V}{R} = \frac{V}{\rho}\frac{C}{\epsilon_0 K}$$

$$\Rightarrow \qquad I = \frac{40}{200}\times\frac{2\times10^{-12}}{8.85\times10^{-12}\times50} = 0.9\,\text{mA}$$ **Ans. (C)**

Sol. 52 The given capacitor can be drawn as equivalent circuit shown below considering it as a combination of three capacitors.

$$C_1 = \frac{(A/2)\varepsilon_0}{d}, \quad C_2 = \frac{K_1(A/2)\varepsilon_0}{d/2} = \frac{k_1A\varepsilon_0}{d}, \quad C_3 = \frac{K_2(A/2)\varepsilon_0}{d/2} = \frac{k_2A\varepsilon_0}{d}$$

The equivalent capacitance of this system is given as

$$C_{eq} = C_1 + \frac{C_3C_3}{C_2+C_3}$$

$$\Rightarrow \qquad C_{eq} = \frac{A\varepsilon_0}{2d} + \frac{\frac{K_1A\varepsilon_0}{d}\times\frac{K_2A\varepsilon_0}{d}}{\frac{K_1A\varepsilon_0}{d}+\frac{K_2A\varepsilon_0}{d}}$$

$$\Rightarrow \qquad C_{eq} = \frac{A\varepsilon_0}{2d} + \frac{A\varepsilon_0}{d}\left(\frac{K_1K_2}{K_1+K_2}\right)$$

$$\Rightarrow \qquad C_{eq} = \frac{A\varepsilon_0}{d}\left(\frac{1}{2}+\frac{K_1K_2}{K_1+K_2}\right)$$ **Ans. (C)**

Sol. 53 Equivalent capacitance of the given system is calculated as

$$C_{eq} = C_1 + \frac{C_2C_3}{C_2+C_3}$$

$$\Rightarrow \qquad C_{eq} = 2 + \frac{6\times12}{6+12} = 6\,\mu\text{F}$$

As C_2 and C_3 are in series, they have same charges and the ratio of charges on C_1 and C_2 are given as

$$\frac{Q_{C_1}}{Q_{C_2}} = \frac{C_1V}{\frac{C_2C_3}{C_2+C_3}V} = \frac{2\times(6+12)}{6\times12} = \frac{1}{2}$$ **Ans. (D)**

Sol. 54 Initially energy stored in capacitor is given as

$$E_i = \frac{1}{2}CV^2 = \frac{1}{2}\times200\times10^{-6}\times(200)^2\,\text{J}$$

After dielectric insertion, capacitance becomes

$$C' = kC = 400\,\mu\text{F}$$

Final energy stored in capacitor is given as

$$E_f = \frac{1}{2}C'V^2 = \frac{1}{2}\times400\times10^{-6}\times(200)^2\,\text{J}$$

Change in energy stored in capacitor is given as

$$\Delta E = \frac{1}{2}\times(400-200)\times10^{-6}\times4\times10^4 = 4\text{J}$$

Ans. (4)

17.1 Current Electricity

Sol. 1

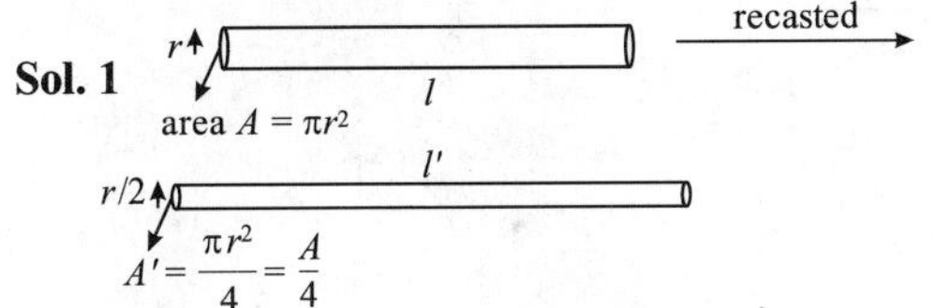

$$R = \rho\frac{l}{A} = 100 \qquad \ldots(1)$$

As volume of material remain same

$$\Rightarrow \quad Al = A'l'$$

$$\Rightarrow \quad l' = 4l$$

New resistance of wire will be

$$R' = \rho\frac{l'}{A'} = 16\rho\frac{l}{A}$$

$$\Rightarrow \quad R' = 1600\,\Omega$$

Ans. (A)

Sol. 2

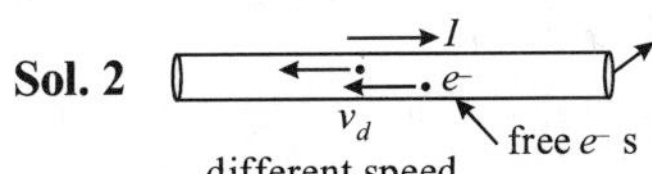

The drift speed of free e^-s (charge carries) is related to current density as

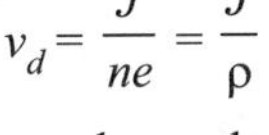

$$v_d = \frac{J}{ne} = \frac{J}{\rho}$$

(As in this case volume charge density (C/m^3) is $\rho = ne$)

$$\Rightarrow \quad v_d = \frac{I}{\rho A} \quad [\text{as } J = \frac{I}{A}]$$

Time the charges take in traveling a distance d is

$$t = \frac{d}{v_d} = \frac{\rho A d}{I}$$

Ans. (C)

Sol. 3 We use $R_T = R_0(1 + \alpha T)$

at room temperature

$$R_T = 100\,\Omega$$

$$\Rightarrow \quad 100 = R_0(1 + \alpha T_R) \qquad \ldots(1)$$

at temperature $T_R + 500°\text{C}$, $R'_T = \frac{220}{2} = 110\,\Omega$

$$\Rightarrow \quad 110 = R_0(1 + \alpha(T_R + 500)) \qquad \ldots(2)$$

$$(2) \div (1) \quad \Rightarrow \quad \frac{110}{100} = \frac{1 + \alpha(T_R + 500)}{1 + \alpha T_R}$$

$$= [1 + \alpha(T_R + 500)]\,(1 + \alpha T_R)^{-1}$$

As α is very small we neglect terms of α^2 which gives

$$\frac{110}{100} = 1 + 500\,\alpha \qquad [RT_1 = RT_0(1 + \alpha\Delta T]$$

$$\Rightarrow \quad \alpha = \frac{10}{500 \times 100}$$

$$= 2 \times 10^{-4}\,°\text{C}^{-1}$$

Ans. (C)

Sol. 4

$$\Rightarrow \quad R = 27 \times 10^3\,\Omega \pm 10\%$$

$$\Rightarrow \quad R = 27\,\text{K}\Omega \pm 10\%$$

Ans. (D)

Sol. 5 Current in wire is given as

$$I = neAv_d$$

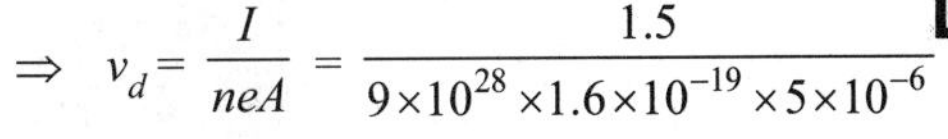

$$\Rightarrow \quad v_d = \frac{I}{neA} = \frac{1.5}{9 \times 10^{28} \times 1.6 \times 10^{-19} \times 5 \times 10^{-6}}$$

$$\Rightarrow \quad v_d = 0.02 \times 10^{-3}\,\text{m/s}$$

Ans. (D)

Sol. 6

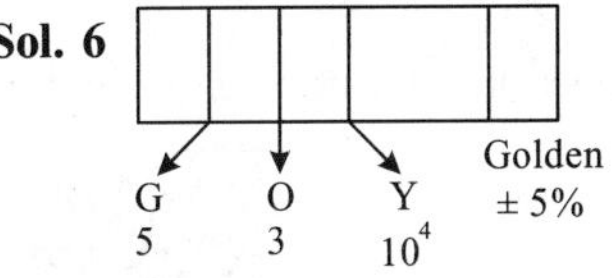

$$\Rightarrow \quad R = 53 \times 10^4 \pm 5\% = 530\,\text{k}\Omega \pm 5\%$$

Ans. (B)

Sol. 7 $P = i^2R$

For i_{max}, R must be minimum

From color coding

$$R = 50 \times 10^2\,\Omega$$

Thus we use $2 = i^2_{max} \times 50 \times 10^2$

$$\Rightarrow \quad i^2_{max} = \frac{2}{5000} = 4 \times 10^{-4}\,\text{A}$$

$$\Rightarrow \quad i_{max} = 20\,\text{mA}$$

Ans. (D)

Sol. 8 When red is replaced with green 1st digit changes to 5 so new resistance will be 500 Ω **Ans. (C)**

Sol. 9 $R = 3\Omega = \frac{\rho l}{A}$

If $l' \to 2l \quad \Rightarrow \quad A' = \frac{A}{2}$ (As volume is constant)

New resistance of wire will be

$$R' = \frac{\rho l'}{A'} = \frac{\rho(2l)}{A/4} = 4\frac{\rho l}{A} = 4R = 4 \times 3 = 12\Omega$$

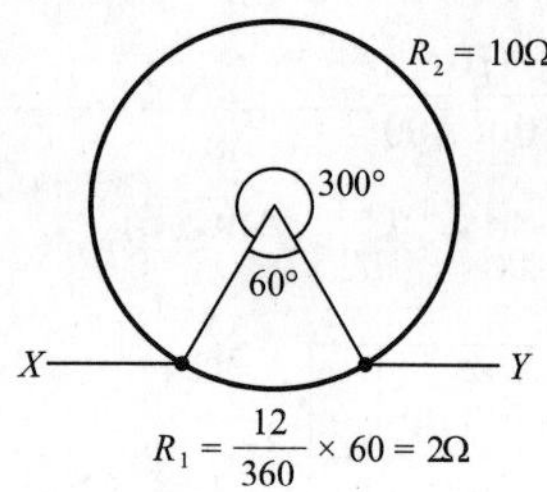

Here $R_{XY} = \frac{R_1 R_2}{R_1 + R_2} = \frac{10 \times 2}{10 + 2}$

$\Rightarrow$ $R_{XY} = \frac{20}{12}\Omega = \frac{5}{3}\Omega$ **Ans. (B)**

Sol. 10 Resistivity of a conductor is given as

$$\rho = \frac{m}{ne^2\tau}$$

$$= \frac{9.1\times10^{-31}}{8.5\times10^{28}\times(1.6\times10^{-19})\times25\times10^{-15}}$$

$\Rightarrow$ $\rho \cong 1.67 \times 10^{-8}\,\Omega\text{-m}$ 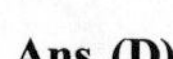 **Ans. (D)**

Sol. 11 The mobility of free electrons in conductor is given as

$$\mu = \frac{v_d}{E}$$

and $E = \rho J$

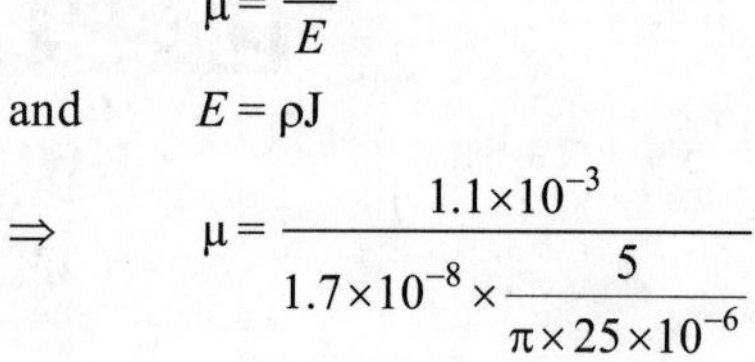

$\Rightarrow$ $\mu = \frac{1.1\times10^{-3}}{1.7\times10^{-8}\times\frac{5}{\pi\times25\times10^{-6}}}$

$\Rightarrow$ $\mu = \frac{1.1\times10^{-3}\times\pi\times25\times10^{-6}}{1.7\times10^{-8}\times5} \approx 1.01\ \text{m}^2/\text{Vs}$ **Ans. (D)**

Sol. 12

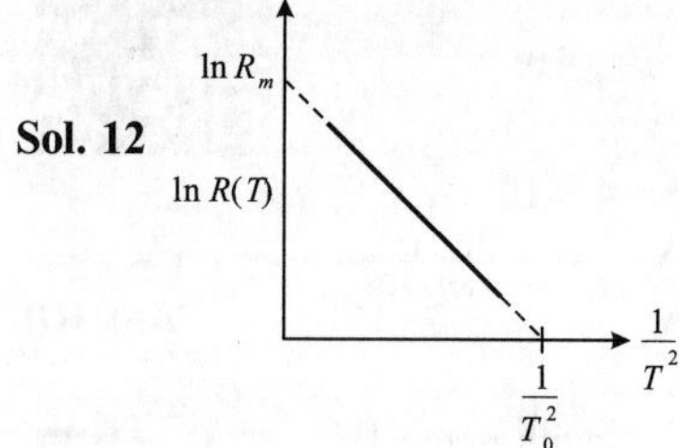

Equation of straight line

$$\ln R(\text{T}) = -\frac{\ln R_m}{(1/T_1^2)} \times \frac{1}{T^2} + \ln R_m$$

$\Rightarrow$ $\ln R(T) = \ln R_m\left(1 - \frac{T_1^2}{T^2}\right)$

$\Rightarrow$ $\ln\left(\frac{R(T)}{R_m}\right) = -\ln \text{R}_\text{m}\left(\frac{T_1^2}{T^2}\right)$

$\Rightarrow$ $R(T) = R_m \cdot e^{-T_1^2/T^2(\ln R_m)} = R_0\, e^{-T_0^2/T^2}$ **Ans. (C)**

Sol. 13 Elemental resistance of shell is

$$dR = \rho.\ \frac{dx}{4\pi x^2}$$

$$dR = \rho \cdot \frac{dx}{4\pi x^2}$$

$$R = \int dR = \frac{\rho}{4\pi}\int_a^b \frac{dx}{x^2}$$

$\Rightarrow$ $R = \frac{\rho}{4\pi}\left[-\frac{1}{x}\right]_a^b$

$$= \frac{\rho}{4\pi}\left[\frac{1}{a} - \frac{1}{b}\right]$$

$$= \frac{\rho(b-a)}{4\pi ab}$$ **Ans. (B)**

Sol. 14 Resistivity of different commonly used metals are given in increasing order below

Silver
Copper
Gold
Aluminium
Magnetism
Tungsten
Zinc
Nickel
Iron
Lead
Mercury
Germanium
Silicon

$\Rightarrow$ $\rho_M > \rho_A > \rho_C$

Ans. (D)

Sol. 15 For the given expression of current, we use

$$i = 20t + 8t^2 = \frac{dq}{dt}$$

Total charge crossed through the wire can be calculated as

$$q = \int idt$$

$\Rightarrow$ $q = \int_0^{15} (20t + 8t^2)dt$

$\Rightarrow$ $q = \left(\frac{20t^2}{2} + \frac{8t^3}{3}\right)_0^{15}$

$\Rightarrow$ $q = 10 \times (15)^2 + \frac{8(15)^3}{3}$

$\Rightarrow$ $q = 2250 + 9000 = 11250\,\text{C}$ **Ans. (D)**

Sol. 16 The electric field in the wire is given as

$$E = 10 \times 10^{-3}\ \frac{V}{m}$$

Current density in wire can be calculated as

$$J = \sigma E = 10 \times 10^{-3} \times 5 \times 10^{7}\ \text{A/m}^2$$

$$J = 5 \times 10^{5}\ \text{A/m}^2$$

If current in wire is i, it can be used as

$$\frac{i}{A} = 5 \times 10^{5}$$

$$\Rightarrow \quad i = 5 \times 10^{5} \times \pi r^2 = 5 \times 10^{5} \times \pi \times (5 \times 10^{-4})^2$$

$$\Rightarrow \quad i = 125\pi \times 10^{-3}\ \text{A} = 125\ \pi\ mA$$

$$\Rightarrow \quad x = 5$$

Ans. (5)

Sol. 17 As described in the question, when length of wire is doubled and cross section area is halved, we have new resistance given as

$$R' = \frac{\rho(2l)}{(A/2)} = \frac{4\rho l}{A}$$

Circuit current is given as

$$i = \frac{V}{R} = \frac{VA}{4\rho l}$$

Ans. (A)

Sol. 18 The effective circuit is shown below

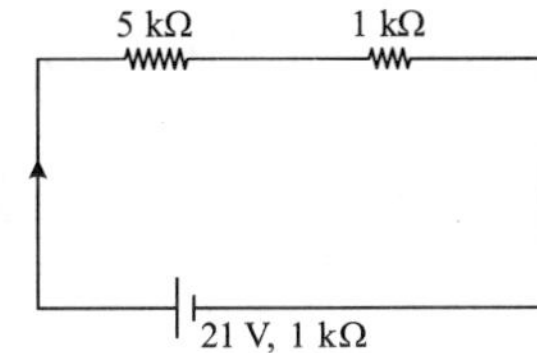

Current in circuit is given as

$$I = \frac{21}{5+1+1} = 3\ \text{mA}$$

Ans. (3)

Sol. 19 Current in wire is given as

$$i = neAv_d$$

$$\Rightarrow \quad 10 = n \times 1.6 \times 10^{-19} \times 5 \times 10^{-6} \times 2 \times 10^{-3}$$

$$\Rightarrow \quad n = 0.625 \times 10^{28} = 625 \times 10^{25}$$

Ans. (B)

Sol. 20 Resistance of conductor is given as

$$R = \frac{\rho l}{A} = \frac{V}{I}$$

$$\Rightarrow \quad \rho = \frac{AV}{Il} = \frac{\pi d^2 V}{4Il} \qquad \left(A = \frac{\pi d^2}{4}\right)$$

Maximum fractional error in resistivity calculation is given as

$$\frac{\Delta\rho}{\rho} = \frac{2\Delta d}{d} + \frac{\Delta V}{V} + \frac{\Delta I}{I} + \frac{\Delta l}{l}$$

$$\Rightarrow \quad \frac{\Delta\rho}{\rho} = 2\left(\frac{0.01}{5.00}\right) + \frac{0.1}{5.0} + \frac{0.01}{2.00} + \frac{0.1}{10.0}$$

$$\Rightarrow \quad \frac{\Delta\rho}{\rho} = 0.004 + 0.02 + 0.005 + 0.01$$

$$\Rightarrow \quad \frac{\Delta\rho}{\rho} = 0.039$$

Maximum percentage error can be given as

$$\%\ \text{error} = \frac{\Delta\rho}{\rho} \times 100 = 0.039 \times 100 = 3.90\%$$

Ans. (A)

Sol. 21 Current passing through the wire can be given as

$$i = JA\cos\theta = \frac{JA}{2}$$

$$\Rightarrow \quad J = \frac{2i}{A}$$

If electric field in wire is E, we use

$$E = \rho J = \frac{2\rho i}{A} = \frac{2 \times 44 \times 10^{-8} \times 5}{4 \times 10^{-2}} = 11 \times 10^{-5}\ \text{V/m}$$

Ans. (B)

Sol. 22 Current in the conductor is given as

$$i = nAev_d$$

$\Rightarrow$ v_d increases as area A decrease

Ans. (A)

Sol. 23 From the given figure voltage at t = 3.2 s can be calculated by linear function as 6 V. So we use, from the below graph

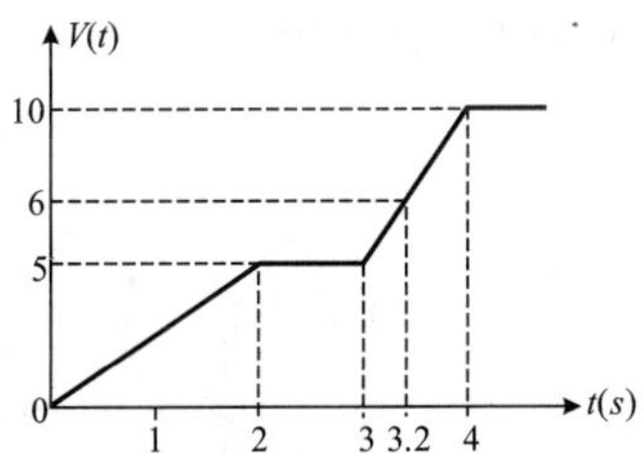

$$\frac{\Delta V}{(3.2-3)} = \frac{5}{1}$$

$$\Rightarrow \quad \Delta V = 1\ \text{V}$$

Thus at $t = 3.2$ s, $V(t) = 5 + 1 = 6$ V

Thus circuit current at this instant is given as

$$I = \frac{6-5}{R} = 1\ \text{A}$$

Ans. (01)

Sol. 24 For parallel combination of wires, we have

$$\frac{R_1 R_2}{R_1 + R_2} = 3\Omega$$

$$\Rightarrow \quad \frac{\rho_1\rho_2}{\rho_1+\rho_2}\times\frac{l}{A}=3\Omega$$

$$\Rightarrow \quad l=3\times A\times\frac{(\rho_1+\rho_2)}{\rho_1\rho_2}\times\frac{3\times\pi(10^{-3})^2\times 63\times10^{-8}}{12\times51\times10^{-16}};$$

$$\Rightarrow \quad l=97\,\text{m}$$ **Ans. (B)**

Sol. 25 From the colour coading of bands on resistor, we can get the below value of color coded resistor given in question.

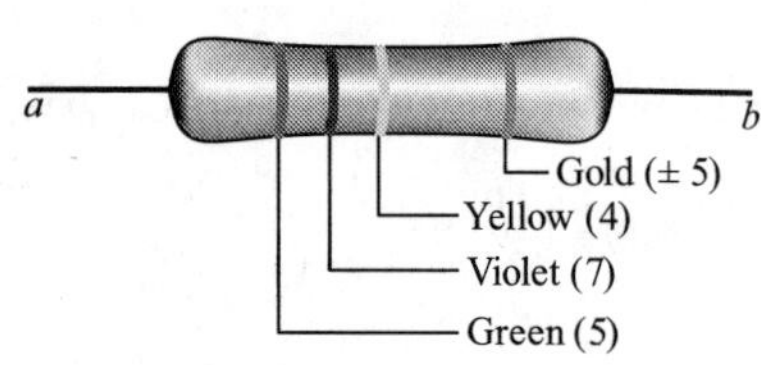

$$R=(7500\pm375)\Omega$$ **Ans. (A)**

17.2 Circuit Analysis

Sol. 1

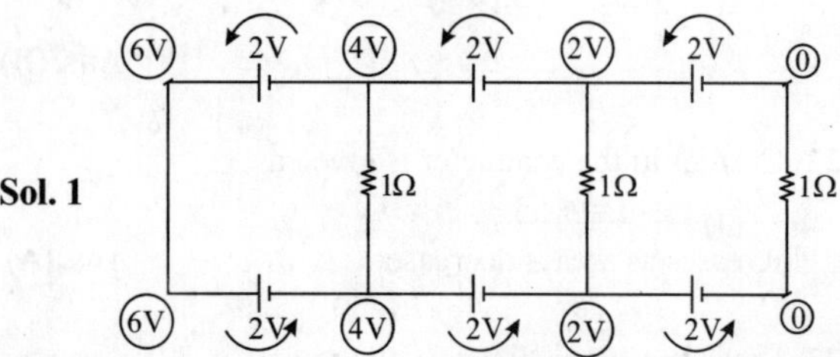

As shown in figure, after distribution of potentials in circuit, potential difference across each resistance is equal to zero.

$$\Rightarrow \quad I_{\text{each }1\Omega}=\frac{V}{R}=0$$ **Ans. (B)**

Sol. 2

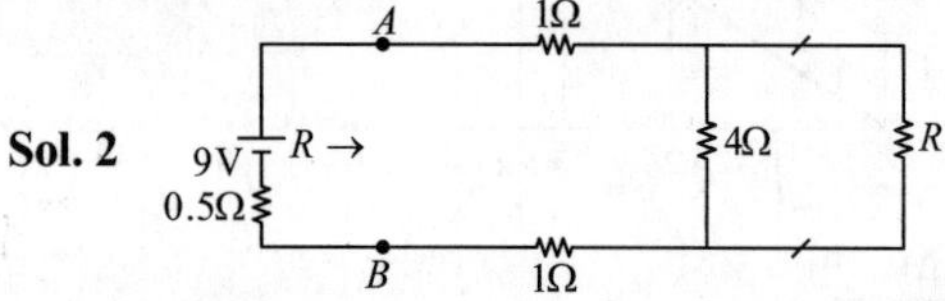

Here R is the equivalent Resistance across terminals A and B

For ∞ grid we use

$$\text{Across } AB \rightarrow R=2+\frac{4R}{4+R}$$

$$\Rightarrow \quad 4R+R^2=8+6R$$

$$\Rightarrow \quad R^2-2R-8=0$$

$$\Rightarrow \quad R^2-2R-8=0$$

$$\Rightarrow \quad R=\frac{2\pm\sqrt{4+32}}{2}=\frac{2\pm6}{2}$$

$$\Rightarrow \quad R=\frac{8}{2}=4\,\Omega$$

$\Rightarrow$

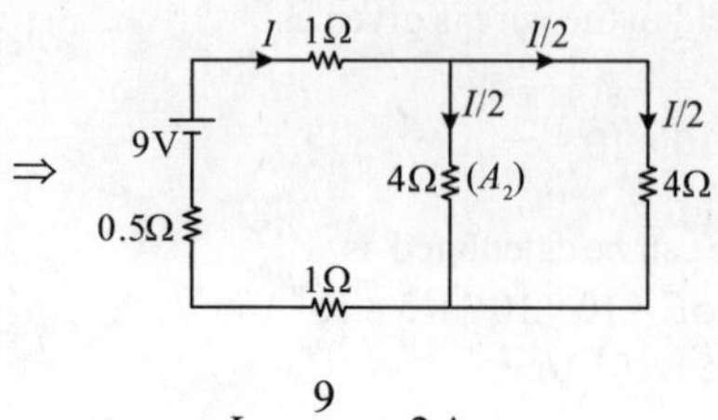

$$I=\frac{9}{4.5}=2\,\text{A}$$

Reading of (A_2) is

$$I_2=\frac{2}{2}=1\,\text{A}$$

In further part of grid current will divide half in 4 Ω and half to next section

$\Rightarrow$ Reading of (A_3) is

$$\frac{I}{4}=\frac{2}{4}=0.5\,\text{A}$$

Reading of voltmeter (V) is

$$=IR=2\times4=8\,\text{V}$$

Here current supplied by battery is

$$I=\frac{9}{0.5+4}=\frac{9}{4.5}=2\,\text{A (Reading of } A_1)$$ **Ans. (A)**

Sol. 3

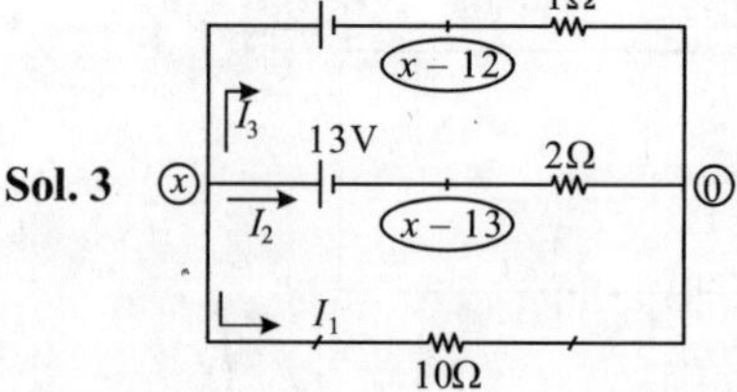

Distributing potentials at junctions of circuit as shown and using KCL for junction with potential x, it gives

$$I_1+I_2+I_3=0$$

$$\Rightarrow \quad \frac{x-0}{10}+\frac{x-13}{2}+\frac{x-12}{1}=0$$

$$\Rightarrow \quad x+5x-65+10x-120=0$$

$$\Rightarrow \quad 16x=185$$

$$\Rightarrow \quad x=11.56\,\text{V}$$ **Ans. (B)**

Sol. 4 Equivalent circuit of above situation is

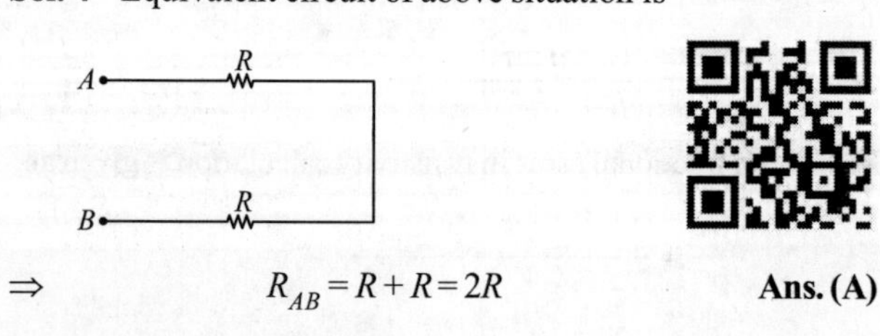

$$\Rightarrow \quad R_{AB}=R+R=2R$$ **Ans. (A)**

Sol. 5

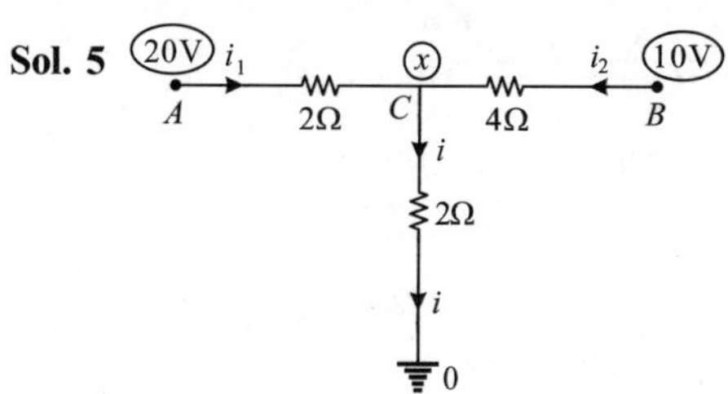

Using KCL equation at C in above circuit, we have

$$\frac{20-x}{2}+\frac{10-x}{4}=\frac{x-0}{2}$$

$\Rightarrow$ $$x=10$$

$\Rightarrow$ $$i=\frac{x}{2}=\frac{10}{2}=5\text{ A}$$ **Ans. (B)**

Sol. 6

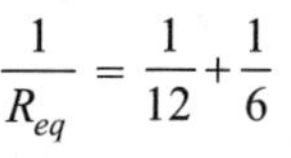

R_{eq} between any two vertex will be

$$\frac{1}{R_{eq}}=\frac{1}{12}+\frac{1}{6}$$

$\Rightarrow$ $$R_{eq}=4\ \Omega$$ **Ans. (C)**

Sol. 7 From the given circuit, we can write

$$i_1=\frac{10}{20}=0.5\text{ A}$$

$$i_2=\frac{0}{20}=0\text{ A}$$

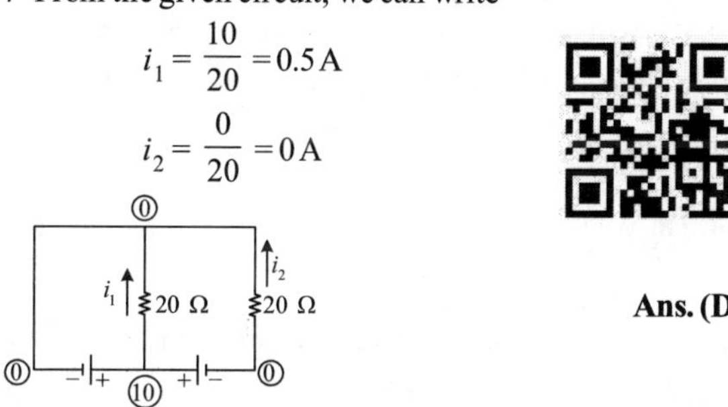

Ans. (D)

Sol. 8 Potential difference across AB will be equal to battery equivalent across CD, given as

$$V_{AB}=V_{CD}=\frac{\frac{E_1}{r_1}+\frac{E_2}{r_2}+\frac{E_3}{r_3}}{\frac{1}{r_1}+\frac{1}{r_2}+\frac{1}{r_3}}=\frac{\frac{1}{1}+\frac{2}{1}+\frac{3}{1}}{\frac{1}{1}+\frac{1}{1}+\frac{1}{1}}$$

$$V_{AB}=\frac{6}{3}=2\text{ V}$$ **Ans. (D)**

Sol. 9 Using KCL, we can write

$$I_3=0.8-0.4=0.4\text{ A}$$

$$I_2=0.4+0.4+0.3$$

$\Rightarrow$ $$I_2=1.1\text{ A}$$

$$I_6=0.4\text{ A}$$ **Ans. (A)**

Sol. 10 Here we consider the three batteries in parallel combination so the potential difference across a and b is given as

$$E_{eq}=\frac{\frac{E_1}{2R_1}+\frac{E_2}{R_2}+\frac{E_3}{2R_1}}{\frac{1}{2R_1}+\frac{1}{R_2}+\frac{1}{2R_1}}$$

$\Rightarrow$ $$E_{eq}=\frac{\frac{2}{2}+\frac{4}{2}+\frac{4}{2}}{\frac{1}{2}+\frac{1}{2}+\frac{1}{2}}$$

$\Rightarrow$ $$E_{eq}=\frac{5}{\frac{3}{2}}=\frac{10}{3}=3.3$$ **Ans. (B)**

Sol. 11 The circuit given is reduced as shown below

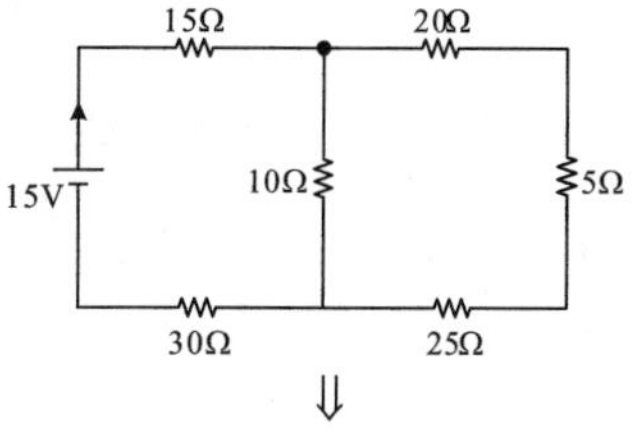

$\Downarrow$

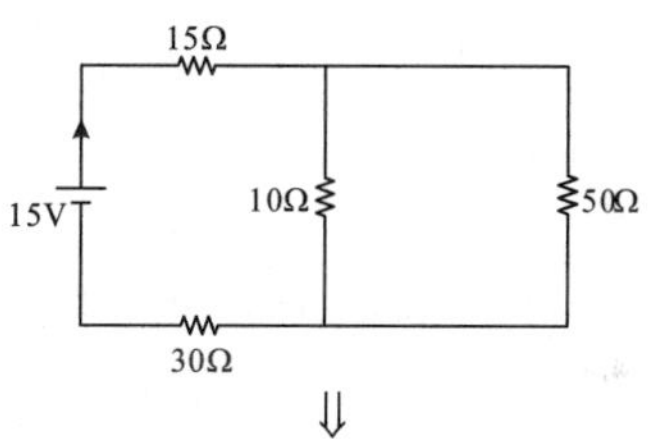

$\Downarrow$

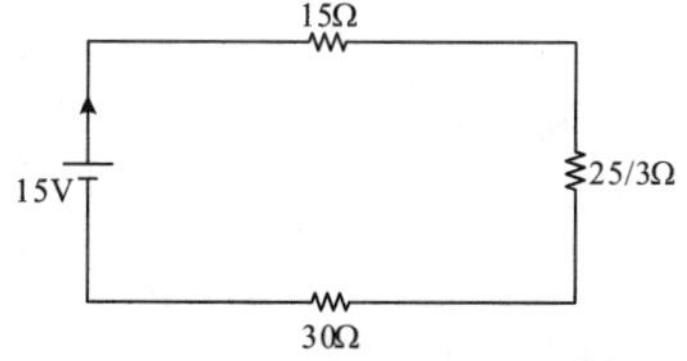

Equivalent resistance across battery is given as

$$R_{eq}=15+\frac{25}{3}+30=\frac{45+25+90}{3}=\frac{160}{3}\Omega$$

Thus current supplied by battery is given as

$$I=\frac{E}{R_{eq}}=\frac{15\times 3}{160}=\frac{9}{32}\text{A}$$ **Ans. (C)**

Sol. 12 The equivalent circuit of the given square is redrawn as shown below

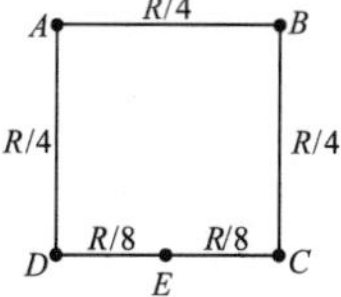

The equivalent resistance across E and C is given as

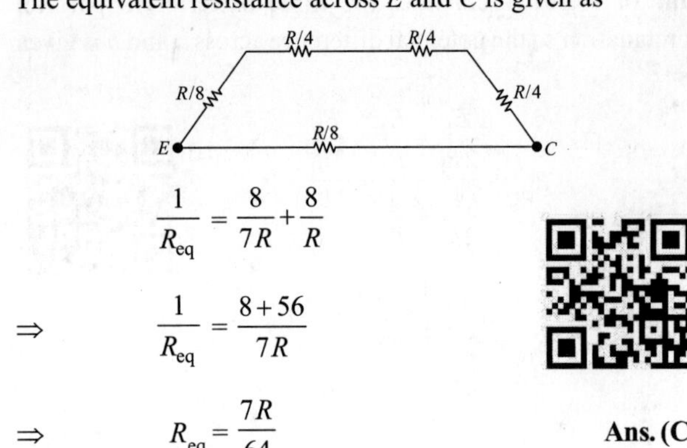

$$\frac{1}{R_{eq}} = \frac{8}{7R} + \frac{8}{R}$$

$$\Rightarrow \quad \frac{1}{R_{eq}} = \frac{8+56}{7R}$$

$$\Rightarrow \quad R_{eq} = \frac{7R}{64}$$

Ans. (C)

Sol. 13 Equivalent resistance of upper branch of circuit is 2.5 Ω so total current in upper branch is 1/2.5 = 0.4A which divides equally in two 1Ω resistances so through each current is 0.2 A.

Ans. (B)

Sol. 14 Figure below shows the circuit for the described situation

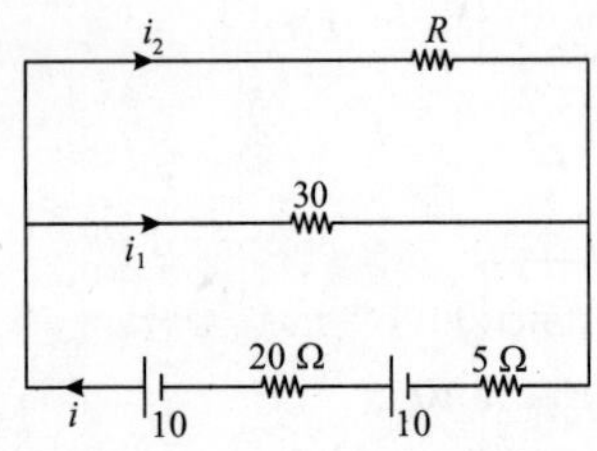

In above circuit, we use potential difference across batteries

$$E_1 = E - ir$$

$$\Rightarrow \quad E_1 = E - ir$$

$$\Rightarrow \quad i = 0.5\,\text{A}$$

and $\quad E_2 = E - ir$

$$\Rightarrow \quad E_2 = 10 - 0.5 \times 5$$

$$\Rightarrow \quad E_2 = 7.5\,\text{V}$$

Using KCL at right most junction, we have

$$0.5 = \frac{7.5}{R} + \frac{7.5}{30}$$

$$R = 30\,\Omega$$

Ans. (30)

Sol. 15 Considering potentials as shown in circuit below we write *KLL* equation for x

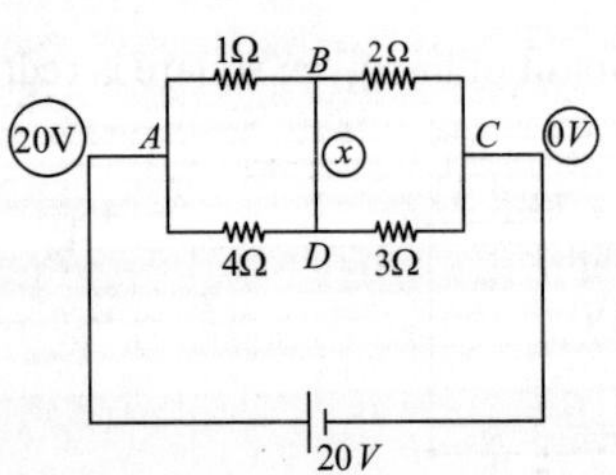

$$\frac{x}{2} + \frac{x}{3} + \frac{x-20}{1} + \frac{x-20}{4} = 0$$

$$\Rightarrow \quad 6x + 4x + 12x - 240 + 3x - 60 = 0$$

$$\Rightarrow \quad 25x = 300$$

$$\Rightarrow \quad x = 12\text{V}$$

Thus current in 1Ω and 2Ω resistance are

$$I_{1\Omega} = \frac{20-12}{1} = 8A$$

and $\quad I_{2\Omega} = \frac{12}{2} = 6A$

Current in branch *BD* is

$$I_{BD} = I_{1\Omega} - I_{2\Omega} = 2A$$

Ans. (B)

Sol. 16 Current in branch of 8Ω resistance is given as

$$i = \frac{V}{R} = \frac{8}{8} = 1\,\text{A}$$

Ans. (D)

Sol. 17 Current in upper & power branches are given as

$$i_1 = \frac{40}{100} = 0.4\,\text{A}$$

and $\quad i_2 = \frac{40}{200} = 0.2\,\text{A}$

We write equation of potential drop from B to D as

$$V_B - 60 \times 0.4 + 110 \times 0.2 = V_D$$

$$\Rightarrow \quad V_B - V_D = 24 - 22 = 2\,\text{V}$$

Ans. (2)

Sol. 18 Writing equation of potential drop A to B, we have,

$$V_A + 1 + 1 \times 2 - 2 = V_B$$

(Current in 2 Ω resistance is 1 A downward)

$$\Rightarrow \quad V_B - V_A = 1\,\text{V}$$

Ans. (A)

Sol. 19 By principle of superposition current in 4Ω resistance is given as

$$I_{10V} = I_{4\Omega} = I_{20V} - I_{10V}$$

$$\Rightarrow \quad I_{10V} = \frac{20}{7 + \frac{20}{7}} \times \frac{10}{14} - \frac{10}{4 + \frac{70}{17}}$$

$$\Rightarrow \quad I_{10V} = \frac{140}{69} \times \frac{10}{14} - \frac{170}{138}$$

$$\Rightarrow \quad I_{10V} = \frac{100}{69} - \frac{170}{138} = \frac{30}{138} \simeq 0.22\,\text{A}$$

Ans. (C)

NOTE : This question can also be solved using KCL or KVL or by Thevenin's analysis. Students are advised to solve this question by all these methods for conceptual clarity.

Sol. 20 The current in circuit is given as

$$I=\frac{6-4}{10}=\frac{1}{5}A$$

By KVL equation in given circuit, we have

$$V_x+4+8\times\frac{1}{5}-V_y=0$$

$$\Rightarrow \qquad V_y-V_x=5.6 \qquad \textbf{Ans. (C)}$$

Sol. 21 For the given parallel combination of resistances, current is divided in inverse ratio of resistance, so we have

$$i_{PQ}=\frac{2}{6}\times 6\text{A} \qquad \textbf{Ans. (2)}$$

Sol. 22 This circuit diagram given in question can also be drawn as shown in figure below as a wheat-stone bridge

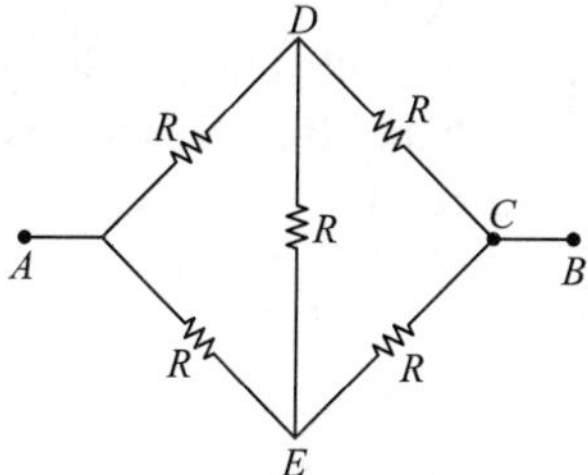

The equivalent resistance of this balanced wheat-stone bridge can be directly given as

$$R_{eq}=\frac{2R\times 2R}{2R+2R}=R \qquad \textbf{Ans. (C)}$$

Sol. 23 Work done by battery in passing a given amount of charge is given as

$$W=Q(\Delta V)=20\times 15=300\text{J} \qquad \textbf{Ans. (300)}$$

Sol. 24 The equivelent resistance of the combinations as described in question are taken as

$$R_1+R_2=s \qquad \ldots(1)$$

$$\frac{R_1R_2}{R_1+R_2}=p \qquad \ldots(2)$$

$$\Rightarrow \qquad R_1R_2=sp$$

$$\Rightarrow \qquad R_1R_2=np^2$$

$$\Rightarrow \qquad R_1+R_2=\frac{nR_1R_2}{(R_1+R_2)}$$

$$\Rightarrow \qquad \frac{(R_1+R_2)^2}{R_1R_2}=n$$

For minimum value of n happens when

$$R_1=R_2=R$$

$$\Rightarrow \qquad n=\frac{(2R)^2}{R^2}=4 \qquad \textbf{Ans. (4)}$$

Sol. 25 The current in circuit is calculated as

$$i=\frac{3E}{R+r_1+r_2}$$

Potential difference across first cell can be calculated as

$$V=2E-ir_1=0$$

$$\Rightarrow \qquad 2E=ir_1$$

$$\Rightarrow \qquad 2E=\frac{3E\times r_1}{R+r_1+r_2}$$

$$\Rightarrow \qquad 2R+2r_1+2r_2=3r_1$$

$$\Rightarrow \qquad R=\frac{r_1}{2}-r_2 \qquad \textbf{Ans. (B)}$$

Sol. 26 To calculate the current in circuit, we distribute the potentials at different junctions of circuit as shown in below figure.

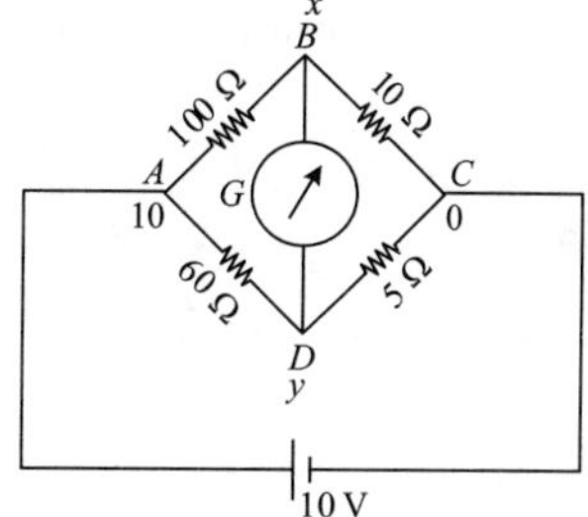

Writing KCL equations for junctions B and D, we have

$$\frac{x-10}{100}+\frac{x-y}{15}+\frac{x-0}{10}=0$$

$$53x-20y=30 \qquad \ldots(1)$$

$$\frac{y-10}{60}+\frac{y-x}{15}+\frac{y-0}{5}=0$$

$$17y-4x=10 \qquad \ldots(2)$$

Solving equations-(1) & (2), we get

$$x=0.865\text{ V}$$

$$y=0.792\text{ V}$$

$$\Delta V=0.073 \; R=15\Omega$$

Current through galvanometer is calculated as

$$i=\frac{V}{R}=\frac{x-y}{R}=\frac{0.073}{15}=4.87\text{ mA}$$

Ans. (C)

Sol. 27 Equivalent resistance of the above circuit can be given as

$$R_{eq_1}=\frac{50\times 20}{70}+10=\frac{170}{7}\Omega$$

Voltage across 10Ω resistance is given as

$$V_1=\left[\frac{170}{\frac{170}{7}}\right]\times 10=70\text{ V} \qquad \textbf{Ans. (70.00)}$$

Sol. 28 In parallel combination, we use

$$R_{net} = \frac{R_1 R_2}{R_1 + R_2}$$

$$\Rightarrow \quad \frac{\rho l}{2A} = \frac{R_1 R_2}{R_1 + R_2}$$

$$\Rightarrow \quad \frac{\rho}{2} = \frac{6\times 3}{6+3} = 2$$

$$\Rightarrow \quad \rho = 4$$ **Ans. (4)**

Sol. 29 Distributing potentials in the circuit shown in figure below, gives

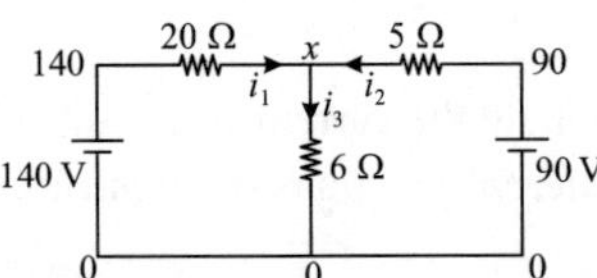

Writing KCL equation for unknown potential x, gives

$$\frac{140-x}{20} + \frac{90-x}{5} + \frac{0-x}{6} = 0$$

$$\Rightarrow \quad x = 60\,\text{V}$$

Thus current in 6Ω resistance is given as

$$i_3 = \frac{x}{6} = \frac{60}{6} = 10\,\text{A}$$ **Ans. (B)**

Sol. 30 When switches are closed the effective circuit across ab can be redrawn as shown in figure below

For this, effective resistance across ab is given as

$$R_{ab} = 10\,\Omega$$ **Ans. (10)**

Sol. 31 The two rods can be considered to be in parallel combination so the effective resistance is given as

$$R_{eq} = \frac{R_1 R_2}{R_1 + R_2} = \frac{\rho_1 \frac{l}{A} \rho_2 \left(\frac{l}{A}\right)}{\rho_1 \left(\frac{l}{A}\right) + \rho_2 \left(\frac{l}{A}\right)}$$

$$\Rightarrow R_{eq} = \frac{l}{A}\left(\frac{\rho_1 \rho_2}{\rho_1 + \rho_2}\right) = \frac{0.25}{3\times 10^{-6}} \left(\frac{1.7\times 10^{-8} \times 2.6\times 10^{-8}}{1.7\times 10^{-8} + 2.6\times 10^{-8}}\right)$$

$$\Rightarrow R_{eq} = 0.085 \times 10^{-2} = 0.858\,\text{m}\Omega$$ **Ans. (A)**

Sol. 32 Figure below shows the circuits as explained in the question.

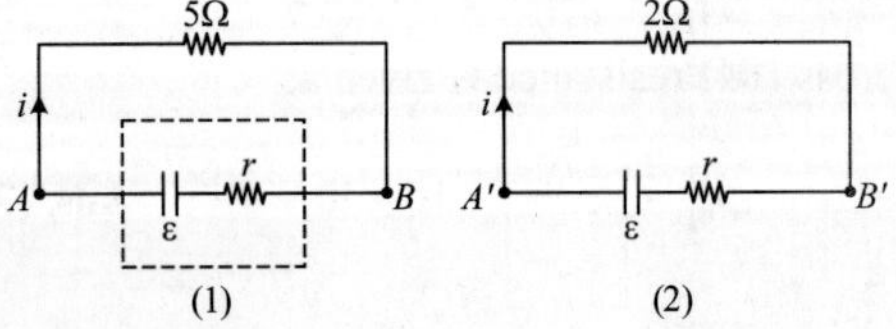

In case-(1), the current in circuit is given as

$$I = \frac{1.25}{5} = 0.25\,\text{A}$$

The potential drop across the load resistance is given as

$$V_{AB} = 5V = \varepsilon - ir$$

$$\Rightarrow \quad 1.25 = \varepsilon - 0.25r \quad \ldots(1)$$

In case-(2), the current in circuit

$$I = \frac{1}{2} = 0.5\,\text{A}$$

$$\Rightarrow \quad V_{AB} = 1 = \varepsilon - 0.5r \quad \ldots(2)$$

Solving equations-(1) & (2), we get

$$\varepsilon = 15\,\text{V}$$ **Ans. (15)**

Sol. 33 The maximum current in circuit is given as

$$i_0 = \frac{20}{10} = 2\text{A}$$

After the current is switched off (battery is removed), current starts decaying through the given inductor so now current as a function of time is given as

$$i = i_0 e^{-Rt/L}$$

At $t = 1$ ms = 0.001 s, we have

$$i = 2\times e^{-\frac{10\times 0.001}{10\times 10^{-3}}} = \frac{2}{2.718} = 0.74\,\text{A}$$

$$\Rightarrow \quad x = 74$$ **Ans. (74)**

Sol. 34 Figure below shows the effective circuit formed for the situation described in the question.

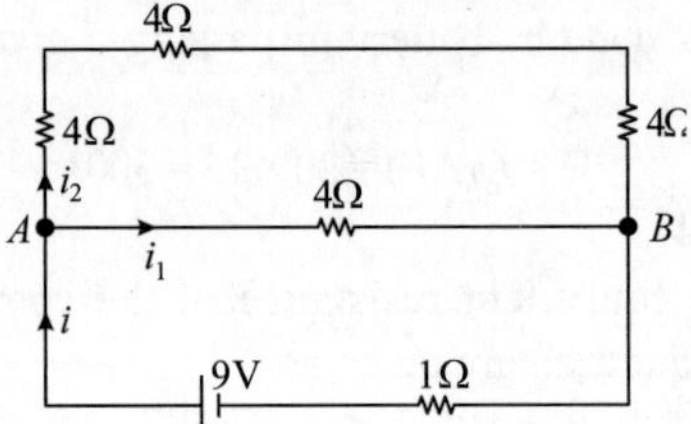

Resistance across battery is given as

$$R_{AB} = \frac{12\times 4}{12+4} = 3\,\Omega$$

Current through battery is given as

$$i = \frac{9}{3+1} = \frac{9}{4}\text{A}$$

Current i_2 is given as

$$i_2 = \frac{r_2}{r_2 + r_1} i = \frac{1}{4}\times\frac{9}{4} = \frac{9}{16};$$

$$\Rightarrow \quad V_{AB} = \frac{9}{2} = 4.5\,\text{V}$$ **Ans. (45)**

Sol. 35 In steady state, current in circuit is given as

$$I = \frac{6V}{6\Omega} = 1\text{ A}$$

Just after T_1 is connected to T_3, current will remain same due to inductor so potential drop across 3 Ω resistor is given as

$$V_{3\Omega} = 3 \times 1 = 3\text{ V}$$ **Ans. (3)**

Sol. 36 Below figures show the calculation of equivalent resistance of the given circuit.

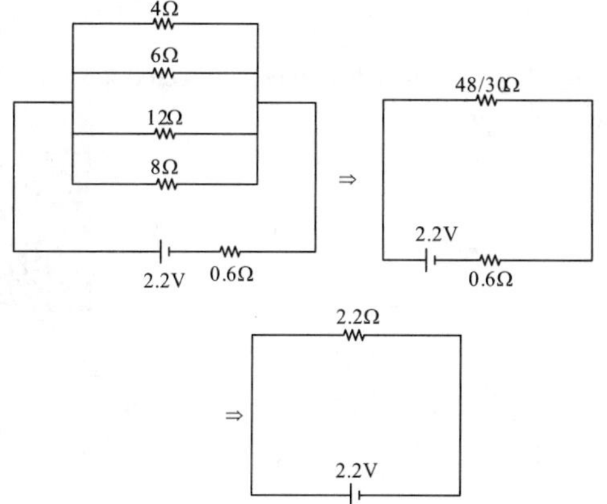

Thus power dissipated in the circuit can be calculated as

$$P = \frac{V^2}{R}, P = \frac{(2.2)^2}{2.2} = 2.2\text{ W}$$ **Ans. (D)**

Sol. 37 For the required equivalent resistance the desired combination is shown in figure below

$$\Rightarrow R_{eq} = \frac{4}{3} + 14 = \frac{46}{3}\Omega$$ **Ans. (A)**

Sol. 38 For series combination of batteries current through effective battery is given as

$$i_1 = \frac{5\varepsilon}{R + 5r}$$

For parallel combination of batteries current through effective battery is given as

$$i_2 = \frac{\varepsilon}{R + \frac{r}{5}}$$

As current in both cases is same, we use

$$i_1 = i_2$$

$$\Rightarrow \frac{5\varepsilon}{R+5r} = \frac{\varepsilon}{R + \frac{r}{5}}$$

$$\Rightarrow R = r = 1\,\Omega$$ **Ans. (B)**

Sol. 39 According to the given conditions, we have

$$i_p = 20 i_a$$

$$\Rightarrow \frac{20}{\left(\frac{10}{n} + 10\right)} = 20\left(\frac{20}{10n + 10}\right)$$

$$\Rightarrow \frac{20n}{10 + 10n} = 20\left(\frac{20}{10n+10}\right)$$

$$n = 20$$ **Ans. (20)**

Sol. 40 After closing the switch, equivalent resistance is given as

$$R_1 = \frac{2R}{3} + \frac{2R}{3} = \frac{4R}{3}$$

When switch is open, the equivalent resistance is given as

$$R_2 = \frac{3R \cdot 3R}{3R + 3R} = \frac{3R}{2}$$

$$\Rightarrow \frac{R_1}{R_2} = \frac{4R}{3} \times \frac{2}{3R} = \frac{8}{9} = \frac{8}{x}$$ **Ans. (9)**

Sol. 41 The equivalent circuit of the given circuit is drawn below

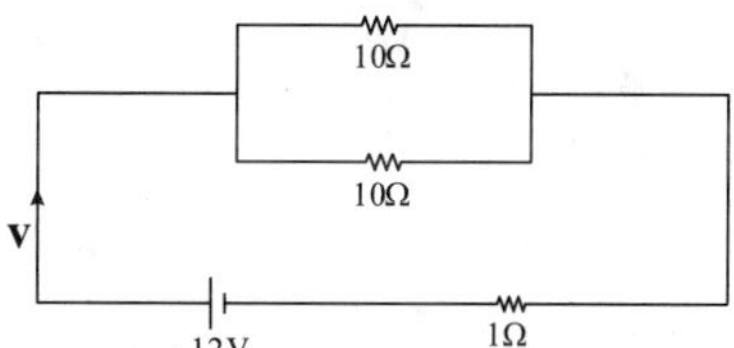

Current in circuit through the battery is given as

$$i = \frac{12}{6} = 2\text{A}$$

Currents in the top and lower branch will be same as both have equal 10 Ω resistance

$$i_1 = i_2 = i/2 = 1\text{ A}$$

Voltage drop across 15 Ω resistance can be calculated by ohms law as

$$V_{15\Omega} = i_1 \times \frac{10 \times 15}{10 + 15} = 1 \times 6 = 6\text{ V}$$ **Ans. (1)**

Sol. 42 Figure below shows the resistances described in the question. When the wire is bent to form a circle then across the diametrically opposite ends there will be two equal resistances of 6 Ω each considered to be connected in parallel as shown. From the equivalent resistance required is equal to 3 Ω.

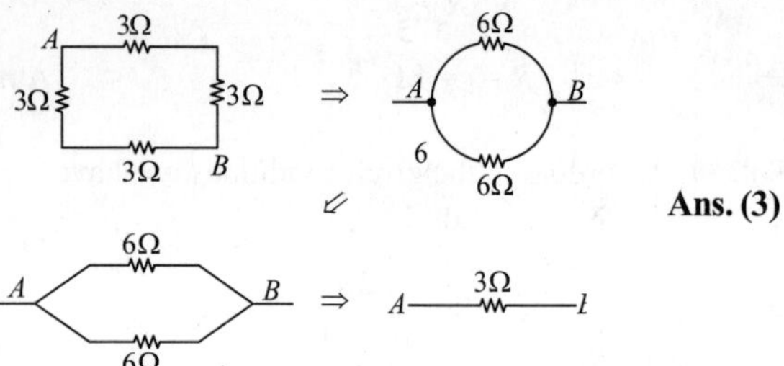

Ans. (3)

Sol. 43 In the given circuit leftmost resistance is short circuited and for right most resistances of 3 Ω in parallel and rest of resistances, we cam find equivalent by using series and parallel analysis so we have

$$R_{AB} = \frac{3\times\frac{3}{2}}{3+\frac{3}{2}} = 1\,\Omega$$ **Ans. (A)**

Sol. 44 The equivalent resistance of the two resistors can be given as

$$\frac{1}{R} = \frac{1}{R_1} + \frac{1}{R_2}$$

For error calculation, we differntiate the above expression as

$$\frac{\Delta R}{R^2} = \frac{\Delta R_1}{R_1^2} + \frac{\Delta R_2}{R_2^2}$$

$$\Rightarrow \quad \Delta R = (R)^2\left[\frac{\Delta R_1}{R_1^2} + \frac{\Delta R_2}{R_2^2}\right]$$

$$\Rightarrow \quad \Delta R = (2)^2\left[\frac{0.8}{4^2} + \frac{0.4}{4^2}\right]$$

$$\Rightarrow \quad \Delta R = [0.2 + 0.1] = 0.3$$

Thus equivalent resistance with error is given as

$$R_{net} = R + \Delta R = 2 \pm 0.3$$ **Ans. (D)**

17.3 Thermal Effects of Current & RC Circuits

Sol. 1

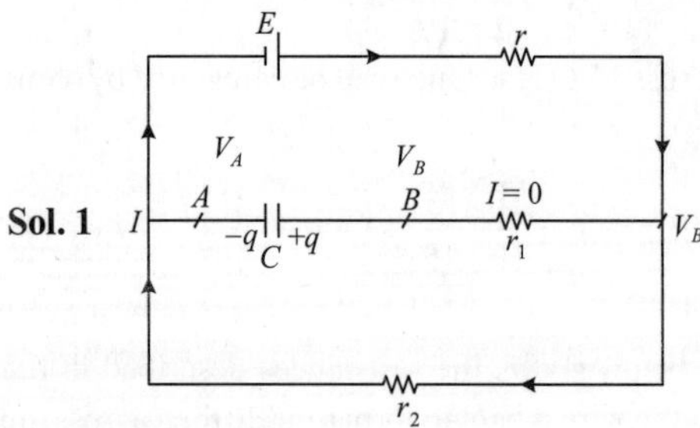

In steady state no current flows through capacitive branch so current only flows in outer loop as shown, and it is given as

$$I = \frac{E}{r + r_2}$$

Potential difference across capacitor is

$$V_B - V_A = Ir_2 = \frac{Er_2}{r + r_2}$$

$\Rightarrow$ Charge on capacitor is

$$q = C(V_B - V_A) = \frac{CEr_2}{r + r_2}$$ **Ans. (A)**

Sol. 2 For a group of resistances connected across a battery the power dissipated is given as

$$P = \frac{V^2}{R_{eq}}$$

$$R_{\text{I}} = 1\,\Omega \rightarrow P_1 = 9\text{ W}$$

$$R_{\text{II}} = 1\,\Omega \,||\, 2\,\Omega \,||\, 2\,\Omega$$

$$P_2 = 9/0.5 = 18\text{ W} = 0.5\,\Omega$$

$$R_{\text{III}} = 1.5\,\Omega \rightarrow P_3 = 9/1.5 = 6\text{ W}$$

$$\Rightarrow \quad P_2 > P_1 > P_3$$ **Ans. (C)**

Sol. 3

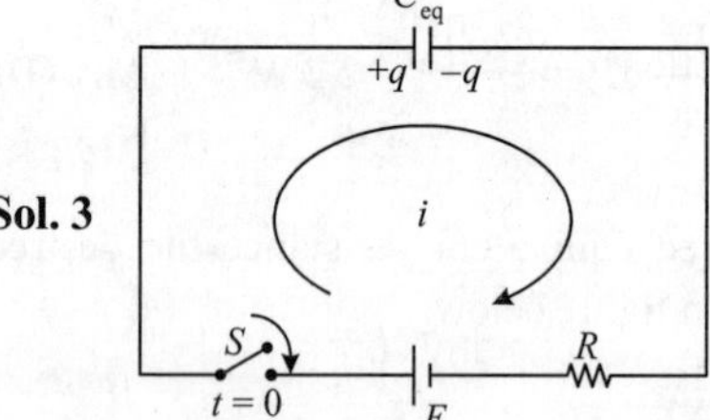

If S is closed at $t = 0$ then charge on capacitor plates as a function of time is given as

$$q = C_q E\,(1 - e^{-t/RC_{eq}})$$

As C_1 and C_2 are in series, charge on their plates must be equal to q only.

C_1 $+q$ $-q$ C_2 $+q$ $-q$

Ans. (A)

Sol. 4

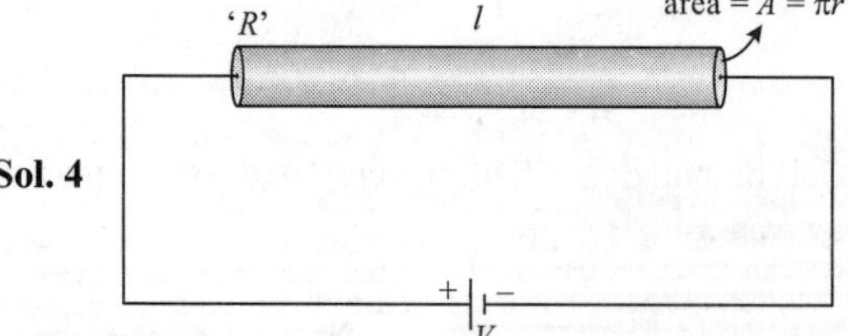

Resistance of wire is

$$R = \frac{\rho l}{A}$$

Thermal power produced in wire

$$P_H = \frac{V^2}{R}$$

when $l \rightarrow l/2$

and $r \rightarrow 2r$

$\Rightarrow$ $A \rightarrow 4A$

$\Rightarrow$ New resistance of wire becomes

$$R' = \frac{\rho(l/2)}{4A} = \frac{1}{8}\frac{\rho l}{A} = \frac{R}{8}$$

Now thermal power produced in wire becomes

$$P'_H = \frac{V^2}{R/8} = \frac{8V^2}{R} = 8P_H$$ **Ans. (A)**

Sol. 5 Power dissipation in resistance is

$$P = I^2R$$

$\Rightarrow$ $4.4 = 4 \times 10^{-6} R$

$\Rightarrow$ $R = 1.1 \times 10^{6}\Omega$

When 11 V power supply is used then

$$P' = \frac{11^2}{R} = \frac{11^2}{1.1}\times 10^{-6} = 11 \times 10^{-5}\text{ W}$$ **Ans. (A)**

Sol. 6 In series combination, equivalent resistance is 2R thus power consumed is $60\text{W} = \frac{\varepsilon^2}{2R}$

In parallel combination, equivalent resistance is $R/2$ thus new power is consumed is

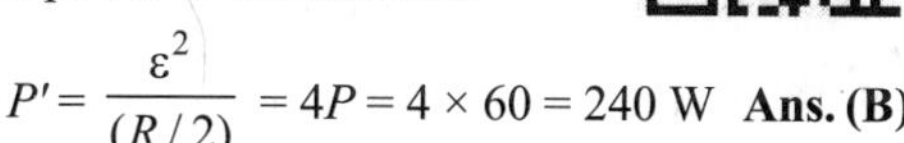

$$P' = \frac{\varepsilon^2}{(R/2)} = 4P = 4 \times 60 = 240\text{ W}$$ **Ans. (B)**

Sol. 7 Resistances of the two bulbs are given as

$$R_1 = \frac{220^2}{25}$$

and $$R_2 = \frac{220^2}{100}$$

When bulbs are connected in series, we use

$$I = \frac{220}{R_1 + R_2}$$

Power consumed by the bulbs are

$$P_1 = i^2 R_1$$

$$\Rightarrow P_1 = \frac{220^2}{\left(\frac{220^2}{25} + \frac{220^2}{100}\right)} \times \frac{220^2}{25}$$

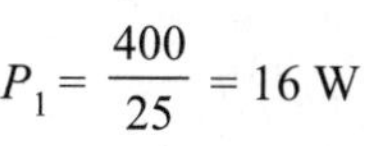

$$\Rightarrow P_1 = \frac{400}{25} = 16\text{ W}$$

and $$P_2 = i^2 R_2 = \frac{220^2}{\left(\frac{220^2}{25} + \frac{220^2}{100}\right)} \times \frac{220^2}{25} = 4\text{ W}$$

Ans. (C)

Sol. 8 Rate of magnetic energy storage in inductor is

$$\frac{d}{dt}\left(\frac{1}{2}LI^2\right) = LIdI$$

$\Rightarrow$ $LIdI = I^2R$

$$LdI = IR$$

We use current as $$I = \frac{E}{R}(1 - e^{-Rt/2})$$

$$\Rightarrow L \times \frac{E}{10}(-e^{-t/2}) \times \frac{-1}{2} = \frac{E}{10}(1 - e^{-t/2}) \times 10$$

$$\Rightarrow e^{-t/2} = 1 - e^{-t/2}$$

$$\Rightarrow t = 2 \ln 2$$ **Ans. (C)**

Sol. 9

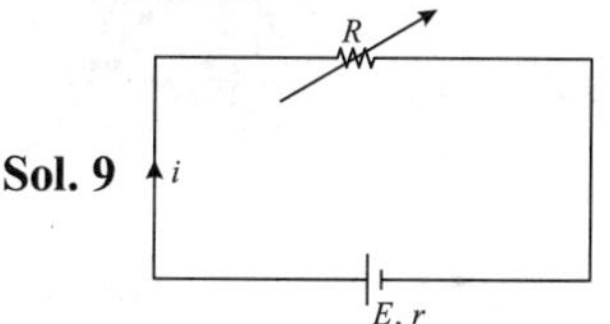

Current in external resistance is given as

$$i = \frac{E}{r + R}$$

Power generated in R is given as

$$P = i^2 R$$

$$P = \frac{E^2 R}{(r+R)^2}$$

For maximum power, we use

$$\frac{dP}{dR} = 0$$

$$\Rightarrow E^2\left[\frac{(r+R)^2 \times 1 - R \times 2(r+R)}{(r+R)^4}\right] = 0$$

$$\Rightarrow r = R$$ **Ans. (D)**

Sol. 10 Applying point potential method of potential distribution in given circuit as shown below

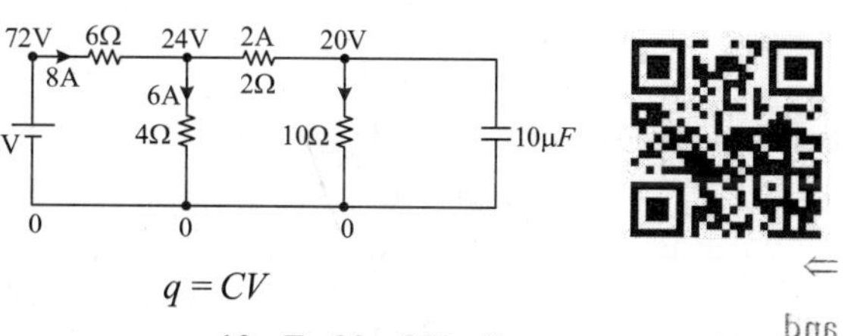

$$q = CV$$

$$q = 10\ \mu F \times 20 = 200\ \mu\text{C}$$ **Ans. (C)**

Sol. 11

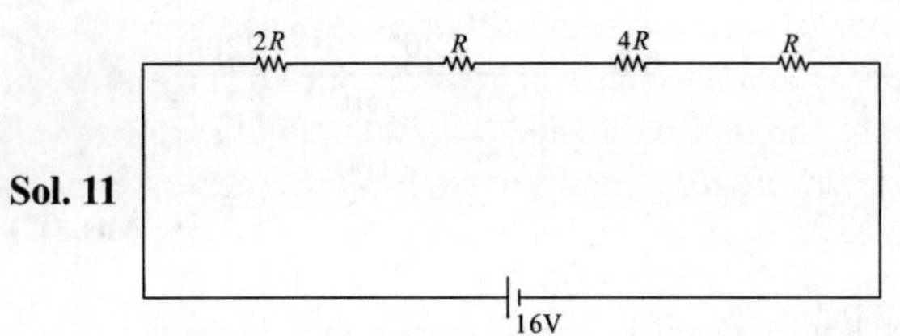

Power consumed by equivalent resistance $8R$ is circuit is

$$P = \frac{16^2}{8R} = 4$$

$$\Rightarrow \quad R = 8\,\Omega \quad \textbf{Ans. (A)}$$

Sol. 12 Total power supplied in time t is

$$Q = P \times t$$

and we use $$Q = mc\Delta T + mL$$

Power supplied by mains is $P = \dfrac{V_{rms}^2}{R}$

Substituting all values, we have

$$4200 \times 80 + 2260 \times 10^3 = \frac{(200)^2}{20} \times t$$

$$\Rightarrow \quad t = 1298 \text{ s}$$

$$\Rightarrow \quad t \simeq 22 \text{ min}$$ **Ans. (B)**

Sol. 13 Total power consumption with current I at full load of building is

$$220I = P = 15 \times 45 + 15 \times 100 + 15 \times 10 + 2 \times 10^3$$

$$\Rightarrow \quad I = \frac{4325}{220} = 19.66$$

$$\Rightarrow \quad I \simeq 20 \text{ A}$$ **Ans. (D)**

Sol. 14 Considering loop currents i_1 and i_2 in circuit as shown in figure

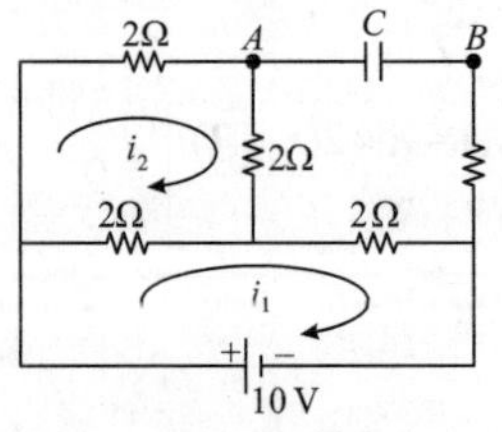

Using KVL equation for i_1 and i_2 gives

$$4i_1 - 2i_2 = 10 \quad \ldots(1)$$

$$6i_2 - 2i_1 = 0 \quad \ldots(2)$$

Solving (1) and (2) gives

$$12i_2 - 2i_2 = 10$$

$$\Rightarrow \quad i_2 = 1\text{A}$$

and $$i_1 = 3\text{A}$$

Potential difference between A and B is given as

$$V_A - 2 \times 1 - 2 \times 3 = V_B$$

$$\Rightarrow \quad V_A - V_B = 8V$$ **Ans. (8)**

Sol. 15 Resistance of an elemental shell of radius x and width dx inside the battery for radial flow of current as shown in figure is given as

$$dr = \frac{\rho dx}{2\pi xl}$$

$$\Rightarrow \quad r = \int dr = \int_a^b \frac{\rho}{2\pi l}\frac{dx}{x} = \frac{\rho}{2\pi l}\ln\left(\frac{b}{a}\right)$$

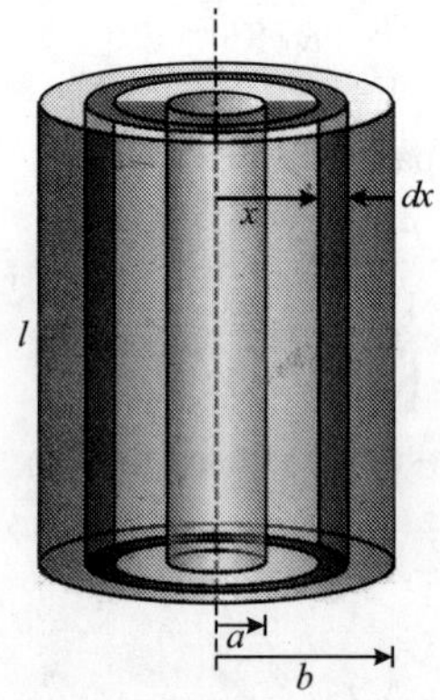

When battery is connected to external resistance R, maximum heat is produced in external resistance when

$$R = r = \frac{\rho}{2\pi l}\ln\left(\frac{b}{a}\right)$$ **Ans. (D)**

Sol. 16 If circuit current is i, we use

$$i^2R = 0.5$$

and $$iR = 2.5$$

Terminal voltage of battery is given as

$$V = E - ir$$

$$\Rightarrow \quad 2.5 = 3 - ir$$

$$\Rightarrow \quad ir = 0.5$$

$$\Rightarrow \quad \frac{R}{r} = 5$$

Power dissipated in r is given as

$$P_r = i^2r$$

Power dissipated in external R is given as

$$P_R = i^2R$$

$$\Rightarrow \quad \frac{P_r}{P_R} = \frac{r}{R} = \frac{1}{5}$$

$$\Rightarrow \quad P_r = \frac{1}{5} \times 0.5 = 0.1 \text{ W}$$ **Ans. (D)**

Sol. 17 Power delivered by transmission line is

$$P = vi = 1000\text{ W}$$

$$\Rightarrow \quad i = \frac{1000}{220} = \frac{50}{11}\text{A}$$

Power loss in line is

$$P_L = i^2R = \left(\frac{50}{11}\right)^2 \times 2 = \frac{5000}{121}\text{W}$$

line efficiency is given as

$$\eta = \frac{1000}{1000 + \frac{5000}{121}} \times 100$$

$$\Rightarrow \quad \eta = \frac{121}{126} \times 100 \simeq 96\%$$ **Ans. (B)**

Sol. 18 When constant voltage is applied across the RC circuit, capacitor charges exponentially and when the voltage drops to zero, it discharges exponentially so most appropriate wave pattern across capacitor matches with option (C) when square wave EMF is applied across RC circuit. **Ans. (C)**

Sol. 19 Thermal energy can be expressed as

$$H = i^2Rt$$

$$\Rightarrow \quad 500 = (1.5)^2 \times R \times 20$$

$$\Rightarrow \quad E = (3)^2 \times R \times 20 = 2000\text{ J}$$ **Ans. (D)**

Sol. 20 Energy dissipated in resistor is given as

$$H = i^2RT$$

$$\Rightarrow \quad R = \frac{Q}{i^2t} = \frac{10 \times 10^{-3}}{4 \times 10^{-6} \times 1} = 2500\,\Omega$$ **Ans. (2500)**

Sol. 21 Just after closing the switch, capacitor can be considered as short circuit so current in circuit can be given as

$$i_0 = \frac{V}{R} = \frac{30/3}{5 \times 10^6} = 2 \times 10^{-6}\text{A}$$ **Ans. (2)**

Sol. 22 Power of bulb is given as

$$P = \frac{V^2}{R_b}$$

$$\Rightarrow \quad 200 = \frac{(100)^2}{R_b}$$

Resistance of bulb is given as

$$R_b = 50\,\Omega$$

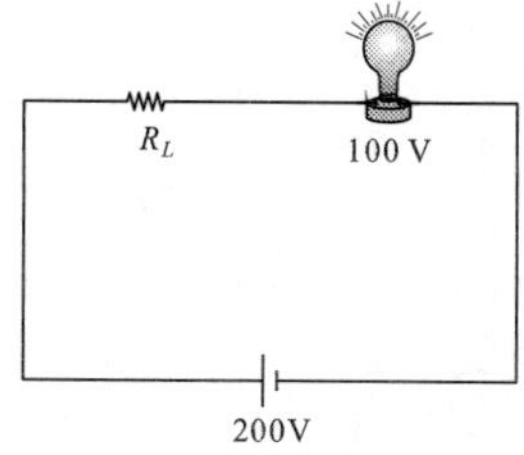

To produce same power in bulb when connected across 200 V supply as shown in figure above as explained in question, the current through bulb should remain same and potential difference across bulb must remain 10 V. Thus current in circuit must be

$$i = \frac{100}{50} = 2\text{A}$$

The voltage across the series resistance now should be $200 - 100 = 100$ V thus resistance R can be given as

$$R_L = \frac{100}{2} = 50\,\Omega$$ **Ans. (50)**

Sol. 23 Resistance of bulb can be calculated as

$$P = \frac{V^2}{R}$$

$$R = \frac{(100)^2}{500} = \frac{100 \times 100}{500} = 20\,\Omega$$

Current through bulb for 500 W power production is calculated as

$$I = \frac{P}{V} = \frac{500}{100} = 5\text{ A}$$

For a supply of 200 V, current 5 A wll flow when total resistance of circuit will be

$$R_T = \frac{200}{5} = 40\,\Omega$$

Thus extra resistance required to be kept in series of bulb is 20 Ω. **Ans. (D)**

Sol. 24 The effective circuit is shown in figure below

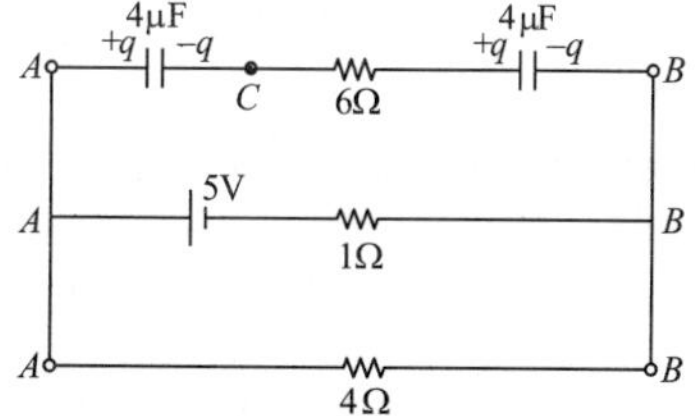

The current in lower loop of circuit is given as

$$i = \frac{5}{4+1} = 1\text{ A}$$

Potential difference across AB is given as

$$V_{AB} = i \times 4 = 4\text{ V}$$

$$\Rightarrow \quad V_{AC} = V_{CB} = 2\text{ V}$$

$$\Rightarrow \quad q \text{ on } 4\mu\text{F} = CV_{AC} = 8\mu\text{C}$$ **Ans. (B)**

Sol. 25 In steady state, the current through capacitor is zero. The potential difference across capacitor is 2 V across the resistance in parallel to which capacitor is connected. So charge on capacitor is given as

$$q = 50 \times 2 = 100\,\mu\text{C}$$ **Ans. (100)**

Sol. 26 Power dissipated in time t is given as

$$P = i^2Rt$$

$$\Rightarrow \quad 192 = 4 \times 4 \times R \times 1$$

$$\Rightarrow \quad R = 12\,\Omega$$

If the current is doubled, amount of energy dissipated is calculated as

$$P' = i'^2Rt'$$

$$\Rightarrow \quad P' = 8 \times 8 \times 12 \times 5 = 3840\text{ J}$$

Ans. (3840)

Sol. 27 The potential difference across capacitor as a function of time is given as

$$V = V(1 - e^{-t/RC})$$

$$\Rightarrow \quad 2 = 20\,(1 - e^{-1\mu S/10C})$$

$$\Rightarrow \quad \frac{1}{10} = (1 - e^{-1 \times 10^{-6}/10C})$$

$$\Rightarrow \quad e^{-1\times10^{-6}/10C} = \frac{9}{10}$$

$$\Rightarrow \quad \frac{1\times10^{-6}}{10C} = ln\,\frac{9}{10}$$

$$\Rightarrow \quad C = \frac{1\times10^{-6}}{10\times\ln\left(\frac{9}{10}\right)} = \frac{10^{-7}}{1.05} = \frac{1}{1.05}\,\mu\text{F} = \frac{100}{105}\,\mu\text{F} = 0.95\,\mu\text{F}$$

Ans. (C)

Sol. 28 When steady state is reached, inductors in circuit behave like short circuit. So all 3 Ω resistance will act in parallel and current supplied by the battery can be given as

$$I = \frac{30}{2+1} = \frac{30}{3} = 10\text{ A}$$

Ans. (C)

Sol. 29 In first case power dissipation is given as

$$P = \frac{V^2}{R}$$

In second case power dissipation in the two parts are given as

$$P_1 = \frac{V^2}{\frac{R}{2}} \text{ and } P_2 = \frac{V^2}{\frac{R}{2}}$$

Total power dissipation in second case is given as

$$P_1 + P_2 = \frac{4V^2}{R}$$

Ratio of final to initial power dissipation is given as

$$\frac{P_1 + P_2}{P} = \frac{\frac{4V^2}{R}}{\frac{V^2}{R}} = 4$$

Ans. (4)

17.4 Electrical Measuring Instruments

Sol. 1 All options (A), (B), (C) and (D) are covered in the theory option (D) is false.

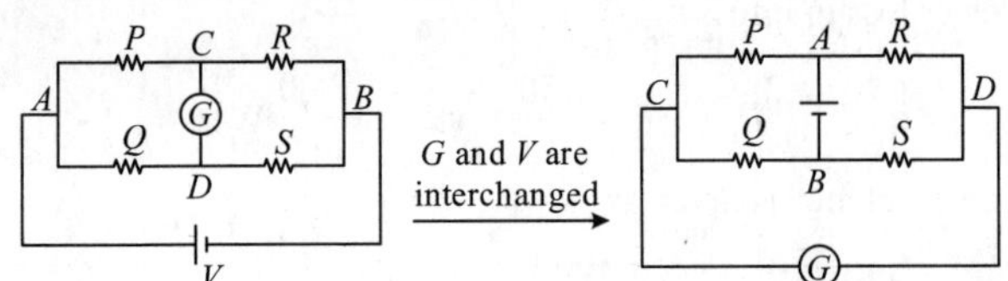

Balancing condition

$$\rightarrow \quad \frac{P}{Q} = \frac{R}{S}$$

Balancing condition

$$\rightarrow \quad \frac{Q}{S} = \frac{P}{R}$$

Option (D) is false. **Ans. (D)**

Sol. 2

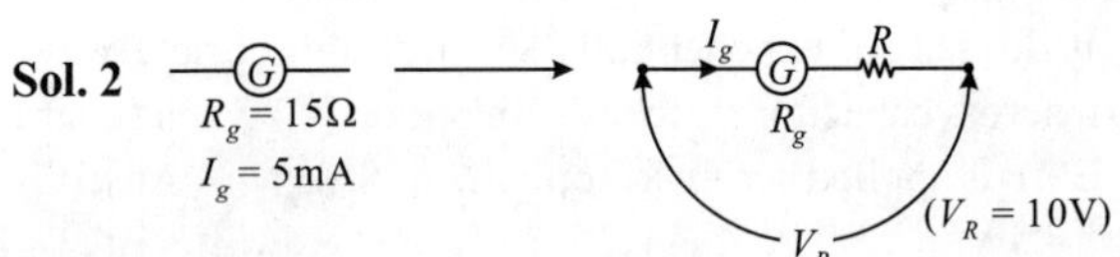

So for full deflection at voltage V_R, we use

$$I_g = \frac{V_R}{R_g + R}$$

$$\Rightarrow \quad R = \frac{V_R}{I_g} - R_g$$

$$\Rightarrow \quad R = \frac{10\times1000}{0.005} - 15 = 2000 - 15 = 1985\,\Omega$$

Ans. (C)

Sol. 3 In first case we use

$$IR_1 = \lambda l, \qquad \ldots(1)$$

where $I \rightarrow$ current flowing through R_1 and R_2 in series

$\lambda \rightarrow$ potential gradient on wire PQ

In second case when K_3 is plugged into points 1 and 3 then R_1 and R_2 in series are balanced across points P and M where

$$PM = l_2$$

$$\Rightarrow \quad I\,(R_1 + R_2) = \lambda l_2 \qquad \ldots(2)$$

$$\frac{(2)}{(1)} \Rightarrow \quad \frac{R_1 + R_2}{R_1} = \frac{l_2}{l_1}$$

$$\Rightarrow \quad 1 + \frac{R_2}{R_1} = \frac{l_2}{l_1}$$

$$\Rightarrow \quad \frac{R_2}{R_1} = \frac{l_2 - l_1}{l_1}$$

$$\Rightarrow \quad \frac{R_1}{R_2} = \frac{l_1}{l_2 - l_1}$$

Ans. (D)

Sol. 4 For meter bridge initially we use

$$AN = 60 \text{ cm}$$

$$NB = 40 \text{ cm}$$

$$\Rightarrow \quad \frac{P}{60} = \frac{Q}{40}$$

$$\Rightarrow \quad \frac{4}{60} = \frac{Q}{40} \quad Q = \frac{8}{3}\,\Omega$$

In next state $P \to 4 + R$ and

$$AN = 80 \text{ cm}$$

$$NB = 20 \text{ cm}$$

$$\Rightarrow \quad \frac{4+R}{80} = \frac{8/3}{20}$$

$$\Rightarrow \quad 12 + 3R = 32$$

$$\Rightarrow \quad 3R = 20 \quad \Rightarrow \quad R = 20/3\,\Omega$$

Ans. (D)

Sol. 5 The internal resistance of cell using potentiometer experiment is given as

$$r = R\left(\frac{l_1}{l_2} - 1\right) = 5 \times \left(\frac{52}{40} - 1\right)$$

$$= 5 \times \frac{12}{40} = \frac{12}{8} = 1.5\,\Omega$$

Alternative : If cell of EMF E is balanced at l_1 length of wire

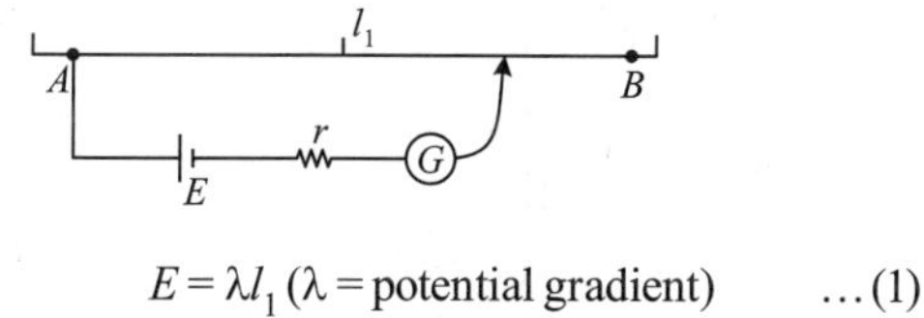

$$\Rightarrow \quad E = \lambda l_1 \; (\lambda = \text{potential gradient}) \quad \ldots(1)$$

When call is shunted with a resistance R

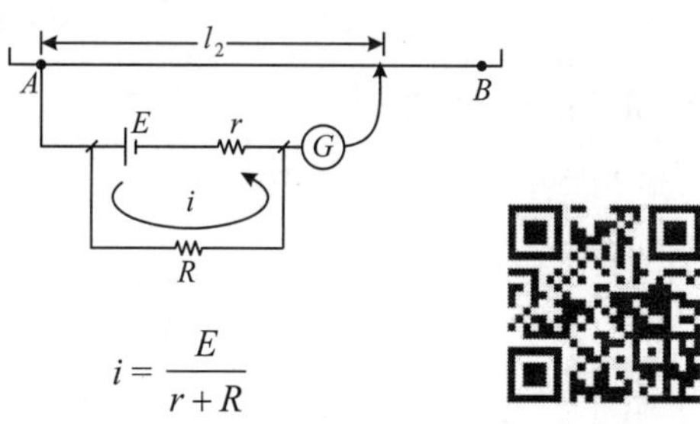

$$i = \frac{E}{r+R}$$

The potential difference across cell terminals after shunting is

$$V_E = iR = \frac{ER}{r+R} = \lambda l_2 \quad \ldots(2)$$

$$\frac{(2)}{(1)} \Rightarrow \quad \frac{R}{r+R} = \frac{l_2}{l_1}$$

$$\Rightarrow \quad r = R\left(\frac{l_1}{l_2} - 1\right) = 1.5\,\Omega$$

Ans. (B)

Sol. 6

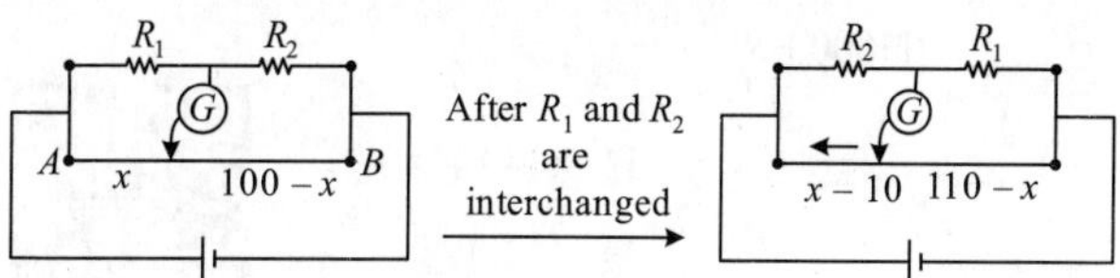

For balanced meter bridge, we use

$$\frac{R_1}{x} = \frac{R_2}{100-x} \quad \ldots(1)$$

$$\frac{R_2}{x-10} = \frac{R_1}{110-x} \quad \ldots(2)$$

$$\frac{(1)}{(2)} \Rightarrow \quad \frac{110-x}{x} = \frac{x-10}{100-x}$$

$$\Rightarrow \quad (110-x)(100-x) = x(x-10)$$

$$\Rightarrow \quad 11000 + x^2 - 210x = x^2 - 10x$$

$$\Rightarrow \quad 200x = 11000$$

$$\Rightarrow \quad x = \frac{110}{2} = 55 \text{ cm}$$

From equation-(1) we have

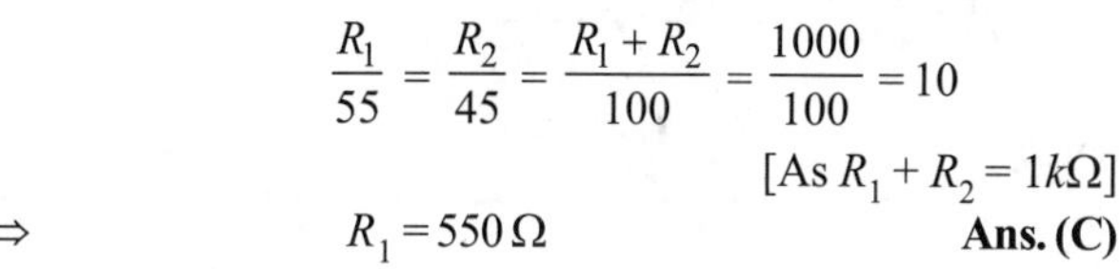

$$\frac{R_1}{55} = \frac{R_2}{45} = \frac{R_1 + R_2}{100} = \frac{1000}{100} = 10$$

[As $R_1 + R_2 = 1k\Omega$]

$$\Rightarrow \quad R_1 = 550\,\Omega$$

Ans. (C)

Sol. 7 When bridge is balanced, we have

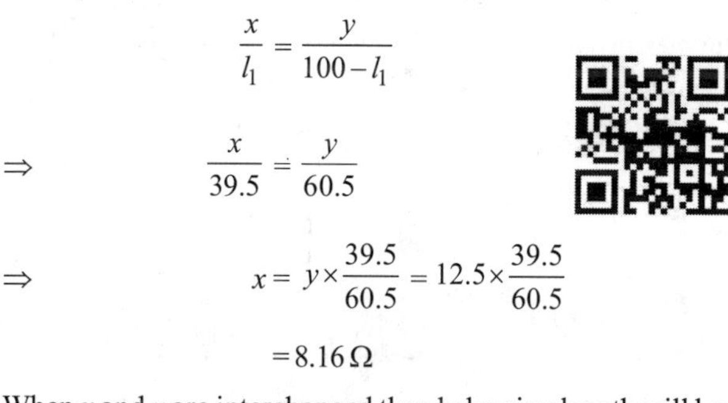

$$\frac{x}{l_1} = \frac{y}{100 - l_1}$$

$$\Rightarrow \quad \frac{x}{39.5} = \frac{y}{60.5}$$

$$\Rightarrow \quad x = y \times \frac{39.5}{60.5} = 12.5 \times \frac{39.5}{60.5}$$

$$= 8.16\,\Omega$$

When x and y are interchanged then balancing length will be

$$l_2 = 100 - l_1$$

$$\Rightarrow \quad l_2 = 100 - 39.5 = 60.5 \text{ cm}$$

Ans. (B)

Sol. 8 **Case-I :** Without using a shunt resistance

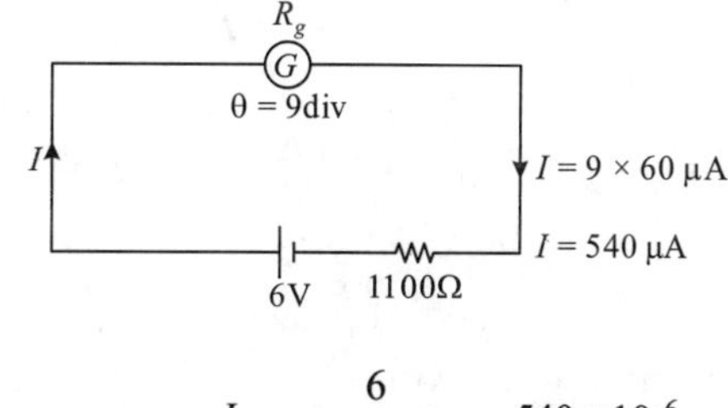

$$\Rightarrow \quad I = \frac{6}{11000 + R_g} = 540 \times 10^{-6}$$

$$\Rightarrow \quad 11000 + R_g = \frac{10^5}{9}$$

$$= 11111.11\,\Omega$$

$$\Rightarrow \quad R_g = 11111.11 - 1000$$

$$= 111.11\,\Omega$$

Case-II : with a shunt resistance

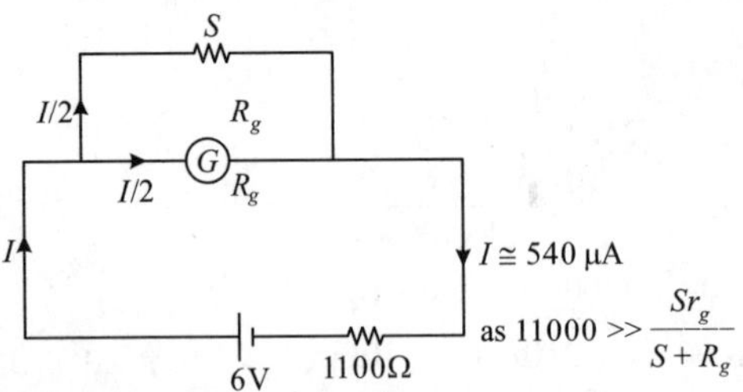

For half deflection in galvanometer we should use

$$S \simeq R_g = 111.11\,\Omega$$ **Ans. (B)**

Sol. 9

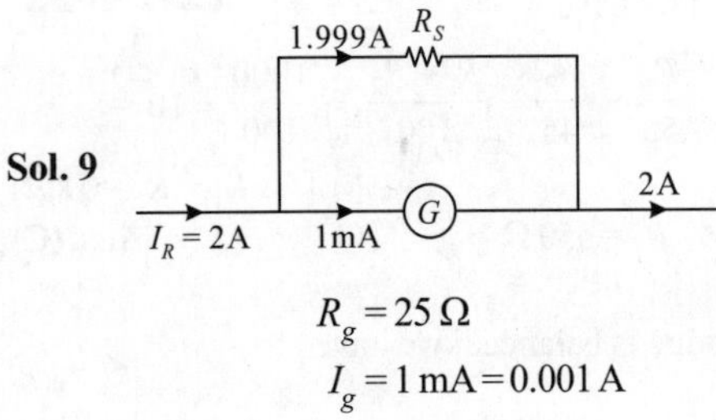

$$R_g = 25\,\Omega$$

$$I_g = 1\text{ mA} = 0.001\text{ A}$$

For making an ammeter of 2A range, we use

$$I_g R_g = (I_R - I_g) R_S$$

$$0.001 \times 25 = 1.999 \times R_S$$

$$\Rightarrow \quad R_S = \frac{25}{1999} \simeq 1.25 \times 10^{-2}\,\Omega$$ **Ans. (D)**

Sol. 10

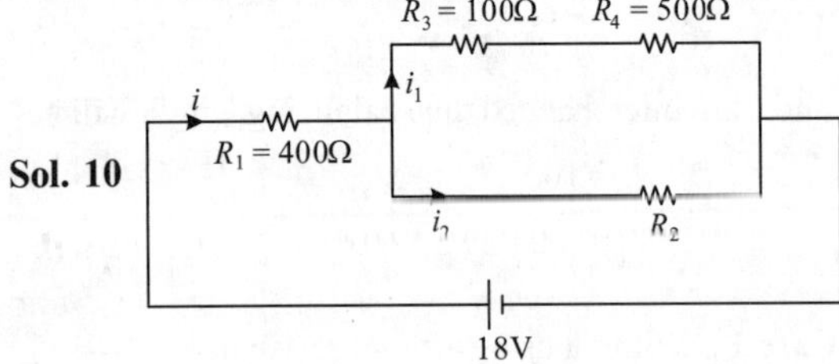

Given that

$$V_4 = 5\text{ V}$$

$$\Rightarrow \quad i_1 = \frac{V_4}{R_4} = 0.01\text{ A}$$

$$\Rightarrow \quad V_3 = i_1 R_3 = 1\text{ V}$$

$$\Rightarrow \quad V_3 + V_4 = 6\text{ V} = V_2$$

$$\Rightarrow \quad V_1 + V_3 + V_4 = 18\text{ V}$$

$$\Rightarrow \quad V_1 = 12\text{ V}$$

$$\Rightarrow \quad i = \frac{V_1}{R_1} = 0.03\text{ A}$$

$$\Rightarrow \quad i - i_1 = 0.02\text{ A}$$

$$\Rightarrow \quad R_2 = \frac{V_2}{i_2} = \frac{6}{0.02} = 300\,\Omega$$ **Ans. (A)**

Sol. 11 Current in primary circuit is

$$i = \frac{\varepsilon}{13r}$$

At no deflection

$$i \times \left(\frac{12r}{L} \times x\right) = \frac{\varepsilon}{2}$$

$$\Rightarrow \quad \frac{\varepsilon}{13r}\left(\frac{12r}{L} \times x\right) = \frac{\varepsilon}{2}$$

$$\Rightarrow \quad x = \frac{13L}{24}$$ **Ans. (D)**

Sol. 12 Using V and I readings measured resistance will be parallel combination of R_V and R, so we use

$$0.95\,R = \frac{R\,R_V}{R + R_V}$$

$$\Rightarrow \quad 0.95\,R^2 + 0.95\,RR_V = RR_V$$

$$\Rightarrow \quad 0.95\,R = 0.05\,R_V$$

$$\Rightarrow \quad R_V = 19 \times 30 = 570\,\Omega$$ **Ans. (B)**

Sol. 13 $R_1 = 32 \times 10 = 320\,\Omega$

For wheat stone bridge, balancing condition is

$$\Rightarrow \quad \frac{R_1}{R_3} = \frac{R_2}{R_4}$$

$$\Rightarrow \quad \frac{320}{R_3} = \frac{80}{40}$$

$$\Rightarrow \quad R_3 = 160\,\Omega$$

Colour code of 160Ω resistance is

(Brown, Blue, Brown) **Ans. (B)**

Sol. 14 Potential gradient with $R_h = 2\Omega$ is

$$\lambda = \left(\frac{6}{2+4}\right) \times \frac{4}{L} = \frac{4}{100}\frac{\text{V}}{\text{cm}}$$

For emf ε_1 let null point be at l cm at J, we use

$$\varepsilon_1 = 0.5\text{ V} = 0.04l \qquad \ldots(1)$$

Now with $R_h = 6\Omega$ new potential gradient is

$$\lambda' = \left(\frac{6}{4+6}\right) \times \frac{4}{L} = \frac{2.4}{100} = 0.024\frac{\text{V}}{\text{cm}}$$

For emf ε_2 null deflection length is same, so we use

$$\varepsilon_2 = 0.024l \qquad \ldots(2)$$

Dividing equation-(1) by (2) we get

$$\frac{0.5}{\varepsilon_2} = \frac{0.04}{0.024}$$

$\Rightarrow \quad \varepsilon_2 = 0.3\text{ V}$ **Ans. (C)**

Sol. 15 At $R = 400\ \Omega$ bridge is balanced so we use

$$\frac{P}{400} = \frac{Q}{X} \quad \ldots(1)$$

On interchanging P and Q for balance $R = 405\ \Omega$. So we use

$$\frac{Q}{405} = \frac{P}{X} \quad \ldots(2)$$

(1) × (2) gives

$$X = \sqrt{400 \times 405} = 402.5\ \Omega$$ **Ans. (D)**

Sol. 16 Given that

$$R_g = 20\ \Omega$$

$$N_{\text{Left}} = N_{\text{Right}} = 30$$

Figure of merit of galvonometer is

$$\text{FOM} = \frac{I}{\phi} = 0.005\text{ A/Div}$$

$$\text{Current sensitivity} = \left(\frac{1}{0.005}\right)\text{Div/A}$$

$$Ig_{\max} = 0.005 \times 30$$

$$= 15 \times 10^{-2} = 0.15\text{ A}$$

For this galvanometer to be used as 15 V range voltmeter, we use

$$15 = 0.15[20 + R]$$

$\Rightarrow \quad 100 = 20 + R$

$\Rightarrow \quad R = 80\ \Omega$ **Ans. (A)**

Sol. 17 Initial balancing condition of bridge is

$$\frac{R_1}{R_2} = \frac{2}{3} \quad \ldots(1)$$

When 10 Ω is connected in series with R_1, we use

$$\frac{R_1 + 10}{R_2} = 1$$

$\Rightarrow \quad R_1 + 10 = R_2 \quad \ldots(2)$

From (1) and (2)

$$\frac{2R_2}{3} + 10 = R_2$$

$\Rightarrow \quad 10 = \frac{R_2}{3}$

$\Rightarrow \quad R_2 = 30\ \Omega$

and $\quad R_1 = 20\ \Omega$

If a resistance R is connected in parallel with $R_1 + 10$ then we use

$$\frac{\frac{30 \times R}{30 + R}}{30} = \frac{2}{3}$$

$\Rightarrow \quad R = 60\ \Omega$ **Ans. (B)**

Sol. 18 In first case galvanometer current is

$$i_g = \frac{E}{220 + R_g} = C\theta_0 \quad \ldots(1)$$

In second case galvanometer current is

$$i_g = \left(\frac{E}{220 + \frac{5R_g}{5 + R_g}}\right) \times \frac{5}{(R_g + 5)} = \frac{C\theta_0}{5}$$

$$\Rightarrow \quad \frac{5E}{225R_g + 1100} = \frac{C\theta_0}{5} \quad \ldots(2)$$

(1)/(2) gives

$$\Rightarrow \quad \frac{225R_g + 1100}{1100 + 5R_g} = 5$$

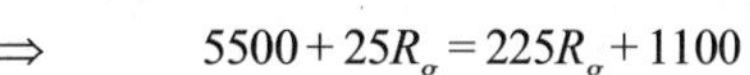

$\Rightarrow \quad 5500 + 25R_g = 225R_g + 1100$

$\Rightarrow \quad 200R_g = 4400$

$\Rightarrow \quad R_g = 22\ \Omega$ **Ans. (D)**

Sol. 19 If resistance of part AP is R_1 and PB is R_2 so for null deflection, we use

$$\frac{R'}{R'} = \frac{R_1}{R_2}$$

$\Rightarrow \quad R_1 = R_2$

On the given wire AB the resistance of an elemental length dl is given as

$$dR = \frac{C}{\sqrt{l}}dl \quad \text{(where } C \to \text{constant)}$$

Thus resistance of length AB is given as

$$R_1 = C\int_0^l l^{-1/2}dl = C(2,\sqrt{l})$$

Similarly of resistance of length PB is given as

$$R_2 = C\int_l^1 l^{-1/2}dl = C(2 - 2\sqrt{l})$$

Using $\quad R_1 = R_2$, we have

$$C_2\sqrt{l} = C(2 - 2\sqrt{l})$$

$\Rightarrow \quad 2\sqrt{l} = 1$

$\Rightarrow \quad \sqrt{l} = \frac{1}{2}$

$\Rightarrow \quad l = \frac{1}{4}\text{m} = 0.25\text{ m}$ **Ans. (A)**

Sol. 20

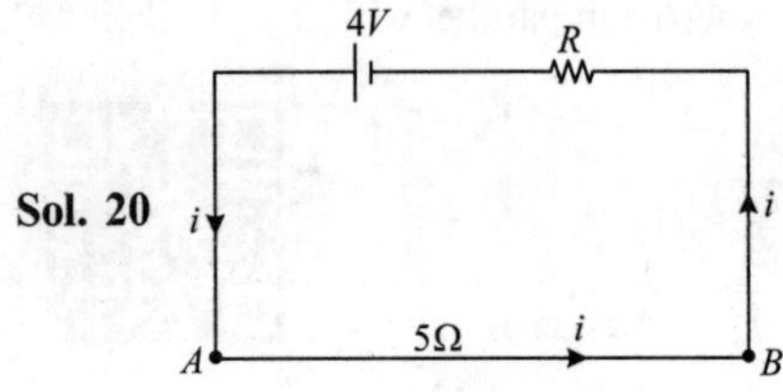

If current flowing in the wire is i, then we use

$$i = \left(\frac{4}{R+5}\right)\text{A}$$

Resistance of 10 cm length of wire is

$$\frac{5}{10} = 0.5\,\Omega$$

Potential difference on 10 cm of wire is

$$5 \times 10^{-3} = \left(\frac{4}{R+5}\right) \times 0.5$$

$$\Rightarrow \quad \frac{4}{R+5} = 10^{-2}$$

$$\Rightarrow \quad R+5 = 400\,\Omega$$

$$\Rightarrow \quad R = 395\,\Omega$$

Ans. (C)

Sol. 21 Full deflection current of galvanometer is

$$I_g = 4 \times 10^{-4} \times 25 = 10^{-2}\text{ A}$$

To convert the galvanometer into a voltmeter, we connect a resistance R in series, such that

$$2.5 = (50+R)\,10^{-2}$$

$$\Rightarrow \quad R = 200\,\Omega$$ **Ans. (C)**

Sol. 22

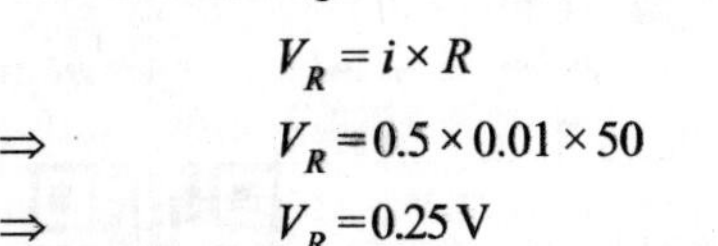

Resistance of wire AB is given as

$$R_{AB} = 400 \times 0.01 = 4\,\Omega$$

Circuit current is

$$i = \frac{1.5+1.5}{1+1+4} = \frac{3}{6} = 0.5\text{A}$$

Voltmeter reading is

$$V_R = i \times R$$

$$\Rightarrow \quad V_R = 0.5 \times 0.01 \times 50$$

$$\Rightarrow \quad V_R = 0.25\text{ V}$$

Ans. (B)

Sol. 23

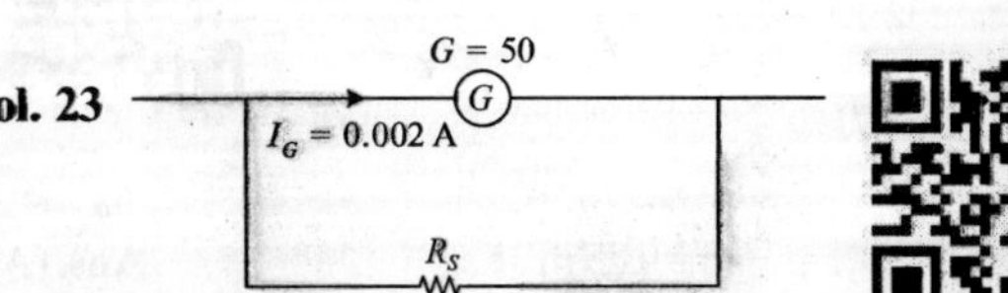

Shunt resistance is calculated as

$$R_S(0.5-0.002) = 50 \times 0.002$$

$$R_S = \frac{50 \times 0.002}{(0.5-0.002)} = \frac{0.1}{0.498} = 0.2$$ **Ans. (A)**

Sol. 24 For balanced state of bridge we use

$$X = \frac{R(100-l)}{l}$$

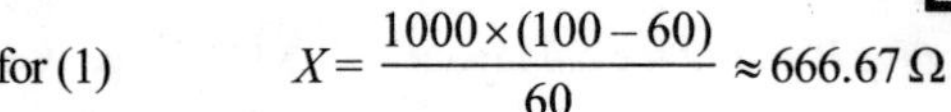

for (1) $$X = \frac{1000 \times (100-60)}{60} \approx 666.67\,\Omega$$

for (2) $$X = \frac{100 \times (100-13)}{13} \approx 669.23\,\Omega$$

for (3) $$X = \frac{10 \times (100-1.5)}{1.5} \approx 656.67\,\Omega$$

for (4) $$X = \frac{1 \times (100-1)}{1} \approx 99\,\Omega$$

So option (A) is completely different so inconsistent with given data.

Ans. (A)

Sol. 25 Equivalent resistance of circuit across battery is

$$R_{\text{eq}} = \frac{15 \times 10}{25} + 2 + 2r$$

$$= 8 + 2r$$

Current in circuit is

$$i = \frac{3}{8+2r}$$

Potential difference across 10Ω resistance is

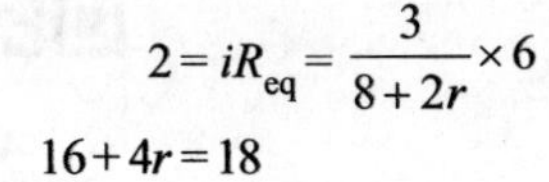

$$2 = iR_{\text{eq}} = \frac{3}{8+2r} \times 6$$

$$\Rightarrow \quad 16 + 4r = 18$$

$$\Rightarrow \quad r = 0.5\,\Omega$$

Ans. (D)

Sol. 26 Full deflection current

$$I_g = 20 \times 10^{-6} \times 50 = 10^{-3}\text{ A} = 1\text{ mA}$$

For 2 V range of voltmeter, we use

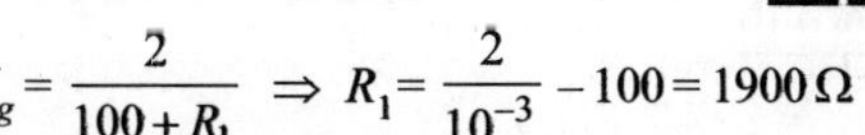

$$I_g = \frac{2}{100+R_1} \Rightarrow R_1 = \frac{2}{10^{-3}} - 100 = 1900\,\Omega$$

For 10 V range of voltmeter, we use

$$I_g = \frac{10}{2000+R_2} \Rightarrow R_2 = \frac{2}{10^{-3}} - 2000 = 8000\,\Omega$$

For 20 V range of voltmeter, we use

$$I_g = \frac{20}{10000+R_3} \Rightarrow R_3 = \frac{2}{10^{-3}} - 10000 = 10000\,\Omega$$

Ans. (D)

Sol. 27 Reading of voltmeter V and current I in circuit (reading of ammeter) can be related as

$$V = E - Ir$$

at $\quad I = 1A,\ V = V_0 \simeq 0$

$\Rightarrow \quad 0 = E - r$

$\Rightarrow \quad E = r$

at $\quad V = 1.5\text{ V}, I = 0$

$\Rightarrow \quad V = E = 1.5\text{ V}$

$\Rightarrow \quad r = 1.5\ \Omega$ **Ans. (C)**

Sol. 28 Internal resistance of cell by potentiometer is given as

$$r = R\left(\frac{l_1 - l_2}{l_2}\right)$$

$\Rightarrow \quad r = 10 \times \dfrac{60}{500} r$

$\Rightarrow \quad r = 1.2\ \Omega = \dfrac{12}{10}\Omega$ **Ans. (12)**

Sol. 29 If λ is potential gradient of potentiometer wire & L is total length of wire, then we use

$$5 = \lambda l$$

$\Rightarrow \quad 5 = \dfrac{\Delta V}{L} l$

$\Rightarrow \quad \Delta V = \dfrac{5 \times L}{l} = 5 \times \dfrac{12}{10} = 6\text{V}$

By Ohm's Law, we use

$$\Delta V = IR$$

$\Rightarrow \quad R = \dfrac{\Delta V}{I} = \dfrac{6}{60 \times 10^{-3}} = 100\ \Omega$ **Ans. (D)**

Sol. 30 For balancing the wheatstone network as shown in figure, we use

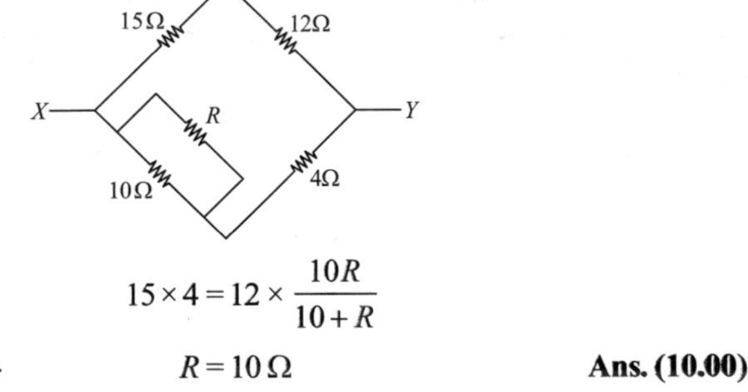

$$15 \times 4 = 12 \times \frac{10R}{10 + R}$$

$\Rightarrow \quad R = 10\ \Omega$ **Ans. (10.00)**

Sol. 31 For the given galvanometer

$$I_g = 1\text{ mA},$$

$$R_g = 100\ \Omega$$

To convert it into a voltmeter, we use shunt resistance R, so we have

$$V = I_g(R + R_g)$$

$\Rightarrow \quad 10 = 1 \times 10^{-3}(R + 100)$

$\Rightarrow \quad R = 9.9\text{ k}\Omega$ **Ans. (A)**

Sol. 32 In balancing condition we use

$$\frac{R}{S} = \frac{25}{75}$$

If R is replaced by a wire half length and one fourth cross-section then resistance becomes $2R$, so we use

$$\frac{2R}{S} = \frac{l'}{100 - l'}$$

$\Rightarrow \quad 2 \times \dfrac{1}{3} = \dfrac{l'}{100 - l'} = 3l' = 200 - 2l'$

$\Rightarrow \quad 5l' = 200$

$\Rightarrow \quad l' = 40$ **Ans. (40.00)**

Sol. 33 If λ be the potential gradient then for balancing length $l = 100 - 49 = 51$ cm, we use

$$E_2 = \lambda l$$

$\Rightarrow \quad 1.05 = \lambda \times 51$

$\Rightarrow \quad \lambda = \dfrac{1.02}{51}$

$\Rightarrow \quad \lambda = 0.02\text{ V/cm}$ **Ans. (A)**

Sol. 34 (A) Is incorrect as there is no connectivity between plates of capacitor so infinite resistance

(B) Is correct because LED works only in forward based mode

(C) For metal wire if considered as a perfect conductor, this is correct

(D) For a resistance both sides resistance value is equal so this is correct. **Ans. (A)**

Sol. 35 Effective circuit is shown below

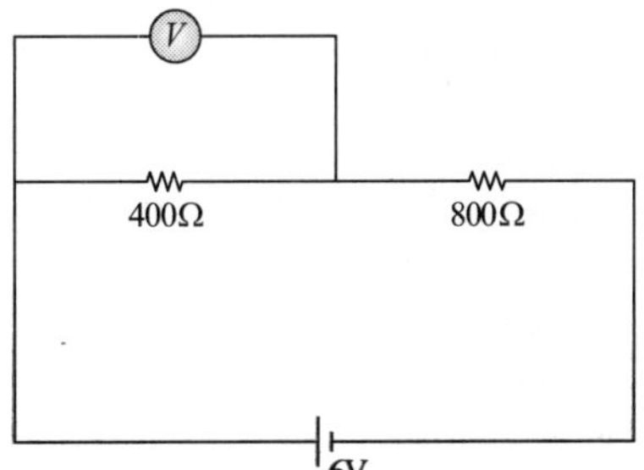

Equivalent resistance across battery is

$$R_{eq} = \frac{400 \times 10000}{10400} + 800 = 1185\ \Omega$$

Potential difference measured is

$$V_{400\Omega} = \frac{6}{1185} \times 400 \cong 1.95\text{ V}$$ **Ans. (B)**

Sol. 36 In first case, we use

$$1 = I_g(G + R_1) \quad ...(1)$$

In second case, we use

$$2 = I_g(G + R_1 + R_2) \quad ...(2)$$

From (1) and (2), we use

$$\frac{G + R_1 + R_2}{G + R_1} = 1$$

$$\Rightarrow \quad 1 + \frac{R_2}{G + R_1} = 2$$

$$\Rightarrow \quad R_2 = G + R_1$$

Ans. (B)

Sol. 37 Figure of Merit of galvanometer is given as

$$C = \frac{i}{\theta}$$

$$\Rightarrow \quad C = \frac{6 \times 10^{-3}}{2} = 3 \times 10^{-3}\,\text{Am}^2$$

Ans. (A)

Sol. 38 For normal operation of circuit ammeter is to be connected in series because it has low resistance and voltmeter is to be connected in parallel as it has high resistance.

Ans. (A)

Sol. 39 Length of potentiometer wire is given as

$$AB = 10\,\text{m}$$

For battery E_1, balancing length is $l_1 = 380$ cm
For battery E_2, balancing length is $l_2 = 760$ cm
For the given value of potential gradient on wire, we can use

$$\frac{E_1}{E_2} = \frac{l_1}{l_2}$$

$$\Rightarrow \quad \frac{E_1}{E_2} = \frac{380}{760} = \frac{1}{2} = \frac{a}{b}$$

$$a = 1 \quad \& \quad b = 2$$

$$a = 1$$

Ans. (1)

Sol. 40 Under balanced state, by Wheatstone bridge condition, we have

$$\frac{12}{6} = \frac{x}{72 - x}$$

$$\Rightarrow \quad x = 48\,\text{cm}$$

Ans. (48)

Sol. 41 At null deflection in galvanometer, the maximum value of unknown *EMF* that can be measured is V_{AB} which is given as

$$V_{AB} = \frac{6}{100 + 20} \times 100 = 5\,\text{V}$$

Ans. (C)

Sol. 42 When the sliding contact is at the middle of the wire, the equivalent circuit of the given potentiometer is given as

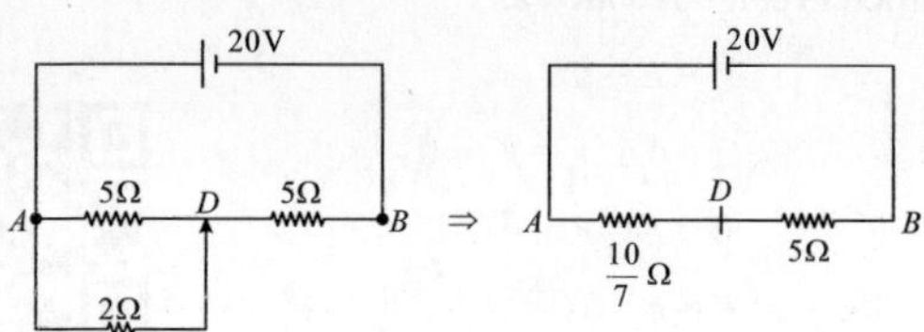

The potential drop across 2 Ω resistance can now be given as

$$V_{2\Omega} = \frac{20}{\frac{10}{7} + 5} \times \frac{10}{7}$$

$$\Rightarrow \quad V_{2\Omega} = \frac{40}{9}\,\text{V}$$

Ans. (A)

Sol. 43 The ratio of secondary circuit emf at the two balancing lengths can be given as

$$\frac{\varepsilon_1}{\varepsilon_1 + \varepsilon_2} = \frac{\lambda \times 250}{\lambda \times 400}$$

$$\Rightarrow \quad 8\varepsilon_1 = 5\varepsilon_1 + 5\varepsilon_2$$

$$\Rightarrow \quad 3\varepsilon_1 = 5\varepsilon_2$$

$$\Rightarrow \quad \frac{\varepsilon_1}{\varepsilon_2} = \frac{5}{3}$$

Ans. (B)

Sol. 44 Current sensitivity of the galvanometer is given as

$$C = \frac{\theta}{I_g}$$

$$\Rightarrow \quad I_g = \frac{\theta}{C} = \frac{25\text{ div}}{2\text{ div/mA}} = 25\text{ mA}$$

Potential difference required for full deflection is 50 mV at which 25 mA current flows so resistance of galvanometer is given as

$$R = \frac{V}{I} = \frac{50\text{mV}}{25\text{mA}} = 2\,\Omega$$

Ans. (B)

Sol. 45 Figure below shows the case described in the question

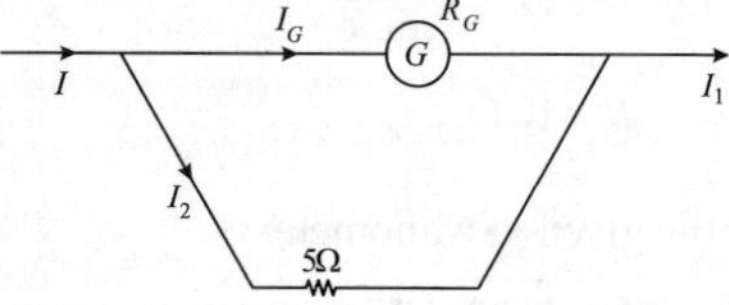

In above circuit galvanometer current which is 2% of the total current, can be given as

$$I_G = \frac{2}{100} \times I = \frac{5}{R_G + 5} \times I$$

$$\Rightarrow \quad \frac{5}{R_G+5}=\frac{1}{50}$$

$$\Rightarrow \quad 250=R_G+5$$

$$\Rightarrow \quad R_G=245\,\Omega$$

Ans. (B)

18.1 Magnetic Effects of Current

Sol. 1

Magnetic induction due to a finite length wire is given as

$$B_0=\frac{\mu_0 I}{4\pi r}(\cos\alpha+\cos\beta)$$

$$\Rightarrow \quad B_0=\frac{\mu_0 I}{4\pi\left(\frac{l}{2\sqrt{3}}\right)}(2\cos 30°) \qquad [\because r=\frac{l}{2\sqrt{3}}]$$

$$\Rightarrow \quad B_0=\frac{\sqrt{3}\times 4\pi\times 10^{-7}\times 1}{2\pi\times 4.5\times 10^{-2}}\times 2\times\frac{\sqrt{3}}{2}$$

Net MI at O is given as

$$B_{net}=3B_0=\frac{3\times 3\times 2\times 10^{-5}}{4.5}$$

$$\Rightarrow \quad B_{net}=4\times 10^{-5}\,\text{Wb/m}^2$$

Ans. (A)

Sol. 2 Magnetic induction due to the two coils at P are in same direction so we have.

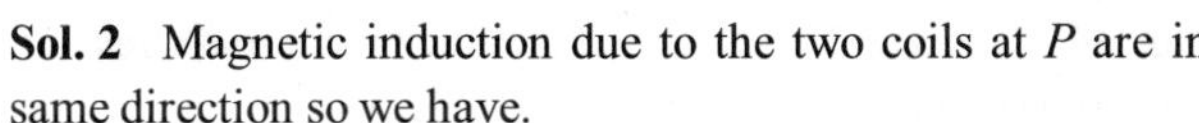

$$B=\frac{\mu_0 I R^2}{2(x^2+R^2)^{3/2}}\times 2\Bigg|_{x=R/2}\times N$$

$$\Rightarrow \quad B=\frac{\mu_0 I R^2 N}{\left(\frac{R^2}{4}+R^2\right)^{3/2}}=\frac{\mu_0 I R^2 N}{\left(\frac{5}{4}\right)^{3/2}R^3}$$

$$\Rightarrow \quad B=\frac{8\mu_0 I N}{(5)^{3/2}R}$$

Ans. (B)

Sol. 3 Magnetic field at O is given as

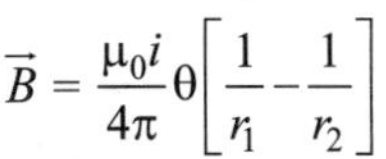

$$\vec{B}=\frac{\mu_0 i}{4\pi}\theta\left[\frac{1}{r_1}-\frac{1}{r_2}\right]$$

$$\Rightarrow \quad \vec{B}=\frac{4\pi\times 10^{-7}}{16}\times 10\times\left[\frac{1}{3\times 10^{-2}}-\frac{1}{5\times 10^{-2}}\right]$$

$$\Rightarrow \quad \vec{B}=\frac{\pi}{3}\times 10^{-5}\,\text{T}$$

Ans. (A)

Sol. 4

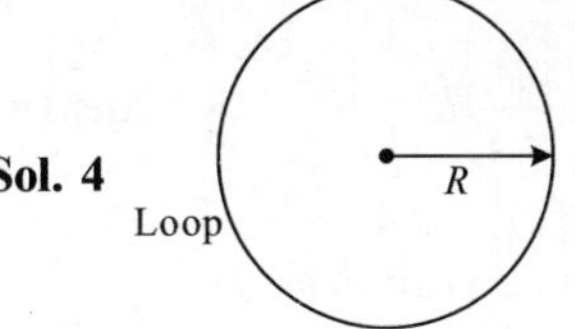

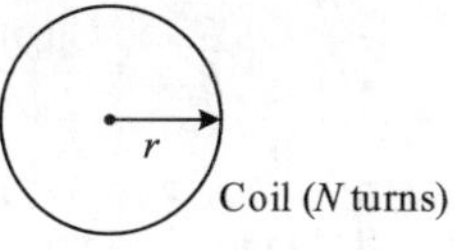

For loop $L=2\pi R$ For coil $L=N\times 2\pi r$

$$\Rightarrow \quad R=Nr$$

Magnetic field at the centres are given as

$$B_L=\frac{\mu_0 i}{2R}$$

and

$$B_C=\frac{\mu_0 N i}{2r}=\frac{\mu_0 N^2 i}{2R}$$

$$\Rightarrow \quad \frac{B_L}{B_C}=\frac{1}{N^2}$$

Ans. (C)

Sol. 5 Magnetic field at 'O' will be only due to 'PS' and 'QN' only, so we have

$$B_0=B_{PS}+B_{QN}$$

Let current in each wire is i so magnetic field at O is given as

$$B_0=\frac{\mu_0 i}{4\pi d}+\frac{\mu_0 i}{4\pi d}$$

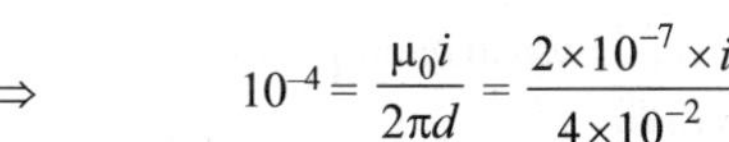

$$\Rightarrow \quad 10^{-4}=\frac{\mu_0 i}{2\pi d}=\frac{2\times 10^{-7}\times i}{4\times 10^{-2}}$$

$$\Rightarrow \quad i=20\,\text{A (into the page)}$$

Ans. (D)

Sol. 6 Dimension of $\sqrt{\frac{\varepsilon_0}{\mu_0}}$

$$[\varepsilon_0]=[M^{-1}L^{-3}T^4A^2]$$

$$[\mu_0]=[MLT^{-2}A^{-2}]$$

Dimensions of

$$\sqrt{\frac{\varepsilon_0}{\mu_0}}=\left[\frac{M^{-1}L^{-3}T^4A^2}{MLT^{-2}A^{-2}}\right]^{\frac{1}{2}}$$

$$=[M^{-2}L^{-4}T^6A^4]^{1/2}$$

$$=[M^{-1}L^{-2}T^3A^2]$$

Ans. (B)

Sol. 7

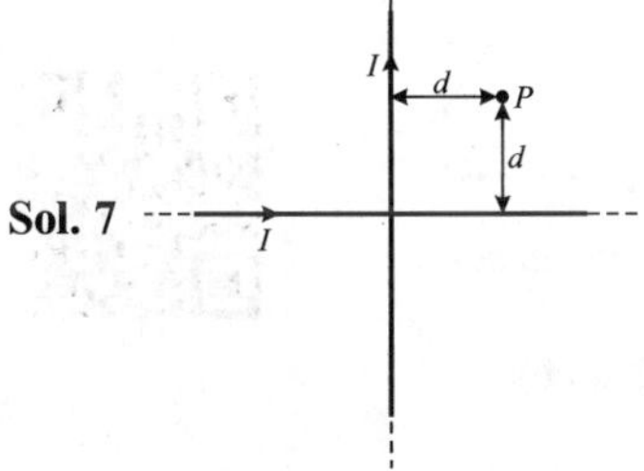

Magnetic field at point P is given as

$$\vec{B}_{\text{net}} = \frac{\mu_0 i}{2\pi d}(-\hat{k}) + \frac{\mu_0 i}{2\pi d}(\hat{k}) = 0 \qquad \textbf{Ans. (A)}$$

Sol. 8 Magnetic field at centroid O is given as

$$B = 3\left[\frac{\mu_0 i}{4\pi r}(\sin 60° + \sin 60°)\right]$$

Here r is given as

$$r = \frac{a}{2\sqrt{3}} = \frac{1}{2\sqrt{3}}$$

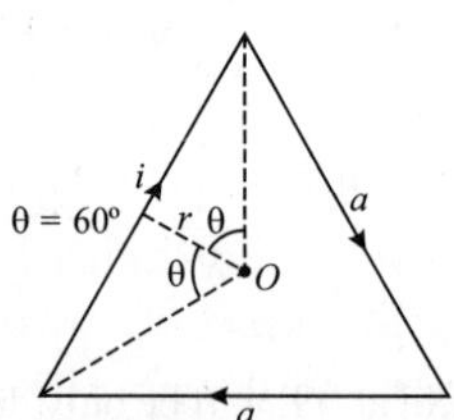

$$\Rightarrow \qquad B = 3\left[\frac{4\pi\times10^{-7}\times10\times2\sqrt{3}}{4\pi\times1}\left[\frac{\sqrt{3}}{2}+\frac{\sqrt{3}}{2}\right]\right]$$

$$\Rightarrow \qquad B = 18\times10^{-6} = 18\,\mu\text{T} \qquad \textbf{Ans. (A)}$$

Sol. 9 Dimensions of given physical quantities are

$$[\varepsilon_0] = \text{M}^{-1}\,\text{L}^{-3}\,\text{T}^{4}\,\text{A}^{2}$$

$$[\mu_0] = \text{M}\,\text{L}\,\text{T}^{-2}\,\text{A}^{-2}$$

$$[R] = \text{M}\,\text{L}^{2}\,\text{T}^{-3}\,\text{A}^{-2}$$

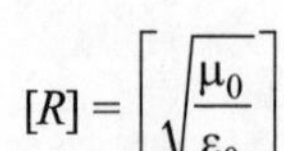

$$\Rightarrow \qquad [R] = \left[\sqrt{\frac{\mu_0}{\varepsilon_0}}\right] \qquad \textbf{Ans. (C)}$$

Sol. 10 Magnetic field due to a rotating ring is given as

$$B = \frac{\mu_0 i}{2R} = \frac{\mu_0 q\omega}{2R2\pi}$$

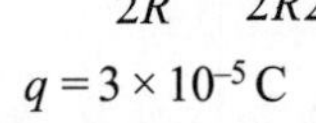

$$\Rightarrow \qquad q = 3\times10^{-5}\,\text{C}$$

Sol. 11

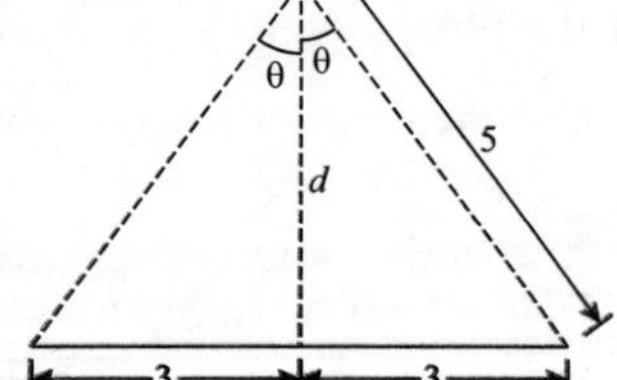

Magnetic field at point P is given as

$$B = \frac{\mu_0 I}{4\pi d}(2\sin\theta)$$

Here $\qquad d = 4$ cm and $\sin\theta = \dfrac{3}{5}$

$$\Rightarrow \qquad B = \frac{10^{-7}\times5}{0.04}\times2\times\frac{3}{5} = 1.5\times10^{-5}\,\text{T} \qquad \textbf{Ans. (D)}$$

Sol. 12 The given expression is of magnetic energy stored per unit volume so its dimensions are $\text{ML}^{-1}\,\text{T}^{-2}$. **Ans. (A)**

Sol. 13 The given wire loop can be considered in three parts A, B and C as shown

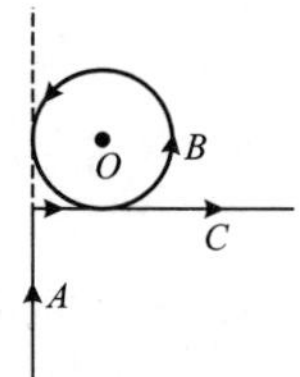

Magnetic field at point O is given as

$$B_0 = B_A + B_B + B_C$$

$$\Rightarrow \qquad B_0 = \frac{\mu_0 I}{4\pi R}(\sin 90° - \sin 45°)\otimes + \frac{\mu_0 I}{2R}\odot$$

$$+ \frac{\mu_0 I}{4\pi R}(\sin 45° + \sin 90°)\odot$$

$$\Rightarrow \qquad B_0 = \frac{\mu_0 I}{2\pi R}(\sin 45° + \pi)$$

$$\Rightarrow \qquad B_0 = \frac{\mu_0 I}{2\pi R}\left(\pi + \frac{1}{\sqrt{2}}\right) \qquad \textbf{Ans. (C)}$$

Sol. 14 If current density in wire is J, then magnitude field due to wire is outside is given as

$$B_1 = \frac{\mu_0 I}{2\pi r} = \frac{\mu_0(J\pi a^2)}{2\pi(2a)} = \frac{\mu_0 J a}{4}$$

Due to wire inside field is given as

$$B_2 = \frac{1}{2}\mu_0 J r = \frac{1}{2}\mu_0 J\left(\frac{a}{3}\right) = \frac{\mu_0 J a}{6}$$

$$\Rightarrow \qquad \frac{B_2}{B_1} = \frac{\mu_0 J a\times4}{\mu_0 J 6a} = \frac{2}{3} \qquad \textbf{Ans. (A)}$$

Sol. 15 In figure shown below, magnetic field due to coil at centre is given as

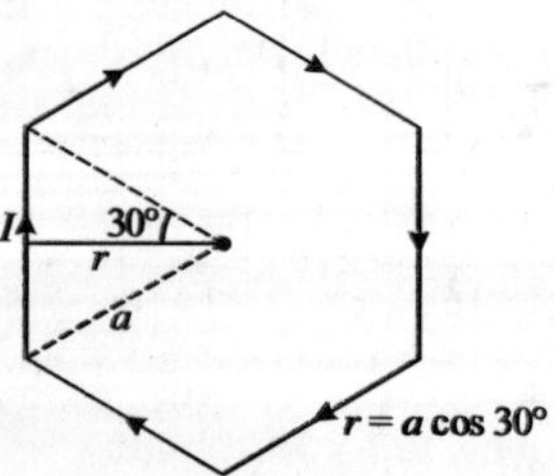

$$B = \frac{6\mu_0 I}{4\pi r}(2\sin 30^\circ)\times 50$$

$$\Rightarrow \quad B = \frac{6\mu_0 I}{4\pi a\cos 30^\circ}\times 2\sin 30^\circ \times 50$$

$$\Rightarrow \quad B = \frac{\mu_0 I}{\pi}\frac{150}{\sqrt{3}a} = \frac{50\sqrt{3}}{0.1}\frac{\mu_0 I}{\pi}$$

$$\Rightarrow \quad B = 500\sqrt{3}\frac{\mu_0 I}{\pi}$$ **Ans. (C)**

Sol. 16 In given figure magnetic fields due to wires A and B are

$$B_A = \frac{\mu_0 I\left(\frac{3\pi}{2}\right)}{4\pi R_A}$$

and $$B_B = \frac{\mu_0 I\left(\frac{5\pi}{3}\right)}{4\pi R_B}$$

$$\Rightarrow \quad \frac{B_A}{B_B} = \frac{I_A(3/2)}{R_A}\times\frac{R_B}{I_B(5/6)}$$

$$\Rightarrow \quad \frac{B_A}{B_B} = \frac{2(3/2)\times 4}{2\times 3\times(5/3)} = \frac{12}{10} = \frac{6}{5}$$ **Ans. (C)**

Sol. 17 The magnetic field on the axis of a current carrying circular coil is is given as

$$B = \frac{\mu_0}{4\pi}\frac{2NIA}{(R^2+x^2)^{3/2}}$$

$$\Rightarrow \quad \frac{B_1}{B_2} = \frac{8}{1} = \left[\frac{R^2+(0.2)^2}{R^2+(0.05)^2}\right]^{3/2}$$

$$\Rightarrow \quad 4[R^2+(0.05)^2] = [R^2+(0.2)^2]$$

$$\Rightarrow \quad 4R^2 - R^2 = (0.2)^2 - 4\times(0.05)^2$$

$$\Rightarrow \quad 3R^2 = 0.3\times 0.1$$

$$\Rightarrow \quad R^2 = (0.1)^2$$

$$\Rightarrow \quad R = 0.1$$ **Ans. (A)**

Sol. 18 As force on a charge by magnetic field is always perpendicular to $\vec{v}$, the work by magnetic force on the point charge is zero. **Ans. (C)**

Sol. 19 The magnetic flux density (or magnetic induction) due to a long solenoid can be given as

$$B = \mu n I = \mu_0\mu_r n I$$

$$\Rightarrow \quad B = 4\pi\times 10^{-7}\times 500\times 1000\times 5 = \pi \text{ Tesla}$$ **Ans. (A)**

Sol. 20 Net magnetic field at point P can be given as

$$B = 2\times B_{\text{wires}} + B_{\text{loop}}$$

$$\Rightarrow \quad B = 2\times\frac{\mu_0 i}{4\pi r} + \frac{\mu_0 i}{2r}\left(\frac{\pi}{2\pi}\right)$$

$$\Rightarrow \quad B = \frac{\mu_0 i}{4\pi r}(2+\pi)$$ **Ans. (B)**

Sol. 21 Magnetic induction B due to a current carrying straight wire of circular cross section inside the body of wire is given as

$$B = \frac{\mu_0 J r}{2}$$

Where current density is given as

$$J = \frac{i}{\pi R^2}$$

Thus in the given curve, slope of straight line is proportional to the current density J and outside the wire magnetic induction varies with inverse relation to distance r. The maximum magnetic induction on surface of wire is given as

$$B_{\max} = \frac{\mu_0 J R}{2} = \frac{\mu_0 i R}{2\pi R^2} = \frac{\mu_0 i}{2\pi R}$$

As $b > a$ so slope of line for second wire is less than that of first wire and maximum magnetic induction will also be more for the wire with smaller radius hence option (B) is most appropriate. **Ans. (B)**

Sol. 22 Figure below shows the vectors of eletric field due to the charges specified in question.

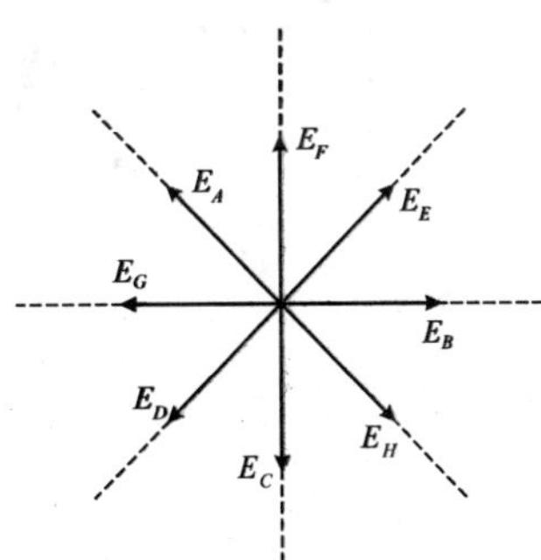

The magnitude of electric fields due to the charges are given as

$$E_A = \frac{kq}{(\sqrt{2}l)^2} = \frac{1}{2}\frac{kq}{l^2}$$

$$E_B = \frac{kq}{(l)^2} = \frac{kq}{l^2}$$

$$E_C = \frac{2kq}{(l)^2} = 2\frac{kq}{l^2}$$

$$E_D = \frac{kq}{(\sqrt{2}l)^2} = \frac{1}{2}\frac{kq}{l^2}$$

$$E_E = \frac{2kq}{(\sqrt{2}l)^2} = \frac{kq}{l^2}$$

$$E_F = \frac{kq}{(l)^2} = \frac{kq}{l^2}$$

$$E_G = \frac{2kq}{(l)^2} = 2\frac{kq}{l^2}$$

$$E_H = \frac{kq}{(\sqrt{2}l)^2} = \frac{1}{2}\frac{kq}{l^2}$$

Calculating the resultant of all fields gives

$$E_{\text{net}} = \sqrt{2}\frac{kq}{l^2} - \frac{kq}{2l^2} = \frac{1}{4\pi\varepsilon_0}\frac{q}{(2l^2)}(2\sqrt{2}-1)$$

Ans. (A)

Sol. 23 Due to a current carrying coil magnetic field at a distance x along its axis is given as

$$B_{\text{axis}} = \frac{\mu_0 i a^2}{2(a^2+r^2)^{3/2}}$$

Magnetic field at centre of the coil is given as

$$B_{\text{centre}} = \frac{\mu_0 i}{2R}$$

Fractional change in magnetic field is given as

$$f = \frac{\frac{\mu_0 i}{2a} - \frac{\mu_0 i a^2}{2(a^2+r^2)^{3/2}}}{\frac{\mu_0 i}{2a}}$$

$$\Rightarrow \quad f = 1 - \frac{1}{\left[1+\left(\frac{r^2}{a^2}\right)\right]^{3/2}} \quad (\text{As } r << a)$$

$$\Rightarrow \quad f \approx 1 - \left[1 - \frac{3}{2}\frac{r^2}{a^2}\right] = \frac{3}{2}\frac{r^2}{a^2}$$

Ans. (B)

Sol. 24 By Ampere's law, we have

$$B_1 \times 2\pi(x) = \frac{\mu_0 I_0 (x)^2}{a^2}$$

and $\quad B_2 \times 2\pi(x) = \mu_0 I_0$

$$\Rightarrow \quad \frac{B_1}{B_2} = \frac{x^2}{a^2}$$

Ans. (D)

Sol. 25 Figure below shows the spiral coil as described in the question. To calculate magnetic field at the center we consider an elemental circular coil of radius x and width dx in it as shown.

The number of turns in this elemental coil is given as

$$dN = \frac{N}{b-a}dx$$

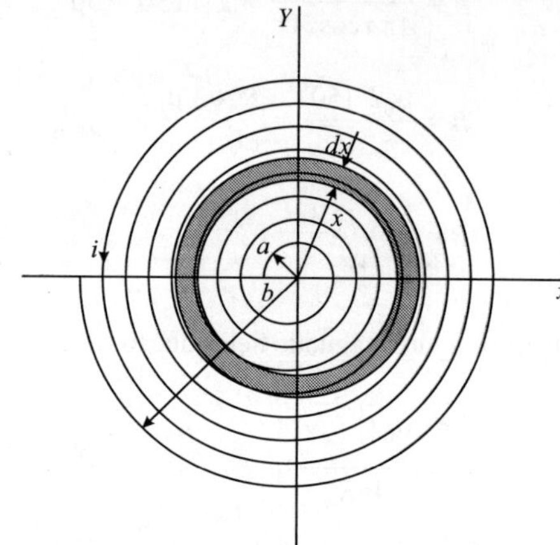

The magnetic induction due to the elemental coil is given as

$$dB = \frac{\mu_0 i}{2x} \times dN$$

Total magnetic induction due to whole coil at center is given as

$$B = \int \frac{\mu_0 dNi}{2x} = \int \frac{\mu_0\left(\frac{N}{b-a}dx\right)i}{2x}$$

$$\Rightarrow \quad B = \frac{\mu_0 Ni}{2(b-a)} \ln\frac{b}{a}$$

Ans. (B)

Sol. 26 Figure below shows the triangular coil as described in the question

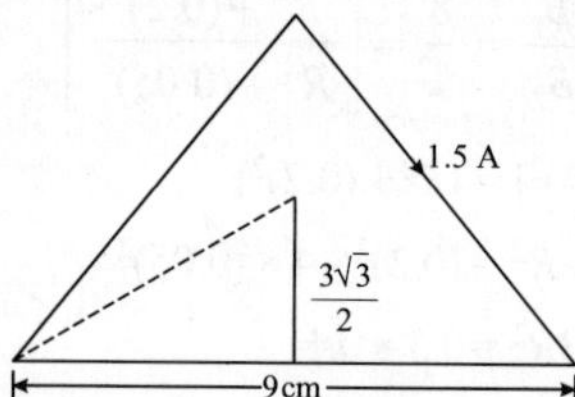

The magnetic field at the centroid of the triangle due to one side of triangle is given as

$$B_1 = \frac{\mu_0 I}{4\pi R}(2\sin 60°) = \frac{4\pi\times10^{-7}\times1.5}{4\pi.\left(\frac{3\sqrt{3}}{2}\right)\times10^{-2}}\times2\times\frac{\sqrt{3}}{2} = 1\times10^{-5}$$

Magnetic field due to whole triangle is given as

$$B_{\text{net}} = 3B_1 = 3\times10^{-5}\,\text{T}$$

As the current is in clockwise manner, direction of field will be into the plane of triangle.

Ans. (D)

Sol. 27 The magnetic field due to the wire at point P is given as

$$B=\frac{\mu_0 I}{4\pi y}(1+\sin\theta_1)+\frac{\mu_0 I}{4\pi x}(1+\sin\theta_2)$$

$$\Rightarrow\; B=\frac{\mu_0 I}{4\pi y}\left(1+\frac{x}{\sqrt{x^2+y^2}}\right)+\frac{\mu_0 I}{4\pi x}\left(1+\frac{y}{\sqrt{x^2+y^2}}\right)$$

$$\Rightarrow\; B=\frac{\mu_0 I}{4\pi x}+\frac{\mu_0 I}{4\pi y}+\frac{\mu_0 I}{4\pi}\left[\frac{x}{y\sqrt{x^2+y^2}}+\frac{y}{x\sqrt{x^2+y^2}}\right]$$

$$\Rightarrow\; B=\frac{\mu_0 I}{4\pi}\left[\frac{1}{x}+\frac{1}{y}\right]+\frac{\mu_0 I}{4\pi}\frac{\sqrt{x^2+y^2}}{xy}$$

$$\Rightarrow\; B=\frac{\mu_0 I}{4\pi}\left[\frac{x+y+\sqrt{x^2+y^2}}{xy}\right]$$ **Ans. (D)**

18.2 Magnetic Force on Moving Charges

Sol. 1 A loop with a current is in stable equilibrium in a uniform magnetic field when its PE is minimum, which is given as

$$U=-\vec{M}\cdot\vec{B}=-MB\cos\theta$$

Stable equilibrium $\Rightarrow \theta=0°$ when $\vec{M}$ is parallel to $\vec{B}$. **Ans. (C)**

Sol. 2 By Lorentz force equation, force on a moving charge in EM field is given as

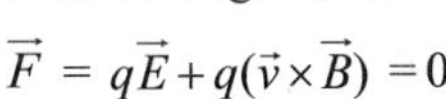

$$\vec{F}=q\vec{E}+q(\vec{v}\times\vec{B})=0$$

[As no force is acting on charge]

$$\vec{E}=-(\vec{v}\times\vec{B})=-v_0(3\hat{i}-\hat{j}+2\hat{k})\times B_0(\hat{i}+2\hat{j}-4\hat{k})$$

$$\vec{E}=-v_0B_0(6\hat{k}+12\hat{j}+\hat{k}+4\hat{k}+2\hat{j}-4\hat{i})$$

$$\vec{E}=-v_0B_0(14\hat{j}+7\hat{k})$$ **Ans. (D)**

Sol. 3

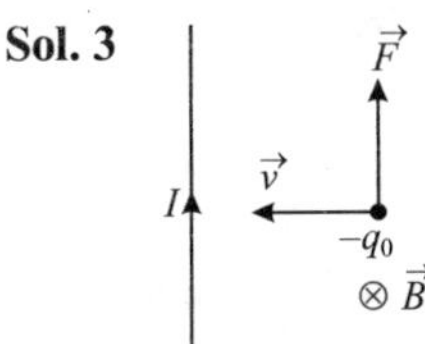

As force $\vec{F}$ and magnetic induction $\vec{B}$ is specified then using Flaming's left hand rule or Right hand Palm Rule we can find the direction of $\vec{v}$ for $-q_0$ charge.

By RHPR fingers point along $\vec{B}$

Palm face vector along force $\vec{F}$

$\Rightarrow$ direction of thumb which is toward right gives the $\vec{v}$ for +ve charge so here as charge is –ve $\Rightarrow$ $\vec{v}$ is leftward. **Ans. (B)**

Sol. 4 Case-I :

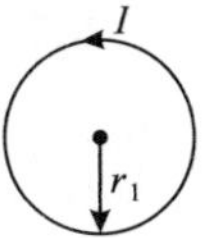

Magnetic moment $m=I\pi r_1^2$

Magnetic induction at centre

$$B_1=\frac{\mu_0 I}{2r_1} \qquad \ldots(1)$$

Case-II : $I\rightarrow$ constant but $m'=2m$

$$m'=I\pi r_2^2=(I\pi r_1^2)$$

$$\Rightarrow\; r_2=\sqrt{2}\,r_1$$

$$B_2=\frac{\mu_0 I}{2r_2}=\frac{\mu_0 I}{2\sqrt{2}r_1} \qquad \ldots(2)$$

By (1)/(2) $\Rightarrow$ $\dfrac{B_1}{B_2}=\sqrt{2}$ **Ans. (C)**

Sol. 5 In a magnetic field for a charge particle in circular motion, radius is given as

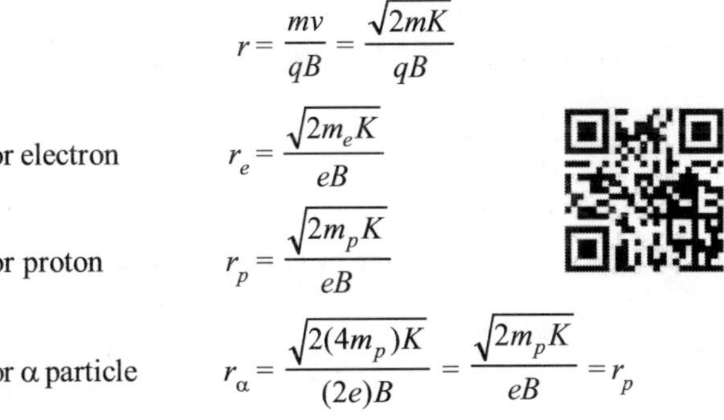

$$r=\frac{mv}{qB}=\frac{\sqrt{2mK}}{qB}$$

for electron $r_e=\dfrac{\sqrt{2m_eK}}{eB}$

for proton $r_p=\dfrac{\sqrt{2m_pK}}{eB}$

for α particle $r_\alpha=\dfrac{\sqrt{2(4m_p)K}}{(2e)B}=\dfrac{\sqrt{2m_pK}}{eB}=r_p$

As $m_e<m_p$

$\Rightarrow$ $r_e<r_p=r_\alpha$ **Ans. (B)**

Sol. 6

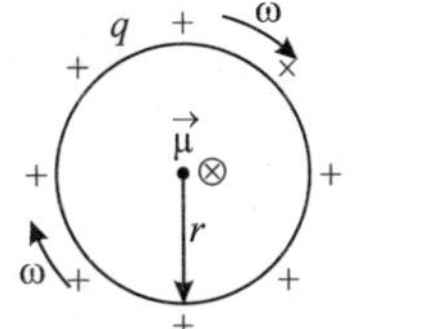

The rotating charged loop behaves like a circular current, which is given as

$$I=\frac{q\omega}{2\pi}$$

⇒ Magnetic moment of this rotating charge is given as

$$\mu = I \times \pi r^2 = \frac{q\omega}{2\pi} \times \pi r^2 = \frac{1}{2}q\omega r^2 \quad \textbf{Ans. (A)}$$

Alternative : We use

$$\frac{\mu}{L} = \frac{q}{2m}$$

$$\Rightarrow \quad \mu = \frac{q}{2m} \times L = \frac{q}{2m} \times mr^2\omega$$

$$\Rightarrow \quad \mu = \frac{1}{2}q\omega r^2 \qquad \textbf{Ans. (A)}$$

Sol. 7 Force on loop due to wire is given as

$$|\vec{F}| = m\left|\frac{dB}{dr}\right|$$

Where m is the dipole moment of loop

$$m = i\pi a^2$$

We use

$$\frac{dB}{dx} = \frac{d}{dx}\left(\frac{\mu_0 I}{2\pi x}\right) = -\frac{\mu_0 I}{2\pi x^2}$$

$$\Rightarrow \quad F = \frac{\mu_0 m I}{2\pi x^2} = \frac{\mu_0 i I \pi a^2}{2\pi x^2}$$

$$\Rightarrow \quad F \propto \frac{a^2}{d^2} \qquad [\text{As } x = d] \quad \textbf{Ans. (C)}$$

According to Newtons III law same force will be exerted by loop on wire.

Sol. 8 In circular motion under magnetic field, we use

$$\frac{mv^2}{R} = qvB$$

$$mv = qBR \qquad \dots(1)$$

As path becomes straight line by electric field, we use

$$qE = qvB$$

$$\Rightarrow \quad E = vB \qquad \dots(2)$$

From equation-(1) & (2)

$$m = \frac{qB^2R}{E}$$

$$\Rightarrow \quad m = 2.0 \times 10^{-24}\,\text{kg} \quad \textbf{Ans. (A)}$$

Sol. 9 Charge on element of with dx on rod revolving in a circle of radius x is given as

$$dq = \lambda dx$$

its magnetic moment is given as

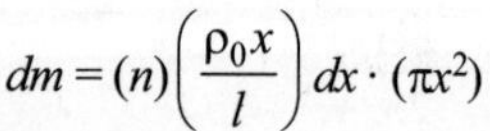

$$dm = (n)\left(\frac{\rho_0 x}{l}\right)dx \cdot (\pi x^2)$$

Total magnetic moment of rod is

$$M = \int dm = \frac{n\rho_0\pi}{l}\cdot\int_0^l x^3 \cdot dx = \frac{n\rho_0\pi}{l}\cdot\left[\frac{l^4}{4}\right]$$

$$\Rightarrow \quad M = \frac{n\rho_0\pi l^3}{4} \qquad \textbf{Ans. (A)}$$

Sol. 10 Radius of circular motion of electron is

$$r = \frac{\sqrt{2mK}}{eB} = \frac{\sqrt{2meV}}{eB}$$

$$\Rightarrow \quad r = \frac{\sqrt{\frac{2m}{e}\cdot V}}{B} = \frac{\sqrt{\frac{2\times 9.1\times 10^{-31}}{1.6\times 10^{-19}}(500)}}{100\times 10^{-3}}$$

$$\Rightarrow \quad r = \frac{\sqrt{\frac{9.1}{0.16}\times 10^{-10}}}{10^{-1}} = \frac{3}{.4}\times 10^{-4}\,\text{m}$$

$$\Rightarrow \quad r = 7.5 \times 10^{-4}\,\text{m} \qquad \textbf{Ans. (A)}$$

Sol. 11

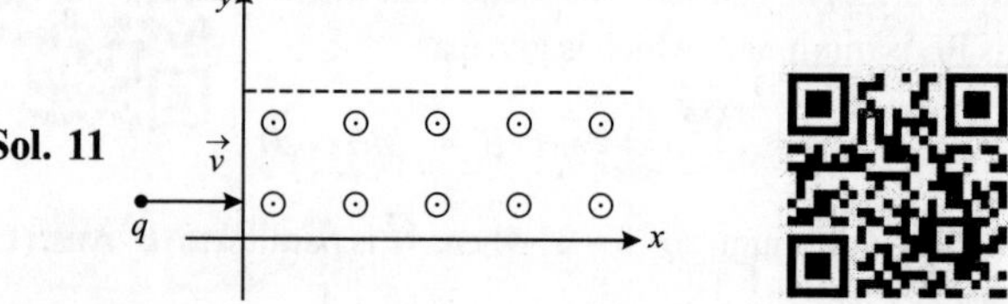

As the initial position of entering particle is not given, acceleration at leaving point cannot be calculated.

Ans. (None)

Sol. 12 Work will only be done against electric force, equal to

$$W = q\vec{E}\cdot\vec{r}$$

$$\Rightarrow \quad W = q(2\hat{i} + 3\hat{j})\cdot(\hat{i} + \hat{j}) = 5q \quad \textbf{Ans. (D)}$$

Sol. 13 Kinetic energy of a charge q accelerated by a potential difference V is

$$k = qV$$

Radius of circular motion in magnetic field is given as

$$r = \sqrt{\frac{2mk}{q_B}} = \sqrt{\frac{2mkV}{q_B}}$$

$$\Rightarrow \quad r \propto \sqrt{\frac{m}{q}}$$

$$\Rightarrow \quad \frac{r_\rho}{r_\alpha} = \frac{1}{\sqrt{2}} \qquad \textbf{Ans. (A)}$$

Sol. 14

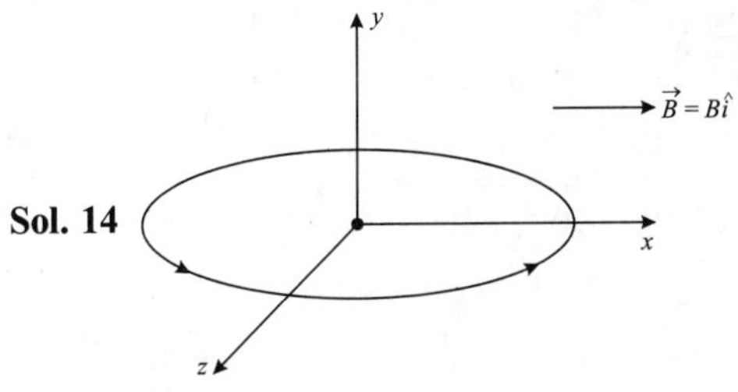

Magnetic moment of coil is given as

$$\vec{M} = NIA\,\hat{j}$$

$$\vec{\tau} = NI(\pi r^2)\hat{j}$$

Torque on loop (coil)

$$= \vec{M}\times\vec{B}$$

$$= NI(\pi r^2)\,B\sin 90^\circ\,(-\hat{k})$$

$$= NI\pi r^2 B(-\hat{k})$$

Ans. (A)

Sol. 15 Given that resitance of galvanometer and shunt are

$$R_g = 50\,\Omega$$

$$R_S = 5000\,\Omega$$

Full deflecttion current is

$$i_g = 4\times 10^{-3}$$

Thus range of voltmeter is given as

$$V_R = i_g(G+S)$$

$$\Rightarrow \quad V_R = 4\times 10^{-3}(50+5000)\,V$$

$$\Rightarrow \quad V_R = 4\times 10^{-3}(5050)\,V$$

$$\Rightarrow \quad V_R = 20.2\,V$$

Ans. (B)

Sol. 16 Force acting on segments of square frame are shown in figure below

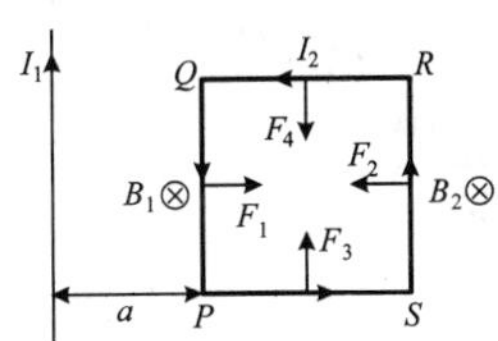

Here F_3 & F_4 cancel each other and force on PQ will be

$$F_1 = I_2 B_1 a$$

$$\Rightarrow \quad F_1 = I_2\left(\frac{\mu_0 I_1}{2\pi a}\right)a$$

$$\Rightarrow \quad F_1 = \frac{\mu_0 I_1 I_2}{2\pi}$$

Force on RS will be

$$F_2 = I_2 B_2 a$$

$$\Rightarrow \quad F_2 = I_2\frac{\mu_0 I_1}{2\pi(2a)}a$$

$$\Rightarrow \quad F_2 = \frac{\mu_0 I_1 I_2}{4\pi}$$

Net force is given as

$$F_{net} = F_1 - F_2 = \frac{\mu_0 I_1 I_2}{4\pi} \text{ repulsion}$$

Ans. (B)

Sol. 17 Torque on coil is given as

$$|\vec{\tau}| = |\vec{M}\times\vec{B}|$$

$$\Rightarrow \quad \tau = NI\times A\times B\times\sin 45^\circ$$

$$\Rightarrow \quad \tau = 100\times 3\times 12.5\times 10^{-4}\times 1\times 0.707$$

$$\Rightarrow \quad \tau = 0.26\,\text{Nm} \simeq 0.27\,\text{Nm}$$

Ans. (B)

Sol. 18 Torque on coil is given as

$$\tau = \vec{M}\times\vec{B}$$

$$C\theta = iNAB$$

$$\Rightarrow \quad 10^{-6}\times\frac{\pi}{180} = 10^{-3}\times 10^{-4}\times 175\times B$$

$$\Rightarrow \quad B = 10^{-3}\,\text{T}$$

Ans. (A)

Sol. 19 Radius of circular motion in magnetic field is given as

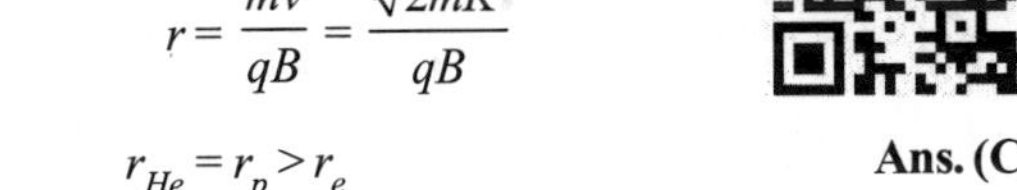

$$r = \frac{mv}{qB} = \frac{\sqrt{2mK}}{qB}$$

$$\Rightarrow \quad r_{He} = r_p > r_e$$

Ans. (C)

Sol. 20

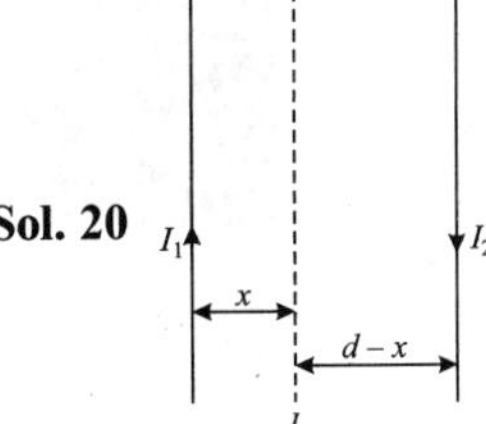

Net force on wire carrying current I per unit length is given as

$$\frac{\mu_0 I_1 I}{2\pi x} + \frac{\mu_0 I_2 I}{2\pi(d-x)} = 0$$

$$\frac{I_1}{x} = \frac{I_2}{x-d}$$

$$\Rightarrow \quad x = \frac{I_1 d}{I_1 - I_2}$$

Ans. (B)

Sol. 21 Given that $I_g = 10^{-4}$ A
and if galvanometer resistance is R_g, we use

$$I_g = \frac{V_R}{R_g + R}$$

$$10^{-4} = \frac{5}{R_g + 2\times10^6}$$

$$\Rightarrow \qquad R_g = 5\times10^4 - 2\times10^6$$

$$= -195\times10^4\,\Omega$$

Practically not possible

None of the options can be chosen. **Ans. (None)**

Sol. 22 When galvanometer is used as an ammeter shunt is used in parallel with galvanometer as shown in figure below

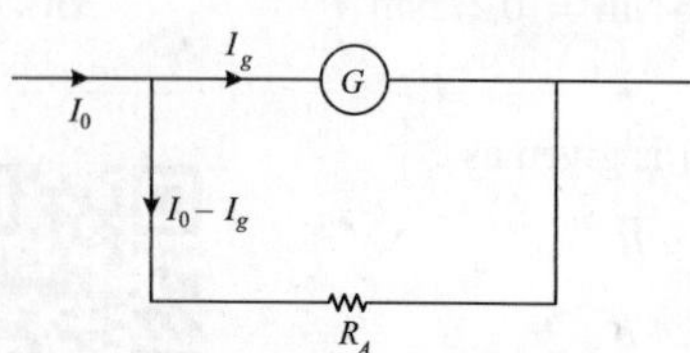

For full deflection, we use

$$I_gG = (I_0 - I_g)R_A$$

$$\Rightarrow \qquad R_A = \left(\frac{I_g}{I_0 - I_g}\right)G$$

When galvanometer is used as a voltmeter, resistance is used in series with galvanometer as shown in figure below

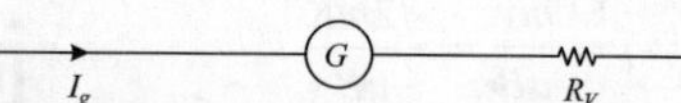

For full deflection, we use

$$I_g(G + R_V) = V = GI_0 \qquad \text{(given } V = GI_0\text{)}$$

$$\Rightarrow \qquad R_V = \frac{(I_0 - I_g)G}{I_g}$$

$$\Rightarrow \qquad R_AR_V = G^2$$

and
$$\frac{R_A}{R_V} = \left(\frac{I_g}{I_0 - I_g}\right)^2$$

Ans. (B)

Sol. 23 Angle of deflection is given as

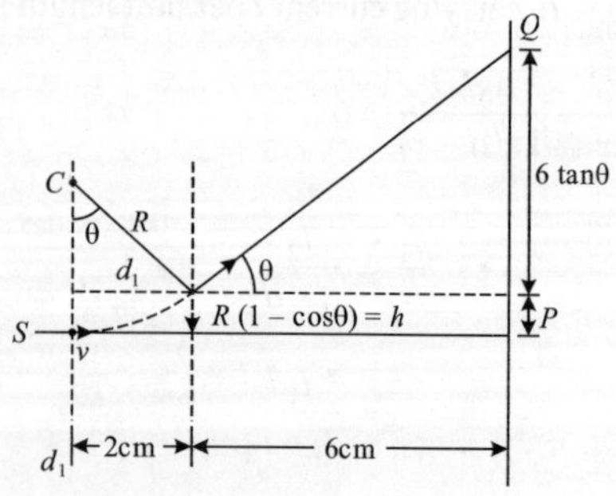

$$\sin\theta = \frac{d_1}{R} = \frac{d_1}{mv/qB} = \frac{qBd_1}{mv} = \frac{qBd_1}{\sqrt{2mE}}$$

$$\Rightarrow \qquad \sin\theta = \frac{1.6\times10^{-19}\times1.5\times10^{-3}\times0.02}{\sqrt{2\times9.1\times10^{-31}\times100\times1.6\times10^{-19}}}$$

$$= \frac{4.8\times10^{-24}}{\sqrt{29}\times10^{-24}} = \frac{4.8}{5.4} \simeq 0.89$$

$$\Rightarrow \qquad \cos\theta = \sqrt{1-(0.89)^2} = \sqrt{0.2} = 0.45$$

$$\Rightarrow \qquad \tan\theta = \frac{0.89}{0.45} \simeq 2$$

$$\Rightarrow \qquad d = PQ = 6\tan\theta + h \cong 12 + h\,\text{cm}$$

Ans. (A)

Sol. 24 We use, after time t

$$(2v_0)^2 = v_0^2 + v_x^2$$

$$\Rightarrow \qquad v_x = \sqrt{3}\,v_0$$

As particle is only accelerated by electric field in X-direction, we use

$$\sqrt{3}\,v_0 = 0 + \frac{qE_0}{m}t$$

$$\Rightarrow \qquad t = \frac{\sqrt{3}v_0m}{qE_0}$$

Ans. (C)

Sol. 25 Acceleration of protons is

$$a = \frac{qvB}{m}$$

$$\Rightarrow \qquad B = \frac{ma}{qv}$$

$$\Rightarrow \qquad B = \frac{ma\sqrt{m}}{\sqrt{2k}}$$

$$\Rightarrow \qquad B = \frac{m^{3/2}a}{e\sqrt{2k}} = \frac{(1.6\times10^{-27})^{3/2}\times10^{12}}{1.6\times10^{-19}\sqrt{2\times1\times10^6\times1.6\times10^{-19}}}$$

$$\Rightarrow \qquad B = 0.71\ \text{mT}$$

Ans. (D)

Sol. 26 As work is only done by EF on charge particle, we use

$$qE(2a) = \frac{1}{2}m(2v)^2 - \frac{1}{2}mv^2$$

$$\Rightarrow \qquad qE(2a) = \frac{3}{2}mv^2$$

$$\Rightarrow \qquad E = \frac{3}{4}\frac{mv^2}{qa}$$

Thus option (A) is correct.

Rate of work done by electric field is

$$P = \vec{F} \cdot \vec{v} = Fv\cos\theta = Fv$$

$$P = qEv$$

$$\Rightarrow \quad P = q\left(\frac{3}{4}\frac{mV^2}{qa}\right)v$$

$$\Rightarrow \quad P = q\left(\frac{3}{4}\frac{mv^3}{qa}\right)$$

$$\Rightarrow \quad P = \frac{3}{4}\frac{mv^3}{a}$$

Thus option (B) is correct.

The angle between electric force and velocity at point Q is 90°, so rate of work done will be zero at Q.

Thus option (C) is correct.

Initial angular momentum about origin is

$$L_i = mva$$

Final angular momentum about origin is

$$L_f = m(2v)(2a)$$

Change in angular momentum $L_f - L_i = 3mva$

Thus option (D) is NOT correct. **Ans. (B)**

Sol. 27 For not to hit the surface of solenoid the radius of electron motion inside must be less than $R/2$, so we use

$$\frac{R}{2} = \frac{mv}{qB}$$

$$\Rightarrow \quad \frac{R}{2} = \frac{mv_{max}}{e(\mu_0 ni)}$$

$$\Rightarrow \quad v_{max} = \frac{e\mu_0 nIR}{2m}$$

Ans. (B)

Sol. 28 Figure below shows the given situation

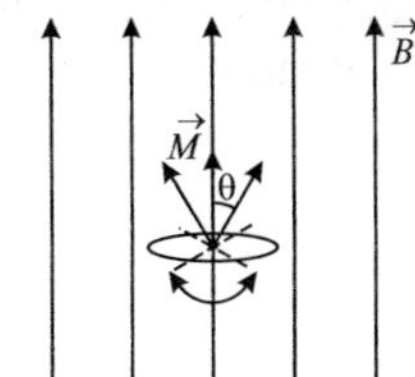

If loop is slightly tilted from equilibrium position (θ = 0°) then restoring torque on it is given as

$$\tau = -MB\sin\theta$$

$$\Rightarrow \quad \tau \simeq -MB\theta$$

Thus restoring angular acceleration is given as

$$\alpha = -\left(\frac{MB}{I}\right)\theta$$

Comparing with $\alpha = \omega^2\theta$, we get

$$\Rightarrow \quad \omega = \sqrt{\frac{MB}{I}} = \sqrt{\frac{(i)(\pi R^2)B}{\left(\frac{mR^2}{2}\right)}}$$

$$\Rightarrow \quad \omega = \sqrt{\frac{2i\pi B}{m}}$$

Thus time period of oscillations is

$$T = \frac{2\pi}{\omega} = \sqrt{\frac{2\pi m}{iB}}$$

Ans. (B)

Sol. 29 Pitch of helix in magnetic field is given as

$$p_H = \frac{2\pi m}{qB}v\cos\theta$$

$$\Rightarrow \quad p_H = \frac{2(3.14)(1.67\times10^{-27})\times4\times10^5\times\cos60^\circ}{(1.69\times10^{-19})(0.3)}$$

$$\Rightarrow \quad p_H = 0.04\,\text{m} = 4\,\text{cm}$$

Ans. (B)

Sol. 30 Acceleration of particle is given as

$$\vec{a} = \frac{-qE}{m}\hat{j}$$

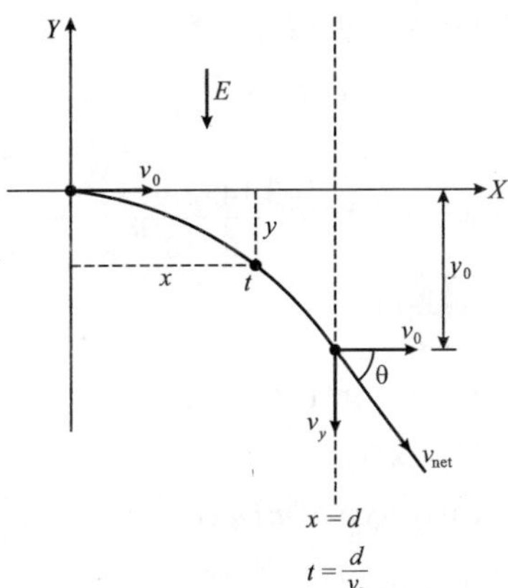

x & y coordinate of particle at time $t = t$ is

$$x = v_0 t \quad \ldots(1)$$

$$y = \frac{1}{2}\left(-\frac{qE}{m}\right)t^2 \quad \ldots(2)$$

From (1) and (2)

$$y = -\frac{qEx^2}{2mv_0^2}$$

At $x = d$, $t = \frac{d}{v}$ $\Rightarrow$ $v_x = v_0$ and $v_y = \frac{qE}{m}\left(\frac{d}{v_0}\right)$

$$\Rightarrow \quad \tan\theta = \frac{v_y}{v_0} = \frac{qEd}{mv_0^2}; \quad y_0 = \frac{-qEd^2}{2mv_0^2}$$

Equation of straight line motion of particle after $x = d$ is

$$y = -(x-d)\tan\theta - y_0$$

$$\Rightarrow \quad y = -x\left(\frac{qEd}{mv_0^2}\right) + \frac{qEd^2}{mv_0^2} - \frac{qEd^2}{2mv_0^2}$$

$$\Rightarrow \quad y = \frac{qEd}{mv_0^2}\left(\frac{d}{2} - x\right)$$ **Ans. (A)**

Sol. 31 For helical motion of proton, time period is given as

$$T = \frac{2\pi m}{qB}$$

Total time taken by proton is

$$t = 10\,\text{T}$$

10 Revolution

Velocity of proton along axis is

$$v_x = v\cos 60° = \frac{v}{2}$$

Thus distance travelled along axis is

$$l = \frac{v}{2}\times t = \frac{v}{2}\times 10\times\frac{2\pi m}{qB}$$

$$\Rightarrow \quad l = 4\times 10^5\times 10\times\frac{3.14\times 1.67\times 10^{-27}}{1.6\times 10^{-19}\times 0.3}\,\text{m}$$

$$\Rightarrow \quad l = 0.439\,\text{m}$$ **Ans. (C)**

Sol. 32 Magnetic moment of loop $ABCDA$ is

$$\vec{M}_1 = abI\,\hat{k}$$

Magnetic moment of loop $ADEFA$ is

$$\vec{M}_2 = abI\,\hat{j}$$

Total magnetic moment of loop $ABCDEAA$ is

$$\vec{M} = \vec{M}_1 + \vec{M}_2$$

$$\Rightarrow \quad \vec{M} = abI(\hat{k} + \hat{j})$$

$$\Rightarrow \quad \vec{M} = \sqrt{2}\,abI\left(\frac{\hat{k}}{\sqrt{2}} + \frac{\hat{j}}{\sqrt{2}}\right)$$ **Ans. (A)**

Sol. 33 Force on particle is given as

$$\vec{F} = q(\vec{v}\times\vec{B})$$

$$\Rightarrow \quad \vec{F} = 1(2\hat{i} + 3\hat{j} + 4\hat{k})\times(5\hat{i} + 3\hat{j} - 6\hat{k})\times 10^{-3}\times 10^{-6}\,\text{N}$$

$$\Rightarrow \quad \vec{F} = (6\hat{k} + 12\hat{j} - 15\hat{k} - 18\hat{i} + 20\hat{j} - 12\hat{i})\times 10^{-9}\,\text{N}$$

$$\Rightarrow \quad \vec{F} = (-30\hat{i} + 32\hat{j} - 9\hat{k})\times 10^{-9}\,\text{N}$$ **Ans. (C)**

Sol. 34 Magnetic torque on coil when it is parallel to magnetic field is

$$\tau = BINA = \tau_{ext}$$

$$\Rightarrow \quad B = \frac{\tau_{ext}}{INA} = \frac{1.5}{0.5\times 500\times 3\times 10^{-4}}$$

$$\Rightarrow \quad B = \frac{100}{5} = 20\,\text{T}$$ **Ans. (20)**

Sol. 35 Work done is rotation of coil by 60° is given as

$$\omega = MB(\cos 30° - \cos 90°) = \frac{1}{2}I\omega^2$$

$$\Rightarrow \quad \omega = \sqrt{\frac{2MB\left(\frac{\sqrt{3}}{2}\right)}{I}} = \sqrt{\frac{20\times 4\times\sqrt{3}}{0.8}}$$

$$\Rightarrow \quad \omega = \sqrt{100\sqrt{3}} = 10(3)^{1/4}\,\text{rad/s}$$

$$\Rightarrow \quad \omega \simeq 13.16\,\text{rad/s}$$

Out of given options closest is option (A). **Ans. (A)**

Sol. 36 Torque on loop is given as

$$\tau = MB = 4a^2I\times\frac{\mu_0 I}{2\pi b}$$

$$\Rightarrow \quad \tau = \frac{2\mu_0 I^2 a^2}{\pi b}$$ **Ans. (A)**

Sol. 37 Magnetic moment of electromagnet is given as

$$M = \mu_r NiA$$

$$\Rightarrow \quad M = \mu_r(nl)iA$$

$$\Rightarrow \quad M = \mu_r niV$$

$$\Rightarrow \quad M = 1000\times 10\times 100\times 0.5\times 10^{-3}$$

$$\Rightarrow \quad M = 500\,\text{Am}^2$$ **Ans. (D)**

Sol. 38 Particle will not hit the screen if its radius of circular motion is less then d, so we use

$$R < d$$

$$\Rightarrow \quad \frac{mv}{qB_0} < d$$

$$\Rightarrow \quad v = \frac{qB_0 d}{m}$$ **Ans. (B)**

Sol. 39 Force on electron due to electric field is given as

$$F_e = eE = 1.6\times 10^{-19}\times 300\times 100\ (\text{Along} - \text{Y-direction})$$

For electron to keep moving along X-direction, net force on it must be zero thus magnetic force must balance electric force so act along +Y-direction so for electron moving in +X-direction, magnetic force is given as

$$\overrightarrow{F_m} = -e(\vec{v} \times \vec{B})$$

For $\overrightarrow{F_m}$ along +Y-direction $\vec{B}$ must be along +Z-direction such that

$$evB = eE$$

$$\Rightarrow \quad B = \frac{E}{v} = \frac{300 \times 100}{6 \times 10^6} = 5 \times 10^{-3}\ \text{T} \qquad \textbf{Ans. (A)}$$

Sol. 40 Velocity vector is given as

$$\vec{v} = v_x\hat{i} + v_y\hat{j} \qquad \ldots(1)$$

Here force on particle is given as

$$\vec{F} = k(v_y\hat{i} + v_x\hat{j}) \qquad \ldots(2)$$

From (1) and (2), if mass of particle is m, we use

$$m\frac{dv_x}{dt} = kv_y$$

and $$m\frac{dv_y}{dt} = kv_x$$

$$\Rightarrow \quad \frac{dv_x}{dv_y} = \frac{v_y}{v_x}$$

$$\Rightarrow \quad v_x dv_x - v_y dv_y = 0$$

$$\Rightarrow \quad v_x^2 - v_y^2 = \text{constant}$$

In option (C) $\vec{v} \times \vec{a} = (v_x\hat{i} + v_y\hat{j}) \times \frac{k}{m}(v_y\hat{i} + v_x\hat{j})$

$$= (v_x^2 - v_y^2) \times \frac{k}{m}\hat{k} = \text{Constant} \qquad \textbf{Ans. (C)}$$

Sol. 41 Radius of circular motion of particle is given as

$$r = \frac{mv}{qB}$$

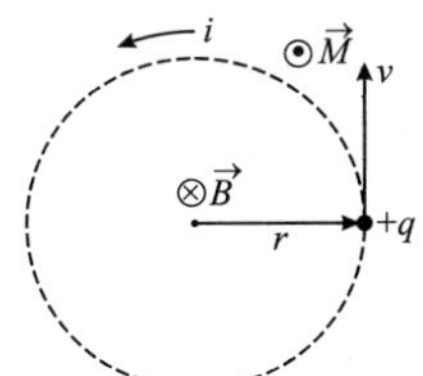

Angular speed of particle is given as

$$\omega = \frac{v}{r} = \frac{qB}{m}$$

Magnetic moment of revolving charge is given as

$$M = i \cdot \pi r^2$$

$$\Rightarrow \quad M = \frac{q\omega}{2\pi} \cdot \pi\left(\frac{mv}{qB}\right)^2 = \frac{1}{2}q\left(\frac{qB}{m}\right)\left(\frac{m^2v^2}{q^2B^2}\right)$$

$$\Rightarrow \quad M = \frac{mv^2}{2B}$$

From figure it is clear that direction of $\vec{M}$ is opposite to $\vec{B}$ so we write

$$\vec{M} = -\frac{mv^2}{2B^2}\vec{B} \qquad \textbf{Ans. (D)}$$

Sol. 42 Figure shows the forces acting on two side wires of square loop due to the long wire placed at (0, b, 0) which is given as

$$F = \frac{\mu_0 I^2(2a)}{2\pi\sqrt{b^2 + a^2}}$$

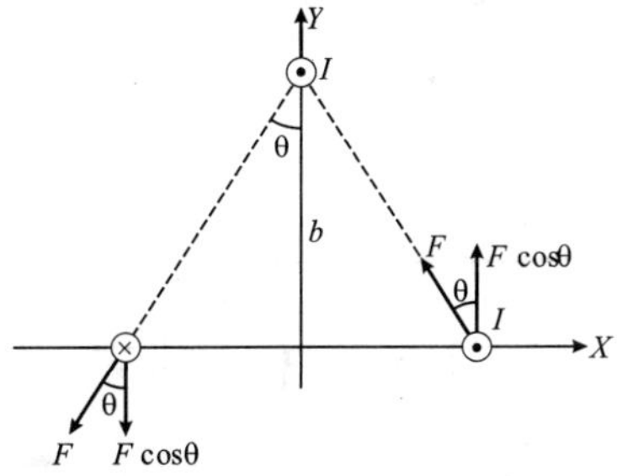

Torque on loop is given as

$$\tau = F\cos\theta \times 2a$$

$$\Rightarrow \quad \tau = \frac{\mu_0 I^2(2a)}{2\pi\sqrt{b^2 + a^2}} \times \frac{b}{\sqrt{b^2 + a^2}} \times 2a$$

$$\Rightarrow \quad \tau = \frac{2\mu_0 I^2 a^2 b}{\pi(b^2 + a^2)} \simeq \frac{2\mu_0 I^2 a^2}{\pi b} (b >> a)$$

For all the option given (A) is most appropriate. **Ans. (A)**

Sol. 43 The magnetic force on a charge moving with some momentum in magnetic field is given as

$$F = q(\vec{v} \times \vec{B}) = \frac{q}{m}(\vec{P} \times \vec{B})$$

$$\Rightarrow \quad F \propto \frac{q}{m}$$

For the three given particles, we can use

$$F_1 : F_2 : F_3 = \frac{q_1}{m_1} : \frac{q_2}{m_2} : \frac{q_3}{m_3} = \frac{e}{m_p} : \frac{e}{2m_p} : \frac{2e}{4m_p}$$

$$\Rightarrow \quad F_1 : F_2 : F_3 = \frac{1}{1} : \frac{1}{2} : \frac{2}{4}$$

$$\Rightarrow \quad F_1 : F_2 : F_3 = 2 : 1 : 1$$

From a given momentum of particle, its speed is given as

$$v \propto \frac{1}{m}$$

$$\Rightarrow \quad v_1 : v_2 : v_3 = \frac{1}{m_p} : \frac{1}{2m_p} : \frac{1}{4m_p}$$

$$\Rightarrow \quad v_1 : v_2 : v_3 = \frac{1}{1} : \frac{1}{2} : \frac{1}{4}$$

$$\Rightarrow \quad v_1 : v_2 : v_3 = 4 : 2 : 1$$

Ans. (B)

Sol. 44 The radius of circular motion of a moving charge in magnetic field is given as

$$r = \frac{mv}{qB} = \frac{p}{qB}$$

As $\frac{m_\alpha}{m_p} = 4$, we have

$$\frac{r_p}{r_\alpha} = \frac{p_p}{q_p}\frac{q_\alpha}{p_\alpha} = \frac{2}{1}$$

$$\Rightarrow \quad \frac{p_p}{p_\alpha} = \frac{2q_p}{q_\alpha} = 2\left(\frac{1}{2}\right)$$

$$\Rightarrow \quad \frac{p_p}{p_\alpha} = 1$$

$$\Rightarrow \quad \frac{K_p}{K_\alpha} = \frac{p_p^2}{p_\alpha^p}\frac{m_\alpha}{m_p} = 4 : 1$$

Ans. (D)

Sol. 45 In magnetic field the radius of circular motion of charge particle is given as

$$r = \frac{mv}{qB} = \frac{\sqrt{2mK}}{qB}$$

$$\Rightarrow \quad r \propto \frac{\sqrt{m}}{q}$$

As $m_\alpha = 2m_d$ and $q_\alpha = 2q_d$ so ratio of radii of their path is give n as

$$\frac{r_d}{r_\alpha} = \frac{\sqrt{m_d}}{q_d} \times \frac{2q_d}{\sqrt{2m_d}} = \sqrt{2}$$

Ans. (B)

Sol. 46 The charge ratio of the ions is given as

$$\frac{Q_1}{Q_2} = \frac{1}{2}$$

The speed ratio is given as

$$\frac{v_1}{v_2} = \frac{2}{3}$$

The radius of circular motion in magnetic field is given as

$$R = \frac{mv}{qB}$$

Thus ratio of radii is given as

$$\Rightarrow \quad \frac{R_1}{R_2} = \frac{v_1}{v_2} \times \frac{Q_2}{Q_1} = \frac{2}{3} \times \frac{2}{1} = \frac{4}{3}$$

Ans. (C)

Sol. 47 If after N revolutions, protons achieve one sixt of the speed of light then we use

$$2\,NeV = \frac{1}{2}mv^2$$

$$\Rightarrow \quad 2N \times 1.6 \times 10^{-19} = \frac{1}{2} \times 1.67 \times 10^{-27} \times (5 \times 10^7)^2$$

$$\Rightarrow \quad N = 543$$

Ans. (543)

Sol. 48 The radius of circular path in magnetic field is given as

$$r = \frac{P}{qB} = \frac{\sqrt{2mk_e}}{qB}$$

$$\Rightarrow \quad r \propto \frac{\sqrt{m}}{q}$$

$$\Rightarrow \quad \frac{r_1}{r_2} = \frac{3}{4}$$

$$\Rightarrow \quad r_2 > r_1$$

Ans. (C)

Sol. 49 At $t = 0$, magnetic fields are given as

$$B \text{ at } \left(0, 0, \frac{\pi}{k}\right) = B_0\left(\frac{\hat{i} + \hat{j}}{\sqrt{2}}\right)\cos(\pi)$$

$$\text{and } B \text{ at } \left(0, 0, \frac{3\pi}{k}\right) = B_0\left(\frac{\hat{i} + \hat{j}}{\sqrt{2}}\right)\cos(3\pi)$$

Force on charged particle q_1 moving at a given velocity is calculated as

$$F_1 = q_1\, \vec{V}_1 \times \vec{B}_1$$

$$\Rightarrow \quad F_1 = 4\pi(0.5c\,\hat{i}) \times \left[-B_0\left(\frac{\hat{i} + \hat{j}}{\sqrt{2}}\right)\right] = \frac{4\pi B_0 c}{2\sqrt{2}}(-\hat{k})$$

Force on charged particle q_2 is given as

$$F_2 = q_2\, \vec{V}_2 \times \vec{B}_2$$

$$\Rightarrow \quad F_2 = 2 \times (0.5c\,\hat{i}) \times \left[-B_0\left(\frac{\hat{i} + \hat{j}}{\sqrt{2}}\right)\right]$$

$$\Rightarrow \quad F_2 = \frac{2\pi B_0 c}{2\sqrt{2}}(-\hat{k})$$

$$\Rightarrow \quad \frac{F_1}{F_2} = 2$$

Ans. (D)

19.1 Classical and Terrestrial Magnetism

Sol. 1 For SHM oscillations of magnetic needle must be with small angular amplitude so that restoring torque on it should be given as

$$\tau = MB\sin\theta \approx MB\theta$$

[For small θ we use $\sin\theta \approx \theta$]

$\Rightarrow$ angular acceleration of needle is given as

$$\alpha = -\frac{\tau}{I} = -\frac{MB}{I}\theta$$

[As $\alpha \propto \theta \Rightarrow$ it executes SHM]

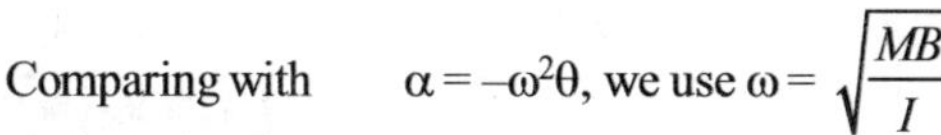

Comparing with $\alpha = -\omega^2\theta$, we use $\omega = \sqrt{\frac{MB}{I}}$

Time period of SHM is

$$T = \frac{2\pi}{\omega} = 2\pi\sqrt{\frac{I}{MB}}$$

$$\Rightarrow \quad T = 2\times 3.14\sqrt{\frac{7.5\times10^{-6}}{6.7\times10^{-2}\times0.1}} \approx 0.665\text{s}$$

For 10 oscillations time taken is $= 0.665 \times 10 = 6.66$s **Ans. (C)**

Sol. 2 Potential energy of a magnetic dipole in constant magnetic field is given as

$$U = -\vec{M}\cdot\vec{B}$$

$$\Rightarrow \quad U = -MB\cos\theta$$

Maximum PE is for $\theta = 180°$ when $U_{max} = -MB(-1) = MB$

Minimum PE is for $\theta = 0°$ when $U_{min} = -MB$

Zero PE is for $\theta = 90°$ when $U = -MB(0) = 0$

and torque on dipole $\vec{\tau} = \vec{M}\times\vec{B}$

$$\Rightarrow \quad \vec{\tau} = MB\sin\theta$$

$$\Rightarrow \quad \tau = MB\sin 180° = 0 \text{ (Min)}$$

$$\Rightarrow \quad \tau = MB\sin 0 = 0 \text{ (Min)}$$

$$\Rightarrow \quad \tau = MB\sin 90° = MB \text{ (Max)}$$

Ans. (C)

Sol. 3 To keep needle horizontal net torque on it must be zero and net magnetic field at the location of dipole is $\sqrt{2}B$

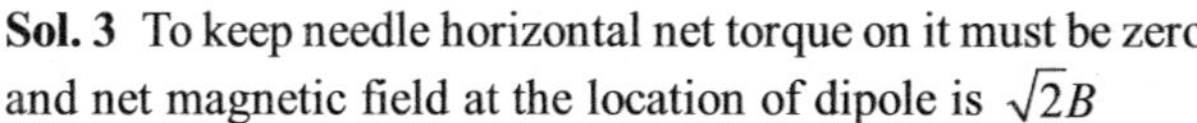

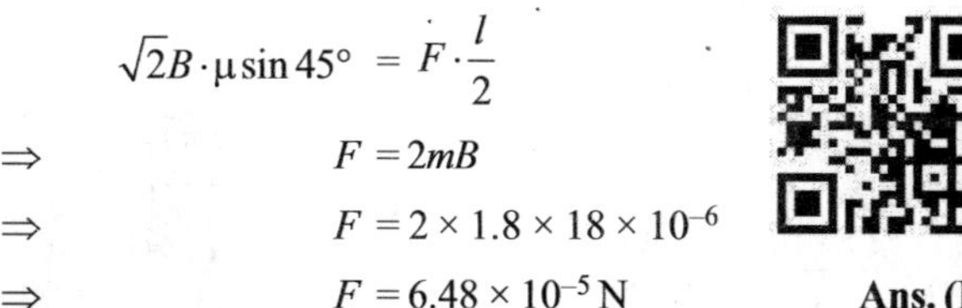

$$\sqrt{2}B\cdot\mu\sin45° = F\cdot\frac{l}{2}$$

$$\Rightarrow \quad F = 2mB$$

$$\Rightarrow \quad F = 2\times1.8\times18\times10^{-6}$$

$$\Rightarrow \quad F = 6.48\times10^{-5}\text{ N}$$

Ans. (B)

Sol. 4 For oscillations of magnetic dipoles, we use

$$T = 2\pi\sqrt{\frac{I}{\mu B}}$$

For hoop, we use

$$T_h = 2\pi\sqrt{\frac{mR^2}{(2\mu)B}}$$

For cylinder, we use

$$T_C = 2\pi\sqrt{\frac{1/2mR^2}{\mu B}} \simeq T_r$$

Ans. (D)

Sol. 5 Magnetic moment of square loop is given as

$$m = NIA = 1\times I\times a^2$$

Here a = side of square

Now, $4a = 2\pi r$

$$r = \frac{2a}{\pi}$$

For circular loop, we use

$$m' = 1\times I\times\pi r^2$$

$$\Rightarrow \quad m' = 1\times I\times\pi\times\left(\frac{2a}{\pi}\right)^2$$

$$m' = \frac{4m}{\pi}$$

Ans. (B)

Sol. 6 For a horizontal compass needle the frequency of small oscillations in earth's magnetic field is given as

$$f = \frac{1}{2\pi}\sqrt{\frac{MB_N}{I}} \text{ (horizontal component of earth's field)}$$

When $\theta_D = 45° \rightarrow B_{H_1} = B_1\cos45°$

$$f_1 = \frac{1}{2\pi}\sqrt{\frac{MB_1\cos45°}{I}} \quad \ldots(1)$$

When $\theta'_D = 30° \rightarrow B_{H_2} = B_2\cos30°$

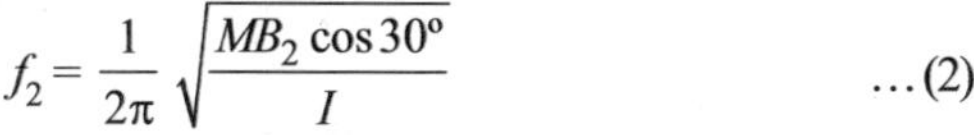

$$f_2 = \frac{1}{2\pi}\sqrt{\frac{MB_2\cos30°}{I}} \quad \ldots(2)$$

$$(1)/(2) \Rightarrow \quad \frac{30}{40} = \sqrt{\frac{B_1/\sqrt{2}}{B_2\sqrt{3}/2}}$$

$$\Rightarrow \quad \frac{B_1}{B_2} = \frac{9}{16}\times\sqrt{\frac{3}{2}} \approx 0.7$$

Ans. (C)

Sol. 7 Stable equilibrium position is at $\theta = 0°$ and unstable equilibrium position is at $\theta = 180°$

Torque on magnet is given as

$$\tau = MB\sin\theta$$

With given values, we use

$$0.018 = M\times0.06\times(10.5)$$

$$\Rightarrow \quad M = \frac{0.018}{0.06\times0.5} = 0.6\text{ A-m}^2$$

Work done in rotation of magnet from 0° to 180° is given as

$$W = MB(\cos 0° - \cos 180°)$$

$$\Rightarrow \quad W = 2MB = 2 \times 0.6 \times 0.06$$

$$\Rightarrow \quad W = 7.2 \times 10^{-2}\,\text{J}$$

Ans. (D)

Sol. 8 Figure below shows the magnetic field due to the bar magnet at the point P where the field of bar magnet nullifies the horizontal component of earth's field.

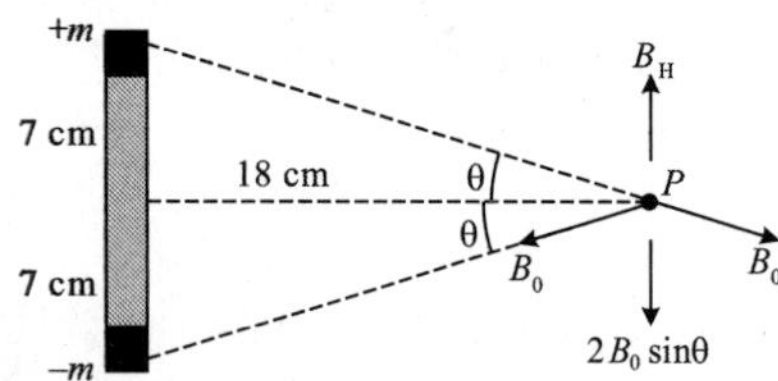

Magnetic induction due to a magnetic pole at point P is given as

$$B_0 = \frac{2\mu_0}{4\pi}\frac{m}{r^2} \times \frac{7}{r} = 0.4 \times 10^{-4}$$

$$\Rightarrow \quad 2 \times 10^{-7} \times \frac{m \times 7}{(7^2 + 18^2)^{3/2}} \times 10^4 = 0.4 \times 10^{-4}$$

$$\Rightarrow \quad m = \frac{4 \times 10^{-2} \times (373)^{3/2}}{14}\ \text{A-m}$$

Magnetic moment of the bar magnet is given as

$$M = md = \text{m} \times \frac{14}{100}$$

$$\Rightarrow \quad M = \frac{0.04 \times (373)^{3/2}}{14} \times \frac{14}{100}$$

$$\Rightarrow \quad M = 4 \times 10^{-4} \times 7203.82 = 2.88\ \text{J/T}$$

Ans. (C)

Sol. 9 Apparent dip angle is related to true dip angle by the relation given as

$$\tan\theta_{AD} = \frac{B\sin\phi}{B\cos\phi\cos\alpha}$$

$$\Rightarrow \quad \tan 45° = \frac{B\sin\phi}{B\cos\phi\cos 30°}$$

$$\Rightarrow \quad \tan\phi = \frac{\sqrt{3}}{2}$$

Ans. (A)

Sol. 10 Apparent dip angle is related to true dip angle by the relation given as

$$\tan\theta_{AD} = \frac{B\sin\phi}{B\cos\phi\cos\alpha} = \frac{\tan\phi}{\cos\alpha}$$

Here we can directly state that true dip angle is alware greater than the apparent dip angle. **Ans. (B)**

19.2 Magnetic Properties of Materials

Sol. 1 For demagnetization of ferromagnetic material it is the coercivity which is given as

$$H = 100\ \text{A/m}$$

For a solenoid $\quad H = nI$

$$\Rightarrow \quad I = \frac{H}{n} = \frac{(100\,\text{A/m})}{(10^5\,\text{turns/m})} = 10^{-3}\,\text{A}$$

$$\Rightarrow \quad I = 1\ \text{mA}$$

Ans. (B)

Sol. 2 Coercivity of material is given as

$$H_c = \frac{B}{\mu_0}$$

$$\Rightarrow \quad H = ni = \frac{N}{l}i = \frac{100}{0.2} \times 5.2\ \text{A/m}$$

$$\Rightarrow \quad H = 2600\ \text{A/m}$$

Ans. (B)

Sol. 3 Magnetic susceptibility of a material is

$$\chi = \frac{I}{H}$$

Where intensity of magnetization is

$$I = \frac{\text{Magnetic moment}}{\text{Volume}}$$

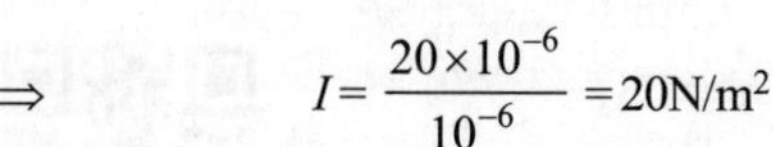

$$\Rightarrow \quad I = \frac{20 \times 10^{-6}}{10^{-6}} = 20\text{N/m}^2$$

$$\Rightarrow \quad \chi = \frac{20}{60 \times 10^{+3}} = \frac{1}{3} \times 10^{-3}$$

$$\Rightarrow \quad \chi = 0.33 \times 10^{-3} = 3.3 \times 10^{-4}$$

Ans. (C)

Sol. 4 We know that susceptibility of a material is given as

$$X \ \alpha\ \frac{1}{T_C}$$

$$\Rightarrow \quad \frac{X_1}{X_2} = \frac{T_{C_2}}{T_{C_1}}$$

$$\Rightarrow \quad \frac{2.8 \times 10^{-4}}{X_2} = \frac{300}{350}$$

$$\Rightarrow \quad X_2 = \frac{2.8 \times 350 \times 10^{-4}}{300}$$

$$\Rightarrow \quad X_2 = 3.266 \times 10^{-4}$$

Ans (C)

Sol. 5

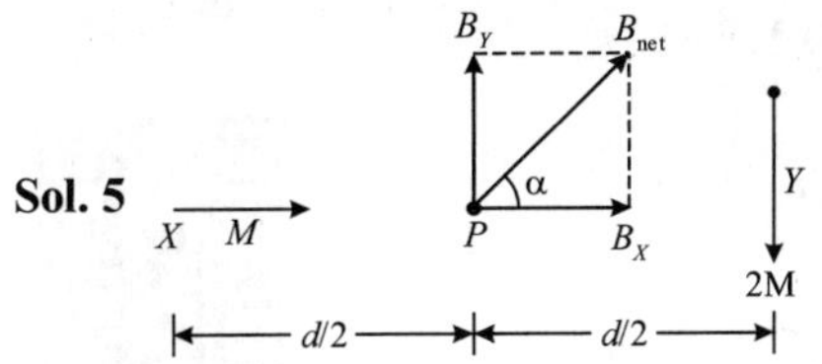

B_X is given as

$$B_X = \frac{2kM_X}{(d/2)^3} = 2\left(\frac{\mu_0}{4\pi}\right)\cdot\frac{M}{(d/2)^3} \qquad (B_X = B_Y)$$

B_Y is given as

$$B_Y = \frac{kM_Y}{(d/2)^3} = \left(\frac{\mu_0}{4\pi}\right)\cdot\frac{2M}{(d/2)^3} \qquad (\alpha = 45^\circ)$$

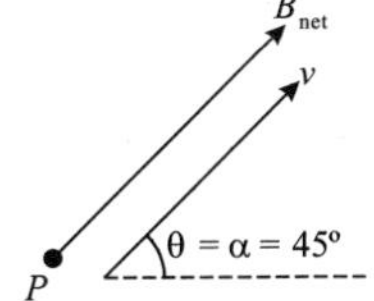

At P, $\vec{B}_{net}$ is || to $\vec{v}$

$\Rightarrow \quad \vec{F}_{mag} = q(\vec{v}\times\vec{B}_{net}) = 0$ **Ans. (D)**

Sol. 6 By definition and from figure, we get

Retentivity = 1.0 T

Coercivity = 50 A/m

Saturation = 1.5 T **Ans. (B)**

Sol. 7 For making of permanent magnets, materials are preferred which have large retentivity and large coercivity. **Ans. (D)**

Sol. 8 The perfect diamagnetic material does not allow external field to enter into it so material inside will not get magnetised. **Ans. (A)**

Sol. 9 By Curies law we use

$$\chi \propto \frac{1}{T}$$

$$\Rightarrow \quad \chi_1 T_1 = \chi_2 T_2$$

$$\Rightarrow \quad \frac{6}{0.4}\times 4 = \frac{I}{0.3}\times 24$$

$$\Rightarrow \quad I = \frac{0.3}{0.4} = 0.75\ \text{A/m}$$ **Ans. (B)**

Sol. 10 Soft ferromagnetic materials are materials which can be magnetised very fast and easily with small magnetic field also and also demagnetised easily by external magnetic field. When external field is applied, the domains experiences a net torque hence change their orientation. **Ans. (B)**

Sol. 11 Domain are the small localized regions of dipoles aligned in same orientation due to coupling of magnetic field by neighbouring dipoles. Out of all the given options, (D) is most appropriate. **Ans. (D)**

Sol. 12 Statements (a) is NOT correct because existance of a charge can be considered as an electric monopole. Statement (b) is NOT correct because magnetic field lines at the ends of a solenoid are not straight but confined as closed loops only. Statement (c) is correct because for a closely wound toroid, magnetic field outside the toroid is zero and all the magnetic field lines are confined within the toroid only. Statement (d) is NOT correct because inside a bar magnet, magnetic lines are considered to be confined within its body only and parallel to each other. Statement (e) is correct because for super conducting materials inside their body the net magnetic field is always zero and they are perfect diamagnetic with relative permeability zero so we have

$$\mu_r = 1 + \chi$$
$$\chi = -1$$ **Ans. (A)**

Sol. 13 Magnetic susceptibility of material is related to its permeability as

$$\mu_r = 1 + \chi$$

$$\Rightarrow \quad \mu_r = 1 + 499 = 500$$

Absolute permeability of the material of rod is given as

$$\mu = \mu_0 \times \mu_r$$

$$\Rightarrow \quad \mu = 500 \times 4\pi \times 10^{-7}$$

$$\Rightarrow \quad \mu = 2\pi \times 10^{-4}\ \text{H/m}$$ **Ans. (D)**

Sol. 14 In a ferromagnetic property of a material depends upon temperature and at high temeperature it looses its ferromagnetic properties hence Statement-I is true and Statement-II is flase. **Ans. (D)**

Sol. 15 (m_1 upward, m_2 to the right, B_1 downward)

$$\vec{\tau} = \vec{M}_2 \times \vec{B}_1$$

$$\tau = M_2 B_1 \sin 90^\circ = 1 \times \frac{\mu_0}{4\pi}\frac{M}{(1)^3}\, 1 = 10^{-7}\ \text{N.m}$$ **Ans. (1)**

Sol. 16 Torque acting on the coil si given as

$$\tau = M \times B$$

$$\Rightarrow \quad \tau = IAB$$

$$\Rightarrow \quad \tau = 0.2 \times \left(\frac{\sqrt{3}a^2}{4}\right) \times 2 \times 10^{-2}$$

$$\Rightarrow \quad \tau = \frac{0.2\times\sqrt{3}\times 10^{-2}\times 2\times 10^{-2}}{4} = \sqrt{3}\times 10^{-5}\ \text{T}$$ **Ans. (3)**

Sol. 17 When the coil is in the form of a square having n_1 turns, we have

$$4an_1 = 24a$$

$$\Rightarrow \quad n_1 = 6 \text{ turns}$$

When coil is in form of an equilateral triangle having n_2 turns, we have

$$3an_2 = 24a$$

$$\Rightarrow \quad n_2 = 8 \text{ turns}$$

Magnetic moment of a coil with current i and n turns having area A is given as

$$M = niA$$

$$\Rightarrow \quad \frac{M_1}{M_2} = \frac{n_1 i A_1}{n_2 i A_2} = \frac{6 \times a^2}{8 \times a^2 \times \frac{\sqrt{3}}{4}} = \frac{\sqrt{3}}{1}$$

$$\Rightarrow \quad \frac{M_2}{M_1} = \frac{1}{\sqrt{3}} = \frac{1}{\sqrt{y}}$$

$$\Rightarrow \quad y = 3$$

Ans. (3)

Sol. 18 Intensity of magnetisation in the material is given as

$$\vec{M} = \frac{\text{magnetic moment } (\vec{m})}{V}$$

$$\Rightarrow \quad \chi \vec{H} = \frac{\vec{m}}{V}$$

$$\Rightarrow \quad \vec{m} = (\mu_r - 1)\, Niv$$

The fractional change in the magnetic moment of core can be calculated as

$$\frac{\Delta m}{m} = \frac{[(750-1)-(500-1)]Niv}{(500-1)Niv} = 250$$

Ans. (250)

Sol. 19 For a diamagnetic material magnetisation and magnetising field are proportional and opposite to each other so option (a) is correct and in diamagnetic materials magnetic suseptibility does not depend on temperature so option (c) is also correct. **Ans. (D)**

20.1 Electromagnetic Induction

Sol. 1 The charge which flows through a coil of resistance R due to change in magnetic flux $\Delta\phi$ is given as

$$q_f = \frac{\Delta\phi}{R} = \text{area under } i\text{-}t \text{ curve}$$

$$\frac{1}{2} \times 10 \times 0.5 = \frac{\Delta\phi}{100}$$

$$\Rightarrow \quad \Delta\phi = 250 \text{ Wb}$$

Ans. (A)

Sol. 2

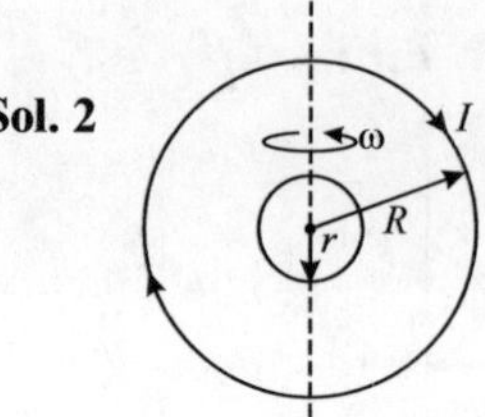

As $r << R$. Magnetic field at centre of larger coil is almost uniform for smaller coil.

The smaller coil is rotating in the magnetic field of larger coil at centre $\left(B_C = \frac{\mu_0 I}{2R}\right)$ so emf induced due to rotation of smaller coil is

$$e = NBA\omega \sin\omega t$$

$$\Rightarrow \quad e = \frac{\mu_0 I}{2R}(\pi r^2)\omega \sin\omega t$$

Ans. (C)

Sol. 3

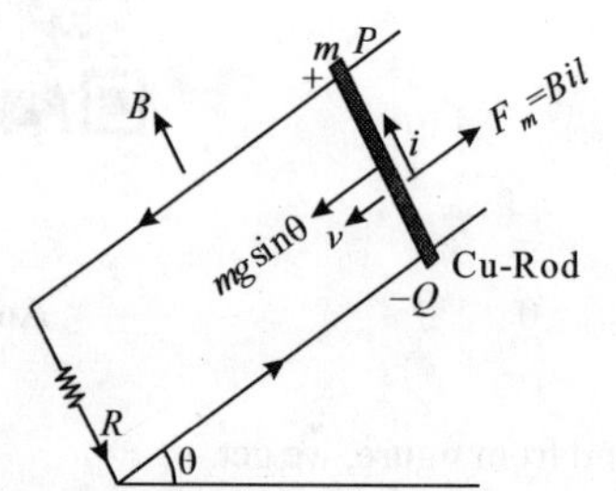

When rod is moving at speed v, the motional emf induced in rod will be

$$e_{PQ} = Bvl$$

$\Rightarrow$ Induced current is

$$i = \frac{e_{PQ}}{R} = \frac{Bvl}{R}$$

Magnetic force on rod due to B is

$$F_m = Bil = \frac{B^2 v l^2}{R}$$

At terminal speed $a_{\text{rod}} = 0$

$$\Rightarrow \quad mg \sin\theta = F_m$$

$$\Rightarrow \quad mg \sin\theta = \frac{B^2 v l^2}{R}$$

$$\Rightarrow \quad v = \frac{mgR \sin\theta}{B^2 l^2}$$

Ans. (B)

Sol. 4

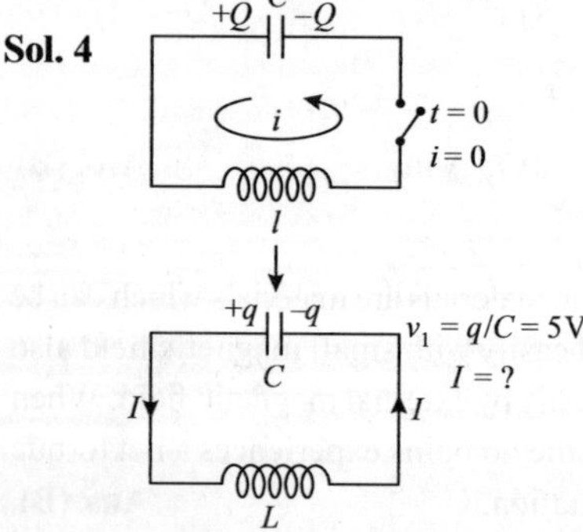

Initial energy in capacitor is

$$E = \frac{1}{2}CV^2 = \frac{1}{2} \times 0.2 \times 10^{-6} \times (10)^2 = 10\,\mu J$$

Energy of circuit (sum of $E_C + E_L$) when potential difference across capacitor is 5 V, is given as

$$E = \frac{1}{2}CV_1^2 + \frac{1}{2}LI^2$$

$$\Rightarrow \quad 10 \times 10^{-6} = \frac{1}{2} \times 0.2 \times 10^{-6} \times (5)^2 + \frac{1}{2} \times 0.5 \times 10^{-3} \times I^2$$

$$\Rightarrow \quad 2.5 \times 10^{-4} I^2 = 7.5 \times 10^{-6} \Rightarrow I^2 = 3 \times 10^{-2}$$

$$\Rightarrow \quad I = \frac{\sqrt{3}}{10}\text{A} = 0.1732\,\text{A}$$ **Ans. (A)**

Sol. 5

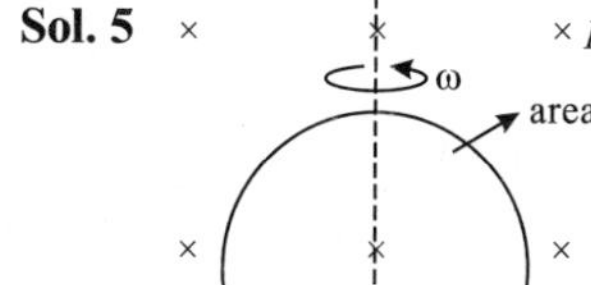

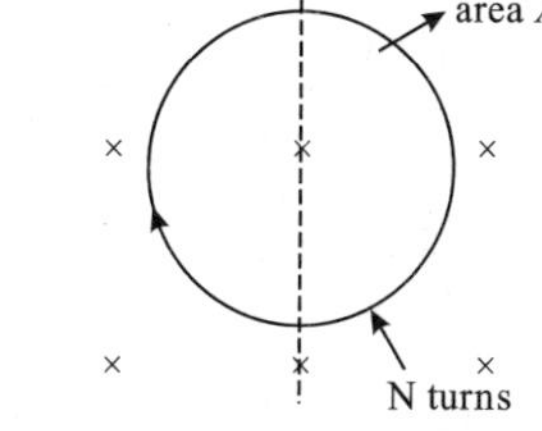

$$\left.\begin{array}{l}\vec{A} \to \otimes \\ \vec{B} \to \otimes\end{array}\right] \Rightarrow \theta = 0^\circ$$

Initial flux through coil $\phi = BAN$

after time t angle between $\vec{A}$ and $\vec{B}$ will be

$$\theta = \omega t$$

$\Rightarrow$ Magnetic flux through coil after time t is given as

$$\phi = (\vec{B} \cdot \vec{A})N = NBA\cos\omega t$$

EMF induced in coil is

$$e = -\frac{d\phi}{dt} = NBA\omega \sin\omega t = e_0 \sin\omega t$$

Here $\quad e_0 = NBA\omega \to$ maximum emf **Ans. (A)**

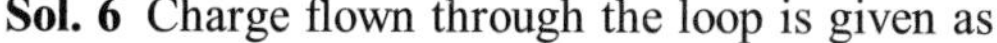

Sol. 6 Charge flown through the loop is given as

$$Q = \frac{\Delta\phi}{R} = \frac{1}{10}A(B_f - B_i)$$

$$\Rightarrow \quad Q = \frac{1}{10} \times 3.5 \times 10^{-3}\left(0.4\sin\frac{\pi}{2} - 0\right)$$

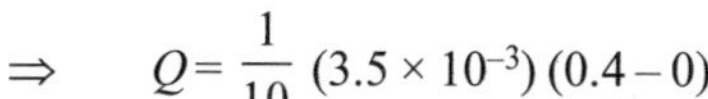

$$\Rightarrow \quad Q = \frac{1}{10}(3.5 \times 10^{-3})(0.4 - 0)$$

$$\Rightarrow \quad Q = 1.4 \times 10^{-4} = 0.14\,\text{mC}$$ **Ans. (A)**

Sol. 7 Potential difference between two faces perpendicular to x-axis will be the motional emf given as

$$e = Bvl = 12\,\text{mV}$$ **Ans. (C)**

Sol. 8 Work done in reversing direction of current is

$$W = \text{M.B.}(\cos\theta_1 - \cos\theta_2)$$

$$\Rightarrow \quad W = 2 \times 10^{-2} \times 1\cos(0.125)$$

$$\Rightarrow \quad W = 0.019\,\text{J}$$

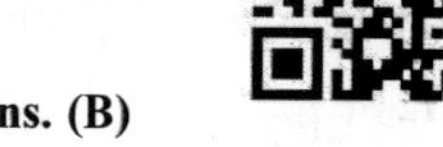

Closest possible option is (B) **Ans. (B)**

Sol. 9 Motional *emf* induced in wire is

$$e = Bvl$$

$$\Rightarrow \quad e = 0.3 \times 10^{-4} \times 5 \times 10$$

$$\Rightarrow \quad e = 1.5 \times 10^{-3}\,\text{V}$$ **Ans (D)**

Sol. 10 For damped oscillation amplitude is given as

$$A = A_0 e^{-t/\gamma}$$

Where $\quad \gamma = \frac{2m}{b}\left(b = \frac{B^2l^2}{R}\right)$

If at $t = t_1$, $\quad A = \frac{A_0}{e}$

$$\Rightarrow \quad \frac{A_0}{e} = A_0 e^{-t/\gamma}$$

$$\Rightarrow \quad \gamma = \frac{2mR}{B^2l^2}$$

$$\Rightarrow \quad \gamma = \frac{2 \times 50 \times 10^{-3} \times 10}{(0.1)^2 \times (0.1)^2} = 10^4\,\text{s}$$

Time period of oscillations

$$\Rightarrow \quad T_0 = 2\pi\sqrt{\frac{m}{k}} = 2\pi\sqrt{\frac{50 \times 10^{-3}}{0.5}} = \frac{2\pi}{\sqrt{10}}\,\text{s}$$

Number of oscillation of strip in t_1 s are

$$N = \frac{t_1}{T_0} = \frac{10^4}{2\pi\sqrt{10}} \sim 5000$$ **Ans. (B)**

Sol. 11 Resistance of Wheatstone bridge is

$$R_{BF} = \frac{4 \times 2}{4 + 2} = \frac{8}{6} = \frac{4}{3}\Omega$$

Equal resistance of circuit is

$$R_{eq} = 1.33 + 1.7 = 3.03\,\Omega \simeq 3\Omega$$

EMF induced in L is

$$e = Bvl = 1 \times 0.05 \times 0.01 = 5 \times 10^{-4}\,\text{V}$$

Circuit current $i = \dfrac{e}{R_{eq}} = \dfrac{5\times10^{-4}}{3} = 1.66 \times 10^{-4}\,\text{A}$

$= 166\,\mu\text{A} \simeq 170\,\mu\text{A}$ **Ans. (B)**

Sol. 12

Magnetic flux (ϕ) through ring is

$$\phi = \pi(R)^2 . B$$

$$\Rightarrow \quad \phi = (\pi R^2)(\mu_0 n I)$$

$$\Rightarrow \quad \phi = (\pi R^2 \mu_0 n I_0)(t - t^2)$$

Induced emf is given as

$$e = \frac{d\phi}{dt}$$

$$\Rightarrow \quad e = (\pi R^2 \mu_0 n I_0)(2t - 1)$$

Induced current is given as

$$I_R = \frac{\pi R^2 \mu_0 n I_0 (2t-1)}{R_{\text{ring}}}$$

In above expressive, we can see that

e and I_R are zero at $t = \dfrac{1}{2} = 0.5$ s

and their sign also changes at $t = 0.5$ s **Ans. (A)**

Sol. 13 Magnetic flux which comes out from the area enclosed by coil is same which goes into the surrounding plane of coil from area outside it. **Ans. (A)**

Sol. 14 Figure shows the loop formed by given corner coordinates

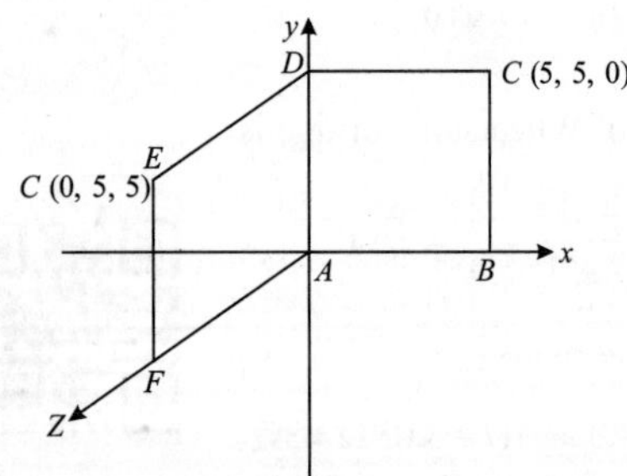

Area vector of loop is given as

$$\vec{A}_{ABCD} = 25\hat{k}$$

$$\vec{A}_{ADEF} = 25\hat{i}$$

$$\vec{A}_{\text{net}} = 25\hat{i} + 25\hat{k}$$

Magnetic flux through loop is given as

$$\phi = \vec{B} \cdot \vec{A}$$

$$\Rightarrow \quad \phi = 25 \times 3 + 25 \times 4$$

$$\Rightarrow \quad \phi = 175\ Wb$$

Ans. (175)

Sol. 15 For a rotating coil in uniform magnetic field.

Induced emf is given as $|e| = BA\omega \sin \omega t$

emf will be maximum at $\omega t = \dfrac{\pi}{2}$

$$\Rightarrow \quad \left(\frac{2\pi}{T}\right)t = \frac{\pi}{2}$$

$$\Rightarrow \quad \left(\frac{2\pi}{10}\right)t = \frac{\pi}{2}$$

$$\Rightarrow \quad t = 2.5\ \text{s}$$

and emf will be minimum at $\omega t = \pi$

$$\Rightarrow \quad \left(\frac{2\pi}{10}\right)t = \pi$$

$$\Rightarrow \quad t = 5\ \text{s}$$

Ans. (D)

Sol. 16 For linear reduction in magnetic field, we use

$$\frac{dB}{dt} = \frac{1000 - 500}{5} = 100\ \text{G/s}$$

$$= 100 \times 10^{-4}\ \text{T/s}$$

Area of loop is

$$A = 16 \times 4 - 4 \times 2 = 56\ \text{cm}^2$$

EMF induced in loop is

$$\varepsilon = \frac{dB}{dt}A = 100 \times 10^{-4} \times 56 \times 10^{-4} = 56\ \mu\text{V}$$

Ans. (C)

Sol. 17 *EMF* induced in a rotating coil in magnetic field is

$$e = BA\omega \sin\omega t$$

Maximum *EMF* is $e_{\max} = BA\omega = BA\left(\dfrac{2\pi}{T}\right)$

$$\Rightarrow \quad e_{\max} = 3 \times 10^{-5} \times 3.14 \times (0.1)^2 \times \frac{2 \times 3.14}{0.4}$$

$$\Rightarrow \quad e_{\max} = \frac{6 \times 10}{4} \times 10^{-6}$$

$$\Rightarrow \quad e_{\max} = 15\ \mu\text{V}$$

Ans. (15)

Sol. 18 *emf* induced in rotating loop is

$$e = NBA\omega \sin \omega t$$

Here $N = 1$

$\Rightarrow$ $e = BA\omega \sin \omega t$

Average power developed in loop is

$$P_{avg} = \left|\frac{e_{avg}^2}{R}\right| = \frac{B^2A^2\omega^2(\sin^2 \omega t)_{avg}}{R}$$

$$\Rightarrow \quad P_{avg} = \frac{B^2A^2\omega^2}{2R} \quad \left[\text{As } (\sin^2 \omega t)_{avg} = \frac{1}{2}\right]$$

$$\Rightarrow \quad P_{avg} = \frac{B^2\omega^2}{2R}(\pi ab)^2 = \frac{\pi^2a^2b^2B^2\omega^2}{2R}$$ **Ans. (A)**

Sol. 19 Resistance of wire is

$$R = \frac{\rho l}{A} = \frac{1.23\times10^{-8}\times30\times10^{-2}}{\pi(2\times10^{-3})^2}$$

EMF induced in loop is given as

$$e = \frac{AdB}{dt} = a^2\frac{dB}{dt} = (7.5\times10^{-2})^2\times0.032$$

Induced current in loop is

$$i = \frac{e}{R} = \frac{(7.5\times10^{-2})^2\times0.032}{1.23\times10^{-8}\times30\times10^{-2}}\times3.14\times(2\times10^{-3})^2$$

$\Rightarrow$ $i \simeq 0.61$ A **Ans. (A)**

Sol. 20 At the time of entering of magnet into coil and at the time of leaving the coil due to flux change in opposite direction emf is induced in opposite sense and when magnet is completely inside flux will not change so reading of galvanometer remain zero. Correct match is with option (C). **Ans. (C)**

Sol. 21 Magnetic field at centre of C_2 is

$$B = \frac{\mu_0 I}{2R_2}\cdot N_2$$

flux linked through coil C_1 due to magnetic field of C_2 is

$$\phi = B.\pi R_1^2N_1 = \frac{\mu_0\pi R_1^2N_1N_2}{2R_2}.(5t^2 - 2t + 3)$$

emf induced in C_1 is given as

$$e = \frac{d\phi}{dt} = \frac{\mu_0\pi R_1^2N_1N_2}{2R_2}\times(10t - 2)$$

$$\Rightarrow \quad e = \frac{4\pi\times10^{-7}\times(0.01)^2\times\pi\times200\times500}{2\times0.2}\times(10-2)$$

$\Rightarrow$ $e = 8\times10^{-4}\,\text{V} = 0.8\,\mu\text{V}$ **Ans. (5.00)**

Sol. 22 Emf induced in sliding conductor is given as

$$e = Bvl$$

$$\Rightarrow \quad e = \frac{\mu_0 I}{2\pi r}vl$$

Induced current in loop is calculated as

$$i = \frac{e}{R} = \frac{\mu_0 Ivl}{2\pi rR}$$ **Ans. (B)**

Sol. 23 Motional EMF induced across the plane wings is given as

$$e = Bvl\sin\theta$$

$$\Rightarrow \quad e = (2.5\times10^{-4}\,\text{T})\left(180\times\frac{5}{18}\,\text{m/s}\right)(10\,\text{m})\sin 60°$$

$\Rightarrow$ $e = 108.25\times10^{-3}$ V **Ans. (D)**

Sol. 24 The effective circuit including motional EMF of the moving rod can be shown in figure below.

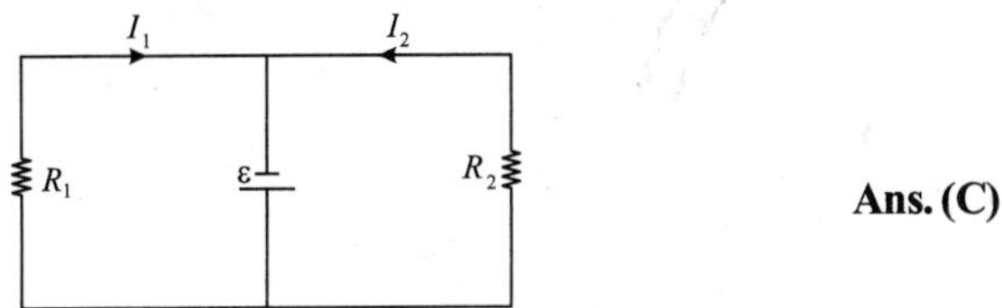

Ans. (C)

Sol. 25 Motional EMF induced in segments parallel to y axes is given as

$$E_1 = \frac{B_0(x+d)}{a}v_0d$$

$$E_2 = \frac{B_0(x)}{a}v_0d$$

The equivalent circuit of the loop is shown in below figure.

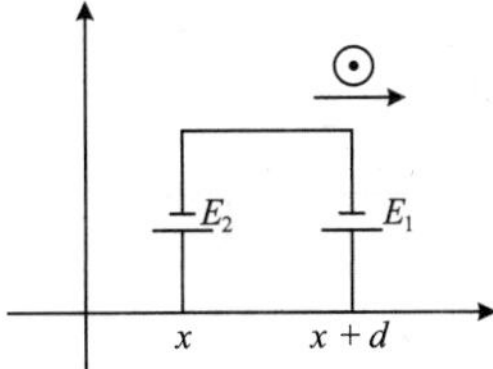

Net loop EMF induced is given as

$$E_{net} = E_1 - E_2$$

$$\Rightarrow \quad E_{net} = \frac{B_0v_0d^2}{a}$$ **Ans. (C)**

Sol. 26 Every elemental part of the wire is pulled by force $Bi(dl)$ and it acts perpendicular to current & magnetic field so it will open up and outward force will be giving it a shape of circle. **Ans. (A)**

Sol. 27 The electric flux passing through the given area is calculated as

$$\phi = \mathbf{B}.\mathbf{A} = BA\cos 0$$

Area enclosing the flux is given as $A = lvt$, thus flux is given as

$$\phi = Blvt$$

Once rod goes at $a > b$ then flux stops changing and maintained at constant value of Blb. When rod comes back and $x < b$ flux start decreasing so graph of variation of flux with time can be drawn as

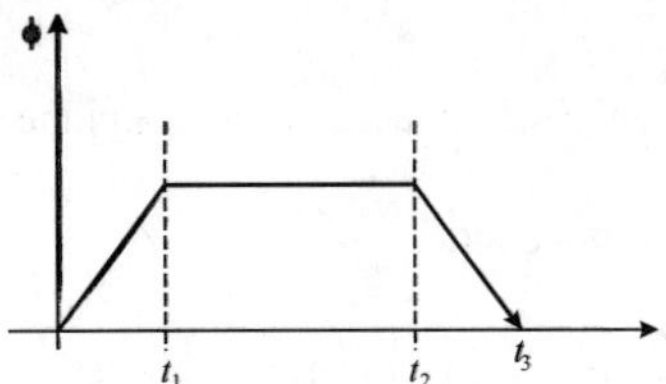

EMF induced can be given as

$$e = \frac{-d\phi}{dt}$$

This can be represented as slope of flux - time curve and *EMF* variation with time can now be plotted as shown in diagram below

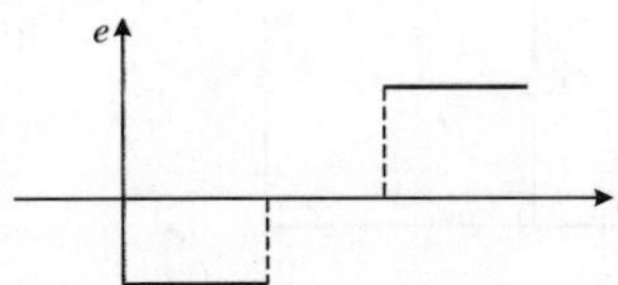

Power dissipated in the circuit is given as e^2/R where resistance is only of rod so resistance of the circuit remain constant thus power is given as

$$P = \frac{B^2 l^2 v^2}{R}$$

Thus power will also remain constant for the duration *EMF* is constant and zero when *EMF* is zero.

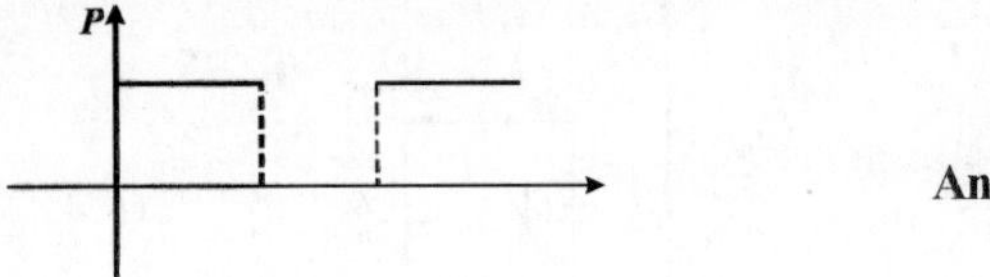

Ans. (D)

Sol. 28 The emf induced in coil can be given as

$$\varepsilon = \left| -A\frac{dB}{dt} \right|$$

$$\Rightarrow \quad \varepsilon = \pi(1)^2 \times \frac{4}{\pi} \times \frac{10^{-3}}{100} \text{ V}$$

$$\Rightarrow \quad \varepsilon = 4 \times 10^{-5} \text{ V}$$

When $B = 0$, $t = 100$ s, the energy dissipated can be directly calculated as emf in coil is constant, given as

$$H = \frac{\varepsilon^2}{R} \times t = \frac{16 \times 10^{-10}}{2 \times 10^{-6}} \times 100 = 8 \times 10^{-2} = 80 \text{ mJ}$$

Ans. (80)

Sol. 29 EMF induced in the loop is given as

$$\varepsilon = \left| -\frac{d\phi}{dt} \right| = (20t + 20) \times 10^{-3}$$

at $t = 5$ s, EMF is calculated as

$$\varepsilon = 120 \times 10^{-3} \text{ V}$$

Current in loop is given as

$$i = \frac{\varepsilon}{R} = \frac{120}{2} \times 10^{-3} = 60 \text{ mA}$$

Ans. (60)

Sol. 30 EMF induced in a rotating coil in uniform magnetic field is given as

$$e = NBA\omega \sin \omega t$$

$$e_{max} = NBA\omega = NB(\pi R^2)\omega$$

$$\Rightarrow \quad e_{max} = 3 \times 10^{-2} \times 5 \times 20 \times \pi \times \frac{64}{10^4}$$

$$\Rightarrow \quad e_{max} = 60 \times 10^{-2} \text{ V}$$

Ans. (60)

Sol. 31 As Magnet is approaches the loop, magnetic flux increases through the loop and rate of increment of flux also increases so emf increases with time. When magnet enters into the loop then at this moment emf is maximum and negative and when magnet is in middle of coil, at this moment emf is zero. Now again flux starts changing however polarity of emf get reversed. When magnet is coming out of the other side from the loop again at this moment emf is maximum and in opposite direction and as it gradually receds away emf becomes zero. Hence optoin (D) is most appropriate for this situation.

Ans. (D)

Sol. 32 Motional emf induced in the ring when half ring enters in the magnetic field (at t=1s) is given as

$$E = Bv(2R)$$

$$\Rightarrow \quad E = 1 \times 1 \times 2 \times 1 = 2 \text{ V}$$

Ans. (B)

Sol. 33 As the current induced is in anticlock wise manner, it can be due to two reasons : (1) magnetic field is inward and increasing with time or (2) magnetic field is outward and decreasing with time. Hence out of given options, (D) is correct.

Ans. (D)

Sol. 34 Motional emf induced in the loop is given as

$$e = Blv$$

Current in circuit is calculated as

$$i = \frac{Blv}{R_{eq}} = 200 \text{ mA}$$

$$\Rightarrow \quad v = \frac{200 \times 10^{-3} \times R_{eq}}{Bl}$$

$$\Rightarrow \quad v = \frac{200 \times 10^{-3} \times 5}{5 \times (0.2)} = 1 \text{ m/s}$$

Ans. (A)

20.2 Self and Mutual Induction

Sol. 1 Transformer efficiency is given as

$$\eta = \frac{P_{out}}{P_{in}} = \frac{V_S I_S}{V_P I_P}$$

$$\Rightarrow \quad 0.9 = \frac{23 \times I_S}{230 \times 5}$$

$$\Rightarrow \quad I_S = 45 \text{ A}$$

Ans. (D)

Sol. 2 Given that

$$L\frac{di}{dt} = 25$$

$$\Rightarrow \quad L \times \frac{15}{1} = 25$$

$$\Rightarrow \quad L = \frac{5}{3} \text{H}$$

Change in energy of inductance is

$$\Delta E = \frac{1}{2} L(I_2^2 - I_1^2)$$

$$\Rightarrow \quad \Delta E = \frac{1}{2} \times \frac{5}{3} \times (25^2 - 10^2) \text{ J}$$

$$\Rightarrow \quad \Delta E = \frac{5}{6} \times 525 = 437.5 \text{ J}$$

Ans. (A)

Sol. 3 From time $t = 0$ to $t = t_0$, current in RL circuit grows and after to current decays.

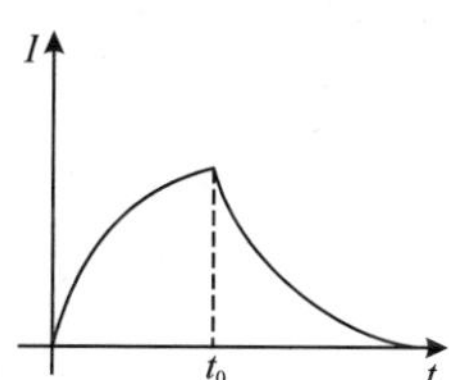

In this case most appropriate option is (B) **Ans. (B)**

Sol. 4 Mutual inductance of the given coils is

$$M = \mu_0 n_1 n_2 \pi r_1^2 l \quad \ldots(1)$$

Self inductance of inner coils is

$$L = \mu_0 n_1^2 \pi r_1^2 l \quad \ldots(2)$$

From (1) and (2)

$$\frac{M}{L} = \frac{n_2}{n_1}$$ **Ans. (D)**

Sol. 5 Self inductance of a solenoid is

$$L = \mu_0 n^2 A l$$

Keeping total length of copper wire same on increasing the frame size to 3 times, the length of solenoid on which terms are wound will become $l/3$ so new self inductance will be

$$L' = \mu_0 n^2 (9A)(l/3) = 3 \text{ L}$$

Ans. (B)

Sol. 6 Ideal inductor will behave like zero resistance long time after switch is closed so equivalent circuit will be

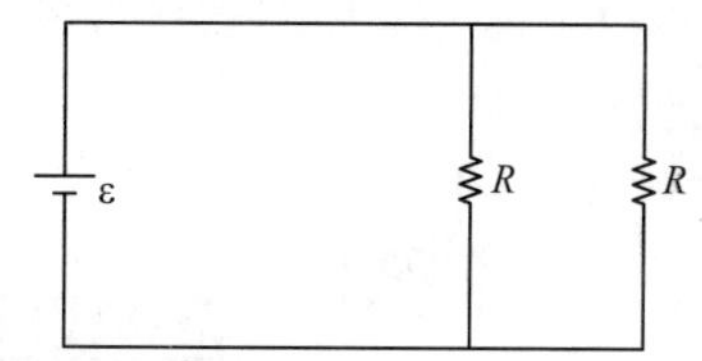

Current through battery is

$$I = \frac{2\varepsilon}{R} = \frac{2 \times 15}{5} = 6 \text{ A}$$

Ans. (A)

Sol. 7 Flux linked with solenoid is given as

$$\phi = NBA = LI$$

$$\Rightarrow \quad N\mu_0 nI\pi R^2 = LI$$

$$\Rightarrow \quad N\mu_0 \left(\frac{N}{l}\right) I\pi R^2 = LI$$

N & R are to be taken constant, so we use

Self inductance $\quad (L) \propto \frac{1}{l} \propto \frac{1}{\text{length}}$ **Ans. (B)**

Sol. 8 If mutual induction between coils P and Q is M, we use

$$\phi_Q = MI_P \quad \ldots(1)$$

and $$\phi_P = MI_Q \quad \ldots(2)$$

$$\Rightarrow \quad \frac{\phi_Q}{\phi_P} = \frac{I_P}{I_Q}$$

$$\Rightarrow \quad \frac{10^{-3}}{\phi_P} = \frac{3}{2}$$

$$\Rightarrow \quad \phi_P = \frac{2}{3} \times 10^{-3} = 0.67 \times 10^{-3} \text{ Wb}$$

Ans. (B)

Sol. 9 Current in solenoid

$$I_{(t)} = kte^{-\alpha t}$$

Flux through solenoid is

$$\phi_{(t)} = B.A = (\mu_0 n I_{(t)}) \times \pi R^2$$

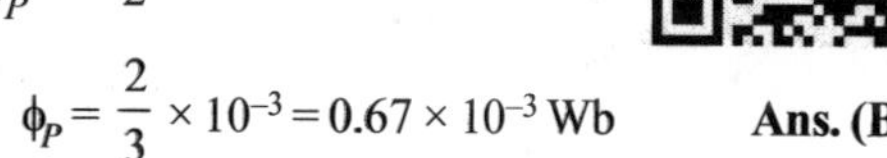

(same flux will pass through coil of radius $2R$)

emf induced in outer coil is

$$e = -\frac{d\phi_{(t)}}{dt}$$

$$\Rightarrow \quad e = -\mu_0 n\pi R^2 k\,[e^{-\alpha t} + t(-\alpha e^{-\alpha t})]$$

$$\Rightarrow \quad e = -Ce^{-\alpha t}(1 - \alpha t)$$

Current induced in coil is

$$I_{coil} = \frac{e}{R_{coil}}$$

$$\Rightarrow \quad I_{coil} = -\frac{C}{R} e^{-\alpha t}(1 - \alpha t)$$

at $t = 0 \rightarrow I_{coil} \rightarrow 0$

at $t = \infty \rightarrow I_{coil} \rightarrow 0$

$I_{coil} \rightarrow -$ ve

Ans. (B)

Sol. 10

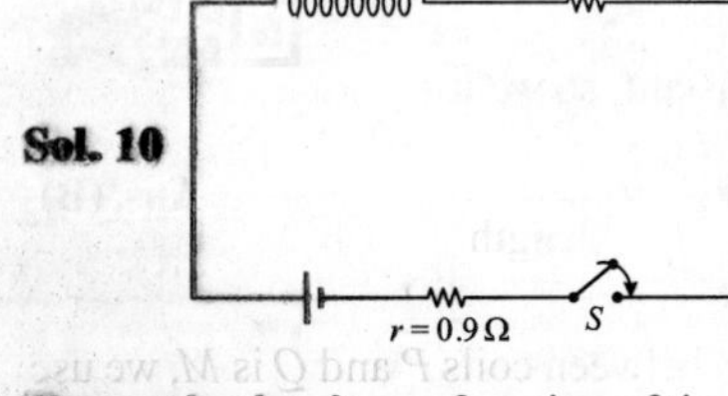

Current in circuit as a function of time is

$$i = i_0(1 - e^{-t/\tau})$$

$$\Rightarrow \quad \frac{80}{100} i_0 = i_0(1 - e^{-t/\tau})$$

$$\Rightarrow \quad 0.8 = 1 - e^{-t/\tau}$$

$$\Rightarrow \quad e^{-t/\tau} = 0.2 = \frac{1}{5}$$

$$\Rightarrow \quad -\frac{t}{\tau} = \ln\left(\frac{1}{5}\right)$$

$$\Rightarrow \quad -\frac{t}{\tau} = -\ln(5)$$

$$\Rightarrow \quad t = \tau.\ln(5)$$

$$\Rightarrow \quad t = \frac{L}{R_{eq}} \cdot \ln(5)$$

$$\Rightarrow \quad t = \frac{10 \times 10^{-3}}{(0.1 + 0.9)} \times 1.6$$

$$\Rightarrow \quad t = 1.6 \times 10^{-2}$$

$$\Rightarrow \quad t = 0.016 \text{ s}$$

Ans. (B)

Sol. 11 Charge flown through battery is given as

$$q = \int I dt$$

$$\Rightarrow \quad q = \int_0^{L/R} \frac{E}{R}\left[1 - e^{\frac{-Rt}{L}}\right] dt$$

$$\Rightarrow \quad q = \frac{EL}{R^2} \frac{1}{e}$$

$$\Rightarrow \quad q = \frac{EL}{2.7 R^2}$$

Ans. (B)

Sol. 12 Growth of current in *RL* circuit is given as

$$i = i_0(1 - e^{-Rt/L})$$

$$\Rightarrow \quad \frac{i_0}{i} = \frac{1}{1 - e^{-2 \times 10^4}}$$

$$\Rightarrow \quad \frac{i_0}{i} \simeq 1$$

Ans. (A)

Sol. 13 For current growth in *LR* circuit, we use

$$i = i_0(1 - e^{-Rt/L}) = i_0(1 - e^{-t/T_C})$$

Charge flow through circuit is given as

$$q = \int_0^{T_C} i dt$$

$$\Rightarrow \quad q = \int_0^{T_C} \frac{\varepsilon}{R}(1 - e^{-t/T_C})$$

$$\Rightarrow \quad q = \frac{\varepsilon}{R}\left(t - \frac{e^{-t/T_C}}{-1/T_C}\right)\Bigg|_0^{T_C}$$

$$\Rightarrow \quad q = \frac{e}{R}(T_C - T_C e^{-1}) - \frac{e}{R}(0 + T_C)$$

$$\Rightarrow \quad q = \frac{e}{R} \times T_C e^{-1}$$

$$\Rightarrow \quad q = \frac{\varepsilon}{R} \times \frac{L}{R} \frac{1}{e}$$

$$\Rightarrow \quad q = \frac{\varepsilon L}{e R^2}$$

Ans. (D)

Sol. 14 Induced voltage in choke coil is given as

$$V = L\left|\frac{di}{dt}\right|$$

$$\Rightarrow \quad L = \frac{V}{\left|\frac{di}{dt}\right|} = \frac{100}{\frac{0.25}{0.025 \times 10^{-3}}} = 10 \text{ mH}$$

Ans. (10)

Sol. 15 Maximum energy stored in inductor is

$$U_{max} = \frac{1}{2} L I_{max}^2$$

In series *L-R* circuit growth of current as a function of time is given as

$$i = I_{max}(1 - e^{-Rt/L})$$

When circuit **current becomes** $\frac{I_{max}}{\sqrt{n}}$ then energy in inductor **becomes** $\left(\frac{1}{n}\right)$ **times maximum value.** So we use

$$\frac{I_{max}}{\sqrt{n}} = I_{max}(1-e^{-Rt/L})$$

$$\Rightarrow \quad e^{-Rt/L} = 1-\frac{1}{\sqrt{n}} = \frac{\sqrt{n}-1}{\sqrt{n}}$$

$$\Rightarrow \quad -\frac{Rt}{L} = \ln\left(\frac{\sqrt{n}-1}{\sqrt{n}}\right)$$

$$\Rightarrow \quad t = \frac{L}{R}\ln\left(\frac{\sqrt{n}}{\sqrt{n}-1}\right)$$

Ans. (C)

Sol. 16 In the given circuit writing equation of potential drop from P to Q, we get

$$V_P - L\frac{di}{dt} - 30 + iR = V_Q$$

$$\Rightarrow \quad V_P - V_Q = 50 \times 10^{-3} \times 10^2 + 30 - 1 \times 2 = 33 \text{ V}$$

NOTE : As current is decreasing in circuit self induced emf in inductor will have high potential end at left side. **Ans. (33)**

Sol. 17 By transformer equation, we use

$$\frac{N_P}{N_S} = \frac{V_P}{V_S}$$

$$\Rightarrow \quad \frac{N_P}{24} = \frac{220}{12}$$

$$\Rightarrow \quad N_P = \frac{220 \times 24}{12} = 440 \text{ turns}$$

Ans. (440)

Sol. 18 Just after the switch is closed, inductor will act like and open circuit or as infinite resistance so the circuit will look like as shown in figure below

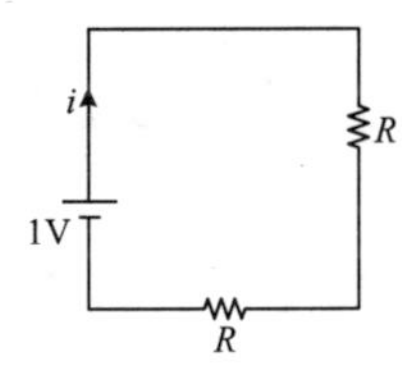

$$i = \frac{9}{R+R} = \frac{9}{4} = 2.25$$

Ans. (A)

Sol. 19 In the given circuit, at $t = 0$, inductor act like an open circuit so current through inductor is zero and equivalent resistance of the circuit across battery is given as

$$R_{eq} = (5+1)\,\|\,(5+4) = \frac{18}{5}\,\Omega$$

Current through battery can now be calculated as

$$i_1 = \frac{E}{18/5} = \frac{5E}{18}$$

At $t = \infty$ i.e. after a long time, inductor act as a short circuit and it behaves like a straight wire and now the circuit can be drawn as shown in figure below

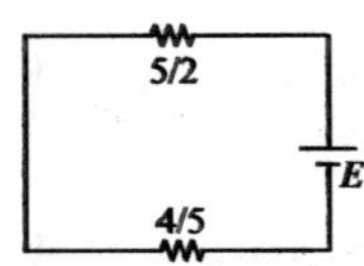

Equivalent resistance of across battery is now given as

$$R_{eq} = (5\,\|\,5) + (4\,\|\,1) = \frac{33}{10}\,\Omega$$

Current through battery can now be calculated as

$$i_2 = \frac{E}{33/10} = \frac{10E}{33}$$

Ans. (A)

Sol. 20 From the circuit described in question the EMF of battery remain equal to the potential difference across the inductor, which gives

$$e = \frac{L\,dI}{dt}$$

Current in circuit at $t = 4$ s can be given as

$$\Rightarrow \quad 3\int_0^4 t\,dt = 2\int_0^I dI$$

$$\Rightarrow \quad \frac{3}{2} \times 16 = 2I$$

$$\Rightarrow \quad I = 12$$

Energy stored in coil can now be given as

$$U = \frac{1}{2}LI^2 = \frac{1}{2} \times 2(12)^2 = 144 \text{ J}$$

Ans. (144)

Sol. 21 Voltage in secondary coil can be calculated from the circuit shown in figure as

$$V_S = \frac{P}{i} = \frac{60}{0.11} = 545.45 \text{ V}$$

As primary coil voltage is $V_P = 220$ V so we have $V_S > V_P$ hence this is a step up transformer. **Ans. (C)**

Sol. 22 As the solenoids are identical, current i will be equally divided in the three solenoids B, C and D and the magnetic field inside the solenoid is directly proportional to the current through solenoids so magnetic field inside these three solenoids will be one third of that of A which will be 1 T. **Ans. (D)**

Sol. 23

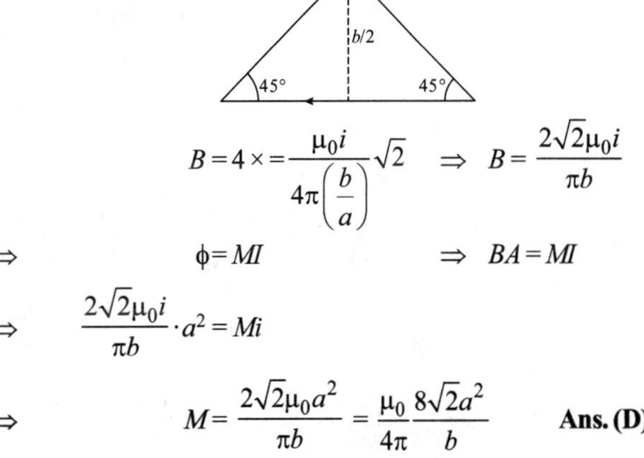

$$B = 4 \times = \frac{\mu_0 i}{4\pi\left(\frac{b}{a}\right)}\sqrt{2} \quad \Rightarrow \quad B = \frac{2\sqrt{2}\mu_0 i}{\pi b}$$

$$\Rightarrow \quad \phi = MI \quad \Rightarrow \quad BA = MI$$

$$\Rightarrow \quad \frac{2\sqrt{2}\mu_0 i}{\pi b} \cdot a^2 = Mi$$

$$\Rightarrow \quad M = \frac{2\sqrt{2}\mu_0 a^2}{\pi b} = \frac{\mu_0}{4\pi}\frac{8\sqrt{2}a^2}{b}$$

Ans. (D)

21.1 Alternating Current

Sol. 1 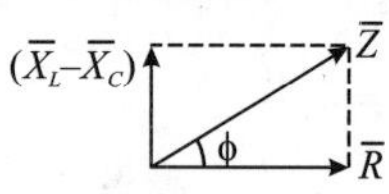

$$R = 5\Omega$$

$$X_L = \omega L = 320 \times 0.025 = 8\,\Omega$$

$$X_C = \frac{1}{\omega C} = \frac{1}{320 \times 10^{-3}} = 3.1\,\Omega$$

Impedance triangle (phasor form)

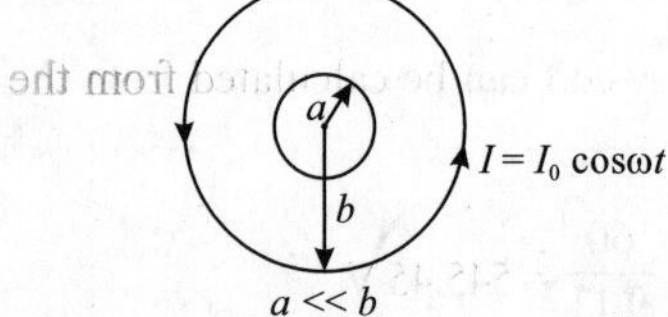

Circuit impedance is given as

$$Z = \sqrt{R^2 + (X_L - X_C)^2}$$

$$Z = \sqrt{(5)^2 + (4.9)^2} = 7\Omega$$

From triangle we have

$$\phi = \tan^{-1}\left(\frac{X_L - X_C}{R}\right) = \tan^{-1}\left(\frac{4.9}{5}\right)$$

$$\Rightarrow \quad \phi \simeq \pi/4 \quad \textbf{Ans. (B)}$$

Sol. 2

Magnetic flux linked with the smaller loop is

$$\phi = \frac{\mu_0 I}{2b} \times \pi a^2 = \frac{\mu_0 \pi a^2}{2b} I_0 \cos\omega t$$

Emf induced in smaller loop is

$$e = -\frac{d\phi}{dt} = \frac{\mu_0 \pi a^2 I_0}{2b} \omega \sin\omega t \quad \textbf{Ans. (A)}$$

Sol. 3 The quality factor for a series RLC circuit is given as

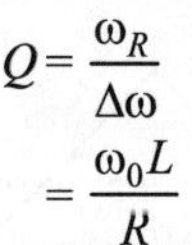

$$Q = \frac{\omega_R}{\Delta\omega}$$

$$= \frac{\omega_0 L}{R}$$

Ans. (A)

Sol. 4 Average power consumed by an AC circuit is given as

$$P_{avg} = e_{rms} i_{rms} \cos\phi$$

Here $e_{rms} = \frac{e_0}{\sqrt{2}} = \frac{100}{\sqrt{2}} V$

and $i_{rms} = \frac{i_0}{\sqrt{2}} = \frac{20}{\sqrt{2}}$ A; $\phi = \pi/4$

$$\Rightarrow \quad P_{avg} = \frac{100}{\sqrt{2}} \times \frac{20}{\sqrt{2}} \times \frac{1}{\sqrt{2}} = \frac{1000}{\sqrt{2}} W$$

Wattless current is given as

$$I_{\omega L} = i_{rms} \sin\phi = \frac{20}{\sqrt{2}} \times \frac{1}{\sqrt{2}} = 10\text{A} \quad \textbf{Ans.(B)}$$

Sol. 5 Power equation of a transformer is given as

$$e_1 i_1 \times \eta = e_2 i_2$$

$$\Rightarrow \quad 2300 \times 5 \times \frac{90}{100} = 230 \times i_2$$

$$\Rightarrow \quad i_2 = 45\text{A} \quad \textbf{Ans. (C)}$$

Sol. 6 Capacitive reactance of circuit is given as

$$x_C = \frac{1}{\omega C} = \frac{1}{100\pi \times 120 \times 10^{-6}}$$

$$\Rightarrow \quad x_C = 26.52\,\Omega$$

Inductive Reactance is

$$x_L = \omega L = 100\pi \times 20 \times 10^{-3} = 2\pi\Omega$$

$$\Rightarrow \quad x_C - x_L = 20.24 \approx 20$$

Phasor inpedance triangle

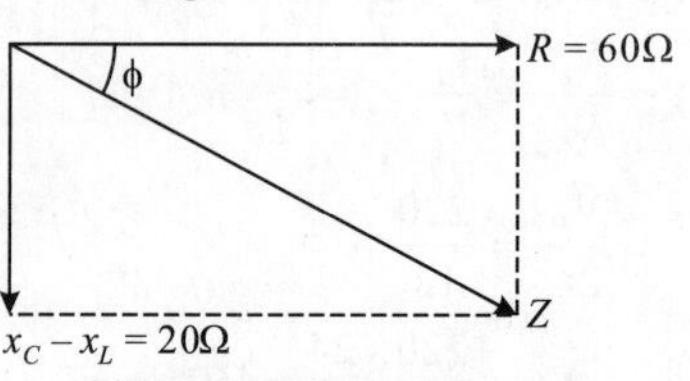

$$\Rightarrow \quad Z = \sqrt{R^2 + (x_C - x_L)^2}$$

$$\Rightarrow \quad Z = 20\sqrt{10}\Omega$$

$$\Rightarrow \quad \cos\phi = \frac{R}{z} = \frac{3}{\sqrt{10}}$$

$$\Rightarrow \quad P_{avg} = \text{VI}\cos\phi \text{ where } I = \frac{V}{Z}$$

$$\Rightarrow \quad P_{avg} = \frac{V^2}{Z}\cos\phi$$

$$\Rightarrow \quad P_{avg} = 8.64\text{ W}$$

Energy spent is

$$Q = P.t = 8.64 \times 60 = 5.18 \times 10^2\text{ J} \quad \textbf{Ans. (D)}$$

Sol. 7 For C-R_2 circuit

$$X_C = \frac{1}{\omega_C} = \frac{2}{10^{-6} \times \sqrt{3} \times 100} = \frac{2 \times 10^4}{\sqrt{3}}\Omega$$

Phase difference of I_2 with AC source emf is given as

$$\tan\theta_1 = \frac{X_C}{R_2} = \frac{2 \times 10^4}{\sqrt{3} \times 20} = \frac{1000}{\sqrt{3}}$$

$\Rightarrow$ θ_1 is close to 90° with current leading emf

For L–R_1 circuit, we use

$$X_L = \omega_L = 100 \times \frac{\sqrt{3}}{10} = 10\sqrt{3}\,\Omega$$

Phase difference of I_1 with AC source emf is given as

$$\tan\theta_2 = \frac{X_L}{R_2} = \frac{10\sqrt{3}}{10} = \sqrt{3}$$

$\Rightarrow$ $\theta_2 = 60°$ with current lagging

Thus phase difference between I_1 and I_2 is given as

$90° + 60° = 150°$ **Ans. (None)**

Sol. 8

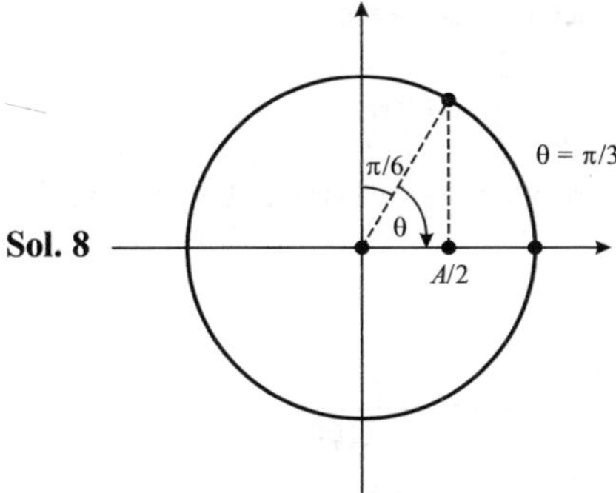

$$V(t) = 220\sin(100\pi t) \text{ volt}$$

Time taken to change phase angle by $\pi/3$ is

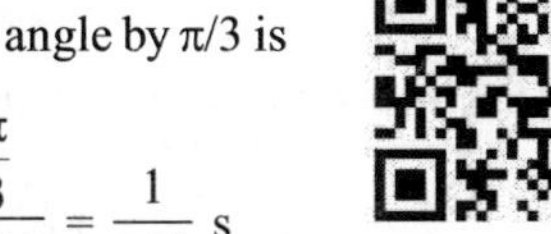

$$t = \frac{\theta}{\omega} = \frac{\frac{\pi}{3}}{100\pi} = \frac{1}{300}\text{ s}$$

$\Rightarrow$ $t = 3.3$ ms **Ans. (C)**

Sol. 9 Given phase difference $\theta = \pi/4$ and $\omega = 100$ rad/s

Phase difference will be 45° when circuit reactance and resitance are equal

$\Rightarrow$ Reactance (X) = Resistance (R) now by checking options

Option (A)

$$R = 1000\,\Omega$$

and $$X_C = \frac{1}{10^{-6}\times 100} = 10^4\Omega$$

Option (B) $R = 10^3\,\Omega$

and $X_L = 10^{-3}\times 100 = 10^{-1}\Omega$

Option (C) $R = 10^3\,\Omega$

and $X_L = 10\times 10^{-3}\times 100 = 1\,\Omega$

Option (D) $R = 10^3\,\Omega$

and $$X_C = \frac{1}{10\times 10^{-6}\times 100} = 10^3\Omega$$

Option (D) matches the given condition. **Ans. (D)**

Sol. 10 Given that for transformer

$$N_P = 300, N_s = 150, P_0 = 2200\text{ W}$$

$$I_s = 10\text{ A}$$

Output power of transformer is

$$P_0 = V_0 I_0$$

$\Rightarrow$ $2200 = V_0 \times 10$

$\Rightarrow$ $V_0 = 220$ V

Using transformer equation, we have

$$\frac{V_i}{V_0} = \frac{N_P}{N_s}$$

$\Rightarrow$ $V_i = 2\times 220 = 440$ V

For an ideal transformer

$$V_i I_i = V_O I_O$$

$\Rightarrow$ $$I_i = \frac{2200}{440} = 5\text{ A}$$ **Ans. (B)**

Sol. 11 For LCR circuit by KVL, we use

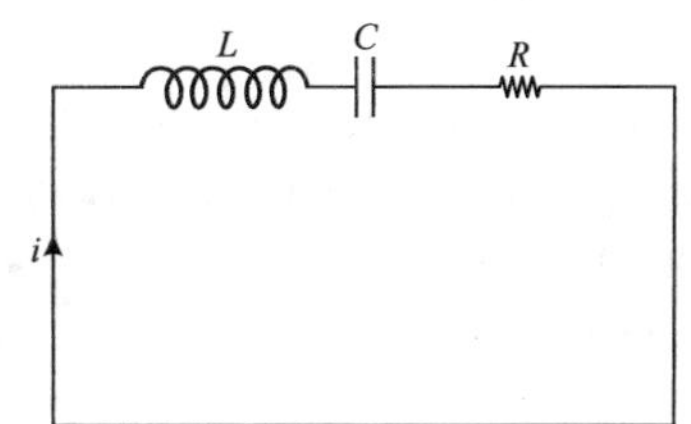

$$-L\frac{di}{dt} - \frac{q}{C} - iR = 0$$

$$L\frac{d^2q}{dt^2} + \frac{1}{C}q + R\frac{dq}{dt} = 0$$

For damped oscillator net force

$$= -kx - bv = ma$$

$$\frac{md^2x}{dt^2} + kx + \frac{bdx}{dt} = 0$$

by comparing; Equivalence is

$$L \to m;\ C \to \frac{1}{k};\ R \to b$$ **Ans. (A)**

Sol. 12 Figure below shows the circuit for situation described in question

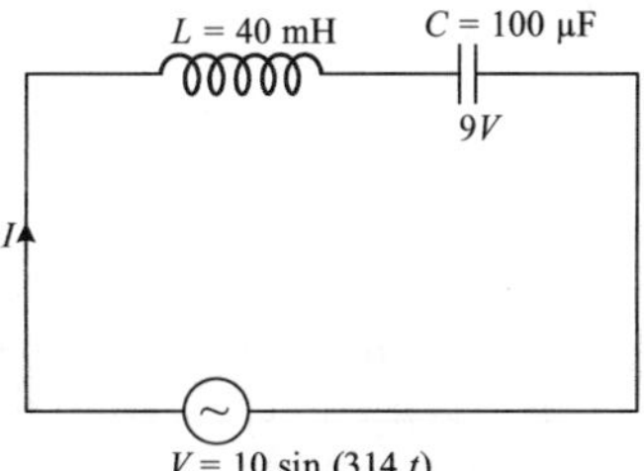

Here reactance of L and C are given as

$$X_L = \omega L = 314\times 10\times 10^{-3} = 12.56\,\Omega$$

$$X_C = \frac{1}{\omega C} = \frac{1}{314\times 100\times 10^{-6}} = \frac{10^4}{314} = 31.84\,\Omega$$

As $\overline{X_C}$ and $\overline{X_L}$ are in opposite phase we use

$$\overline{Z} = \overline{X_C} - \overline{X_L}$$

$$Z = 31.84 - 12.56 = 19.28\,\Omega$$

Peak current in circuit is given as

$$I = \frac{10}{19.28} = 0.52\,\text{A}$$

As $X_C > X_L$, in this case current will lease voltage by $\pi/2$ so it is given as

$$I = 0.52\sin\left(314t + \frac{\pi}{2}\right) \qquad \textbf{Ans. (A)}$$

Sol. 13 As due to inductance voltage lead current by 45° that means resistance of coil is equal to that of inductive reactance as shown in phasor diagram below

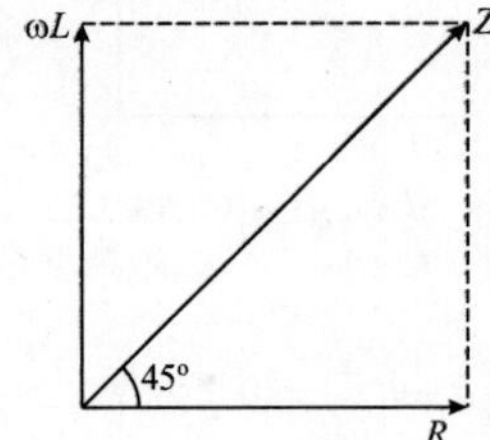

We use

$$R = X_L = \omega L \text{ and } Z = \sqrt{R^2 + \omega^2 L^2}$$

$$\Rightarrow \quad L = \frac{R}{\omega} = \frac{100}{2000\pi} = \frac{1}{20\times 3.14}$$

$$\Rightarrow \quad L = 1.6\times 10^{-2}\,\text{H}$$

Most appropriate, among given answers is (A) **Ans. (A)**

Sol. 14 Given circuit impedance is

100Ω 0.1803H 10μF

20V, 750Hz

$$Z = \sqrt{R^2 + (X_L - X_C)^2}$$

$$\Rightarrow \quad Z = \sqrt{(100)^2 + \left(2\pi nL - \frac{1}{2\pi nL}\right)^2}$$

$$\Rightarrow \quad Z = \sqrt{(100)^2 + \left[2\times 3.14\times 750\times 0.1803 - \frac{1}{2\times 3.14\times 750\times 10\times 10^{-6}}\right]^2}$$

$$\Rightarrow \quad Z \simeq 834\,\Omega$$

Current in circuit is given as

$$i = \frac{e}{Z} = \frac{20}{834}A$$

Thermal power produced in resistance is

$$P_{\text{th}} = i^2 R = \left(\frac{20}{834}\right)^2 \times 100\,\text{W}$$

In time t if temperature of resistance is raised by 10°C, we use

$$\theta = ms\Delta\theta$$

$$\Rightarrow \quad P_{\text{th}} t = ms\Delta\theta$$

$$\Rightarrow \quad t = \frac{2\times 10}{\left(\frac{20}{834}\right)^2 \times 100} = \frac{(834)^2}{20\times 100}$$

$$\Rightarrow \quad t \cong 348\text{ s} \qquad \textbf{Ans. (C)}$$

Sol. 15 Power dissipated in circuit is given as

$$P = \left(\frac{e^2}{Z}\right)\cos\phi$$

$$\Rightarrow \quad 400 = \frac{(250)^2}{Z}\times 0.8$$

$$\Rightarrow \quad Z = 125\,\Omega$$

Capacitance required to change power factor to unity is such that

$$\frac{1}{\omega C} = \omega L = Z\sin\phi$$

$$\Rightarrow \quad C = \frac{1}{\omega Z\sin\phi}$$

$$\Rightarrow \quad C = \frac{1}{2\pi(50)\times 125\times 0.6}$$

$$\Rightarrow \quad C = \frac{\left(\frac{1}{2500}\right)}{3\pi}F = \frac{n}{3\pi}\,\mu\text{F}$$

$$\Rightarrow \quad n = \frac{10^6}{2500} = 400 \qquad \textbf{Ans. (400)}$$

Sol. 16 Qality factor of resonance circuit is given as

$$Q = \omega_0\frac{L}{R} = (2\pi f)\frac{L}{R}$$

$$\Rightarrow \quad Q = 2\pi\times 10^7\times\frac{2\times 10^{-4}}{6.28} = 2\times 10^3 = 2000$$

Ans. (2000)

Sol. 17 (a) Magnetron is a source of microwave frequencies.
(b) Vibration of atoms and molecules causes them to emit infrared frequency.
(c) Radioactive decay of a nucleus causes it to emit radioactive radions including alpha, beta and gamma rays.
(d) Inner shall electrons in heavy elements during transition emits high frequency radiations of the order of X-rays.

Ans. (B)

Sol. 18 At resonance the power of LCR circuit is given as

$$P=\frac{V^2}{R}$$

$$\Rightarrow \quad R=\frac{V^2}{P}=\frac{(120)^2}{16}=900\,\Omega \qquad \textbf{Ans. (900)}$$

Sol. 19 Current through $60\,\Omega$ resistance in the circuit given in question is calculated as

$$i_R=\frac{15}{60}=\frac{1}{4}\text{A}$$

Thus capacitor current is also the same, so we have

$$i_C=\frac{1}{4}\text{A}$$

Potential difference across capacitor is given as

$$V_C=IX_C$$

$$\Rightarrow \quad 10=\frac{1}{4}\times\frac{1}{\omega C}$$

$$\Rightarrow \quad C=\frac{1}{40\omega}=\frac{1}{4000}=250\,\mu\text{F}$$

Now, current through 40Ω resistance is given as

$$i_{R'}=\frac{20}{40}=\frac{1}{2}\text{A}$$

Thus current through inductor is given as

$$i_L=\frac{1}{2}-\frac{1}{4}=\frac{1}{4}\text{A}$$

Potential difference across inductor is given as

$$V_L=IX_L=\frac{1}{4}\times\omega L$$

$$\Rightarrow \quad 20=\frac{1}{4}\times 100\times L$$

$$\Rightarrow \quad L=0.8\,\text{H} \qquad \textbf{Ans. (D)}$$

Sol. 20 As the phase difference between current and voltage is 45°, we can use

$$\tan 45^\circ=\frac{1}{\omega CR}=\frac{\omega L}{R}$$

$$\Rightarrow \quad X_L=X_C$$

At resonance, we have

$$i=\frac{V}{R}=\frac{220}{110}=2\text{A} \qquad \textbf{Ans. (A)}$$

Sol. 21 (a) Rectifier is used to convert from AC to DC.
(b) Stabilizers are used for stable (constant) output voltage when input voltage varies with time.
(c) Transformers are used to Step - up or step - down ac voltages.
(d) Filters are used to remove ripples in the rectified output voltage from a rectifier. **Ans. (A)**

Sol. 22 For the given AC current, RMS value can be calculated by taking square root of average of squares, given as

$$i=i_1\sin\omega t+i_2\sin(\omega t+90)$$

$$i^2=i_1^2\sin^2\omega t+i_2^2\cos^2\omega t+i_1i_2\sin\omega t\cos\omega t$$

$$<i^2>=i_1^2\left(\frac{1}{2}\right)+i_2^2\left(\frac{1}{2}\right)+0$$

$$i_{RMS}=\sqrt{<i^2>}=\sqrt{\frac{i_1^2+i_2^2}{2}} \qquad \textbf{Ans. (C)}$$

Sol. 23 The quality factor of a series LCR resonant circuit is given as

$$Q=\frac{X_L}{R}=\frac{\omega L}{R}=\frac{1}{\sqrt{LC}}\frac{L}{R}=\frac{\sqrt{L}}{R\sqrt{C}}$$

When the inductance is increased by two fold and resistance decreased by two fold, the new quality factor becomes

$$Q'=\frac{\sqrt{2L}}{\left(\frac{R}{2}\right)\sqrt{C}}=2\sqrt{2}Q=2\sqrt{2}(100)$$

$$\Rightarrow \quad Q'=282.84 \qquad \textbf{Ans. (283)}$$

Sol. 24 Impedance of the circuit is given as

$$Z=\sqrt{(X_L-X_C)^2+R^2}$$

Inductive and capacitive reactance of circuit are calculated as

$$X_L=\omega L=100\times 100\times 10^{-3}=10\,\Omega$$

and

$$X_C=\frac{1}{\omega C}=\frac{1}{100\times 100\times 10^{-6}}=10\,\Omega$$

$$\Rightarrow \quad Z=\sqrt{(10-100)^2+R^2}=\sqrt{90^2+120^2}$$

$$\Rightarrow \quad Z=150\,\Omega$$

Peak current in circuit can be calculated as

$$i_{\text{peak}}=\frac{e}{Z}=\frac{30}{150}=\frac{1}{5}=0.2\,\text{A}$$

Resonant frequency of circuit is given by

$$\omega = \frac{1}{\sqrt{LC}}$$

$$\Rightarrow \quad f = \frac{1}{2\pi\sqrt{LC}} = \frac{1}{2\pi\sqrt{100\times10^{-3}\times100\times10^{-6}}}$$

$$\Rightarrow \quad f = \frac{100\sqrt{10}}{2\pi} = \frac{100\pi}{2\pi} = 50\text{ Hz}$$ **Ans. (D)**

Sol. 25 At resonance power dissipated is given as

$$P = \frac{(V_{rms})^2}{R}$$

$$P = \frac{(250/\sqrt{2})^2}{8} = 3906.25\text{ W} \approx 4\text{ kW}$$

Ans. (4)

Sol. 26 Hot wire ammeter gives rms value of the current. So it is calculated for the given current as

$$I_{rms} = \frac{I_0}{\sqrt{2}} = \sqrt{\frac{I_1^2 + I_2^2}{2}}$$ **Ans. (B)**

Sol. 27 All options can be chosen by definition using phasor analysis for an AC circuit in which an emf is applied across pure resistance, inductor and capacitor. **Ans. (D)**

Sol. 28 Inductive reactance in purely inductive circuit is given as

$$X_L = \omega L$$

Thus circuit current is given as

$$i = \frac{e_0}{\omega L}$$

So on reducing the frequency to half, reactance will get halved and current gets doubled. **Ans. (B)**

Sol. 29 Bandwidth of a series LCR circuit is given as

$$w = R/L$$

$$\Rightarrow \quad w \propto R$$

Thus on increasing resistance, bandwidth will increase. Resonance frequency does not depend on resistance so it will remain same. Quality factor of series LCR circuit is given as

$$Q = \frac{X_L}{R} = \frac{\omega L}{R}$$

Thus on increasing resistnace, quality factor will decrease.

Ans. (A)

Sol. 30 For an ideal AC source, sinusoidal current is given as

$$i = i_0\cos(\omega t)$$

Here $i = i_0$ at $t = 0$ and $i = \frac{i_0}{\sqrt{2}}$ occurs at $\omega t = \frac{\pi}{4}$ which gives

$$t = \frac{\pi}{4\omega} = \frac{\pi}{4(2\pi f)} = \frac{1}{8f}$$

$$\Rightarrow \quad t = \frac{1}{400} = 2.5\text{ ms}$$ **Ans. (A)**

Sol. 31 The power factor of the circuit is given as

$$\cos\phi = \frac{R}{Z} \quad \ldots(1)$$

The circuit impedence in series LCR circuit is given as

$$Z = \sqrt{R^2 + (X_L - X_C)^2} \quad \ldots(2)$$

$$\Rightarrow \quad Z = \sqrt{6^2 + (10-4)^2}$$

$$\Rightarrow \quad Z = 6\sqrt{2}\ \Omega$$

The power factor is calculated as

$$\cos\phi = \frac{6}{6\sqrt{2}} = \frac{1}{\sqrt{2}}$$ **Ans. (C)**

Sol. 32 Magnetic energy stored in inductor carrying a current i is given as

$$U = \frac{1}{2}Li^2$$

This energy reaches 25% of its maximum value when current reaches half of maximum value, which gives

$$i = \frac{i_0}{2}$$

For growth of current in LR circuit, current is given as

$$i = i_0(1 - R^{-Rt/L})$$

If at $t = t_1$, $i = \frac{i_0}{2}$, from above expression, we have

$$t_1 = \frac{L}{R}\ln 2$$ **Ans. (C)**

Sol. 33 The phasor diagrams of voltage and current in circuit are given as

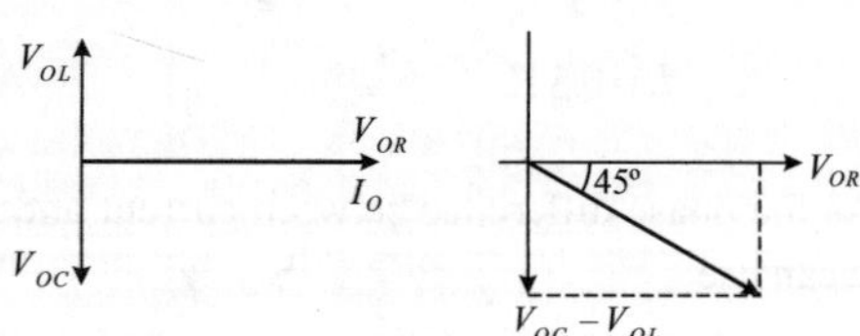

$$\Rightarrow \quad \tan 45° = \frac{V_{OC} - V_{OL}}{V_{OR}} = \frac{X_C - X_L}{R}$$

$$\Rightarrow \qquad R = \frac{1}{\omega C} - \omega L$$

$$1\Omega = \frac{1}{300C} - 30 \times 10^{-3} \times 300$$

$$\Rightarrow \qquad C = \frac{1}{3} \times 10^{-3}\,\text{F}$$ **Ans. (3)**

Sol. 34 The angular frequency of the supply is given as

$$\omega = 2\pi f = 100\,\pi$$

Resistance & reactances of the given circuit are calculated as

$$R = 100\,\Omega$$

$$X_L = \omega L = \frac{100\pi \times 0.5}{1000} = \frac{\pi}{200}\Omega$$

$$X_C = \frac{1}{\omega C} = \frac{10^{12}}{100\pi \times 0.1} = \frac{10^{11}}{\pi}\Omega$$

As $X_C >> X_L$ & $X_C >> R$ so the circuit impedence is dominently capacitive.

Then $$\tan\phi = \frac{|X_L - X_C|}{R} \approx \infty; \phi = 90^\circ$$ **Ans. (B)**

Sol. 35 As circuit is given in resonance so we use

$$X_L = X_C$$

$$\Rightarrow \qquad Z = R$$

and $i_{rms} = V/Z = V/R$

Power at resonance is given as

$$P = i^2_{rms} R$$

$$\Rightarrow \qquad P = \frac{V^2}{R} = \frac{250 \times 250}{5} = 12500\,\text{J/s} = 125 \times 10^2\,\text{W}$$

Ans. (125)

Sol. 36 In case of a pure capacitive circuit, current leads voltage by an angle $\pi/2$ in phase. **Ans. (D)**

Sol. 37 At resonant frequency in a circuit, emf is in same phase with current and the resonant frequency is given as

$$\omega_R = \frac{1}{\sqrt{LC}}$$

At resonant frequency the capacitive reactance of circuit becomes equal to inductive reactance of circuit and current in circuit becomes maximum, so we use

$$\omega L = \frac{1}{\omega C}$$

If inductive reactance is more current will lag behind emf in phase and when capacitive reactance is more current will lead emf in phase. **Ans. (B)**

Sol. 38 The angular frequency of the AC source is given as

$$\omega = 2\pi f = 100\,\pi\,\text{rad/s},$$

The AC emf can be given in standard form as

$$e = e_0 \sin\left(\omega t + \frac{\pi}{2}\right)$$

At rms value of emf, we use

$$\cos\omega t = \frac{1}{\sqrt{2}}$$

$$\Rightarrow \qquad \omega t = \frac{\pi}{4}$$

$$\Rightarrow \qquad t = \frac{\pi/4}{100\pi} = \frac{1}{400}\,\text{s} = \frac{1000}{400}\,\text{ms} = 2.5$$ **Ans. (D)**

Sol. 39 Given that per cycle average power dissipation in both the circuits is same, so we have

$$P_1 = P_2$$

$$\left(\frac{V^2}{R}\right)_1 = \left(\frac{V^2}{Z}\right)_2$$

$$\Rightarrow \qquad R = Z$$

$$\Rightarrow \qquad R = \sqrt{\left(\omega L - \frac{1}{\omega C}\right)^2 + R^2}$$

$$\Rightarrow \qquad 5 = \sqrt{\left(\omega L - \frac{1}{\omega C}\right)^2 + R^2}$$

$$\Rightarrow \qquad 25 = \left[\omega(0.1) - \frac{1}{\omega(40 \times 10^{-6})}\right]^2 + 25;$$

$$\Rightarrow \qquad \omega^2 = (0.1) = \frac{1}{40 \times 10^{-6}}$$

$$\Rightarrow \qquad \omega^2 = \frac{1}{4} \times 106$$

$$\Rightarrow \qquad \omega = 500\,\text{rad/s}$$ **Ans. (500)**

Sol. 40 Angular frequency of the source is given as

$$\omega = 2\pi f = 100\,\pi$$

$$\Rightarrow \qquad X_L = \omega L = 100 \times \frac{22}{7} \times 0.07 = 22\,\Omega$$

Circuit impedance can be given as

$$Z = \sqrt{R^2 + x_L^2}$$

$$\Rightarrow \qquad Z = \sqrt{(12)^2 + (22)^2} \approx 25.00$$

Circuit current is given as

$$I = \frac{220}{25} = 8.8$$

Phase angle between current and voltage can now be calculated as

$$\tan\phi = 22/12 = 11/6$$

$$\Rightarrow \quad \phi = \tan^{-1}\left(\frac{11}{6}\right) \qquad \textbf{Ans. (C)}$$

Sol. 41 Given angular frequncy of source is

$$\omega = 2\pi f = 120\,\pi$$

In resonance condition, we use

$$\omega L = \frac{1}{\omega C}$$

$$\Rightarrow \quad L = \frac{1}{\omega^2 C} = \frac{1}{(120\pi)^2 \times 0.1\times 10^{-6}}$$

$$\Rightarrow \quad L = 70.3\ \text{H} \qquad \textbf{Ans. (D)}$$

Sol. 42 As the give inductor is storing energy as well as dissipating heat, it acts like a *RL* circuit shown below.

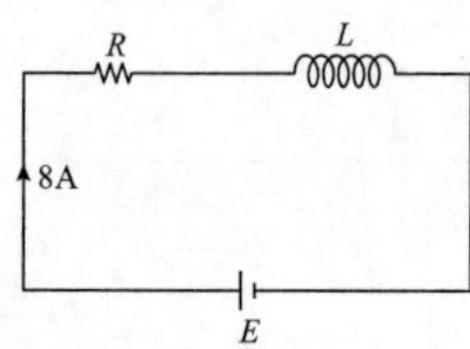

The energy stored in inductor is given as

$$U = \frac{1}{2}LI^2$$

$$\Rightarrow \quad 64 = \frac{1}{2}.L.(8)^2$$

$$\Rightarrow \quad L = 2\ \text{H}$$

Power dissipated in coil resistance is given as

$$P = I^2R$$

$$\Rightarrow \quad 640 = (8)^2.R$$

$$\Rightarrow \quad R = 100$$

Time constant of the circuit is calculated as

$$\tau = \frac{L}{R} = \frac{2}{10} = 0.2 \qquad \textbf{Ans. (D)}$$

Sol. 43 At resonance of series *LCR* circuit, we use

$$\omega = 2\pi f = 2\pi\times 50 = 100\pi$$

and $$X_C = X_L$$

$$\Rightarrow \quad \frac{1}{\omega C} = \omega L$$

$$\Rightarrow \quad \frac{1}{\omega C} = 250\pi$$

$$\Rightarrow \quad C = \frac{1}{100\pi\times 250\pi} = 4\ \mu\text{F} \qquad \textbf{Ans. (A)}$$

Sol. 44 The impedence of upper branch in circuit is given as

$$Z_C = \sqrt{\left(\frac{1}{\omega C}\right)^2 + R^2}$$

$$\Rightarrow \quad Z_C = \sqrt{\left(\frac{1}{100\times 100\times 10^{-6}}\right)^2 + 100^2}$$

$$\Rightarrow \quad Z_C = \sqrt{100^2 + 100^2} = 100\sqrt{2}\ \Omega$$

The impedence of lower branch in circuit is given as

$$\Rightarrow \quad Z_L = \sqrt{(\omega L)^2 + R^2} = \sqrt{(100\times 0.5)^2 + 50^2}$$

$$\Rightarrow \quad Z_L = 50\sqrt{2}\ \Omega$$

Current in the two branches of circuit can be calculated as

$$i_L = \frac{200}{Z_L} = \frac{200}{50\sqrt{2}} = \frac{4}{\sqrt{2}} = 2\sqrt{2}\ \text{A}$$

$$\Rightarrow \quad i_C = \frac{200}{Z_C} = \frac{200}{100\sqrt{2}} = \sqrt{2}\ \text{A}$$

Power factors of the two branches are given as

$$\cos\phi_1 = \frac{100}{100\sqrt{2}} = \frac{1}{\sqrt{2}},\ \phi_1 = 45^\circ$$

and $$\cos\phi_2 = \frac{50}{50\sqrt{2}} = \frac{1}{\sqrt{2}},\ \phi_2 = 45^\circ$$

Due to one branch current leads by 45° and due to other branch it lags by 45° so the two currents are at 90° phase difference hence current through source is given as

$$I = \sqrt{I_C^2 + I_L^2} = \sqrt{8+2} = \sqrt{10}A$$

Ans. (None)

Sol. 45 Square of the given current is written as

$$i^2 = 42\sin^2\left(\frac{4t}{T}\right) + 100 + 2\sqrt{42}\sin\left(\frac{4t}{T}\right)$$

Taking mean of each term of squares, gives

$$<i^2> = \frac{42}{2} + 100$$

Calculating root of mean square gives

$$i_{\text{rms}} = \sqrt{<i^2>} = \sqrt{121} = 11\ \text{A} \qquad \textbf{Ans. (11)}$$

Sol. 46 Power factor of circuit is given as

$$PF_{\text{initial}} = \frac{R}{\sqrt{9R^2 + R^2}} = \frac{1}{\sqrt{10}}$$

Final power factor of circuit is given as

$$PF_{\text{final}} = \frac{R}{\sqrt{(3-2)^2R^2 + R^2}} = \frac{1}{\sqrt{2}}$$

$$\Rightarrow \quad \frac{PF_{\text{final}}}{PF_{\text{initial}}} = \sqrt{5} \qquad \textbf{Ans. (1)}$$

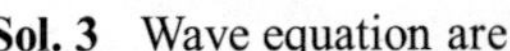

Sol. 47 As $X_L = X_C$ circuit will be in resonance and circuit impedance can be given as

$$Z = \sqrt{(X_L - X_C)^2 + R^2}$$

$$\Rightarrow \quad Z = R \qquad \textbf{Ans. (B)}$$

5Sol.48 At very high frequencies i.e. at $\omega \to \infty$, capacitive reactance can be taken as zero and inductive reactance can be considered as open circuit so the effective circuit can be drawn as shown in figure below

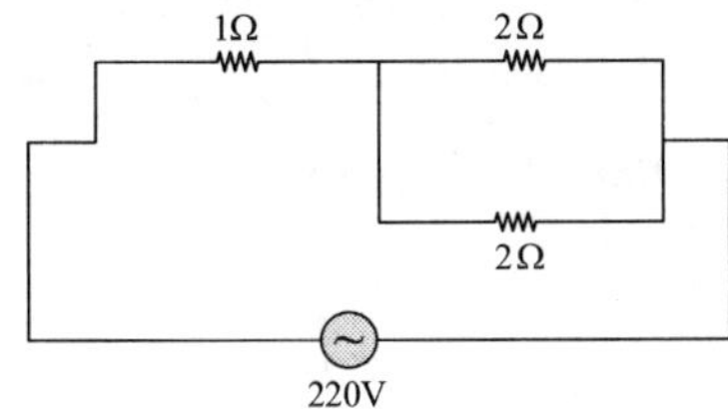

Thus equivalent resistance in this case can be calculated as

$$Z = 1 + \frac{2\times 2}{2+2} = 2\Omega \qquad \textbf{Ans. (2)}$$

22.1 Electromagnetic Wave

Sol. 1 Electric field amplitude is related to B_0 with speed of EM wave as

$$c = E_0/B_0$$

$$\Rightarrow \quad E_0 = cB_0$$

As wave is travelling along $-X$ direction so vector $\vec{E}\times\vec{B}$ must be along the direction of propagation of EM wave i.e. along $-\hat{i}$ direction. So here Electric field can be given as

$$\vec{E} = E_0 \sin(kx + \omega t)\,\hat{k}$$

$$\Rightarrow \quad \vec{E} = cB_0 \sin(kx + \omega t)\,\hat{k} \qquad \textbf{Ans. (A)}$$

Sol. 2 As we know for an EM wave we use

$$\frac{\partial B}{\partial t} = -\frac{\partial E}{\partial z}$$

$$\Rightarrow \quad \frac{\partial B}{\partial t} = -(-2E_0 k \sin kz \cos\omega t)$$

$$\Rightarrow \quad \partial B = 2E_0 k \sin kz \cos\omega t\,\partial t$$

$$\Rightarrow \quad \int \partial B = 2E_0 k \sin kz \int \cos\omega t\,\partial t$$

$$\Rightarrow \quad B = 2E_0 k \sin kz \left(\frac{\sin\omega t}{\omega}\right)$$

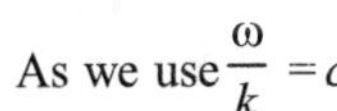

As we use $\dfrac{\omega}{k} = c$

$$\Rightarrow \quad B = \frac{2E_0}{c} \sin kz \sin\omega t \qquad \textbf{Ans. (C)}$$

Sol. 3 Wave equation are

$$E_1 = E_{01} \cos\left(\frac{\omega z}{c} - \omega t\right) \qquad \dots(1)$$

[We use $\omega = 2\pi\nu$ and $\omega = kv$ or $v = \omega/k$]

$$E_2 = E_{02} \cos\left(\frac{2\omega z}{c} - \omega t\right) \qquad \dots(2)$$

$$\Rightarrow \quad E_2 = E_{02} \cos\left(\frac{\omega z}{c/2} - \omega t\right) \qquad [v = c \text{ in air}]$$

In air speed of EM wave is $v_{air} = c$

In medium speed of EM wave is $v_m = c/2$

The speed of EM wave can be written as

$$v = \frac{1}{\sqrt{\mu \in}} = \frac{1}{\sqrt{\mu_0 \in_0 \mu_r \in_r}}$$

As here medium is non magnetic $\Rightarrow \mu_{r_1} = \mu_{r_2} = 1$

In this case we use

$$\frac{v_{air}}{v_m} = \frac{c}{c/2} = \frac{\frac{1}{\sqrt{\in_{r_1}}}}{\frac{1}{\sqrt{\in_{r_2}}}} = \sqrt{\frac{\in_{r_2}}{\in_{r_1}}}$$

$$\Rightarrow \quad \frac{\in_{r_2}}{\in_{r_1}} = 4 \Rightarrow \frac{\in_{r_1}}{\in_{r_2}} = \frac{1}{4} \qquad \textbf{Ans. (C)}$$

Sol. 4 If $\vec{B}$ is the magnetic field then it should be such that $\vec{E}\times\vec{B}$ is along the direction of propagation of EM wave i.e. parallel to $\left(\dfrac{\hat{i}+\hat{j}}{\sqrt{2}}\right)$

taking cross product $\hat{k}\times\left(\dfrac{\hat{i}-\hat{j}}{\sqrt{2}}\right) = \dfrac{\hat{j}-(-\hat{i})}{\sqrt{2}} = \left(\dfrac{\hat{i}+\hat{j}}{\sqrt{2}}\right)$

So option (A) and (B) here are resulting $\vec{B}$ such the direction of wave propagation is along the given direction.

here $\vec{r}$ is the wave displacement along direction of propagation

So $\qquad \vec{r} = r\hat{r} = r\left(\dfrac{\hat{i}+\hat{j}}{\sqrt{2}}\right)$

By substituting $\vec{r}$ in (A) and (B) we get

$$\text{(A)} \quad \frac{E_0}{c}\left(\frac{\hat{i}-\hat{j}}{\sqrt{2}}\right)\cos\left[\underbrace{10^4\left(\frac{\hat{i}-\hat{j}}{\sqrt{2}}\right)\cdot r\left(\frac{\hat{i}+\hat{j}}{\sqrt{2}}\right)}_{0} - \overset{\omega=2\pi\nu}{\overbrace{3\times 10^{12}t}}\right]$$

So (A) represent on oscillating field not the travelling field.

$$\text{(B)} \quad \frac{E_0}{c}\left(\frac{\hat{i}-\hat{j}}{\sqrt{2}}\right)\cos\left[10^4\left(\frac{\hat{i}-\hat{j}}{\sqrt{2}}\right)\cdot r\left(\frac{\hat{i}+\hat{j}}{\sqrt{2}}\right) - 3\times 10^{12}t\right]$$

$$\Rightarrow \quad \frac{E_0}{c}\left(\frac{\hat{i}-\hat{j}}{\sqrt{2}}\right)\cos[10^4 r - 3\times 10^{12}t] \text{ travelling wave equation.}$$

Ans. (B)

Sol. 5 For an EM wave $\frac{E}{B} = c \Rightarrow |\vec{B}| = \frac{E}{c}$

If I is intensity then energy density in EM wave is given as

$$\frac{1}{2} \in_0 E_0^{\,2} = \frac{I}{c}$$

$$\Rightarrow \qquad E_0 = \sqrt{\frac{2I}{\in_0 c}}$$

We also know that direction of propagation of EM wave is along vector $\vec{E} \times \vec{B}$ which should be along $\hat{j}$ here. **Ans. (C)**

Sol. 6 As z_1 and z_2 are two neighbouring points where EF = 0

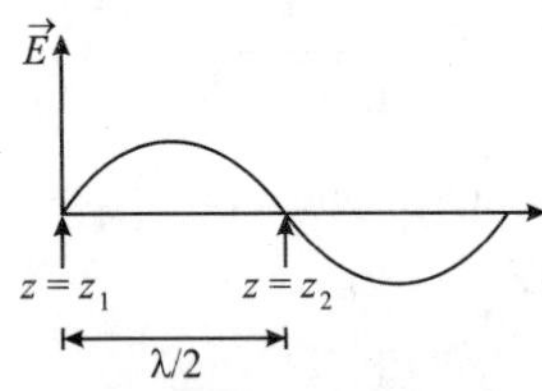

As we have $|(z_2 - z_1)| = \frac{\lambda}{2}$

or $\qquad \lambda = c/\nu$

$$\Rightarrow \qquad |z_2 - z_1| = \frac{c}{2\nu}$$

$$\Rightarrow \qquad \nu = \frac{c}{2|z_2 - z_1|} = \frac{3\times10^8}{2|z_2 - z_1|}$$

$$\Rightarrow \qquad \nu = \frac{1.5\times10^8}{|z_2 - z_1|}$$ **Ans. (C)**

Sol. 7 In EM wave E, B and C are related as

$$B = \frac{E}{C} = \frac{6.3}{3\times10^8} = 2.1 \times 10^{-8}\,\text{T}$$

Direction vectors $\hat{E}$, $\hat{B}$ and $\hat{C}$ are related as

$$\Rightarrow \qquad \hat{E} \times \hat{B} = \hat{C}$$

$$\Rightarrow \qquad \hat{j} \times \hat{B} = \hat{i}$$

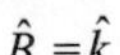

$$\Rightarrow \qquad \hat{B} = \hat{k}$$ **Ans. (C)**

Sol. 8 Average energy density of magnetic field,

$u_B = \frac{B_0^2}{2\mu_0}$, B_0 is maximum value of magnetic field.

Average energy density of electric field,

$$u_E = \frac{\in_0 E_0^2}{2}$$

Now, $\qquad E_0 = cB_0$, and $c^2 = \frac{1}{\mu_0 \in_0}$

$$u_E = \frac{\in_0}{2} \times c^2 B_0^2$$

$$\Rightarrow \qquad u_E = \frac{\in_0}{2} \times \frac{1}{\mu_0 \in_0} \times B_0^2 = \frac{B_0^2}{2\mu_0} = u_B$$

$$\Rightarrow \qquad u_E = u_B$$

As energy density of electric & magnetic field is same, energy associated with equal volume will be equal. **Ans. (C)**

Sol. 9 Electric and magnetic field amplitudes are related as

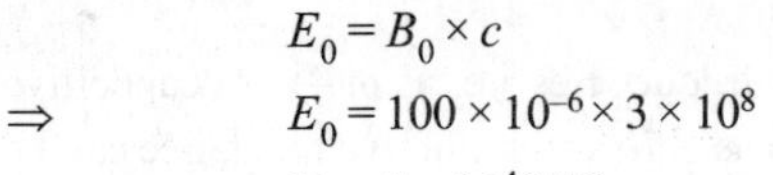

$$E_0 = B_0 \times c$$

$$\Rightarrow \qquad E_0 = 100 \times 10^{-6} \times 3 \times 10^8$$

$$\Rightarrow \qquad E_0 = 3 \times 10^4\,\text{N/C}$$ **Ans. (D)**

Sol. 10 The magnetic field magnitude is given as

$$B = \frac{E}{c} = \frac{10}{c}$$

We write EF at $t = 0$ as

$$\vec{E}(x, y) = (10\,\hat{j}) \cos [(6\,\hat{i} + 8\,\hat{k}) \cdot (x\,\hat{i} + z\,\hat{k})]$$

$$= (10\,\hat{j}) \cos [\vec{k} \cdot \vec{r}]$$

Here $\vec{k} = 6\,\hat{i} + 8\,\hat{k}$ is the direction along which wave propagation

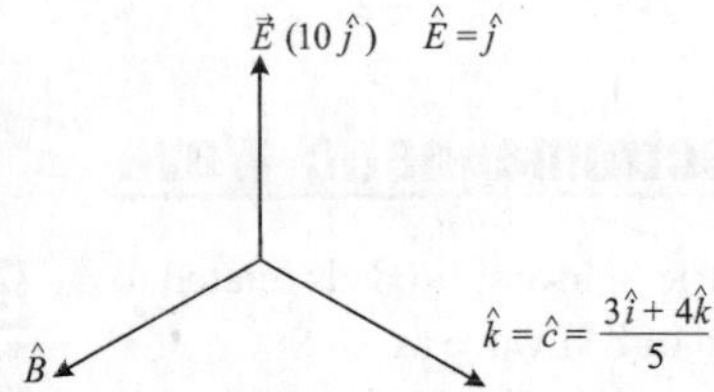

Direction vector of MF is given as

$$\hat{B} = \hat{c} \times \hat{E} = \frac{-4\hat{i} + 3\hat{k}}{5}$$

MF vector is given as

$$\vec{B} = \frac{10}{c}\left(\frac{-4\hat{i} + 3\hat{j}}{5}\right) \cos [\vec{k} \cdot \vec{r} - \omega t]$$

$$\Rightarrow \qquad \vec{B} = \frac{10}{c}\left[\frac{3\hat{i} - 4\hat{i}}{5}\right] \cos [6x + 8z - 10ct]$$ **Ans. (B)**

Sol. 11 Speed of EM wave in free space is

$$c = \frac{1}{\sqrt{\mu_0 \in_0}}$$

In a medium speed is given as

$$v = \frac{1}{\sqrt{k \in_0 \mu_0}} \quad \text{[For transparent medium } \mu_r \approx \mu_0]$$

$$\Rightarrow \qquad \frac{c}{v} = \sqrt{k} = n \qquad \text{(refractive index of medium)}$$

EM wave intensity is given as

$$I = \frac{1}{2} \in_0 E_0^2 c = \frac{1}{2} \in_0 k E^2 v$$

$$\Rightarrow \qquad E_0^2 c = kE^2 v$$

$$\Rightarrow \quad \frac{E_0^2}{E^2} = \frac{kv}{c} = \frac{n^2}{n}$$

$$\Rightarrow \quad \frac{E_0}{E} = \sqrt{n}$$

For magnetic field we can use

$$\frac{B_0^2 c}{2\mu_0} = \frac{B^2 v}{2\mu_0}$$

$$\Rightarrow \quad \frac{B_0}{B} = \frac{1}{\sqrt{n}}$$ **Ans. (B)**

Sol. 12 Intensity of electromagnetic wave is given by

$$I = \frac{\text{Power}}{\text{Area}} = \frac{1}{2}\varepsilon_0 E_0^2 c$$

$$\Rightarrow \quad I = \frac{27\times10^{-3}}{10\times10^{-6}} = \frac{1}{2}\times 9\times10^{-12}\times E_0^2 \times 3\times10^8$$

$$\Rightarrow \quad E_0 = \sqrt{2}\times10^3 \text{ kV/m}$$

$\Rightarrow \quad E_0 = 1.4\,\text{kV/m}$ **Ans. (C)**

Sol. 13 Power of refracted transmitted light is 96% so, we have

$$P_{\text{refracted}} = \frac{96}{100}P_I$$

$$\Rightarrow \quad K_2 A_t^2 = \frac{96}{100}K_1 A_i^2$$

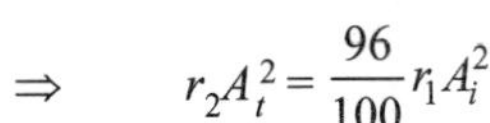

$$\Rightarrow \quad r_2 A_t^2 = \frac{96}{100} r_1 A_i^2$$

$$\Rightarrow \quad A_t^2 = \frac{96}{100}\times\frac{1}{\frac{3}{2}}\times(30)^2$$ [Given that $A_i = 30$ V/m]

$$\Rightarrow \quad A_t = \sqrt{\frac{64}{100}\times(30)^2} = 24\,\text{V/m}$$ **Ans. (B)**

Sol. 14 Intensity of electromagnetic wave is given as

$$I = \varepsilon_0 c E_{\text{rms}}^2$$

and $\quad E_{\text{rms}} = cB_{\text{rms}}$

$$\Rightarrow \quad I = \varepsilon_0 c^3 B_{\text{rms}}^2$$

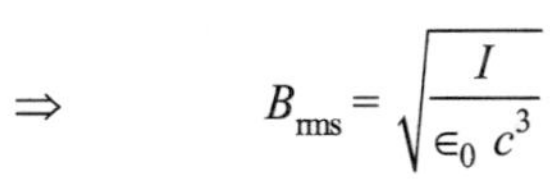

$$\Rightarrow \quad B_{\text{rms}} = \sqrt{\frac{I}{\epsilon_0 c^3}}$$

$\Rightarrow \quad B_{\text{rms}} \approx 10^{-4}$ **Ans. (B)**

Sol. 15 The direction of propogation of an *EM* wave is direction of $\vec{E}\times\vec{B}$. Thus from given information

$$\hat{i} = \hat{j}\times\hat{B}$$

$$\Rightarrow \quad \hat{B} = \hat{k}$$

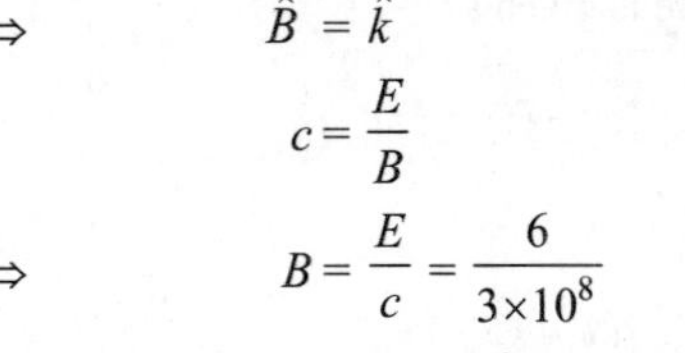

$$c = \frac{E}{B}$$

$$\Rightarrow \quad B = \frac{E}{c} = \frac{6}{3\times10^8}$$

$B = 2\times10^{-8}\,\text{T}$ Along *z* direction **Ans. (C)**

Sol. 16 If we use that direction of light propagation will be along $\vec{E}\times\vec{B}$. Then (D) option is correct

Detailed solution is as following

magnitude of $\; E = CB$

$$E = 3\times10^8\times1.6\times10^{-6}$$

$$E = 4.8\times10^2$$

$\vec{E}$ and $\vec{B}$ are perpendicular to each other

$$\Rightarrow \quad \vec{E}\cdot\vec{B} = 0$$

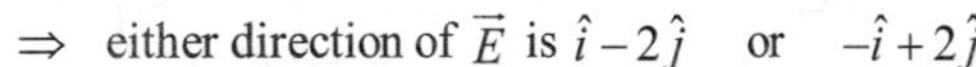

$\Rightarrow$ either direction of $\vec{E}$ is $\hat{i}-2\hat{j}$ or $-\hat{i}+2\hat{j}$

From given option

Also wave propagation direction is parallel to $\vec{E}\times\vec{B}$ which is $-\hat{k}$

$\Rightarrow \quad \vec{E}$ is along $(-\hat{i}+2\hat{j})$ **Ans. (D)**

Sol. 17 Maximum Electric field is given as

$$E = (B)(c)$$

$$\vec{E}_0 = (3\times10^{-5})c\,(-\hat{j})$$

$$\vec{E}_1 = (2\times10^{-6})c\,(-\hat{i})$$

Maximum force

$$\vec{F}_{0\max} = q\vec{E} = qc(-3\times10^{-5}\hat{j} - 2\times10^{-6}\hat{i})$$

$$|\vec{F}_{0\max}| = 10^{-4}\times3\times10^8\sqrt{(3\times10^{-5})^2+(2\times10^{-6})^2}$$

$$= 0.9\,\text{N}$$

$$F_{\text{rms}} = \frac{F_{0\max}}{\sqrt{2}} = 0.63\,\text{N (approx)}$$ **Ans. (D)**

Sol. 18 Direction of $\vec{E}\times\vec{B}$ is along the propogaion of EM wave

$$\vec{E}\times\vec{B} \parallel \vec{v} \qquad \ldots(1)$$

Given that wave is propagating along positive *z*-axis and $\vec{E}$ along positive *x*-axis. Thus from Maxwell's equation we have

$$\vec{\nabla}\times\vec{E} = -\frac{\partial B}{\partial t}$$

$$\Rightarrow \quad \frac{\partial E}{\partial z} = -\frac{\partial B}{dt}$$

and $\quad B_0 = \dfrac{E_0}{c}$ **Ans. (B)**

Sol. 19 Electric field wave is given as

$$\vec{E} = E_0 \hat{n} \sin[\omega t + (6y - 8z)]$$

Comparing with equation

$$\vec{E} = E_0 \hat{n} \sin[\omega t - \vec{s} \cdot \vec{r}]$$

Where $\vec{s}$ is along direction of propagation and

$$\vec{r} = x\hat{i} + y\hat{i} + z\hat{k}$$

$$\vec{s} \cdot \vec{r} = -6y + 8z$$

$$\Rightarrow \quad \vec{s} = -6\hat{i} + 8\hat{k}$$

Unit vector along direction of propagation as

$$\hat{s} = \frac{\vec{s}}{|\vec{s}|} = \frac{-6\hat{j} + 8\hat{k}}{10} = \frac{-3\hat{j} + 4\hat{k}}{5}$$ **Ans. (C)**

Sol. 20 Magnetic field when electromagnetic wave propagates in $+z$ direction so general equation of magnetic field is

$$B = B_0 \sin(kz - \omega t)$$

Where $B_0 = \dfrac{E_0}{C} = \dfrac{60}{3\times10^8} = 2\times10^{-7}\,\text{T}$

We use $k = \dfrac{2\pi}{\lambda} = 0.5\times10^3\,\text{m}^{-1}$

and $\omega = 2\pi f = 1.5\times10^{11}\,\text{rad/s}$ **Ans. (C)**

Sol. 21 Given that $\vec{B}$ is along $\hat{j}$ and wave is travelling in X-direction so using

$$\vec{E}\times\vec{B} = \vec{C} = -\hat{i}$$

$\Rightarrow$ $\vec{E}$ is along $\hat{k}$ and its magnitude is given as

$$E = 3\times10^{-8}\times3\times10^{8} = 9\,\text{V/m}$$ **Ans. (A)**

Sol. 22 Force on a charge particle is given as

$$\vec{F} = q(\vec{E} + \vec{v}\times\vec{B})$$

Electric field at location of charge is

$$\vec{E} = E_0\left(\frac{\hat{i}+\hat{j}}{\sqrt{2}}\right)\cos\pi = -E_0\left(\frac{\hat{i}+\hat{j}}{\sqrt{2}}\right)$$

For *EM* wave, we use

$$\vec{E}\times\vec{B} = \vec{c}$$

$$\Rightarrow \quad +E_0\left(\frac{\hat{i}+\hat{j}}{\sqrt{2}}\right)\times\vec{B} = c\hat{k}$$

$$\Rightarrow \quad \vec{B} = -\left(\frac{\hat{i}-\hat{j}}{\sqrt{2}}\right)\frac{E_0}{c}$$

Thus force on charge is given as

$$\vec{F} = q\left(-E_0\left(\frac{(\hat{i}+\hat{j})}{\sqrt{2}}\right) - v_0\hat{k}\times\left(\frac{(\hat{i}-\hat{j})}{\sqrt{2}}\right)\frac{E_0}{c}\right)$$

$$\Rightarrow \quad \vec{F} \simeq -qE_0\left(\frac{\hat{i}+\hat{j}}{\sqrt{2}}\right) \quad [\text{As } v_0 << c]$$

$\Rightarrow$ F is antiparallel to $\dfrac{\hat{i}+\hat{j}}{\sqrt{2}}$ **Ans. (C)**

Sol. 23 For an electromagnetic wave, we use

$$\vec{E} = \vec{B}\times\vec{c}$$

$$\Rightarrow \quad \vec{E} = (5\times10^{-8}\,\hat{j})\times(3\times10^{8}\,\hat{k})$$

$$\Rightarrow \quad \vec{E} = 15\hat{i}\ \text{V/m}$$ **Ans. (B)**

Sol. 24 Electric field vector is given as

$$\vec{E}_1 = E_0\hat{j}\cos(\omega t - kx)$$

Its corresponding magnetic field will be

$$\vec{B}_1 = \frac{E_0}{c}\hat{k}\cos(\omega t - kx)$$

and for second wave corresponding fields are

$$\vec{E}_2 = E_0\hat{k}\cos(\omega t - ky)$$

$$\vec{B}_2 = \frac{E_0}{c}\hat{i}\cos(\omega t - ky)$$

Net force on charge particle is given as

$$\Rightarrow \vec{F} = q\vec{E}_1 + q\vec{E}_2 + q\vec{v}\times B_1 + q\vec{v}\times B_2$$

$$\Rightarrow \vec{F} = qE_0\hat{j} + qE_0\hat{k} + q(0.8c\,\hat{j})\times\left(\frac{E_0}{c}\hat{k}\right) + q(0.8c\,\hat{j})\times\left(\frac{E_0}{c}\hat{i}\right)$$

$$\Rightarrow \vec{F} = qE_0\hat{j} + qE_0\hat{k} + 0.8qE_0\hat{i} - 0.8qE_0\hat{k}$$

$$\Rightarrow \vec{F} = qE_0[0.8\hat{i} + 1\hat{j} + 0.2\hat{k}]$$ **Ans. (C)**

Sol. 25 Direction of polarisation of wave is given as

$$\hat{E} = \hat{k}$$

Direction of propagation

$$\hat{c} = \hat{E}\times\hat{B} = \frac{\hat{i}+\hat{j}}{\sqrt{2}}$$

$$\Rightarrow \quad \hat{B} = \frac{\hat{i}-\hat{j}}{\sqrt{2}}$$ **Ans. (A)**

Sol. 26 Energy density of electromagnetic waves is given as

$$u_m = \frac{B_0^2}{2\mu_0}$$

$$\Rightarrow B_0 = \sqrt{2\mu_0 u_m} = \sqrt{2\times4\pi\times10^{-7}\times1.02\times10^{-8}}$$

$$\cong \sqrt{256\times10^{-16}}$$

$$= 16\times10^{-8}\,\text{T} = 160\,\text{nT}$$ **Ans. (B)**

Sol. 27 Unit vectors along electric and magnetic fields are

$$\hat{E} = \hat{k}$$

$$\hat{B} = \frac{\vec{B}}{|B|} = \frac{2\hat{i} - 2\hat{j}}{2\sqrt{2}}$$

$$\Rightarrow \quad \hat{B} = \frac{1}{\sqrt{2}}(\hat{i} - \hat{j})$$

Direction of wave propagation is given as

$$\hat{c} = \hat{E} \times \hat{B}$$

$$\Rightarrow \quad \hat{c} = \hat{k} \times \left[\frac{1}{\sqrt{2}}(\hat{i} - \hat{j})\right]$$

$$\Rightarrow \quad \hat{c} = \frac{1}{\sqrt{2}}(\hat{k} \times \hat{i} - \hat{k} \times \hat{j})$$

$$\Rightarrow \quad \hat{c} = \frac{1}{\sqrt{2}}(\hat{i} + \hat{j})$$

Ans. (A)

Sol. 28 Direction of speed in an electromagnetic wave is given as

$$\hat{c} = \hat{E} \times \hat{B}$$

(given) $-\hat{j}$ ↓, $\hat{i}$ (given) ↓, $-\hat{k}$ (analysed)

Amplitudes of electric field and magnetic fields are related as

$$E_0 = cB_0 = 3 \times 10^8 \times 3 \times 10^{-8} = 9 \text{ V/m}$$

As wave is propagating in $-y$-direction direction of $\vec{E}$ must be along $-\hat{k}$ so we have

$$\vec{E} = -9 \sin(200\pi(y + ct))\,\hat{k}$$

Ans. (B)

Sol. 29 At any instance magnetic field magnitude in electromagnetic wave is given as

$$B_0 = \frac{E_0}{c} = E_0\sqrt{\mu_0 \in_0}$$

Direction of wave propagation is given by Poynting vector

$$\vec{S} = \frac{1}{\mu_0}(\vec{E} \times \vec{B})$$

$$\hat{S} = \hat{E} \times \hat{B}$$

↑ ↑ ↓

$\hat{i}_{(\text{given})}$ $\hat{j}_{(\text{given})}$ $\Rightarrow \hat{k}_{(\text{analysed})}$

Thus magnetic field is given as

$$B = B_0 \cos(\omega t - kx)\,\hat{k}$$

at $t = 0$ $\quad B = E_0\sqrt{\mu_0 \in_0} \cos kx\hat{k}$

Ans. (C)

Sol. 30 Electromagnetic spectrum is given as

$\lambda_{\text{increasing}}$ ↑

Radio waves		few cm to few m
Micro waves		0.01 mm to few cm
IR	~	8000 Å – 10^5 Å
Visible	~	4000 Å – 8000 Å
UV	~	100 Å – 4000 Å
X-Rays	~	0.1 Å – 100 Å

Ans. (C)

Sol. 31 From wave equation, we can see the direction of wave propagation is $+\hat{k}$ so we use

$$\hat{k} = \hat{E} \times \hat{B}$$

Unit vector along $\vec{E}$ is given as

$$\hat{E} = \frac{1}{\sqrt{2}}(\hat{i} + \hat{j})$$

$$\Rightarrow \quad \hat{k} = \frac{1}{2}(\hat{i} + \hat{j}) \times \hat{B}$$

$$\Rightarrow \quad \hat{B} = \frac{1}{\sqrt{2}}(-\hat{i} + \hat{j})$$

Equation of magnetic field wave is given as

$$\vec{B} = \frac{E_0}{C}(-\hat{x} + \hat{y})\sin(kz - \omega t)$$

Ans. (B)

Sol. 32 Magnetic field amplitude in *EM* wave is given as

$$B_0 = \frac{E_0}{C} = \frac{30}{1.5 \times 10^7} \times 5 \times 10^{-2} = 10^{-7} \text{ T}$$

Maximum force on electron is given as

$$F_{max} = evB_0 = 1.6 \times 10^{-19} \times 0.1 \times 3 \times 10^8 \times 10^{-7}$$

$$\Rightarrow \quad F_{max} = 4.8 \times 10^{-19} \text{ N}$$

Ans. (B)

Sol. 33 *EM* radiation spectrum is given below in decreasing order of wavelength

$\lambda_{\text{decreasing}}$ ↓

Radio waves	–	few cm to several meters
Micro waves	–	$10^{-2} - 10^{-5}$ m
IR	–	8000 Å – 10^{-5} m
Visible	–	4000 – 8000 A
UV	–	100 – 4000 Å
X-Rays	–	0.01 – 100 Å
Gama rays	–	10^{-5} to 10^{-3} Å

Ans. (D)

Sol. 34 *Q*-factor of a series *RLC* circuit is given by

$$Q = \frac{1}{R}\sqrt{\frac{L}{C}} = \frac{1}{100}\sqrt{\frac{80 \times 10^{-3}}{2 \times 10^{-6}}}$$

$$\Rightarrow \quad Q = \frac{1}{100}\sqrt{40 \times 10^3}$$

$$\Rightarrow \quad Q = \frac{200}{100} = 2$$

Ans. (B)

Sol. 35 Intensity of EM wave is given as

$$I = \epsilon_0 E^2 c$$

$$\Rightarrow \quad E^2 = \frac{I}{\epsilon_0\, c}$$

$$\Rightarrow \quad E^2 = \frac{315}{\pi\,\epsilon_0} \times \frac{1}{3\times10^8}$$

$$\Rightarrow \quad E^2 = \frac{315\times4\times9\times10^9}{3\times10^8} = 315\times120$$

$$\Rightarrow \quad E = \sqrt{315\times120} \simeq 194.4$$

Ans. (194)

Sol. 36 For given electromagnetic wave electric field amplitude is given as

$$E_0 = cB_0 = 3\times10^8\times1.2\times10^{-7} = 36 \text{ V/m}$$

As wave is travelling in X-direction, we use $\hat{c} = -\hat{i}$
From Poynting vector direction, we use

$$\hat{c} = \hat{E}\times\vec{B}$$

$$-\hat{i} = \hat{E}\times\hat{k}$$

Thus $\hat{E} = -\hat{j}$ so instantaneous electric field is given as

$$\vec{E} = -36\sin(0.5\times10^3 x + 1.5\times10^{11} t)\,\hat{j} \text{ V/m}$$

Ans. (B)

Sol. 37 Frequency of the EM wave given as

$$f = 5 \text{ GHz} = 5\times10^9 \text{ Hz}$$

Relative permittivity of medium is $\epsilon_r = 2$
and Relative permeability is also $\mu_r = 2$
Speed of light in a medium in terms of the speed of light in vacuum is given as

$$v = \frac{c}{\sqrt{\mu_r\,\epsilon_r}} = \frac{3\times10^8}{\sqrt{4}} = \frac{30\times10^7}{2} \text{ m/s}$$

$$\Rightarrow \quad v = 15\times10^7 \text{ m/s}$$

Ans. (15)

Sol. 38 Wavelength in vacuum is given as

$$\lambda = \frac{c}{f} = \frac{3\times10^8}{3\times10^9} = 0.1 \text{ m}$$

Wavelength in medium is given as

$$\lambda_m = \frac{0.1}{\mu}$$

Where refractive index m of the medium in terms of medium permittivity is given as (considering non-magnetic medium for which relative permeability of medium is taken to be unity)

$$\mu = \sqrt{\varepsilon_r} = \sqrt{2.25} = 1.5$$

$$\Rightarrow \quad \lambda_m = \frac{0.1}{1.5} = \frac{1}{15} m = 6.67 \text{ cm}$$

$$\Rightarrow \quad \lambda_m = 667\times10^{-2} \text{ cm}$$

Ans. (667)

Sol. 39 Average intensity of radiation in terms of electric field is given as

$$I = \frac{1}{2} c\,\epsilon_0 E_0^2$$

Intensity at a distance 10 m from the source of 8 W power with 50% efficiency is given as

$$I = \frac{8}{4\pi\times10^2}$$

$$\Rightarrow \quad \frac{8}{4\pi\times10^2} = \frac{1}{2}\times c\times\frac{1}{\mu_0 c^2}\times E_0^2$$

$$\Rightarrow \quad E_0 = \frac{2}{10}\times\sqrt{\frac{\mu_0 c}{\pi}}$$

$$\Rightarrow \quad x = 2$$

Ans. (2)

Sol. 40 Average intensity of radiation in terms of peak electric field is given as

$$I_{avg} = \frac{1}{2}\varepsilon_0 E_0^2 c$$

$$\frac{1.25}{100}\times\frac{1000}{4\pi(2)^2} = \frac{1}{2}\times8.85\times10^{-12}\times3\times10^8\times E_0^2$$

$$E_0^2 = 187.4$$

$$\Rightarrow \quad E_0 = 13.689 = 136.89\times10^{-1} \text{ V/m}$$

$$\Rightarrow \quad x = 136.89$$

Rounding off to nearest integer $x = 137$. **Ans. (137)**

Sol. 41 EM wave is travelling towards $+\hat{y}$ direction so electric field is given as

$$\vec{E} = \vec{B}\times\vec{C} = (8\times10^{-8}\hat{z})\times(3\times10^8\hat{y})$$

$$\Rightarrow \quad \vec{E} = -24\hat{x} \text{ V/m}$$

Ans. (A)

Sol. 42 In EMW, Average energy density due to electric (U_e) and magnetic (U_m) fields is same. **Ans. (A)**

Sol. 43 The intensity of bulb at a distance 3 m from it is given as

$$I = \frac{100}{4\pi\times3^2}$$

$$\Rightarrow \quad c\epsilon_0 E^2 = \frac{100}{4\pi\times3^2}$$

$$\Rightarrow \quad c\epsilon_0\left(\sqrt{\frac{x}{5}}E\right)^2 = \frac{60}{4\pi\times3^2}$$

$$\Rightarrow \quad \frac{x}{5} = \frac{3}{5}$$

$$\Rightarrow \quad x = 3$$

Ans. (3)

Sol. 44 Conduction current density in seawater is given as

$$J_c = \frac{E}{\rho} = \frac{V}{\rho d}$$

Displacement current density can be given as

$$J_d = \frac{1}{A}\frac{dq}{dt} = \frac{C}{A}\frac{dV_c}{dt} = \frac{\in}{d}\frac{dV_c}{dt}$$

$$\Rightarrow \quad \frac{V_0 \sin 2\pi ft}{\rho d} = 10^x \times \frac{80\varepsilon_0}{d} V_0(2\pi f)\cos 2\pi ft$$

$$\Rightarrow \quad \tan\left(2\pi \times \frac{900}{800}\right) = 10^x \times \frac{40}{9\times10^9} \times 900$$

$$\Rightarrow \quad x = 6$$ **Ans. (6)**

Sol. 45 Electric field magnitude of EM wave is given as

$$E = Bc = 6 \text{ V/m}$$

Direction of propagation of wave is parallel to the vector $(\vec{E}\times\vec{B})$.

As the wave is travelling along x-direction and magnetic field is along z-direction hence the electric field vector is given as

$$\vec{E} = 6\hat{j} \text{ V/m}$$ **Ans. (C)**

Sol. 46 For an electromagnetic wave direction of field vectors is perpendicular to the direction of propagation. As wave is propagating along y-direction, field vectors must have components only in x and z directions. **Ans. (C)**

Sol. 47 Displacement current through capacitor plates can be given as

$$I_d = \varepsilon_0 \frac{d\phi_E}{dt}$$

Flux can be calculated as

$$\phi_E = \frac{V(t)A}{d} = \frac{20\sin 100\pi t \cdot 1}{2\times10^{-3}} = 10^4 \sin 100\pi t$$

$$\Rightarrow \quad I_d = \varepsilon_0 \frac{d}{dt}(10^4 \sin 100\pi t)$$

$$\Rightarrow \quad I_d = 8.85\times10^{-12}\times10^4\times100\pi \cos 10\pi t$$

$$\Rightarrow \quad I_d = 27.79\ \mu\text{A} \cos 100\,\pi t$$ **Ans. (C)**

Sol. 48 The direction of propagation of electromagnetic wave in always along the direction of vector $\vec{E}\times\vec{B}$. The unit vector in this direction is given as

$$\hat{n} = \frac{\vec{E}\times\vec{B}}{|\vec{E}\times\vec{B}|}$$

$$\Rightarrow \quad \hat{n} = \frac{E_0\hat{i}\times B_l\hat{k}}{E_0 B_0 \sin 90^\circ} = \hat{i}\times\hat{k} = -\hat{j}$$ **Ans. (A)**

Sol. 49 Average intensity of sunlight is related to the peak electric field in radiation as

$$\frac{1}{2}\varepsilon_0 E_0^2 c = I$$

$$\Rightarrow \quad E_0^2 = \frac{2I}{\varepsilon_0 C}$$

$$\Rightarrow \quad E_0 = \sqrt{\frac{2I}{\varepsilon_0 C}}$$

Peak value of magnetic field can now be given as

$$B_0 = \frac{E_0}{C} = \frac{1}{C}\sqrt{\frac{2I}{\varepsilon_0 C}} = 2.77\times10^{-8}\text{ T}$$ **Ans. (C)**

Sol. 50 The frequency of the electromagnetic wave is given as

$$2\pi f = 5.3\times10^3$$

$$\Rightarrow \quad f = \frac{5.6\times10^3}{2\pi} = \frac{5.6\times10^3}{2\times3.14} = 891.7\text{ Hz}$$

Wavelength of the electromagnetic wave is given as

$$\lambda = \frac{c}{f} = \frac{3\times10^8}{891.7} = 3.36\times10^5\text{ m}$$

Reflecting wave is in opposite direction considered to have same amplitude is given by the equation

$$E = 3.1\cos[(1.8)z + (5.4\times106)t]\,\hat{i}\text{ N/C}$$

Ans. (B)

Sol. 51 The velocity of light depends upon the electric permittivity & mangetic permeability of the dielectric medium, given as

$$v = \frac{c}{\sqrt{\mu_r\varepsilon_r}} = 3.33\times10^7\text{ m/s}$$ **Ans. (A)**

Sol. 52 Magnetic field amplitude due to light beam is calculated as

$$B_0 = \frac{E_0}{c} = \frac{800}{3\times10^8}\text{ V/m}$$

$$\Rightarrow \quad F_m = qvB_0 = \frac{1.6\times10^{-19}\times3\times10^7\times800}{3\times10^8}$$

$$\Rightarrow \quad F_m = 12.8\times10^{-18}\text{ N}$$ **Ans. (C)**

Sol. 53 Speed of electromagnetic wave is given as

$$v = \frac{\omega}{k} = \frac{10^{10}}{50} = 2\times10^8 = \frac{2}{3}c$$ **Ans. (A)**

Sol. 54 Magnetic field at a point on electromagnetic wave in space **is** given as

$$B=\frac{E}{c}=\frac{6}{3\times10^{8}}=2\times10^{-8}\,\text{T}$$ **Ans. (2)**

Sol. 55 Energy density in electromagnetic wave is given as

$$u_e=\frac{1}{2}\epsilon_0 E^2$$

Energy confined in the given volume can be calculated as

$$\text{Energy}=\frac{1}{2}\epsilon_0 E^2\times\text{volume}$$

$$\Rightarrow\quad 5.5\times10^{-12}=\frac{1}{2}\times8.8\times10^{-12}\times50^2\times V$$

$$\Rightarrow\quad V=500\,\text{cm}^3$$ **Ans. (500)**

Sol. 56 Wave speed is given as

$$v=\frac{\omega}{x}=\frac{20\times10^{10}}{20\times10^{2}}=10^8\,\text{m/s}$$

Refractive index of the wave is given as

$$\mu=\frac{c}{v}=\frac{3\times10^8}{10^8}=3=\sqrt{\mu_r\varepsilon_r}$$

$$\Rightarrow\quad \varepsilon_r=9$$ **Ans. (D)**

23.1 Geometrical Optics-I Reflection of Light

Sol. 1 As object oscillates $\Rightarrow$ image will also oscillate so (a) is correct.

For the two ext positions of object at $u_1=8$ cm and $u_2=12$ cm the corresponding image location are – (for $f=5$ cm)

For left position of object

$$u_1=8\,\text{cm}\,(+ve)$$

$$f=5\,\text{cm}\,(+ve)$$

$$\Rightarrow\quad \frac{1}{f}=\frac{1}{v_1}+\frac{1}{u_r}$$

$$\Rightarrow\quad v_1=\frac{u_1 f}{u_1-f}=\frac{8\times5}{8-5}=+\frac{40}{3}\,\text{cm}$$

For right position of object

$$u_2=12\,\text{cm}$$

$$f=+5\text{cm}$$

$$\Rightarrow\quad \frac{1}{f}=\frac{1}{v_2}+\frac{1}{u_2}$$

$$\Rightarrow\quad v_2=\frac{u_2 f}{u_2-f}=\frac{12\times5}{12-5}+\frac{60}{7}\text{cm}$$

As v_1 and v_2 are not same so the extreme position in oscillation not symmetric about image mean position which is

$$x_0=10\,\text{cm}\,(2f)$$

$\Rightarrow$ Option (c) is correct

The distance between extreme positions of image is

$$\Delta S=v_1-v_2$$

$$=\frac{40}{3}-\frac{60}{7}=\frac{280-180}{21}=\frac{100}{21}\text{cm}$$

$\Rightarrow$ Option (d) is correct **Ans. (C)**

Sol. 2

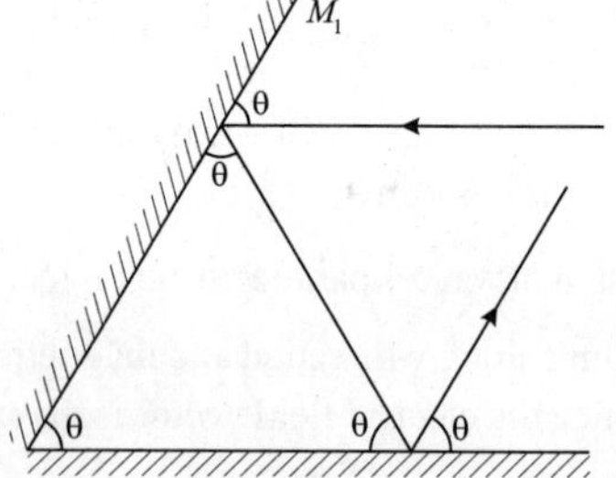

Assuming angles between two mirrors be θ as per geometry, a triangle sum of angles of is

$$3\theta=180°$$

$$\Rightarrow\quad \theta=60°$$ **Ans. (D)**

Sol. 3

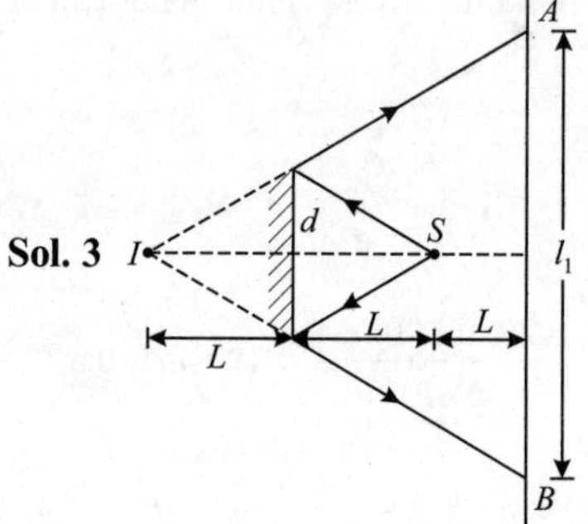

Field of view of image is $l_1(=AB)$, given as

$$\frac{l_1}{3L}=\frac{d}{L}$$

$$\Rightarrow\quad l_1=3d$$ **Ans. (A)**

Sol. 4 Magnification for spherical mirror is given as

$$m=\frac{f}{f-u}$$

$$\Rightarrow\quad 5=\frac{-40}{-40-u}$$

$$\Rightarrow\quad u=-32\,\text{cm}$$ **Ans. (D)**

Sol. 5 Light from particle P will be reflected at mirror.

So for mirror formula, we use

$$u = -5\text{ cm},$$

$$f = -\frac{R}{2} = -20\text{ cm}$$

Using mirror formula, we get

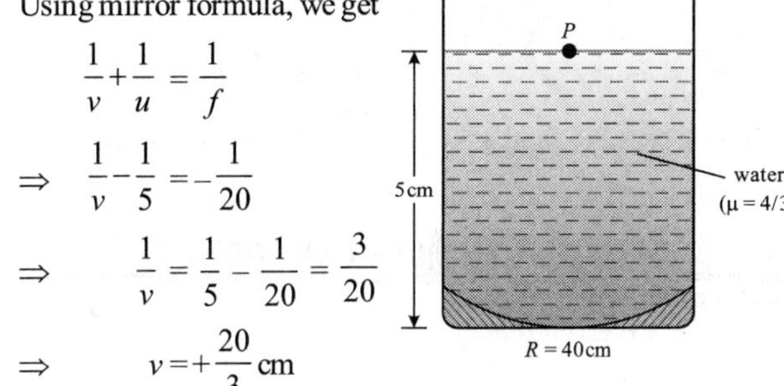

$$\frac{1}{v}+\frac{1}{u} = \frac{1}{f}$$

$$\Rightarrow \quad \frac{1}{v}-\frac{1}{5} = -\frac{1}{20}$$

$$\Rightarrow \quad \frac{1}{v} = \frac{1}{5}-\frac{1}{20} = \frac{3}{20}$$

$$\Rightarrow \quad v = +\frac{20}{3}\text{ cm}$$

This image will act as object for light getting refracted at water surface so, object distance is taken as

$$d = 5+\frac{20}{3} = \frac{35}{3}\text{ cm}$$

Light rays from this object will refract at plane surface (water level) so after refraction, final image will form at

$$d' = d\left(\frac{\mu_2}{\mu_1}\right)$$

$$\Rightarrow \quad d' = \left(\frac{35}{3}\right)\left(\frac{1}{4/3}\right)$$

$$\Rightarrow \quad d' = \frac{35}{4} = 8.75\text{cm}$$

$$\Rightarrow \quad d' \approx 8.8\text{ cm}$$

Ans. (A)

Sol. 6 By mirror formula, we have

$$\frac{1}{v}+\frac{1}{u} = \frac{1}{f}$$

At focus $\quad m = \infty$ and $x = f$

At centre $\quad m = -1$ and $x = 2f$

Thus option (B) is most appropriate. **Ans. (B)**

Sol. 7 Using mirror formula, we have

$$f = -4\text{ cm}$$

$$u = -10\text{ cm}$$

Substituting values in formula

$$\frac{1}{v}+\frac{1}{u} = \frac{1}{f}$$

$$\Rightarrow \quad \frac{1}{v}+\left(\frac{1}{-10}\right) = \frac{1}{-4}$$

$$\Rightarrow \quad \frac{1}{v} = \frac{1}{10}-\frac{1}{4}$$

$$\Rightarrow \quad \frac{1}{v} = \frac{4-10}{40}$$

$$\Rightarrow \quad v = \frac{40}{-6}\text{ cm}$$

$$\Rightarrow \quad v = \frac{-20}{3}\text{ cm}$$

$$m = -\frac{v}{u} = \frac{-\left(-\frac{20}{3}\right)}{-10} = \frac{-2}{3}$$

$\Rightarrow$ Image will be real, inverted and diminished. **Ans. (D)**

Sol. 8 For the given situation, object and image distances are given as

$$u = 30\text{ cm}$$

$$v = 10\text{ cm}$$

$$\text{Object speed} = 9\text{ cm/s}$$

Considering object motion along axis, we use

$$\text{Image speed} = \left|\frac{v}{u}\right|^2 \times \text{Object speed}$$

$$v_{\text{image}} = \left(\frac{1}{3}\right)^2 \times 9 = 1\text{ cm/s}$$

Ans. (1)

Sol. 9 For convex mirror, focus is behind the mirror which is taken as positive side so we use

$$f = +\frac{r}{2}$$

Ans. (A)

Sol. 10 As orientation is same image must be virtual and as image is smaller than object, the mirror has to be convex.

Ans. (B)

Sol. 11 The unit vectors along incident ray and reflected ray can be written from the figure shown below as

$$\vec{a} = \sin\theta\hat{i} - \cos\theta\hat{j}$$

$$\vec{b} = \sin\theta\hat{i} + \cos\theta\hat{j}$$

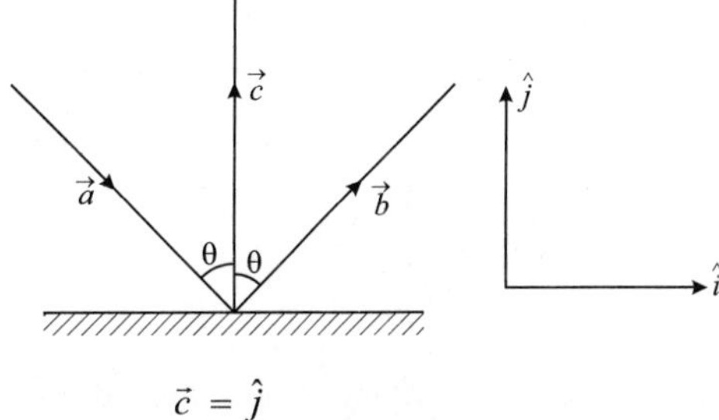

$$\vec{c} = \hat{j}$$

$$\vec{a} - 2(\vec{a}\cdot\vec{c})\vec{c} = \sin\theta\hat{i} + \cos\theta\hat{j}$$

From all the options given (B) is correct. **Ans. (B)**

Sol. 12 The field of view of mirror can be drawn by reflected rays as shown in figure below.

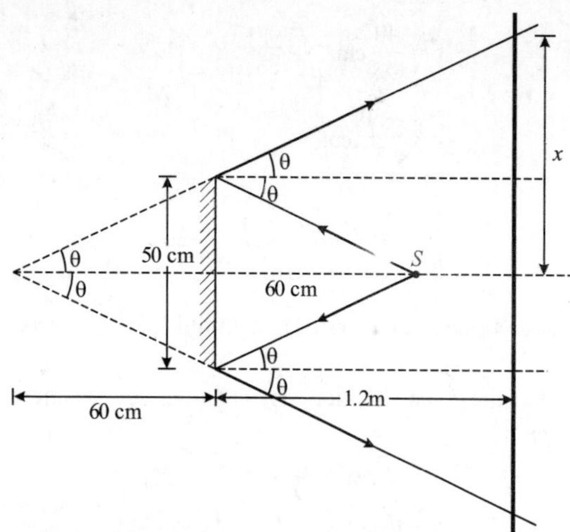

In the above figure, we use

$$\tan\theta = \frac{25}{60} = \frac{x}{180}$$

$\Rightarrow$ $x = 75$ cm

Thus distance between extreme points is given as

$$d = 2x = 2 \times 75 = 150 \text{ cm}$$ **Ans. (150)**

Sol. 13 Magnification of the mirror, is given a s

$$m = \frac{f}{f-u} = \frac{10}{10-(-190)}$$

$$\Rightarrow \quad m = \frac{10}{200} = \frac{1}{20}$$

Velocity of image in terms of object velocity is given as

$$V_I = -m^2V_0 = -\left(\frac{1}{20}\right)^2 \times 40$$

$$\Rightarrow \quad V_I = -0.1 \text{ m/s}$$ **Ans. (C)**

Sol. 14 Using Newton's formula, we have

$$xy = f^2$$

$$\Rightarrow \quad (f+d_1)(f-d_2) = f^2$$

$$\Rightarrow \quad f = \frac{d_1d_2}{d_1-d_2}$$

$$\Rightarrow \quad R = 2f = \frac{2d_1d_2}{d_1-d_2}$$ **Ans. (D)**

Sol. 15 Figure below shows the images formed by the two mirrors. As mirrors are at right angle total three images are produced but due to direct reflection by the mirrors images I_1 and I_2 are produced.

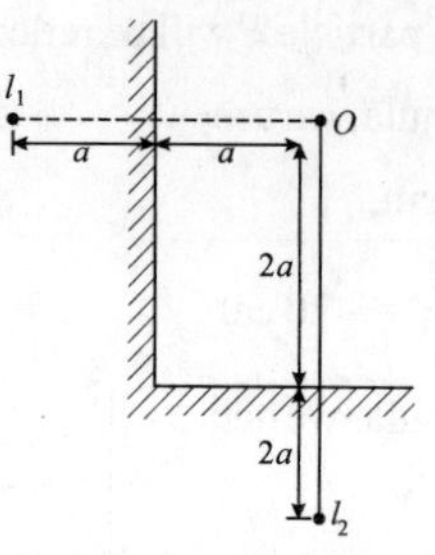

Distance between I_1 and $I_2 = \sqrt{(4a)^2+(2a)^2} \approx 4.6a$. **Ans. (A)**

23.2 Geometrical Optics-II Refraction of Light

Sol. 1 We are given that

$$n_{12} = \frac{n_1}{n_2} = \frac{1}{\sin\theta_c} \qquad (n_1 > n_2)$$

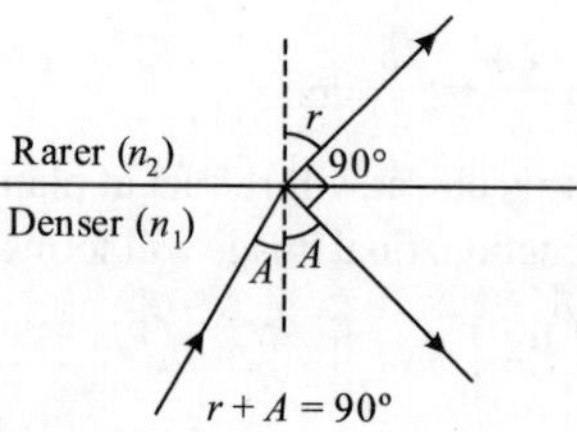

By Snells' law we have

$$n_1 \sin A = n_2 \sin r$$

$$\frac{\sin A}{\sin\theta_c} = \sin(90° - A) = \cos A$$

$$\Rightarrow \quad \tan A = \sin\theta_c$$

$$\Rightarrow \quad A = \tan^{-1}(\sin\theta_c)$$ **Ans. (D)**

Sol. 2

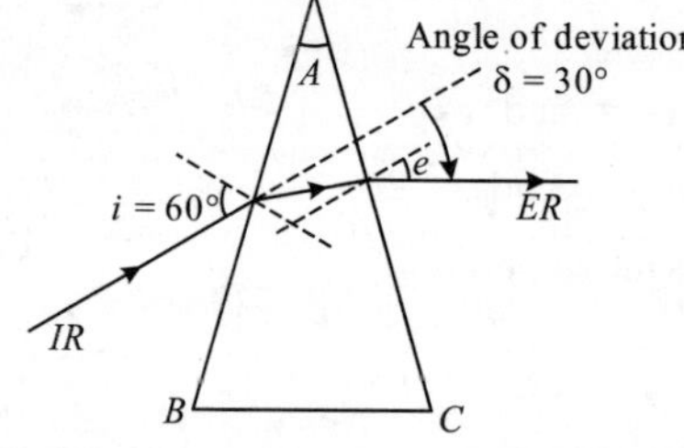

For a trihedral prism we know

$$\delta = i + e - A$$

$$\Rightarrow \quad e = \delta + A - i = 30° + 30° - 60° = 0°$$

$\Rightarrow$ Emergent ray comes out of prism perpendicular to face AC.

Ans. (C)

Sol. 3 Using refraction formula at spherical surface, we have

$$\frac{\mu_2}{v} - \frac{\mu_1}{u} = \frac{\mu_2 - \mu_1}{R}$$

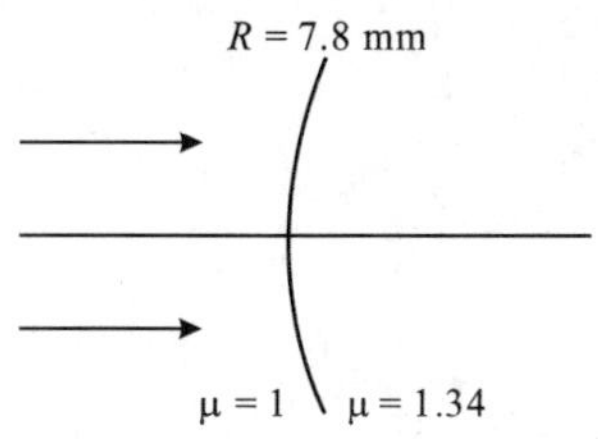

$$\Rightarrow \quad \frac{1.34}{v} - \frac{1}{\infty} = \frac{1.34-1}{7.8}$$

$$\Rightarrow \quad v = 30.7\,\text{mm}$$

Ans. (C)

Sol. 4 For minimum deviations by a prism, we use

$$i = e$$

and $$r_1 = r_2 = \frac{A}{2} = 30°$$

by Snell's law, we use

$$\mu_{air} \sin i = \mu_p \sin r$$

$$\Rightarrow \quad 1 \times \sin i = \sqrt{3} \times \frac{1}{2} = \frac{\sqrt{3}}{2}$$

$$\Rightarrow \quad i = 60°$$

Ans. (D)

Sol. 5 By Snell's law, we use

$$1.\sin 40° = 1.31 \sin\theta_2$$

$$\Rightarrow \quad \sin\theta_2 = \frac{0.64}{1.31} = \frac{64}{131}$$

$$\Rightarrow \quad \tan\theta_2 = \frac{64}{\sqrt{(131)^2-(64)^2}} = \frac{64}{\sqrt{13065}} \simeq 0.49 = \frac{d}{x}$$

$$\Rightarrow \quad x = \frac{d}{0.49}$$

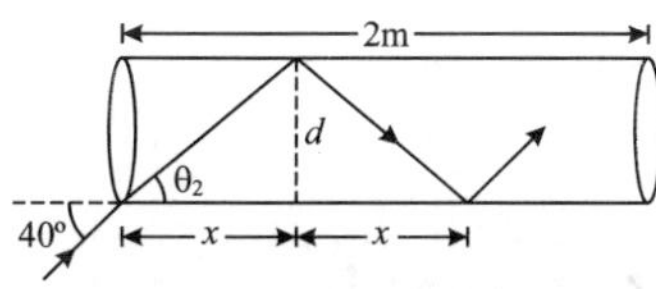

Total number of reflections in fiber are

$$N = \frac{2 \times 0.49}{20 \times 10^{-6}} = 55991$$

Closes to 55000

Ans. (A)

Sol. 6 From Snell's law

$$1.\sin 60° = \mu \sin 30°$$

$$\Rightarrow \quad \mu = \sqrt{3}$$

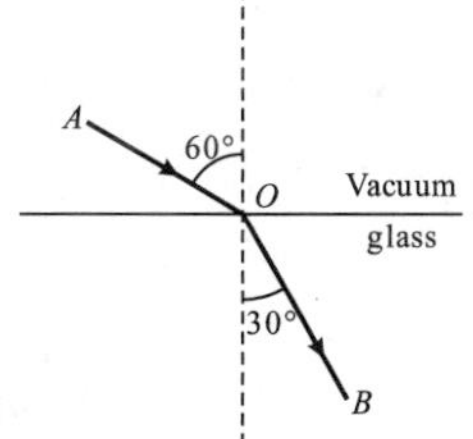

Optical path length of light ray from A to B is

$$OA = \mu(OB)$$

$$= \frac{a}{\cos 60°} + \sqrt{3}\frac{b}{\cos 30°}$$

$$= 2a + 2b$$

Ans. (A)

Sol. 7

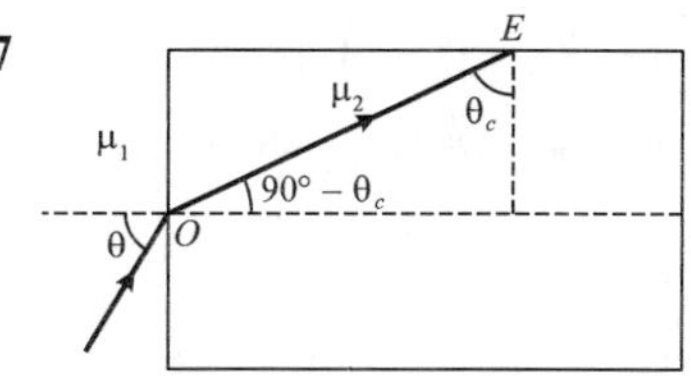

For total internal reflection at point E, we use

$$\sin\theta_c = \frac{\mu_1}{\mu_2}$$

For refraction at point O, we use

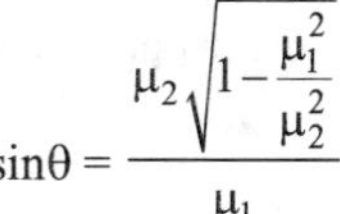

$$\mu_1 \sin\theta = \mu_2 \sin(90° - \theta_c)$$

$$\Rightarrow \quad \sin\theta = \frac{\mu_2\sqrt{1-\frac{\mu_1^2}{\mu_2^2}}}{\mu_1}$$

$$\Rightarrow \quad \theta = \sin^{-1}\sqrt{\frac{\mu_2^2-\mu_1^2}{\mu_1^2}}$$

For total internal reflection to take place at point E, we use

$$\theta < \sin^{-1}\sqrt{\frac{\mu_2^2}{\mu_1^2}-1}$$

Ans. (B)

Sol. 8 Critical angle for a medium is given as

$$\sin\theta_C = \frac{1}{\mu} = \frac{1}{\sqrt{3 \times 4/3}}$$

[where Refractive index $\mu = \sqrt{\mu_0 \varepsilon_0}$]

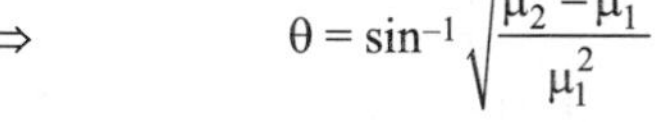

$$\theta_C = 30°$$

Ans. (D)

Sol. 9 Figure shows the situation described in question

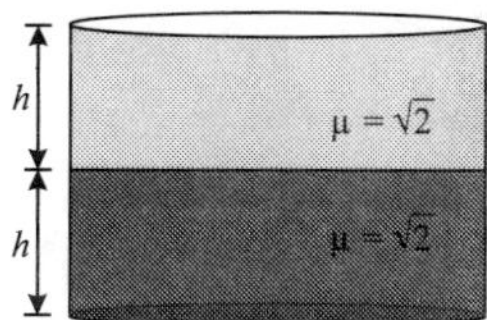

For normal viewing, apparent depth of vessel is given as

$$h_{app} = \frac{h}{\mu_1} + \frac{h}{\mu_2}$$

$$\Rightarrow \quad h_{app} = \frac{h}{\sqrt{2}} + \frac{h}{2\sqrt{2}}$$

$$\Rightarrow \quad h_{app} = \frac{3}{2\sqrt{2}}h = \frac{3}{4}\sqrt{2}\,h$$

Ans. (B)

Sol. 10

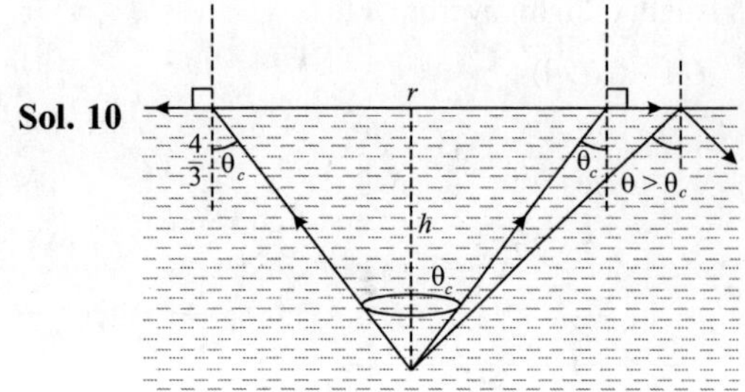

Critical angle for air-water interface is given as

$$\frac{4}{3}\sin\theta_c = 1 \sin 90^\circ$$

$$\sin\theta_c = \frac{3}{4}$$

Solid angle of cone from which light escapes out from water surface is

$$\Omega = 2\pi(1-\cos\theta)$$

$$\Rightarrow \quad \Omega = 2\pi\left(1-\frac{\sqrt{7}}{4}\right)$$

Fraction of light energy which escapes out from water is

$$P_{esc} = \frac{P}{4\pi}\times\Omega$$

$$\Rightarrow \quad f = \frac{P_{esc}}{P} = \frac{1}{2}\left(1-\frac{\sqrt{7}}{4}\right)$$

$$\Rightarrow \quad f = \frac{0.33}{2} = 0.17$$

$\Rightarrow$ % fraction = 17% **Ans. (A)**

Sol. 11

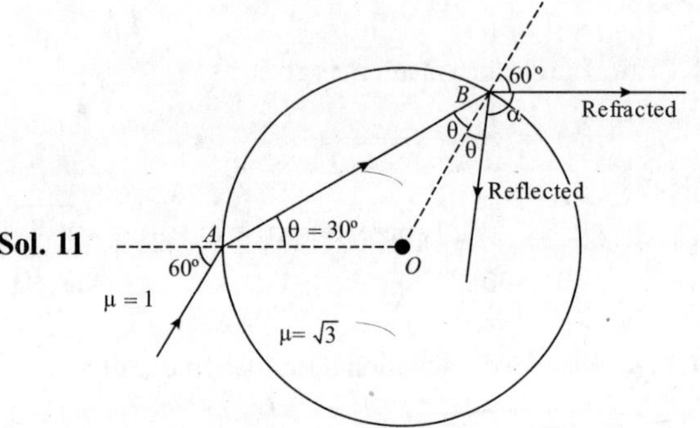

Using Snell's law at point A, we have

$$1\times\sin 60^\circ = \sqrt{3}\times\sin\theta$$

$$\Rightarrow \quad \frac{\sqrt{3}}{2} = \sqrt{3}\times\sin\theta$$

$$\Rightarrow \quad \sin\theta = \frac{1}{2}$$

$$\Rightarrow \quad \theta = 30^\circ$$

At point B, we have

$$\theta + 60^\circ + \alpha = 180^\circ$$

$$\Rightarrow \quad 30^\circ + 60^\circ + \alpha = 180^\circ$$

$$\Rightarrow \quad \alpha = 90^\circ$$

Ans. (90)

Sol. 12 From figure, we have

$$\tan r = \frac{15}{30} = \frac{1}{2}$$

$$\Rightarrow \quad \sin r = \frac{1}{\sqrt{5}}$$

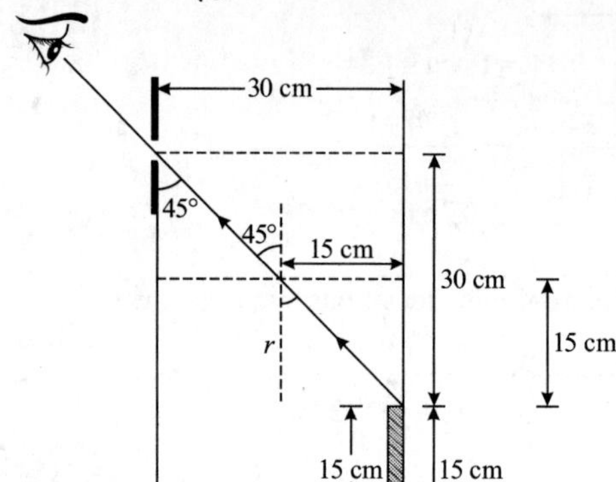

By smell's law, we have

$$\mu \sin r = \sin 45^\circ$$

$$\Rightarrow \quad \mu = \frac{\frac{1}{\sqrt{2}}}{\frac{1}{\sqrt{5}}} = \sqrt{\frac{5}{2}} = 1.581$$

Given value of

$$\mu = \frac{N}{100} = 1.581$$

$$\Rightarrow \quad N = 158.1$$

Ans. (158)

Sol. 13 Angle of minimum deviation for small angled prism is given as

$$\delta_{min} = (\mu-1)A$$
$$= (1.5-1)1$$
$$= 0.5^\circ$$

$$\delta_{min} = \frac{N}{10} = 0.5^\circ$$

$$\Rightarrow \quad N = 5$$

Ans. (5)

Sol. 14 By lens makers formula, we use

$$P = (\mu-1)\left[\frac{2}{R}\right] \quad \ldots(1)$$

For another plano convex lens with power $1.5P$, we use

$$1.5P = (\mu-1)\left[\frac{1}{R'}\right] \quad \ldots(2)$$

From (1) and (2), we have

$$1.5 = \frac{1/R'}{2/R}$$

$$\Rightarrow \quad R' = \frac{R}{2\times 1.5} = \frac{R}{3}$$

Ans. (D)

Sol. 15 By lens makers formula, we have

$$\frac{1}{f}=\left[\frac{\mu_L}{\mu_S}-1\right]\left[\frac{1}{R_1}-\frac{1}{R_2}\right]$$

If $\mu_L=\mu_S$, it gives $f=\infty$. **Ans. (B)**

Sol. 16 Using the construction diagram below for the lens, the radius of lens can be related with its thickness and radius of curvature as

$$R^2=r^2+(R-t)^2$$

$$\Rightarrow \quad R^2=r^2+R^2+t^2-2Rt \quad \ldots(1)$$

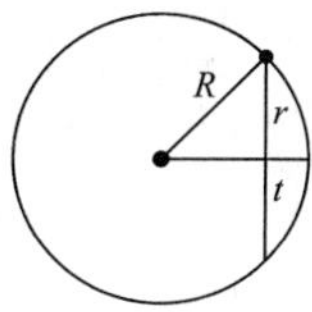

For small thickness neglecting t^2 in equation-(1), we get

$$R=\frac{r^2}{2t}$$

Using lens maker's formula, we have

$$\frac{1}{f}=(\mu-1)\left(\frac{1}{R}-\frac{1}{\infty}\right)=\frac{\mu-1}{R}$$

$$\Rightarrow \quad f=\frac{R}{\mu-1}=\frac{r^2}{2t(\mu-1)}$$

$$\Rightarrow \quad f=\frac{(3\times10^{-2})^2}{2\times3\times10^{-3}\times\left(\frac{3}{2}-1\right)}$$

$$\Rightarrow \quad f=\frac{9\times10^{-4}}{6\times10^{-3}\times1}\times2=0.3\text{ m}=30\text{ cm}$$ **Ans. (D)**

Sol. 17 As wavelength of light in medium is given as

$$\lambda_m=\frac{\lambda_a}{\mu}$$

Thus refractive index of the medium is given as

$$\mu=\frac{3}{2}$$

Using refraction formula for the surface, we have

$$\frac{\mu}{v}-\frac{1}{u}=\frac{\mu-1}{R}$$

Given that image is located at 10 cm behind the surface and it is 2/3 of the object distance so object distance can be taken as 15 cm in from of the lens. Substituting $v=+10$ cm and $u=-15$ cm in above formula, we have

$$\Rightarrow \quad \frac{3}{2\times10}+\frac{1}{15}=\frac{\frac{3}{2}-1}{R}$$

$$\Rightarrow \quad R=\frac{30}{13}\text{ cm}$$

$$\Rightarrow \quad x=30$$ **Ans. (30)**

Sol. 18 In the situation described in question we consider that the given right angled prism is an isoceles one, so the angles at which light will incident on face PR is 45° as shown below.

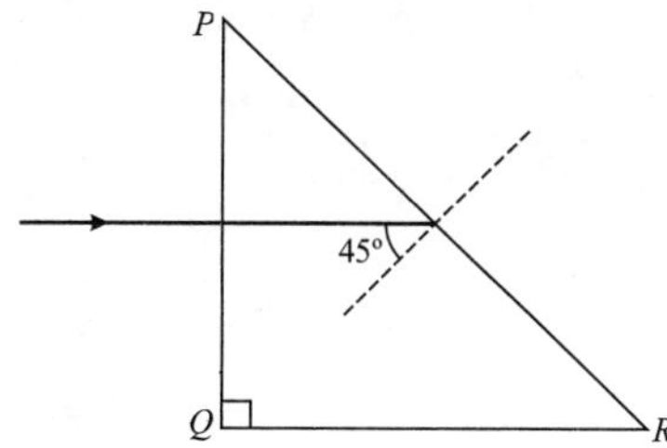

Critical angle θ_C for the three light rays is given as

$$\theta_C=\sin^{-1}\left(\frac{1}{\mu}\right)$$

If $\theta_C\geq45°$ for any ray then this light ray will pass through the prism.

$$(\theta_C)_{\text{Red}}=\sin^{-1}\left(\frac{1}{1.27}\right)=51.94°$$

Thus red light will pass

$$(\theta_C)_{\text{Green}}=\sin^{-1}\left(\frac{1}{1.42}\right)=44.76°$$

Thus green light will not pass

$$(\theta_C)_{\text{Blue}}=\sin^{-1}\left(\frac{1}{1.49}\right)=42.15°$$

Thus blue light will not pass. So only red light will pass through the prism face PR. **Ans. (B)**

Sol. 19 Using refraction formula for image formation, we have

$$\frac{n_2}{v}-\frac{n_1}{u}=\frac{n_2-n_1}{R}$$

$$\Rightarrow \quad \frac{1.4}{v}-\frac{1.25}{-40}=\frac{1.4-1.25}{(-25)}$$

$$\Rightarrow \quad v=-37.58\text{ cm}$$ **Ans. (D)**

Sol. 20 In figure, we can see that

$$i=r\ \&\ r'=90-r$$

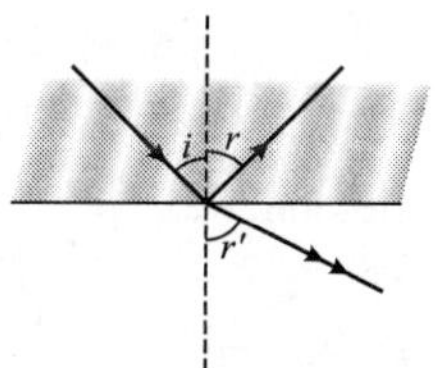

Critical angle is given as

$$\theta_C = \sin^{-1}\left(\frac{1}{\mu}\right) = \sin^{-1}\left(\frac{\sin i}{\sin r'}\right)$$

$$\Rightarrow \quad \theta_C = \sin^{-1}\left(\frac{\sin i}{\sin r}\right) = \sin^{-1}(\tan r)$$ **Ans. (A)**

Sol. 21 The situation described in question is shown in figure below. In this state the Prism angle is given as

$$A = \frac{i}{2} + \frac{i}{2} = i$$

From Snell's law we use

$$1 \sin i = \mu \sin \frac{i}{2}$$

$$\Rightarrow \quad 2\sin\frac{i}{2}\cos\frac{i}{2} = \sqrt{3}\sin\frac{i}{2}$$

$$\Rightarrow \quad \cos\frac{i}{2} = \frac{\sqrt{3}}{2}$$

$$\Rightarrow \quad \frac{i}{2} = 30^\circ$$

$$\Rightarrow \quad i = 60^\circ$$

$$\Rightarrow \quad A = i = 60^\circ$$ **Ans. (60)**

A
i
i/2
i/2
i
μ = √3

Sol. 22 Critical angle for diamond-air interface is given as

$$\theta_C = \left(\frac{1}{2.42}\right) = 24.4^\circ$$

Given Incident angle 30° is more than critical angle so laser will suffer total internal reflection and will not get refracted.

Ans. (C)

Sol. 23 By Snell's law, we use

$$1 \times \sin\theta = \frac{4}{3}\sin(90 - C)$$

$$\Rightarrow \quad \sin\theta = \frac{4}{3}\cos C$$

For critical angle, we use

$$\sin C = \frac{1}{\mu} = \frac{5}{4}$$

$$\Rightarrow \quad \cos C = \frac{\sqrt{7}}{4}$$

$$\Rightarrow \quad \sin\theta = \frac{4}{3} \times \frac{\sqrt{7}}{4} = \frac{\sqrt{7}}{3}$$

For Total Internal Reflection, we have

$$\sin\theta \le \frac{\sqrt{7}}{3} \text{ as } \theta = \sin^{-1}\frac{\sqrt{7}}{3}$$ **Ans. (C)**

Sol. 24 Angle of deviation for the light ray passing through the prism first decreases with increase in angle of incidence from 90° as for grazing incidence and grazing emergence the deviation is maximum and when light ray passes through the prism symmetrically, it is minimum thus option (C) only matches with this analysis. **Ans. (C)**

Sol. 25 As it is specified that in initial siutation, image coincides with the object this indecases that Image oby lens must be on centre of curvature of mirror. Hence, distance between object and image after removing mirror = 12 + 8 + 30 = 50 cm.

Ans. (50)

Sol. 26 For first refraction, using lens formula, we have

$$\frac{1}{v} - \frac{1}{u} = \frac{1}{f}$$

$$\Rightarrow \quad \frac{1}{v} - \frac{1}{-30} = \frac{1}{10}$$

$$\Rightarrow \quad \frac{1}{v} = \frac{1}{10} - \frac{1}{30}$$

$$\Rightarrow \quad \frac{1}{v} = \frac{3-1}{30};\ v = 15 \text{ cm}$$

For second refraction we use $u = +10$ cm and again by using lens formula, we have

$$\frac{1}{v} - \frac{1}{10} = \frac{1}{-10}$$

$$\Rightarrow \quad \frac{1}{v} = \frac{-1}{10} + \frac{1}{10} \qquad v = \infty$$

For third refraction as light rays incident parallel to principal axis, final image is formed at focal point of the third lens so we have $v = +30$ cm from third lens. **Ans. (B)**

Sol. 27 By lens maker's formula for plano convex and plano concave lens, we have

$$\frac{1}{f_1} = (\mu_1 - 1)\left[\frac{1}{R}\right]$$

and $$\frac{1}{f_1} = (\mu_2 - 1)\left[\frac{1}{-R}\right]$$

$$\Rightarrow \quad \frac{1}{f_{eq}} = \frac{1}{f_1} + \frac{1}{f_2} = \frac{(\mu_1 - 1)}{R} - \frac{(\mu_2 - 1)}{R}$$

$$\Rightarrow \quad \frac{1}{f_{eq}} = \frac{(\mu_1 - \mu_2)}{R}$$

$$\Rightarrow \quad \frac{R}{f_{eq}} = \mu_1 - \mu_2$$ **Ans. (D)**

Sol. 28 de-Broglie wavelength of electron is given as

$$\lambda_D = \frac{h}{\sqrt{2m\left(\frac{3}{2}kT\right)}}$$

$$\Rightarrow \quad \lambda_D = \frac{h}{\sqrt{3mkT}}$$

$$\Rightarrow \quad \lambda_D = \frac{6.6\times10^{-34}}{\sqrt{3\times9\times10^{-31}\times1.38\times10^{-23}\times300}}$$

$$\Rightarrow \quad \lambda_D = \frac{6.6\times10^{-34}}{105.72\times10^{-27}} = 62\ \text{Å} = 6.2\ \text{nm}$$

Ans. (A)

Sol. 29 Height of water as observed by observer is given as

$$h' = \frac{H}{\mu_w} = \frac{H}{(4/3)} = \frac{3H}{4}$$

Height of air as observed by observer = 17.5 – H
Acoording to the given condition, we use

$$\frac{3H}{4} = 17.5 - H$$

$$\Rightarrow \quad H = 10\ \text{cm}$$

Ans. (B)

23.3 Geometrical Optics-III Thin Lenses

Sol. 1

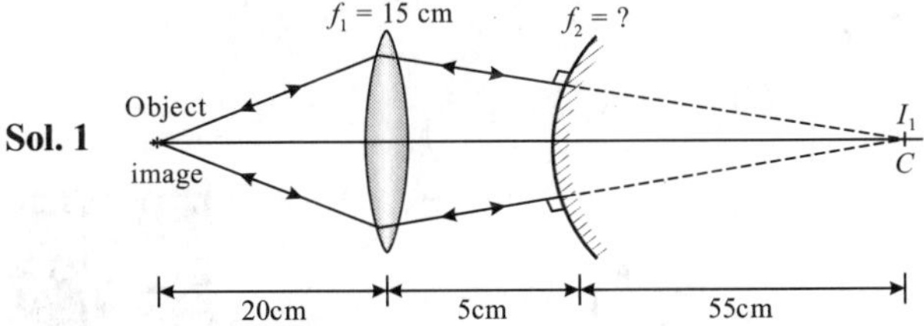

For refraction through lens we use lens formula for which we use

$$u = -20\ \text{cm}$$

$$f_1 = +15\ \text{cm}$$

$$\Rightarrow \quad \frac{1}{f} = \frac{1}{v} - \frac{1}{u}$$

$$\Rightarrow \quad \frac{1}{v} = \frac{1}{15} - \frac{1}{20} = \frac{1}{60}$$

$$\Rightarrow \quad v = 60\ \text{cm}$$

If image is produced on object then it can happen light rays retrace the path of incident ray after reflection or IR at mirror falls normally. For this I_1 must be at centre of curvature of mirror

$$\Rightarrow \quad R = 55\ \text{cm}$$

$$\Rightarrow \quad f_2 = \frac{R}{2} = 27.5\ \text{cm}$$

Ans. (A)

Sol. 2 If μ be the refractive index of lens and R be its radius of curvature then focal length of a thin plano convex lens is given by lens makers formula as

$$f = \frac{R}{(\mu-1)} \qquad \left[\text{As } \frac{1}{f} = (\mu-1)\left[\frac{1}{\underset{R}{R_1}} - \frac{1}{\underset{\infty}{R_2}}\right]\right]$$

Case-I : (Figure-A)

$$f_{\text{mirror}} = \infty$$

$$f_L = \frac{R}{\mu-1}$$

Effective focal length of lens system is given as

$$\frac{1}{f_{eq}} = \frac{2}{f_L} + \frac{1}{f_M}$$

$$\Rightarrow \quad f_{eq} = \frac{f_L}{2} = \frac{R}{2(\mu-1)}$$

$$\Rightarrow \quad \frac{R}{2(\mu-1)} = 28\ \text{cm} \qquad \ldots(1)$$

Case-II : (Figure-B)

$$f_M = \frac{R}{2}$$

$$f_L = \frac{R}{2(\mu-1)}$$

Equivalent focal length of lens system is given as

$$\frac{1}{f'_{eq}} = \frac{2}{f_C} + \frac{1}{f_M}$$

$$\Rightarrow \quad \frac{1}{f'_{eq}} = \frac{2(\mu-1)}{R} + \frac{2}{R} = \frac{24}{R}$$

$$\Rightarrow \quad f'_{eq} = \frac{R}{2\mu} = 10\ \text{cm} \qquad \ldots(2)$$

We use $\frac{(1)}{(2)}$

$$\Rightarrow \quad \frac{\mu}{\mu-1} = \frac{28}{10} = 2.8$$

$$\Rightarrow \quad \mu = 2.8\mu - 2.8$$

$$\Rightarrow \quad \mu = \frac{2.8}{1.8} = \frac{14}{9} = 1.55$$

Ans. (B)

Sol. 3

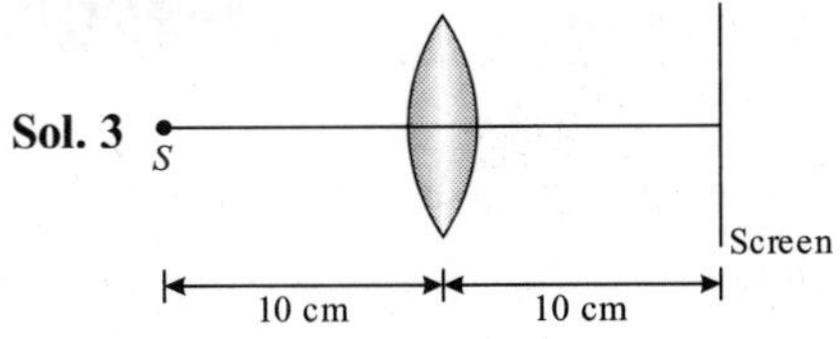

By lens formula, we have

$$\frac{1}{v}-\frac{1}{u}=\frac{1}{f}$$

$$\Rightarrow \quad \frac{1}{10}-\frac{1}{-10}=\frac{1}{f}$$

$$\Rightarrow \quad f=5\text{ cm}$$

Shift due to slab $= t\left(1-\frac{1}{\mu}\right)$ in the direction of incident ray

$$\Rightarrow \quad S=1.5\left(1-\frac{2}{3}\right)=0.5$$

Again using lens formula, we have

$$\frac{1}{v}-\frac{1}{-9.5}=\frac{1}{5}$$

$$\Rightarrow \quad \frac{1}{v}=\frac{1}{5}-\frac{2}{19}=\frac{9}{95}$$

$$\Rightarrow \quad v=\frac{95}{9}=10.55\text{ cm}$$

Ans. (A)

Sol. 4 By lens maker's formula, we have

$$\frac{1}{2f_2}=\frac{1}{f_1}=(\mu_1-1)\left(\frac{1}{\infty}-\frac{1}{-R}\right)$$

and $$\frac{1}{f_2}=(\mu_2-1)\left(\frac{1}{-R}-\frac{1}{\infty}\right)$$

$$\Rightarrow \quad \frac{(\mu_1-1)}{R}=\frac{(\mu_2-1)}{2R}$$

$$2\mu_2-\mu_1=1$$

Ans. (B)

Sol. 5 From lens formula, we have

$$\frac{1}{v}-\frac{1}{u}=\frac{1}{f}$$

$$\frac{1}{v}-\frac{1}{(-20)}=\frac{1}{(.3)}=\frac{10}{3}$$

$$\frac{1}{v}=\frac{10}{3}-\frac{1}{20}$$

$$\frac{1}{v}=\frac{197}{60};\ v=\frac{60}{197}\text{ m}$$

Lateral magnification by lens is

$$m=\left(\frac{v}{u}\right)=\frac{\left(\frac{60}{197}\right)}{20}=\frac{3}{197}$$

Velocity of image w.r.t. to lens is given by

$$v_I=m^2v_O$$

Direction of velocity of image is same as that of object

As $\quad v_O=5\text{ m/s}$

$$\Rightarrow \quad v_I=\left(\frac{60\times 1}{197\times 20}\right)^2(5)$$

$$\Rightarrow \quad v_I=1.16\times 10^{-3}\text{ m/s toward lens}$$

Ans. (C)

Sol. 6 For first lens, using lens formula, we have

$$\frac{1}{v}-\frac{1}{-20}=\frac{1}{5}$$

$$v=\frac{20}{3}\text{ cm}$$

For second lens, using lens formula, we have

$$u_2=\frac{20}{3}-2=\frac{14}{3} \quad \text{(object distance)}$$

$$\Rightarrow \quad \frac{1}{v_2}-\frac{1}{\frac{14}{3}}=\frac{1}{-5}$$

$$\Rightarrow \quad v_2=70\text{ cm}$$

Ans. (D)

Sol. 7 Using lens marker's formula, new focal length of lens will be given as

$$\frac{1}{f}=(\mu_{rel}-1)\left(\frac{1}{R_1}-\frac{1}{R_2}\right)$$

As focal length of lens will change hence image disappears from the screen. **Ans. (A)**

Sol. 8

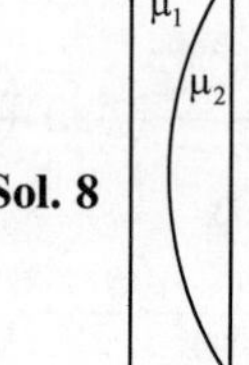

Focal length of this combination is given as

$$\frac{1}{f}=\frac{1}{f_1}+\frac{1}{f_2}=\frac{1-\mu_1}{R}+\frac{\mu_2-1}{R}=\frac{\mu_2-\mu_1}{R}$$

Ans. (C)

Sol. 9

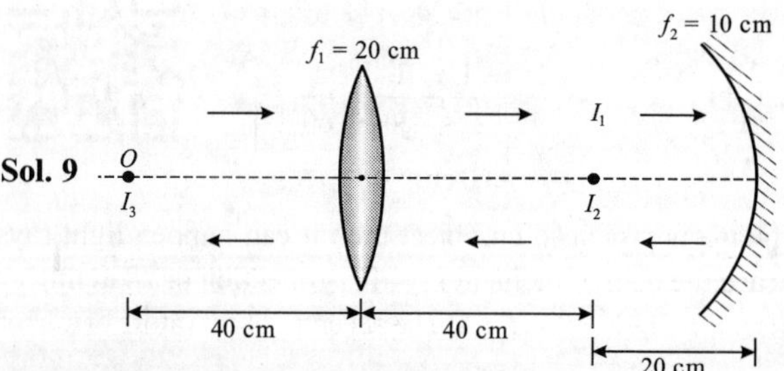

After all refractions and reflection final image is produced at object i.e. 40 cm from lens and of same size.

If I_2 is considered final image after one refraction and one reflection. Then **Ans. (None)** Safe and most appropriate.

Sol. 10

By lens formula we use

$$u = -30\text{ cm}$$

$$f = +0\,20\text{ cm}$$

$$\Rightarrow \quad \frac{1}{v} + \frac{1}{30} = \frac{1}{20}$$

$$\Rightarrow \quad V = +60\text{ cm}$$

I_1 will be producing image at same position in cancave mirror if it is centre of curvature

$$\Rightarrow \quad R_m = 20\text{ cm}$$

$$f_m = \frac{R_m}{2} = 10\text{ cm}$$

A cancave mirror produces a virtual image for object to be located from P to F.

$$u_{max} = f_m = 10\text{ cm}$$ **Ans. (B)**

Sol. 11 Initially we use

$$f_L = \frac{R}{2(\mu - 1)} R$$

We also use $\frac{1}{f_{eq}} = \frac{1}{f_L} + \frac{1}{\infty} + \frac{1}{f_L} = \frac{2}{f_L}$

$$\Rightarrow \quad f_L = 2f_{eq} = 18\text{ cm} = R$$

After insertion of liquid, the focal length of plano concave lens is $f_{L_2} = \frac{R}{\mu_1 - 1} = \frac{18}{\mu_1 - 1}$ (–ve) as diverging lens

equal focal length for the combination now is

$$\frac{1}{f'_{eq}} = \frac{2}{f_L} + \frac{2}{f_{L_2}}$$ [here we use $2f'_{eq} = 27$ cm]

$$\frac{1}{27/2} = \frac{2}{18} - \frac{2(\mu_1 - 1)}{18}$$

$$\Rightarrow \quad -\frac{\mu_1 - 1}{18} = \frac{1}{27} - \frac{1}{18} = \frac{2-3}{54} = -\frac{1}{54}$$

$$\Rightarrow \quad \mu_1 - 1 = \frac{1}{3}$$

$$\Rightarrow \quad \mu_1 = 1 + \frac{1}{3} = \frac{4}{3}$$ **Ans. (B)**

Sol. 12

In first case we use

$$m = \frac{f}{x_1 - f} = 2$$

$$\Rightarrow \quad f = 2x_1 - 2f$$

$$\Rightarrow \quad x_1 = \frac{3f}{2} \quad \ldots(1)$$

In second case we use

$$m = \frac{f}{x_2 - f} = -2$$

$$\Rightarrow \quad f = -2x_2 + 2f$$

$$\Rightarrow \quad x_2 = \frac{f}{2} \quad \ldots(2)$$

$$(1)/(2) \Rightarrow \quad \frac{x_1}{x_2} = \frac{3}{1}$$ **Ans. (D)**

Sol. 13 By lens maker's formulas, focal lengths of the two lenses are

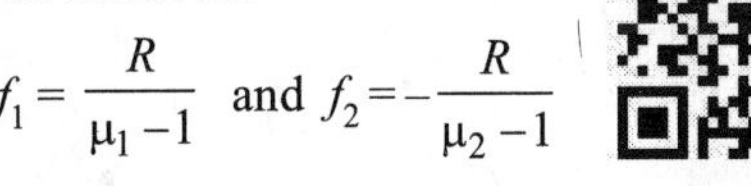

$$f_1 = \frac{R}{\mu_1 - 1} \text{ and } f_2 = -\frac{R}{\mu_2 - 1}$$

For combination, we use

$$\frac{1}{f_{eq}} = \frac{1}{f_1} + \frac{1}{f_2} = \frac{\mu_1 - 1}{R} - \frac{\mu_2 - 1}{R} = \frac{\mu_1 - \mu_2}{R}$$

$$\Rightarrow \quad f_{eq} = \frac{R}{\mu_1 - \mu_2}$$ **Ans. (D)**

Sol. 14 By lens formula, we have

$$\frac{1}{v} - \frac{1}{u} = \frac{1}{f}$$

$$\Rightarrow \quad 1 - \frac{v}{u} = \frac{v}{f}$$

$$\Rightarrow \quad 1 - m = \frac{v}{f}$$

$$\Rightarrow \quad m = 1 - \frac{v}{f}$$

At $\quad v = a,$

$$m_1 = 1 - \frac{a}{f}$$

At $\quad v = a + b,$

$$m_2 = 1 - \frac{a+b}{f}$$

$$m_2 - m_1 = c = \left(1 - \frac{a+b}{f}\right) - \left(1 - \frac{a}{f}\right)$$

In magnitude $\quad c = \frac{b}{f}$

$$f = \frac{b}{c}$$

Ans. (D)

Sol. 15 Using Lens Maker's Formula, we have

$$\frac{1}{f} = \left(\frac{\mu_2}{\mu_1} - 1\right)\left(\frac{1}{R_1} - \frac{1}{R_2}\right)$$

In air, we use

$$\frac{1}{f} = \left(\frac{1.5}{1} - 1\right)\left(\frac{1}{R_1} - \frac{1}{R_2}\right) \quad \ldots(1)$$

In liquid, we use

$$\frac{1}{f_1} = \left(\frac{1.5}{1.42} - 1\right)\left(\frac{1}{R_1} - \frac{1}{R_2}\right) \quad \ldots(2)$$

$$\Rightarrow \quad \frac{f_1}{f} = \frac{0.5}{0.056} = 8.93 \approx 9$$

Ans. (C)

Sol. 16 By use of Lens-Maker's formula, we have

$$\frac{1}{f} = (\mu - 1)\left(\frac{1}{R_1} - \frac{1}{R_2}\right)$$

$$\Rightarrow \quad \frac{1}{f} = (1.5 - 1)\left(\frac{1}{30} - 0\right)$$

$$\Rightarrow \quad f = 60 \text{ cm}$$

Ans. (60)

Sol. 17 Using displacement method on optical bench we have

$$f = \frac{D^2 - d^2}{4D}$$

$$\Rightarrow \quad f = \frac{100^2 \quad 40^2}{4(100)} = 21 \text{ cm}$$

Power of lens is given as

$$P = \frac{1}{f} = \frac{100}{21} D = \left(\frac{N}{100}\right)_D$$

$$\Rightarrow \quad N = \frac{10000}{21} \simeq 476$$

Ans. (476)

Sol. 18 By lens formula, we have

$$\frac{1}{v} - \frac{1}{u} = \frac{1}{f}$$

For $\quad v = u$ (both positive)

This happens only when object is at pole

For + *ve* u, *ve* is given as

$$\frac{1}{v} = \frac{1}{f} + \frac{1}{u}$$

Thus $|v| < |u|$ so the $u - v$ curve must lie below $u = v$ reference line and at $u \to \infty$, $v = f$ thus option (D) is correct. **Ans. (D)**

Sol. 19 For lens refraction, we use its image produced is I_1 at position v from lens, given as

$$u = -1 \text{ m}$$
$$f = +0.5 \text{ m}$$

$$\Rightarrow \quad v = \frac{uf}{u+f} = \frac{-1 \times 0.5}{-0.5} = 1 \text{ m}$$

Image I_1 produced by convex lens above will act like object for plane mirror so in mirror its image I_2 is produced at a distance 1 m behind the mirror. (Refer ray diagram given below)

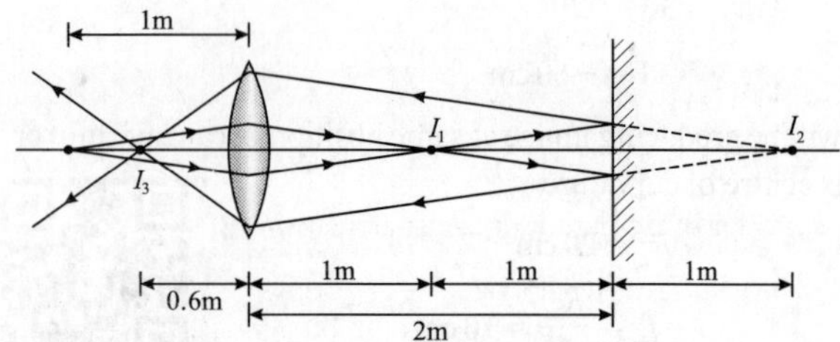

For lens this image I_2 will act as object so again for lens refraction we use

$$u = +3 \text{ m}$$
$$f = -0.5 \text{ m}$$

$$\Rightarrow \quad v = \frac{uf}{u+f} = \frac{3 \times -0.5}{2.5} = -\frac{3}{5} = -0.6 \text{ m}$$

This final image is produced to the left of lens at 0.6 m and real. So correct answer is option-(C) **Ans. (C)**

Sol. 20 By lens formula, we have

$$\frac{1}{f} = \frac{1}{v} - \frac{1}{u}$$

Using $u = -f$ and $f = -f$, we have

$$\frac{1}{-f} = \frac{1}{v} - \frac{1}{-f}$$

$$\Rightarrow \quad \frac{-1}{f} = \frac{1}{f} - \frac{1}{v}$$

$$\Rightarrow \quad \frac{-2}{f} = \frac{1}{v}$$

$$\Rightarrow \quad v = -\frac{f}{2}$$

Magnification is given as

$$m = v/u = \frac{-f}{2(-f)} = \frac{1}{2} \qquad \textbf{Ans. (A)}$$

23.4 Dispersion of Light

Sol. 1 For doublet lenses at a separation d and with focal lengths f_1 and f_2 the equivalent focal length is given as

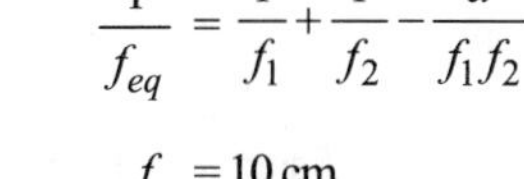

$$\frac{1}{f_{eq}} = \frac{1}{f_1} + \frac{1}{f_2} - \frac{d}{f_1 f_2}$$

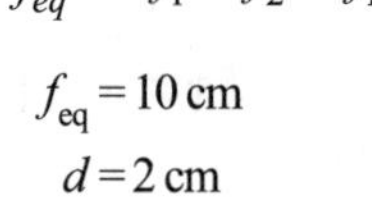

Here $f_{eq} = 10$ cm

and $d = 2$ cm

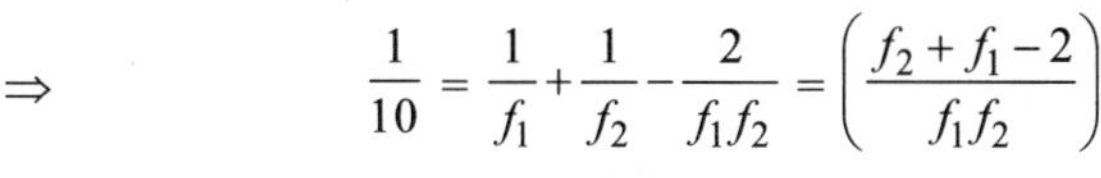

$$\Rightarrow \quad \frac{1}{10} = \frac{1}{f_1} + \frac{1}{f_2} - \frac{2}{f_1 f_2} = \left(\frac{f_2 + f_1 - 2}{f_1 f_2}\right)$$

For option (A) $f_1 = 18$ cm & $f_2 = 20$ cm

$$\Rightarrow \quad \frac{1}{10} = \frac{20+18-2}{18\times 20} = \frac{36}{18\times 20} = \frac{1}{10}$$

$$\Rightarrow \quad \text{LHS} = \text{RHS} \qquad \textbf{Ans. (A)}$$

Sol. 2 As for thin prism, minimum deviation is

$$D_m = (\mu - 1)\text{A}$$

So on increasing the wavelength, μ decreases & hence D_m decreases. **Ans. (B)**

Sol. 3 Deviation is minimum in a prism when :

$i = e$, $r_1 = r_2$ and refracted ray inside prism is parallel to base of prism.

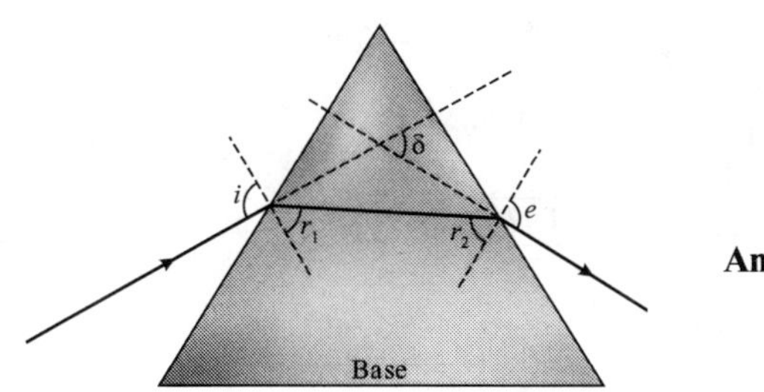

Ans. (A)

Sol. 4 For Achromatic combination of prisms, we use

$$\theta_{net} = 0$$

$$\Rightarrow \quad \theta_1 - \theta_2 = 0$$

$$\Rightarrow \quad \theta_1 = \theta_2$$

$$\Rightarrow \quad \omega_1\delta_1 = \omega_2\delta_2$$

Total deviation is given as

$$\delta_{net} = \delta_1 - \delta_2 = 2^o$$

$$\Rightarrow \quad \delta_1 - \frac{\omega_1\delta_1}{\omega_2} = 2^o$$

$$\Rightarrow \quad \delta_1\left(1 - \frac{\omega_1}{\omega_2}\right) = 2^o$$

$$\Rightarrow \quad \delta_1\left(1 - \frac{2}{3}\right) = 2^o$$

$$\Rightarrow \quad \delta_1 = 6^o$$

For small angle prism, deviation of light is given as

$$\delta_1 = (\mu_1 - 1)A_1$$

$$\Rightarrow \quad 6^o = (1.5 - 1)A_1$$

$$\Rightarrow \quad A_1 = 12^o \qquad \textbf{Ans. (12)}$$

Sol. 5 The refractive index of prism is given as

$$\mu = \frac{\sin\left(\frac{A+\delta_m}{2}\right)}{\sin\left(\frac{A}{2}\right)}$$

$$\Rightarrow \quad \mu = \frac{\sin A}{\sin A/2}$$

$$\Rightarrow \quad \mu = 2\cos\frac{A}{2}$$

$$\Rightarrow \quad A = 2\cos^{-1}\left(\frac{\mu}{2}\right) \qquad \textbf{Ans. (D)}$$

Sol. 6 For without deviation passing at interface, we must have same refractive index on the two sides of the interface, so we use

$$n_1 = n_2$$

$$\Rightarrow \quad 1.2 + \frac{1.8\times 10^{-14}}{\lambda^2} = 1.45 + \frac{1.8\times 10^{-14}}{\lambda^2}$$

$$\Rightarrow \quad \frac{(10.8 - 1.8)\times 10^{-14}}{\lambda^2} = 1.45 - 1.2 = 0.25$$

$$\Rightarrow \quad \lambda = 600 \text{ nm} \qquad \textbf{Ans. (600)}$$

Sol. 7 Minimum deviation of light through the prism is given as

$$\delta_{min} = 2i - A$$

As per given condition in question, we use

$$A = 2i - A$$

Refractive index of prism in terms of minimum deviation angle is given as

$$\mu = \frac{\sin\left(\frac{\delta_{min} + A}{2}\right)}{\sin\left(\frac{A}{2}\right)} = \frac{\sin(A)}{\sin\left(\frac{A}{2}\right)} = \frac{\sin 60°}{\sin 30°}$$

$$\Rightarrow \quad \mu = \sqrt{3}$$

Velocity of light in prism is given as

$$v = \frac{c}{\mu} = \frac{3\times 10^8}{\sqrt{3}} \text{ m/s}$$

Time taken by light to cover the distance AP is given as

$$t=\frac{AP}{V}=\frac{5\sqrt{3}\times10^{-3}}{\left(\frac{3\times10^{8}}{\sqrt{3}}\right)}=5\times10^{-10}\,\text{s}$$ **Ans. (5)**

23.5 Optical Instruments

Sol. 1 Limit of resolution of telescope is given as

$$\theta=\frac{1.22\lambda}{D}$$

$$\Rightarrow \quad \theta=\frac{1.22\times500\times10^{-9}}{200\times10^{-2}}$$

$$\Rightarrow \quad \theta=305\times10^{-9}\,\text{rad}$$ **Ans. (A)**

Sol. 2 Limit of resolution for objective of telescope is

$$\theta=\frac{1.22\lambda}{d}$$

$$\Rightarrow \quad \theta=\frac{1.22\times600\times10^{-9}}{250\times10^{-2}}$$

$$\Rightarrow \quad \theta=2.9\times10^{-7}\,\text{rad}$$ **Ans. (C)**

Sol. 3 Numerical aperature of the microscope is given as

$$NA=\frac{0.61\lambda}{d}$$

Where d = minimum sparation between two points to be seen as distinct

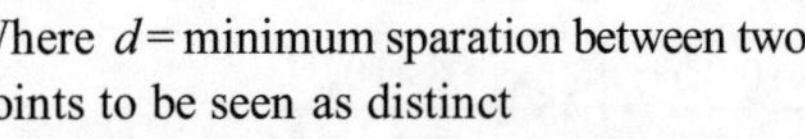

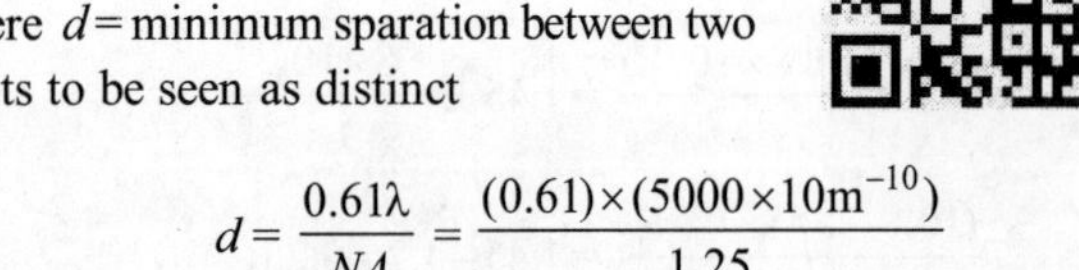

$$\Rightarrow \quad d=\frac{0.61\lambda}{NA}=\frac{(0.61)\times(5000\times10\text{m}^{-10})}{1.25}$$

$$\Rightarrow \quad d=2.4\times10^{-7}\,\text{m}$$

$$\Rightarrow \quad d=0.24\,\mu\text{m}$$ **Ans. (A)**

Sol. 4 Magnification of compound microscope is

$$m=\frac{LD}{f_e\times f_0}=\frac{150\times250}{f_e\times25}=375$$

$$\Rightarrow \quad f_e=20\,\text{mm}$$ **Ans. (A)**

Sol. 5 Tube length is taken as

$$L=f_0+f_e=60\,\text{cm}$$

and $$M=\frac{f_0}{f_e}=5$$

$$\Rightarrow \quad f_0=5f_e$$

$$\Rightarrow \quad 6f_e=60\,\text{cm}$$

$$\Rightarrow \quad f_e=10\,\text{cm}$$ **Ans. (D)**

Sol. 6 If the distance between objects on moon is l and D is diameter of telescope and d = distance between earth and moon, then we use

$$\theta=\frac{1.22\lambda}{D}$$

$$\Rightarrow \quad \frac{l}{d}=\frac{1.22\lambda}{D}$$

$$\Rightarrow \quad l=\frac{1.22\times(5500\times10^{-10})\times(4\times10^{8})}{5}=53.68\,\text{m}$$

Here most appropriate option is (C) **Ans. (C)**

Sol. 7 If object is placed at a distance x from objective, its image is produced at for objective

$$\frac{1}{v}+\frac{1}{x}=1$$

$$\Rightarrow \quad v=\frac{x}{x-1}$$

and $$m_0=\left|\frac{v}{u}\right|=\frac{1}{x-1}$$

For eye-piece

$$\frac{1}{-25}-\frac{1}{-\left(20-\frac{x}{x-1}\right)}=\frac{1}{f_e}$$

and $$m_e=\frac{25}{20-\frac{x}{x-1}}$$

Given that

$$\left(\frac{1}{x-1}\right)\times\left(\frac{25}{20-\frac{x}{x-1}}\right)=100$$

Solving we get $$x=\frac{81}{76}\,\text{cm}$$

For eye piece lens formula, we use

$$-\frac{1}{25}-\frac{1}{20-\frac{81/76}{81/76-1}}=\frac{1}{f_e}$$

Solving we get

$$f_e\simeq4.5\,\text{cm}$$

Answer can be 4 or 5. **Ans. (5)**

Sol. 8 Given focal lengths of objective and eyepiece are

$$f_0=1\,\text{cm}$$

$$f_e=5\,\text{cm}$$

$$l=10\,\text{cm (Tube length)}$$

For final image at infinity (for least strain on eye), image produced by objective must be at focal point of eyepiece i.e. 5 cm from it. For objective we use

$$f_0=+1\,\text{cm}$$

$V_0 = +5\text{ cm}$

$$\Rightarrow \quad u = \frac{V_0 f_0}{V_0 - f_0} = \frac{5\times1}{4} = 1.25\text{ cm} = \frac{N}{40}$$

$\Rightarrow \quad N = 50$ **Ans. (50)**

Sol. 9 In case of a simple microscope the line of vision of object and image is same for eye so assertion is correct and magnification of simple microscope is also greater than one so both A and R are true but R is not the correct explanation of A. **Ans. (C)**

Sol. 10 For a person if far away objects are not clear or blurry then problem is Myopia. When objects are looking distorted then it is due to Astigmatism. **Ans. (B)**

Sol. 11 Resolving power of a microscope is reated to wavelength used as

$$RP \propto \frac{1}{\lambda}$$

de-Broglie wavelength of the protons is given as

$$\lambda = \frac{h}{P} = \frac{h}{mv}$$

$$\Rightarrow \quad RP \propto \frac{mv}{h}$$

$$\Rightarrow \quad RP \propto m$$

Mass of proton is 1837 times that of an electron. **Ans. (A)**

Sol. 12 Magnification of microscopic lens is given as

$$m = 1 + \frac{D}{f_0}$$

$$\Rightarrow \quad 6 = 1 + \frac{25}{f_0}$$

$$\Rightarrow \quad f_0 = 5\text{ cm}$$

Magnification of compound microscope when image is formed at infinity, is given as

$$m = \frac{l}{f_0} \times \frac{D}{f_e}$$

$$\Rightarrow \quad 12 = \frac{60}{5} \times \frac{25}{f_e}$$

$f_e = 25\text{ cm}$ **Ans. (25)**

24.1 Interference of Light

Sol. 1 In *YDSE* the distance of nth bright fringe from central maximum is given as

$$x_{nB} = \frac{n\lambda D}{d}$$

Where $D = 1.5\text{m}$

$d = 0.5 \times 10^{-3}\text{ m}$

If n^{th} bright fringe of λ, coincides with n_2^{th} bright fringe of λ_2 then we use

$$x_{n_1B} = x_{n_2B}$$

$$\Rightarrow \quad \frac{n_1\lambda_1 D}{d} = \frac{n_2\lambda_2 D}{d}$$

$$\Rightarrow \quad \frac{n_1}{n_2} = \frac{\lambda_2}{\lambda_1} = \frac{520}{560} = \frac{4}{5}$$

Last distance from central maxima of coinciding bright fringe is

$$x_{n_1B} = \frac{4\times650\times10^{-9}\times1.5}{0.5\times10^{-3}}$$

$n_1 = 4$

$n_2 = 5$

$\Rightarrow \quad x_{n_1B} = 7.8 \times 10^{-3}\text{m}$ **Ans. (D)**

Sol. 2 Ratio of maximum to minimum intensities of interference is given as

$$\frac{I_{max}}{I_{min}} = 16$$

If I_1 and I_2 are intensities of compoment waves, we have

$$\Rightarrow \quad \frac{I_{max}}{I_{min}} = \frac{\sqrt{I_1}+\sqrt{I_2}}{\sqrt{I_1}-\sqrt{I_2}} = 4$$

$$\Rightarrow \quad \sqrt{\frac{I_1}{I_2}} = \frac{5}{3}$$

$$\Rightarrow \quad \frac{I_1}{I_2} = \frac{25}{9}$$

Ans. (B)

Sol. 3

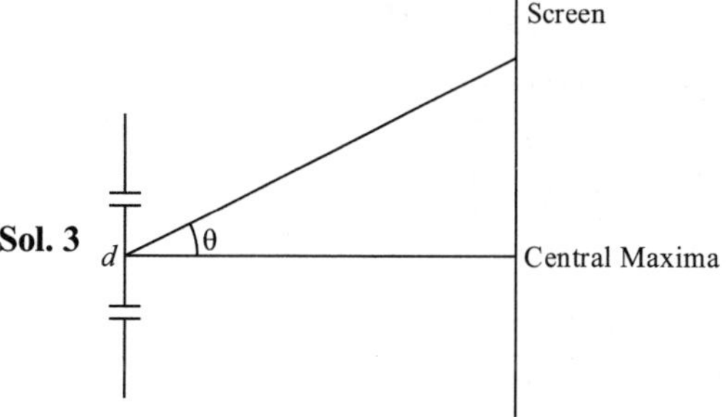

Path difference for angular position θ is

$$d \sin\theta = n\lambda$$

$$0.32 \times 10^{-3} \sin(30°) = n \times 500 \times 10^{-9}$$

$$n = 320$$

Thus including central maxima there will be 641 fringes seen in the given region. **Ans. (B)**

Sol. 4 Path difference

$$\Delta = d\sin\theta \approx d\theta$$

$$\Rightarrow \quad \Delta = 0.1\times\frac{1}{40}\text{ mm} = 2500\text{ nm}$$

On bright fringe, path difference must be integral multiple of λ.

$$\Rightarrow \quad 2500 = n\lambda_1 = m\lambda_2$$

$$\Rightarrow \quad \lambda_1 = 625,$$

$$\lambda_2 = 500 \text{ (from } m=5 \text{ and } n=4)$$

Ans. (B)

Sol. 5 For first minima in front of one slit, path difference at point P is

$$\sqrt{5}d - 2d = \frac{\lambda}{2}$$

$$\Rightarrow \quad d = \frac{\lambda}{2(\sqrt{5}-2)}$$

Ans. (D)

Sol. 6 Given that path difference at the point

$$\Delta x = \frac{\lambda}{8}$$

Phase difference at this point is

$$\Delta\phi = \frac{(2\pi)}{\lambda}\times\frac{\lambda}{8} = \frac{\pi}{4}$$

Intensity at this point is

$$I = I_0\cos^2\left(\frac{\Delta\phi}{2}\right) = I_0\cos^2\left(\frac{\pi}{8}\right)$$

$$\Rightarrow \quad \frac{I}{I_0} = \cos^2\left(\frac{\pi}{8}\right) = \frac{1+\cos\left(\frac{\pi}{4}\right)}{2}$$

$$\Rightarrow \quad \frac{I}{I_0} = \frac{1+0.707}{2} = 0.85$$

Ans. (C)

Sol. 7 For diffraction, location of 1st minima

$$y_1 = \frac{D\lambda}{b} = 0.2469\,D\lambda$$

location of 2nd minima is

$$y_2 = \frac{2D\lambda}{b} = 0.4938D\lambda$$

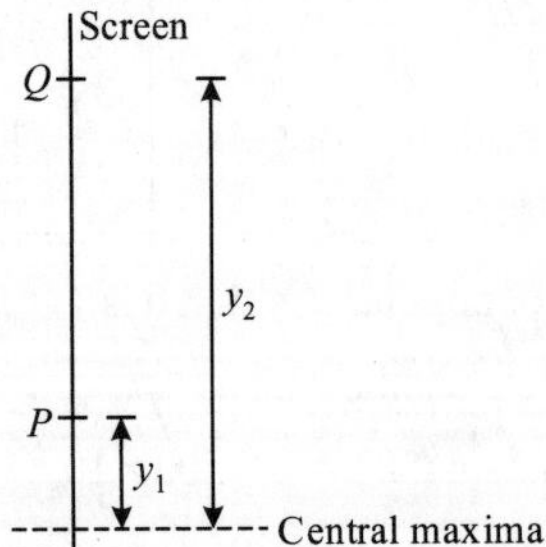

Now for interference

Path difference at point P on screen is

$$\frac{dy}{D} = 4.8\lambda$$

Path difference at point Q on screen is

$$\frac{dy}{D} = 9.6\lambda$$

So orders of maxima in between P & Q is are 5, 6, 7, 8, 9

Thus 5 bright fringes are present between P & Q. **Ans. (D)**

Sol. 8 Given that ratio of amplitudes is

$$\frac{a_1}{a_2} = \frac{1}{3}$$

Ratio of intensities of the waves can be given as

$$\frac{I_1}{I_2} = \left(\frac{a_1}{a_2}\right)^2 = \frac{1}{9}$$

Ratio of maximum to minimum intensities is given as

$$\frac{I_{max}}{I_{min}} = \left(\frac{\sqrt{I_1}+\sqrt{I_2}}{\sqrt{I_1}-\sqrt{I_2}}\right)^2 = \left(\frac{1+3}{1-3}\right)^2 = 4$$

Ans. (A)

Sol. 9 If we introduce a thin sheet in path of light from one of the slits then path difference introduced is

$$\Delta = t(\mu - 1)$$

If fringe pattern displaces by x, we use

$$\Delta = \frac{xd}{D}\left(\frac{\lambda D}{d}\right) = n\lambda$$

$$\Rightarrow \quad t(\mu-1) = n\lambda$$

$$\Rightarrow \quad t = \frac{n\lambda}{\mu-1}$$

Ans. (None)

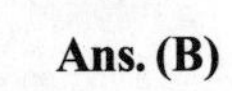

Sol. 10 If I_1 and I_2 are light intensities from the two slits, we use

$$I_2 = I_0$$

$$\Rightarrow \quad I_1 = 4I_0$$

We use $$\frac{I_{max}}{I_{min}} = \left[\frac{\sqrt{I_1}+\sqrt{I_2}}{\sqrt{I_1}-\sqrt{I_2}}\right]^2$$

$$\Rightarrow \quad \frac{I_{max}}{I_{min}} = \left[\frac{2\sqrt{I_0}+\sqrt{I_0}}{2\sqrt{I_0}-\sqrt{I_0}}\right]^2 = \frac{9}{1}$$

Ans. (B)

Sol. 11

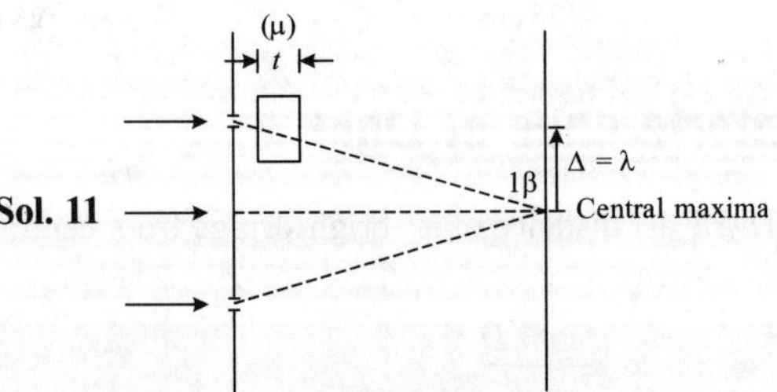

Path difference due to introducing a thin film is given as

$$\Delta = t(\mu - 1)$$

As shift of central maxima is by one fringe width

$$\Rightarrow \quad \Delta = \lambda$$

$$t(\mu - 1) = \lambda$$

$$\Rightarrow \quad t = \frac{\lambda}{\mu - 1}$$ **Ans. (D)**

Sol. 12 Fringe width in YDSE is given as,

$$\beta = \frac{D\lambda}{d}$$

$$\Rightarrow \quad \beta = \frac{1.5 \times 589 \times 10^{-9}}{0.15 \times 10^{-3}} \text{ m}$$

$$\Rightarrow \quad \beta = 5.9 \times 10^{-3} \text{m}$$

$$\Rightarrow \quad \beta = 5.9 \text{mm}$$ **Ans. (B)**

Sol. 13 Intensity of light on YDSE screen is given as

$$I = I_0 \cos^2\left(\frac{\phi}{2}\right)$$

$$\frac{I}{I_0} = \cos^2\left(\frac{\phi}{2}\right)$$

Phase difference for the given point is

$$\phi = \frac{2\pi}{\lambda} \times \frac{\lambda}{8} = \frac{\pi}{4}$$

$$\Rightarrow \quad \frac{I}{I_0} = \cos^2\left(\frac{\pi}{8}\right)$$

$$\Rightarrow \quad \frac{I}{I_0} = 0.853$$ **Ans. (D)**

Sol. 14 If amplitude of each wave is considered as A then for resulting amplitude, we use the figure below

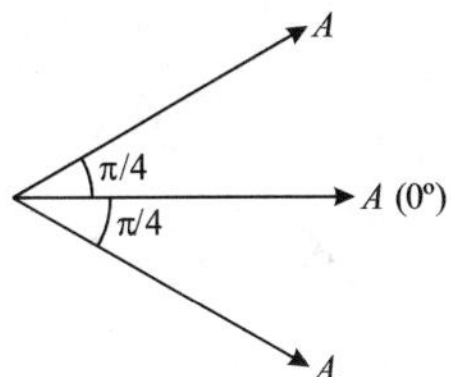

Resulting amplitude is given as

$$R = A + 2\left(\frac{A}{\sqrt{2}}\right) = A(\sqrt{2} + 1)$$

As intensity is proportional to square of amplitude, we use

$$I_0 = kA^2 \text{ and } I_R = kR^2$$

$$\Rightarrow \quad \frac{I_R}{I_0} = (\sqrt{2} + 1)^2$$

$$\Rightarrow \quad I_R \cong 5.8\, I_0$$ **Ans. (A)**

Sol. 15 The length of portion on screen used for 15 fringes is

$$l = 15B = 15\frac{\lambda_0 D}{d}$$

When wavelength changes to λ, 10 fringes are observed on same position of screen, so we use

$$l = 10\frac{\lambda D}{d} = \frac{15\lambda_0 D}{d}$$

$$\Rightarrow \quad l = \frac{15\lambda_0}{10} = \frac{3}{2} \times 500 = 750 \text{ nm}$$ **Ans. (750)**

Sol. 16 Distance of nth bright fringe from centre is

$$Y = \frac{nD\lambda}{d}$$

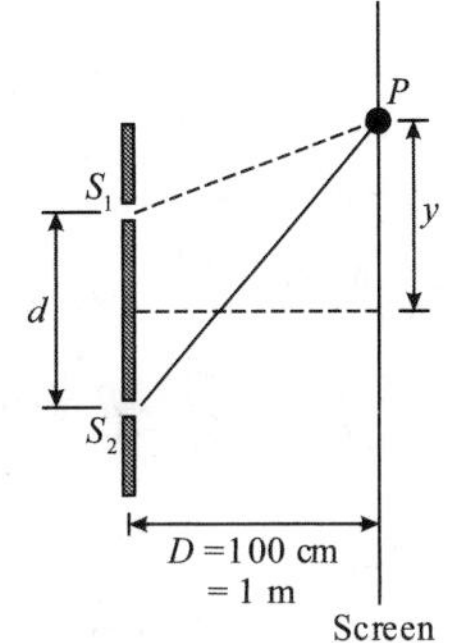

$$\Rightarrow \quad n = \frac{Yd}{D\lambda} = \frac{1.27 \times 10^{-3} \times 10^{-3}}{1 \times 632.8 \times 10^{-9}} = 2$$

For bright fringe path difference is given as

$$\Delta x = n\lambda$$

$$\Rightarrow \quad \Delta x = 2 \times 632.8 \text{ nm}$$

$$\Rightarrow \quad \Delta x = 1265.6 \text{ nm}$$

$$\Rightarrow \quad \Delta x = 1.27\ \mu\text{m}$$ **Ans. (A)**

Sol. 17 If N fringes are there in screen segment of length l and fringe width is β, we use

$$N\beta = l$$

$$\Rightarrow \quad N\left(\frac{\lambda D}{d}\right) = l$$

For first wavelength $\lambda_1 = 700$ nm, we use

$$\frac{N_1\lambda_1 D}{d} = l$$

For second wavelength $\lambda_2 = 400$ nm, we use

$$\frac{N_2\lambda_2 D}{d} = l$$

$$\Rightarrow \quad N_1\lambda_1 = N_2\lambda_2$$

$$\Rightarrow \quad 16 \times 700 = N_2 \times 400$$

$$\Rightarrow \quad N_2 = 28$$ **Ans. (A)**

Sol. 18 In YDSE setup given data is

$$d = 0.05 \times 10^{-3}\,\text{m}$$

and $\lambda = 5 \times 10^{-7}\,\text{m}$

Angular fringe width is given as

$$\theta = \frac{\beta}{D} = \frac{\lambda D/d}{D} = \frac{\lambda}{d}\ \text{rad}$$

In degrees it is given as

$$\theta = \frac{\lambda}{d} \times \frac{180}{\pi}\ \text{degree}$$

$$\Rightarrow \quad \theta = \frac{5\times10^{-7}\times180}{5\times10^{-5}\times3.14} \simeq 0.57^\circ$$ **Ans. (D)**

Sol. 19 Optical path difference between two waves is given as

$$\Delta x = n_1L_1 - n_2L_2$$

Phase difference is given as

$$\Delta\phi = \frac{2\pi}{\lambda}\Delta k = \frac{2\pi}{\lambda}(n_1L_1 - n_2L_2)$$ **Ans. (A)**

Sol. 20 At initial position

$$S_1L = \sqrt{2^2 + \left(\frac{3}{2}\right)^2} = \frac{5}{2}\,\text{m}$$

and $S_2L = 2\,\text{m}$

Path difference at initial position is given as

$$\Delta = S_1L - S_2L = 2.5 - 2 = 0.5\,\text{m}$$

Here $\Delta = \frac{\lambda}{2}$

Thus intensity is minimum. When L moves away to new position L', next maximum is obtained when

$$\Delta = \lambda = 1\,\text{m}$$

$$\Rightarrow \quad d = S_1L' = S_2L' + 1 = 3\,\text{m}$$ **Ans. (B)**

Sol. 21 As P is 90° ahead in phase so path lead of P w.r. to Q is

$$\Delta = \frac{\lambda}{2\pi} \times \frac{\pi}{2} = \frac{\lambda}{4} = \frac{20}{4} = 5\,\text{m}$$

For point A, path difference in waves from P and Q is

$$\Delta_A = 5 - 5 = 0$$

Thus at A intensity is maximum I_0

For point B, path difference in waves from P and Q is

$$\Delta = 5\,\text{m}$$

$$\Rightarrow \text{Phase difference } \phi = \frac{2\pi}{\lambda}\Delta = \frac{2\pi}{20} \times 5 = \frac{\pi}{2}$$

Thus intensity at B is

$$I_B = I_0\cos^2\frac{\phi}{2} = I_0\cos^2\frac{\phi}{4} = \frac{I_0}{2}$$

For point C, path difference in waves from P and Q is

$$\Delta = 5 + 5 = 10\,\text{m}$$

As $\Delta = \frac{\lambda}{2}$ so intensity will be minimum at C due to distractive interference, so we have

$$\Rightarrow \quad I_A : I_B : I_C = I_0 : \frac{I_0}{2} : 0 = 2 : 1 : 0$$ **Ans. (D)**

Sol. 22 The point on screen where path difference is λ is the point of bright fringe so at a general point resulting light intensity is given as

$$I_R = K\cdot\cos^2\frac{\phi}{2}$$

At the point where path difference is $\frac{\lambda}{6}$, phase difference is given as

$$\phi = \frac{2\pi}{\lambda} \times \frac{\lambda}{6} = \frac{\pi}{3}$$

Thus intensity at this point is given as

$$\Rightarrow \quad I = \frac{3K}{4} = \frac{nK}{12}$$

$$\Rightarrow \quad n = 9$$ **Ans. (9)**

Sol. 23 Amplitude of light wave is directly proportional to the width of slit in YDSE setup so we use

$$A_2 = 3A_1$$

$$\Rightarrow \quad \frac{I_{max}}{I_{min}} = \left(\frac{\sqrt{I_1} + \sqrt{I_2}}{|\sqrt{I_1} - \sqrt{I_2}|}\right)^2$$

As light intensity is directly proportional to square of amplitude, we use

$$\frac{I_{max}}{I_{min}} = \left(\frac{A_1 + A_2}{|A_1 - A_2|}\right)^2$$

$$\Rightarrow \quad \frac{I_{max}}{I_{min}} = \left(\frac{A_1 + 3A_1}{|A_1 - 3A_1|}\right)^2$$

$$\Rightarrow \quad \frac{I_{max}}{I_{min}} = \left(\frac{4A_1}{2A_1}\right)^2 = 4 : 1$$ **Ans. (B)**

Sol. 24 Fringe width in YDSE setup is given as

$$\beta = \frac{\lambda.D}{d}$$

As $\lambda_R > \lambda_V \Rightarrow \beta_R > \beta_V$

Thus fringe pattern will shrink on changing color from red to violet. **Ans. (D)**

Sol. 25 The ratio of component light intensities is given as

$$\frac{I_1}{I_2}=2x$$

Maximum and minimum intensities due to interference are given as

$$I_{max}=(\sqrt{I_1}+\sqrt{I_2})^2$$

and $$I_{min}=(\sqrt{I_1}-\sqrt{I_2})^2$$

$$\Rightarrow \quad \frac{I_{max}-I_{min}}{I_{max}+I_{min}}=\frac{2\sqrt{I_1I_2}}{I_1+I_2}$$

$$\Rightarrow \quad \frac{I_{max}-I_{min}}{I_{max}+I_{min}}=\frac{2\sqrt{I_1/I_2}}{\frac{I_1}{I_2}+1}=\frac{2\sqrt{2x}}{2x+1}$$ **Ans. (A)**

Sol. 26 Fringe width in YDSE pattern is given as

$$\beta=\frac{\lambda D}{d}=\frac{500\times10^{-9}\times1}{2\times10^{-3}}$$

$$\Rightarrow \quad \beta=\frac{5}{2}\times10^{-4}\,\text{m}=2.5\times10^{-1}\,\text{mm}$$

$$\Rightarrow \quad \beta=0.25\,\text{mm}$$ **Ans. (A)**

Sol. 27 Fringe withd in YDSE setup is given as

$$\beta=\frac{\lambda D}{d}$$

$$\Rightarrow \quad \lambda=\frac{\beta d}{D}=\frac{6\times10^{-3}\times10^{-3}}{10}$$

$$\Rightarrow \quad \lambda=6\times10^{-7}\,\text{m}=600\times10^{-9}\,\text{m}=600\,\text{nm}$$ **Ans. (600)**

Sol. 28 In *YDSE* setup, fringe width in interference pattern is given as

$$\beta=\frac{D\lambda}{d}$$

$$\Rightarrow \quad \beta=\frac{D\lambda}{(d_0+a_0\sin wt)}$$

The difference between the largest and smallest fringe width can be calculated as

$$\beta_{max}-\beta_{min}=\frac{D\lambda}{d_0-a_0}-\frac{D\lambda}{d_0+a_0}$$

$$\Rightarrow \quad \beta_{max}-\beta_{min}=D\lambda\left[\frac{d_0+a_0-d_0+a_0}{d_0^2+a_0^2}\right]=\frac{2\lambda Da_0}{(d_0^2-a_0^2)}$$ **Ans. (A)**

Sol. 29 Distance of first red and violet fringe from central white fringe is given as

$$y_1=\frac{\lambda_r D}{d}$$

and $$y_2=\frac{\lambda_v D}{d}$$

$$\Rightarrow \quad \lambda_r-\lambda_v=(y_1-y_2)\frac{d}{D}=(3.5-2)\,\text{mm}\times\frac{0.3\text{mm}}{1.5\text{m}}=300\,\text{nm}$$

Ans. (300)

Sol. 30 As $I_1=I_2=I_0$, the maximum intensity at the point of interference is given as

$$I_{max}=(\sqrt{I_1}+\sqrt{I_2})^2=4I_1$$ **Ans. (C)**

Sol. 31 In YDSE setup intensity of light is proportional to the slit width. The ratio of maximum to minimum intensity is given as

$$\frac{I_{max}}{I_{min}}=\frac{(\sqrt{I_1}+\sqrt{I_2})^2}{(\sqrt{I_1}-\sqrt{I_2})^2}=\left[\frac{\sqrt{9I}+\sqrt{I}}{\sqrt{9I}-\sqrt{I}}\right]^2$$

$$\Rightarrow \quad \frac{I_{max}}{I_{min}}=\left[\frac{(3+1)\sqrt{I}}{(3-1)\sqrt{I}}\right]^2=4=\frac{x}{4}$$

$$\Rightarrow \quad x=16$$ **Ans. (16)**

24.2 Diffraction of Light

Sol. 1

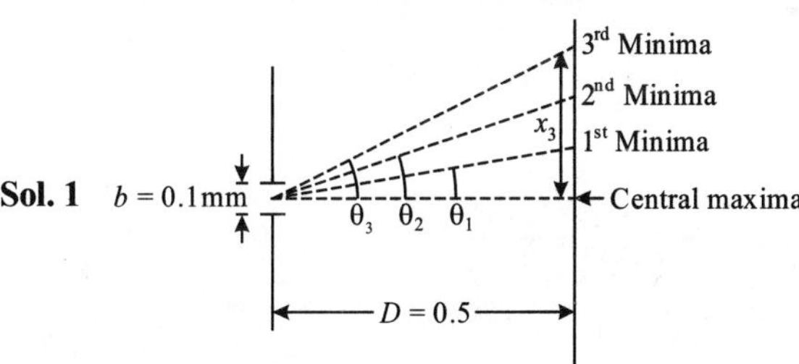

The angular separation of nth minima from screen centre (central maxima) is given as

$$b\sin\theta_n=n\lambda$$

from 3rd dark band ($n=3$) we have

$$\theta_3\simeq\sin\theta_3=\frac{3\lambda}{b}\quad[\text{as }\theta\text{ is }v\text{ small}]$$

distance of 3rd minima from O is

$$x_3=D\theta_3=0.5\times\frac{3\times6000\times10^{-10}}{0.1\times10^{-3}}$$

$$\Rightarrow \quad x_3=9\times10^{-3}\,\text{m}=9\,\text{mm}$$ **Ans. (B)**

Sol. 2

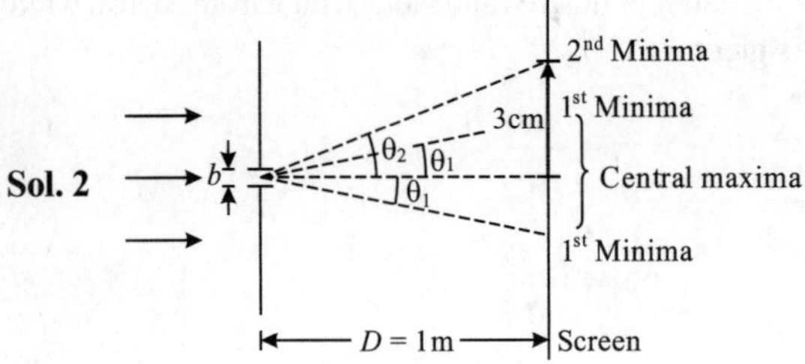

For first minima in diffraction pattern we use

$$b \sin \theta = n\lambda \qquad \ldots (1)\ [n=1]$$

The angular position of second minimum from screen centre is given as

$$\tan \theta_2 \simeq \sin \theta_2 \simeq \theta_2 = \frac{3\text{cm}}{100\text{cm}}$$

$$\Rightarrow \qquad \theta_2 = \frac{3\text{cm}}{1\text{cm}} = 0.03$$

From equation (1), for 2nd minima, we use

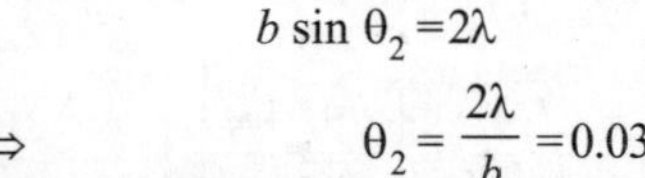

$$b \sin \theta_2 = 2\lambda$$

$$\Rightarrow \qquad \theta_2 = \frac{2\lambda}{b} = 0.03$$

Angular width of central maxima in diffraction pattern is given as

$$2\theta_1 = \frac{2\lambda}{b} = 0.03$$

Width of central maxima is

$$w = 2\theta_1 D = 0.03 \times 1\text{ m} = 0.03\text{ m} = 3\text{ cm}$$

Ans. (B)

Sol. 3

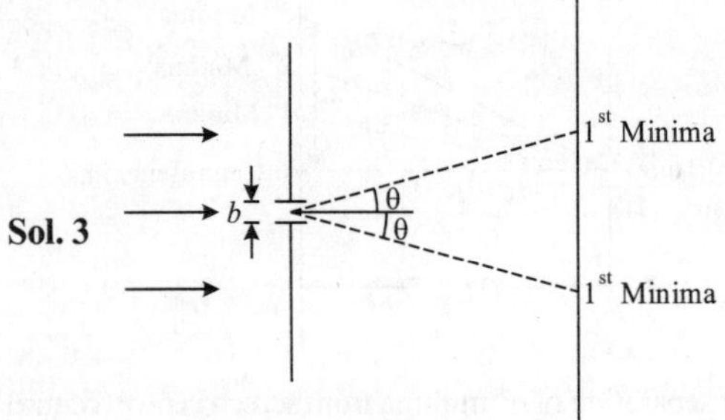

Angular position of first minima

$$b \sin \theta = \lambda$$

$$\Rightarrow \qquad \sin \theta = \frac{\lambda}{b}$$

So wavelength of light used

$$\lambda = b \sin 30° = \frac{b}{2}$$

given $\qquad 2\theta = 60°$

$$\Rightarrow \qquad \theta = 30°$$

If similar two slits are used in *YDSE* at a separation '*d*' then we have

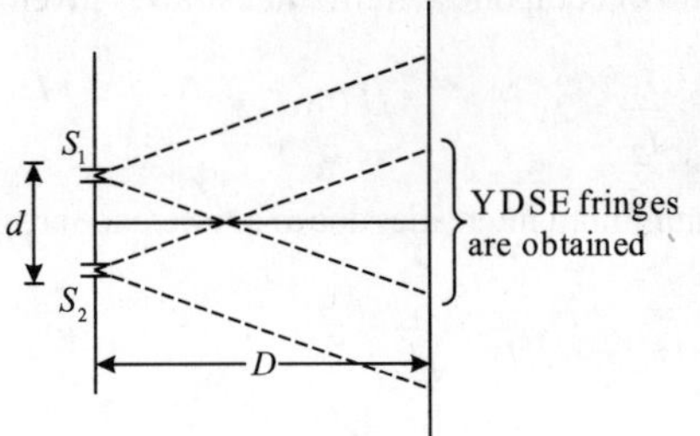

Fringe width in *YDSE* interference pattern is given as

$$b = \frac{\lambda D}{d} = \frac{bD}{2d}$$

$\Rightarrow$ Slit separation

$$d = \frac{bD}{2\beta} = \frac{10^{-6} \times 0.5}{2 \times 0.01} = 25 \times 10^{-6}\text{m}$$

$$\Rightarrow \qquad d = 25\ \mu\text{m}$$

Ans. (A)

Sol. 4

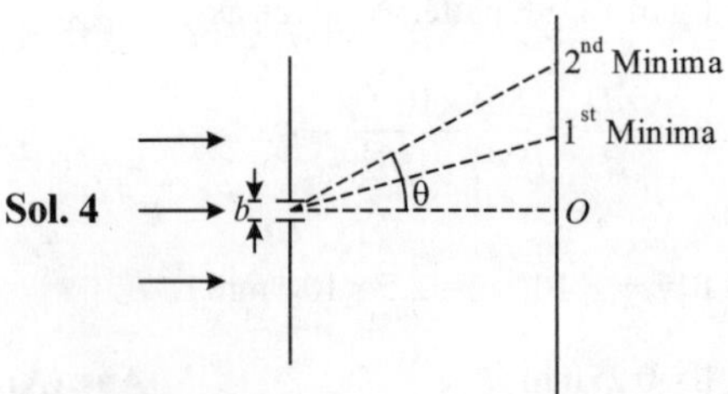

If at angular position θ from central maxima we get n^{th} minima then we use

$$b \sin \theta = n\lambda$$

So for 2nd minima we use $n = 2$

$$\Rightarrow \qquad b \sin \theta = 2\lambda$$

$$\Rightarrow \qquad \sin \theta = \frac{2\lambda}{b} = \frac{2 \times 550 \times 10^{-9}}{22 \times 10^{-7}} = 50 \times 10^{-2}$$

$$\Rightarrow \qquad \sin \theta = 0.5$$

$$\Rightarrow \qquad \theta = \frac{\pi}{6}$$

Ans. (D)

Sol. 5 For second minima, we use

$$b \sin \theta_2 = 2\lambda$$

$$\Rightarrow \qquad \sin 60° = \frac{2\lambda}{b}$$

$$\Rightarrow \qquad \frac{\lambda}{b} = \sqrt{\frac{3}{4}}$$

Now for first minima, we use

$$\sin \theta_1 = \frac{\lambda}{b} = \frac{\sqrt{3}}{4}$$

$$\Rightarrow \qquad \theta_1 \simeq 25°$$

Ans. (D)

Sol. 6 For n^{th} diffraction minima, we use

$$b \sin\theta = n\lambda$$

$$\Rightarrow \quad \sin\theta = \frac{n\lambda}{b} \le 1 \quad \text{(for maximum limit)}$$

$$\Rightarrow \quad n \le \frac{b}{\lambda} = \frac{6\times10^{-5}}{6\times10^{-7}} = 100$$

Thus maximum number of minima on either side are 99 so total minima on screen are

$$99 \times 2 = 198$$ **Ans. (198)**

Sol. 7 First intensity maximum occurs at an angle θ, given as

$$2d \sin\theta = \lambda = \frac{h}{\sqrt{2mE}}$$

$$\Rightarrow \quad E = \frac{h^2}{8md^2 \sin^2\theta} \times \frac{1}{e} eV$$

$$\Rightarrow \quad E = \frac{(6.64\times10^{-34})^2}{8\times9.1\times10^{-31}\times(10^{-10})^2\times\frac{3}{4}\times1.6\times10^{-19}}$$

$$\Rightarrow \quad E \cong 50.47$$ **Ans. (50)**

Sol. 8 For a pin hole of diameter a, the angular width of disc produced is given as

$$\sin\theta = \frac{m\lambda}{a}$$

When a increases, θ decreases due to which the area of disc decreases so intensity will increases. **Ans. (A)**

24.3 Polarization of Light

Sol. 1

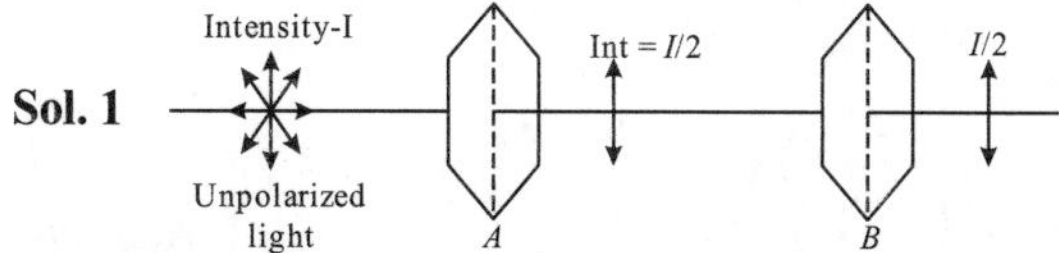

Initially when A and B are present then light intensity does not change on passing through $B \Rightarrow$ Axis of B is parallel to axis of A.

After inserting polarizer C, intensity after B becomes $I/8$

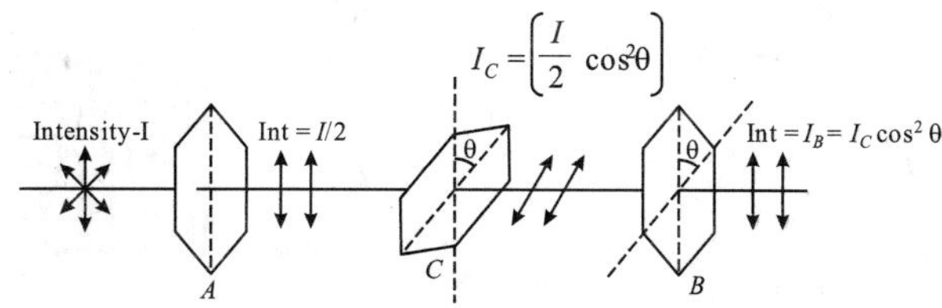

By Malus law $\quad I_C = \frac{I}{2}\cos^2\theta$

and $\quad I_B = I_C\cos^2\theta = \frac{I}{2}\cos^4\theta$

Given that finally $\quad I_B = \frac{I}{8}$

$$\Rightarrow \quad \frac{I}{8} = \frac{I}{2}\cos^4\theta$$

$$\Rightarrow \quad \cos^4\theta = \frac{1}{4}$$

$$\Rightarrow \quad \cos\theta = \frac{1}{\sqrt{2}}$$

$$\Rightarrow \quad \theta = 45°$$ **Ans. (C)**

Sol. 2

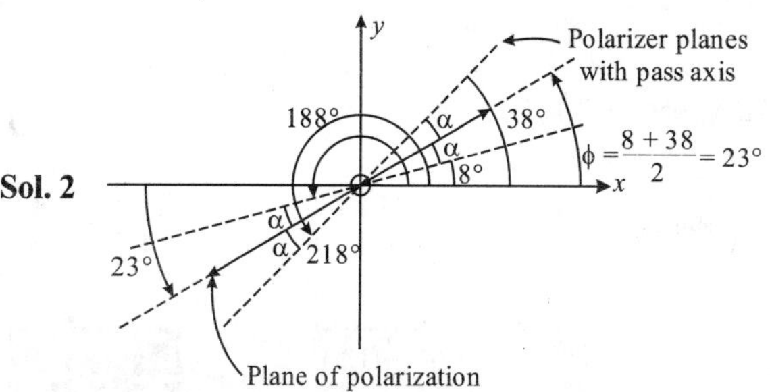

As the light intensity after passing through polarizer is same for the given pass axis (polarization axis)

$\Rightarrow$ Plane of polarization must be at same angle with the given direction of pass axis.

$$I_T = I\cos^2\alpha$$

Values of $\quad \phi = 23°$

or $\quad 180 + 23 = 203°$ **Ans. (A)**

Sol. 3 Case-I :

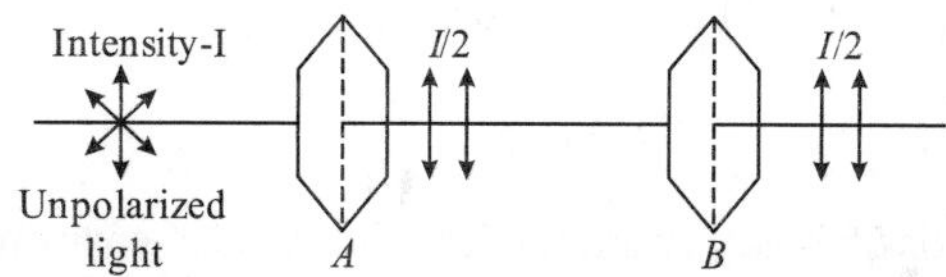

As light intensity does not change after passing through polarizer B

$\Rightarrow$ Axis of B is parallel to axis of A

Case-II :

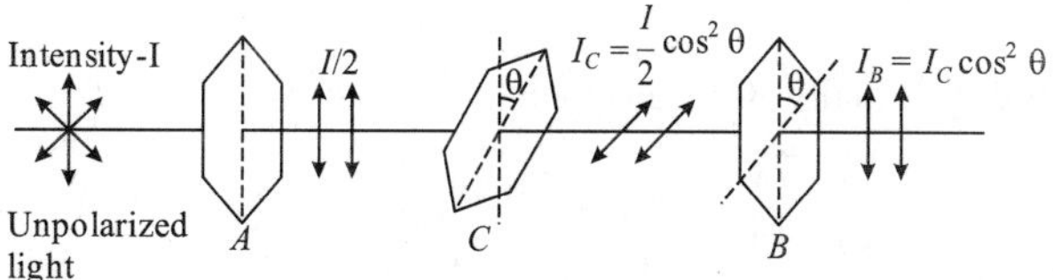

By Malus law have

$$I_C = \frac{I}{2}\cos^2\theta$$

and $\quad I_B = I_C\cos^2\theta = \frac{I}{2}\cos^4\theta$

Given that $I_B = \frac{I}{3}$

$$\Rightarrow \quad \frac{I}{3} = \frac{I}{2}\cos^4\theta$$

$$\Rightarrow \quad \cos\theta = \left(\frac{2}{3}\right)^{1/4}$$

Ans. (A)

Sol. 4 For light to never be reflected as polarized, incident angle on liquid glass interface must be less then brewster's angle so we use

$$\theta_{CL} < i_b$$

$$\Rightarrow \quad \sin\theta_c < \sin i_b$$

Here we use $\left[\tan i_b = \frac{1.5}{\mu}\right]$

$$\Rightarrow \quad \sin i_b = \frac{1.5}{\sqrt{\mu^2 + (1.5)^2}}$$

$$\Rightarrow \quad \frac{1}{\mu} < \frac{1.5}{\sqrt{\mu^2 + (1.5)^2}}$$

$$\Rightarrow \quad \sqrt{\mu^2 + (1.5)^2} < 1.5 \times \mu$$

$$\Rightarrow \quad \mu^2 + (1.5)^2 < (\mu \times 1.5)^2$$

$$\Rightarrow \quad \mu > \frac{3}{\sqrt{5}}$$

Ans. (A)

Sol. 5 Since unpolarised light falls on P_1, intensity of light transmitted from P_1 is given as

$$I_1 = I_0/2$$

Pass axis of P_2 is at an angle of 30° with P_1 so intensity of light transmitted from P_2 is given by Malus law as

$$I_2 = \frac{I_0}{2}\cos^2 30° = \frac{3I_0}{8}$$

Pass axis of P_3 is at an angle of 60° with P_2 so intensity of light transmitted from P_3 is given as

$$I_3 = \frac{3I_0}{8}\cos^2 60° = \frac{3I_0}{32}$$

$$\Rightarrow \quad \left(\frac{I_0}{I}\right) - \frac{32}{3} - 10.67$$

Ans. (D)

Sol. 6 Initially 10% light comes out so we use

$$\frac{I_0}{10} = \frac{I_0}{2} \times \cos^2\theta$$

$$\Rightarrow \quad \cos\theta = \frac{1}{\sqrt{5}}$$

$$\Rightarrow \quad \theta = 63.44°$$

Thus rotation angle is $\theta_R = 90 - 63.44° = 26.56°$

Out of given options, closest is (A) **Ans. (A)**

Sol. 7 Total power of light incident on polarizer is

$$P = IA = 3.3 \times 3 \times 10^{-4} = 9.9 \times 10^{-4}\,\text{W}$$

Average light intensity passes through polarizer is $I/2$ so we use amount of light energy passing through it in one revolution is

$$E = \frac{I}{2} \times A \times t = \frac{1}{2} \times 9.9 \times 10^{-4} \times \frac{2\pi}{31.4}$$

$$\Rightarrow \quad E = 4.95 \times 10^{-4} \times 0.2$$

$$\Rightarrow \quad E = 0.99 \times 10^{-4}$$

$$\Rightarrow \quad E \simeq 1 \times 10^{-4}\,\text{J}$$

Ans. (C)

Sol. 8 The intensity of emerging light from analyzer after it is rotated by 30° is given as

$$I = \frac{I_0}{2}\cos^2 30°$$

$$\Rightarrow \quad I = 100\left(\frac{\sqrt{3}}{2}\right)^2 = 100 \times \frac{3}{4} = 75$$

Ans. (75)

Sol. 9 The situation described in question is shown in figure below

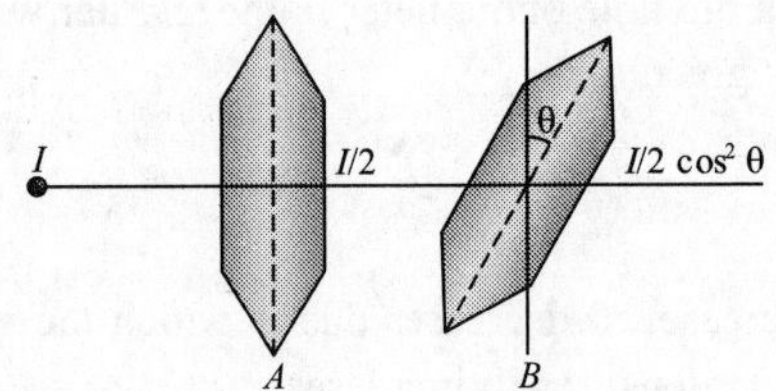

The intensity of light after polarizer A becomes $I/2$ then after passing through polarizer B it is given by Malus's law as

$$I' = \frac{I}{2}\cos^2\theta$$

$$\Rightarrow \quad \frac{3I}{8} = \frac{I}{2}\cos^2\theta$$

$$\Rightarrow \quad \theta = 30°$$

Ans. (30)

25.1 Photoelectric Effect

Sol. 1 By photo electric effect equation we use

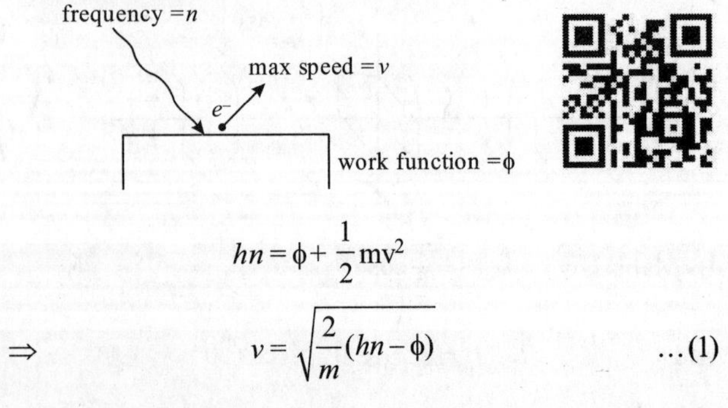

$$hn = \phi + \frac{1}{2}mv^2$$

$$\Rightarrow \quad v = \sqrt{\frac{2}{m}(hn - \phi)} \qquad \ldots(1)$$

If incident radiation frequency is increased to $3n \Rightarrow$ The speed of ejected fastest electron is given as

$$\Rightarrow \quad v' = \sqrt{\frac{2}{m}(3hn - \phi)}$$

$$\Rightarrow \quad v' > \sqrt{3}v$$

Ans. (C)

Sol. 2 $\delta t = 60\text{ms}$ $\lambda = 6 \times 10^{-7}$ m P_{watt}

Pulse energy $E = \rho\delta t$

Energy of one photon in pulse $E_{ph} = \dfrac{hc}{\lambda}$

Number of photons in light pulse are

$$N = \frac{E}{E_{ph}} = \frac{E\lambda}{hc} = \frac{\rho\delta t\lambda}{hc}$$

$$\Rightarrow \quad N = \frac{0.5\times10^3 \times 60\times10^{-3} \times 6\times10^{-7}}{6.62\times10^{-34} \times 3\times10^8}$$

$$\Rightarrow \quad N \simeq \frac{10^{-7}\times10}{10^{-34}\times10^8} = 10^{20}$$

Ans. (A)

Sol. 3 We use for velocity of ejected photoelectrons

$$\frac{hc}{\lambda_1} = \phi + \frac{1}{2}m(2v)^2 \quad \dots(1)$$

and
$$\frac{hc}{\lambda_2} = \phi + \frac{1}{2}mv^2 \quad \dots(2)$$

$$\frac{(1)}{(2)} \Rightarrow \frac{\frac{hc}{\lambda_1} - \phi}{\frac{hc}{\lambda_2} - \phi} = 4$$

$$\Rightarrow \quad \frac{hc}{\lambda_1} - \phi = \frac{4hc}{\lambda_2} - 4\phi$$

$$\Rightarrow \quad \frac{4hc}{\lambda_2} - \frac{hc}{\lambda_1} = 3\phi$$

$$\Rightarrow \quad \phi = \frac{1}{3}hc\left(\frac{4}{\lambda_2} - \frac{1}{\lambda_1}\right)$$

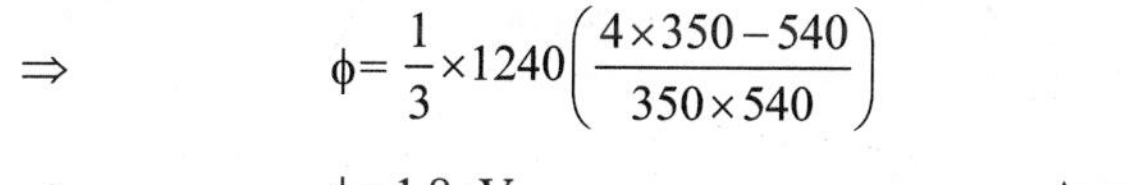

$$\Rightarrow \quad \phi = \frac{1}{3}\times1240\left(\frac{4\times350 - 540}{350\times540}\right)$$

$$\Rightarrow \quad \phi = 1.8 \text{ eV}$$

Ans. (A)

Sol. 4 $B = B_0\sin(\pi\times10^7c)t + B_0\sin(2\pi\times10^7c)t$ since there are two EM waves with different frequency, to get maximum kinetic energy we take the photon with higher frequency

$$B_1 = B_0\sin(\pi\times10^7c)t$$

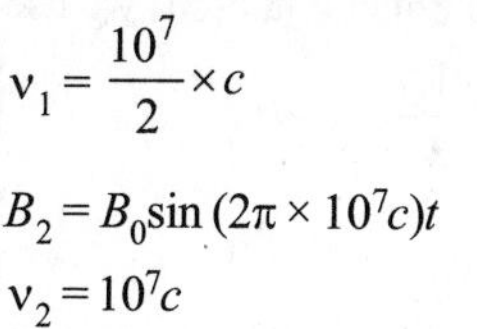

$$\nu_1 = \frac{10^7}{2}\times c$$

$$B_2 = B_0\sin(2\pi\times10^7c)t$$

$$\nu_2 = 10^7c$$

Where c is speed of light

As $\nu_2 > \nu_1$ so KE of photoelectron will be maximum for photon of higher energy $\nu_2 = 10^7c$ Hz

We use

$$h\nu = \phi + KE_{\text{max}} \quad \dots(1)$$

Energy of photon incident is

$$E_{\text{ph}} = h\nu = 6.6\times10^{-34}\times10^7\times3\times10^9$$

$$\Rightarrow \quad E_{\text{ph}} = 6.6\times3\times10^{-19}\text{ J}$$

$$\Rightarrow \quad E_{\text{ph}} = \frac{6.6\times3\times10^{-19}}{1.6\times10^{-19}}\text{eV} = 12.375\text{ eV}$$

from (1)

$$KE_{\text{max}} = E_{\text{ph}} - \phi$$

$$\Rightarrow \quad KE_{\text{max}} = 12.375 - 4.7$$

$$= 7.675\text{ eV} \approx 7.7\text{ eV}$$

Ans. (A)

Sol. 5 Energy incident on plate per second is

$$E = IA$$

Number of electrons emitted per second are

$$n = \frac{E\times0.1}{10\text{eV}} = \frac{16\times10^{-3}\times1\times10^{-4}\times0.1}{10\times1.6\times10^{-19}}$$

$$= 10^{11} \text{ electrons/sec}$$

Maximum KE of ejected electrons is

$KE = 10 - 5 = 5$ eV **Ans. (D)**

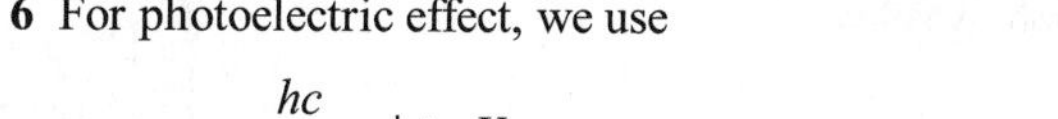

Sol. 6 For photoelectric effect, we use

$$\frac{hc}{\lambda_1} = \phi + eV_1 \quad \dots(\text{i})$$

and
$$\frac{hc}{\lambda_2} = \phi + eV_2 \quad \dots(\text{ii})$$

(i) – (ii)

$$hc\left(\frac{1}{\lambda_1} - \frac{1}{\lambda_2}\right) = e(V_1 - V_2)$$

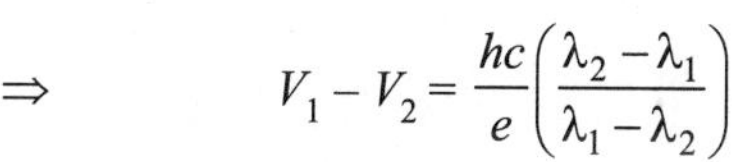

$$\Rightarrow \quad V_1 - V_2 = \frac{hc}{e}\left(\frac{\lambda_2 - \lambda_1}{\lambda_1 - \lambda_2}\right)$$

$$\Rightarrow \quad \Delta V = (1240\text{ nm} - V)\frac{100\text{nm}}{300\text{nm}\times400\text{nm}}$$

$$\Rightarrow \quad \Delta V = 1\text{ V}$$

Ans. (B)

Sol. 7 By photoelectric effect equation, we use

$$h_{\nu} = \phi + \frac{eV_0}{2} \quad \ldots (1)$$

and $$\frac{h\nu}{2} = \phi + eV_0 \quad \ldots (2)$$

Above equations cannot holds so given data is incorrect.

Ans. (None)

Sol. 8 Frequency of the incident light is

$$\nu = \frac{\omega}{2\pi} = \frac{2\pi \times 6 \times 10^{14}}{2\pi} = 6 \times 10^{14}\,\text{Hz}$$

Light wavelength is

$$\lambda = \frac{c}{\nu} = \frac{3\times10^8}{6\times10^{14}} = 5000\,\text{Å}$$

Photo energy $E = \dfrac{12375}{5000} = 2.475\,\text{eV}$

KE (max) of ejected photoelectrons is

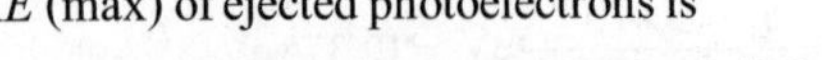

$$KE_{\max} = E - \phi = 2.475 - 2 = 0.475\,\text{eV}$$

Stopping potential

$$V = \frac{KE_{\max}}{e} = 0.475\,\text{V} \simeq 0.48\,\text{V}$$ **Ans. (A)**

Sol. 9 Max *KE* of ejected photoelectrons is given as

$$K_{\max} = \frac{hc}{\lambda} - \phi$$

$$\Rightarrow \quad K_{\max} = \frac{hc}{\lambda} - \frac{hc}{\lambda_0} = \frac{hc}{260\text{ nm}} - \frac{hc}{380\text{ nm}}$$

$$\Rightarrow \quad K_{\max} = 1237\left[\frac{1}{260} - \frac{1}{380}\right]$$

$$= 1237\left[\frac{380-260}{3800\times260}\right]\text{eV} \simeq 1.5\,\text{eV}$$ **Ans. (A)**

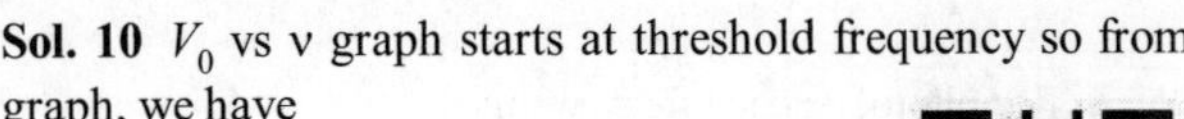

Sol. 10 V_0 vs ν graph starts at threshold frequency so from graph, we have

$$\Rightarrow \quad \nu_{\text{th}} = 4\times10^{14}\,\text{Hz}$$

So work function for this metal is given as

$$\phi = h\nu_{\text{th}} = \frac{6.63\times10^{-34}\times4\times10^{14}}{1.6\times10^{-19}}\,\text{eV}$$

$$\Rightarrow \quad \phi = 1.657\,\text{eV} \simeq 1.66\,\text{eV}$$ **Ans. (C)**

Sol. 11 Total photons incident on surface are

$$n = \frac{I\lambda}{hc}\times A$$

$$\Rightarrow \quad n = \frac{6.4\times10^{-5}\times1}{\frac{1240}{310}\times1.6\times10^{-19}} = \frac{6.4\times10^{-5}\times310\times1}{1240\times1.6\times10^{-19}}$$

$$\Rightarrow \quad n = 10^{14}\text{ photons/sec}$$

Number of electrons emitted from surface per second are

$$N_e = 10^{14}\times10^{-3} = 10^{11}\text{ per/sec}$$

$$\Rightarrow \quad x = 11$$ **Ans. (11)**

Sol. 12 Given that $\lambda_B = 2\lambda_A$

$$\Rightarrow \quad \frac{h}{\sqrt{2T_Bm}} = \frac{2h}{\sqrt{2T_Am}}$$

$$\Rightarrow \quad T_A = 4T_B \quad \ldots (1)$$

Also given that

$$T_B = (T_A - 1.5)\,\text{eV} \quad \ldots (2)$$

$$\Rightarrow \quad 3T_B = 1.5\,\text{eV}$$

$$\Rightarrow \quad T_B = 0.5\,\text{eV}$$

and $$0.5\,\text{eV} = 4.5\,\text{eV} - \phi_B$$

$$\Rightarrow \quad \phi_B = 4\,\text{eV}$$ **Ans. (C)**

Sol. 13 Using dimensions of given physical quantities, we have

$$[V_0] = [h]^x[c]^y[G]^z[A]^w$$

$$\Rightarrow \quad \frac{\text{ML}^2\text{T}^{-2}}{\text{AT}} = (\text{ML}^2\text{T}^{-1})^x(\text{LT}^{-1})^y(\text{M}^{-1}\text{L}^3\text{T}^{-2})^zA^w$$

Comparing dimensions at LHS and RHS, we have

$\Rightarrow$ $w = -1, x = 0, z = 1$ and $y = 5$

$$\Rightarrow \quad [V_0] = h^0c^5G^{-1}A^{-1}$$

So Bonus. **Ans. (Bonus)**

Sol. 14 If the metal surface work function is ϕ then maximum kinetic energy of ejected electrons will be

$$\text{KE}_{\max} = \frac{hc}{\lambda} - \phi$$

In magnetic field, radius of circular motion will be

$$R = \frac{mv}{qB} = \frac{\sqrt{2mKE_{\max}}}{qB} = \frac{\sqrt{2m\left(\frac{hc}{\lambda} - \phi\right)}}{qB}$$

$$\Rightarrow \quad \frac{R^2q^2B^2}{2m} = \frac{hc}{\lambda} - \phi$$

$$\Rightarrow \quad \phi = \frac{hc}{\lambda} - \frac{R_{\max}^2q^2B^2}{2m} = 1.0899\,\text{eV} \approx 1.1\,\text{eV}$$ **Ans. (B)**

Sol. 15 We use for given data

$$\frac{hc}{\lambda} = \frac{hc}{\lambda_{\text{th}}} + eV \quad \ldots (1)$$

and $$\frac{hc}{3\lambda} = \frac{hc}{\lambda_{\text{th}}} + \frac{eV}{4} \quad \ldots (2)$$

$4 \times (2) - (1)$ gives

$$\frac{1}{3}\frac{hc}{\lambda} = \frac{3hc}{\lambda_{th}}$$

$$\Rightarrow \quad \lambda_{th} = 9\lambda$$

$$\Rightarrow \quad n = 9$$ **Ans. (9)**

Sol. 16 By photoelectric effect equation, we use

$$\frac{hc}{\lambda} = \phi + k_m$$

We use $$\frac{hc}{500\text{ nm}} = \phi + k_m \quad \ldots(1)$$

and $$\frac{hc}{200\text{ nm}} = \phi + 3k_m \quad \ldots(2)$$

from (1) and (2) we get

$$3 = \frac{\frac{hc}{200\text{ nm}} - \phi}{\frac{hc}{500\text{ nm}} - \phi}$$

We use $$hc = 1243.1\text{ eV} - \text{nm}$$

$$\Rightarrow \quad 3 \times \frac{1243.1}{500} - 3\phi = \frac{1243.1}{200} - \phi$$

$$\Rightarrow \quad 2\phi = 1243.1\left[\frac{3}{500} - \frac{1}{200}\right]$$

$$\Rightarrow \quad \phi = \frac{1243.1}{1000}[6-5]$$

$$\Rightarrow \quad \phi = \frac{1.243}{2} = 0.612\text{ eV}$$ **Ans. (A)**

Sol. 17 From the given graph, we use

$$hv = \phi + eV_0$$

At point B, $V_0 = 0$,

$$v = 5.5 \times 10^{14}\text{ Hz}$$

$$\Rightarrow \quad h(5.5 \times 10^{14}) = \phi$$

$$\Rightarrow \quad \phi = \frac{6.62 \times 10^{-34} \times 5.5 \times 10^{14}}{1.6 \times 10^{-19}}$$

$$\Rightarrow \quad \phi = 2.27\text{ eV}$$ **Ans. (A)**

Sol. 18 With variation in intensity we consider change in photon flux at constant frequency so the graph does not change. **Ans. (C)**

Sol. 19 For emitted photoelectrons, we use

$$E_1 = \phi + k_1$$

and $$E_2 = \phi + k_2$$

As $$\frac{v_1}{v_2} = 2$$

$$\Rightarrow \quad \frac{k_1}{k_2} = \frac{v_1^2}{v_2^2} = 4$$

Given that $$\frac{k_1}{k_2} = 4 = \frac{4-\phi}{2.5-\phi}$$

$$\Rightarrow \quad 10 - 4\phi = 4 - \phi$$

$$\Rightarrow \quad \phi = 2\text{ eV}$$ **Ans. (2)**

Sol. 20 By photoelectric effect equation, we have

$$\Rightarrow \quad \frac{hc}{\lambda} = \phi + eV_s$$

$$\Rightarrow \quad \frac{1240}{491} = \phi + 0.71 \quad \ldots(1)$$

$$\Rightarrow \quad \frac{1240}{\lambda} = \phi + 1.43 \quad \ldots(2)$$

$$\Rightarrow \quad \lambda = 382\text{ nm}$$ **Ans. (A)**

Sol. 21 By photoelectric effect equation, we have

$$KE_{max} = h\nu - \phi$$

$$\Rightarrow \quad \frac{1}{2}mv^2 = h\nu - \phi$$

$$\Rightarrow \quad v = \sqrt{\frac{2(h\nu - \phi)}{m}}$$

If ν_1 and ν_2 are the frequencies of the incident radiations, it is given that

$$h\nu_1 = 2\phi$$

and $$h\nu_2 = 10\phi$$

$$\Rightarrow \quad \frac{v_1}{v_2} = \sqrt{\frac{h\nu_1 - \phi}{h\nu_2 - \phi}}$$

$$\Rightarrow \quad \frac{v_1}{v_2} = \sqrt{\frac{2\phi - \phi}{10\phi - \phi}} = \frac{1}{3}$$ **Ans. (1)**

Sol. 22 Stopping potential changes linearly with frequency of incident radiation and it does not depend on any other listed physical quantitites. **Ans. (D)**

Sol. 23 Force due to radiation pressure on the surface is calculated as

$$F = \frac{IA}{C}$$

$$\Rightarrow \quad I = \frac{FC}{A} = \frac{2.5 \times 10^{-6} \times 3 \times 10^8}{30} = 25\text{ W/cm}^2$$ **Ans. (25)**

Sol. 24 By photoelectric effect equation in the two cases, we use

$$\frac{1}{2}mv_1^2 = hf_1 - \phi$$

and $$\frac{1}{2}mv_2^2 = hf_2 - \phi$$

$$\Rightarrow \quad v_1^2 - v_2^2 = \frac{2h}{m}[f_1 - f_2]$$ **Ans. (A)**

Sol. 25 Energy of radiation from hydrogen atom is given as

$$E_p = 13.6\left[\frac{1}{R_1^2} - \frac{1}{R_2^2}\right]\text{eV} = 13.6\left[\frac{1}{4} - \frac{1}{9}\right]\text{eV}$$

$$\Rightarrow \quad E_p = 1.89\,\text{eV}$$

Speed of electrons moving in circular path in magnetic field is given as

$$v = \frac{RqB}{m}$$

$$\Rightarrow \quad v = \frac{7\times10^{-3}\times1.6\times10^{-19}\times5\times10^{-4}}{9.1\times10^{-31}} = 6.15\times10^5$$

Maximum kinetic energy of electrons is given as

$$KE_{max} = \frac{1}{2}mv^2$$

$$\Rightarrow \quad KE_{max} = \frac{1}{2}\times\frac{9.1\times10^{-31}\times(6.15\times10^5)}{1.6\times10^{-19}}\text{eV} = 1.075\,\text{eV}$$

For Gold plate, we use

$$\phi = E_p - KE_{max}$$

$$\Rightarrow \quad \phi = 1.89 - 1.075;\ \phi = 0.82\,\text{eV}$$ **Ans. (A)**

Sol. 26 By photoelectric effect equation, we have

$$KE = h\nu - \phi$$

$$\Rightarrow \quad eV = \frac{hc}{\lambda} - \phi$$

For first case when stopping potential is $3V_0$, we have

$$e(3V_0) = \frac{hc}{\lambda} - \phi \quad \ldots(1)$$

For second case when stopping potential is $2V_0$, we have

$$e(2V_0) = \frac{hc}{2\lambda} - \phi \quad \ldots(2)$$

From equations-(1) and (2), we use

$$\phi = \frac{hc}{4\lambda} = \frac{hc}{\lambda_{th}}$$

$$\Rightarrow \quad \frac{hc}{4\lambda} = \frac{hc}{\lambda_{th}}$$

$$\Rightarrow \quad \lambda_{th} = 4\lambda$$ **Ans. (4)**

Sol. 27 By photoelectric effect equation, we have

$$KE = h\nu - \phi$$

$$\Rightarrow \quad eV = \frac{hc}{\lambda} - \phi$$

For first case, we use

$$4.8 = \frac{hc}{\lambda_0} - \phi \quad \ldots(1)$$

For second case, we use

$$1.6 = \frac{hc}{2\lambda_0} - \phi \quad \ldots(2)$$

From equation (1) and (2)

$$\phi = \frac{hc}{4\lambda_0}$$

For threshold wavelength λ_{th} we have

$$\phi = \frac{hc}{\lambda_{th}}$$

$$\Rightarrow \quad \frac{hc}{4\lambda_0} = \frac{hc}{\lambda_{th}}$$

$$\Rightarrow \quad \lambda_{th} = 4\lambda_0$$ **Ans. (B)**

Sol. 28 Using photoelectric effect equation, we have

$$\frac{hc}{\lambda} = \phi + KE_{max}$$

$$\Rightarrow \quad \frac{1240}{500} = 1.25 + KE_{max}$$

$$\Rightarrow \quad KE_{max} = 1.23\,\text{eV}$$

Radius of circular motion of electron in magnetic field is given as

$$R = \frac{mv}{qB} = \frac{\sqrt{2mKE}}{qB}$$

$$\Rightarrow \quad B = \frac{\sqrt{2mKE}}{qR}$$

$$\Rightarrow \quad B = \frac{\sqrt{2\times9.1\times10^{-31}\times1.23\times1.6\times10^{-19}}}{1.6\times10^{-19}\times30\times10^{-2}}$$

$$\Rightarrow \quad B = 0.125\times10^{-4} = 125\times10^{-7}$$ **Ans. (125)**

Sol. 29 Initial kinetic energy of electron is given as 3 eV. After the proton captures the electron, hydrogen atom in second excited state is formed i.e. in $n = 3$. In this state final energy of electron is – 1.51 eV. Thus the energy taken out by photon is given as

$$E_p = 3 + 1.51 = 4.51\,\text{eV}$$

Work function of photosensititve metal is given as

$$\phi = \frac{12431}{4000} = 3.1\text{ eV}$$

By photoelectric effect equation, the maximum kinetic energy of ejected electron is given as

$$K_{max} = E_P - \phi$$

$$\Rightarrow \quad K_{max} = 4.51 - \frac{12431}{4000}$$

$$\Rightarrow \quad K_{max} = 4.51 - 3.1 = 1.41\text{ eV}$$

Ans. (B)

Sol. 30 Photoelectric effect equation is written as

$$eV_0 = \frac{hc}{\lambda} - \phi$$

For the two cases, we have

$$eV_1 = \frac{hc}{280nm} - \phi$$

and $$eV_2 = \frac{hc}{400\text{nm}} - \phi$$

$$\Rightarrow \quad e(V_1 - V_2) = 1240\left(\frac{120}{280 \times 400}\right) = 1.3\text{ eV}$$

Ans. (C)

Sol. 31 The amplitude of electric field

$$E_0 = 200\text{ V/m}$$

Intensity of wave is given as

$$I = \frac{1}{2} e_0 E_0^2 . c$$

Radiation pressure due to wave is given as

$$P = \frac{2I}{c} = \left(\frac{2}{c}\right)\left(\frac{1}{2}\varepsilon_0 E_0^2 c\right) = e_0 E_0^2$$

$$\Rightarrow \quad P = 8.85 \times 10^{-12} \times 200^2$$

$$\Rightarrow \quad P = 8.85 \times 10^{-8} \times 4 = \frac{354}{10^9}$$

Ans. (1)

Sol. 32 The number of photons incident per second on a surface depends on the intensity of light as

$$N = \frac{IA}{h\nu}$$

Thus with increase in light intensity, number of photons increases but it does not change the kinetic energy of photoelectrons as frequency is not changing. **Ans. (A)**

Sol. 33 From photoelectric effect equation, we have

$$eV_s = \frac{hc}{\lambda} - \phi$$

$$\Rightarrow \quad e(V' - V) = hc\left(\frac{1}{\lambda'} - \frac{1}{\lambda}\right)$$

$$\Rightarrow \quad V' = \frac{hc}{e}\left(\frac{1}{\lambda'} - \frac{1}{\lambda}\right) + V$$

$$\Rightarrow \quad V' = 1240\left(\frac{1}{474} - \frac{1}{670}\right) + 0.48 = 1.25\text{ eV}$$

Ans. (C)

Sol. 34 Using conservation of energy, we have

$$K_e = E_{H\text{-atom}} + E_{\text{photon}}$$

$$\Rightarrow \quad 2.6 = \frac{-13.6}{4} + h\nu$$

$$\Rightarrow \quad h\nu = 6\text{ eV}$$

$$\Rightarrow \quad \nu = \frac{6 \times 1.6 \times 10^{-19}}{6.626 \times 10^{-34}}$$

$$\Rightarrow \quad \nu = 1.45 \times 10^{15}\text{ Hz} = 1.45 \times 10^{9}\text{ MHz}$$

Ans. (B)

25.2 Matter Waves and X-Ray

Sol. 1

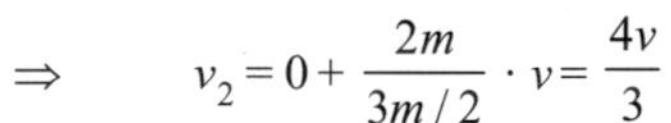

For head-on electric collision we use

$$v_1 = \left(\frac{m_1 - m_2}{m_1 + m_2}\right)u_1 + \left(\frac{2m_2}{m_1 + m_2}\right)u_2$$

$$\Rightarrow \quad v_1 = \frac{m/2}{3m/2} \cdot v + 0 = \frac{v}{3}$$

$$\Rightarrow \quad v_2 = \left(\frac{m_1 - m_1}{m_1 + m_2}\right)u_2 + \left(\frac{2m_1}{m_1 + m_2}\right)u_1$$

$$\Rightarrow \quad v_2 = 0 + \frac{2m}{3m/2} \cdot v = \frac{4v}{3}$$

After collision de-Broglie wavelength of A and B are

$$\Rightarrow \quad \lambda_A = \frac{h}{mv_1} \text{ and } l_B = \frac{h}{\left(\frac{m}{2}\right)v_2} = \frac{3h}{2mv}$$

$$\Rightarrow \quad \frac{\lambda_A}{\lambda_B} = \frac{3h/mv}{3h/2mv} = 2$$

Ans. (D)

Sol. 2 For continuous X-Ray the cut of wavelength λ_{min} is related to the voltage applied in discharge tube V as

$$\lambda_{min} = \frac{hc}{eV} = \frac{12431}{V}\text{ Å}$$

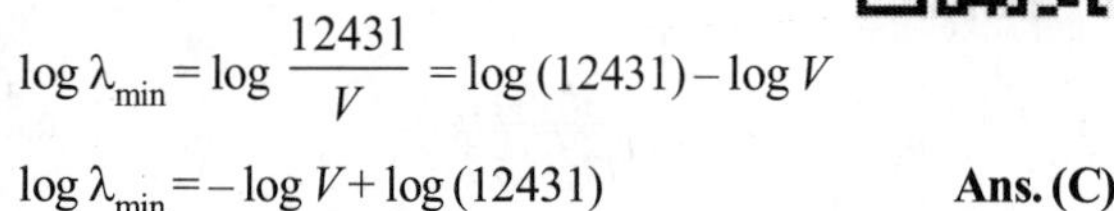

Taking log on both sides

$$\log \lambda_{min} = \log\frac{12431}{V} = \log(12431) - \log V$$

$$\Rightarrow \quad \log \lambda_{min} = -\log V + \log(12431)$$

Ans. (C)

Sol. 3 It is given that de-Broglie wavelength for proton and α-particle are equal

$$\Rightarrow \quad \lambda_P = \lambda_\alpha$$

$$\Rightarrow \quad \frac{h}{m_\rho v_\rho} = \frac{h}{m_\alpha v_\alpha}$$

$$\Rightarrow \quad \frac{v_\rho}{v_\alpha} = \frac{m_\alpha}{m_\rho} = \frac{4m_\rho}{m_\rho} = \frac{4}{1} \quad [\text{As } m_\alpha = 4m_\rho]$$ **Ans. (C)**

Sol. 4

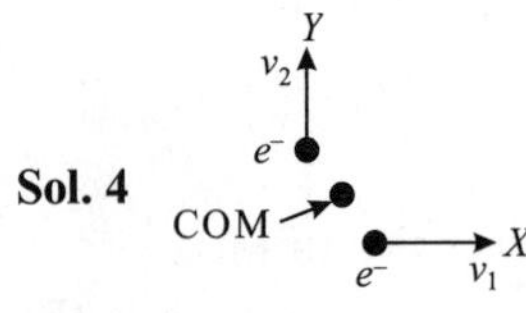

As $\vec{v}_1 = v_1\hat{i} = \frac{h}{m\lambda_1}\hat{i}$

$$\vec{v}_2 = v_2\hat{i} = \frac{h}{m\lambda_2}\hat{j}$$

We also use

$$\lambda_1 = \frac{h}{mv_1} \text{ and } \lambda_2 = \frac{h}{mv_2}$$

Vector speed of COM of this two electron system is given as

$$\vec{V_C} = \frac{m\vec{v}_1 + m\vec{v}_2}{m+m} = \frac{1}{2}(\vec{v}_1 + \vec{v}_2)$$

$$\Rightarrow \quad \vec{V_C} = \frac{1}{2}\frac{h}{m}\left(\frac{1}{\lambda_1}\hat{i} + \frac{1}{\lambda_2}\hat{j}\right)$$

If reference frame of COM, velocity of first electron travelling in X-direction is given as

$$\vec{V}_{1C} = \vec{v}_1 - \vec{v}_2$$

$$\Rightarrow \quad \vec{V}_{1C} = \frac{h}{m\lambda_1}\hat{i} - \left(\frac{h}{2m\lambda_1}\hat{i} + \frac{h}{2m\lambda_2}\hat{j}\right)$$

$$\Rightarrow \quad \vec{V}_{1C} = \frac{h}{2m\lambda_1}\hat{i} - \frac{h}{2m\lambda_2}\hat{j}$$

$$\Rightarrow \quad |\vec{V}_{1C}| = \frac{h}{2m}\sqrt{\frac{1}{\lambda_1^2 + \frac{1}{\lambda_2^2}}}$$

[speed of first electron in frame of COM]

de-Broglie wavelength of first electron in COM frame is given as

$$\lambda_{1C} = \frac{h}{mv_{1C}} = \frac{h}{m} \times \frac{1}{\frac{h}{2m}\sqrt{\frac{1}{\lambda^2} + \frac{1}{\lambda_2^2}}}$$

$$\Rightarrow \quad \lambda_{1C} = \frac{2\lambda_1\lambda_2}{\sqrt{\lambda_1^2 + \lambda_2^2}}$$ **Ans. (C)**

Sol. 5 By Bohr's II Postulate we have

$$mvr = \frac{nh}{2\pi} \Rightarrow mv = \frac{nh}{2\pi r}$$

de-Broglie wavelength of electron in second excited state ($n = 3$), we use

$$\lambda = \frac{h}{mv} = \frac{2\pi h}{n} = \frac{2\pi}{n}(0.529\, n^2)\ [\text{as } z = 1]$$

$$\Rightarrow \quad \lambda = 1.06\,\pi n \Rightarrow \lambda \propto n$$

As $n = 3$ for λ_B and $n = 1$ for λ_G so we use

$$\frac{\lambda_B}{\lambda_G} = \frac{3}{1}$$

$$\Rightarrow \quad \lambda_B = 3\lambda_G$$ **Ans. (D)**

Sol. 6 deBroglie wavelength of electrons is

$$\lambda = \frac{h}{p} \qquad [\lambda = 7.5 \times 10^{-12}]$$

$$\Rightarrow \quad p = \frac{h}{\lambda}$$

Kinetic energy of electrons is

$$KE = \frac{p^2}{2m} = \frac{(h/\lambda)^2}{2m} = \frac{\left\{\frac{6.6\times10^{-34}}{7.5\times10^{-12}}\right\}}{2\times9.1\times10^{-31}}\text{ J}$$

$$\Rightarrow \quad KE = 25\text{ keV}$$ **Ans. (C)**

Sol. 7 Given that de-Broglie wavelength of electron is

$$\lambda_e = \frac{h}{mv} = 10^{-3}\left(\frac{3\times10^8}{6\times10^{14}}\right)$$

$$\Rightarrow \quad v = \frac{6.63\times10^{-34}\times6\times10^{14}}{9.1\times10^{-31}\times3\times10^5}$$

$$\Rightarrow \quad v = 1.45\times10^6\text{ m/s}$$ **Ans. (A)**

Sol. 8 De-Broglie wavelength of an accelerated charge particle is given as

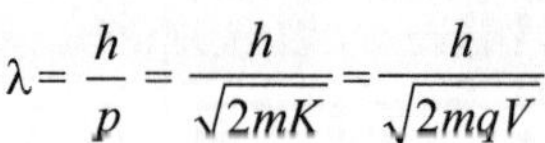

$$\lambda = \frac{h}{p} = \frac{h}{\sqrt{2mK}} = \frac{h}{\sqrt{2mqV}}$$

$$\Rightarrow \quad \frac{\lambda_A}{\lambda_B} = \frac{\sqrt{2m_Bq_BV_B}}{\sqrt{2m_Aq_AV_A}}\sqrt{\frac{4m\times q\times2500}{m\times q\times50}} = 2\sqrt{50}$$

$$\Rightarrow \quad \frac{\lambda_A}{\lambda_B} = 2\times7.07 = 14.14$$ **Ans. (B)**

Sol. 9 Momentum of the two particle are given as

$$\frac{h}{\lambda_1} = p_1 \qquad p_2 = \frac{h}{\lambda_2}$$

$$\vec{p}_1 = \frac{h}{\lambda_1}\hat{i}$$

and $$\vec{p}_2 = \frac{h}{\lambda_2}\hat{j}$$

Using momentum conservation before and after collision we have

$$\vec{p} = \vec{p}_1 + \vec{p}_2$$

$$\vec{p} = \frac{h}{\lambda_1}\hat{i} + \frac{h}{\lambda_2}\hat{j}$$

$$|\vec{p}| = \sqrt{\left(\frac{h}{\lambda_1}\right)^2 + \left(\frac{h}{\lambda_2}\right)^2}$$

de-Broglie wavelength of final particle λ is given as

$$\frac{h}{\lambda} = \sqrt{\left(\frac{h}{\lambda_1}\right)^2 + \left(\frac{h}{\lambda_2}\right)^2}$$

$$\Rightarrow \quad \frac{1}{\lambda^2} = \frac{1}{\lambda_1^2} + \frac{1}{\lambda_2^2} \quad \textbf{Ans. (D)}$$

Sol. 10

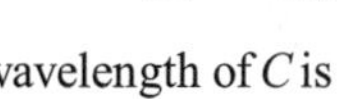

By conservation of momentum

$$mv_0 = \frac{m}{2}v - \frac{mv}{4} = \frac{mv}{4}$$

$$\Rightarrow \quad v = 4v_0$$

De-Broglie wavelength of B is

$$\lambda_B = \frac{2h}{mv} = \frac{2h}{4mv_0} = \frac{\lambda_A}{2}$$

De-Broglie wavelength of C is

$$\lambda_C = \frac{4h}{mv} = \frac{4h}{4mv_0} = \lambda_A \quad \textbf{Ans. (D)}$$

Sol. 11

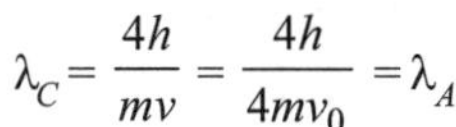

By momentum conservation before and after collision, we have

$$p_x - p_y = p_P$$

$$\Rightarrow \quad \frac{h}{\lambda_x} - \frac{h}{\lambda_y} = \frac{h}{\lambda_P}$$

$$\Rightarrow \quad \lambda_P = \frac{\lambda_x \lambda_y}{|\lambda_y - \lambda_x|} \quad \textbf{Ans. (C)}$$

Sol. 12

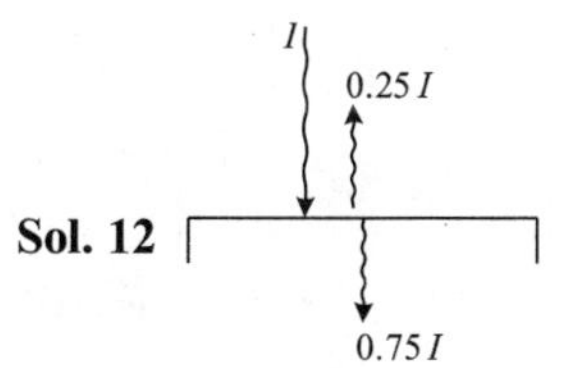

Pressure on surface due to incident, reflected and absorbed radiation is

$$p = 0.25\left[\frac{2I}{c}\right] + 0.75\left[\frac{I}{c}\right]$$

$$= 1.25\frac{I}{c} = 1.25 \times \frac{50}{3\times10^8} = 20.83 \times 10^{-8}\,\text{N} \quad \textbf{Ans. (D)}$$

Sol. 13 Radiation pressure on surface is

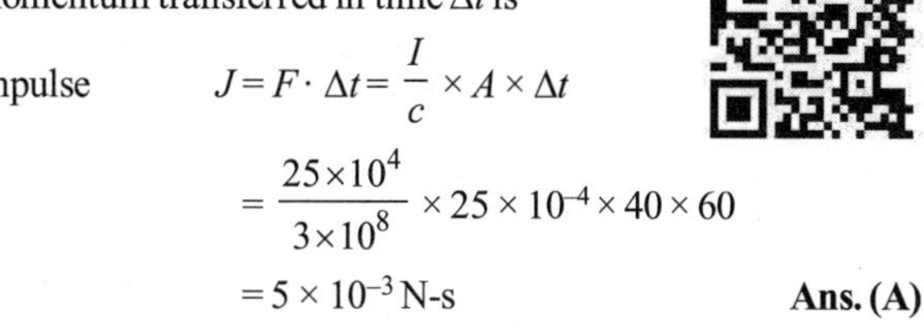

$$p = \frac{I}{c}$$

Momentum transferred in time Δt is

Impulse $$J = F \cdot \Delta t = \frac{I}{c} \times A \times \Delta t$$

$$= \frac{25\times10^4}{3\times10^8} \times 25 \times 10^{-4} \times 40 \times 60$$

$$= 5 \times 10^{-3}\,\text{N-s} \quad \textbf{Ans. (A)}$$

Sol. 14 From an energy source of power P_{watt}, number of photons emitted/sec are

$$n = \frac{P\lambda}{hc}$$

$$= \frac{2\times10^{-3}\times500\times10^{-9}}{6.6\times10^{-34}\times3\times10^8} \simeq 1.5 \times 10^{16}\,\text{ph/s} \quad \textbf{Ans. (B)}$$

Sol. 15 Photon energy is given as

$$E = \frac{hc}{\lambda_{\text{photon}}} \qquad \ldots(1)$$

and $$\lambda_{\text{electron}} = \frac{h}{\sqrt{2mE}} \qquad \ldots(2)$$

From (1) and (2) we have

$$\Rightarrow \quad \frac{\lambda_{\text{electron}}}{\lambda_{\text{photon}}} = \frac{1}{c}\left(\frac{E}{2m}\right)^{1/2} \quad \textbf{Ans. (B)}$$

Sol. 16 Using de-Broglie hypothesis, wavelength of electron is given as

$$\lambda = \frac{h}{mv}$$

$$\Rightarrow \quad \lambda_0 = \frac{h}{m\sqrt{2}v_0} \quad [\text{As } v = \sqrt{2}v_0] \qquad \ldots(1)$$

After time t, we use

$$\lambda' = \frac{h}{\sqrt{v_0^2 + v_0^2 + \left(\frac{eE_0t}{m}\right)^2}}$$

$$\Rightarrow \quad \lambda' = \frac{h}{m\sqrt{2v_0^2 + \frac{e^2E_0^2t^2}{m^2}}} \qquad \ldots(2)$$

From (1) and (2), we use

$$\Rightarrow \quad \lambda' = \frac{\lambda_0}{\sqrt{1+\dfrac{e^2E_0^2t^2}{2m^2v_0^2}}} \qquad \textbf{Ans. (C)}$$

Sol. 17 de-Broglie wavelength of particle of kinetic energy E is given as

$$\lambda = \frac{h}{\sqrt{2mE}} \qquad \ldots(1)$$

and
$$\frac{\lambda}{2} = \frac{h}{\sqrt{2m(E+\Delta E)}} \qquad \ldots(2)$$

From (1) and (2), we have

$$\frac{E+\Delta E}{E} = 4$$

$$\Rightarrow \quad \Delta E = 3E \qquad \textbf{Ans. (C)}$$

Sol. 18 Acceleration of electron due to electric field is

$$a = \frac{eE}{m}$$

Speed of electron after time t is given as

$$v = u + at = \left(\frac{eE}{m}\right)t$$

de Broglie wavelength of electron is

$$\lambda = \frac{h}{mv} = \frac{h}{eEt}$$

$$\Rightarrow \quad \frac{d\lambda}{dt} = -\frac{h}{|e|Et^2} \qquad \textbf{Ans. (A)}$$

Sol. 19 deBroglie wavelength of particle is given as

$$\lambda_d = \frac{h}{p} = \frac{h}{mv}$$

$$\Rightarrow \quad \lambda_{dp} = \frac{h}{m(5v)} \qquad \ldots(1)$$

$$\Rightarrow \quad \lambda_{de} = \frac{h}{m_e v} \qquad \ldots(2)$$

Given that

$$\frac{\lambda_{dp}}{\lambda_{de}} = 1.878\times10^{-4}$$

$$\Rightarrow \quad m = \frac{m_e}{5\times1.878\times10^{-4}}$$

$$\Rightarrow \quad m = \frac{9.1\times10^{-31}}{5\times1.878\times10^{-4}}$$

$$\Rightarrow \quad m = 9.7\times10^{-28}\text{ kg} \qquad \textbf{Ans. (D)}$$

Sol. 20 Number of photons emitted per second are given as

$$N_1 = \frac{P\lambda_1}{hc}$$

and
$$N_2 = \frac{P\lambda_2}{hc}$$

$$\Rightarrow \quad \frac{N_1}{N_2} = \frac{\lambda_1}{\lambda_2} = \frac{1}{500} \qquad \textbf{Ans. (A)}$$

Sol. 21 Before Collision

$m/2$ (A) → V_0 ; $m/3$ (B)

After Collision

A → V_A ; B → V_B

By momentum conservation, we use

$$\frac{m}{2}V_0 = \frac{m}{2}V_A + \frac{m}{3}V_B$$

$$\Rightarrow \quad V_0 = V_A + \frac{2V_B}{3} \qquad \ldots(1)$$

As collision is elastic

$$V_0 = V_B - V_A \qquad \ldots(2)$$

Solving (1) & (2), we get

$$V_A = \frac{V_0}{5} \text{ and } V_B = \frac{6V_0}{5}$$

De-Broglie wavelength of A before collision

$$\lambda_{DA} = \frac{2h}{mV_0}$$

De-Broglie wavelength A after collision

$$\lambda'_{DA} = \frac{h}{mV_A} = \frac{10h}{mV_0}$$

$$\Rightarrow \quad \Delta\lambda = \lambda'_{DA} - \lambda_{DA} = \frac{8h}{mv_0} = 4\lambda_{DA} = 4\lambda_0 \qquad \textbf{Ans. (A)}$$

Sol. 22 For a particle of kinetic energy E, de-Broglie wavelength is given as

$$\lambda = \frac{h}{p} = \frac{h}{\sqrt{2mE}}$$

$$\Rightarrow \quad \lambda \propto \frac{1}{\sqrt{m}}$$

As $\quad m_{He^{++}} > m_p > m_e$

$$\Rightarrow \quad \lambda_{He^{++}} < \lambda_p < \lambda_e \qquad \textbf{Ans. (D)}$$

Sol. 23 RMS speed of molecules is given as

$$v_{rms} = \sqrt{\frac{3RT}{M}} = \sqrt{\frac{3KT}{m}}$$

Here $M = mN_{AV}$ where m is mass of one molecule and $R = k \cdot N_{AV}$

de-Broglie wavelength of molecules is given as

$$\lambda = \frac{h}{mv}$$

$$\Rightarrow \quad \lambda = \frac{h}{m\sqrt{\frac{3KT}{m}}}$$

$$\Rightarrow \quad \lambda = \frac{h}{\sqrt{3KTm}}$$

$$\Rightarrow \quad \lambda = \frac{6.63\times10^{-34}}{\sqrt{3\times1.38\times10^{-23}\times400\times\left(\frac{28\times10^{-3}}{6.023\times10^{-23}}\right)}}$$

$$\Rightarrow \quad \lambda = \frac{6.63\times10^{-11}}{2.77} = 2.39\times10^{-11}\,\text{m}$$

$\Rightarrow \quad \lambda \simeq 0.24$ Å **Ans. (B)**

Sol. 24 If linear momentum of photons are equal then it can be considered that wavelengths are also equal because wavelength is inversely proportional to momentum so Statement-I is true. On decreasing wavelength, momentum and energy of photon increases so Statement-II is false. **Ans. (C)**

Sol. 25 The shortest wavelength of the photon produced in X-Ray tube can be given as

$$\lambda_{min} = \frac{1240}{\Delta V}\text{(nm)}$$

$$\Rightarrow \quad \lambda_{min} = \frac{1240}{1.24\times10^6} = 10^{-3}\text{nm}$$ **Ans. (C)**

Sol. 26 de-Broglie wavelength of a moving particle is given as

$$\lambda = \frac{h}{mv}$$

Given that $\lambda_P = \lambda_\alpha$

$\Rightarrow \quad m_P v_P = m_\alpha v_\alpha$

$\Rightarrow \quad m_P v_P = 4m_p v_\alpha$ (As $m_\alpha = 4m_P$)

$\Rightarrow \quad \frac{v_P}{v_\alpha} = 4$ **Ans. (D)**

Sol. 27 de-Broglie wavelength of particles can be given as

$$\lambda = \frac{h}{p} = \frac{h}{\sqrt{2mqV}}$$

$$\Rightarrow \quad \frac{\lambda_p}{\lambda_\alpha} = \sqrt{\frac{m_\alpha q_\alpha}{m_p q_p}} = \sqrt{\frac{4m_p\times2e}{m_p\times e}}$$

$\Rightarrow \quad \frac{\lambda_p}{\lambda_\alpha} = \sqrt{8} = 2\sqrt{2} == 2\times1.4 = 2.8$ **Ans. (C)**

Sol. 28 The ratio of de-Broglie wavelengths of electron and proton is given as

$$\frac{\lambda_e}{\lambda_p} = \frac{\frac{h}{m_e v}}{\frac{h}{m_p v}} = 1836$$ **Ans. (C)**

Sol. 29 Considering energy of X-Ray beam when equated to the energy of the given mass, we get

$$\frac{hc}{\lambda} = mc^2$$

$\Rightarrow \quad m = \frac{h}{c\lambda}$ **Ans. (10)**

Sol. 30 Resolving power $\propto \frac{1}{\lambda}$

As the wavelength of electron is much less than visible light, its resolving power will be much more. **Ans. (D)**

Sol. 31

(ΔE) Releases when photon going from $n = 5$ to $n = \Delta E = (13.6 - 0.54)\,\text{eV} = 13.06\,\text{eV}$

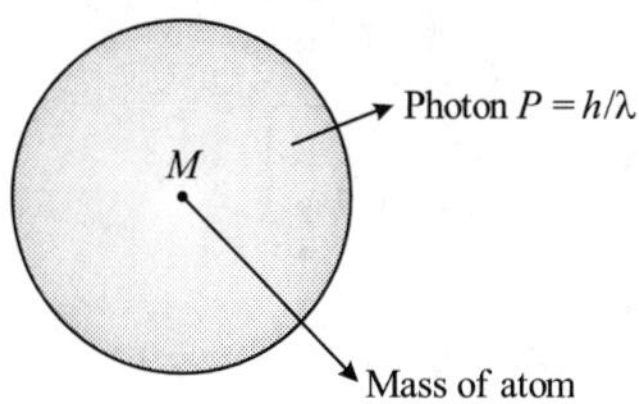

$P_i = P_f$ (By linear momentum conservation)

$$0 = \frac{h}{\lambda} - Mv = V_{Recoil} = \frac{h}{\lambda M} \quad \ldots\text{(i)}$$

& $$\Delta E = \frac{hc}{\lambda} = \frac{hc}{\lambda M}\times M$$

$\Rightarrow \quad McV_{Recoil}$

$$V_{Recoil} = \frac{\Delta E}{Mc} = \frac{13.06\times1.6\times10^{-19}}{1.67\times10^{-27}\times3\times10^8}$$

$= 4.17$ m/sec **Ans. (C)**

Sol. 32 The de-Broglie wavelength of a charge particle is given as

$$\lambda = \frac{h}{mv} = \frac{h}{\sqrt{2mK}} = \frac{h}{\sqrt{2mqV}}$$

$$\Rightarrow \quad \frac{\lambda_1}{\lambda_2} = \sqrt{\frac{m_2}{m_1}}$$

$$\Rightarrow \quad \frac{\lambda_e}{\lambda_P} = \sqrt{\frac{m_P}{m_e}} = \sqrt{1831.4} = 42.79$$ **Ans. (D)**

Sol. 33 Red light and blue light have different wavelength and different frequency in space as both travel at same speed. **Ans. (A)**

Sol. 34 de-Broglie wavelength of electron with kinetic energy E is given as

$$\lambda_1 = \frac{h}{\sqrt{2mE}}$$

Wavelength of photon is given as

$$\lambda_2 = \frac{hc}{E}$$

$$\Rightarrow \quad \frac{\lambda_1}{\lambda_2} = \frac{1}{c}\left(\frac{E}{2m}\right)^{1/2}$$ **Ans. (B)**

Sol. 35 de-Broglie wavelength for the particle travelling at momentum p is given as

$$\lambda = \frac{h}{p}$$

The ratio of de-Broglie wavelengths of particle to that of electron is given as

$$\frac{\lambda_p}{\lambda_e} = \frac{p_e}{p_p} = \frac{m_e v_e}{m_p v_p}$$

$$\Rightarrow \quad 2 = \frac{m_e}{m_p}\left(\frac{v_e}{4v_e}\right)$$

$$\Rightarrow \quad m_p = \frac{m_e}{8}$$ **Ans. (D)**

Sol. 36 De-broglie wavelength of electron can be given as

$$\lambda = \frac{h}{P}$$

$$\Rightarrow \quad P = \frac{h}{\lambda}$$

Kinetic energy of electron in the tube can be given as

$$\Rightarrow \quad E = \frac{P^2}{2m} = \frac{h^2}{2m\lambda^2}$$

Cut-off wavelength of emitted X-Ray can be calculated as

$$E = \frac{hc}{\lambda}$$

$$\Rightarrow \quad \frac{h^2}{2m\lambda^2} = \frac{hc}{\lambda}$$

$$\Rightarrow \quad \lambda = \frac{2mc\lambda^2}{h}$$ **Ans. (D)**

Sol. 37 de-Broglie wavelength associated with the charged particle accelerated through a given potential difference is given as

$$\lambda = \frac{h}{p} = \frac{h}{\sqrt{2mqV}}$$

As it is given that potential difference is same for both electron and proton, so we have

$$\Rightarrow \quad \frac{\lambda_e}{\lambda_p} = \sqrt{\frac{m_p}{m_e}}$$ **Ans. (D)**

Sol. 38 de-Broglie wavelengths of the particles is given as

$$\lambda = \frac{h}{P}$$

As momentum is same for both the particles, we use

$$\frac{\lambda_{2m}}{\lambda_m} = \frac{1}{1}$$ **Ans. (C)**

Sol. 39 The de-Broglie wavelength of a particle with kinetic energy K is given as

$$\lambda = \frac{h}{\sqrt{2mK}}$$

$$\Rightarrow \quad \lambda \propto \frac{1}{\sqrt{m}}$$

As $\quad m_\alpha > m_p > m_e,$

$$\Rightarrow \quad \lambda_e > \lambda_p > \lambda_\alpha$$ **Ans. (D)**

Sol. 40 de-Broglie wavelength of the particle is given as

$$\lambda = \frac{h}{p}$$

Thus kinetic energy of photon is related to its momentum as

$$K_p = mc^2 = pc \quad \ldots(1)$$

Kinetic energy of electron is related to its momentum as

$$K_e = \frac{1}{2}mv^2 = \frac{pv}{2} \quad \ldots(2)$$

$$\Rightarrow \quad \frac{K_e}{K_p} = \frac{pv/2}{pc} = \frac{v}{2c}$$ **Ans. (C)**

Sol. 41 Particle deBroglie wavelength is given as

$$\lambda_{particle} = \frac{h}{p} = \frac{h}{9.1\times10^{-31}\times10^{6}}$$

Wavelength of photon is given as

$$\lambda_{photon} = \frac{h}{p} = \frac{h}{10^{-27}}$$

$\Rightarrow$ $\lambda_{photon}/\lambda_{particle} = 910$ **Ans. (910)**

Sol. 42 From the given data, we have

$$K_\alpha = \frac{hc}{\lambda} = 0.071\times10^{-9} = 17.49\ \text{eV}$$

Energy difference between L and K lines can be given as

$$E_K - E_L = 17.49\ \text{eV}$$

$\Rightarrow$ $E_L = 27.5 - 17.49 \approx 10\ \text{eV}$ **Ans. (10)**

Sol. 43 Momentum of a particle is inversely proportional to its de-Broglie wavelength so its energy is related to wavelength as

$$E \propto \frac{1}{\lambda^2}$$

$\Rightarrow$ $E = \dfrac{k}{\lambda^2}$

If de-Broglie wavlength reduces to 75% of initial value then final energy will become

$$E_1 = \frac{16k}{9\lambda^2},$$

Thus extra energy required will be

$\Rightarrow$ $\Delta E = \dfrac{7}{9}E$ **Ans. (B)**

Sol. 44 From photoelectric effect equation, we have

$$eV_s = \frac{hc}{\lambda} - \phi$$

$\Rightarrow$ $e(V' - V) = hc\left(\dfrac{1}{\lambda'} - \dfrac{1}{\lambda}\right)$

$\Rightarrow$ $V' = \dfrac{hc}{e}\left(\dfrac{1}{\lambda'} - \dfrac{1}{\lambda}\right) + V$

$\Rightarrow$ $V' = 1240\left(\dfrac{1}{474} - \dfrac{1}{670}\right) + 0.48 = 1.25\ \text{eV}$

Ans. (C)

Sol. 45 de-Broglie wavelength of particles is given as

$$\lambda_{deb} = \frac{h}{p} = \frac{h}{\sqrt{2mK}}$$

If λ is same then P will be same so we have

$$K \propto 1/m$$

As $m_p > m_e$

$\Rightarrow$ $K_p < K_e$ **Ans. (A)**

26.1 Atomic Structure

Sol. 1 For first transition, we use

$$\frac{hc}{\lambda_1} = -E - (-2E) = E$$

$\Rightarrow$ $\lambda_1 = \dfrac{hc}{E}$ ...(1)

For second transition, we use

$$\frac{hc}{\lambda_2} = -E - \left(-\frac{4E}{3}\right) = \frac{E}{3}$$

$\Rightarrow$ $\lambda_2 = \dfrac{3hc}{E}$ ...(2)

From equation-(1) and (2)

$$\frac{\lambda_1}{\lambda_2} = \frac{hc/E}{3hc/E} = \frac{1}{3}$$ **Ans. (B)**

Sol. 2 Due to motion of electron in n^{th} orbit, the magnetic induction at centre is given by Biot Savart's law as

$$B = \frac{\mu_0}{4\pi}\frac{qv\sin\theta}{r^2} = \frac{\mu_0}{4\pi}\cdot\frac{ev_n}{r_n^2}$$

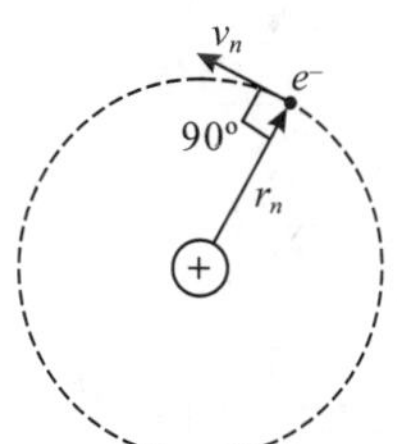

As for n^{th} orbit – for $Z = 1$

$$v_n = \frac{2\pi ke^2}{nh}$$

$$r_n = \frac{n^2h^2}{4\pi^2ke^2m}$$

$\Rightarrow$ $B_n = \dfrac{\mu_0}{4\pi}e\times\dfrac{2\pi ke^2}{nh}\times\left(\dfrac{4\pi^2ke^2m}{n^2h^2}\right)^2$

$\Rightarrow$ $B_n \propto n^{-5}$ **Ans. (B)**

Sol. 3 From Bohr's II Postulate we use

$$mvr = \frac{h}{2\pi}\ [n=1]$$ ...(1)

Acceleration of electron in first orbit of H-atom is :

$$a = \frac{v^2}{r} \quad \ldots(2)$$

From equation-(1) we use

$$v = \frac{h}{2\pi mr}$$

From equaton-(2) we have

$$a = \left(\frac{h}{2\pi mr}\right)^2 \times \frac{1}{r} = \frac{h^2}{4\pi^2 m^2 r^3}$$

Ans. (C)

Sol. 4 Series limit of Lyman series is corresponding to transition $\infty \to 1$

Here by using Rydberg's formula, we have

$$\nu_L = \frac{c}{\lambda_{\infty\to 1}} = c\left[RZ^2\left(\frac{1}{1^2} - \frac{1}{\infty^2}\right)\right]$$

$$\Rightarrow \quad \nu_L = RcZ^2$$

For Pfund series limit is corresponding to transition $\infty \to 5$

So we use

$$\nu_F = \frac{c}{\lambda_{\infty\to 5}} = c\left[RZ^2\left(\frac{1}{5^2} - \frac{1}{\infty^2}\right)\right]$$

$$\Rightarrow \quad \nu_F = RcZ^2\left(\frac{1}{25}\right) = \frac{\nu_L}{25}$$

Ans. (D)

Sol. 5 For transition from n^{th} state to ground state emitted wavelength is given as

$$\frac{1}{\Lambda_n} = RZ^2\left[\frac{1}{1^2} - \frac{1}{n^2}\right]$$

As $n >> 1$

$$\Lambda_n = \frac{1}{RZ^2}\left(1 - \frac{1}{n^2}\right)^{-1}$$

$$\Lambda_n = \frac{1}{RZ^2}\left(1 + \frac{1}{n^2}\right) \quad \ldots(1)$$

[As $\frac{1}{n^2} << 1$ we use $(1+x)^b \simeq (1+xb)$]

For n^{th} state de-Broglie wavelength of e^- is given as :

$$\lambda_n = \frac{h}{p_n} = \frac{h}{mv}$$

By angular momentum quantization postulate of Bohr model, we use

$$mvr = \frac{nh}{2\pi}$$

$$\Rightarrow \quad \lambda_n = \frac{h}{mv}$$

$$= \frac{2\pi}{n} \times (0.529 \times n^2) \simeq 1.06\,\pi n \quad \ldots(2)$$

From equation-(1) we have

$$\Lambda_n = \frac{1}{RZ^2} + \frac{1}{RZ^2 n^2}$$

$$\Rightarrow \quad \Lambda_n = \frac{1}{RZ^2} + \frac{(1.06\pi)^2}{RZ^2} \cdot \frac{1}{\lambda_n^2}$$

$$\Rightarrow \quad \Lambda_n = A + \frac{B}{\lambda_n^2}$$

Ans. (A)

Sol. 6 Radius of nth orbit for Muonic atom is

$$r_\mu = \frac{n^2h^2}{4\pi^2 kZe^2 m_\mu} \simeq \frac{n^2h^2}{4\pi^2 kZe^2 m_e} \times \frac{1}{200} = \frac{r_e}{200}$$

[As $m_\mu = 200\, m_e$]

$\Rightarrow$ Option (a) is correct.

Speed of muon in n^{th} orbit can be given as

$$v_\mu = \frac{2\pi kZe^2}{nh}$$

[It is independent of mass of orbiting particle]

$$\Rightarrow \quad v_\mu = v_e$$

$\Rightarrow$ Option (b) is NOT correct.

The ionization energy in Bohr model is given as

$$E_\mu = \frac{2\pi^2 k^2 Z^2 e^4 m_\mu}{n^2h^2}$$

$$\simeq \frac{2\pi^2 k^2 Z^2 e^4 m_e}{n^2h^2} \times 200 = 200\, E_e$$

$\Rightarrow$ Option (c) is correct.

Momentum of muon in atom is given as :

$$P_\mu = m_\mu v_\mu = 200\, m_e \cdot v_e = 200\, P_e$$

$\Rightarrow$ Option (d) is correct. **Ans. (D)**

Sol. 7 $\text{He} \xrightarrow{E_1} \text{He}^{+} \xrightarrow{E_2} \text{He}^{+2}$

Total energy required to completely ionize He atom is $(E_1 + E_2)$.

As He^+ is a hydrogenic atom (single electron system) so E_2 in the *IP* for He^+ which is given by Bohr model as

$$E_2 = 13.6\, Z^2 = 13.6 \times (2)^2 = 13.6 \times 4 = 54.4 \text{ eV}$$

It is given that

$$E_2 = 2.2\, E_1$$

$$\Rightarrow \quad E_1 = \frac{54.4}{2.2} \times \frac{272}{11} = 24.72 \text{ eV}$$

Total required energy is $E_1 + E_2 = 54.4 + 24.72 = 79.12$ eV

Ans. (B)

Sol. 8 Energy of photon $= \dfrac{12500}{980} = 12.75$ eV

So electron will excite to $n = 4$ state

As radius of nth orbit $r_n \propto n^2$ so the radius of atom will be $16a_0$. **Ans. (D)**

Sol. 9 For $M \rightarrow L$ shell transition, we use

$$\frac{1}{\lambda} = K\left(\frac{1}{2^2} - \frac{1}{3^2}\right) = \frac{K \times 5}{36} \quad \ldots(1)$$

For $N \rightarrow L$ shell transition, we use

$$\frac{1}{\lambda'} = K\left(\frac{1}{2^2} - \frac{1}{4^2}\right) = \frac{K \times 3}{16} \quad \ldots(2)$$

From (1) and (2)

$$\Rightarrow \quad \lambda' = \frac{20}{27}\lambda$$ **Ans. (C)**

Sol. 10 For the given circular orbit, we use

$$F = \frac{dU}{dr} = kr = \frac{mv^2}{r} \quad \ldots(1)$$

and $$mvr = \frac{nh}{2\pi} \quad \ldots(2)$$

$$\Rightarrow \quad r^2 \propto n$$

$$\Rightarrow \quad r \propto \sqrt{n}$$

Energy of particle in n^{th} energy level is

$$E = \frac{1}{2}kr^2 + \frac{1}{2}mv^2$$

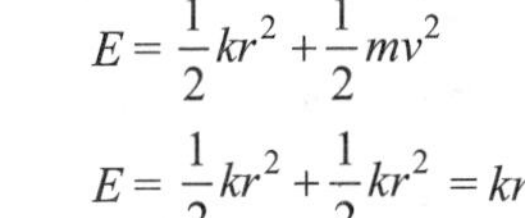

$$\Rightarrow \quad E = \frac{1}{2}kr^2 + \frac{1}{2}kr^2 = kr^2$$

$$\Rightarrow \quad E \propto n$$ **Ans. (D)**

Sol. 11 The energy lost by electrons in excitation of mercury atoms is

$$\Delta E = 5.6 - 0.7 = 4.9 \text{ eV}$$

Minimum wavelength (maximum energy) of photons emitted by mercury atoms in de-excitation is given as

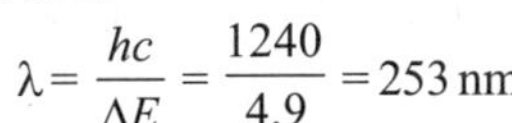

$$\lambda = \frac{hc}{\Delta E} = \frac{1240}{4.9} = 253 \text{ nm}$$ **Ans. (C)**

Sol. 12 Energy released for transition $n = 2$ to $n = 1$ of hydrogen atom is given as

$$E = 13.6\, Z^2\left(\frac{1}{n_1^2} - \frac{1}{n_2^2}\right)$$

For $Z = 1$, $n_1 = 1$, $n_2 = 2$, we get

$$E = 13.6 \times 1 \times \left(\frac{1}{1^2} - \frac{1}{2^2}\right)$$

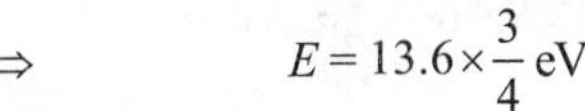

$$\Rightarrow \quad E = 13.6 \times \frac{3}{4} \text{ eV}$$

For He$^+$ ion $z = 2$

(A) $n = 1$ to $n = 4$

$$E = 13.6 \times 2^2 \times \left(\frac{1}{1^2} - \frac{1}{4^2}\right) = 13.6 \times \frac{15}{4} \text{ eV}$$

(B) $n = 2$ to $n = 4$

$$E = 13.6 \times 2^2 \times \left(\frac{1}{2^2} - \frac{1}{4^2}\right) = 13.6 \times \frac{3}{4} \text{ eV}$$

(C) $n = 2$ to $n = 5$

$$E = 13.6 \times 2^2 \times \left(\frac{1}{2^2} - \frac{1}{5^2}\right) = 13.6 \times \frac{21}{25} \text{ eV}$$

(D) $n = 2$ to $n = 3$

$$E = 13.6 \times 2^2 \times \left(\frac{1}{2^2} - \frac{1}{3^2}\right) = 13.6 \times \frac{5}{9} \text{ eV}$$

Energy required for transition of He^+ for $n = 2$ to $n = 4$ matches exactly with energy released in transition of H for $n = 2$ to $n = 1$ **Ans. (B)**

Sol. 13 Using Rydberg's formula for first line of Balmer series, we have

$$\frac{1}{660} = R\left(\frac{1}{2^2} - \frac{1}{3^2}\right) = \frac{5R}{36} \quad \ldots(1)$$

For second line of Balmer series, we have

$$\frac{1}{\lambda} = R\left(\frac{1}{2^2} - \frac{1}{4^2}\right) = \frac{3R}{16} \quad \ldots(2)$$

From equations-(1) and (2), we use

$$\frac{\lambda}{660} = \frac{5 \times 16}{36 \times 3}$$

$$\Rightarrow \quad \lambda = \frac{4400}{9} = 488.88 \simeq 488.9 \text{ nm}$$ **Ans. (C)**

Sol. 14 Energy levels in Hydrogen like atom are given as

$$E = -13.6\, \frac{Z^2}{n^2} \text{ eV}$$

As He$^+$ is 1st excited state, we use

$$z = 2, n = 2$$

$$\Rightarrow \quad E = -13.6 \text{ eV}$$

As total energy of He$^+$ in 1st excited state is –13.6 eV, its ionisation energy will be +13.6 eV. **Ans. (B)**

Sol. 15 For 6 spectral lines emission, excitation must be upto $n = 4$

$\Rightarrow$ Energy required for transition from $n = 1$ to $n = 4$ is

$$\Delta E = E_4 - E_1$$

$$= (-0.85\,z^2) - (-13.6\,z^2)\,\text{eV}$$
$$= 12.75\,z^2\,\text{eV}$$

Wavelength of radiation of this energy

$$12.75 \times 9 = \frac{hc}{\lambda}$$

or $$\lambda = \frac{hc}{9 \times 12.75e} = \frac{12431}{9 \times 12.75}\,\text{Å} \simeq 108\,\text{Å}$$ **Ans. (C)**

Sol. 16

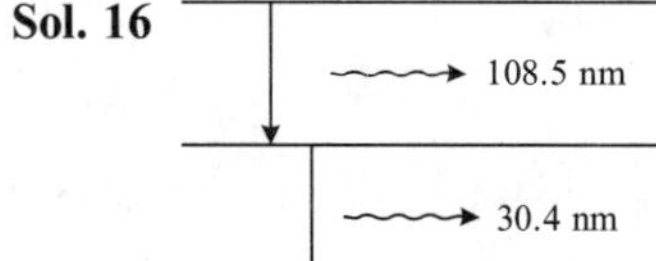

$$\Delta E_1 = \frac{1240}{108.5}\,\text{eV} = 11.73\,\text{eV}$$
$$\Delta E_2 = \frac{1240}{30.4}\,\text{eV} = 40.79\,\text{eV}$$
$$= 52.22\,\text{eV}$$
$$E_n - E_1 = 52.22\,\text{eV}$$

For He^+, we use $z = 2$

$$\left(-\frac{13.6z^2}{n^2}\right) - (-13.6z^2) = 52.22\,\text{eV}$$
$$54.4 - \frac{54.4}{n^2} = 52.22\,\text{eV}$$

$$\Rightarrow \quad 54.4 - 52.22 = \frac{54.4}{n^2}$$
$$\Rightarrow \quad n^2 = \frac{54.4}{2.18} = 24.95 \simeq 25$$
$$\Rightarrow \quad n = 5$$ **Ans. (A)**

Sol. 17 By Rydberg's formula

$$\frac{1}{\lambda} = R\left(\frac{1}{n_1^2} - \frac{1}{n_2^2}\right)$$

For $n = 4 \rightarrow n = 3$ transition

$$\frac{1}{\lambda_1} = R\left[\frac{1}{3^2} - \frac{1}{4^2}\right] = \frac{16-9}{144}R$$
$$\Rightarrow \quad \lambda_1 = \frac{144}{7R} \quad \ldots(1)$$

For $n = 3 \rightarrow n = 2$ transition

$$\frac{1}{\lambda_2} = R\left[\frac{1}{2^2} - \frac{1}{3^2}\right] = \frac{9-4}{36}R$$
$$\Rightarrow \quad \lambda_2 = \frac{36}{5R} \quad \ldots(2)$$

(1)/(2) gives $$\frac{\lambda_1}{\lambda_2} = \frac{144/7}{36/5} = \frac{20}{7}$$ **Ans. (D)**

Sol. 18 De-Broglie wavelength is given as

$$\lambda = \frac{h}{mv} = \frac{hr}{mvr}$$

By Bohr's II Postulate, we use

$$mvr = \frac{hr}{2\pi}$$
$$\Rightarrow \quad \lambda = \frac{hr}{(hr/2\pi)} = \frac{2\pi r}{n} \quad [n = 3]$$
$$\Rightarrow \quad \lambda = \frac{2 \times 3.14 \times 4.65 \times 10^{-10}}{3}$$
$$\Rightarrow \quad \lambda \simeq 9.7\,\text{Å}$$ **Ans. (C)**

Sol. 19 Time period of revolution of electron in n^{th} orbit is given as

$$T \propto \frac{n^3}{Z^2}$$
$$\Rightarrow \quad \frac{T_2}{T_1} = \frac{(2)^3}{(1)^3} = 8$$
$$\Rightarrow \quad T_2 = 8 \times 1.6 \times 10^{-16}\,\text{s}$$

Frequency in $n = 2$ orbit is

$$f_2 = \frac{1}{T_2} = \frac{10^{16}}{8 \times 1.6}$$
$$\approx 7.8 \times 10^{14}\,\text{Hz}$$ **Ans. (C)**

Sol. 20 For scattering experiment we use number of α particles at a specific angle are inversely proportional to fourth power of sine of scattering angle. Thus most appropriate option is (C). **Ans. (C)**

Sol. 21 Using Rydberg's formula, we use

$$\frac{1}{\lambda} = R_H\left(\frac{1}{2^2} - \frac{1}{n_2^2}\right) \quad \text{[For Balmer Series]}$$

For first and second line of Balmer Series, we have

$$\frac{\lambda_2}{\lambda_1} = \frac{\left(\frac{1}{2^2} - \frac{1}{3^2}\right)}{\left(\frac{1}{2^2} - \frac{1}{4^2}\right)}$$
$$\Rightarrow \quad \frac{\lambda_2}{6561} = \frac{5/36}{3/16}$$
$$\Rightarrow \quad \lambda_2 = \frac{20}{27} \times 6561$$
$$\Rightarrow \quad \lambda_2 = 4860\,\text{Å}$$
$$\Rightarrow \quad \lambda_2 = 486\,\text{nm}$$ **Ans. (486)**

Sol. 22 Ionization energy of H-like ion with atomic number z is given as

$$IP = 13.6 Z^2 = 9 \times 13.6 \text{ eV}$$

$$\Rightarrow \quad Z = 3$$

Using Rydberg formula, we use

$$\frac{1}{\lambda} = RZ^2\left(\frac{1}{1^2} - \frac{1}{3^2}\right) = 1.09 \times 10^7 \times 9 \times \frac{8}{9}$$

$$\Rightarrow \quad \lambda = 11.4 \text{ nm}$$

Ans. (D)

Sol. 23 For very high value of n, transition between adjoining orbits causes emission of radiation of frequency equal to revolution frequency of electron in n^{th} orbit which is given as

$$f_n = \frac{v_n}{2\pi r_n} = \frac{\frac{2\pi kZe^2}{nh}}{\frac{n^2h^2}{2\pi kZe^2 m}}$$

$$\Rightarrow \quad f_n \propto \frac{Z^2}{n^3}$$

$$\Rightarrow \quad f_n \propto \frac{1}{n^3}$$

Ans. (B)

Sol. 24 For lyman series first & last time wavelengths are given as

$$\frac{1}{\lambda_1} = R\left[\frac{1}{1^2} - \frac{1}{2^2}\right] = \frac{3R}{4}$$

$$\frac{1}{\lambda_L} = R\left[\frac{1}{1^2} - \frac{1}{\infty^2}\right] = R$$

$$\Delta\lambda_L = \lambda_1 - \lambda_L = \frac{4}{3R} - \frac{1}{R} = \frac{1}{3R} = 304 \text{ Å}$$

For paschen series we use

$$\frac{1}{\lambda_1} = R\left[\frac{1}{3^2} - \frac{1}{4^2}\right] = \frac{7R}{144}$$

$$\Rightarrow \quad \frac{1}{\lambda_L} = R\left[\frac{1}{3^2} - \frac{1}{\infty^2}\right] = \frac{R}{9}$$

$$\Rightarrow \quad \Delta\lambda_p = \lambda_1 - \lambda_L = \frac{144}{7R} - \frac{9}{R} = \frac{81}{7R}$$

$$\Rightarrow \quad \Delta\lambda_p = \frac{81}{7} \times 304 \times 3 = 10553.14 \text{ Å}$$

Ans. (10553)

Sol. 25 If initial KE of particle is E_1 and KE of H-atom after collision is E_2 we use

$$E_1 - 10.2 \text{ eV} = E_2$$

By conservation of momentum we use

$$\sqrt{2m_P E_1} = \sqrt{2m_H E_2}$$

$$m_P E_1 = m_H(E_1 - 10.2)$$

$$\Rightarrow \quad E_1 = \frac{10.2 \times m_H}{m_H - m_P} eV$$

$$\Rightarrow \quad E_1 = \frac{10.2 \times 10^9}{10^9 - 2 \times 10^8} eV$$

$$\Rightarrow \quad E_1 = \frac{10.2}{0.8} = 12.75\, eV = \frac{N}{4}$$

$$\Rightarrow \quad N = 51$$

Ans. (51)

Sol. 26 Series limit of Lymen series is corresponding to the transision from infinite to $n = 1$ which is denoted by A. Third member of Balmer series is corresponding to $n = 5$ to $n = 2$ which is denoted by B. Second member of Paschen series is corresponding to $n = 5$ to $n = 3$ which is denoted by C. **Ans. (A)**

Sol. 27 Out of the given options, maximum energy difference is there in transition from $n = 2$ to $n = 1$. **Ans. (A)**

Sol. 28 For the given transition energy difference is 10.2 eV and corresponding wavelength is given as

$$\lambda = \frac{12431}{10.2} = 1218.72 \text{ A} = 121.8 \text{ nm}$$

Ans. (A)

Sol. 29 By Rydberg's formula, we have

$$\frac{1}{\lambda_1} = R\left[\frac{1}{1^2} - \frac{1}{4^2}\right]$$

$$\frac{1}{\lambda_2} = R\left[\frac{1}{3^2} - \frac{1}{4^2}\right]$$

$$\Rightarrow \quad \frac{\lambda_1}{\lambda_2} = \frac{\left[\frac{1}{9} - \frac{1}{16}\right]}{\left[1 - \frac{1}{16}\right]} = \frac{7}{9 \times 15} = \frac{7}{135}$$

Ans. (D)

Sol. 30 For 1st line of Balmer series, its wavelength is given as

$$\frac{1}{\lambda_1} = Rz^2\left(\frac{1}{2^2} - \frac{1}{3^2}\right)$$

$$\Rightarrow \quad \frac{1}{\lambda_1} = Rz^2 \frac{5}{36} \quad \ldots(1)$$

For 3rd line of Balmer series, its wavelength is given as

$$\frac{1}{\lambda_3} = Rz^2\left(\frac{1}{2^2} - \frac{1}{5^2}\right)$$

$$\Rightarrow \quad \frac{1}{\lambda_3} = Rz^2 \frac{21}{100} \quad \ldots(2)$$

From equations-(1) and (2), we have

$$\frac{\lambda_1}{\lambda_3} = \frac{21}{100} \times \frac{36}{5} = 1.512 = 15.12 \times 10^{-1}$$

$$\Rightarrow \quad x \approx 15$$

Ans. (15)

Sol. 31 Velocity of electron in n^{th} energy level of hydrogen atom is given as

$$v = \frac{2\pi kZe^2}{nh}$$

$$\Rightarrow \quad v \propto \frac{1}{n}$$ **Ans. (B)**

Sol. 32 Energy of hydrogen atom in ground state is given as

$$E = -13.6\, Z^2/n^2$$

For $Z = 1$ & $n = 1$, we have

$$E = -13.6 \times \frac{1^2}{1^2} = -13.6 \text{ eV}$$

For single ionized carbon atom having same energy, we use

$$E = -13.6\, \frac{Z^2}{n^2} = -13.6 \text{ eV}$$

$$\Rightarrow \quad n^2 = 6^2$$

$$\Rightarrow \quad n = 6$$ **Ans. (B)**

Sol. 33 As a fact it is studied from hydrogen spectrum that Lyman series lies in UV region, Balmer series lies in visible region and and next 3 lies in infrared region. **Ans. (D)**

Sol. 34 For hydrogen like atoms ionization potential is related to mass of revolving particle as

$$E \propto m$$

Thus ionization potential of muon, we use

$$E_\mu = 13.6 \times \frac{(\text{Mass}_\mu)eV}{(\text{Mass}_e)}$$

$$\Rightarrow \quad E_\mu = 13.6 \times 207 \text{ eV} = 2815.2 \text{ eV}$$ **Ans. (B)**

Sol. 35 Energy band gap of the PN junction can be calculated by using the light wavelength which triggers the electron-hole pair generation, given as

$$E = \frac{hc}{\lambda} = \frac{1224 - \text{nm}}{e \times 612\text{nm}} = 2 \text{ eV}$$ **Ans. (C)**

Sol. 36 Average current due to electron revolution is given as

$$i = \frac{q}{t} = \frac{q}{2\pi r / v}$$

$$\Rightarrow \quad i = \frac{1.6 \times 10^{-19} \times 2.2 \times 10^6}{2\pi \times 0.5 \times 10^{-10}}$$

$$\Rightarrow \quad i = \frac{1.6 \times 10^{-19} \times 2.2 \times 10^6}{2 \times \frac{22}{7} \times 0.5 \times 10^{-10}}$$

$$\Rightarrow \quad i = 112 \times 10^{-2} \text{ mA}$$ **Ans. (112)**

Sol. 37 Statement-(1) is correct as atom of each element has various fixed energy levels so thats why emit characteristics spectrum

Statement-(2) is correct, as Bohr's model shows the motion of electrons revolution in well define orbits called energy levels

Statement-(3) is wrong, as density nucleus almost remain constant and that is the basis of defining fermi radius.

Statement-(4) is wrong, as neutrons can decay in proton, electron and an antineutrino

Statement-(5) is Correct, as unstable nuclei shows radioactivity.

Ans. (A)

Sol. 38 If the thickness of air column is taken as l, difference in number of waves through air and vacuum columns are given as

$$n_a - n_v = 1$$

$$\Rightarrow \quad l\left(\frac{1}{\lambda_a} - \frac{1}{\lambda_v}\right) = 1$$

$$\Rightarrow \quad l\left(\frac{1.0003}{6000} - \frac{1}{6000}\right) = 1$$

$$\Rightarrow \quad l = 2 \text{ mm}$$ **Ans. (2)**

Sol. 39 Photon energy radiated from hydrogen like ion for transion of electron between orbits, we have

$$h\upsilon = E_0 Z^2 \left[\frac{1}{n_1^2} - \frac{1}{n_2^2}\right]$$

$$\Rightarrow \quad \upsilon \; \alpha \left[\frac{1}{n_1^2} - \frac{1}{n_2^2}\right]$$

$$\Rightarrow \quad \frac{2.19 \times 10^{15}}{\upsilon} = \frac{\left[\frac{1}{1} - \frac{1}{9}\right]}{\left[\frac{1}{1} - \frac{1}{4}\right]}$$

$$\Rightarrow \quad \upsilon = 2.19 \times 10^{15}\, \frac{3}{4} \times \frac{9}{8} = 2.46 \times 10^{15} \text{ Hz}$$

Ans. (B)

Sol. 40 Number of spectral lines emitted when atoms of hydrogen gas are excited to nth energy level, are given as

$$N = {}^6C_2 = 15$$ **Ans. (15)**

27.1 Radioactivity

Sol. 1 After time t, number of undecayed nuclei of A is given by radioactive decay equation as

$$N_A = N_0 (2)^{-t/T} \quad \ldots(1)$$

[where $N_0 \rightarrow$ number of nuclei of A at $t = 0$]

Number of nuclei of B at time t are

$$N_B = N_0 - N_A = N_0 (1 - (2)^{-t/T}) \quad \ldots(2)$$

$$(2)\div(1) \Rightarrow \frac{N_B}{N_A} = \frac{N_0[1-(2)^{-t/T}]}{N_0(2)^{-t/T}} = 0.3$$

$$\Rightarrow 1-(2)^{-t/T} = 0.3\,(2)^{-t/T}$$

$$\Rightarrow 1.3\,(2)^{-t/T} = 1$$

$$\Rightarrow 2^{t/T} = 1.3$$

$$\Rightarrow t = \frac{T\log(1.3)}{\log(2)}$$

Ans. (D)

Sol. 2 At $t=0$, Activity of sample is $A_{C_0} = 0.8\ \mu Ci$

$$= 0.8\times10^{-6}\times3.7\times10^{10}\ \text{dps}$$

At $t = 10$hrs activity/cm^3 is given as AC/cm^3 = 300 dpm

$$= \frac{300}{60} = 5\ \text{dps}$$

If V is the volume of animal's blood then total activity at

$t = 10$ hr is $A_C = 5V$ dps

Using radioactive decay equation, we have

$$A_C = A_{C_0}e^{-\lambda t}$$

$$5V = 0.8\times3.7\times10^4\times0.84$$

$$V = \frac{0.8\times3.7\times10^4\times0.84}{5}\ \text{cm}^2$$

$$\Rightarrow V = 4.97\times10^{-3}\ \text{m}^3 \simeq 5\ \text{ltr}$$

Ans. (D)

Sol. 3 Activity of a sample

$$A_C = \lambda N$$

If S_1 has N_0 nuclei then S_2 will be having $\frac{N_0}{2}$ nuclei at that instant, so we use

$$A_{CS_1} = \lambda_1 N_0 \quad \ldots(1)$$

and

$$A_{CS_2} = \lambda_2\left(\frac{N_0}{2}\right) \quad \ldots(2)$$

$$(1)\div(2) \Rightarrow \frac{5}{10} = \frac{\lambda_1}{\lambda_2/2}$$

$$\Rightarrow \frac{\lambda_1}{\lambda_2} = \frac{1}{4} = \frac{T_2}{T_1}$$

$$\Rightarrow T_1 = 4T_2$$

So half life of S_1 is 4 times that of S_2 **Ans. (D)**

Sol. 4 For RA material A if there ae $2N_0$ atoms then, we use

$$A_{C_A} = \lambda_A(2N_0) \quad \ldots(1)$$

For material B, number of atoms are N_0 so, we use

$$A_{C_B} = \lambda_B(N_0) \quad \ldots(2)$$

$$(1)/(2) \quad \frac{A_{C_A}}{A_{C_B}} = \frac{2\lambda_A}{\lambda_B} = \frac{10}{20}$$

$$\Rightarrow 4\lambda_A = \lambda_B$$

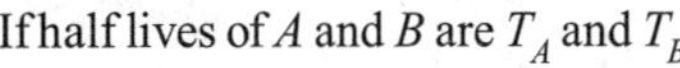

If half lives of A and B are T_A and T_B

$$\Rightarrow \frac{T_A}{T_B} = \frac{\lambda_B}{\lambda_A} = 4$$

Ans. (A)

Sol. 5 Activities of A and B after time t are given as

$$R_A = A_{COA}e^{-\lambda_A t} \quad \left[\text{Here } \lambda_A = \frac{\ln(2)}{T_A} = \frac{\ln(2)}{\ln(2)} = 1\right]$$

$$R_B = A_{COB}e^{-\lambda_B t}$$

$$\Rightarrow \frac{R_B}{R_A} = \frac{A_{COB}e^{-\lambda_B t}}{A_{COA}e^{-\lambda_A t}} = e^{-(\lambda_B-\lambda_A)t} = e^{-3t}$$

$$\Rightarrow \lambda_B - \lambda_A = 3$$

$$\Rightarrow \lambda_B = 3+1 = 4$$

$$\Rightarrow T_B = \frac{\ln(2)}{\lambda_B} = \frac{\ln(2)}{4}$$

Ans. (C)

Sol. 6 By radioactive decay law, we use

$$A = A_0(2)^{-t/T}$$

at $t=0$, $A_O = 1600$ cps

at $t = 8$s, $A_C = 100$ cps

$$\Rightarrow 100 = 1600\,(2)^{-8/T}\ \text{half life}$$

$$\Rightarrow 2^{8/T} = 2^4$$

$$\Rightarrow \frac{8}{T} = 4$$

$$\Rightarrow T = 2\text{s}$$

At $t = 6$s, we use

$$A = 1600\,(2)^{-6/2} = \frac{1600}{2^3} = \frac{1600}{8}$$

$$= 200\ \text{cps}$$

Ans. (C)

Sol. 7 If we write nulcear equation for α and β decay

$${}^{232}_{90}\text{Th} \longrightarrow {}^{208}_{78}Y + 6\,{}^{4}_{2}\alpha$$

$${}^{208}_{78}Y \longrightarrow {}^{208}_{82}X + 4\,{}^{0}_{-1}\beta$$

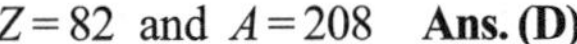

$Z = 82$ and $A = 208$ **Ans. (D)**

Sol. 8 By using RA decay law, we have

For $A \leftarrow N = N_0e^{-\lambda t} \rightarrow$ For A

$$N_A = N_0e^{-10\lambda t} \quad \ldots(1)$$

$$N_B = N_0e^{-\lambda t} \quad \ldots(2)$$

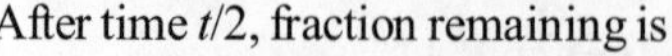

$(1)/(2) \Rightarrow \quad \dfrac{N_A}{N_B} = e^{-9\lambda t} = \dfrac{1}{e}$

$\Rightarrow \quad 9\lambda t = 1$

$\Rightarrow \quad t = \dfrac{1}{9\lambda}$ **Ans. (B)**

Sol. 9 By *RA* decay equation for *A* and *B*, number of nuclei after time *t* are

$$N_A = N_0 e^{-5\lambda t}$$

and $\quad N_B = N_0 e^{-\lambda t}$

$\Rightarrow \quad \dfrac{N_A}{N_B} = \dfrac{e^{-5\lambda t}}{e^{-\lambda t}} = \dfrac{1}{e^2}$

$\Rightarrow \quad e^{-4\lambda t} = e^{-2}$

$\Rightarrow \quad 4\lambda t = 2$

$\Rightarrow \quad t = \dfrac{1}{2\lambda}$ **Ans. (C)**

Sol. 10 By radioactive decay equation, we use

For $A \rightarrow \quad N_A = N_0 2^{-t/T_A}$

Number of decayed nuclei

$$N_{DA} = N_0 - N_A$$

$$= N_0 - N_0 2^{-6} = N_0\left(1 - \frac{1}{64}\right) = \frac{63}{64}N_0$$

For $B \rightarrow \quad N_B = N_0 2^{-t/T_B}$

Number of decayed nuclei

$$N_{DB} = N_0 - N_B$$

$$= N_0 - N_0 2^{-3} = N_0\left(1 - \frac{1}{8}\right) = \frac{7}{8}N_0$$

$\dfrac{N_{DA}}{N_{DB}} = \dfrac{63/64N_0}{7/8N_0} = \dfrac{9}{8}$ **Ans. (A)**

Sol. 11 From radioactive decay law, we use

$$A = A_0 2^{-t/T}$$

$\Rightarrow \quad 500 = 700\, 2^{-t/T}$

$\Rightarrow \quad 0.7 \simeq 2^{-t/T}$

$\Rightarrow \quad 2^{-1/2} \simeq 2^{-t/T}$

$\Rightarrow \quad T \simeq 2t = 2 \times 30 = 60\,\text{m}$ **Ans. (D)**

Sol. 12 By radiative decay equation, we use

$$N = N_0 e^{-\lambda t}$$

Given that $\quad \dfrac{N}{N_0} = \dfrac{9}{16} = e^{-\lambda t}$

After time *t*/2, fraction remaining is

$\left(\dfrac{N}{N_0}\right)_{t/2} = e^{-\lambda t/2} = \sqrt{e^{-\lambda t}} = \sqrt{\dfrac{9}{16}} = \dfrac{3}{4}$ **Ans. (A)**

Sol. 13 By radioactive decay equation, we have

$$R = R_0 e^{-\lambda t}$$

$\Rightarrow \quad \ln R = \ln R_0 - \lambda t$

For element *A* we use

$$\lambda_A = \frac{6}{10}$$

$\Rightarrow \quad T_A = \dfrac{10}{6}\ln 2$

For element *B* we use

$$\lambda_B = \frac{6}{5}$$

$\Rightarrow \quad T_B = \dfrac{5\ln 2}{6}$

for element *C* we use

$$\lambda_C = \frac{2}{5}$$

$\Rightarrow \quad T_C = \dfrac{5\ln 2}{2}$

$\Rightarrow \quad T_A : T_B : T_C = \dfrac{10}{6} : \dfrac{5}{6} : \dfrac{15}{6} :: 2 : 1 : 3$ **Ans. (C)**

Sol. 14 For simultaneous decay modes of nucleus, effective decay constant is given as

$$A \xrightarrow{\lambda_1} B, \quad A \xrightarrow{\lambda_2} C$$

$$\lambda_{eff} = \lambda_1 + \lambda_2$$

$\Rightarrow \quad \dfrac{\ln 2}{T_{eff}} = \dfrac{\ln 2}{T_1} + \dfrac{\ln 2}{T_2}$

$\Rightarrow \quad T_{eff} = \dfrac{T_1T_2}{T_1 + T_2} = \dfrac{100 \times 10}{100 + 10} = \dfrac{100}{11} = 9.09\,\text{s}$

Ans. (A)

Sol. 15 If half life of *X* is *t*, than that of *Y* will be 2*t*. After three half lives of *Y* i.e. after a time 6*t* the number of nuclei of both becomes equal so we use

$$N_1' = N_2'$$

$\Rightarrow \quad N_1 e^{-\lambda_1 6t} = N_2 e^{-\lambda_2 6t}$

$\Rightarrow \quad \dfrac{N_1}{N_2} = e^{(\lambda_1 - \lambda_2)6t} = e^{\ln 2\left(\frac{1}{t} - \frac{1}{2t}\right) \times 6t} = e^{(\ln 2) \times 3}$

$\Rightarrow \quad \dfrac{N_1}{N_2} = e^{\ln 8} = 8$ **Ans. (A)**

Sol. 16 If initial activity of the sample be A_0 then by radioactive decay law, after time t_1 and t_2, we have

$$A = A_0 e^{-\lambda t_1} \qquad \ldots(1)$$

and
$$\frac{A}{5} = A_0 e^{-\lambda t_2} \qquad \ldots(2)$$

From above equations-(1) and (2), we have

$$5 = e^{\lambda(t_2 - t_1)}$$

$$\Rightarrow \quad \lambda = \frac{\ln 5}{t_2 - t_1} = \frac{1}{\tau}$$

$$\Rightarrow \quad \tau = \frac{t_2 - t_1}{\ln 5}$$

Ans. (A)

Sol. 17 The activity is given as

$$A = \lambda N \qquad \left[t_{1/2} = \frac{\ln 2}{\lambda}\right]$$

where
$$N = nN_A$$

$$\Rightarrow \quad N = \left(\frac{1.5 \times 10^{-3}}{198}\right) N_A$$

$$\Rightarrow \quad A = \left(\frac{\ln 2}{t_{1/2}}\right) N$$

As 1 Ci = 3.7×10^{10} Bq, calculating for activity we get

$$A = 365 \text{ Ci}$$

Thus closes matching option is (B). **Ans. (B)**

Sol. 18 By radioactive deay law, for first case we have

$$N_1 = N_0 e^{-\lambda t_1}$$

$$\Rightarrow \quad \frac{N_1}{N_0} = e^{-\lambda t_1}$$

$$\Rightarrow \quad 0.67 = e^{-\lambda t_1}$$

$$\Rightarrow \quad \ln(0.67) = -\lambda t_1 \qquad \ldots(1)$$

For second case, we use

$$N_2 = N_0 e^{-\lambda t_2}$$

$$\Rightarrow \quad \frac{N_2}{N_0} = e^{-\lambda t_2}$$

$$\Rightarrow \quad 0.33 = e^{-\lambda t_2}$$

$$\Rightarrow \quad \ln(0.33) = -\lambda t_2 \qquad \ldots(2)$$

From equations-(1) annd (2), we have

$$\ln(0.67) - \ln(0.33) = \lambda t_1 - \lambda t_2$$

$$\Rightarrow \quad \lambda(t_1 - t_2) = \ln\left(\frac{0.67}{0.33}\right)$$

$$\Rightarrow \quad \lambda(t_1 - t_2) \cong \ln 2$$

$$\Rightarrow \quad t_1 - t_2 = \frac{\ln 2}{\lambda} = t_{1/2} = 20 \text{ min}$$

Ans. (B)

Sol. 19 For independent decay processes in a radioactive sample, the effective decay constant is given as

$$\lambda_{\text{eff}} = \lambda_1 + \lambda_2$$

$$\Rightarrow \quad \frac{1}{T_{1/2}} = \frac{1}{T_{1/2}^{(1)}} + \frac{1}{T_{1/2}^{(2)}}$$

$$\Rightarrow \quad T_{1/2} = \frac{T_{1/2}^{(1)} T_{1/2}^{(2)}}{T_{1/2}^{(1)} + T_{1/2}^{(2)}}$$

Ans. (C)

Sol. 20 For simultaneous decay modes of the element X, we can write the total decay rate of the element as

1400 year → A

X

700 year → B

$$-\frac{dN_x}{dt} = \lambda_1 N_x + \lambda_2 N_x$$

$$\Rightarrow \quad \int_N^{N/3} \frac{dN_x}{N_x} = \int_0^t (\lambda_1 + \lambda_2) dt$$

$$\Rightarrow \quad \ln 3 = \left(\frac{\ln 2}{1400} + \frac{\ln 2}{700}\right) t$$

$$\Rightarrow \quad t = \frac{\ln 3}{\ln 2} \times \frac{1400}{3} = 740 \text{ year}$$

Ans. (C)

Sol. 21 By radioactive decay law, the activity of substance after time t is given as

$$R = R_0 e^{-\lambda t}$$

$$\ln R = \ln R_0^{-\lambda t}$$

Slope of curve from the given graph is calculated as

$$-\lambda = \frac{-6}{40}$$

$$\Rightarrow \quad \lambda = \frac{3}{20}$$

$$\Rightarrow \quad t_{1/2} = \frac{\ln 2}{\lambda} = \frac{\ln 2}{3} \times 20 = \frac{20}{3} \ln 2 = 4.62 \text{ s}$$

Ans. (A)

Sol. 22 As activity is decayed to one sixteenth of its initial value means total time passed is four half lives. Thus the half life of the substance is 80/4 = 20 days. **Ans. (3)**

Sol. 23 By Radioactive decay law, we have

$$N = N_0 e^{-\lambda t}$$

$$\Rightarrow \quad \frac{3N_0}{4} = N_0 e^{-\lambda t_1}$$

$$\Rightarrow \quad t_1 = \frac{\ln\left(\frac{4}{3}\right)}{\lambda}$$

$$\Rightarrow \quad t_2 = \frac{\ln 2}{\lambda}$$

$$\Rightarrow \quad t_2 - t_1 = \frac{\ln 2}{\lambda} - \frac{\ln\left(\frac{4}{3}\right)}{\lambda} = \frac{\ln\left(\frac{3}{2}\right)}{\lambda}$$ **Ans. (A)**

Sol. 24 Number of nuclei in 2mg of gold are calculated as

$$N = \frac{m}{M} . N_A = \frac{2\times10^{-3}}{198} \times 6.02 \times 10^{23}$$

Activity of the substance is given as

$$A_0 = \lambda N = \frac{0.693}{3\times27\times60\times60} \times \frac{2\times10^{-3}}{198} \times 6.02\times10^{23}$$

$$\Rightarrow \quad A_0 = 1618\times10^{-8}\times10^{18} = 16.18\times10^{12}$$ **Ans. (D)**

Sol. 25 By radioactive decay law, we use

$$A = A_0 e^{-\lambda t}$$

After half life, we have

$$A/2 = A_0 e^{-\lambda T}$$

$$\Rightarrow \quad \frac{1}{2} = e^{-\lambda T} \qquad \ldots(1)$$

Given that half life is 30 years, so we use

$$1/8 = e^{-\lambda 30} \qquad \ldots(2)$$

Solving (1) and (2)

$$e^{-3\lambda T} = e^{-\lambda 30}$$

$$\Rightarrow \quad T = 10 \text{ Yrs}$$ **Ans. (10)**

Sol. 26 By radioactive decay law, we have

$$N = N_0 e^{-\lambda t}$$

Number of nuclei decayed upto time t are given as

$$N_d = N - N_0$$

$$\Rightarrow \quad N_d = N_0(1 - e^{-\lambda t})$$

The fraction f is now given as

$$f = \frac{N_d}{N_0}(1 - e^{-\lambda t})$$

$$\Rightarrow \quad \frac{df}{dt} = \lambda e^{-\lambda t}$$ **Ans. (A)**

Sol. 27 During discharging of capacitor, charge on capacitor is given

$$q = q_0 e^{\frac{-t}{RC}} \qquad \ldots(1)$$

$$A = A_0 e^{-\lambda t} \qquad \ldots(2)$$

Dividing equations-(1) & (2), we get

$$\frac{q}{A} = \frac{q_0}{A_0} \frac{e^{\frac{-t}{RC}}}{e^{-\lambda t}}$$

$$\Rightarrow \quad -\lambda t = -\frac{-t}{RC}$$

$$\Rightarrow \quad RC = 1/\lambda$$

$$\Rightarrow \quad R = \frac{30\times10^{-3}}{200\times10^{-6}} = 150\,\Omega$$ **Ans. (150)**

Sol. 28 By radioactive decay law for element A we have

$$N_A = N_{0A} e^{-\lambda t}$$

Rate at which nuclie of B are being produced is given as

$$\frac{dN_B}{dt} = \lambda N_A - \lambda N_B$$

$$\Rightarrow \quad \frac{dN_B}{dt} = \lambda N_{0A} e^{-\lambda t} - \lambda N_B$$

$$\Rightarrow \quad \frac{dN_B}{dt} + \lambda N_B = \lambda N_{0A} e^{-\lambda t}$$

$$\Rightarrow \quad N_B = N_{0B}(1 - 2\lambda t e^{-\lambda t})$$

$$\Rightarrow \quad \frac{N_B(t)}{N_B(0)} = 1 - 2\lambda t e^{-\lambda t}$$

Out of given option, best is matching with option (A).

Ans. (A)

Sol. 29 By radioactive decay law, we have

$$N = N_0 e^{-\left(\frac{\ln 2}{T}\right)t}$$

$$\Rightarrow \quad N = 10^{10} e^{-0.35} = 7\times10^9$$ **Ans. (B)**

Sol. 30 Equation below shows the disintegration in the radioactive series from elements A to element C.

$$A \xrightarrow{\lambda_1} B \xrightarrow{\lambda_2} C$$

Accumulation rate of nuclei of B is given by the below expression

$$\frac{dN_B}{dt} = \lambda_1 N_1 - \lambda_2 N_2$$

As time passes decay rate of A decreases and decay rate of B increases as quantity of B is increasing. In such situation after a period of time when the quantity of B becomes maximum it will be in radioactive equilibrium and after this point quantity of B starts decreasing. Hence out of given option only (D) is most appropriate. **Ans. (D)**

Sol. 31 From the given condition, we have

$$\frac{0.693}{\lambda_x} = \frac{1}{\lambda_y}$$

$$\Rightarrow \quad \lambda_x = 0.693\,\lambda_y$$

$$\Rightarrow \quad \lambda_x < \lambda_y$$ **Ans. (D)**

27.2 Nuclear Physics

Sol. 1 The nuclear reaction is

$${}^{2}_{1}\text{H} + {}^{2}_{1}\text{H} \longrightarrow {}^{4}_{2}\text{He} + \Delta E$$

The released energy is equal to difference in total binding energies of product in this reaction.

Total BE of the reaction $E_R = 4 \times 1.1 = 4.4$ MeV

Total BE of the reaction $E_P = 7 \times 4 = 28$ MeV

Energy released (Q-value) $\Delta E = E_P - E_R = 28 - 4.4 = 23.6$ MeV

Ans. (C)

Sol. 2 Mass m produces energy $E = mc^2$

Power of reactor $P = 10^9\ W$

Mass of fuel consumed per second is $m' = \dfrac{P}{c^2} = \dfrac{10^9}{(3\times10^8)^2}$ kg

Mass of fuel consumed per hour is $m = m' \times 3600$

$$\Rightarrow \quad m = \frac{3600\times10^9}{8\times10^{16}} = 4\times10^{-5}\text{ kg}$$

$$\Rightarrow \quad m = 4\times10^{-2}\text{ gm}$$ **Ans. (C)**

Sol. 3 We consider an element X

$X^* \longrightarrow X + E_r$ $\quad(\lambda_N \sim 0.1$ Å or less ≈ 0.01Å$)$

(E_r = Energy of the order of some MeV)

For atom X when its e^- transit from $n = 2$ to $n = 1$

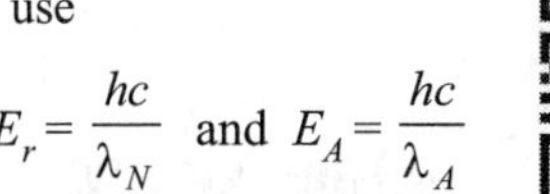

$$\frac{1}{\lambda_A} = R(z-\sigma)^2\left[\frac{1}{1^2} - \frac{1}{2^2}\right]$$

Energy is of the order of few eV only.

For photons we can use

$$E_r = \frac{hc}{\lambda_N} \text{ and } E_A = \frac{hc}{\lambda_A}$$

$$\Rightarrow \quad \frac{\lambda_N}{\lambda_A} = \frac{E_A}{E_r} = \frac{\text{order of } eV}{\text{order of } MeV} \simeq \frac{1}{10^6} = 10^{-6}$$ **Ans. (A)**

Sol. 4 Formula for fermi radius is $R = R_0(A)^{1/3}$

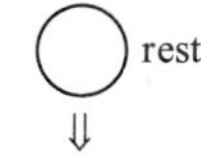

⇓

$8v \leftarrow m_1 \qquad m_2 \rightarrow 27v$

By conservation of linear momentum we use

$$m_1(8v) = m_2(27v)$$

$$8m_1 = 27m_2$$

If R_1 and R_2 are the radii of nuclei of masses m_1 and m_2, we have

$$\frac{R_1}{R_2} = \left(\frac{m_1}{m_2}\right)^{1/3}$$

$$\Rightarrow \quad \frac{R_1}{R_2} = \left(\frac{27}{8}\right)^{1/3} = \frac{3}{2}$$ **Ans. (C)**

Sol. 5 For the given nuclear reaction energy released is

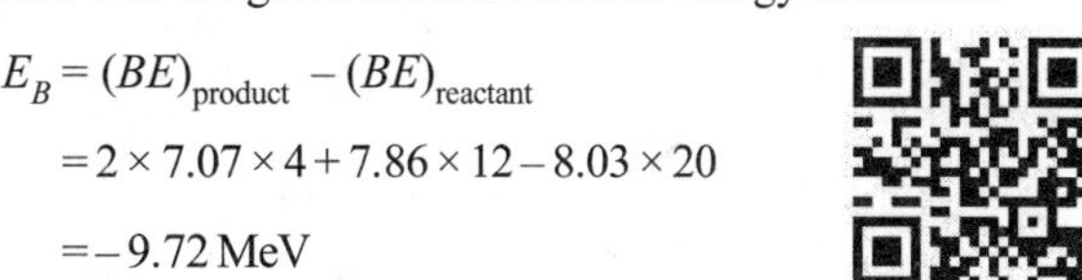

$$E_B = (BE)_{\text{product}} - (BE)_{\text{reactant}}$$

$$= 2\times7.07\times4 + 7.86\times12 - 8.03\times20$$

$$= -9.72\text{ MeV}$$

Thus minimum 9.72 MeV or more energy to be supplied. Thus option (C) is most appropriate **Ans. (C)**

Sol. 6 Mass densities of all nuclei are same so their ratio is 1 **Ans. (A)**

Sol. 7 Power output is energy released per second. Total energy released in 30 days is given as

$$E = 200\text{ MeV} \times \left[\frac{2\times6.03\times10^{23}}{235\times10^{-3}}\right]$$

Power output of reactor is given as

$$\Rightarrow \quad P = \frac{E}{T} = \frac{200\times10^6\times1.6\times10^{-19}}{30\times24\times86400} \times \frac{2\times6.023\times10^{23}}{235\times10^{-3}}$$

$$\cong 63.2\times10^6\text{ W}$$ **Ans. (B)**

Sol. 8 Mass density of nucleus is given as

$$\rho = \frac{m}{V}$$

$$\Rightarrow \quad \rho = \frac{m}{V} = \frac{A}{\frac{4}{3}\pi R^3} = \frac{3}{4\pi r^3}$$

$$\Rightarrow \quad \rho = 2.3\times10^{17}\text{ kg/m}^3$$ **Ans. (C)**

Sol. 9 Binding energy of ${}^{120}_{50}\text{S}_\text{n}$ given as

$$\Delta E = \Delta mc^2$$

where $\quad \Delta m = 50(1.00783) + 70(1.00867) - 119.902199$

$\Rightarrow \quad \Delta E = \Delta m \times 931 = 1020.56$ MeV

BE per nucleon is given as

$$(BE)_n = \frac{1020.56}{120} = 8.5\text{ MeV}$$ **Ans. (A)**

Sol. 10 Mass defect of proton capture reaction is given as

$${}^{7}_{3}\text{Li} + {}^{1}_{1}\text{H} \rightarrow 2\,{}^{4}_{2}\text{He}$$

$$\Delta m = m_{Li} + m_H - 2M_{He}$$

$$\Rightarrow \quad \Delta m = [(7.016 + 1.0079) - 2 \times 4.0026] = 0.0187\,u$$

Energy released by 20 gm of Li is given as

$$E = \frac{\Delta mc^2}{m_{Li}} \times 20$$

$$\Rightarrow \quad E = \frac{0.0187 \times 1.6 \times 10^{-19} \times 10^9}{7.016 \times 1.6 \times 10^{-24}} \times 20\text{ J}$$

$$\Rightarrow \quad E = 5.33 \times 10^{12}\text{ J}$$

$$\Rightarrow \quad E = \frac{5.33 \times 10^{12}}{1000 \times 3600} = 1.48 \times 10^6\text{ kWh}$$ **Ans. (B)**

Sol. 11 Only in option-(A) masses of reactants is more than that of product so this is only possible reaction. **Ans. (A)**

Sol. 12 It is possible only inside the nucleus for a proton to decay into a neutron otherwise it is not possible as mass of a free proton is less than that of a free neutron. **Ans. (B)**

Sol. 13 When nucleus emits the photon it also recoils, say with a recoil speed v to conserve momentum. By conservation of linear momentum, we use

$$Mv = \frac{h}{\lambda} = \left(\frac{h\nu}{c}\right)$$

Total loss of energy of nucleus is given as

$$E = \frac{1}{2}Mv^2 + h\nu = \frac{1}{2}\frac{p^2}{M} + h\nu = \frac{1}{2M}\left(\frac{h\nu}{c}\right)^2 + h\nu$$

$$\Rightarrow \quad E = h\nu\left(1 + \frac{h\nu}{2Mc^2}\right)$$ **Ans. (B)**

Sol. 14 As already explained in Physics Galaxy Vol. 4 about the kinetic energy of α-particle as

$$k_\alpha = \left(\frac{A-4}{A}\right)Q = 5.38\text{ MeV}$$ **Ans. (C)**

Sol. 15 Binding energy of Al nucleus is given as

$$BE = \Delta mC^2$$

Mass defect is calculated as

$$\Delta m = [13 \times 1.00726 + 14 \times 1.00866 - 27.18846]$$

$$\Delta m = 0.02716 = 27.16 \times 10^{-3}\,u \approx 27\,u$$

$$\Rightarrow \quad BE = 27 \times 10^{-3}\text{J}$$ **Ans. (27)**

28.1 Semiconductor

Sol. 1 Conductivity of a semiconductor is given as

$$\sigma = n_e\mu_e e + n_h\mu_h e$$

$$\Rightarrow \quad \sigma = (n_e\mu_e + n_h\mu_h)e$$

$$\Rightarrow \quad \sigma = [5 \times 10^{18} \times 2.0 + 5 \times 10^{19} \times 0.01] \times 1.6 \times 10^{-19}$$

$$\Rightarrow \quad \sigma = [10^{19} + 0.05 \times 10^{19}] \times 1.6 \times 10^{-19}$$

$$\Rightarrow \quad \sigma = 1.05 \times 1.6 = 1.68\,(\Omega\text{-m})^{-1}$$ **Ans. (A)**

Sol. 2 Resistivity of a semiconductor is given as

$$\rho = \frac{1}{\sigma} = \frac{1}{ne\mu_e}$$

$$\Rightarrow \quad \rho = \frac{1}{10^{19} \times 1.6 \times 10^{-19} \times 1.6}$$

$$\Rightarrow \quad \rho = 0.4\,\Omega\text{m}$$ **Ans. (B)**

Sol. 3 By fundamental theory it is evident that on increasing doping concentration the fermi level of p type semiconductor goes down due to acceptor impurity and that of n type semiconduct goes up due to donor impurity. **Ans. (A)**

Sol. 4 For in extrinsic semi-conductor by law of mass action in thermal equilibrium, we use

$$n_e n_h = n_i^2$$

$$\Rightarrow \quad n_e = \frac{n_i^2}{n_h} = \frac{(1.5 \times 10^{16})^2}{4.5 \times 10^{22}} = 5 \times 10^9\text{ m}^{-3}$$ **Ans. (5)**

Sol. 5 Collector current in circuit is given as

$$I_C = \frac{0.6}{1000}$$

Using $\beta = \dfrac{I_C}{I_B}$, we have

$$\Rightarrow \quad I_B = \frac{0.6}{1000 \times 24} = 25\mu\text{A}$$ **Ans. (25)**

Sol. 6 Pentavalent material have excess free electrons so free electron density increases but overall semiconductor always remain neutral. **Ans. (D)**

28.2 PN Junction and Diodes

Sol. 1 Forward bias resistance is given as

$$r_{FB} = \left.\frac{\Delta V}{\Delta I}\right|_{FB} = \frac{0.8 - 0.7}{(20 - 10) \times 10^{-3}}$$

$$\Rightarrow \quad r_{FB} = \frac{0.1}{10 \times 10^{-3}} = 10\,\Omega$$

Reverse bias resistance is given as

$$r_{RB} = \left.\frac{\Delta V}{\Delta I}\right|_{RB} = \frac{10 - 0}{(1 - 0) \times 10^{-6}} = 10^7\,\Omega$$

$$\Rightarrow \quad \frac{r_{FB}}{r_{RB}} = \frac{10}{10^7} = 10^{-6}$$ **Ans. (B)**

Sol. 2 As silicon is forward biased, it will behave like an emf of 0.7 V which is the barrier potential for a Si diode in Forward bias. So equivalent circuit here is

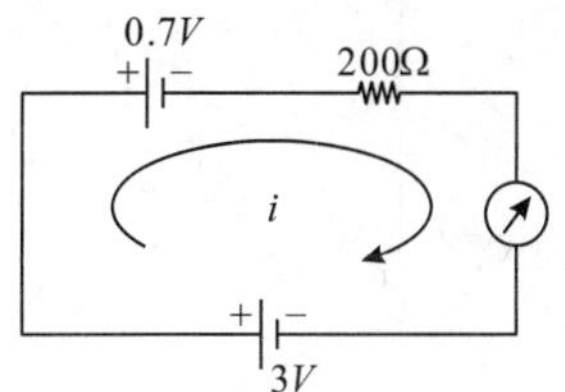

Current in circuit is

$$I = \frac{3-0.7}{200} = \frac{2.3}{200} = 0.00115\text{ A}$$

$\Rightarrow$ $I = 11.5\text{ mA}$ **Ans. (C)**

Sol. 3 In the given circuit across terminals A and B the potential difference due to 15 V battery (if Zener is not present) is given as

$$V_{AB} = \frac{15}{2000} \times 1500 = \frac{225}{20} = 11.25\text{ V}$$

So here $V_{AB} > V_Z \Rightarrow$ Zener will operate in breakdown region and maintain V_Z across terminals A and B.

Under the state of breakdown the equilibrium circuit will be

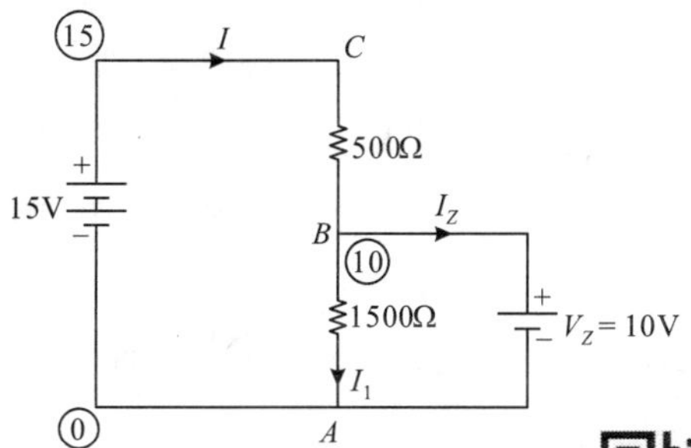

Current in 1500 Ω resistance is

$$I_1 = \frac{10}{1500} = \frac{1}{150}\text{ A}$$

Current in 500 Ω resistance is

$$I = \frac{15-10}{500} = \frac{5}{500} = \frac{1}{100}\text{ A}$$

Current through zener diode is

$$I_Z = I - I_1 = \frac{1}{100} - \frac{1}{150} = \frac{3-2}{300}$$

$\Rightarrow$ $I_Z = \frac{1}{300} = 3.33\text{ mA}$ **Ans. (B)**

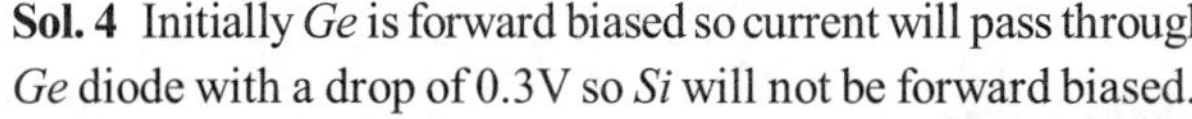

Sol. 4 Initially *Ge* is forward biased so current will pass through *Ge* diode with a drop of 0.3V so *Si* will not be forward biased.

If "*Ge*" is reversed then current will flow through "*Si*" diode hence an effective drop of (0.7 – 0.3) = 0.4V is observed. **Ans. (C)**

Sol. 5 Assuming zener diode doesn't breakdown, current in circuit

$$= \frac{120}{15000} = 8\text{ mA}$$

$\Rightarrow$ The voltage drop across diode = 80 V > 50 V.

$\Rightarrow$ The diode will breakdown and maintain 50 V across 10 kΩ resistance as shown below

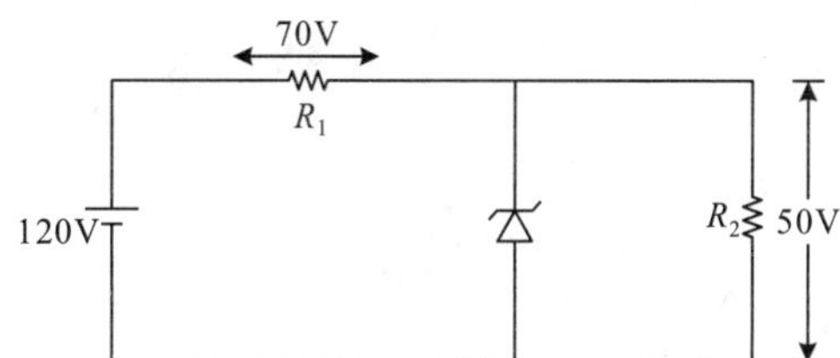

$\Rightarrow$ Current is $R_1 = \frac{70}{5000} = 14\text{ mA}$

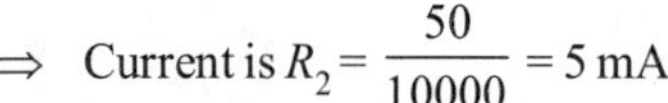

$\Rightarrow$ Current is $R_2 = \frac{50}{10000} = 5\text{ mA}$

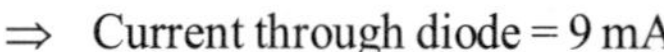

$\Rightarrow$ Current through diode = 9 mA **Ans. (D)**

Sol. 6 As voltage across zener diode must be less than 10V so it will not work in breakdown region, & its resistance will be infinite & current through it = 0. **Ans. (D)**

Sol. 7 Here diode D_2 is reverse biased and D_1 is forward biased so current flows in outer loop only, given as

$$I = \frac{6}{100+150+50} = \frac{6}{300}$$

$\Rightarrow$ $I = 0.02\text{ A}$ **Ans. (B)**

Sol. 8

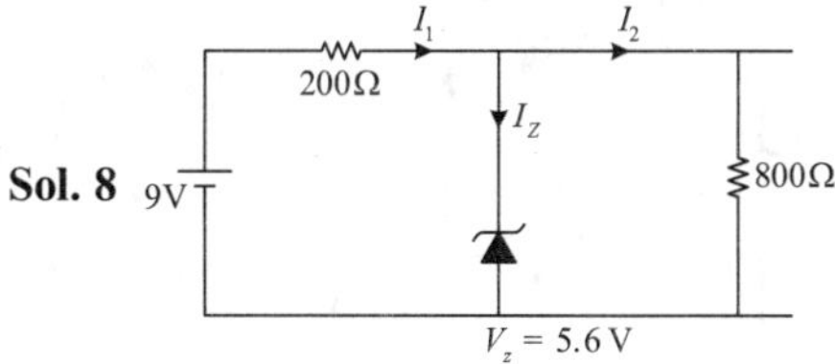

In first loop

$$9V = 200\,I_1 + 5.6$$

$\Rightarrow$ $I_1 = \frac{3.4}{200} = 17\text{ mA}$

In second loop

$$I_2 = \frac{5.6}{800} = 7\text{ mA}$$

Here $I_Z = I_1 - I_2$

$\Rightarrow$ $I_Z = 17 - 7 = 10\text{ mA}$ **Ans. (C)**

Sol. 9 At $V = 8V$

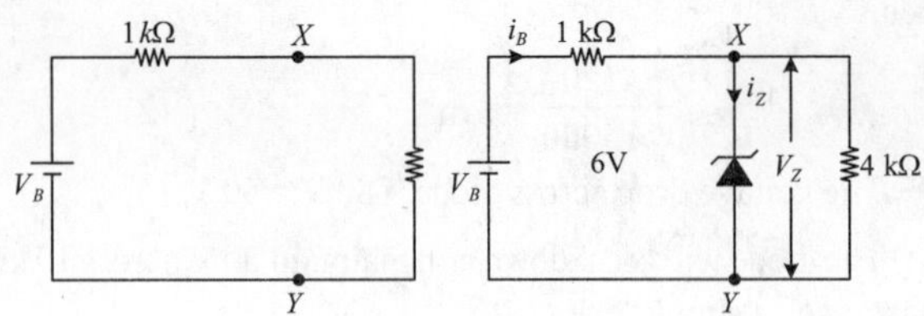

$$V_{XY} = \frac{V_B}{5} \times 4 = \frac{32}{5} = 6.4\text{ V} > \frac{V_Z}{6\text{V}}$$

At $V = 16$ V

$$V_{XY} = \frac{16 \times 4}{5} = \frac{64}{5} = 12.8\text{ V} > 6\text{ V}$$

In breakdown state

$$V_{XY} = 6\text{ V}$$

$$\Rightarrow \quad i_L = \frac{6}{4\text{k}} = 1.5 \times 10^{-3}\text{ A}$$

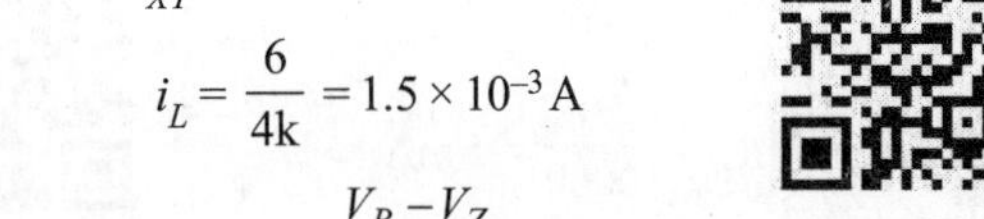

$\Rightarrow$ Battery current $i_B = \dfrac{V_B - V_Z}{1000}$

At 8 V $\rightarrow$ $\quad i_B = \dfrac{8-6}{1000} = 2 \times 10^{-3}$ A

$$i_Z = i_B - i_L = (2 - 1.5) \times 10^{-3} = 0.5\text{ mA}$$

At 16 V $\rightarrow$ $\quad i'_B = \dfrac{16-6}{1000} = 10 \times 10^{-3}$ A

$$i_Z = i'_B - i_L = (10\ 1.5) \times 10^{-3} = 8.5\text{ mA}$$ **Ans. (B)**

Sol. 10 Current I_2 will be maximum when input voltage is also maximum at 16 V

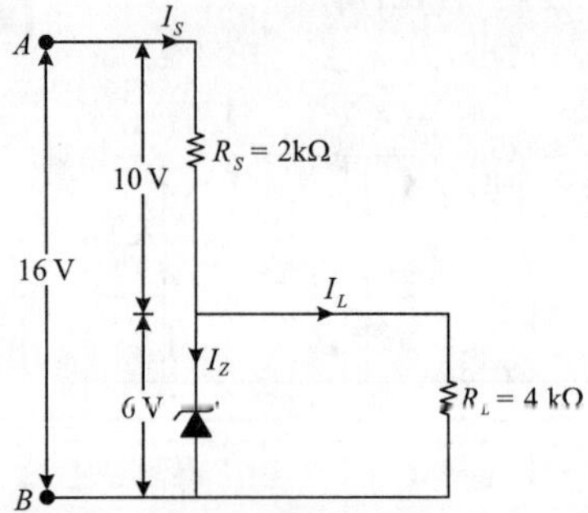

At voltage across R_L is maintained at $6V$

$$\Rightarrow \quad I_L = \frac{6}{4000} = 1.5\text{ mA}$$

Current through R_S is

$$I_S = \frac{10}{2000} = 5\text{ mA}$$

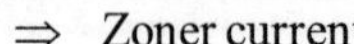

$\Rightarrow$ Zoner current

$$I_Z = I_S - I_L = 5 - 1.5 = 3.5\text{ mA}$$ **Ans. (B)**

Sol. 11 As diode is forword based, we use circuit as

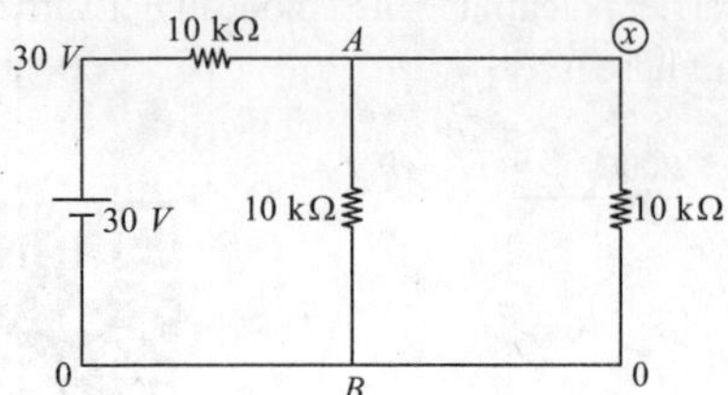

Using KCL equation for x, we have

$$\frac{30-x}{10} + \frac{0-x}{10} + \frac{0-x}{10} = 0$$

$$3 = \frac{3x}{10}$$

$$x = 10\text{ V}$$ **Ans. (B)**

Sol. 12 In the figure shown we can if diode D_1 is forward biased potential of point A will become

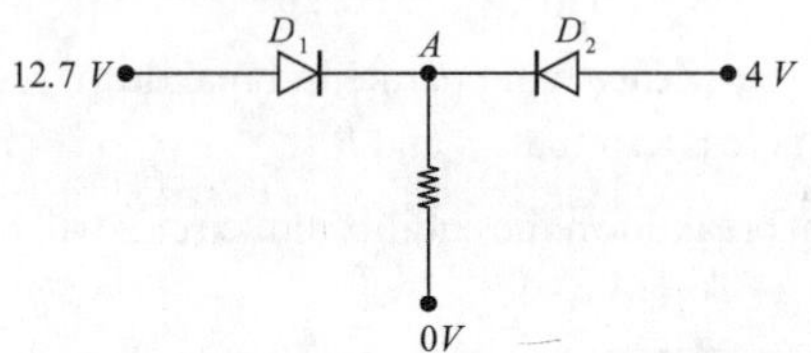

$$V_A = 12.7 - 0.7 = 12\text{V}$$

and diode D_2 will be reverse biased and behave like open circuit. **Ans. (12)**

Sol. 13 In given circuit both the diodes will be reverse biased so effective circuit is shown below

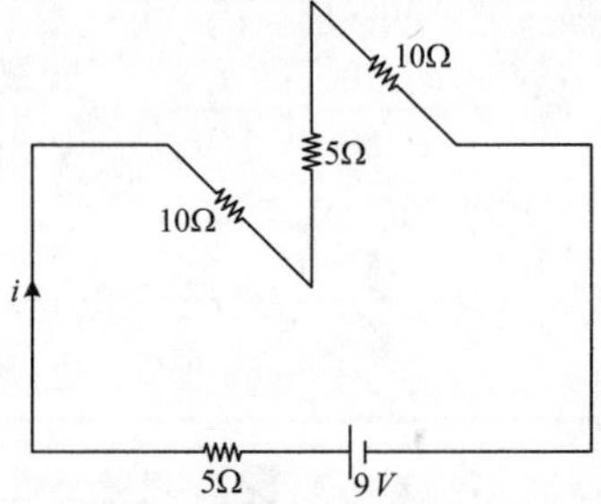

Thus current in circuit is given as

$$i = \frac{9}{(5+10+5+10)} = \frac{9}{30}\text{ A}$$ **Ans. (C)**

Sol. 14 Diode in circuit of capacitor A will remain reverse biased so no current flows. Thus charge on capacitor A will remain constant at

$$Q_A = CV$$

Discharge current flows in circuit of capacitor B so charge capacitor is given as

$$\Rightarrow \quad q = CVe^{-\frac{t}{RC}}$$

at $t = CR$, charge is given as

$$Q_B = CVe^{-1} = \frac{CV}{e}$$ **Ans. (A)**

Sol. 15 For the given circuit potential difference across 400 Ω resistance is 4V, so current in circuit is

$$I = \frac{4}{400} = \frac{1}{100} A$$

Power dissipated in each diode is given as

$$P_d = VI$$

$$\Rightarrow \quad P_d = 4 \times \frac{1}{100} W$$

$$\Rightarrow \quad P_d = 40 \times 10^{-3} \text{mW}$$ **Ans. (40)**

Sol. 16 Circuit of described situation is shown below

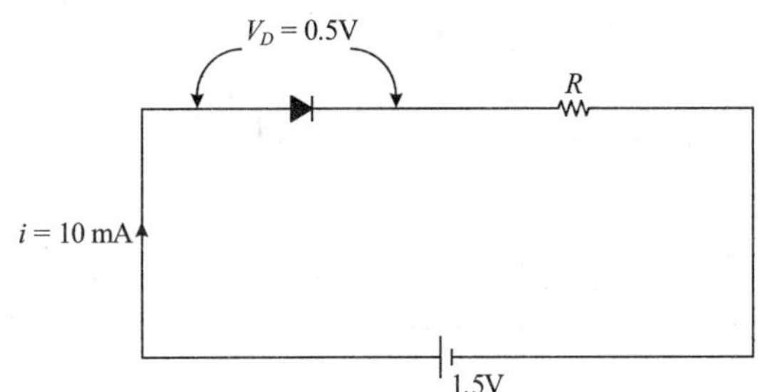

Using KVL equation in loop gives

$$1.5 - 0.5 - iR = 0$$

$$\Rightarrow \quad R = \frac{1}{i} \times \frac{1}{10 \times 10^{-3}} = 100\,\Omega$$ **Ans. (A)**

Sol. 17 Energy band gap is given as

$$\Delta E = \frac{hc}{\lambda} = \frac{19.89 \times 10^{-26}}{400 \times 10^{-9} \times 1.6 \times 10^{-19}}$$

$$\Rightarrow \quad \Delta E = 3.107 \text{ eV}$$ **Ans. (C)**

Sol. 18 As there are two zener diodes in opposite sense in both cycles one will breakdown and cut off voltage at 6 V so option (B) is correct. **Ans. (B)**

Sol. 19 For photodiode on increasing reverse voltage, current increases then saturation arises at a specific intensity of light falling on it.

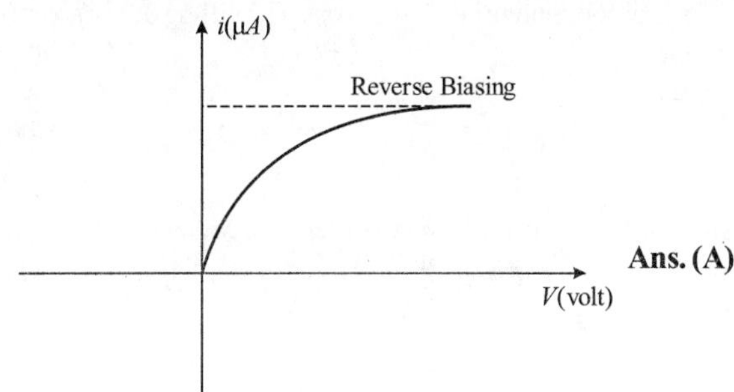

Ans. (A)

Sol. 20 When input voltage increases from 0, breakdown in 4V zener take place when $V_{in} = 4V$ but when $V_{in} > 4V$ then also V_{out} increases because of potential drop across 100Ω resistance and when $V_{in} = 6V$ or more breakdown in 6V zener occurs and V_{out} remain constant at 6V now on lord. **Ans. (D)**

Sol. 21 During $t = 0$ to $t = 5$ μC charge on capacitor grows and becomes constant for $t = 5$ μC to $t = 10$ μC then again grows with time so most appropriate option is (A) **Ans. (A)**

Sol. 22 In the figure shown current through 2kΩ resistance can be directly given by zener voltage as

$$I = \frac{5}{2 \times 10^3} = 2.5 \times 10^{-3} \text{A}$$

$$I = 25 \times 10^{-4} \text{A}$$ **Ans. (25)**

Sol. 23 Zener diode is constructed with high doping concentration and have a thin depletion layer. **Ans. (B)**

Sol. 24 Back to back connections of diodes will not act like a transistor because its working require different dopin concentrations in the emitter, base and collector along with different sizes of these regions so Statement-I is false. For a transistor the current amplification factor is defined as

$$\beta = \frac{i_c}{i_b}$$

Thus Statement-II is true. **Ans. (B)**

Sol. 25 After connection of battery as described in question, circuit is shown in figure below

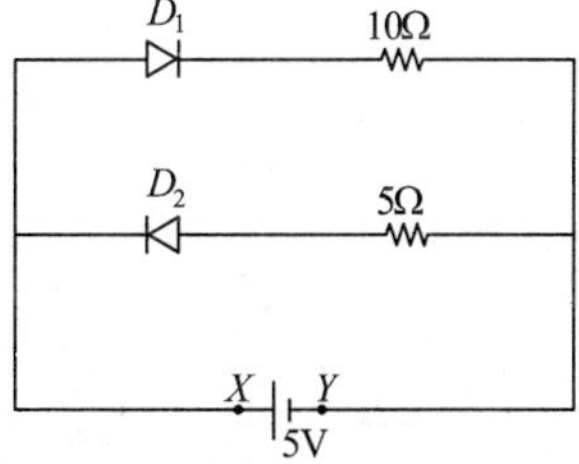

In this only D_1 will be forward biased and conduct and for

silicon diode, potential drop across D_1 will be 0.7 V so current is given as

$$I = \frac{5-0.7}{10} = 0.43\text{ A}$$ **Ans. (C)**

Sol. 26 From the energy gap given, the wavelength of emitted light due to transition of electrons in depletion layer can be calculated as

$$\lambda = \frac{12431}{1.9} \times 10^{-10} = 654\text{ nm}$$

The color corresponding to this wavelength is red. **Ans. (C)**

Sol. 27 After connection of battery as described in question, circuit is shown in figure below

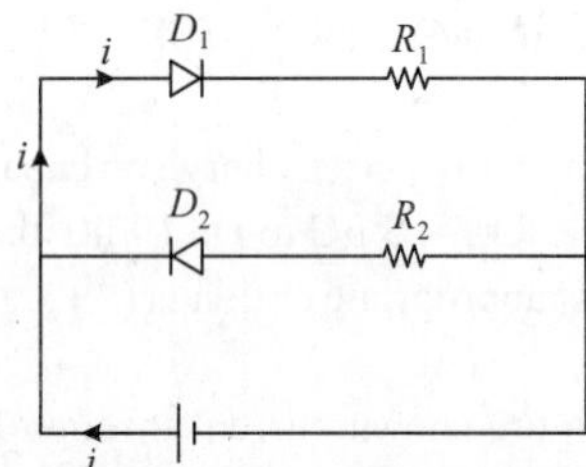

In this circuit D_1 will be forward biased and D_2 will be revers biased so no current through D_2 and R_2. Here current through the battery is calculated as

$$i = \frac{6}{120+130+50} = \frac{6}{300} = 20\text{ mA}$$ **Ans. (20)**

Sol. 28 Under breakdown conditions current through 4000 Ω resistance can be given as

$$i = \frac{60}{4000}\text{ A}$$

Current through 5000 Ω resistance is given as

$$i_1 = \frac{30}{5000}\text{ A}$$

Current through the diode can now be calculated as

$$i_z = i - i_1 = \frac{60}{4000} - \frac{30}{5000} = \frac{9}{1000} = 9\text{ mA}$$ **Ans. (9)**

Sol. 29 Figure below shows the current distibution in the given circuit.

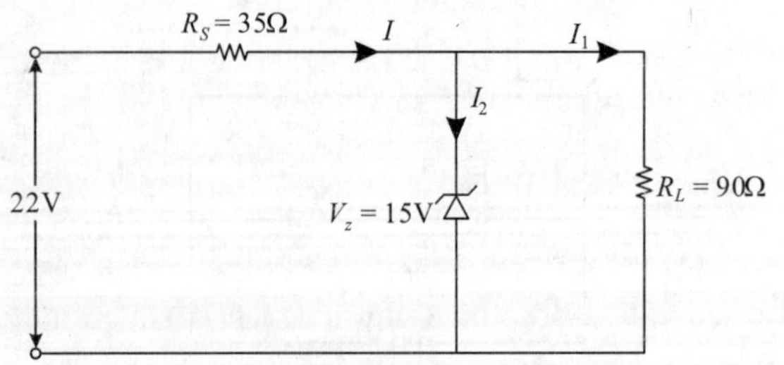

Voltage across R_S is calculated as

$$V_S = 22 - 15 = 7\text{ V}$$

Current through R_S is calculated as

$$I = \frac{7}{35} = \frac{1}{5}\text{ A}$$

Current through 90Ω resistance is calculated as

$$I_2 = \frac{15}{90} = \frac{1}{6}\text{ A}$$

Current through the zener diode is calculated as

$$I_1 = \frac{1}{5} - \frac{1}{6} = \frac{1}{30}\text{ A}$$

Power dissipated across zener diode is calculated as

$$P = VI$$

$$\Rightarrow \quad P = 15 \times \frac{1}{30} = 0.5 = 5 \times 10^{-1}\text{ W}$$ **Ans. (5)**

Sol. 30 In the circuit shown current through battery can be calculated as

$$I = \frac{100-50}{1000} = 50\text{ mA}$$

Current through 2000Ω resistance is calculated as

$$I' = \frac{50}{2000} = 0.025\text{ mA}$$

$$\Rightarrow \quad I_z = I - I' = 50\text{ mA} - 25\text{ mA} = 25\text{ mA}$$ **Ans. (B)**

Sol. 31 For a PN junction diode,dynamic resistance in forward biasing is given as

$$r_D = \frac{\Delta V}{\Delta I}$$

$$\Rightarrow \quad r_D = \frac{0.75-0.65}{(5-1)\times 10^{-3}} = 25$$ **Ans. (25)**

Sol. 32 For maximum potential drop across the diode through potential divider, the voltage across protective resistance will be 20 – 8 = 12 V and the effective resistance across the zener diode can be given as

$$R_D = \frac{V_Z^2}{P} = \frac{64}{0.5} = 128\,\Omega$$

So the current through the branch is given as

$$i = \frac{20}{128+R_P}$$

$$\Rightarrow \quad 12 = \frac{20}{128+R_P} \times R_P$$

$$\Rightarrow \quad 1536 = 8\,R_P$$

$$\Rightarrow \quad R_P = 192\,\Omega$$ **Ans. (192)**

Sol. 33 From the circuit given in question, as current through zener is 90 mA, we use

$$\frac{50-5}{R} \geq 90\text{ mA}$$

$\Rightarrow \quad R \le \frac{45}{90} \times 10^3$

$\Rightarrow \quad R \le 500\,\Omega$ **Ans. (500)**

Sol. 34 The currents in the various branches

$$i_1 = \frac{14}{1} = 14\text{ mA}$$

$$i_3 = \frac{10}{5} = 2\text{ mA}$$

$$i_2 = i_1 - i_3 = 12\text{ mA}$$

Power consume in zerer diode is given as

Pavg $= V \times i = 0.12$ W **Ans. (120)**

Sol. 35 Power consumption in diode can be written as

$$P = Vi$$

$\Rightarrow \quad i = \frac{P}{V} = \frac{2}{10} = 0.2\text{ mA}$

For $V_S = 14$ V, the voltage drop across R_S can be given as

$$V_R = 14 - 10 = 4\text{ V}$$

By Ohm's law for R_S, we have

$$V = iR_s$$

$\Rightarrow \quad R_S = \frac{4}{0.2 \times 10^{-2}} = 20\text{ m}\Omega$ **Ans. (20)**

Sol. 36 In the given circuit current will flow through the diode when it is in forward biased state. The potential difference across the diode $\Delta V = (10 \sin\omega t - 3)$ V and ΔV across the diode will be zero, when diode is in reverse biased state. For this case the correct waveform is drawn in figure below.

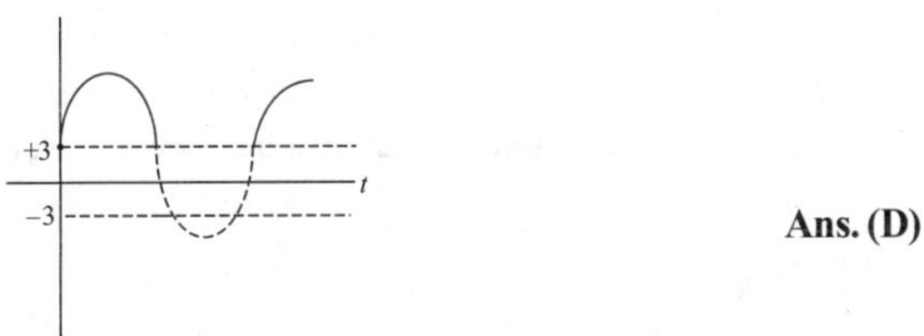

Ans. (D)

Sol. 37 A capacitor has low reactance in AC circuit but act as an open circuit for dc so it short circuits the pulsating voltage and allows direct current to pass hence statement-I is true. Similarly an inductor has high reactance in AC circuit but act as short circuit for dc so it is connected in series to allow dc to pass through it and blocks ac hence statement-II is also true. **Ans. (C)**

Sol. 38 Equivalent resistance of the given circuit with forward biased diodes to be considered short circuited and reverse biased diods as open circuit, is calculated as

$$R_{eq} = \frac{160}{2} + 20$$

$\Rightarrow \quad R_{eq} = 80 + 20$

$\Rightarrow \quad R_{eq} = 100$

$\Rightarrow \quad i = \frac{200}{100} = 2\text{A}$ **Ans. (A)**

28.3 Transistors

Sol. 1 Common emitter current gain $\beta = 69$

$\Rightarrow \quad \alpha = \frac{\beta}{1+\beta} = \frac{69}{70}$

we use $\quad \alpha = \frac{I_C}{I_E}$

$\Rightarrow \quad I_C = \alpha I_E = \frac{69}{70} \times 7.0 = 6.9\text{ mA}$ **Ans. (B)**

Sol. 2

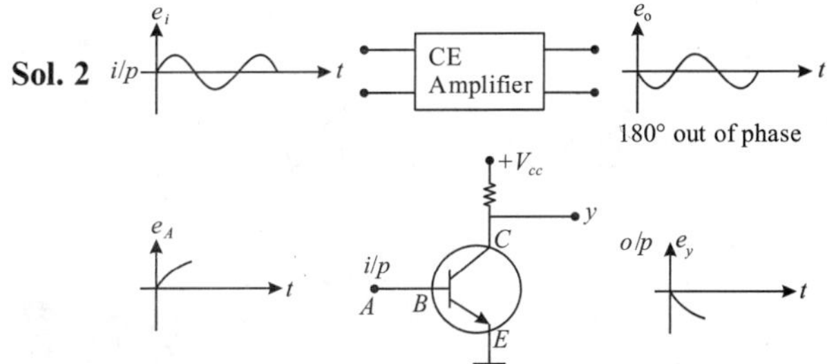

Behaves like a NOT gate (inverter)

Ans. (B)

Sol. 3

V_{cc}, R_L, $\Delta V_{out} = \Delta I_C R_L$, I_C, C, B, I_B, V_{CE}, V_{BE}, E, I_E, $\Delta V_{BE} = \Delta I_B \cdot R_{BE}$, ΔV_i

Current gain $\quad \beta = \frac{\Delta I_C}{\Delta I_B}$

Voltage gain $\quad A_V = \frac{\Delta V_{out}}{\Delta V_{in}} = \frac{\Delta I_C R_L}{\Delta I_B R_{BE}}$

$\Rightarrow \quad A_V = \beta \frac{R_L}{R_{BE}}$

Power gain = Voltage gain × Current gain

$A_P = \beta^2 \frac{R_L}{R_{BE}}$ **Ans. (D)**

Sol. 4 At saturation, CE voltage becomes zero, so we have

$$V_{CE}=0$$

$$\Rightarrow \quad V_{CE}=V_{CC}-I_C R_C$$

$$\Rightarrow \quad I_C=\frac{V_{CC}}{R_C}=5\times10^{-3}\,\text{A}$$

Given that $\beta_{dc}=200$

$$\Rightarrow \quad \beta_{dc}=\frac{I_C}{I_B}$$

$$\Rightarrow \quad I_B=\frac{5\times10^{-3}}{200}$$

$$\Rightarrow \quad I_B=25\,\mu\text{A}$$

At input side of transistor, we use

$$V_{BB}=I_B R_B+V_{BE}$$

$$\Rightarrow \quad V_{BB}=(25\mu\text{A})(100k\Omega)+1V$$

$$\Rightarrow \quad V_{BB}=3.5V$$

Ans (B)

Sol. 5 At saturation

$$V_{CE}=0$$

$$\Rightarrow \quad i_C=\frac{V_{CC}}{R_C}=\frac{10}{1000}0.01\,\text{A}$$

De current gain of transistor is

$$\beta=\frac{i_C}{i_B}$$

$$\Rightarrow \quad i_B=\frac{i_C}{\beta}=\frac{0.01}{250}=4\times10^{-5}\,\text{A}=40\,\mu\text{A}$$ **Ans. (C)**

Sol. 6 Input current

$$i_b=15\,\mu\text{A}$$

Output current

$$i_c=3\,\text{mA}$$

Output resistance

$$R_L=10\,\text{mV}$$

Input voltage

$$V_{BE}=10\,\text{mv}$$

Input resistance

$$r_{in}=\frac{V_{BE}}{i_b}$$

$$\Rightarrow \quad r_{in}=\frac{10\times10^{-3}}{15\times10^{-6}}=\frac{2000}{3}=0.67\,\text{k}\Omega$$

Output voltage

$$v_{out}=i_c R_L=3\,\text{V}$$

Voltage gain

$$A_V=\frac{V_{out}}{V_{in}}=\frac{3}{10\times10^{-3}}=300$$ **Ans. (D)**

Sol. 7 Power gain in dB is given as

$$10\log_{10}=\left(\frac{P_{out}}{P_{in}}\right)=60$$

$$\Rightarrow \quad \frac{P_{out}}{P_{in}}=\text{P}_{gain}=60^6$$

We use $$P_{gain}=\beta^2\times\frac{R_{out}}{R_{in}}$$

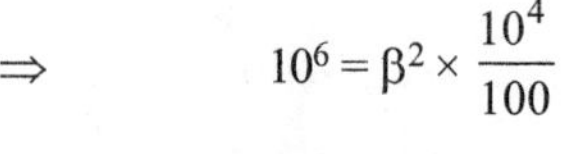

$$\Rightarrow \quad 10^6=\beta^2\times\frac{10^4}{100}$$

$$\Rightarrow \quad \beta^2=10^4$$

$$\Rightarrow \quad \beta=100$$ **Ans. (D)**

Sol. 8 Voltage gain of a transistor is

$$V_{gain}=\left(\frac{\Delta I_C}{\Delta I_B}\right)\times\frac{R_{out}}{R_{in}}$$

$$=\left(\frac{5\times10^{-3}}{100\times10^{-6}}\right)\times\frac{100\times10^3}{100}=5\times10^4$$

Power gain $$P_{gain}=\frac{\Delta I_c}{\Delta I_b}\times V_{gain}$$

$$=\frac{5\times10^{-3}}{100\times10^{-6}}\times5\times10^4=2.5\times10^6$$ **Ans. (C)**

Sol. 9 In figure shown if any switch is closed base of transistor will become forward biased and output potential difference becomes zero so truth table is as shown below which is for

$$Y=\bar{A}\cdot\bar{B}$$

A	B	Y
0	0	1
1	0	0
0	1	0
1	1	0

Ans. (C)

Sol. 10 From the given graph at $I_C=4$ mA, we calculate ΔI_B and ΔI_C as

$$\Delta I_B=(30-20)=10\,\mu\text{A}$$

$$\Delta I_C=(4.5-3)\,\text{mA}=1.5\,\text{mA}$$

β_{ac} is given as

$$\beta_{ac}=\frac{\Delta I_C}{\Delta I_B}=\frac{1.5\text{mA}}{10\mu\text{A}}=150$$

$$\Rightarrow \quad \beta_{ac}=150$$ **Ans. (150)**

Sol. 11 For a transistor current through terminals are related as

$$I_\varepsilon = I_C + I_B$$

$$\Rightarrow \quad \Delta I_\varepsilon = \Delta I_C + \Delta I_B$$

$$4\,\text{mA} = 3.5\,\text{mA} + \Delta I_B$$

$$\Rightarrow \quad \Delta I_B = 0.5\,\text{mA}$$

Transistor constant β is given as

$$\beta = \frac{\Delta I_C}{\Delta I_B} = \frac{3.5}{0.5} = 7$$ **Ans. (A)**

Sol. 12 For the two positions of the object, magnification is same for the convex lens so we use

For $u = -10$cm

$$m = \frac{f}{-10+f} \quad \ldots(1)$$

For $u = -20$ cm

$$-m = \frac{f}{-20+f} \quad \ldots(2)$$

From equations-(1) & (2), we have

$$-1 = \frac{f-20}{f-10}$$

$$\Rightarrow \quad 10 - f = f - 20$$

$$\Rightarrow \quad 30 = 2f$$

$$\Rightarrow \quad f = 15\,\text{cm}$$ **Ans. (15)**

Sol. 13 For the common emmitter amplifier the power gain can be expressed as

$$G_P = \beta^2 \times \frac{R_0}{R_i}$$

$$\Rightarrow \quad 10^6 = \beta^2 \times \frac{R_0}{R_i}$$

$$\Rightarrow \quad 10^6 = \beta^2 \times \frac{10^4}{10^2}$$

$$\Rightarrow \quad \beta^2 = 10^4 = 100$$ **Ans. (100)**

Sol. 14 The currents in transistor terminals are related to transistor constants are given as

$$\alpha = \frac{I_C}{I_E}, \beta = \frac{I_C}{I_B}$$

We use

$$I_E = I_B + I_C$$

$$\Rightarrow \quad \alpha = \frac{I_C}{I_B + I_C} = \frac{1}{\frac{I_B}{I_C}+1}$$

$$\Rightarrow \quad \alpha = \frac{1}{\frac{1}{\beta}+1} = \frac{\beta}{1+\beta}$$ **Ans. (D)**

Sol. 15 Current gain from the figure is given as

$$\beta = \frac{\Delta I_C}{\Delta I_B} = \frac{2\times 10^{-3}}{10\times 10^{-6}}$$

$$\beta = \frac{1}{5} \times 10^3 = 2 \times 10^2 = 200$$ **Ans. (200)**

Sol. 16 For amplification action transistors to be used in active region. **Ans. (B)**

Sol. 17 In the circuit given, we can analyze that diodes are forward biased so these will act as short circuit and whole of battery current will pass through short circuited wire and hence input current to transistor will be zero so output current will also remain zero hence the output voltage will be maintained at 5V. **Ans. (5)**

Sol. 18 Currents in transistor terminals are related **as**

$$I_E = I_B + I_C$$

$$\Rightarrow \quad \frac{I_E}{I_C} = \frac{I_B}{I_C} + 1$$

$$\Rightarrow \quad \frac{1}{\alpha} = \frac{1}{\beta} + 1$$

$$\Rightarrow \quad \beta = \frac{\alpha}{1-\alpha}$$ **Ans. (C)**

28.4 Logic Gates

Sol. 1

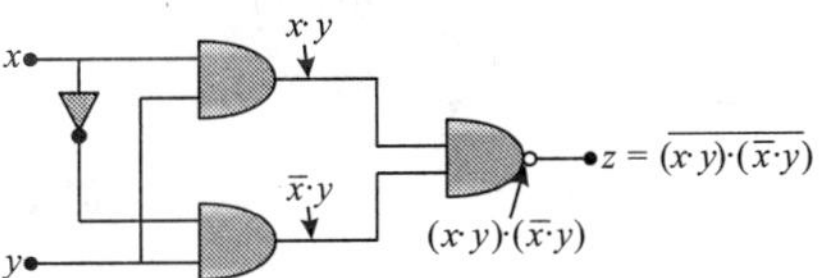

We calculate z for cases

$x = 0, 0, 1, 1$

$y = 0, 1, 0, 1$

output $\quad z = \overline{(x.y).(\bar{x}.y)}$

At $x = 0, y = 0 \Rightarrow \quad z = \overline{0\cdot 0} = 1$

At $x = 0, y = 1 \Rightarrow \quad z = \overline{0\cdot 1} = 1$

At $x = 1, y = 0 \Rightarrow \quad z = \overline{0\cdot 0} = 1$

At $x = 1, y = 1 \Rightarrow \quad z = \overline{1\cdot 0} = 1$

Ans. (C)

Sol. 2 $\quad P = \bar{X} + Y$

$$Q = \overline{\bar{Y}\cdot X} = Y + \bar{X}$$

$$\text{Output} = \overline{P+Q}$$

To make output '1' $P + Q$ must be '0'

So, $Y = 0$ and $X = 1$ **Ans. (D)**

Sol. 3 The boolean output of given circuit is shown in figure below

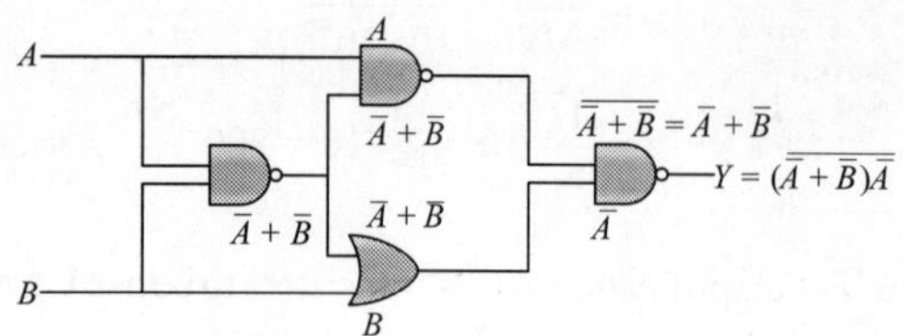

The output is simplified as

$$Y=\overline{(\overline{A}+\overline{B})\overline{A}}$$

$\Rightarrow$ $$Y=\overline{\overline{A}+\overline{A}B}$$

$\Rightarrow$ $$Y=A(\overline{\overline{A}B})$$

$\Rightarrow$ $$Y=A(A+\overline{B})$$

$\Rightarrow$ $$Y=A+A\overline{B}=A\overline{B}$$

Ans. (B)

Sol. 4 Truth table of the given logic circuit is written as

A	B	Equivalent
0	0	0
0	1	1
1	0	1
1	1	1

This is the table for 'OR' gate **Ans. (A)**

Sol. 5

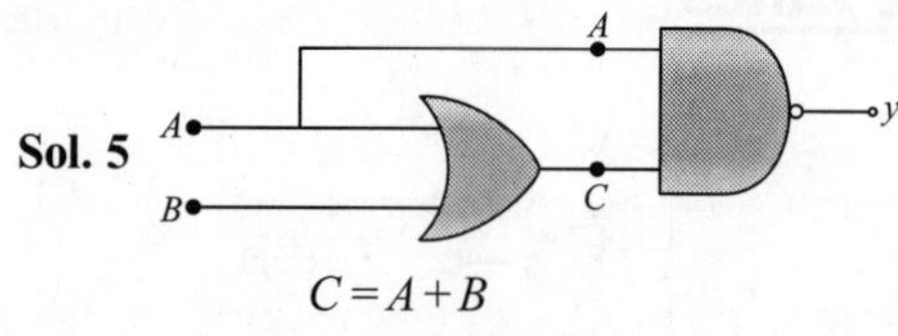

$$C=A+B$$

and $$y=\overline{A.C}$$

Truth table for this circuit is given as

A	B	$C=(A+B)$	$A.C.$	$y=\overline{A.C}$
0	0	0	0	1
0	1	1	0	1
1	0	1	1	0
1	1	1	1	0

Ans. (A)

Sol. 6 Reversible operation is only possible in one input, one output systems so only option is (B). **Ans. (B)**

Sol. 7 For the given circuit boolean expression of output Y is given as

$$Y=\overline{\overline{AB}.A}$$

$\Rightarrow$ $$Y=\overline{\overline{AB}}+\overline{A}$$

$\Rightarrow$ $$Y=0+0$$

$\Rightarrow$ $$Y=0$$

Ans. (B)

Sol. 8 In below figure intermediate outputs of gates are mentioned as C, D and E

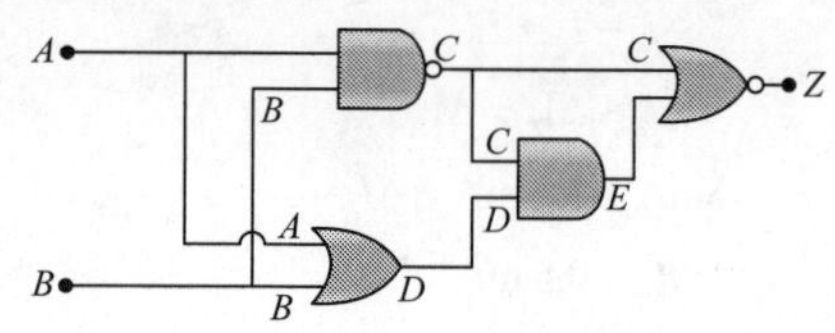

Relation between output z and inputs A and B are deduced as

$$Z=\overline{(C+E)}$$

$\Rightarrow$ $$Z=\overline{(C+CD)}$$

$\Rightarrow$ $$Z=\overline{(C(1+D))}=\overline{C}$$

$\Rightarrow$ $$Z=\overline{\overline{AB}}=AB$$

Thus truth table for given circuit is made as shown in figure

A	B	Z
0	0	0
0	1	0
1	0	0
1	1	1

Ans. (C)

Sol. 9 The truth table of given circuit is shown below

A	B	C	
0	0	0	0
1	0	0	0
0	1	0	0
0	0	1	0
1	1	0	0
1	0	1	0
0	1	1	0
1	1	1	1

This corresponds to AND gate. **Ans. (A)**

Sol. 10 Truth table of given logic circuit is given as

A	B	Y
0	0	0
0	1	1
1	0	1
1	1	1

For the given input wavelength correct output wavelength is given as

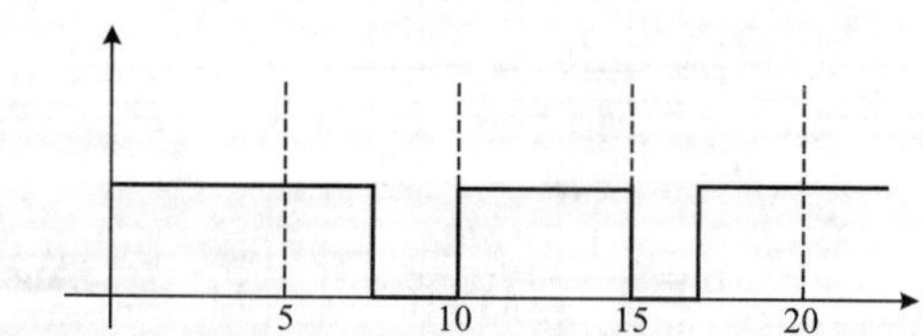

None of given option is correct. **Ans. (None)**

Sol. 11 The Boolean expression for output of the given logic circuit can be written as

$$\overline{A+\overline{B}}=C$$

Simplifying the expression, we get

$$C=\overline{A}\cdot B$$

This expression is realized by the circuit shown in option (D).

Ans. (D)

Sol. 12 The boolean expression for the given logic circuit is written as

$$y=\overline{(A\overline{B}+\overline{A}B)}$$

$$\Rightarrow \quad y=\overline{A\overline{B}}\cdot\overline{\overline{A}B}$$

$$\Rightarrow \quad y=(\overline{A}+B)\cdot(A+\overline{B})$$

$$\Rightarrow \quad y=\overline{A}.A+\overline{A}.\overline{B}+A.B+B.\overline{B}$$

$$\Rightarrow \quad y=AB+\overline{A}\,\overline{B}$$

The truth table for this boolean expression is written below

A	B	$Y=AB+\overline{A}\overline{B}$
0	0	1
0	1	0
1	0	0
1	1	1

Ans. (C)

Sol. 13 The boolean expression of the given logic circuit is written as

$$Y=\overline{\overline{A}+B}=A\cdot\overline{B}$$

Ans. (A)

Sol. 14 The flow of inputs through the logic circuit is shown in the diagram below with voltage values (1 for high and 0 for low) are shown at every terminal of circuit.

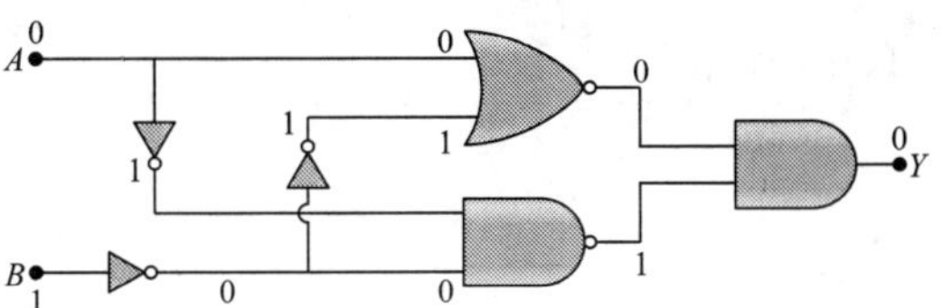

Hence the output value in this circuit is 0. **Ans. (0)**

Sol. 15 The boolean expresson for the final output of the logic cirucit given in question is written as

$$Y=\overline{A}.\overline{B}=\overline{A+B}$$

This represents a NOR gate. **Ans. (A)**

Sol. 16 Boolean expression of the given logic circuit can be expressed as shown in figure below which represents a NAND gate.

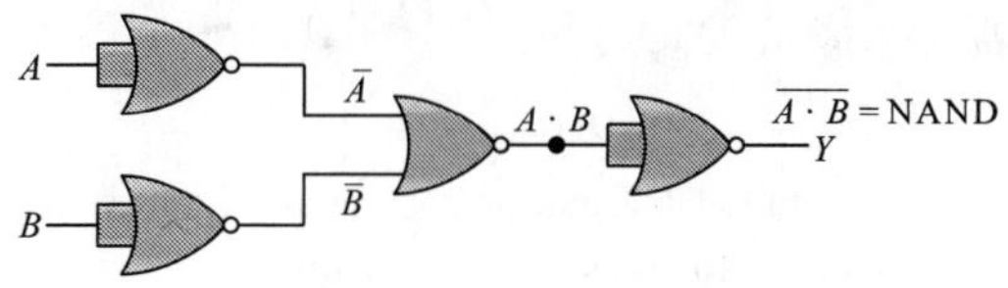

Ans. (B)

Sol. 17 The boolean expression of the given logic circuit is given as

$$Y=\overline{AB}.(A+B)=(\overline{A}+\overline{B}).(A+B)=\overline{A}B+A\overline{B}$$

This is the output of XOR gate. **Ans. (D)**

Sol. 18 The given logic circuit output can be expressed as boolean expression

$$Y=\overline{\overline{A}.\overline{B}}=A+B$$

This indicates the output of *OR* gate. **Ans. (A)**

Sol. 19 The boolean expression of the given logic circuit can be written as

$$Y=AB+\overline{B}$$

Ans. (C)

Sol. 20 The boolean expression for the given logic circuit is expressed as

$$Y=\overline{A+B}$$

Above expression represents NOR gate. **Ans. (A)**

Sol. 21 The boolean expression for the output of the given logic circuit can be written using the diagram shown below

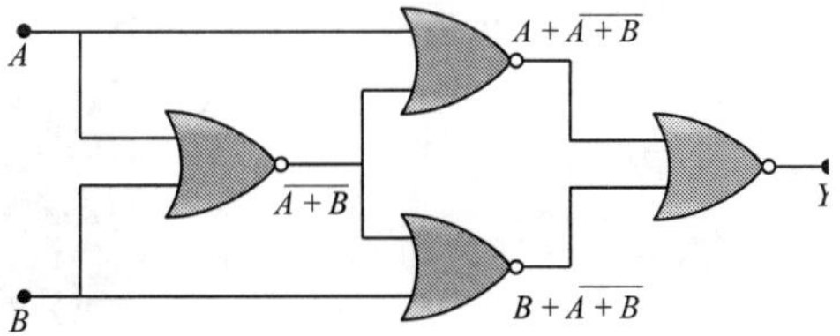

The boolean expression of output Y is written as

$$y=\overline{\overline{(A+\overline{A+B})+(B+\overline{A+B})}}$$

$$y=(A+\overline{A+B})\cdot(B+\overline{A+B})$$

According to the above expression, truth table can be drawn as

A	B	Y
0	0	1
1	1	1
0	1	0
1	0	0

Ans. (A)

Sol. 22 For the given logic circuit the boolean expression for the output Y is given as

$$Y = \overline{\overline{AB}.(A+B)} = \overline{A+B}$$

This is equivalent to NOR gate so cooresponding truth table of this circuit can be drawn as shown in below

A	B	Y
0	0	1
0	1	1
1	0	1
1	1	0

Ans. (D)

Sol. 23 In first circuit either of inputs A or B goes high, will make the output high so it is behaving like an OR gate and in second circuit, due to npn transistor as the input B goes high, base of transistor allows conduction and due to which output goes low and vice versa so this is behaving like a NOT gate. **Ans. (B)**

29.1 Communication Systems

Sol. 1 For resonating condition in antena for transmission of EM wave the antenna length should be of the order of λ. Power transmission in EM waves is inversely proportional to λ^2

$$\Rightarrow \quad P_{eff} \propto \frac{1}{\lambda^2}$$

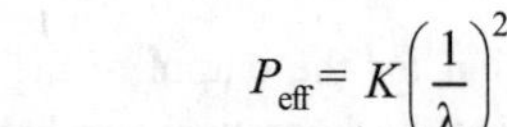

$$\Rightarrow \quad P_{eff} = K\left(\frac{1}{\lambda}\right)^2$$

Ans. (A)

Sol. 2 Maximum distance upto which signal can be broadcasted is

$$d_{max} = \sqrt{2Rh_T} + \sqrt{2Rh_R}$$

$$\Rightarrow \quad d_{max} = \sqrt{2\times 6.4\times 10^6}\left[\sqrt{140}+\sqrt{40}\right]$$

$$\Rightarrow \quad d_{max} = 65\,\text{km}$$

Ans. (D)

Sol. 3 As range of transmission on ground is

$$R \propto \sqrt{h}$$

$\Rightarrow$ To double the range h should be made 4 times. **Ans. (B)**

Sol. 4 Range of transmission in LOS communication is given as

$$\text{Range} = \sqrt{2Rh_T} + \sqrt{2Rh_R}$$

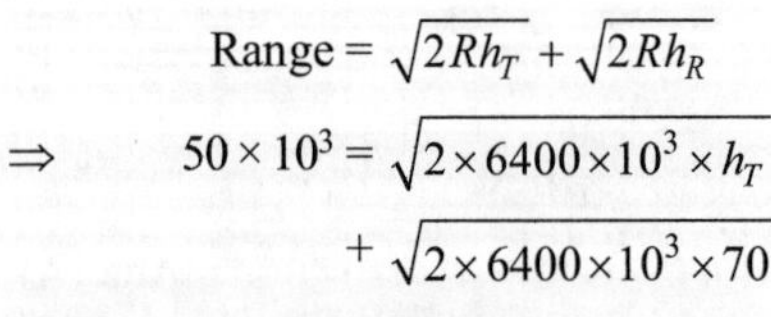

$$\Rightarrow \quad 50\times 10^3 = \sqrt{2\times 6400\times 10^3\times h_T} + \sqrt{2\times 6400\times 10^3\times 70}$$

$$\Rightarrow \quad h_T = 32\,\text{m}$$

Ans. (C)

Sol. 5 The given question is based on conceptual facts. Optical fibre communication uses infrared light for signal transmission, radar uses radio waves, sonar uses ultrasound and mobile phones use microwaves for their signal transmission. **Ans. (D)**

Sol. 6 As it is given that sound level decreases by 5dB every km so sound level decreased in 20 km will be 100 dB. So we use

$$L_2 - L_1 = 10\log_{10}\frac{I_2}{I_1}$$

$$\Rightarrow \quad -100 = 10\log_{10}\frac{I_2}{I_1}$$

$$\Rightarrow \quad \frac{I_1}{I_2} = 10^{10}$$

$$\Rightarrow \quad I_2 = 10^{-10} I_1$$

$$\Rightarrow \quad P_2 = 10^{-10} P_1$$

$$\Rightarrow \quad P_2 = 10^{-10}\,\text{W}$$

$$\Rightarrow \quad P_1 = 10^{-10}\times 100 = 10^{-8}\,\text{W}$$

$$\Rightarrow \quad x = 8$$

Ans. (8)

Sol. 7 Resonant frequency of the circuit is given as

$$\omega_0 = \frac{1}{\sqrt{LC}}$$

$$\Rightarrow \quad 2\pi f_0 = \frac{1}{\sqrt{LC}}$$

$$\Rightarrow \quad 2\pi\frac{c}{\lambda} = \frac{1}{\sqrt{LC}}$$

$$\Rightarrow \quad 4\pi^2\frac{c^2}{\lambda^2} = \frac{1}{LC}$$

$$\Rightarrow \quad \frac{4\times 10\times (3\times 10^8)^2}{(960)^2} = \frac{1}{L\times 2.56\times 10^{-6}}$$

$$\Rightarrow \quad \frac{1}{L} = \frac{4\times 10\times 9\times 10^{16}\times 2.56\times 10^{-6}}{960\times 960}$$

$$\Rightarrow \quad L = 10\times 10^{-8}\,\text{H}$$

Ans. (10)

Sol. 8 Signal Bandwidth is for AM broadcast is given as

$BW = 2\times$ maximum frequency at modulating signal

$= 2\times 5\text{kHz}$

$= 10\,\text{kHz}$

Number of stations accommodated using this bandwidth in 90kHz total bandwidth are calculated as

$$N = \frac{90}{10} = 9$$

Ans. (9)

Sol. 9 Length of antenna tower is taken one fourth of the wavelength of transmission signal so we use

$$L = 25\text{m} = \frac{\lambda}{4}$$

$$\Rightarrow \quad \lambda = 100\,\text{m} \qquad \textbf{Ans. (D)}$$

Sol. 10 For line of sight communication the distance between antennas is given as

$$D = 2\sqrt{2Rh}$$

$$\Rightarrow \quad h = \frac{D^2}{8R} = \frac{45^2}{8\times 6400}\,\text{km} \cong 39.55\,\text{m} \qquad \textbf{Ans. (B)}$$

Sol. 11 Ground distance between antennas is given as

$$d = \sqrt{2Rh}$$

where $A = \pi d^2$, so we have

$$A = \pi 2Rh = 3.14\times 2\times 6400 \times \frac{30}{1000}$$

$$\Rightarrow \quad A = 1205.76\,\text{km}^2 \sim 1206\,\text{km}^2 \qquad \textbf{Ans. (1206)}$$

Sol. 12 Bandwidth of the modulated signal is given as

$$w = 2f_m$$

Angular frequency of the signal is given as

$$\omega_m = 1.57\times 10^8 = 2\pi f_m$$

$$\Rightarrow \quad w = 2f_m = \frac{10^8}{2}\,\text{Hz} = 50\text{ MHz} \qquad \textbf{Ans. (D)}$$

Sol. 13 In order of atmospheric layers from bottom we have Troposphere, stralospherre, Mesospehre, Thermosphere.

Ans. (A)

Sol. 14 Antenna range in first case is calculated as

$$R_1 = \sqrt{2Rh}$$

Antenna range in nsecond case is calculated as

$$R_2 = \sqrt{2Rh} + \sqrt{2Rh'}$$

Where $h = 20\,\text{m}$ & $h' = 5\,\text{m}$

Percentage increase in range can be given as

$$\Delta = \frac{R_2 - R_1}{R_1}\times 100\%$$

$$\Rightarrow \quad \Delta = \frac{\sqrt{2Rh'}}{\sqrt{2Rh}}\times 100\%$$

$$\Rightarrow \quad \Delta = \frac{\sqrt{5}}{\sqrt{20}}\times 100\% = 50\% \qquad \textbf{Ans. (50)}$$

Sol. 15 Transmitting distance is given as

$$d = \sqrt{2Rh}$$

$$\Rightarrow \quad h = \frac{d^2}{2R} = \frac{150\times 150\times 10^6}{2\times 6.5\times 10^6} = 1730.7 \approx 1731\,\text{m}$$

Population covered can be given as

$$p = 2\pi Rh \times \text{Population density}$$

$$\Rightarrow \quad p = 2\pi\times 6.5\times 10^6\times 1730.7\times 2000\times 10^{-6} \approx 1413\times 10^5$$

Ans. (A)

Sol. 16 Band width of the modulated signal is given as

$$BW = 2f_m = 2\times 10^5\,\text{Hz} = 200\,\text{kHz} \qquad \textbf{Ans. (B)}$$

Sol. 17 From the given data, we can use

$$\frac{a}{10} = \frac{b}{10} = \frac{\mu A_c}{2}$$

$$\Rightarrow \quad \frac{a}{b} = 1 \qquad \textbf{Ans. (1)}$$

Sol. 18 Range between transmitting and receiving antenna is given as

$$R = \sqrt{2Rh_1} + \sqrt{2Rh_2} = 57.28\,\text{km} \qquad \textbf{Ans. (B)}$$

Sol. 19 Range on ground by line of sight is given as

$$R = \sqrt{2Rh_1} + \sqrt{2Rh_2}$$

$$\Rightarrow \quad R = \sqrt{2\times 6400\times 320} + \sqrt{2\times 6400\times 2000}$$

$$\Rightarrow \quad R \approx 224\,\text{km} \qquad \textbf{Ans. (224)}$$

Sol. 20 For better range and directivity height of antenna should be proportional to the wavelength of signal to be transmitted. The higher the antenna, better is the range and for and optimum antenna its height should be equal to its wavelength or longer than that so option (A) is most appropriate out of all given options. **Ans. (A)**

Sol. 21 Sum of heights of transmitting and receiving antennas is given as

$$h_r + h_t = 160$$

Range of LOS communication is given as

$$R = \sqrt{2R}\left(\sqrt{h_r} + \sqrt{h_t}\right)$$

$$\Rightarrow \quad R = \sqrt{2R}\left(\sqrt{h_r} + \sqrt{160 - h_r}\right)$$

For maximum range of LOS communication, we have

$$\frac{dR}{dh_r} = 0 = \sqrt{2}\,R\left(\frac{1}{2\sqrt{h_r}} - \frac{1}{2\sqrt{160 - h_r}}\right) = 0$$

$$\Rightarrow \quad h_r = h_t = 80\,\text{m}$$

$$\Rightarrow \quad R = \sqrt{2R}\left(\sqrt{80} + \sqrt{80}\right)$$

$$\Rightarrow \quad R = 2\sqrt{2\times 6400\times 10^3\times 80} = 64\,\text{km}$$

Ans. (64)

Sol. 22 Total bandwidth in terms of number of channels and modulating frequency can be calculated as

Band width = 2 × n × highest modulating frequency

$$\Rightarrow \quad 6\,\text{MHz} = 2\times n\times 6\,\text{KHz}$$

$$\Rightarrow \quad n = \frac{6\times10^6}{2\times6\times10^3} = 500 \qquad \textbf{Ans. (500)}$$

29.2 Modulation and Demodulation

Sol. 1 If carrier frequency is ω_c than side band frequencies as

lower side band $\omega_1 = \omega_c - \omega_m$

upper side band $\omega_2 = \omega_c + \omega_m$

in transmission of modulated signal all frequencies between $\omega_1 \leftrightarrow \omega_2$ are transmitted.

Out of given option ω_m is NOT contained in modulated wave. **Ans. (C)**

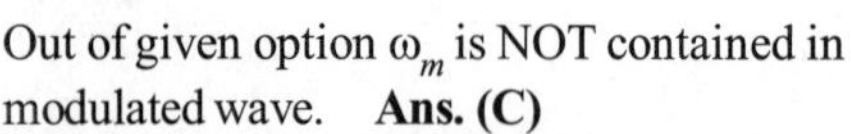

Sol. 2 Modulation index

$$m = \frac{V_{0m}}{V_{0c}} = \frac{5}{25} = \frac{1}{5} = 0.2$$

Modulation frequency f_m = 20 kHz

Carrier frequency f_c = 1.2 MHz = 1200 kHz

Side band frequencies as

$$f_{LSB} = f_c - f_m = 1200 - 20 = 1180\,\text{kHz}$$

and $$f_{USB} = f_c + f_m = 1200 + 20 = 1220\,\text{kHz}$$

Ans. (D)

Sol. 3 Modulation Index

$$m = \frac{V_{0m}}{V_{0c}}$$

$$\Rightarrow \quad V_{0m} = m\times V_{0c}$$

$$\Rightarrow \quad V_{0m} = 0.8\times14 = 11.2\,\text{V} \qquad \textbf{Ans. (A)}$$

Sol. 4 Total transmission band width available is

$$= \frac{10}{100}\times10\times10^9\,\text{Hz} = 10^9\,\text{Hz}$$

Each channel requires and individual Band width = 5 × 10³ Hz

If the total Band width can accommodate N channels then we use

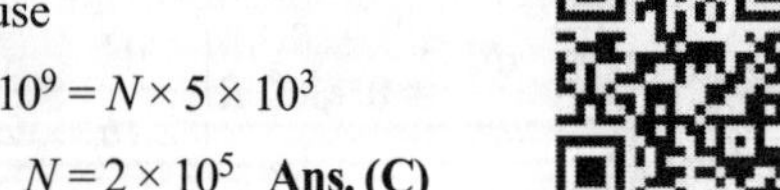

$$10^9 = N\times5\times10^3$$

$$\Rightarrow \quad N = 2\times10^5 \quad \textbf{Ans. (C)}$$

Sol. 5 The frequency of carrier wave is given as

$$f_c = \frac{1}{2\pi\sqrt{LC}}$$

$$\Rightarrow \quad f_c = \frac{1}{2\pi\sqrt{49\times10^{-6}\times2.5\times10^{-9}}}$$

$$\Rightarrow \quad f_c = \frac{1}{2\pi\times7\times5\times10^{-8}} = \frac{10^8}{70\pi}$$

$$f_c = 454.54\,\text{kHz}$$

The frequency range used for transmission of this signal is

$$\Delta f = 2f_m$$

$$f_1 \leftrightarrow f_2 = f_c \pm f_m$$

$$\Rightarrow \quad f_1 = f_c - f_m = 454.54 - 12 = 442.54\,\text{kHz}$$

$$\Rightarrow \quad f_2 = f_c + f_m = 454.54 + 12 = 466.54\,\text{kHz} \qquad \textbf{Ans. (C)}$$

Sol. 6 Total bandwidth available for transmission is

$$\Delta BW = 300\times10^3\,\text{Hz}$$

Band width of each channel is $CBW = 2f_m = 2\times15\times10^3$

$$\Rightarrow \quad CBW = 30\times10^3\,\text{Hz}$$

If N are the number of channels accommodated for transmission, we have

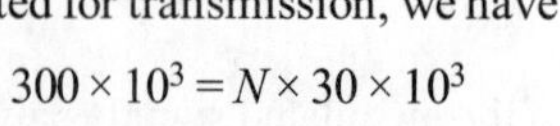

$$300\times10^3 = N\times30\times10^3$$

$$\Rightarrow \quad N = 10 \quad \textbf{Ans. (B)}$$

Sol. 7 Frequency of source is given as

$$f = \frac{3\times10^8}{8\times10^{-7}} = \frac{30}{8}\times10^{14}\,\text{Hz}$$

$$\Rightarrow \quad f = 3.75\times10^{14}\,\text{Hz}$$

$$1\%\text{ of } f = 0.0375\times10^{14}\,\text{Hz}$$

$$\Rightarrow \quad \Delta f = 3.75\times10^{12}\,\text{Hz} = 3.75\times10^6\,\text{MHz}$$

$\Rightarrow$ Number of channels

$$= \frac{3.75\times10^6}{6} = 6.25\times10^5 \qquad \textbf{Ans. (D)}$$

Sol. 8 $$f_{\text{carrier}} = \frac{250}{0.1}\,\text{kHz}$$

$\Rightarrow$ Range of signal = 2250 Hz to 2750 Hz

In given options; for 2000 kHz

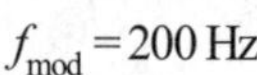

$$f_{\text{mod}} = 200\,\text{Hz}$$

So its range is 1800 Hz to 2200 Hz

Which is not overlapping with the existing range. **Ans. (B)**

Sol. 9 The given signal can be written as

$$V(t) = 10\sin B + \frac{3}{2}[2\cos A\sin B]$$

$$\Rightarrow \quad V(t) = 10\sin B + 10 + \frac{3}{2}\,[\sin(A+B) - \sin(A-\text{B})]$$

$$\Rightarrow \quad V(t) = 10 \sin B\,(5.5 \times 10^5\, t) + \frac{3}{2}\,[\sin(57.2 \times 10^4\, t)]$$

$$- \sin(52.8 \times 10^4\, t)$$

Here $\omega_1 = 57.2 \times 10^4 = 2\pi f_1$

$$\Rightarrow \quad f_1 = \frac{57.2\times10^4}{2\times\left(\frac{22}{7}\right)} = 9.1 \times 10^4$$

$$\Rightarrow \quad f_1 \simeq 91\text{kHz}$$

and $\omega_2 = 52.8 \times 10^4 t = 2\pi f_2$

$$\Rightarrow \quad f_2 = \frac{52.8\times10^4}{2\times\left(\frac{22}{7}\right)}$$

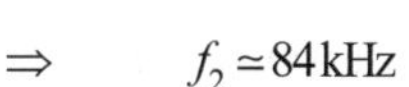

$$\Rightarrow \quad f_2 \simeq 84\,\text{kHz}$$

Ans. (C)

Sol. 10 Analysis of graph shows

(1) Amplitude varies between

$8 - 10\,V$ or 9 ± 1

(2) Two time period are

100 μs (modulating wave) & 8 μs (carrier wave)

Thus modulated signal is $\left[9 \pm 1 \sin\left(\frac{2\pi t}{T_1}\right)\right] \sin\left(\frac{2\pi t}{T_2}\right)$

$= [9 \pm 1 \sin(2\pi \times 10^4 t)] \sin(2.5\pi \times 10^5\, t)$ **Ans. (D)**

Sol. 11 Given that by modulation carrier wave amplitude changes maximum upto 160 V, so we use

$$A_m + A_c = 160$$

$$\Rightarrow \quad A_m + 100 = 160$$

$$\Rightarrow \quad A_m = 60$$

Modulation index is

$$\mu = \frac{A_m}{A_c} = \frac{60}{100}$$

$$\Rightarrow \quad \mu = 0.6$$

Ans. (A)

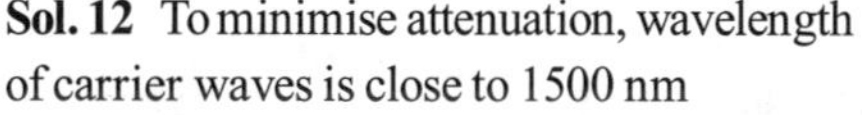

Sol. 12 To minimise attenuation, wavelength of carrier waves is close to 1500 nm

Ans. (D)

Sol. 13 AM signal contains one carrier frequency and two (ω_0) side band frequencies

$$\text{USB} = \omega_0 + \omega$$

$$\text{LSB} = \omega_0 - \omega$$

Ans. (B)

Sol. 14 The physical size of antenna of reciver and transmitter both directly proportional to the wavelength of carrier wave so both are inversely proportional to carrier frequency. **Ans. (C)**

Sol. 15 Modulation inxex

$$= \frac{V_{0m}}{V_{0c}} = \frac{100}{400} = 0.25$$

USB frequency $= f_c + f_m$

LSB frequency $= f_c - f_m$

USB – LSB $= 2f_m = 2 \times 100 \times 10^6 = 2 \times 10^8$ Hz **Ans. (D)**

Sol. 16 Modulation index

$$m = \frac{A_m}{A_c} = \frac{2}{4} = 0.5$$

Carrier frequency

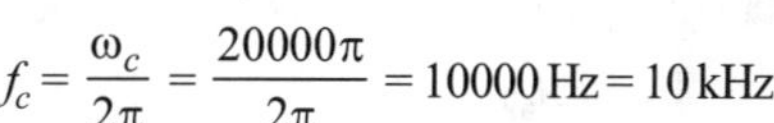

$$f_c = \frac{\omega_c}{2\pi} = \frac{20000\pi}{2\pi} = 10000\,\text{Hz} = 10\,\text{kHz}$$

Modulating frequency

$$f_m = \frac{\omega_m}{2\pi} = \frac{2000\pi}{2\pi} = 1000\,\text{Hz} = 1\,\text{kHz}$$

LSB frequency $= f_c - f_m = 10 - 1 = 9$ kHz **Ans. (A)**

Sol. 17 Modulated signed voltage is given as

$$V_m = 5\,(1 + 0.6 \cos(6280t) \sin(2\pi \times 10^4 t)$$

$$V_m = 5 \sin(2\pi \times 10^4 t) + 3 \cos(6280t) \sin(2\pi \times 10^4 t)$$

In above wave maximum amplitude is

$$V_{max} = 5 + 3 = 8\,\text{V}$$

$$V_{min} = 5 - 3 = 2\,\text{V}$$

Ans. (None)

Sol. 18 Percentage modulation in AM is given as

$$p = \frac{Am}{Ac} \times 100$$

$$\Rightarrow \quad p = \frac{20}{80} \times 100 = 25\%$$

Ans. (25)

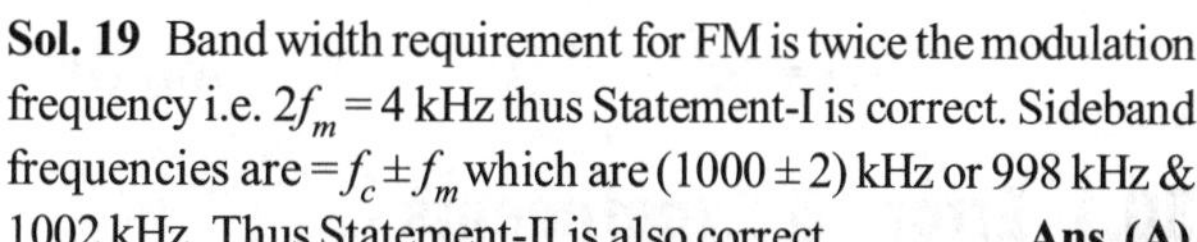

Sol. 19 Band width requirement for FM is twice the modulation frequency i.e. $2f_m = 4$ kHz thus Statement-I is correct. Sideband frequencies are $= f_c \pm f_m$ which are (1000 ± 2) kHz or 998 kHz & 1002 kHz. Thus Statement-II is also correct. **Ans. (A)**

Sol. 20 The wavelength of signal in air will be at carrier frequency only so it is given as

$$\lambda = \frac{c}{f_c}$$

Ans. (D)

Sol. 21 Modulation index for the modulated wave is given as

$$m=\frac{A_{\max}-A_{\min}}{A_{\max}+A_{\min}}$$

$$m=\frac{16-8}{16+8}=\frac{8}{24}=\frac{1}{3}=0.33$$

$\Rightarrow \quad x\times 10^{-2}=0.33$

$\Rightarrow \quad x=33$ **Ans. (33)**

Sol. 22 Maximum voltage in the transmitting signal can be given as

$$V_{\max}=A_C+A_m$$

$\Rightarrow \quad 200=160+A_m$

$\Rightarrow \quad A_m=40\text{ V}$ **Ans. (40)**

Sol. 23 Modulation index of the wave is calculated as

$$\mu=\frac{A_m}{A_c}=\frac{20}{20}=1$$ **Ans. (1)**

Sol. 24 Modulation index of AM wave is given as

$$\mu=\frac{A_{\max}-A_{\min}}{A_{\max}+A_{\min}}$$

$$\mu=\frac{12-3}{12+3}=0.6$$

$\Rightarrow \quad x=1$ **Ans. (1)**

Sol. 25 The amplitude modelated wave is generally represented as

$$C_m=Ac\left[1+\frac{A_m}{A_C}\sin 2\pi f_m t\right]$$

Comparing with the given signal, we have

$$2\pi f_m=12560$$

$\Rightarrow \quad f_m=2000\text{ Hz}=2\text{ KHz}$ **Ans. (2)**

Sol. 26 Maximum and minimum amplitude of modulated signal are given as

$$A_{\max}=A_c+A_m=250+150=400$$

and $\quad A_{\min}=A_c-A_m=250-150=100$

$\Rightarrow \quad \frac{A_{\min}}{A_{\max}}=\frac{100}{400}=\frac{1}{4}=\frac{50}{200}$

$\Rightarrow \quad x=200$ **Ans. (200)**

30.1 Errors & Experiments

Sol. 1 For the given relation of surface tension, the relative % error in its measurement is given as

$$\frac{dT}{T}=\frac{dr}{r}+\frac{dh}{h}=\frac{dD}{D}+\frac{dh}{h} \qquad \ldots(1)$$

In measurement of diameter and water rise the least count of readings are

$$dr=\frac{dD}{2}=\frac{0.01\times10^{-2}}{2}\text{ m}$$

and $\quad dh=0.01\times10^{-2}\text{ m}$

from equation-(1) we use

$$\%\text{ error}=\frac{dT}{T}\times 100=\left(\frac{dD}{D}+\frac{dh}{h}\right)\times 100$$

$$\Rightarrow \quad \frac{dT}{T}\times 100=\left(\frac{0.01\times10^{-2}}{1.25\times10^{-2}}+\frac{0.01\times10^{-2}}{1.45\times10^{-2}}\right)\times 100$$

$$\Rightarrow \quad \frac{dT}{T}\times 100=\frac{1}{1.25}+\frac{1}{1.45}=0.8+0.69\%$$

$$=1.49\%$$ **Ans. (D)**

Sol. 2 From the given expression the measurement of P gives its relative maximum % error as

$$\frac{dP}{P}=\frac{1}{2}\left(\frac{da}{a}\right)+2\left(\frac{db}{b}\right)+3\left(\frac{dc}{c}\right)+4\left(\frac{dd}{d}\right)$$

$$\Rightarrow \%\text{ error in }P=\frac{1}{2}\times 2+2\times 1+3\times 3+4\times 5$$

$$=1+2+9+20=32\%$$ **Ans. (C)**

Sol. 3 Expression for A is

$$A=\frac{P^3Q^3}{\sqrt{RS}}$$

Maximum relative % error in A is given as

$$\frac{dA}{A}\times 100\%=\left(3\frac{dP}{P}+2\frac{dQ}{Q}+\frac{1}{2}\frac{dR}{R}+\frac{dS}{S}\right)\times 100\%$$

$$\Rightarrow \quad \%\text{ error in }A=3\times 0.5+2\times 1+\frac{1}{2}\times 3+1.5$$

$$=1.5+2+1.5+1.5=6.5\%$$ **Ans. (D)**

Sol. 4 5 revolution displacement of 0.25 cm

$$\Rightarrow \text{ 1 main scale division}=\frac{0.25}{5}=0.05\text{ cm}$$

$$\text{least count of Screw gauge}=\frac{\text{1 main scale division}}{\text{no. of circular scale division}}$$

$$=\frac{0.05}{100}=5\times10^{-4}\text{ cm}$$

Reading for thickness of wire = 4 MS + 30 CS

$$\text{wire thickness}=4\times 0.05+30\times 5\times 10^{-4}\text{ cm}$$

$$=0.2+0.015=0.215\text{ cm}$$ **Ans. (D)**

Sol. 5 If R is radius of a sphere then

its surface area $S = 4\pi R^2$

$$\Rightarrow \quad \frac{dS}{S} = \frac{2dR}{R} = \alpha$$

$$\Rightarrow \quad \frac{dR}{R} = \frac{\alpha}{2}$$

its volume $V = \frac{4}{3}\pi R^3$

$$\Rightarrow \quad \frac{dV}{V} = \frac{3dR}{R} = 3\left(\frac{\alpha}{2}\right) = \frac{3}{2}\alpha$$

Ans. (C)

Sol. 6 Using Kepler's law we have

$$T^2 = \frac{4\pi^2}{GM}R^3$$

taking log on both sides and differentiating, gives

$$2\ln T = \ln\left(\frac{4\pi^2}{GM}\cdot R^3\right)$$

$$\Rightarrow \quad 2\ln T = \ln\left(\frac{4\pi^2 R^3}{G}\right) - \ln M$$

$$\Rightarrow \quad 2\frac{dT}{T} = \frac{dM}{M}$$

$$\Rightarrow \quad \frac{dM}{M} = 2\times 10^{-2}$$

Ans. (C)

Sol. 7 Using mass and length of a cube, its density is given as

$$\rho = \frac{\text{Mass}}{\text{Volume}} = \frac{M}{L^3}$$

relative error (maximum) in density calculation is given as

$$\frac{\Delta\rho}{\rho} = \frac{\Delta M}{M} + 3\frac{\Delta L}{L}$$

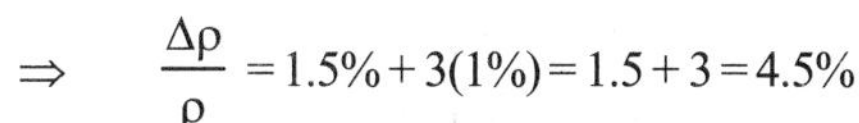

$$\Rightarrow \quad \frac{\Delta\rho}{\rho} = 1.5\% + 3(1\%) = 1.5 + 3 = 4.5\%$$

Ans. (C)

Sol. 8 Resistance of a conductor is given as

$$R = \frac{\rho l}{A} \text{ and volume } (V) = Al$$

$$R = \frac{\rho l^2}{V}$$

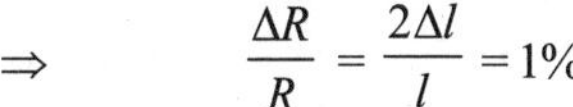

$$\Rightarrow \quad \frac{\Delta R}{R} = \frac{2\Delta l}{l} = 1\%$$

Ans. (C)

Sol. 9 Least count of screw gauge is given as

$$LC = \frac{\text{Pitch}}{\text{No. of division}}$$

$$\Rightarrow \quad LC = 0.5\times 10^{-2}\,\text{mm}$$

Zero error is given as

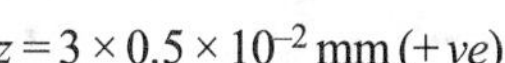

$$z = 3\times 0.5\times 10^{-2}\,\text{mm}\,(+ve)$$

$$\Rightarrow \quad z = 1.5\times 10^{-2}\,\text{mm} = 0.015\,\text{mm}$$

Thickness of sheet measured by screw gauge is

$$t = MSR + CSR - ZE\ (+ve \text{ zero error})$$

$$\Rightarrow \quad t = 5.5\,\text{mm} + (48\times 0.5\times 10^{-2}) - 0.015$$

$$\Rightarrow \quad t = 5.5 + 0.24 - 0.015 = 5.725\,\text{mm}$$

Ans. (B)

Sol. 10 For volume of cylinder, relative error is given as

$$\frac{\Delta V}{V} = 2\frac{\Delta d}{d} + \frac{\Delta h}{h} = 2\left(\frac{0.1}{12.6}\right) + \frac{0.1}{34.2} \simeq 0.016$$

$$V = 34.2\times\frac{\pi}{4}\times(12.6)^2$$

$$\Rightarrow \quad V = 4262.23\,\text{cm}^3$$

and $\Delta V = 0.016\times 4262.23 = 68.19\,\text{cm}^3$

Volume in 3 SD is given as

$$V = 4260 \pm 68\,\text{cm}^3$$

Out of given options most appropriate is (A). **Ans. (A)**

Sol. 11 Least count of a screw gauge is given as

$$\text{Least count} = \frac{\text{Pitch}}{\text{Number of division on circular scale}}$$

$$\Rightarrow \quad 5\times 10^{-6} = \frac{10^{-3}}{N}$$

$$\Rightarrow \quad N = 200$$

Ans. (C)

Sol. 12 Time of one oscillation measured is

$$T = \frac{30\,\text{sec}}{20}$$

Maximum possible error in measurement (least count) is

$$\Delta T = \frac{1}{20}\,\text{sec}$$

Length of pendulum is given as

$$L = 55\,\text{cm}$$

$$\Delta L = 1\,\text{mm} = 0.1\,\text{cm}$$

Value of g is calculated as

$$g = \frac{4\pi^2 L}{T^2}$$

Percentage error in measurement of g is

$$\frac{\Delta g}{g}\times 100\% = \left(\frac{\Delta L}{L}+\frac{2\Delta T}{T}\right)100\%$$

$$= \left(\frac{0.1}{55}+\frac{2\left(\frac{1}{20}\right)}{\frac{30}{20}}\right)100\% \simeq 6.8\% \quad \textbf{Ans. (D)}$$

Sol. 13 Density of cube is given as

$$\rho = \frac{m}{v} \qquad [v = l^3]$$

Percentage error or fractional error in density measurement is

$$\frac{\Delta\rho}{\rho} = \frac{\Delta m}{m} + 3\frac{\Delta l}{l}$$

$$\Rightarrow \quad \Delta\rho = \left[\frac{0.10}{10.00}+3\left(\frac{0.01}{0.10}\right)\right]\times\frac{10.00}{(0.10)^3}\ \text{kg/m}^3$$

$$\Rightarrow \quad \Delta\rho = [0.01+0.3]\times 10^4\ \text{kg/m}^3$$

$$\Rightarrow \quad \Delta\rho = 0.31\times 10^4\ \text{kg/m}^3 \quad \textbf{Ans. (None)}$$

Sol. 14 Total Area of seven squares is

$$S = A_1 + A_2 + \ldots A_7$$

$$\Rightarrow \quad S = A + A + \ldots 7 \text{ times}$$

$$\Rightarrow \quad S = 37.03\ \text{cm}^2$$

Addition of 7 terms all having 2 terms beyond decimal, so final answer must have 2 terms beyond decimal (as per rules of significant digits.) **Ans. (C)**

Sol. 15 Time period of a simple pendulum is given as

$$T = 2\pi\sqrt{\frac{l}{g}}$$

$$\Rightarrow \quad g = \frac{4\pi^2 l}{T^2}$$

Fractional error is measurement of g is

$$\frac{\Delta g}{g} = \frac{\Delta l}{l} + \frac{2\Delta T}{T}$$

$$\Rightarrow \quad \frac{\Delta g}{g} = \frac{0.1}{25} + \frac{2\times 1}{50} = 0.044$$

Thus percentage error or accuracy is given as

$$\frac{\Delta g}{g} = 4.4\% \quad \textbf{Ans. (C)}$$

Sol. 16 As by on six rotation, reading of main scale changes by 3mm, so main scale least count is 0.5 mm. Number of division on circular scale are 50 so least count of the screw gauge will be

$$LC = \frac{0.5}{50}\ \text{mm} = 0.001\ \text{cm} \quad \textbf{Ans. (B)}$$

Sol. 17 For the given set of physical quantities we use

$$A_1 + B_1 + C_1 = 24.36 + 0.0724 + 256.2$$
$$= 280.6324$$
$$= 280.6$$
$$A_2 + B_2 + C_2 = 24.44 + 16.082 + 240.2$$
$$= 280.722$$
$$= 280.7$$
$$A_3 + B_3 + C_3 = 25.2 + 19.2812 + 236.183$$
$$= 280.6642$$
$$= 280.7$$
$$A_4 + B_4 + C_4 = 25 + 236.191 + 19.5$$
$$= 280.691$$
$$= 281$$

$$A_4 + B_4 + C_4 > A_3 + B_3 + C_3 = A_2 + B_2 + C_2 > A_1 + B_1 + C_1$$

Out of given options none is correct. **Ans. (Bonus)**

Sol. 18 Zero error in scale is

$$Z = 0 + 0.01\times 7 = 0.07\ \text{cm}$$

Reading of vernier caliper for length of cylinder is

$$R = 3.1 + 0.01\times 4 - 0.07 = 3.07\ \text{cm} \quad \textbf{Ans. (D)}$$

Sol. 19 Pitch of screw gauge $p = 0.1$ cm

No of divisions on circular scale

$$N = 50$$

least count of screw gauge is

$$L = \frac{p}{N} = 0.002\ \text{cm}$$

The result of measurement by this screw gauge must be in multiples of L, so here result would be 2.124 cm. **Ans. (D)**

Sol. 20 Percentage error in measurement of z is given as

$$\frac{\Delta z}{z} = \frac{2\Delta a}{a} + \frac{2}{3}\frac{\Delta b}{b} + \frac{1}{2}\frac{\Delta c}{c} + \frac{3\Delta d}{d}$$

$$\Rightarrow \quad \frac{z}{\Delta z} = 2\times 2 + \frac{2}{3}\times 15 + \frac{1}{2}\times 4 + 3\times 2.5\%$$

$$\Rightarrow \quad \frac{\Delta z}{z} = 14.5\% \quad \textbf{Ans. (B)}$$

Sol. 21 Pitch of circular scale = 0.5 mm

Number of divisions of CS = 50

$$\text{Least count} = \frac{0.5}{50}\text{ mm}$$
$$= 0.01\text{ mm}$$
$$= 0.01 \times 10^{-3}\text{ m}$$
$$= 10 \times 10^{-6}\text{ m} = 40\ \mu\text{m}$$

Zero error $z = 4 \times 0.01 = 0.04$ mm (+ve) **Ans. (B)**

Sol. 22 Fractional error in density measurement is given as

$$\frac{d\rho}{\rho} = \frac{dm}{m} + \frac{3dD}{D} \qquad \left[\text{As } \rho = \frac{m}{\frac{4}{3}\pi\left(\frac{D}{2}\right)^3}\right]$$

We use

$$\frac{d\rho}{\rho} = \times 100 = 6 + 3 \times 1.5 = 10.5\% = \frac{x}{100}$$

$\Rightarrow \quad x = 1050$ **Ans. (1050)**

Sol. 23 As all readings are given in 3 significant digits, final results must be in 3 significant digits only so option (C) can be only correct answer. **Ans. (C)**

Sol. 24 The least count of screw guage given is calculated as

$$LC = \frac{1\text{mm}}{100} = 0.01\text{mm}$$

Zero error in the given screw guage is given as

$$z = +8 \times LC = +0.08\text{ mm}$$

True reading for diameter of wire is given as

$$TR = (1\text{ mm} + 72 \times LC) - z$$
$$\Rightarrow \quad TR = (1\text{ mm} + 72 \times 0.01\text{ mm}) - 0.08\text{ mm}$$
$$\Rightarrow \quad TR = 1.72\text{ mm} - 0.08\text{ mm} = 1.64\text{ mm}$$

Radius of wire is given as

$$r = \frac{1.64}{2} = 0.82\text{ mm}$$ **Ans. (D)**

Sol. 25 As volume of wire remains constant while stretching so we use

$$A_0 l_0 = A_1 l_1$$
$$\Rightarrow \quad A_1 = \frac{l_0 A_0}{l_1}$$

Resistance of wire can be given as

$$R = \frac{\rho l}{A}$$

If R_0 and R_1 are the initial and final resistances of wire, we use

$$\frac{R_0}{R_1} = \frac{l_0 / A_0}{l_1 / A_1}$$
$$\Rightarrow \quad \frac{1}{R_1} = \frac{l_0}{A_0}\left(\frac{l_0 A_0}{l_1 \times l_1}\right)$$
$$\Rightarrow \quad R_1 = \frac{l_1^2}{l_0^2} = 1.5625\ \Omega$$

Thus percentage change in resistance is given as

$$p = \frac{R_1 - R_0}{R_0} \times 100\%$$
$$\Rightarrow \quad p = \frac{1.5625 - 1}{1} \times 100\% = 56.25\%$$ **Ans. (D)**

Sol. 26 The percentage error in value of resistance R is given as

$$\frac{\Delta R}{R} \times 100 = \frac{\Delta V}{V} \times 100 + \frac{\Delta I}{I} \times 100$$
$$\frac{\Delta R}{R} \times 100 = \frac{2}{50} \times 100 + \frac{0.2}{20} \times 100$$
$$\frac{\Delta R}{R} \times 100 = 4 + 1 = 5\%$$ **Ans. (5)**

Sol. 27 Positive zero error is to be subtracted from the final reading. As main scale reading is 8.5 cm and vernier coincidence is 6, so final reading of vernier scale reading is given as

$$\text{VSR} = 6 \times 0.01 = 0.06\text{ cm}$$

Thus final reading is calculated as

$$\text{MSR} + \text{VSR} - \text{ZE} = 8.5 + 0.06 - 0.02 = 8.54\text{ cm}$$ **Ans. (B)**

Sol. 28 By the expression of time period of a simple pendulum, value of g is given as

$$g = \frac{4\pi^2 l}{T^2}$$

Fractional error in value of g is given as

$$\frac{\Delta g}{g} = \frac{\Delta l}{l} + 2\frac{\Delta T}{T} = \frac{0.1}{10} + 2\left(\frac{1}{\frac{200}{0.5}}\right)$$
$$\Rightarrow \quad \frac{\Delta g}{g} = \frac{1}{100} + \frac{1}{50}$$

Percentage error in value of g is

$$\Rightarrow \quad \frac{\Delta g}{g} \times 100 = 3\%$$ **Ans. (B)**

Sol. 29 Volume of sphere is expressed as

$$V = \frac{4}{3}\pi r^3$$

Fractional error in volume can be written as

$$\frac{dV}{V} = 3\frac{dr}{r}$$

Percentage error in volume is given as

$$\frac{dV}{V} \times 100\% = \frac{3 \times 0.85}{7.5} \times 100\% = 34\% \qquad \textbf{Ans. (34)}$$

Sol. 30 Percentage error in calculation of g can be written as

$$\frac{\Delta g}{g} = \left(\frac{\Delta l}{l} + \frac{2\Delta T}{T}\right) \times 100\,\% \qquad \textbf{Ans. (1)}$$

Sol. 31 The difference between the two student reading for any set of experiment should be equal to approximately the difference between zero error in the two instruments. From the figure, it is calculated as

$$= 5 - (-8) = 13 \text{ divisions} \qquad \textbf{Ans. (13)}$$

Sol. 32 The magnetic permeability of the medium is related to susceptibility of medium is given as

$$\mu_r = 1 + \chi$$

$$\Delta\mu_r = \Delta\chi$$

As $\quad B \propto \mu r$

$\Rightarrow \quad B = k\mu_r \qquad$ (Here k is a constant)

Percentage change in magnetic field is given as

$$\frac{\Delta B}{B} \times 100 = \frac{k(\Delta\mu_r)}{k\mu_r} \times 100 = \frac{2.2 \times 10^{-5}}{1} \times 100 = 0.022$$

$\Rightarrow \quad x = 22 \qquad$ **Ans. (22)**

Sol. 33 Pitch of screw gauge is the distance travelled on main scale in one rotation which is 1 mm thus least count is given as

$$LC = 1\text{ mm}/50 = 0.02\text{ mm}$$

Thus Statement-1 is wrong. Statement-2 is right **Ans. (B)**

Sol. 34 Percentage error in calculation of y is given as

$$\%y = 2(\%m) + 4(\%r) + x(\%g) + \frac{3}{2}(\%l)$$

$\Rightarrow \quad 18 = 2 \times 1 + 4 \times 0.5 + x(p) + \frac{3}{2} \times 4$

$\Rightarrow \quad 18 = 2 + 2 + x(p) + 6$

$\Rightarrow \quad x(p) = 8$

Out of given options only option (C) can be correct. **Ans. (C)**

Sol. 35 Time period of simple pendulum is given as

$$T = 2\pi\sqrt{\frac{l}{g}}$$

For percentage error calculation, we have

$$\frac{\Delta g}{g} + \frac{2\Delta T}{T} = \frac{\Delta L}{L}$$

$\Rightarrow \quad 4 + 2 \times 3 = 10\%$

$\Rightarrow \quad E = \frac{1}{2}m\omega^2A^2 = \frac{1}{2}m\frac{g}{l}A^2$

$\Rightarrow \quad E = \frac{\Delta g}{g} + \frac{\Delta L}{L} = 4 + 10 = 14\% \qquad$ **Ans. (14)**

* * * * *

NOTES

NOTES